Materials Under Extreme Conditions

Materials Under Extreme Conditions

Recent Trends and Future Prospects

Edited By

A.K. Tyagi

S. Banerjee

Bhabha Atomic Research Centre, Mumbai, India

ELSEVIER elsevier.com

Elsevier
Radarweg 29, PO Box 211, 1000 AE Amsterdam, Netherlands
The Boulevard, Langford Lane, Kidlington, Oxford OX5 1GB, United Kingdom
50 Hampshire Street, 5th Floor, Cambridge, MA 02139, United States

Notices

Knowledge and best practice in this field are constantly changing. As new research and experience broaden our understanding, changes in research methods, professional practices, or medical treatment may become necessary.

Practitioners and researchers must always rely on their own experience and knowledge in evaluating and using any information, methods, compounds, or experiments described herein. In using such information or methods they should be mindful of their own safety and the safety of others, including parties for whom they have a professional responsibility.

To the fullest extent of the law, neither the Publisher nor the authors, contributors, or editors, assume any liability for any injury and/or damage to persons or property as a matter of products liability, negligence or otherwise, or from any use or operation of any methods, products, instructions, or ideas contained in the material herein.

Library of Congress Cataloging-in-Publication Data

A catalog record for this book is available from the Library of Congress

British Library Cataloguing-in-Publication Data

A catalogue record for this book is available from the British Library

ISBN: 978-0-12-801300-7

Publisher: John Fedor
Acquisition Editor: Anita Koch
Editorial Project Manager: Amy Clark
Production Project Manager: Maria Bernard
Cover Designer: Greg Harris

Typeset by TNQ Books and Journals

Contents

List of Contributors

S.N. Achary
Bhabha Atomic Research Centre, Mumbai, India

A. Ananthanarayanan
Bhabha Atomic Research Centre, Mumbai, India

V. Arora
Raja Ramanna Centre for Advanced Technology, Indore, India

S. Banerjee
Bhabha Atomic Research Centre, Mumbai, India

Y.K. Bhardwaj
Bhabha Atomic Research Centre, Mumbai, India

J.A. Chakera
Raja Ramanna Centre for Advanced Technology, Indore, India

S.S. Chopade
Bhabha Atomic Research Centre, Mumbai, India

G.K. Dey
Bhabha Atomic Research Centre, Mumbai, India

K.A. Dubey
Bhabha Atomic Research Centre, Mumbai, India

Nandini Garg
Bhabha Atomic Research Centre, Mumbai, India

M. Goswami
Bhabha Atomic Research Centre, Mumbai, India

V. Grover
Bhabha Atomic Research Centre, Mumbai, India

P.D. Gupta
Raja Ramanna Centre for Advanced Technology, Indore, India

Satish C. Gupta
Bhabha Atomic Research Centre, Mumbai, India

D. Harish
Bhabha Atomic Research Centre, Mumbai, India

P.A. Hassan
Bhabha Atomic Research Centre, Mumbai, India

D. Jain
Bhabha Atomic Research Centre, Mumbai, India

M. Joseph
Indira Gandhi Centre for Atomic Research, Kalpakkam, India

K.D. Joshi
Bhabha Atomic Research Centre, Mumbai, India

U. Kamachi Mudali
Indira Gandhi Centre for Atomic Research, Kalpakkam, India

R. Kapoor
Bhabha Atomic Research Centre, Mumbai, India

R. Kar
Bhabha Atomic Research Centre, Mumbai, India

G.P. Kothiyal
Bhabha Atomic Research Centre, Mumbai, India

N. Manoj
Bhabha Atomic Research Centre, Mumbai, India

P. Manoravi
Indira Gandhi Centre for Atomic Research, Kalpakkam, India

R. Mishra
Bhabha Atomic Research Centre, Mumbai, India

P.A. Naik
Raja Ramanna Centre for Advanced Technology, Indore, India

R.S. Ningthoujam
Bhabha Atomic Research Centre, Mumbai, India

J. Nuwad
Bhabha Atomic Research Centre, Mumbai, India

M.K. Patel
Bhabha Atomic Research Centre, Mumbai, India

D.S. Patil
Formerly at Bhabha Atomic Research Centre, Mumbai, India

Y.B.S.R. Prasad
Raja Ramanna Centre for Advanced Technology, Indore, India

M. Roy
Bhabha Atomic Research Centre, Mumbai, India

S.B. Roy
Raja Ramanna Centre for Advanced Technology, Indore, India

N.K. Sarkar
Bhabha Atomic Research Centre, Mumbai, India

P. Sengupta
Bhabha Atomic Research Centre, Mumbai, India

P. Sharma
Bhabha Atomic Research Centre, Mumbai, India

Surinder M. Sharma
Bhabha Atomic Research Centre, Mumbai, India

N. Sivakumar
Indira Gandhi Centre for Atomic Research, Kalpakkam, India

V. Sudarsan
Bhabha Atomic Research Centre, Mumbai, India

R. Tewari
Bhabha Atomic Research Centre, Mumbai, India

A.K. Tyagi
Bhabha Atomic Research Centre, Mumbai, India

R.K. Vatsa
Bhabha Atomic Research Centre, Mumbai, India

M. Vijayalakshmi
Indira Gandhi Centre for Atomic Research, Kalpakkam, India

B. Vishwanadh
Bhabha Atomic Research Centre, Mumbai, India

About the Editors

Dr. A. K. Tyagi obtained his MSc (Chemistry) degree in 1985 from Meerut University, Meerut, India, and joined the 29th batch of BARC Training School, Mumbai, India, in the same year. After completing a one-year orientation course, he joined the Chemistry Division, Bhabha Atomic Research Center (BARC), Mumbai, in 1986. Presently, he heads the Solid State Chemistry Section of the Chemistry Division, BARC, and is a professor of chemistry at Homi Bhabha National Institute (HBNI), Mumbai. In addition, he is Academic Dean of Chemical Sciences, HBNI, Mumbai. His research interests are in the field of chemistry of materials, which includes functional materials, nanomaterials, nuclear materials, hybrid materials, and structure-property correlation.

Dr. Tyagi was awarded a PhD by Mumbai University in 1992. He was at Max-Planck Institute for Solid State Research, Stuttgart, Germany, for postdoctoral research during 1995–96 on a Max-Planck Fellowship. In addition, he has also visited the Institute of Superior Technology, Portugal; Institute for Chemical Process and Environmental Technology, Ottawa, Canada; Dalhousie University, Halifax, Canada; Moscow State University, Moscow, Russia; Institute for Materials, Nantes, France; University of Malay, Malaysia; National Institute of Materials Science, Tsukuba, Japan; National University of Singapore, Singapore; Royal Institute of Technology, Stockholm, Sweden; Rice University, Houston, USA; Tongji University, Shanghai, China; University of Valencia, Valencia, Spain; Weizmann Institute of Science, Israel; University of Queensland, Brisbane, Australia; US-Air Force Research Lab, Dayton, USA; and Institute for Studies of Nanostructured Materials, Palermo, Italy.

In recognition of his significant contributions to the field of chemistry of materials, Dr. Tyagi has been conferred with several awards, including the Dr. Lakshmi Award by the Indian Association of Solid State Chemists and Allied Scientists (2001); Rheometric Scientific-Indian Thermal Analysis Society Award (2002); Gold Medal of Indian Nuclear Society (2003); Materials Research Society of India's Medal (2005); Chemical Research Society of India's Medal (2006); DAE-Homi Bhabha Science and Technology Award (2006); IANCAS-Dr. Tarun Datta Memorial Award (2007); Rajib Goyal Prize in Chemical Sciences (2007); RD Desai Memorial Award from Indian Chemical Society (2009); DAE-SRC Outstanding Research Investigator Award (2010); CRSI-Prof. CNR Rao National Prize in Chemical Sciences (2012); ISCB Award for Excellence in Chemical Sciences (2013); MRSI-ICSC Materials Science Senior Award (2014); Coastal Chemical Research Society's Award (2014); and Platinum Jubilee Lecture Award in Materials Science from Indian Science Congress Association (2015). He is an elected fellow of the Indian Academy of Sciences; National Academy of Sciences India; Maharashtra Academy of Sciences; Royal Society of Chemistry, UK, and Asia Pacific Academy of Materials.

Dr. S. Banerjee is a DAE-Homi Bhabha Chair Professor at Bhabha Atomic Research Center (BARC), Mumbai, India. Presently he is also Chancellor of Homi Bhabha National Institute, DAE and Central University of Kashmir, Srinagar. He received his BTech (Metallurgical Engineering) from Indian Institute of Technology, Kharagpur, in 1967. He joined the erstwhile Metallurgy Division, BARC, Mumbai, in 1968 after graduating from the 11th batch of BARC Training School, Mumbai. He obtained his PhD in 1974 from IIT, Kharagpur. Dr. Banerjee occupied several important positions during his career spanning more than 40 years, such as Head, Metallurgy Division, BARC; Director, Materials Group, BARC; Director, BARC; and Chairman, Atomic Energy Commission. He has done pioneering work in the field of martensitic transformations, rapid solidification, omega transformation, quasicrystalline solids, and shape-memory alloys.

Dr. Banerjee was a senior visiting fellow at the University of Sussex, UK; Humboldt Foundation Fellow and Awardee at Max-Planck Institute for Metallforschung, Stuttgart, and Forschungszentrum, Juelich, Germany; and a visiting professor at Ohio State University, Columbus, USA. He currently holds the position of courtesy faculty as Professor of Materials Science and Engineering at the University of North Texas, Denton, USA.

In recognition of his seminal contribution to the field of materials science, Dr. Banerjee has been conferred with many prestigious national and international awards including the INSA Young Scientist Award (1976); National Metallurgists' Day Award (1981); Acta Metallurgica Outstanding Paper Award (1984); Shanti Swaroop Bhatnagar Prize in Engineering Sciences (1989); GD Birla Gold Medal of the Indian Institute of Metals (1997); INSA Prize for Materials Science; MRSI-Superconductivity and Material Science Prize (2003); Indian Nuclear Society Award (2003); Alexander von Humboldt Research Award (2004); Prof. Brahm Prakash Memorial Medal (2004) from INSA; Padma Shri from Government of India (2005); Distinguished Materials Scientist Award from MRSI (2008); Indian Science Congress Association's Excellence in Science and Technology Award (2009); Ram Mohun Puraskar of Rammohun Mission (2010); CNR Rao Prize in Advanced Materials (2011); M N Saha Birth Centenary Award from the Indian Science Congress Association (2012); Presidential Citation of American Nuclear Society (2012); Indian Institute of Metals Award (2012); W. J. Kroll Zirconium Medal Award from ASTM (2013); and Robert Cahn Memorial Award (2016). He has been conferred with Doctor of Science (Honoris Causa) degrees from 11 universities and institutions including University of Calcutta and Indian Institute of Technology, Kharagpur. He is an elected fellow of the Indian Academy of Sciences, National Academy of Sciences India, Indian National Science Academy, Indian National Academy of Engineering, Third World Academy of Sciences, and International Nuclear Energy Academy.

Preface

Today's advanced technologies demand performance of materials under extreme conditions, such as elevated temperatures, high pressure, hostile chemical environments, high radiation fields, and high magnetic and electric fields. Understanding materials' behavior under extreme conditions is therefore essential for designing high-tech equipment and processes. The relations of materials' behavior to the chemical bonding of its constituent elements, crystallography in different-length scales, are of great significance in understanding and even predicting the performance of materials under extreme conditions. There have been significant developments in a wide variety of engineering materials that include oxide and nonoxide ceramics, multicomponent alloys and intermetallics, glasses (both oxide and metallic), carbon-based materials, polymers, and composites. Several new processing techniques for design and fabrication of these materials have also evolved recently. Because of the diversity of these materials and their applications, it is difficult to obtain information from a single source. In view of this, it was felt worthwhile to compile a book on "Materials Under Extreme Conditions" to fill the gap in the literature.

This book provides a kaleidoscopic view of materials under extreme conditions. The book is subdivided into 22 chapters, and each chapter is dedicated to a particular class of materials or phenomenon. Each chapter has been appropriately structured to provide an essence of the physical basis of resistance to a given condition, materials synthesis and processing, characterization, properties, and applications. The concepts of crystallographic structures and chemical bonding in imparting the resistance to extreme conditions are discussed. The book can be viewed from two different angles, viz., from the specific service condition or the specific materials. Various types of extreme conditions, such as high pressure (both static and dynamic) to ultrahigh vacuum, high temperature, harsh chemically corrosive environments to high energy and intense radiation fields, and extremely high electric and magnetic fields, and their implications on materials response are discussed in different chapters. Two chapters are devoted to radiation-stable materials. In addition, there are some dedicated chapters for extreme conditions generated during a specific function, such as severe plastic deformation and intense laser irradiation. Dedicated chapters highlighting the noncoventional processing condition phenomena, such as natural glasses under harsh environments, clusters under extreme ionization conditions, and cold plasma processing, are also presented.

This book is aimed at providing a review of recent developments and the scope of future directions in research not only for beginners but also for practicing scientists and researchers in multidisciplinary areas, including physics, chemistry, material sciences, and engineering.

New concepts are introduced and the most recent literature is cited. The editors are immensely thankful to all the authors for their rich contributions, without which it was unthinkable to conceive this book. Due efforts have been taken to make the book as free of errors as possible. Despite this, some errors might have crept in. We shall be thankful to the vigilant readers for bringing such unintentional errors to our notice. We do hope that our efforts will be of use to both beginners and the experienced researchers in the field.

A.K. Tyagi
S. Banerjee

CHAPTER

1 MATERIAL STUDIES AT HIGH PRESSURE

Surinder M. Sharma, Nandini Garg
Bhabha Atomic Research Centre, Mumbai, India

CHAPTER OUTLINE

Materials Under Extreme Conditions. http://dx.doi.org/10.1016/B978-0-12-801300-7.00001-2

1. INTRODUCTION

Pressure and temperature are two important thermodynamic variables that can bring about several changes in condensed matter. Application of pressure brings about larger changes at very low pressures compared to temperatures as the change in Gibbs free energy of elements on application of pressure is ~10 kJ/mol/GPa, much larger compared to 0.1 kJ/mol/°C as a function of temperature [1]. (Commonly used units of pressure are related to each other as: atmospheric pressure: bar = 750.064 mm of Hg = 10^5 Pa = 0.986923 atm = 1 kg/cm^2, psi = 14.50373 bar, 1 GPa = 10 kbar and 1 Mbar = 100 GPa.) In nature pressures vary over almost 63 orders of magnitude, i.e., from that of hydrogen in intergalactic space (~10^{-32} bar) to that at the center of neutron stars (~10^{31} bar). Further, in nature pressures could be static like the pressure at the center of the earth or at the bottom of an ocean, or dynamic like that caused by meteoric impacts witnessed recently in Russia or an earthquake causing the disastrous tsunami. Study of matter at extreme conditions helps in understanding several natural phenomena, synthesize new materials, has applications for energy [2] and defense technologies, provides insights into reaction chemistry of energetic materials, etc. Therefore high pressure investigations span a broad range of scientific disciplines (e.g., material science chemistry, physics, biology, and earth and planetary sciences, etc.).

When a material is compressed, its constituent atoms are brought closer, causing an increase in the overlap of electronic orbitals, resulting in changes in the hybridization of the outer electronic orbitals, bringing about the modifications in the nature of interatomic bonding and band structure. Significant increase in overlap of orbitals also manifests itself in terms of stronger repulsive forces, particularly between the nonbonded atoms. These changes propel the materials to evolve to new states having different physical properties, including structure, modifications occurring continuously or discontinuously. Understanding these alterations at extreme conditions provides insight into the nature of phase transformations, molecular to nonmolecular transitions [3], crystal to amorphous transitions [4], amorphous to amorphous transitions [5], chemical changes [6], increasing/lowering of melting temperature [7], insulator to metal transitions [8], valence transitions [9], Lifshitz transitions [10], changes in the dielectric function [11], refractive index [12–14] determination of the equation of state etc. [15,16] High-pressure studies also help to understand how the micro- and nanostructural aspects, such as crystallite size, dislocations, voids, and grain boundaries evolve on application of pressure [17].

Knowledge of high-pressure behavior of minerals like zircon can be very useful in determining peak pressures of meteoric impacts [18]. In fact, high-pressure studies of many minerals like pyrochlores, zircons, or glasses can throw light on their usefulness as nuclear encapsulants [19]. Application of pressure has also found its use in the chemical industry as there are several reactions that are accelerated on application of pressure, for example, the production of ammonia by the Habers process. Pressure plays an important role even in the food preservation industry and the health sector. Calcium hydroxyapatite is an important bioceramic and is the main constituent of teeth and bone tissues of human beings. Also, monocalcium monohydrate is widely used in bone and teeth cements, and it is known that masticatory stresses can be quite high. High-pressure studies on these cements and bioceramics help the understanding of their hardness, elastic moduli, and behavior under shear stresses, etc. [20,21].

By varying the temperature and pressure simultaneously it is possible to simulate the conditions deep down in the earth or outer planet/super-earth interiors, or changes on the earth's surface, due to meteor impacts. Thus, knowledge of the myriad changes that take place in materials under these conditions helps unravel the mysteries of the planetary interiors or events that took place several millions of years ago. The pressure on the surface of the earth is just 10^{-4} GPa and increases at the rate of 1 GPa/30 Km toward the center of earth where the pressure is ~350 GPa. The formation of postperovskite phase in $MgSiO_3$ at 125 GPa, 2500K [22], and the iron-rich postperovskite phase at 130 GPa, 3000K has helped understanding of the seismic ultralow velocity zones of earth's core-mantle boundary [23]. Tateno et al. have shown that iron that exists in the bcc phase at ambient conditions transforms to the hcp phase and is elastically highly anisotropic at 377 GPa, 5700K, the conditions corresponding to the inner core of our earth. These studies could hold the key to the understanding of the seismic anisotropies of our earth [24]. Studies of the rheology of geophysical systems as a function of pressure can help in understanding the deep earthquakes [25]. Recent laser shock studies on the insulator magnesium oxide, an important constituent of the earth's mantle, have shown for the first time that it transforms to a new high-pressure (CsCl) phase at 0.36 TPa and 50,000K, before melting to a metallic liquid at 0.6 TPa. Since these temperature and pressure conditions may exist in the deep mantles of super-earth planets, MgO may exist in the CsCl form or as a metallic liquid on these planets. This shows that even the oxide-rich regions inside terrestrial planets could be electrically conductive enabling a magnetic field producing dynamo action [26]. Knowledge about the carbon to diamond transformation at high temperature and pressures, and the recent laser shock experiments on diamond that showed that it was stable until ~9000K and 600−1100 GPa [27], suggest that carbon in the diamond form could be present deep inside the giant planets like Uranus and Neptune. Several studies have also indicated that a considerable mass of carbon-rich planets with lower concentrations of oxygen could be made up of diamonds [28,29].

Studies at high pressure and low temperature have helped in finding new superconductors [30] with higher T_c or for studying magnetic ordering in materials [31]. The highest T_c cuprate superconductor ($HgBa_2Ca_2Cu_3O_8$, $T_c = 135K$) found so far has been obtained at a pressure of 30 GPa [32]. However, recently Drozdov et al. have shown that a noncuprate material like sulfur hydride superconducts at $T_c = 203K$, 90 GPa [33]. CsI, an ionic insulator at ambient conditions, has been shown to metalize at 115 GPa and eventually superconduct at 2K, 180 GPa [34]. It is interesting to observe that though alpha uranium and the high-pressure orthorhombic phase of cerium have similar crystal structure but different electronic configuration (one is 5f and the other is 4f), their high-pressure behavior is strikingly similar, with cerium metal showing a strong electron-phonon coupling associated with the superconducting transition at 4 GPa [35]. Though there are several properties that are enhanced on application of pressure, there are others like ferroelectricity, ferromagnetism, superconductivity in conventional elemental superconductors, charge density waves, etc. that are suppressed. However, recent computations and experiments have shown that lead titanate displays reentrant ferroelectricity at high pressures, which is quantum in nature. Kornev et al. have explained that the overlap of titanium and oxygen orbitals is energetically more demanding and hence to reduce this overlap the ferroelectric behavior sets in at high pressure [36,37]. Recent experiments by Sun et al. have reported the reemergence of superconductivity in FeAs at 12.5 GPa accompanied with an increase in the T_c to 48K [38]. This reemergence has not yet been fully understood, and these investigators believe that it could be associated with quantum critical transition, magnetism, superstructure, vacancy ordering, or spin fluctuation. Application of pressure could lead to a transfer of electrons from valence to conduction band or within the conduction band itself leading to a change in the physical properties or pressure-induced valence transitions. For example, the increase of interband absorption in rubidium and cesium at high pressure was found to be due to the s-d transfer of the conduction electrons [9] and the valence discontinuity and closing of gap of Yb and Sm monochalcogenides at high pressure was due to the filling of the lowest unoccupied $5d\ t_{2g}$ orbital by the 4f electrons [39]. Different gases have been shown to exhibit either an insulator to metal transition (oxygen, xenon, iodine) [40–42] or molecular to nonmolecular transition (carbon dioxide and nitrogen) [43,44] at high pressure. Recently, based on experimental evidence [8], xenon fluoride was termed as a novel energy material as it undergoes both insulator to metal and molecular to nonmolecular transition at high pressure resulting in a large amount of reversible conversion of mechanical to chemical energy. However, subsequent theoretical calculations do not support the transition to a nonmolecular solid and rather suggest a self-dissociation to an ionic solid [45]. Theoretical calculations have predicted that between 200 and 400 GPa molecular hydrogen with paired protons would become conducting whereas between 400 and 600 GPa it would undergo a dissociative transition to a metallic phase [46–48]. Shock experiments by Nellis et al. [49] have shown that hydrogen transforms to a metallic diatomic fluid at 140 GPa, 3000K. However, though recently Eremet et al. [50] claimed to have formed metallic hydrogen in the diamond anvil cell at 260 GPa, several other investigators do not agree with their results and hence hydrogen metallization yet remains to be unambiguously observed in the static high-pressure experiments [51–53].

Most of industrial or laboratory-based synthetic chemistry occurs in the condensed phase. Extreme P-T conditions can alter the kinetics, chemical reactivity, elemental partitioning, diffusion coefficients, mechanical/chemical properties of condensed matter. In fact, even for understanding the behavior of energetic materials under impact and detonation conditions it is important to understand their reaction chemistry at high pressure and high temperature (HPHT). Knowledge of equation of state, kinetics, and thermodynamic phase diagram at HPHT can help synthesize new and exotic materials. This can improve the reaction yield or can also help in lowering the synthesis temperature of materials. In addition, many metastable phases can be obtained under HPHT condition. In the early 1950s and 1960s Bundy et al. demonstrated the synthesis of diamond and B28 (a new phase of boron) [54–56] at high pressure and high temperature in a large volume press. Subsequently, graphite to diamond transformation has also been observed under dynamic pressures. It has also been claimed that methane can be synthesized in the diamond anvil cell using FeO, $CaCO_3$ (calcite), and water at pressures between 5 and 11 GPa and temperatures ranging from 500 to 1500°C, which are the pressures and temperatures found in the upper mantle region of earth [57–59]. Recently Zhang et al. have shown for the first time that sodium chloride, which is known to exist with the rock salt structure at ambient conditions, can form compounds with different stoichiometries at high pressure, high temperature [6]. A unique carbon-substituted cristobalite form of silica ($C_{0.6(1)}Si_{0.4(1)}O_2$) has been synthesized by Santoro et al. at $P = 16–22$ GPa, $T > 4000$K by reacting carbon dioxide and silica [60]. This opens up a new oxide chemistry at high pressure and high temperature where carbon can replace some of the silicon atoms in silicates thus making the synthesis of new stronger and lighter materials possible.

Nowadays many scientists are involved in the HPHT synthesis of novel hydrogen storage materials, super hard materials, high T_c superconductors, etc. [61–63] Cerium and aluminum have been shown to form a substitutional alloy at high pressure, despite the fact that the atomic radii and electronegativities of both these metals do not satisfy the Hume Rothery rules for alloy formation at ambient conditions [64]. Cubic boron nitride, which finds wide industrial usage owing to its high thermal and chemical stability (bulk modulus comparable to diamond but hardness half that of diamond) is routinely synthesized at HPHT [65]. High-purity bulk nanostructured cubic boron nitride having a hardness almost double that of bulk cubic boron nitride has been synthesized at 20 GPa and 1770K from graphite-like BN [66]. Sumiya et al. synthesized nanopolycrystalline diamond from graphite at 15 GPa, 2300°C and showed that it was harder than diamond [67]. However, recently Tian et al. have shown that a non-diamond material, i.e., nanotwinned cubic boron nitride, which was synthesized from nanospheres of orthorhombic cubic boron nitride at high pressure, high temperature, is harder than diamond [68].

There have been several studies [69,70] of high-pressure-initiated photodisassociation and chemical reactions. In fact, many of these studies have shown that photoactivated molecular species act as a catalyst for several reactions and trigger them at much lower pressure and temperature conditions, attainable even with the large volume presses. At ambient pressure 10.5 eV energy is required to dissociate the triple bond in N_2 and CO, whereas in the presence of water and on application of pressures <0.5 GPa a mere 3.5 eV energy is sufficient to trigger the dissociation. This has been explained to be due to the dissociation of water into hydrogen and active OH radical [71]. Cepattali et al. have shown that ethanol, which is stable up to 1.7 GPa, dissociates into hydrogen, ethane, and methane at pressures as low as 7 MPa when irradiated with 350 nm radiation. They showed that both dissociation and its products depend on the pressure of the experiment and the wavelength of radiation [72]. Recently the same group has also shown that less reactive red phosphorous reacts with water and forms reaction products at 0.6 GPa when exposed to near UV radiation [73].

It has been observed that as material grain sizes are decreased to the nanometer range, hardness increases initially but starts decreasing below a critical size of 10–20 nm. It has been proposed that at the onset of the inverse Hall–Petch effect, there is a crossover from dislocation-mediated plastic deformation to grain boundary sliding, grain rotation, or dislocations created at the grain boundaries [74,75]. Recent nonhydrostatic radial X-ray diffraction studies on nanocrystalline nickel have shown that at high pressure even 3 nm particles develop texture due to preferential deformation on slip planes leading to crystal rotations, indicating that dislocation glides are still active. These studies suggest that the deformation crossover at the onset of the Hall–Petch effect may shift to smaller particle sizes on application of pressure [17].

It is not necessary that only very high pressures can bring about changes in materials. In soft matter macromolecular assemblies that include proteins, biomembranes, surfactants, lipids, etc., a very small pressure of the order of a few kbar can bring about changes in the ensemble of conformational substates of a macromolecular soft matter assembly [76]. It was shown as early as 1914 by Bridgman that a pressure of 5 kbar could coagulate egg white [77]. There are several processes where the application of a few bars of pressure increases the rate of reaction or increases the reaction yield. For example, pressure enhances decarboxylation, lactonization, and hydration reactions. It has also been observed that subjecting pyruvic acid to high pressure enhances the yield of the cyclic aromatic compounds [78]. Since enzyme activity can be controlled with pressure, enzyme catalysis at high pressure can be used to synthesize fine chemicals, pharmaceuticals, or to produce modified proteins of medical or pharmaceutical interest. Inactivation of several bacterial strains, viruses, etc. at high pressure can be used to sterilize fragile biopharmaceuticals or medical compounds that are sensitive to heat. Antigenic sites of proteins, viruses, and killed bacteria get unmasked due to protein unfolding at high pressure. This enhances their immunogenic properties thus making them useful for preparation of new vaccines [79]. Since water freezes at ~ −20°C at 2 kbar pressure (critical point), cells, animal tissues, blood cells, organs for transplant, and so forth can be stored at subzero temperatures without freezing [80–84]. As in the pressure range of biotechnological interest, pressure has, in general, no effect on covalent bonds, and several natural compounds such as flavors, aromas, dyes, or pharmacologically active molecules do not get destroyed on application of pressure. However, as pressure results in inactivation of pathogens, spoilage microorganisms, and food-spoiling enzymes, high-pressure processing is widely used for food processing and preservation. It has been demonstrated that the shelf life of milk can be increased on application of pressure. It has been observed that even at pressures ~ 6–8 kbar several provitamins and vitamins like β-carotene, thiamin, riboflavin, folic acid, ascorbate, etc. are stable. Several studies on biomolecules have shown that proteins are the most sensitive to high pressure [79].

Since food is packaged and then subjected to pressure for preservation it is necessary to understand the high-pressure behavior of the packaging material also. High-pressure studies on one such biodegradable polymer, i.e., polyactide, suggest that these polymers can be used for the development of food packaging material [85].

2. EXPERIMENTAL AND THEORETICAL METHODS TO STUDY MATERIALS AT HIGH PRESSURE

The nature of the pressure response depends upon the peak pressures as well as the strain rates. The strain rates can be as slow as 10^{-17}–10^{-14} s^{-1} (in the case of movement of tectonic plates inside the earth) or as rapid as 10^{5}–10^{7} s^{-1} (materials under shock). Several experimental and theoretical methods have been developed to study the behavior of materials on application of both static and dynamic pressures. For theoretical understanding several methods can be used, such as classical molecular dynamics (CMD), ab initio molecular dynamics (AIMD), Monte Carlo calculations, local density approximation (LDA) based various first principles methods, etc. These theoretical methods are useful when one wants to study the changes in materials at the atomistic level, or for prediction of new phases in different materials at high pressures or temperatures. In many experiments the exact pressure temperature conditions are very critical and are difficult to attain in the laboratory. In such cases also it is possible to model these experiments at extreme conditions and carry out theoretical calculations including CMD or ab initio MD to obtain even the detailed trajectories of the atoms as the new states evolve.

On the experimental side, there are several techniques that can be used to achieve the required pressures and strain rates between 10^{-4} and 10^{-2} s^{-1} (in the case of static pressures) or 10^{4}–10^{11} s^{-1} (in the case of dynamic pressures). Recently there have been efforts to extend the studies to the unexplored 10^{-2}–10^{4} s^{-1} strain rate regime, by the use of a dynamic diamond anvil cell [86]. These authors have shown that water that has a solid liquid phase line at 0.9 GPa can be supercompressed to 1.8 GPa using the dynamic diamond anvil cells (DAC). Dynamic pressures can be achieved with the help of a high-speed projectile from a gas gun, explosives, or with the help of lasers. With two-stage gas guns the maximum pressure that can be reached is ~700 GPa. However, with the use of pulsed high-power lasers the pressures can be in the TPa regime and the temperatures can reach 100,000K. Static high pressures are routinely generated with the help of DAC or large volume presses. Though the pressure limits of large volume presses is ~30 GPa, with tungsten carbide as second-stage anvils [87], the pressure limit can be raised to 100 GPa with sintered diamond anvils [88]. In the DAC pressures up to ~100 GPa are easily and routinely achievable in most of the laboratories across the world. With appropriate modification (beveling) of anvils, pressures ~350 GPa have been achieved. However, recent experiments by Dubrovinsky et al. using microball nanodiamond anvils have pushed the static pressures attainable in the laboratory to ~ 1 TPa [89,90a, 90b]. Temperatures up to 6000K can be attained by heating with continuous-wave lasers. However, recent advancements in pulse laser heating technique has enabled heating to temperatures greater than 12,000K in the DAC [91]. Generally the high temperature facilitates diffusion, but the short time of the laser pulse helps to restrict the diffusion of mobile gases like hydrogen. Also, the short duration of the pulse prevents heating of the gasket and parts of the cell, further reducing diffusion into the anvils and gasket, helping containment and experiments on chemically reactive and mobile gases [92].

2.1 DIAMOND ANVIL CELLS

In the rest of this article the primary focus will be on studies of materials under static high pressures. Most of these studies employ DAC-based techniques. A DAC is a simple device, with which we can compress a tiny sample (~100 μg or less) between the culets (ϕ ~ 300–400 μm) of two gem-quality diamonds. A thin gasket (~200–300 μm thick), typically formed of metals and alloys, is preindented with the anvils and then a small hole is made at the center of the diamond indentation, as shown schematically in Fig. 1.1. The small hole in the gasket between the diamond anvil surfaces serves as the sample chamber, and, for high-pressure experiments it is filled with the sample material to be studied, the pressure marker, and a fluid medium for applying hydrostatic pressure around the sample. Force is applied to the piston diamond along the optical axis so that the diamonds are squeezed together against the gasket, sealing the sample chamber. The sample is suspended in the medium and does not directly bridge diamonds. Diamonds make ideal anvils for achieving high pressures, not only because it is the hardest material known with very high thermal conductivity but also because it is an excellent window for observing the sample in situ. Except for the two-phonon infrared absorption band at about 0.2–0.5 eV and interband electronic transitions in the soft X-ray region, the diamonds are transparent to electromagnetic radiation extending from the far infrared to X-ray and γ-ray energies [93–96]. And finally, there is the advantage that the high-pressure diamond-window (or diamond-anvil) cell is very small, typically weighing no more than a few kilograms even for the highest-pressure models.

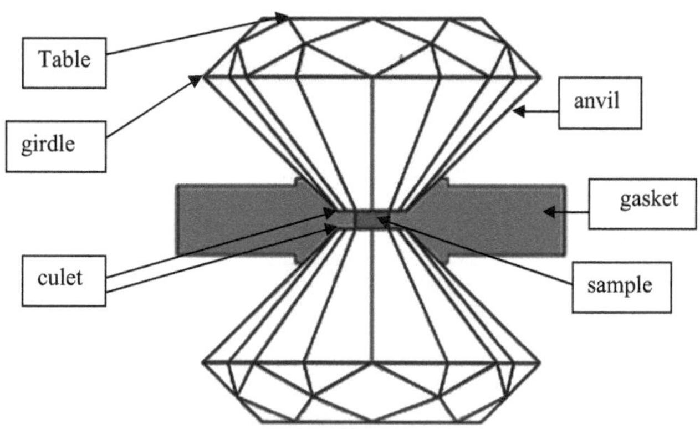

FIGURE 1.1

A schematic of the diamond anvils, as used in a diamond anvil cell.

Therefore, a sample can be examined by many techniques while under pressure, and can be easily transported from a spectrometer in one laboratory to an X-ray source in another. Several diamond anvil cells (shown in Fig. 1.2) have been developed by us to carry out high-pressure, powder, and single crystal diffraction studies; IR studies; and low-temperature high-pressure studies. These cells are capable of achieving pressures of ~100 GPa. For single-crystal studies and also for increasing the Q range of X-ray diffraction measurements, some of the cells are equipped with beryllium or cubic boron–nitride backing plates, the former fabricated in-house. Apart from this we have also developed diamond anvil cells for transport properties measurements, low-temperature magnetic studies, and Brillouin scattering studies.

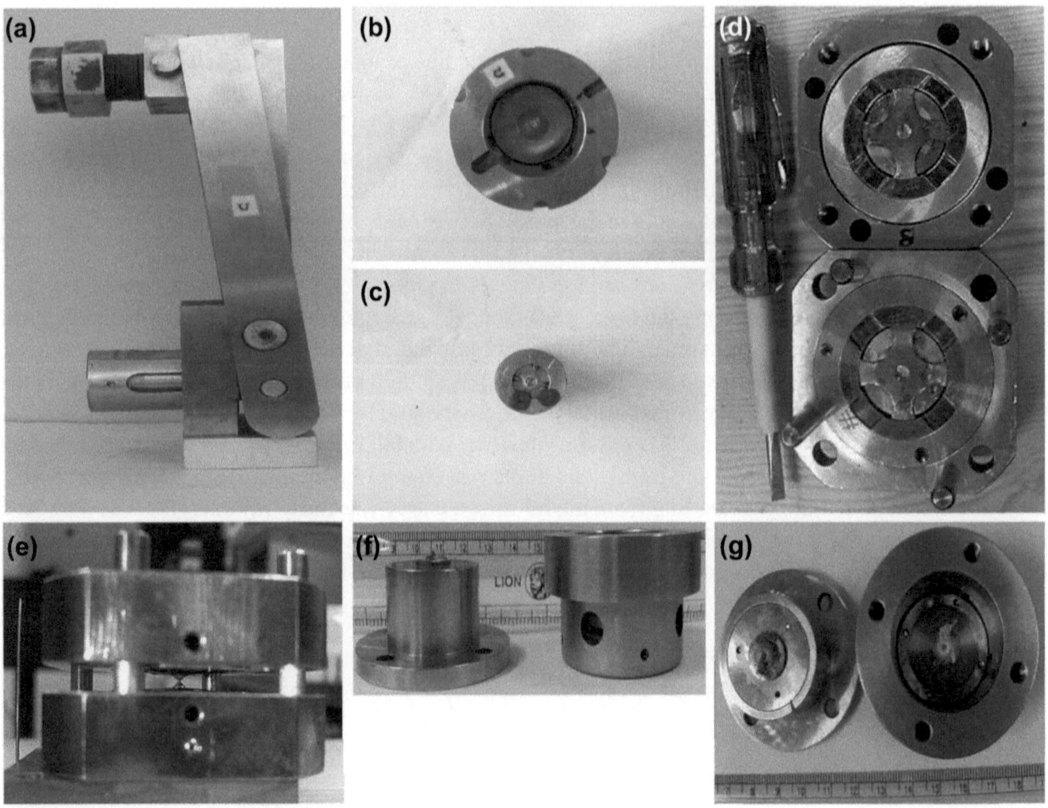

FIGURE 1.2

The different kinds of diamond anvil cells developed in our laboratory to carry out high-pressure X-ray diffraction studies. (a) Side view of a modified Mao Bell (MMB) kind of cell with the lever arm; (b) cylinder of the MMB cell; (c) piston of the MMB cell; (d) diamond anvil cell with beryllium/cubic boron–nitride backing plates for single crystal X-ray diffraction with (e) as the side view of this cell; (f) side view of a new Mao Bell kind of cell equipped with a beryllium/cubic boron–nitride backing plate; and (g) showing piston and cylinder of this cell.

2.2 PREPARATION OF DIAMOND ANVIL CELLS FOR EXPERIMENTS

While preparing the samples for carrying out the high-pressure experiments, it has been observed that size of the samples, or the method in which they are prepared, may also affect the sequence of structural changes under high pressures. While in some cases the phenomenon may depend upon the particle size [97,98] or shape [99], in others the process of reduction of particle size by grinding or mechanical milling, etc. may also cause structural changes (e.g., silver cobalt cyanide, alpha quartz, Ti, etc.) [100–102].

As mentioned before, a pressure transmitting fluid is used in DAC-based experiments so that the sample is preferably under hydrostatic pressure or at least under homogeneous stress, free of any differential stress and consequent shear strain. To achieve this, the pressure transmitting media are so chosen that they ideally exhibit hydrostatic behavior, i.e., have no shear strength. Apart from this the medium should not dissolve the sample being studied. In the case of microporous or mesoporous inorganic samples like zeolites, it should not enter into the tunnels or structural cages of the sample, as otherwise it would modify their high-pressure behavior [103–105]. These considerations have led to the usage of a wide range of media like soft solids, condensed gases, mixture of alcohols, fluorocarbons, and inert silicone oils.

Inhomogeneous strains due to nonhydrostatic stresses result in broadening of characteristic (diffraction or spectroscopic) peaks from the sample. This results in a reduction of the signal-to-noise ratio of the measured signal. Nonhydrostatic media can also modify the relative evolution of the cell parameters of the crystalline samples with pressure [106]. Nonhydrostatic stresses can adversely affect the results of equation of state or elasticity studies. By coupling to the spontaneous strain, these stresses can also promote or suppress phase transitions [107,108] and/or lead to amorphization of crystalline samples [109,110]. Therefore the decision about the pressure transmitting medium must be made on the bases of requirements of the experiments and also on the tolerable pressure gradients of the measured quantity.

It has been observed that mixtures of liquids can be taken to higher pressures than the solidification pressures of individual liquids. The viscosity of these fluids rises with pressure, and eventually many liquids transform to a glassy state above a critical pressure. It has been recently observed that the amorphous phase formed in 4:1 methanol-ethanol and 16:3:1 methanol-ethanol-water are quite stiff above 10 GPa having shear wave velocities of 3.3 km/s at ~1 GHz [111]. Klotz et al. based on measurement of pressure gradient from several ruby spheres, have shown that in terms of requisites of pressure media, nitrogen is comparable to neon until 10 GPa, and in principle Argon can also be used up to ~18 GPa [112]. In general, they have shown that for the medium not to affect the results significantly the pressure gradient (ΔP) should be smaller than $B\varepsilon$ where **B** is the bulk modulus and ε is the resolution of the experimental technique used in measuring any property. The pressure limits of hydrostaticity of several media are given in Table 1.1. Liquid and solid media can be loaded in the sample chamber at ambient conditions. However, gases need either cryogenic loading or compressed gas loading, using specialized apparatus.

Accurate determination of pressure is one of the most important issues in high-pressure experiments. Pressure can be determined using the absolute/primary scale or the secondary scale. Usually the primary pressure scale is used to generate the secondary scale, which is used for determination of the pressure in DAC-based experiments. The absolute scale can be determined using the piston cylinder [124], hugoniot of materials determined by carrying out shock wave measurements [125,126], or by

Table 1.1 Some Pressure Transmitting Media and Their Pressure Ranges

Medium	Hydrostatic Limit (GPa)	Pressure Range up to Which Behavior is Nearly Hydrostatic	References
Methanol:ethanol (4:1)	10.5	20	Klotz et al. [112]
	9.8		Angel et al. [111]
	10.4		Piermarini et al. [113]
Methanol:ethanol:water (16:3:1)	10.5	20	Klotz et al. [112]
	10.5		Angel et al. [111]
	14.5		Fujishiro et al. [114]
Isopropyl alcohol	4.2		Angel et al. [111]
	4.3		Piermarini et al. [113]
Glycerine:Water (3:2)	5.3		Sidorov et al. [115]
Petroleum ether	6		Barnet and Bosco [116]
Pentane:isopentane (1:1)	7.4		Piermarini et al. [113]
Pentane:isopentane (1:5)	7.5		Klotz et al. [112]
Methanol	8.6		Piermarini et al. [113]
Glycerol	1.4		Angel et al. [111]
Silicone oil, viscosity 0.65 cSt, 37 cSt	0.9		Angel et al. [111]
Polydimethyl-siloxane oil (Rhodorsil 47V1000) Viscosity 0.97 Pa s at 25°C	2.5		Klotz et al. [112]
FC84-FC87 Fluorinert (1:1) [(C8 F18)n (C8 F16 O)m]	2.3		Klotz et al. [112]
Daphne 7474 (2,2,8,8 tetra-alkylsilane and silicone oil) (Mostly used for magnetic measurements)	4 (20°) 6.5 (100°)		Klotz et al. [112]
Argon	2.0	18	Klotz et al. [112]
	1.9	1.9	Angel et al. [111]
	1.2	9	Bell and Mao [117]
Nitrogen	12		Klotz et al. [112]
	3.0	3.0	Angel et al. [111]
	2.4	13	Le Sar et al. [118]
Hydrogen	5.7	177	Mao and Bell [119]
Deuterium	5.3		Mao and Bell [119]
Neon	15	20	Klotz et al. [112]
	4.7	16	Bell and Mao [117]
Xenon		30	Liebenberg [120]
		55	Asaumi and Ruoff [121]
Helium	23		Klotz et al. [112]
	11.8	60–70	Bell and Mao [117]
			Eremets [122]
Oxygen	5.9		Nicol et al. [123]

direct measurements of elasticity and density [127,128]. In the megabar region the primary scale has been established by determining the shock hugoniot of Au, Ag, W, etc. With this scale we get an accuracy of 1% in the measurement [126] of pressure. However, since a shocked sample is compressed uniaxially and numerous defects could be introduced into the sample, the comparison of a shock compressed and a hydrostatically compressed sample is not ideal. The accuracy can be improved to better than 1% by using the absolute pressure scale developed by Smith et al. [129] and further extended to higher pressures by several other experimentalists [130−132]. This formulation does not need shock wave data and uses a combination of volume V and compressibility to extract the true thermodynamic pressure by integration of the isothermal bulk modulus determined accurately by ultrasonic measurements, Brillouin scattering, or laser-induced phonon spectroscopy (also called impulsive stimulated scattering). Recently Holzapfel has also shown that thermodynamically deduced isotherms determined theoretically or using the above methods give a better equation of state (EOS) when compared to the shock wave−reduced isotherms (SWRI) [133]. He has argued that EOS determined using SWRI does not take into account the anharmonicity, dispersion, and temperature dependence in the mode-Grueneisen parameters, creation of defects, melting and correspondingly lower temperatures in stronger shocks, differences in the models required for the different regions along the hugoniot curve corresponding to solid states, mixed regions, and fluid final states and corrections for deviatoric stresses [133]. Subsequent to this study, Jin et al. have proposed an EOS of gold and platinum up to 300 GPa, 3000K based on experimental shock Hugoniot and thermodynamic data [134].

Secondary pressure standards can be continuous or fixed. They are obtained by calibrating any pressure-dependent variable with the primary pressure scale. The fixed standards are usually used to calibrate the large volume presses. The reproducible phase transitions, of materials like bismuth, barium, etc., are used to construct a practical pressure scale. Large changes in electrical resistance or volume at the phase transitions can be measured and hence can be used to give the pressure. In IR absorption experiments the pressure behavior of vibrational modes of quartz [135], $BaSO_4$ [136], or the antisymmetric N-O stretch of nitrate in sodium bromide matrix [137] are also used as a secondary pressure standard. For Raman studies, the pressure-induced shift of 1331 cm^{-1} Raman active mode measured from the tip of the diamond could be used [138]. A Raman shift from a chip of isotopically enriched C_{13} has also been used as a pressure marker [139,140]. The most commonly used continuous secondary standards in the diamond anvil cells are the "fluorescence scale" (mostly of ruby) or the equation of state of some materials like Au, Pt, Ag, NaCl, etc. The choice of pressure standard is based on the requirements of an experiment. For example, if we are using a large volume press the ruby fluorescence technique cannot be used, whereas for optical measurements the diamond anvil cell fluorescence techniques would be the best option.

Though there are calibrants like rare earth element doped materials that show relatively higher spectral changes with pressure and are relatively more stable than ruby at high temperatures, the most popular fluorescence sensor is ruby (Cr^{3+} doped $\alpha{-}Al_2O_3$). Ruby has a strong luminescence doublet at $R_1 = 694.2$ nm and $R_2 = 692.8$ nm, which red shift on application of pressure [141−144]. The $R_1{-}R_2$ line separation also allows us to know whether the sample environment is hydrostatic or non-hydrostatic. $R_1{-}R_2$ separation increases for uniaxial strain along **a**-axis of ruby while it decreases for the strain along the **c**-axis [145]. Calibration of this standard with the equation of state of NaCl showed that it had a linear dependence of 0.365 nm/GPa. However, later this pressure scale was extended to the megabar region by Mao et al. and others [146−153] and very recently it was further extended to

300 GPa by Kishchenko et al. [154] The latest calibration represents the P(λ) dependence from 0 to 300 GPa by a single equation and has been shown to be consistent with room temperature isotherms of Al, Au, Cu, Pt, Ta, and W and does not contradict the recent EOSs of diamond [155]. Pressure measurements with the ruby scale can be monitored to within 10 MPa [112]. However, recently Doherty et al. have demonstrated that using the optical and spin resonance of the negatively charged vacancy of diamond pressure changes within 1 MPa can be monitored [156].

Apart from these standards, many times when samples get opaque at high pressures the first-order Raman spectra of diamond anvils can be used as an indicator of the peak pressures. Akahama et al. have calibrated the high-frequency edge of the Raman band of the diamond anvil up to pressures of 310 GPa with the equation of state of platinum [138]. The high-frequency edge essentially gives the pressure at the center of the culet. Recently this method of pressure determination has been extended to high-temperature, high-pressure studies (800K, 106 GPa) [157]. For X-ray diffraction studies in situ pressure measurements can be made by using internal diffraction standards mentioned above. With the help of finite element calculations, Conil et al. have shown that the Poisson's ratio of the internal pressure standards should be ~0.4 and their Young's modulus should be ~12% less than that of the sample [158]. If the difference in the Young's modulus of the sample and the pressure standard is larger than 12%, the difference in pressure on the sample and the pressure standard is large at high pressures. There are several standards like CsCl, NaCl [159,160], Al, Au, Pt [161], etc. However, the most commonly used pressure standards for ultrahigh-pressure studies are Au, Pt, Cu, etc.

3. DIAGNOSTICS

Once a sample is loaded in a diamond anvil cell there are several in situ measurements that can be carried out. These include transport property measurements, spectroscopic measurements like IR, optical, or X-ray absorption; Raman, Brillouin, or X-ray inelastic scattering; Mossabauer, X-ray diffraction from powder or single crystals, etc. Due to the excellent thermal stability of the diamond anvils, temperatures also can be varied from mili-kelvin to several kilo-kelvin. Since the sample size used in the diamond anvil cell is ~picoliter, intense sources are required for carrying out experiments at high pressure, such as lasers and synchrotrons. For example, Raman measurements are carried out using confocal micro-Raman (spot size a few microns), and most of the X-ray diffraction investigations at synchrotrons employ X-rays, focused down to a few microns. However, with the availability of third-generation synchrotron sources one can focus the X-rays down to <5 nm [162,163]. Thus in the diamond anvil cell, a very tiny amount of sample can be subjected to pressures >100 GPa range, and subtle details in the structures like weak superstructures etc. can be reliably obtained.

3.1 HIGH-PRESSURE DIFFRACTION

Diffraction studies can be carried out using both X-rays and neutrons on single crystals as well as powders. These studies help in identifying structural changes across the phase transformations in crystalline and amorphous materials, determining the atomic positions in a crystal or the radial distribution functions in a glass. For crystalline solids one can determine the unit cell precisely and hence the comprehensive thermodynamic equation of state. Single-crystal diffraction can help obtain the structure more accurately. However, since a large portion of the reciprocal space is not accessible due to the geometrical constraints enforced by the shape of DACs, analysis of high-pressure diffraction data is somewhat nontrivial.

Neutron measurements complement the X-ray studies particularly for low Z elements and for magnetic materials. However, the flux of the neutron sources is several orders of magnitude smaller than the available flux of photons from the third-generation X-ray sources. Until the last decade, high-pressure neutron diffraction studies were limited to ~30 GPa using miniature Paris Edinburgh cell and up to 50 GPa using the small Kurchatov LLB cells [164]. However, Tulk et al. have recently studied ice VII up to pressures of 80 GPa using the high- pressure diffractometer SNAP at the Oakridge neutron spallation source [165]. The high-pressure studies in these small cells have been made possible by focusing the neutrons with the help of the Ni-Ti super-mirrors [166]. However, limited flux available and the low count rates resulting from small samples have prevented detailed structural studies at very high pressures and most of the studies at pressures larger than 25 GPa have mainly concentrated on identification of the magnetic ordering in different structures.

3.2 HYDROSTATIC AND NONHYDROSTATIC STRESSES AND RADIAL X-RAY DIFFRACTION

A material responds elastically until the applied stress exceeds the yield stress, beyond which it yields to a plastic flow. It is not necessary that a plastic flow is accompanied by a phase transition. However, elastic instability can lead to phase transitions like melting, amorphization, or polymorphism. It has been shown recently that all different formulations for the stability criterion of a crystal lattice essentially stem from the fact that the internal free energy of the crystal should display convexity on application of an external load [167]. Since elastic constants also get modified on application of pressure, mechanical stability of crystals at high pressure is governed by the modified Born elastic stability criteria based on a combination of elastic stiffness constants, representing the bulk, shear and spinodal instability. Elastic instability at high pressures could also be due to nonhydrostatic stresses. During nonhydrostatic application the nondiagonal components of the stress matrix, i.e., the shear components come into play. Deviatoric stress tensor is obtained by subtracting the hydrostatic stress ($1/3\ (\Sigma\sigma_{ij}\ \delta_{ij})$ where σ_{ij} are the components of the stress tensor) from the total stress. The shear strength and yield stress of a material is decided by the off-diagonal components of the stress tensor. In the diamond anvil cell the extent of nonhydrostaticity can be determined with the help of radial X-ray diffraction. The relationship between elasticity and plasticity of a material can be determined from their Pughs ratio (shear modulus (G)/bulk modulus (B)). Brittleness and hardness of a material are proportional to the Pughs ratio and ductility and flow are inversely proportional to this ratio. From this ratio it can be seen that a high bulk modulus alone does not ensure that the material is hard [168].

3.3 HIGH-PRESSURE SPECTROSCOPY AND INELASTIC SCATTERING

The dynamics of atomic motion of matter can be studied by inelastic scattering techniques. Inelastic optical (Raman, Brillouin), neutron. and X-ray scattering probe the elementary excitations such as, phonons magnons and plasmons. With inelastic X-ray scattering, apart from the advantage of being able to use small samples for high-pressure studies one can also determine the dispersion relation of the acoustic modes, which are difficult or not possible with neutron inelastic scattering. However, until a few years ago these studies were very few, as phonon excitations are of the order of a few meV and the probe photon energies are of the order of KeV. So the resolution required for the detection of inelastic X-ray scattering is of the order of $\Delta E/E = 10^{-7}$. But with the advent of third-generation high-flux synchrotrons, microfocusing optics and high-resolution energy analyzers, inelastic X-ray scattering (IXS) measurements are now being routinely carried out. In a recent inelastic scattering study on cerium, Krisch et al. have determined the phonon dispersion curve across the famous isostructural γ to α transition in cerium [169]. By comparing their dispersion curves with those determined from calculations, they have shown that ab initio calculations in the γ phase should treat the 4f electrons as localized whereas in the α phase they should be treated as a part of the valence band [169].

X-ray absorption or emission spectroscopy like XAS, XES, XANES, and EXAFS gives information regarding the structural and the magnetic interactions of the atom with its neighbors in the vicinity of the absorption edge. This is especially very effective for glasses and disordered solids where X-ray diffraction may not give a clear picture about the coordination environment of different species in a material. XMCD gives spin-resolved electronic density of states and local magnetic structure information. Recent XMCD studies on colossal magnetoresistance manganites have shown that it undergoes an orbitally disordered ferromagnetic to an orbitally ordered antiferromagnetic phase transition at 23 GPa [170].

3.4 HIGH-PRESSURE TRANSPORT PROPERTY MEASUREMENTS

Measurements of electrical resistivity and thermoelectric power using conventional four-probe method have helped to discover a variety of phenomena up to a few tens of GPa over several decades. With the advent of the designer anvils resistivity studies have been successfully carried out up to pressures of 180 GPa [171]. There are specially designed nonmagnetic diamond anvil cells that can be used with closed cycle refrigerator to study the magnetic behavior and presence of superconductivity in materials at low temperatures and high pressures. Nuclear magnetic resonance and electron spin resonance studies also can be carried out at high pressures.

3.5 HIGH-PRESSURE 3D TOMOGRAPHY

With third-generation synchrotrons where the beam sizes can be reduced down to a few nanometers and with the high-resolution transmission X-ray microscopes, which enable a spatial resolution of 30 nm, 3D tomographic studies have become possible. This requires collection of images across 0–180 rotation of the object. However, even if the panoramic DAC is used this full rotation is not possible. Still, with the help of an iterative reconstruction algorithm, Wang et al. [172] have shown how the 3D shape of tin sample changes across the beta tin to bct phase transition at high pressure. Volume changes have also been determined by this method for noncrystalline materials like Se glass [173] and thus this method can be effectively used to determine the EOS of glasses and melts up to megabar pressures.

4. SOME CASE STUDIES

As discussed in the previous paragraphs there are a variety of phenomena occurring at high pressures and correspondingly several techniques have been developed to study them. Before describing a few representative studies, many of those from our own laboratory, it is worth mentioning that at BARC, many in-house facilities are available for carrying out both experimental and theoretical high-pressure studies. A dedicated high-pressure beamline for X-ray diffraction was set up at BL-11 at Indus-2, 2.5 GeV synchrotron. This beamline has both energy-dispersive as well as angle-resolved X-ray diffraction facilities [174]. For spectroscopic measurements a confocal micro-Raman facility has been set up at Trombay with spatial resolution of ~5 nm. A triple spectrometer has been installed to carry out Raman measurements up to 10 cm^{-1}, and a Brillouin scattering facility that employs JRS scientific instrument (3 × 3) pass tandem Fabry–Perot interferometer has also been established. High-pressure infrared absorption and reflection studies can be carried out on our Bruker 80V–based FTS system installed at the Indus-1 beamline at RRCAT, Indore, India. High-pressure, high-temperature measurements in the diamond anvil cell can be carried out with laser-heated DAC facility, built using a 100 W fiber CW laser (1064A). Electrical resistance measurements have been carried out in our laboratory up to ~2K and ~25 GPa using a Syassen–Holzapfel kind of DAC in an optical cryostat [175]. For studies related to magnetic and superconducting state of materials at high pressure, a squid magnetometer–based facility has been set up. Computational studies are carried out on dedicated Pluto and other parallel computing systems at BARC, Trombay. Computations involving classical molecular dynamics and first principles calculations are carried out using a variety of software codes like DLPOLY (CMD), Wien2k, VASP, quantum-expresso, etc.

4.1 MECHANISM OF PHASE TRANSITIONS UNDER HIGH PRESSURE

Phase transitions under pressure are ubiquitous. In general, parent-daughter phases may be connected to each other either through total reconstruction or smooth distortions of motifs brought about through correlated atomic movements. Generally reconstructive phase transitions are diffusion controlled, necessitating breaking/making of old/new bonds and are therefore very sluggish. In contrast, displacive phase transitions are brought about through correlated atomic movements, without any drastic reconstruction, resulting in a smooth evolution of the new structure. This is the reason why these displacive transitions have fast kinetics. Generally reconstructive phase changes are also first order (displaying discontinuity in first-order derivative of free energy), while displacive are either of second order (associated with discontinuities of properties representing second derivatives of free energy) or weakly first order, usually associated with emergence and evolution of an order parameter. Of course, phase changes may also be nonstructural, such as superconducting, or ferroic, etc., which have their own characteristic signatures. As structural changes are far more abundant, much space in this article has been allocated to these studies.

For structural phase transitions the knowledge of the path of transformation not only provides insight into how the atoms rearrange themselves but may also help in designing new materials or understanding different materials having the same structure, etc. In an earlier high-pressure study on silicon [176], it was shown that primitive hexagonal phase is related to the β-tin phase through correlated atomic movements, appropriate of a soft mode $\frac{1}{2}(101)_{\gamma}$. The geometric orthohexagonal construction used by us also implied intermediate orthorhombic state, later identified experimentally as Imma phase [177]. In a similar study on CdHg alloys [178], we showed that body-centered tetragonal phase is related to the high-pressure hexagonal close-packed (hcp) phase through an orthorhombic state, which was again discovered later [179]. A similar analysis on shock-driven phase transition in CdS has helped establish the existence of an intermediate phase [180,181]. We will discuss some of our more recent studies in the next sections.

4.1.1 Nanozircon-Structured Chromates

Zircon ($ZrSiO_4$) is an important mineral found in the igneous rocks and sediments of the earth's crust [182], and it transforms from the zircon (S.G. $I4_1/amd$, $Z=4$) to the reidite (scheelite $I4_1/a$, $Z=4$) form at ~23 GPa. This transformation was suggested to be brought about by shear deformations followed by small displacement of the atoms and rotation of the SiO_4 tetrahedra [18]. However, based on Raman results, Jayaraman et al. [183] suggested that this transformation involves a substantial rearrangement of the cations and the anions, both in length and angles, and is not as simplistic as proposed by Kusaba et al. [18] As per the mechanism of Kusaba et al. [18], zircon to scheelite phase transformation proceeds through a shear in the basal plane such that the angle of intersection (AXB shown in Fig. 1.3b), between (100) and (010) direction, changes abruptly from 90° to 115°. However, if the transition is not abrupt and this angle changes smoothly during this transformation, the space group of the intermediate states cannot have tetragonal symmetry and the unit cell will essentially satisfy the monoclinic symmetry. In fact, in a recent theoretical calculation on $ZrSiO_4$, based on group-subgroup analysis, it has been argued that the transformation path of the zircon to scheelite phase transformation is $I4_1/amd \rightarrow Fddd \rightarrow C2/c \rightarrow I4_1/a$ [184].

If the zircon and scheelite structures shown in Fig. 1.4 are compared, it appears that the zircon to scheelite transition may be brought about by uniform compression along the a and b axes and an elongation along the c axis. However, Florez et al. have calculated the energy barrier heights through first principles calculations and have shown that the energy barrier in the direct transformation of tetragonal to tetragonal phase is 236 kJ/mol compared to 80 kJ/mol when the path has an intermediate transient monoclinic state [185]. This monoclinic phase was experimentally observed by us for the first time in zircon-structured chromates.

It was observed that some of the diffraction peaks like (321), (312), and (332) showed unusual broadening and others like (101), (200), and (202) showed very little change (Fig. 1.5). Earlier studies by Tolbert et al. had shown that in silicon this kind of change in FWHM of the diffraction peaks can be

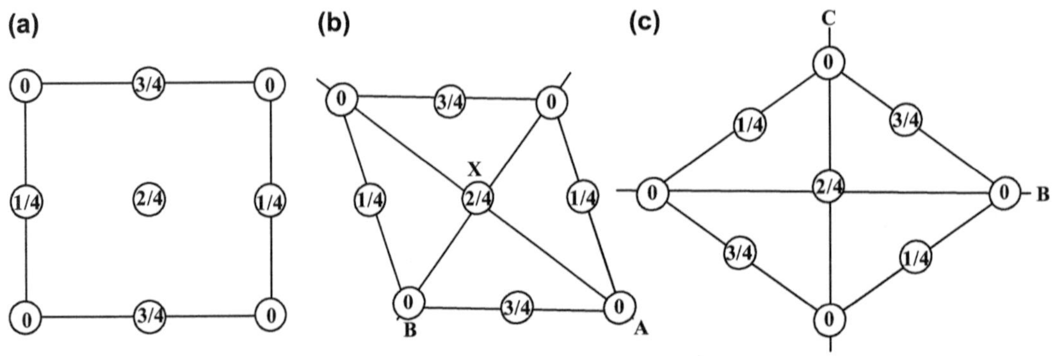

FIGURE 1.3

These three figures give the projection of the zircon and scheelite phase. (a) Down the c-axis of zircon-structured materials, (b) down the c-axis of distorted zircon-structured material in the intermediate state where it is 10% denser than the zircon phase, and (c) projection along the a-direction of the distorted zircon.

Reproduced from K. Kusaba, T. Yagi, M. Kikuchi, Y. Syono, J. Phys. Chem. Solids 47 (1986) 675 with copyright permission from Elsevier.

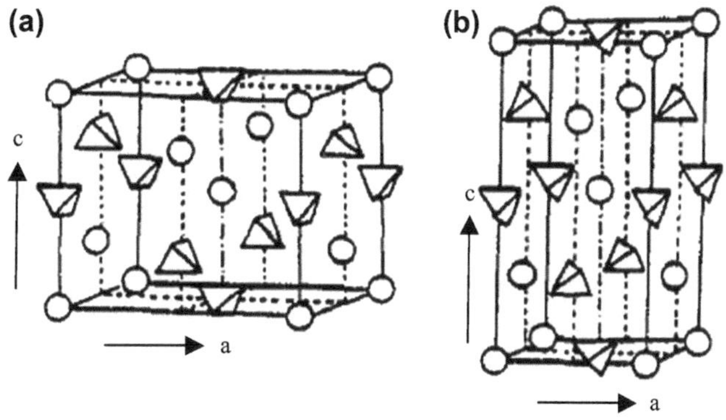

FIGURE 1.4

Crystal structure of (a) zircon and (b) scheelite.

Reproduced from K. Kusaba, T. Yagi, M. Kikuchi, Y. Syono, J. Phys. Chem. Solids 47 (1986) 675 with copyright permission from Elsevier.

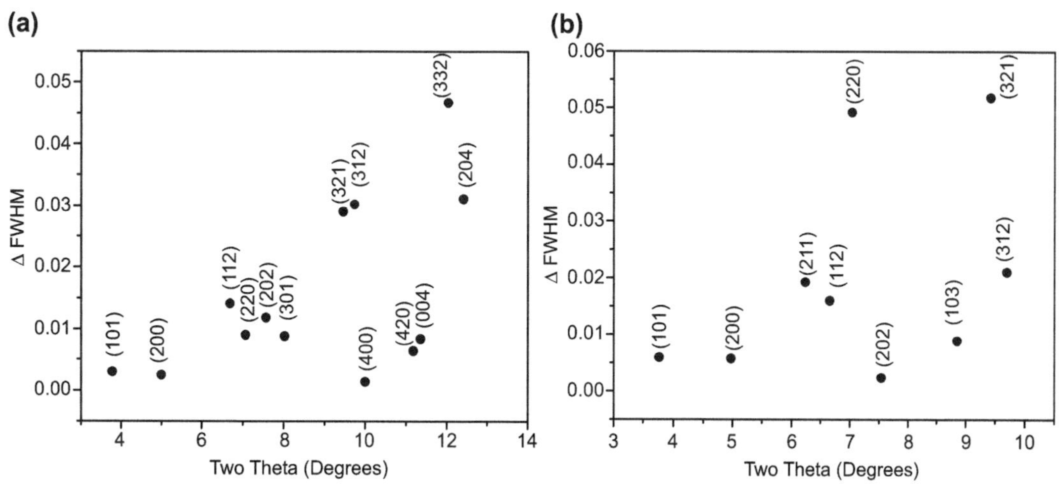

FIGURE 1.5

The increase in FWHM of some of the diffraction peaks of (a) $YCrO_4$ at 4.6 GPa and (b) $HoCrO_4$ at 6.5 GPa. The FWHM of the (200) diffraction peak did not increase as the difference between the "a" and "b" cell constants in the monoclinic phase is ~0.01%.

Reproduced from A.K. Mishra, N. Garg, K.K. Pandey, K.V. Shanavas, A.K. Tyagi, S.M. Sharma, Phys. Rev. B 81 (2010) 104109 with copyright permission from APS.

assigned to shape change of the crystal [186]. However, in this case it does not imply either shape change or simple strain broadening. With the help of Rietveld fitting and group-subgroup analysis the observed specific broadening could be rationalized by the presence of an intermediate monoclinic phase (C2/c) [98]. The structure of this monoclinic phase (MP) was determined to be similar to that of the zircon phase (ZP) except for a slight rotation of the chromate tetrahedra as shown in Fig. 1.6 and a change in the

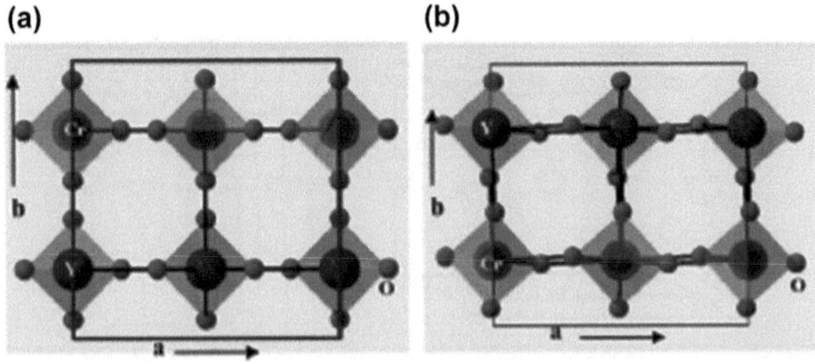

FIGURE 1.6

(a) Zircon and (b) monoclinic structure of $YCrO_4$ as determined from the diffraction data. The γ angle is 90.4. The chromium, yttrium, and oxygen atoms have been marked as Cr (gray), Y (blue (dark gray in print versions)), and O (red (light gray in print versions)), respectively.

Reproduced from A.K. Mishra, N. Garg, K.K. Pandey, K.V. Shanavas, A.K. Tyagi, S.M. Sharma, Phys. Rev. B 81 (2010) 104109 with copyright permission from APS.

gamma angle from 90 to 90.44 degree. This intermediate phase was similar to the monoclinic phase predicted by first principle calculations of Florez et al. [185]. Our first principles calculations also support the presence of this monoclinic phase. Thus by understanding the path of transformation we were able to detect the intermediate monoclinic phase in the zircon to scheelite transformation [98].

4.1.2 Alpha Cristobalite

Cristobalite is a high-temperature polymorph of silica and is often found in volcanic rocks. Experimentally it has been observed that the structures of the high-pressure phases of alpha cristobalite are very sensitive to the degree of nonhydrostatic stresses supported by the fluid medium, whereas in simulations it has been found that they are dependent upon the interatomic potentials or the number of unit cells used for the simulation [187]. Experiments have shown that alpha cristobalite transforms to lower symmetry phases or amorphizes at high pressure [188–193]. However, calculations have shown that it transforms to a stishovite phase, either directly or via an intermediate Cmcm phase, which has a mixed coordination of 4 and 6 [194–198]. Ab initio studies by Klug et al. observed a direct transformation from cristobalite-SiO_2 to stishovite [199]. However, subsequent theoretical calculations (both ab initio and CMD) of Huang et al. showed that at high pressure alpha cristobalite ($P4_12_12$) transforms to $C222_1$ to X1 ($P4_12_12$) to stishovite to $CaCl_2$ structure [200]. The absence of Cmcm phase, in their classical molecular dynamics simulation studies, could be due to a difference in the interatomic potential, choice of number of unit cells in the macrocell, or could be due to its transient nature as discussed later.

As observed in earlier classical MD simulations, our studies also [187] show that the Cmcm phase (high-pressure phase of alpha cristobalite) can be stabilized with the TTAM [194,195] potentials. When further pressurized this Cmcm phase transforms to the six coordinated stishovite phase. In addition, our studies also show that with the BKS [201] potentials the intermediate Cmcm phase is not stabilized, and even with the TTAM potentials this intermediate phase can be skipped if the pressure is raised in somewhat larger steps of 5 GPa. Our results showed that at 15 GPa, for the first 11 ps the structure was similar to α-cristobalite (Fig. 1.7). However, at 13 ps the structure is similar to Cmcm,

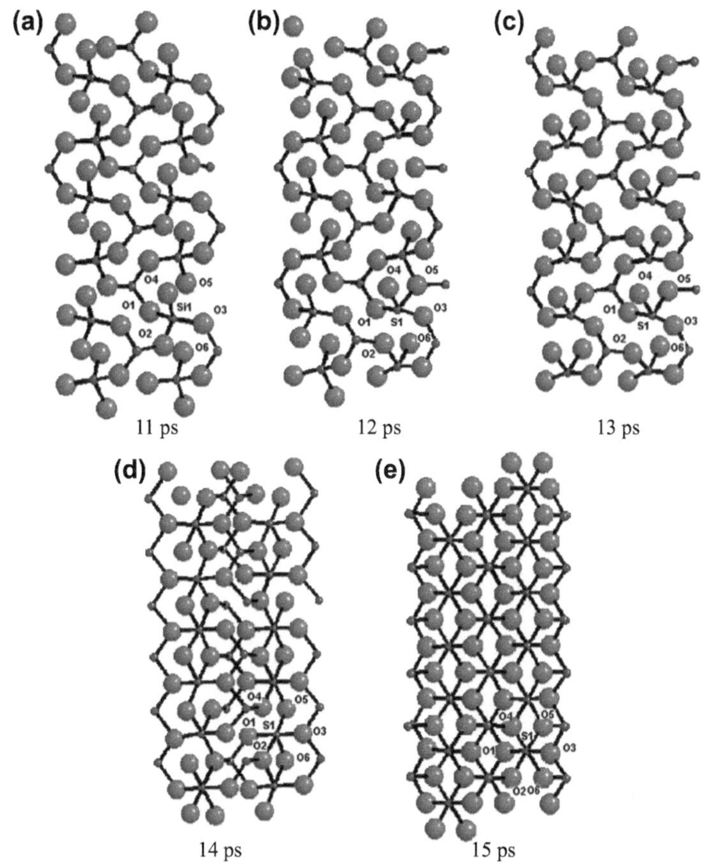

FIGURE 1.7

(a–e) Structural evolution of alpha cristobalite to stishovite transformation. In this figure we can see that at 12 ps new bond is formed between S1 (silicon) and O5 (oxygen) and simultaneously the bond between S1 and O2 breaks resulting in the Cmcm structure at 13 ps. The O2 and O6 coordinate bond to the S1 atom at 14 ps, and the structure finally relaxes to an almost perfect stishovite structure by 15 ps.

Reproduced with permission from N. Garg, S.M. Sharma, J. Phys. Condens. Matter 19 (2007) 456201. Copyright (2007), IOP Publishing. All rights reserved.

which evolves to the stishovite structure at 15 ps. In this context, we observed that the only difference between different pair potentials is that the Cmcm phase may or may not have a range of stability. This may explain why Huang et al. [200] did not observe the Cmcm phase.

Structural evolution shows that no diffusion is necessary to bring about a transformation from α-cristobalite to stishovite phase through the Cmcm phase. It can be seen from Fig. 1.7 that the structural change at high pressure is brought about by a slight local atomic rearrangement. These studies show that the transformation of the α-cristobalite phase to the high-pressure stishovite phase is always via an intermediate orthorhombic Cmcm phase irrespective of whether this intermediate phase stabilizes or not [187].

Our high-pressure studies on isostructural low-cristobalite form of $AlPO_4$ (c-$AlPO_4$) indicate that under nonhydrostatic conditions c-$AlPO_4$ initially transforms to a monoclinic phase, which then transforms to the Cmcm phase via an intermediate, disordered structure [202]. In contrast, X-ray diffraction measurements made under hydrostatic conditions show that the ambient structure transforms directly to the Cmcm phase, consistent with the classical molecular dynamics simulations and also with our simulations on isostructural c-SiO_2 [187,194]. This shows that there is a similarity in the high-pressure behavior of these cristobalite structured oxides. It is possible that the Cmcm phase is a transient phase in c-SiO_2 and hence has not been observed so far experimentally. However, as discussed in the earlier section it may be possible to capture the transient intermediate phases by carrying out studies at high-intensity, high-resolution sources or by changing the range of stability of the high-pressure phases by using nanoparticles.

4.1.3 Scandium Tungstate

Several materials that have framework structures exhibit negative thermal expansion, i.e., their volume decreases on increase of temperature. It has been observed that in these structures the volume compression is brought about by not just the rotation of the tetrahedra but also the compression and/or the distortion of the polyhedra. These compounds crystallize in the orthorhombic space group *Pnca* with four formula units per crystallographic unit cell and are made up of corner-shared WO_4 tetrahedron and AO_6 octahedron (A = Sc, Al) as shown in Fig. 1.8.

Our studies on scandium tungstate [$Sc_2(WO_4)_3$] have shown that it transforms to the monoclinic phase at $\sim 0.6 \pm 0.3$ GPa. This was established by observing the changes (arrows indicate the diffraction peaks of the new high-pressure phase) in the diffraction patterns as shown following (Fig. 1.9) [4].

With the help of Rietveld analysis we determined that the high-pressure phase (m1) has a $P2_1/a$ symmetry. This symmetry change is similar to the one observed in the ferroelastic phase transition at low temperature [203,204]. The cell constants of the new high-pressure phase were found to be $a_{m1} = 16.03$ (2) Å, $b_{m1} = 9.4$ Å, $c_{m1} = 18.6$ Å, and $\beta = 124.89°$ (3). The new phase can be related to

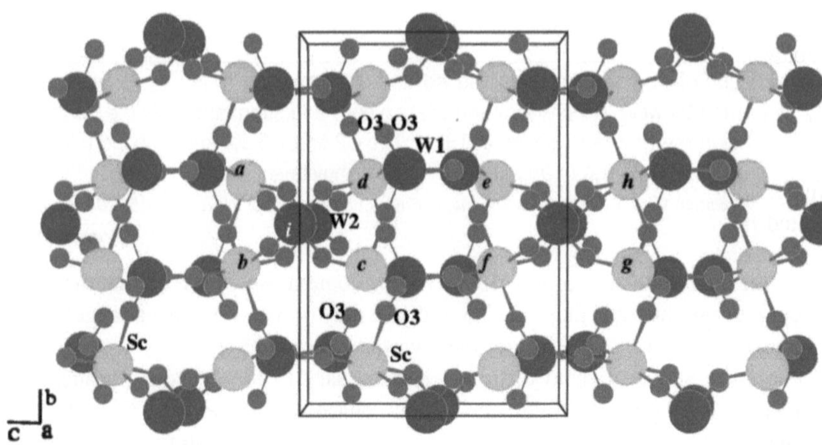

FIGURE 1.8

Orthorhombic structure of $Sc_2(WO_4)_3$ compound. The unit cell is shown by the *blue solid lines* (gray in print versions).

Reproduced from N. Garg, C. Murli, A.K. Tyagi, S.M. Sharma, Phys. Rev. B 72 (2005) 064106 with copyright permission from APS.

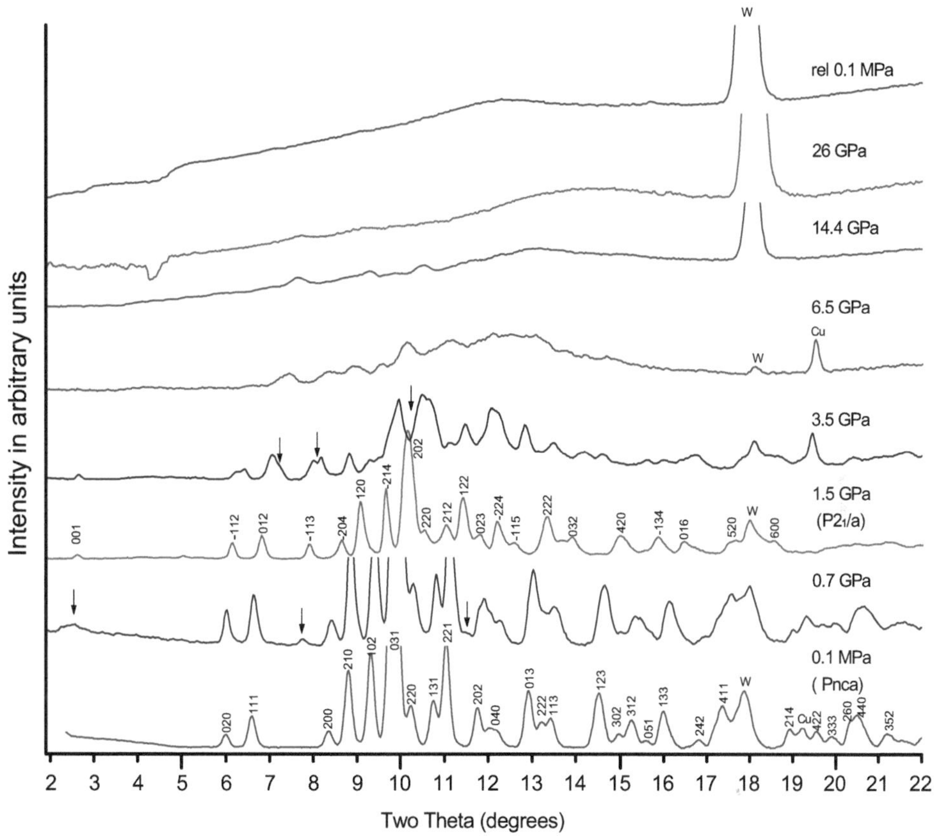

FIGURE 1.9

Background-subtracted X-ray diffraction patterns of $Sc_2(WO_4)_3$ at different pressures. The W and Cu represent diffraction lines from the tungsten gasket and copper pressure marker, respectively.

Reproduced from N. Garg, C. Murli, A.K. Tyagi, S.M. Sharma, Phys. Rev. B 72 (2005) with copyright permission from APS.

the initial orthorhombic phase through cell doubling along the ***c***-axis. The cell constants of these two phases are related by the following relations:

$$a_{\mathrm{m1}} \approx -b_{\mathrm{o}} + c_{\mathrm{o}} \tag{1.1}$$

$$b_{\mathrm{m1}} \approx -a_{\mathrm{o}} \tag{1.2}$$

$$c_{\mathrm{m1}} \approx -2\,c_{\mathrm{o}} \tag{1.3}$$

where "m" stands for monoclinic and "o" stands for orthorhombic.

Looking at the structure of the first high-pressure phase (Fig. 1.10) it can be seen that the ***ac*** plane of this phase is equivalent to the ***bc*** plane of the orthorhombic phase (Fig. 1.8). Rotation of the WO_4 tetrahedra at ***i***, and elongation of the four ScO_6 octahedra at sites ***a***, ***b***, ***c***, and ***d*** along the [011] direction of the orthorhombic cell are responsible for the transformation to the monoclinic structure. In this process the distance between scandium atoms at sites ***a*** & ***c*** and ***e*** & ***g*** are no longer equal resulting in doubling of the unit cell along the $\boldsymbol{c_o}$ axis.

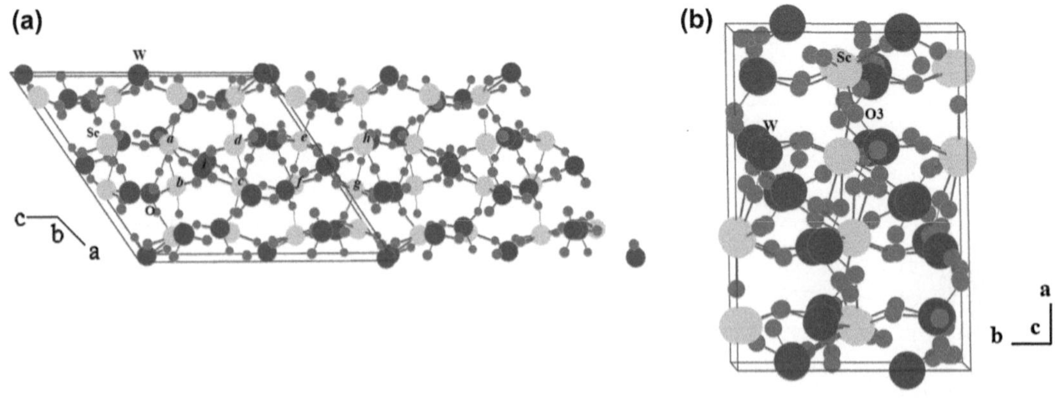

FIGURE 1.10

View of the $Sc_2(WO_4)_3$ structure in the first monoclinic phase ($P2_1/a$) down the (a) *b*-axis (b) *c*-axis.

Reproduced from N. Garg, C. Murli, A.K. Tyagi, S.M. Sharma, Phys. Rev. B 72 (2005) with copyright permission from APS.

A second phase transition to a monoclinic phase (m2) of $P2_1/a$ symmetry distinct from that of the first high-pressure phase was also observed. There is a substantial rearrangement of the atoms accompanied by an expansion of the $\boldsymbol{b}_{m1}$ axis and a considerable collapse of the $\boldsymbol{a}_{m1}$ axis, along with a slight decrease in the β angle. The observed pressure-induced variation of the fractional coordinates implies increasing bending of WO_4 tetrahedra at the O_3 linkages as the pressure is increased from ~0.6 GPa. Since layers of tetrahedra and octahedra are connected only by the oxygen atom O_3, it is likely that sudden collapse of the $\boldsymbol{a}_{m1}$ axis and increase in $\boldsymbol{b}_{m1}$ axis at 1.9 GPa is due to a discontinuous jump in the tetrahedral deformations beyond some limiting value of bending.

4.2 PRESSURE-INDUCED AMORPHIZATION

Pressure-induced amorphization has been observed in a variety of materials. It has been explained as a metastable state that arises due to kinetic frustration of the system to reach a higher density crystalline phase [205]. Several tetrahedrally coordinated solids have been shown to become amorphous under pressure. Amorphization implies loss of long-range translational order in the crystal, and the most common experimental methodologies used to characterize this phenomenon are based on diffraction (X-ray, neutron) and spectroscopic (Raman, IR, EXAFS, etc.) techniques. However, since different experimental probes give information over different length scales, it has been observed that materials that are X-ray amorphous may not show amorphization when probed by Raman spectroscopy, as the length scale of the latter is almost one to two orders of magnitude smaller than the X-ray diffraction length scales. Several times it has been observed that even when a material shows features implying an amorphous state, it may actually not be amorphous. For example, if the probing source is weak and/or is not collimated, or the sample environment is nonhydrostatic, or the sample is nanocrystalline or poorly crystallized, we may conclude that the material has amorphized. Confirmation of an amorphous phase with only one technique may also be misleading. For example, if a high-pressure phase does not have any Raman active modes then interpretation of the Raman spectra alone may lead to incorrect conclusions. To elucidate some of these points we will next discuss a few examples.

4.2.1 Berlinite

Due to their geophysical relevance and substantial natural abundance, tetrahedral framework structures have been extensively investigated under high pressure. The high-pressure behavior of α-$AlPO_4$ (berlinite), having a tetrahedral framework, was claimed to be extraordinary [206] as it was supposed to exhibit the memory glass effect. A material is termed memory glass if its high-pressure amorphous phase reverts back to the original single crystal with the same orientations as the starting crystal [206]. Loss of sharp X-ray diffraction peaks (collected with the lab X-ray source) [206–208] around 12–18 GPa was inferred as amorphization of berlinite. This high-pressure behavior was also supported by subsequent Raman, IR, and Brillouin studies [206,209,210]. Several theoretical investigations were inspired by these exciting experimental results. Energy minimization and lattice dynamical calculations proposed the dynamical mechanism of pressure-induced amorphization, whereas classical molecular dynamics (MD) calculations [211–214] showed anisotropic structural disordering of $AlPO_4$ above 30 GPa. Tse and Klug [213] tried to explain the memory effect in berlinite by showing that the spiraling AlO_4 and PO_4 tetrahedra distort and rotate into empty spaces in the crystal on increasing pressure resulting in higher Al-O coordination and unwind on release of pressure. On comparing the enthalpies of both the quartz and Cmcm phase of $AlPO_4$, Chaplot and Sikka [211] concluded that the latter was stable beyond 12 GPa, implying that berlinite $AlPO_4$ could undergo a crystal-to-crystal phase transition at high pressure. However, though the molecular dynamical calculations could not observe the Cmcm phase in $AlPO_4$, experimental studies on isostructural α-$FePO_4$ and α-$GaPO_4$ showed that they undergo crystal-to-crystal phase transitions and transform to the Cmcm phase at higher pressures [215–217]. Our MD simulations on $AlPO_4$ [218] showed agreement with the Raman measurements of Gillet et al. [219] and indicated that at ~15 GPa the structure of $AlPO_4$ was not amorphous and translational order persisted even up to ~60 GPa.

Our X-ray diffraction studies [220,221] carried out at Spring8 showed that sharp Bragg diffraction peaks exist even at 40 GPa contrary to earlier experimental results of Kruger et al. [206]. Moreover, the weak diffraction peaks that emerged (shown in Fig. 1.11) beyond 11 GPa could be indexed to the Cmcm phase. Due to kinetic impedance, these new peaks grew marginally and had a large FWHM. This indicates that the new Cmcm phase is poorly crystallized, i.e., the particle size of the new phase may be very small, and also since the growth of the new phase is not homogeneous the sample may develop local strains (insert in Fig. 1.11). The existence of poor crystallinity of the Cmcm phase (Fig. 1.11) also explains why the earlier X-ray diffraction studies done with the laboratory X-ray sources [206,207] interpreted this crystalline–crystalline phase transformation in terms of pressure-induced amorphization. Earlier Raman results of Gillet et al. also can be rationalized with the presence of the high-pressure Cmcm phase as the weak Raman bands that were observed at $P > 14$ GPa are very similar to that of $CuCrO_4$, which exists in the Cmcm phase at ambient conditions [222].

The presence of a crystalline orthorhombic phase beyond ~13 GPa provides a more easily acceptable rationale for the explanation of the observed elastic anisotropy of the high-pressure phase in Brillouin scattering results [210]. Also, though the total energy calculations had supported the stability of Cmcm phase at high pressures, this phase could not be dynamically reached in any of the classical MD calculations carried out so far [218]. This feature, along with our experimental results, suggest the possibility of large kinetic resistance for the α-phase to Cmcm phase transformation. In fact, recent studies by Pellicer Pores confirm this suggestion as laser annealing of the sample at ~14 GPa resulted in the emergence of very sharp peaks of the Cmcm phase [223].

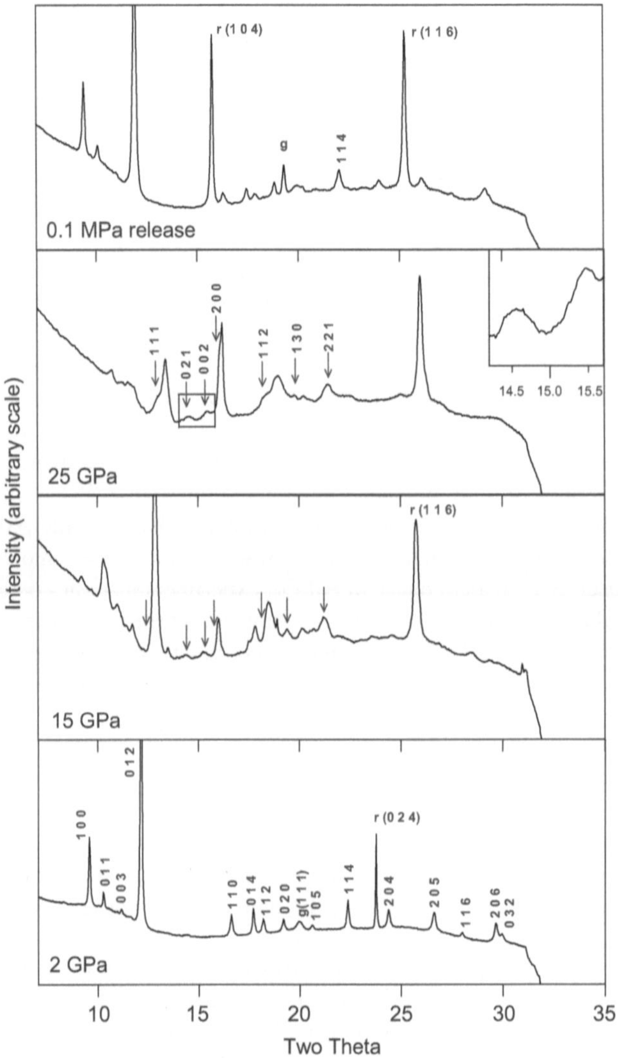

FIGURE 1.11

Diffraction pattern of $AlPO_4$ (a) at different pressures.

Reproduced from S.M. Sharma, N. Garg, S.K. Sikka. Phys. Rev. B 62 (2000) 8824 with copyright permission from APS.

4.2.2 Silicon

Silicon has been extensively studied under different thermodynamic conditions, both by experiments and theory, as it has immense technological importance. In particular, under pressure its structures have been shown to evolve to denser and higher coordinated forms [186,224–230]. For example, in the bulk crystalline form, the initial cubic diamond phase of silicon is known to transform to the β-tin structure at 11 GPa and then to Imma phase at 13 GPa followed by primitive hexagonal (ph), orthorhombic (Cmca), hexagonal close-packed (hcp), and face-centered cubic (fcc) phases at 16, 38, 42, and 78 GPa, respectively. In contrast, the cubic diamond phase of nanocrystalline Si ($\sim$50 nm) transforms directly to the primitive hexagonal (ph) form at 22 GPa, bypassing β-tin and Imma phases observed in the bulk [186]. Moreover, this transformation is thought to be associated with the shape change of nanocrystallites. Qualitatively, the nonrealization of intermediate phases has been explained in terms of increased stability field of the diamond phase in the nano form [186].

Earlier experimental studies on nanocrystalline porous silicon (π- Si) claimed that the initial cubic phase transforms to a high-density amorphous (HDA) phase at $\sim$14 GPa, which on release of pressure converts to a low-density amorphous (LDA) structure [231]. These observations were rationalized in terms of pressure induced pseudo-melting, implied by the negative Clapeyron slope for Si. The pressure-induced amorphization (PIA) of nanocrystalline porous silicon was concluded primarily based on Raman measurements. This, in principle, does not rule out the possibility of the transformation to a phase that may not have a Raman active mode, such as in primitive hexagonal [186].

Our Raman spectra [232] of nanocrystalline porous silicon shows that the Raman active peak of the cubic phase vanishes at high pressure (Fig. 1.12b), as observed by Deb et al. [231] and interpreted as an indicator of amorphization of silicon. However, our X-ray diffraction experiments (at Spring8) on the same sample showed that it did not amorphize even up to pressures of $\sim$36 GPa. On the contrary (Fig. 1.12a), several new diffraction peaks were observed at $\sim$20 GPa indicating a phase transformation. This phase could be indexed to the eight coordinated primitive hexagonal phase thus explaining why earlier authors had misinterpreted the Raman results as indicative of amorphization of silicon. Thus the loss of Raman spectra at high pressure must be verified by other experimental techniques before being termed as amorphous.

4.2.3 Yttrium Chromate and Holomium Chromate

Several compounds like zircon, TiO_2, etc. are known to be poor glass formers. However, bulk $ZrSiO_4$ is known to amorphize when irradiated with heavy ions [233], and compounds like TiO_2, etc. amorphize only in the nano form [234]. To understand the effect of high pressure on the structural stability of zircon structured nanocrystalline holomium and yttrium chromate, we carried out high-pressure X-ray diffraction experiments.

The diffraction patterns of $YCrO_4$ and $HoCrO_4$ at a few representative pressures are shown in Fig. 1.13. The diffraction patterns of both these compounds show that the diffraction peaks from the zircon phase show broadening with pressure, and beyond 10 GPa these diffraction peaks become very broad and are no longer discernible, indicating that the sample may have become disordered. These diffraction patterns seem to suggest that the kinetics in nanocrystalline samples are such that the scheelite phase does not crystallize even up to $\sim$10 GPa. However, our Raman studies indicate an irreversible crystal-to-crystal phase transformation whereas the X-ray diffraction experiments suggest that the high-pressure phase is amorphous. How do we reconcile with such contrasting results? There could be two explanations. (a) Since in the Raman spectra translational modes of the chromate tetrahedra are not observable and the only new modes that are observable are weak and broad, we can hypothesize that there is some sort of orientational disorder of the chromate tetrahedra that is leading to long-range disorder as observed in the X-ray diffraction experiments. As Raman spectra probes shorter length scales it is not surprising that we could observe the internal modes of the chromate tetrahedra. (b) A second explanation to these results could be that the new high-pressure phase has a small crystallite size (result

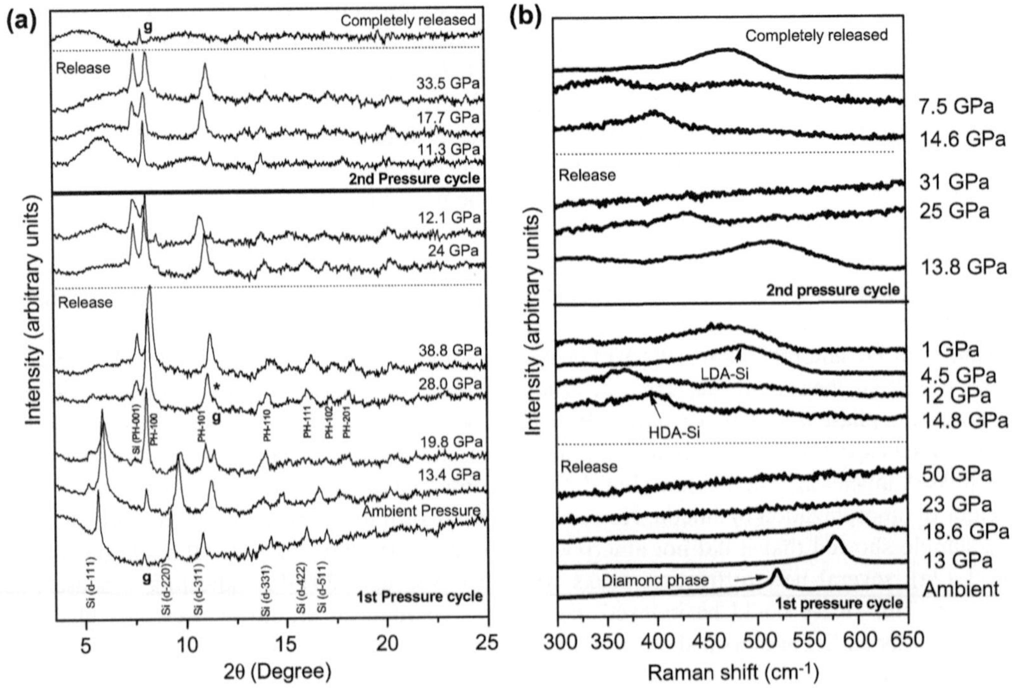

FIGURE 1.12

(a) The X-ray diffraction patterns of π-Si in the compression and decompression cycles. The background of the empty gasketed cell was subtracted from the diffraction patterns. The pattern at ambient pressure shows all the diffraction peaks of the cubic diamond phase of silicon. The tungsten gasket peaks have been labeled as "g." The diffraction peaks of the high-pressure phase have been indexed with the primitive hexagonal ("ph") structure. Calculated diffraction patterns of the "ph" silicon phase and the gasket peaks are marked as "gasket" and "Ph-Si." (b) The Raman spectra of nano π-Si at different pressures. At ambient pressure the asymmetric Raman peak of the diamond Si–Si stretch mode can be seen at 519.54 cm^{-1}. At 23 GPa the Raman mode of diamond phase is not visible and we get a flat background as "ph" phase does not have any optical phonon. On release of pressure, at ~15 GPa, the broad band centered at 392 cm^{-1} was assigned to the high-density amorphous phase (HDA). (Emergence of HDA instead of beta-tin phase implies that the amorphous phase in the present case is due to kinetics preference.) The low-density amorphous phase was visible at ~4.5 GPa, characterized by a hump at 484 cm^{-1}.

Reproduced from N. Garg, K.K. Pandey, K.V. Shanavas, C.A. Betty, S.M. Sharma, Phys. Rev. B 83 (2011) 115202 with copyright permission from APS.

of multiple nucleation sites of new phase). It is known that when solid–solid phase transitions are accompanied by a volume change, single crystals fragment into much smaller crystallites [235].

In fact a careful inspection of the diffraction data reveals that the intensity of the diffraction peaks starts diminishing beyond 5 GPa in case of $HoCrO_4$ and beyond 7 GPa in case of $YCrO_4$, which indicates that the zircon phase does become unstable at these pressures. However, these studies did not help us in ascertaining whether the high-pressure phase was disordered or poorly crystalline. To

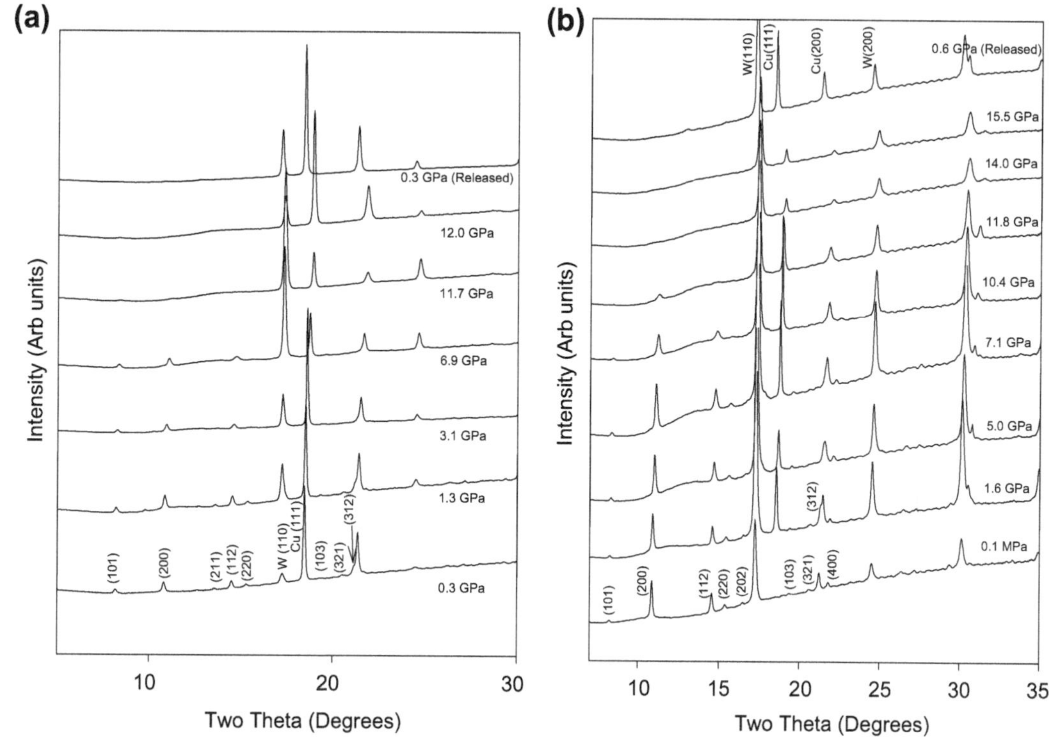

FIGURE 1.13

Diffraction pattern of zircon-structured nanocrystalline yttrium chromate and holomium chromate collected at the Elettra synchrotron source.

resolve this issue we carried out experiments at a higher energy synchrotron source (Spring8) and found that the high-pressure phase was not amorphous and zircon-structured chromates had transformed to the scheelite phase [98]. These studies show that in the case of zircon-structured nanochromates one could have misinterpreted the weak high-pressure phase as amorphous if experiments had not been carried out with a brighter source.

4.2.4 Reduced Graphene Oxide

Graphene has been shown to have remarkable, physical, optical, and electronic properties [236–240]. High-pressure studies on a few layer graphene (up to 50 GPa) under both hydrostatic and nonhydrostatic conditions have shown that it loses long-range order when the interlayer distance reaches a critical value of ~2.8 Å [241]. These conclusions were based on the loss of the only observable (002) diffraction peak of graphene, as also observed in graphite by Mao et al. [242]. Clark et al. explained the high-pressure disorder by invoking buckling and fracturing of the graphene layers due to enhanced interlayer interaction and attributed it to partial conversion of interlayer π (sp2) to intralayer σ (sp3) bond, as was observed [241]. Observability and recovery of the G Raman band at the highest pressure and on pressure release was attributed to disorder different from that observed in graphite and carbon nanotubes. Recent studies on reduced graphene oxide [243] showed that since these samples are essentially layered compounds, due to texture and usage of the Bragg Brentano geometry, the (100) diffraction peak (essentially diffraction from in-plane hexagonal ordering of carbon atoms) was not observed in earlier studies. These studies also show that though reduced graphene oxide disorders reversibly along the (002) direction, its structure remains intact in the basal plane. Based on the analysis of the peak shape of the (100) diffraction peak it was concluded [243] that the a-b basal plane had puckered regions that flattened irreversibly on application of pressure. This example shows that in layered compounds, if we rely on only X-ray diffraction results, then texture of the sample may lead us to conclude complete disorder when in reality the material may have lost long-range order in only one dimension.

4.3 NEGATIVE THERMAL EXPANSION AND AMORPHIZATION

Several materials that have open structures and small shear moduli display negative thermal expansion. In these materials the low-frequency transverse vibrations that can be excited at low temperatures, and have a negative Gruneisen parameter, are responsible for NTE. In many framework structures that are made up of rigid tetrahedral and octahedral networks, linked by a corner oxygen atom, the rigid unit modes or the quasi-rigid modes have a role to play in bringing about negative thermal expansion in these materials. Due to the presence of a large number of low-frequency transverse vibrations, it is easy to destabilize the structure of an NTE material at very low pressures. The question has been asked several times whether NTE is responsible for the amorphization of these materials at high pressure. Our studies show that NTE materials are prone to destabilize at very low pressures and undergo phase transitions. However, it is not necessary that the high-pressure phase has to be an amorphous phase. We elucidate it here with a few examples.

4.3.1 Scandium Tungstate

Earlier studies have shown that while some NTE materials like $Sc_2(MoO_4)_3$ etc. [203,244] undergo several pressure-induced phase transitions prior to becoming amorphous at high pressures, some others, like $Y_2(WO_4)_3$, $Lu_2(WO_4)_3$ [245,246], amorphize directly from their initial structure. It has also been suggested that pressure-induced amorphization in these materials maybe related to the transverse modes causing NTE [247]. X-ray diffraction studies on recovered samples of $Sc_2(WO_4)_3$ had shown that it amorphizes partially when pressurized to 4 GPa, but amorphizes completely and irreversibly when pressurized to ~8 GPa [248].

Our high-pressure Raman and X-ray diffraction studies on scandium tungstate show that this compound undergoes structural phase transitions around $\sim 0.6 \pm 0.3$ and ~1.6 GPa [4]. At still higher pressures, at ~6.5 GPa this compound becomes disordered, which progressively and irreversibly grows into an amorphous phase. The presence of broad bands at 473 and 955 cm^{-1} and softening of the higher band frequency at high pressures (Fig. 1.14), indicates that the constituent polyhedra are highly distorted, and the effective W-O coordination may be higher. The structural changes up to ~1.6 GPa were found to be reversible, and can be rationalized in terms of the monoclinic phase being related to the initial phase through small rotations of the WO_4 tetrahedra, which springs back on release of pressure. In $Sc_2(WO_4)_3$, the second monoclinic phase is a precursor to the amorphous phase, as also observed in aluminum tungstate and scandium molybdate [203,244]. Though the thermal expansion behavior of the high-pressure monoclinic phase of $Sc_2(WO_4)_3$ has not been studied, Evans and Mary [204] have shown that the low-temperature monoclinic phase of scandium molybdate has a positive

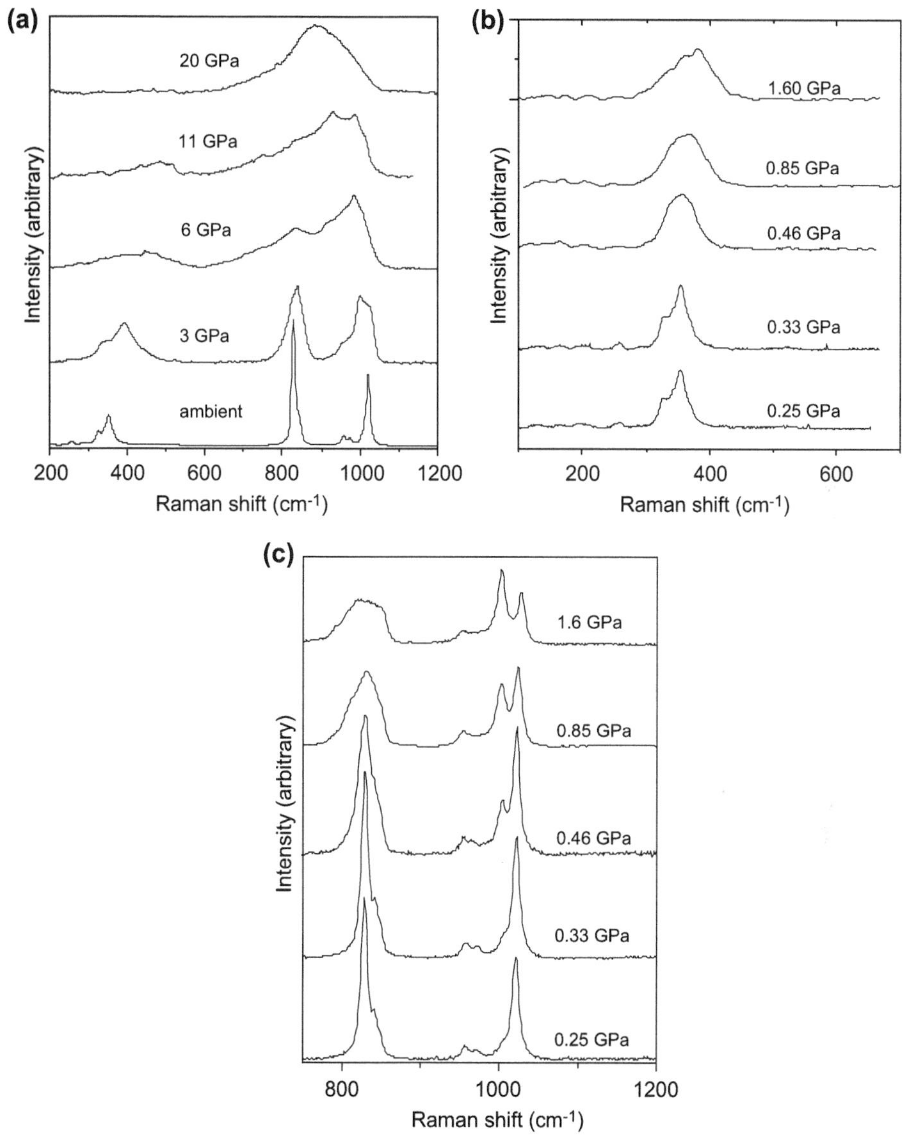

FIGURE 1.14

Background subtracted Raman modes of $Sc_2(WO_4)_3$ at different pressures. (a) Overall variations in the Raman spectra with pressure, (b & c) the Raman spectra at low pressures in the range 200–600 and 700–1100 cm^{-1}, respectively.

Reproduced from N. Garg, C. Murli, A.K. Tyagi, S.M. Sharma, Phys. Rev. B 72 (2005) 064106 with copyright permission from APS.

thermal expansion. If the behavior is the same for the high-pressure phase too (due to the structural similarity), then it is doubtful that in this compound the pressure-induced amorphization is related to negative thermal expansion behavior. The proposed relation between NTE and pressure-induced amorphization has also been debated in a recent paper [249]. Purely structurally, we note that in the higher density monoclinic phase, the reduced compressibility implies an increase in the repulsive forces between various polyhedra. Hence on further increase of pressure the polyhedra would try to rearrange so as to minimize the free energy. In most cases the pressure-induced amorphization is due to the inability of the structure to transform to a stabler crystalline phase due to poor kinetics. It is likely that the present case is similar and structural frustration arises from the inability to cross the energy barrier to the higher pressure-ordered phase and the structure locally deforms to minimize the energy.

4.3.2 $Ag_3[Co(CN)_6]$

Silver hexacyanocobaltate $Ag_3[Co(CN)_6]$ shows colossal anisotropic negative thermal expansion with a CTE much larger than that of many NTE materials including ZrW_2O_8 [250,251]. $Ag_3[Co(CN)_6]$ is a framework material in which the Co-CN-Ag-NC-Co linkages are not constrained. Its crystal structure consists of alternating layers of Ag^+ and $[Co(CN)_6]^{3-}$ ions, stacked perpendicular to the "*c*" axis of the trigonal $P\bar{3}\,1m$ unit cell. In the silver layer the atoms are arranged at the vertices of a Kagome lattice, and the octahedral $[Co(CN)_6]^{3-}$ ions are positioned above and below the hexagonal Kagome holes. This compound behaves like a garden fencing where a contraction along *c*-direction is coupled to an expansion along *a*-direction. The compound also exhibits a negative linear compression (NLC) [251]. It is understood that the flexible framework structure along with the Ag...Ag argentophillic interactions give rise to such an unusual NTE effect [252,253]. Goodwin et al. carried out high-pressure studies on silver cobalto cyanide and observed that it transformed to a monoclinic phase at very low pressures of 0.19 GPa [251]. They also observed that this high-pressure phase also shows weak negative thermal expansion behavior. Subsequent Raman studies showed that it amorphized above 13 GPa, thus indicating that it was in the NTE state when it amorphized [254].

Our FTIR studies on $Ag_3[Co(CN)_6]$ show that it undergoes an irreversible phase transition at 0.3 GPa to a monoclinic phase [102]. We have observed that this phase transition is very sensitive to shear stresses and can also be induced by the small shear stresses introduced while grinding this sample. Our studies, however, showed that this monoclinic phase is recoverable even after it had been pressurized to ~5 GPa. X-ray diffraction studies carried out in very fine pressure steps (Fig. 1.15), however, show that the first-phase transition occurs at very low pressure viz., ~0.1 GPa, and it is partly reversible when pressure is released from 1.5 GPa [255]. We also observed that prior to amorphization at 21 GPa $Ag_3[Co(CN)_6]$ undergoes a second phase transition. Its structure and NTE behavior have not been determined so far. So it yet remains to be ascertained in this material whether NTE is responsible for the amorphization of the second high pressure phase.

4.3.3 $Zn(CN)_2$

Cyanide framework compounds are known to exhibit negative thermal expansion [111,256–258]. Zinc cyanide, which crystallizes with a cubic structure, has a doubly interpenetrated nonporous framework of two diamondoids made up of tetrahedrally coordinated Zn ions bridged by disordered CN anions. It is known to exhibit a large isotropic negative thermal expansion behavior over an extended temperature range. Neutron diffraction studies (up to 1 GPa) of Chapman et al. on this framework compound have shown that it has a negative K_0' implying softening of the material with

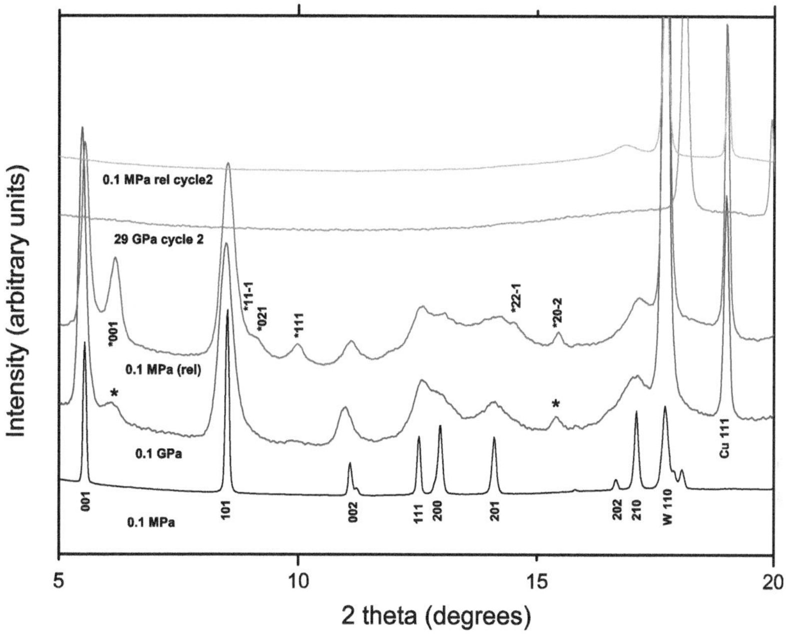

FIGURE 1.15

Diffraction pattern of $Ag_3(Co(CN)_6)$ at different pressures. The diffraction peaks of the ambient and monoclinic phase (*) have been indexed. These data show that the ambient phase is partly recovered when released from ~1.5 GPa but remains amorphous when released from 28 GPa.
To be published in Ref. [255].

pressure. They have also shown that this softening was accompanied by an increase in the NTE behavior of this compound [259]. Ravindran et al. showed that several Raman modes of $Zn(CN)_2$ have a negative Gruneisen parameter, and at high pressures it shows disordering ~ 1.6 GPa [260]. Our studies have shown that the high-pressure behavior of $Zn(CN)_2$ depends upon the hydrostaticity of the pressure transmitter surrounding the compressed sample. In the case of hydrostatic environment (methanol-ethanol-water mixture 16:3:1) disordering was observed ~9 GPa with eventual amorphization at ~11 GPa preceded by two crystal-to-crystal phase transitions, i.e., cubic (Pn3m) -orthorhombic ($Pmc2_1$) -cubic-II ($P4_232$). In contrast, in the nonhydrostatic case a poorly crystallized phase was observed at low pressures with eventual disordering at ~11.7 GPa. The disordered phase obtained under nonhydrostatic conditions was distinct from that obtained under hydrostatic conditions [261]. It is possible that in the hydrostatic case the liquid medium may have penetrated the framework, destabilizing it at much lower pressures. Recently Lapidus et al. have shown that apart from the degree of hydrostaticity even the type of pressure medium alters the high-pressure behavior of zinc cyanide [262]. They have shown that when no fluid or large molecule fluids like isopropanol, ethanol, or Fluorinert were used, zinc cyanide transformed to an orthorhombic (Pbca) phase. However, when small molecule fluids like water, methanol, methanol-ethanol-water were used, it transformed to Fd3m, lonsdaleite P63/mmc, and pyrite-like Pa3 structures, respectively. Another interesting observation of this study was an intercalation of the pressure transmitting fluid into the framework of zinc cyanide, thus increasing its volume with pressure. These authors could retrieve these intercalated phases of zinc cyanide. Since this material is highly sensitive to both the pressure transmitting fluid and hydrostaticity of the pressure environment, it is not surprising that the structure of the high-pressure phase obtained by Lapidus et al. for the methanol-ethanol-water pressure transmitting medium is distinct from that determined by us [262].

4.4 RECRYSTALLIZATION OF AMORPHOUS PHASE

Crystallization of a liquid or an amorphous solid is a complex nonequilibrium process and is fundamental to a variety of materials. It basically involves nucleation and growth of crystallites, which could either be a random nucleation or preferential nucleation when in contact with a crystalline substrate. As many times specifically amorphous materials are required in several industries like, electronic, pharma, etc., it is important to understand the different stimuli that may trigger crystallization and thus change the properties of the material. Several studies have shown that the crystallization process is affected by presence of impurities, the annealing temperatures, external force fields, ion beam radiation, or applied stress. High viscosity of the amorphous phase at room temperature inhibits structural relaxations thus slowing down the process of crystallization. The amorphous to crystal phase transition mostly involves large volume changes and substantial movement and rearrangement of the constituent atoms. Since densities increase at high pressure, one would expect that crystallization would be suppressed because of steric constraints. Though many studies have shown this effect, there are several other studies that have shown that pressure facilitates crystallization by reducing the temperature of crystallization. As long-range diffusion is significantly suppressed at high pressure, crystallization of only those amorphous materials can take place at high pressure where it takes only a local atomic rearrangement to bring about crystallization. Hence, only polymorphous crystallization, where there is no compositional difference between parent and daughter phases, may be feasible under high pressure. In this context we will discuss the crystallization of silicon in both the bulk and the nano form.

4.4.1 Silicon

Compared to crystalline silicon, the high-pressure studies on amorphous silicon are very few. These can also be categorized into two types, polyamorphic phase transitions or pressure induced crystallization. Crystallization of amorphous silicon at high pressure was observed several years ago. In one of the early reports, Minomura et al. [263] had shown that amorphous silicon transformed to a mixture of β-tin and BC8 phases at ~17 GPa. Subsequent high-pressure diffraction studies by Imai et al. [264] claimed that amorphous silicon crystallized reversibly to β-tin structure. However, this conclusion was based on only two very weak and broad diffraction peaks.

Our X-ray diffraction studies (Fig. 1.16) on both bulk amorphous silicon and nano amorphous silicon (prepared by pressurizing porous silicon having particle size ~4 nm) carried out at SPring8 show that they crystallize to a primitive hexagonal (**ph**) structure at 14 and 18 GPa, respectively [232,265,266]. It is well known that the crystalline phases obtained from amorphous materials at high pressure have a large number of defects and voids and their presence is likely to affect the sequence as well as the pressure of

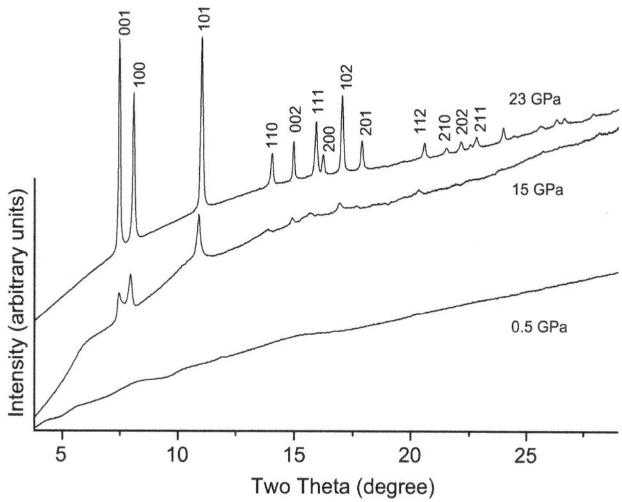

FIGURE 1.16

The high-pressure diffraction pattern of amorphous silicon collected at Spring8. The diffraction peaks in the high-pressure phase have been indexed to the "ph" phase.

Reproduced from N. Garg, K.K. Pandey, K.V. Shanavas, C.A. Betty, S.M. Sharma, Phys. Rev. B 83 (2011) 115202 with copyright permission from APS.

phase transformations [267]. We have observed that amorphous silicon directly transforms to the **ph** phase, skipping the β-Sn and Imma phase on increase of pressure, but transforms to these phases on slow release of pressure as observed in bulk silicon. We have explained the preferred crystallization of a-Si to **ph** phase in the pressure region where other crystalline phases of Si, viz., β-Sn and Imma, are also stable, by invoking the thermodynamic model based on random nucleation and growth process. Our studies show that both the critical diameter for nucleation and the nucleation work are lower for the "ph" phase beyond 5 GPa, resulting in a higher nucleation rate for its preferential crystallization [265].

4.5 AMORPHOUS TO AMORPHOUS PHASE TRANSITION

Study of polyamorphic transitions in liquids and solids is gaining a lot of importance as it helps us to synthesize and understand new glassy and liquid states, which have different physical and electronic properties. Amorphous to amorphous transitions (AAT) are possible only if both the metastable amorphous phases are not close to the stability region of their respective crystalline forms. AAT has been observed in several oxide glasses, elemental semiconductors, ice, etc. The structure of the different amorphous phases has been studied by molecular dynamic simulations, EXAFS, or pair distribution function using X-rays. The change in structure can be continuous or abrupt. For example, in ice it has been observed that the LDA to HDA transformation is abrupt. However, in $GeSe_2$ it has been seen that it transforms gradually from edge-shared tetrahedra to a more open corner-shared tetrahedra accompanied by a decrease in rigidity to a higher coordinated glass. The transformation of ice from LDA to HDA in the high-pressure cryo-cooled protein crystals is a boon for protein crystallographers as it reduces the mosaicity of the crystals due to the higher density of the HDA phase [268].

4.5.1 Silicon

High-pressure polymorphism of elemental semiconductors like Si and Ge in all the forms, i.e., crystalline, amorphous, porous, and liquid, has been extensively studied both experimentally as well as theoretically. Macmillan et al. [225] have shown that amorphous silicon undergoes an LDA to HDA phase transition at high pressures. This transition has been explained within the framework of liquid–liquid transition in the supercooled state in analogy with water. There are some theoretical studies that have shown the possibility of a denser VHDA form of Si, which is 10% denser than the HDA and has a coordination of 8–9. However, experimentally this VHDA form of silicon is still elusive [269,270].

Our recent XRD and Raman results on porous silicon have shown that unlike earlier studies porous silicon does not amorphize at high pressure but mimics the high-pressure behavior of nanocrystalline silicon. It transforms to the "ph" phase ~23 GPa and amorphizes only on release of pressure. However, on repressurizing this LDA phase of silicon, amorphous porous silicon (a-ps), was found to transform to the HDA phase, and beyond 19 GPa the Raman signal was lost indicating that it may have transformed to the "ph" phase. On release of pressure the amorphous phase was retrieved. An interesting part of this study is that this transformation is reproducible even after two pressure cycles [232,265].

We have observed that bulk "a-Si" also transforms from LDA to HDA on application of pressure. On further pressurizing HDA silicon, it transforms partially, to a poorly crystallized "ph" phase. This further transforms to a Cmca phase. Also depending on the rate of release of pressure and time spent in the stability regime of the different crystalline phases, bulk amorphous silicon crystallizes or transforms to the LDA phase (shown in Fig. 1.17). This behavior is distinct from the high-pressure behavior of "a-ps," which is essentially in the nano form.

4.5.2 GeO_2

Network forming glasses like silica, BeF_2, amorphous GeO_2, etc. have been extensively studied both theoretically and experimentally, as they are important materials used in industry and also have significant geophysical relevance. Understanding the structural changes of disordered materials at high pressure is a nontrivial task. These oxide glasses have been shown to undergo low-density to high-density amorphous phase transitions. Though, almost two decades ago classical molecular dynamical calculations had shown that silica (amorphous SiO_2) undergoes a phase transition from four coordinated to six coordinated state via an intermediate five coordinated state [214,271], only recent [272–274] X-ray diffraction and O *K*-edge resonant Raman scattering studies have confirmed its presence. These studies have also shown the Si-O coordination in amorphous SiO_2 increases to values larger than 6 at pressures ~ 140 GPa.

Since GeO_2 is similar to SiO_2, and earlier studies, based on structural considerations suggest that it is a high-pressure analog of SiO_2, one can expect that its high-pressure behavior would be similar to SiO_2, but at much lower pressures. There have been several high-pressure studies on amorphous GeO_2 that have shown that the densification, and concomitant continuous coordination changes, from 4 to 5 to 6 are associated with an amorphous to amorphous transition. However, X-ray and neutron diffraction studies of Gutherie et al. [275] claimed that they had observed an intermediate state between 6 and 10 GPa where the average Ge-O coordination was a constant equal to 5. This is at variance with SiO_2 where it had been observed that Si-O coordinations do not change abruptly from 4 to 6 but continuously evolve from 4 to 6 via an intermediate 5 coordinated phase [5].

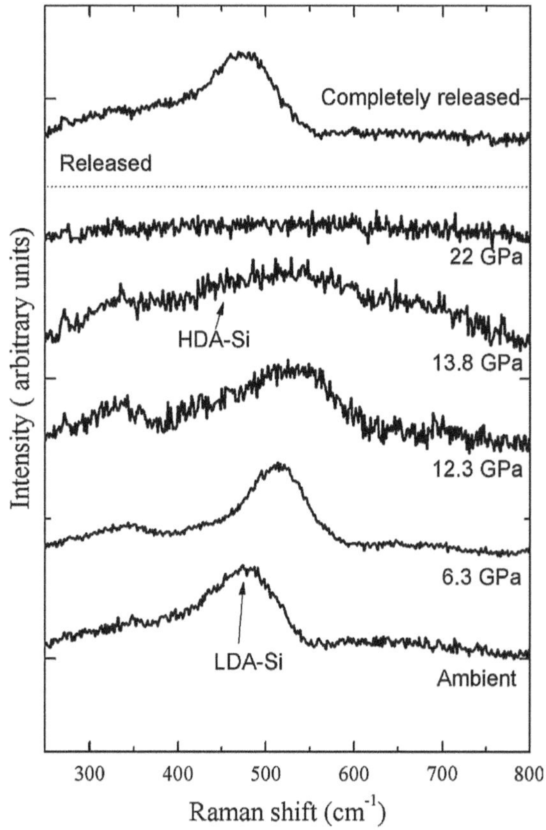

FIGURE 1.17

Raman spectra of bulk a-Si at a few pressures when pressure was released abruptly.

Figure reproduced with permission from K.K. Pandey, N. Garg, K.V. Shanavas, S.M. Sharma, S.K. Sikka, J. Appl. Phys. 109 (2011) 113511. Copyright (2011), AIP Publishing LLC.

Using the rescaled pair potentials of Oeffner and Elliot, we simulated the high-pressure behavior of amorphous GeO_2 with the help of classical molecular dynamics simulations [5,276]. This potential is known to reproduce the vibrational density of states of both the quartz and rutile phases of GeO_2. Our studies show that amorphous GeO_2 continuously evolves from a four to almost six coordinated state at 30 GPa (shown in Fig. 1.18). These results were in agreement with that of Micoulat [277]. However, these are at variance with experimental results where it was shown that at 15 GPa itself, amorphous GeO_2 is fully 6 coordinated. Our studies also showed that even though we find that ~50% of the Ge-O atoms are 5 coordinated below 7 GPa there is no signature of a pure pentahedral state below this pressure. To rule out the effect of deviatoric stresses in the experiments we also carried out simulations in the presence of nonhydrostatic stresses and found that it also did not aid the formation of a pure pentahedral state. Our studies also showed that similar to amorphous SiO_2 even vitreous GeO_2 remains in a densified state on release of pressure. It is heartening to observe that our studies spurred

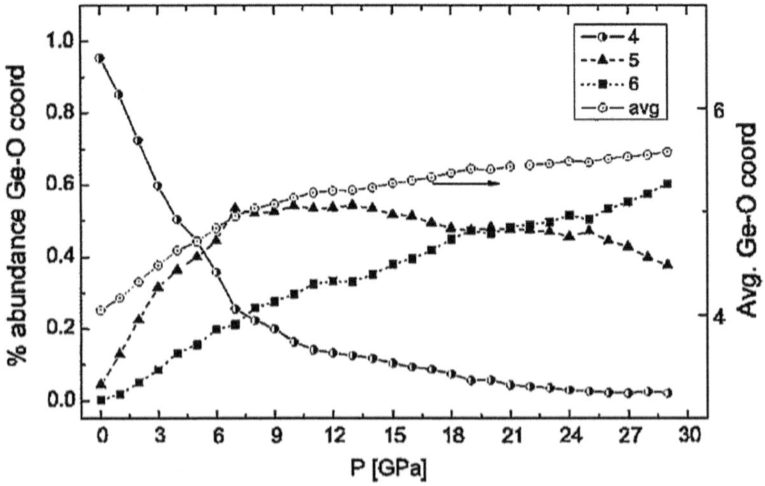

FIGURE 1.18

Average coordination of Ge-O and variation of abundance of Ge-O coordination as a function of pressure.

Reproduced from K.V. Shanavas, N. Garg, S.M. Sharma, Phys. Rev. B 73 (2006) 094120 with copyright permission from APS.

experimental investigations on vitreous GeO_2. EXAFS experiments of Baldini et al. have shown that indeed the transformation to six coordinated Ge-O is gradual and is complete only after 30 GPa. They have also ruled out the possibility of a pure pentahedral state. Their studies also support the densification of amorphous GeO_2 on release of pressure [278]. Recent neutron diffraction studies and molecular dynamics simulations using pair potentials that also incorporate dipole polarization effects are also in agreement with our studies and have shown that in amorphous GeO_2 the GeO_4 tetrahedra are replaced by GeO_5 square pyramids, and at still higher pressures the tetrahedra are also converted to GeO_6. These studies have also shown that to preserve the stoichiometry the four- to fivefold conversion of Ge-O coordination is associated with oxygen become three coordinated [279].

4.6 DISSOCIATION UNDER PRESSURE

4.6.1 Lanthanum Hafnate

It was mentioned earlier that under pressure the amorphous state arises primarily due to kinetics frustration. Mostly the daughter equilibrium phase is a denser crystalline phase. However, it can also happen that a dissociation leads to a denser phase with lower free energy [205]. This may have implications on the usefulness of some materials for some specific applications. For example, compounds having the pyrochlore lattice, mostly of $A_2B_2O_7$ type (where the B cation is transition metal element Zr, Ti, etc.) are known to retain considerable amounts of actinides and lanthanides and hence are promising candidates for fixation of nuclear waste. Materials to be used for nuclear encapsulants should have high chemical durability, high solubility for actinides and other radionuclides, radiation resistance, chemical durability in water, long-term chemical and mechanical durability under sudden thermodynamic changes such as under earthquakes, etc., retention of trace elements because of low diffusivity, etc. Several pyrochlores like $Gd_2Zr_2O_7$ and $Er_2Zr_2O_7$, have been found to be radiation resistant and hence are attractive candidates for the immobilization of actinides.

However, there are several studies that have shown that pyrochlores amorphize directly or via an intermediate monoclinic phase at high pressures. An example is gadolinium zirconate (a strong candidate for immobilization of nuclear waste), which has been found to transform to a monoclinic phase at pressures as low as ~3 GPa [281]. Lanthanum hafnate having a pyrochlore structure is also one of the potential compounds that can be used in the nuclear waste disposal (due to the presence of Hf as a neutron absorber). Our X-ray diffraction and Raman spectroscopic studies supported by our first principles calculations show that lanthanum hafnate dissociates into its constituent oxides (HfO_2 and La_2O_3) at high pressure (shown in Fig. 1.19) [280]. These studies showed that the resulting oxides are disordered at high pressure and retain this disordered state on release of pressure. However, with time this amorphous state partially transforms to low-pressure crystalline forms of HfO_2 and La_2O_3. This is the first time that a pyrochlore has been shown to decompose at high pressure. Our studies

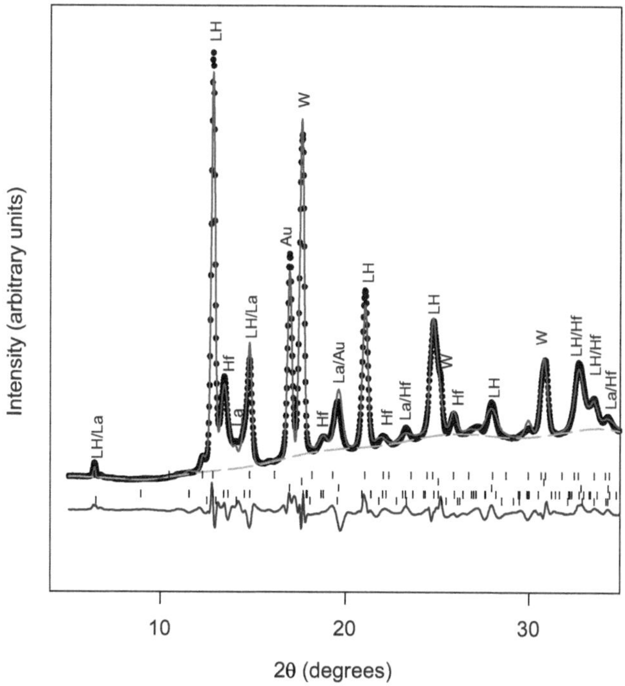

FIGURE 1.19

Rietveld fits to the recorded diffraction pattern of lanthanum hafnate (LH) at 26 GPa (red (dark gray in print versions)). The *green dashed line* (light gray in print versions) shows the subtracted background and vertical bars give the expected positions of the diffraction peaks from the sample (LH), the gasket (w), and the pressure marker (Au). The difference in the calculated and experimental diffraction pattern is given at the bottom of the graph (blue (gray in print versions)). The diffraction peaks from lanthanum hafnate, hafnium oxide (orthorhombic-II) and lanthanum oxide, gold and tungsten are marked as LH, Hf, La, Au, and W, respectively.

Reproduced from N. Garg, K.K. Pandey, C. Murli, K.V. Shanavas, B.P. Mandal, A.K. Tyagi, S.M. Sharma, Phys. Rev. B 77 (2008) 214105 with copyright permission from APS.

indicate that lanthanum hafnate and other compounds showing similar behavior may not be a suitable nuclear encapsulant and should encourage careful studies on other materials of this class.

4.7 MOLECULAR SOLIDS

Physical properties of minerals or molecular solids are a direct consequence of bonding within these minerals or molecular solids. Differences in interatomic forces along different directions may result in anisotropic physical properties. Out of the different types of bonding in molecular solids, biological molecules, and hydrous substances, hydrogen bond is the most important directional intermolecular interaction. Despite the fact that it is a weak interaction, it is responsible for the different properties of these compounds. For example, the high heat of vaporization, strong surface tension, high specific heat, and nearly universal solvent properties of water can be attributed to the formation of hydrogen bonds.

A hydrogen bond is defined as an attractive interaction between the more electropositive hydrogen atom of a group X−H (mostly covalently bonded and called the acceptor) and an atom or a group of atoms A (donor), in the same or different molecule(s). Long hydrogen bonds are weak and are characterized by a double-well proton transfer potential and on reducing the distance between the acceptor and donor this evolves into a single-well potential where the hydrogen atom sits at the center of the acceptor and donor forming a symmetric hydrogen bond [282−287]. A symmetric hydrogen bond facilitates proton migration due to reduction in potential barrier and hence is of great importance in the study of enzymatic reactions and hydrogenous materials [288]. Hydrogen bonds have been widely studied due to their importance in novel functional materials, biology, proteins, DNA, crystal engineering, and supramolecular assemblies [289−291]. Recent studies have shown that the ordering of the hydrogen bond at low temperatures in metal organic hybrids is responsible for its ferroelectric properties [292].

Application of pressure varies bond strengths, facilitates molecular reorientations, thus enabling close packing and different structural rearrangements, leading to different conformational, structural, and ionic forms of molecular solids. In the case of hydrogen bonded solids it can result in formation, strengthening or weakening of hydrogen bonds, proton disorder, symmetrization, etc. [293−296] It can also lead to saturation of the unsaturated bonds of molecular solids (having high carbon to hydrogen ratio) leading to polymerization [297].

Our studies have shown that the intramolecular hydrogen bond in glycine lithium sulfate strengthens at high pressures. This leads to a transformation of the zwitterionic glycine to the neutral form of glycine. At higher pressures strengthening of the intersheet hydrogen bonding interactions leads to a reversible noncrystalline polymeric phase of glycine lithium sulfate [298]. Acrylamide (C_3H_5NO), a monomer of an important polymer polyacrylamide, is a hydrogen-bonded amide. It has been shown to irreversibly polymerize at high pressure through the reconstruction of the N−H—O hydrogen bonds [297]. However, imidazole, which is rich in unsaturated bonds, did not show clear signatures of polymerization even when pressurized to 20 GPa, in contrast to its derivative carnosine, which polymerized at very low pressures (2.8 GPa) [295,296].

Bis(glycinium)oxalate, an important complex of the simplest amino acid glycine, possesses a center of symmetry and has a very strong O−H—O hydrogen bond (H—O = 1.3 Å) close to the symmetrization limit. This hydrogen bond shows further strengthening at small pressures and changes discontinuously at the phase transformation across 1.6 GPa. This along with a reformation of N−H—O hydrogen bonds leads to a reorientation of the molecular units of bis glycinium oxalate into ordered layers at this pressure [299]. In this process, the compound loses its center of inversion and a new, noncentrosymmetric phase is formed [300]. Our recent studies have shown that in oxalic acid dehydrate the simplest dicarboxylic acid there is an onset of proton migration at very low pressures resulting in the formation of dianionic oxalate and hydronium ion motifs [294a].

5. FUTURE PROSPECTS

This article demonstrates the usefulness of high pressure as a thermodynamic variable to realize states that otherwise may not be accessible. In doing so it provides unique possibilities of testing the theoretical understanding of materials. There is considerable amount of work that is being pursued in the context of hydrogen storage. Metallization of hydrogen and realization of its elemental state at high pressures continues to be elusive and is still a challenging problem, which may be settled in years to come. A significant motivation for the high-pressure studies is also its relevance to gain deeper insights into the states of materials deep inside earth and planetary objects. An issue of the water budget on earth and in its mantle is of vital importance and might provide new geophysical insights. In addition, the high-pressure studies are now being pursued by using several techniques simultaneously. This, as shown through several examples in this chapter, may provide self-consistent explanation of divergent results and hence a unified understanding.

ACKNOWLEDGMENTS

Authors acknowledge Dr. R. Chidambaram and Dr. S. K. Sikka for their guidance and support through several decades of work in this field. We also acknowledge several colleagues, particularly, Drs. Chitra Murli, S. Karmakar, H. K. Poswal, A. K. Mishra, K. Shanavas, and K. K. Pandey, who have been coworkers in several studies presented here.

REFERENCES

[1] J.V. Badding, L.J. Parker, D.C. Nesting, J. Solid State Chem. 117 (1995) 229.
[2] Y. Song, Phys. Chem. Chem. Phys. 15 (2013) 14524.
[3] M. Santoro, F. Gorelli, J. Haines, O. Cambon, C. Levelut, G. Garbarino, PNAS 108 (2011) 7689.
[4] N. Garg, C. Murli, A.K. Tyagi, S.M. Sharma, Phys. Rev. B 72 (2005) 064106.
[5] K.V. Shanavas, N. Garg, S.M. Sharma, Phys. Rev. B 73 (2006) 094120.
[6] W. Zhang, A.R. Oganov, A.F. Goncharov, Q. Zhu, S.E. Boulfelfel, A.O. Lyakhov, E. Stavrou, M. Somayazulu, V.B. Prakapenka, Z. Konôpková, Science 342 (2013) 1502.
[7] E. Gregoryanz, O. Degtyareva, M. Somayazulu, R.J. Hemley, H.-K. Mao, Phys. Rev. Lett. 94 (2005) 185502.
[8] M. Kim, M. Debessai, C.-S. Yoo, Nat. Chem. 2 (2010) 784.
[9] H. Tups, K. Takemura, K. Syassen, Phys. Rev. Lett. 49 (1982) 1776.
[10] W. Sun, S. Chakraborty, R. Ahuja, Appl. Phys. Lett. 103 (2013) 251901.
[11] T. Honda, T. Aoyama, J.S. White, T. Strassle, L. Keller, M. Kenzelmann, F. Honda, A. Miyake, K. Shimizu, Y. Wakabayashi, T. Kimura, Phys. Rev. B 89 (2014) 104405.

[12] K. Vedam, E.D.D. Schmidt, Phys. Rev. 146 (1966) 548.
[13] D. Pan, Q. Wan, G. Galli, Nat. Commun. 5 (2014) 1.
[14] A.R. Go, F. Kaess, J.S. Reparaz, M.I. Alonso, M. Garriga, G. Callsen, M.R. Wagner, A. Hoffmann, Z. Sitar, Phys. Rev. B 90 (2014) 045208.
[15] K.K. Pandey, J. Gyanchandani, M. Somayazulu, G.K. Dey, S.M. Sharma, S.K. Sikka, J. Appl. Phys. 115 (2014) 233513.
[16] V. Panchal, N. Garg, S.N. Achary, A.K. Tyagi, S.M. Sharma, J. Phys. Condens. Matter 18 (2006) 8241.
[17] B. Chen, K. Lutker, S.V. Raju, J. Yan, W. Kanitpanyacharoen, S.Y. Jialin Lei, H.-R. Wenk, H.-K. Mao, Q. Williams, Science 338 (2012) 1448.
[18] K. Kusaba, T. Yagi, M. Kikuchi, Y. Syono, J. Phys. Chem. Solids 47 (1986) 675.
[19] L.G. Liu, Earth Planet. Sci. Lett. 44 (1979) 390.
[20] J. Xu, D.F.R. Gilson, I.S. Butler, Spectrochim. Acta A 54 (1998) 1869.
[21] J. Xu, D.F.R. Gilson, I.S. Butler, I. Stangel, J. Biomed. Mater. Res. 30 (1996) 239.
[22] M. Murakami, K. Hirose, K. Kawamura, N. Sata, Y. Ohishi, Science 304 (2004) 855.
[23] W.L. Mao, H.-K. Mao, W. Sturhahn, J. Zhao, V.B. Prakapenka, Y. Meng, J. Shu, Y. Fei, R.J. Hemley, Science 312 (2006) 564.
[24] S. Tateno, K. Hirose, Y. Ohishi, Y. Tatsumi, Science 330 (2010) 359.
[25] W.B. Durham, D.J. Weidner, S. Karato, Y. Wang, New developments in deformation experiments at high pressure Reviews, in: Mineralogy and Geochemistry: Plastic Deformation of Minerals and Rocks, vol. 51, Mineralogical Society of America, Washington, DC, 2002, p. 21.
[26] R.S. McWilliams, D.K. Spaulding, J.H. Eggert, P.M. Celliers, D.G. Hicks, R.F. Smith, G.W. Collins, R. Jeanloz, Science 338 (2012) 1330.
[27] J.H. Eggert, D.G. Hicks, P.M. Celliers, D.K. Bradley, R.S. McWilliams, R. Jeanloz, J.E. Miller, T.R. Boehly, G.W. Collins, Nat. Phys. Lett. 6 (2010) 40.
[28] C.T. Unterborn, J.E. Kabbes, J.S. Pigott, D.M. Reaman, W.R. Panero, Astrophys. J. 793 (2014) 124.
[29] N. Madhusudhan, K.K.M. Lee, O. Mousis, Astrophys. J. Lett. 759 (2012) L40.
[30] J. Nagamatsu, N. Nakagawa, T. Muranaka, Y. Zenitani, J. Akimitsu, Nature 410 (2001) 63.
[31] I. Mirebeau, I.N. Goncharenko, P. Cadavez-Peres, S.T. Bramwell, M.J.P. Gingras, J.S. Gardnerk, Nature 420 (2002) 54.
[32] A. Schilling, M. Cantoni, J.D. Guo, H.R. Ott, Nature 363 (1993) 56.
[33] A.P. Drozdov, M.I. Eremets, I.A. Troyan, V. Ksenofontov, S.I. Shylin, Nature 525 (2015) 73.
[34] M.I. Eremets, K. Shimizu, T. Kobayashi, K. Amaya, Science 281 (1998) 1333.
[35] I. Loa, E.I. Isaev, M.I. McMahon, D.Y. Kim, B. Johansson, A. Bosak, M. Krisch, Phys. Rev. Lett. 108 (2012) 045502.
[36] P.-E. Janolin, P. Bouvier, J. Kreisel, P.A. Thomas, I.A. Kornev, L. Bellaiche, W. Crichton, M. Hanfland, B. Dkhil, Phys. Rev. Lett. 101 (2008) 237601.
[37] I.A. Kornev, L. Bellaiche, P. Bouvier, P.-E. Janolin, B. Dkhil, J. Kreisel, Phys. Rev. Lett. 95 (2005) 196804.
[38] L. Sun, X.-J. Chen, J. Guo, P. Gao, Q.-Z. Huang, H. Wang, M. Fang, X. Chen, G. Chen, Q. Wu, C. Zhang, D. Gu, X. Dong, L. Wang, K. Yang, A. Li, X. Dai, H.-K. Mao, Z. Zhaoe, Nature 483 (2012) 67.
[39] I. Jarrige, H. Yamaoka, J.-P. Rueff, J.-F. Lin, M. Taguchi, N. Hiraoka, H. Ishii, K.D. Tsuei, K. Imura, T. Matsumura, A. Ochiai, H.S. Suzuki, A. Kotani, Phys. Rev. B 87 (2013) 115107.
[40] M. Bastea, A.C. Mitchell, W.J. Nellis, Phys. Rev. Lett. 86 (2001) 3108.
[41] R. Reichlin, K.E. Brister, A.K. McMahan, M. Ross, S. Martin, Y.K. Vohra, A.L. Ruoff, Phys. Rev. Lett. 62 (1989) 669.
[42] F. Siringo, R. Pucci, N.H. March, Phys. Rev. B 37 (1988) 2491.
[43] M.I. Eremets, R.J. Hemley, H.K. Mao, E. Gregoryanz, Nature 411 (2001) 170.

[44] J. Sun, D.D. Klug, R. Martoňák, J.A. Montoya, M.-S. Lee, S. Scandolo, E. Tosattid, PNAS 106 (2009) 6077.
[45] D. Kurzydzowski, P. Zaleski-Ejgierd, W. Grochala, R. Hoffmann, Inorg. Chem. 50 (2011) 3832.
[46] D.E. Ramaker, L. Kumar, F.E. Harris, Phys. Rev. Lett. 34 (1975) 812.
[47] C. Friedli, N.W. Ashcroft, Phys. Rev. B 16 (1977) 662.
[48] P. Cudazzo, G. Profeta, A. Sanna, A. Floris, A. Continenza, S. Massidda, E.K.U. Gross, Phys. Rev. Lett. 100 (2008) 257001.
[49] S.T. Weir, A.C. Mitchell, W.J. Nellis, Phys. Rev. Lett. 76 (1996) 1860.
[50] M.I. Eremets, I.A. Troyan, Nat. Mater. 10 (2011) 927.
[51] G. Mazzola, S. Yunoki, S. Sorella, Nat. Commun. 5 (2014) 3487.
[52] R.E. Cohen, I.I. Naumov, R.J. Hemley, PNAS 110 (2013) 13757.
[53] S. Azadi, T.D. Kühne, JETP Lett. 95 (2012) 449.
[54] R.H. Wentorf, Science 147 (1965) 49.
[55] H.P. Bovenkerk, F.P. Bundy, R.M. Chrenko, P.J. Codella, H.M. Strong, R.H. Wentorf, Nature 365 (1993) 19.
[56] F.P. Bundy, H.T. Hall, H.M. Strong, R.H. Wentorf, Nature 176 (1955) 51.
[57] A. Sharma, G.D. Cody, R.J. Hemley, Energy Fuel 23 (2009) 5571.
[58] J.Y. Chen, L.J. Jin, J.P. Dong, H.F. Zheng, G.Y. Liu, Chin. Chem. Lett. 19 (2008) 475.
[59] H.P. Scott, R.J. Hemley, H.-K. Mao, D.R. Herschbach, L.E. Fried, W.M. Howard, S. Bastea, Nature 101 (2004) 14023.
[60] M. Santoro, F.A. Gorelli, R. Bini, A. Salamat, G. Garbarino, C. Levelut, O. Cambon, J. Haines, Nat. Commun. 5 (2014) 3761.
[61] S. Wang, H. Mao, X. Chen, W. Mao, Proc. Natl. Acad. Sci. USA 106 (2009) 14763.
[62] W. Mao, H. Mao, Proc. Natl. Acad. Sci. USA 101 (2004) 708.
[63] V.V. Brazhkin, High Pressure Res. 27 (2007) 333.
[64] Q.-S. Zeng, Y. Ding, W.L. Mao, W. Luo, A. Blomqvist, R. Ahuja, W. Yang, J. Shuh, S.V. Sinogeiking, Y. Mengg, D.L. Brewei, J.-Z. Jianga, H.-K. Mao, PNAS 106 (2009) 2515.
[65] H.R.J. Wentorf, Chem. Phys. 26 (1957) 956.
[66] V.L. Solozhenko, O.O. Kurakevych, Y. Le Godec, Adv. Mater. 24 (2012) 1540.
[67] H. Sumiya, T. Irfune, Sci. Tech. Rev. 66 (2008) 85.
[68] Y. Tian, B. Xu, D. Yu, Y. Ma, Y. Wang, Y. Jiang, W. Hu, C. Tang, Y. Gao, K. Luo, Z. Zhao, Li-M. Wang, B. Wen, J. He, Z. Liu, Nature 493 (2013) 385.
[69] R. Bini, Laser-assisted high-pressure chemical reactions, Acc. Chem. Res. 37 (2004) 95.
[70] M. Citroni, M. Ceppatelli, R. Bini, V. Schettino, J. Phys. Chem. B 111 (2007) 3910.
[71] M. Ceppatelli, R. Bini, V. Schettino, PNAS 106 (2009) 11454.
[72] M. Ceppatelli, S. Fanetti, M. Citroni, R. Bini, J. Phys. Chem. B 114 (2010) 15437.
[73] M. Ceppatelli, R. Bini, M. Caporali, M. Peruzzini, Angew. Chem. Int. Ed. 52 (2013) 2313.
[74] J. Schiøtz, K.W. Jacobsen, Science 301 (2003) 1357.
[75] J. Schiøtz, F.D. Di Tolla, K.W. Jacobsen, Nature 391 (1998) 561.
[76] S.M. Gruner, in: A. Katrusiak, P. Macmillan (Eds.), High Pressure Crystallography, 2004, p. 543.
[77] P.W. Bridgman, J. Biol. Chem. 19 (1914) 511.
[78] R.M. Hazen, N. Boctor, J.A. Brandes, G.D. Cody, R.J. Hemley, A. Sharma Jr., H.S. Yoder, J. Phys. Condens. Matter 14 (2002) 11489.
[79] P. Masson, C. Tonello, C. Balny, J. Biomed. Biotechnol. 1 (2001) 85.
[80] F. Abe, C. Kato, K. Horikoshi, Trends Microbiol. 7 (1999) 447.
[81] V. Mozhaev, K. Heremans, J. Frank, P. Masson, C. Balny, Trends Biotechnol. 12 (1994) 493.
[82] H. Ludwig, in: Advances in High Pressure Bioscience and Biotechnology, Springer-Verlag, Berlin, 1999.
[83] R. Hayashi, C. Balny, in: High Pressure Bioscience and Biotechnology, Elsevier, Amsterdam, 1996.

[84] C. Balny, R. Hayashi, K. Heremans, P. Masson, in: High Pressure and Biotechnology, John Libbey Eurotex, London & Paris, 1992.
[85] J. Ahmed, S.K. Varshney, J.-X. Zhang, H.S. Ramaswamy, J. Food Eng. 93 (2009) 308.
[86] W.J. Evans, C.-S. Yoo, G.W. Lee, H. Cynn, M.J. Lipp, K. Visbeck, Rev. Sci. Instrum. 78 (2007) 073904.
[87] T. Katsura, K. Funakoshi, A. Kubo, N. Nishiyama, Y. Tange, Y. Sueda, T. Kubo, W. Utsumi, Phys. Earth Planet. Inter. 143 (2004) 497.
[88] E. Ito, D. Yamazaki, T. Yoshino, H. Fukui, S. Zhai, A. Shatzkiy, T. Katsura, Y. Tange, K. Funakoshi, Earth Planet. Sci. Lett. 293 (2010) 84.
[89] L. Dubrovinsky, N. Dubrovinskia, V.B. Prapenka, A.M. Abakumov, Nat. Commun. 3 (2012) 1163.
[90] (a) N. Dubrovinskaia, et al., Sci. Adv. (2016) e1600341;
(b) L. Dubrovinsky, N. Dubrovinskaia, E. Bykova, M. Bykov, V. Prakapenka, C. Prescher, K. Glazyrin, H.-P. Liermann, M. Hanfland, M. Ekholm, Q. Feng, L.V. Pourovskii, M.I. Katsnelson, J.M. Wills, I.A. Abrikosov, Nature 525 (2015) 226.
[91] A.F. Goncharov, J.A. Montoya, N. Subramanian, V.V. Struzhkin, A. Kolesnikov, M. Somayazulu, R.J. Hemley, J. Synchrotron. Rad. 16 (2009) 769.
[92] S. Deemyad, I.F. Silvera, Phys. Rev. Lett. 100 (2008) 155701.
[93] R.J. Hemley, P.M. Bell, H.K. Mao, Science 237 (1987) 605.
[94] R.J. Hemley, H.-K. Mao, G. Shen, J. Badro, P. Gillet, M. Hanfland, D. Häusermann, Science 276 (1997) 1242.
[95] R.J. Hemley, Rigaku J. 6 (1989) 2.
[96] T. Takahashi, W.A. Bassett, Science 145 (1964) 483.
[97] V. Swamy, A. Kuznetsov, L.S. Dubrovinsky, P.F. McMillan, V.B. Prakapenka, G. Shen, B.C. Muddle, Phys. Rev. B 96 (2006) 135702.
[98] A.K. Mishra, N. Garg, K.K. Pandey, K.V. Shanavas, A.K. Tyagi, S.M. Sharma, Phys. Rev. B 81 (2010) 104109.
[99] M.I. McMahon, R.J. Nelmes, Phys. Rev. Lett. 78 (1997) 3884.
[100] X.S. Wei, W. Xu, K. Xia, Scripta Mater. 93 (2014) 32.
[101] K.J. Kingma, C. Meade, R.J. Hemley, H.K. Mao, D.R. Veblen, Science 259 (1993) 666.
[102] P. Ghalsasi, H.K. Poswal, R. Tripathi, N. Garg, M.N. Deo, S.M. Sharma, J. Phys. Conf. Ser. 377 (2012) 012009.
[103] Y. Lee, T. Vogt, J.A. Hriljac, J.B. Parise, J.C. Hanson, S.J. Kimk, Nature (London) 420 (2002) 485.
[104] R.M. Hazen, L.W. Finger, J. Appl. Phys. 56 (1984) 1838.
[105] R.M. Hazen, Science 219 (1983) 1065.
[106] T. Kenichi, Phys. Rev. B 60 (1999) 6171.
[107] R. Resel, M. Oehzelt, K. Shimizu, A. Nakayama, K. Takemura, Solid State Commun. 129 (2004) 103.
[108] D.L. Decker, S. Petersen, D. Debray, M. Lambert, Phys. Rev. B 19 (1979) 3552.
[109] D. Machon, V.P. Dmitriev, P. Bouvier, P.N. Timonin, V.B. Shirokov, H.-P. Weber, Phys. Rev. B 68 (2003) 144104.
[110] J. Haines, J.M. Lé ger, F. Gorelli, M. Hanfland, Phys. Rev. Lett. 87 (2001) 15503.
[111] R.J. Angel, M. Bujak, J. Zhao, G. Diego Gattac, S.D. Jacobsen, J. App. Crys. 40 (2007) 26.
[112] S. Klotz, J.-C. Chervin, P. Munsch, G. Le Marchand, J. Phys. D: Appl. Phys. 42 (2009) 075413.
[113] G.J. Piermarini, S. Block, J.D. Barnett, J. Appl. Phys. 44 (1973) 5377.
[114] I. Fujishiro, G.J. Piermarini, S. Block, R.G. Munro, C.M. Backman, T. Johannison, L. Tegner (Eds.), High Pressure in Research and Industry, 1981, p. 608. Proc. 8th AIRAPT Conf. Uppsala.
[115] V.A. Sidorov, O.B. Tsiok, Fizika I Tekhnika Vysokikh Davlenii 34 (1991) 74.
[116] J.D. Barnett, C.D. Bosco, J. Appl. Phys. 40 (1969) 3144.
[117] P.M. Bell, H.K. Mao, Carnegi Inst. Wash. Year Book 80 (1981) 404.

[118] S.R. Le, S.A. Ekberg, L.H. Jones, R.L. Mills, L.A. Schwalbe, D. Schiferl, Solid State Comm. 32 (1979) 131.
[119] H.K. Mao, P.M. Bell, Science 103 (1979) 1004.
[120] D.H. Lienbenberg, Phys. Lett. A 73 (1979) 74.
[121] K. Asaumi, A.L. Ruoff, Phys. Rev. B 33 (1986) 5633.
[122] M.I. Eremets, High Pressure Experimental Methods, Oxford university press, Oxford, New York, Tokyo, 1996.
[123] M. Nicol, K.R. Hirsch, W.B. Holzapfel, Chem. Phys. Lett. 68 (1979) 49.
[124] G.N. Peggs, High Pressure-High Temp. 12 (1980) 1.
[125] G.E. Duvall, R.A. Graham, Rev. Mod. Phys. 49 (1977) 523.
[126] J.C. Jamieson, J.N. Fritz, M.H. Manghnani, in: S. Akimoto, M.H. Manghnani (Eds.), High Pressure Research in Geophysics, D. Reidel, Boston, 1982, p. 27.
[127] J. Zaug, E. Abramson, J.M. Brown, L.J. Slutsky, in: Y. Syono, M.H. Manghnani (Eds.), High Pressure Research: Applications to Earth and Planetary Sciences, 1992, p. 157.
[128] J.M. Brown, L.J. Slutsky, K.A. Nelson, L.T. Cheng, Science 241 (1989) 65.
[129] A.W. Smith, A.W. Lawson, J. Chem. Phys. 22 (1954) 351.
[130] S.M. Dorfman, V.B. Prakapenka, Y. Meng, T.S. Duffy, J. Geophys. Res. 117 (2012) B08210.
[131] B. Li, K. Woody, J. Kung, J. Geophys. Res. 111 (2006) B11206.
[132] C.S. Zha, H.K. Mao, R.J. Hemley, Proc. Natl. Acad. Sci. USA 97 (2000) 13494.
[133] W.B. Holzapfel, High Pressure Res. 30 (2010) 372.
[134] K. Jin, Q. Wu, H. Geng, X. Li, L. Cai, X. Zhou, High Pressure Res. 31 (2011) 560.
[135] P.T.T. Wong, D.J. Moffat, F.L. Baudais, Appl. Spect. 39 (1985) 733.
[136] P.T.T. Wong, D.J. Moffat, Appl. Spect. 43 (1989) 1279.
[137] D.D. Klug, E. Whalley, Rev. Sci. Instrum. 54 (1983) 1205.
[138] Y. Akahamaa, H. Kawamura, J. Appl. Phys. 100 (2006) 043516.
[139] W. Qiu, P.A. Baker, N. Velisavljevic, Y.K. Vohra, S.T. Weir, J. Appl. Phys. 99 (2006) 064906.
[140] D. Schiferl, M. Nicol, J.M. Zaug, S.K. Sharma, T.F. Cooney, S.-Y. Wang, T.R. Anthony, J.F. Fleischer, J. Appl. Phys. 82 (1997) 3256.
[141] H.K. Mao, P.M. Bell, J.W. Shaner, D.J. Steinberg, J. Appl. Phys. 49 (1978) 3276.
[142] R.G. Munro, G.J. Piermarini, S. Block, W.B. Holzapfel, J. Appl. Phys. 57 (1985) 165.
[143] G.J. Piermarini, S. Block, J.D. Barnett, R.A. Forman, J. Appl. Phys. 46 (1975) 2774.
[144] R.A. Forman, G.J. Piermarini, J.D. Barnett, S. Block, Science 176 (1972) 284.
[145] S.M. Sharma, Y.M. Gupta, Phys. Rev. B 43 (1991) 879.
[146] K. Syassen, High Pressure Res. 28 (2008) 75.
[147] A.D. Chijioke, W.J. Nellis, A. Soldatov, I.F. Silvera, J. Appl. Phys. 98 (2005) 114905.
[148] I.F. Silvera, A.D. Chijioke, W.J. Nellis, A. Soldatov, J. Temp. Phys. Stat. Sol. 244 (2007) 460.
[149] J.H. Eggert, F. Moshry, W.J. Evans, K.A. Goettel, I.F. Silvera, Phys. Rev. B 44 (1991) 7202.
[150] J.H. Eggert, K.A. Goettel, I.F. Silvera, Phys. Rev. B 40 (1989) 5724.
[151] R.J. Hemley, C.S. Zha, A.P. Jephcoat, H.K. Mao, L.W. Finger, Phys. Rev. B 39 (1989) 11820.
[152] H.K. Mao, J. Xu, P.M. Bell, J. Geophys. Res. 91B (1986) 4673.
[153] J.A. Xu, H.K. Mao, P.M. Bell, Science 232 (1986) 1404.
[154] K.V. Khishchenko, P.R. Levashov, I.V. Lomonosov, Shock Wave Database, 2001. http://teos.ficp.ac.ru/rusbank.
[155] P.I. Dorogokupets, A.R. Oganov, Phys. Rev. B 75 (2007) 024115.
[156] M.W. Doherty, V.V. Struzhkin, D.A. Simpson, L.P. McGuinness, Y. Meng, A. Stacey, T.J. Karle, R.J. Hemley, N.B. Manson, L.C.L. Hollenberg, S. Prawer, Phys. Rev. Lett. 112 (2014) 047601.
[157] S. Ono, K. Mibe, Y. Ohishi, J. Appl. Phys. 116 (2014) 053517.
[158] C. Nathalie, A. Kavner, J. Phys. Condens. Matter 18 (2006) S1039.

[159] D.L. Decker, J. Appl. Phys. 42 (1971) 3239.
[160] K. Takemura, A. Dewaele, Phys. Rev. B 78 (2008) 104119.
[161] A. Dewaele, L.P. Loubeyre, M. Mezouar, Phys. Rev. B 70 (2004) 094112.
[162] H.C. Kang, J. Maser, G.B. Stephenson, C. Liu, R. Conley, A.T. Macrander, S. Vogt, Phys. Rev. Lett. 96 (2006) 127401.
[163] F. Döring, A.L. Robisch, C. Eberl, M. Osterhoff, A. Ruhlandt, T. Liese, F. Schlenkrich, S. Hoffmann, M. Bartels, T. Salditt, H.U. Krebs, Opt. Express 21 (2013) 19311.
[164] I.N. Goncharenko, J.M. Mignot, G. Andre, O.A. Lavrova, I. Mirebeau, V.A. Somenkov, High Pressure Res. 14 (1995) 41.
[165] Web site of Oak Ridge National Laboratory.
[166] D. Braithwate, I.N. Goncharenko, J.M. Mignot, A. Ochiai, O. Vog, Europhys. Lett. 35 (1996) 121.
[167] H. Wang, J. Mo Li, J. Phys. Condens. Matter 24 (2012) 245402.
[168] S. Kamran, K. Chen, C. Liang, Phys. Rev. B 77 (2008) 094109.
[169] M. Krischa, D.L. Farberb, R. Xud, D. Antonangelib, C.M. Aracneb, A. Berauda, T.-C. Chiangd, J. Zarestkyg, D.Y. Kimh, E.I. Isaevh, R. Ahuja, B. Johanssonh, PNAS 108 (2011) 9342.
[170] Y. Ding, D. Haskel, Y.-C. Tseng, E. Kaneshita, M. van Veenendaal, J.F. Mitchell, S.V. Sinogeikin, V. Prakapenka, H.-K. Mao, Phys. Rev. Lett. 102 (2009) 237201.
[171] S.T. Weir, J. Akella, C.A. Ruddle, Y.K. Vohra, S.A. Catledge, Appl. Phys. Lett. 77 (2000) 3400.
[172] J. Wang, W. Yang, S. Wang, X. Xiao, F. De Carlo, Y. Liu, W.L. Mao, J. Appl. Phys. 111 (2012) 112626.
[173] H. Liu, L. Wang, X. Xiao, F. De Carlo, J. Feng, H.-K. Mao, R.J. Hemley, PNAS 105 (2008) 13229.
[174] K.K. Pandey, H.K. Poswal, A.K. Mishra, A. Dwivedi, R. Vasanthi, N. Garg, S.M. Sharma, Pramana J. Phys. 80 (2013) 607.
[175] P.S. Malavi, S. Karmakar, N.N. Patel, H. Bhatt, S.M. Sharma, J. Phys. Condens. Matter 26 (2014) 125701.
[176] S.M. Sharma, S.K. Sikka, J. Phys. Chem. Solids 46 (1985) 477.
[177] M.I. McMahan, R.J. Nelmes, Phys. Rev. B 47 (1993) 8337.
[178] S.M. Sharma, H. Sankaran, S.K. Sikka, R. Chidambaram, Phys. Rev. B 36 (1987) 7730.
[179] O. Schulte, W.B. Holzapfel, Phys. Lett. A 131 (1998) 38.
[180] S.M. Sharma, Y.M. Gupta, Phys. Rev. B 58 (1998) 5964.
[181] M.D. Knudson, Y.M. Gupta, Phys. Rev. Lett. 81 (1998) 2938.
[182] A. Navrotsky, Rev. Mineral. Geochem. 44 (2001) 73.
[183] A. Jayaraman, G.A. Kourouklis, G.P. Espinosa, A.S. Cooper, L.G. Van Uitert, J. Phys. Chem. Solids 48 (1987) 755.
[184] M.B. Smirnov, A.P. Mirgorodsky, V.Y. Kazimirov, R. Guinebretière, Phys. Rev. B 78 (2008) 094109.
[185] M. Flórez, J. Contreras-García, J.M. Recio, M. Marqués, Phys. Rev. B 79 (2009) 104101.
[186] S.H. Tolbert, A.B. Herhold, L.E. Brus, A.P. Alivisatos, Phys. Rev. Lett. 76 (1996) 4384.
[187] N. Garg, S.M. Sharma, J. Phys. Condens. Matter 19 (2007) 456201.
[188] A.J. Gratz, L.D. DeLoach, T.M. Clough, W.J. Nellis, Science 259 (1993) 663.
[189] D.C. Palmer, L.W. Finger, Amer. Miner. 79 (1994) 1.
[190] D.C. Palmer, R.J. Hemley, C.T. Prewitt, Phys. Chem. Miner. 21 (1994) 481.
[191] M. Yamakata, T. Yagi, Technical report, ISSP University of Tokyo, 1996.
[192] R.T. Downs, D.C. Palmer, Amer. Miner. 79 (1994) 9.
[193] Y. Tsuchida, T. Yagi, Nature 347 (1990) 267.
[194] S. Tsuneyuki, Y. Matsui, H. Aoki, T. Tsukada, Nature 339 (1989) 209.
[195] S. Tsuneyuki, M. Tsukada, H. Aoki, Y. Matsui, Phys. Rev. Lett. 61 (1988) 869.
[196] J.S. Tse, D.D. Klug, J. Chem. Phys. 95 (1991) 9176.
[197] N.R. Keskar, J.R. Chelikowsky, Phys. Rev. B 46 (1992) 1.

[198] M. Dove, M. Tucker, S. Redfern, M. Craig, K. Trachenko, W. Marshall, D. Keen, ISIS Annual Report 1999–2000, 2000, p. 32.
[199] D.D. Klug, R. Rousseau, K. Uehara, M. Bernasconi, Y. Le Page, J.S. Tse, Phys. Rev. B 63 (2001) 104106.
[200] L. Huang, M. Durandurudu, J. Kieffer, Nat. Miner. 5 (2006) 977.
[201] B.W.H. van Beest, G.J. Kramer, R.A. van Santen, Phys. Rev. Lett. 64 (1990) 1955.
[202] H.K. Poswal, N. Garg, M. Somayazulu, S.M. Sharma, Amer. Miner. 98 (2013) 285.
[203] W. Paraguassu, M. Maczka, A.G. Souza Filho, P.T.C. Freire, J. Mendes Filho, F.E.A. Melo, L. Macalik, L. Gerward, J. Staun Olsen, A. Waskowska, J. Hanuza, Phys. Rev. B 69 (2004) 094111.
[204] J.S.O. Evans, T.A. Mary, Int. J. Inorg. Mater. 2 (2000) 143.
[205] S.M. Sharma, S.K. Sikka, Prog. Mater. Sci. 40 (1996) 1.
[206] M.B. Kruger, R. Jeanloz, Science 249 (1990) 647.
[207] H. Sankaran, S.M. Sharma, S.K. Sikka, R. Chidambaram, Pramana J. Phys. 35 (1990) 177.
[208] M.S. Somayazulu, N. Garg, S.M. Sharma, S.K. Sikka, Pramana J. Phys. 43 (1994) 1.
[209] A. Jayaraman, D.L. Wood, R.G. Maines, Phys. Rev. B 35 (1987) 8316.
[210] A. Polian, M. Grimsditch, E. Philippot, Phys. Rev. Lett. 71 (1993).
[211] S.L. Chaplot, S.K. Sikka, Phys. Rev. B 47 (1993) 5710.
[212] B. Vessal, Trans. Am. Cryst. Assoc. 27 (1991) 37.
[213] J.S. Tse, D.D. Klug, Science 255 (1992) 1559.
[214] J.S. Tse, D.D. Klug, Phys. Rev. Lett. 67 (1991) 3559.
[215] M.P. Pasternak, G.K. Rozenberg, E. Milher, M. Amanowicz, T. Zhou, U. Schwarz, K. Syassen, R.D. Taylor, M. Hanfland, Phys. Rev. Lett. 79 (1997) 4409.
[216] C. Murli, S.M. Sharma, S.K. Kulshreshtha, S.K. Sikka, Pramana J. Phys. 49 (1997) 285.
[217] J. Badro, J.P. Itie, A. Polian, Eur. Phys. J. B1 (1998) 265.
[218] N. Garg, S.M. Sharma, J. Phys. Condens. Matter 12 (2000) 375.
[219] P. Gillet, J. Badro, B. Varrel, P.F. McMillan, Phys. Rev. B 51 (1995) 11262.
[220] S.M. Sharma, N. Garg, S.K. Sikka, J. Phys. Condens. Matter 12 (2000) 6683.
[221] S.M. Sharma, N. Garg, S.K. Sikka, Phys. Rev. B 62 (2000) 8824.
[222] E.J. Baran, Spectrochim. Acta A 50 (1994) 2385.
[223] J. Pellicer-Porres, A.M. Saitta, A. Polian, J.P. Itie, M. Hanfland, Nat. Mater. 6 (2007) 698.
[224] H. Olijynk, S.K. Sikka, W.B. Holzaphel, Phys. Lett. A 103A (1984) 137.
[225] P.F. McMillan, M. Wilson, M.C. Wilding, D. Daisenberger, M. Mezouar, G.N. Greaves, J. Phys. Condens. Matter 19 (2007) 415101.
[226] J.C. Jamieson, Science 139 (1963) 762.
[227] R.H. Wentorf, J.H. Kasper, Science 139 (1963) 338.
[228] S.H. Tolbert, A.P. Alivisatos, Annu. Rev. Phys. Chem. 46 (1995) 595.
[229] H.K. Poswal, N. Garg, S.M. Sharma, E. Busetto, S.K. Sikka, G. Gundiah, F.L. Deepak, C.N.R. Rao, J. Nanosci. Nanotech. X (2005) 1.
[230] V. Domnich, Y. Gogotsi, Rev. Adv. Mater. Sci. 3 (2002) 1.
[231] S.K. Deb, M.C. Wilding, M. Somayazulu, P.F. McMillan, Nature 414 (2001).
[232] N. Garg, K.K. Pandey, K.V. Shanavas, C.A. Betty, S.M. Sharma, Phys. Rev. B 83 (2011) 115202.
[233] W.J. Weber, R.C. Ewing, L.M. Wang, J. Mater. Res. 9 (1994) 688.
[234] V. Swamy, A. Kuznetsov, L.S. Dubrovinsky, P.F. McMillan, V.B. Prakapenka, G. Shen, B.C. Muddle, Phys. Rev. Lett. 96 (2006) 135702.
[235] A. San-Miguel, Chem. Soc. Rev. 35 (2006) 876.
[236] A. Geim, K. Novoselov, Nat. Mater. 6 (2007) 183.
[237] Y. Zhu, S. Murali, W. Cai, X. Li, Ji W. Suk, J.R. Potts, R.S. Ruoff, Adv. Mater. 22 (2010) 3906.
[238] K. Novoselov, Science 306 (2004) 666.

[239] K. Novoselov, Two-dimensional atomic crystals, Proc. Natl. Acad. Sci. USA 102 (2005) 10451.
[240] V. Tung, M.J. Allen, Y. Yang, R. Kaner, Nat. Nanotechnol. 4 (2009) 25.
[241] S. Clark, K.-J. Jeon, J.-Y. Chen, C.-S. Yoo, Solid State Commun. 154 (2013) 15.
[242] W. Mao, H.K. Mao, P.J. Eng, T.P. Trainor, M. Newville, C.-C. Kao, D.L. Heinz, J. Shu, Y. Meng, R.J. Hemley, Science 302 (2003) 425.
[243] K.K. Pandey, H.K. Poswal, M.N. Deo, S.M. Sharma, K.S. Vasu, A.K. Sood, Carbon 70 (2014) 199.
[244] N. Garg, V. Panchal, A.K. Tyagi, S.M. Sharma, J. Solid State Chem. 178 (2005) 998.
[245] S. Karmakar, S.K. Deb, A.K. Tyagi, S.M. Sharma, J. Solid State Chem. 177 (2004) 4087.
[246] H. Liu, A. Secco, N. Imanaka, G. Adachi, Solid State Commun. 121 (2002) 177.
[247] C.A. Perottani, J.A.H. da Jornada, Science 280 (1998) 886.
[248] R.A. Secco, H. Liu, N. Imanaka, G. Adachi, J. Mat. Sci. Lett. 20 (2001) 1339.
[249] S.K. Sikka, J. Phys. Condens. Matter 16 (2004) S1033.
[250] A.L. Goodwin, M. Calleja, M.J. Conterio, M.T. Dove, J.S.O. Evans, D.A. Keen, L. Peters, M.G. Tucker, Science 319 (2008) 794.
[251] A.L. Goodwin, D.A. Keen, M.G. Tucker, Proc. Natl. Acad. Sci. USA 48 (2008) 18708.
[252] A.L. Goodwin, D.A. Keen, M.G. Tucker, M.T. Dove, L. Peters, J.S.O. Evans, J. Am. Chem. Soc. 130 (2008) 9660.
[253] M.J. Conterio, A.L. Goodwin, M.G. Tucker, D.A. Keen, M.T. Dove, L. Peters, J.S.O. Evans, J. Phys. Conden. Matter 20 (2008) 255225.
[254] R. Rao, S.N. Achary, A.K. Tyagi, T. Sakuntala, Phys. Rev. B 84 (2011) 054107.
[255] N. Garg, P. Ghalsasi, S.M. Sharma (To be published).
[256] D. Williams, D.E. Parting, F.J. Lincoln, J. Kouvetakis, M. O'Keeffe, J. Solid State Chem. 134 (1997) 164.
[257] J.W. Zwanziger, Phys. Rev. B 76 (2007) 52102.
[258] B.F. Hoskins, R. Robson, J. Am. Chem. Soc. 112 (1990) 1546.
[259] K.W. Chapman, J. Peter, Chupas. Am. Chem. Soc. 129 (2007) 10090.
[260] N.V.C. Shekar, T.R. Ravindran, P.C. Sahu, A.K. Arora, Indian J. Phys. 83 (2009) 1289.
[261] H.K. Poswal, A.K. Tyagi, A. Lausi, S.K. Deb, S.M. Sharma, J. Solid State Chem. 182 (2009) 136.
[262] S.H. Lapidus, G.J. Halder, P.J. Chupas, K.W. Chapman, J. Am. Chem. Soc. 135 (2013) 7621.
[263] S. Minomura, Semicond. Semimetals 21A (1984) 273.
[264] M. Imai, T. Mitamura, K. Yaoita, K. Tsuji, High Pressure Res. 15 (1996) 167.
[265] K.K. Pandey, N. Garg, K.V. Shanavas, S.M. Sharma, S.K. Sikka, J. Appl. Phys. 109 (2011) 113511.
[266] K.V. Shanavas, K.K. Pandey, N. Garg, S.M. Sharma, J. Appl. Phys. 111 (2012) 063509.
[267] A.D. Cicco, A. Congeduti, F. Coppari, J.C. Chervin, F. Baudelet, A. Polian, Phys. Rev. B 78 (2008) 033309.
[268] C.U. Kim, Yi-F. Chen, M.W. Tate, S.M. Gruner, J. Appl. Cryst. 41 (2008) 1.
[269] M. Durandurdu, D.A. Drabold, Phys. Rev. B 64 (2001) 014101.
[270] T. Morishita, Phys. Rev. Lett. 93 (2004) 055503.
[271] M.S. Somayazulu, S.M. Sharma, N. Garg, S.L. Chaplot, S.K. Sikka, J. Phys. Condens. Matter 5 (1993) 6345.
[272] T. Sato, N. Funamori, Phys. Rev. Lett. 101 (2008) 255502.
[273] T. Sato, N. Funamori, Phys. Rev. B 82 (2010) 184102.
[274] M. Wu, Y. Liang, J.-Z. Jiang, J.S. Tse, Sci. Rep. 2 (2012) 398.
[275] M. Guthrie, C.A. Tulk, C.J. Benmore, J. Xu, J.L. Yarger, D.D. Klug, J.S. Tse, H. Mao, R.J. Hemley, Phys. Rev. Lett. 93 (2004) 115502.
[276] R.D. Oeffner, S.R. Elliot, Phys. Rev. B 58 (1998) 14791.
[277] M.J. Micoulaut, J. Phys. Condens. Matter 16 (2004) L131.
[278] M. Baldini, G. Aquilanti, H.-K. Mao, W. Yang, G. Shen, S. Pascarelli, W.L. Mao, Phys. Rev. B 81 (2010) 024201.

[279] K. Wezka, P.S. Salmon, A. Zeidler, D.A.J. Whittaker, J.W.E. Drewitt, S. Klotz, H.E. Fischer, D. Marrocchelli, J. Phys. Condens. Matter 24 (2012) 502101.
[280] N. Garg, K.K. Pandey, C. Murli, K.V. Shanavas, B.P. Mandal, A.K. Tyagi, S.M. Sharma, Phys. Rev. B 77 (2008) 214105.
[281] N.R.S. Kumar, N.V.C. Shekar, P.C. Sahu, Solid State Commun. 147 (2008) 357.
[282] W.C. Hamilton, J.A. Ibers, Hydrogen Bonding in Solids, Benjamin, New York, 1968.
[283] G.A. Jaffery, An Introduction to Hydrogen Bonding, Oxford University Press, Oxford, 1997.
[284] T. Steiner, Angew. Chem. Int. Ed. 41 (2002) 48.
[285] G.C. Pimentel, A.L. McClellan, The Hydrogen Bond, W.H. Freeman, San Francisco, 1960.
[286] E. Arunan, Curr. Sci. 92 (2007) 17.
[287] S.K. Sikka, S.M. Sharma, Phase Transitions 81 (2008) 907.
[288] H. Ishikita, K.J.R. Saito, Soc. Interface 11 (2014) 20130518.
[289] R.J. Wojtecki, M. Meador, S.J. Rowan, Nat. Mater. 10 (2011) 14.
[290] J. Lewinski, J. Zachara, I. Justyniak, M. Dranka, Coord. Chem. Rev. 249 (2005) 1185.
[291] T. Aida, E.W. Meijer, S.I. Stupp, Science 335 (2012) 813.
[292] Y. Tian, A. Stroppa, Y. Chai, L. Yan, S. Wang, P. Barone, S. Picozzi, Y. Sun, Sci. Rep. 4 (2014) 6062.
[293] M. Benoit, D. Marx, M. Parrinello, Lett. Nat. 392 (1998) 258.
[294] (a) H. Bhatt, A.K. Mishra, C. Murli, A.K. Verma, N. Garg, M.N. Deo, and S.M. Sharma, Phys. Chem. Chem. Phys. 18 (2016) 8065−8074;
(b) H. Bhatt, C. Murli, A.K. Mishra, A.K. Verma, N. Garg, M.N. Deo, R. Chitra, and S.M. Sharma, J. Phys. Chem. B 120 (4) (2016) pp. 851−859.
[295] C. Murli, A.K. Mishra, S. Thomas, S.M. Sharma, J. Phys. Chem. B 116 (2012) 4671.
[296] B.B. Sharma, A.K. Verma, S. Thomas, C. Murli, S.M. Sharma, J. Phys. Chem. B 119 (2015) 372.
[297] B.B. Sharma, C. Murli, S.M. Sharma, J. Raman Spectrosc. 44 (2013) 785.
[298] A.K. Mishra, C. Murli, A.K. Verma, Y. Song, M.R.S. Kumar, S.M. Sharma, J. Phys. Chem. A 117 (2013).
[299] A.K. Mishra, C. Murli, N. Garg, R. Chitra, S.M. Sharma, J. Phys. Chem. B 114 (2010) 17084.
[300] H. Bhatt, C. Murli, N. Garg, M.N. Deo, R. Chitra, R.R. Choudhury, S.M. Sharma, Chem. Phys. Lett. 532 (2012) 57.

CHAPTER

MATERIALS UNDER SHOCK WAVES

2

Satish C. Gupta, K.D. Joshi, S. Banerjee
Bhabha Atomic Research Centre, Mumbai, India

CHAPTER OUTLINE

Materials Under Extreme Conditions. http://dx.doi.org/10.1016/B978-0-12-801300-7.00002-4

1. INTRODUCTION

The studies on the response of materials under dynamic compressions that create extreme conditions of pressure, temperature, and strain rates are of great interest from the viewpoint of technological applications and basic sciences. Such studies provide knowledge of the pressure-induced phase transitions, equation of state, and dynamic deformation behavior of materials, which are vital inputs to the computer codes for simulation of many situations such as reactor accidents, meteorite impact on Earth, and inertial confinement fusion. Besides, these studies are essential for unraveling the processes occurring at the Earth and planetary interiors and variety of astrophysical phenomena. In addition, these are important for synthesis of materials such as centimeter-sized nanopolycrystalline diamonds, which show superior mechanical properties than the natural diamonds and have been synthesized using high-pressure, high-temperature techniques [1]. On the other hand, a new physics insight is derived from high-pressure measurements, since solids can be compressed to more than twice their initial densities [2], which can significantly alter their electronic structure and can cause in them unexpected pressure-induced changes such as the metallization of oxygen at 96 GPa [3], conversion of sodium to a transparent material at 200 GPa [4], amorphization of germanium dioxide at 10 GPa [5], and transition of shock-compressed diamond to metallic fluid above 1 TPa and 10^4K [6].

High-pressure and high-temperature conditions can be achieved in materials using quasistatic or dynamic methods. In quasistatic method, material is compressed to steady-state high pressure using diamond anvil cell (DAC) or large volume press, and the pressurized material is heated to high temperatures by laser or resistive heating. The pressure and temperature can be independently varied in such experiments. The present-day techniques can achieve up to 640 GPa [7] pressure and a few thousand Kelvin temperatures [8]. The material under high-pressure, high-temperature conditions is characterized using X-ray diffraction [9] or spectroscopic techniques [10].

In dynamic methods, the state of high pressures and temperature is achieved at high strain rates by launching a dynamic compression wave—a shock wave or a ramp wave—into material either by impacting the material by a flying plate moving at high velocity, where impedance of the flying plate and its velocity determine the peak pressure in target materials, or by a rapid deposition of energy in target materials using explosives [11], lasers [12], or particle beams [13], where the rate of the energy deposition decides the peak pressure of the propagating wave. Using dynamic compression methods in laboratory, pressures of several hundred GPa, temperature of several thousand Kelvin, and strain rates exceeding 10^9/s have been achieved; and these are limited only by the capability to launch the flyer plates to higher velocities or by the availability of powerful sources providing higher rates of energy deposition. In dynamic methods, the pressure, temperature, and the strain rates are interlinked and can only be varied independently in a limited way.

In solids, a planar shock wave introduces a high strain rate, uniaxial strain compression, which lasts for a few microseconds or less, and is always accompanied by shear. For a given uniaxial strain, the shear stress supported by a material increases with increasing strain rate, and above 10^9/s it can be as large as the ideal shear strength of the material. The relaxation of shear stress with wave propagation accompanies the development of inelastic deformation in the compressed material [14–16]. The presence of shear and high strain rate in dynamic compression could alter the mechanism and kinetics of phase transitions observed in static pressure conditions. For instance, many phase transitions that need many hours for completion under static pressure can occur within a few microseconds under shock compression [17]; sometimes, a phase transition occurring under shock wave loading may not be seen on static compression. Because of very short duration of the experiment, the microscopic measurements like X-ray diffraction and Raman scattering on dynamically compressed materials are very difficult, nevertheless these are being carried out in a few laboratories [18,19]. Generally, the response of material is studied from the measurements of velocity history of its free surface.

Let us consider first some basic concepts of dynamic compression. Then experimental techniques in Section 3 will be presented. Some examples of experimental investigations on shock wave–induced response of zirconium and a zirconium–niobium alloy, and mechanical properties of aluminum alloys at high strain rates are presented in Section 4. The theoretical methods are discussed in Section 5; and examples of theoretical studies are given in Section 6 on phase transitions in carbides of thorium and uranium and also shock-induced melting in bismuth. Finally, in Section 7, the future directions for the scope of research in the field of materials response under dynamic compression are discussed.

2. DYNAMIC COMPRESSION PROCESS

The dynamic compression techniques are different from the static compression techniques in the manner that the pressure is generated in the specimen, the nature of stress state achieved in it, and the diagnostic tools used for examining the compressed state. In contrast to the static technique, where the specimen is slowly pressurized (in about a fraction of a second or more) to a uniform hydrostatic pressure by the surrounding pressurizing medium, in dynamic technique the specimen is compressed very rapidly by introducing a compression pulse from one side of the specimen. As the compression pulse propagates through the specimen, different parts of the specimen feel the uniaxial-strain compression and the resulting stress at different times. At any location inside the specimen, the stress rises in a very short time (hundreds of nanoseconds or smaller) to the maximum value and then falls back to zero; the duration of compression at any location may be a few microseconds or smaller. Because of such short duration of the compression in dynamic techniques, it is very difficult to employ the conventional microscopic measurement techniques like X-ray diffraction and Raman scattering, which are popular in static-pressure studies; and most of the time material response under dynamic compression is inferred from the macroscopic measurements like the compression wave velocity and the material velocity histories.

For achieving dynamic compression, a pressure pulse can be launched into the material by irradiating it with laser beam of suitable temporal profile. As the pressure pulse propagates into the material, it gradually changes its shape, and ultimately it travels with a rapidly rising shock front, where pressure increases from zero to the maximum, followed by a gradual release. The formation of the shock front takes place irrespective of the initial profile of the pressure pulse.

To understand dynamic compression process, let us consider a high-amplitude pressure disturbance propagating in one dimension in an ideal fluid. At a given time t_0, let the disturbance traveling to the right be represented by the pressure profile in Fig. 2.1.

The disturbance consists of two parts. In the first, the loading part, where the pressure P, and all other flow variables like density ρ, specific volume V, internal energy E, temperature T, and particle velocity u increase from point 0 on the profile with their values in the undisturbed medium gradually through point 1 and point 2 to final point 3. In the second, the unloading part, pressure decreases gradually through point 4 to point 5 to final value P_0, at point 6.

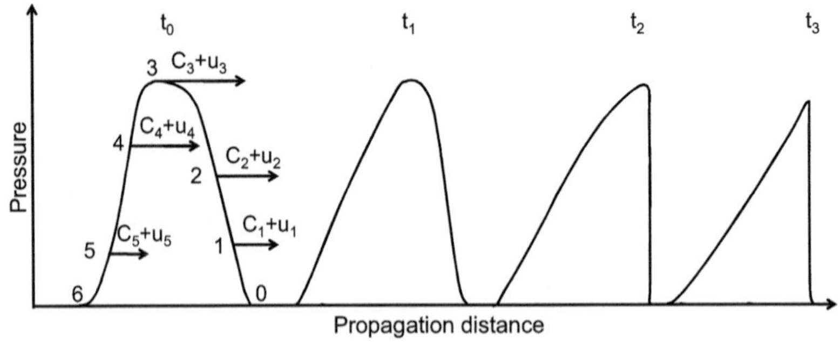

FIGURE 2.1

Propagation of a dynamic compression wave: changes in the shape of a given disturbance at time t_0 are shown at various times. The lengths of the arrows indicate the velocity of the point in the profile; this velocity increases with pressure. The rising part of the wave steepens with time, and at some time t_2 attains minimum width that does not change afterward with time. Thereafter, this compression wave moves as a front with shock velocity.

Any pressure point on the profile moves with the sound velocity plus the material velocity $C + u$. Assuming that the disturbance is occurring during a few μs timescale, so that the process is adiabatic, and assuming that the strain rates are such that dissipation effects due to viscosity and thermal conductivity are negligible, the process of loading and unloading is along the isentrope for undisturbed medium (entropy S remains constant). The disturbance compresses the material from the initial state (P_0, V_0) along the isentrope to state (P, V_S), and the decompression brings it back to the initial state. The sound velocity at any point on the profile is therefore given by $C = \sqrt{(\partial P/\partial \rho)_S}$.

Since both the rising and the falling part of the pressure disturbance are simple waves, the particle velocity along the profile is given by the following differential equation [20],

$$du = \frac{dP}{\rho C} \tag{2.1}$$

or, by the Riemann integral

$$u - u_0 = \int_{P_0}^{P} \frac{dP}{\rho C} \tag{2.2}$$

Since the compression process is nonlinear, and compressibility decreases with increasing compression for normal material, the velocity of sound at higher pressure (higher compression) is higher; so that $C_3 + u_3 > C_2 + u_2 > C_1 + u_1 > C_0$; similarly, $C_3 + u_3 > C_4 + u_4 > C_5 + u_5 > C_6 + u_6$. Therefore, as the wave propagates, the compression wave becomes steeper and steeper, whereas the release wave spreads more and more.

The steepening of the compression wave is opposed by the spreading due to the irreversible effects of viscosity and thermal conductivity, which become important when the strain rates becomes very large ($\sim 10^4 s^{-1}$). On further wave propagation, the steepening of the wave due to nonlinearity is balanced by spreading of wave due to viscosity and thermal conductivity, with the result that the compression wave travels with a constant width as a shock front with a shock velocity U. Shock front moves supersonically with respect to the undisturbed medium in front of it, but subsonically with respect to the compressed medium behind it. On further propagation through the material, the pressure pulse attenuates and the amplitude of the pressure pulse decreases since the peak of the release wave always travels faster than the shock front. In other words, the pressure disturbance propagates as an unsteady shock wave.

Across the shock front, the irreversible effects cause the entropy and temperature of the medium to increase to S_H and T_H. The release wave, however, unloads isentropically, but along the isentrope ($S_H =$ constant) passing through the compressed state (P, V). On complete unloading, the medium releases to state (P_0, V_{0H}), which is slightly expanded compared to the initial state. The entropy and temperature of the released state are more than that of the initial undisturbed medium.

The determination of material properties in the compressed state involves solving differential equations of hydrodynamics. For unsteady shock waves, the solutions of these equations are very complicated; however, for a steady shock wave, which connects the undisturbed state to the uniformly shocked state, these equations assume simpler form amenable to easy solution. Therefore, dynamic experiments are designed to propagate steady shock through the sample material.

For steady shock wave, which connects the undisturbed state to the uniformly shocked state, the conservation of mass momentum and energy relate the dynamical variables across the shock front as follows [21]:

$$\rho_0 U = \rho(U - u) \tag{2.3}$$

$$P - P_0 = \rho_0 U u \tag{2.4}$$

$$E - E_0 = \frac{1}{2}(P + P_0)(V_0 - V) \tag{2.5}$$

Eqs. (2.3)–(2.5) are known as the Rankine–Hugoniot jump conditions, or simply jump conditions.

Using Eqs. (2.3) and (2.4), the shock velocity and particle velocity can be written in terms of pressure and specific volume as:

$$U = V_0 \sqrt{\frac{(P - P_0)}{(V_0 - V)}} \tag{2.6}$$

$$u = \sqrt{(P - P_0)(V_0 - V)} \tag{2.7}$$

The locus of all the thermodynamic states achievable by propagating a single shock into a material in a given initial (P_0, V_0) is known as Hugoniot (Fig. 2.2). The loading path in the pressure volume space is a straight line, termed as Rayleigh line. The magnitude of the slope of the Rayleigh line as seen from Eq. (2.6) is equal to U^2/V_0^2. For material undergoing no phase transformation or elastic–plastic transition, the shock velocity increases monotonically with increasing shock pressure.

Shock Hugoniot of a material is different from its isotherm or its isentrope. An isotherm displays the thermodynamic path followed during compression or decompression of a material at constant temperature, implying that the states along the isotherm can be achieved from one another by compression or decompression process. Experimental isotherm of a material is derived from quasi-static loading experiments in DAC. An isentrope is, similarly, a thermodynamic path achieved in an adiabatic and reversible, hence, isentropic compression process, for example, the path followed during dynamic loading and unloading by smoothly rising or releasing wave, as shown in Fig. 2.1,

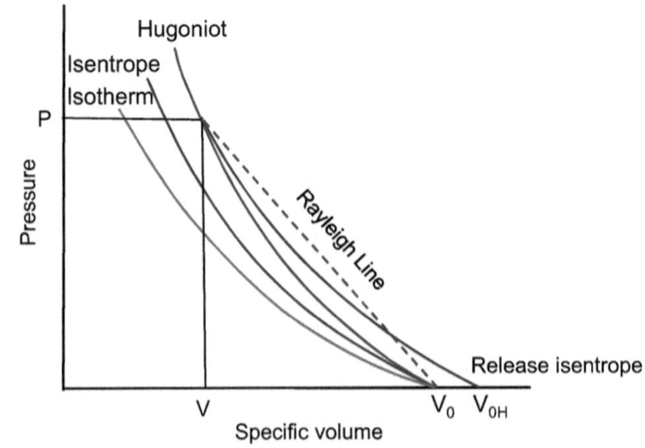

FIGURE 2.2

Schematic representation of shock Hugoniot in *P-V* plot. The Hugoniot curve lies above the isentrope, which lies above the isotherm. Hugoniot does not represent the loading or unloading path; it represents the locus of (P, V) states attainable from initial (P_0, V_0) by a single shock.

before time t_2. Generally, the isentrope is determined from the dynamic ramp wave compression experiments. Temperature does rise along the isentrope, which contributes to the thermal pressure; therefore, as shown in Fig. 2.2, the isentrope lies above the isotherm. A Hugoniot, on the other hand, does not represent a thermodynamic path, although it displays the locus of all the thermodynamic states achievable by propagating a single shock into a material in a given initial (P_0, V_0) state. As stated earlier, shock compression process is adiabatic but irreversible, and irreversible heating increases entropy and temperature. For a given compression, the rise in temperature is much higher and the corresponding thermal pressure is much higher than that in the isentropic compression. Therefore, the Hugoniot lies above the isentrope. The isentrope and the Hugoniot passing through the ambient state of material are termed as the principal isentrope and principal Hugoniot, respectively.

Unlike that in ideal fluids, where a planar dynamic compression wave generates isotropic states of strains and stress, in isotropic solids, such a wave introduces a uniaxial strain, so that the longitudinal stress is larger than the transverse stress. If the difference between the two stress components exceeds the elastic limit, the longitudinal stress relaxes to the value permitted by the dynamic strength, as the wave propagates. The compression wave splits in two parts; the first one compresses the material up to the elastic limit and travels with longitudinal sound speed, and the second one compresses up to the peak stress.

For dynamic compression in solids, it can be shown that Eqs. (2.1)–(2.7) are valid if pressure P is replaced by longitudinal stress and sound speed by longitudinal sound speed. In shock wave literature, the terms *shock pressure* and *shock stress* are used interchangeably, since the deviatoric stresses (a few GPa) are much smaller than the peak shock stress.

Shock Hugoniot is measured from a series of shock wave experiments, each compressing material from an initial state (P_0, V_0) with different magnitude of pressure P to different final volume V. Since U and u are related through Eqs. (2.3) and (2.4), measurements of any two of these quantities suffice to

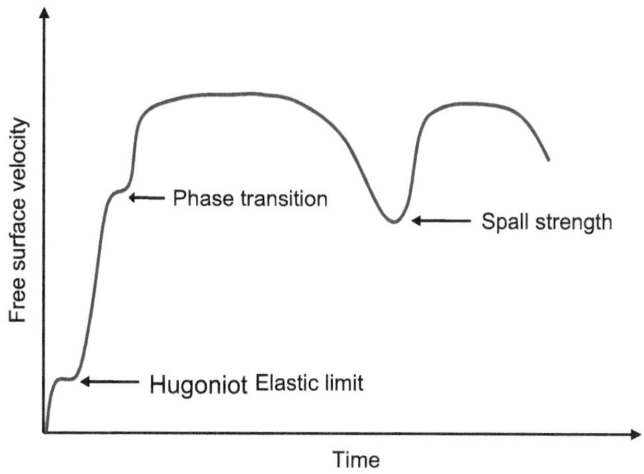

FIGURE 2.3

Schematic of a free-surface velocity history for a material undergoing elastic–plastic transition and phase transformation during shock wave loading, and undergoing spall fracture during unloading.

measure the Hugoniot. Generally, measurements of U and u are carried out. Experimentally, the relation U and u is observed to be linear for a normal material; but any deviation from linearity or a break in it indicates an elastic–plastic transition or a phase transition. Using jump conditions, the $U - u$ relation can be converted to $P - u$ representation of Hugoniot.

More detailed information about the shock wave response of the material is derived from the measurement of the particle velocity history $u(t)$ or the free surface velocity history $u_{fs}(t)$ of the material. In normal materials the particle velocity (and the corresponding peak pressure) rises to the peak value in a single jump; but, any feature in the rising part of the particle velocity history, as shown in Fig. 2.3, represents elastic-to-plastic transition or a phase transition.

2.1 INTERACTIONS OF WAVES

For generation and measurements of shock waves, it is important to understand how the waves interact with each other and with interfaces between two media. The wave interactions are analyzed using *x-t* diagram and *P-u* diagram.

An *x-t* diagram provides the location of the pressure disturbance in space and time coordinates. For illustration, consider propagation of a one-dimensional square wave pulse of amplitude P_1 in forward direction as shown in Fig. 2.4a.

Since the loading part of the wave is a steady shock and it moves with constant speed, in *x-t* diagram (Fig. 2.4b) the location of the shock front is represented by a straight line whose slope equals the inverse of the shock speed of amplitude P_1. All the (x, t) coordinates below this line correspond to the undisturbed state of the medium. The unloading part of the wave, which spreads as it moves through the material, is represented in *x-t* diagram by collection of diverging straight lines, known as rarefaction fan. The slope of each line in rarefaction fan equals the inverse of the propagation speed of corresponding pressure amplitude. The lowest line corresponds to propagation of the maximum

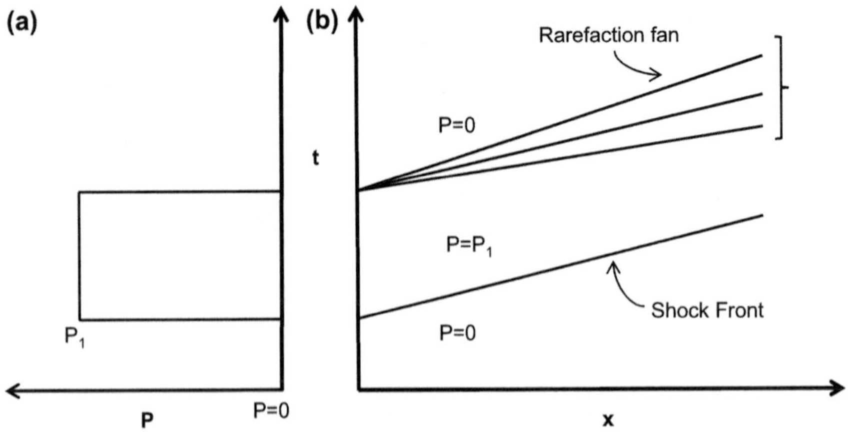

FIGURE 2.4

x-t diagram for propagation of a rectangular pulse of P_1 amplitude in a medium.

pressure of the release wave, and the uppermost line corresponds to the minimum pressure of the release wave. For all (x, t) coordinates lying between the lowest line of the rarefaction fan and the line representing shock propagation, the pressure in the medium is P_1. The pressure falls gradually across the rarefaction fan, and finally it becomes zero for all the (x, t) coordinates lying above the uppermost line of the rarefaction fan.

Conservation laws restrict the dynamic states that may be achieved due to wave propagation in a medium; additionally, P and u are constrained to be continuous at the interface of any two media. The analysis of the wave interactions is more convenient if the allowable states are represented in the P-u plane. For instance the curve OA in P-u plane (Fig. 2.5) is the plot of Eq. (2.4), and it gives all the (P, u) states reachable by a single forward facing shock wave in a medium at rest. It may be noted that there is one-to-one mapping between this curve and the Hugoniot curve in the P-V plane; in other words, this is an alternative representation of the Hugoniot. Similarly, the (P, u) states reachable by a single

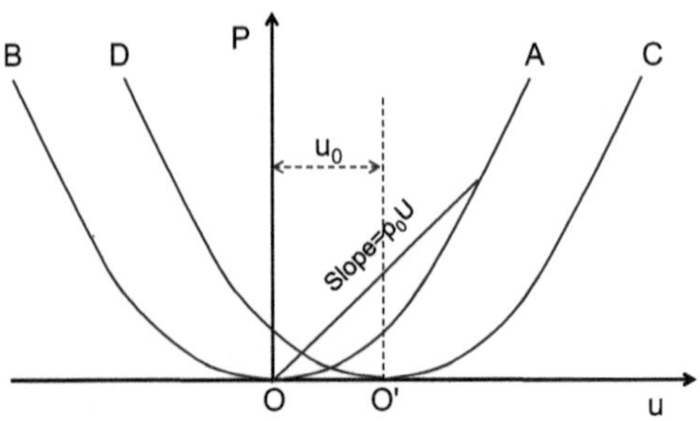

FIGURE 2.5

Representation of Hugoniot in P-u plane.

backward-facing shock wave in a medium at rest are given by the curve OB, which is the reflection of the curve OA about vertical axis passing through $u = 0$.

In the case of a shock wave propagating in the medium moving with uniform velocity u_0, the conservation of momentum [Eq. (2.4)] can be written as:

$$P - P_0 = \rho_0(U - u_0)(u - u_0) \tag{2.8}$$

So that the allowable states in P-u plot are represented by curve O′C, which is parallel to curve OA and passing through $u = u_0$. Similarly, for the backward-facing shock wave in the moving medium, the achievable states are represented by curve O′D, which is parallel to curve OB. Under a few approximations, all these curves also represent the allowable states achievable by a pressure releasing wave.

Now, a family of curves parallel to OA and other family parallel to curve OB can be drawn. For a medium in a particular state (P_1, u_1), a forward-facing disturbance can take the medium to any (P, u) state lying only on the curve parallel to OA and passing through (P_1, u_1). Similarly, a backward-facing wave into medium in state (P_1, u_1) can take it to any (P, u) state lying only on the curve parallel to OB and passing through (P_1, u_1).

2.2 GENERATION OF SHOCK WAVES

In most of the experimental techniques, shock waves are generated by impacting a flying plate onto a stationary target plate. On impact, a planar flat top shock wave is produced in target plate as illustrated in Fig. 2.6.

Consider the impact of a stationary target plate B with thickness x_b by a parallel flying plate of material A with thickness x_a and moving with velocity u_a, as shown in Fig. 2.6a. For this impact, the x-t diagram and the P-u diagram are shown in Fig. 2.6b and c, respectively. Upon impact, a forward shock propagates into the plate B and a backward shock into the plate A. These shock waves, which are represented by straight line in x-t diagram, load the material A and B to state (P_1, u_1), which can be determined from the P-u diagram.

As shown in the P-u diagram, the achievable (P, u) states in plate B, which is initially at state (0, 0), lie on the curve OB, the Hugoniot of material B; whereas the states in plate A lie on curve OA′, which is a reflection of the Hugoniot of material A passing through (0, u_a). Since the (P, u) state at the contact surface of the two plates must be same, the only permissible state is given by (P_1, u_1) lying at the intersection of the curves OA′ and OB.

With the passage of time, the backward-facing shock wave in the plate A is reflected from its rear surface as a forward-facing rarefaction wave, which is represented by a rarefaction fan in the x-t diagram. The propagation of rarefaction wave into the plate B releases the pressure. The pressure history at any location x is given in Fig. 2.6d. It is clear from the previous analysis that the amplitude P_1 of the shock wave depends on the Hugoniot of the plate materials; the shock amplitude increases with increase in velocity of flying plate; also, the width of the shock wave increases with the thickness of the flying plate.

In most of the laboratory methods, shock waves are generated by plate impact method. Various laboratory methods for shock wave generation are explosive systems [11], lasers [12], pulsed electron or ion beams [13], various guns (like gas gun [22], propellant gun [23], two stage gun [24], electric gun [25], and rail gun [26]), and magnetic flux compression [27].

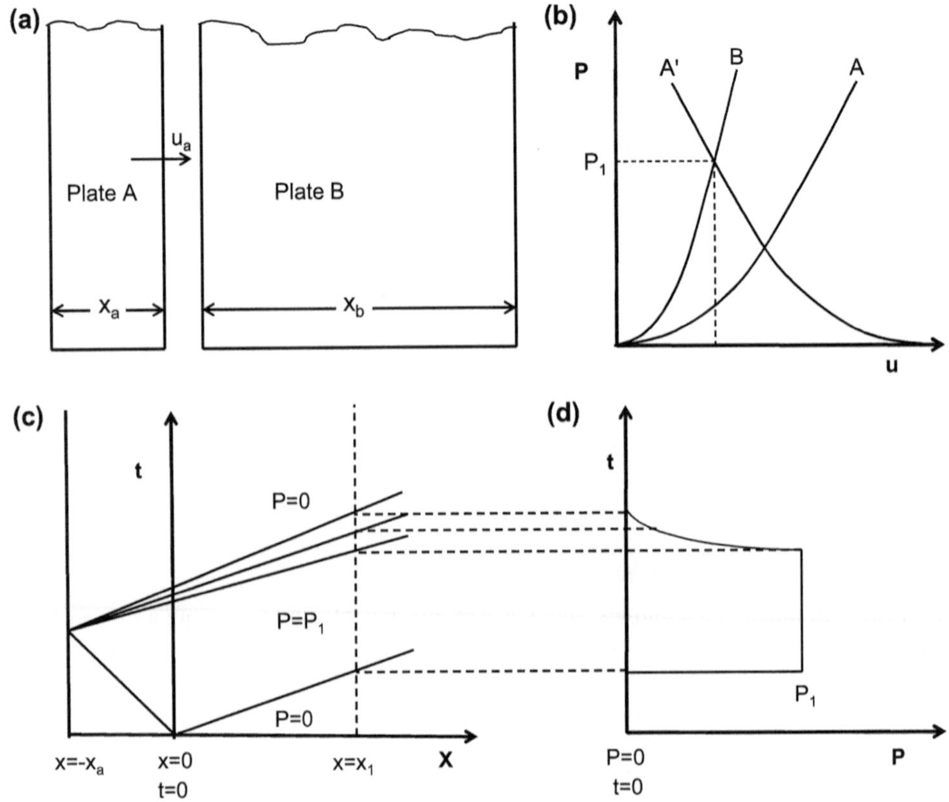

FIGURE 2.6

(a) Impact of target plate by a parallel flying plate, (b) *P-u* diagram for the impact, (c) *x-t* diagram for the wave propagation, and (d) the pressure profile at a point x_1 in the target.

Dynamic compression of solids in isentropic manner has also been achieved by ramp wave loading with laser beam by suitably tailoring the temporal profile of the incident laser [28], or by using the magnetic flux compression method [27], or by impacting the target with a graded impedance flyer plate in gun experiments [29]. The stress-density states along the isentrope are determined by recording the particle velocity histories at two thicknesses of the sample, then measuring longitudinal sound speeds at all particle velocities, and finally using Eq. (2.2). Such experiments are complementary to the shock wave experiments, as these explore the different regions of the pressure-temperature space. Unlike that in shock compression experiments, where only one datum is determined in one experiment, in isentropic compression experiment, full isentrope is generated in a single experiment.

2.3 NEGATIVE PRESSURE AND SPALLING

The suitably designed shock compression experiments can generate not only the high compressive stresses but also the high tensile stresses in materials [30]. The tensile stresses can be large enough to cause fracture during submicrosecond timescale. Such experiments provide opportunity to study fracture response of materials at high strain rates of ~10^4/s and higher. The generation of high tensile stresses and occurrence of fracture in solids in shock wave experiments is explained next.

Fig. 2.7 shows the state of the material during shock loading and unloading in a target plate impacted by a thinner flying plate of same material. In *x-t* diagram (Fig. 2.7a), the *x* is the distance from the impact surface and *t* is the time after the impact. Upon impact of the flyer plate onto the stationary target, the right-going shock wave in target takes the material from state 0 to state 1 as shown in the *P-u* diagram (Fig. 2.7b), and its propagation is represented by the line OB in the *x-t* diagram. Similarly, the left-going shock wave in flyer plate takes the material from state 3 to state 1 in the *P-u* diagram and its propagation is represented by the line OA in the *x-t* diagram. The reflection of the shock wave in target at free surface produces a rarefaction wave represented by a rarefaction fan with C_{-h} and C_{-t} as the characteristic of head and tail, respectively. This rarefaction wave progressively unloads the material

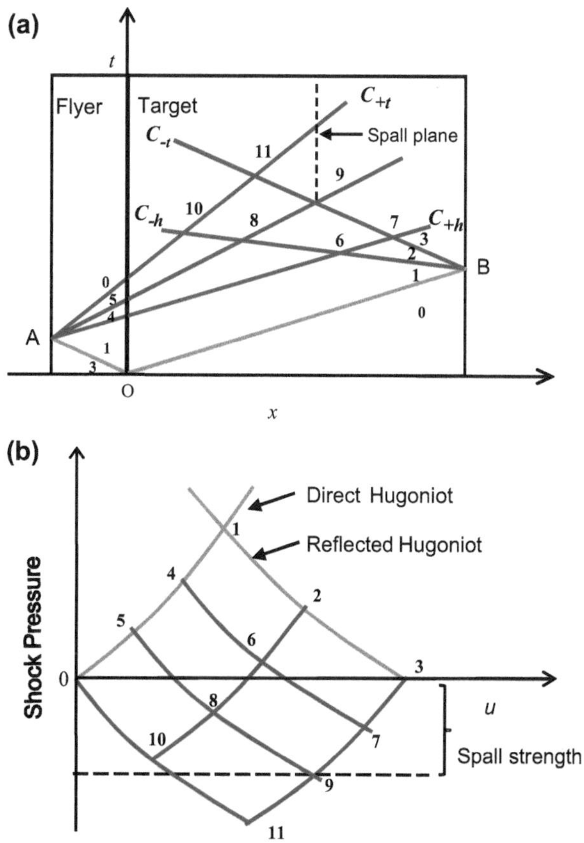

FIGURE 2.7

Generation of tensile stresses in material through plate impact experiment. The propagation of shock wave and release waves is shown in (a) *x-t* diagram and (b) in *P-u* diagram. The tensile stress corresponding to state "9" is the maximum stress that solid can withstand before spall fracture.

from state 1 to 3. Similarly, the reflection of the shock wave in the flyer at free surface results in a rarefaction wave, the propagation of which is represented by the family of C_+ characteristics; this rarefaction wave releases the pressure from state 1 to 0. The interactions of the right-going and left-going rarefaction waves result in various states in target material represented by 6, 7, 8, 9, 10, and 11. It is clear that the interactions of these rarefaction waves results in negative pressure (tensile stress) states in the target material (for example states 7, 8, 9, 10, and 11 in *P*-*u* diagram). When the negative pressure exceeds the strength of the material, creation of new surface occurs parallel to the free surface. This process is called spalling, and tensile stress at which the fracture of the material starts is known as spall strength (shown by point 9 in *P*-*u* diagram). The spalling of the material is manifested in the velocity history of the free surface (Fig. 2.3) as fall in the free surface velocity followed by an increase that results from the recompression of the spalled layer of the target material.

3. EXPERIMENTAL TECHNIQUES

A typical laboratory involved in shock wave research comprises various techniques (gas gun, propellant gun, electric gun, rail gun, etc.) for generation of shock waves in materials, and associated diagnostics (recording of pressure history, shock velocity, and particle velocity history, radiation pyrometry) to examine material response during a few microsecond shock duration with a few nanosecond time resolution. Some of these techniques are presented here.

3.1 GAS GUN

The gas gun is capable of accelerating the projectiles carrying a disc of about 5 mm thickness to velocities up to 1.2 km/s and causing a parallel impact on a target [22]. Depending on the impedances of the impactor and the target material, shock pressure up to 40 GPa can be generated.

A typical gas gun consists of four main parts: breech, barrel, target chamber, and catcher tank (Fig. 2.8). The breech is a cylindrical vessel, which can hold about 25 L of gas at high pressure up to 400 bars. The barrel is 3 m long with a smooth bore of 62.8 mm diameter. To accelerate a projectile to

FIGURE 2.8

Photograph of a typical gas gun.

a desired velocity, first the projectile is inserted into the breech, then the breech is filled with gas (nitrogen or helium) up to the required pressure, and after that this gas is released behind the projectile within a short time of a few milliseconds. Because of the differential pressure across it, the projectile accelerates while moving through the barrel. As the projectile emerges out of the barrel, a flat disc attached to front of the projectile impacts the parallel target plate of sample material mounted at the muzzle end of the barrel. Sometimes, a gun is designed to have a 3-mm-wide and 2-mm-deep slot machined at its inner diameter. A key fitted to the projectile moves along the slot in the barrel and prevents the rotation of the projectile during acceleration. Such motion of the projectile can be used for the impact of the inclined flying plate onto the parallel target plate, which causes propagating a shear wave following the compression wave in the sample target. So, it is possible to measure the shear wave velocity in the compressed state along with the compressive wave velocity in these experiments.

The target chamber is mounted at the muzzle end of the barrel. The target chamber houses the instrumented target and is fitted with feedthrough for transporting the electrical and optical signals using cables from inside the chamber to outside it and then to the recording instruments inside the control room. The target chamber and the barrel are evacuated to 10 millitorr vacuum before experiment. The catcher tank is used for arresting the target and projectile debris after the impact has taken place.

The gun is operated from a remote control room, where the data acquisition system is located. By selecting the gas and its initial pressure in the breech, a projectile can be accelerated precisely to the desired velocity. When the projectile hits the target material, a shock wave is propagated into the target material, and different shock parameters like pressure, shock wave velocity, and particle velocity are measured by different diagnostic techniques.

3.2 PROPELLANT GUN

Propellant guns, in general, are capable of accelerating projectiles to velocities much higher than achievable from single-stage gas guns. A typical propellant gun can accelerate a 500-g projectile, 40 mm in diameter, up to a velocity of about 2 km/s, which upon impact can generate shock pressure of 80 GPa in the target material.

Similar to the gas gun, the propellant gun consists of the breech, barrel, target chamber, and catcher system (Fig. 2.9). The only difference is that in propellant gun the propulsion of the projectile is done

FIGURE 2.9

Photograph of a typical propellant gun.

by a propellant. The projectile is attached to the nose of the cartridge filled with double-based solid propellant. For accelerating the projectile, the cartridge is inserted into the breach, and the burning of its propellant is triggered from the electric signal controlled remotely. On triggering, the propellant burns and turns into high-temperature, high-pressure gas, which acts behind the projectile and accelerates it, as it moves through the barrel. Finally, the projectile impacts the target mounted at the muzzle end inside the evacuated target chamber, which is permanently attached to the muzzle of the gun. Upon impact, shock parameters from the compression of the target material are measured and recorded on the data acquisition system located in the control room. After this, the moving target and the projectile debris move toward the catcher tank, which is attached to the target chamber. The moving debris enters the catcher tank by rupturing the Mylar diaphragm that initially separates the target chamber from the catcher tank, and is finally absorbed inside the catcher tank.

The projectile velocity is varied by varying the mass of the projectile, or the quantity of the propellant mass inside the cartridge, or the nature of the propellant, or by varying all of these parameters.

3.3 ELECTRIC GUN

An electric gun is a small device suitable for performing shock wave experiments on small samples. A typical electric gun is capable of accelerating about 100-μm-thick flyers of about 6 mm diameter up to 6.3 km/s, and generating up to 70 GPa shock pressures in materials like tantalum [25].

The electric gun consists of an exploding foil assembly connected to a fast capacitor bank through a triggered spark gap switch (Fig. 2.10). The exploding foil assembly is mainly a metallic foil about 25 μm thick, which rests on a 2-mm-thick Perspex tamper with a thin dielectric sheet in between. On the upper side of the metallic foil is attached a 100—150-μm-thick dielectric sheet that acts as flyer,

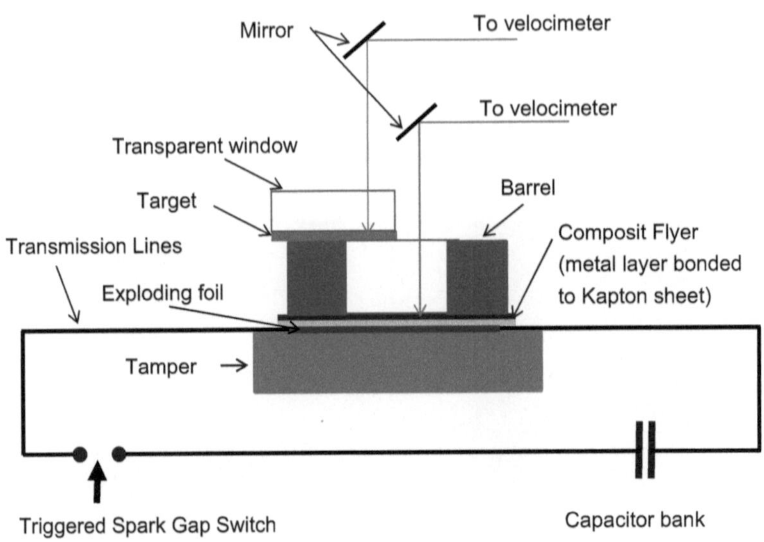

FIGURE 2.10

A schematic diagram of an electric gun.

which is pressed from above by a 2–6-mm-long steel barrel with 6–8-mm-inner diameter. The barrel is held in position by a spring-loaded assembly. If a metallic flyer is required, a thin sheet of metal is bonded over the upper dielectric sheet. The gun uses 11-μF capacitor with maximum charging voltage of 40 kV, a triggered rail-gap switch, and parallel plate transmission line. The exploding foil assembly is enclosed in a cylindrical chamber, which is fitted with a port for transporting the electrical and optical signals.

The gun is fired by discharging the capacitor bank in the electric circuit, which induces a large current through the foil in about a few μs. The large current through the foil rapidly heats the foil and increases its thermal pressure. The rapid buildup of pressure in a very short time leads to rapid expansion of the foil (generally termed as explosion of the foil), which pushes the dielectric sheet (flyer) above it, punches it out through the barrel, and accelerates it to high velocities. The flyer impacts the target material kept at the end of the barrel. Upon impact a shock is generated in the target. The response of the material to the shock loading is determined from the measurements of the flying plate velocity and the velocity history of the free surface of the target using Fabry–Pérot interferometers.

The velocity of the flyer can be varied by changing the electric gun parameters, such as the charging voltage of the capacitor bank, rise time of the current, material and mass of the flyer, and barrel length.

3.4 PARTICLE VELOCITY MEASUREMENT

The response of a dynamically compressed material is examined by taking measurements of particle velocity histories at a few chosen locations along the direction of shock wave propagation. The record of the arrival times of the wave at these locations yields shock velocity. The particle velocity and the shock velocity are then used for deriving pressure and density using the jump conditions [Eqs. (2.3) and (2.4)].

The measurement of particle velocity histories is generally carried out using techniques based on optical interferometer. The widely used techniques are described in the following section.

3.4.1 Velocity Interferometer System for Any Reflector (VISAR)

Velocity interferometer system for any reflector (VISAR) is a modified wide-angle Michelson interferometer that is used for measuring the velocity history of a moving surface. The measurement is made by recording the interference from laser light signals reflected from the moving surface at two instants of time differing by a few nanoseconds. The velocity history is derived directly from the interference history.

The schematic of a typical VISAR system [31] and its photograph are shown in Figs. 2.11 and 2.12, respectively. As shown in Fig. 2.11, a beam of 532-nm wavelength from a diode pumped Nd: Vanadate laser is incident on the moving target through an optical fiber of core diameter 125 μm. The light scattered from the target surface is collected by a 1000-μm-thick fiber and transported to the beam expander, which expands and collimates the beam to 20-mm-diameter size. This beam of light enters the interferometer, the two legs of which are designed such that their apparent lengths from detector point of view are same, but the time of arrival of light signal from one leg is delayed compared to the other beam by a few ns. This is achieved by having an air gap with mirror M_1 in one leg of the interferometer, but a glass etalon within air gap along with mirror M_2 in the other leg. With such design of interferometer, a good interference pattern is possible with even diffuse light. The fringe frequency

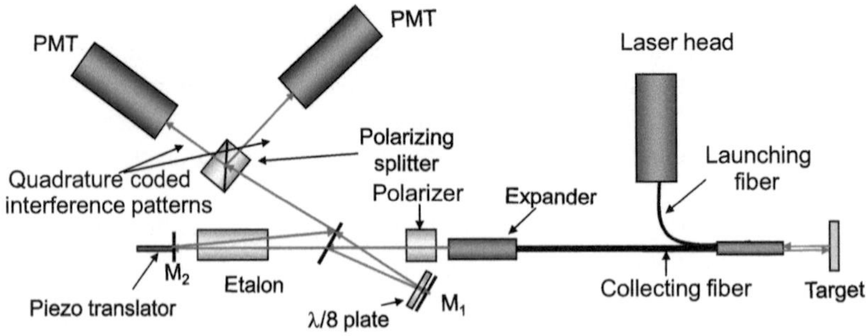

FIGURE 2.11

The schematic diagram of a VISAR system.

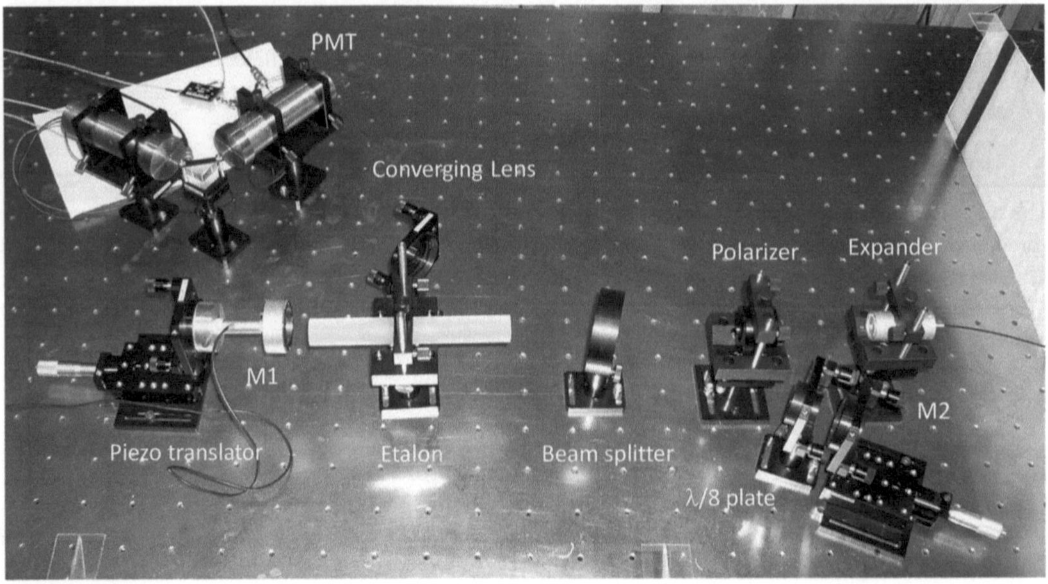

FIGURE 2.12

Photograph of a typical VISAR system.

in the interference pattern relates to the rate of change of the velocity of the moving target; the rate of velocity change can be converted to the velocity history.

In order to distinguish between the acceleration and deceleration of the target, quadrature coding is used employing polarized light. Though the laser light is initially polarized, the polarization is destroyed after scattering from the target surface. Therefore, a polarizer is placed after the beam expander in such a way that the axis of the polarization is kept at 45 degree from vertical, so that polarized output can be considered as linear combination of vertically polarized component (P) and horizontally polarized component (S).

A beam splitter splits the polarized light into two parts, one of which travels through an eighth wave retardation plate placed in one of the interferometer legs and other travels through the other leg. After traveling the eighth wave plate twice (onward and return), the P component light is retarded by 90 degree phase with respect to the S component. When the light from the two legs recombine at the beam splitter, two interference patterns 90 degree out of phase are formed. These patterns are separated by a polarizing beam splitter and recorded separately using photomultiplier tubes.

The free surface velocity is determined from fringe count using the following expression [32]:

$$u_{fs}(t) = \frac{\lambda_0}{2\tau} \times F(t)$$

where $F(t)$ is the measured fringe count, λ_0 is the wavelength of the laser, and τ is the optical delay introduced by etalon.

3.4.2 Fabry–Pérot Velocimeter

For accelerating a flyer plate, an electric gun uses discharge of a capacitor bank, which is likely to generate a large electromagnetic pulse causing interference in the data acquisition system. This entails the use of complete optical system for data acquisition, which is immune to the electromagnetic interference. Therefore, shock wave experiment with an electric gun requires the use of a Fabry–Pérot velocimeter [33], which is basically an interferometer that measures the time-varying Doppler shift in the light reflected from the moving surface. The typical capability of the instrument is measuring velocity of 0.2–10 km/s with time resolution of 0.2 km/s.

A schematic diagram of the Fabry–Pérot velocimeter is given in Fig. 2.13. A laser light of 530 nm and 5 MHz line width from a 350 mW Nd: vanadate laser is incident on the surface, the velocity of which needs to be measured. The light reflected from the target is collected by a mirror and collimated by a lens in the direction of a cylindrical lens, which converts the circular light beam to a line beam.

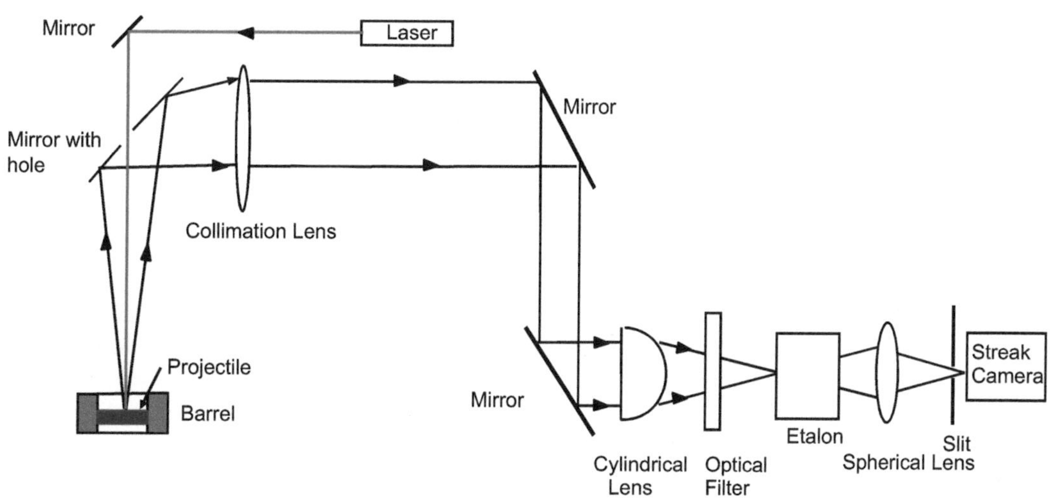

FIGURE 2.13

Schematic diagram of Fabry–Pérot velocimeter.

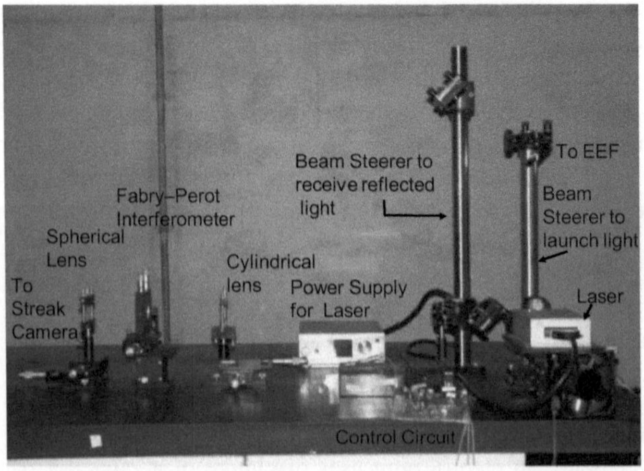

FIGURE 2.14

Photograph of a Fabry–Pérot velocimeter.

The light beam interferes in the Fabry–Pérot etalon and produces an interference pattern. Since the incident beam is not circular but a line, the interference pattern consists of dots in a straight line instead of usual circular pattern. This interference fringe pattern is focused by a spherical lens onto a streak camera and recorded. The streak camera record provides the changes in the fringe spacing and emergence of new fringes as a function of time resulting from the projectile motion.

The instantaneous velocity $u(t)$ of the projectile is derived from the following expression:

$$u(t) = \frac{c\lambda}{4d\mu}\left(n + \frac{D_{0t}^2 - D_0^2}{D_1^2 - D_0^2}\right)$$

where c is the speed of light in air and λ is the wavelength of laser light; μ and d are the refractive index and thickness of etalon; D_0 and D_1 are the diameters of the innermost fringe and the one next to it, when the projectile is at rest; and D_{0t} is the diameter of the innermost fringe at instant t, whereas n is the number of fringes emerging from the center. A setup of the Fabry–Pérot velocimeter is given in Fig. 2.14.

4. CASE STUDIES

In this section, we present two experimental investigations: (1) determination of mechanical properties of aluminum alloys at high strain rates, and (2) shock wave–induced omega-phase formation in zirconium and zirconium–niobium alloy.

4.1 HIGH STRAIN RATE MECHANICAL PROPERTIES OF AL ALLOYS, AL2014-T4 AND AL2024-T4

Aluminum alloys, which display high strength-to-weight ratio and good fracture toughness, are widely used for industrial applications like armors, automobile, and aerospace. For analyzing response of impulsive loading for such applications, the mechanical properties of these alloys under high strain rates become crucial. Gas gun experiments, which shock load and unload material with strain rate of 10^4/s–10^8/s, are best suited for measurement of such properties. Suitably designed gas gun experiment can generate not only compressive stresses but also high tensile stresses at high strain rate, so that both the dynamic yield strength and the spall strength (dynamic fracture strength) can be measured. In the following, typical measurements on two aluminum alloys, Al2014-T4 and Al2024-T4, are presented.

Three samples of both the alloys are prepared in form of disc, 8 mm thick and 57 mm in diameter. Each sample is impacted by a 5-mm-thick parallel flying plate of same material and same diameter but accelerated to the different chosen velocity in the gas gun. The velocity of the flyer plate is measured before the impact. Upon impact of the flyer onto the target, a planar forward-moving shock wave propagates in the target and a backward-moving shock wave in the flyer plate. The shock wave that compresses target material beyond elastic limit splits into two waves, the first one loading the material up to the elastic limit, and the second following it, plastically deforming the material and loading it to the peak stress. The shock wave reflects from the free surface as a backward-moving release wave. Similarly, the backward-moving shock wave reflects from the rear free surface of the flyer plate as forward-moving release wave. Since the target is thicker than the flyer plate, the two release waves intersect within the target plate and generate large tensile stress in it. In case the tensile stress exceeds the dynamic fracture strength known as spall strength, the target material undergoes dynamic fracture. This phenomenon is also known as spalling of the target.

The velocity history of the free surface of the target is derived from the recorded fringe pattern from VISAR instrument. A typical measured fringe pattern for the impact velocity of 0.37 mm/μs and the corresponding velocity history of the target free surface are given in Fig. 2.15. The velocity history shows initial jump associated with the Hugoniot elastic limit or the dynamic yielding of the material, then increase to the maximum value related to the peak loading pressure, followed by a fall in the velocity corresponding to the spalling of the target, and again increase in velocity due to recompression of the spalled target.

The Hugoniot elastic limit σ_{HEL}, the dynamic yield strength Y, and spall strength σ_s are determined from this profile as follows [34–36]:

$$\sigma_{HEL} = \frac{1}{2} u_H \rho_0 c_l$$

$$Y = \sigma_{HEL} \frac{(1-2\nu)}{(1-\nu)}$$

$$\sigma_s = \frac{1}{2} \Delta u_{fs} \rho_0 c_b$$

where u_H is free surface velocity at σ_{HEL}; the Δu_{fs} is pullback velocity of free surface defined as $u_{fs} - u_{\mathrm{m}}$ with u_{fs} as the peak free surface velocity and u_m as the free surface velocity just ahead of spall pulse; ρ_0, c_b, c_l and ν are the initial density, bulk sound speed, longitudinal sound speed, and Poisson ratio, respectively.

Three experiments each on Al2014-T4 and Al2024-T4 have been performed. The free surface velocity histories measured using VISAR instruments are plotted in Figs. 2.16 and 2.17. For determining the mechanical properties from the velocity histories, the Hugoniot data and the values of c_b, c_l, and ν are taken from available literature [37]. The mechanical properties determined from this study such as spall strength, Hugoniot elastic limit, and strain rates are listed in Table 2.1.

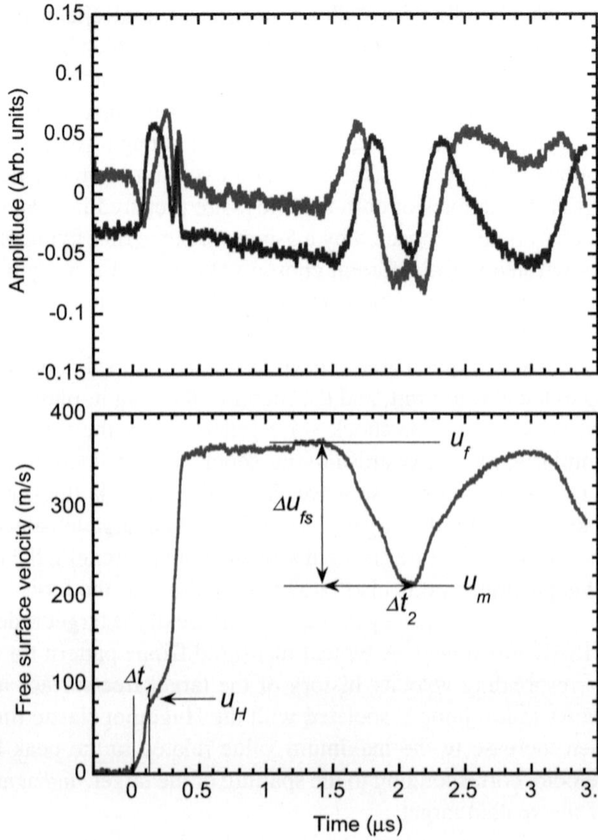

FIGURE 2.15

The recorded fringe pattern from VISAR and free surface velocity history of aluminum alloy, which displays features corresponding to different phenomena.

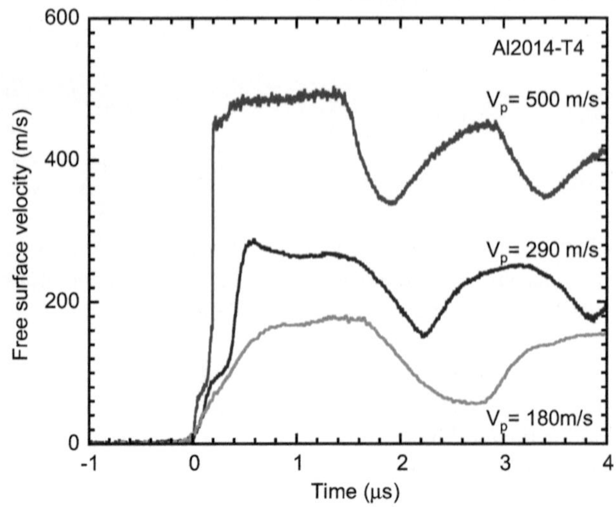

FIGURE 2.16

Free surface velocity history of Al2014-T4 alloy at different impact velocities.

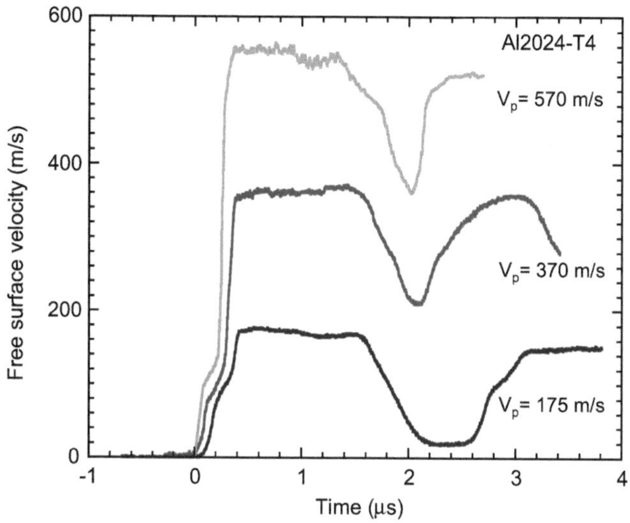

FIGURE 2.17

Free surface velocity history of Al2024-T4 alloy at different impact velocities.

Table 2.1 Mechanical Properties of Aluminum Alloys During Dynamic Failure Process Occurring at High Strain Rates

Shot No.	v_p (m/s)	u_{fs} (m/s)	P (GPa)	σ_s (GPa)	σ_H(GPa)	$\partial\varepsilon_c/\partial t(\times 10^4 s^{-1})$	$\partial\varepsilon_t/\partial t(\times 10^4 s^{-1})$
Al2014-T4							
1.	500	484	3.6	1.11	0.79	9.4	2.4
2.	290	286	2.1	0.97	0.77	3.6	1.6
3.	180	175	1.3	0.87			1.4
Al2024-T4							
1.	570	560	4.2	1.43	0.84	9.9	2.5
2.	370	366	2.6	1.16	0.79	4.6	2.0
3.	175	169	1.3	1.11	0.79	3.0	1.9

Table 2.1 shows that the spall strength value of 0.87, 0.97, and 1.11 GPa for Al2014-T4 at strain rates of 1.4×10^4/s, 1.6×10^4/s, and 2.4×10^4/s increase monotonically with increasing strain rate, and all these values are significantly higher than the quasistatic value of 0.511 GPa. On the other hand, the dynamic yield strength values 0.77 and 0.79 GPa at strain rates of 3.6×10^4/s and 9.4×10^4/s are almost the same. The dynamic yield strength values of 0.395 and 0.400 GPa at these strain rates are higher than the quasistatic values of 0.355 GPa. The spall strength values of Al2024-T4 also show increasing trend with increasing strain rates. For this material, the spall strength values at strain rates of 1.9×10^4/s, 2.0×10^4/s, and 2.5×10^4/s are 1.11, 1.16, and 1.43 GPa, respectively. Again, all values are higher than the quasistatic value of 0.469 GPa. As also seen for Al2014-T4, the dynamic yield

strength values for Al2024-T4 are higher than the quasistatic value of 0.324 GPa and also show the same trend of increase with strain rate.

The strain rate hardening of material against tensile fracture has been reported by other researchers also [30,38,39]. The reason for the strain rate hardening can be understood in terms of reduced tendency of microvoid nucleation at higher strain rates. Above strain rates of about 10^9/s it becomes equal to the ideal fracture strength of the material [16].

4.2 SHOCK WAVE–INDUCED OMEGA PHASE FORMATION IN Zr AND Zr–Nb ALLOYS

Zirconium and Zr–Nb alloys, because of their low-neutron-absorption cross-section, are used in pressure tubes of coolant channels for pressurized heavy water reactors. The understanding of the formation of omega phase in these materials is important as the presence of this phase embrittles them and has detrimental effects on their mechanical properties.

Zr metal occurs in hcp (α) phase at ambient conditions of temperature and pressure. It is well established that under application of 2.2 GPa pressure, α phase irreversibly transforms to ω (hexagonal) phase, which upon further compression to 30 GPa transforms to bcc (β) phase [40–42]. The structural sequence $\alpha \rightarrow \omega \rightarrow \beta$ has been associated with the pressure-induced increase in *d* electron population due to *sp* to *d* electron transfer [43]. The transformation of α phase to ω phase rather than to the expected more-symmetric β structure (the structure of Nb having one more electron than in Zr) has been explained by theoretical calculations, which suggest the lowering of electronic energy of ω structure relative to that of β structure because of lifting of degeneracy of electronic states near Fermi level [44]. Since ω phase can be retained in a metastable state in pressure-treated sample, this is amenable to examination using electron microscopy.

For examining the mode of formation of ω phase under static pressure and shock-loading techniques, electron microscopy examinations have been carried out on zirconium samples recovered after pressurization. For static-pressure treatment, zirconium samples have been compressed to 6.8 GPa for 24 h using tungsten carbide Bridgman anvils in a hydraulic press. The quasi-hydrostatic environment around the sample has been produced by using talc discs. In dynamic loading, a zirconium sample sandwiched between two stainless steel plates has been compressed to 11.8 GPa by a reverberating shock wave of a few μs duration in a recovery fixture using gas gun. Thin foils of pressure-treated samples have been prepared, polished and suitably thinned, and examined using a JEOL 2000FX electron microscope.

Fig. 2.18 shows typical selected area diffraction (SAD) pattern taken from coexistence region of α and ω phases for pressure-treated Zr samples. The orientation relationship (OR) between α and ω phases from the SAD pattern (Fig. 2.18a) for static-compressed sample is found to be $\left[7\ \ \bar{2}\ \ \bar{5}\ \ 3\right]_\alpha \parallel \left[1\ \ \bar{2}\ \ 1\ \ \bar{3}\right]_\omega$, whereas the OR determined from one of the SAD pattern (Fig. 2.18b) of shock-treated sample is detected to be $\left[1\ \ 0\ \ 1\ \ 0\right]_\alpha \parallel \left[1\ \ 1\ \ \bar{2}\ \ \bar{3}\right]_\omega$ [45]. It can be shown that these two ORs are equivalent as both can be satisfied by the same correspondence matrix [46]; therefore, the lattice correspondences between the two structures in both the pressure loading conditions are the same. Moreover, these ORs are equivalent to the one derived by Usikov and Zilberstein [47], who proposed the transformation mechanism of α to ω phase via an intermediate β phase.

Although the ORs between α and ω phases for Zr samples pressurized in the two loading conditions are equivalent, the morphology displayed by micrographs (Fig. 2.19) for the two cases is significantly

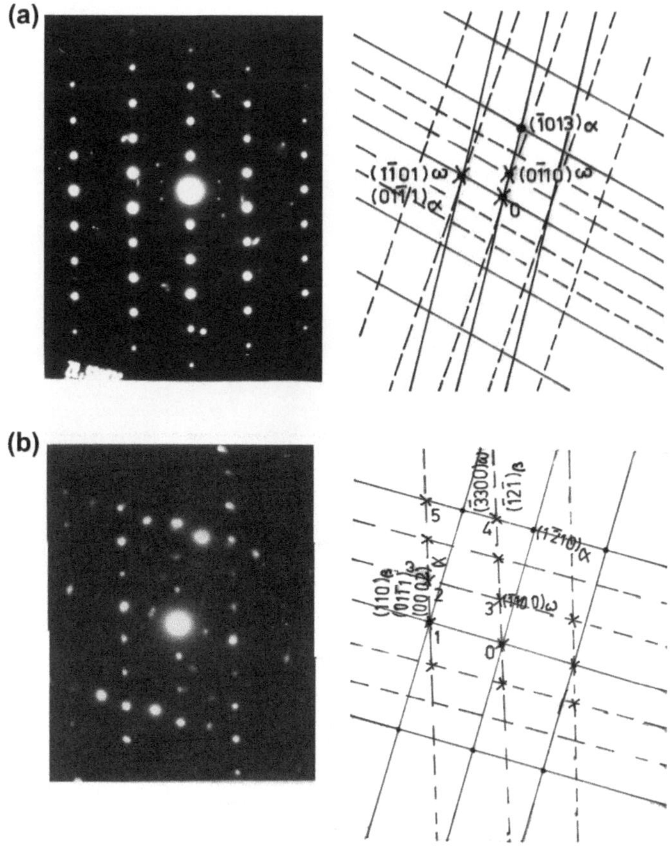

FIGURE 2.18

SAD patterns from transmission electron microscopy (TEM) measurements on (a) static-compressed Zr sample and (b) shock-loaded Zr sample.

different. The static-compressed sample shows ω grains of irregular shape and sizes, whereas the shock-treated sample displays the formation of large plates of ω phase from the α phase. The plate morphology of the ω phase is characteristic of a martensitic transformation in which shear plays a dominant role under shock compression [48]. The grain morphology of ω phase from the static-compressed sample corresponds to nucleation and growth mechanism in which transformation proceeds by nucleation of small ω particles, which grow due to overpressure and large pressure soaking time of 24 h. The particle morphology has also been observed during quenching of Zr-Nb alloys from high-temperature β-phase field. This implies that the mechanism of phase transformation of the ω phase under dynamic compression is different than under the static-loading conditions.

The pronounced effect of shear stresses present in the shock wave loading on the occurrence and mechanism of phase transitions is brought out by the studies on formation of ω phase on pressure treatment of β-stabilized Zr20%Nb alloy [48]. In this study, samples of Zr20%Nb, recovered after

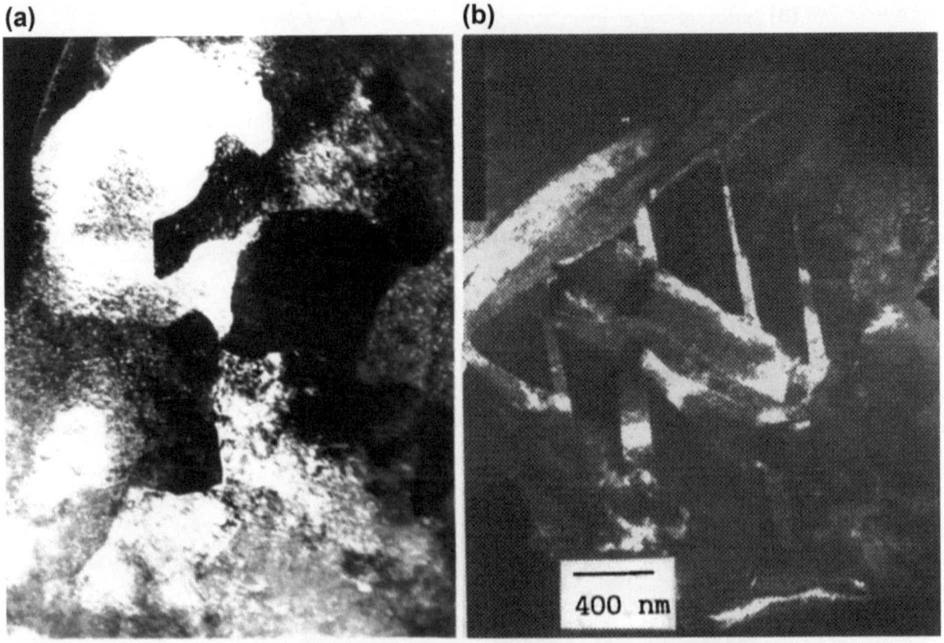

FIGURE 2.19

Dark field micrograph of transformation corresponding to SAD patterns given in Fig. 2.18; (a) for static-compressed Zr sample and (b) shock-loaded Zr sample.

pressurization up to 8 GPa under quasi-hydrostatic condition for 96 hrs, and examined using X-ray diffraction and transmission electron microscopy (TEM), did not show any ω-phase. Since ω-phase, if formed at high pressure, does not revert back to the parent phase, the absence of ω-phase in the static-compressed samples indicates that $\beta \rightarrow \omega$ transition could not take place under applied static-pressure conditions. On the other hand, TEM examination of similar specimens recovered after shock loading to 15 GPa for a few μs revealed presence of platelike second phase, which was identified from the SAD patterns.

The platelike morphology of the ω phase in shock-treated material is in stark contrast to the fine-particle morphology observed in the $\beta \rightarrow \omega$ transformation caused by quenching of high-temperature β phase of Zr–Nb alloys. The fine-particle morphology of the ω phase has been explained by considering formation of ω from the β phase by collapse of two adjacent (222) planes in the bcc structure to an intermediate position, leaving the next plane undisturbed, and again collapsing the next two planes to their intermediate position, and so on. Since there are four equivalent $<111>$ directions, four equivalent variants of ω are possible. The platelike morphology, however, arises due to formation of ω from β phase by application of homogeneous shear on $\{112\}$ planes along $<111>$ directions plus some atomic shuffles [17]. Since any $\{112\}$ plane contains only one $<111>$ direction, a single orientation variant can be formed parallel to this plane. In shock wave loading, this $\{112\}$ plane will correspond to the direction of maximum resolved shear stress [17].

The studies on both Zr and Zr—Nb alloys illustrate significant influence of the rapid rate of loading and the presence of shear stress in the shock compression process on the occurrences and the mechanism of phase transitions.

5. THEORETICAL METHODS

For determination of complete equation of a state of material, we need to obtain the Helmholtz free energy $F(V, T)$, which under adiabatic approximation can be decomposed into three parts,

$$F(V,T) = E_C(V,0) + \Delta F_T(V,T) + \Delta F_e(V,T) \tag{2.9}$$

where the first term is the total energy at volume V with frozen nuclei of the solid at $T = 0$, the second term and the third term, respectively, represent free energy contributions from lattice vibrations and electronic excitations at temperature T.

The total energy $E_C(V, 0)$ of solid can be accurately calculated using ab initio electronic band structure methods based on density functional formalism [49], such as ab initio pseudopotential method [50], linearized augmented plane wave method [51], linear muffin tin orbital method [52], full-potential LMTO method [53], and full-potential LAPW methods [54], and so on. These methods essentially differ in terms of the choice of the basis functions used for constructing the wave function and in terms of the form of the potential in the single electron Kohn—Sham equation. Some standard software packages are available for carrying out band structure calculations, which include commercial packages, Vienna Ab initio Simulation Package, VASP [55], and WIEN2K [56], and also open source package Quantum Espresso [57] and ABINIT [58].

The free energy contribution from lattice vibration arises due to excitation of phonons, the frequencies of which generally depend on V and T. Various available ab initio methods based on density functional perturbation theory can provide phonon frequencies within quasi-harmonic approximation (QHA), which account for volume dependence of phonon frequencies (implicit anharmonicity), but not the temperature dependence (explicit anharmonicity). Generally, explicit anharmonicity is small but can be very large in some cases like for solids undergoing phase transition with softening of phonon modes, or at temperatures close to melting, or for class of materials showing negative thermal expansion. The complete lattice free energy calculations can be performed only by molecular dynamics calculations, which are very time-consuming and need large computer resources.

Under QHA, $\Delta F_T(V,T)$ and the vibrational entropy $\Delta S_T(V,T)$ can be expressed as

$$\Delta F_T(V,T) = \sum_{q,n} \frac{1}{2}\hbar\omega_n(q,V) + k_B T \sum_{q,n} \ln\left[1 - exp\left(\frac{-\hbar\omega_n(q,V)}{k_B T}\right)\right] \tag{2.10}$$

$$\Delta S_T(V,T) = -k_B \sum_{q,n} \left[\ln\left\{1 - exp\left(\frac{-\hbar\omega_n(q,V)}{k_B T}\right)\right\} + \frac{\hbar\omega_n(q,V)}{k_B T}\left\{exp\left(\frac{\hbar\omega_n(q,V)}{k_B T}\right) - 1\right\}^{-1}\right] \tag{2.11}$$

with $\omega_n(q, V)$ being the frequency of the n-th phonon band at the point q in the Brillouin zone, and $\hbar$ and k_B are, respectively, the Plank constant and the Boltzmann constant.

For estimating the free energy contribution from electronic excitation, electronic energy levels are first obtained from band structure calculations at volume V at $T = 0$, (since energy levels are not much affected by thermal nuclear motion) and then populating the energy levels according to Fermi–Dirac distribution at temperature T, so that $\Delta F_E(V,T)$ can be expressed [59] as

$$\Delta F_e(V,T) = -\frac{1}{6}\pi^2 k_B^2 N(E_f) T^2 \tag{2.12}$$

where $N(E_f)$ is the electronic density of states at Fermi level.

The Helmholtz free energy is calculated as a function of unit cell volume and temperature; then at any temperature the equilibrium volume $V_0(T)$, the pressure P, and the bulk modulus B are determined from the following relations:

$$\left.\frac{\partial F}{\partial V}\right|_{V_0} = 0 \tag{2.13}$$

$$P = -\frac{\partial F}{\partial V} \tag{2.14}$$

$$B = -V\frac{\partial P}{\partial V}. \tag{2.15}$$

5.1 CALCULATION OF EQUATION OF STATE

The pressure–volume relation, generally known as equation of state, depends on the loading path, which can be isothermal, isentropic, or shock wave loading. These relations are derived as follows.

5.1.1 Calculation of Isotherm and Isentrope

The common output of a quasistatic high-pressure experiment is 300K isotherm, which can be used for testing the theoretical calculations. The 300K isotherm can be calculated using Eqs. (2.9)–(2.14), since the current theoretical methods can calculate accurate total energy of the frozen lattice, and accurate phonon dispersion relations in QHA; the electronic excitation energy being very small (of the order of 10^{-2} mRy/atom) is generally neglected.

In order to determine the isentrope, we need to first calculate temperature rise along the isentropic compression. For this purpose, the entropy $S(T, V)$ is expressed by the following thermodynamic relation:

$$TdS = T\left(\frac{\partial S}{\partial T}\right)_V dT + T\left(\frac{\partial S}{\partial V}\right)_T dV \tag{2.16}$$

which using Maxwell relations reduces to

$$TdS = C_V dT + T\left(\frac{\partial P}{\partial T}\right)_V dV \tag{2.17}$$

where C_V is specific heat at constant volume. For isentropic compression, $dS = 0$, so that Eq. (2.17) implies

$$\frac{dT}{T} = -\frac{\gamma(V)}{V} dV \tag{2.18}$$

Since

$$\frac{1}{C_V}\left(\frac{\partial P}{\partial T}\right)_V = \left(\frac{\partial P}{\partial E}\right)_V = -\frac{\gamma(V)}{V}dV \tag{2.19}$$

the thermal Grüneisen parameter $\gamma(V)$ can be derived from

$$\gamma(V) = \frac{\alpha VB}{C_V} \tag{2.20}$$

where thermal expansivity (α) is given by the knowledge of $V_0(T)$ from Eq. (2.13), C_V from the knowledge of vibration entropy from Eq. (2.11), and B from Eq. (2.15).

Integration of Eq. (2.18) provides the temperature T_S reached on isentropic compression to volume V of material initially at volume V_0 and temperature T_0.

$$\ln\left(\frac{T_S}{T_0}\right) = -\int_{V_0}^{V}\frac{\gamma(V)}{V}dV \tag{2.21}$$

The volume V_0 is generally taken as the theoretical equilibrium volume at temperature T_0 as 300K. Once the temperature T_S is known, $P(V, T_S)$ is obtained using Eqs. (2.9)–(2.14).

5.1.2 Calculation of Hugoniot

At a given compression under shock loading, the temperature is not known and needs to be determined along with the pressure. For this, the total energy $E(V, T)$ and pressure $P(V, T)$ at given temperature T and volume V is expressed as [60]:

$$E(V,T) = E_C(V) + \Delta E_T(V,T) + \Delta E_e(V,T) \tag{2.22}$$

$$P(V,T) = -\frac{\partial E_C}{\partial V} + \gamma\frac{\Delta E_T}{V} + \gamma_e\frac{\Delta E_e}{V} \tag{2.23}$$

where ΔE_T is vibrational energy of the ions including zero point vibration energy, which is determined from lattice dynamics calculations. ΔE_e is thermal electronic excitation energy and γ_e is the electronic Grüneisen parameter, which use the following expressions:

$$\Delta E_e = \frac{1}{2}\beta T^2 \tag{2.24}$$

$$\beta = \frac{1}{3}\pi^2 k_B^2 N(E_f) \tag{2.25}$$

$$\gamma_e = \frac{\partial \ln \beta}{\partial \ln V} \tag{2.26}$$

The shock Hugoniot is determined using Eqs. (2.22) and (2.23) in conjunction with the Rankine–Hugoniot relation [Eq. (2.5)], which can be written in the following form:

$$E(V,T_H) - E(V_0,T_0) = \frac{1}{2}[P(V,T_H) + P(V_0,T_0)][V_0 - V] \tag{2.27}$$

Here T_H and T_0 are the temperatures of the shocked and unshocked state, respectively.

The solution of Eq. (2.27), obtained by substituting the expression for $E(V, T)$ and $P(V, T)$ from Eqs. (2.22) and (2.23), yields the Hugoniot temperature T_H, which can then be used in Eq. (2.23) to obtain the Hugoniot pressure.

5.2 PHASE TRANSITIONS

At a given temperature and pressure, a crystal adopts the structure with minimum Gibbs free energy. At high temperatures, the analysis of the stability of various crystal structures becomes very expensive since it requires ab initio calculations of lattice vibrations at various compressions. Also, in case of materials where the explicit anharmonic contribution becomes large [61], for example, for materials Cr, Mo, and W showing directional character of bonding, and in case of materials where phase transformations are accompanied by phonon softening, for example, in Zr and Ti, the calculations based on QHA are not very accurate. So, generally the structure stability analysis is carried out at 0K only. At 0K, however, the structure stability analysis has been very successful [62].

To determine the stable crystal structures at various pressures, comparison of Gibbs free energies (enthalpies at 0K) of all the plausible structures as function of compression is carried out. The probable structures are generally guessed from the use of principle of corresponding states, which suggests similar phase diagrams for substances containing elements in the same group of the periodic table. The candidate structures are also derived from symmetry considerations, which are based on the analysis of symmetry systematics of pressure-induced phase transitions [63,64]. For each plausible structure, the enthalpy of the solid at a given volume is computed by calculating the total energy (E_C) using Eq. (2.9), and pressure obtained from the negative volume derivative of E_C. The differences between zero point energy of various structures, being very small ($\sim$about 0.1 mRy/atom), are normally neglected. Such methods can determine the most stable structure from the chosen set of candidate structures but cannot by itself discover the structure with lowest enthalpy.

Prediction of most stable structures from ab initio methods, without any knowledge of probable structures, is very difficult because it requires locating the global minimum of the free energy surface in configuration space of all structural variables. However, with the advent of advances in computing power and progress in the searching methodologies, good progress has been made for structure prediction. These methodologies include AIRSS [65,66], USPEX [67,68], and metadynamics [69,70]. The AIRSS method is based on the search of stable structure by relaxing from randomly chosen initial configurations and has been used for predicting high-pressure phases of silane [66]. The method is expected to work for up to 10 atoms per unit cell. USPEX uses a self-improving evolutionary algorithm in conjunction with optimization using density functional band structure methods for locating step by step the best structure. Metadynamics, on the other hand, uses density functional molecular dynamics calculations to explore free energy dependence on configurational space going from one local minimum to another through lowest energy barrier with proper distortion of the unit cell and finally discovers the stable structure. Despite the claims of the successes of these methods, it is difficult to ensure the discovery of the global minimum free energy state of the system.

For some studies, we have adopted a hybrid approach in which we determine the lowest enthalpy structures at a chosen number of compressions using USPEX algorithm and use these structures as the probable structures for comparing their enthalpies at all the pressures. Once a structure with lowest free energy is found, detailed calculations of its phonon frequencies are carried out to ensure the dynamic stability of this structure.

6. THEORETICAL STUDIES

Here we present two examples of the use of theoretical techniques for prediction of phase transitions in materials and for determination of Hugoniot of materials.

6.1 THEORETICAL INVESTIGATIONS ON CARBIDES OF URANIUM AND THORIUM

A good understanding of the high-pressure behavior of carbides of uranium and thorium is important since these are used as nuclear fuels in fast breeder reactors [71–73]. The studies are also interesting as the itinerancy of 5f electrons in actinide carbides is expected to have influence on the cohesive properties of these materials.

Under application of pressure in a material, various energy bands broaden and shift relative to each other, which play an important role in deciding the physical properties and the stability of a particular structure. The studies on the structural stability of these carbides have been carried out by performing detailed ab initio static-lattice and lattice-dynamic calculations. These studies have reproduced the experimentally observed phase transition in uranium carbide (UC) and determined the structural parameters of its high-pressure phase, and have predicted new high-pressure phase transitions in thorium carbide (ThC). The results from these studies are presented in the following discussion.

6.1.1 Phase Transition in UC

The elemental uranium has electronic configuration $\{[Xe]4f^{14}5d^{10}6s^2\}7s^26p^66d^15f^3$ and its monocarbide occurs as solid in B1 phase at ambient conditions. The high-pressure energy dispersive X-ray diffraction experiments by Olsen et al. [74] showed that this phase changed to a new structure around 27 GPa. The diffraction peaks from new phase corresponded to body-centered orthorhombic (bco) structure; however, two diffraction peaks were unexplained. Therefore, authors termed the structure as pseudo bco orthorhombic, which is a distortion of body-centered tetragonal cell of the fcc lattice. The atomic coordinates in the new structure, however, could not be determined.

With the objective to better understand the high-pressure phase transition in UC and also the mechanism of this phase transition, Sahoo et al. [75] have carried out detailed theoretical analyses of the structure stability and the dynamic stability under high pressures. The calculations have been performed using plane wave pseudopotential method [57]. All the calculations are done for nonmagnetic case, since their initial calculations for B1 phase show 1.36 meV/formula unit lower energy for nonmagnetic case than that of antiferromagnetic case.

The calculations have been performed at various volumes starting with orthorhombic structure with space group Pmmn. The relation of this structure with the B1 and bco structures is given in Fig. 2.20. In terms of symmetry, Pmmn structure is a subgroup of bco structure with space group Immm; and both Pmmn and Immm are subgroups of B1 structure with space group Fm3m. For Pmmn structure $a \neq b \neq c$, and U and C atoms are located, respectively, at 2a $(0\ 0\ z_1, \frac{1}{2}\ \frac{1}{2}\ -z_1)$ and 2b $(0\ \frac{1}{2}\ z_2, \frac{1}{2}\ 0\ -z_2)$; this structure becomes bco for $z_1 = z_2 = \frac{1}{4}$ and becomes B1 structure for $c = a = b/\sqrt{2}$ with $z_1 = z_2 = \frac{1}{4}$. At each volume, the cell constants a, b, c and internal parameter z_1, z_2 are optimized by minimizing the total energy.

More details of the calculations are given elsewhere [75]; here, we have briefly given the results.

Optimization of cell constants and internal parameters of the Pmmn phase show that for compressions up to 23 GPa, $b/a = \sqrt{2}$; $c/a = 1$; $z_1 = z_2 = \frac{1}{4}$, which correctly corresponds to the

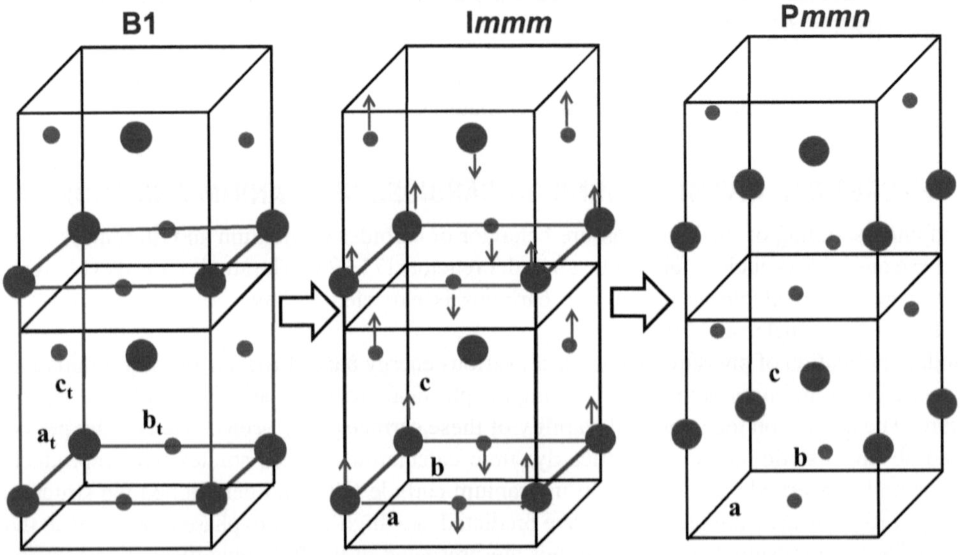

FIGURE 2.20

Relationship between B1 (NaCl Type), Immm, and Pmmn structure of UC. The large circles represent U atoms while the small circles show C atoms. For B1 phase (represented as body-centered tetragonal cell) the $c = a = b/\sqrt{2}$ with position coordinates of U and C as (0 0 ¼) and (½ ½ ¼), respectively. For Immm phase, $a \neq b \neq c$ and position coordinates of U and C atoms are (0 0 ¼) and (½ ½ ¼), respectively. For Pmmn phase, $a \neq b \neq c$ with positions of U and C atoms as (0 0 z_1, ½ ½ $-z_1$) and (0 ½ z_2, ½ 0 $-z_2$).

stabilization of the B1 phase. Beyond 23 GPa, b/a and c/a increase gradually and settle, respectively, at about 1.48 and 1.2 around 26 GPa; these theoretically optimized axial ratios of 1.55 and 1.15 agree with experimental data (Olsen et al. [74]) within ~7.6%. Moreover, z_1 increases gradually and settles at 0.31, whereas z_2 decreases gradually and settles at 0.15.

In order to examine the relative stability of B1 and orthorhombic structure under pressure, we have compared the enthalpies of these structures as function of pressure (Fig. 2.21). We find up to pressure of 20 GPa that the orthorhombic structure relaxes with respect to variations in axial ratios and the internal coordinates of U and C atoms to the B1 structure; however, beyond 20 GPa the orthorhombic structure is lower in enthalpy up to 40 GPa, the highest pressure of the present calculations. This comparison suggests B1 to orthorhombic transition in UC at 20 GPa.

The elastic constants of the B1 phase computed theoretically with increasing pressure are plotted in Fig. 2.22. The plot shows that the elastic constants C_{11} and C_{12} increase monotonically and their difference $C_{11}-C_{12}$ increases with increasing pressure; also, the bulk modulus increases with pressure. The elastic constant C_{44}, however, decreases with increasing pressure and becomes negative beyond 23 GPa. The negative values of C_{44} beyond 23 GPa show that B1 phase becomes unstable because of its instability for shear deformation. This implies that the B1 to orthorhombic phase is caused by shear instability.

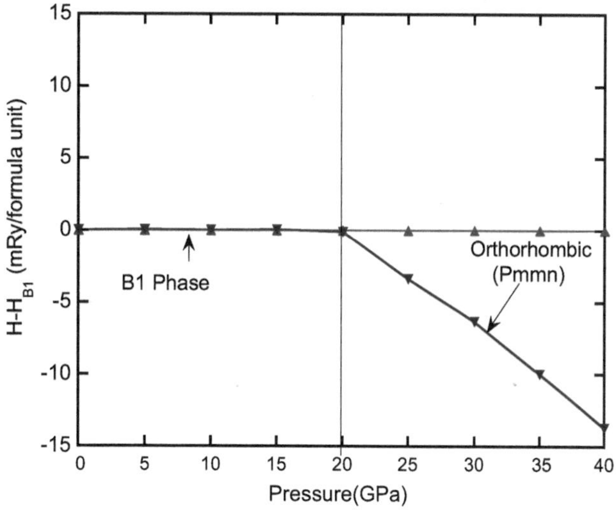

FIGURE 2.21

Enthalpies of Pmmn phase relative to that of B1 phase for UC. The B1 → Pmmn transition in UC takes place at ~20 GPa.

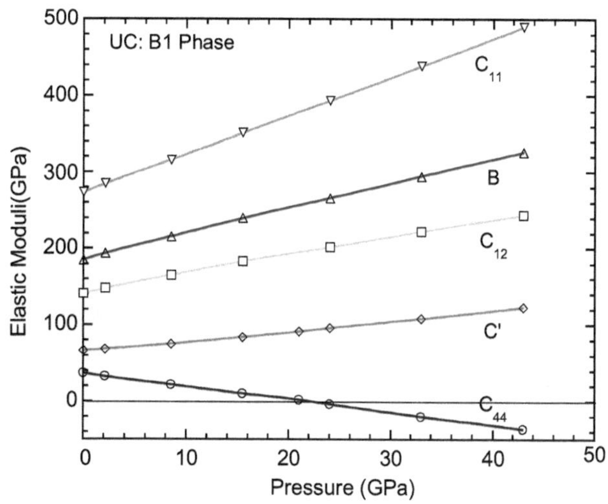

FIGURE 2.22

Elastic constants and elastic moduli of UC in B1 phase as a function of pressure. The C_{44} shear modulus fails around the B1 → Pmmn transition pressure.

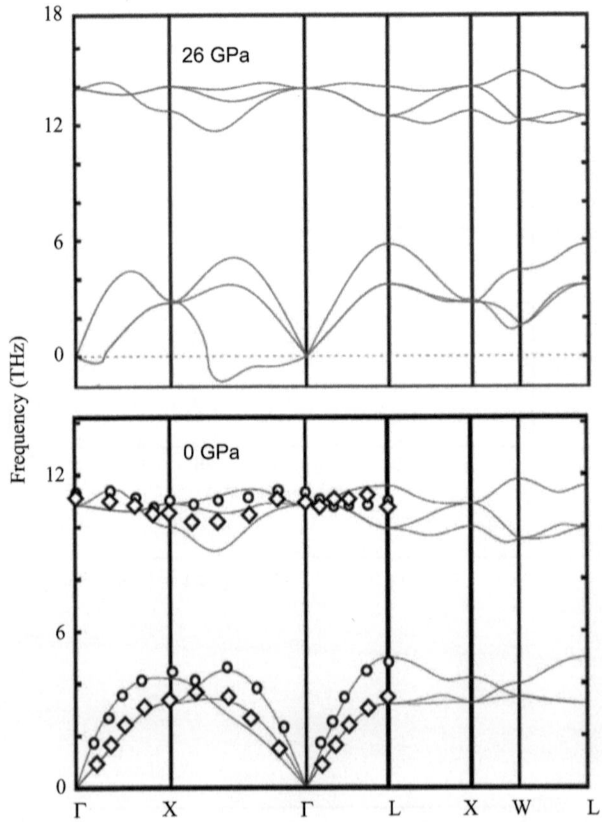

FIGURE 2.23

The phonon spectrum of UC in B1 phase at ambient conditions and at ~26 GPa.

For understanding the dynamical stability of B1 phase in terms the phonon dynamics, the phonon dispersion relations of the B1 structure at ambient conditions and 26 GPa, pressure above the transition pressure, are calculated and displayed in Fig. 2.23. The calculated dispersion relations agree very well with the available experimental measurements [76]. The phonon frequencies throughout the Brillouin zone are positive real, in conformity with the dynamic stability of the B1 phase, which is stable at ambient conditions. The plot of the phonon spectra at 26 GPa show that the frequencies of the transverse acoustic phonons are imaginary near the zone center along the Γ-X direction of the Brillouin zone. These imaginary frequencies of the zone center phonons suggest the dynamical instability of B1 phase at this pressure.

The instability of the long-wavelength phonons around phase transition is consistent with elastic instability of the B1 phase caused by the shear failure. Therefore, analysis of the elastic instability and the dynamic instability suggests that the B1 to Pmmn transition is driven by failure under shear.

6.1.2 Phase Transitions in ThC

The high-pressure experimental studies on thorium monocarbide, which occur in B1 phase at ambient conditions, are limited. Although diamond cell experiments have been performed on ThC sub-stoichiometric samples with carbon deficiencies of above 20%, which show no structural phase transitions up to 50 GPa [77,78], these experiments are not representative of the behavior of ideal thorium monocarbide. Since the previously available theoretical studies dealt with physical and electronic properties of ThC only at the ground state, Sahoo et al. [79] carried out analysis on the structure stability and the prediction of phase transitions in ThC on application of pressure up to 250 GPa. The results of the analysis are briefly presented here.

As mentioned in Section 5.2, the structural stability analysis requires the investigation of the static lattice structure stability, which involves comparisons of enthalpies of probable structures at all the pressure values. The stable structure at the given pressure is indicated by the minimum enthalpies structure. The static lattice analysis must be followed by ensuring the dynamic stability of minimum enthalpy structure so obtained.

Considering that B1 (NaCl-type structure) and B2 (CsCl-type structure) are probable structures, we analyzed the relative stability of B1 phase and B2 phase as a function of pressure. We find that B1 phase has lower enthalpy up to 35 GPa, beyond which B2 phase has lower enthalpy, suggesting the stability of B2 phase above 35 GPa. The lattice vibration calculations, on the other hand, reveal that B1 phase is dynamically unstable above 36 GPa, whereas the B2 phase is dynamically unstable below 50 GPa; thus, in 36–50 GPa pressure range, neither B1 nor B2 phase is stable. These calculations brought out the inadequacy of the choice of B1 and B2 phases as the only probable structures.

For discovering the stable structure, ideally one should employ the evolutionary structure search method, like USPEX Code [80], which can predict stable as well as metastable structures at given pressures. Such calculations, however, are very time-consuming and require large computer resources. In order to reduce the requirement of computer resources, a hybrid approach is used. In this approach, a set of probable structures is found by determining the least enthalpy structures at a number of pressure values using evolutionary structure search method in conjunction with a standard band structure method, and then the enthalpies of these structures are calculated and compared at all the pressures using standard band structure methods.

For ThC, we search the lowest enthalpies structures at pressures 0, 25, 60, and 225 GPa using USPEX Code [80], which for structure relaxation utilizes density functional theory–based projector augmented wave method executed in VASP Code. Using this method, at 0 GPa, B1 structure is found to be the least enthalpy structure. At 25 GPa, the least enthalpy structure is associated with primitive orthorhombic structure having space group Pnma. The structure has eight atoms in the cell with four atoms of Th and four atoms of C occupying the 4c Wyckoff positions (International Table of Crystallography, Vol. 2) (x,1/4,z) (−x+1/2,3/4,z+1/2) (−x,3/4,−z) and (x+1/2,1/4,−z+1/2). On further increasing the pressure to 60 GPa, the least enthalpy is exhibited by a 16-atom structure with space group Cmcm. In the unit cell, the eight Th atoms occupy two inequivalent 4c sites (0,y1,1/4) (0,−y1,3/4) (1/2,1/2+y1,1/4), (1/2,1/2−y1,3/4) and (0,y2,1/4) (0,−y2,3/4) (1/2,1/2+y2,1/4), (1/2,1/2−y2,3/4), whereas the eight C atoms are located at 8f sites. Finally, at 225 GPa, B2 structure is found to have the lowest enthalpy.

After discovering the least enthalpy structures as B1, Pnma, Cmcm, and B2 phases, the enthalpy of these structures are calculated and compared at all the pressures up to 250 GPa. For calculations of

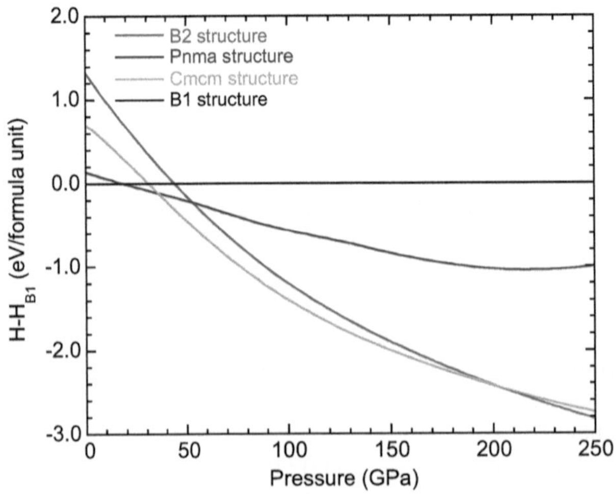

FIGURE 2.24

Theoretically calculated enthalpies of ThC in B1, Pnma, Cmcm, and B2 structure as a function of pressure. The plotted enthalpies are relative to that of B1 phase.

enthalpies at each pressure, the axial ratios and the internal parameters are optimized for Pnma and Cmcm structures. The comparison of the enthalpies as a function of pressure shown in Fig. 2.24 suggests that the B1 phase, stable at ambient conditions, transform to Pnma orthorhombic structure at 19 GPa. On increasing the pressure, Pnma phase transforms to Cmcm phase at 36 GPa. At still higher pressures, the Cmcm phase transforms to B2 structure at 200 GPa.

For Pnma structure in the pressure range 19—36 GPa, the optimized values of structural parameters do not change; these optimized axial ratios are $b/a = 0.593$, $c/a = 0.998$ and the internal parameters are $x = 0.250$ and $z = 0.916$ for Th, and $x = 0.998$ and $z = 0.247$ for C.

For Cmcm structures at 36 GPa, we find the optimized values of $b/a = 3.662$, $c/a = 1.631$ and $y1 = 0.458$, $y2 = 0.742$ for Th, and $y = 0.895$, $z = 0.891$ for C. On increase of pressure up to 200 GPa (the transition pressure) the b/a and c/a gradually increase to 0.384 and 0.684, respectively, but the internal parameters remain fixed.

In order to confirm the dynamic stability of the four structures in their static-lattice stability range of pressure, the phonon dispersion relations of these structures have been determined. Fig. 2.25 displays the plots of phonon dispersion relations for B1 structure at 0 GPa, Pnma structure at 25 GPa, Cmcm phase at 45 GPa, and B2 phase at 225 GPa. All these plots show positive phonon frequencies indicating the dynamical stability of these structures. These observations substantiate the findings from the static-lattice calculations.

Thus, it is predicted that ThC will undergo three structural phase transitions on application of pressure; these transitions are B1 to Pnma at 19 GPa, Pnma to Cmcm at 36 GPa, and Cmcm to B2 at 200 GPa. This study on ThC clearly brings out the necessity of carrying out dynamical stability analysis in addition to static-lattice analysis for theoretical determination of phase transitions in materials under high pressures.

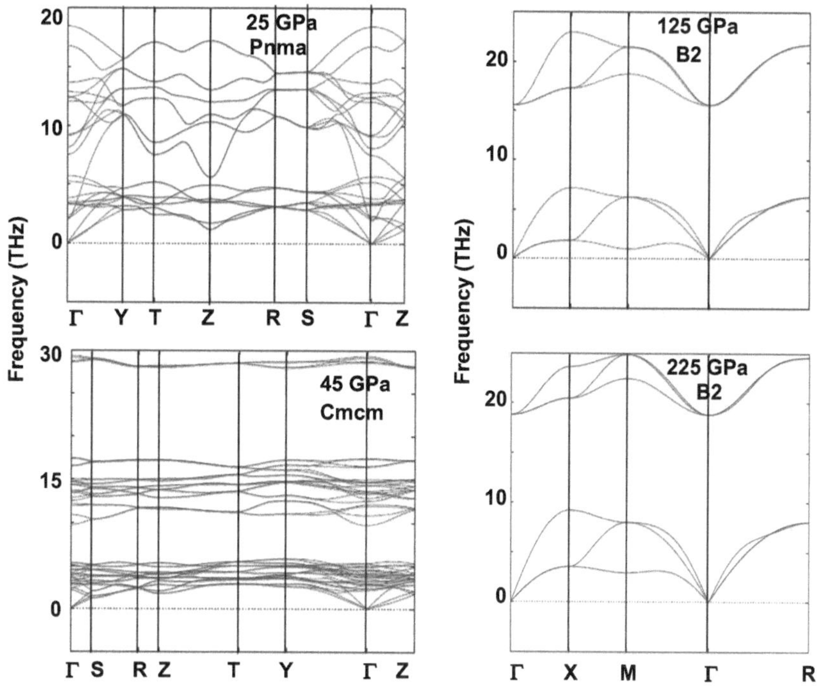

FIGURE 2.25

The phonon spectra of Pnma, Cmcm, and B2 structures at high pressures.

6.2 SHOCK HUGONIOT AND DYNAMIC STABILITY OF BI IN BCC PHASE

Bismuth has the highest atomic number among the nonradioactive elements and is less toxic than its neighboring elements lead, arsenic, and antimony; therefore, it is a potential candidate for replacement of toxic lead in many applications. High-pressure behavior of Bi has been studied under static and dynamic high-pressure conditions. The static high-pressure experiments [81] on Bi have reported 300K isotherm up to 222 GPa, and it has been shown to undergo a sequence of phase transformations: Phase I(rhombohedral) → Phase II (monoclinic) → Phase III (bct) → Phase V (bcc) occurring at 2.6, 2.8, and 7.7 GPa, respectively. Also, high-pressure, high-temperature measurements have revealed Phase IV (orthorhombic) structure at 2.4 GPa and 453K [82]. Under shock compression, the experimental Hugoniot of Bi is available up to 125 GPa [83]. Moreover, the shock-induced melting of Bi has been studied by three groups. Blanco et al. [84] used time-resolved radiation pyrometry and found Bi to melt in the pressure range 30–45 GPa (and corresponding temperatures range 2300–2800K). Another study (Sebban et al. [85]) employing radiation pyrometry detected shock melting of Bi in pressure range 18–28 GPa (and corresponding temperatures range 1250–1700K). In yet another study conducted recently, the shock Hugoniot of Bi has been measured and also shock melting of Bi has been examined from the sound velocity measurements behind the shock front [86]. This study reported onset of melting at 17 GPa and its completion at 28 GPa. Shock-induced melting of Bi has been analyzed theoretically also. Pelissier and Wetta [87], using empirical model potential and statistical mechanics approximations, determined shock melting to occur in 27–28 GPa range. Also, using empirical calculations, Cox [88] reported melting of Bi in 16–22 GPa shock pressure range.

Although many experimental studies at high pressures are available on Bi, the theoretical investigations are only a few; even theoretical studies addressing the shock-induced melting are empirical. In order to provide better understanding of high-pressure behavior of Bi, Mukherjee et al. [89] have carried out ab initio calculations on this material. These authors have constructed its high-pressure equation of state, determined its shock-induced melting pressure, and examined the dynamic stability of the bcc phase at high pressures. The results for this study are briefly presented here.

The 0K isotherm of the bcc Bi has been determined by first calculating the total energy at various volumes using VASP package [55], and then evaluating the negative volume derivative of total energy at various compressions. The bulk modulus as a function of pressure is derived from 0K isotherm, and the other two independent elastic moduli $C' = (C_{11} - C_{12})/2$ and C_{44} at various volumes are calculated from the second derivative of the deformation energy of the strained lattice with respect to deformation strains [89].

Next, the 300K isotherm is constructed by adding thermal contribution to the 0K isotherm. The thermal contribution at a given compression is estimated within Debye approximation, where the Debye temperature is calculated from the estimated elastic constants at that compression. Also, from the knowledge of Debye temperature as a function of volume, the Gruneisen parameter at a particular volume is evaluated as negative derivative of the logarithm of Debye temperature with respect to logarithm of volume. Finally, the shock Hugoniot is constructed using the Mie–Gruneisen equation of state model and ensuring that the pressure and volume satisfy the Rankin–Hugoniot relation.

This procedure provides the Hugoniot, which corresponds to the locus of all end *P-V-T* states reachable by a single shock from the bcc phase of Bi initially at zero pressure. For comparison with experimental Hugoniot, which is measured by shock loading of Bi in rhombohedral phase at ambient condition, the theoretical Hugoniot determined before has been converted to Hugoniot of Bi at volume initially corresponding to 0 GPa in rhombohedral phase.

The theoretical isotherm and Hugoniot are compared with the available experimental data in Fig. 2.26. The theoretical isotherm agrees better with the experimental data of Akahama et al. [81] than the data of Liu et al. [90]. Also, the agreement of the theoretical Hugoniot is very good with the experimental measurements [83,86].

Further, the variation of melting temperature of Bi with pressure has been determined. Fig. 2.27 displays the melting curve along with the theoretical Hugoniot in the pressure-temperature plane. The figure also shows the measured shock temperatures of Bi by two research groups. It is clear from the figure that temperatures measured by Blanco et al. [84] are much higher than those measured by Sebban et al. [85], and also much higher than the theoretical calculations. Our theoretical temperatures are closer to the temperatures reported by Sebban et al. The shock pressure and temperature indicated by the intersection of theoretical melting line with the theoretical Hugoniot are 23 GPa and 1333K. These values are close to the experimental pressure value of 18–28 GPa reported by Sebban et al. [85] and Tan et al. [86], and the temperature values reported by Sebban et al. [85].

It is found that the shock temperatures are very sensitive to the initial density of the samples. For example, if the Hugoniot for material is recalculated with 2.3% lower initial density (this is the difference between the calculated and experimental densities), the Hugoniot temperature turns out to

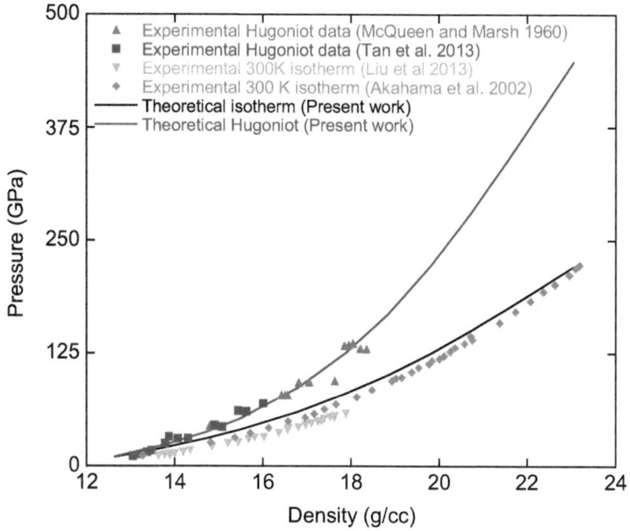

FIGURE 2.26

Comparison of theoretical shock Hugoniot and 300 K isotherm of bcc bismuth with available experimental data.

be higher. As shown in Fig. 2.27, the intersection of the melting line with the recalculated Hugoniot gives the shock melting pressure of 19 GPa in much better agreement with the experimental measurements.

After finding the good agreement of the theoretical equation of state and the shock melting pressures with the experimental data, the phonon dispersion relations have been calculated for Bi at high pressures, the experimental data for which are not available. Some of the phonon frequencies are found to be imaginary below 8 GPa pressure, which is the pressure near the bct to B2 phase transition,

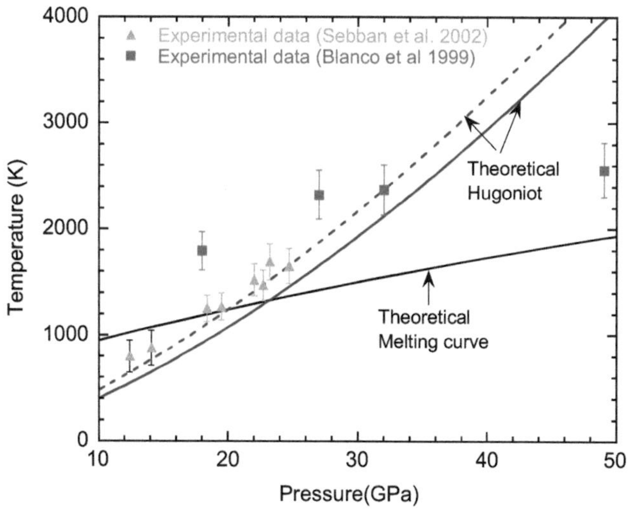

FIGURE 2.27

Theoretical melting curve and shock Hugoniot of bcc bismuth. Also plotted are the pressure-temperature data from radiation pyrometry experiments.

correctly suggesting the dynamic instability of the B2 phase below 8 GPa. At pressures above 8 GPa, however, B2 phase remains dynamically stable as frequencies of all the phonon modes are positive: the frequencies of longitudinal and transverse phonon modes at high pressures at H,P, and N points of the Brillouin zone are plotted in (Fig. 2.28a). As expected, all the frequencies increase monotonically with pressure because of the increase in the stiffness constants with increasing pressure. The plot of the complete dispersion relations for Bi at 200 GPa is given in (Fig. 2.28b).

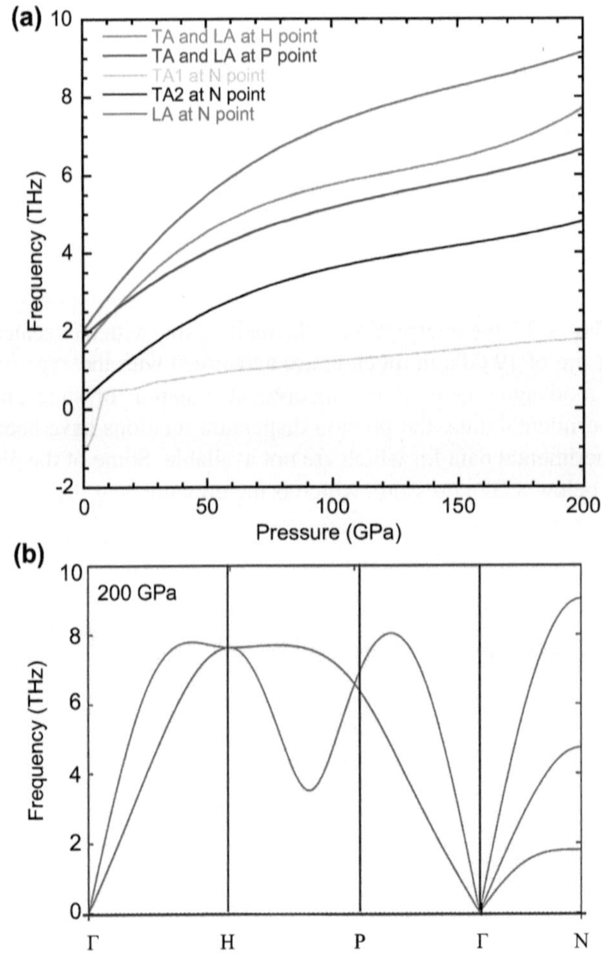

FIGURE 2.28

(a) The pressure variation of frequency of various acoustic phonon modes corresponding to different high-symmetry points of Brillouin zone in bcc bismuth. On vertical axis, imaginary values are plotted below zero, and (b) the phonon spectrum of bcc phase of bismuth at ~200 GPa.

7. SUMMARY AND DISCUSSION

Laboratory experimental methods have been described for studying the changes taking place in the materials subjected to shock wave loading. These methods include a gas gun, a propellant gun, and an electric gun for generation of shock waves, and a VISAR interferometer and Fabry–Pérot velocimeter for examining the response of the shocked materials. To illustrate response of materials to shock wave loading, two examples have been discussed. These examples deal with the determination of the high strain-rate mechanical properties of aluminum alloys, and study on the mechanism of phase transition in zirconium and zirconium–molybdenum alloy. Theoretical methods have also been discussed for ab initio calculations of equation of state of crystalline solid and for determination of stable structure at given pressure. Two examples have been presented showing calculations of equation of state for bismuth and for structural stability analysis in carbides of uranium and thorium.

In recent years many noteworthy developments have taken place. With the advent of powerful laser source at the National Ignition Facility, solids have been dynamically compressed in shock-less manner to about 5 TPa pressures without material temperature exceeding melting point [28]. New experiments have been possible for probing material in the increased region of stress-density-temperature space by various combinations of the static, shock wave, and ramp wave loadings. For instance, shock wave compression of a material, precompressed to different static pressures in a DAC, can create much higher pressure (10–100 TPa) states off the principal Hugoniot [91]. Also, compression by multiple shocks or by ramp wave following a shock wave can generate almost all the stress-temperature states lying between the principal Hugoniot and the principal isentrope.

The integration of a propellant gun with beam line at Advanced Photon Source (APS) synchrotron is a recent development, which facilitates microscopic measurements to study response of materials at atomic levels. By establishing synchronization of individual pulses separated by about 150 ns period from the APS synchrotron, and sensitive X-ray detectors with the impact event, the observation of high-resolution, time-resolved diffraction from a single crystal dynamically compressed in a gun has been demonstrated [14].

All these developments will provide exciting opportunities for studying the response of materials subjected to conditions in the much wider range of pressure, temperature, and strain rates.

REFERENCES

[1] T. Irifune, A. Kurio, S. Sakamoto, T. Inoue, H. Sumiya, Nature 421 (2003) 599.
[2] Y. Ping, et al., Phys. Rev. Lett. 111 (2013) 065501.
[3] Y. Akahama, H. Kawamura, D. Hausermann, M. Hanfland, O. Shimomura, Phys. Rev. Lett. 74 (1995) 4690.
[4] Y. Ma, et al., Nature 458 (2009) 182.
[5] N. Suresh, G. Jyoti, S.C. Gupta, S.K. Sikka, J. Appl. Phys. 76 (1994) 1530.
[6] J.H. Eggert, et al., Nat. Phys. 6 (2010) 40.
[7] L. Dubrovinsky, N. Dubrovinskaia, V.B. Prakapenka, A.M. Abakumov, Nat. Commun. 2160 (2012), http://dx.doi.org/10.1038/ncomms.
[8] J.F. Lin, M. Santoro, V.V. Struzhkin, H.K. Mao, R.J. Hemley, Rev. Sci. Instrum. 75 (2004) 3302.
[9] D.H. Bilderback, P. Elleaume, E. Weckert, J. Phys. B: Mol. Opt. Phys. 38 (2005) S773.
[10] J.H. Nguyen, M.B. Kruger, R. Jeanloz, Phys. Rev. Lett. 78 (1997) 1936.
[11] L.V. Al'tshuler, et al., J. Appl. Mech. Tech. Phys. 22 (1981) 145.

[12] L.J. Perkins, R. Betti, K.N. LaFortune, W.H. Williams, Phys. Rev. Lett. 103 (2009) 045004.
[13] G.I. Kanel, et al., Int. J. Impact Eng. 22 (1999) 421.
[14] Y.M. Gupta, et al., Rev. Sci. Instrum. 83 (2012) 123905.
[15] R.F. Smith, et al., Phys. Rev. Lett. 98 (2007) 065701.
[16] S.I. Ashitkov, M.B. Agranat, G.I. Kanel, P.S. Komarov, V.E. Fortov, JETP Lett. 92 (2010) 516–520.
[17] G.K. Dey, et al., Acta Mater. 52 (2004) 5243–5254.
[18] Y.M. Gupta, et al., Rev. Sci. Instrum. 70 (1999) 4008.
[19] J. Hawreliak, et al., Phys. Rev. B 74 (2006) 184107.
[20] G.R. Fowles, J. Appl. Phys. 31 (1960) 655.
[21] G.E. Duvall, G.R. Fowles, in: R.S. Bradley (Ed.), High Pressure Physics and Chemistry, vol. 2, Academic, New York, 1963, p. 2.
[22] S.C. Gupta, et al., in: S.C. Schmidt, R.D. Dick, J.W. Forbes, D.G. Tasker (Eds.), Shock Compression of Condensed Matter-1991, 1992, p. 839.
[23] J.R. Asay, L.C. Chhabildas, L.M. Barker, Sandia National Laboratories Report No. SAND85-2009, 1985.
[24] M.D. Furnish, L.C. Chhabildas, W.D. Reinhart, Int. J. Impact Eng. 23 (1999) 261.
[25] T.C. Kaushik, L.V. Kulkarni, S.K.H. Auluck, IEEE Trans. Plasma Sci. 30 (2002) 2133.
[26] J. Osher, et al., Int. J. Impact Eng. 10 (1990) 439.
[27] M.D. Knudson, et al., J. Appl. Phys. 94 (2003) 4420.
[28] R.F. Smith, et al., Nature 511 (2014) 330.
[29] L.C. Chhabildas, J.R. Asay, L.M. Barker, SAND88–0306 Report No. Sandia National Laboratory Report, 1988.
[30] E. Moshe, et al., Appl. Phys. Lett. 76 (2000) 1555.
[31] K.D. Joshi, A.S. Rav, S.C. Gupta, S. Banerjee, J. Phys. Conf. Ser. 215 (2010) 012149.
[32] L.M. Barker, R.E. Hollenbach, J. Appl. Phys. 1972 (1972) 4669.
[33] A.K. Sexana, T.C. Kaushik, A.M. Rawool, L.V. Kulkarni, S.C. Gupta, Bhabha Atomic Research Center Report No. BARC/2007/E/013, 2007.
[34] G.I. Kanel, et al., J. Appl. Phys. 74 (12) (1993) 7162.
[35] E.B. Zaretsky, G.I. Kanel, S.V. Razorenov, K. Baumung, Int. J. Impact Eng. 31 (2005) 41.
[36] K.D. Joshi, A. Rav, S. Amit, P.C. Das, S.C. Gupta, AIP Conf. Proc. 1665 (2015) 060024.
[37] S.P. Marsh, LASL Shock Hugoniot Data1980, University of California Press, 1980.
[38] G.I. Kanel, S.V. Razorenov, K. Baumung, J. Singer, J. Appl. Phys. 90 (2001) 136.
[39] X. Chen, S.K. Assay, S.K. Dwivedi, D.P. Field, J. Appl. Phys. 99 (2006) 023528.
[40] S.K. Sikka, Y.K. Vohra, R. Chidambaram, Prog. Mater. Sci. 27 (1982) 245.
[41] H. Xia, S.J. Duclos, A.L. Ruoff, Y.K. Vohra, Phys. Rev. Lett. 64 (1990) 204.
[42] J.C. Jamieson, Science 140 (1963) 72.
[43] J.S. Gyanchandani, S.C. Gupta, S.K. Sikka, R. Chidambaram, High Press. Res. 4 (1990) 472.
[44] S.C. Gupta, K.D. Joshi, S. Banerjee, Metall. Mater. Trans. A 39 (2008) 1593.
[45] G. Jyoti, et al., in: Proceedings of the Symposium Zirconium-2002 [Zirc-02], 372, 2002.
[46] G. Jyoti, et al., Phil. Mag. Lett. 75 (5) (1997) 291–300.
[47] M.P. Usikov, V.A. Zilbershtein, Phisica Stat. Solidi 19 (1973) 53.
[48] G.K. Dey, et al., Phil. Mag. Lett. 82 (6) (2002) 333–340.
[49] W. Kohn, L.J. Sham, Phys. Rev. A 1133 (1965).
[50] M.L. Cohen, Phys. Scrip. T1 (1982) 5.
[51] O.K. Anderson, Phys. Rev. B 12 (1975) 3060.
[52] H.L. Skriver, LMTO Method, Springer, Berlin, 1984.
[53] S.Y. Savrasov, D.Y. Savrasov, Phys. Rev. B 46 (1992) 12181.
[54] P. Blaha, K. Schwarz, P. Sorantin, S.B. Trickey, Comput. Phys. Commun. 59 (1990) 399–415.

[55] J. Hafner, G. Kresse, D. Vogtenhuber, M. Marsman. http://www.vasp.at.
[56] P. Blaha, K. Schwarz, G. Madsen, D. Kvasnicka, J. Luitz. http://www.wien2k at 2013.
[57] A. Baroni, et al. http://www.quantum-expresso.org.
[58] http://www.abinit.org.
[59] D.C. Wallace, Statistical Physics of Crystals and Liquids, World Scientific, Singapore, 2002.
[60] B.K. Godwal, R. Jeanloz, Phys. Rev. B 40 (1989) 7501.
[61] D.C. Wallace, Phys. Rev. B 58 (1998) 15433–15439.
[62] M.T. Yin, M.L. Cohen, Phys. Rev. Lett. 45 (1980) 1004.
[63] S.C. Gupta, R. Chidambaram, High Press. Res. 12 (1994) 51–70.
[64] S.C. Gupta, Proc. Shock Compression Condens. Matter-1991 157–163 (1992).
[65] C.J. Pickard, R.J. Needs, J. Phys. Condens. Matter 23 (2011) 053201.
[66] C.J. Pickard, R.J. Needs, Phys. Rev. Lett. 97 (2006) 045041–045044.
[67] A.R. Oganov, C.W. Glass, S. Ono, Earth Planet. Sci. Lett. 241 (2006) 95–103.
[68] C.W. Glass, A.R. Oganov, N. Hensen, Comp. Phys. Commun. 175 (2006) 713–720.
[69] R. Martonak, A. Laio, M. Parrinello, Phys. Rev. Lett. 90 (2003) 075503.
[70] R. Martonak, et al., Z. Krist 220 (2005) 489–498.
[71] D. Petti, D. Craford, N. Chauvin, MRS Bull. 34 (2009) 40–45.
[72] D.C. Craford, D.L. Porter, S.L. Hayes, J. Nucl. Mater. 371 (2007) 202–231.
[73] I.R. Shein, K.I. Shein, A.L. Ivanovskii, J. Nucl. Mater. 353 (2006) 19.
[74] J.S. Olsen, L. Gerward, U. Benedict, J.P. Itie, K. Richter, J. Less Common Metals 121 (1986) 445–453.
[75] B.D. Sahoo, K.D. Joshi, S.C. Gupta, J. Nucl. Mater. 437 (2013) 81–86.
[76] J.A. Jackman, T.M. Holden, W.J.I. Buyers, Phys. Rev. B 33 (1986) 7144–7153.
[77] J. Gerward, J.S. Olsen, U. Benedict, J.-R. Itie, J.C. Spirlet, J. Appl. Crystallogr. 19 (1986) 308.
[78] L. Gerward, J.S. Olsen, U. Benedict, H. Luo, J. Less-Common Metals 161 (1990) L11.
[79] B.D. Sahoo, K.D. Joshi, S.C. Gupta, J. Appl. Phys. 117 (2015) 185903.
[80] A.R. Oganov, C.W. Glass, J. Chem. Phys. 124 (2006) 244704.
[81] Y. Akahama, H. Kawamura, A.K. Singh, J. Appl. Phys. 92 (2002) 5892.
[82] R.E. Duff, F.S. Minshall, Phys. Rev. 108 (1957) 1207.
[83] R.G. McQueen, S.P. Marsh, J. Appl. Phys. 31 (7) (1960) 1253.
[84] E. Blanco, J.M. Mexmain, P. Chapron, Shock Waves 9 (1999) 209.
[85] P.D. Sebban, D.B. Holtkamp, J.L. Pelissier, J. Taboury, A. Rouyer, Shock Waves 11 (2002) 385.
[86] Y. Tan, et al., J. Appl. Phys. 113 (2013) 093509.
[87] J.L. Pelissier, N. Wetta, Physica A 289 (2001) 459.
[88] G.A. Cox, in: M. Elert, M.D. Furnish, R. Chau, N. Holmes, J. Nguyen (Eds.), Shock Compression of Condensed Matter, 151, AIP, 2007.
[89] D. Mukherjee, B.D. Sahoo, K.D. Joshi, S.C. Gupta, J. Appl. Phys. 115 (2014) 053702.
[90] L. Liu, et al., Phys. Status Solid B 250 (2013) 1398.
[91] R. Jeanloz, et al., PANS 104 (22) (2007) 9172–9177.

[45] [illegible] Y. [illegible] Vollmer [illegible] M. [illegible]
[46] [illegible]
[47] [illegible]
[illegible]
[49] [illegible] Wallace, [illegible] of Crystals and [illegible] Vol [illegible] 2006.
[50] D.C. [illegible] Phys. Rev. B 40 (1989) 7497.
[51] D.C. Wallace, Phys. Rev. B [illegible] (1996) 154 [illegible] 5930.
[52] M. [illegible] Phys. Rev. Lett. 43 (1980) 1946.
[53] S.C. Gupta, [illegible] High Press. Res. 12 (1994) [illegible]
[54] S.C. Gupta, [illegible] Compression [illegible] M. [illegible]
[55] C.J. Pickard, [illegible] Needs, [illegible] 23 (2011) 053201.
[56] C.J. Pickard, R.J. Needs, Phys. Rev. Lett. 97 (2006) 045504.
[57] A.R. Oganov, C.W. Glass, [illegible] Earth Planet. Sci. Lett. 241 (2006) 95–103.
[58] C.W. Glass, A.R. Oganov, N. Hansen, Comput. Phys. Commun. 175 (2006) 713–720.
[59] R. Martoňák, A. Laio, M. Parrinello, Phys. Rev. Lett. 90 (2003) 075503.
[60] R. Martoňák, et al., [illegible] (2005) 488–493.
[61] D. Petti, D. Crawford, [illegible] MRS Bull. 34 (2009) 40–45.
[62] D.C. Crawford, D.L. Porter, S.L. Hayes, J. Nucl. Mater. 371 (2007) 232–242.
[63] L.C. Shein, K. Ashuri, A.L. Ivanovskii, J. Nucl. Mater. 353 (2006) 19.
[64] T.S. Olsen, L. Gerward, U. Benedict, J.P. Itie, [illegible] J. Less Common Metals 121 (1986) 445–453.
[65] [illegible] Sahoo, K.D. Joshi, S.C. Gupta, J. Nucl. Mater. 437 (2013) 81–[illegible]
[66] J.A. Jackman, T.M. Holden, W.J.L. Buyers, Phys. Rev. B 33 (1986) 7144–7153.
[67] L. Gerward, J.S. Olsen, U. Benedict, J.-P. Itié, J.C. Spirlet, J. Appl. Crystallogr. 19 (1986) 308.
[68] L. Gerward, J.S. Olsen, U. Benedict, H. Luo, J. Less Common Metals 161 (1990) L11.
[69] B.D. Sahoo, K.D. Joshi, S.C. Gupta, J. Appl. Phys. [illegible] (2013) [illegible]
[70] A.R. Oganov, C.W. Glass, J. Chem. Phys. 124 (2006) 244704.
[71] S. [illegible] H. [illegible] A.K. Singh, J. Appl. Phys. 91 (2002) [illegible]
[72] R.E. Duff, F.S. Minshall, Phys. Rev. 108 (1957) 1207.
[73] R.G. McQueen, S.P. Marsh, J. Appl. Phys. 31 (7) (1960) 1253.
[74] B. Bleton, J.M. Mozamis, [illegible] Shock Waves 7 (1997) [illegible]
[75] J.D. [illegible] D.B. [illegible] J. [illegible] Phys. [illegible]
[76] Y. [illegible] J. Appl. Phys. [illegible]
[77] J.L. Feldman, [illegible] Physica A [illegible]
[78] [illegible] M. [illegible] R. [illegible] N. [illegible] L. Nguyen [illegible] Shock Compression of Condensed Matter, [illegible] AIP, [illegible]
[79] [illegible] Mukherjee, B.D. Sahoo, K.D. Joshi, S.C. Gupta, J. Appl. Phys. [illegible]
[80] [illegible] et al., Phys. Status Sol. [illegible] (2011) [illegible]
[81] R. Jeanloz, et al., PNAS 104 (22) (2007) [illegible]

CHAPTER

MATERIALS FOR HOSTILE CORROSIVE ENVIRONMENTS

3

U. Kamachi Mudali
Indira Gandhi Centre for Atomic Research, Kalpakkam, India

CHAPTER OUTLINE

Materials Under Extreme Conditions. http://dx.doi.org/10.1016/B978-0-12-801300-7.00003-6

1. INTRODUCTION

The science of corrosion prevention and control is highly complex in hostile corrosive and radioactive environments, exacerbated by the fact that corrosion assumes different forms and is affected by numerous factors. Most metals and alloys have inherent corrosion behavior, and no material can be used universally in every operating condition of various environments. In nuclear power plants (NPPs) based on fast breeder reactor technology, there are several aggressive environments like liquid sodium in the primary heat exchanger, steam water in the secondary heat exchanger, seawater in cooling water systems, nitric acid for aqueous reprocessing, and molten chloride for pyrochemical reprocessing [1–3]. Liquid metals (i.e., Na, Pb, Pb-Bi, etc.) have been studied since the early development of fission energy as reactor core coolants for fast reactors; liquid sodium is used as the coolant in fast breeder reactors (FBRs) since it meets all the requirements of high-thermal conductivity, high-heat-transfer coefficients, low-vapor pressure, high-boiling point, low cost, and excellent compatibility with structural materials for fast reactor coolant (Fig. 3.1). In its pure form, devoid of any impurities, there are no corrosion issues with respect to liquid sodium [3,4]. However, in the presence of certain

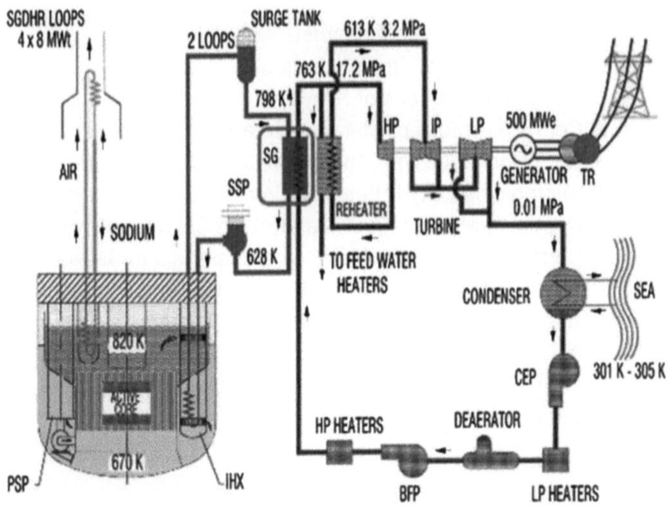

FIGURE 3.1

Typical flow diagram of prototype fast breeder reactor.

impurities like oxygen, carbon, etc., the materials coming in contact with sodium can get affected adversely. This can lead to a reduction in the wall thickness of components resulting in poor load-bearing capacity and less heat-transfer efficiency, and hence is an important issue to be considered. In addition, the corrosion reactions and mass transfer issues in sodium and the effects of sodium environment on the mechanical properties related to sodium-cooled FBRs need more elaborate investigation. Thereby, the materials selection is to be made based on the specific service conditions and the material-environment interaction issues.

The steam generator (SG) materials of NPPs are exposed to extreme corrosive environments of steam and water in the shell side. Thereby, the important requirements of SG heat-transfer tubing are as follows [5,6]:

1. Resistance to stress corrosion cracking (SCC) in aggressive primary water and secondary side environments at 330°C.
2. Material availability.
3. Acceptable heat-transfer characteristics.
4. Manufacturability of thin-walled tubes up to the length of 35 m.
5. Weldability.
6. Inspectability by standard eddy current techniques and ultrasonic techniques.

Furthermore, the SG materials are exposed to the typical operating condition in a pressurized water reactor (PWR) on the secondary shell side up to a pressure of 6.2 MPa, inlet temperature of 493 K, and outlet temperature of 548 K. Severe corrosion problems of SG tubes led to the shutdown of many PWRs in the late 1970s. Regular inspection of the steam generator tubes, monitoring of water chemistry, and selection of corrosion-resistant tube materials helped to solve the problems in many other PWRs. In addition, commercial exploitation of FBR technology is held in the hands of steam generator designer and manufacturer. Of the various issues that need to be overcome in fast reactor technology, the most difficult one is related to SGs that have high-pressure water or steam on one side and liquid sodium on other sides [1,5]. This is because in sodium-cooled FBRs the consequence of a leak, namely the generation of large quantities of hydrogen and an extremely aggressive environment due to sodium water reaction demands excellent design and maintenance of the integrity of heat exchanger tubes. The fundamental starting point to design a high-integrity system rests on the choice of material that forms the sodium-containment boundary, the heart of an SG: the heat-exchanger tube, 2–3 mm of highly stressed material that separate sodium and water [1,3,6].

In the extremely corrosive seawater environment, the materials showed various forms of corrosion such as crevice corrosion, pitting corrosion, and microbiologically influenced corrosion (MIC). It has been estimated that nearly 20% of the total corrosion cost is due to MIC [7,8]. MIC is often seen as a pitting attack that is generally associated with the presence of surface-associated microbial communities embedded in a bioinorganic matrix, referred to as a biofilm.

The degradation of materials in a hostile corrosive nitric acid environment in nuclear fuel reprocessing plants is an important issue, and corrosion research plays a significant role to support the reliability of a plutonium uranium redox extraction (PUREX) reprocessing plant [1–3,9–11]. The reprocessing by PUREX process of high burnup, high Pu containing spent fuel from fast breeder reactors involves various stages that employ head-end treatment involving chemical or mechanical decladding followed by dissolution of fuel in nitric acid, feed clarification, and chemical conditioning of the solution for solvent extraction [11,12]. Additionally, during different stages of the spent fuel dissolution by PUREX process, nitric acid is the main medium used in various concentrations from dilute (1–4 M) to concentrated (10–14 M); temperatures ranging from room temperature (solvent extraction) to intermediate (warm, raffinate waste storage tanks) to boiling temperature (dissolver, evaporator); and including uranium, transuranium elements (TRUs), and powerful oxidizers (Ce^{4+}, Ru^{6+}, Fe^{4+}, Cr^{6+}, Pu^{6+}, etc.) from the fission products. Furthermore, materials are also exposed to the intense radiation field, particularly gamma, and radioactive contamination of surfaces that can cause serious corrosion problems [1,10,11]. Thereby, the operating conditions of high temperatures (boiling condition), and a severely corrosive environment of highly oxidizing nitric acid environments (up to 14 M HNO_3) for the dissolution of the spent fuel pose a risk leading to materials properties degradation or even catastrophic plant failures [2,10–12].

The pyrochemical reprocessing is a refining process of spent metallic fuel and uses high-temperature oxidation–reduction reactions in nonaqueous media to separate the actinides. Since the pyrochemical reprocessing utilizes molten LiCl-KCl salt and molten uranium in an electrorefiner and cathode processor operating at 500–1300°C, high radiation along with fission products (Ru, Pd, Rh, and lanthanides, etc.), chloride salt and reactive metals such as uranium at high temperatures, results in extremely hostile corrosive environment. In order to overcome the corrosion of materials in extreme environment of molten chloride and molten uranium of high and corrosive temperature, efforts have been made to develop compatible materials, fabrication technology, and coating useful for various unit operations like salt preparation, electrorefining, and cathode processing in pyrochemical reprocessing is being actively explored [3,9,12].

The various components of the nuclear industry are designed with the objective of zero incident failures. Hence, the choice of materials for various extreme/hostile environments like seawater, high-temperature/pressure steam, liquid sodium in FBRs, aqueous and pyrochemical reprocessing plant to resist the degradation processes (e.g., corrosion, stress, mechanical, radiation effects, etc.) is a significant issue that must be closely examined. Furthermore, material requirements for such environments are a complicated process involving many parameters such as high corrosion resistance, mechanical properties, ease in fabrication, availability, cost, etc. In reactor core and spent fuel reprocessing, the presence of intense radiation is an additional factor of concern [10,11]. Furthermore, the safety requirements in hostile corrosive nuclear environments demand a high level of materials reliability, and hence the key to any successful design for long life is the proper selection of materials. This chapter briefly highlights the challenges, and the approaches adopted toward the development of special corrosion-resistant materials (i.e., Cr-Mo, nickel alloys, amorphous metals, nitric acid grade SS, Ti and Zr and their alloys, etc.), coating technology, i.e., double-oxide coatings, thermal barrier coatings, superhydrophobic coatings, self-healing repair coatings, antifouling coatings, etc. and corrosion monitoring for highly corrosive and hostile environments, particularly toward development of FBRs and their associated fuel cycle technologies.

2. CORROSION BY LIQUID SODIUM

The general principles that govern the corrosion processes by liquid sodium are different from that in aqueous environments. Unlike aqueous corrosion, which is electrochemical in nature, requiring a passive film breakdown, and occurs over a wide range of temperatures, sodium corrosion is a high-temperature, complex phenomenon depending on a number of factors, the most important of which is the presence of impurities, which will be elaborated in the later part of this chapter. Corrosion of alloys by liquid sodium is often a combination of several different processes. High-temperature sodium with dissolved carbon in a system of different alloys of construction is an extreme condition that the materials have to endure during operation.

The corrosion processes that take place in sodium–stainless steel systems at high temperatures are [13–15]:

- *General corrosion*: This occurs due to complete loss of material by dissolution and leads to a reduction in wall thickness. This process is dependent on the amount of oxygen present at the interface in the system.
- *Formation of ferrite layer*: Preferential leaching of nickel results in the transformation of austenite to ferrite layer of inferior properties. The average concentration of Cr and Ni in the ferrite layer is 6–8 wt% and about 2–4 wt%, respectively [14].
- *Carburization/Decarburization*: These processes are diffusion controlled, and the depth of the layer carburized/decarburized depends on the duration of exposure, the temperature, and carbon potential of sodium at the interface.
- *Precipitation of carbides* in the matrix brought about by thermal effects as well as carbon transfer through sodium.

2.1 MECHANISMS OF SODIUM CORROSION

- **Dissolution**: Dissolution of elements is one of the basic mechanisms of sodium corrosion, which is highly dependent on temperature. The dissolution of the metal depends on the ratio of the chemical activities of the solute in the alloy and the liquid sodium. Dissolution occurs in sodium loops if this ratio acts as a driving force for dissolution and the driving force for dissolution is increased by compound formation resulting in a higher dissolution and higher corrosion rates.
- **Exchange reactions of nonmetallic elements between alloys and liquid sodium**: The chemical activity of oxygen in sodium influences corrosion mechanism. The exchange of oxygen often causes the formation of stable oxide layers on the alloy surface thus decreasing the corrosion rates.
- **Mass Transfer**: Liquid sodium can dissolve some of the constituents of stainless steel [15] from one region and deposit them in another region depending on the temperature gradient or composition gradient across the materials in the sodium system. Generally, elements from the materials in the hot sections are removed, transported by flowing sodium, and are deposited in the cooler sections. Such mass transfer of metallic elements is not only closely related to solubility but is also influenced by the presence of nonmetallic impurities such as oxygen and carbon in sodium [13,16]. Mass transfer not only affects the corrosion rates but also affects the flow rates of sodium in the cooler sections where there is deposition of sodium compounds.

2.2 ROLE OF IMPURITIES IN SODIUM CORROSION

- **Carbon**: Being an interstitial element, carbon can easily diffuse in and out of stainless steel depending on its thermodynamic activity in sodium resulting in carburization/decarburization of the stainless steel. In a nonisothermal system, made of entirely one type of material, the direction of transport would be from the high-temperature region to the low-temperature region. In a system where the materials are different, the carbon transport depends on the relative carbon activities of the steels at their respective temperatures of operation [15,16].

- **Oxygen**: The presence of oxygen in sodium is highly deleterious. The solubility of oxygen in sodium varies from 1 ppm at 120°C to 200 ppm at 500°C. Dissolved oxygen can combine with sodium and with the transition metals [4,14] in the alloy in contact and can form oxides at very low levels of oxygen in sodium. Ternary compounds of other transition metals like iron are formed only if the oxygen content is very high [15,16].
- **Hydrogen**: Hydrogen is usually present in alkali metals as a result of their reactions with moisture. The solubility of hydrogen in sodium is less than oxygen, and hence, the hydride precipitates even at higher temperatures. The exchange of hydrogen between alkali metals and alloys can cause embrittlement of materials.

Because of their importance in sodium corrosion and technology, the above said impurities are kept well under control in the sodium circuit. The dissolved oxygen and hydrogen level in sodium is maintained by cold trap method [15,16].

2.3 FACTORS INFLUENCING SODIUM CORROSION

Corrosion by liquid sodium is a combination of many processes, which depend on a number of factors such as solubility of metals in sodium, solubility of nonmetals in sodium, thermophysical condition of the material, temperature of exposure, downstream effect, axial rate of heating of sodium, mechanical stress on materials, temperature differential in the loop, velocity of sodium, etc. [16–19]. Among the previously mentioned factors, the solubility of metals and nonmetals in sodium and temperature are the main parameters that determine the corrosion and mass transfer phenomenon in sodium systems. In liquid sodium, solubility is highly dependent on temperature. The major elements of importance are iron, nickel, chromium, molybdenum, and manganese, which get dissolved in sodium depending on the temperature of operation. For instance, based on solubility expression, the typical value of solubility of nickel is 1.3, 1.9, and 2.06 ppm at 673, 773, and 873 K, respectively [4,13,19]. Similarly, for other elements, the solubility increases with an increase in temperature. As a result of solubility effects, leaching of elements takes place in combination with oxygen thereby forming corrosion products. The most commonly encountered corrosion products in the sodium system are $NaCrO_2$ and Na_4FeO_3 [13,16]. However, the leaching of nickel is independent of oxygen levels in the sodium system [16,17]. Accordingly, the solubility of oxygen is more than 1000 ppm at 500°C and that of carbon is about 30 ppm at 700°C [13]. The solubility of nitrogen is extremely less, far less than 1 ppb even at 700°C [4]. The solubility of carbon is in the range of 1 ppm at 500°C to 30 ppm at 700°C [13]. Structural materials can get carburized or decarburized depending on the amount of carbon activity, which is a function of the soluble carbon present in sodium.

2.4 MATERIAL TESTING AND EVALUATION IN MOLTEN SODIUM

To understand corrosion of materials in sodium, data on sodium corrosion is generated by exposing materials of interest in either isothermal or nonisothermal systems or loops. The operating loops may be monometallic or bimetallic depending on the nature of studies embarked on. In monometallic systems, mass transfer occurs due to temperature or chemical potential gradient across the materials. In a nonisothermal system, when the materials are of the same composition, carbon transfer would be from a high-temperature to a low-temperature region; when the materials are different, the direction of carbon transfer would depend on the local carbon activity between the material and sodium [18]. Carburization data evolved on American Iron and Steel Institute (AISI) type 316LN stainless steel (SS) and modified 9Cr-1Mo steel retrieved after 16,000 h of sodium exposure at 525°C from a bimetallic sodium loop facility in our laboratory showed that both the materials got carburized up to a depth of 100 and 70 μm, respectively, from the surface [19]. Sodium used in the studies contained 25 wppm of carbon out of which 2 wppm are in solution. Hence, its carbon activity would be higher than 316LN SS. Takushi et al. have reported carburization of modified 9Cr-1Mo steel with carbon concentrations between 0.01 and 0.10 ppm exposed to sodium for 200,000 h [20]. It is also reported that carburization has not affected the tensile strength of the material. In our studies, there was no significant change in the tensile properties of this material; however, there was a reduction in impact energy. 316LN SS showed a reduction in ductility of 34%, 60%, and 39% for specimen after exposure for 16,000, 30,000, and 40,000 h at 525°C as shown in Tables 3.1, 3.2, and 3.3, respectively. This was due to carburization and thermal-aging effects by secondary phase precipitation like sigma, which is an embrittling phase. With respect to strength, there were no significant changes between the three durations of exposure. Natesan et al. [21] have reported the influence of sodium exposure on the tensile strength of 316 SS and titanium modified 316 SS due to carburization.

Table 3.1 Tensile Test Results of 316LN SS Exposed to Flowing Sodium for 16,000 h at 798 K

Material Condition	YS MPa	UTS MPa	UE %	TE %
Mill-annealed	324	627	60	66
Thermally aged	324	632	58	66
Sodium exposed	356	657	39	43

Table 3.2 Tensile Test Results of 316LN SS Exposed to Flowing Sodium for 30,000 h at 798 K

Material Condition	YS MPa	UTS MPa	UE %	TE %
Mill-annealed	324	627	60	66
Thermally aged	303	577	53	61
Sodium exposed	373	576	20	24

Table 3.3 Tensile Test Results of 316LN SS Exposed to Flowing Sodium for 40,000 h at 798 K

Material Condition	YS MPa	UTS MPa	UE %	TE %
Mill-annealed	324	627	60	66
Thermally aged	285	576	59	69
Sodium exposed	343	606	35	40

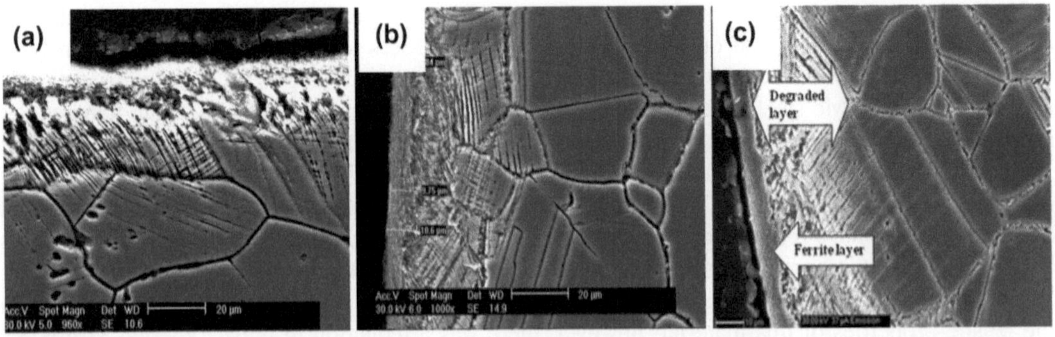

FIGURE 3.2

Scanning electron microscope micrographs showing the microstructures of sodium-exposed 316LN SS after exposure for (a) 16,000 h, (b) 30,000 h, and (c) 40,000 h at 525°C.

Micrographs of 316LN SS exposed to flowing sodium for 16,000, 30,000, and 40,000 h examined under a scanning electron microscope (FEI XL ESEM) with energy dispersive spectroscopic (EDS) attachments are shown in Fig. 3.2a–c. The microstructures for all the three durations revealed the presence of a surface-modified layer of ferrite to about 5–10 μm thickness [16]. Generation of a modified surface layer has been observed and reported in earlier studies [21,22] in sodium-exposed-type 316 and 316LN SS in bimetallic loop operating and simulating the primary and secondary circuit of reactor conditions in terms of temperature, velocity, area ratios, and material sequence.

Microchemical changes obtained from EDS spectra for every 2 μm up to a depth of 40 μm, across the cross-section of the specimen, indicated dissolution of nickel and chromium from the surface of 316LN SS for specimens exposed to sodium for 16,000, 30,000, and 40,000 h. The nickel and chromium contents in the modified ferrite layer were less than 5 wt% each (Fig. 3.3). Generally, the ferrite layer contains 1–4 wt% nickel and 5–7 wt% chromium [22]. Natesan et al. [21] reported 20–30-μm-thick ferrite layers, which attained a steady state after the initial corrosion period. However, modified 9Cr-1Mo steel neither showed dissolution nor deposition of any element at the surface. Precipitate coarsening due to long-term thermal aging could be observed.

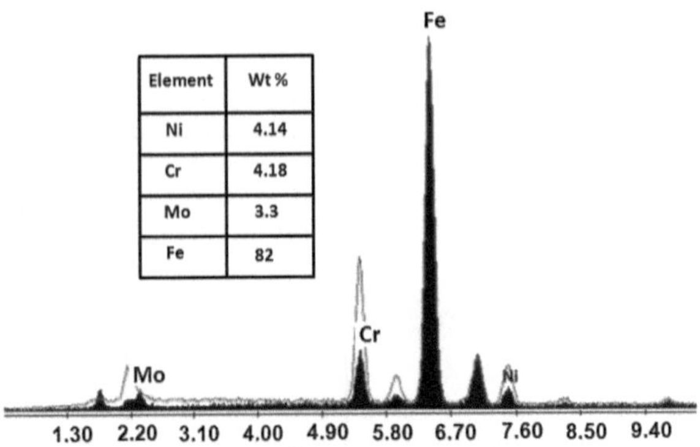

FIGURE 3.3

Energy dispersive spectroscopic spectrum of sodium exposed 316LN SS taken at a distance of 2 μm from the edge, showing leaching of Ni and Cr.

3. MATERIALS FOR THE HOSTILE CORROSIVE ENVIRONMENTS IN STEAM WATER ENVIRONMENTS OF NUCLEAR POWER PLANTS

The function of steam generators is to exchange heat produced in the core of a reactor in a nuclear power plant with water and convert the same into steam. In pressurized water reactors (PWRs), this component with the largest surface area is located between the primary and secondary cooling loops. Each reactor has multiple steam generators (SGs) and tubes in SGs varying from 3000 to 15,000 numbers. They also provide a barrier between the radioactivity in the primary heat transport system and the rest of the plant. Hence, the integrity of these tubes assumes great significance [5]. However, these SG materials are exposed to extreme corrosive environments of steam and water in the shell side. The typical operating condition in a PWR on the secondary shell side is a pressure of 6.2 MPa, inlet temperature of 493 K, and outlet temperature of 548 K. Severe corrosion problems led to the shutdown of many PWRs in the late 1970s. Regular inspection of the steam generator tubes, monitoring of water chemistry, and selection of corrosion-resistant tube materials helped to solve the problems in many other PWRs.

3.1 MATERIAL SELECTION FOR STEAM WATER ENVIRONMENTS IN NPPS

Initially, carbon steel was used as pipe material in the secondary side and then was replaced by copper with high-thermal conductivity. Ferritic stainless steel is selected for turbine blades. Austenitic SS (18% Cr–8% Ni) was used only for a few PWR plants. Russian VVERs still use austenitic stainless steels as SG material. Subsequently, based on superior strength and steam oxidation resistance, the nickel alloys were selected. Thereby, several materials were used as heat-transfer tubing in light water reactor steam generators, including Type 316 stainless steels, Alloy 600 MA (mill annealed), Alloy 600 TT (thermally treated), Alloy 800 modified, and Alloy 690 TT. Alloy 600 MA (Ni-16Cr-9Fe) was susceptible to stress corrosion cracking and hence the need for new alloys was felt [5,6]. In the mid-1980s, Alloy 690 TT (Ni-30Cr-10Fe) was selected as the best SG tubing material with an alternative of Alloy 800mod (Fe-33Ni-22Cr).

Persistent issues of corrosion degradation in the inner diameter of the tubes like primary water stress corrosion cracking and in outer diameter intergranular cracking and stress corrosion cracking led to the improvement of metallurgical characteristics of Alloy 600. Alloy 600 was exposed to temperatures of 700–725°C to dissolve all carbides, and by rapid cooling sensitization was avoided. Thus thermally treated Alloy 600 showed SCC resistance. Further effort to improve SG heat-transfer tubing led to the qualification and selection of thermally treated Alloy 690 TT. Weld filler metals (Alloys 52 and 152) for Alloy 690 TT have been developed, qualified, and used.

The tube-manufacturing process includes melting by air or vacuum induction melting and argon-oxygen or vacuum oxygen decarburization followed by electro slag remelting to produce a chemically homogeneous composition and low levels of metallic inclusions. The ingots are hot worked into round bars in a blooming mill or forge shop and tube hollows are made by hot extrusion while outer diameter and tube thickness are controlled by cold pilgering and intermediate mill annealing. One of the most critical processes is final mill annealing at approximately 1080°C minimum to control carbon content and grain size (ASTM No. 5-8). Rapid cooling to room temperature retains carbon in solid solution. Tubes are straightened and polished by belt grinding. The tube defects are tested by nondestructive techniques like ultrasonic tests and eddy current tests. Then the straight tubes are given special thermal treatment in vacuum at temperatures of 715–735°C for 10 h. During the heat treatment, carbides are precipitated at grain boundaries depleting matrix of super saturation of carbon, and holding time helps to diffuse chromium back to depleted regions. For recirculating steam generators, small-radius U-bend tubes are stress relieved in the same vacuum furnace used for thermal treatment at the same temperature, but for a minimum period of 2 h.

3.2 STEAM GENERATOR CORROSION PROBLEMS

The majority of the SG corrosion problems were experienced in the secondary side and Fig. 3.4 summarizes the worldwide SG corrosion experience [6]. All PWR plants started phosphate chemistry with concentrations of 2–6 mg/kg for high-purity plants and 40–50 mg/kg for plants with frequent condenser leaks. With this phosphate treatment, wastage corrosion was experienced, resulting in SG leaks.

In the absence of phosphate treatment along with condenser leaks, tube denting was the main problem. Carbon steel–drilled hole tube support plate (TSP) supports SG tubes, and it is very sensitive to impurities and denting corrosion. Severe corrosion of carbon steel support plates led to the accumulation of corrosion products in the crevice between the support plates and tubes leading to tube denting. For denting to occur, acidic and oxidizing conditions are required. Condenser leaks lead to chloride ingress and acidic conditions. Oxidizing conditions are favored by air ingress or presence of copper ions on SG tube deposits.

The phosphate treatment was shifted to all volatile treatments (AVT) by replacing all copper bearing materials in the secondary side. Denting stopped with high AVT chemistry. Further improvements of secondary side water chemistry were done by maintaining condenser integrity that avoids impurity and oxygen ingression. The tube support was changed to SS material. In older plants with CS-supported SG tubes, phosphate treatment denting was avoided by boric acid treatment (BAT).

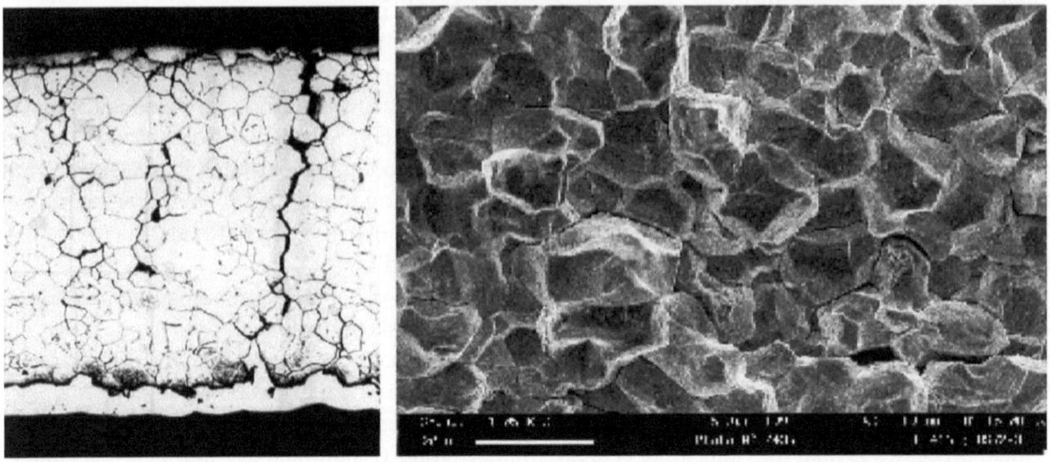

FIGURE 3.4

Stress corrosion cracking in a steam generator tube – Alloy 600 [26].

Reproduced with the permission of Elsevier.

Pitting corrosion of SG tubes happened due to the accumulation of chlorides beneath the deposits. This was avoided by eliminating condenser leaks. Secondary side stress corrosion cracking occurred under stress conditions in corrosive media, very caustic pH >9 or very acidic conditions pH <4. Stress damages the protective oxide layer, and SCC can be initiated. Several mill-annealed Alloy 600 tubes and few thermally treated Alloy 600 tubes suffered from stress corrosion cracking. Alloy 800 with a Ti/C ratio of 12 exhibits resistance to SCC. SCC mechanism in nickel alloys includes two models: hydrogen embrittlement (HE) model and internal oxidation model. HE model is based on the high concentration of H in a region ahead of crack tip leading to weakening of atomic bonds (strength) of material. Internal oxidation (IO) model hypothesizes faster oxidation at the stressed crack tip where preferential oxidation of Cr leaves Ni-rich substrate underneath the crack tip filled with voluminous corrosion products. Both models predict faster crack propagation.

The corrosion products also affected the heat-transfer properties by the phenomenon of SG tube fouling. This fouling behavior strongly depended upon surface chemistry and on the amine used for pH control. Oxidizing conditions favor the formation of hematite, and reducing conditions favor the formation of magnetite. All the degradation problems can be controlled by maintaining cleanliness of SG tubes. The main objective of secondary side water chemistry is to build and maintain protective layers on CS. Oxygen control during startup and shutdown operations is essential to minimize corrosion product ingress into SG tubes. In systems with high medium velocities, flow-assisted corrosion (FAC) can happen.

The main source of corrosion products in the system is the FAC. If the protective layers on CS pipes dissolve in an iron unsaturated fluid medium at the metal-fluid interface, or if iron ions released from the CS surfaces are immediately removed by a high flow, the protective layers cannot be built, and this results in the FAC degradation of CSs. FAC can occur in both one-phase flows, like in condensate, and also in two-phase (wet steam) flows. Visual pattern of FAC is different depending on under which flow condition it was caused. FAC caused by one-phase flow shows “horseshoe pits” whereas it has “tiger stripping” pattern if it is produced under two-phase flow conditions in wet steam [23].

3.3 EQUIPMENT FAILURES IN STEAM GENERATORS

Different equipment failures in steam generators are described next [24].

Deaerator Cracking: A deaerator is a device widely used to remove oxygen up to 7 ppb and other dissolved gases like carbon dioxide completely from feedwater to steam-generating boilers. Environmentally assisted fatigue cracking of deaerators has been reported at the head-to-shell weld below the water level in the storage compartment. Identifications of cracks are done by wet fluorescent magnetic particle testing. Various steps adopted to minimize this problem involve stress relieving of welds and minimizing thermal and mechanical stress during operation.

Feedwater line erosion: High-velocity water or water/steam mixtures cause erosion corrosion at hairpin bends and in feedwater lines. The mixture of steam and water will cause thinning of elbows and leave a characteristic print of reverse horseshoe. A sequence of bends can increase the local velocity even when moderate average velocities occur. Hence, in addition to water chemistry control, the design and operational changes are also required to control erosion corrosion.

Economizer tubes: Water tube economizers are subjected to serious damage by oxygen pitting at the inlet, and tube weld seams. Effective deaeration by fast-acting oxygen scavengers like sodium sulfite can prevent this problem. Another problem that can occur under deposits is caustic gouging due to under-deposit accumulation of caustic soda.

Failure due to overheating: Thin-lipped ruptures of super heater tubes happen when the deposit restricts the flow; the steam flow becomes insufficient leading to rapid heating of tubes, etc. Wrinkled, bulged external surface and a final thick-lipped fissure indicated long-term creep failure due to repetitive scale formation, causing overheating and bulging of tubes.

Failures due to corrosion: Where tubes are rolled into sheets, drums and headers caustic embrittlement occurred due to stress and frees NaOH in water. Addition of sodium nitrate in an appropriate ratio to NaOH inhibits this stress corrosion cracking problem at pressures less than 900 psig. Above this pressure phosphate/pH control is the usual internal treatment.

Straight tubes, shell-and-tube waste heat boilers undergo transgranular cracking due to uneven distribution of stresses by hot gases on the face of tubes and also known as corrosion fatigue failures. Design features that allow steam pockets to form are known to cause cyclic overheating and quenching.

3.4 CHEMISTRY CONTROL OF A STEAM/WATER CIRCUIT

The following main water chemistry-related factors can destroy the protective oxide layers and cause SG tubing corrosion [24]:

pH: Extremely high or low local pH values can dissolve the oxide layers and accelerate different types of corrosion.

Electro Chemical Potential (ECP): ECP is a measure of the strength of the oxidizing or reducing conditions. Local ECP affects strongly the occurrence and rate of the corrosion.

Specific species: Certain specific impurities, like sulfur, chlorides, lead, etc., accelerate tubing corrosion either indirectly by changing the local pH and ECP in the crevices or beneath deposits, or they directly react with the protective oxide layers on the tubing surface.

High AVT or high pH ammonia chemistry can stop FAC in two-phase flow stream systems. The source of ammonia is hydrazine, which thermally decomposes at temperatures above 180°C, producing ammonia. Hydrazine is continuously injected in condensate system for oxygen control. The chemistry conditions achieved are ammonia: 5–15 mg/kg (ppm), hydrazine: 20–150 μg/kg (ppb), pH values: >9.8 up to 10.1, and specific conductivity values: 15–35 μS/cm. Elimination of impurities necessitated the adoption of condensate polishing systems for condenser leaks and steam generator blow down for SG bulk water concentration. CPS consists of a series of ion-exchange resin (IER) beds to eliminate impurity transfer to SG tubes. Corrosion testing programs involve long-term exposures to all reasonable and highly faulted chemical environments, with varying metallurgical features like carbon concentration, final mill annealing, and thermally treating temperatures.

3.5 MATERIALS FOR STEAM WATER ENVIRONMENTS IN FBRS

3.5.1 Choice for Steam Generator Material

The prime requirement for any potential material for SG is that it should exhibit adequate mechanical properties when exposed to the design conditions for the designed life of the operating conditions. The optimum core outlet temperature for fast reactors is in the range of 500–550°C, and the secondary sodium temperature is approximately 20–30°C less [25]. Thus SG materials are working in the creep regime during operation. By providing appropriate wall thickness the decarburization problem in the sodium side can be accommodated. Along with creep, fatigue properties must be well defined under high- and low-cycle conditions and effect of fretting also needs to be considered. Availability of design codes or adequate data to allow sufficient high-temperature analysis to validate the design is needed.

The second area of importance for the designer is the fabrication [25]. High inspection standards are a safeguard against faults in complete systems. The rejection standard of ferritic tubes is based upon a V notch of 5% tube wall thickness for ultrasonic examination and 0.5–0.75 mm diameter through thickness holes on eddy current testing. Machining, rolling, or bending and welding procedures on ferritic materials demand strict specifications in chemical composition like well-defined carbon and silicon levels, manganese or sulfur ratios, and trace element levels. Autogenous welding procedures are employed for fabrication of heat exchanger materials and limiting the number of material melts can reduce cast-to-cast variation. Electroslag refining and vacuum remelting techniques provide cleaner steels.

The final area of concern for selection of material is the environmental performance [25]. SG tubes contain a boiling steam and water mixture at pressures up to 180 bars with controlled water chemistry to reduce waterside corrosion. However, in seawater-cooled stations full flow condensates polishing are mandatory. System startup and shutdown variations need to be considered. In the steam generators with heat flux, as high as 650 kW/m^2 tube wall dry out lead to thermal cycling and salt concentration in the last vestiges of water film left on the wall. SG materials need tolerance to faulty water chemistry arising from chloride excursions due to condenser leakage, sulfate slip from regeneration of polishing plant, and oxygen ingress due to faulty deaeration. The material should be amenable to chemical cleaning to avoid spallation, to remove oxidation, and to remove magnetite formation. Spallation increases corrosion products, oxidation increases concentration of chemical solutions, and magnetite formation causes pressure drop.

3.5.2 Candidate Materials for Fast Reactor SG

At the temperature proposed for the fast reactor SG, choice of materials is narrowed down to low-chromium steels and 300 series austenitic steels [25]. The candidate materials considered were 2.25Cr-1Mo, 2.25Cr-Mo-Nb, 9Cr-1Mo, 12 Cr, Alloy 800, and Type 316 SS. Though Type 316 SS has excellent mechanical properties and fabricability, its susceptibility to stress corrosion cracking under faulty water chemistry conditions puts this material into the risk category. Though superior to 300 series stainless steel by its improved resistance to wastage by small water leakages and also to chloride stress corrosion (Fig. 3.4) [26], Alloy 800 is rejected on the grounds of loss of ductility and also not being completely immune to chloride/caustic SCC [27]. Since the 1960s, 2.25Cr-1Mo unstabilized steel was used and is a well-known structural material fabricated in all shapes and thicknesses and is considered as the favorite SG material for FBRs. Though it suffers decarburization in sodium environments, it can be used for tube plates, pressure vessels, and support members where section thickness is high. However, application above 500°C needs to be taken care of due to low stress to rupture value in high temperatures. Use of this material is limited to economizer and evaporator sections. This material is not preferred for tubes with defined thickness and tube support grids that are in close contact with tubes to avoid decarburization and fretting problem. The stabilized variety with Nb can be used as tube material.

Since SG will be subjected to high stress at an elevated temperature, it was decided to increase chromium content with preferred varieties of 12 and 9Cr steels. The strongest ferritic steels are the 12Cr Mo steels, which are used as super-heater tubes in fossil-fired boilers. However, due to difficulty in forming and welding and making forgings or plates, 9Cr-1Mo was preferred as tubing. The allowable stresses for normalized and tempered 9Cr-1Mo steel are sufficiently high at the operating temperature to allow wall thickness between 2 and 3.5 mm with 15-mm-bore diameter for 2,50,000 h life [25]. This steel readily allows use of ultrasonic and eddy current inspection both during manufacture and for in-service monitoring. With proper heat treatment, this material shows excellent resistance to waterside corrosion, especially chloride/caustic SCC.

The most difficult environment is encountered in the super-heater steam header because of high pressure (170 bars) and high temperature (515°C) of the steam. Since conventional 9Cr-1Mo steel would not allow stress higher than for 2.25Cr-1Mo steel, modified 9Cr steel, 9Cr-2Mo stabilized with Nb and V (EM 12) was selected for the super-heater section.

3.5.3 Testing the Material for Fast Reactor SG

After the steam generator fabrication, the unit is submitted to hydraulic pressure test at 1.5 times design pressure. The absence of excessive embrittlement is assessed from the ductile–brittle transition curve obtained from Charpy impact testing. A lower shelf energy absorption of 68J, temperature of 27°C or lower and upper shelf energy absorption higher than 101 J is highly desirable [28]. Exposure studies to flowing liquid, two-phase water, high-pressure and high-temperature steam, and sodium for long duration (17,000 h) are also desirable.

The feed water can be polluted by chloride during condenser leakage and sodium hydroxide by the sodium–water reaction. Even in the absence of leakage a localized high concentration of Na or Cl can occur on the tube surface with a low concentration of oxygen and normal stress. Hence, the following testing procedures are recommended.

1. To test stressed ferritic steel specimen in water at saturation temperature (350–360°C) with a 100 ppm Cl concentration for 3000 h.
2. To test stressed specimens in caustic solutions at 0%, 0.1%, and 1% NaOH at 350°C for 5000 h.

During these tests, neither excessive corrosion nor cracking should be observed.

3.5.4 Unresolved Concerns

Though decarburization in sodium is a major problem, proper heat treatments can take care of this problem. Low-temperature (450°C) stress corrosion is a potential problem and should be explored to minimize the problem [29]. SG tube material is also prone to pitting corrosion under some defined conditions. Rusting is another problem when not in use. Weldability and postweld treatment processes need to be controlled. Strict maintenance of water chemistry during operation, standby, layup, partial power conditions, and during an in-service inspection of components is a must to avoid pitting corrosion.

4. MATERIALS IN SEAWATER ENVIRONMENTS

Marine environment influences the degradation of the structural materials, due to the aggressiveness and complexity of the seawater. Dissolved oxygen and chloride ions in seawater play the imminent role in passivity and breakdown, respectively, on metals and alloys [30]. The degree of corrosion in marine environments is dependent on many parameters, including temperature, dissolved oxygen content, salinity, water chemistry, pH, biofouling, pollution/contamination, galvanic interactions, fluid velocity characteristics and mode, alloy composition, alloy surface films, geometry, surface roughness, and heat-transfer rate.

4.1 EXTREME CONDITIONS OF SEAWATER

Seawater environment has been divided into five zones, namely, subsoil, continuously submerged, tidal, splash zone above high tidal, and an atmospheric zone [31]. The corrosion behavior of metals and alloys differs from one zone to another. In the splash zone, the stainless steels show satisfactory performance while the carbon and low-alloy steels undergo corrosion.

The oceanic zone based on the temperature is subdivided into the epipelagic (40°C to −3°C), mesopelagic (4–5°C), and bathypelagic zones (0–6°C). Water temperature has long been recognized as an important influencing factor, largely based on theoretical considerations and laboratory observations of corrosion. Temperature affects corrosion kinetics, oxygen concentration, and diffusion. It is also an important factor in biological activity. As a general rule, corrosion reaction rate in the seawater increases as the temperature is increased. Dexter [32] indicated the corrosion rates in tropical surface waters are about twice those in the polar regions or in deep water. Corrosion rates of stainless steel alloys are usually higher in shallow seawater, where temperatures are warmer than in the colder deep ocean [33]. As the water temperature increases, the oxygen solubility will decrease, which will influence the increase in reaction kinetics and oxygen diffusivity. The corrosion rate is also proportional to dissolved oxygen concentration. At an ordinary temperature in neutral or near neutral water, dissolved oxygen is necessary for appreciable corrosion. According to Behrens et al. [34], corrosion rates of copper alloys in a seawater increase significantly with an increase in oxygen concentration.

Seawater pH is buffered and varied from 7.5 to 8.3. This variation in pH does not affect the corrosion of most metals and alloys, except aluminum. The relative velocity between a metal surface and its environment is a major factor in determining the corrosion rate of the metal. Titanium alloys and Ni-Cr-Mo alloys form very tenacious and protective surface oxide films and impart excellent corrosion resistance at all velocities. Most stainless steels and nickel alloys form protective surface films and provide excellent corrosion resistance at high and intermediate velocities. Though copper-based alloys exhibited excellent corrosion resistance at low velocities, they get corroded by erosion-corrosion in the high and intermediate velocities due to the stripping of the protective oxide layers from the metal surface [35].

Anaerobic mud in the seabed harbor sulfate-reducing bacteria (SRB) and cause microbiologically influenced corrosion (MIC). Sulfide ions, the metabolite of SRB metabolism, undergo transformations, which decrease the protectiveness of the metal surfaces. The initial stage in the SRB corrosion process produces mackinawite (FeS) that is protective on the steel surface. However, in the later stages, the film becomes less stable during the SRB proliferation and transforms into corrosive pyrrhotite (FeS_2) [36].

Ship construction materials undergo various forms of corrosion, and different corrosive environments also exist inside ship cargo tanks. Corrosion also causes considerable damage to marine steel infrastructures, such as offshore oil installations and pipeline systems, leading to revenue losses. Marine steel facilities, such as the marine platform, endure tremendous static and cyclic stress due to the ocean waves. The sea-mud zone is the place where the steel facilities endure the most severe stresses. The movement of sea bottom sediment can cause erosion corrosion.

4.2 CORROSION ISSUES OF METALLIC MATERIALS IN SEAWATER

All forms of corrosion have been exhibited by the materials used for seawater environment. Interactions with marine environments can result in various corrosion modes (uniform corrosion, localized corrosion or pitting, crevice corrosion, galvanic corrosion, intergranular corrosion,

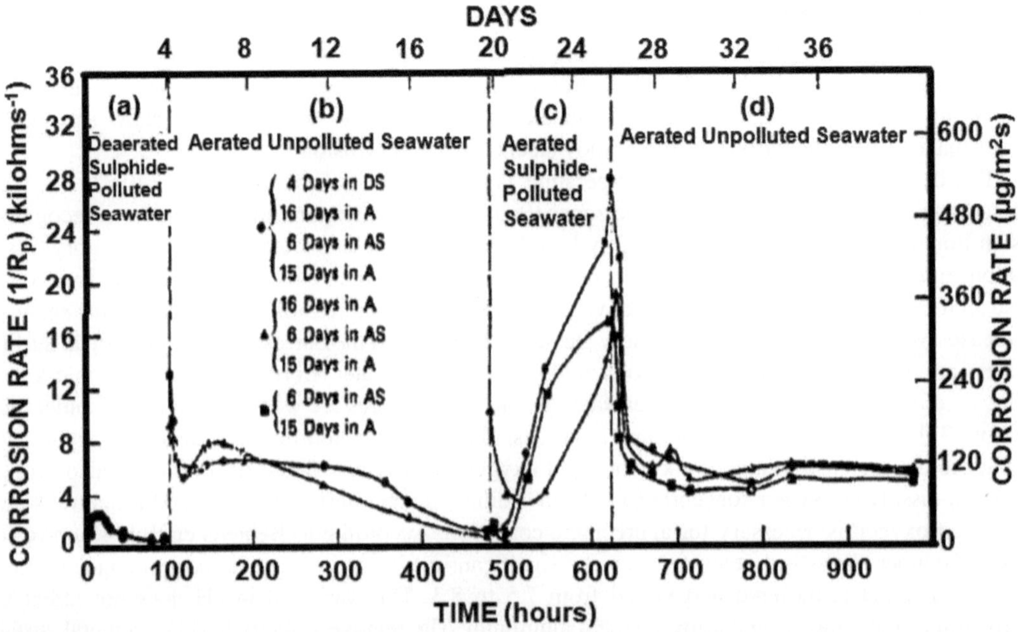

FIGURE 3.5

Corrosion rates for 90Cu-10Ni alloy in flowing seawater showing the effect of exposure history and seawater chemistry (pH = 8.1 ± 0.2, temperature = 23.2 ± 1.3°C, Reynolds number ~44,000, area = 11.04 cm^2) [37].

Reproduced with the permission of Elsevier.

dealloying, erosion corrosion, environmentally assisted cracking, high-temperature corrosion, and microbiologically influenced corrosion). High-alloy stainless steel covers both the superaustenitic and super duplex alloys 316 stainless steel (UNS S31600), 22% Cr duplex stainless steel (UNS S39225), and Alloy 825 (UNS N08825), all of which are susceptible to crevice corrosion in seawater at an ambient temperature. Corrosion rate of copper alloys used in condenser tube applications in seawater has been given in Fig. 3.5 [37]. Copper alloys in seawater do not have a truly passive layer but rely for protection on a very tenacious oxide layer, which limits the rate of metal dissolution, and a thin film beneath limiting the cathodic efficiency. Copper alloy coupled with stainless steel increases the corrosion rate by providing a more active cathode surface.

Anderson and Ross [38] found that the austenitic steel grades performed much better than martensitic and ferritic grades. The Ni-, Cu-, and P-alloyed steels were found to be much more resistant than carbon steel in the splash zone [39]. Furthermore, it was found that Mn, P, and Al had an appreciable influence on corrosion rates of low-carbon steels under tidal exposure. Melchers [40] showed that low-carbon mild steel undergoes general or uniform corrosion in average seawater temperature with higher corrosion rates in warm surface waters than in cold deep waters. High-oxygen content favors passive film formation on stainless steel and aluminum. However, in fully aerated seawater, surface deposits on the passive film-forming alloys create oxygen concentration cells leading to pitting and crevice corrosion. Valve components made of Ni-Cu alloys face both pitting and crevice corrosion in seawater as shown in Fig. 3.6 [35].

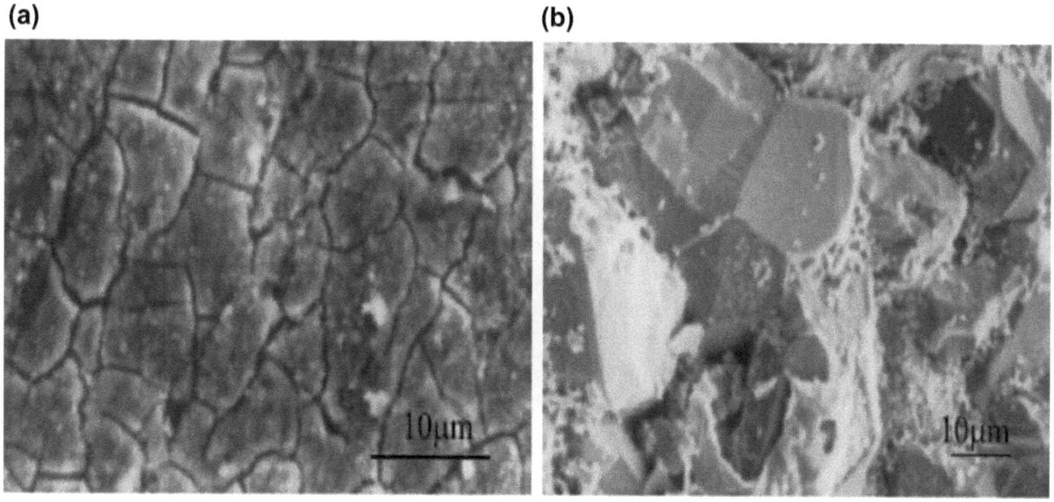

FIGURE 3.6

Localized corrosion of 70Cu-30Ni alloy immersed in seawater for 3 months [35].

Reproduced with the permission of Elsevier.

4.3 CORROSION ISSUES OF CONCRETE MATERIALS IN SEAWATER

In the cooling water systems of nuclear reactors there are many concrete structures in the form of tanks, pillars, and reservoirs that come in contact with seawater. Concrete undergoes several reactions when exposed to aggressive seawater environments. Carbon dioxide dissolved in seawater can reduce the alkalinity of concrete by carbonation. Sulfate in seawater can cause sulfate attack, and chlorides can penetrate concrete leading to reinforcement corrosion. This will be aggravated by attachment of marine organisms like barnacles and mollusks on the concrete surface and biofilm formation by microbes. Acid metabolites secreted by barnacles, sea urchins, and mollusks cause boreholes in concrete and lead to pitting corrosion. Ammonium carbonate by some mollusks and H_2S by anaerobic bacteria attack weak and permeable concrete, which in turn cause pitting in the embedded steel. Hydrostatic pressure and extreme temperature accelerate the material deterioration of both coastal and offshore structures [41–43]. The collective effects of oxygen, H_2S, and acidification of seawater by *Thiobacillus* species corrode mild steel in concrete structures [44]. According to the American Concrete Institute (ACI) Report 222 [45], corrosion of metals in concrete under some conditions need a chloride content of as little as 0.15% by weight of cement to initiate corrosion of embedded steel in concrete in the presence of oxygen and moisture. Thus, higher chloride content in seawater provides very aggressive environments to concrete structures.

4.4 MICROBIOLOGICALLY INFLUENCED CORROSION OF MATERIALS IN SEAWATER ENVIRONMENTS

Microorganisms and macroorganisms in seawater can accelerate corrosion of materials exposed in seawater environments. Microorganisms formed a biofilm, and anaerobic niches may exist within the biofilm in a fully oxygenated system or oxygen may be formed in an anoxic environment due to bacterially mediated reactions [46]. The immersion of stainless steels, as well as any kind of material, in natural seawater induces the development of a microbial film called biofilm, which is established after 1–3 weeks. The activity of the microorganisms in the biofilm causes the electrochemical open circuit potential (OCP) of the stainless steel to increase. This increase is called potential ennoblement that will increase the risk of crevice corrosion or pitting. Pits have been observed in SS pipes, especially in weld-affected zones, and have led to leaks in some instances. Biofilm formation in water resistance and wastewater conduits as well as a condenser and heat-exchange tubes causes increased power consumption for pumped systems or reduced capacity in gravity systems and energy losses [47]. Macroorganisms like barnacles and mussels can attach on stainless steels and create oxygen concentration cells leading to crevice corrosion.

Microfouling due to biofilms and macrofouling can affect heat transfer in the seawater cooling systems. Biomineralization by microbes can enhance deposit formation on material surfaces and increase the aggressivity of the fouling problems.

4.5 CORROSION MONITORING OF MATERIALS IN SEAWATER ENVIRONMENTS

Corrosion of materials used in seawater is monitored by various electrochemical techniques in addition to weight-loss measurements. Corrosion monitoring techniques in marine environment also include simulated service test, rotating disc/spindle test, rotating drum test, and multivelocity jet test. Some modeling procedures are used to evaluate corrosion of reinforcements in concrete structures and fatigue and corrosion under combined loading. Simulation of the erosion of the coatings on the hull is an important tool in evaluating the efficacy of novel formulations. Simulated service test is used to determine the corrosion rate of materials in simulated conditions of distillation plants or ships' seawater systems. Rotating disc/spindle test is cheap and convenient to determine the effect of velocity on the corrosion rate of metal disc specimens rotating in a tank of the liquid of interest. Rotating-drum test or dynamic testing (ASTM D 4938) describes the use of high-velocity seawater flowing through a channel with coated panels to simulate erosion of the coatings as the ship travels through the water [48]. Multivelocity jet test is also used to monitor erosion-corrosion in coupon specimens that are positioned in a pipe parallel to the flow direction. This test has an advantage in that the velocity of the liquid over the specimen surface can be controlled and measured with considerable precision. According to ASTM standard D-3623 [49] for testing marine antifouling coatings, the coated panels are submerged in a heavily fouled marine environment, typically port areas, and left for periods of time to determine the degree of resistance provided by the test coatings against attachment of hard (barnacles, oysters) and soft (algae, seaweeds, sponges, etc.) fouling. Mathematical models of corrosion degradation processes are developed to predict corrosion of marine materials. Mathematical model [50] represents marine aerosol behavior in a marine atmosphere zone to monitor atmospheric corrosion of metals and corrosion of reinforcements in concrete structures. The model has the distance from the sea and wind speed as the major variables and takes into account changes in marine aerosol due to the settlement of salts when being transported inland. This model is also a useful tool to indicate the salinity level expected at different distances from sea in the marine atmosphere zone. A model proposed by Garbatov [51] assesses the reliability of longitudinal members of a ship hull as a result of fatigue and corrosion under combined loading. The model allows for the existence of multiple cracks, and it accounts for the crack growth process. Mathematical models drew failure because of many difficulties found during designing such experiments is the dependence of corrosion on many environmental factors [52].

4.6 CORROSION PROTECTION METHODS OF MATERIALS IN SEAWATER

Corrosion protection starts from the selection of appropriate materials for usage and operations in the aggressive marine environment. In seawater process industries, Type 316 SS are the candidate materials for the construction of evaporators, distiller pipes, pump valves, and condensers.

Though these alloys are stable under dynamic or flow conditions and virtually showed no corrosion even on very long exposures, these alloys are prone to localized corrosion in the presence of chloride ions and under static or stagnant conditions. Stainless steels were replaced with high-chromium-containing Ni-base alloys (Cr-Ni-Mo) to combat localized corrosion. The main disadvantage is the cost of Ni-base alloys, which has restricted its use in process industries. The Cr, Mo, and N contents of the alloy greatly influence their pitting and crevice corrosion behavior, and Cr and Mo have a synergic effect in enhancing the pitting resistance of steel in seawater [33].

The pitting resistance equivalent (PREN) represented by the PREN $= \%$ Cr $+ 3.2\%$ Mo $+ 16\%$ N equation for Cr, Mo, and N containing alloys with above 38 is supposed to provide resistance to marine corrosion [53,54]. Crevice corrosion of highly alloyed SS (PREN > 40) exposed to seawater, once initiated, can propagate at temperature far lower than the initiation temperature. Repassivation properties of the material are also the important factors to be taken into account in material selection and corrosion control [33].

Seawater piping and a variety of applications in the marine heat-exchanger tube are being constructed with iron-modified copper-nickel alloys (30% Ni or 10% Ni) and aluminum brass (78Cu-20Zn-2Al), which can withstand the action of flowing seawater. Titanium and its alloys are very stable in seawater and, therefore, are finding increasing application. The addition of antifouling agents to the seawater (chlorination) will be required since biocompatibility makes it prone to biofouling. Widespread use of glass-reinforced plastic (GRP) vessels is adopted due to excellent marine corrosion resistance. Traditionally, the composite material used most frequently has been E-glass in a matrix of polyester or epoxy resin. E-glass is less strong and slightly less stiff than the other commonly available glass (known as R-glass in Europe, S2-glass in the United States, and T-glass in Japan) but is markedly cheaper [55]. Marine structures, including ships hulls and submarines, are protected by cathodic protection from corrosion. Cathodic protection can be obtained either by using sacrificial anodes or by using an impressed current cathodic protection (ICCP) system [56]. Corrosion in heat exchanger tubing is avoided by coating water boxes made of cast iron with coal tar epoxy, neoprene, or glass flake and by changing the tube plate to high-alloy stainless steels [57].

The most common method of controlling biofilm accumulation is periodic or continuous chlorination. Biofouling in the cooling water pipelines is inhibited by chlorination, which oxidizes biofilm causing disruption and partial removal. Mechanical cleaning such as sponge ball cleaning can physically remove portions of the attached film. Several coatings are developed that can provide antibacterial activity [58], superhydrophobicity [59], and self-cleaning properties [60] to control corrosion and biofouling of materials in marine environments.

5. CORROSION ISSUES OF MATERIALS FOR NITRIC ACID CORROSIVE ENVIRONMENTS

As part of corrosion research, extensive studies need to be carried out for selecting the most suitable materials for fabrication of equipment for hostile and highly oxidizing nitric acid applications in spent nuclear fuel reprocessing plants. The constructional materials most commonly used in reprocessing plants are austenitic grades of stainless steel (SS) along with smaller quantities of refractory metals like titanium and zirconium, and its alloys [1,2,11,12]. Studies have also been carried out to use valve metals (Ti, Zr, Hf, Nb, and Ta) and its alloys (Ti–5%Ta, Ti–5%Ta–1.8%Nb, etc.) as alternate to austenitic stainless steel for use in nitric acid application [1,2,61,62].

5.1 STAINLESS STEELS FOR NITRIC ACID APPLICATION

AISI type 304L SS is extensively used for fabrication of vessels, tanks, piping, and equipment in reprocessing plants. However, in highly oxidizing nitric acid environment, SSs are susceptible to severe intergranular corrosion (IGC), even if the steels are not sensitized [11,12,63]. Most SSs, except certain high chromium types, exhibit an average corrosion rate of 0.13 mm per year in boiling 65% HNO_3 [11,64,65]. However, several incidences of failures of components made of AISI type 304L SS have been reported in spent nuclear fuel reprocessing plants attributed to [11,12,62,63] (1) transpassive corrosion, (2) end-grain attack (in forged SS), (3) IGC due to segregation of impurities at grain boundaries, and (4) selective corrosion of welds [61,62]. Close control of chemical composition, steel-making practices, and processing parameters are essential to achieve consistently good intergranular corrosion resistance for SS in nitric acid. These efforts have identified extra-low-carbon and restricted levels of C, Si, P, S, and Mo of nitric acid grade (NAG) SS with (1) controlled chemical composition of alloying elements, (2) modified microstructures leading to the elimination of weaker sites for passive film breakdown and dissolution, and (3) enhanced passive film stability against transpassive dissolution [11,12,66]. AISI type 304L SS and its improved varieties with extra-low-carbon and restricted levels of C, Si, P, S, and Mo (AISI type 304ELC), with Nb (AISI type 347), with Ti (AISI type 321), or AISI type 310L SS and its equivalent with Nb stabilization and high Cr content, have been used in several reprocessing plants in the world [11,64,67,68].

5.1.1 *Corrosion Evaluation of SS for Nitric Acid Service*

In general, the corrosion evaluation of conventional AISI type 304L SS material for the reprocessing plants is evaluated as per ASTM A262 practice C (HUEY test). The test is also recommended by the US Nuclear Regulatory Commission (USNRC) for evaluating the materials intended for nitric acid service in reprocessing plants. It is an accelerated test involving boiling nitric acid of 65% nitric acid, and the experiments are conducted by immersing the test coupon for five times of 48 h periods (240 h) in the solution. The limitations of the ASTM test motivated the design of a dynamic nitric acid loop (NAL) of 400 L capacity made of AISI type 304L SS, with flowing 6 M HNO_3 (inactive conditions) at different temperatures for evaluating the corrosion performance of materials over long operating periods (Fig. 3.7a) [11,12]. The corrosion resistance of the AISI type 304L SS samples was evaluated in this loop in different metallurgical conditions, namely, solution annealed, alloy composition; cold worked, sensitized (675°C/1 h), and welded conditions, used in demonstration fast reactor reprocessing plant (DFRP) along with different flow velocities (up to 1.55 m/s) and temperatures (40°C, 60°C, 80°C, and vapor phase) up to 10,000 h exposure. The average corrosion rate in the boiling nitric acid was 14.2 mpy (Fig. 3.7b) for the DFRP 304L SS specimen compared to 6.6 mpy for the NAG 304L SS, which is still within the acceptable limit. The observed corrosion rates were much lower at temperatures of 80°C (0.4 mpy) and 60°C (0.03 mpy). The acid analysis of the loop showed a significant increase in the quantity of the elements with Fe ($\approx$3586 ppm), Cr ($\approx$553 ppm), and Ni

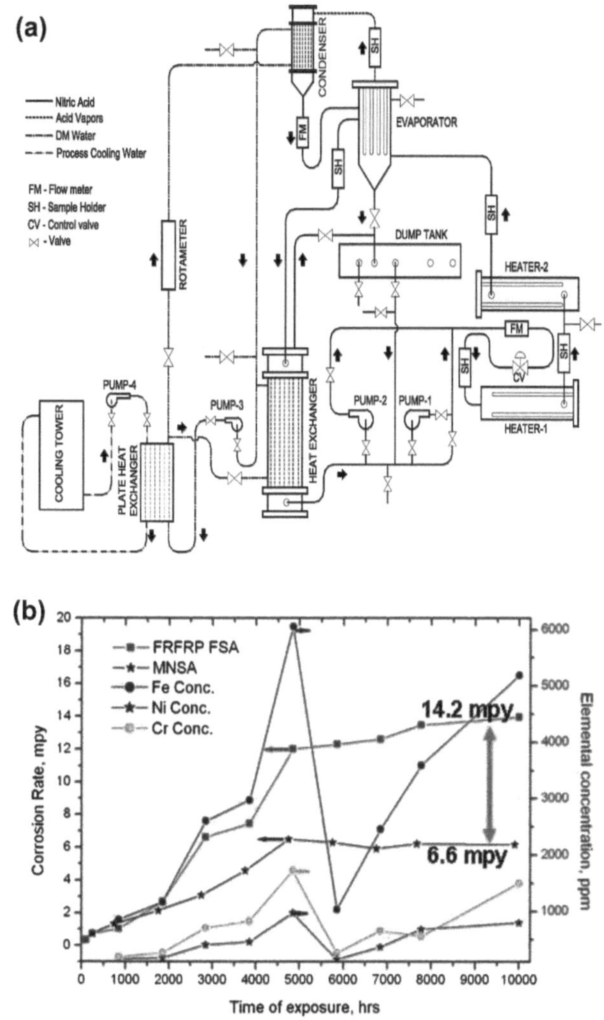

FIGURE 3.7

(a) Schematic flow sheet showing the various parts of dynamic nitric acid loop, and (b) Corrosion rates of 304L SS and NAG SS after 10,000 h exposure in NAL [10].

Reproduced with the permission of Scientific Publishers.

(≈680 ppm) [11,68]. Direct correlation was observed between the corrosion rates and the corrosion product concentration in the loop acid. The long-term corrosion rate data over such a long period of exposure (up to 10,000 h) would provide acceptable corrosion rate of AISI type 304L SS to be used for various applications in nitric acid service. The availability of such data will be helpful in predicting the remnant life of the components used in the reprocessing plants by developing suitable analytical and modeling tools of the corrosion processes. An artificial neural network (ANN) model for the corrosion process was developed, using the large data bank obtained for 10,000 h of exposure of the samples. The model predicted that the average thickness loss up to 50% thickness provides a typical life of 6 years for a 3-mm-thick and 11 years for a 5-mm-thick component; however, for NAG 304L SS the life would be more than double that of conventional 304L SS [69].

The applicability of high nitrogen steel (HNS) of type 304L SS for nitric acid service has been recently explored for highly corrosive nitric acid service [11,68]. The IGC susceptibility of the 304LN SS with varying nitrogen contents in cast, hot rolled + heat treated (wrought), and welded condition was evaluated along with commercial and nitric acid grade 304L SS. As shown in Fig. 3.8a and b, the corrosion rate of 304LN SS with various nitrogen contents (0.132% N, 0.193% N, and 0.401% N) indicated no significant difference in corrosion rate values in wrought and as-received conditions. These results clearly indicated that nitrogen containing (up to 0.1% N) 304LN1 SS is better than commercial 304L SS and higher nitrogen 304LN2 and 304LN3 SS. Further, the corrosion rate of 304LN welds deposit and weldment was found to be marginally higher than the base metal. The higher corrosion resistance of 304LN SS filler wire was attributed to the synergistic effect of high N, Cr, and

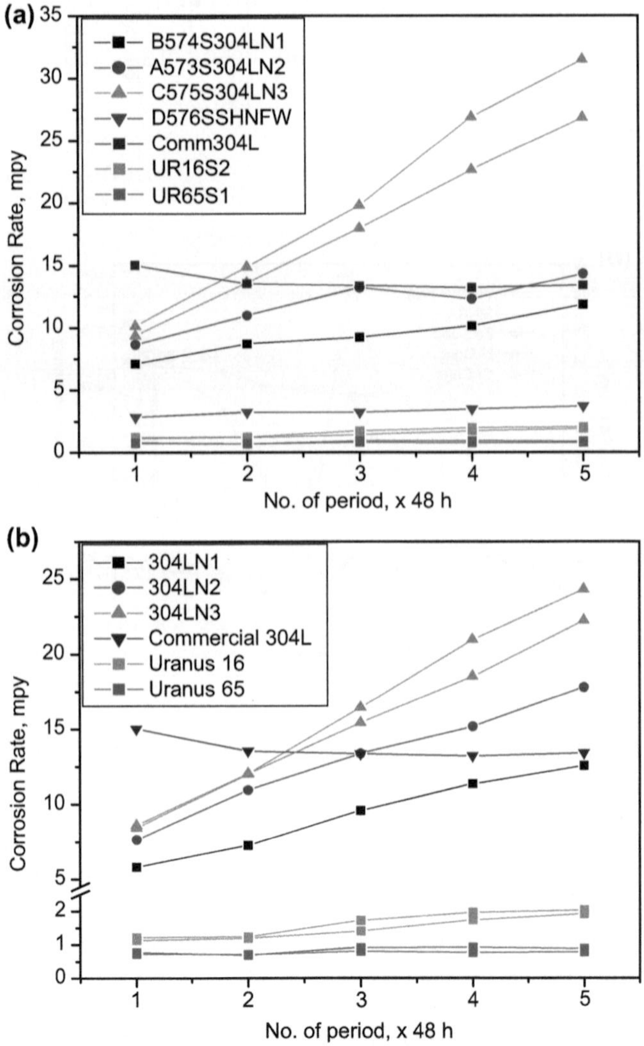

FIGURE 3.8

Corrosion rate of 304LN SS with various nitrogen contents as per ASTM Practice-C test in (a) cast and (b) wrought condition. (304L stainless steels of other grades are also presented for comparison purposes), (a) Hot Forged and (b) Hot Rolled [68].

Reproduced with the permission of VRK Publisher, Chennai.

Ni content [11,68]. However, the average corrosion rates of all the SS were below 18 mpy, which are acceptable for reprocessing plant applications.

5.2 TITANIUM AND ITS ALLOYS FOR NITRIC ACID SERVICE

High-performance NAG SS of higher Cr and Ni content and controlled chemical compositions of impurity elements even undergoes severe IGC in higher temperature and high concentrations of nitric acid [11,12,68]. Titanium with a stable and wide passive range can be used in 11.5 M boiling nitric acid, and titanium exhibits excellent resistance to even recirculating nitric acid process streams. However, significant corrosion in hot and pure nitric acid solutions and vapor condensates are observed [64,70,71]. A new alloy with the addition of refractory metals like Ta and Nb can overcome such corrosion problems attributed to similar metal ion size, low solubility, and superior oxide film in HNO_3 [11,71]. Hence Ti, Ti-5%Ta, Ti-5%Ta-1.8%Nb, Ti-4%Nb-4%Zr, etc. were explored as alternate materials of construction for such highly corrosive environments [11,61].

The three-phase corrosion rate results of Ti-5%Ta-1.8%Nb alloy in boiling 11.5 M HNO_3 are shown in Fig. 3.9. Specimens of the as-received alloy were heat treated in a range of 1100−1273 K and both air-cooled and water quenched to understand their corrosion behavior (Fig. 3.11), and the results indicated that heat treatment at 1273 K followed by water quenching resulted in the best corrosion resistance with very low corrosion rates (<1 mpy). This is attributed to small amounts of beta phases (<5%) in alpha matrix of the alloy that is beneficial to dissolve iron and provide homogeneous distribution of alloying elements in order to achieve low-corrosion rate. The welded specimen exhibited low-corrosion rates (<1 mpy) for this alloy in liquid, vapor, and condensate conditions of boiling 11.5 M HNO_3 [11,66,71]. Although possessing the required corrosion resistance, the presence of tantalum in Ti-5%Ta-1.8%Nb alloy renders it rather expensive and, hence, an attempt was made to develop economic alternatives by replacing the expensive and scarce tantalum with zirconium, without affecting the corrosion properties. Alloys belonging to Ti-Nb-Zr ternary alloys that will possess high-corrosion resistance in all the three phases, and of low cost, compared to the currently used alloys have also been developed for this purpose [72,73]. Newly developed near-alpha Ti-4%Nb-4%Zr alloy possessed superior corrosion resistance in all three phases and excellent weldability compared to conventional alloys used for nitric acid application in spent nuclear reprocessing plants. Further, the corrosion resistance of the beta heat-treated Ti-4%Nb-4% Zr alloy was superior when compared to the sample heat-treated in the alpha + beta phase [72]. Similarly, corrosion resistance of Ti-4%Nb-4%Zr alloy was ~90 times greater than CP titanium in the liquid phase, ~12 times greater in the vapor phase, and ~2 times greater in the condensate phase. The corrosion rate of Ti-4Nb-4Zr in condensate phase is similar to Ti-5%Ta-1.8%Nb alloy.

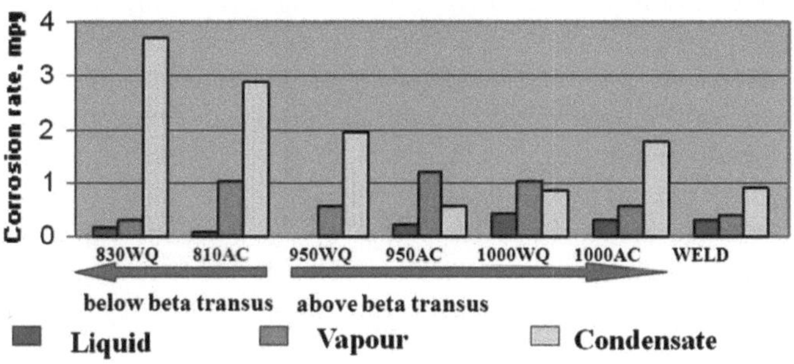

FIGURE 3.9

Corrosion rates of TiTaNb alloy specimens in heat-treated as well as welded conditions after three-phase corrosion test in 11.5 M HNO_3.

5.3 CORROSION OF ZIRCONIUM AND ITS ALLOYS FOR NITRIC ACID SERVICE

Zirconium and its alloys possess high-corrosion resistance in severely corrosive nitric acid conditions attributed to stable and an adherent ZrO_2 film formed on their surface. Zirconium's corrosion rate is typically below 1 mpy up to 70% HNO_3 and temperatures up to 260°C [74–76]. In addition, zirconium does not have complicated microstructures like SS and, hence, plates, sheets, tubes, and pipes have similar corrosion resistance without intergranular corrosion. After significant R&D, zirconium has been used in La Hague extension facilities for about 80 tons weight and 5500-m-long piping (30 years plant life) [11]. Rotating dissolvers, acid recovery evaporator and distillation column, oxalic mother liquor evaporator and heat exchangers, vitrification dust scrubber and liquid waste treatment reactor are some of the equipment employed within 3 M–12 M HNO_3 in the presence of oxidizing ions Cr (VI), Fe (III), Pu (IV), insoluble platinoids, and sulfides [77]. The corrosion resistance of zirconium and its alloys in nitric acid is not affected by concentration, temperature, and applied potential [16,18]. However, concerns associated with the application of Zircaloy are cost, fabrication, and stress corrosion cracking (SCC) behavior. SCC does not occur in Zr in boiling HNO_3 at concentrations less than 70%, whereas SCC is initiated in 6–94% HNO_3 when a potential is applied above the critical potential for SCC, attributed to anisotropic nature of the hcp structure; thus, crystallographic texture control through rolling process has been established to overcome the SCC susceptibility [74,75,78]. Also, new Zr-15%Ti alloy is known to be immune to SCC under applied potential conditions in hot HNO_3, attributed to form a stable passive film composed of ZrO_2 and TiO_2 [78]. In addition, Japanese experience at Rokkasho plant simulated studies showed that Zr exhibited excellent corrosion resistance and did not have the susceptibility of SCC in the simulated dissolver solution at small-scale mock-up test even after 7 years study (50,000 h) [79]. Studies were carried out on the corrosion behavior of zircaloy and zircaloy-4 in the wrought and welded form in 11.5 M HNO_3 [73,76]. Three-phase corrosion results indicated that zircaloy-4 exhibits superior corrosion resistance in both wrought and welded condition in comparison with other materials (Fig. 3.10). Similarly, zircaloy-4-based high-temperature corrosion testing system of 10 L capacity was made with provisions for testing under liquid, vapor, and condensate conditions [66,70]. The system has been operated for 15,000 h with 11.5 M nitric acid. Studies conducted so far indicated corrosion rate values below 1 mpy for zircaloy-4 samples. U-bend SCC specimens of zircaloy exposed for 12,000 h also did not show any tendency for cracking in boiling concentrated nitric acid.

Based on the results obtained, a prototype dissolver vessel similar to the design of the operating dissolver at CORAL plant was manufactured for long-term corrosion evaluation under simulated dissolver conditions (SDSs). Zirconium-702, Zr-4, autoclaved Zr-4, and commercial pure titanium (CP-Ti) exhibited low-corrosion rates of 0.003–0.009 mpy in the as-received and welded conditions exposed to the boiling-liquid phase of the dissolver solution for 2500 h (Fig. 3.11) [80]. Thereby insignificant corrosion rates for candidate materials both in as-received and weld conditions of zirconium-702 and zircaloy-4 were obtained in liquid and vapor phase of the simulated dissolver environment. Long-term corrosion under plant-simulated conditions through 10,000 h of operation are in progress. In the reprocessing plant, the zirconium dissolver needs to be connected to the rest of process vessels and pipings made up of SS 304L. Friction welding of these dissimilar welds was not considered, as the bend ductility is poor when compared to the values obtained for the explosive welding of dissimilar joints. Thus, solid-state joining processes such as explosive and friction joining are employed to join Zr-4 and CP-Ti to SS 304L. When the corrosion rates of the explosively welded

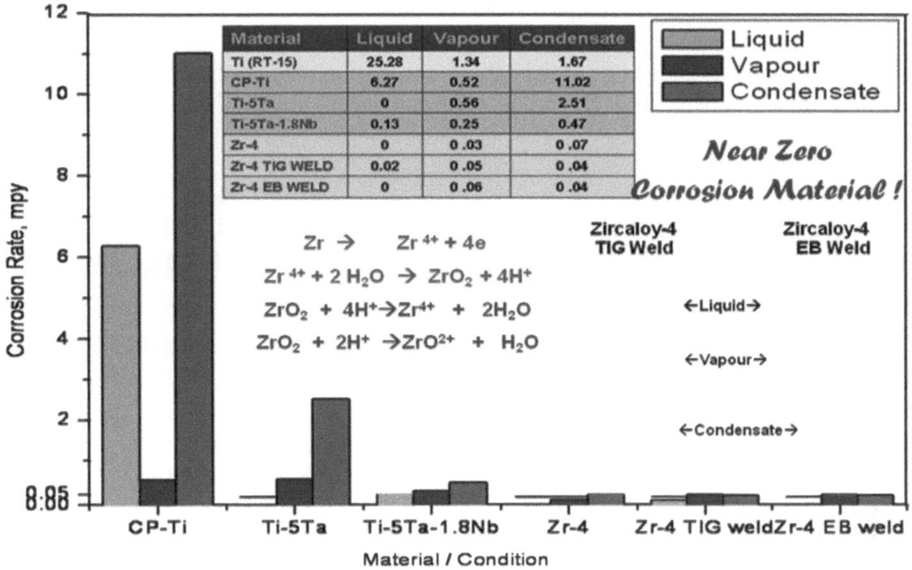

FIGURE 3.10

Three-phase corrosion studies Zr alloys in comparison with Ti alloys in boiling 11.5 M HNO_3.

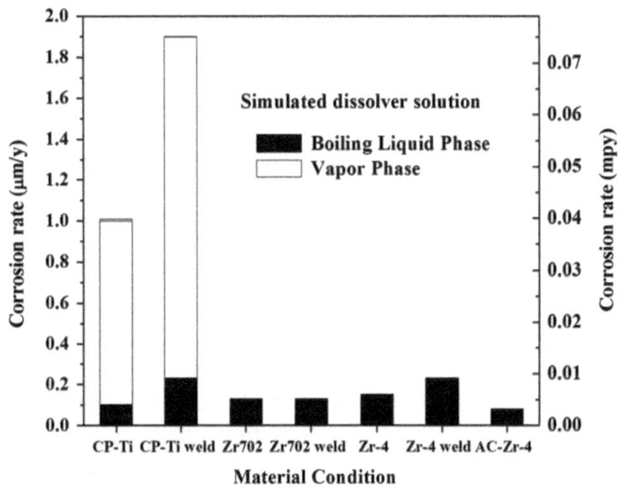

FIGURE 3.11

Average corrosion rates for all the candidate materials exposed for 2500 h in the boiling liquid and vapor phases of simulated dissolver conditions (SDSs) [80].

Reproduced with the permission of American Nuclear Society, Illinois.

Zr-4/SS 304L and CP-Ti/SS 304L samples were evaluated in SDS, for a longer exposure period of 1000 h, Zr-4/SS 304L dissimilar weld samples failed. A code of practice was recommended for the safe operation of plant that transfer of dissolver solution through the dissimilar weld joints should be done only at room temperature and dissimilar welds should be located far away from the dissolver vessel to minimize vapor phase attack.

6. PYROCHEMICAL REPROCESSING OF SPENT METALLIC FUEL AND DURABILITY OF MATERIALS IN MOLTEN SALT AND URANIUM ENVIRONMENTS

The pyrochemical reprocessing involves electrorefining of spent fuel (i.e., metallic or oxide) at high temperature (500–900°C) by chemical processing using molten salt (i.e., LiCl-KCl, LiCl-KCl-CsCl, KCl-NaCl-$MgCl_2$, LiCl-KCl-$MgCl_2$, CsCl-NaCl, LiF-CaF_2, etc.) [1,2,9,12].The methods use high-temperature oxidation–reduction reactions in nonaqueous media to separate the actinides, U and Pu, from fission products [2,4,81]. These methods exploit the differences in the volatilities or thermodynamic stabilities of the compounds of actinides and fission product to achieve the separation. The advantages of pyrochemical reprocessing include the following: (1) short cooling time and direct refining of spent fuels from metal to metal (U-Pu-Zr alloy) and MOX to MOX, (2) small/compact equipment systems based on electrorefining or extraction from the molten salt phase into liquid metal, and (3) simple, minimum steps and single pot ideology with no irradiated fuel transport, and the process could yield an economical nuclear fuel cycle for FBR [81,82]. Hence, colocated with the reactor itself, (4) recycling of transuranics (TRUs)—reduces high-level waste (HLW), also in the solid form, (5) all actinides can be recovered together, and hence, there is no need to develop additional partitioning processes, (6) reprocessing short-cooled fuels, direct waste processing and no liquid low-level waste streams, and (7) intrinsic proliferation-resistance (no product purification) [1–3,81].

6.1 CORROSION OF MATERIALS IN MOLTEN LICL-KCL SALT MEDIUM

The pyrochemical reprocessing technologies in engineering-scale with actual spent fuels and/or surrogates are evaluated in many countries, especially in the United States (Integral fast reactor, IFR), Russia (Research Institute of Atomic Reactors, RIAR), Japan, Korea, and India. This technology with intrinsic nuclear proliferation resistance is regarded as one of the most promising nuclear fuel cycle technologies of the next generation [3,81,82]. In India, the U-Pu-Zr metallic-fuel is proposed for future fast breeder reactors. A flow sheet for laboratory pyrochemical reprocessing of spent metallic fuel is shown in Fig. 3.12, which involves four major operations [81,83]: (1) shearing of the cladded fuel pins with a pin chopper, (2) recovery of U, Pu, actinides from the spent fuel in an electrorefiner, (3) distillation of the entrained salt/cadmium from the electrodeposit in a cathode processor, and (4) casting of the product from the cathode processor into fuel rod in a casting furnace. The major step of pyrochemical reprocessing is an electrorefining step that employs molten salt electrochemical process. Thereby, the corrosion resistance to the chloride/fluoride salts at high temperature has been recognized as a key issue in the choice of structural material. Molten chloride/fluoride salts are quite corrosive at high temperatures with corrosion being driven by thermodynamics, impurity effects, and temperature gradients [2,3,82]. However, the corrosivity of molten chloride salts can be controlled by the oxygen

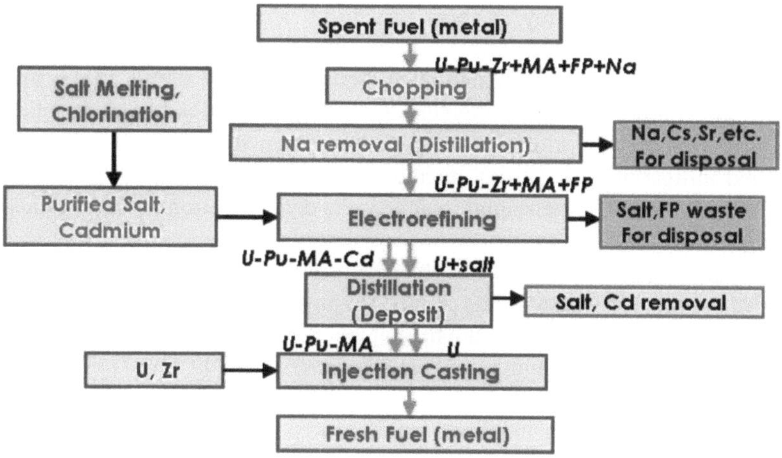

FIGURE 3.12

Flowchart of pyrochemical reprocessing of spent metallic fuel.

and water adsorbed in the salt [82,84]. Furthermore, the corrosion rate also depends on other factors including oxidants, salt flow rate, and galvanic coupling, etc. [84] in pyrochemical environment, since pyrochemical reprocessing utilizes molten LiCl-KCl salt and molten uranium in electrorefiner and cathode processor operating at 500−1300°C; high radiation along with fission products (Ru, Pd, Rh and lanthanides, etc.), chloride salt and reactive metals such as uranium at high temperatures results in extremely hostile corrosive environment.

In order to overcome the corrosion of materials in extreme environment of molten chloride and molten uranium at high temperature, efforts have been made to develop compatible materials, fabrication technology and coating useful for various unit operations like salt preparation, electrorefining, and cathode processing in pyrochemical reprocessing [4,9,12,85−89].

6.2 DEVELOPMENT OF CORROSION-RESISTANT MATERIALS AND COATINGS FOR MOLTEN LICL-KCL SALT MEDIUM

The selection of materials of construction for pyrochemical reprocessing plants is challenging because of the extreme environments, i.e., high radiation, highly corrosive salt, reactive metals, and high temperature [12,85,87]. Furthermore, in any high-temperature application, corrosion is one of the primary concerns. The corrosion/oxidation products of metals by molten fluoride and chloride salts tend to be completely soluble in the corroding media, as passivation is precluded [88,90]. Thereby, materials for pyrochemical processes using molten salts must have adequate corrosion resistance and good response to high-dose radiation. Hostile and aggressive conditions demand structural materials and components with high-temperature strength, dissolution resistance, high-creep rupture resistance, thermal stability, etc. [12,90]. Furthermore, several aspects of molten salt research are extremely challenging for a variety of reasons: toxicity and safety associated with high-temperature liquids and surfaces, limited research activities in fluorides/chlorides molten salts, most data from the 1950s to 1970s; extreme purity requirements of salt, high radiation, etc. [9,12,90]. Hence, the goal is to look into a wide range of materials and alloy systems for the most widespread characterization possible. Different materials such as Cr-Mo steels, 347, 410, 316L SS, dissolution resistance (SiC/SiC composite, TZM, Mo-30W, etc.), high-temperature refractory alloy, Ni-based alloys, and high-density graphite have been proposed for the equipment and components for salt purification and electrorefining operations in molten chloride salt [1,2,9,91,92]. Extensive studies have been carried out for understanding and mitigation of corrosion degradation and qualification of structural materials and coatings in molten salts, which are essential issues for the demonstration of the technical feasibility of critical systems [1,2,9,10,93].

The corrosion databases most relevant for chloride salts of different materials for comparisons are displayed in Table 3.4 [86,87]. These results show that chromium content in the alloy does not seem to be an important factor. The corrosion rates are high at low temperature (~550°C) and similar to results of fluoride salts in contact with stainless steels and Inconel at ~800°C [87]. The corrosion rates of different materials exposed in molten 2CsCl-NaCl salt with a surface temperature of 200°C in the vapor phase showed that Hastelloys exhibited better corrosion resistance, especially Hastelloy C-22 (under 0.01 mm/year) in both molten salt and vapor phases (Fig. 3.13). The difference in corrosion rates between Hastelloy C-22 and X was attributed to the difference in the molybdenum content in the alloys [88]. Similarly, Takeuchi et al. [89] investigated the corrosion resistance of ceramic materials in the extreme/severe corrosion environment and evaluated the effects of chlorine, oxygen, carbon, and fission products in RIAR molten salt environment. The corrosion rate of ceramic materials that show good corrosion resistance (0.1 mm/year), are silicon nitride (Si_3N_4), alumina (Al_2O_3), beryllia (BeO), mullite ($Al_6Si_2O_{13}$), and cordierite ($Mg_2Al_3(AlSi_5O_{18})$) that also show better corrosion resistance than pyrographite (Fig. 3.14). However, the corrosion resistance of zirconia (ZrO_2) is not so excellent in molten salt (about 0.5 mm/year) because of impurity as CaO.

The molten salt corrosion test assembly employed for molten salt corrosion studies is shown in Fig. 3.15. The candidate materials such as 9Cr-1Mo, 2.25Cr-1Mo, 316L SS, INCONEL-600, INCONEL-625, INCONEL-690, low-density graphite, high-density graphite, glassy carbon, and

Table 3.4 Summary of Brookhaven Loop Corrosion Tests for Molten Chloride Salts [86,87]

Loop[a]	Loop Material	% Cr–Ni–Mo in Fe Alloy	Duration (h)	T_{max} (°C)	ΔT (°C)	Corrosion Rate (μm/year)
Tests With LiCl-KCl Eutectic Salt						
TCL-F	347 SS	17.5–1.4–0.2	5500	575	155	12
TCL-L1	410 SS	12.4–0.2–0.1	2200	570	160	50
TCL-L3	2.25Cr–1Mo	2.25–0–1	697	550	150	High[b]
Tests With 30NaCl–20KCl–50MgCl$_2$ Eutectic Salt (mol%)						
TCL-L5	347SS	17.5–11.4–0.2	2467	500	45	93
TCL-L6	410SS	12.4–0.2–0.1	3971	494	42	79
FCL-M1	347SS	17.5–11.4–0.2	1034	520	0	31
FCL-M2	347SS	17.5–11.4–0.2	656	515	0	256

[a] *TCL refers to thermal convection loop, FCL refers to a forced convection loop.*
[b] *No specimen corrosion depth was reported, but salt analysis showed 0.11% iron.*

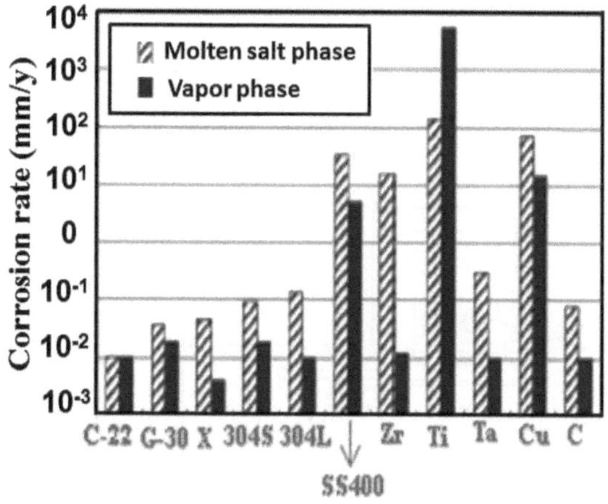

FIGURE 3.13

Corrosion rate of materials at 473 K in molten salt of 2CsCl-NaCl (vapor phase: 393 K) with chlorination [88].

Reproduced with the permission of Elsevier.

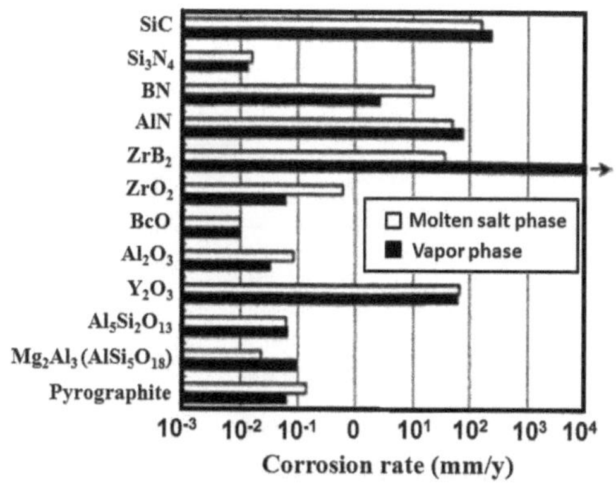

FIGURE 3.14

Corrosion rate of ceramic materials in molten NaCl–KCl salt under Cl_2 bubbling at 750°C for 24 h [89].

Reproduced with the permission of Elsevier.

pyrolytic graphite were investigated and exposed to the molten LiCl-KCl salt at 600°C for various durations under argon atmosphere. The weight of the samples before and after corrosion testing was recorded, and the percentage weight change in the samples was calculated (Fig. 3.16). Thermal cycling studies were carried out on the coated samples to determine the integrity of the coating.

Fig. 3.16a and b shows the percentage weight loss of uncoated and coated samples exposed to molten LiCl-KCl salt at 600°C for different time periods. Weight loss results indicated that the corrosion resistance of the materials increased in the following order: 9Cr-1Mo < 2.25Cr-1 Mo < Ni base alloys < PSZ coating [9]. The surface morphology of the uncoated samples revealed the corrosion product region rich in Cr and deficient in Ni, while the porous-attacked region was rich in Ni and depleted in Cr [94–96]. This indicated that preferential leaching of Cr occurred from the surface resulting into the formation of Cr-rich corrosion products. Low- and high-density graphite exhibited

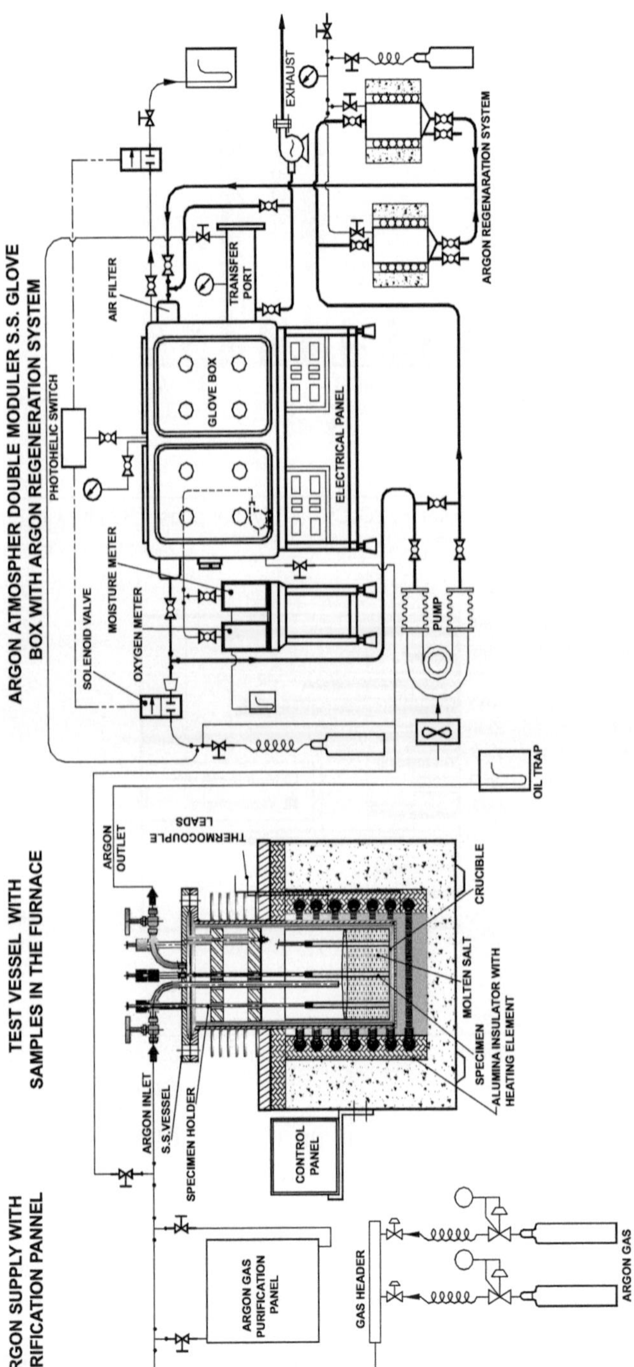

FIGURE 3.15

Molten salt corrosion test assembly (MOSTA) with glove box, gas purification system, and furnace with test vessel.

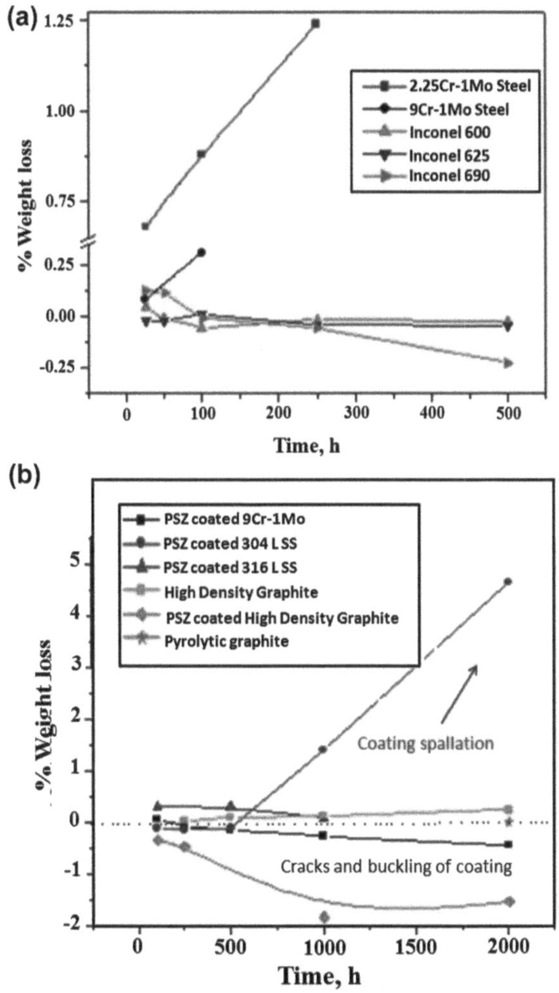

FIGURE 3.16

Percentage weight loss of samples exposed for different time periods to molten LiCl-KCl salt under UHP argon atmosphere (a) uncoated samples and (b) coated samples.

weight loss after exposure to molten salt at 600°C for 2000 h, while glassy carbon (9.2×10^{-4}%) and pyrolytic graphite (3.3×10^{-4}%) showed negligible weight loss [96]. The significant degradation of low-density and high-density graphite compared to glassy carbon and pyrolytic graphite could be attributed to the presence of the significant amount of open pores present in the low-density graphite and high-density graphite as shown in Fig. 3.17a–d, respectively [96]. The excellent corrosion resistance of glassy carbon and pyrolytic graphite is also evident from the micrographs of Fig. 3.17e–h, respectively.

The surface morphology of as-received glassy carbon exhibited smooth and shiny appearance; while pyrolytic graphite exhibited rough morphology, however, the surfaces were dense. The corrosion-tested surfaces of glassy carbon and pyrolytic graphite exhibited similar appearance as that of as-received samples [96].

Fig. 3.18a and b showed the scanning electron microscope micrograph of the partially stabilized zirconia (PSZ)-coated 316L SS in the as-sprayed and molten salt exposed condition, respectively [91]. It is evident from the surface morphology that there is no significant degradation of the PSZ-coated

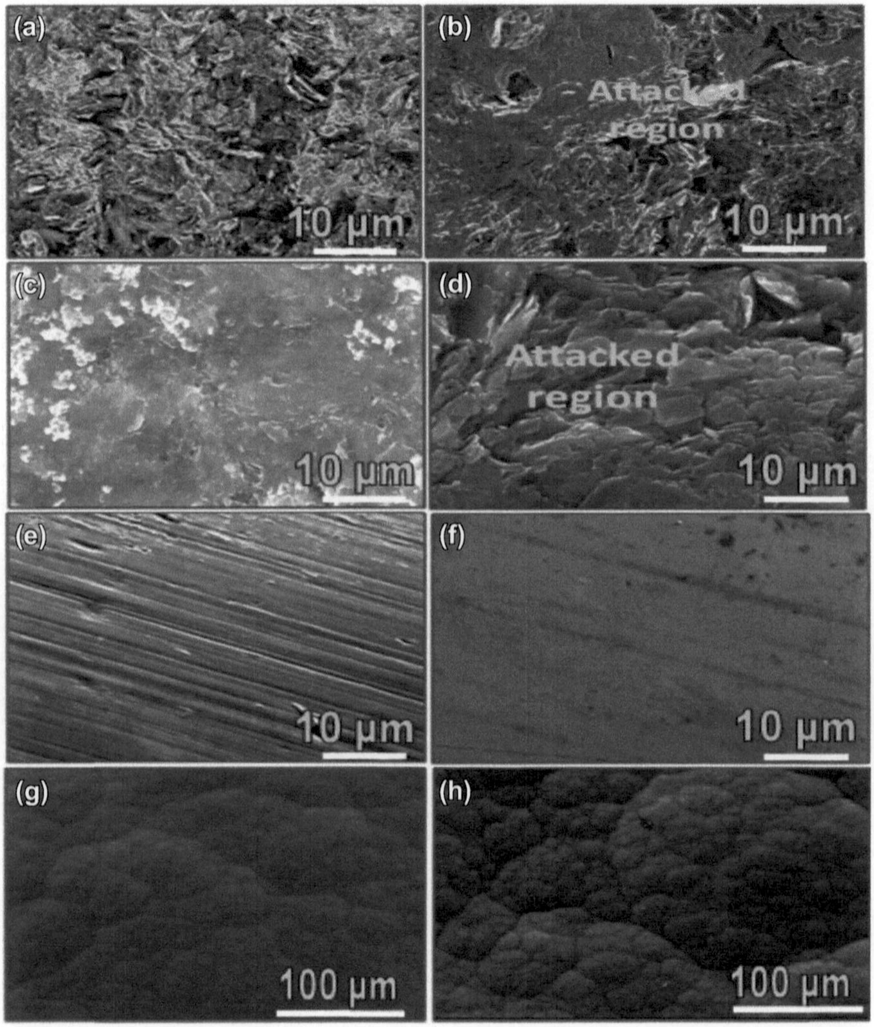

FIGURE 3.17

Scanning electron microscope micrograph of unexposed and exposed low-density graphite (a and b), high density graphite (c and d), glassy carbon (e and f), and pyrolytic graphite (g and h) samples in molten LiCl-KCl salt for 2000 h.

surface. Cross-section microstructural examination revealed that the salt did not penetrate into the coating to attack the substrate. Similarly, the surface morphology of the yttria stabilized zirconia (YSZ) coating on high-density graphite exposed for 2000 h in molten salt exhibited insignificant attack. However, the mismatch in the thermal expansion coefficient between the metallic bond coat and graphite substrate resulted in the formation of microcracks and subsequent failure of the coating [96,97].

Thermal cycling studies carried out on PSZ-coated 9Cr-1Mo at different temperatures in the air (650°C, 850°C, and 950°C) did not show any spallation even after 500 cycles at 650°C. However, on

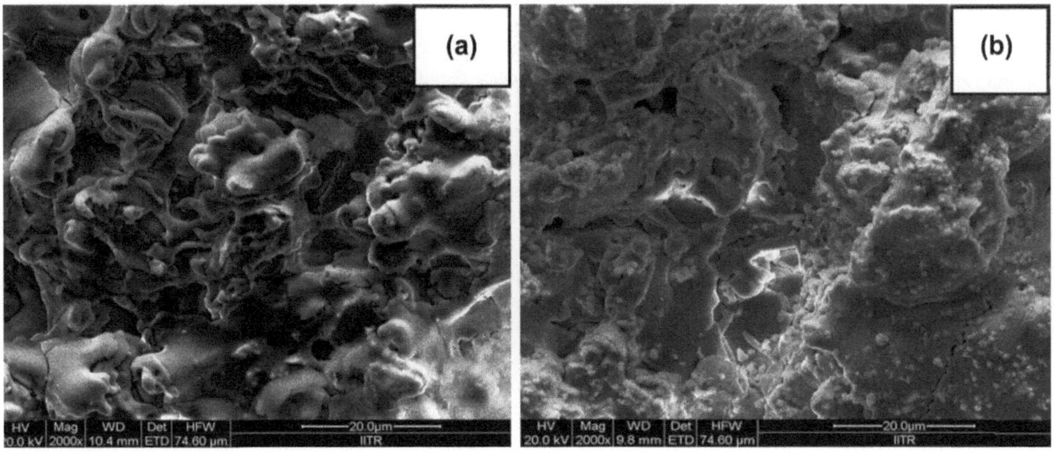

FIGURE 3.18

Surface morphology of yttria stabilized zirconia coating (a) as-sprayed and (b) molten salt exposed condition [90].

Reproduced with the permission of Elsevier.

increasing the temperature to 850°C, cracking and spallation of the coating occurred. Since the salt purification and electrorefining operations are carried out at 500°C under ultrahigh pure argon atmosphere, these coatings could exhibit excellent durability [91].

6.3 MOLTEN URANIUM MEDIUM

For the distillation of the entrained LiCl-KCl salt and cadmium from the uranium, electrodeposit obtained from electrorefining process, graphite crucible is usually employed in the cathode processor. To avoid the undesirable reaction of the graphite crucible with molten uranium, protective coating needs to be provided inside the graphite crucible [92–94]. Various coatings on graphite crucibles were proposed and their performance assessments in contact with molten uranium were made. Transition metal nitride coatings on graphite were initially pursued, and uranium melting experiments were carried out on TiN, Ti-Si-N, ZrN, and HfN coated on high-density graphite crucibles at 1300°C for 20 min [93]. The poor oxidation resistance of the nitride coatings observed in our studies restricted the use of these coatings at such high temperatures. Partially stabilized zirconia and yttria were considered as suitable candidate materials and studies were carried out.

Plasma-sprayed YSZ-coated high-density graphite crucibles were tested by melting uranium at 1400°C for 20 min. The YSZ-coated sample in as-sprayed condition and after uranium melting is shown in Fig. 3.19 [94]. As shown in the figure, no detachment of coating was observed after uranium melting, indicating good performance of the coating.

Similarly, yttria-coated high-density graphite also showed good performance during uranium melting experiments. Plasma-sprayed YSZ and yttria coatings deposited on high-density graphite were subjected to thermal cycling studies at 1000°C and 1100°C under vacuum. Thermal cycling studies carried out at 1000°C on YSZ-coated high-density graphite failed during the 11th cycle and yttria-coated high-density graphite failed during the 36th cycle. Thermal cycling studies carried out at

FIGURE 3.19

Photograph of yttria stabilized zirconia coating on high-density graphite in as-sprayed and after uranium melting [94].

Reproduced with the permission of Elsevier.

1100°C on YSZ-coated high-density graphite failed during the 10th cycle while yttria-coated high-density graphite failed during the 20th cycle. These studies indicate that yttria coating offers better inertness and durability for high-temperature uranium melting applications. Further studies are in progress to establish coating technology for uranium coating up to 1500°C using SiC interlayer.

7. CONCLUSIONS

The phenomenon of corrosion is as old as the history of metals, and it has been looked at as a menace that destroys the beauty of objects and shortens their life. The NPPs operating under extreme conditions, i.e., liquid sodium, steam water, seawater, aqueous reprocessing using nitric acid, and pyrochemical reprocessing of molten chloride environments employed in nuclear industry face failures associated to corrosion. The science of corrosion prevention and control is highly complex in hostile corrosive and radioactive environments, exacerbated by the fact that corrosion takes many forms and is affected by numerous factors. Each metal and alloy has unique and inherent corrosion behavior, and no single material is totally immune to corrosion or can be used in every operating condition. Hence, the choice of materials for NPP and the reprocessing plant is a complicated process involving many parameters such as high-corrosion resistance, mechanical properties, ease in fabrication, availability, cost, etc. The presence of intense radiation, particularly gamma, and of radioactive contamination of surfaces, usually alpha radiation–emitting particles, during the dissolution of spent fuel is an additional factor of concern. Furthermore, the safety requirements in hostile corrosive nuclear environments demand a high level of materials reliability and hence the key to any successful design for the long life of any NPP is the proper selection of materials. The NPPs are designed with the objective of zero incident failures as leakages could cost huge economy loss in shutdown, repair, inspection, and replacements. Hence, the environmental degradation process (e.g., corrosion, stress, radiation effects, etc.) is significant, and that must be closely examined. This chapter highlighted the challenges, and the approaches adopted toward the development of new corrosion-resistance materials, coating technology, simulation of material degradation, and corrosion monitoring in highly hostile corrosive environment for handling equipment for FBRs and their associated fuel cycle technologies. Development of special stainless steels (i.e., Cr-Mo, nickel alloys, nitric acid grade SS, etc.), Ti and zirconium and their alloys, double-oxide coatings for electrolytic dissolver, dissimilar joints, noble metal coating, thermal barrier coatings, and antifouling coatings, etc. for highly corrosive and hostile environments (i.e., liquid sodium in primary heat exchanger, steam water in secondary heat exchanger, seawater in cooling water systems, nitric acid for aqueous reprocessing, and molten chloride for pyrochemical reprocessing) are discussed.

ACKNOWLEDGMENTS

The author thanks Dr. Rani P. George and Dr. S. Ningshen of Corrosion Science and Technology Group, IGCAR for their help in the preparation of manuscript.

REFERENCES

[1] B. Raj, Energy Procedia 7 (2011) 186.
[2] U.K. Mudali, A. Ravishankar, S. Ningshen, G. Suresh, R. Sole, K. Thyagarajan, Energy Procedia 7 (2011) 468.
[3] J.J. Laidler, J.E. Battles, W.E. Miller, J.P. Ackerman, E.L. Carls, Prog. Nucl. Energy 31 (1997) 131−140.
[4] S.P. Awasthi, H.U. Borgstedt, J. Nucl. Mater. 116 (1983) 103−111.
[5] D.L. Harrod, R.E. Gold, R.J. Jacko, Alloy optimization for PWR steam generator heat-transfer tubing, JOM (July 2001) 14−17.
[6] Proceedings: Steam Generator Sludge Management Workshop, Arizona, Report TR114854, Electric Power Research Institute, California, USA, April 2000.
[7] E. Heitz, H.C. Flemming, W. Sand (Eds.), Microbially Influenced Corrosion of Materials, Springer-Verlag GmbH & Co. KG, Berlin, Germany, 1996.
[8] I.B. Beech, J. Sunner, Biocorrosion: towards understanding interactions between biofilms and metals, Curr. Opin. Biotechnol. 15 (2004) 181.
[9] A.R. Shankar, K. Thyagarajan, U.K. Mudali, Corrosion 69 (2013) 655.
[10] U.K. Mudali, Austenitic stainless steels for back end of nuclear fuel cycle, Adv. Mater. Res. 794 (2013) 530−538.
[11] B. Raj, U.K. Mudali, Materials development and corrosion problems in nuclear fuel reprocessing plants, Prog. Nucl. Energy 48 (2006) 283−380.
[12] T. Jayakumar, U.K. Mudali, in: K.L. Ramakumar, S.C. Parida, S.M. Mukerjee, R.V. Pai, S.K. Rakshit (Eds.), Proceedings of Conference on Molten Salts in Nuclear Technology, 2013, pp. 122−133.
[13] S. Rajendran Pillai, H.S. Khatak, in: H.S. Khatak, B. Raj (Eds.), Corrosion of Austenitic Stainless Steels: Mechanisms, Mitigation and Monitoring, Narosa Publishing, New Delhi, 2002, pp. 241−264 (n.13).
[14] B.H. Kolster, J. Nucl. Mater. 55 (1975) 155−168.
[15] T. Gnanasekaran, C.K. Mathews, J. Nucl. Mater. 140 (1986) 202−213.
[16] N.S. Bharasi, M.G. Pujar, K. Thyagarajan, C. Mallika, U.K. Mudali, A. Dhaul, et al., Metall. Mater. Trans. A 46 (2015) 6065−6080.
[17] A.M. Azad, O.M. Sreedharan, J.B. Gnanamoorthy, J. Nucl. Mater. 144 (1987) 94.
[18] C.K. Mathews, T. Gnanasekaran, S.R. Pillai, Trans. IIM 40 (1987) 89−98 (n.18).
[19] N.S. Bharasi, K. Thyagarajan, H. Shaikh, M. Radhika, A.K. Balamurugan, S. Venugopal, et al., Metall. Mater. Trans. A 43A (2012) 561−571.

[20] T. Ito, S. Kato, M. Aoki, J. Nucl. Sci. Technol. 29 (1992) 367–377.
[21] K. Natesan, M. Li, O.K. Chopra, S. Majumdar, J. Nucl. Mater. 392 (2009) 243–248.
[22] S.R. Pillai, H.S. Khatak, J.B. Gnanamoorthy, Mater.Trans. 39 (1998) 370–377.
[23] Investigation into flow-accelerated corrosion at low temperatures, in: Plant Experience-PWRs 1013474, Final Report, Electric Power Research Institute, California, USA, November 2006. Galvanic Corrosion Dealloying Corrosion Velocity Phenomena – Corrosion and Corrosion Control in LWRs.
[24] GE Power and Water, Water and Process Technologies. (Handbook of Industrial Water Treatment, Chapter 14 Boiler System Failure), General Electric Company, 1997–2012.
[25] C. Wilby, J. Walters, Material choices for the commercial fast reactor steam generators, in: S.F. Pugh, E.A. Little (Eds.), Ferritic Steels for Fast Reactor Steam Generators, Proceedings of an International Conference held by the British Nuclear Society, vol. 1, 30 May to 2 June 1977, pp. 40–49. London.
[26] F. Cattant, D. Crusset, D. Féron, Mater. Today 11 (10) (2008) 32.
[27] G.J. Bignold, Review of water side performance of 9Cr1Mo boiler tube material, in: S.F. Pugh, E.A. Little (Eds.), Ferritic Steels for Fast Reactor Steam Generators, Proceedings of an International Conference by the British Nuclear Society, vol. 2, 30 May to 2 June 1977, pp. 342–345. London.
[28] P. Patriarca, S.D. Harkness, J.M. Duke, L.R. Cooper, Nucl. Technol. 28 (1976) 516–536.
[29] C.N. Spalaris, K.D. Challenger, D. Dutina, P. Ring, Sodium heated steam generators: near term and projected information needs – ferritic steels, in: S.F. Pugh, E.A. Little (Eds.), Ferritic Steels for Fast Reactor Steam Generators, Proceedings of an International Conference by the British Nuclear Society, vol. 1, 30 May to 2 June 1977, pp. 55–63. London.
[30] A.S. Al-Fozan, A.U. Malik, Desalination 228 (2008) 61.
[31] F.L. LaQue, Marine Corrosion and Prevention, Wiley, New York, 1975.
[32] S.C. Dexter, Metals handbook, in: ninth ed., in: J.R. Davis, et al. (Eds.), Corrosion, vol. 13, ASM International Handbook Committee, Metals Park, Ohio, 1989.
[33] A.U. Malik, S. Ahmad, I. Andijani, S. A1-Fouzan, Desalination 123 (1999) 205.
[34] H.C. Behrens, F.D. Martin, O. Osborn, L. Rice, W.B. Russell, C.F. Schrieber, et al., Sea Water Corrosion Test Program, Research and Development Progress Report No. 417, U.S. Department of the Interior, Office of Saline Water, 1969. Contract No. 14-01-0001-1090.
[35] X. Zhu, T. Lei, Characteristics and formation of corrosion product films 70Cu-30Ni alloy in seawater, Corros. Sci. 44 (2002) 67–79.
[36] X. Zhao, J. Duan, B. Hou, Y.F. Cheng, Can. Metall. Q. 53 (October 2014) 450–454.
[37] L.E. Eiselstein, B.C. Syrett, S.S. Wing, R.D. Caligiuri, The accelerated corrosion of Cu-Ni alloys in sulphide-polluted seawater: mechanism no. 2, Corros. Sci. 23 (1983) 223–239.
[38] D.B. Anderson, R.W. Ross Jr., in: Proceedings of the 4th International Congress on Marine Corrosion and Fouling, France, 1976, p. 461.
[39] C.P. Larrablee, Corrosion 14 (1958) 501.
[40] R.E. Melchers, Corrosion 61 (2005) 895.
[41] F.M. Lea, The Chemistry of Cement and Concrete, third ed., Chemical Publishing Co., New York, 1971.
[42] B.C. Gerwick, Construction of Offshore Structures, Wiley-Interscience, 1986.
[43] P.K. Mehta, Concrete in the Marine Environment, Elsevier Applied Science, London and New York, 1991.
[44] M. Eashwar, G. Subramanian, P. Chandrasekaran, G. Subramanian, K. Balakrishman, Corrosion 49 (1993) 108–113.
[45] ACI Committee 222, Protection of Metals in Concrete against Corrosion, ACI 222R-01, American Concrete Institute, Farmington Hills, MI, 2001, p. 41.
[46] M. Stipanicev, F. Turcu, L. Esnault, O. Rosas, R. Basseguy, M. Sztyler, et al., Bioelectrochemistry 97 (2014) 76.
[47] W.G. Characklis, Bioengineering Report, Fouling Biofilm Development: A Process Analysis Biotechnology and Bioengineering, 1981.

[48] American Society of Testing Materials, Standard Test Method for Erosion Testing of Antifouling Paints Using High Velocity Water, D 4938−89 (re-approved), 1996.
[49] American Society of Testing Materials, Standard Test Method for Testing Antifouling Panels in Shallow Submergence ASTM D3623-78a, re-approved, 2012.
[50] G.R. Meira, C. Andrade, C. Alonso, I.J. Padaratz, J.C. Borba, Corros. Sci. 50 (2008) 2724.
[51] C.G. Soares, Y. Garbatov, J. Const. Steel Res. 52 (1999) 93.
[52] C.G. Soares, P.K. Das (Eds.), Analysis and Design of Marine Structures, Proceedings of Marstruct 2009, the 2nd International Conference on Marine Structures, Lisbon, Portugal, 2009.
[53] S. Feliu, J.C. Galván, S. Feliu Jr., J.M. Bastidas, J. Simancas, M. Morcillo, et al., Corros. Sci. 35 (1993) 1351.
[54] C. Andrade, M. Keddam, X.R. Nóvoa, M.C. Pérez, C.M. Rangel, H. Takenouti, Electrochim. Acta 46 (2001) 3905.
[55] D.W. Chalmers, Mar. Struct. 7 (1994) 441.
[56] I. Gurrappa, J. Mat. Proc. Technol. 166 (2005) 256.
[57] R. Francis, Br. Corros. J. 29 (1994) 53.
[58] J. Gopal, R.P. George, P. Muraleedharan, H.S. Khatak, Biofouling 20 (2004) 167−175.
[59] P.V. Mahalakshmi, S.C. Vanithakumari, J. Gopal, U. Kamachi Mudali, B. Raj, Curr. Sci. 101 (2011) 1328−1336.
[60] A. Chenan, S. Ramya, R.P. George, U.K. Mudali, Corrosion 70 (2014) 496−511.
[61] U.K. Mudali, R.K. Dayal, G.B. Gnanamoorthy, J. Nucl. Mater. 203 (1993) 73−82.
[62] G.O.H. Whillock, S.E. Worthington, Corrosion in nitric acid, in: T.J.A. Richardson, B.R.A. Cottis, R. Lindsay, S. Lyon, D.J.D. Scantlebury, H. Stott, M. Graham (Eds.), Shreir's Corrosion, vol. 2, Elsevier Ltd., Amsterdam, 2010, pp. 1250−1269.
[63] B.J. Mincher, Degradation issues in aqueous reprocessing systems, Comp. Nucl. Mater. 5 (2012) 367−388.
[64] K. Kapoor, V. Kain, T. Gopalkrishna, T. Sanyal, P.K. De, J. Nucl. Mater. 322 (2003) 36−44.
[65] S. Ningshen, U.K. Mudali, S. Ramya, B. Raj, Corros. Sci. 53 (2011) 64−70.
[66] U.K. Mudali, A. Ravishankar, S. Ningshen, G. Suresh, R. Sole, K. Thyagarajan, Energy Procedia 7 (2011) 468−473.
[67] S. Ningshen, U.K. Mudali, B. Raj, Corros. Sci. 51 (2009) 322−329.
[68] U.K. Mudali, High nitrogen type 304L stainless steels for nitric acid service, in: U.K. Mudali, M.G. Pujar, B. Raj (Eds.), Proceedings of International Conference on High Nitrogen Steels (HNS-2012), VRK publishers, Chennai, India, 2014.
[69] R. Sole, U.K. Mudali, Life prediction of 304L stainless steel component using ANN modeling of long term corrosion data, in: Proceedings of the International Conference on Structural Integrity (ICONS 2014), Kalpakkam, 2014.
[70] U.K. Mudali, R.K. Dayal, J.B. Gnanamoorthy, J. Mat. Eng. Perf. 4 (1995) 214−220 (n. 79).
[71] A.R. Shankar, R.K. Dayal, R. Balasubramaniam, V.R. Raju, R. Mythili, S. Saroja, et al., J. Nucl. Mater. 372 (2008) 277−284.
[72] G. Manivasagam, V. Anbarasan, U.K. Mudali, B. Raj, Metal. Mat. Trans. A 42 (2011) 2685−2695.
[73] A.R. Shankar, U.K. Mudali, Trans. IIM 62 (2009) 545−549.
[74] T. Furuya, H. Satoh, K. Shimogori, A. Aoshima, S. Takeda, Corrosion resistance of zirconium and titanium alloy in HNO_3 solutions, in: Fuel Reprocess, Waste Manage, Proceedings of the American Nuclear Society International Topical Meeting, vol. 1, 1984, pp. 249−258.
[75] H. Nagano, H. Kajimura, Corros. Sci. 38 (1996) 781−791.
[76] A.R. Shankar, V.R. Raju, M.N. Rao, U.K. Mudali, H.S. Khatak, B. Raj, Corros. Sci. 49 (2007) 3527−3538.
[77] Te-L. Yau, Zirconium for nitric acid solutions, in: Fourth ASTM Symposium on Titanium and Zirconium in Industrial Applications, Philadelphia, PA, 10−11 October, 1984, pp. 57−68.
[78] H. Nagano, H. Kajimura, K. Yamanaka, Mater. Sci. Eng. A 198 (1995) 127−134.

[79] C. Kato, T. Motooka, M. Yamamoto, Global 2007, Advanced Nuclear Fuel Cycles and Systems, Boise, Idaho, USA, Sept. 2007, pp. 245–250.

[80] J. Jayaraj, K. Thyagarajan, C. Mallika, U.K. Mudali, Nucl. Technol. 191 (2015) 38–70.

[81] K. Nagarajan, B.P. Reddy, S. Ghosh, G. Ravisankar, K.S. Mohandas, U.K. Mudali, et al., Energy Procedia 7 (2011) 431.

[82] J.P. Glatz, Compr. Nucl. Mater. 5 (2012) 343–366.

[83] B.P. Reddy, K. Nagarajan, in: K.L. Ramakumar, S.C. Parida, S.M. Mukerjee, R.V. Pai, S.K. Rakshit (Eds.), Proceedings of Conference on Molten Salts in Nuclear Technology, 2013, pp. 56–61.

[84] M. Hofmeister, L. Klein, H. Miran, R. Rettig, S. Virtanen, R.F. Singer, Corros. Sci. 90 (2015) 46–53.

[85] B. Raj, U.K. Mudali, T. Jayakumar, K.V. Kasiviswanathan, R. Natarajan, Sadhana 25 (2000) 519.

[86] H. Susskind, Corrosion Studies for a Fused Salt-liquid Metal Extraction Process for the Liquid Metal Fuel Reactor, BNL-585, Brookhaven National Laboratory, Brookhaven, NY, 1960 (n.59).

[87] D.F. Williams, Assessment of Candidate Molten Salt Coolants for the NGNP/NHI Heat-transfer Loop, Report No: ORNL/TM-2006/69, June 2006.

[88] M. Takeuchi, Y. Arai, T. Kase, Y. Nakajima, J. Nucl. Mater. 432 (2013) 35–41.

[89] M. Takeuchi, T. Kato, K. Hanada, T. Koizumi, S. Aose, J. Phys. Chem. Solids 66 (2005) 521–525.

[90] A.R. Shankar, U.K. Mudali, R. Sole, H.S. Khatak, B. Raj, J. Nucl. Mater. 372 (2008) 226–232.

[91] U.K. Mudali, A.R. Shankar, S. Ningshen, Plasma sprayed thermal barrier coatings for molten chloride environments, in: T.S. Sudarshan, P. Nylén (Eds.), Surface Modification Technologies XXV, Valardoc Publishers, 2011, pp. 143–152.

[92] C.E. Holcombe, J.G. Banker, Metall. Trans. B 9 (1978) 317–319.

[93] A.R. Shankar, B.P. Reddy, V. Chawla, M.J. Preyanga, R. Chandra, U.K. Mudali, Surf. Coat. Technol. 204 (2010) 3214–3221.

[94] J. Sure, Ch J. Rao, P. Venkatesh, B.P. Reddy, C. Mallika, U.K. Mudali, Ceram. Int. 40 (2014) 6509–6523.

[95] A.R. Shankar, U.K. Mudali, Mater. Corros. 59 (2008) 878–882.

[96] J. Sure, A.R. Shankar, S. Ramya, C. Mallika, U.K. Mudali, Carbon 67 (2014) 643–655.

[97] J. Sure, A.R. Shankar, S. Ramya, U.K. Mudali, Ceram. Int. 38 (2012) 2803–2812.

CHAPTER

MATERIALS FOR HOSTILE CHEMICAL ENVIRONMENTS

4

V. Sudarsan
Bhabha Atomic Research Centre, Mumbai, India

CHAPTER OUTLINE

Materials Under Extreme Conditions. http://dx.doi.org/10.1016/B978-0-12-801300-7.00004-8

1. INTRODUCTION

Evolution of mankind in the universe and development of materials are two important phenomena that go hand in hand. It is a reality that materials play an important role in the life of all creatures in the world. Materials can be either living or nonliving. Existence and normal functioning of living and nonliving materials heavily depend on the local environment. This local or surrounding environment can be favorable or friendly, noninteracting or dangerous or hostile or unfavorable. Favorable environments are always advisable for the long-term survival and functioning of any materials. However, it is not always possible to have a favorable or noninteracting environment. Due to rapid industrialization, specific applications as well as the increased production requirements, many materials are subjected to drastic or hostile environments that are detrimental to long-term performance of materials. Hostile environment can be temperature, pressures, and climatic or chemical in nature. In this chapter, the hostile environment created by chemicals in different forms (solid, liquid, and gas) and the materials that can withstand such hostile environments are discussed in detail. Initially different types of hostile chemical environments are discussed. This will be followed by materials that can withstand such hostile chemical environments.

2. DIFFERENT TYPES OF HOSTILE ENVIRONMENTS

Hostile chemical environments are classified into different types mainly based on the chemical nature of materials. As a large number of materials/systems exist in nature, depending upon the properties and origin of the materials (systems) a hostile environment for one may be different compared to another. Hostile environments can also be called as corrosive environments; commonly encountered hostile environments are created by acidic, alkaline and saline media, corrosive gases like fluorine, chlorine,

H_2S, CO, etc. Vapors of volatile acids, oxides of sulfur (SO_x), nitrogen (NO_x) etc., also create corrosive environments. Hostile environments can also be site or location specific. For example, tailpipe regions of automobiles, pipelines of petrochemical and fertilizer plants, oil and natural gas carrying containers/vessels/pipelines, drainage/wastewater transporting pipes etc. are locations or sites wherein corrosive environments of different types can be present. In addition to this, high temperature and pressure in the previous situations add to the extent of hostility (corrosive nature) of chemical environment surrounding the material. Acids and alkalies are the two classes of compounds/reagents generally encountered in any plant that deal with chemicals, desalination, petroleum refining, etc. Acids can be generally classified as corrosive or noncorrosive and/or oxidizing or reducing depending up on the nature of the structural materials used for constructing or making the plant. Similarly, there exist materials that are sensitive to alkali or basic environments. In the following section, the nature of chemically hostile environment and the materials used in such environments are described in detail.

3. ACID-RESISTANT MATERIALS

Many industrial processes involve handling acids. Commonly encountered acids are inorganic in nature and include HCl, HNO_3, H_2SO_4, HF etc. It is necessary that exposure of materials (both living and nonliving) to these acids used in the industry should be kept as minimal as possible. In nonliving materials, depending up on the nature of material, corrosion process will be initiated. Acid-proof and acid-resistant materials are the materials that are stable toward destructive action of acids. They include both metallic and nonmetallic-based materials. Metallic-based acid-proof materials include alloy steels, Ti alloys, metals such as Ni, Al, Cu, Zr, Ta, Nb, etc. Actual environmental parameters like temperature, pressure, nature of the surrounding anions, concentration of acid etc. also decide the corrosion properties of such materials. For strongly oxidizing acidic environments, metals that form a stable oxide layer on the surface are ideal to prevent corrosion. Examples include Al, Ti, Cr, etc. and their alloys. In reducing acid environments, the metal used must be thermodynamically stable. Examples of such materials include Cu, Ni etc. and their alloys, for example, the Monel metal. (It is an alloy of Ni and Cu with minor amounts of Fe, C, Si, and Mn). Complex alloys are required for weakly oxidizing and reducing environments. Alloying steels with metals like Ni, Mo, Cu etc. is one of the options available for this. Such processes invariably increase the thermodynamic stability of the material.

Acid-resistant nonmetallic materials have been mainly divided into two categories depending upon whether they are organic or inorganic in nature. Acid-proof inorganic materials include glass, certain ceramics, asbestos, and concrete, whereas materials like polyvinyl chloride, polyethylene, polypropylene, Teflon, polycarbonates etc. belong to the category of organic acid-resistant materials. Most of the herein-mentioned materials will be discussed in the latter sections of this chapter.

4. ALKALI-RESISTANT MATERIALS

Like the behavior of materials with different acids, alkali-resistant properties of materials depend on a variety of parameters like composition of the materials, particle size, microstructure, concentration of alkali, temperature, pressure, duration of interaction, etc. Metals like Mo, W, V, etc., are stable in alkaline media. Also, generally metals like Cu, Ag, and Au are inert toward solutions with high-pH values. However, at high temperatures and in melting conditions they tend to react with alkali. Zirconium (Zr) and hafnium (Hf) are exceptions to this and are found to be stable even in presence of melts of alkali metals. Many inorganic oxides are stable under alkaline mediums. They include compounds like ThO_2, CeO_2, Cr_2O_3, etc. Oxides of Al and Zr are stable under alkaline medium at room temperature. However, they react with alkali upon heating at higher temperatures. Reaction of oxides of aluminum with alkali metal hydroxide is governed by the following chemical equation.

$$Al_2O_3 \cdot nH_2O + 2NaOH \rightarrow 2NaAlO_2 + (n+1)H_2O$$

$$NaAlO_2 + 2H_2O \rightarrow Al(OH)_3 + NaOH$$

Containers such as glass bottles and vessels are unstable for long-term storage of alkalies. Borosilicate glasses undergo leaching under alkaline environments and cations get leached out from the glass. Addition of oxides like ZrO_2, TiO_2, etc. are found to improve alkali-resistant nature of the glass samples. Many nitrides such as GaN, BN, Si_3N_4, etc. are highly alkali resistant. In the case of organic materials, polymers such as polyethylene, polypropylene, etc., do not react with alkalies under cold or hot conditions. Some of the polymers such as polyvinylchloride (PVC) and polystyrene exhibit high resistance against cold alkalies but undergo reaction under hot conditions. Conventionally, materials like compositionally modified stainless steel, borosilicate glasses, oxide ceramics, paints, plastics, etc. are extensively used for storing hostile or corrosive chemicals as they are quite inert to corrosive materials like acids, alkalis, etc. For example, corrosive acids like HCl, HNO_3, H_2SO_4, etc. are kept in borosilicate glass as it is quite inert to such chemicals. Alloying iron (Fe) with carbon and other variety elements produces different types of steels, which can withstand different chemical environments. Paints, plastics, etc. are used to cover different metallic structures/structural components so that the chemicals do not reach the surface of the metal, thereby preventing corrosion. These aspects will be discussed at a later stage of this chapter. Other than acidic and alkaline environments, corrosive gases such as H_2S, chlorine, HF, oxides of sulfur (SO_x) and nitrogen (NO_x), etc. create corrosive environments and affect materials. This aspect is discussed briefly in the following section.

5. MATERIALS RESISTANT TO CORROSIVE GASES (H_2S, SO_X, NO_X, ETC.)

Automation has become very common in many industries. This not only reduces the workforce but also increases productivity and quality of the materials. For automation of different processes, microprocessor-based devices are generally used. These devices are surface mounted on different boards and used for process control applications. The gas hydrogen sulfide (H_2S) is the main culprit that corrodes such surface-mounted devices. Corrosion of the previously mentioned type arises due to the presence of silver resistor in many of the surface-mounted devices. Silver metal is mainly loaded with ceramic and an organic material like epoxide in many of the device control boards. Such materials under industrial conditions (in H_2S environment) react with H_2S to form silver sulfide. Silver sulfide being semiconductor/insulator creates problems with contact leading to short circuits. This results in failure of the automation system and associated increase in downtime of the unit. A similar effect has been also observed with copper-containing devices, even though the affinity of Cu toward H_2S is relatively less as compared to silver. This type of corrosion phenomenon is observed in oil-based industries, gas refineries, sewage disposal and wastewater plants, etc. where H_2S gas is generated in large amounts. It should also be taken into consideration that the solubility of H_2S in certain solvents is relatively high, which invariably leads to extended or accelerated corrosion of Cu and Ag with H_2S.

In addition to H_2S, corrosive and hostile environments created by other gases are also responsible for the failure and nonfunctioning of different types of materials. Typical examples in this category are oxides of sulfur and its compounds with carbon and organic molecules. Oxides of sulfur are corrosive in nature and are generated in the atmosphere either by natural phenomena or from chemical/metallurgical plants. In many of the metallurgical plants, sulfur-containing ores (pyrites) are used for the extraction of metal. Among different oxides of sulfur, sulfur dioxide (SO_2) is the most common. SO_2 can react with construction materials (building stones), which are mainly based on carbonates. Further, they can also react with metals in presence of oxygen to form different sulfates leading to corrosion of metal surface. A concentration of less than 1 ppm of SO_2 gas is sufficient for such corrosion process to take place with metals like iron, aluminum, etc. In presence of moisture, oxides of sulfur form acids that are detrimental to papers, fabrics, and electrical equipment. For human beings they create breathing problems when inhaled in amounts more than 1 ppm. SO_2 is also hostile to plants as it interferes with photosynthesis and destroys plant cells. Similar to oxides of sulfur, CO and oxides of nitrogen (NO_x) are also toxic, corrosive, and hostile to materials and environment. Nitrogen oxides are generated in the automobile exhaust pipes and while burning fossil fuels in powerplants. Like oxides of sulfur, NO_x creates acid rain, which spoils our ecosystem, posing a major threat to plants, fisheries, and human beings. This also causes erosion of nutrients from soils, making sand unusable for cultivation or growth of trees.

Protective polymer coating over different components can reduce the reactivity of metals like Cu and Ag toward H_2S. To a certain extent this can also prevent the reaction of electronic components with oxides of sulfur, nitrogen, etc. For this purpose, acrylic-, epoxy-, silicone-, polyurethane-, paraxylene-based polymers or resins can be used. Each one of the resins mentioned herein is briefly described in the latter section of this chapter. Conventionally, chemical resistance of such materials/polymers is evaluated based on certain standard test procedures approved by the American Society for Testing and Materials (ASTM). The tests ASTM D543, ASTM D3912-10, etc. [1,2] are examples of such standard procedures to check the chemical resistance or compatibility of a material under a particular chemical environment. In such procedures the chemical environment similar to the actual environment encountered by the material is simulated prior to the testing experiments. During the above tests, changes in weight, dimensions, appearance, strength, etc. will be measured for various exposure times, temperatures, other reaction conditions, etc.

6. METHODS TO MONITOR SURFACE MODIFICATIONS OCCURRING WITH MATERIALS DUE TO HOSTILE ENVIRONMENTS

Under hostile chemical environments the mostly affected part of a material is its surface. Hence it is necessary to monitor the surface of such materials by different techniques. Scanning electron microscopy (SEM), X-ray photoelectron spectroscopy (XPS), atomic force microscopy (AFM), etc. are surface-probing techniques that are widely used to monitor the changes occurring on the surface of a material. In this chapter, the results from SEM and AFM techniques have been used to monitor the surface of materials when subjected to hostile chemical environments. In the following section, brief descriptions of SEM and AFM techniques are given.

6.1 SCANNING ELECTRON MICROSCOPIC TECHNIQUE

When a beam of electrons interacts with a solid surface, electrons get scattered both elastically and inelastically. In elastic scattering there is change in direction without change in energy, whereas in inelastic scattering there is change in both direction and energy. These two phenomena result in the emission of secondary electrons, backscattered electrons and transmission of electrons. Imaging secondary and backscattered electrons forms the basis of SEMtechnique[3].

In a scanning electron microscope, a well-focused electron beam is generated and allowed to fall on sample. The beam is scanned over the sample surface by two pairs of electromagnetic deflection coils. The signals generated from the sample in the form of secondary and backscattered electrons are detected and fed to a synchronously scanned cathode ray tube (CRT), and thus, the specimen image is displayed on CRT screen. Changes in the darkness/brightness represent changes of a particular property within the scanned area of the specimen. For SEM analysis sample should withstand vacuums of the order of 10^{-6} torr or better and also it should be electrically conducting. Major advantage of SEM experiments is that the technique can throw light on the nature of grains existing on the surface and surface topography. This information combined with energy dispersive X-ray spectroscopy analysis results will help one to understand chemical reaction occurring with material and in turn the hostility of the chemical environment. Representative SEM images obtained from a sample of sodium borosilicate glass before and after corrosion and the information obtained from the images are discussed in this chapter at the appropriate places.

6.2 ATOMIC FORCE MICROSCOPIC TECHNIQUE

AFM measures topography of thin film surfaces of any type of material. The tip or the force probe is located within few angstroms from the sample surface [4]. AFM images are recorded by moving a fine tip attached to a cantilever across the surface of sample. The deflections that occur in the tip, due to the interaction forces between the tip and sample surface, can be measured by focusing a laser beam onto a cantilever and detecting the reflected light from the cantilever using a position sensitive detector. Schematic diagram showing the principle of atomic force imaging is shown in Fig. 4.1a. As the tip is moved over the surface, a feedback mechanism is employed to ensure that the piezoelectric motor maintains a constant tip force or height above the sample surface. The tip movements normal to the surface are digitally recorded and can be processed and displayed in three dimensions by a computer. This technique has a lateral (X-Y) resolution of 1–5 nm with height resolution of ~1 Å. AFM is typically used to obtain a three-dimensional surface image or to determine the surface roughness of thin films and crystal grains. AFM images can be recorded in mainly two modes. The first one is the contact mode and the second one the semicontact/taping mode, which are used for imaging the samples. The schematic representation of the different types of forces acting between tip and sample in the two methods of imaging is shown in Fig. 4.1b and described below.

In the contact mode, force between the probe tip and substrate is repulsive, and it is within the range of 10^{-8} to 10^{-7} N. The force is set by pushing a cantilever against the sample surface. The contact mode can obtain a higher atomic resolution than the other modes, but it may damage a soft material due to excessive tracking forces applied from the probe on sample. Unlike the other modes, frictional and adhesive forces will affect the image. The method is ideal for hard surfaces, thin films of inorganic materials, etc.

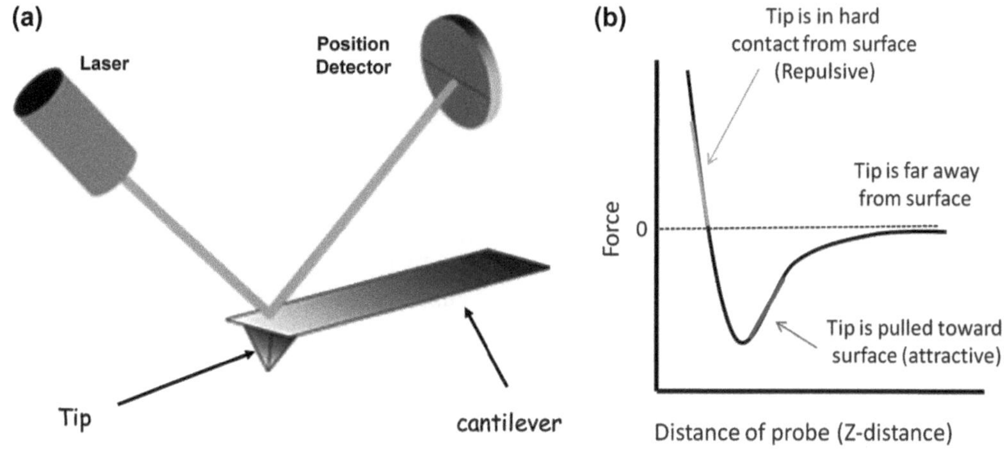

FIGURE 4.1

(a) Principle of atomic force microscope (AFM) imaging, (b) variation of interaction force versus distance between the AFM tip and substrate.

In the noncontact or the tapping mode, the main interaction force between the tip and the substrate is attractive due to van der Waals force and it is in the range of 10^{-10} to 10^{-12} N. The tip in this mode is generally 20–150 nm above the sample surface. In this mode, the cantilever oscillates in the attractive region and its oscillation frequency gets modulated depending on the sample surface features. The modulation can be both in the amplitude (magnitude) as well as in phase and the images can be recorded in both the modes. There are distinct advantages in both modes. The resolution in these modes is limited by the interactions with the surrounding environment. AFM technique gives valuable information regarding surface roughness, which is a measure of the reaction occurring at the surface of an exposed material, and this can be correlated with reactivity of local environment around the material.

As mentioned earlier, steels are important category of structural materials that have gained wide acceptance and applicability for various industrial applications. Hence, it is essential to understand different types of steel and how they behave in hostile chemical environments. This aspect is addressed in the following section.

7. STAINLESS STEEL IN HOSTILE CHEMICAL ENVIRONMENTS

Steels are widely employed for various industrial and domestic applications due to their extraordinary strength and corrosion resistance behavior under different chemical and physical environments. In the following section different aspects of steels, which are used as a corrosion resistant materials for various applications, are discussed in detail.

Steel is an alloy of Fe with carbon. In most of the alloys carbon content is of the order of 2 wt% and they act as hindrance to the free movement of dislocations in Fe lattice, thereby hardening the material. The carbon steel is susceptible to corrosion when exposed to air and moisture due to formation of an iron oxide layer. This iron oxide layer has a higher volume than metal and gets removed from the surface (cracks get generated at the surface), thereby exposing the surface to further corrosion or reaction with the surrounding hostile chemical environment. Alloying Fe with Cr leads to the formation of stainless steel. The layer of chromium oxide formed on the surface of stainless steel prevents oxygen diffusion towards the surface, and this helps to prevent further corrosion of material. Steel is the material that is extensively used for making pipelines and reaction vessels for petrochemical refineries, chemical, fertilizer, desalination plants, etc. Depending upon the nature of process involved, different types of steels are selected. Such steels are made by varying the concentration of Fe and carbon, along with elements like Cr, Mn, Ni, etc. that are present in the alloy as minor components. Iron exists in three crystallographic modifications, namely, the α-, γ-, and δ-forms. In the α-form, Fe exists in body-centered cubic (BCC) structure and is stable below 910°C. The γ-form is characterized by Fe in face centered cubic structure with stability in the range of 910–1400°C. The δ -form is stable over the temperature range of 1400–1534°C and also has BCC structure like the α form. Among the three different forms of Fe, the γ -form has the maximum solubility of C with slightly more than 2 at %. (Phase diagram corresponding to Fe-C system is shown in Fig. 4.2). Upon melting carbon and Fe at temperatures above 1534°C and cooling it over the temperature range of 1400–1800°C, carbon gets dissolved in γ-form of Fe and this phase is known by the name austenite. Further cooling to a temperature less than 723°C leads to the phase separation of alloy as the γ-form is unstable below 723°C. The α-form of Fe along with iron carbide (Fe_3C) is formed during this process. However, cooling at very fast rate leads to the stabilization of γ-form of solid solution, and C atoms in Fe are retained in their locations. Such type of steel is known as martensitic steel and is much harder compared to the austenitic steels. This is clear from the phase diagram of Fe-C shown in Fig. 4.2 [5].

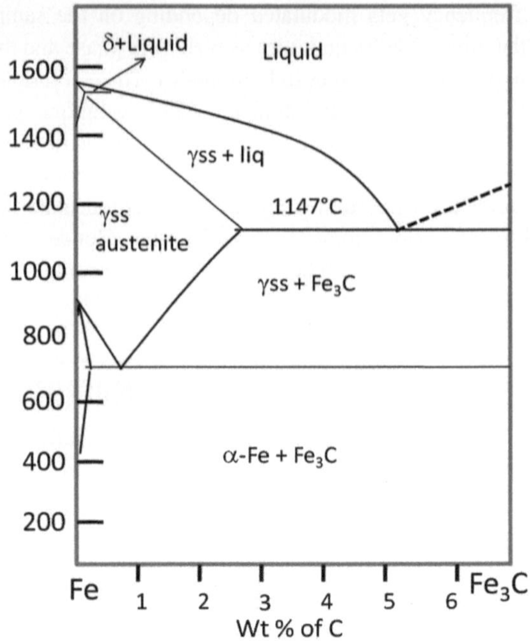

FIGURE 4.2

Schematic representation of Fe-C phase diagram.

8. DIFFERENT TYPES OF STEELS

Before going into the details of corrosion-resistant properties of such materials, it is necessary to know what are the different types of stainless steel available and how their properties change with composition. Steel can be broadly classified into five types depending up on its constituents (additives) and end use. They are known as ferritic, austenitic, martensitic, duplex, and precipitation-hardening steels. Ferritic steels are alloys of Fe and Cr containing usually less than 0.10% carbon. Sometimes it is also incorporated with Mo to improve its resistance to corrosion with seawater. Its mechanical strength is, however, found to be poor. Austenitic steel is an alloy of Fe with Cr and containing nickel, manganese and nitrogen. Cr present in the steel is responsible for Cr_2O_3 film formation on the surface, which prevents corrosion of steel. Further it has high mechanical strength and is extensively used in pipelines in desalination plants. The third category of steel is the martensitic steel and is based on Cr and carbon levels as high as 1%. They offer moderate corrosion resistance with high mechanical strength. Duplex steel belongs to the fourth category of steel and can be considered as a material formed from equal amounts of ferritic and austenitic steels. Such steels have extreme resistance to corrosion by chemical environments. Steel undergoes extensive corrosion in the environment of Cl^- ions and such type of corrosion is generally known as pitting and or crevice corrosion. This can be prevented by adding Mo, N, etc. in the steel. The last category of steel is the precipitation-hardening steels, produced by adding elements such as Al, Cu, and niobium to any kind of steel. They have extremely high mechanical strength and their corrosion resistance is comparable to that of austenitic steels.

9. CLASSIFICATION OF STEEL FOR ENGINEERING USES

The Society for Automotive Engineers International (SAE International), in association with the American Iron and Steel Institute, classified steels according to their composition, mechanical properties, and applications. A few examples are given in Table 4.1.

Depending up on the actual end use, steel of particular type is chosen for fabrication. For example, stainless steels such as 316L, 317L, and 904L are commonly used in seawater reverse osmosis plants. Chemical modifications occurring on the surface of such steel can be monitored by SEM and AFM techniques. AFM images of stainless steel surface before and after exposing to aqueous medium containing chloride ions are given in Fig. 4.3. In the AFM technique, in addition to the observed images, evaluation of surface roughness can give valuable information regarding the extent of corrosion occurring on the steel surface. A brief description of the surface roughness is also given below. From the images it is clear that the pristine sample has nearly smooth surface with certain marks and dips (Fig. 4.3a). Upon exposure to corrosive environments (water containing chloride ions), hills and valleys are developed as can be seen from the image shown in Fig. 4.3b. Surface roughness values have been evaluated from the images and the results are described next.

10. SURFACE ROUGHNESS

Profile surface roughness is a parameter that strongly depends on the reaction/corrosion occurring at the surface. Profile surface roughness can be expressed either as average surface roughness (R_a) or as

Table 4.1 Examples of a Few Types of Stainless Steel With Their SAE Designation and Important Constituents

Type (SAE Designation)	Name of Steel and Application	Composition
102	Austenitic steel for general applications	Plain carbon steel with maximum 1% Mn
200 series (201&202)	Austenitic structure (FCC structure), general-purpose steel. The material can be hardened by cold working	0.15% carbon, ~16% chromium, and sufficient nickel and/or manganese
300 series 301, 302, 303, 304, 304L, 308, 309, 310, 316, 316L, 316Ti	Austenitic steel containing chromium and nickel. Better resistance to corrosion, good weldability	Composition: 18%Cr and ~ 8–10% Ni depending upon the type of steel. Some of them, like 316, contain Mo to improve the corrosion resistance
400 series, 401, 405, 408, 409, 410, 416	Ferritic and martensitic steel containing Cr. Poor corrosion resistance, easy weldability, cutlery, and decorative applications	Types like 404 have 11%Cr and 8%Ni. Some of them contain S (like 416), which improves the machinability
500 series	Alloys of Cr and act as a good heat resistant material	In addition to other components, around 0.4–0.6% Mo is also present. Improved corrosion resistance
600 series 601, 610, 614, 630, 635 etc.	Martensitic steels. The types 630–635 belong to precipitation-hardening steels	Type such as 630 contain 3–5% Cu and 0.15–0.45% Ta

root mean square surface roughness (R_q). Both are the measure of the fluctuations on the surface of a material. They can be expressed by the following equations [4,6].

$$R_a = \frac{1}{n}\sum_{i=1}^{n}|y_i| \tag{4.1}$$

$$R_q = \sqrt{\frac{1}{n}\sum_{i=1}^{i=n}y_i^2} \tag{4.2}$$

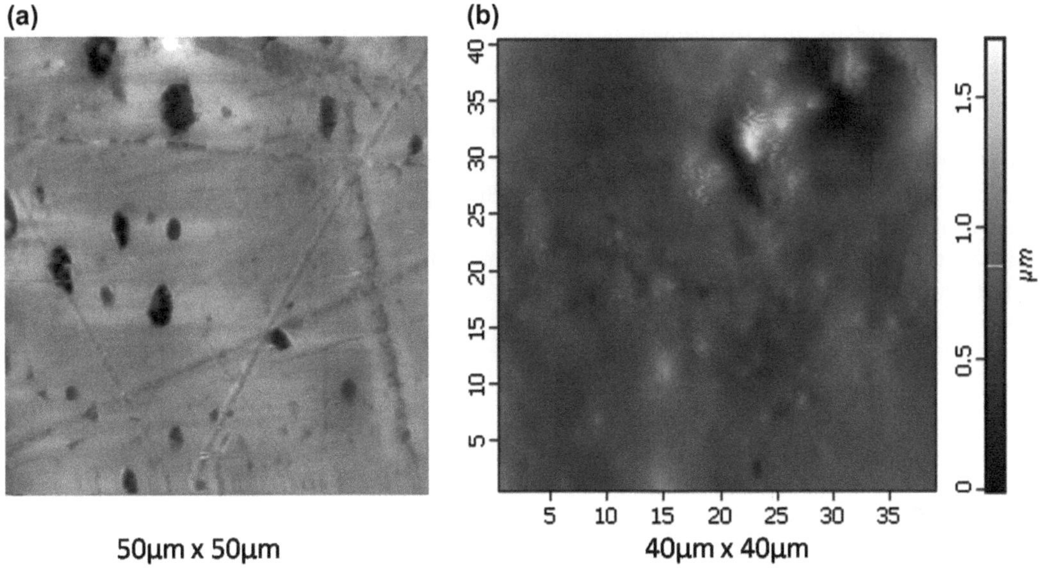

FIGURE 4.3

Atomic force microscopy images of stainless steel rod subjected to corrosion in water containing chloride ions: (a) before corrosion and (b) after corrosion.

In the expression $i = 1$ to n represents the number of lines scanned and y_i is a measure of the surface fluctuations. In the images shown in Fig. 4.3, R_a is around 46 nm with R_q of 78 nm. Upon corrosion the corresponding values change to 100 and 146 nm, respectively. In general, R_q is larger than R_a. In addition to steels discussed previously, there is another category of inorganic materials known as the refractory materials. These materials are in general thermally very stable and resistant to a wide variety of chemical environments. In the following section different types of inorganic refractory materials that are commonly encountered are discussed.

11. REFRACTORY MATERIALS

Materials or compounds that are highly resistant to chemical environments even at high temperatures (more than 1000°C) are generally called refractory materials. They are very essential components in glass and cement plants, iron and steel industries, etc. They usually have very high melting points. Examples of refractory materials include boron nitride (BN), boron carbide (B_4C), silicon nitride (Si_3N_4), silicon carbide (SiC), niobium boride (NbB_2), etc. All of these compounds have melting temperatures more than 2000°C. Also, nitrides such as hafnium nitride (HfN), thallium nitride (TaN), etc. have melting temperatures more than 3000°C, and they are also used as refractory materials in certain specific applications. Another important property of such materials is that they have very high resistance toward oxidation. For example, Si_3N_4 can be heated in air up to 1400°C while retaining its chemical composition. This material finds wide application as bearings and blades in motors, which are subjected to high temperatures. Similarly, SiC can withstand still higher temperatures of the order of ~1700°C compared to Si_3N_4. However, during heating a protective layer of SiO_2 film formation occurs on SiC surface. In a similar way, a protective coating of SiO_2 on $MoSi_2$ (molybdenum disilicide) facilitate, the use of $MoSi_2$ as a candidate for heating element in furnaces. Alloying Si_3N_4 with Al and oxygen leads to the formation of silicon aluminum nitrides (SIALONS) generally represented as $(Si,Al)_3(ON)_4$. These are oxynitride phases containing both $SiO_4(N)$ and $AlO_4(N)$ tetrahedra and have very high chemical resistance even at higher temperatures. In many applications they are found to show better performance characteristics than that of Si_3N_4 phase.

As many refractory materials are used in metallurgical plants, it is essential to know how these refractory materials react with slag during the extraction of different metals. This aspect is discussed in the following section.

12. REACTION OF REFRACTORY MATERIALS WITH SLAG

Slag is usually defined as the stony waste matter separated during the refining process of an ore. Slag can react with container materials thereby reducing lifetime of plants. For example, in blast furnace for Fe and steel production, containers should be resistant to reaction of molten iron oxides. For this purpose silica bricks are generally used. Silica is advantageous because under atmospheric conditions, Fe_2O_3-SiO_2/Fe_3O_4-SiO_2 system has a large immiscible region as is clear from the phase diagram in Fig. 4.4 [5].

13. POZZOLANS

Pozzolans are a different type of refractory materials. They are naturally available and consist of very fine particles of siliceous and aluminous materials that in presence of water react with $Ca(OH)_2$ to form cementitious materials [7]. The extent of pozzolanic activity is measured with respect to the extent of reactivity of fine dispersion of pozzolan with $Ca(OH)_2$ and water. The reaction mainly occurs due to the very fine nature of silica or alumina particles. Pozzolans are generally added to Portland cement to improve the quality of cement. The improvement in quality is taking place due to the pozzolanic

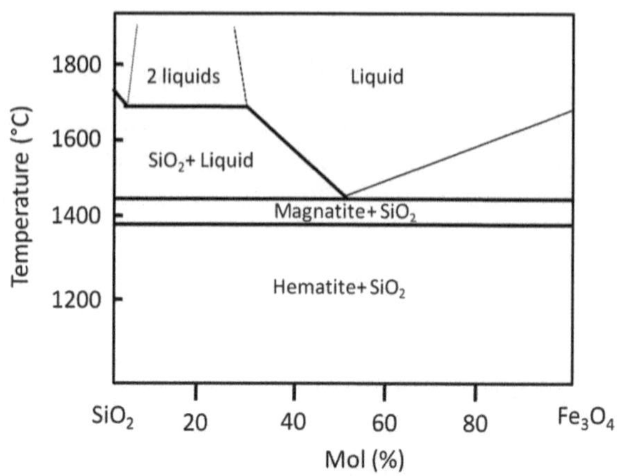

FIGURE 4.4

Schematic representation of Fe_3O_4-SiO_2 phase diagram.

reaction between $Ca(OH)_2$ and silica in the presence of water to form calcium silicate hydrate (C–S–H). This occupies pores formed in cement resulting in reduction in the number density of available pores and in turn reducing the permeability of binder in pores and subsequent interaction with the structural materials. Reducing the binder permeability in pores also reduces the incorporation of harmful ions like chloride and carbonates, thereby reducing the extent of corrosion. Further, sulfate attack is prevented due to the consumption of $Ca(OH)_2$ in the material. The just-mentioned action of pozzolan is of quite significance as it reduces the corrosion of iron in concrete mixtures thereby increasing the life of concrete structures. In conventional concrete, alkaline Portland cement is mainly used. This cement upon drying leads to the development of pores, voids, and cracks. Through these voids and cracks acidic/alkaline/neutral solution can migrate and can damage the structure. Similar to pozzolanic materials, organic resins such as polyester, epoxy, etc., once added, eliminate pores and voids, and help to improve the resistance to medium surrounding the materials. Hydrophobic nature of polymers makes them a better candidate against hostile environments [8].

14. TUNGSTEN CARBIDE

Phase diagram of tungsten carbide (WC) system is very important from the technological point of view. There exist three binary compounds (corresponding phase diagram is shown in Fig. 4.5) in the WC system [9]. They are $W_2C(\beta)$, $WC_{1-x}(\gamma)$, and $WC(\delta)$ phases as can be seen from Fig. 4.5. In the commercial-cemented carbides (compounds) the δ form is mainly present. In addition to this there

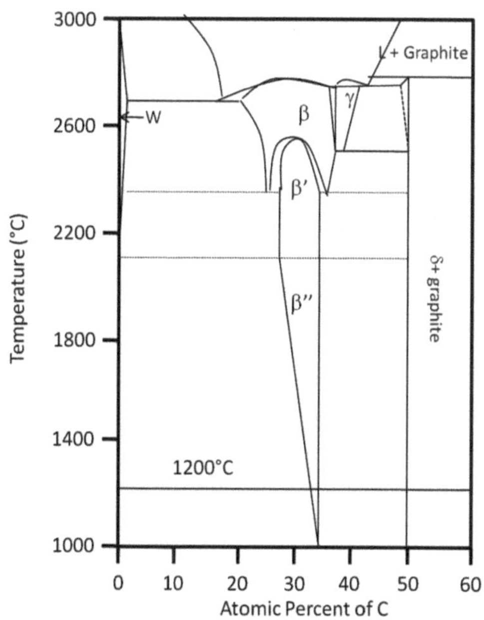

FIGURE 4.5

Schematic representation of phase diagram of WC system.

exist also three temperature-dependent modifications denoted as β, β′, and β″, which are described in detail next (also shown in Fig. 4.5).

The phase W_2C can be considered as hexagonal close-packed (HCP) array of W atoms with half of the octahedral holes occupied by C atoms. The close-packed array is usually distorted and depending upon the extent of order (disorder) there exist phases like β, β′, and β″. The β phase is normally formed at temperatures of the order of 2400°C and above (Fig. 4.5). Upon heating in air, W_2C can get oxidized at a temperature above 500°C. It is resistant to most of the acids but reacts with conc. HNO_3 as well as HNO_3-HF mixtures. Such strong oxidizing agents convert W-C to W^{6+} species (in solution) and CO_2 gas. Upon dilution, W^{6+} undergoes hydrolysis and precipitates as yellow-colored $WO_3 \cdot xH_2O$. Halogens like fluorine and chlorine gas attack W_2C, the former occurs at room temperature and the latter at temperatures above 200°C. The other compound WC_{1-x} is the high-temperature cubic form of tungsten carbide and undergoes eutectoid decomposition at around 2530–2535°C to β-W_2C and δ-WC.

WC is a very hard material with very high chemical resistance. It exists in two crystalline modifications, namely the low-temperature alpha form with hexagonal unit cell and a high-temperature β-form with cubic NaCl type structure. It is one of the hardest materials but its hardness is lower with respect to that of diamond and boron nitride. It is prepared by the reaction of W with C over the temperature range of 1400–2000°C. It can also be prepared by the reaction of WCl_6 or WF_6 and hydrogen in presence of CH_4 or methanol at elevated temperatures. The corresponding reactions are given following [10]:

$$WCl_6 + H_2 + CH_4 \rightarrow WC + 6HCl \text{ (at a temperature of 670°C)}$$

$$WF_6 + H_2 + CH_3OH \rightarrow WC + 6HF + H_2O \text{ (at a temperature of 350°C)}$$

WC up on heating above 500–600°C results in its oxidation and under this condition (temperature more than 400°C), it reacts with chlorine and fluorine gas. Also, at high temperatures it reacts with sodium carbonate to form sodium tungstate. It is resistant to acids like H_2SO_4, HCl, etc., however, with concentrated HNO_3 and HNO_3-HF mixtures, it readily reacts and gets dissolved at slightly higher temperature. The reaction is found to be highly exothermic. Mixture of H_2SO_4 and H_2O_2 also slowly acts with WC, leading to its dissolution [10].

WC reacts with molten metals of both nickel and cobalt and phase diagram of W-Co system shows the compound formation between WC and Co [11].

15. SILICON CARBIDE

Silicon carbide is also known by the name carborundum and is a widely used abrasive material. It finds applications in car brakes, car clutches, ceramic plates, components in turbines, ball bearings, hot gas flow liners, heat exchange materials, etc. It is generally prepared by the reaction of very fine particles of silica with carbon powder over the temperature range of 1600–2500°C. It can also be prepared in the form of films by the chemical vapor deposition of organic polymers like poly(methylsilyne) in an inert atmosphere [12–15]. Silicon carbide exists in a variety of crystalline forms and exhibits property known as polytypism. The most commonly encountered SiC phase is the alpha form and has an HCP array of silicon atoms with half of the tetrahedral holes occupied by carbon atoms. The alpha form of SiC is formed at a temperature above 1700°C. The cubic form of SiC, known as the β-form, is formed at temperatures less than 1700°C. The alpha form is employed in most of the technological applications and the β-form has application as catalyst due to its higher surface area compared to the alpha form. Up to 800°C it is not attacked by acids, alkalies, and molten salts. Silicon carbide can be used for applications that involve temperatures up to 1600°C, even though above 1200°C, a film of SiO_2 is formed over of SiC, which acts as a protective layer for the material. The material has exceptionally high shock resistance and this has been attributed to the high thermal conductivity (0.25 W/cm-C at 200°C) and low thermal expansion coefficient ($3.8 \times 10^{-6}\ K^{-1}$ at 200°C) [16].

16. BORON NITRIDE

Boron nitride is a refractory material with very high chemical resistance. It is insoluble in acids like HCl, HNO_3, H_2SO_4, etc. but soluble in molten liquids of LiOH, KOH, NaOH, etc. Hence, the above compounds in general are used for etching boron nitride (BN). Boron nitride exists in a variety of crystalline modifications and among them the cubic and hexagonal forms are the most common. The hexagonal form has properties similar to graphite, and the cubic form has properties similar to diamond. Boron is three coordinated in the hexagonal form whereas it is tetrahedrally coordinated in the case of cubic BN (Fig. 4.6).

There also exists an amorphous form of BN that exhibits properties similar to that of amorphous carbon. Heating BN in oxygen or air at high temperatures (~1300°C) leads to the formation of surface boron oxide, which protects BN from further corrosion [17]. Reactivity of cubic and hexagonal form of BN is similar. Elements like Fe, Co, and Ni (which are components in different types of steels) react with BN only at temperatures above 1400°C. Reactive and low-melting metals like Ga and Sn do not wet BN until 1100°C. With silica it forms borosilicate network, which is the main structural unit present in borosilicate glass. With Al_2O_3 no reaction occurs even after heating around 1300°C.

Boron nitride can be prepared by a variety of methods like heating B_2O_3 or H_3BO_3 in ammonia, urea, etc., at elevated temperatures of the order of 900–1000°C and above. The reactions are the following [18]:

$$B_2O_3 + 2\ NH_3 \rightarrow 2\ BN + 3H_2O$$

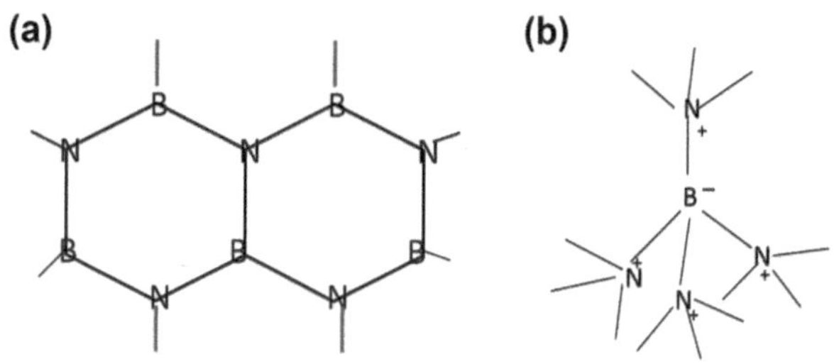

FIGURE 4.6

Coordination environment around B and nitrogen in hexagonal (a) and cubic (b) forms of BN.

$$B(OH)_3 + NH_3 \rightarrow BN + 3H_2O$$

$$B_2O_3 + CO(NH_2)_2 \rightarrow 2\ BN + CO_2 + 2H_2O.$$

BN upon red heating reacts with steam, producing NH_3 and H_3BO_3 (reverse reaction involved in the preparation of BN, mentioned previously). It is also attacked by fluorine gas and HF as can be seen from the reactions given here [19]:

$$BN + 3F_2 \rightarrow 2BF_3 + N_2$$

$$BN + 4HF \rightarrow NH_4BF_4$$

17. GALLIUM NITRIDE

Gallium nitride is a III-V semiconductor and has very high chemical resistance to corrosive environments. Strong bond existing between Ga and nitrogen is responsible for the corrosion resistance properties the compound. Bulk gallium nitride is a direct band gap semiconductor (band gap = 3.4 eV) having wurtzite type structure and is the material used for making light-emitting devices that can withstand corrosive environments. Gallium nitride is prepared by the reaction of Ga_2O_3 with NH_3 at elevated temperatures of the order of 1000°C.

$$Ga_2O_3 + 2\ NH_3 \rightarrow 2\ GaN + 3H_2O$$

It can be also prepared by the chemical vapor deposition of organometallic compounds containing Ga and nitrogen atoms. Corrosive acids and alkaline environments do not have any effect on GaN phase as revealed by X-ray diffraction and electrical conductivity measurements.

18. BOROSILICATE GLASS

Glasses in different forms like fibers, sheets etc. are widely used in various construction applications. Glasses in general are susceptible to attack by both acids and alkalies (at lower and higher pH values). In basic solutions the alteration or corrosion rate is relatively higher when compared to acidic solutions. During corrosion of glass multiple steps are involved, namely the diffusion of network-modifying ions like Na^+, hydrolysis of glass network, formation of alteration products that may or may not protect the glass surface. If the surface area and density of glass and the leachate concentration in solution are known, normalized leach rate or release rate can be calculated based on Eq. (4.3) given here [20]:

$$NR_x = \frac{C_x}{(f_x)\cdot(SA/V)\cdot(t)} \tag{4.3}$$

where C_x is concentration of element "x" in the solution with the unit "g/L," and "f_x" is the fraction of element "x" in the unleached glass. The term "SA" represents surface area of glass powder (cm^2) and V represents leachate volume (L). The term "t" represents the duration of leaching. Magnitude of NR_x can be used as a measure of the extent of leaching that can take place with the glass samples. Smaller alkali metal ions like Na and Li leach out immediately from the glass matrix. In certain cases, a secondary phase like sodium silicate also gets leached out from the glass, thereby changing composition of the residual glass. Extent of the herein-mentioned types of alterations occurring with glass samples can be determined from concentrations of Na or B species in solution. From the amount of Na and B present in leached solution, fraction of altered glass after leaching can be calculated based on the following expression, Eq. (4.4) [21]:

$$\text{Altered glass fraction (AG)} = \frac{m_{\text{Boron}}^{\text{leached}}}{m_0 \cdot f_B} \tag{4.4}$$

where "$m_{\text{Boron}}^{\text{leached}}$" is the mass of boron in leached solution, "m_0" is the initial mass of glass powder (in grams), and "f_B" is boron mass fraction in the glass (a unitless parameter). Oxides like TiO_2, ZrO_2, etc. are known to improve stability and reduce leaching properties of glass samples. Glass corrosion in alkaline medium takes place due to the attack of OH groups on positively charged silicon species. In acidic medium, protons attack nonbridging oxygen atoms. Both processes lead to weakening of glass network. To check the fact that the environment attacks glass, surface of the glass needs to be probed at regular intervals after exposing the materials to corrosive environments. SEM and AFM (whose principles of functioning are described earlier in this chapter) can be effectively used for this purpose. SEM image of typical sodium borosilicate glass with composition $[(SiO_2)_{0.482}(B_2O_3)_{0.244}(Na_2O)_{0.220}(BaO)_{0.054}]$ subjected to leaching for 700 days in mineral water is shown in Fig. 4.7.

Pristine sodium borosilicate glass is characterized by homogeneous distribution of different components as can be seen from Fig. 4.7a. No grains or phase separation could be seen in the image

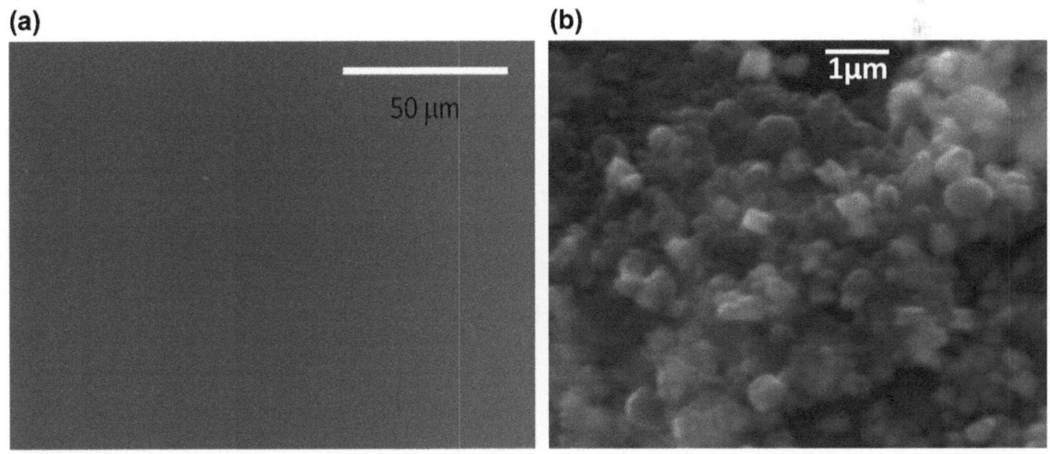

FIGURE 4.7

Scanning electron microscope images corresponding to (a) unleached sodium borosilicate and (b) sodium borosilicate glass subjected to leaching for more than 700 days.

confirming the existence of single-phase isotropic glassy phase. Unlike that for leached sample, aggregated particles with varying sizes over the range of 200–600 nm could be seen in the image (Fig. 4.7b). The results suggest the phase separation from glass or the precipitation of crystalline phase form leached solution. From the energy dispersive X-ray analysis technique, it is confirmed that the segregated phase is silica rich in nature. Thus the SEM results confirm the interaction of environment with glass and subsequent modification of the glass surface.

19. DIAMOND-BASED MATERIALS

Diamond is the hardest material known to humankind. The material can withstand highly corrosive and harsh chemical environments. Graphite and diamond are two commonly found allotropes of carbon in nature. Graphite is the thermodynamically stable form of carbon whereas the diamond phase is only kinetically stable. Diamond is characterized by many desirable physical properties, which include high thermal conductivity, extremely low thermal expansion, very high chemical resistance, etc. Values for the previously mentioned parameters are included in Table 4.2. Diamond films for the protective coating of different components in hostile chemical environments are mainly prepared by the chemical vapor deposition of alkanes like CH_4 in presence of hydrogen. A schematic diagram for the coating or deposition of diamond films on substrates is given in Fig. 4.8.

Even though diamond has many superior properties compared to other materials, use of diamond for industrial applications is limited due high cost involved in making bulk and thin films of diamond.

Table 4.2 Values Corresponding to Certain Important Properties of Diamond

Property	Value
Mechanical hardness	90 GPa
Thermal conductivity	2×10^3 W/mK
Thermal expansion coefficient	1×10^{-6}K
Room temperature resistivity	10^{13} Ω cm

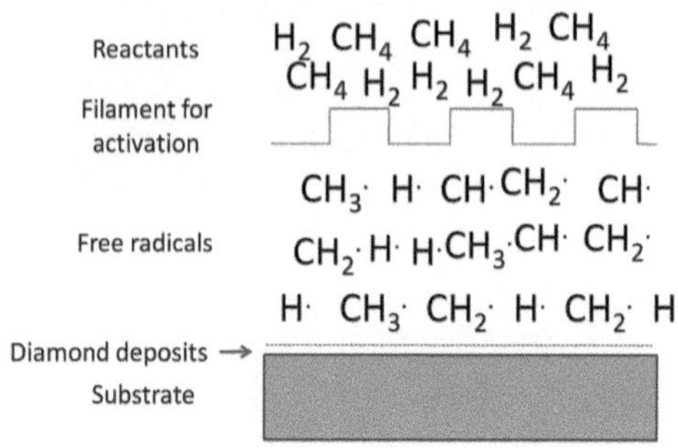

FIGURE 4.8

Schematic diagram for the chemical vapor deposition (CVD) of diamond films on substrates.

Natural diamond is only obtained from the mines of a few African countries and hence it is quite costly. However, with the advent of chemical vapor deposition technique diamond films can be easily grown and coated over different substrates at a much cheaper costs. Coating of diamond films over mechanical tools is widely used to reduce wear and tear as well as to improve the mechanical strength of tools mainly used for cutting nonferrous alloys. In the case of ferrous alloys, diamond phase has a tendency to react with such alloys. Due to its extreme resistance against chemical attacks, diamond film-based detectors find extensive application in radioactive reprocessing plants, where one routinely encounter high concentration of nitric acid. The insulating diamond can be converted to semiconductor material by doping with elements such as B during its growth. Generally during diamond synthesis, a thermodynamically more stable form of carbon, namely graphite, also forms. This can be removed by treating the product with chromic acid.

20. RESINS AND POLYMERS FOR HOSTILE CHEMICAL ENVIRONMENTS

In addition to the inorganic materials mentioned before, organic materials also play major roles for protecting different materials against hostile environments. Certain resins and polymers have been extensively used for chemically hostile environments, such as acidic and alkaline medium, corrosive gases, etc. Resins are mainly compounds whereas polymers are macromolecular in nature. There is a large difference in the molecular weights between resins and polymers. Similarity between resin and polymer is the existence of chainlike structure in both the materials, however the chain length is smaller in resins compared to polymers. Polymers many times are called "plastics" and they usually contain other materials (also called "filler materials") in addition to polymers. Plastics can be mainly of two types: thermoplastics and thermosetting plastics. The former can be deformed easily on heating and brought back to original condition. The latter, thermosetting plastics, once molded cannot be softened by heating. Polyoxybenzylmethylenglycolanhydride, known by the name Bakelite, is an example of a thermosetting plastic.

Under corrosive environments, chemical nature of the polymer can change in addition to its microstructure and texture. In other words, the supramolecular structure can change with the extent of hostility of surrounding chemical environment. Such changes can result in decrease in molecular weight, reduction in tensile strength, reduction in thermal stability, introduction of functional groups, colors, etc. with the polymers. In certain cases, crystallization of different phases also occurs, which can significantly change the transparency of polymers and associated optical properties. Thus it is clear that polymer needs to be characterized by different techniques to understand changes brought about by the chemical environment. In the following section, first resins that are commonly used for hostile environments will be discussed, and this will be followed by discussion of polymers.

21. ACRYLIC RESINS

Acrylic resins are materials derived from acrylic acid (structural formula given following) or methacrylic acid and its derivatives. The simplest example is polymethacrylate (PMA) and it can be represented by the formula given here:

$CH_2{=}CH{-}COOH$(acrylic acid)	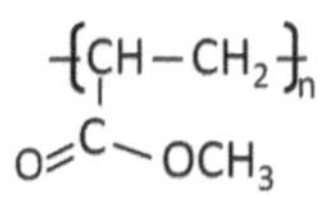 Polymethyl acrylate(PMA)

Polymethyl acrylate is a polymer resin and is highly hydrophobic and water resistant in nature. Such polymers are generally prepared by the thermal polymerization of a reaction involving a monomer using an initiator like Azo isobutyl nitrate (AIBN). PMA is soluble in dimethyl sulfoxide and is unstable against alkaline environment. A similar compound, methyl methacrylate, a derivative of acrylic acid, $CH_2{=}C(CH_3){-}COOCH_3$, upon polymerization leads to the formation of polymethyl methacrylate (PMMA) with structural formula:

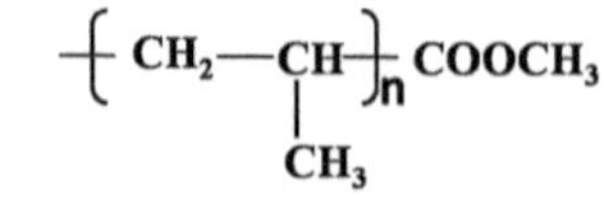

The material has a variety of trade names such as Plexiglas, Acrylite, Lucite, Perspex, etc. In appearance it looks like silica glass, however, this material is machinable and shatter resistant and is also known as organic glass. With suitable modification during the synthesis conditions including temperature, the working temperature range of PMMA can be varied over the range of 85−165°C. At temperatures above 460°C, the material undergoes ignition producing large volumes of mainly CO_2 and water [22]. Due to the presence of hydrolyzable $-COOCH_3$ group, the material has a tendency to undergo degradation in alkaline and acidic environments. Also it can undergo swelling by absorbing organic solvents. However, its stability is much better than that of other polymers like polystyrene and polyethylene (polythene) [23].

22. EPOXY RESINS

Epoxy resins are polymers containing epoxide group. They are also known by the name glycidyl or oxirane-containing polymers. The resin provides a tough protective coating with excellent hardness. Starting materials for making epoxide-based resins are alcohols, phenols, polyols, and other petroleum-derived products. Examples of such materials include resins like Bisphenol A fumerate, Bisphenol F epoxy, Novolac epoxy, aliphatic epoxy, etc. Epoxide content in such resins is an important parameter that decides the properties of resins. It is expressed by the parameter epoxide number and is defined as the number of epoxide equivalents in 1 kg of resin. Normally epoxide resins are cured with materials known as hardeners prior to its applications. The amount of hardener required depends on the epoxide number of the resin. Hardeners or curatives are the molecules containing active hydrogen atoms and include molecules like alcohols, phenols, amines, thiols, etc. Epoxy resins themselves, when allowed to react in presence of a Lewis base (catalyst), homopolymer-containing ether bridges are formed. Such materials have very high chemical resistance. Another important epoxide resin is bisphenol A fumerate resin, which is prepared by the reaction of oxide diether of bisphenol A and fumeric acid [24]. The aromatic ring moiety offers rigidity to the system, thereby increasing the thermal stability. Another advantage is that the number of interchain ester groups have been reduced in the resin and this significantly improves the resistance to hydrolysis and chemical reaction. The chemical structure of Bisphenol fumerate A polyester resin in given here:

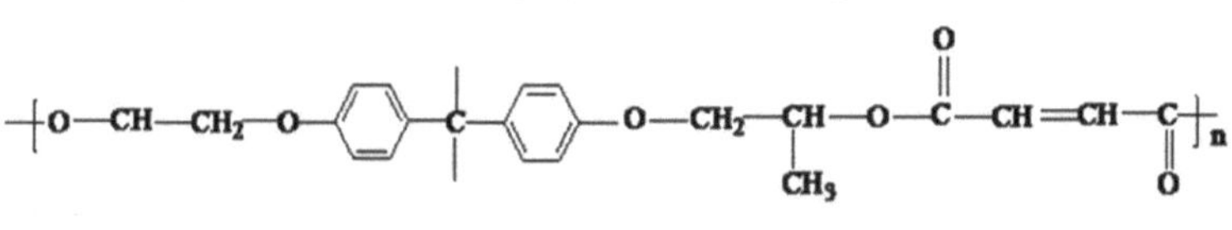

Table 4.3 Common Chemicals and Gases Including the Conditions in Which They Are Resistant to Bisphenol Fumerate A Resin

Chemicals	Concentration and Temperature
Hydrochloric acid	10%, 93°C
Phosphoric acid	80%, 93°C
Sulfuric acid	30%, 93°C
Trichloroacetic acid	50%, 93°C
Chloroacetic acid	15%, 93°C
Chromic acid	5%, 93°C
KOH, $Ca(OH)_2$	25%, 71°C
NH_4OH	5%, 71°C
CO_2, CO, Cl_2, SO_2 gases	93°C

These resins undergo hydrolysis under highly concentrated acidic and alkaline conditions. The following chemicals given in Table 4.3 can be handled by Bisphenol fumerate A polyester resin under mentioned conditions in the same table. Bisphenol fumerate A resins are extensively used for storing dilute sulfuric acid. Carbon steel cannot be used for this purpose as it is highly corrosive for dilute sulfuric acid. Commercially available DIOR 382 is Bisphenol A fumerate resin [24] and is widely used all over the world for storing dilute sulfuric acid.

23. SILICONE RESINS

Silicones or silicon resins are organosilicon compounds (alternatively known as organosiloxanes) and contain both Si−O and Si−C linkages in the material. Such materials can also be considered as organic-inorganic hybrids, the organic nature arising from Si−C bond and inorganic nature comes from the Si−O−Si linkages. The organic groups linked to Si decide properties of the resin. Further, they can also be mixed with organic resin thereby modifying their properties. Certain organosilicon polymer with high oxygen permeability is used as contact lenses for myopia patients. Organosilicon resins are also used in antifriction protection for transparent plastics, glass, and lens. They have wide application in weather-resistant exterior coating after mixing with concrete. The starting materials for many of the silicone resins are chlorosilanes and ethoxy or methoxy silicates. The latter silicate reagents require acid or alkali as catalyst for hydrolysis. Initially hydrolysis occurs followed by condensation of OH groups to form more cross-linked structures. Complete condensation generally does not occur, and around 2−6 wt% silanol groups (Si−OH linkages) remain with the resin even after the reaction. A schematic diagram of the molecular structure of silicon resin is given in Fig. 4.9.

The Si−O−Si groups are quite stable in acidic environments. However, in presence of hydroxyl ions, Si can undergo further hydrolysis and facilitate formation of soluble silicates.

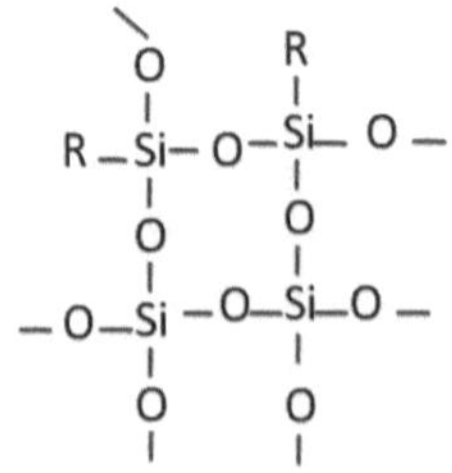

FIGURE 4.9

Schematic representation of structural units present in silicone resin.

24. BUNA N

This is a polymer called polybutadiene and is generated by the polymerization of butadiene monomer. The monomer butadiene can be produced from acetylene or ethanol, and the reaction with ethanol as the starting materials is given in Fig. 4.10.

The polymerization reaction requires an initiator, and initially Na metal (suspended in mineral oil) was used as the initiator, hence the name Buna N [25] (it is also called Buna-N rubber). With the advent of modern technology, the reaction is carried out with Ziegler–Natta catalyst or Li metal. It is a polymer resistant to oils, solvents, etc., and has the property of hydraulic fluid resistance. However, it is sensitive to highly polar organic solvents such as acetone, methyl ethyl ketone, chlorinated hydrocarbons, ozone, or nitro hydrocarbons. The operating temperature can be −54 to 135°C.

25. ETHYLENE PROPYLENEDIENE MONOMER RUBBER

These are a category of rubber containing polymethylene-type saturated chains. It can be also considered as a copolymer consisting of three different components, namely ethylene, propylene, and a diene. The last component in these materials, diene, is in the concentration range of 2.5–12%. Peroxide curing of this polymer leads to higher cross-link density with these samples. It can withstand

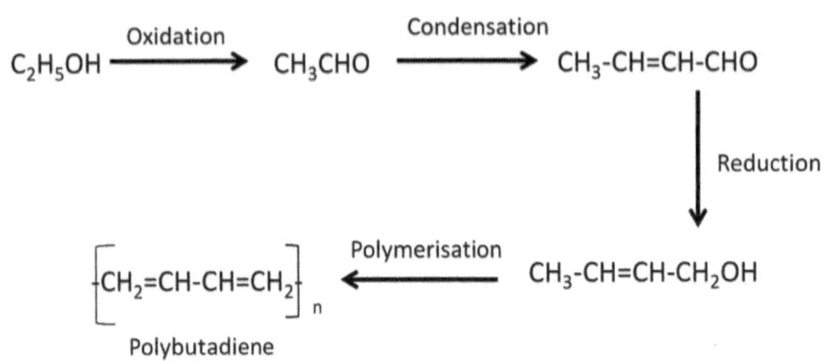

FIGURE 4.10

Schematic diagram for the preparation of polybutadiene.

to a maximum service temperature of 150°C and is resistant to ozone, steam, polar compounds, and weather changes. This polymer offers high resistance to tear and abrasion. The material is highly resistant to dilute acids and alkalies but not stable for petroleum and strong acids or alkalies.

26. FLUOROCARBON ELASTOMERS

These materials belong to the category of fluoroelastomers and is a family comprising of copolymers of fluorine-containing monomers like hexafluoropropylene, vinylidene fluoride, etc. Around 66–70% fluorine content is present in most of the polymers. Some of the polymers in this category are known by the commercial names Viton A, Viton B, etc. Such polymers are compatible to a broad spectrum of chemicals. They are resistant to strong mineral acids, salt solutions, chlorinated hydrocarbons, etc. Stability of such polymers depends on the components constituting the polymer. These types of materials are widely used for making O-rings, gaskets, etc., which are to be employed in very corrosive environments like conc. HNO_3, HCl, etc.

27. STYRENE-BUTADIENE-BASED POLYMERS

These materials are compatible with water, alcohols, weak acids, etc. They do not react with mineral oils, petroleum greases, and fuels. Clothes used in hostile environment are usually coated with a layer of such resins. The commercially available resin, Tyvek, is nothing but copolymer of styrene butadiene with vinylidene chloride. To improve the extent of protection, Tyvek is usually coated with polyethylene or other polymers.

28. PLASTICS THAT CAN SURVIVE HOSTILE ENVIRONMENTS

Plastics are materials that are generally made from organic molecules by polymerization reaction. They are very widely employed for various applications due to their easy method of preparation of and processability. Some of the polymers like polyvinylidene difluoride (PVDF), polyether ether ketone (PEEK), and polyvinyl chloride (PVC) are known to have very high resistance toward acids and other corrosive liquids. The polymer PVDF is a highly chemical-resistant fluorine-containing polymer. It is prepared by the polymerization of vinylidene difluoride. The material has very high resistance for different types of hot concentrated acids with varying concentrations. Another important acid resistant polymer is PEEK. It also has very high chemical and creep resistance and is widely used in medical implants.

29. DESIGNING FIBERS OF POLYMERS WITH RESISTANCE TO HOSTILE CHEMICAL ENVIRONMENTS

While making or designing polymers that can withstand corrosive chemical environments the following approaches are generally followed. Materials with strong chemical bonds are usually stable under hostile environments. Similarly, absence of reactive side groups, polymer backbone with no hydrolyzable group etc. facilitate the chemical inertness of a fiber or polymer. In certain situations a hostile chemical environment can arise at high temperatures. Hence a material that is chemically inert must be also thermally stable. For example, polymers like polyethylene and polypropylenes are chemically resistant under ambient conditions. However, at slightly higher temperatures above 50°C, they undergo reaction in oxidizing environment. It is observed that generally halogenated polymers and polymer-containing aromatic rings have chemical resistance and thermal stabilities and find extensive applications in protective clothing, wet and dry/hot gas and liquid filtrations, conveyer belts, etc.

30. CHLORINE AND FLUORINATED POLYMERS

Such polymers belong to the category of chlorinated and fluorinated polymers, and examples include polyvinylidene chloride (PVDC), polytetrafluoroethylene (PTFE) etc. PVDC as such has high chemical resistance. It is obtained by the free radical polymerization of vinylidene chloride monomer. The monomer is obtained by heating trichloroethylene above 400°C.

$$CHCl_2\text{-}CH_2Cl \xrightarrow[-\,HCl]{400^{\circ}C} CH_2 = CCl_2 \xrightarrow[\text{Emulsion polymerisation}]{} \left[CH\text{-}CCl_2 \right]_n$$

This polymer is extensively used for making fibers and is resistant to many organic solvents. However, its application is limited due to its poor processability. To improve the processability of PVDC, copolymers with monomers such as acrylonitrile, methyl acrylate, etc. are used. For example, a copolymer of vinylidene chloride with vinyl chloride gives fibers that are resistant to acids, alkalis, hydrocarbons, etc., up to a temperature of 100°C. Limiting oxygen index value is a measure of the corrosion resistance of materials. The value is ~60% for this material. Such materials (particularly based on PVDC) have also exhibited excellent weather resistance and are widely used in vehicle seat covers, furniture in parks, public vehicles, etc.

Limiting oxygen index [26] is a parameter that decides the thermal stability of a polymer. It is defined as the minimum amount of oxygen (in percentage) that can support combustion of a particular polymer. It can be easily evaluated under laboratory condition using a mixture of nitrogen and oxygen. While burning polymer in oxygen nitrogen mixture, the amount of oxygen is reduced to such a level that below that level no burning occurs and above which the burning sustains. There are standard tests [27] that can determine the limit of oxygen index values, for example, ASTM D2863.

31. FLUORINATED POLYMERS OR FIBERS

Fluorinated polymer fibers have very high chemical resistance and thermal stability. Among the different polymers in this category, polytetrafluoroethylene (PTFE) is most widely encountered. Fluorinated polymers/fibers also include polyvinyl fluoride (PVF), polyvinylidene fluoride (PVDF), etc. Fluorinated acetylene polymers have added stability over PVF and PVDF with respect to acidic environments like more than 90% sulfuric acid, 70% nitric acid, or comparable alkaline environments.

31.1 TEFLON

PTFE is known by the trade name Teflon. It can be represented by the following structural formula.

$$\left[CF_2\text{-}CF_2 \right]_n$$

The monomer for this polymer, tetrafluoroethylene, can be prepared by different methods like dechlorination of dichlorotetrafluoroethane, decarboxylation of sodium perfluoropropionate, pyrolysis of chlorofluoromethane, etc. Polymerization of tetrafluoroethylene (monomer) is carried out in presence of redox initiators like $H_2O_2/FeSO_4$. It is also possible to carry out this reaction with oxygen as the initiator. PTFE is a linear polymer with helical structure and is highly crystalline in nature with a melting point of around 330°C. It is observed that the mechanical properties of Teflon remain the same over the temperature range of −100–350°C. Barring molten alkali metals, Teflon is quite inert toward, all organic solvents, strong acids and alkalies, etc. Unlike other polymers that can be processed by melting followed by molding, Teflon cannot be processed by that method due to the very slow melt flow rate. Compaction of powder followed by sintering at high temperatures is generally followed to make Teflon into different shapes according to the requirements. Due to the extremely high chemical resistance of Teflon as discussed earlier, it is extensively used for making pump valves in chemical plants, petrol or diesel tanks, etc. Teflon also finds applications in different types of clothes for using in highly acidic and alkaline environments.

Other aliphatic fluorine-containing polymers include PVF and PVDF. PVF can be represented as $-(CH_2-CHF-)_n-$. Monomer corresponding to PVF can be prepared by the reaction of HF with acetylene. Polymerization reaction can be initiated by benzoyl peroxide.

$$CH \equiv CH + HF \longrightarrow CH_2{=}CHF \xrightarrow{\text{Benzoyl peroxide}} \left[CH_2\text{-}CHF \right]_n$$

Its chemical resistance resembles that of Teflon and it is used as a glazing material over cemented surfaces in building industry. Surface coating of such materials is found to have better corrosion-resistance property than paints. Such coatings are stable below 150°C. PVDF is a polymer similar to PVF with slightly higher temperature stability (170°C). It can be represented as $-(CH_2-CF_2-)_n-$. Its resistance to chemical environment is similar to that of PVF.

31.2 FLUORINATED ETHYLENE PROPYLENE (FEP)

This polymer is similar to PTFE but softer than PTFE with a melting temperature of around 260°C. It is a copolymer of tetrafluoroethylene and hexafluoropropylene. Fluorinated ethylene propylene (FEP) is insoluble in many solvents and is prepared by free radical polymerization of tetrafluoroethylene and hexafluoropropylene in appropriate amounts. It is represented by the structural formula:

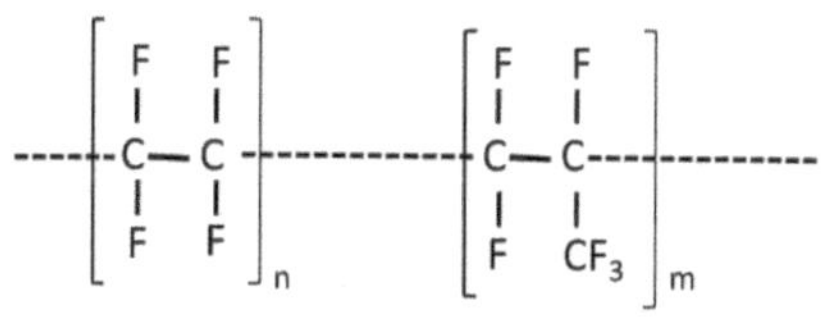

Its corrosion and chemical resistance properties are similar to that of PTFE. However, the stability of FEP is better when exposed to detergents. A comparison of the different properties of herein-mentioned polymers are given in Table 4.4.

32. CHEMICALLY RESISTANT FIBERS BASED ON AROMATIC RING-CONTAINING POLYMERS

Aromatic ring–containing polymers is another category of chemically resistant and thermally stable polymers. Examples of such polymers include polyphenylene oxide (PPO), polyphenylene sulfide

Table 4.4 Important Properties of Representative Chemically Resistant Polymers

Name of the Polymer	Melting Temperature (°C)	Glass Transition Temperature (°C)	Maximum Service Temperature (°C)	Limit of Oxygen Index (LOI) (%)
FEP	241	149	180	48
PVDF	156	100	149	44
PVF	170	100	150	23
PTFE	347	177	290	98
PVDC	171	–	115	60

(PPS), PEEK, etc. Molecular structure of these polymers is given following. It is generally observed that melting of such polymers occurs before their decomposition. They have got a wide range of temperature stability, namely 60–260°C. These materials are, however, susceptible to attack by concentrated sulfuric and nitric acids.

32.1 POLYPHENYLENE SULFIDES

PPSs are an interesting class of compounds and can be represented by the structural formula given here:

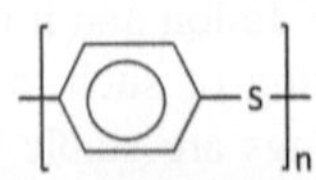

These types of polymers have excellent chemical resistance with respect to nonoxidizing acids and hot concentrated alkalis. The material has got considerable stability against hydrolysis along with high strength-to-weight ratio with easy machinability and is a potential candidate for making pressure-release valves in chemical industries, protective coating, flow meters, etc. It is a semicrystalline solid with a melting point of around 285°C [28].

However, under long-term exposure to oxidizing conditions (like conc. HNO_3), sulfur undergoes oxidation to form SO_x species. With organic solvents it shows different extent of stability depending up on the nature of solvents. PPS is prepared by the reaction para dichlorobenzene with Na_2S in the solvent *N*-methyl-2-pyrrolidone [28]:

$$n\text{ClC}_6\text{H}_4\text{Cl} + n\text{Na}_2\text{S} \rightarrow [\text{C}_6\text{H}_4\text{S}]_n + 2n\ \text{NaCl}$$

Like inorganic semiconductors, such polymers can be doped to improve their conductivity.

32.2 POLYPHENYLENE OXIDE

PPO is poly(*p*-phenylene oxide) and is represented by the structural formula given following. It is also known by the name polyphenylene ether. For the purpose of improving its machinability, it is blended with polystyrene, polyamide etc. [29].

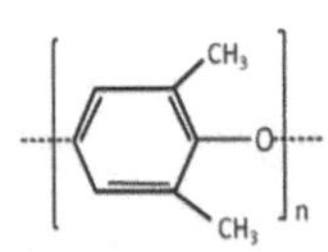

It is prepared by the polymerization of naturally occurring phenol like syringic acid in presence of enzymes like peroxidase and laccase. Such polymers have got high-heat resistance and dimensional stability and are used for making structural and automotive parts. It is also widely used as a low-temperature air separation membrane to generate nitrogen from air.

32.3 POLYETHER ETHER KETONES

PEEK is a polymer used for a variety of engineering applications. The materials can be used for the previously mentioned application due to its high thermal stability and chemical resistance at elevated temperatures. PEEK polymers are prepared by the dialkylation of bisphenol followed by step growth polymerization. It is schematically shown here:

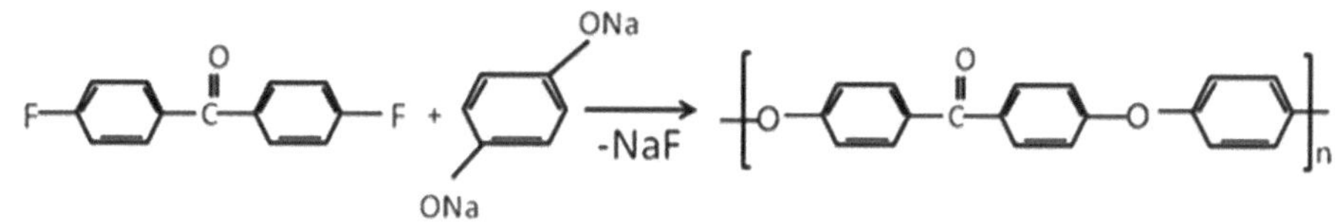

The material can withstand a temperature of around 200°C and is resistant to attack by organic and aqueous environments. It is attacked by H_2SO_4 and a fine powder of polymer can be dissolved in concentrated H_2SO_4. It is also attacked by halogenated solvents, hydrocarbons, strong Bronsted and Lewis acids, etc. It has wide application as bearings, piston parts, HPLC columns, compressor valves, etc. The material is highly resistant to biodegradation and is used as bioimplants [28,30,31].

32.4 POLY (ETHER IMIDE) FIBERS

Polyether imide is a polymer similar to that of PEEK, but with lower impact strength compared to PEEK. The family of PEI polymers is known by the commercial name Ultem resins and such materials have high solvent and thermal resistance. Their stability in alkaline environment is relatively poor. They can withstand temperatures as high as 190°C and have reasonably good mechanical properties. The material finds extensive application in hot gas filtering under corrosive chemical environments. In addition to this, such polymers have excellent resistance toward dilute acids and alkalies.

33. COMPOSITE MATERIALS

Fiber-reinforced plastics (FRPs) or fiber-reinforced polymer belongs to the category of materials produced by incorporating fibers like glass, carbon, paper, wood, etc. in a polymer matrix based on epoxy, vinyl, etc. Such composite materials are extensively employed as container materials or storage tanks in chemical plants. FRPs are currently replacing the conventionally used stainless steel, premium alloys, carbon steels, etc. They also possess additional advantages like light weight and high strength-to-weight ratios, the only negative point being the high manufacturing costs involved during their development. For example, terephthalic resin with higher extent of cross-linking is widely used for the underground storage of gasoline. Many pickling solutions contain high concentration of chloride ions and FRPs are the ideal candidate as container materials for such solutions compared to metals or alloys. FRPs are also used for constructing pipelines of pulp bleach and sewage plants, etc. in municipalities and towns. H_2S and H_2SO_4 are two chemical components present in sewage and drainage pipes. Concrete pipes are severely affected by these reagents, thereby reducing the lifetime of pipes and degrading its performance. In chloralkali industry toxic and corrosive chemicals like Cl_2 gas,

$NaOH$, $NaClO_3$, etc., are used. FRPs are the ideal materials used for developing pipelines and joints in the just mentioned industries.

33.1 TEREPHTHALIC RESINS AS STORAGE TANK MATERIALS

Resins and polymers based on terephthalic acids are prepared by the polymerization of the monomer ethylene terephthalate. Copolymerization with the polymer polyethylene leads to the formation of polyethylterephthalate (PET), which can be represented by the structural formula:

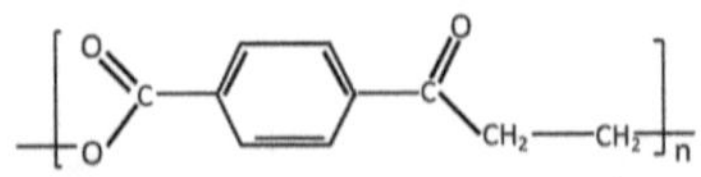

PET can be prepared from starting materials like ethylene glycol and dimethyl terephthalate ($C_6H_4(CO_2CH_3)_2$) or terephthalic acid. The reaction takes place over the temperature of 150–200°C.

33.2 CORROSION OR DEGRADATION OF PET

PET undergoes corrosion by processes like thermal oxidation, hydrolytic attack, etc. During thermal oxidation, color of the material changes. Hydrolytic attack leads to breaking of bonds and results in reduced molecular weight, formation of compounds like acetaldehyde, etc. Change in color is due to the formation of different functional groups (chromophores) upon heat treatment at higher temperatures. This prevents the use of such materials for optical applications. Copolymerization with monomers like isophthalic acid can reduce the extent of such degradations by increased cross-linking and reducing the extent of monomer formation [32].

33.3 MATERIALS FOR CONSTRUCTING PICKLING AND GALVANIZING TANKS

Pickling and galvanizing plants are very essential in metallurgical and mineral industries. Pickling plants are generally used to remove contaminants, impurities, stains, rust, etc. from materials like copper, aluminum, and precious and noble metals. Materials for such applications should be strong and must be highly resistant to acid corrosion. Materials should resist the corrosion introduced by species like Cl^-, SO_4^{2-} etc., present in the medium or environment. Materials that are generally used for such purposes are polypropylene, high-density polyethylene (HDPE), polypropylene homopolymer (PPHP), etc. Molecular structure of these polymers are given here:

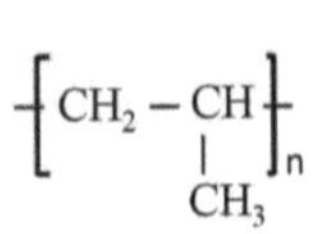	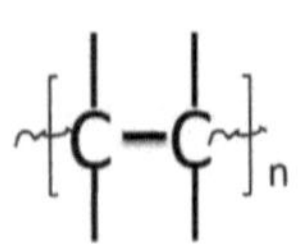	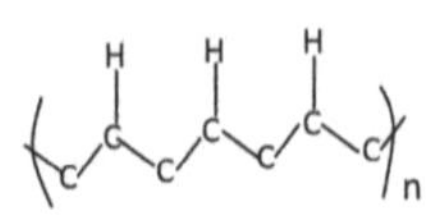
Polypropylene	high density polyethylene	Polypropylene homo-polymer

HDPE has reduced extent of branching compared to low-density polyethylene (LDPE). Little branching in HDPE leads to increased intermolecular interactions and associated increase in properties like tensile strength, rigidity, etc. Similarly, PPHP has more strength compared to the corresponding copolymer.

Table 4.5 Strength of Representative Polymer/Fibers After Exposing to 7 Days in Steam

Name of the Polymer	% Strength Retained				
Temperature	100°C	150°C	200°C	250°C	300°C
PEEK	100	100	100	95	80
PPS	100	100	90	0	Destroyed
meta-ara mid	100	95	0	0	0
PET	90	0	0	0	Destroyed

33.4 AROMATIC POLYAMIDES (ARAMIDS)

Aramids or aromatic polyamides are extremely strong materials and are prepared by the reaction of amine group with carboxylic acid halide group. A typical example is given here:

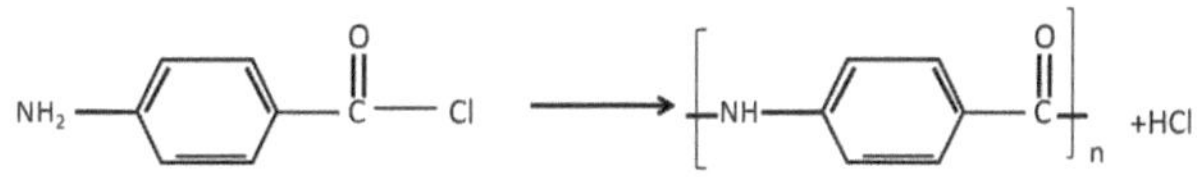

There is a wide variety of aramids that are formed by the copolymerization with different monomers. They have very interesting properties like good abrasion resistance, inert to many organic solvents, nonconducting, low flammability, etc. They are sensitive toward acids and ultraviolet radiations. Percentage strengths retained by some of the previously mentioned polymers/fibers/composites after subjecting them to aggravated test conditions like treatment with steam are given in Table 4.5.

34. CONCLUSIONS

In this chapter different types of chemically hostile environments and their influence on the material properties are discussed in detail. Brief descriptions are provided on various types of inorganic and organic materials that can withstand such environments. A variety of materials that are currently used for withstanding hostile chemical environments and their properties are also discussed in this chapter.

REFERENCES

[1] ASTM D3912–95, Standard Test Method for Chemical Resistance of Coatings and Linings for Use in Nuclear Power Plants, ASTM International, West Conshohocken, PA, 2001.

[2] ASTM D543-95, Standard Practices for Evaluating the Resistance of Plastics to Chemical Reagents, ASTM International, West Conshohocken, PA, 2001.

[3] W. Zhou, Z.L. Wang, Scanning Microscopy for Nanotechnology: Techniques and Applications, Springer Science and Business Media, New York, 2006.

[4] R. Wiesendanger, Scanning Probe Microscopy and Spectroscopy: Methods and Applications, Cambridge University Press, Cambridge, UK, 1994.

[5] R. Anthony, West, Solid State Chemistry and its Applications, John Wiley & Sons (Asia) Pvt. Ltd, Singapore, 2003.

[6] K.C. Khulbe, C.Y. Feng, T. Matsuura, Synthetic Polymer Membranes: Characterization by Atomic Force Microscopy (Chapter 5), Springer, Berlin, Heidelberg, 2008. http://dx.doi.org/10.1007/978-3-540-73994-4.

[7] P.K. Mehta, Natural Pozzolans: Supplementary Cementing Materials in Concrete, 86, CANMET Special Publication, 1987, pp. 1–33.

[8] T. Chappex, K. Scrivener, Alkali fixation of C-S-H in blended cement pastes and its relation to alkali silica reaction, Cem. Concr. Res. 42 (2012) 049.
[9] E. Lassner, W.D. Schubert, Tungsten: Properties, Chemistry, Technology of the Elements, Alloys and Chemical Compounds, Springer Science and Business Media, New York, 1999, p. 139.
[10] H. Tulhoff, "Carbides. Metal Like Carbides of Industrial Importance," Ullmann's Encyclopedia of Industrial Chemistry (Electronic Edition), Wiley-VCH Verlag, Weinheim, 2000.
[11] E. Peter, W. Lengauer, Carbides: Transition Metal Solid State Chemistry, Encyclopedia of Inorganic Chemistry, John Wiley & Sons, USA, 1994.
[12] Y.S. Park, SiC Materials and Devices, Academic Press, 1998, pp. 20–60.
[13] M.W. Pitcher, S.J. Joray, P.A. Bianconi, Smooth continuous films of stoichiometric silicon carbide from poly(methylsilyne), Adv. Mater. 16 (2004) 706.
[14] A.R. Bunsell, A. Piant, J. Mater. Sci. 41 (2006) (2006) 823.
[15] R.M. Laine, F. Babonneau, Preceramic polymer routes to silicon carbide, Chem. Mater. 5 (1993) 260.
[16] Y.S. Touloukian (Ed.), Thermophysical Properties of High Temperature Solid Materials, vol. 5, Part 1, MacMillan Co, New York, 1967, pp. 123–129.
[17] G. Leichtfried, et al., 13.5 Properties of diamond and cubic boron nitride, in: P. Beiss, et al. (Eds.), Landolt-börnstein – Group VIII Advanced Materials and Technologies: Powder Metallurgy Data. Refractory, Hard and Intermetallic Materials, 2A2, Springer, Berlin, 2002, pp. 118–139.
[18] S. Rudolph, Boron nitride (BN), Am. Ceram. Soc. Bull. 79 (2000) 50.
[19] R.B. Heslop, P.L. Robinson, Inorganic Chemistry: A Guide to Advanced Study, Elsevier Publishing Company, Amsterdam, 1967.
[20] ASTM-C-1285-02, Standard Test Methods for Determining Chemical Durability of Nuclear, Hazardous, and Mixed Waste Glasses and Multiphase Glass Ceramics: The Product Consistency Test (PCT), ASTM International, West Conshohocken, PA, 2002.
[21] N. Rajmohan, P. Frugier, S. Gin, Composition effects on synthetic glass alteration mechanisms: Part 1. Experiments, Chem. Geol. 279 (2010) 106.
[22] W.R. Zeng, S.F. Li, W.K. Chow, Preliminary studies on burning behavior of polymethylmethacrylate (PMMA), J. Fire Sci. 20 (2002) 297.
[23] M. Ezrin, Plastics Failure Guide: Cause and Prevention, Hanser Verlag, Munich, 1996, p. 168.
[24] P.A. Schweitzer, Corrosion of Polymers and Elastomers, second ed., CRC Press, Taylor and Francis Group, Boca Raton, London, New York, 2006, p. 166.
[25] V.R. Gowariker, N.V. Viswanathan, J. Sreedahr, Polymer Science, New Age International (P) Limited, Publishers, New Delhi, 2001.
[26] N.G. Mc Crum, C.P. Buckley, C.B. Bucknall, Principles of Polymer Engineering, second ed., Oxford University Press, Canada, 1998.
[27] D. ASTM, Standard Test Method for Measuring the Minimum Oxygen Concentration to Support Candle-Like Combustion of Plastics (Oxygen Index), ASTM International, West Conshohocken, PA, 2013, pp. 2863–2913.
[28] J. Bussink, D. Parker, H.T. van de Grampel, G.W. Wheatley, E.-U. Dorf, E. Ostlinning, K. Reinking, Polymers, high-temperature, in: Ullmann's Encyclopedia of Industrial Chemistry, Wiley-VCH, Weinheim, 2002.
[29] H. Uyama, R. Ikeda, S. Yaguchi, S. Kobayashi, Polymers from renewable resources, ACS Symp. Ser. 764 (2001) 113.
[30] D. Kemmish, Update on the Technology and Applications of Poly Aryl Ether Ketones, 2010, ISBN 978-1-84735-408-2. Materials properties data.
[31] M. Lauzon, Diversified Plastics Inc., PEEK Playing Role in Space Probe, Crain Communications Inc., 2012.
[32] H. Köpnick, M. Schmidt, W. Brügging, J. Rüter, W. Kaminsky, Polyesters, in: Ullmann's Encyclopedia of Industrial Chemistry, A21, Wiley-VCH, Weinheim, 2005, pp. 233–238.

CHAPTER

HIGH PERFORMANCE POLYMER NANOCOMPOSITES FOR STRUCTURAL APPLICATIONS

5

K.A. Dubey, P.A. Hassan, Y.K. Bhardwaj
Bhabha Atomic Research Centre, Mumbai, India

CHAPTER OUTLINE

Materials Under Extreme Conditions. http://dx.doi.org/10.1016/B978-0-12-801300-7.00005-X

1. INTRODUCTION

Almost all structural applications demand materials with high strength, stiffness, and toughness [1,2]. Metals, ceramics, and their composites are therefore the conventional choices for such applications. However, they are unable to meet the requirements of light weight, low cost, easy processability, and corrosion resistance. For construction, infrastructure, wind energy, automobile, and space applications, light weightiness is the foremost requirement, as it significantly affects aerodynamics, cost, and energy economy [3–7]. Polymers excel in terms of low density, easy processability, and corrosion resistance; however, their stiffness and strength are much less than that needed for structural applications. Polymer composites circumvent the inherent limitations of polymers, metals, and ceramics through a combinatorial approach, synergizing the best properties of different ingredients [6,8–11].

Nanosize particles have considerably higher surface area than their microcounterparts, allowing a profound increase in the polymer-nanoparticle interfacial area and associated interactions. Such interactions facilitate efficient load transfer and high reinforcement at much lower filler loading. High degree of reinforcement has been achieved with 0.5–2% filler loading, whereas conventional filler requires 30–70% loading to achieve similar strengths. As a result, polymer nanocomposites are envisaged to be light weight and high strength alternatives of traditional microcomposites. However, with a decrease in size, the dispersion and distribution of filler in the polymer matrix becomes difficult [12]; moreover, intrinsic properties of polymer and nanoparticles play a crucial role in limiting the micromechanics of nanocomposites [13–16]. In the following sections, properties of polymers and fillers commonly used/proposed for structural applications, the micromechanical theory of reinforcement, strategies for interface modification and nanocomposite synthesis, and some recent advances in highly reinforced polymer nanocomposites have been discussed. Synergistic interactions between fillers that allow significant enhancement in mechanical properties at lower volume fractions of nanoparticles have also been discussed.

2. NANOFILLERS

Polymer nanocomposites refer to a two-phase system in which nanofillers are dispersed in a continuous polymer phase. Synthesis and physicomechanical characteristics of some of the commonly employed nanofillers are discussed next.

2.1 CARBON NANOTUBES

Carbon nanotubes (CNTs) are allotropes of carbon, having a tubular structure with exceptional mechanical and electrical properties [17–19]. Their diameters are of the order of few nm and length can be up to several mm, translating into extremely high aspect ratio. High intrinsic strength together with high aspect ratio makes CNTs as one of the most promising material for structural applications [18–20]. CNTs can be visualized as seamlessly rolled graphene sheets. They are commonly categorized as single-walled nanotube (SWNT), double-walled, and multiwalled nanotube (MWNT). Elastic modulus and density of nanotubes are listed in Table 5.1. It can be seen that mechanical properties of CNTs are considerably higher than conventionally used structural materials. Several theoretical and indirect experimental estimates have been made for the mechanical characteristics of CNTs; however, recently mechanical properties of MWNTs have been measured directly through stress-strain measurements using an electron microscope [21–24]. The elastic modulus was reported to be as 0.27–0.95 TPa, strengths 10–52 GPa, elongation at break ~5%, and toughness of ~770 J/g. Modulus of 1 TPa and strength of 100 GPa are now commonly accepted for MWNTs. SWNTs have higher strength than MWNTs as concentric cylinders in MWNTs undergo shear-induced sliding. SWNTs have superior mechanical properties, but their dispersion in polymers is generally more difficult than the dispersion of MWNTs.

Arc discharge, laser ablation, and chemical vapor deposition are frequently used for the synthesis of nanotubes [25,26]. The arc discharge procedure used for CNT synthesis is similar to the one used for fullerene synthesis. The procedure involves the production of arc discharge between two high-purity graphite electrodes in an inert atmosphere. After arc discharge for optimum time, nanotubes get deposited on the cathode. This process generally leads to the formation of MWNTs but SWNT can also be produced if metal catalyst such as Fe, Co, Ni, Y, or Mo are used. The characteristics of nanotubes formed depend on various processing conditions such as the type of gas, catalyst used, and system and plasma configurations. In the laser ablation technique, a high-power laser is used to vaporize carbon from a graphite target at high temperature. Almost all lasers are used for the ablation though Nd; YAG and CO_2 lasers are most common. Both MWNTs and SWNTs can be produced by this technique. In order to

Table 5.1 Elastic Modulus and Density of Some Common Reinforcing Fillers

Nanofiller	Density (g/cc)	Modulus (TPa)
SWNT	1.3–1.5	1–5
MWNT	1.8–2	~1
Graphene	1.5–2.0	0.5–1
Nanocellulose	1.5–1.6	0.15
Fiber	1.35–1.55	0.03–0.08
Boron nitride (BN)	1.37	1.18
Steel	7.85	0.19–0.2
Kevlar	1.44	0.07–0.1
Carbon fiber	1.58	0.125–0.18
E glass	2.66	0.03–0.04

generate SWNTs, metal particles as catalysts must be added to the graphite targets. Carbon nanotubes produced by laser ablation are purer, crystalline, and have a narrow distribution of diameters. Chemical vapor deposition (CVD) is presently considered to be the most economically viable process for large scale synthesis. In CVD, thermal decomposition of a hydrocarbon vapor is achieved in the presence of a metal catalyst. The process is easy to control and yields high-purity nanotubes [25].

2.2 GRAPHENE

Graphene has almost all the advantages of CNTs; in addition, it has easier synthesis and lower cost. Graphene is a two-dimensional one-atom thick (<1 nm) sheet of carbon with a hexagonal conjugated structure. Due to the presence of in-plane σ-bonds and out-of-plane π-bonds, the graphene sheet has exceptionally high mechanical strength, making it the strongest material ever discovered [27,28]. Single-layered graphene exhibits Young's modulus of $\sim$1100 GPa and the tensile strength of 130 GPa [29]. Due to the presence of delocalized electrons in the completely conjugated two-dimensional network, the thermal and electrical in-plane conductivity of graphene are remarkable [30].

Graphene can be synthesized by both top-down and bottom-up approaches [30]. CVD and epitaxial growth have been extensively used to synthesize graphene; though upscalability and cost effectiveness of these processes are lower in comparison to top-down approach. The epitaxial growth of graphene has been used to grow graphene films on a single-crystal substrate under ultrahigh vacuum at high temperature [31]. Unzipping of CNTs and the arc discharge have also been explored extensively for production of graphene. In top-down approach, graphene is generated from natural graphite through mechanical peeling, exfoliation, or delamination. Hummer's process is the easiest and most effective process for the production of graphene. The process involves oxidation using $KMnO_4$ and H_2SO_4 followed by thermal expansion. Many refined versions of Hummer's processes are now available. However, all these processes involve oxidation of graphene, resulting in the formation of graphene oxide with defects in the matrix (Fig. 5.1). As a result, the electrical conductivity of the graphene is severely compromised. However, graphene oxides can be reduced to graphene using hydrazine or metal halides. Graphene intercalated compound (GIC) or expandable graphene are also available, in which graphite is intercalated using different reagents [32,33].

2.3 INORGANIC FILLERS

Inorganic nanoparticles possess unique electronic, optical, magnetic, and mechanical properties and can be synthesized by physical or chemical approaches [34,35]. Physical methods generally involve vapor deposition and mechanical milling whereas chemical routes can be hydrothermal, sonochemical, solvothermal, reverse micelles, sol–gel, flame spray pyrolysis, organometallic synthesis, microwave method, thermal evaporation, and mechanochemical synthesis. In terms of mechanical reinforcement, silica, MoS_2, boron nitride and nanoclays are of special importance. Nanoclays are generally derived from naturally occurring clays such as montmorillonite, bentonite, kaolinite, hectorite, and halloysite [36]. Among them, plate-like montmorillonite and tubular halloysite are the common nanoclays used for polymer nanocomposite synthesis. Like graphene, montmorillonite nanoclays are also nanosize sheets, having lateral dimensions varying from a few nm to several microns. Unlike graphene, though, the interlayer interaction between layers is electrostatic. Isomorphic substitution of an atom within the layer by other metal atom leads to the generation of negative charge that is counterbalanced by

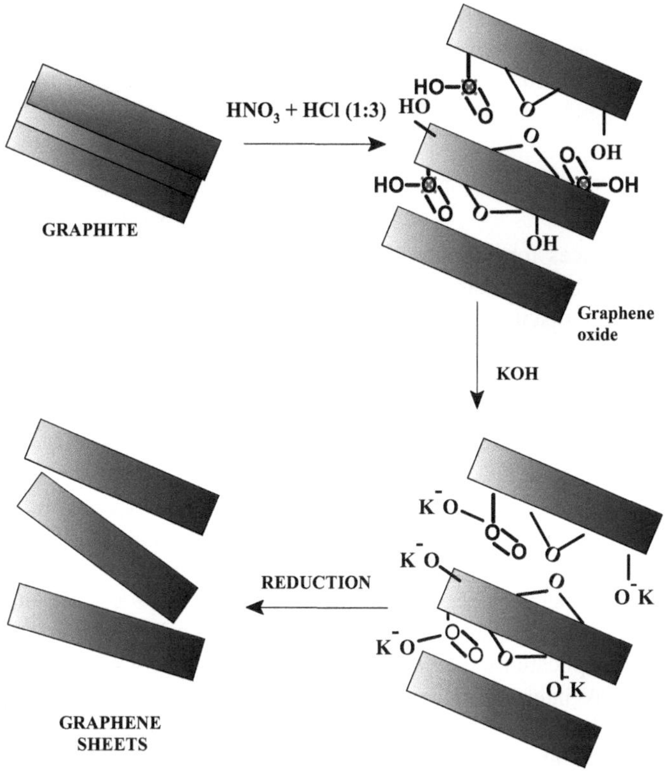

FIGURE 5.1

Preparation of graphene from graphite.

hydrated alkali or alkaline earth cations. The presence of weak forces between clay layers makes their separation possible with the action of suitable surfactant or polymer. Halloysite is also a naturally occurring aluminosilicate that is used for polymer nanocomposite synthesis. Boron nitride nanotubes (BNTs) are also promising filler for polymer reinforcement owing to their high elastic modulus comparable to CNT [37,38]. BNTs are structural analogues of CNTs where C atoms are substituted by alternating B and N atoms. MoS_2 and the related WS_2 nanotubes have also been explored and they offer a combination of strength and flexibility [39,40].

2.4 NATURAL FILLERS

Flax, sisal, cotton, coir, ramie, jute, and bamboo fibers are widely used in reinforcing polymer composites [41,42]. Their availability, good mechanical properties, easy processability, low cost, low density, and biodegradability makes them an attractive choice. However, because of their natural origin, their mechanical characteristics and density vary significantly with the source of origin. These fibers are generally of microscale diameter, and, therefore, do not show high surface area effect on the polymer reinforcement, as is shown by nanosize fillers. Nanocrystalline cellulose (NCC) is a recent development that possesses advantages because of nanoscale dimensions [43,44]. Nanocrystalline cellulose is generally synthesized by acid hydrolysis of native cellulose, and the properties of final product markedly depend upon reaction time, temperature, and acid concentration. NCCs are rigid rod-like crystals with a diameter in the range of 10–20 nm and lengths of a few hundred nanometers. NCC has high specific strength and modulus making it a promising reinforcing agent for polymers [43].

3. POLYMERS

Polymers are considered to be exciting matrices for structural applications, by virtue of their light weight, easy processability, and long-term stability. A large variety of polymers is used in the preparation of nanocomposites, some of which are discussed in the next sections.

3.1 ELASTOMERS

An elastomeric matrix offers high toughness and excellent elongation at break along with good tensile strength. Such matrices are widely used in automobile, space, wire and cable, sports, and nuclear industry. Elastomers have high molecular weight, amorphous structure, and weak inter and intermolecular forces. Their glass transition temperature (T_g) is considerable below room temperature; therefore, at room temperature segmental motions are possible, making them soft and flexible. However, reversibility of deformation demands covalent or ionic linkages between polymer chains. This can be achieved by using peroxides, metal oxides, sulfur, ionic moieties, or by radiation cross-linking. The elastomers are saturated or unsaturated and are used for a variety of applications such as tyres, conveyor belts, vibration dampeners, and insulating structures. Due to weak intermolecular bonding, elastomers need further reinforcement even after cross-linking to attain desired mechanical properties. Carbon black, fumed silica, clay, talc, and mica are commonly used fillers for elastomer reinforcement. However, conventional fillers demand high loading, resulting in increased weight and hysteresis of the composite. High filler loading also makes the processability difficult. Therefore, there has been considerable interest in using nanofillers for their reinforcement.

3.2 THERMOPLASTICS

Thermoplastics have unique virtue of processability. These polymers become soft or melt over a specified temperature range and solidify again on cooling. Unlike elastomers, these polymers do not require cross-linking due to the presence of crystalline domains or polar intermolecular interactions. The presence of crystalline and amorphous domains allows a wide range of thermal and mechanical properties, opacity, and permeability. The T_g of thermoplastic can be above or below the room temperature. These polymers can be extruded, injection molded, compression molded, and transfer molded reversibly. Nanoparticle additions have been reported to significantly improve properties like flame retardancy, durability, and elastic modulus of thermoplastics such as acrylic ABS, nylon, polybenzimidazole, polyethylene, polystyrene, polyvinyl chloride, and Teflon.

3.3 THERMOSETS

Thermosetting polymers are highly cross-linked polymers and cannot be recycled or reprocessed. Prior to cross-linking they can be monomers/oligomer (liquid) or molten polymer filled in a mold of predefined shape and size, where in situ cross-linking takes place. Using suitable initiator, a sensitizer or radiation, monomers/oligomer (liquid)/molten polymer transform into a highly rigid three-dimensional structure. Epoxy, polyurethane, and acrylate resins are widely used as thermosetting polymers. They contain unsaturation or highly strained groups that can be cross-linked at room temperature or at high temperature by aromatic/aliphatic amines or anhydrides or by high energy or UV radiation. Generally, high-temperature cured thermosets have superior physicomechanical properties than room temperature cured systems. A variety of fillers are used to improve modulus of thermosetting polymers. Glass fiber, carbon fiber, and natural fibers are the traditional choices and several commercial applications have been developed using them. Recently there has been a considerable interest in using nanofillers as reinforcing agents for thermosets. Due to their high strength and flexibility, thin yet strong structural objects can be made by using suitable thermoset-filler compositions. However, unlike thermoplastic nanocomposites, thermoset nanocomposites cannot be recycled and are less chemically resistant.

4. PREPARATION OF POLYMER NANOCOMPOSITES

Homogeneous dispersion and distribution of filler in the polymer matrix are essential to realize the reinforcing effect of the filler. However, as the size of the particle reduces from micro- to nanodomain, the surface energy increases many folds, leading to agglomeration of nanoparticles. This demands energy intensive procedures to achieve desired distribution or exfoliation. Solvent mixing, melt mixing, and in situ polymerization are some of the most common routes for nanocomposite synthesis.

4.1 SOLVENT MIXING

In this approach, fillers are dispersed in a polymer solution and composites are formed by evaporation or precipitation or phase inversion. As mentioned earlier, nanofiller tends to agglomerate due to high surface energy. Sonication is generally done to break these aggregates. Sonication-assisted dispersion has been ascribed to the cavitation induced by ultrasonic waves, which can increase localized temperature up to 5000 K and pressure up to 500 atm [45]. This route is simple and provides good molecular-level mixing but process efficiency is low as polymer solubility is limited. Moreover, in the case of nonaqueous dispersions, the cost, regeneration, and efficiency of the mixing process are also to be considered.

4.2 EMULSION (LATEX) MIXING

This approach is more environmental friendly and can be used for polymers that are difficult to dissolve. Polymers are emulsified using suitable surfactant and solvent followed by addition of nanoparticles to the emulsion. Films are cast in a mold followed by evaporation of the solvent, leading to the formation of polymer embedded with nanoparticles as a film.

4.3 MELT MIXING

This is a solvent-free approach in which fillers are incorporated into the molten polymer and can be effectively used for thermoplastics and elastomers. Good distributive and dispersive mixing can be achieved due to the high shear force generated during melt compounding. This shear force breaks down the filler agglomerates. Extrusion and internal compounding are frequently used in melt compounding. Industrial-scale applicability of such processes is quite good. However, the processing gets difficult with the increase in filler loading as melt viscosity increases significantly.

4.4 IN SITU POLYMERIZATION

Polymer melts or solutions have high viscosity; therefore dispersion of fillers demands high agitation to get uniformly distributed in the polymer matrix. This problem can be circumvented to a great extent if fillers are dispersed in the monomer and then polymerized. High filler loading can be achieved through this route. For thermosetting polymers, this is the only approach of nanocomposites synthesis, as thermosets cannot be melted or solubilized in the solvent.

4.5 IN SITU NANOPARTICLE SYNTHESIS

In this approach, nanoparticles are synthesized in a polymer solution. As a result, the distribution of nanoparticles in the polymer matrix is more homogeneous. This approach can be applied to nanoparticles synthesized through sol–gel or solvent-based routes. This leads to the formation of interpenetrating networks of inorganic-organic phases. Sol–gel hydrolysis and condensation of a precursor, such as tetraethoxysilane, tetra butyl titanate, or aluminum isopropoxide can be carried out.

4.6 EXFOLIATION

Exfoliation is important for clays and graphene, as their precursors have layered structures held together through ionic and van der Waals forces, respectively [1,46]. Exfoliation involves separation of platelets from one another. In the case of clay-based polymer composites, three types of morphologies, viz. microcomposites, intercalated and exfoliated are reported. These morphologies depend on the extent of integrally expansion. It is assumed that intercalation is a prestep for exfoliation. Since the process of intercalation by polymer chain involves a significant decrease in entropy, it is assumed that intercalation is a result of strong interactions between polymer chains and clay. However, layer–layer delamination has also been reported under high shear conditions. Graphene, on the other hand, demands a completely different mechanism for exfoliation and dispersion as graphite does not bear any net charge between layers [47]. Nonetheless, it can easily be intercalated by various molecules, metal complexes, and salts. If exfoliated graphene (EG)/GIC are exposed to high temperature, the intercalated compounds decompose into gaseous products, leading to the development of sudden pressure in the *c*-axis direction. This rapid pressure shock, results in expansion of graphite. The expansion potential depends on the intercalation compound, graphite particle size, and nature of the thermal shock.

5. CHARACTERIZATION OF NANOCOMPOSITES

5.1 MORPHOLOGICAL CHARACTERIZATION

Analysis of the microstructure of nanocomposite is an important aspect to arrive at the structure-property relationship. Various techniques have been widely used in the characterization of the composites. Commonly used techniques are electron microscopy, small-angle X-ray scattering (SAXS), and X-ray diffraction (XRD).

Scanning electron microscopy (SEM) provides information about the surface features of composites. For SEM, the surface of the material should be electrically conductive to prevent charge accumulation on the surface. This necessitates coating of polymeric material with an ultrathin conductive material. SEM in combination with energy dispersive spectroscopy (EDS) can be used to identify the degree of dispersion of nanoparticles in composites. Direct observation of fractured surfaces by SEM is a valuable tool to understand the dispersion characteristics. Recently, scanning probe microscopic techniques such as scanning tunneling microscopy (STM), atomic force microscopy (AFM), etc. have also become important tools for nanostructure imaging. In these techniques, the interaction between a sharp tip and surface of the specimen is monitored for surface profiling. This method can provide surface structural and electronic information, at atomic-scale resolution (in some cases). AFM can be operated either in contact or tapping mode (noncontact). In tapping mode, the cantilever vibrates above the sample surface at a certain frequency and the amplitude and phase of oscillation can be used to reconstruct the image. This way it is possible to reduce the shear effects associated with tip movement. The phase images can be used to differentiate the electronic properties of the sample surface and hence generate contrast between the polymer and nanoparticles. Transmission electron microscopy (TEM) allows direct observation of internal structure, the spatial distribution of different phases, defects and crystal structure of nanoparticles, etc. at very high resolution. However, this requires necessary skill to take images at different magnifications and at different places to obtain a representative and statistically averaged information.

Due to ease of availability of wide-angle, XRD instrument it is commonly used for nanocomposite characterization. It can provide information about internal structure of the composites and even kinetics of intercalation can be investigated. For layered nanocomposites, the peak corresponding to platelet separation can be clearly seen in the XRD pattern and the peak position shifts to lower angles as *d*-spacing increases. To characterize structures larger than 10 Å, small-angle X-ray scattering is employed, which measures the scattering intensity in the range of 0.02 to 5 degrees. The scattering intensity gives information about the variation in the electron density of the specimen, which in turn depends on the nanofiller shape, polymer conformation, and interactions.

In the case of nanocomposites, one is usually interested in the microstructure of the filler itself. For a dilute dispersion of fillers, i.e., the typical interaction distances are smaller than the interparticle distance between nanoparticles, the absolute scattering intensity [$I(q)$] can be written as [48]:

$$I(q) = \phi \cdot (\Delta\rho)^2 \cdot V \cdot P(q) \tag{5.1}$$

where ϕ is the volume fraction of filler, $\Delta\rho$ is the electron density difference between filler and polymer, V is the geometrical volume of nanoparticle, and $P(q)$ is known as the particle form factor, which depends on the size and shape of the nanofiller. Here, q is the magnitude of scattering vector given by $q = (4\pi/\lambda)\sin(\theta/2)$, λ being the wavelength of X-rays and θ the scattering angle.

For a spherical particle of radius R, $P(q)$ is given by

$$P(q) = \left[3 \cdot \frac{\sin(qR) - qR\cos(qR)}{(qR)^3}\right]^2 \tag{5.2}$$

For a thin rod of length L, the form factor can be given as

$$P(q) = \frac{2}{x}\int_0^x \frac{\sin v}{v} dv - \left(\frac{\sin(x/2)}{x/2}\right)^2 \tag{5.3}$$

where $x = qL$.

For a randomly coiled macromolecule, the factor will be

$$P(q) = \frac{2}{u^2}\left(e^{-u} + u - 1\right) \tag{5.4}$$

where $u = (qR_g)^2$ and R_g is the radius of gyration. Thus by fitting the experimentally observed $I(q)$ with appropriate equations, one can get information about the geometry and dimension of the particle. It has been shown by Guinier that in the small-angle limit, i.e., for $qR_g << 1$, the radius of gyration of the object can be obtained, irrespective of its shape, using the following relation:

$$P(q)_{q \to o} = 1 - \frac{q^2 R_g^2}{3} \tag{5.5}$$

SAXS can be more informative and quantitative; however, this has not been used to the extent XRD has been used, due to a limited number of instruments and experience in interpretation.

5.2 MECHANICAL PROPERTIES

Mechanical properties of polymer micro-, meso-, or nanocomposites mainly depend on the intrinsic characteristics of the polymer matrix and filler, the aspect ratio of filler, orientation, and packing density [13,14,16,19,20,24]. Polymer filler interface also plays a critical role in determining mechanical properties of the composites [49]. Due to the additive nature of the properties, several models have been proposed to predict mechanical properties of polymer composites. Most of the work, however, is confined to fiber-reinforced composites, and the reinforcement theory for nanofillers is not yet fully established. Still, it is proposed that established micromechanical models can be applied to nanocomposites with a great deal of success. Among several models, the rule of mixture model, the Halpin–Tsai model and the Neilsen's model are most frequently used [50,51].

Elastic modulus (E_c) of the polymer composites in the simplest form can be described using following equations.

$$E_c = E_m V_m + E_f V_f \tag{5.6}$$

E_m and E_f are the moduli of the matrix and filler, respectively, and V_m and V_f are their respective volume fractions. It is assumed that the filler distribution is isotropic and the filler covers the full length of the matrix. It is also assumed that the bonding between filler and polymer matrix is strong and under applied stress, and filler and polymer are equally strained. However, in practice, most of the composites depart from these ideal conditions, and, therefore, different efficiency factors have been incorporated into the previous equation. If reinforcement length and orientation efficiency factors are considered, the equation is modified as:

$$E_c = E_m V_m + E_f \eta_l \eta_o V_f \tag{5.7}$$

where η_l, η_o are length (for asymmetric fillers) and orientation efficiency factors. The length efficiency factor (η_l) can be defined using Cox's shear lag model as follows

$$\eta_l = 1 - \left[\frac{\text{Tanh}(\alpha \cdot l/d)}{\alpha \cdot l/d}\right] \tag{5.8}$$

where α is defined as:

$$\alpha = \sqrt{\frac{-3E_m}{2E_f \ln V_f}} \tag{5.9}$$

Another important concept in determining length efficiency factor is the critical length of the filler, which is essential for the stress transfer from the matrix to the filler. Stress transfer is expected to increase as the length-to-diameter ratio (*l/d*) of the filler increases since the surface area is expected to increase with an increase in the length of a fixed diameter. It is, therefore, possible that at a particular length, transferred stress would be higher than the tensile strength of the filler and may lead to the breakage of the filler. This length is generally defined as a critical length (l_c) for a matrix-filler system. Subcritical length will lead to inefficient stress transfer and reinforcement and length efficiency factor approaches 1 as *l/d* increases [52].

Orientation efficiency factor (η_o) is generally considered to be 1 for completely aligned fillers, 3/8 for fillers with planer alignment, and 1/5 for randomly oriented filler. It can be calculated using the following relation:

$$\eta_o = \sum_n \alpha_n \cos^4 \theta_n \tag{5.10}$$

where θ_n is the angle with respect to the axis of applied stress. The orientation distribution of anisotropic fillers is determined by the flow pattern and shear forces developed during processing. The anisotropic orientation of asymmetric nanoparticles is observed in extrusion, spinning, and injection molding, and even in compression molding [49].

A porosity and filler area (dimension) correction factor can also be introduced into the previous equation to take care of porosity generated during composite formation and to take into consideration the fact that dimension of all fillers is not equal. The modified equation can be presented as:

$$E_c = [E_m V_m + E_f \eta_l \eta_o k V_f] \cdot (1 - V_p)^2 \tag{5.11}$$

where k and V_p are dimensional and porosity factors respectively.

Halpin–Tsai model predicts modulus of polymer composites with oriented composites using the following relation [50]:

$$E_c = E_m \frac{[1 + (2l/d)\eta V_f]}{1 - \eta V_f} \tag{5.12}$$

where η is defined as

$$\eta = \frac{\frac{E_f}{E_m} - 1}{\frac{E_f}{E_m} + 1} \tag{5.13}$$

For random orientation, the efficiency factors η_l and η_t are introduced. The model can be further refined as:

$$E_c = E_m\left[\frac{3}{8}\cdot\frac{1+(2l/d)\eta_l V_f}{1-\eta_l V_f}+\frac{5}{8}\cdot\frac{1+2\eta_t V_f}{1-\eta_t V_f}\right] \tag{5.14}$$

where

$$\eta_l = \frac{\frac{E_f}{E_m}-1}{\frac{E_f}{E_m}+2\frac{l}{d}} \tag{5.15}$$

and

$$\eta_t = \frac{\frac{E_f}{E_m}-1}{\frac{E_f}{E_m}+2} \tag{5.16}$$

These models consider aspect ratio as well as filler volume fractions and predict almost linear increase in the modulus whereas some models such as Guth have power functions [53]. In case of symmetrical particles Nielsen's model is generally employed; it can be described as

$$\frac{E_c}{E_m} = \frac{1+ABV_f}{1-\psi BV_f} \tag{5.17}$$

$$A = k_E - 1 \tag{5.18}$$

$$B = \frac{\frac{E_f}{E_m}-1}{\frac{E_f}{E_m}+A} \tag{5.19}$$

$$\psi \cong 1+\frac{1-\phi_m}{\phi_m^2}V_f \tag{5.20}$$

where E_f, E_c, E_m are the elastic modulus of the filler, polymer composite, and polymer matrix respectively, and k_E is the Einstein coefficient. The constant A takes care of the aspect ratio and orientation of the filler, and the factor B addresses the relative difference in the modulus of the two components. ϕ_m is related to the packing density of the filler and ψ is associated with the filler volume fraction. A modified ψ equation (Eq. 5.21) has also been used quite frequently.

$$\psi = 1+\frac{V_m}{\phi_m}[\phi_m V_f + (1-\phi_m)V_m] \tag{5.21}$$

6. TAILORING MORPHOLOGY AND INTERFACE OF POLYMER NANOCOMPOSITES

6.1 INTERFACE MODIFICATION

The polymer-filler interface plays an important role in deciding the mechanical properties of polymer composites. Both agglomerations of filler and efficient stress transfer from the matrix to filler depend on interfacial characteristics. As the particle size decreases, the tendency of agglomeration increases and the extent of agglomeration depends on the forces between filler particles and the forces between polymer and filler surfaces. These agglomerated domains can act as a crack propagation site and lead to the failure under stress. Electrostatic interactions, hydrogen bonding, van der Waals interactions, mechanical interlocking, and geometrical synergies play important roles in achieving good dispersion of the filler in the polymer matrix and in efficient stress transfer. If no attempts are made to modify the filler or polymer surface to improve, the wetting characteristics of the interface, the dispersion of the filler will depend mainly on shear and elongation forces exerted during mixing. The adhesive forces and shear forces can be correlated using equation

$$\frac{F^a}{F^s} = k\frac{W_{ab}}{\eta\gamma R} \tag{5.22}$$

where F^a and F^s are adhesive and shear forces, respectively. W_{ab} is work of adhesion, R is effective radius, γ is the shear rate, and η is the melt viscosity [47]. It is evident from the equation that high shear rate can easily improve the dispersion of a filler. However, there are good chances that higher shear might induce degradation of the polymer matrix; therefore, shear stress cannot be increased beyond a certain limit. Use of surfactants can reduce W_{ab}. The increasing radius can improve dispersion but it will decrease reinforcement as predicted by aforementioned micromechanical theories. In cases where interfacial interactions are poor, it is possible that filler will be pulled out of the polymer matrix and the shear strength of the interface takes predominance in determining mechanical properties. Shear strength can be expressed using the following relation [49,52,54]:

$$\sigma_c = \left(1 + \frac{2b}{d}\right)[(\sigma_s l/d) - (1 + 2b/d)\sigma_m]V_f + \sigma_m \tag{5.23}$$

where σ_c is the composite strength, b is the thickness of interfacial region, l is length, d is the diameter, V_f is the volume fraction of the filler, σ_s is the shear strength of the interface, and σ_m is the strength of the matrix. It is evident from the previous relation that interface characteristic plays a crucial role in determining mechanical properties of the composites. The common methodology for interface modification includes covalent and noncovalent functionalization of the filler surface. The covalent approach involves attachment of a functional group or grafting a suitable polymer. Noncovalent approaches involve the use of surfactants or high shear processing conditions. Treatment with surfactant significantly reduces the surface energy of the filler due to adsorption. Selection of the suitable surfactant and optimum quantity is, however, essential to improve filler dispersion in a polymer matrix. The noncovalent modification is easier, though it leads to a reduction in the yield strength. Another important factor that might influence mechanical properties of the interface is the crystallization kinetics of the polymer. Interfacial modification improves the stiffness of interface, but it might also induce imperfect crystallization. Therefore, the optimum degree of functionalization is essential to get the desired reinforcement. Some of the functionalization approaches commonly used to modify nanofillers are described here.

6.2 CNT MODIFICATION

Carbon nanotubes have a high aspect ratio (~1000) and exceptionally high mechanical properties (modulus); therefore, there has been considerable work on exploring their applications for structural applications. Even though the surface of CNTs is hydrophobic, they show poor dispersion even in nonpolar polymers. This is attributed to the high surface energy and strong van der Waals interactions between nanotubes. As a result, CNTs form good dispersion only in a limited number of solvents such as dimethylformamide (DMF), tetrahydrofuran (THF), and *N*-Methylpyrrolidone (NMP) [47]. Nanotubes can be functionalized through covalent and noncovalent routes. Noncovalent routes are preferred if electrical applications are of interest whereas covalent route is preferred if high mechanical reinforcement is needed.

6.3 NONCOVALENT FUNCTIONALIZATION

Noncovalent functionalization of nanotube surface can be achieved by surfactants or by polymer wrapping. A surfactant with suitable hydrophilic and hydrophobic segments can be used for partial or complete wetting of nanotubes surface. Nanotubes surface can also be modified by wrapping in a suitable polymer. CNTs have been dispersed using cationic, anionic, and neutral surfactants [55]. The quality of dispersion depends on the alkyl chain length and the function group present in the surfactant. A variety of common ionic surfactants such as sodium dodecylbenzenesulfonate (SDBS), sodium dodecyl sulfonate (SDS), tetradecyltrimethylammonium bromide (TTAB), sodium cholate (SC), and lauryl dodecyl sulfonate (LDS) have been explored for these applications. Surfactant molecules get absorbed onto nanotubes surface through hydrophobic segments, and in the case of ionic surfactants, ion gets dissociated from the head groups and acts as a counterion, forming an electric double layer as shown in Fig. 5.2 [47]. Zeta potential associated with this electric double layer depends on the magnitude and sign of the charge and decides the quality of dispersion. The dispersion quality of some commonly used surfactant-CNT dispersions varies as SDS > LDS > SDBS > TTAB > SC. Nonionic surfactants especially Triton X-100, Tween-80, Tween-60, and Tween-20 were also found to be effective in dispersing nanotubes. However, since these surfactants are nonionic, the mechanism of surface modification and disaggregation is attributed to entropic factors and is found to be dependent on the molecular weight of the surfactant. As the higher molecular weight affects diffusion of surfactant molecule within nanotubes aggregates, it can also adversely affect dispersion. The presence of an unsaturated carbon bond or heteroatom is also expected to further facilitate the dispersion, as π-stacking is possible between nanotubes surface and functional groups with unsaturation, a benzene ring, or hetero atom [47,49,55].

Wrapping of nanotubes with a suitable polymer is another way to improve CNT dispersion [56]. In this approach, CNT-polymer forms a supramolecular complex. SWNTs have been modified by encapsulating them within cross-linked amphiphilic poly(styrene)-block-poly(acrylic acid) copolymer micelles. The mechanism proposed for better dispersion by polymer-wrapped nanotubes is entropic repulsion between polymer segments of polymer-decorated tubes. The reversible wrapping of SWNTs with linear polymers such as polyvinyl pyrrolidone (PVP) was found to be thermodynamically driven by the elimination of a hydrophobic interface between the tubes and the aqueous medium [57]. The diameter of CNTs is generally of the order of nm whereas length can be up to a few micrometers; therefore considerable optimization is needed to find out a suitable surfactant and critical micelle concentration and filler loading should be carefully considered to achieve homogeneous and stable dispersion.

6.4 COVALENT FUNCTIONALIZATION

Though noncovalent modification has advantages in context of electrical applications, its load transfer efficacy is quite low due to poor interfacial characteristics. Covalent functionalization involves a

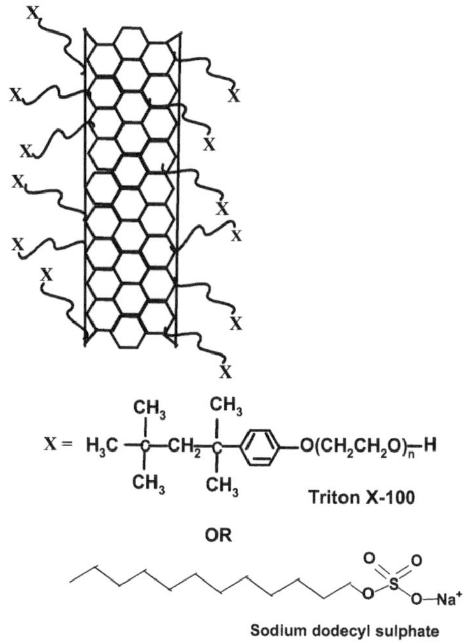

FIGURE 5.2

Schematic representation of surfactant adsorbed on CNT surface.

chemical reaction on CNTs surface, and, therefore, hybridization of the C-atom changes from sp^2 to sp^3, adversely affecting the electrical conductivity. During CNT synthesis and purification, several functional groups and defects are generated on CNT surface. In covalent functionalization, these groups play a critical role, as a defect-free CNT is too resistant to any chemical reaction. Hydroxyl and carboxyl groups generated during synthesis or purification (Fig. 5.3) can be converted to different functionality by esterification, amidization, nucleophilic coupling reactions, cycloaddition, and

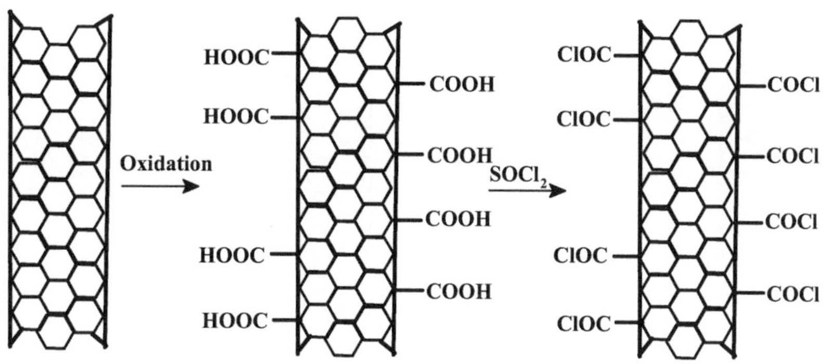

FIGURE 5.3

Oxidation of CNT and conversion of carboxylic acids into acid chlorides.

different sonochemical reactions [12,29,55]. Fluorination of CNTs is also used quite often. Fluorinated CNTs can be subsequently treated with suitable Grignard reagent or alkyl lithium compounds to get different functionalities [55].

Polymer chains can also be coupled with CNT surfaces using "grafting-to" and "grafting-from" approaches. In grafting-to approach, a polymer with functional end group is attached to the nanotube surface through amidization, esterification, or radical coupling. It is a two-step approach; the first step involves the synthesis of a polymer with specific molecular weight and reactive groups or radicals. In the next step, the polymer chain is attached to the surface of nanotubes by conventional reactions. Many polymers have been successfully attached to CNTs by this method; however, it is difficult to achieve good grafting extent, as the reactivity of functional groups is limited and polymer chains possess stearic hindrance. On the other hand, in grafting-from approach, the polymer is generated through in situ polymerization in the presence of CNTs with a reactive functional group. The "grafting-from" approach, therefore, involves the growth of polymer chain from the CNT surface. The polymerization is initiated by the chemical species immobilized on the CNT sidewalls or edges; it is therefore also termed as surface-initiated polymerization. Grafting-from, therefore, is more efficient and versatile. Atom transfer radical polymerization (ATRP), reversible addition−fragmentation chain transfer (RAFT), ring opening polymerization (ROP), free radical polymerization, cationic/anionic polymerization, condensation polymerization, reduction/oxidation polymerization, and metallocene catalysis polymerization can be used for "grafting-from" approach. Electrochemical grafting, nitroxide-mediated radical polymerization have also been used to graft polymer chains onto CNTs surface [55]. Fig. 5.4 shows radiation induced grafting of styrene on CNT, which can be used as such or the grafted styrene ring can be further sulfonated by chemical treatment.

6.5 GRAPHENE MODIFICATION

Graphene functionalization can also be done through noncovalent and covalent routes and also strongly depends on the surface oxidation and defects generated during synthesis and purification. The number of graphene layers stacked together also varies from process to process. The most commonly used procedure of graphene synthesis is Hummer's process in which graphene oxide is produced first and then reduced to graphene. Graphene oxide contains hydroxyl, epoxide, carbonyl, and carboxyl groups; therefore, it shows good dispersion in an aqueous medium (Fig. 5.5) [47]. Graphene,

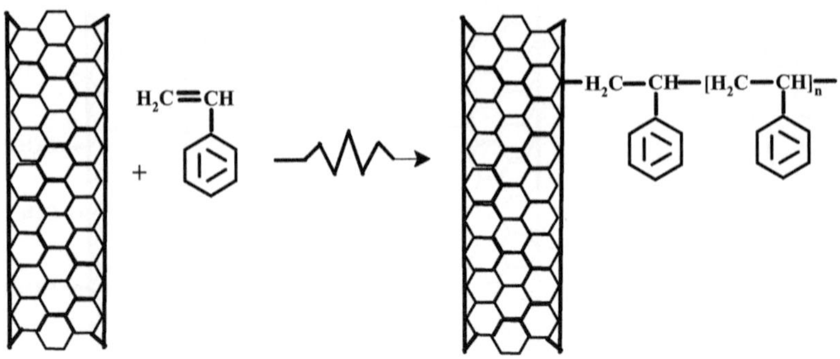

FIGURE 5.4

Radiation-induced grafting of styrene on CNT.

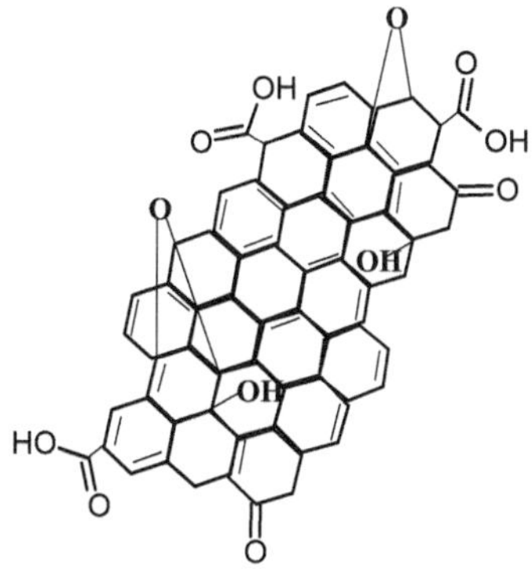

FIGURE 5.5

Different oxy group generated in graphene oxide.

depending on the exfoliation procedure, can be dispersed with the aid of a range of surfactants. Surfactants are especially useful in liquid phase exfoliation of graphene, enabling exfoliation as well as stabilization of graphene platelets. Graphene can be homogeneously dispersed using SDS, SDBS, LDS, cetyl trimethylammonium bromide, TTAB, SC, sodium deoxycholate, sodium taurodeoxycholate, IGEPAL CO-890, Triton X-100, Tween-20, and Tween-80. In a recent study, acetone has been used to modify the graphene surface. Graphene was sonicated in acetone, and it has been proposed that compounds with keto group (O=C) tend to adsorb and react with defective sites at GO sheet [58]. Liu et al. exploited π-π interactions and modified graphene surface by poly(*N*-isopropyl acrylamide) (PNIPAAm) using pyrene-terminated PNIPAAm in water [59]. Different organic amines, porphyrins, poly-L-lysine (water), PVA, alkyl lithium reagents, isocyanates, and diisocyanate compounds have been used for graphene exfoliation. Electrochemical synthesis and functionalization of graphene have also been reported using ionic liquids. In addition, through cycloaddition of nitrenes onto −C=C− double bonds, a number of different organic compounds, such as azido phenylalanine and azido trimethyl silane, have been attached to graphene [30].

6.6 INORGANIC/METALLIC NANOPARTICLE MODIFICATION

Inorganic nanoparticles can be synthesized by chemical or physical routes, and their surface characteristics strongly depend on the synthesis route [10]. Silane coupling agents are the most frequently used surface modifiers for inorganic nanoparticles. The mechanism of functionalization involves hydrolysis of silane followed by adhesion at the surface of nanoparticles through its hydrophilic end (Fig. 5.6). The silanes with different functional groups have been extensively used for modification of a wide variety of silicates, aluminates, and titanates. The surface of nanoparticles may also be modified using metal alkoxides, epoxides, and isocyanates. Grafting-to and grafting-from approaches have also been used to modify the surface of inorganic nanoparticles. The in situ polymerization is also useful in disaggregation and dispersion of inorganic filler particles, as the monomer being a smaller molecule can easily occupy the space within agglomerates and on polymerization size expansion takes place leading to the disaggregation of agglomerates [60–62]. Fig. 5.7 shows agglomeration of unmodified and dispersal of surface modified nanoparticles in parent polymer matrix.

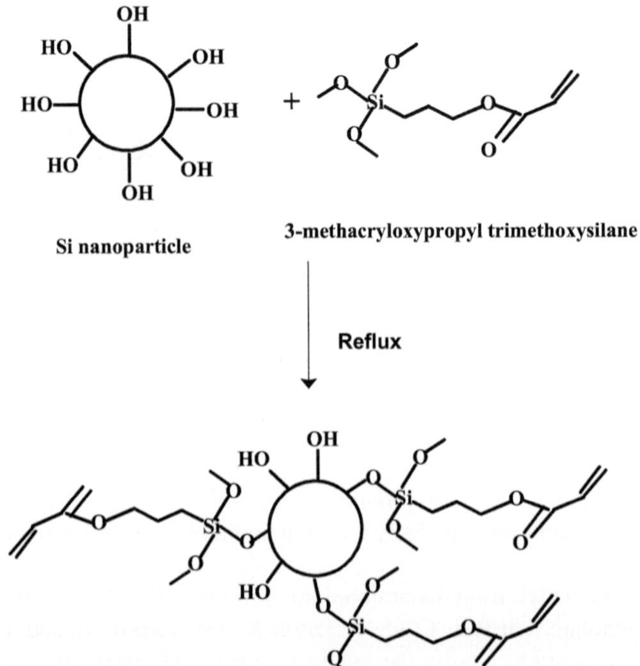

FIGURE 5.6

Modification of Si nanoparticles to make its surface hydrophobic.

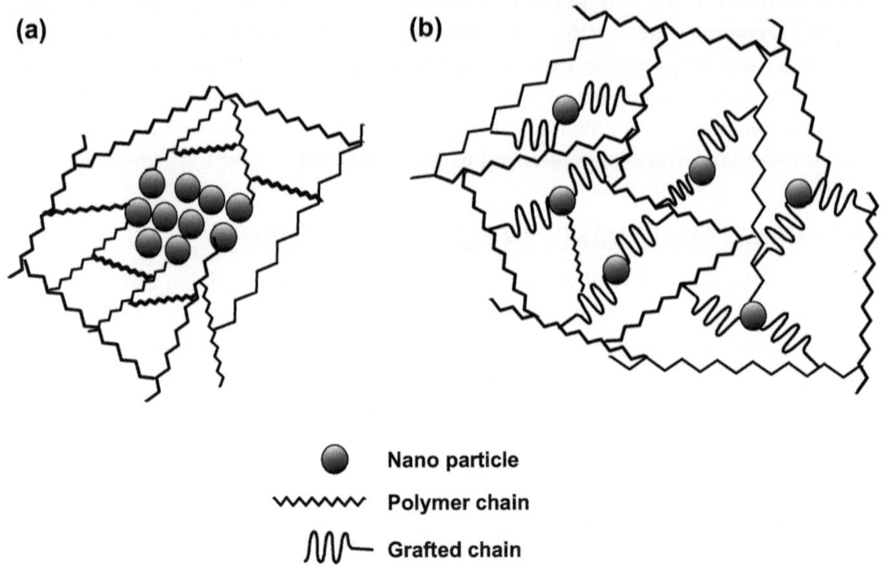

FIGURE 5.7

(a) Agglomeration of nanoparticles and (b) dispersion of grafted nanoparticles in polymer matrix.

6.7 NATURAL FILLERS

Most of the natural fillers, mainly fibers, are derived from plant sources and are hydrophilic therefore show poor compatibility with hydrophobic polymers. They also have a tendency to absorb moisture, which adversely affects the mechanical properties of the fiber as well as interfacial interaction. Both physical and chemical modifications are used in improving surface characteristics of natural fillers. The physical modification involves plasma, heat, or mechanical treatment to remove the impurities from the surface and vary the roughness of the filler surface. Chemical treatment using silanes and isocyanates can be used to modify the natural fiber filler surface. However, it is often argued that if fiber length is greater than the critical length, surface modification is not essential, as it might add to the process cost and decrease the mechanical characteristics of the filler [52].

7. HIGH-MODULUS POLYMER NANOCOMPOSITES: RECENT DEVELOPMENTS

7.1 NANOCARBON/POLYMER COMPOSITES

Nanocarbon black (NCB) has been extensively used for the reinforcement of polymers for the last several decades and is perhaps the first nanomaterial to be used for polymer reinforcement; whereas, graphene, carbon nanotubes, and nanodiamonds are the recent entrants. Following are some of the interesting cases wherein high-reinforcing effects are demonstrated by nanocarbon in elastomeric, thermoplastic, and thermosetting matrices.

7.1.1 Elastomeric Composites

Reinforcement of elastomers is needed for almost all applications, and the amount of filler needed to achieve desired mechanical properties is very high. Nanofillers, therefore, are of great importance as they are envisaged to yield high reinforcement at relatively lower volume fractions. In addition to the reinforcement by filler, elastomers need a significant amount of cross-link density to achieve desired physicomechanical characteristics. Dubey et al. have investigated a number of multiphasic elastomers matrices cross-linked by high-energy radiation. It was demonstrated that MWNT reinforces the system considerably and in some cases acts synergistically with radiation-induced cross-linking, translating in a multifold increase in elastic modulus and other properties of interest. Fig. 5.8A–D shows Charlesby–Pinner plot, mechanical modeling, dynamic mechanical analysis, and thermal analysis of SBR/EPDM/MWNT nanocomposites [63]. A significant increase in the Young's modulus was observed as MWNT content increased from 1 to 3 wt%. This behavior was attributed to the strong mechanical interlocking among nanotubes. The experimental results were compared using many micromechanical models. Additive law, Guth model, and Halpin-Tsai (HT) equations are used to predict the bulk properties of the carbon nanotube composites. The MWNT diameter, length, and Young's modulus of 20 nm, 50 μm, and 400 GPa were used for reinforcing efficiency calculations. Interestingly, with an increase in MWNT content the reinforcing efficiency, which was defined as dE_c/dV_f, increased considerably (Table 5.2). It was also reported that with the addition of MWNT, both gel fraction and cross-link density increase. The cross-linking density followed the order $SENT_{00} < SENT_{05} < SENT_1 < SENT_3 < SENT_5$ [subscript indicates % MWNT in SBR: EPDM (50:50) blend], in the entire dose range studied, indicating the predominantly cross-linking behavior of nanocomposites on irradiation and enhancement in the cross-linking density with the incorporation of

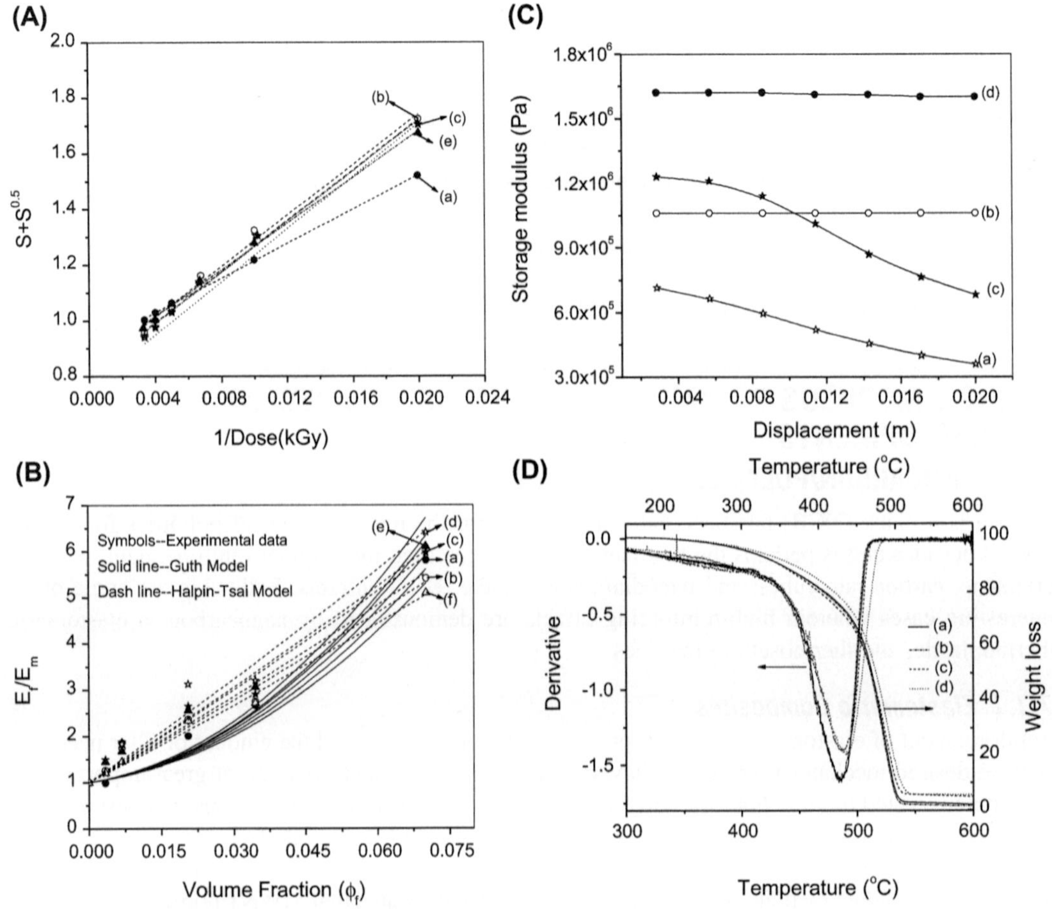

FIGURE 5.8

(A) Charlesby–Pinner plot for different nanocompositions (a) ES_{NT00} (b) ES_{NT05}(c) ES_{NT1} (d) ES_{NT3} (e) ES_{NT5} (B) Halpin–Tsai and Guth model fittings for nanocomposites irradiated at different doses: (a) 50 kGy (b) 100 kGy (c) 150 kGy (d) 200 kGy (e) 250 kGy (f) 300 kGy (C) Strain amplitude dependence of storage modulus for irradiated (dose = 200 kGy) and unirradiated nanocomposites (a) ES_{NT0} (b) irradiated ES_{NT0} (c) ES_{NT3} (d) irradiated ES_{NT3} (D) thermograms of irradiated and unirradiated (a) ES_{NT0} (b) irradiated ES_{NT0} (c) ES_{NT3} (d) irradiated ES_{NT3}: subscript indicates % MWNT in SBR: EPDM (50:50) blend.

Reproduced with permission of John Wiley and Sons [63].

MWNT. Thus, the enhancement in the cross-linking yield on MWNT incorporation was attributed to the formation of multiionization spurs yielding Y-cross-linking and also increased the possibility of spur overlap due to reduced free volume. Polymer-filler interactions in SBR/EPDM/MWNT nanocomposites were assessed by Kraus plot. The negative slope of nanocomposites implied strong polymer-filler interaction and reinforcement of the matrix. The C value for the system was found to be 4.03, which suggested high reinforcement of the matrix. Not only the mechanical properties and

Table 5.2 Aspect Ratio Values for Different SBR/EPDM/MWNT Nanocomposites Derived From Model Fitting

Model		Dose (kGy)					
		50	100	150	200	250	300
Halpin Tsai	(l/d)	29.58	28.29	31.93	35.58	32.76	26.43
	SE	2.04	1.29	1.73	2.14	1.23	0.79
Guth	(l/d)	22.07	21.29	22.75	24.00	23.09	20.48
	SE	0.91	1.20	1.37	1.72	1.26	1.11
dE_c/dV_f	(dE_c/dV_f)	5.50	7.02	9.04	10.45	10.77	10.58
	SE	0.58	0.49	0.75	0.86	0.63	0.50

SE, *Standard error.*

cross-linking density of the composites improved but significant improvement was also observed in the dynamic mechanical and thermal properties. Storage modulus (E') improved substantially both with the addition of MWNT and with radiation treatment (Fig. 5.8B). It was concluded that with the radiation treatment, the mobility of polymer chains close to the MWNT surface is restricted, preventing the structural breakdown during dynamic motions [63]. The decomposition temperature of the nanocomposite increased with the addition of nanotubes. However, this improvement was observed only for MWNT-filled SBR/EPDM system (Fig. 5.8D). The observed improvement in thermal stability of the nanocomposite was attributed to the barrier effect rendered by MWNT agglomerates, minimizing the oxygen ingress into the matrix. In another multiphase MWNT-based nanocomposite, similar results were observed. In PCR/EPDM/MWNT nanocomposites, high reinforcement was displayed by MWNTs (Table 5.3) [64]. The observed manifold increase in the modulus of parent matrices on the addition of nanofillers has been attributed to the inherent defect-free morphology of the nanofillers. Radiation processing of these nanocomposites further affects the mechanical properties due to cross-linking/degradation of the polymer matrix. Different mechanical models were used to

Table 5.3 Aspect Ratio Values for Different PCR/EPDM/MWNT Nanocomposites Derived From Model Fitting

Model		Dose (kGy)					
		50	100	150	200	250	300
Halpin Tsai	(l/d)	8.80	7.21	6.39	10.21	9.70	10.03
	SE	0.69	0.32	0.21	0.59	0.57	0.31
Guth	(l/d)	10.33	9.24	8.70	11.30	11.01	11.21
	SE	0.75	0.52	0.20	0.62	0.52	0.46
dE_c/dV_f	(dE_c/dV_f)	2.42	3.07	3.57	5.66	5.65	7.12
	SE	0.19	0.13	0.17	0.44	0.48	0.29

SE, *Standard error.*

explain the observed results. The additive rule and Guth model were found to be inadequate in explaining the modulus behavior of the nanocomposites as the values calculated were much higher than the experimental and Halpin–Kardos values. The deviation of experimental results from the above-mentioned models indicates that assumptions, such as homogeneous distribution and efficient transfer of tensile load to the MWNTs, made to derive these equations do not hold good for these nanocomposites. Thus, these models were modified by applying boundary conditions for the elastomers systems and by varying apparent aspect ratio of MWNTs in the elastomer matrix. The increase in the magnitude of dE_c/dV_f (a measure of reinforcement) with radiation indicated that there is an improvement in load transfer effect on irradiation and an interface link between MWNTs and elastomer matrix is established after radiation treatment.

7.1.2 Thermoplastic Composites

Thermoplastic polymer composites have advantages of easy processability and reusability. However, in addition to having a strong dependence on interface and filler characteristic, their mechanical properties might also be affected due to nanofiller-induced crystallization of polymer matrix. Low-density polyethylene (LDPE) is among widely used thermoplastics, and it has been reported that MWNTs affect its crystallization kinetics. In a multiphasic LDPE/PCR/EPDM composite prepared by melt compounding, reinforcing effects of MWNTs were reported [65]. The X-ray diffractograms of the composites are shown in Fig. 5.9A. The incorporation of nanotubes into LDPE/EVA/PCR blends did not affect the crystalline structure of LDPE, as the nanocomposites also exhibited an orthorhombic structure typical of LDPE and

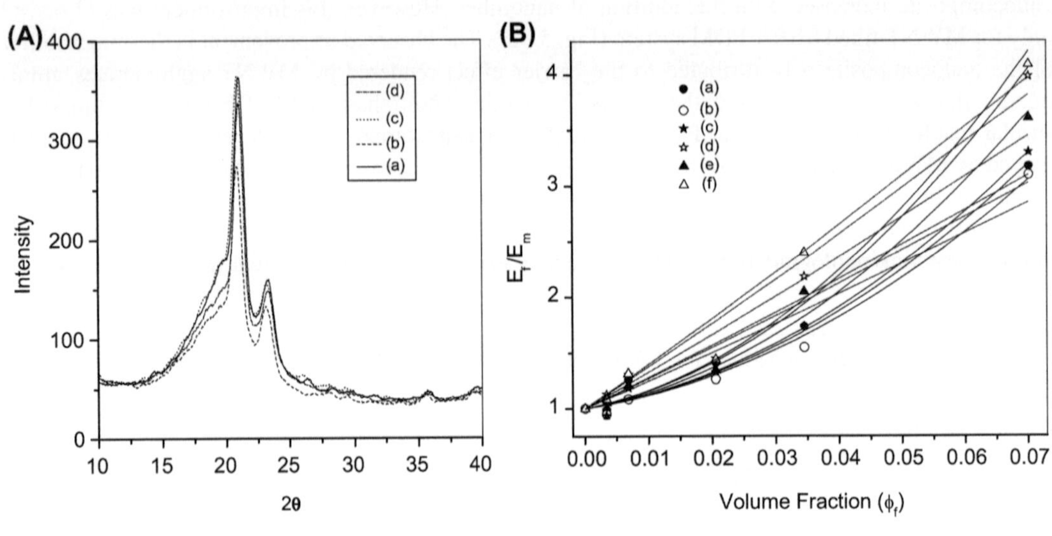

FIGURE 5.9

(A) X-ray diffraction patterns of blends (a) PEP_{NT0} (b) PEP_{NT05} irradiated (dose = 200 kGy) (c) PEP_{NT3} (d) PEP_{NT3} irradiated (dose = 200 kGy); The subscript in sample indicates % MWNT in LDPE/EVA/PCR (40:40:20): (b) Halpin–Tsai and Guth model fittings for nanocomposites [LDPE/EVA/PCR (40:40:20) with different volume fraction of MWNT] irradiated to different doses (a) 50 kGy, (b) 100 kGy, (c) 150 kGy, (d) 200 kGy, (e) 250 kGy, (f) 300 kGy.

Reproduced with permission of Springer [65].

no significant shift in the peak position was observed (Table 5.4). However, Young's modulus of the composites increased with the increase in MWNT content and with an increase in radiation dose. A significant increase in the Young's modulus was observed when the MWNT content increased from 0.5 to 1 wt%. It may be noted that this fraction was considerably lower than for elastomers described in the previous section. This suggests the crystalline domain of thermoplastics plays a crucial role in determining reinforcement efficacy of the filler. The reinforcement increased with an increase in radiation dose up to 200 kGy, which indicates that there are a good load transfer and an interface link between MWNTs and elastomer matrix after radiation treatment Fig. 5.9B. The storage modulus of the ternary blends decreases with increase in EVA component and deviates significantly from both series and parallel model results. The bulk density of the system also showed a positive deviation, indicating good EVA/PCR interaction in the intermediate composition range.

NCB has also been reported to reinforce thermoplastic polymers. In polycaprolactam/EVA/NCB nanocomposites significant reinforcement reflected as improvement in mechanical properties is reported, and the mechanism of reinforcement was investigated by SAXS and WAXRD measurements (Fig. 5.10) [66]. Micromechanical modeling and physical properties of the composites indicated high susceptibility of mechanical and physical properties to the interfacial changes (Fig. 5.11). SAXS studies indicated surface fractal structures in ternary composites. Crystallographic parameter studies of polycaprolactam/ EVA blend using wide-angle X-ray diffraction and differential scanning calorimetry (DSC) established that introduction of NCB did not affect these parameters.

It is well established that for many thermoplastics introduction of nanofillers leads to higher modulus of nanocomposites; however, in the majority of the cases a significant reduction in the toughness is reported. In a recent study, it has been demonstrated that with the addition of compatible elastomers, a significant improvement in the toughness can also be achieved. Polypropylene/LDPE blends were reinforced by NCB. A third component, EPDM, (an elastomer), was added to improve elongation at break and toughness of the composites. Marked improvement in elongation at break for 30 wt% NCB-filled LDPE/PP (1:1) on the addition of 10 wt% EPDM addition was observed [67]. DSC studies revealed good interactions between LDPE and EPDM phases, and there was no reduction in the crystallinity of either LDPE or PP phases up to 50% EPDM loading. This improvement in flexibility and elongation at break was attributed to the percolation of carbon black in different phases and the role of EPDM as a compatibility modifier of LDPE and PP phases (Fig. 5.12).

Table 5.4 XRD Results of LDPE/EVA/PCR/MWNT Nanocomposites

[a]Sample	X_c (%)	θ_{110} (degrees)	θ_{200} (degrees)	d_{110} (Å)	l_{110} (Å)	d_{200} (Å)	l_{200} (Å)
PEP_{NT0}	37.60	21.05	23.39	4.25	167.01	3.84	130.62
PEP_{NT0} (irradiated)	36.80	20.89	23.17	4.23	167.49	3.81	129.68
PEP_{NT3}	37.89	20.97	23.41	4.26	165.69	3.84	128.93
PEP_{NT3} (irradiated)	37.53	20.91	23.35	4.21	166.70	3.78	130.76

[a]*The subscript in sample indicates % MWNT in LDPE/EVA/PCR (40:40:20).*

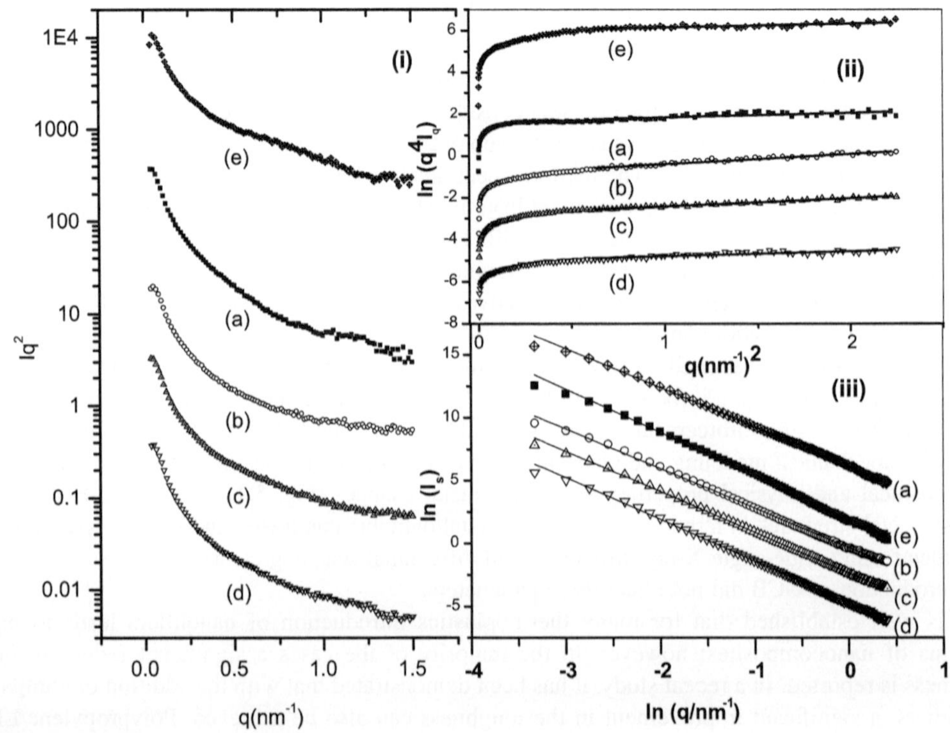

FIGURE 5.10

Small-angle X-ray patterns of polymer composites (i) variation of Iq^2 with q (ii) variation of $\ln(q^4I_q)$ with q^2 (iii) liner fitting of the variation of $\ln(I_q)$ with $\ln(q)$ (a) EVA/CB 30 wt% (b) EVA/polycaprolactam (3:1)/CB 30 wt % (c) EVA/polycaprolactam (1:1)/CB 30 wt% (d) EVA/polycaprolactam (1:3)/CB 30 wt% (e) Polycaprolactam/ CB 30 wt%.

Reproduced with permission of John Wiley and Sons [66].

7.1.3 Thermosetting Composites

Thermosetting polymers generally involve monomer/oligomers in the first step of synthesis that are cross-linked to a three-dimensional network. As a result, dispersion and distribution of filler are relatively easier. There have been extensive studies on the reinforcement of thermosetting matrices, mainly epoxies using different nanocarbons [54,68–71]. Wang et al. recently used in situ functionalization method to improve dispersibility of SWNTs in an epoxy matrix. It was shown that at 1 wt% loading of modified SWNTs, the tensile strength increased by 40% and elastic modulus by 60%. These results were ascribed to the increased load transfer efficacy due to in situ–grafted SWNTs [72]. Feng et al. used nanotubes to modify carbon fibers using aerosol-assisted chemical vapor deposition. They demonstrated that CNT-assisted modification leads to around three times increase in surface area of carbon fiber. It was claimed that the interfacial shear strength of composite reinforced by modified carbon fibers is around 94% higher [73]. On the other hand, Khare et al. used amidoamine functionalized carbon nanotubes to reinforce epoxy matrix and demonstrated the enhancement in the

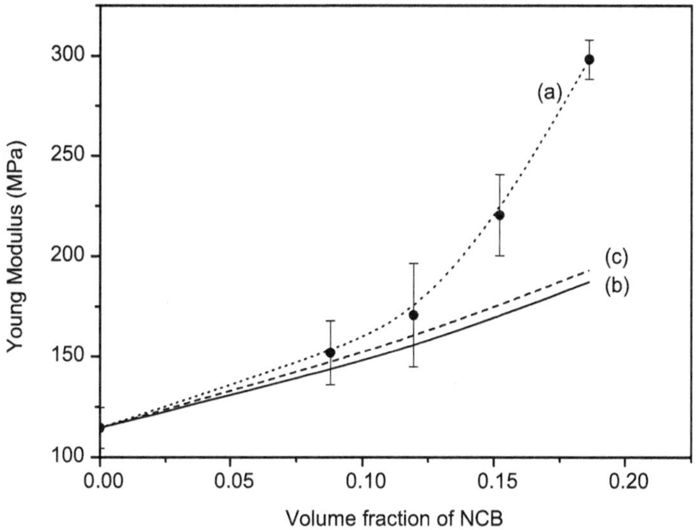

FIGURE 5.11

Variation in the Young's modulus of EVA/polycaprolactam (50:50) with NCB volume fraction (a) experimental data (b) Nielsen's model fit (c) Nielsen's model fit with modified.

Reproduced with permission of John Wiley and Sons [66].

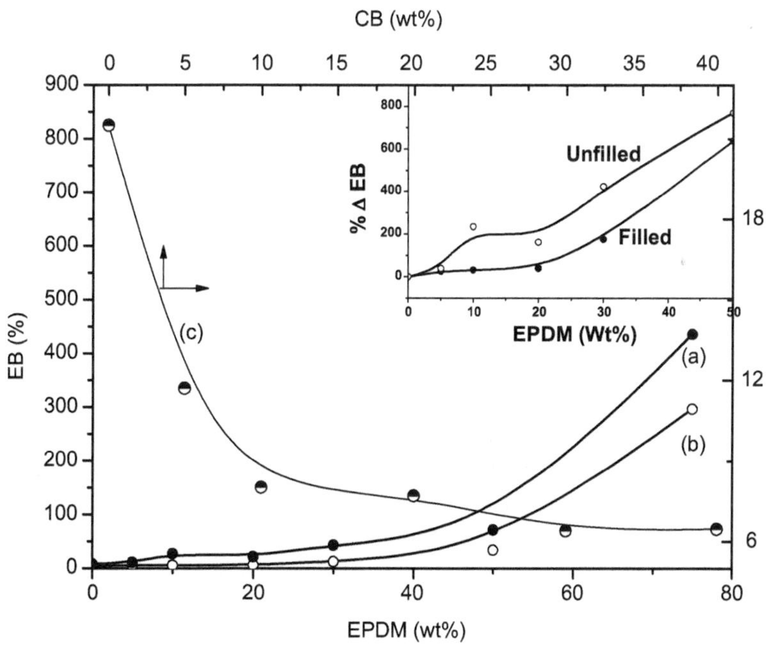

FIGURE 5.12

Elongation at break of (a) PE/PP (1:1) blend with variation in EPDM content (b) PE/PP (1:1) blends, CB 30 wt% with variation in EPDM content (c) PE/PP/EPDM (4:4:1) blend with variation in CB fraction. Inset: relative increase in EB for (a) PE/PP (1:1) blend with variation in EPDM content (b) PE/PP (1:1) blends, CB 30 wt% with variation in EPDM content.

Reproduced with permission of Taylor & Francis [67].

mechanical and thermal properties [11]. Ma et al. used silane functionalization to improve interfacial adhesion between epoxy and carbon nanotubes. Their results showed significantly enhanced mechanical and thermal properties [68].

7.1.4 Inorganic Filler, Natural Fiber and Other Polymer Nanocomposites

A variety of inorganic fillers has been used for developing polymer composites for different applications ranging from optoelectronic to high-energy radiation shielding. Flexible lead-free PDMS-Bi_2O_3 composites for radiation shielding were developed through internal compounding [74]. Polymer-filler interaction, physicomechanical characteristics of the composites, and extensively high filler loading (up to 70%) have been reported. Nielsen model was applied to understand the micromechanics of the system. These composites had high elastic modulus (Fig. 5.13A) and were demonstrated to attenuate soft gamma rays effectively (Fig. 5.13B). Thermal stability of the composites also improved and onset temperature for the second step of PDMS degradation increased from 451°C to 491°C as loading of Bi_2O_3 increased from 0 to 70 wt%. In another interesting study, PTFE micropowder (PTFEMP) generated by degrading PTFE ribbons using gamma irradiation in the presence of CCl_4 was used as a filler to reinforce silicone rubber. The elastic modulus of the composites increased significantly on the addition of PTFE (Fig. 5.14A) [75]. Polymer-filler interaction, XRD, and FTIR investigations revealed complex changes in composites on compounding. PTFEMP not only improved mechanical properties but also the coefficient of friction (Fig. 5.14B). It has been proposed that during cross-linking of PDMS on irradiation, interfacial bonding takes place between PTFEMP and PDMS, significantly improving load transfer mechanism (Fig. 5.15). In another study, $CaCO_3$ was used to reinforce polypropylene. The composites were prepared by melt compounding and ~500% increase in the fracture toughness was reported. By using melt compounding again, polyethylene terephthalate/alumina composites were made, and it was shown that wear resistance considerably improves with the loading of nanoalumina. Boron nitride nanosheets were recently used to reinforce polyvinylchloride matrices. It was demonstrated that even at 200% strain profound improvement in mechanical properties can be achieved.

Natural fillers, mainly fibers, have also been used as reinforcing fillers. In recent years, there has been a lot of interest in nanocrystalline cellulose as filler due to its high aspect ratio. Significant improvement in the mechanical properties has been reported in polymer nanocrystalline cellulose composites [43]. Gindl et al. used gluconacetobacter xylinus to synthesize cellulose sheets and reinforced cellulose acetate butyrate (CAB) with it [9]. It was shown that 0.1–0.3 volume fraction of cellulose increases the CAB yield strength by 50–130 MPa. It was also reported that reorientation of cellulosic crystals during cyclic tensile strength measurements affects elastic modulus of the composites. In a very interesting study, graphene-reinforced nanocellulose nanopaper was prepared and its successful application was demonstrated for highly stretchable strain sensors [76].

7.2 SYNERGISTIC INTERACTIONS IN HYBRID FILLERS

There have been several reports wherein exceptional synergistic effect has been observed between/among fillers in the composites [77–80]. As a result, with the use of dual or multifiller significant improvement in elastic modulus was reported. Synergistic effects were found between organic-inorganic

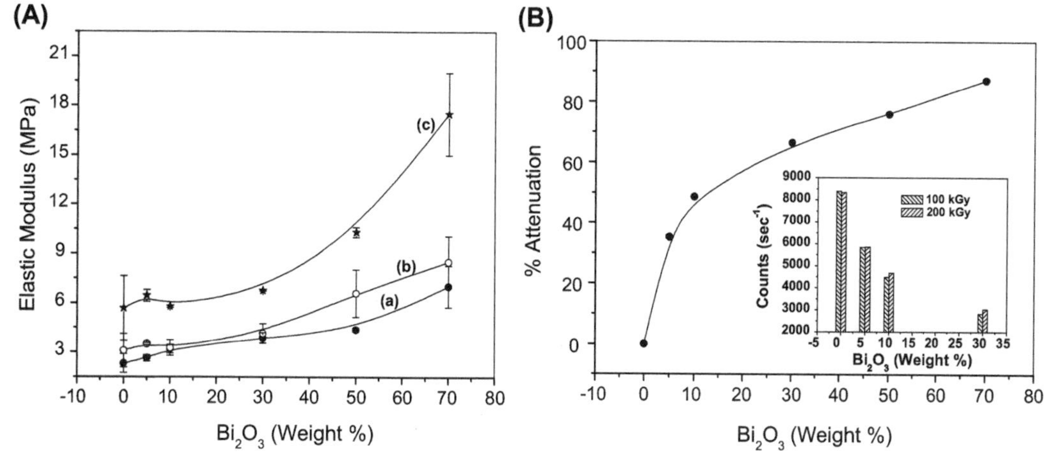

FIGURE 5.13

(A) Variation in elastic modulus of composites with absorbed dose (a) 100 kGy (b) 200 kGy (c) 400 kGy (B) Variation in attenuation of different composites irradiated to different doses. Inset: Variation in counts with Bi_2O_3 (wt%) at two different radiation doses.

Reproduced with permission of John Wiley and Sons [74].

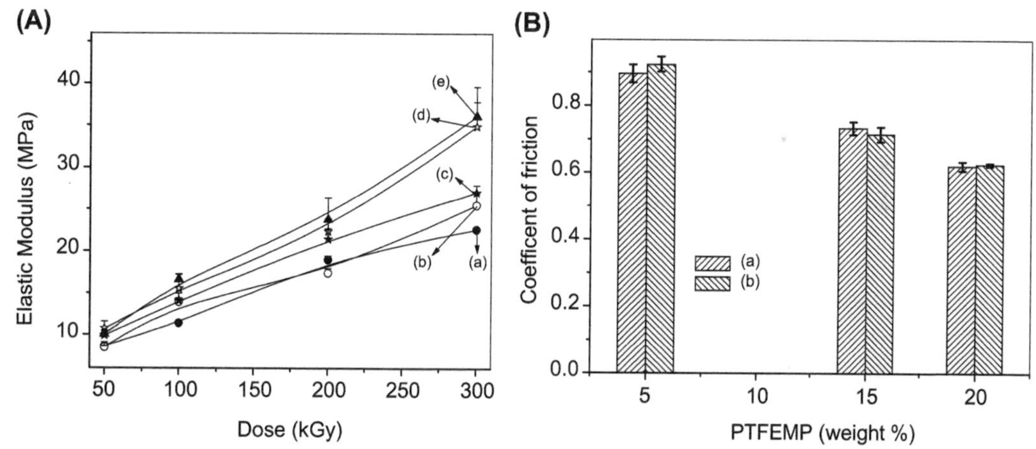

FIGURE 5.14

(A) Variation in elastic modulus of composites (a) $\mathbf{C}_0^0$ (b) $\mathbf{C}_5^0$ (c) $\mathbf{C}_{10}^0$ (d) $\mathbf{C}_{15}^0$ (e) $\mathbf{C}_{20}^0$ [superscript shows the radiation dose and subscript shows wt% of polytetrafluroethylene microfine powder (PTFEMP)] (B) Coefficient of friction PDMS-PTFEMP composites (a) static (b) dynamic.

Reproduced with permission of Taylor & Francis [75].

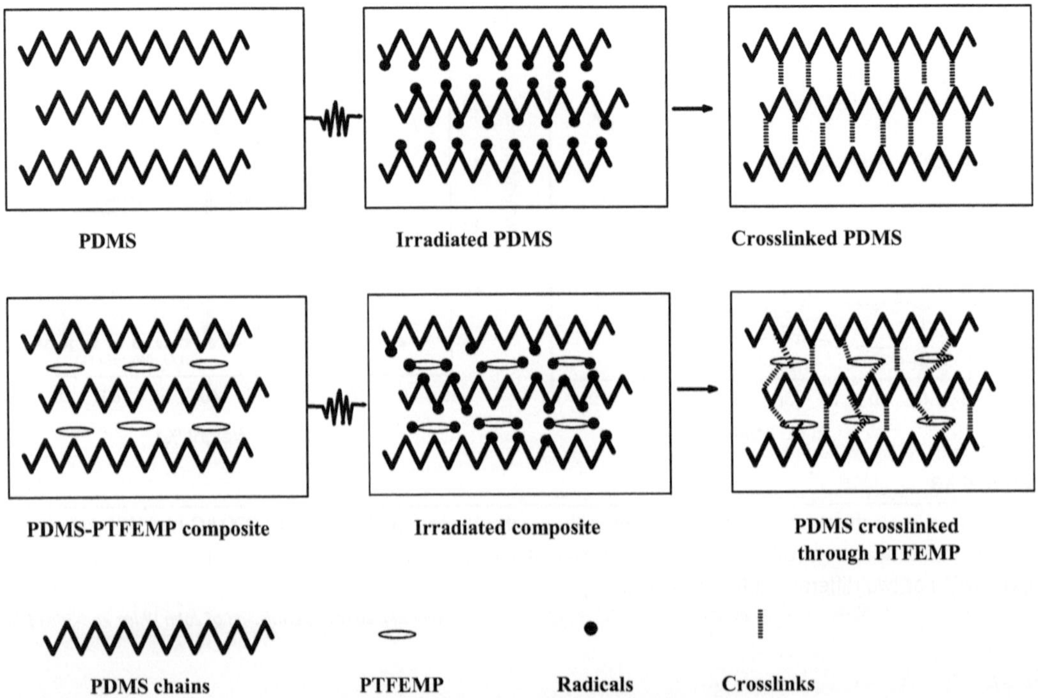

FIGURE 5.15

Schematic showing radiation-induced cross-linking of PDMS and PDMS-PTFEMP composite.

Reproduced with permission of Taylor & Francis [75].

dual filler in EVA composites [81]. Mechanical properties of composites containing hybrid fillers exhibited high-synergic interactions between fillers. The synergy between organoclay and PTFEMP in EVA matrix has been plotted in Fig. 5.16. This synergy may be the cumulative effect of synergistic interaction between organoclay and PTFEMP and formation of PTFEMP network within the amorphous domain of EVA at such a high PTFEMP weight fraction. Micromechanical models also suggested synergistic interactions and predicted lower values than experimentally observed values. X-ray diffractograms of different systems were investigated. The intensity of the EVA's characteristics peaks was lower in ternary composites than it was in binary EVA/organoclay nanocomposites. EVA/organoclay (10%) nanocomposites also showed characteristic EVA peaks and there was no shift in peak position. However, the ratio of organoclay and EVA peak intensity (I_{110}/I_{001}) significantly decreased (from 2.5 to 1.4), highlighting the fact that, in ternary composites, the crystalline structure of EVA was largely intact, however, a substantial decrease in the organoclay crystalline structure took place. This was attributed to the polar synergistic interaction between vinyl acetate-organoclay under a high-shear condition in presence of PTFEMP. In $EV_{0.05}^{0.50}$ (5 wt% of clay in EVA: PTFE; 50:50) nanocomposites, significant reduction in the peak intensity was not observed but a shift in the peak position was observed, increasing $d(001)$ spacing from 1.9 to 2.1 nm (10.5%). This observation indicated that extent of organoclay intercalation improves with the addition of PTFEMP. As around 10% intergallery expansion was

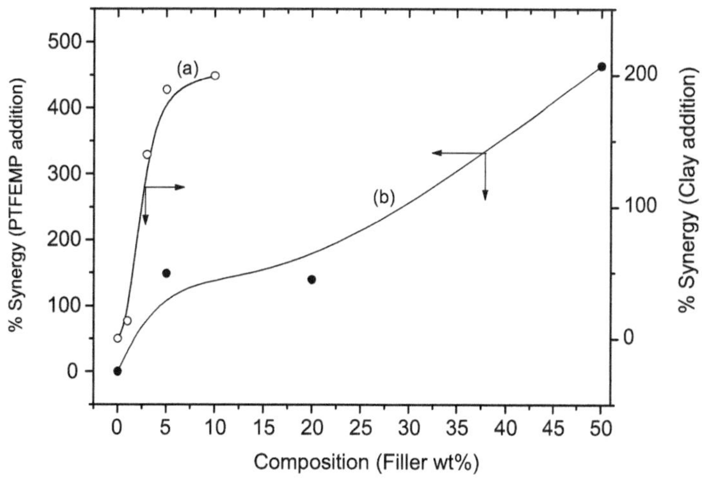

FIGURE 5.16

Synergy between PTFEMP and nanoclays (a) EVA/PTFEMP(15%) composites containing different wt% of nanoclays (b) EVA/Clay(5%) composites containing different wt% of PTFEMP.

Reproduced with permission of Elsevier [81].

observed only in $EV_{0.05}^{0.50}$ composite, it was concluded that delamination of the organoclay has been facilitated in the presence of PTFEMP as shown in (Fig. 5.17). It was debated that though there was only 10% increase in the *d*-spacing, this increase may induce a substantial effect on the viscous flow of polymer chains during tensile elongation, as both organoclay and PTFEMP are going to be present in the amorphous region of the polymer and will improve the stiffness of the matrix. Synergistic effect of thin-walled carbon nanotubes (TWNTs) and reduced graphene oxide (RGO) on laser-induced shape recovery performance was observed in a polyurethane matrix [80]. They observed enhanced shape memory and recovery was highest for 7/3 of TWNT/RGO ratio and a 1 wt% of nanocarbon loading (Fig. 5.18). The elastic modulus and electrical conductivity of the composites also showed TWNT/RGO-based synergistic effect. Prasad et al. demonstrated exceptionally enhanced elastic modulus with the use of two different nanocarbons in polyvinyl alcohol. They used nanoindentation technique to evaluate mechanical properties and demonstrated extraordinary synergy, resulting in 400% increase in mechanical properties in comparison to composites containing single nanocarbon [82]. In an epoxy matrix, synergistic interactions were also observed between graphene nanoplates (GnPs) and carbon nanotubes [8]. Interestingly, they used GnPs of two different sizes, 5 and 25 mm, to understand the role of aspect ratio in such interactions. It was found that in the hybrid samples CNTs align themselves on the GnP flakes to form networked structures. It was also found that CNT-GnP ratio significantly affects the reinforcement. Shin et al. have represented an excellent example of synergy between graphene oxide and carbon nanotubes [83]. They combined carbon nanotubes and reduced graphene oxide flakes in solution-spun PVA polymer fibers, resulting in the gravimetric toughness of 1000 J/g, which is almost 12 times higher than that of Kevlar. Interestingly these fibers are weaveable, making the development of different structures possible (Fig. 5.19).

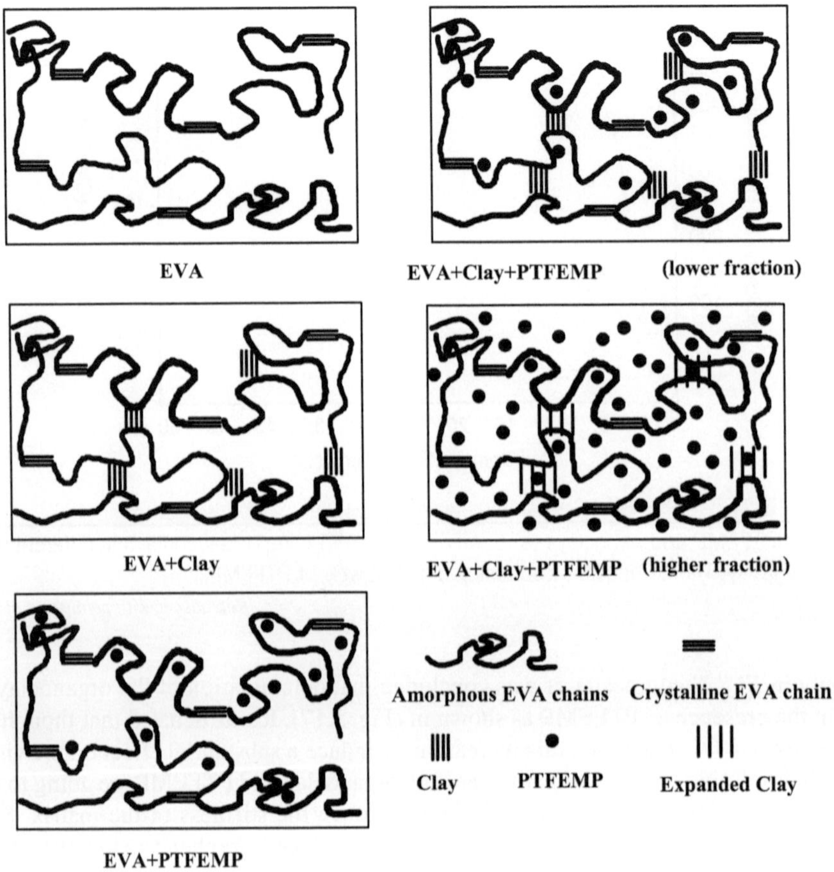

FIGURE 5.17

Schematic showing delamination of clay at higher PTFEMP fraction for EVA-PTFEMP-Clay composites.

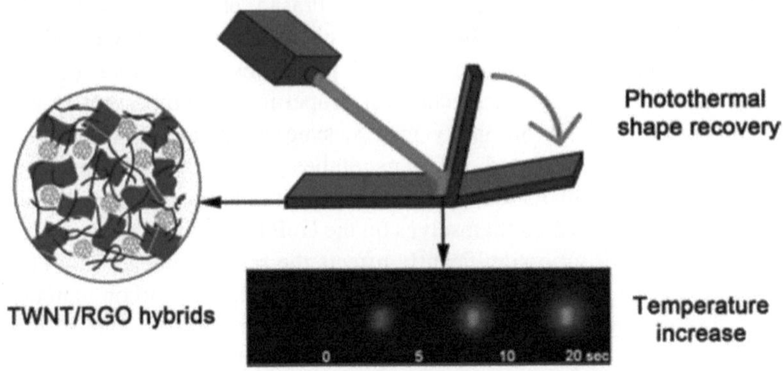

FIGURE 5.18

Schematic showing enhanced shape memory in TWBT/RGO hybrids.

Reproduced with permission of Elsevier [80].

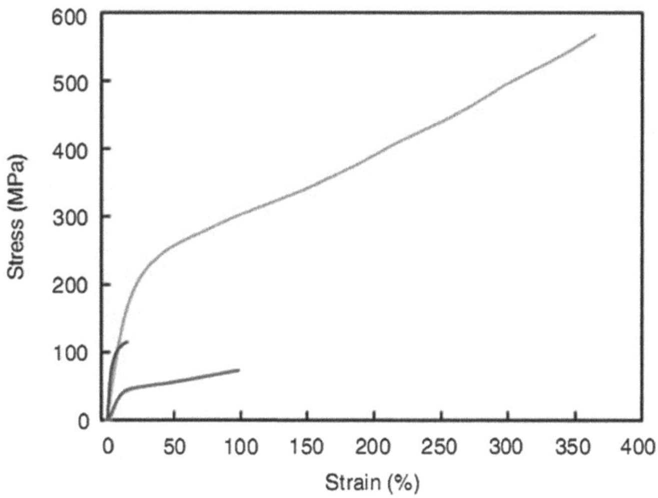

FIGURE 5.19

Enhancement in toughness of PVA fiber due to synergistic effect of fillers.

Reproduced with permission of Nature Publishing Group [83].

8. CONCLUSIONS

Polymer nanocomposites show tremendous potential for use in structural applications. High strength, high stiffness along with low cost, weight and process ease can be achieved through the optimum selection of nanofiller and polymer matrix and with the careful tailoring of composition and interface. Thermoplastics, elastomers, and thermosets all can be reinforced by nanofillers. Nanocarbons, nanoclays, and biosource-derived fillers show significant reinforcement effect in different polymer matrices. Micromechanics of the system strongly depends on the filler characteristics, orientation, and packing. Different surface modification approaches, covalent and noncovalent, can be used to modify the interface and dispersion of the filler in a polymer matrix. Use of hybrid filler was found to yield synergistic effect in several cases, leading to an exceptionally high elastic modulus of the nanocomposites. In spite of several remarkable properties of nanocomposites, still it has not been possible to achieve the reinforcement levels predicted by simulations and theories of polymer nanocomposites. Future developments in this area should involve efficient surface modification and dispersion techniques to overcome limitations posed by agglomeration and poor interfacial shear strength. Innovative strategies for the in situ synthesis of polymer nanocomposites and facile synthesis of tailored nanoparticles are going to be the focus of research in this area.

LIST OF ACRONYMS AND ABBREVIATIONS

AFM Atomic force microscopy
Bi_2O_3 Bismuth oxide
BN Boron nitride
BNT Born nitride tubes
DMF Dimethyl formamide
CNTs Carbon nanotubes
CTAB Cetyl trimethylammonium bromide
CVD Chemical vapor deposition

EG Exfoliated graphene
EVA Ethylene-vinyl acetate
EPDM Ethylene propylene diene monomer
GIC Graphene intercalated compound
GO Graphene oxide
LDPE Low-density polyethylene
LDS Lauryl dodecyl sulfonate
MoS_2 Molybdenum disulfide
MWNT Multiwalled nanotube
NCB Nanocarbon black
NCC Nanocrystalline cellulose
NMP N-Methylpyrrolidone
PCR Polychloroprene
PDMS Polydimethyl siloxane
PVA Poly vinyl alcohol
SAXS Small-angle X-ray scattering
SBR Styrene butadiene rubber
SC Sodium cholate
SDBS Sodium dodecylbenzenesulfonate
SDS Sodium dodecyl sulfonate
SEM Scanning electron microscopy
SENT SBR-EPDM-MWNT nanocomposite
STM Scanning tunneling microscopy
SWNT Single-walled nanotube
TDOC Sodium taurodeoxycholate
THF Tetrahydrofuran
TTAB Tetradecyltrimethylammonium bromide
WS_2 Tungsten sulfide
XRD X-ray diffraction

REFERENCES

[1] S. Sinha Ray, M. Okamoto, Polymer/layered silicate nanocomposites: a review from preparation to processing, Prog. Polym. Sci. 28 (2003) 1539–1641, http://dx.doi.org/10.1016/j.progpolymsci.2003.08.002.

[2] A.A. Azeez, K.Y. Rhee, S.J. Park, D. Hui, Epoxy clay nanocomposites – processing, properties and applications: a review, Compos. B Eng. 45 (2013) 308–320, http://dx.doi.org/10.1016/j.compositesb.2012.04.012.

[3] P.C. Ma, Y. Zhang, Perspectives of carbon nanotubes/polymer nanocomposites for wind blade materials, Renew. Sustain. Energy Rev. 30 (2014) 651–660, http://dx.doi.org/10.1016/j.rser.2013.11.008.

[4] A.D. Sahin, Progress and recent trends in wind energy, Prog. Energy Combust. Sci. 30 (2004) 501–543, http://dx.doi.org/10.1016/j.pecs.2004.04.001.

[5] D. Wu, F. Zhang, P. Liu, X. Feng, Two-dimensional nanocomposites based on chemically modified graphene, Chemistry 17 (2011) 10804–10812, http://dx.doi.org/10.1002/chem.201101333.

[6] D. Feldman, Polymer nanocomposites in building, construction, J. Macromol. Sci. A 51 (2014) 203–209, http://dx.doi.org/10.1080/10601325.2014.871948.

[7] H. Lu, Z. Chen, C. Ma, Bioinspired approaches for optimizing the strength and toughness of graphene-based polymer nanocomposites, J. Mater. Chem. 22 (2012) 16182, http://dx.doi.org/10.1039/c2jm32763e.

[8] S. Chatterjee, F. Nafezarefi, N.H. Tai, L. Schlagenhauf, F.A. Nüesch, B.T.T. Chu, Size and synergy effects of nanofiller hybrids including graphene nanoplatelets and carbon nanotubes in mechanical properties of epoxy composites, Carbon 50 (2012) 5380–5386, http://dx.doi.org/10.1016/j.carbon.2012.07.021.

[9] W. Gindl, J. Keckes, Tensile properties of cellulose acetate butyrate composites reinforced with bacterial cellulose, Compos. Sci. Technol. 64 (2004) 2407–2413, http://dx.doi.org/10.1016/j.compscitech.2004.05.001.

[10] S. Kango, S. Kalia, A. Celli, J. Njuguna, Y. Habibi, R. Kumar, Surface modification of inorganic nanoparticles for development of organic–inorganic nanocomposites—a review, Prog. Polym. Sci. 38 (2013) 1232–1261, http://dx.doi.org/10.1016/j.progpolymsci.2013.02.003.

[11] K.S. Khare, F. Khabaz, R. Khare, Effect of carbon nanotube functionalization on mechanical and thermal properties of cross-linked epoxy-carbon nanotube nanocomposites: role of strengthening the interfacial interactions, ACS Appl. Mater. Interfaces (2014) 6098–6110, http://dx.doi.org/10.1021/am405317x.

[12] M.E. Mackay, A. Tuteja, P.M. Duxbury, C.J. Hawker, B. Van Horn, Z. Guan, et al., General strategies for nanoparticle dispersion, Science 311 (2006) 1740–1743, http://dx.doi.org/10.1126/science.1122225.

[13] E. Kontou, Micromechanics model for particulate composites, Mech. Mater. 39 (2007) 702–709, http://dx.doi.org/10.1016/j.mechmat.2006.12.001.

[14] X.Y. Ji, Y.-P. Cao, X.-Q. Feng, Micromechanics prediction of the effective elastic moduli of graphene sheet-reinforced polymer nanocomposites, Model. Simul. Mater. Sci. Eng. 18 (2010) 045005, http://dx.doi.org/10.1088/0965-0393/18/4/045005.

[15] B. Burks, J. Middleton, M. Kumosa, Micromechanics modeling of fatigue failure mechanisms in a hybrid polymer matrix composite, Compos. Sci. Technol. 72 (2012) 1863–1869, http://dx.doi.org/10.1016/j.compscitech.2012.07.017.

[16] G.M. Odegard, T.C. Clancy, T.S. Gates, Modeling of the mechanical properties of nanoparticle/polymer composites, Polymer 46 (2005) 553–562, http://dx.doi.org/10.1016/j.polymer.2004.11.022.

[17] P.M. Ajayan, Nanotubes from carbon, Chem. Rev. 99 (1999) 1787–1800, http://dx.doi.org/10.1021/cr970102g.

[18] M.S. Dresselhaus, G. Dresselhaus, R. Saito, Physics of carbon nanotubes, Carbon 33 (1995) 883–891, http://dx.doi.org/10.1016/0008-6223(95)00017-8.

[19] A. Kis, A. Zettl, Nanomechanics of carbon nanotubes, Philos. Trans. A. Math. Phys. Eng. Sci. 366 (2008) 1591–1611, http://dx.doi.org/10.1098/rsta.2007.2174.

[20] M. Moniruzzaman, K.I. Winey, Polymer nanocomposites containing carbon nanotubes, Macromolecules 39 (2006) 5194–5205, http://dx.doi.org/10.1021/ma060733p.

[21] X. Wei, Q. Chen, L. Peng, R. Cui, Y. Li, In situ measurements on individual thin carbon nanotubes using nanomanipulators inside a scanning electron microscope, Ultramicroscopy 110 (2010) 182–189, http://dx.doi.org/10.1016/j.ultramic.2009.11.007.

[22] A. Kis, G. Csányi, J.P. Salvetat, T.N. Lee, E. Couteau, A.J. Kulik, et al., Reinforcement of single-walled carbon nanotube bundles by intertube bridging, Nat. Mater. 3 (2004) 153–157, http://dx.doi.org/10.1038/nmat1076.

[23] T.W. Tombler, C. Zhou, L. Alexseyev, J. Kong, H. Dai, L. Liu, et al., Reversible electromechanical characteristics of carbon nanotubes under local-probe manipulation, Nature 405 (2000) 769–772, http://dx.doi.org/10.1038/35015519.

[24] M. Yu, O. Lourie, M.J. Dyer, K. Moloni, T.F. Kelly, R.S. Ruoff, Strength and breaking mechanism of multiwalled carbon nanotubes under tensile load, Science 287 (2000) 637–640, http://dx.doi.org/10.1126/science.287.5453.637.

[25] J. Prasek, J. Drbohlavova, J. Chomoucka, J. Hubalek, O. Jasek, V. Adam, et al., Methods for carbon nanotubes synthesis – review, J. Mater. Chem. 21 (2011) 15872–15884, http://dx.doi.org/10.1039/c1jm12254a.

[26] W. Li, S. Xie, L. Qian, B. Chang, B. Zou, W. Zhou, et al., Large-scale synthesis of aligned carbon nanotubes, Science 274 (1996) 1701–1703.

[27] A.K. Geim, K.S. Novoselov, The rise of graphene, Nat. Mater. 6 (2007) 183–191, http://dx.doi.org/10.1038/nmat1849.

[28] A.K. Geim, Graphene: status and prospects, Science 324 (2009) 1530–1534, http://dx.doi.org/10.1126/science.1158877.

[29] X. Sun, H. Sun, H. Li, H. Peng, Developing polymer composite materials: carbon nanotubes or graphene? Adv. Mater. 25 (2013) 5153–5176, http://dx.doi.org/10.1002/adma.201301926.
[30] X. Huang, Z. Yin, S. Wu, X. Qi, Q. He, Q. Zhang, et al., Graphene-based materials: synthesis, characterization, properties, and applications, Small 7 (2011) 1876–1902, http://dx.doi.org/10.1002/smll.201002009.
[31] W.A. de Heer, C. Berger, X. Wu, P.N. First, E.H. Conrad, X. Li, et al., Epitaxial graphene, Solid State Commun. 143 (2007) 92–100, http://dx.doi.org/10.1016/j.ssc.2007.04.023.
[32] C. Nethravathi, M. Rajamathi, N. Ravishankar, L. Basit, C. Felser, Synthesis of graphene oxide-intercalated α-hydroxides by metathesis and their decomposition to graphene/metal oxide composites, Carbon 48 (2010) 4343–4350, http://dx.doi.org/10.1016/j.carbon.2010.07.047.
[33] C. Nethravathi, J.T. Rajamathi, N. Ravishankar, C. Shivakumara, M. Rajamathi, Graphite oxide-intercalated anionic clay and its decomposition to graphene-inorganic material nanocomposites, Langmuir 24 (2008) 8240–8244, http://dx.doi.org/10.1021/la8000027.
[34] C.N.R. Rao, H.S.S. Ramakrishna Matte, R. Voggu, A. Govindaraj, Recent progress in the synthesis of inorganic nanoparticles, Dalton Trans. 41 (2012) 5089–5120, http://dx.doi.org/10.1039/c2dt12266a.
[35] R. Krahne, G. Morello, A. Figuerola, C. George, S. Deka, L. Manna, Physical properties of elongated inorganic nanoparticles, Phys. Rep. 501 (2011) 75–221, http://dx.doi.org/10.1016/j.physrep.2011.01.001.
[36] F. Uddin, Clays, nanoclays, and montmorillonite minerals, Metallur. Mater. Trans. A 39 (2008) 2804–2814, http://dx.doi.org/10.1007/s11661-008-9603-5.
[37] N.G. Chopra, A. Zettl, Measurement of the elastic modulus of a multi-wall boron nitride nanotube, Solid State Commun. 105 (1998) 297–300, http://dx.doi.org/10.1016/S0038-1098(97)10125-9.
[38] D. Golberg, Y. Bando, Y. Huang, T. Terao, Boron nitride nanotubes and nanosheets, ACS Nano 4 (2010) 2979–2993, http://dx.doi.org/10.1021/nn1006495.
[39] H.S.S. Ramakrishna Matte, A. Gomathi, A.K. Manna, J. Dattaray, R. Datta, S.K. Pati, et al., MoS_2 and WS_2 analogues of graphene, Angew. Chem. – Int. Ed. Engl. 49 (2010) 4059–4062, http://dx.doi.org/10.1002/anie.201000009.
[40] Z. Li-na, W. Cheng-biao, W. Hai-dou, X. Bin-shi, Z. Da-ming, L. Jia-jun, Microstructure and tribological properties of WS_2/MoS_2 multilayer films, Appl. Surf. Sci. 258 (2012) 1944–1948, http://dx.doi.org/10.1016/j.apsusc.2011.06.148.
[41] E. Galli, Natural organic fillers, in: Plastics Compounds, 1982.
[42] V. Dhawan, S. Singh, I. Singh, Effect of natural fillers on mechanical properties of GFRP composites, J. Compos. 2013 (2013) 1–8, http://dx.doi.org/10.1155/2013/792620.
[43] E. Lam, K.B. Male, J.H. Chong, A.C.W. Leung, J.H.T. Luong, Applications of functionalized and nanoparticle-modified nanocrystalline cellulose, Trends Biotechnol. 30 (2012) 283–290, http://dx.doi.org/10.1016/j.tibtech.2012.02.001.
[44] L. Brinchi, F. Cotana, E. Fortunati, J.M. Kenny, Production of nanocrystalline cellulose from lignocellulosic biomass: technology and applications, Carbohydr. Polym. 94 (2013) 154–169, http://dx.doi.org/10.1016/j.carbpol.2013.01.033.
[45] A. Gedanken, Using sonochemistry for the fabrication of nanomaterials, Ultrason. Sonochem. 11 (2) (2004) 47–55.
[46] J.M. Tour, Scaling up exfoliation, Nat. Mater. 13 (2014) 545–546, http://dx.doi.org/10.1038/nmat3961.
[47] E.E. Tkalya, M. Ghislandi, G. de With, C.E. Koning, The use of surfactants for dispersing carbon nanotubes and graphene to make conductive nanocomposites, Curr. Opin. Colloid Interface Sci. 17 (2012) 225–232, http://dx.doi.org/10.1016/j.cocis.2012.03.001.
[48] L. Boldon, F. Laliberte, L. Liu, Review of the fundamental theories behind small angle X-ray scattering, molecular dynamics simulations, and relevant integrated application, Nano Rev. 6 (2015) 25661, http://dx.doi.org/10.3402/nano.v6.25661.

[49] J. Móczó, B. Pukánszky, Polymer micro and nanocomposites: structure, interactions, properties, J. Ind. Eng. Chem. 14 (2008) 535–563, http://dx.doi.org/10.1016/j.jiec.2008.06.011.
[50] J.C.H. Affdl, J.L. Kardos, The Halpin-Tsai equations: a review, Polym. Eng. Sci. 16 (1976) 344–352, http://dx.doi.org/10.1002/pen.760160512.
[51] A.P. Chatterjee, A model for the elastic moduli of three-dimensional fiber networks and nanocomposites, J. Appl. Phys. 100 (2006), http://dx.doi.org/10.1063/1.2336088.
[52] D.U. Shah, Developing plant fibre composites for structural applications by optimising composite parameters: a critical review, J. Mater. Sci. 48 (2013) 6083–6107, http://dx.doi.org/10.1007/s10853-013-7458-7.
[53] X.L. Ji, J.K. Jing, W. Jiang, B.Z. Jiang, Tensile modulus of polymer nanocomposites, Polym. Eng. Sci. 42 (2002) 983–993, http://dx.doi.org/10.1002/pen.11007.
[54] K. Yang, M. Gu, Y. Guo, X. Pan, G. Mu, Effects of carbon nanotube functionalization on the mechanical and thermal properties of epoxy composites, Carbon 47 (2009) 1723–1737, http://dx.doi.org/10.1016/j.carbon.2009.02.029.
[55] N.G. Sahoo, S. Rana, J.W. Cho, L. Li, S.H. Chan, Polymer nanocomposites based on functionalized carbon nanotubes, Prog. Polym. Sci. 35 (2010) 837–867, http://dx.doi.org/10.1016/j.progpolymsci.2010.03.002.
[56] W. Yi, A. Malkovskiy, Q. Chu, A.P. Sokolov, M.L. Colon, M. Meador, et al., Wrapping of single-walled carbon nanotubes by a pi-conjugated polymer: the role of polymer conformation-controlled size selectivity, J. Phys. Chem. B 112 (2008) 12263–12269, http://dx.doi.org/10.1021/jp804083n.
[57] M.J. O'Connell, P. Boul, L.M. Ericson, C. Huffman, Y. Wang, E. Haroz, et al., Reversible water-solubilization of single-walled carbon nanotubes by polymer wrapping, Chem. Phys. Lett. 342 (2001) 265–271, http://dx.doi.org/10.1016/S0009-2614(01)00490-0.
[58] J. Rafiee, M.A. Rafiee, Z.Z. Yu, N. Koratkar, Superhydrophobic to superhydrophilic wetting control in graphene films, Adv. Mater. 22 (2010) 2151–2154, http://dx.doi.org/10.1002/adma.200903696.
[59] J. Liu, W. Yang, L. Tao, D. Li, C. Boyer, T.P. Davis, Thermosensitive graphene nanocomposites formed using pyrene-terminal polymers made by RAFT polymerization, J. Polym. Sci. A Polym. Chem. 48 (2010) 425–433, http://dx.doi.org/10.1002/pola.23802.
[60] N.G. Portney, K. Singh, S. Chaudhary, G. Destito, A. Schneemann, M. Manchester, et al., Organic and inorganic nanoparticle hybrids, Langmuir 21 (2005) 2098–2103, http://dx.doi.org/10.1021/la047525f.
[61] T. Hanemann, D.V. Szabó, Polymer-nanoparticle composites: from synthesis to modern applications, Materials (Basel) 3 (2010) 3468–3517, http://dx.doi.org/10.3390/ma3063468.
[62] R.P. Bagwe, L.R. Hilliard, W. Tan, Surface modification of silica nanoparticles to reduce aggregation and nonspecific binding, Langmuir 22 (2006) 4357–4362, http://dx.doi.org/10.1021/la052797j.
[63] K.A. Dubey, Y.K. Bhardwaj, C.V. Chaudhari, N.K. Goel, S. Sabharwal, K. Rajkumar, et al., Radiation effects on styrene-butadiene-ethylene-propylene diene monomer-multiple walled carbon nanotube nanocomposites: vulcanization and characterization, Polym. Adv. Technol. 22 (2011) 1888–1897, http://dx.doi.org/10.1002/pat.1688.
[64] K.A. Dubey, Y.K. Bhardwaj, K. Rajkumar, L. Panicker, C.V. Chaudhari, S.K. Chakraborty, et al., Polychloroprene rubber/ethylene-propylene diene monomer/multiple walled carbon nanotube nanocomposites: synergistic effects of radiation crosslinking and MWNT addition, J. Polym. Res. 19 (2012) 9876, http://dx.doi.org/10.1007/s10965-012-9876-7.
[65] K.A. Dubey, Y.K. Bhardwaj, C.V. Chaudhari, S. Sabharwal, LDPE/EVA/PCR/MWNT nanocomposites: radiation crosslinking and physicomechanical characteristics, Polym. Compos. 32 (2011) 737–746, http://dx.doi.org/10.1002/pc.21092.
[66] S.K. Sinha, K.A. Dubey, V. Grover, P. Sastry, Y.K. Bhardwaj, L. Panicker, et al., Selective interfacial localization in conducting polycaprolactam/ethylene vinyl acetate/carbon black composites, Polym. Compos 34 (2013) 912–919, http://dx.doi.org/10.1002/pc.22497.

[67] K.A. Dubey, S.K. Sinha, Y.K. Bhardwaj, L. Panicker, L. Varshney, Carbon black-filled PE/PP/EPDM blends: phase selective localization of carbon black and EPDM-induced phase stabilization, Polym. Plast. Technol. Eng. 53 (2014) 442–450, http://dx.doi.org/10.1080/03602559.2013.844832.
[68] P.C. Ma, J.K. Kim, B.Z. Tang, Effects of silane functionalization on the properties of carbon nanotube/epoxy nanocomposites, Compos. Sci. Technol. 67 (2007) 2965–2972, http://dx.doi.org/10.1016/j.compscitech.2007.05.006.
[69] L. Guadagno, B. De Vivo, A. Di Bartolomeo, P. Lamberti, A. Sorrentino, V. Tucci, et al., Effect of functionalization on the thermo-mechanical and electrical behavior of multi-wall carbon nanotube/epoxy composites, Carbon 49 (2011) 1919–1930, http://dx.doi.org/10.1016/j.carbon.2011.01.017.
[70] A. Montazeri, The effect of functionalization on the viscoelastic behavior of multi-wall carbon nanotube/epoxy composites, Mater. Des. 45 (2013) 510–517, http://dx.doi.org/10.1016/j.matdes.2012.09.013.
[71] A. Garg, S.B. Sinnott, Effect of chemical functionalization on the mechanical properties of carbon nanotubes, Chem. Phys. Lett. 295 (1998) 273–278, http://dx.doi.org/10.1016/S0009-2614(98)00969-5.
[72] S. Wang, R. Liang, B. Wang, C. Zhang, Reinforcing polymer composites with epoxide-grafted carbon nanotubes, Nanotechnology 19 (2008) 085710, http://dx.doi.org/10.1088/0957-4484/19/8/085710.
[73] F. An, C. Lu, Y. Li, J. Guo, X. Lu, H. Lu, et al., Preparation and characterization of carbon nanotube-hybridized carbon fiber to reinforce epoxy composite, Mater. Des. 33 (2012) 197–202, http://dx.doi.org/10.1016/j.matdes.2011.07.027.
[74] K.A. Dubey, C.V. Chaudhari, S.K. Suman, N. Raje, R.K. Mondal, V. Grover, et al., Synthesis of flexible polymeric shielding materials for soft gamma rays: physicomechanical and attenuation characteristics of radiation crosslinked polydimethylsiloxane/Bi_2O_3 composites, Polym. Compos. (2014), http://dx.doi.org/10.1002/pc.23232.
[75] S. Majji, K.A. Dubey, R.K. Mondal, Y.K. Bhardwaj, S. Acharya, Development of electron beam cross-linked PDMS/PTFEM composites with low coefficient of friction and high elastic modulus, Polym. Plast. Technol. Eng. 53 (2014) 435–441, http://dx.doi.org/10.1080/03602559.2013.844831.
[76] C. Yan, J. Wang, W. Kang, M. Cui, X. Wang, C.Y. Foo, et al., Highly stretchable piezoresistive graphene-nanocellulose nanopaper for strain sensors, Adv. Mater. 26 (2014) 2022–2027, http://dx.doi.org/10.1002/adma.201304742.
[77] Y. Tang, J. Gou, Synergistic effect on electrical conductivity of few-layer graphene/multi-walled carbon nanotube paper, Mater. Lett. 64 (2010) 2513–2516, http://dx.doi.org/10.1016/j.matlet.2010.08.035.
[78] E.S. Kim, G. Hwang, M. Gamal El-Din, Y. Liu, Development of nanosilver and multi-walled carbon nanotubes thin-film nanocomposite membrane for enhanced water treatment, J. Memb. Sci. 394–395 (2012) 37–48, http://dx.doi.org/10.1016/j.memsci.2011.11.041.
[79] W. Zhang, T. Yang, D. Huang, K. Jiao, G. Li, Synergistic effects of nano-ZnO/multi-walled carbon nanotubes/chitosan nanocomposite membrane for the sensitive detection of sequence-specific of PAT gene and PCR amplification of NOS gene, J. Memb. Sci. 325 (2008) 245–251, http://dx.doi.org/10.1016/j.memsci.2008.07.038.
[80] D.H. Yi, H.J. Yoo, S.S. Mahapatra, Y.A. Kim, J.W. Cho, The synergistic effect of the combined thin multi-walled carbon nanotubes and reduced graphene oxides on photothermally actuated shape memory polyurethane composites, J. Colloid Interface Sci. 432 (2014) 128–134, http://dx.doi.org/10.1016/j.jcis.2014.06.060.
[81] K.A. Dubey, S. Majji, S.K. Sinha, Y.K. Bhardwaj, S. Acharya, C.V. Chaudhari, et al., Synergetic effects of radiolytically degraded PTFE microparticles and organo-clay in PTFE-reinforced ethylene vinyl acetate composites, Mater. Chem. Phys. 143 (2013) 149–154, http://dx.doi.org/10.1016/j.matchemphys.2013.08.039.
[82] K.E. Prasad, B. Das, U. Maitra, U. Ramamurty, C.N.R. Rao, Extraordinary synergy in the mechanical properties of polymer matrix composites reinforced with 2 nanocarbons, Proc. Natl. Acad. Sci. U. S. A. 106 (2009) 13186–13189, http://dx.doi.org/10.1073/pnas.0905844106.
[83] M.K. Shin, B. Lee, S.H. Kim, J.A. Lee, G.M. Spinks, S. Gambhir, et al., Synergistic toughening of composite fibres by self-alignment of reduced graphene oxide and carbon nanotubes, Nat. Commun. 3 (2012) 650, http://dx.doi.org/10.1038/ncomms1661.

CHAPTER

GLASSES AND GLASS-CERAMICS FOR VACUUM AND HIGH-TEMPERATURE APPLICATIONS

6

A. Ananthanarayanan, M. Goswami, G.P. Kothiyal
Bhabha Atomic Research Centre, Mumbai, India

CHAPTER OUTLINE

Materials Under Extreme Conditions. http://dx.doi.org/10.1016/B978-0-12-801300-7.00006-1

1. INTRODUCTION

Glasses and glass-ceramics are widely used for a variety of scientific, technological, industrial, and general-purpose day-to-day applications owing to a combination of special inherent properties such as chemical bonding, electrical and thermal insulation, chemical inertness, wettability, etc., which in turn then play an important role in various application, including vacuum technology [1–13]. Among the various technological applications, joining of glass to metal, or other types of glasses are quite important and also technically demanding. Vacuum glass sealed-off devices like electric bulbs, flash lamps, hollow cathode lamps, electronic valves, cathode ray tubes, X-ray tubes, TV tubes, etc., are some of the prominent examples [14,15]. Further, in many chemical plants and scientific laboratories glass tubings, flasks, and gadgets like stop cocks, joints, etc., for vacuum lines are employed. The aim of this chapter is to acquaint readers with the applications of glasses and glass-ceramics for vacuum/hermetic (air tight) sealing/high-pressure sealing/high-temperature sealing and introduce them to material selection criteria in terms of compositions, heat treatments, and metal treatments to enable good glass (or glass-ceramic) to metal bonding, vacuum compatibility, and other required properties [9,16,17]. By no means though is the application of glass for preparing coatings, or devices a recent occurrence. As such, glasses have been used for rendering the surface of ceramic vessels shiny, decorative, and impervious by ancient civilizations, perhaps first by the Egyptians, in around 1500 BC [18,19].

In the present context, glasses and glass-ceramics, owing to improved manufacturing processes are supplementing and sometimes replacing conventional forms of hermetic sealing using epoxies, elastomers, solders or brazes. In part, this is due to the ability of glasses and glass-ceramics to withstand corrosive environments and/or harsh environments such as those encountered in solid oxide fuel cell stacks (operating temperature $>700^{\circ}C$) and ultra-high vacuum ($<10^{-6}$ mbar) [1,2,5,17].

In order to ensure satisfactory material performance under rigorous operating environments, a thorough understanding and judicious optimization of structure property correlations is required. In view of this, we will briefly present a discussion on the nature of the glassy state, the glass transition, and the crystallization of glasses. We will also present a functional understanding of some of the characterization techniques that allow us to optimize the properties of these materials to ensure optimal seal performance.

Since this chapter is mainly focused on hermetic/pressure sealing applications, we need an understanding of what we imply by the term "vacuum." Industrially, high vacuum is simply another way of describing a region of low pressure, usually the vacuum chamber, separated from a region of high pressure, usually normal atmosphere [20,21]. Glasses (and glass-ceramics) are used to provide seals that allow monitoring of pressure or conditions in the vacuum chamber while still providing pressure isolation [22]. As a converse, glasses and glass-ceramics are also useful in seals that separate a high pressure ambient from atmospheric pressure, and we will also describe such seals later in this chapter [23].

In general, glasses and glass-ceramics exhibit the properties listed following, which enable them to be used as sealants for high-vacuum environments or to isolate extreme conditions such as high pressure, low pressure, or high temperature [24]:

1. Glasses and glass-ceramics are chemically durable and are resistant to most corrosive environments. However, an exception must be made for highly alkaline environments, which are highly deleterious to the longevity of even the most durable silicate glasses.

2. Glasses can be shaped easily and applied in the molten or semimolten state. In situ crystallization will subsequently enable the glass to be ceramized to yield glass-ceramics.
3. Glass and glass-ceramics show good bonding with a range of materials and most importantly with most of the major alloys that are used in vacuum technology.
4. The thermal expansion coefficient (TEC) of glasses can be optimized for various applications by suitably tailoring the composition [25,26]. The same can be achieved in glass-ceramics by a judicious choice of heat treatment protocols in addition to composition optimization [1,5,22,23,25].
5. Glass and glass-ceramics have very low outgassing rates. As a result, they are useful even in high or ultrahigh vacuum applications, where plastics and polymers are limited in their utility [27]. In this regard, there also exist special materials, viz., machinable glass-ceramic compositions that retain all the advantages of glass-ceramics in terms of vacuum compatibility, the ability to be shaped by machining methods, viz., drilling, grinding, or turning. We will describe these materials subsequently.

We will proceed to describe briefly the structure of glasses, glass formation, and ceramization of glasses to acquaint nonspecialist readers, so that further descriptions of processes and materials can be understood easily.

2. GLASS AND GLASS-CERAMICS AND RELATED PROPERTIES

2.1 GLASSES

In this section, a brief overview of glasses, focusing on glass formation and network model attributed to Zachariasen, is discussed [28]. These aspects are chosen since they will be invoked during materials selection and process refinement during fabrication of glass-ceramic-to-metal seals. Let us begin by asking a seemingly simple question: What is glass?

The popular and reasonably correct answer is that a glass is a supercooled liquid that has lost its ability to flow. It follows from this basic idea that a glass is essentially a solid in terms of its shear modulus and rigidity, with the atomic arrangement more akin to a liquid [29,30]. This implies that the long-range order of crystalline materials is absent in glasses. In addition, glass material must show a glass transition behavior. One of the ways, and by far the most popular way, to get glass is to quench a melt from high temperature [31,32]. Consequently, the method is called the melt-quench method. Immediately, it follows that the glassy state obtained is not unique. In addition to the chemical composition, the thermal history of the melt also influences the glassy state realized [29,33]. Since this chapter deals with glasses for sealing (vacuum/high pressure) applications and therefore mainly focuses on oxide glasses, certain terminology used commonly in relation to glasses is elucidated. The main constituents/backbone of glass, are known as glass formers (SiO_2, P_2O_5 or B_2O_3), which can form glass networks [34]. Intermediate oxides such as Al_2O_3, PbO, etc., cannot form glass networks by themselves but can participate in glass network formation or act as network modifiers depending upon the conditions in the glass network. Finally, there are network modifiers (Na_2O, Li_2O, BaO, CaO, SrO) among others that break up the glass network and reduce network connectivity [34]. Modifiers are commonly added to glasses in order to control melt rheology and influence the glass transition temperature and enhance the crystallization tendency of glasses. An artist's conception of glass structure, superimposed on a micrograph of a silica sheet of two-atom thickness, is shown in Fig. 6.1 exemplifying an open random network [35].

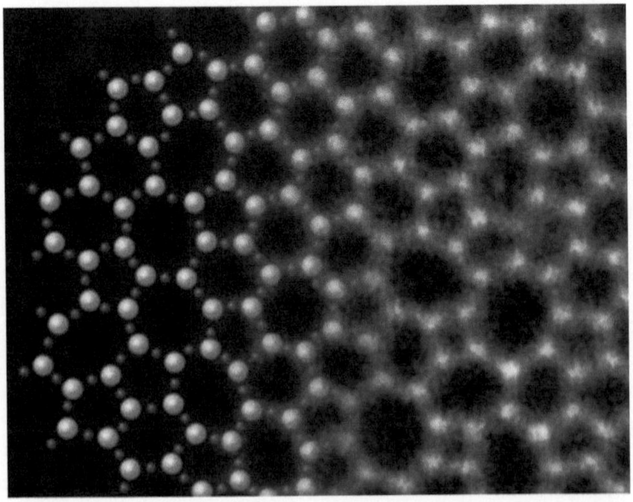

FIGURE 6.1

Random network of glass superimposed on a micrograph of a two-atom-thick silica sheet as perceived by an artist.

Reprinted with permission from P.Y. Huang, S. Kurasch, A. Srivastava, V. Skakalova, J. Kotakoski, A.V. Krasheninnikov, R. Hovden, Q. Mao, J.C. Meyer, J. Smet, D.A. Muller, U. Kaiser, Direct imaging of a two-dimensional silica glass on Graphene, Nano Lett. 12 (2012) 1081–1086. Copyright (2016) American Chemical Society.

The classification of oxides mentioned previously, into glass formers, intermediate oxides, and network modifiers is based in large part on the studies of Zachariasen in the 1920s. He formulated the following rules for glass formation in a general oxide A_mO_n [28]:

1. A metal atom can at most be linked to two oxygen atoms.
2. The oxygen coordination around A (metal ion) must be small (three or four).
3. The cation polyhedra share corners, not edges or faces.
4. At least three corners of the cation-tetrahedra are shared.

These rules are based only upon the structure of the amorphous materials and can be considered as the pinnacle of the structural models of glass formation. With a better understanding of glassy materials, it has become clear that glass formation in many nonoxide or metallic glasses cannot be understood solely based upon structural models [7,36–41]. One of the reasons is that there are also thermodynamic and kinetic phenomena underlying the glass transition [42]. These aspects will be briefly presented in the following sections.

2.1.1 The Glass Transition Enigma

A glass can be viewed as a liquid, in which a huge slowing down of the diffusive motion of the particles has destroyed its ability to flow on experimental time scales [31,34,39]. The slowing down is expressed through the relaxation time, that is, generally speaking, the characteristic time at which the slowest measurable processes relax to equilibrium [43,44]. Cooling down from the liquid phase, the slow degrees of freedom of the glass former are no longer accessible and the viscosity of the undercooled melt grows several orders of magnitude over a relatively small temperature interval [44]. As a result, in the cooling process, from some point on, the time effectively spent at a certain temperature is not enough to attain equilibrium [45]. The system is said to have driven out of equilibrium, which is at the origin of the glass transition. Of course, the preparation process, indeed, plays a significant role to get a glass out of a liquid, thus avoiding the crystallization of the substance. To better understand this, let us first track the change in specific volume of a melt as the temperature is lowered as shown in Fig. 6.2.

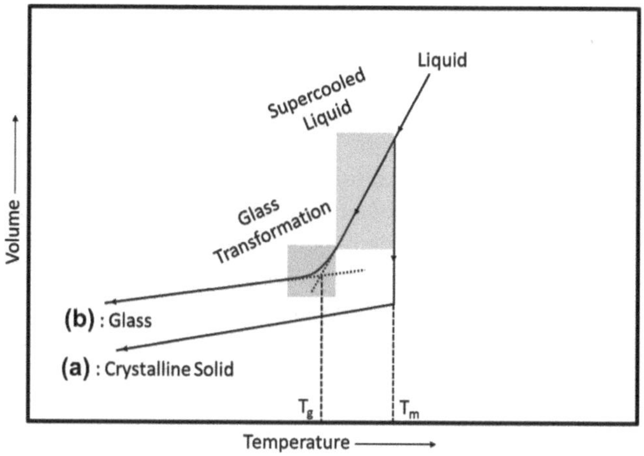

FIGURE 6.2

Variation in specific volume of a liquid cooled at different rates. (a) Represents the solidification of a liquid to give crystalline solid, while (b) represents glass formation.

With a reduction in temperature, the specific volume decreases linearly with temperature. As the temperature reaches the melting point, T_m, two possibilities exist:

1. Crystallization: This process is manifested by a discontinuous change in the specific volume at T_m. Since the change in specific volume with temperature is discontinuous, the crystallization phenomenon is an example of a first-order transition.
2. Supercooling: In this process, the liquid state is maintained below T_m. The liquid state that persists below T_m is known as the supercooled liquid (region marked blue in Fig. 6.2).

The formation of the supercooled liquid can be thought as the first step in glass formation. As the supercooled liquid cools further, its viscosity increases rapidly. A stage is reached where the viscosity of the liquid becomes so large that it can be considered to be a solid. When the material solidifies, the slope of the specific volume versus temperature curve becomes nearly equal to solid-like as shown in Fig. 6.2.

The region over which the slope of the specific volume versus temperature curve changes from liquid-like to solid-like is the glass transition region. Without loss of generality, the specific volume can be replaced with entropy S or enthalpy H in Fig. 6.2. The glass transition is not a single temperature but a region (orange region in Fig. 6.2), so for practical purposes, it is customary to define the temperature at the intersection point of the tangents drawn to the liquid and glass curves as the glass transition temperature, T_g.

Even at present there is no clear consensus on the nature and the physics underlying the glass transition phenomenon. Interested readers are referred to the bibliography for a more detailed discussion of the physics and nature of the glass transition phenomenon [7,29,40,41,43,45]. From a glass sealant point of view, it is important to note that above the glass transition, the glass is no longer solid but liquid-like. At these temperatures, the glass can deform under an applied load and dimensional tolerances of glass pieces can be lost. Further, with the rapid increase in $\alpha(T)$, the TEC in the vicinity of T_g can lead to integrity problems with glass-to-metal seals and cause the seals to delaminate. One of the important considerations in sealing is to obtain a glass that can wet the metal it is being sealed with, but at the same time have a high enough T_g to avoid creep under service conditions. This will be discussed in greater detail in the section on glass sealing and wetting of metals.

Before closing this section, we mentioned that the scientific interest into the structure of glass manifests itself in a singularly useful property of glasses, viscosity. Running with the premise of the simplified free volume model [46], as temperature increases, the free volume in the glass increases, which makes it easier for the glass to flow [47,48]. In other words, the changing glass structure in the vicinity of T_g is important for the rapid decrease in viscosity of the glass from very high solid-like values below T_g to low, more moderate, values just above T_g, and finally transitioning to liquid-like values near the fusion temperature [46,48]. In context of glasses and the fact that they are a continuation of the liquid state into the solid state, it is hard to define a melting temperature akin to that of a crystalline solid. Instead, when the viscosity of the glass-forming liquid is 10 Pa s, it is considered to be a glass melt. The ability to change the viscosity of a glass by changing the temperature is the prime mover for glass blowing techniques, for glass-ceramic formation, and also sealing with metals, joining of different glasses, and annealing.

The viscosity of some glass formers increases rapidly and exponentially with temperature, the archetype being SiO_2. Such glass formers are known as strong glass formers [49]. In these, the amorphous states are separated by large energy barriers [50]. Strong glass formers display Arrhenius behavior in their temperature versus viscosity curves. The Arrhenius relation is given in Eq. (6.1) [51,52]:

$$\text{Viscosity}\eta = \eta_0 \exp\left(\frac{A}{T}\right) \tag{6.1}$$

In this expression η_0 is a constant while A is the activation energy for viscous flow.

In fragile glass formers such as fluorides, the viscosity versus temperature relation follows the Vogel–Fulcher–Tammann–Hesse law, as given in Eq. (6.2) [53,54]:

$$\eta = \eta_0 \exp\left(\frac{B}{T - T_0}\right) \tag{6.2}$$

where, η_0, B, and T_0 are constants.

The fragility of amorphous materials is also characterized by the Doremus ratio [55,56] as expressed in Eq. (6.3):

$$R_D = \frac{Q_H}{Q_L} \tag{6.3}$$

where, Q_L is the viscosity at low temperature (in the glassy state) and Q_H is the viscosity at high temperature. For a fragile glass former $R_D - 1 \geq 1$ and for strong glass formers, $R_D - 1 < 1$.

In closing, we must reflect that although all the glass-based industries in the world rely upon the structural changes and atomic scale rearrangement taking place in glasses near T_g to render glasses workable in operations such as shaping, blowing, molding, annealing, glazing, sealing, and joining, a complete understanding (kinetic and thermodynamic) of the glass transition phenomenon remains elusive.

2.1.2 Crystallization of Glasses and the Time-Temperature-Transformation Diagram

In the discussions about glasses, it has been emphasized that it is essential to traverse the region around the glass transition temperature T_g rapidly to preclude crystallization in the glasses. Clearly, if the

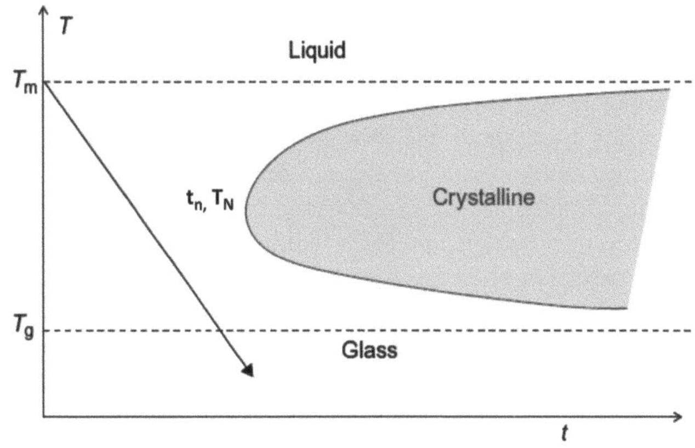

FIGURE 6.3

Schematic of time-temperature-transformation (TTT) diagram for one thermal history (*arrow*).

Adapted from J.L.M. Rupp, B. Scherrer, N. Schäuble, L.J. Gauckler, Time–temperature–transformation (TTT) diagrams for crystallization of metal oxide thin films, Adv. Funct. Mater. 20 (2010) 2807–2814.

cooling rate is slow enough, the melt will reorganize into a crystalline solid and glass formation will not be possible. To predict the lowest possible rate at which a glass melt may be cooled to yield a glass, a time-temperature-transformation (TTT) plot, also known as a TTT diagram, is drawn as shown in Fig. 6.3.

On the y-axis the temperature, and on the x-axis the time required for crystallization, are plotted. At very high temperature, well above the melting temperature, T_m, crystalline solid cannot form in the melt. Consequently, the time required for the formation of a crystalline solid is infinitely large. As the temperature of the melt decreases below T_m, the probability of nucleating the crystalline solid increases, leading to a decrease in the time required for nucleating the crystalline solid in the melt. Simultaneously, the viscosity of the melt also increases upon cooling below T_m, which increases the time required for nucleation of the crystalline solid.

At a certain temperature, the time required for crystallization of a crystalline nucleus is the least. This point is the nose of the shaded region in Fig. 6.3. The temperature corresponding to this point is the crystallization temperature, T_N, and time required is t_n. Upon cooling below the crystallization temperature, viscosity begins to dominate and time required for nucleating a crystalline state increases rapidly. By the time the temperature falls to T_g, the glass former has been solidified and time required for nucleating the crystalline state is again effectively infinite.

The region to the right of the curve (shaded gray) in Fig. 6.3 represents the region where crystallization can occur. During glass preparation, any cooling trajectory must not pass through the shaded region to avoid crystallization. The TTT diagram can also be used to understand the critical cooling rate to obtain a glass. The "critical cooling" rate is the lowest cooling rate that still allows glass formation. On the TTT diagram, the critical cooling trajectory is tangential to the TTT curve at the "nose." In other words, the critical cooling trajectory intersects the TTT curve at coordinates (T_c, t_c). Assuming the glass melt is cooled from the melting temperature T_m, the critical cooling rate is described by the following expression:

$$\frac{dT}{dt_c} = \frac{T_m - T_n}{t_n} \tag{6.4}$$

2.2 GLASS-CERAMICS

While uncontrolled crystallization of a glass during cooling is undesirable, controlled crystallization of glasses can be used to yield an intrinsic composite of glass and crystalline material known as glass-ceramics [17,22]. Glass-ceramics are materials that blend the best properties of glasses and ceramics in the same material. In general they exhibit superior chemical durability and mechanical strength as compared to the glasses they are derived from. In addition, they offer very good dimensional tolerance, the ability to be produced in the shape and form required, and also excellent surface finish. Unlike conventional ceramics, they also offer the possibility of zero or nearly zero porosity, an important consideration for hermetic sealing [3,4,17,22,23]. Crystallization of a glass to yield glass-ceramics involves a nucleation step and then a growth of crystallites (crystallization step). During nucleation, nuclei of the crystalline phase or of an initiator phase are formed in the glass. Then during crystallization, the major crystalline phase grows on/from the nuclei. The readers are referred to the bibliography for a more detailed discussion on nucleation and crystallization of glasses [58–61].

2.2.1 Nucleation

We begin with the classical theory of nucleation, with a description of what we mean by a nucleus. A nucleus is an entity that belongs to a different phase but is in unstable equilibrium with respect to the supersaturated, surrounding parent phase. Let us now proceed by saying that the nucleus is a solid phase in a liquid, such as the melt. When the atoms of the melt come together to form a nucleus, a new surface is created between the emerging nucleus and the embedding liquid, and the creation of this surface requires energy. The phase transformation from liquid to crystal provides the energy necessary for the creation of the new surface. Let us now define w_c as the energy necessary to form a particle that then grows into a crystallite; w_c is the critical energy below which the atoms aggregate and then separate. Seen in this light, w_c can be thought to be a barrier to crystallization, and we will develop expressions for w_c in case of homogeneous and heterogeneous nucleation. Further, the atoms must reach the growing crystallite by overcoming a diffusion barrier, which is expressed as $\Delta g'$. The nucleation rate will therefore be an expression consisting of two terms. One term involves w_c and the other involves $\Delta g'$. The nucleation rate can be expressed as:

$$n(T) = n\upsilon_E \exp\left(-\frac{w_c}{k_B T}\right)\exp\left(-\frac{\Delta g'}{k_B T}\right) = n\upsilon_E \exp\left(-\frac{Nw_c}{RT}\right)\exp\left(-\frac{\Delta G'}{RT}\right) \tag{6.5}$$

where, n is the concentration of atoms, υ_E is the frequency of atomic transitions in the liquid (of the order of $10^{13}\ \mathrm{s}^{-1}$), T is the temperature, and N is Avogadro's number. The first term $\exp\left(-\frac{Nw_c}{RT}\right)$ is the thermal activation related to an energy barrier w_c for a particle to form. The second term $\exp\left(-\frac{\Delta G'}{RT}\right)$ is diffusion term, which denotes the energy required by the atoms to diffuse through the melt and reach the solid nucleus.

In the limiting case of the melt containing only one type of atom, the atoms required for further growth of the crystalline nucleus are readily available in the immediate vicinity. In such a case, diffusion of atoms over long distances is superfluous. Consequently, the diffusion barrier appearing in the expression for nucleation rate is the barrier against viscous flow of the melt.

With this simplified, but essentially realistic, discussion, we can proceed to the cases of homogeneous nucleation and heterogeneous nucleation.

2.2.1.1 Homogeneous Nucleation

As the name suggests, homogeneous nucleation is the spontaneous creation of a solid nucleus in a liquid. There are now two contributions to the energy of the particle:

1. The energy of formation of the crystal from the liquid, w_f.
2. The interface energy, w_i.

The former can be expressed as:

$$w_f = \frac{V}{V_m}\Delta G_v \tag{6.6}$$

This expression is obtained under the assumption that the nucleus is spherical with a radius r. ΔG_v denotes the change in free energy of the liquid-to-crystal transformation. V_m denotes the molar volume of the crystal.

The interfacial energy term can be expressed as:

$$w_i = 4\pi r^2 \varsigma_{sl} \tag{6.7}$$

In this expression, ς_{sl} represents the interfacial energy between the solid and the liquid. Based on these expressions, if we plot the free energies as a function of the particle size, it is evident that $w = w_i + w_f$ shows a maximum at:

$$r_c = \frac{2\varsigma_{sl} V_m}{\Delta G_v} \tag{6.8}$$

The maximum free energy is then expressed as:

$$w_c = \frac{16\pi \varsigma_{sl}^3 V_m^2}{\Delta G_v} \tag{6.9}$$

The maximum in energy w_c sets a limit on size r_c. When an aggregate reaches the critical size, it will develop further and grow. Below the critical size, the atoms aggregate and then separate again.

With the activation term expressed, let us now focus on the diffusion term. For this we invoke the Stoke–Einstein equation. The diffusion coefficient can now be written as:

$$D = \frac{k_B T}{3\pi a_0 \eta} = \upsilon_E a_0^2 \exp\left(-\frac{\Delta G}{RT}\right) \tag{6.10}$$

In this expression, a_0 is the atomic jump distance (which is assumed to be twice the particle size) and η represents the viscosity.

Using this, we get the expression for nucleation rate as:

$$n(T) = n\upsilon_E \exp\left(-\frac{w_c}{k_B T}\right)\exp\left(-\frac{\Delta g'}{k_B T}\right) = n\upsilon_E \exp\left(-\frac{N w_c}{RT}\right)\exp\left(-\frac{\Delta G'}{RT}\right)$$
$$= n\frac{k_B T}{3\pi a_0 \eta}\exp\left(-\frac{N w_c}{RT}\right) \tag{6.11}$$

From this expression, an important conclusion can be obtained. We observe that there is balance between the contributions of decreasing w_c (caused by an increase in ΔG_v with decreasing temperature, with $\Delta G_v = 0$ at the liquidus temperature T_L) and an increase in viscosity with decreasing temperature.

2.2.1.2 Heterogeneous Nucleation

Heterogeneous nucleation is the case when the crystallization is aided by the presence of impurities. The impurities that aid crystallization are termed as catalyzers. In case of heterogeneous nucleation, there are three interfaces to be considered: the liquid catalyzer interface, the solid catalyzer interface, and the solid liquid interface. The mechanical equilibrium of the solid growing on the catalyzer is expressed by the Young–Dupree equation [62]:

$$\varsigma_{cl} = \varsigma_{cs} + \varsigma_{sl} \cos\theta \tag{6.12}$$

where, θ is the contact angle between solid and liquid interface.

To find the surface of the particle, we have to consider the area of the particle in contact with the liquid, S_{sl} and the area in contact with the catalyzer, S_{cs}.

$$S = S_{sl} + S_{cs} \tag{6.13}$$

where,

$$S_{sl} = \int_0^{\theta} 2\pi r \sin\theta' \, d\theta'$$

also,

$$S_{cs} = \pi r^2 \sin^2\theta = \pi r^2 (1 - \cos^2\theta)$$

The volume of the solid nucleus can then be expressed as:

$$V = \int_0^{\theta} \pi \left(r \sin\theta' \right)^2 r \sin\theta' \, d\theta' = \frac{2}{3}\pi r^3 \left(1 - \frac{3}{2}\cos\theta + \frac{1}{2}\cos^3\theta \right) \tag{6.14}$$

Then, expressing the energy of the particle as a sum of surface and volume contributions,

$$w_f = \frac{V}{V_m}\Delta G_v = \frac{\frac{2}{3}\pi r^3 \left(1 - \frac{3}{2}\cos\theta + \frac{1}{2}\cos^3\theta \right)}{V_m} \tag{6.15}$$

and,

$$w_i = \varsigma_{sl}S_{sl} + \varsigma_{cs}S_{cs} - \varsigma_{cl}S_{cs} = \varsigma_{sl}S_{sl} + (\varsigma_{cs} - \varsigma_{cl})S_{cs} \tag{6.16}$$

Invoking the expression for mechanical equilibrium,

$$w_i = \varsigma_{sl}S_{sl} + (\varsigma_{sl}\cos\theta)S_{cs} = \varsigma_{sl}\pi r^2 (2 - 3\cos\theta + \cos^3\theta) \tag{6.17}$$

therefore,

$$w = w_i + w_f = \left(1 - \frac{3}{2}\cos\theta \frac{1}{2}\cos^3\theta\right)\left\{\frac{\frac{2}{3}\pi r^3 \Delta G_v}{V_m} + 2\varsigma_{sl}\pi r^2\right\} \tag{6.18}$$

We can then define the critical radius for heterogeneous nucleation as:

$$r_c = 2\frac{\varsigma_{sl} V_m}{\Delta G_v} \tag{6.19}$$

Further,

$$w_{c\ net} = \frac{1}{2}\left(1 - \frac{3}{2}\cos\theta + \frac{1}{2}\cos^3\theta\right) w_{c\ hom} = \left(\frac{1}{2} - \frac{3}{4}\cos\theta + \frac{1}{4}\cos^3\theta\right) w_{c\ hom} \tag{6.20}$$

Eq. (6.20) shows that even for large contact angle, θ, the presence of a catalyzer can greatly reduce the barrier against nucleation. This equation shows why heterogeneous nucleation is important in the manufacture of glass-ceramic components and parts. In addition, heterogeneous nucleation results in larger particle size as indicated by the next equation:

$$\frac{r_{het}}{r_{hom}} = \left(\frac{1}{2} - \frac{3}{4}\cos\theta + \frac{1}{4}\cos^3\theta\right)^{-1/3} \tag{6.21}$$

It may be mentioned in passing that the glass-to-metal interface can often act as a nucleating site for crystallization [5,22].

2.2.2 Crystal Growth

Once nucleation takes place, crystal growth occurs by diffusion of atoms to the interface between the glass and growing crystalline regions. At the liquidus temperature, it is intuitively clear that the crystal growth rate tends to zero. As the temperature reduces, the growth rate becomes increasingly favorable from a thermodynamic standpoint. At the same time, melt viscosity increases with reducing temperature and serves as a kinetic barrier against crystallization. Therefore, there is a temperature where equilibrium is attained between thermodynamic and kinetic factors, where maximum growth rate is obtained.

During the fabrication of glass-ceramics, the heat treatment temperature is tuned to obtain the required crystalline phase fraction and maintain an optimum crystal/glass ratio in the glass-ceramic.

2.3 METHODS OF GLASS AND GLASS-CERAMIC FORMATION

We begin this section by mentioning that there are many specific methods of glass formation [58,63]. Since we are focusing on glass sealing, we will describe the melt-quench technique, which allows large-scale glass formation and sealing [31]. In the melt-quench technique, the chemicals used for glass formation are melted in a suitable crucible (alumina, platinum, etc.) by heating in a furnace. The melt is then held at high temperature to remove gases formed by the decomposition of nitrates and carbonates in the precursors. Melt homogeneity can be ensured by periodic stirring, particularly if the melt is viscous. Agitation of the melt by stirring also allows removal of gas bubbles. The melt is then poured out into graphite or metallic molds to shape the glass piece. Cooling rates of the order of several hundred degrees per second can be realized by this method, and it can also be adapted for continuous processing. Another advantage is that the amount of glass prepared can be easily scaled up using larger furnaces and crucibles.

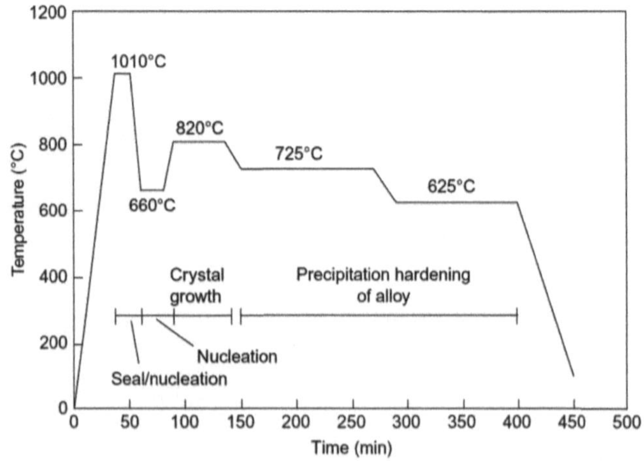

FIGURE 6.4

Typical heat treatment schedule for sealing lithium aluminum silicate glasses to steel.

Adapted from I.W. Donald, Glass to Metal Seals, Society of Glass Technology, London, 2009.

Once glass is formed, it may be crystallized in situ using a controlled heat treatment. Typical heat treatment profile for crystallization is shown in Fig. 6.4.

In terms of glass formation, other techniques such as sol–gel, chemical vapor deposition, glow discharge decomposition, radiation bombardment, etc. exist, and we direct the interested reader to the bibliography for further reading about these techniques [58,63,64].

3. WETTABILITY AND JOINING OF METALS USING GLASSES

Oxide glasses find application in sealing and coating of metals. In these applications, wetting of the metal by the glass is of great importance. In this section we shall detail some of the basic concepts involved in wetting and also consider the particular effectiveness of glass in wetting the metal. Wetting is a surface phenomenon; therefore consider a flat, perfectly smooth, nondeformable, and chemically homogeneous surface in contact with a nonreactive liquid phase in the presence of the vapor phase. If the liquid does not completely cover the solid, the liquid surface will intersect the solid surface at a "contact angle" θ. The equilibrium value of θ, used to define the wetting behavior of the liquid, obeys the classical equation of Young [61]. This is given by Eq. (6.22) [62]:

$$\cos \theta_Y = \frac{\sigma_{SV} - \sigma_{SL}}{\sigma_{LV}} \tag{6.22}$$

where, σ_{SV}, σ_{SL}, and σ_{LV} are the interfacial free energies. A contact angle of less than 90 degrees will identify a wetting liquid, while a greater value will identify a nonwetting liquid. If the contact angle is 0 degree, the liquid will be considered to be perfectly wetting. Eq. (6.22) can be easily derived by calculating the variation of the surface free energy F_s of the system caused by a small displacement δz of the S/L/V triple line, usually hereafter referred to as TL. In Fig. 6.5 the solid surface is vertical and TL is perpendicular to the plane of the figure and assumed to be a straight line, rendering the problem a two-dimensional one. Thus, the total length of TL is constant during its displacement, as in the case of a meniscus formed on a vertical plate. Moreover, for the radius R of the TL region considered in this calculation (Fig. 6.5) for metallic and covalent ceramics, this range is roughly

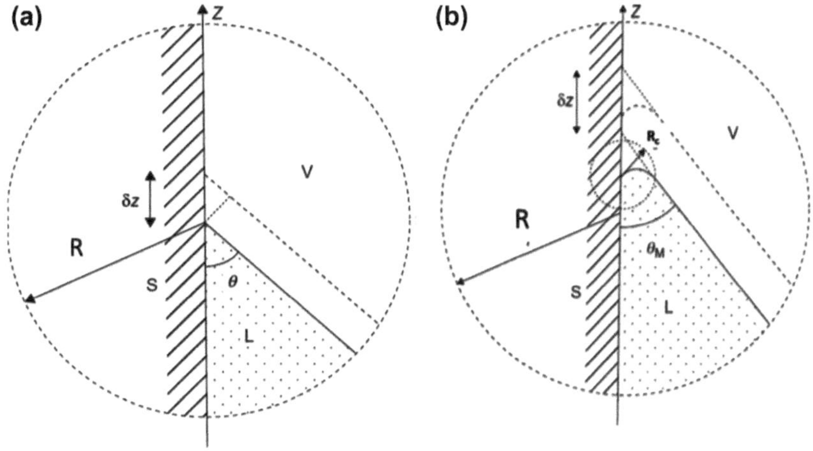

FIGURE 6.5

Schematic figure showing solid, liquid, and vapor interface for (a) macroscopic wetting, while (b) shows that the microscopic contact angle is different from the macroscopic contact angle.

Adapted from N. Eusthopoulos, M.G. Nicholas, B. Drevet, Wettability at High Temperatures, Oxford, 1999.

1–2 nm, while for bodies with long-range van der Waals interactions it can attain 10 nm, which is much larger than the range of atomic (or molecular) interactions in the system.

However, the radius R must be small compared to a characteristic dimension of the liquid, and the region of radius R, the intersection of the L/V surface with the plane of Fig. 6.5a, is assumed to be essentially a straight line. With these assumptions, the variation of interfacial free energy per unit length of TL, resulting from a small linear displacement δz of TL, is:

$$F_s(z + \delta z) - F_s(z) = \delta F_s = (\sigma_{SL} - \sigma_{SV})\delta z + \cos(\theta)\sigma_{LV}\delta z \tag{6.23}$$

The equilibrium condition $d(\delta F_s)/d(\delta z) = 0$ leads to Eq. (6.23).

In practice, the geometry in a core region around TL, i.e., inside a sphere of radius R_c, which is of the order of the range of atomic interactions in the system, can be different from that shown in Fig. 6.5b.

Due to long-range interactions, the energy of an atom lying on a given interface inside this core region is different from the energy of an atom at the same interface far from TL. Thus, the three relevant interfacial energies close to and far from TL are different and this difference increases with the range of atomic interactions. Consequently, the local contact angle, as far as it can be defined at this nanometric scale, may be different from the macroscopic contact angle θ_M. This is schematically shown in Fig. 6.5b.

When the drop size becomes of the order of magnitude of the radius R_c of the core region defined in Fig. 6.5a, the relevant contact angle is no longer the Young contact angle but the microscopic contact angle. The relation between these two angles has been discussed using the concept of line energy (or line tension). During wetting, the increase δ_R of the drop base radius leads to an increase of the TL length (Fig. 6.5b) and the free energy associated with the increase of the TL length must be taken into account in the interracial energy calculation. The triple line can be treated in phenomenological manner as an equilibrium line defect with a specific excess free energy z (per unit length). To some extent, it can be compared to the step energy on faceted crystals used to describe crystal growth or changes of crystal surface energy with crystallographic orientation in the case of vicinal surfaces.

The variation of interfacial energy during wetting can then be expressed as:

$$\delta F_s = 2\pi R(\sigma_{SL} - \sigma_{SV})\delta R + 2\pi R \cos(\theta)\sigma_{LV}\delta R + 2\pi\tau\delta R \quad (6.24)$$

Applying the equilibrium condition $d(\delta F_s)/d(\delta R) = 0$,

$$\cos\theta_{eq} = \cos\theta_Y - \frac{\tau}{R\sigma_{LV}} \quad (6.25)$$

The corrective term $\tau(R\sigma_{LV})^{-1}$ becomes less significant as the drop size increases to the order of 100 nm. Therefore for the typical glass sealing configurations, the Young–Dupree equation is valid. The picture presented herein is rather elementary and does not delve into the details of surface imperfections such as roughness of varying types. For a detailed analysis of wetting of such nonideal surfaces, refer to Eusthopoulos et al. [10].

It is very important to note that good wetting need not imply a thermodynamically strong interface. Good wetting simply implies that the interfacial bond is nearly as strong as the cohesive force in the liquid. As mentioned earlier, a contact angle less than 90 degrees implies wetting while one greater than that implies that the liquid does not wet the solid.

Glasses are random network structures, as a result of which the number of bonds broken per unit area during surface creation is lower than in the case of crystalline solids. Further, the open structure of glasses readily allows minor structural rearrangements allowing the minimization of the energy penalty incurred during surface creation. As a result of the structural features of glasses outlined before, glasses readily wet metals and oxide ceramics as well. To ensure better bonding between the glass and a metal, the metal may be oxidized. The presence of an oxide layer on the metal greatly enhances glass adherence due to the similarity of the binding force between the glass and the oxide layer and the cohesive force operational within the glass itself. Of course, reaction between glasses and the surface can change the contact dynamics greatly. Oxidization of the metal must also be judiciously optimized. Poor oxidation can lead to improper wetting and poor seal integrity. On the other hand, a very thick oxide layer may peel off, compromising the seal. The oxidation and sealing parameters must be carefully optimized depending upon the glass and the metal being used.

3.1 THEORIES OF GLASS-TO-METAL BONDING

The bonding of glass to metal is a very important consideration for glass/glass-ceramic-to-metal seal formation. There exist many different theories of glass-to-metal bonding [8,10,16,65,66].

3.1.1 Mechanical Bonding

These early theories of glass-to-metal bonding assumed a mechanical reason for the glass/glass-ceramic being bonded to the metal substrate. There are two major mechanical theories of glass-to-metal bonding.

3.1.1.1 Dendrite Theory

Various early workers in the field of glass-to-metal sealing such as King, Healy, and Andrews or Harrison observed the existence of dendrites of a metallic phase such as Co or an intermetallic phase such as Co-Fe in the interfacial regions [8].

$$CoO(glass) + Fe(Metal) \rightarrow Co(dendrites) + FeO(glass) \left[\Delta G = -43\ kJ \cdot mol^{-1}\right] \quad (6.26)$$

They suggested that the dendrites hold the coating in place by acting as anchor points between the metal substrate and the glass/glass-ceramic. This theory is also known as the lock-key theory of glass-to-metal bonding [8].

3.1.1.2 Electrolytic Theory

The electrolytic theory, put forth by Staley and Dietzel, and collated by Donald [8], proposed that the metal substrate is roughened during coating/intimate contact with the sealing glass owing to galvanic corrosion. Similar to the dendritic theory, if CoO-containing glass is applied onto an Fe substrate, then metallic cobalt precipitates. However, the formation of the precipitates was attributed to galvanic corrosion of the substrate, which causes dissolution of Fe into the glass, leaving behind a pitted and rough surface. As the glass melts during sealing, it flows into these pits and attaches itself to the substrate by means of the mechanical lock and key mechanism. The reaction sequence for the same is given as:

$$2Co(precipitate) + O_2(atmospheric) \rightarrow 2Co^{2+} + 2O^{2-} \quad (6.27)$$

$$Co^{2+} + 2e \rightarrow Co \quad (6.28)$$

$$Fe \rightarrow Fe^{2+} + 2e^- \quad (6.29)$$

However, mechanical bonding theories met with opposition when Harrison and coworkers observed that the bonding strength was not greatly influenced by the ease of reduction of the metal oxide constituent of the glass [8,66]. Thus, if CoO is replaced by an oxide that is more easily reducible, we would expect it to create more roughness in the iron surface, which must in turn improve bonding. However, no such improvement in bonding was observed, which has led to mechanical bonding theories falling out of favor.

3.1.2 Chemical Bonding

Some of the earliest critics of the mechanical bonding models were King, Tripp, and Ducksworth [8,16,66], who investigated the nature of glass bonding to an iron substrate. They observed an increase in the bonding strength of the glass to the metal if the glass contains glass formers such as SiO_2 or B_2O_3, while modifiers such as Na_2O reduced bonding tendency. They concluded that strong bonding was encouraged if the metal-to-glass interface region was saturated in a metal oxide that could not be reduced by the presence of the metal. In case of Fe substrate, the relevant oxide is FeO.

Similar results were observed for a variety of metals, which led to the conclusion that the bonding between a glass and a metal must occur via an interfacial zone where the nature of bonding changes from metallic in the metal to ionocovalent in the glass. Therefore, it has been suggested that metal ions will dissolve in the glass near the glass-metal interface and saturate it completely, leaving metal ions at the interface, which then helps bonding. As a result, it can be stated that the easiest method to ensure glass-to-metal bonding is to preoxidize the metal, creating an oxide layer on the metal surface. This oxide layer then dissolves in the glass, leading to a saturation of metal ions in the glass, which in turn helps glass-to-metal bonding. Indeed, in most glass/glass-ceramic-to-metal bonding/sealing activities, the metal components are carefully preoxidized to allow the formation of an oxide layer, which helps bonding.

However, this does not mean that an oxide layer on the metal is mandatory for the glass-to-metal bonding. Pask and Tomsia have shown that it is indeed possible to achieve good bonding even between a clean substrate and glass [16]. They suggested that the glass metal interface can become saturated with metal ions by means of suitable redox reactions at the sealing temperature, after which glass-to-metal bonding proceeds as described previously. Even in case of reactions where the change in Gibbs free energy for the formation of an oxide layer is positive, changes in sealing conditions can help improve bonding and even allow bonding where it is unfavorable under ambient conditions.

Indeed, Pask and Tomsia studied the bonding between a clean Fe substrate and sodium silicate glass. They found that bonding is enhanced by either low partial pressure of Na, low activity of FeO, or high activity of Na_2O. This implies that although bonding is not favored at ambient conditions, it can occur at low pressure. However, sealing operations at low pressure, particularly for glass prepared under ambient pressure, can be delicate owing to extensive bubbling in the glass.

The chemical bonding theory shows that it is not only the nature of the interface but the sealing conditions that play a dominant role in the integrity of glass-to-metal and glass-ceramic-to-metal seals.

3.1.3 Adherence Promoters

As discussed previously, it is essential that during sealing the glass-to-metal interface be saturated with the ions of the metal. Therefore, certain easily reducible oxides such as CoO or NiO can enhance the efficacy of interfacial bonding. The steps by which this happens are listed here:

1. When a glass reacts with the metal at the interface during sealing, the glass dissolves the metal oxide layer on the surface of the metal substrate.
2. When the oxide layer is completely consumed, the glass will come into contact with the metal substrate.
3. The adherence promoters in the glass will be reduced by the metal substrate, allowing a further supply of substrate metal oxide, which helps maintain interfacial saturation during the sealing operation.

In actual sealing, it is usually not possible to maintain interfacial saturation at all times during the sealing operation. Therefore, if the entire oxide layer is not dissolved in the glass, then bonding occurs via the interfacial oxide layer. On the other hand, if sealing time exceeds the dissolution time for the surface oxide layer, then the interface characteristics are influenced only by the nature of redox reactions at the interface.

It may be mentioned that the exact nature of the oxide layer and how quickly it will dissolve in the glass must be optimized to achieve good bonding between a chosen pair of glass and metal [8,16]. It may be emphasized that a very thick oxide layer on the metal must be avoided as this will lead to metal oxide saturation deep into the glass at the interface, which may deleteriously impact properties such as TEC of the glass/glass-ceramic.

4. TYPES OF SEALS AND THEIR CONFIGURATION

We now discuss the various types of seals, such as expansion-matched seals and compressive seals. Historically, soldered seals and mechanical joints are also two other types of glass-to-metal seals but they have been largely superseded in applications by direct matched seals or unmatched/compressive seals [8,16]. Seals are also classified based on their geometries as bead seals, strip seals, tubular seals, etc. They can also be classified based on the glasses used (hard or soft) or the metal used. For this chapter we will classify seals as matched seals and unmatched seals. The latter type includes compressive seals and also housekeeper seals.

4.1 EXPANSION MATCHED SEALS

The first type of glass-to-metal or indeed glass-ceramic-to-metal seals are the expansion matched seals. As the name suggests, these are seals where the TEC of the glass/glass-ceramic matches closely with the metal with which sealing is carried out. For best sealing performance, the TEC matching should be seen close to set temperature of glass. In some specialized vacuum/pressure applications, we need to join two glasses of widely different TECs such as quartz (TEC $\sim 5 \times 10^{-6}/°C$) with borosilicate glass(TEC $\sim 50 \times 10^{-6}/°C$). For this purpose, the strategy of graded expansion glasses is used to minimize TEC mismatch. In case of glass-ceramics being used, the initial composition and the phase emergence are important in determining the TEC. Typical examples of matched seals are lithium zinc silicate (LZS) glass-ceramics to SS321 or Cu, sodium aluminium phosphate glass to aluminum, etc. [5,8,58]. Photographs of LZS glass-ceramic-to-metal seals with and without center pins are shown in Fig. 6.6. It is interesting to observe that LZS glass-ceramic is capable of simultaneously forming matched seals with SS-321 and Cu as shown in Fig. 6.6. This is exploited in the seal shown in

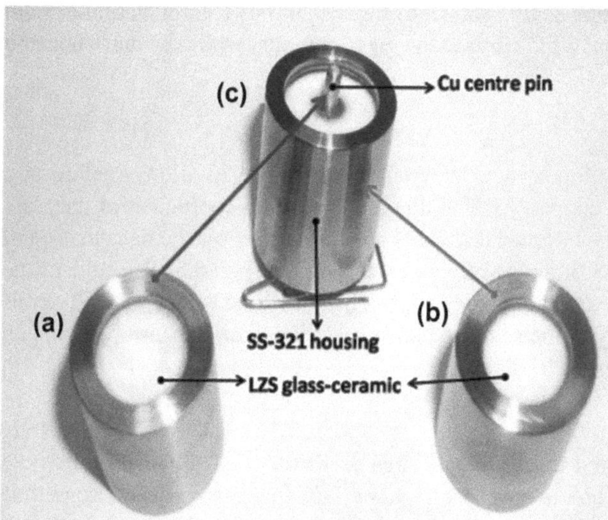

FIGURE 6.6

Photograph of lithium zinc silicate (LZS) glass-ceramics-to-SS-321/Cu seals, with and without central pin.

Fig. 6.6b, where LZS glass-ceramic is used to form a matched seal with the SS-321 outer housing and the copper central pin. For high-temperature sealing, such as that in solid oxide fuel cells, expansion matched seals are preferred since such seals maintain integrity even under repeated thermal cycles [1]. As with all seals, the integrity of these seals depends upon management of stresses at the glass-to-metal interface [1,21]. It is of course preferred that the glass be under no stress or a compressive stress, as glasses and indeed glass-ceramics show a tendency to fail by crack opening modes under tensile stress. The stress levels must be low enough to prevent the chances of static fatigue [6,9].

4.2 UNMATCHED SEALS

In this section, we will discuss three types of seals: (1) compressive seal, (2) Housekeeper seal, and (3) graded seal.

4.2.1 Compressive Seals

Compressive seals are designs where the expansion coefficient of the metal housing exceeds that of the glass/glass-ceramic. Therefore, upon cooling the assembly, compressive stresses are generated at the glass-to-metal interface, which is responsible for the integrity of the seal. Seals of the compressive type are seldom used in applications involving extensive thermal cycling. However, such seals are used commonly in applications where high pressure isolation is desired. For example, lead silicate–based single or multipin seals are used as level sensor feedthroughs in pressure vessels [67]. In such applications, a pressure tolerance of the order of 70 MPa is called for. More recently, lead silicate glasses have been replaced by lead- free compositions in view of the health implications of working with PbO powders. These materials are barium borosilicate–based glasses and provide similar performance to lead-based glasses [2,68].

Lithium aluminum silicate (LAS) glass-ceramics are also candidates for such sealing applications because the glass-ceramic offers even higher pressure tolerance to the order of 70 MPa [22]. A photograph of an LAS glass-ceramic-to-metal seal is shown in Fig. 6.7. It has a combination of both matched and unmatched seal. Central pin has TEC close to the glass-ceramic, while the outer housing has value much higher than the sealing glass-ceramic.

4.2.2 Housekeeper Seal

A special type of unmatched seal first developed by G.W. Housekeeper in 1923 allows sealing of a glass tube to highly ductile metals such as copper [8]. The thermal expansion coefficient of the glass does not match that of the glass tube, but the interfacial stresses are relieved by plastic deformation of the metal tube. A Housekeeper seal consists of a tube of copper as shown in Fig. 6.8. The edge of the tube is thinned and a bead of glass is applied to the thinned edge. The glass tube is then bonded to this beaded edge. As evident from the geometry of the seal, any stress resulting from expansion mismatch is relieved by plastic deformation of the metal [69].

4.2.3 Graded Seals

In glass-to-metal sealing, it may be required to use metals such as Kovar (Fig. 6.9) with a thermal expansion coefficient of $5 \times 10^{-6\circ}$/C with quartz glass, which has a TEC that is an order of magnitude smaller, $0.5 \times 10^{-6\circ}$/C. In such a case, a graded seal is fabricated in which the low TEC quartz glass is

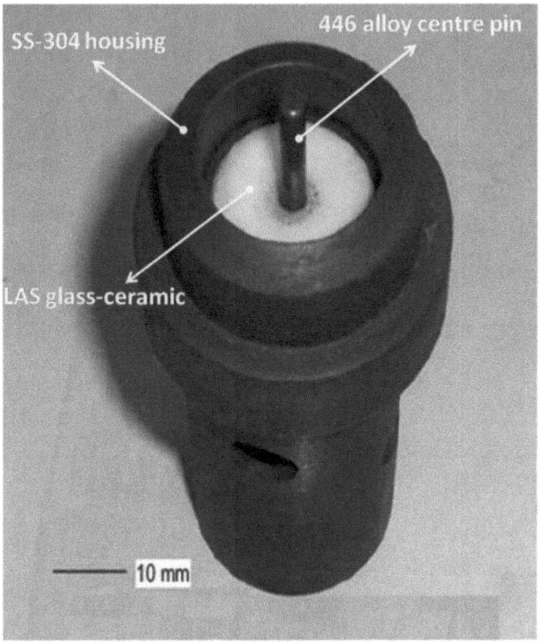

FIGURE 6.7

Photographs of cylindrical geometry lithium aluminum silicate (LAS) compressive seal.

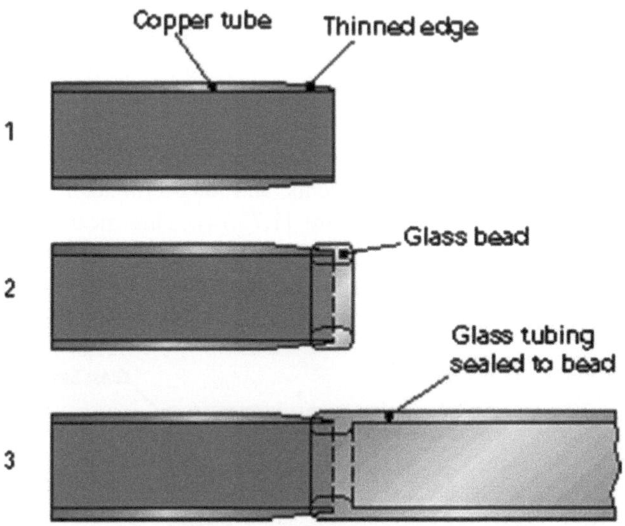

FIGURE 6.8

Schematic of a Housekeeper seal.

Adapted from http://www.twiglobal.com.

attached to the comparatively high TEC metal through a series of intermediate TEC glasses. This connecting glass region allows the thermal expansion coefficient to vary smoothly from metal to quartz glass. The stress in the transition region from metal to quartz is optimized such that it is well below the crack initiation stress of the glass. Fig. 6.8 depicts a graded seal fabricated in our laboratory.

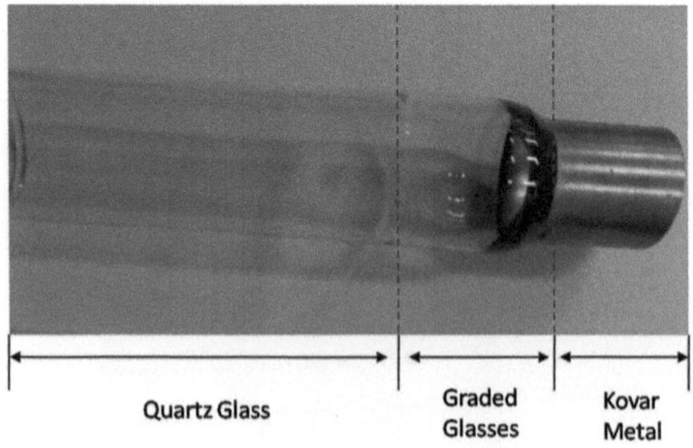

FIGURE 6.9

Graded seals for joining quartz glass with Kovar metal [70].

5. SEALANTS FOR HIGH TEMPERATURE (SOLID OXIDE FUEL CELL SEALANTS)

Solid oxide fuel cells (SOFCs) are energy conversion devices that convert chemical energy in fuels into electrical energy without the intermediate step of conversion to heat. Presently, there are two SOFC geometries that are commonly implemented, the tubular type and the planar type (*p*-SOFC) [6] (Fig. 6.10).

Various types of seals have been proposed for SOFCs, and one of the techniques is the use of glasses/glass-ceramics to form a rigid seal. Since SOFC operation requires continuous exposure of the sealant to high temperature, glass seals used would, most likely, devitrify in situ. Therefore glass-ceramics would be preferable due to their enhanced stability and durability, both chemical and mechanical [71,72].

The glass-ceramic used for sealing must show good matching of thermal expansion coefficient with other SOFC components, low reactivity with other fuel cell components, and long-term stability at elevated temperature in either oxidizing or wet-reducing atmosphere [1,73,74]. This means that normally phosphate- and borate-based sealants can be excluded [75–78]. As a result, most of the glass-ceramic sealants reported so far have been based on the alkaline earth aluminosilicate system such as barium aluminosilicate due to good thermal expansion matching and the ability to wet

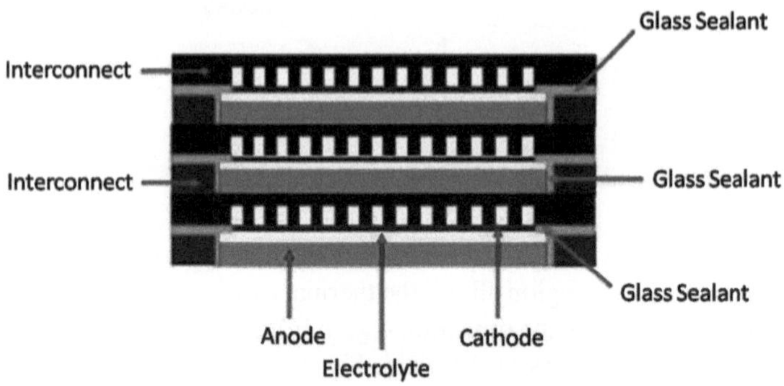

FIGURE 6.10

Exploded view of a planar solid oxide fuel cell (SOFC) stack.

materials such as the interconnect allowing good sealing [71,74,79–83]. While the chemical durability of these glass-ceramics is generally good, one significant problem is reaction with chromium-containing ferritic stainless steels such as Crofer-22 APU used as interconnects, leading to the formation of chromate species ($BaCrO_4$ in case of barium aluminosilicate) at the interface in the presence of air [84]. The large TEC ($18–20 \times 10^{-6}$°/C) of this phase can lead to cracking of the seal [84]. Therefore, many efforts are devoted to the optimization of alkaline earth aluminosilicate glass-ceramics compositions by the addition of a small quantity of oxides that may help in optimizing TEC, surface tension, viscosity, wettability, sintering, crystallization, and thermal stability among others. Our aim is to demonstrate that P_2O_5 can be advantageously added to sealing compositions. Indeed, we have already shown that P_2O_5 containing aluminosilicate glasses may present interesting properties as sealing glass-ceramics [23,85]. Moreover, P_2O_5 has been known to interact with Cr-containing steels, leading to the formation of Cr_2O_3 at the interface, which aids binding [86]. In addition, P_2O_5 has been widely employed in various silicate glasses to aid crystallization [23,85,87]. P_2O_5 has been well recognized as a nucleating agent in various silicate glasses, and different mechanisms are reported for the nucleation action [23,85,88,89]. As a result, P_2O_5 incorporation could be potentially beneficial for improving interfacial properties and phase formation. However, rapid crystallization is deleterious to the sintering and flow properties of any glass and can potentially impede sealing [90,91].

Therefore, to get greater insight on the use of P_2O_5 and reduce its adverse effects, a study is of interest to ascertain the optimum quantity of P_2O_5 in the BaO-CaO-Al_2O_3-B_2O_3-SiO_2 (BCABS) system. This system has been chosen since it has already demonstrated acceptable sealing performance [74,79–82], which may however be further optimized. In order to seal glass-ceramics with SOFC interconnect materials, a combination of studies into thermophysical properties is essential [2,21,68,93,94]. For example, Sharma et al. [92] have added phosphate species in the form of $Ba_3(PO_4)_2$, to barium-strontium-aluminosilicate glasses in order to avoid the scavenging of modifier cations for charge compensation, which increases the sealing temperature [93]. In addition, they have added V_2O_5, to reduce the maximum shrinkage temperature from 764 to 738°C. The sintering/shrinkage curves from the studies of Sharma et al. [92] are reproduced in Fig. 6.11. Also of concern in

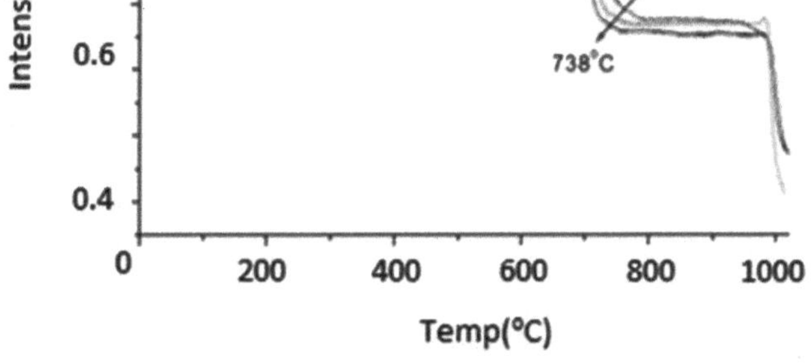

FIGURE 6.11

Typical sintering/shrinkage curves that are useful in determining softening and sealing temperatures of sealing glasses. This shows reduction in the maximum shrinkage temperature with addition of V_2O_5.

Reprinted from K. Sharma, G.P. Kothiyal, L. Montagne, F.O. Mear, B. Revel, Synergic effect of V_2O_5 and P_2O_5 on the sealing properties of barium-strontium-alumino-silicate glass/glass-ceramics, Int. J. Hydrog. Energy 38 (2013) 15542–15552. Copyright (2016), with permission from Elsevier.

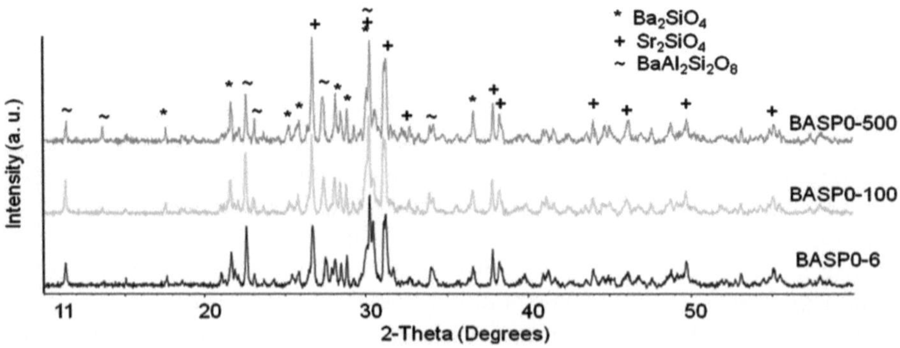

FIGURE 6.12

X-ray diffractogram of barium-strontium-aluminosilicate glass heat treated at 800°C for 6, 100, and 500 h (bottom to top).

SOFC applications is phase stability at elevated temperatures and migration of chromium from the metal into the glass/glass-ceramic [9,68,92,93].

The use of in situ/ex situ X-ray diffraction (XRD) allows a clear insight into phase emergence and also shows that no further phase emergence/modification may result, leading to stable thermomechanical properties. Fig. 6.12 shows phase emergence in a barium-strontium-aluminosilicate glass for 6, 100, and 500 h of heat treatment at 800°C. Note that the phases are stable and there is no significant change in the phases with increasing holding time. The thermophysical properties of this sealant are stable.

Further structural insights are possible using solid-state nuclear magnetic resonance (NMR) studies to better understand short-range order as this will affect sealing and therefore material performance under the extreme operating conditions of an SOFC. ^{29}Si MAS-NMR (magic angle spinning nuclear magnetic resonance) spectra of some barium calcium borosilicates are presented in Fig. 6.13. An

FIGURE 6.13

^{29}Si MAS-NMR spectra of BCABS-0P to 5P glasses. Note the increasing network polymerization with P_2O_5 concentration from 0 to 5 wt% [93].

Reprinted from A. Ananthanarayanan, L. Montagne, B. Revel, G.P. Kothiyal, Interaction studies between Crofer-22APU alloy and P_2O_5 containing barium calcium alumino-borosilicate (BCABS) sealant glass-ceramics, AIP Conf. Proc. 1313 (2009) 370–372. Copyright (2016), with permission from Elsevier.

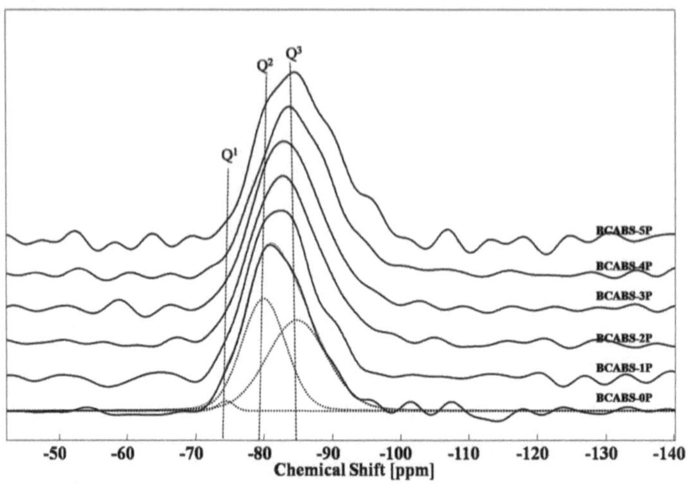

increase in network polymerization with increasing phosphorus content is observed in the figure. This indicates that sealing temperature will increase progressively with incorporation of P_2O_5 as discussed previously.

Testing sealants for high-temperature SOFC applications is also quite demanding. While the sandwich-type seal is quite common, ensuring optimum wetting between the metal plates and the sealant is quite difficult. As a result, a grooved seal testing assembly as shown in Fig. 6.14 can be used. The groove is filled with glass powder to be tested and the flange is then placed on top of the powder to ensure optimum sealing.

This set-up also allows testing of seal performance for extended time durations. Such studies helped to ascertain that Cr migration at the glass-ceramic-to-metal interface was not excessive despite holding at 800°C for over 1500 h as shown in Fig. 6.15 [92].

6. PREPARATION OF GLASS-TO-METAL SEALS

In the formation of a hermetic seal, the property of glass melts to wet metal is exploited. As mentioned earlier, glasses are random network structures, as a result of which the number of bonds broken per unit area during surface creation is lower than in the case of crystalline solids. Further, the open structure of glasses readily allows minor structural rearrangements, which in turn minimize the energy penalty incurred during surface creation. As a result of this, glasses readily wet metals and oxide ceramics as well [58].

6.1 PRETREATMENT OF METAL PARTS AND THEIR ASSEMBLY

In order to enhance the wetting of metal by the glass, the metal surface is treated suitably before the sealing process. The steps involved in the fabrication of a glass-to-metal seal, including the

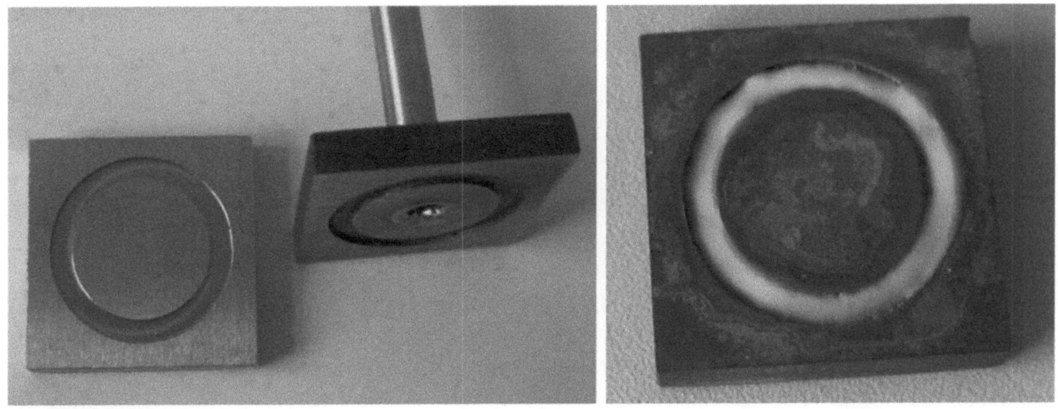

FIGURE 6.14

(a) Haynes metal with groove used for sealing experiments and (b) gilling of groove with glass.

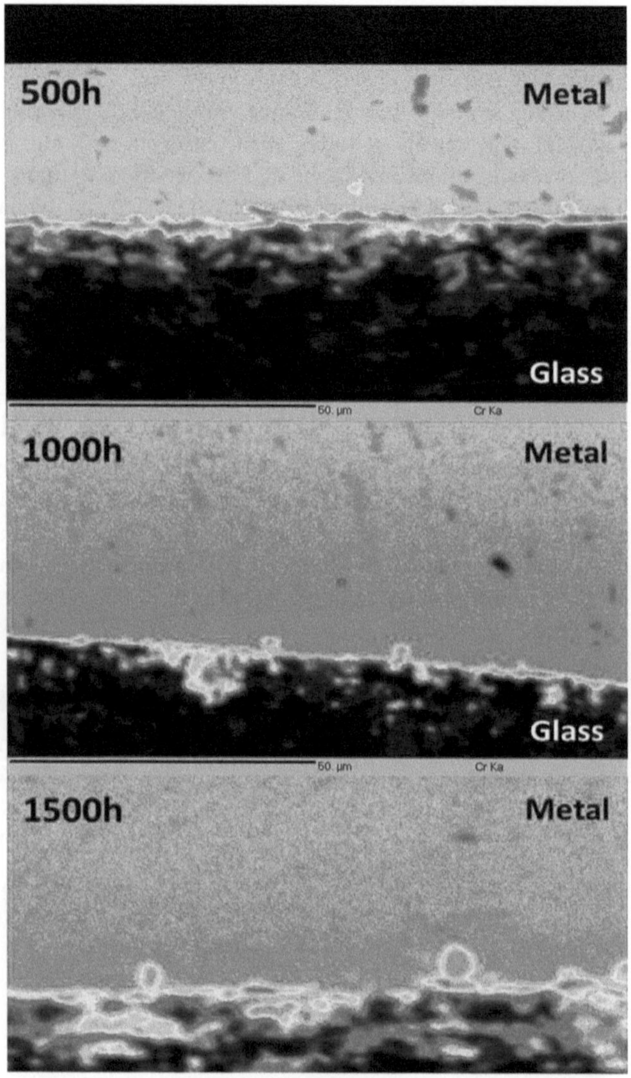

FIGURE 6.15

Chromium distribution at the glass-metal (Crofer 22 APU alloy) interface. Regions of high chromium concentration are colored orange. Even upon holding at 800°C for up to 1500 h, the chromium distribution at the interface does not change significantly, indicating stable sealant performance.

Reprinted from K. Sharma, G.P. Kothiyal, L. Montagne, F.O. Mear, B. Revel, Synergic effect of V_2O_5 and P_2O_5 on the sealing properties of barium-strontium-alumino-silicate glass/glass-ceramics, Int. J. Hydrog. Energy 38 (2013) 15542–15552. Copyright (2016), with permission from Elsevier.

pretreatment, are outlined in Fig. 6.16a. In order to ensure that the metal parts are completely degreased, they must be rinsed and cleaned in soap water and alcohol or some other suitable organic solvent. The parts were then rinsed in ultrasonic bath in trichloroethylene followed by rinsing in alcohol before drying. They were then heat-treated in dry followed by wet H_2 environment to allow for the formation of an oxide layer on the surface of the metal. As mentioned earlier, the presence of an oxide layer on the metal greatly enhances glass adherence due to the similarity of the binding force between the glass and the oxide layer and the cohesive force operational within the glass itself. The heat treatment was carried out at least 25°C above the sealing temperature to ensure that adsorbed

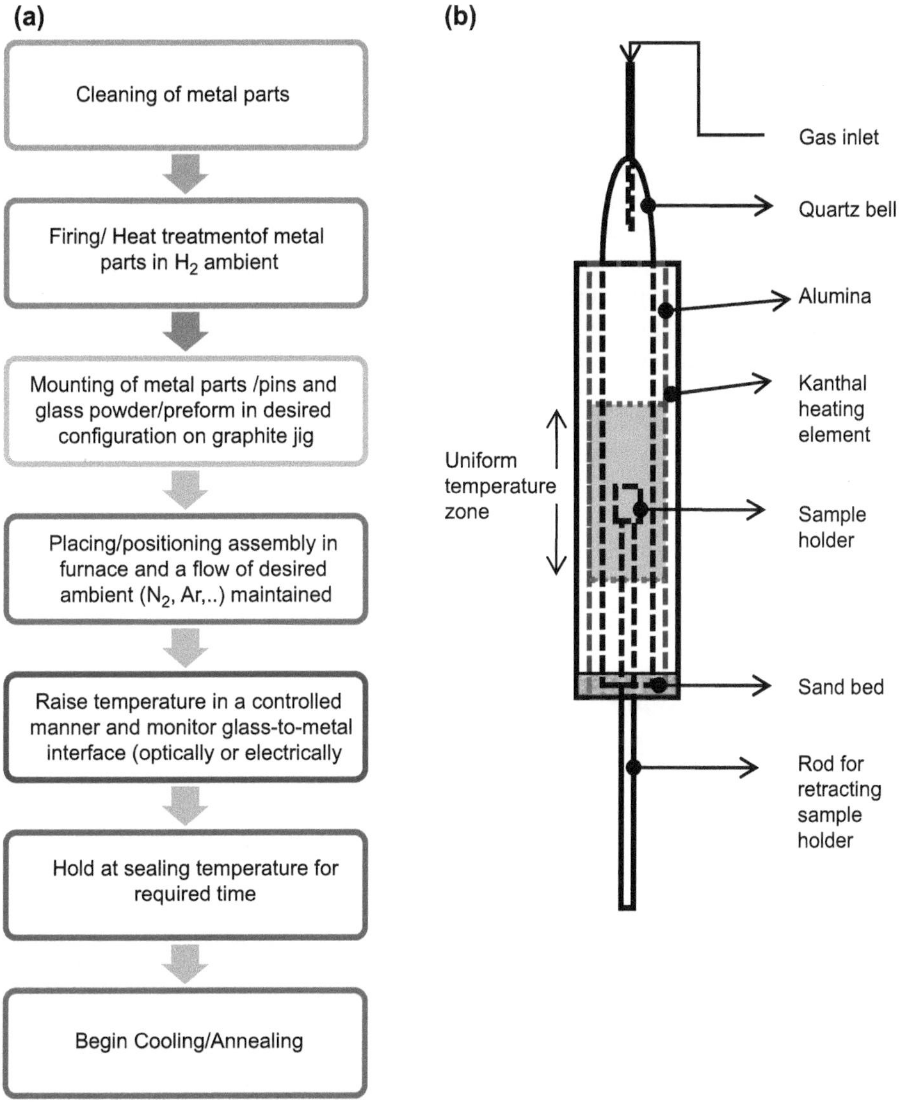

FIGURE 6.16

(a) Flow chart involved in seal fabrication and (b) furnace arrangement for fabrication of glass-ceramic to metal seal.

gases do not create bubbles during sealing. The heat-treated parts like housing/casings, pins, etc., were assembled in a desired configuration. For this, the parts of the seal such as the outer housing and pins and the glass powder/frit/preform must be assembled on a graphite jig, which has suitable grooves and holes machined in it for anchoring the housing and the pins. Graphite is chosen since molten glass does not wet graphite. Since graphite will oxidize under atmospheric heating, seal fabrication is carried out under a protective ambient of either argon or nitrogen.

The assembly was then transferred to a controlled ambient furnace for sealing and ceramization. It may be mentioned that the optimization of oxide layer thickness formed during sealing is a very important factor in seal integrity, which may be done occasionally with the help of suitable thickness measurement techniques. Poor oxidation can lead to improper wetting and poor seal integrity. On the other hand, a very thick oxide layer may peel off, compromising the seal.

The seal assembly is mounted securely on a suitable holder/pedestal, which is introduced into the uniform temperature zone of the furnace, as depicted in Fig. 6.16b. The assembly is housed within a suitable enclosure/a bell jar—type quartz tube with a provision for maintaining an inert gas flow/ambient to prevent degradation of the graphite jigs used. The upper part of the bell jar that projects out of the furnace can be thermally insulated with ZrO_2 wool blankets to minimize heat loss (not shown). Once the assembly is securely mounted and flow of inert gas maintained, the temperature of furnace is raised in a controlled manner as decided on the basis of prior DTA and XRD results. Heating may be carried out by employing a resistance or inductive heater. The glass-to-metal interface is also closely monitored either by optical or thermal devices to decide the dwell time and temperature.

Such an assembly has been used in our laboratory for the fabrication of different types of seals including LAS glass-ceramic-to-metal (SS304) compressive seals and LZS glass-ceramic-to-metal (SS321)) expansion matched seals (Figs. 6.6 and 6.7) [5,22,23].

6.2 QUALIFICATION OF SEALS

6.2.1 Qualification of Seals for Hermeticity

One of the important qualifications of a seal is hermeticity or leak lightness of interface. A leak means an unintended crack, hole, or porosity in an enveloping wall or a joint, which allows fluids and gases to escape out from a closed medium [20]. The basic functions of leak detection are the localization and size measurement of leaks in sealed systems.

6.2.1.1 Low-Pressure Testing

Among the various leak detection techniques, mass spectrometry—based techniques are the most sensitive. In helium (He) leak testing, the test part is connected to a leak detector. If a leak is present, He penetrates through the leak and its partial pressure is sensed. This is displayed in terms of flow rate. The most important subsystem of the leak detector is the mass spectrometer. In simple terms, at low pressures, the gases in the spectrometer are ionized by bombardment with energetic electrons. Different ions have different charge-to-mass ratio (q/m). In the presence of a magnetic field perpendicular to the direction of ion velocity, the ions move in circular orbits. Invoking the laws of circular motion, the radius of the ion path is:

$$r = \sqrt{\frac{mv}{qB}} \tag{6.30}$$

where, m is the mass of the ion; v its velocity, which depends upon the field applied; q the charge on the ion; and B is the applied magnetic field.

In the present case, the He ions are collected and the partial pressure is calculated. For He leak detection, magnetic sector leak detectors are preferred as gas sensors. These operate in the pressure range of $\sim 10^{-6}$–10^{-8} mbar, which is achieved by means of an integral vacuum system. The auxiliary vacuum pump required for rough pumping the tested equipment is either incorporated or can be attached via suitable connection. Fig. 6.17 shows the schematic diagram of a He mass spectrometer leak detector.

Vacuum in the spectrometer tube is maintained by means of a diffusion pump backed by a mechanical (rotary) pump or by a turbomolecular pump, avoiding a cold trap. The presence of the diffusion pump necessitates the use of a cold trap to condense oil vapors and prevent their entry into the spectrometer tube.

For measurement of leaks the seal is mounted on the test port and evacuated to about 10^{-2}–10^{-3} mbar by the auxiliary vacuum system, and then it is further evacuated by the vacuum system of leak detector. Once suitable vacuum is attained at the sample, it is isolated from the roughing pump by means of the roughing valve. The test valve is then opened to connect the sample to the spectrometer and He is sprayed on the sample. The presence of a leak allows the He to enter the spectrometer allowing its detection. Leak rate is determined by comparing with the standard leaks. Once a sample has been tested, the test valve is closed and the sample is vented to air by means of the vent valve. The sample is then removed.

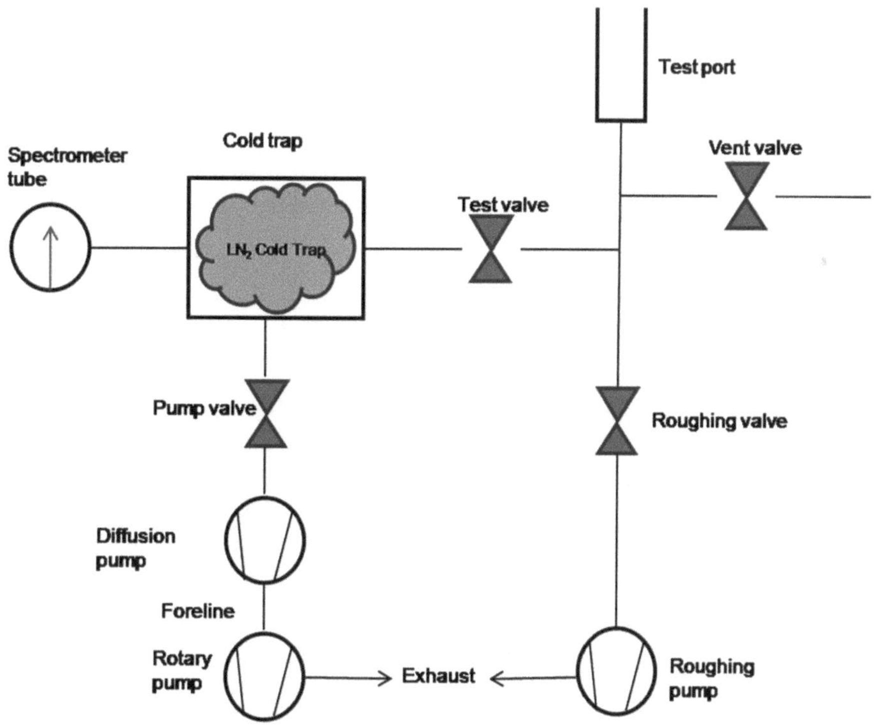

FIGURE 6.17

Schematic layout of a helium (He) mass spectrometer leak detector.

6.2.1.2 Pressure Testing

In some applications, in addition to hermeticity, the seal must also be capable of withstanding high pressure. To check the pressure compatibility of the seal, pressure testing is also carried out.

6.2.1.2.1 Conventional Method. In one of the methods pressure can be applied by either using the seal to isolate very low pressure from atmospheric pressure or by applying pressure higher than atmosphere to the seal and checking the isolation of the high pressure from the ambient. In our experiments, oil was used to hydraulically apply pressure to the seal. If the seal is compromised, then there would be a drop in pressure as oil leaks out from the crack. The test using overpressure is also a good indicator of hermeticity of a given seal. A schematic diagram of the pressure-testing instrument is shown in Fig. 6.18. The seal to be tested was mounted securely on the pressure chamber. Teflon O-rings were used to secure the seal on the lower flange. To hold the seal in place, another flange was attached on top of the first one as shown in Fig. 6.18. Since the pressure in the chamber could be as high as 100 MPa, the upper flange was attached using heavy-duty nuts and bolts. The oil drain valve was then closed and oil from the tank was pumped into the chamber. The oil being pumped was under pressure, and to prevent reverse flow, a nonreturn valve was employed. As the oil was pumped, line pressure and chamber pressure both increased. Once the required pressure was reached, pumping was stopped. The line downstream of the nonreturn valve and the chamber were at the same pressure. If any leak exists in the test piece such as a crack, then the oil will leak out through

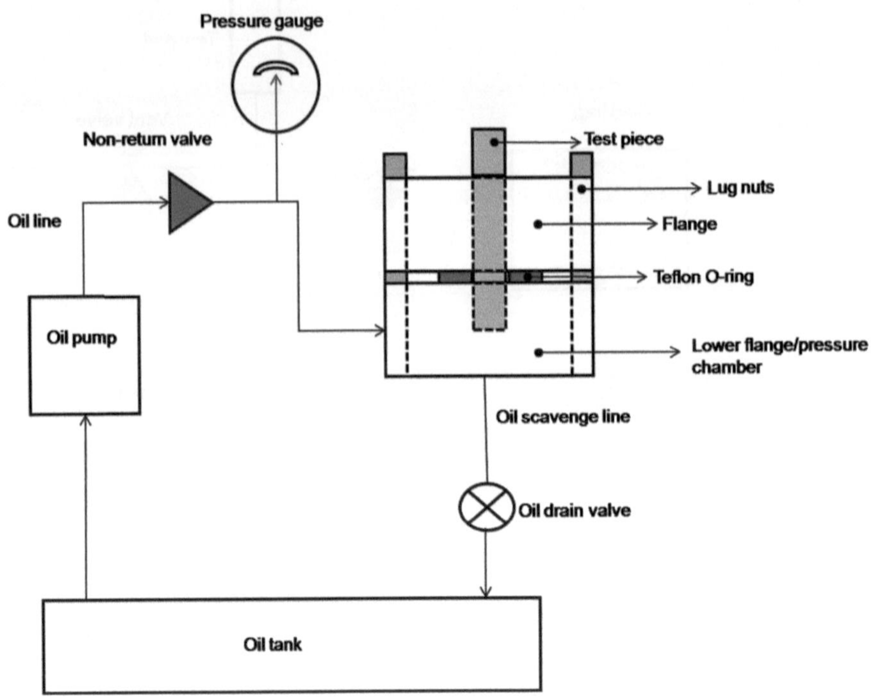

FIGURE 6.18

Schematic diagram of pressure-testing equipment.

the seal and cause a drop in the pressure registered on the gauge. The test piece (the seal) is held at high pressure for prolonged durations (30 min to 2 h) by this method. To remove the test piece, pressure is released by opening the drain valve that allows the oil to return to its tank. Subsequent to this test, the seal was retested on the He mass spectrometer for hermeticity, to ensure that no cracks had been opened by the overpressure test cycle.

6.2.1.2.2 Differential Pressure Measurement System. Seals may also be tested for hermeticity using pressure difference technique even at high operating temperatures such as needed for SOFC sealants as shown in Fig. 6.19. The seals are pressurized with argon gas with a pressure difference of, say, 200 mbar. The pressure is then monitored by pressure gauges, which indicates the air tightness at a fixed temperature. The benefit of this set-up is that the leak tightness at elevated temperatures can also be ascertained, which is very important for applications such as solid oxide fuel cells.

6.2.2 Strength and Mechanical Properties of Glass-to-Metal Seals

Whatever be the type of seal, it is in essence a joint between dissimilar materials. The dissimilarity of materials implies a difference in elastic properties, thermal expansion behavior, etc. At present there is little documented work determining the mechanical strength of glass or indeed glass-ceramic to metal interfaces, and this is an area receiving increasing attention, both experimentally and particularly using computational techniques, which we intend to highlight in this section.

Part of the problem lies in the nature of precipitated phases and their morphology at the interface. An example is the reaction layer in case of LZS glass-ceramics sealed with SS321 shown in Fig. 6.20. We observe dendritic crystals extending from the interface into the crystalline region. Clearly, the various phase morphologies and chemical changes at the interface will affect seal durability.

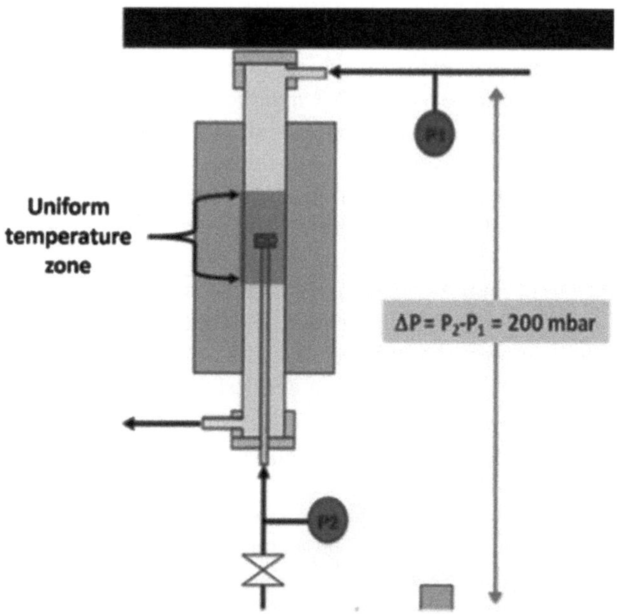

FIGURE 6.19

Schematic of differential pressure leak tightness measurement.

Reprinted from K. Sharma, G.P. Kothiyal, L. Montagne, F.O. Mear, B. Revel, Synergic effect of V_2O_5 and P_2O_5 on the sealing properties of barium-strontium-alumino-silicate glass/glass-ceramics, Int. J. Hydrog. Energy 38 (2013) 15542–15552. Copyright (2016), with permission from Elsevier.

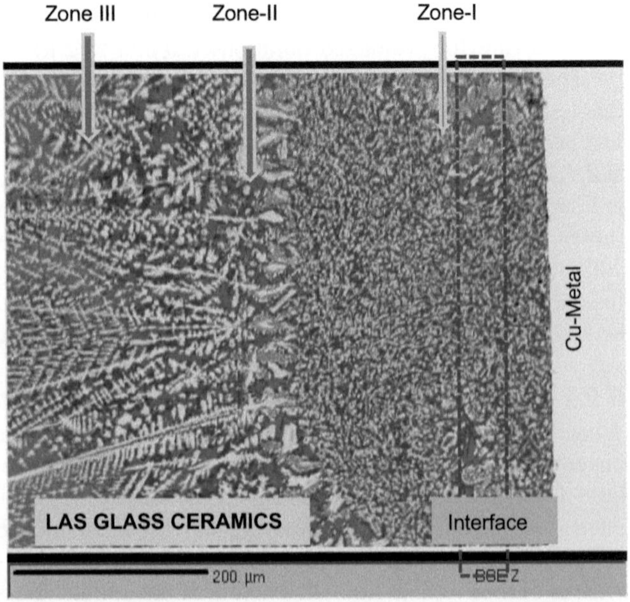

FIGURE 6.20

The interface between lithium zinc silicate (LZS) glass-ceramic and copper, showing three distinct microstructures in regions marked I, II, and III [5].

Reprinted from B. Indrajit Sharma, M. Goswami, P. Sengupta, V.K. Shrikhande, G.B. Kale, G.P. Kothiyal, Study on some thermo-physical properties in Li_2O-ZnO-SiO_2 glass-ceramics, Mater. Lett. 58 (2004) 2423–2428. Copyright (2016), with permission from Elsevier.

Further, the seal is subjected to thermal cycles and also pressure cycles; or it may be under static pressure, all of which leads to fatiguing of the seal. Returning to the matter of stresses on seals and joints, the ambiguity of the nature of the joints leads most researchers and industries to use residual stress determined by various methods and the thermal expansion mismatch for understanding the stresses on seals in service [8].

However, the emerging explosives and armament industry, which makes extensive use of glass-to-metal and more recently glass-ceramic-to-metal seals in cartridge-actuated detonators (CADs). CAD designers are increasingly resorting to computer-aided modeling of the stresses at the glass-to-metal interface using finite element analysis techniques. Such sophisticated techniques allow an understanding of the behavior of the seal during explosive initiation under shock. Indeed, the CAD glass seal is perhaps one of the most extreme environment applications of glasses discussed in this chapter.

The most common technique for this is the use of finite element analysis methods using 2-D or 3-D grids. A considerable amount of work exists in this domain, but it is beyond the scope of this chapter to delve into the mathematical specifics of the finite element analysis method [95].

A schematic figure showing a CAD-type assembly for air bag initiation is shown in Fig. 6.21.

This geometry was analyzed using the closed-form analysis method, and with some assumptions about material and interface behavior, a stress distribution profile as shown in Fig. 6.22 was observed [97].

Clearly, the seal exhibits a large tangential stress component, which is tensile in nature, which may lead to cracking of the glass in the tangential direction. It may be mentioned in passing that the radial stress component leads to circumferential cracking, while the tangential components cause axial cracks [98].

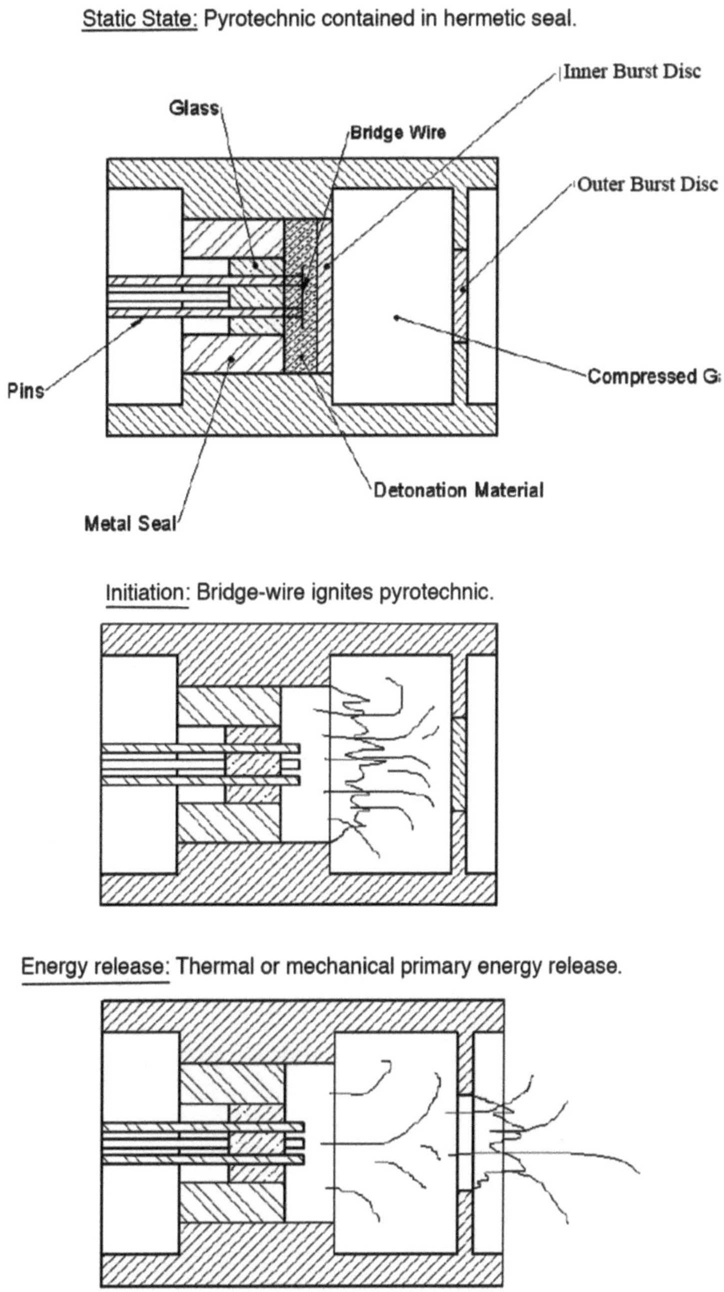

FIGURE 6.21

Cartridge-actuated detonator (CAD) assembly and sequence of failure.

Adapted from W. Horton, Empirical Study: Failure of Glass-to-Metal Seals During Shock Loading, Materials Science and Engineering, Clemson University, South Carolina, 2011, pp. 7.

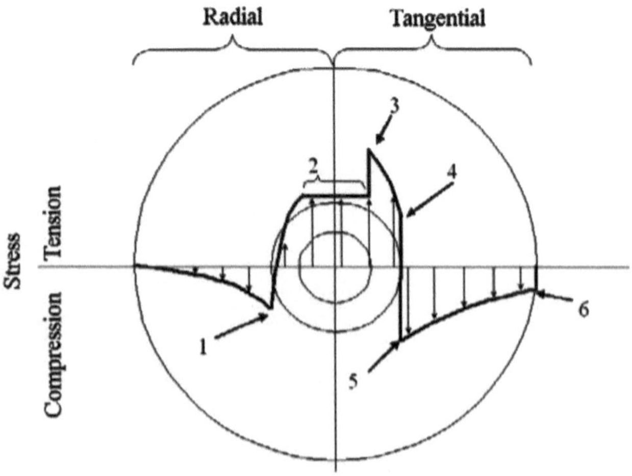

FIGURE 6.22

Stress distribution profile in the cross-section of a cartridge-actuated detonator (CAD) used for air bag initiation. Note the high stress profile in the vicinity of the central pin marked as region 3, which is most vulnerable to crack initiation.

Adapted from K.K. Rink, Fundamental considerations concerning the detection of gross leaks in bridge-wire initiators, Int. J. Veh. Saf. 1 (2006) 253–266.

A more-detailed finite element analysis by the same group allowed the determination of the expansion coefficients of the pin, which will leave the glass under compressive stress to avoid crack opening as shown in Fig. 6.23.

This example makes it quite clear that the stresses at the glass/glass-ceramic-to-metal interface must be well known and modeled. We envisage a growing interest among technologists and engineers to understand stresses in glass seals depending upon the nature of the interfaces.

7. MACHINABLE GLASS-CERAMICS

In many of the vacuum applications of glass/glass-ceramics, we may require materials with very low outgassing rates and high voltage compatibility to be used as spacers, lugs, nuts and bolts, plates, isolators, seals, etc. The jigs/gadgets with multiple pins or intricate pin layouts may be required. In such applications, it may be possible to fabricate such components using glass-ceramics, which are

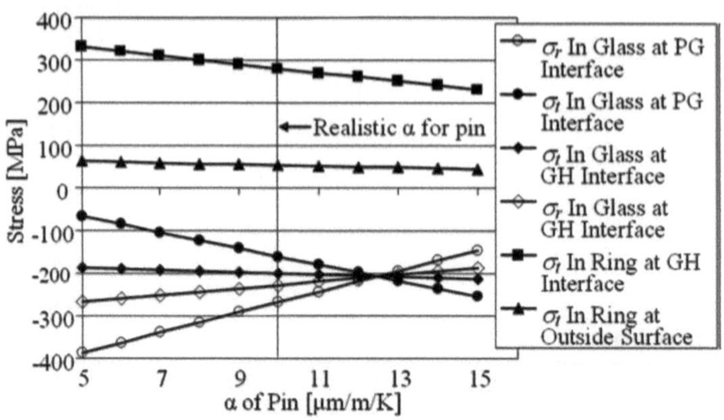

FIGURE 6.23

Effect of thermal expansion coefficient (TEC) of central pin on the stress distribution in the seal.

Adapted from W. Horton, Empirical Study: Failure of Glass-to-Metal Seals During Shock Loading, Materials Science and Engineering, Clemson University, South Carolina, 2011, pp. 7.

machinable. These glass-ceramics contain fluorophlogopite mica phase, which allows the glass-ceramic to be drilled, turned, and machined using conventional carbide tools.

Fabrication of components with precise dimensions demonstrates preparation of good machinable quality magnesium aluminosilicate glass-ceramics. The increase in Mg content favors the formation of magnesium silicate phase while increase in fluorine (MgF_2) content improves the machinability through the formation of more fluorophlogopite phase and modifying the microstructure [12]. In addition, the thermomechanical properties, i.e., the glass transition temperature, average thermal expansion coefficient, microhardness, etc., are also found to depend upon the content of MgF_2 and MgO affecting the residual glassy network and nature and concentration of resultant crystalline phases [11,99].

Based on the MAS-NMR study of the base glasses it is inferred that a change in alumina contents affects their network thereby affecting their physical properties such as average thermal expansion coefficient, microhardness, and glass transition temperature [13]. With increasing Al_2O_3 concentration there is systematic decrease in the values of thermal expansion coefficient and increase in the values of glass transition temperature and microhardness. Unlike for glass-ceramic samples, the variation of these parameters is not monotonic in nature and is governed by the relative concentration of different crystalline phases and the nature of residual glass and its extent [11]. Based on ^{29}Si NMR studies, it has been shown that with increasing Al_2O_3 contents, the Q^2 and Q^3 structural units are partially transformed to Q^3 structural units, and for higher concentration of Al_2O_3, the phase separation occurs and crystalline Mg_2SiO_4 with $\delta \approx -61.5$ ppm is formed (Fig. 6.24). For glass-ceramics samples, the crystalline phases formed for varying concentration of alumina have been identified as $MgSiO_3$, fluorophlogopite, and Mg_2SiO_4 based on powder X-ray diffraction and NMR spectroscopic measurements [11,12]. The thermomechanical characteristics such as glass transition temperature, microhardness, and average thermal expansion coefficient of base glasses with varying concentration of Al_2O_3 have been explained in terms of the structural features of these glasses, which have been

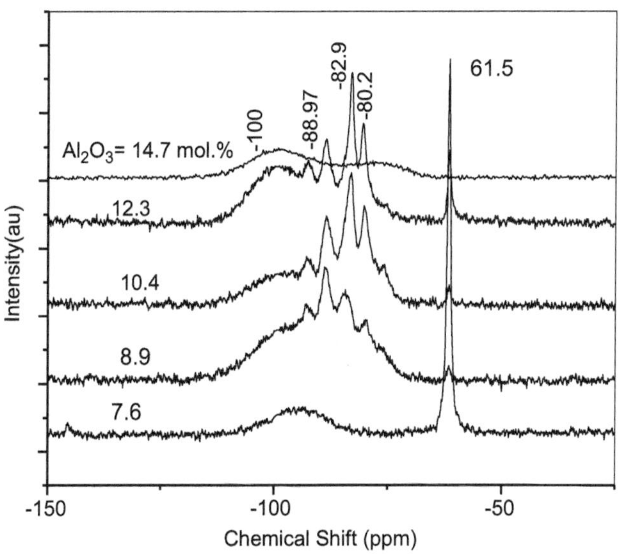

FIGURE 6.24

^{29}Si MAS NMR patterns for of glass-ceramics samples.

Adapted from M. Goswami, G.P. Kothiyal, V. Sudarsan, S.K. Kulshreshtha, The effects of alumina content on the thermomechanical and structural properties of machineable magnesium alumino-silicate glasses and their glass ceramics, Glass Technol. 46 (2005) 341–346.

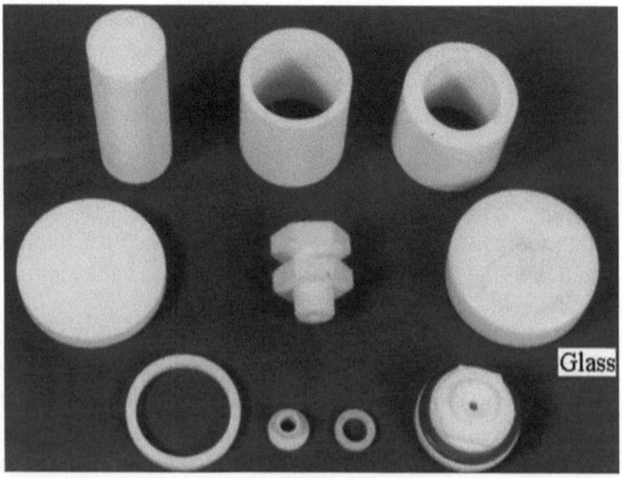

FIGURE 6.25

Components fabricated in our laboratory from machinable glass-ceramics.

elucidated by ^{27}Al and ^{29}Si NMR investigations. For glass-ceramics the variation of these physical properties with increasing concentration of Al_2O_3 is not systematic due to the varying nature of the crystalline phases present in the residual glass matrix and their relative concentration.

The dielectric behavior is seen to depend markedly on the presence of crystalline phases and microstructures of the sample. The sample having pure fluorophlogopite phase responsible for machinability shows low dissipation factor and is less affected by temperature [100]. Further improvements in dielectric properties are possible using techniques such as ion exchange [101].

The combination of properties such as low outgassing rates, high strength, machinability, and high insulation properties means that many different components can be fabricated using machinable glass-ceramics. A selection of high-density, extremely low-porosity machinable glass-ceramic parts is shown in Fig. 6.25, which depicts some of the components fabricated in our laboratory using machinable glass-ceramics.

8. FUTURE PERSPECTIVES

Glasses and glass-ceramics find wide application in fields where extreme conditions are routine. For the chapter, we have restricted our discussion to mainly sealing. It is expected that the reader has gained an understanding that analysis of glass-to-metal interfaces remains an open and challenging area of research. A concerted push, both experimentally to understand the microstructure at the interface and on the computational fronts, is needed to make analysis of glass-to-metal interfaces possible and useful. This remains an industrially relevant area of research, one to which there will be further addition as the computational capacity at the disposal of researchers continues to grow.

Glass-ceramics with their tunable thermal expansion and higher durability are also expected to find increasing use in sealing applications under extreme conditions. The use of multiple techniques to better characterize and understand these materials will be required to ensure good material performance. While alternatives are becoming available, glasses and glass-ceramics remain important materials in energy, electronics, and chemical industries to name a few, and we anticipate them to maintain this importance well into the future.

REFERENCES

[1] J.W. Fergus, Sealants for solid oxide fuel cells, J. Power Sources 147 (2005) 46–57.

[2] K.D. Meinhardt, D.S. Kim, Y.S. Chou, K.S. Weil, Synthesis and properties of a barium aluminosilicate solid oxide fuel cell glass-ceramic sealant, J. Power Sources 182 (2008) 188–196.

[3] R.D. Rawlings, J.P. Wu, A.R. Boccaccini, Glass-ceramics: their production from wastes. A review, J. Mater. Sci. 41 (2006) 733–761.

[4] R.R. Tummala, Ceramic and glass-ceramic packaging in the 1990s, J. Am. Ceram. Soc. 74 (1991) 895–908.

[5] B. Indrajit Sharma, M. Goswami, P. Sengupta, V.K. Shrikhande, G.B. Kale, G.P. Kothiyal, Study on some thermo-physical properties in Li_2O-ZnO-SiO_2 glass-ceramics, Mater. Lett. 58 (2004) 2423–2428.

[6] A. Buonomano, F. Calise, M.D. da Accadia, A. Palombo, M. Vicidomini, Hybrid solid oxide fuel cells and gas turbine systems for combined heat and power: a review, Appl. Energy 156 (2015) 32–85.

[7] H.S. Chen, Thermodynamic considerations on the formation and stability of metallic glasses, Acta Metall. 22 (1974) 1505–1511.

[8] I.W. Donald, Glass to Metal Seals, Society of Glass Technology, London, 2009.

[9] I.W. Donald, B.L. Metcalfe, L.A. Gerrard, Interfacial reactions in glass–ceramic-to-metal seals, J. Am. Ceram. Soc. 91 (2008) 715–720.

[10] N. Eusthopoulos, M.G. Nicholas, B. Drevet, Wettability at High Temperatures, Oxford, 1999.

[11] M. Goswami, G.P. Kothiyal, V. Sudarsan, S.K. Kulshreshtha, The effects of alumina content on the thermomechanical and structural properties of machineable magnesium alumino-silicate glasses and their glass ceramics, Glass Technol. 46 (2005) 341–346.

[12] M. Goswami, A. Sarkar, T. Mirza, V.K. Shrikhande, Sangeeta, K.R. Gurumurthy, G.P. Kothiyal, Study of some thermal and mechanical properties of magnesium aluminium silicate glass ceramic, Ceram. Int. 28 (2002) 585–592.

[13] M. Goswami, A. Sarkar, B.I. Sharma, V.K. Shrikhande, G.P. Kothiyal, Effect of alumina concentration on thermal and structural properties of MAS glass and glass-ceramics, J. Therm. Anal. Calorim. 78 (2004) 699–705.

[14] M.R. Werner, W.R. Fahrner, Review on materials, microsensors, systems and devices for high-temperature and harsh-environment applications, IEEE Trans. Ind. Electron. 48 (2001) 249–257.

[15] T.C. Wolfe, T.J. Vickers, Optimization of pulsing conditions for hollow cathode lamps for atomic fluorescence spectrometry, Appl. Spectrosc. 32 (1978) 265–268.

[16] J.A. Pask, R.M. Fulrath, Fundamentals of glass-to-metal bonding: VIII, nature of wetting and adherence, J. Am. Ceram. Soc. 45 (1962) 592–596.

[17] A. Arvind, R. Kumar, M.N. Deo, V.K. Shrikhande, G.P. Kothiyal, Preparation, structural and thermomechanical properties of lithium aluminum silicate glass-ceramics, Ceram. Int. 35 (2009) 1661–1666.

[18] A.L. Wilson, Elemental analysis of pottery in the study of its provenance: a review, J. Archeol. Sci. 5 (1978) 219–236.

[19] T. Rehren, A review of factors affecting the composition of early Egyptian glasses and faience: alkali and alkali earth oxides, J. Archeol. Sci. 35 (2008) 1345–1354.

[20] J.F. O'Hanlon, A User's Guide to Vacuum Technology, third ed., John Wiley and Sons, New Jersey, 2005.

[21] S.M. Gross, T. Koppitz, J. Remmel, J.-B. Bouche, U. Reisgen, Joining properties of a composite glass-ceramic sealant, Fuel Cells Bull. 2006 (2006) 12–15.

[22] A. Ananthanarayanan, R. Kumar, S. Bhattacharya, V.K. Shrikhande, G.P. Kothiyal, Some properties of lithium aluminium silicate (LAS) glass-ceramics used in glass-ceramic to metal compressive seal for vacuum applications, J. Phys. Conf. Ser. 114 (2008) 012042.

[23] A. Ananthanarayanan, G.P. Kothiyal, L. Montagne, B. Revel, MAS-NMR investigations of the crystallization behaviour of lithium aluminum silicate (LAS) glasses containing P_2O_5 and TiO_2 nucleants, J. Solid State Chem. 183 (2007) 1416–1422.
[24] R. Shukla, N. Mencinger, Moisture evolution analysis of sealing glasses used in integrated circuit packaging, J. Electron. Mater. 14 (1985) 461–471.
[25] J.-M. Wu, S.-P. Hwang, Effects of (B_2O_3, P_2O_5) additives on microstructural development and phase-transformation kinetics of stoichiometric cordierite glasses, J. Am. Ceram. Soc. 83 (2000) 1259–1265.
[26] Y.-M. Sung, The effect of additives on the crystallization and sintering of $2MgO\text{-}2Al_2O_3\text{-}5SiO_2$ glass-ceramics, J. Mater. Sci. 31 (1996) 5421–5427.
[27] The Complete Book on Glass and Ceramics Technology, NIIR Board of Consultants and Engineers, Asia Pacific Business Press Inc., New Delhi, 2005.
[28] W.H. Zachariasen, The atomic arrangement in glasses, J. Am. Chem. Soc. 54 (1932) 3841–3851.
[29] P.G. Debenedetti, F.H. Stillinger, Supercooled liquids and the glass transition, Nature 410 (2001) 259–267.
[30] V.V. Hoang, T. Odagaki, Glass formation and thermodynamics of supercooled monatomic liquids, J. Phys. Chem. B 115 (2011) 6946–6956.
[31] T.R. Anantharaman, C. Suryanarayana, Review: a decade of quenching from the melt, J. Mater. Sci. 6 (1971) 1111–1135.
[32] M. Subhadra, P. Kistaiah, Infrared and Raman spectroscopic studies of alkali bismuth borate glasses: evidence of mixed alkali effect, Vibrational Spectrosc. 62 (2012) 23–27.
[33] J. Habasaki, A. Ueda, Molecular dynamics study of one-component soft-core system: thermodynamic properties in the supercooled liquid and glassy states, J. Chem. Phys. 138 (2013) 144503.
[34] A.K. Varshneya, Fundamental of Inorganic Glasses, Academic Press, London, 1994.
[35] P.Y. Huang, S. Kurasch, A. Srivastava, V. Skakalova, J. Kotakoski, A.V. Krasheninnikov, R. Hovden, Q. Mao, J.C. Meyer, J. Smet, D.A. Muller, U. Kaiser, Direct imaging of a two-dimensional silica glass on Graphene, Nano Lett. 12 (2012) 1081–1086.
[36] X.H. Lin, W.L. Johnson, Formation of Zr–Ti–Cu–Ni–Al bulk metallic glasses, J. Appl. Phys. 78 (1995) 6514–6519.
[37] C.C. Hays, C.P. Kim, W.L. Johnson, Microstructure controlled shear band pattern formation and enhanced plasticity of bulk metallic glasses containing in situ formed ductile phase dendrite dispersions, Phys. Rev. Lett. 84 (2000) 2901–2904.
[38] Z.P. Lu, C.T. Liu, W.D. Porter, Role of yttrium in glass formation of Fe-based bulk metallic glasses, Appl. Phys. Lett. 83 (2003) 2581–2583.
[39] Z.P. Lu, C.T. Liu, A new glass-forming ability criterion for bulk metallic glasses, Acta Mater. 50 (2002) 3501–3512.
[40] A. Feltz, H. Aust, A. Blayer, Glass formation and properties of chalcogenide systems XXVI: permittivity and the structure of glasses As_xSe_{1-x} and Ge_xSe_{1-x}, J. Non-Cryst. Solids 55 (1983) 179–190.
[41] X. Feng, W.J. Bresser, P. Boolchand, Direct evidence for stiffness threshold in chalcogenide glasses, Phys. Rev. Lett. 78 (1997) 4422–4425.
[42] H. Tanaka, Relation between thermodynamics and kinetics of glass-forming liquids, Phys. Rev. Lett. 90 (2003) 055701.
[43] V.N. Novikov, A.P. Sokolov, Poisson's ratio and the fragility of glass-forming liquids, Nature 431 (2004) 961–963.
[44] H. Tanaka, T. Kawasaki, H. Shintani, K. Watanabe, Critical-like behaviour of glass-forming liquids, Nat. Mater. 9 (2010) 324–331.
[45] E.W. Fischer, Light scattering and dielectric studies on glass forming liquids, Physica A: Stat. Mech. Appl. 201 (1993) 183–206.

[46] D. Turnbull, M.H. Cohen, Free-volume model of the amorphous phase: glass transition, J. Chem. Phys. 34 (1961) 120–125.
[47] W.L. Johnson, J. Lu, M.D. Demetriou, Deformation and flow in bulk metallic glasses and deeply undercooled glass forming liquids—a self consistent dynamic free volume model, Intermetallics 10 (2002) 1039–1046.
[48] D. Turnbull, M.H. Cohen, On the free-volume model of the Liquid-glass transition, J. Chem. Phys. 52 (1970) 3038–3041.
[49] A.P. Sokolov, E. Rassler, A. Kisliuk, D. Quitmann, Dynamics of strong and fragile glass formers: differences and correlation with low-temperature properties, Phys. Rev. Lett. 71 (1993) 2062–2065.
[50] M.I. Ojovan, W.E. Lee, Connectivity and glass transition in disordered oxide systems, J. Non-Cryst. Solids 356 (2010) 2534–2540.
[51] E. Rassler, A.P. Sokolov, The dynamics of strong and fragile glass formers, Chem. Geol. 128 (1996) 143–153.
[52] R. Boehmer, K.L. Ngai, C.A. Angell, D.J. Plazek, Nonexponential relaxations in strong and fragile glass formers, J. Chem. Phys. 99 (1993) 4201–4209.
[53] T. Rouxel, Elastic properties and short-to medium-range order in glasses, J. Am. Ceram. Soc. 90 (2007) 3019–3039.
[54] J.C. Dyre, The glass transition and elastic models of glass-forming liquids, Rev. Mod. Phys. 78 (2006) 953–972.
[55] M.D. Ediger, Spatially heterogeneous dynamics in supercooled liquids, Annu. Rev. Phys. Chem. 51 (2000) 99–128.
[56] S.P. Das, Mode-coupling theory and the glass transition in supercooled liquids, Rev. Mod. Phys. 76 (2004) 785–851.
[57] J.L.M. Rupp, B. Scherrer, N. Schäuble, L.J. Gauckler, Time–temperature–transformation (TTT) diagrams for crystallization of metal oxide thin films, Adv. Funct. Mater. 20 (2010) 2807–2814.
[58] G.P. Kothiyal, A. Ananthanarayanan, G.K. Dey, S.B.K. Tyagi, 9-Glass and Glass-Ceramics, Functional Materials, Elsevier, London, 2012, pp. 323–386.
[59] R. Casasola, J.M. Rincon, M. Romero, Glass-ceramic glazes for ceramic tiles: a review, J. Mater. Sci. 47 (2012) 553–582.
[60] K. Thieme, C. Russel, Nucleation and growth kinetics and phase analysis in zirconia-containing lithium disilicate glass, J. Mater. Sci. 50 (2014) 1488–1499.
[61] E. Apel, C. van t' Hoen, V. Rheinberger, W. Holland, Influence of ZrO_2 on the crystallization and properties of lithium disilicate glass-ceramics derived from a multi-component system, J. Eur. Ceram. Soc. 27 (2007) 1571–1577.
[62] M.E. Schrader, Young-Dupre revisited, Langmuir 11 (1995) 3585–3589.
[63] J.E. Shelby, Principles of glass formation, introduction to glass science and technology (2), Royal Soc. Chem. (2005) 7–25.
[64] J. van Wonterghem, S. Morup, S.W. Charles, S. Wells, J. Villadsen, Formation of a metallic glass by thermal decomposition of $Fe(CO)_5$, Phys. Rev. Lett. 55 (1985) 410–413.
[65] J.A. Davidson, Joining ceramics glass and metals, Mater. Manuf. Process. 8 (1993) 267–268.
[66] I.W. Donald, Preparation, properties and chemistry of glass- and glass-ceramic-to-metal seals and coatings, J. Mater. Sci. 28 (1993) 2841–2886.
[67] V.K. Shrikhande, V. Sudarsan, G.P. Kothiyal, S.K. Kulshreshtha, ^{29}Si MAS NMR and microhardness studies of some lead silicate glasses with and without modifiers, J. Non-Cryst. Solids 283 (2001) 18–26.
[68] J.-H. Hsu, C.-W. Kim, R.K. Brow, J. Szabo, R. Crouch, R. Baird, An alkali-free barium borosilicate viscous sealing glass for solid oxide fuel cells, J. Power Sources 270 (2014) 14–20.
[69] http://www.twiglobal.com.

[70] R.G. Barge, G.P. Kothiyal, Graded Seals to Join Quartz Glass to Kovar Metal, Unpublished Work, 2006.
[71] K.L. Ley, M. Krumpelt, R. Kumar, J.H. Meiser, I. Bloom, Glass-ceramic sealants for solid oxide fuel cells: Part I. Physical properties, J. Mater. Res. 11 (1996) 1489–1493.
[72] http://www.mo-sci.com/sealing-glass.
[73] A. Atkinson, B. Sun, Residual stress and thermal cycling of planar solid oxide fuel cells, Mater. Sci. Technol. 23 (2007) 1135–1143.
[74] S. Ghosh, A.D. Sharma, P. Kundu, S. Mahanty, R.N. Basu, Development and characterizations of $BaO-CaO-Al_2O_3-SiO_2$ glass–ceramic sealants for intermediate temperature solid oxide fuel cell application, J. Non-Cryst. Solids 354 (2008) 4081–4088.
[75] R.N. Singh, Sealing technology for solid oxide fuel cells (SOFC), Int. J. Appl. Cerm. Technol. 4 (2007) 134–144.
[76] K. Kendall, S.C. Singhal, N.Q. Minh, Cell and stack design, in: S.C. Singhal, K. Kendall (Eds.), High Temperature Solid Oxide Fuel Cells, Elsevier, Amsterdam, 2003, pp. 197–225.
[77] P.H. Larsen, P.F. James, Chemical stability of $MgO/CaO/Cr_2O_3-Al_2O_3-B_2O_3$–phosphate glasses in solid oxide fuel cell environment, J. Mater. Sci. 33 (1998) 2499–2507.
[78] T. Zhang, W.G. Fahrenholtz, S.T. Reis, R.K. Brow, Borate volatility from SOFC sealing glasses, J. Am. Ceram. Soc. 91 (2008) 2564–2569.
[79] K.D. Meinhardt, D.S. Kim, Y.S. Chou, K.S. Weil, Synthesis and properties of a barium aluminosilicate solid oxide fuel cell glass–ceramic sealant, J. Power Sources 182 (2008) 188–196.
[80] J. Piao, K. Sun, N. Zhang, X. Chen, D. Zhou, Sealing glass of barium-calcium-aluminosilicate system for solid oxide fuel cells, J. Rare Earths 25 (2007) 434–438.
[81] F. Smeacetto, A. Chrysanthou, M. Salvo, Z. Zhang, M. Ferraris, Performance and testing of glass-ceramic sealant used to join anode-supported-electrolyte to Crofer-22APU in planar solid oxide fuel cells, J. Power Sources 190 (2009) 402–407.
[82] M.K. Mahapatra, K. Lu, Glass-based seals for solid oxide fuel and electrolyzer cells—a review, Mater. Sci. Eng: R: Rep. 67 (2010) 65–85.
[83] S. Ghosh, P. Kundu, A.D. Sharma, R.N. Basu, H.S. Maiti, Microstructure and property evaluation of barium aluminosilicate glass-ceramic sealant for anode-supported solid oxide fuel cell, J. Eur. Ceram. Soc. 28 (2008) 69–76.
[84] Z. Yang, K.D. Meinhardt, J.W. Stevenson, Chemical compatibility of barium-calcium-aluminosilicate-based sealing glasses with the ferritic stainless steel interconnect in SOFCs, J. Electrochem. Soc. 150 (2003) A1095–A1101.
[85] M. Goswami, G.P. Kothiyal, L. Montagne, L. Delevoye, MAS-NMR study of lithium zinc silicate glasses and glass-ceramics with various ZnO content, J. Solid State Chem. 181 (2008) 269–275.
[86] I.W. Donald, Glass-to-Metal Seals, Society of Glass Technology, Sheffield, 2009.
[87] T.J. Headley, R.J. Loehman, Crystallization of a glass-ceramics by epitaxial growth, J. Am. Ceram. Soc. 67 (1984) 354–361.
[88] S.M. Carr, K.N. Subramanian, Spherulitic crystal growth in P_2O_5-nucleated lead silicate glasses, J. Cryst. Growth 60 (1982) 307–312.
[89] W. Holland, V. Rheinberger, E. Apel, C. Ritzberger, H. Eckert, C. Monster, Mechanisms of nucleation and crystallisation in high strength glass-ceramics Glass Technology – European, J. Glass Sci. Technol. Part B 48 (2007) 97–102.
[90] C. Lara, M.J. Pascaul, M.O. Prado, A. Duran, Sintering of glasses in the system $RO-Al_2O_3-BaO-SiO_2$ (R=Ca, Mg, Zn) studied by hot-stage microscopy, Solid State Ion. 170 (2004) 201–208.
[91] A. Goel, D.U. Tulyaganov, A.M. Ferrari, E.R. Shaaban, A. Prange, F. Bondioli, J.M.F. Ferreira, Structure, sintering, and crystallization kinetics of alkaline-earth aluminosilicate glass-ceramic sealants for solid oxide fuel cells, J. Am. Ceram. Soc. 93 (2010) 830–837.

[92] K. Sharma, G.P. Kothiyal, L. Montagne, F.O. Mear, B. Revel, Synergic effect of V_2O_5 and P_2O_5 on the sealing properties of barium-strontium-alumino-silicate glass/glass-ceramics, Int. J. Hydrog. Energy 38 (2013) 15542–15552.
[93] A. Ananthanarayanan, L. Montagne, B. Revel, G.P. Kothiyal, Interaction studies between Crofer-22APU alloy and P_2O_5 containing barium calcium alumino-borosilicate (BCABS) sealant glass-ceramics, AIP Conf. Proc. 1313 (2009) 370–372.
[94] G. Kreysa, K.-i. Ota, R. Savinell, S. Singhal, Solid Oxide Fuel Cells, History, Encyclopedia of Applied Electrochemistry, Springer, New York, 2014, pp. 2008–2018.
[95] K. Ho-Le, Finite element mesh generation methods: a review and classification, Computer-Aided Des. 20 (1988) 27–38.
[96] W. Horton, Empirical Study: Failure of Glass-to-Metal Seals During Shock Loading, Materials Science and Engineering, Clemson University, South Carolina, 2011, p. 7.
[97] K.K. Rink, Fundamental considerations concerning the detection of gross leaks in bridge-wire initiators, Int. J. Veh. Saf. 1 (2006) 253–266.
[98] D.Q. Lei, Z.F. Wang, Z.J. Wang, Effects of geometry and material properties on the residual stress of glass-to-metal seals in solar receiver tubes, Energy Procedia 49 (2014) 418–427.
[99] C. Patzig, T. Haeche, M. Dittmer, C. Raessel, Temporal evolution of crystallization in MgO-Al_2O_3-SiO_2-ZrO_2 glass ceramics, Crystal Growth Des. 12 (2012) 2059–2067.
[100] M.S. Jogad, M. Goswami, A. Sarkar, T. Mirza, L.A. Udachan, G.P. Kothiyal, Dielectric measurement on magnesium aluminum silicate glass-ceramics prepared by different routes, Mater. Lett. 57 (2002) 619–627.
[101] P. Zhai, X.-P. Ma, N. Zheng, G.-J. Zhang, Surface flashover performance of ion-exchanged machineable glass ceramics material in vacuum, Mater. Lett. 77 (2012) 71–73.

CHAPTER

NATURAL GLASSES UNDER EXTREME CONDITIONS

7

P. Sengupta
Bhabha Atomic Research Centre, Mumbai, India

CHAPTER OUTLINE

1. INTRODUCTION

Earth is the largest museum of materials produced under extreme conditions. If we take a look inside the Earth, from crust (continental crust: ~25–40 km; oceanic crust: ~5 km) through mantle (upper mantle: ~650 km; lower mantle: ~2235 km) to the core (liquid outer core: ~2270 km; solid inner core: ~1216 km), we will find a series of discontinuities, each of which is marked with phase transformations due to changes in pressure-temperature conditions. For example, the Mohorovičić

Materials Under Extreme Conditions. http://dx.doi.org/10.1016/B978-0-12-801300-7.00007-3

discontinuity (Moho or M discontinuity), between crust and upper mantle, is identified with the disappearance of feldspar. Similarly, disappearance of shear seismic waves (S-waves) at the lower mantle—outer core boundary indicates the presence of liquid below this interface. With further increase in depth, pressure increases and the outer core liquid undergoes a phase change to form the solid inner core. In addition to these traditional structural units, phase transitions also take place from olivine to a spinel structure at a depth of 400 km, and silicates to a perovskite structure at 670 km, respectively. Revelations of all these discontinuities were actually made by studying seismic P- and S-wave velocities, which indicated that all such phase transformations do influence material properties significantly. These discontinuities are therefore boundaries separating chemically and mineralogically distinct zones. Of course phase identification at such great depths cannot be done with certainty but a reasonable prediction can be made through analyzing data obtained for seismic velocities, mass, and moment of inertia of Earth, and planetary and meteoritic abundances of elements.

Coming back to Earth's surface, the most common signature of "extreme conditions" is perhaps preserved in the form of meteoritic craters. Beside extraterrestrial impacts, several other natural processes leave imprints of "extreme conditions" on Earth's surface, the prominent ones being volcanism, seismicity, lightning strike, etc. All these processes have been known for some time now and their products are being studied in depth using sophisticated electron microscopic-microanalytical, diffraction, and spectroscopic techniques. These techniques help(ed) us in deciphering minute variations of interatomic distances, bond lengths, and angles from the smallest possible samples. The scientific understanding that has emerged out of these investigations states that pressure as an independent variable can influence chemical processes/chemistry of materials significantly under extreme conditions. Similar observations also have been made from other branches of physical and chemical sciences but the uniqueness of geological science is that the pressure regime dealt here is much higher than 1 GPa and involves solid materials only. In other cases all types of matters, e.g., solid, liquid, and gas, are studied.

Needless to say, the wide range of pressure-temperature conditions witnessed inside the Earth as well as on its surface lead to a large variety of chemical phenomenon through bringing atoms closer to each other within crystalline frameworks, redistributions of internal energy, and substantial overlapping of electron distributions leading to different properties. It is highly possible that the crystal/mineral phases and their properties witnessed in nature are highly deviated from results obtained in laboratory due to wide variation in extremes conditions as well as timescales. Examples include "shock metamorphism" on the Earth's crust as the pressures (>10 GPa) and stress rates (10 GPa/10^{-9} s) encountered are phenomenally high and the time durations are exceptionally small ($\sim 10^{-6}$s) [1]. Such phase transformation phenomena and associated mechanisms have been extensively studied by high-pressure physicists whereas the effects on transformed materials are mostly observed by mineralogists and metamorphic petrologists. Therefore a thorough comprehension of the subject "materials behavior under extreme conditions" merits multidisciplinary approach; and over the last 50 years extensive experimental mineralogy/petrology, mantle mineralogy, meteorology, and geophysical techniques have been focused toward this. Such investigations not only help us to speculate about material properties under situations far from equilibrium but also to offer logical solutions for many catastrophic natural situations like earthquakes, tsunamis, meteoritic impacts, etc. Microstructural and microchemical analyses of shock-metamorphosed rocks have revealed many interesting features including the possibilities of extreme variations in pressure and temperature conditions within a single mineral grain itself [2]. Experimentalists have reproduced such observations from nature in the laboratory using projectiles launched by a gun or explosives; and computational experts have attempted to explain all such similar localized phenomena using various simulation codes and methods. Of course, loading histories of laboratory samples are quite different from natural counterparts in terms of pressure pulses ($\sim$0.1 ms vs. a few seconds). Material scientists have taken these inputs to synthesize high-strength materials for various peaceful applications. The newest addition to the list of useful applications is perhaps extending the database to carry out long-term performance assessment of nuclear materials under extreme geological conditions, and this chapter is written from this viewpoint only.

In the current scenario, "energy" plays a crucial role behind the social and economic growth of any country. In India, a three-stages nuclear energy program based on closed nuclear fuel cycle is considered as an important component for present, as well as futuristic, energy mixtures. Among the various scientific and technological challenges, success of closed nuclear fuel cycles depends on effective isolation of high-level nuclear wastes from the biosphere over a considerably long time span (as long as one million years)! Concentration and containment of high-level nuclear waste within silicate (borosilicate, aluminosilicate)-based inert glass matrices placed inside multiple barrier systems of deep geological repository is considered as the most feasible methodology to achieve this. However, validation of such a concept needs long-term performance assessments of the host glass matrices, which is very difficult to achieve even using accelerated experimental studies in laboratories. In fact, such timescale is much outside than the experiences gathered so far by mankind on material synthesis and its property evaluation. On the other hand, reliability prediction is a fundamental step toward establishing credibility of a given robust system. To address this issue, natural analogue studies. i.e., similar processes or materials exposed to comparable environments over geological timescale, are considered as a useful methodology [3].

Considering the performance assessment as one of the major aspects, "natural glasses" have been studied to understand the responses of amorphous materials to various harsh geological processes, especially those important for nuclear waste isolation. This also aptly fits into the focal theme of the present edited volume because natural glasses do form under extreme conditions of temperature, cooling rates, pressure, and compositional variations; and they also survive through the extremities encountered in the geological environment. The interdependence between performance assessments and safety guidelines stems from the possibilities of any toxicity releases from the host amorphous matrices over long-term exposure to harsh service conditions. Therefore, a special focus has been given to "leaching of natural glasses." A brief description on various natural glasses is given next. However, before getting into the topic it felt prudent to discuss a little bit on "glass" in general, especially for a refresher in the subject. Experienced learners can skip this section and directly go to the one on natural glasses.

2. GLASS

Our first familiarity with glassy materials is probably linked to our childhood experiences of tasting chocolate, lollypops, sugar candy, etc. In scientific parlance, the former two are examples of rigid blocks of glassy materials whereas the later one belongs to fibrous variety. Structurally, "glass" is an amorphous material having short-range ordering (SRO), i.e., it exhibits symmetrical arrangements in first- or second-nearest neighbors of a given atom. Schematic diagrams showing SRO and long-range ordering (LRO; characteristic of crystalline materials) are shown in Fig. 7.1a and b, respectively. First,

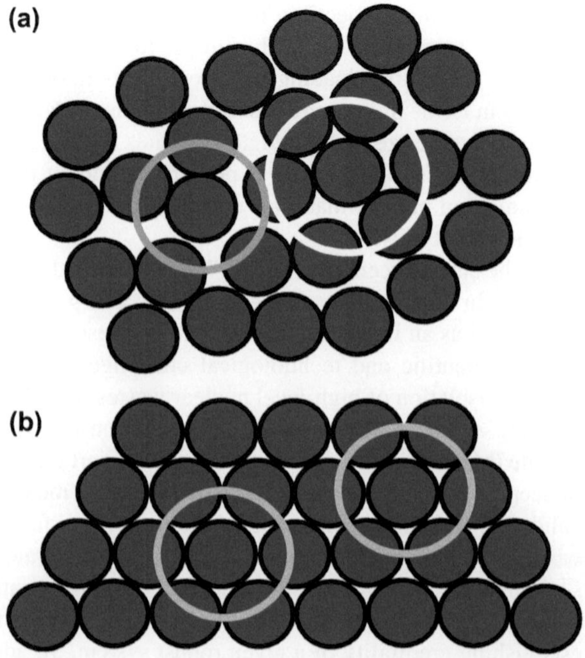

FIGURE 7.1

Schematic diagram showing (a) amorphous network with short-range ordering (SRO) and (b) crystalline network with long-range ordering (LRO) structures. Fivefold and sixfold symmetry in amorphous material is indicated by yellow (*gray in print version*) and white circles respectively. Green circle (*light gray in print version*) in crystalline matrix is indicative of LRO.

Fig. 7.1a shows localized fivefold (marked by yellow circle) and sixfold (marked by white circle) ordering within an amorphous structure while Fig. 7.1b exhibits sixfold symmetry all along the crystalline material. According to american society for testing and materials (ASTM) definition, glass is "an inorganic product of fusion which has cooled to a rigid condition without crystallizing."

This definition, although it satisfies most of the plant-scale vitrification processes, has many limitations. For example, amorphous materials can be prepared by sol–gel process, chemical vapor deposition technique, etc. without involving "fusion stage" within respective synthesis procedures. Similarly on the other hand, noncrystalline sugar derived from body fluid (cytosol) by arctic grasses, wood frogs (*Rana sylvatica*) and flatworms (*Turbellaria*) to survive life-threatening chilling winter, spider-made carbon soot, etc., although structurally amorphous, according to ASTM do not qualify to be "glasses" since they are not inorganic in nature. Recognizing these shortcomings, many scientists have tried to redefine glass and the most popular among them is "any material, inorganic, organic, or metallic formed by any technique, which exhibits glass transformation behavior is a glass" [4].

Glasses are thus microscopically disordered solid forms of matter. Constituents of glass are broadly classified as network formers, modifiers, and intermediates on the basis of "field strength (F)." "F" is a ratio of valence state of a given cation (Z) and square of its ionic radius (a^2). Network formers are those cations whose F value varies within ~1.5–2, and whose molten oxides can form glasses during cooling at moderate quenching rate. Glass formers that are commonly present within natural glasses are given in Table 7.1. Among the various network formers, the predominant one is Si^{4+}, which forms tetrahedron with four oxygen anions (O^{2-}). During cooling of silicate melts, such SiO_4 tetrahedron can go on polymerizing and create various network structures. In the simplest case, two Si–O tetrahedrons

Table 7.1 Classification of Glass Constituents

Element	Valence Z	Coordination Number (CN)	Field Strength F	Bond Strength (kJ/mol)	Function in Glass
Si	4	4	1.57	443	Network formers: F 1.5–2.0
P	5	4	2.1	368–464	
Al	3	6	1.04	304	Intermediates: F ~0.5–1.0
	3	4	0.96	335–423	
Fe	3	6	0.84	224–284	
	3	4	0.85		
Fe	2	6	0.43		
Mg	2	4	0.53		
	2	6	0.45	155	
Ca	2	8	0.33	134	Network modifiers: F 0.1–0.4
Na	1	6	0.19	84	
K	1	8	0.13	54	

can join together to form a structural unit called a "double tetrahedron." In this case the O-ion common to both tetrahedrons is called "bridging oxygen" and others are called "nonbridging oxygens." To indicate the nature of silicate network polymerization, usually a descriptive term Q^n is used where n signifies the number of bridging oxygens. Therefore for "double tetrahedron" structural units the descriptive term would be Q^1. Like this the extent of silicate polymerization can vary between Q^0 and Q^4. In natural glasses all these varieties of silicate polymerizations are possible. There are, however, certain elements that can modify such polymerization and are commonly known as "network modifiers." F values for modifiers are low and for those that are present within natural glasses it varies within 0.1–0.4 (Table 7.1). Besides network formers and modifiers, there are certain cations like Mg^{2+}, Fe^{2+}, and Al^{3+} that are known to play the roles of both formers and modifiers within natural glasses. These cations are therefore referred to as "intermediates" and their F value ranges between ~0.5 and 1.0.

Typically, glass state is reached by a molten mass through rapid cooling. This is explained in Fig. 7.2 where a small volume of melt is envisioned to have achieved equilibrium at a high temperature "A." As the melt cools, kinetic energy of its constituent particles (ions, molecules, clusters) get continuously reduced, volume shrinks, and solubility limit for a particular component decreases. When the rate of cooling is slow, binding energy gradually takes over kinetic energy leading to ordering of constituents and crystal formation. Time required for such ordering depends on transport rate of constituents from its current position to energetically preferred points at crystal nodes. In the present diagram the said cooling melt can witness such a favorable situation at Point B when an infinitely small amount of crystals (with melting point T_m) is in thermodynamic equilibrium with the parental liquid. However, for considerable amount of crystallization, the parental melt should undergo some amount of undercooling, such that (1) an adequate number of nuclei are present in the melt, and (2) its growth to form a crystal is possible. Crystallization from melt is associated with discontinuous jump in enthalpy, entropy, and volume; therefore it is a "first-order transition" type.

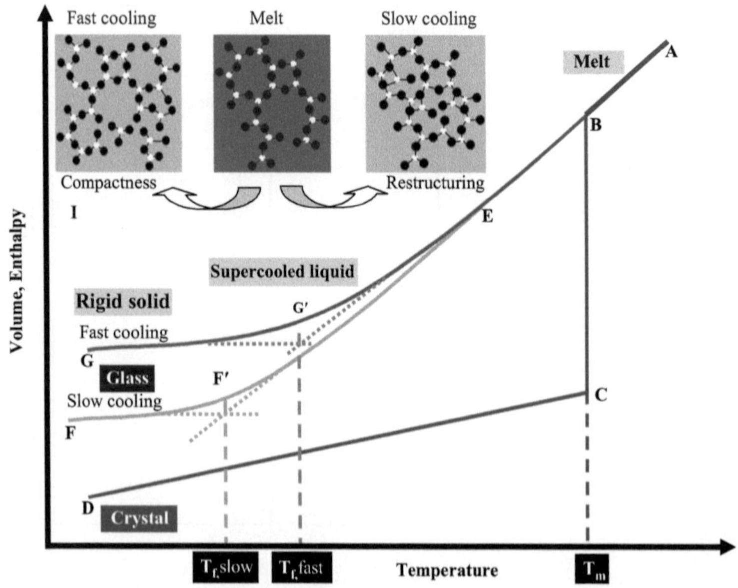

FIGURE 7.2

Glass formation.

On the other hand, if the system undergoes cooling at a faster rate, then the melt may surpass the crystallization step and move along the line BF'. All through the journey from B, the melt structure rearranges itself continuously into a smaller volume and its viscosity increases rapidly. Continuous withdrawal of energy drives the melt system to a situation where the constituents are no longer able to rearrange themselves fast enough to match the volume characteristic of that particular temperature. At this juncture (F′), the "liquid line of descent" assumes a new trend and becomes a near-straight line (FF′), often roughly parallel to CD. The material at this stage (FF′) behaves like a solid and it is said to have attained "glassy state," a stage at which any material inherits an amorphous structure within it. Detail of the amorphous structure depends on rate of cooling and therefore the network of a glass at point G (cooled fast) is very likely to be different from that of F (cooled slowly). Structure of the slowly cooled glass is expected to be more (low volume/more ordered) polymerized and ordered than that noted within fast-cooled glass (see inset I, Fig. 7.2). The smooth curve portion of the liquid line descent between the onset of departure from the supercooled liquid line and reaching the rigid condition is called the "glass transition region." Across this region, viscosity of the melt changes drastically from $\sim 10^8$ Pa s ($=10^9$ Poise) at higher temperature domain to $\sim 10^{15}$ Pa s or more at the lower temperature part. Usually T_g is defined as the temperature at which the viscosity exceeds 10^{13} P. Around the same temperature steep changes in thermal properties like specific heat, thermal expansion, etc. also take place. Intersection between extrapolated glass line and supercooled liquid line is often considered to be characteristic of the "cooling rate" for a specific system and it is termed the fictive temperature (T_f). Slowly cooled glass has lower T_f compared to quickly cooled ones.

3. NATURAL GLASSES OF VOLCANIC ORIGIN

Like in laboratories, glasses are produced in nature due to rapid cooling of silicate melts called magma. Such natural glasses are commonly referred to as volcanic glasses and are found on Earth's surface, both on landmasses as well as on ocean bottom. Their linkage to "extreme conditions" is due to fast undercooling. On the basis of bulk compositions, volcanic rocks can be classified as basic, intermediate, and acidic type. Within basic volcanic rocks, like basalt (Fig. 7.3), subordinate amounts of natural glasses occur in the groundmass whereas in acid lavas it can be the major constituent as well. Examples include acidic rocks like obsidians and pitchstones, each of which can be entirely made up of glasses (Fig. 7.4). The reason behind such wide variation in glass content is essentially due to differences in bulk compositions, which in turn define viscosity of the magma and therefore play crucial roles in crystal nucleation

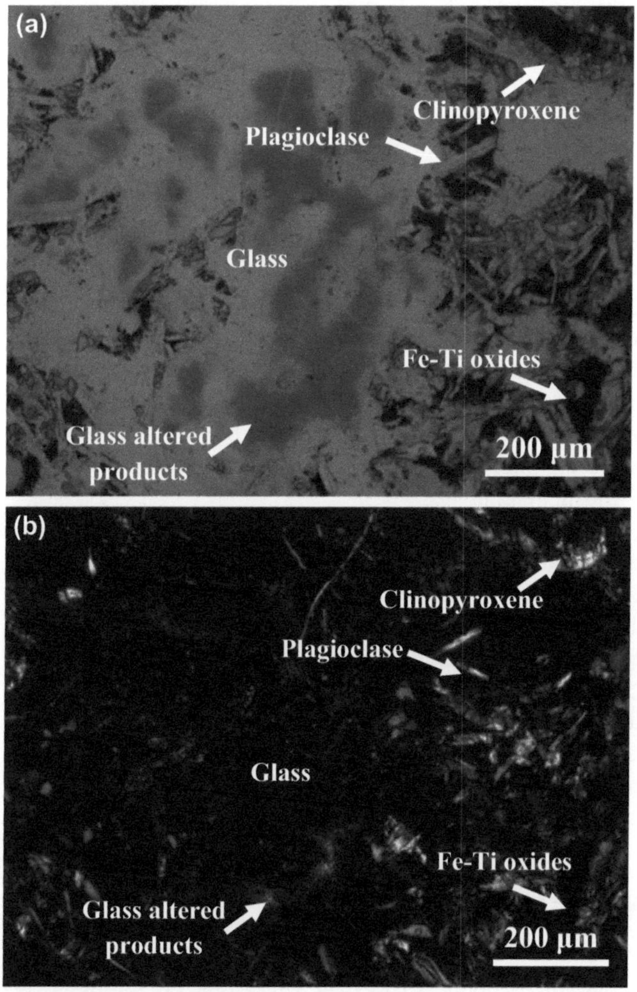

FIGURE 7.3

Optical microscope images showing distributions of natural glass and other mineralogical constituents of basalt under (a) plane-polarized and (b) cross-polarized light conditions.

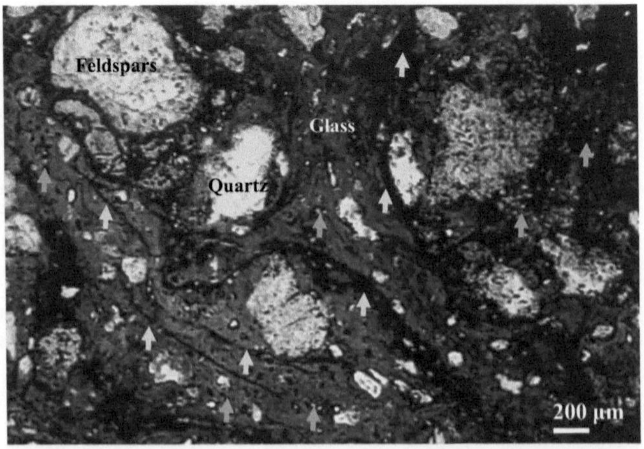

FIGURE 7.4

Optical microscope image (plane-polarized light) showing distributions of natural glass and other mineralogical constituents of rhyolite.

and growth within it upon its extrusion on the Earth's surface. Higher polymerization within silicate network enhances viscosity and therefore favors glass formation. This explains why natural glass is more common in rhyolite than in basalt as the average SiO_2 content in the former is much more than the latter. In hand specimen or under optical microscope, natural glasses exhibit characteristic curved, partially concentratic "conchoidal" fracture systems (very similar to beer glass fractures), which are also known as perlitic structures (when caused due to hydration induced expansion).

Representative microstructure of a basaltic rock containing natural glass is shown in Fig. 7.3. It can be seen from the plane polarized (Fig. 7.3a) and crossed polarized (Fig. 7.3b) optical images that the crystalline groundmass of the rock is constituted of plagioclase feldspar laths, clinopyroxene and Fe–Ti oxides within which colorless glasses are randomly distributed. Being isotropic in nature, glasses always appear dark under cross-polarized light (Fig. 7.3b). The gray domains observed on the glasses (Fig. 7.3a) are the fine-grained alteration products that are most likely to be chlorite in this case. Such alteration is more common within ocean floor basalts and the usual secondary phases are saponite, smectite, chlorite, celadonite, philipsite, etc.

Typical microstructure of rhyolite rock is shown in Fig. 7.4. The irregular domains (gray) are natural glass, which has included irregular-shaped quartz and feldspars grains within it. Note the flow structures (marked by white arrows; Fig. 7.4) within the glass domains, often defined by the parallel alignment of various constituent fabric elements like tiny dark and white inclusions (marked by light gray arrows; Fig. 7.4), plagioclase lathes, etc. Under higher magnifications, these are found to be needlelike crystals of feldspars, quartz, and Fe–Ti oxides. Often these needles orient themselves radially around scattered nuclei and build up spherical bodies called spherulites. At times such spherulites exhibit cross-cutting relations with flow structures indicating that they have grown after flow of the host magma had ceased. Obsidian is a variety of the "rhyolite group of rocks" that is partially to completely amorphous, i.e., typically free of crystals.

4. NATURAL GLASSES OF NONVOLCANIC ORIGIN

Besides volcanic glasses, amorphous materials are also produced in nature under various other extreme conditions like (1) meteoritic impact, (2) seismic events, (3) nuclear explosion, (4) lightning, (5) radiation damage, etc. Brief notes on such nonvolcanic natural glasses are given following. The major difference between these two situations of extreme conditions is in kinetics. Both volcanic and nonvolcanic glasses refer to nonequilibrium situations but the departure from equilibrium condition is maximum in the case of nonvolcanic glasses. Some typical features of nonvolcanic glasses are they lack phenocrysts as they are the product of superheated conditions and presence of extensive deformation features and high-pressure phases. However, the said high-pressure condition exists momentarily and therefore only diffusionless polymorphic transformations can take place.

4.1 ORIGIN: METEORITIC IMPACT

Meteoritic craters are often characterized with amorphous materials formed due to sudden energy deposition on country rocks through imposition of shock waves. These lead to successive events of vaporization and melting followed by crater formation. Glasses so produced due to rapid quenching of the melts are known as "impact glasses" and are heterogeneous in composition and microstructures (Fig. 7.5a) [48]. The extent of inhomogeneities is essentially dependent on target rocks and magnitude of shock pressures [5]. Texturally impact glasses [6–9] can be classified as (1) diaplectic and (2) normal varieties, where the former usually preserve parental mineralogical (even twin lamellae)/petrological signatures and the latter one (normal glasses) are characterized by features like flow structures, distorted crystal boundaries, vesiculation, etc. Owing to the extremely high pressure (*P*)–temperature (*T*) conditions, phases like coesite, stishovite, maskelynite, etc. are often present within impact glasses [10,11].

"Rock glasses" are another variety of amorphous materials often seen within craters. These are compositionally similar with host rocks and usually occur as melt ejecta, coherent melt layer, dikes, and veins, with mineral/lithic clasts exhibiting varied degrees of (1) deformation features (undulose extinction, kinking of cleavage, etc.), (2) flow structures, (3) vesiculations, etc.

Tektites are another variety of glasses that are found within craters. They are about centimeter sized, usually rounded, natural glass bodies formed by melting and quenching of terrestrial surficial sediments due to hypervelocity meteorite impact [12–19]. Cooling rates of tektites are ($<10^4$ K/min), much faster than for massive volcanic glasses ($<10^{-4}$ K/min) [20]. They are (1) chemically homogeneous (on a scale of 10–100 μm), reduced (valence state) in nature, almost water-free, (2) exhibit bright colors, peculiar shapes (commonly spherical-symmetric shaped, or fragments thereof), wide range of surficial ornamentations (contain lechatelierite and bubbles), and (3) are typically restricted in space and time (Cenozoic). A majority of tektites are distributed over four regions known as (1) North American (age: ~34.9 Ma old; related with Chesapeake Bay impact crater) [21,22], (2) Central European (~14.5 Ma; Ries crater) [14,23,24], (3) Ivory Coast (~1.07 Ma; Bosumtwi crater) [25,26], and (4) Australasian (~0.77 Ma; source crater not yet identified) [27] strewn fields. Based on shapes, tektites can be classified as normal/splash type, aerodynamic shaped, and layered (Muong Nong) type [28]. The current idea suggests that upon meteoritic impact on the Earth, the terrestrial rocks are melted and ejected into the upper atmosphere (hundreds of kilometers above) at hypervelocities. After a few minutes the partially molten mass rained on the Earth's surface due to gravitational pull, and it is during this period of traverse through outer space or vacuum that tektites assumed their peculiar morphologies due to differential ablation, rotation, etching, etc.

The North American tektites can be broadly categorized as bediasites (usually black colored; named after Bedias Indians of Grimes County, Texas) and the georgiaites (usually green color) found

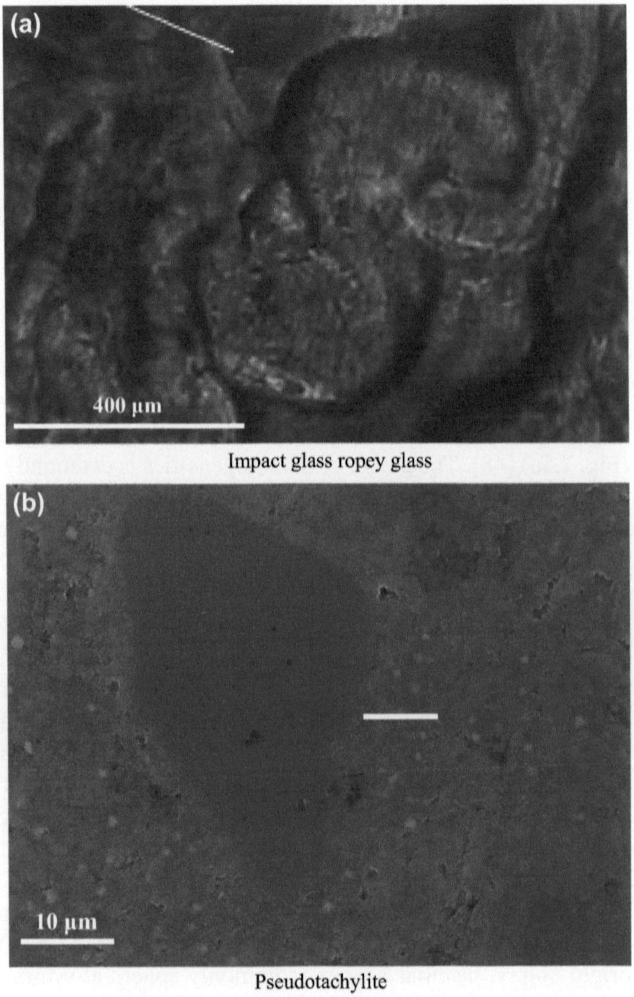

FIGURE 7.5

Microstructure of selected nonvolcanic natural glasses.

in East-Central Georgia. Compositionally both the varieties are different with the latter being more siliceous and containing a smaller amount of aluminum and iron.

Moldavites are bright green, transparent tektites (a few centimeters in size) commonly found in Germany, the Czech Republic, and Austria. In general they are siliceous in nature, with few varieties that occur as pure silica glass called "lechatelierite." Often thin layers of silica-rich and silica-poor glass alternate, giving rise to "schlieren" structure, preserving the flow patterns within the melt before quenching.

The Australasian-strewn field is the largest among the four with an estimated total tektite mass of 2000 megatons. Based on geographic locations, tektites are subdivided as australites, indochinites, philippinites, thailandites, malaysianites, tibetanites, and rizalites. Within each such subgroup, there exist compositional variations. There is another interesting variety of tektite called "Muong Nong type," which occurs as layered blocks, often containing relict minerals. Major element chemistry and age suggest that Muong Nong tektites are similar to other Australasian tektites but are more primitive and less altered. Possibly all these varieties of Australasian tektites were formed in the same event, but the Muong Nong ones were heated up to lower temperature and traveled only a short distance from the crater.

Despite a global consensus on "meteorite impact," origin for tektites, the detail mechanism is not yet understood. Departure of tektite compositions from its precursors is being explained on the basis of selective volatilization and vapor fractionation, incomplete mixing of melts, weathering, etc. Examples include depletion of volatile elements within (1) Muong Nong tektites compared to Australasian tektites, (2) Moldavites [14,29], etc. Reduced iron state of tektites is most commonly explained by superheating associated with the process [30].

4.2 ORIGIN: SEISMIC EVENTS

Pseudotachylites are the best-known example of natural glasses produced due to seismic events. Earthquake causes rubbing of one rock mass against another along a fault plane, which leads to development of frictional heat causing melting of rocks. The melt so produced gets channelized along the major/minor fault planes and adjacent cracked domains and thereby undergoes rapid cooling, leading to glass formation.

Pseudotachylites are more common in quartz-bearing rocks than others, probably more because of mechanical reasons than for compositional ones. Usually, relative to country rocks, pseudotachylite contains more of granitic melt components like potash and silica than others like soda and lime. In the field, pseudotachylite can be distinguished from cataclasite or breccia by the lack of any transitional zones of decreasing brittle deformation intensity toward the wall rock. Usually it is characterized as dark matrix material with minor inclusions of mineral or wall rock fragments. Under the microscope pseudotachylites show presence of glass, vesicles, amygdules, microlites, and flow structures that indicate frictional melt generation [31]. A representative microstructure of pseudotachylite is shown in Fig. 7.5b where random distribution of quartz clasts within glassy matrix is noted. In a separate investigation [32], microchemical analyses of glass, microlite, and crystallite with proximity to quartz clasts showed compositional variations indicating chemical reactions between the clasts and melts. Presence of quartz and feldspar clasts, and complete absence of mafic fragments in most pseudotachylites, indicate "selective" melting of minerals as and when the temperature generated during frictional heating is reached the melting temperature of individual minerals [33–35]. At times, pseudotachylite exhibits compositional banding, which may or may not conform country rock outline.

4.3 ORIGIN: NUCLEAR EXPLOSION

Civil underground nuclear explosions for various peaceful applications such as (1) creation of underground storage for liquid hydrocarbons, (2) recovery of oil from bituminous shale or sandstone through in situ cracking, (3) extinguishing fires in oil and gas wells, and (4) destroying potentially dangerous near-Earth objects like asteroids, etc., have led to formation of heterogeneous glassy materials within wide varieties of host rocks (acid volcanic rocks, alluvium, salt, dolomite, granite, sandstone, etc.). Usually such detonations are carried out at depths sufficient enough to contain the explosion and its debris. Such debris samples are usually collected through drilling the collapsed explosion cavities. Glasses are very commonly found within the debris samples, which are physically and compositionally heterogeneous in nature [36]. Occasionally the glasses are decorated with iron-enriched domains, referred to as "marble cake." In the Rainier event (September 19, 1957), which was a "contained" underground nuclear explosion test (yield: 1.7 kt; produced temperatures and pressures of 1×10^6 K and 7 Mbar, respectively) carried out in a tuffaceous country rock, three different glasses were found. These were (1) highly radioactive dark-colored dense glasses, (2) less radioactive, vesicular light-colored glasses, and (3) highly radioactive dense glasses with appearances varying from blebs, globules, coatings, etc.

Much before the Rainer event, the Trinity test (July 16, 1945) was carried out, which generated a crater with a 1.2 m depth. Country rock of the site was an arkosic sand containing dominantly quartz, k-feldspar, albite along with some minor amounts of muscovite, amphibole, and calcite. Post detonation, the entire area developed a layer of glass, popularly known as trinitite, which contained only quartz from the precursor, and all other minerals melted. Broadly, the trinitite glass layer can be divided into four types: (1) differentiated pancake trinitite with glassy surface and solidified globules/beads at the bottom, (2) red trinitite, (3) scoriaceous green trinitite fragments, and (4) trinitite beads and dumbbells. Red coloration of trinitite is due to copper that is present inside the glass as a part of Fe-Pb-Cu metallic chondrules. Green trinitite fragments were a major component of the glazed layer and compositionally (variable Ca and Fe contents) plus structurally they were heterogeneous in nature. Rounded quartz grains were the only crystalline phase present within the amorphous matrix. It is believed that the heterogeneity is due to partial melting of quartz and other minerals, and also due to lack of melt mixing. Trinitite beads and dumbbells are believed to have formed from materials present within post-detonation rising cloud of hot gases, and were showered on the Earth as melt droplets.

4.4 ORIGIN: LIGHTNING

During thunderstorms in the rainy season there is always charge imbalance between atmosphere and ground, which gets partially released through electron flow between the two media in the form of "cloud-to-ground" lightning. Estimates show that a single lightning event can dissipate as high as 10^9 J [37,38], thereby raising the surrounding air temperature instantaneously to a very high value ($\sim 10^5$ K; [39]). When such lightning strikes the Earth's crust, it deposits a huge amount of energy on a very short timescale (a few tens of microseconds). As a result the target gets heated up at an enormously high rate ($\geq$1000 K/s) and the temperature can go even beyond $\sim$3000 K (or even higher, $\sim$10,000–30,000 K, in the lightning channel) over a period of less than 1 ms [40]. A lightning event can therefore be considered as a point source in space-time domain. Such an instantaneous extreme condition can induce extensive physical, chemical, and morphological changes of the target rocks through its melting and vaporization. Incineration of any organic material present within the rock takes place immediately [41]. The lightning process continues until it encounters some conductive layer like aquifer, surface water or matter with high-moisture content (e.g., plant roots), etc. Usually the path of lighting on Earth's surface is marked with heterogeneous tubular (often branching) porous mass, commonly known as "fulgurite." Such masses are found to be very resistant to natural weathering on the geological timescale and are therefore extensively studied for "paleoenvironment reconstruction" [42].

Fulgurite morphology depends on several factors like target composition, striking energy, multiplicity of lightning, time duration, etc. The interiors are made up of several cavities, mostly decorated with thin glassy material containing unaltered grains embedded within it (Fig. 7.6). Often the glasses

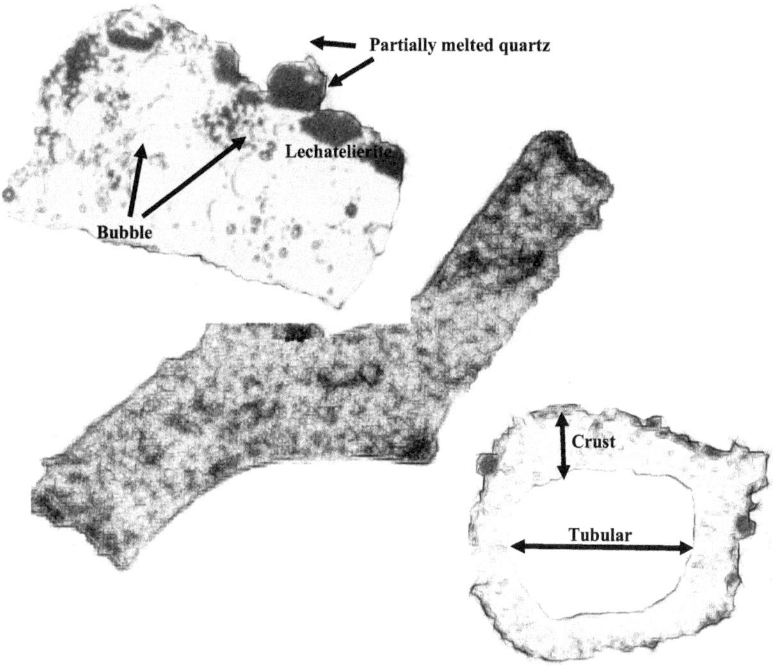

FIGURE 7.6

Some important physical features of fulgurite.

are vesicular in nature with elongated vesicles oriented normal to the central tube. On the exterior side, fulgurites are mostly constituted of partially melted/baked target rock components (high-temperature refractory materials) with numerous "blister-like" features. At times acicular texture resulting from partially melted quartz grains and fine needles of glass is also seen on the external crust of fulgurites. Based on morphology and petrology, Pasek et al. [43] identified five different types of fulgurites: sand- (type I), clay- (type II), caliche- (type III), rock- (type IV), and droplet fulgurites (type V; brown to green glasses). According to them [43] glass domains are maximum in type I (95−100%) and type V (90−100%) fulgurites and minimum within type III (10−50%). In type II and IV, glass content varies between 30% and 90% and 50−90% respectively. Lechatelierite, which contains 90−99 wt% SiO_2 glass with or without gas bubbles, constitutes the major component of the glass domains. It has, however, very recently been shown that the glass composition within a given fulgurite can vary from domain to domain. For example, Crespo et al. [44] reported significant microstructural and compositional variations among axial (homogeneous) and radial facies (heterogeneous) of a fulgurite developed on granitic rock.

Among the various types, rock fulgurites exhibit a number of interesting features. Usually they occur as thin coats of black glass (few tens of microns thick; covering several square centimeters; [40,42]), containing fine-grained anisotropic minerals, carbonaceous matters (if any), and vesicles. Under transmission electron microscope, some of these anisotropic minerals like quartz exhibit planar deformation lamellae [44−46], very similar to those reported from "shock metamorphosed rocks."

Recently comparative strain analyses on Sahara fulgurite and host rock quartz crystals using neutron diffraction technique (plus Rietveld refinement) identified lattice parameter variations that can be indicative of shock pressure as high as 22–30 GPa. X-ray computed tomography examinations revealed that the stressed quartz crystals were mostly restricted on the outer shell domains of Sahara fulgurite [47]. Despite such increasing evidence of shock pressure–induced phase transformations within fulgurite, the exact reason is not yet clearly understood. It is probable that very energetic lightning strikes can raise pressure inside channels and thereby accelerate materials radially outward. Such motions, however, get resistance due to the strength of surrounding soil and therefore build up the pressure instantaneously at the fulgurite/host matrix interfaces [49].

4.5 ORIGIN: RADIATION DAMAGE

Natural glasses are also very commonly produced within minerals due to radiation damage. Usually accessory minerals like zircon, monazite, apatite, etc. incorporate actinides during their crystallization and growth. With time the radionuclides (uranium, thorium, etc.) undergo decay and emit alpha and beta particles. These particles, either partially or totally, break the regular lattice network of the adjacent matrix (exhibited by pleochroic halo, discoloration) and impart aperiodicity through amorphization while preserving the external crystal habit. In mineralogical parlance this natural process of transforming a structurally ordered crystalline state into a special state of aggregation resembling solid colloids is known as metamictization, and the host mineral is called metamictic mineral. Usually metamictic minerals exhibit conchoidal fracture and isotropic optical properties. Density, refractive indices, and birefringence are also found to decrease with increase in radiation damage. On heating in the range of 400–800°C (sometimes to 1000°C), metamict minerals often get back to an ordered state, with properties of the initial crystalline mineral.

Structural studies have revealed that metamictic disintegration is common in minerals having weakly bonded cationic and anionic groups (Zr, Th, U, REE, and others with Si, Nb, Ta, Ti, and other elements). This is why minerals such as zircon, thorite, orthite, gadolinite, pyrochlore, samarskite, euxenite, niobotantalates, etc. preserve the signature of radiation damage very commonly.

5. LEACHING OF GLASS

Besides formation under extreme conditions, natural glasses experience unusually long interaction with surface water, which renders compositional alterations at the water–glass interfaces. As mentioned earlier, elemental profiling of such natural glasses is very important from the "long-term performance assessment of nuclear waste glasses" point of view. Within geological repository system, the easiest pathway for radionuclides' release is through interaction between nuclear waste glass and groundwater. Various hypotheses have been put forward to explain glass corrosion, and the most favored is the Doremus model [50]. According to this model glass hydration is the result of ionic interdiffusion between the alkali ions of the glass with hydronium ions from solution. The evidence that supports interdiffusion model includes (1) depletion of alkali ions within hydrated natural glasses, (2) experimental results with technical glasses showing amount of ions diffused out of the amorphous matrix corroborates well with ion-depletion profiles measured in the hydrated domains, and (3) the fact that glasses that hydrate are ionic conductors whereas glasses without mobile ions do not generally hydrate (such as SiO_2). In the following section leaching of natural glasses over long (geological) timescale and under accelerated conditions are described separately. However, for the benefit of less-experienced researchers a small description on the leaching mechanisms of glasses is also given.

5.1 LEACHING MECHANISMS OF GLASSES

"Leaching of glasses" takes place in phase-wise manner [51]. The process starts with entry of water into the glass structure and exchange of H^+ ions with alkali ions [52]. It is noted that independent of temperature and pH, about three hydrogen atoms enter the glass structure for every alkali ion (R^+) released [53]. In fact, the partial molar volumes of the released alkali ions govern the number of water molecules entering the glass together with the H^+ ions necessary for charge compensation [54]. This can be explained through the following chemical reaction

$$\equiv \text{SiOR} + \left(\text{H}^+ + \text{OH}^-\right) \rightarrow \equiv \text{Si} - \text{OH(silanol group)} + \text{ROH}$$

or

$$\equiv \text{Si} - \text{O}^-\text{R}^+ + 2\left(\text{H}^+ + \text{OH}^-\right) \rightarrow \equiv \text{Si} - \text{O}^-\text{H}_3\text{O}^+ + \text{ROH}$$

Since the leaching process is interdiffusion controlled, the concentration profile of R in glass surface would be solutions of Fick's second law of diffusion with appropriate boundary conditions, and the amount of R in the leachate would increase as a function of $(\text{time})^{1/2}$. The interdiffusion model is convenient because it can be transformed to an equivalent simple diffusion problem with a concentration-dependent diffusion coefficient, for which standard methods of solution exist and expressed as

$$D = D_\text{H} D_\text{A} / (C_\text{H} D_\text{H} + C_\text{A} D_\text{A})$$

where, D_H and D_A are the diffusion coefficients of hydronium and the alkali ion, and C_H and C_A are their respective concentrations.

Hence, if mobile alkali ions (such as Na^+) interdiffuse with hydronium, one can immediately predict the shape of the hydration profile. The rate of hydration can be calculated if the diffusion coefficients of Na^+ and H_3O^+ are known.

In the next stage of leaching, proton on the Si–OH group (silanol group) exchanges sites with an inner R^+. In other words, the proton diffuses inward and the R^+ moves outward. The Si–OH ultimately combines with an Si–OR to form a bridging bond ≡Si–O–Si≡, or the "siloxane" group.

$$\equiv \text{Si} - \text{OR} + \text{HO} - \text{Si} \equiv \; \rightarrow \; \equiv \text{Si} - \text{O} - \text{Si} \equiv + \text{ROH}$$

or

$$\equiv \text{Si} - \text{O}^-\text{H}_3\text{O}^+ + \text{R}^+\text{O}^- - \text{Si} \equiv \; \rightarrow \; \equiv \text{Si} - \text{O} - \text{Si} \equiv + \text{ROH} + \text{H}_2\text{O}$$

With time, the H_2O molecule gets retained within the glass structure, whereas ROH is able to gradually move outward [55].

However, the situation may change drastically if the surrounding medium is basic in nature. A completely different reaction occurs where OH^- group in a basic solution of ROH breaks the network:

$$\equiv \text{Si} - \text{O} - \text{Si} \equiv + \left(\text{R}^+ + \text{OH}^-\right) \rightarrow \equiv \text{Si} - \text{OH} + \text{RO} - \text{Si} \equiv$$

The bridge is broken to form a more open gel-like structure on the surface. With time, more of the R^+ OH^- molecules are able to break up more of the bridging oxygen bonds, resulting in a gradual dissolution of the network. The surface appears corroded or etched. The rate of such a corrosion reaction or etching is generally linear with time. When the dissolution is uniform, the ratio of the various species in the attacking medium is the same as that in the glass; this process is known as "congruent dissolution."

Under neutral condition, ion exchange, glass hydration, and dissolution of hydrated glass network take place in parallel. With time, diffusion and dissolution processes operate at the same rates, thereby establishing a steady state and nearly congruent glass dissolution [56,57]. Recently, water diffusion coefficients under silica-saturated conditions have been estimated to be smaller than 10^{-20} m^2/s at 90°C, indicating solid-state diffusion [54]. The fastest ion exchange reactions involve the alkali ions, and the activation energy varies between 100 and 120 kJ mol^{-1}, while slower rates are characteristic of Ca, Mg, and Al and breaking of Si–O–Si bonds is the slowest process.

With increasing reaction, more and more dissolved silica is resorbed onto the surface, initiating the formation of a surface gel layer. The gel layer's density increases when silica concentrations within solution rise [58]. Contrary to hydrated glass surface, gel layer is not part of the glass matrix, rather it is an entirely restructured, compositionally different, highly porous, amorphous reaction product [59]. It can also contain alteration products like oxyhydrates, powellite, clays, and zeolites. Following gel layer formation, the reactive sites are no longer available at the natural glass surface and therefore it acts as barrier to the transport of dissolved silica from active surface sites at the glass–gel interface to the solution. A typical image of an altered glass surface showing different domains is given in Fig. 7.7. The Si Kα X-ray line scan clearly indicates widths of the different domains.

Leaching of natural glasses also depends on pH conditions as it can vary from extremely acidic to basic. Under mild acidic condition of pH <7, alkali ions get preferentially leached from the glass matrix through exchange of hydronium ions. However, in extreme acidic condition this is not the case and instead of hydronium proton gets exchanged for alkali ions. Since the volumes of hydronium ions and protons are not same, natural glasses exposed to extreme acidic conditions usually develop cracks. The proposed reaction is:

$$\equiv Si - ONa + H^+ \rightarrow \equiv Si - OH + Na^+$$

Apparently, the exchange of Na^+ in the glass with the smaller sized proton creates tensile stress in the surface, which leads to cracking. On the other hand in basic media, where pH >7, the dominant leaching mechanism is network dissolution and therefore besides alkali ions, alkaline earths also get released in the leachant.

6. ALTERATION OF NATURAL GLASSES UNDER EXTREME GEOLOGICAL CONDITIONS

6.1 BASALTIC GLASSES

Earlier investigations showed that dredged ocean floor basaltic glasses (constant static water/natural glass interaction over millions of years under low temperature of ~3°C) collected from beneath sediment cover (silica-saturated pore water conditions) alter at a much lower rate (~10^{-3} times) than those on the sediment surface (silica undersaturated conditions) [60]. Accumulations of dissolved silica in bulk solution as well as within pore water of gel layers reduce reaction rates with time. Such "saturation effect" implies that local equilibrium is reached between dissolved silica and

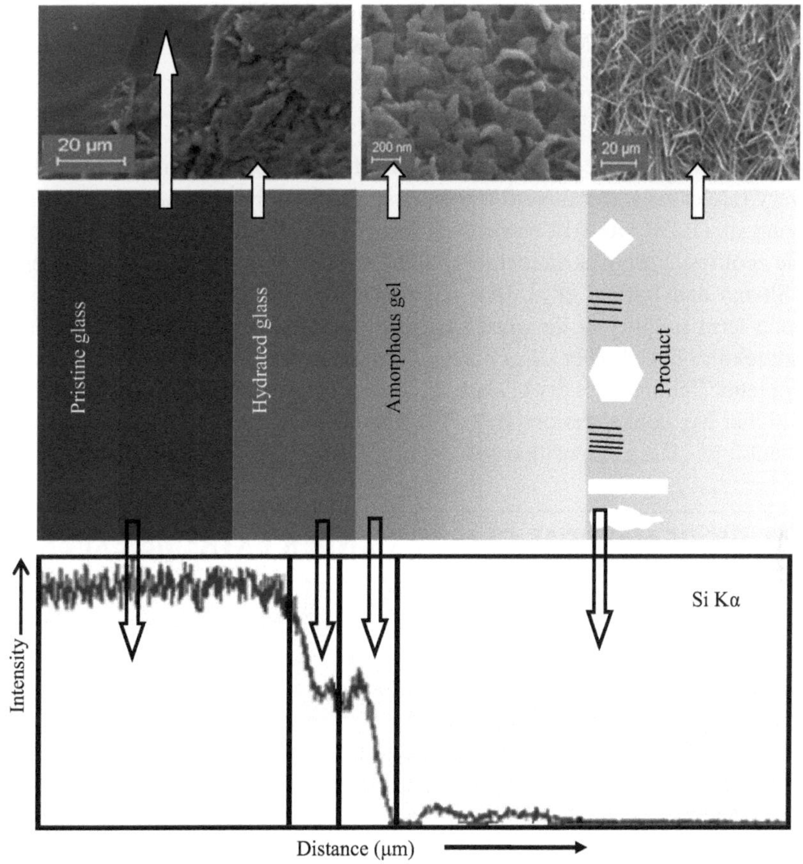

FIGURE 7.7

Typical image of an altered glass surface showing different domains. The Si Kα X-ray line scan clearly indicates widths of the different domains.

alkali-depleted basaltic glass surface [61]. In both cases, silica-saturated or unsaturated situations, similar alteration products like palagonite, analcime, and mat of crystal rosettes (probably clay) can be seen. All such mineralogical changes corroborate well with observed compositional changes within glass matrices, i.e., (1) depletion in SiO_2, (2) approximately constant CaO and Al_2O_3 concentrations, and (3) building up of TiO_2, FeO, MnO, and Na_2O concentrations within surface layers.

6.2 IMPACT GLASSES

Very little information is available on leaching of impact glasses. Koeberl and Sigurdsson [62] have carried out some in-depth analysis on alterations of Haiti impact glasses along with translucent light brown–gray glass shard with high Si, K content. In the case of black glass spherules, the amorphous particles are partially to completely rimmed with smectite. Compositionally, smectite differs from black glasses in its higher contents of Mg, Fe, Zn, and Br and its lower abundance for most other elements, especially Na, K, Ca, Mn, Cr, Co, Rb, Sr, Cs, Th, and U.

6.3 TEKTITES

Tektites generally do not produce any alteration products on exposure to environmental moisture [63], however, in very rare cases some alteration been observed [64]. For example, North American tektite fragments found at DSDP site 612 sometimes have rims of hydrated glass. Lovering et al. [65] observed some zeolites in grooves on tektites, but there is no proof that these are real tektite alteration products. El-Shamy and Barkatt et al. [66–68] found that the tektite composition is an important factor—the long-term stability of the glass against dissolution requires a silica content in excess of 67 mol%; most tektites have higher silica contents, although there are a few Australasian tektites with lower silica contents. Tektite glass dissolves more slowly in seawater, which seems to have a buffering effect due to higher Mg concentrations [68–70]. Barkatt et al. (1984) found that the speed of dissolution is a function of glass composition [67].

7. ALTERATION OF NATURAL GLASSES UNDER EXTREME EXPERIMENTAL CONDITIONS

All the various mechanisms discussed earlier can be verified in laboratory using pellet or powder samples. The latter experiments require shorter periods of time because of the much larger exposed surface area. The experimental choice depends largely upon the intended application. For example, since the glass constituents are likely to affect the attacking medium characteristics, the experiments can be performed in closed (fixed quantity) medium or in open (flowing) medium where the fluid is continually replenished. In closed-medium experiments, pH of the leachant increases with time because of the extraction of alkali, which encourages network dissolution. Closed-medium experiments also suffer from the additional possibility that reaction products are deposited on the glass surface, acting as barriers for further reaction. Thus, the relationship of the amount of material extracted with time is rarely constant. The use of gentle shake (provided by a shaker bath) to reduce the deposit problem is often recommended.

To understand leaching mechanisms within natural glasses, accelerated tests are commonly being done where small pieces of glasses are exposed to static or dynamic water bath of distilled water at 90°C for different lengths of timescales. After experiments, cross-sections of samples are quantitatively studied for hydrogen profiles using ^{15}N method, which relies on the resonant nuclear reaction ^{1}H (^{15}N, $\alpha\gamma$) ^{12}C. The measured hydrogen profile is expected to have contributions from all alkali ions (D_A) present within a given natural glass, and influence of each individual cation is difficult to identify.

7.1 OBSIDIAN GLASS

Laursen and Lanford studied Mahogany obsidian (USA) samples exposed to water bath up to 348 h, and for reference purposes they also included soda-lime glasses in the experiments [71]. Comparison of the results showed that diffusion coefficients of alkali ions (Na^+ and K^+) in obsidian glass were at least one order smaller in magnitude with respect to soda-lime glasses. However, one should exercise caution here as obsidian and soda-lime glasses are compositionally widely different. Obsidian has much less total alkali than the examined soda-lime glass (~15.9% Na_2O). Such compositional variations have significant influence in elemental diffusion as D_{Na} decreases rapidly with rise in Na^+ concentrations in glasses. Further, alkali ions like K^+ are less mobile than Na^+, and hence D_A is expected to be smaller than D_{Na}. Extrapolation of experimentally measured D_{Na} in obsidian glass (temperature range 350–850°C) to 90°C (present hydration condition) gives a rough estimation of 5×10^{-18} m^2/s. On the altered surface perlite is found to form, which changed the mechanical properties significantly.

7.2 TEKTITE GLASS

Rauch et al. [72] examined hydrogen concentration profiles of polished and etched slices (8–15 mm diameter, 1–2 mm thick) of an indochinite tektite exposed to water (in 23 mL Teflon containers filled with 15 mL distilled water within a stainless steel vessel) at 125–250°C (vessels were put for various times in an oven set at different temperatures) for up to 24 days using ^{15}N nuclear reaction technique. The growth kinetics of hydrated layers with time and temperature (up to several hundred nanometers) indicated activation energy of 52 kJ/mol. The diffusion coefficient at 200°C is 6×10^{-20} m^2/s.

In a separate study with fractured tektite surfaces, warm water (90°C) interaction led to formation of very thin (<100 nm) hydration layers whose thickness did not increase monotonically with time. In another investigation under similar experimental conditions [67], Na- and K-depleted leach layer was only 25 nm thick after 10 days and did not increase further. This probably indicates matrix dissolution and explains the absence of tektite hydration. However, measurements on old tektite surfaces lying on the soil for ~1 million years revealed hydration layers in which the H content decreased exponentially from the surface [73]. During natural glass/water interactions, chemistry of water bath is also found to be changed. For example, when the experiment was carried out at 140°C, the water bath pH of 5.9 changed to 6.2 after 1 day, to 6.9 after 4 days, and to 7.7 after 16 days. After the treatments, the samples (but not the Teflon containers) were found to be covered with thin loosely attached, very brittle reddish layers whose thickness increased (tens to few hundreds of nm) with rise in experimental temperature and time. All H (or H^+) profiles had the same characteristic shape—the H concentration was highest at the surface with concentrations of (3–5) × 10^{21} H/cm^3, and decreased gradually following error function (erfc) pattern. Scanning electron micrograph of the flake showed the layer to have a spongelike structure composed of small needles with typical dimensions of a few μm in length and fractions of a μm in width. Micrographs of sample surface areas from which flakes had been removed showed that the surface were completely smooth with no trace of residues from the layer material. The only noticeable features were sparsely distributed groups of so-called etch pits. It appears that these layers are very likely due to precipitation of initially dissolved glass constituents that had come into the water bath by congruent dissolution of the surface layers of the samples. It appears less likely that the layers were generated on the samples by leaching of glass constituents because they had only very weak adhesion to the samples. This deduction is supported by the fact that no residues from the layer material were found on the surfaces.

H profiles of hydrated tektites are quite different from that found for hydrated soda-lime silicate glasses and obsidian glasses. For the latter set of glasses the H concentration is roughly constant throughout the hydrated layer and decreases rather steeply in the boundary zone to the pristine glass. Such variations in results are probably due to different interdiffusing hydrogen-bearing species (H^+, H_3O^+, H_2O) for substitutions of alkali ions. By contrast, the error function-like shape of the tektite profiles points to the interdiffusion of water molecules as the dominant process for the formation of the hydrated layers. If the same mechanism as in other natural glasses would hold for tektite glass, the experimentally found H concentration values of about 4×10^{21} H/cm^3 would be connected with leaching of at least 13×10^{21} alkali/cm^3. However, the RBS measurement showed at most 0.4×10^{21} alkali/cm^3 have been leached.

The calculated D values define a constant activation energy for the corresponding temperature range, with a value of 52 kJ/mol. Upon extrapolation to $T = 90°C$ the diffusion coefficient comes in the order of 10^{-21} m^2/s. This data corroborates well with the observed hydrated layers of a few tens of nanometers within tektites. The absolute values of diffusion coefficients found are smaller by about two orders of magnitude than those for the diffusion of water into vitreous silica and for the hydration/leaching process of soda-lime silicate glasses. The activation energy of 52 kJ/mol is typical for silica-rich materials, being in the range spanned by the activation energy for soda-lime silicate glass and obsidian of about 80 kJ/mol at 60–90°C and for vitreous silica, 40 kJ/mol at 90–400°C. Also the maximum H concentrations of (3–5) $\times 10^{21}$ H/cm^3 are between those for soda-lime silicate glass, about 2×10^{22} H/cm^3 and for vitreous silica, about 5×10^{20} H/cm^3 and similar to those for obsidian.

8. CONCLUSIONS AND FUTURE DIRECTIONS

From the previously mentioned microstructural and compositional heterogeneities within natural glasses it can be concluded that extreme conditions do provide an alternative activation technique for ions/molecules/clusters to overcome the energy barrier for certain chemical reactions that cannot be achieved under normal pressure-temperature conditions. As the natural glasses are highly viscous in nature and compositionally extremely complicated, they can easily preserve the products of such extraordinary chemical reactions very easily. However, so far not many detailed investigations have been done on natural glasses produced under extreme conditions. There are many interesting observations made from natural glasses that can be studied in the future using usual experimental and characterization approaches coupled with improved computational capabilities. A few such interesting topics are outlined following:

1. The most important aspect that evolves from studying phase relations under extreme conditions is that surfaces and interfaces of phases in multicomponent systems are structurally and compositionally dynamic in nature. The interdependence between the two actually drives the reaction mechanisms and rates. Examples include shock-induced congruent melting and devolatilization of hydrosilicates like biotite, muscovite, amphiboles, etc. and generation of homogeneous glasses. These glasses are quite different from those obtained through melting hydrous silicates under natural and/or experimental static conditions. It appears that homogeneous glasses are uniquely obtained in shock loading of hydrous silicates, but why this so is yet to be understood.
2. From leaching studies a number of parallel and sequential chemical reactions have been identified. The rate-controlling process within this overall mechanistic reaction network changes with time and with geochemical boundary conditions. This adds complexity to the analysis of the experimental data, since different rate-controlling processes may be invoked to explain experimentally observed behavior or field observations. This also establishes the need for natural analogue studies in deciding experimental protocols keeping in mind the ultimate reaction

products. It may be noted that saturation effects coupled to steadily decreasing water diffusion rates can provide effective stabilization of the glass for many millions of years, even in the event of permanent contact with very slow-flowing, stagnant deep groundwater.

3. It is interesting to note that there are significant departures in leaching mechanisms for tektites and other natural glasses, the reasons for which are still not clearly understood. An answer to the fundamental question of why ionic exchange and interdiffusion mechanisms are not active in tektite glass corrosion is yet to be found out.

Finally, it may appear that this review has considerable variation in the level of information. This is so because not many intense studies have been done on the subject. However, the subject upholds significant promises for future advancements. Information gathered here is expected to be useful in decision making at various stages of nuclear waste disposal programs and may also constitute the basis of safety concepts and their assessments.

ACKNOWLEDGMENTS

Continuous encouragement received from Dr. Srikumar Banerjee and Dr. A.K. Tyagi during preparation of the manuscript is gratefully acknowledged. Scientific thoughts and explanations that were generated through discussions with Dr. G.K. Dey and Dr. Pooja Mahadik are highly appreciated.

REFERENCES

[1] P. Lambert, M.A. Lange, Glasses produced by shock melting and devolatilization of hydrous silicates, J. Non–Cryst. Solids 67 (1984) 521–542.
[2] S.W. Kieffer, P.P. Phakey, J.M. Christie, Shock processes in porous quartzite: transmission electron microscope observations and theory, Contrib. Mineral. Petrol. 59 (1976) 41–93.
[3] M.J. Apted, Natural analogues for predicting the reliability of the engineered barrier system for high level waste, J. Geochem. Explor. 46 (1992) 35–62.
[4] J.E. Shelby, Introduction to Glass Science and Technology, second ed., Royal Society of Chemistry, 2009, p. 285.
[5] D. Stoffler, Glasses formed by hypervelocity impact, J. Non–Cryst. Solids (1984) 465–502.
[6] W.V. Engelhardt, W.J. Arndt, D. Stoffler, W.F. Muller, H. Jezorkowski, R.A. Gubser, Diaplektische Gläser in den Breccien des Ries von Nördlingen als Anzeichen für Stosswellenmetamorphose, Contrib. Mineral. Petrol. 15 (1967) 93–102.
[7] W.V. Engelhardt, D. Stoffler, in: B.M. French, N.M. Short (Eds.), Shock Metamorphism of Natural Materials, Mono Book Corp, Baltimore, 1968, p. 159.
[8] E.C.T. Chao, Shock effects of certain rock-forming minerals, Science 156 (1967) 192–202.
[9] J.A. O'Keefe, Natural glasses, J. Non–Cryst. Solids 67 (1984) 1–17.
[10] G. Tschermak, Sitz Ber Akad Wiss 55 (1872) 122.
[11] T.E. Bunch, A.J. Cohen, M.R. Dence, Shock induced structural disorder in plagioclase and quartz, in: B.M. French, N.M. Short (Eds.), Shock Metamorphism of Natural Materials, Mono Book Corp, Baltimore, 1968, pp. 59–81.
[12] S.R. Taylor, Consequences for tektite composition of an origin by meteoritic splash, Geochim. Cosmochim. Acta 26 (1962) 915–920.
[13] B.P. Glass, Tektites and microtektites: key facts and inferences, Tectonophysics 171 (1990) 393–404.

[14] W.V. Engelhardt, E. Luft, T. Arndt, H. Schoch, W. Weiskirchner, Origin of moldavites, Geochim. Cosmochim. Acta 51 (1987) 1425–1443.
[15] H. Singurdsson, S. D'Hondt, M.A. Arthur, T.J. Bralower, J.C. Zachos, M.V. Fosses, JET Channel, Glass from the cretaceous/tertiary boundary in Haiti, Nature 349 (1991a) 482–487.
[16] H. Singurdsson, P. Bonte, L. Turpin, M. Steinberg, P. Pradel, C. Jehanno, et al., Geochemical constraints on source region of cretaceous/tertiary impact glasses, Nature 353 (1991) 839–842.
[17] J. Smit, Meteorite impact extinctions and the cretaceous–tertiary boundary, Geol. Mijnbouw 69 (1990) 187–204.
[18] B.P. Glass, J. Wu, Coesite and shocked quartz discovered in the Australasian and North American microtektite layers, Geology 21 (1993) 435–438.
[19] K.T. Howard, S. Bunopas, C. Burrett, M.D. Norman, The 770 Ka tektite producing impact event: evidence for distal environmental effects in NE Thailand (abstract #1308), in: 31st Lunar and Planetary Science Conference. CD-ROM, 2000.
[20] G. Heide, B. Müller, G. Kloess, D. Moseler, G.H. Frischat, Structural classification of natural non-crystalline silicates, J. Non-Cryst. Solids 323 (2003) 68–71.
[21] D. Storzer, G.A. Wagner, Fission track ages of North American tektites, Earth Planet. Sci. Lett. 10 (1971) 435–440.
[22] C. Koeberl, C.W. Poga, W.U. Reimold, D. Brandt, Impact origin of Chesapeake Bay structure and the source of North American tektites, Science 271 (1996) 1263–1266.
[23] W.H. Schwarz, H.J. Lippolt, Coeval argon-40/argon-39 ages of moldavites from the Bohemian and Lusatian strewn fields, Meteorit. Planet. Sci. 37 (2002) 1757–1763.
[24] M.A. Laurenzi, G. Bigazzi, M.L. Balestrieri, V. Bouska, $^{40}Ar/^{39}Ar$ probe dating of the Central European tektite-producing impact event, Meteorit. Planet. Sci. 38 (2003) 887–893.
[25] C. Koeberl, R.J. Bottomley, B.P. Glass, D. Storzer, Geochemistry and age of Ivory Coast tektites and microtektites, Geochim. Cosmochim. Acta 61 (1997) 1745–1772.
[26] C. Koeberl, W.U. Reimold, J.D. Blum, C.P. Chamberlain, Petrology and geochemistry of target rocks from the Bosumtwi impact structure, Ghana, and comparison with Ivory Coast tektites, Geochim. Cosmochim. Acta 62 (1998) 2179–2196.
[27] G.A. Izett, J.D. Obradovich, Laser-fusion $^{40}Ar/^{39}Ar$ ages of Australasian tektites (abstract), Lunar Planet. Sci. XXIII (1992) 593–594.
[28] C. Koeberl, Geochemistry and origin of Muong Nong-type tektites, Geochim. Cosmochim. Acta 56 (1992) 1033–1064.
[29] W.V. Engelhardt, C. Berthold, T. Wenzel, T. Dehner, Chemistry, small-scale inhomogeneity, and formation of moldavites as condensates from sands vaporized by the Ries impact, Geochim. Cosmochim. Acta 69 (2005) 5611–5626.
[30] P. Jakes, S. Sen, K. Matsuishi, A.M. Reid, E.A. King, I. Casanova, Silicate melts at super liquidus temperatures: reduction and volatilization, Lunar Planet. Sci. XXIII (1992) 599–600.
[31] C.D. Rowe, J.C. Moore, F. Meneghini, A.W. McKeirnan, Large–scale pseudotachylytes and fluidized cataclasites from an ancient subduction thrust fault, Geology 33 (2005) 937–940.
[32] L.B. Goodwin, Controls on pseudotachylyte formation during tectonic exhumation in the South Mountains metamorphic core complex, Arizona (special publications), Geol. Soc. London 154 (1999) 325–342.
[33] R.L. Maddock, Effects of lithology, cataclasis and melting on the composition of fault generated pseudotachylytes in Lewisian gneiss, Scotland, Tectonophysics 204 (1992) 261–278.
[34] J.F. Magloughlin, J.G. Spray, Frictional melting processes and products in geological materials: introduction and discussion, Tectonophysics 204 (1992) 197–204.
[35] A.M. Boullier, T. Ohtani, K. Fujimoto, H. Ito, M. Dubios, Fluid inclusions in pseudotachylytes from the Nojima fault, Japan, J. Geophys. Res. B 106 (2001) 21965–21977.

[36] L. Schwartz, A. Piwinskii, F. Ryerson, H. Tewes, W. Beiriger, Glass produced by underground nuclear explosions, J. Non-Cryst. Solids 67 (1984) 559−591.

[37] G.N. Eby, R. Hermes, N. Charnley, J.A. Smoliga, Trinitite − the atomic rock, Geol. Today 26 (2010) 181−186.

[38] W.J. Borucki, W.L. Chameides, Lightning − estimates of the rates of energy dissipation and nitrogen fixation, Rev. Geophys. Space Phys. 22 (1984) 363−372.

[39] M.A. Uman, W.H. Beasley, J.A. Tiller, Y. Lin, E.P. Krider, C.D. Weidmann, et al., An unusual lightning flash at Kennedy Space Center, Science 201 (1978) 9−16.

[40] R. Grapes, S.H. Müller, Lightning − strike fusion of gabbro and formation of magnetite − bearing fulgurite, Comone di Blumone, Adamello, Western Alps, Italy, Mineral. Petrol. 99 (2009) 67−74.

[41] J. Knight, S.W. Grab, Lightning as a geomorphic agent on mountain summits: evidence from southern Africa, Geomorphology (2013). http://dx.doi.org/10.1016/j.geomorph.2013.07.029.

[42] R. Navarro-Gonzalez, S.A. Mahan, A.K. Singhvi, R. Navarro-Aceves, J.L. Rajot, C.P. McKay, et al., Paleoecology reconstruction from trapped gases in a fulgurite from the late Pleistocene of the Libyan Desert, Geology 35 (2007) 171−174.

[43] M.A. Pasek, K. Block, V. Pasek, Fulgurite morphology: a classification scheme and clues to formation, Contrib. Mineral. Petrol. 164 (2012) 477−492.

[44] T.M. Crespo, R.P. Lozano Fernandez, R. Gonzalez Laguna, The fulgurite of Torre de Moncorvo (Portugal): description and analysis of the glass, Eur. J. Mineral. 21 (2009) 783−794.

[45] G.S. Collins, H.J. Melosh, M.A. Pasek, Can lightning strikes produce shocked quartz?, in: 43rd Lunar and Planetary Science Conference, 2012, 1160.

[46] R. Giere, W. Wimmenauer, H. Muller−Sigmund, R. Wirth, G.R. Lumpkin, K.L. Smith, Lightning − induced shock lamellae in quartz, Am. Mineral. 100 (2015) 1645−1648.

[47] M. Ende, S. Schorr, G. Kloess, A. Franz, M. Tovar, Shocked Quartz in Sahara Fulgurite, P22−P04, DMG, Salzburg, 2011, pp. 146−147.

[48] A.A. Sheffer, Chemical Reduction of Silicates by Meteorite Impacts and Lightning Strikes (Ph.D. dissertation), The University of Arizona, 2007.

[49] B.J. Saikia, G. Parthasarathy, R.R. Borah, Distribution of microcrystalline quartz in glassy fulgurites from Garuamukh and Kimin, India, J. Appl. Math. Phys. 3 (2015) 1343−1351.

[50] R.H. Doremus, Interdiffusion of hydrogen and alkali ions in a glass surface, J. Non−Cryst. Solids 19 (1975) 137−144.

[51] B. Grambow, Nuclear waste glasses − how durable? Elements 2 (2006) 357−364.

[52] R. Hellmann, J.C. Dran, M.G. Della, The albite − water system: Part III. Characterization of leached and hydrogen enriched layers formed at 300°C using MeV ion beam techniques, Geochim. Cosmochim. Acta 61 (1997) 1575−1594.

[53] W.A. Lanford, ^{15}N hydrogen profiling: scientific applications, Nucl. Instrum. Methods 149 (1978) 1−8.

[54] K. Ferrand, A. Abdelouas, B. Grambow, Water diffusion in the simulated French nuclear waste glass SON 68 contacting silica rich solutions: experimental and modelling, J. Nucl. Mater. 355 (2006) 54−67.

[55] B. Grambow, R. Müller, First order dissolution rate law and the role of surface layers in glass performance assessment, J. Nucl. Mater. 298 (2001) 112−124.

[56] D. Perret, J.L. Crovisier, P. Stille, G. Shields, U. Mäder, T. Advocat, et al., Thermodynamic stability of waste glasses compared to leaching behaviour, Appl. Geochem. 18 (2003) 1165−1184.

[57] E.H. Oelkers, General kinetic description of multioxide silicate mineral and glass dissolution, Geochim. Cosmochim. Acta 65 (2001) 3703−3719.

[58] D. Rebiscoul, A. Van der Lee, F. Rieutord, F. Né, O. Spalla, E. El-Mansouri, et al., Morphological evolution of alteration layers formed during nuclear glass alteration: new evidence of a gel as a diffusive barrier, J. Nucl. Mater. 326 (2004) 9−18.

[59] I. Munier, J.L. Crovisier, B. Grambow, B. Fritz, A. Clément, Modelling the alteration gel composition of simplified borosilicate glasses by precipitation of an ideal solid solution in equilibrium with the leachant, J. Nucl. Mater. 324 (2004) 97–115.
[60] W. Lutze, G. Malow, R.C. Ewing, M.J. Jercinovic, K. Keil, Alteration of basalt glasses: implications for modelling the long-term stability of nuclear waste glasses, Nature 314 (1985) 252–255.
[61] B. Grambow, A general rate equation for nuclear waste glass corrosion, Mater. Res. Soc. Symp. Proc. 932 (1985) 55–66.
[62] C. Koeberl, H. Sigurdsson, Geochemistry of impact glasses from the K/T boundary in Haiti: Relation to smectites and a new type of glass, Geochim. Cosmochim. Acta 56 (1992) 2113–2129.
[63] P.H. LaMarche, F. Rauch, W.A. Lanford, Reaction between water and tektite glass, J. Non–Cryst. Solids 67 (1984) 361–369.
[64] J.J. Mazer, J.K. Bates, J.P. Bradley, C.M. Stevenson, Alteration of tektite to form weathering products, Nature 357 (1992) 573–576.
[65] J.F. Lovering, B. Mason, G.E. Williams, D.H. McColl, Stratigraphical evidence for the terrestrial age of australites, J. Geol. Soc. Australia 18 (1972) 409–418.
[66] T.M. El-Shamy, The chemical durability of K_2O-CaO-MgO-SiO_2 glasses, Phys. Chem. Glasses 14 (1973) 1–5.
[67] A. Barkatt, M.S. Boulos, A.L. Barkatt, W. Sousanpour, M.A. Bo-Roomand, P.B. MaCedo, et al., The chemical durability of tektites – a laboratory study and correlation with long-term corrosion behavior, Geochim. Cosmochim. Acta 48 (1984) 361–371.
[68] A. Barkatt, E.E. Saad, R. Adiga, W. Sousanpour, A.L. Barkatt, S. Alterescu, Leaching of microtektite glass compositions in seawater, Adv. Ceram. 20 (1986) 681–687.
[69] A. Barkatt, E.E. Saad, R. Adiga, W. Sousanpour, A.L. Barkatt, M.A. Adel-Hadadi, et al., Leaching natural and nuclear waste glasses in seawater, Appl. Geochem. 4 (1989) 593–603.
[70] B.P. Glass, Solution of natural glasses in the geological environment, Adv. Ceram. 20 (1986) 723–732.
[71] T. Laursen, W.A. Lanford, Hydration of obsidian, Nature 276 (1978) 153–156.
[72] F. Rauch, J.E. Ericson, W. Wagner, C. Grimm-Leimsner, R.P. Livi, C. Shi, et al., Hydration of tektite glass, J. Non-Crsyt. Solids 144 (1992) 224–230.
[73] R.R. Lee, D.A. Leich, T.A. Tombrello, J.E. Ericson, I. Friedman, Obsidian hydration profile measurements using a nuclear reaction technique, Nature 250 (1974) 44–47.

CHAPTER

PROTECTIVE HARD COATINGS FOR TRIBOLOGICAL APPLICATIONS

8

M. Roy
Bhabha Atomic Research Centre, Mumbai, India

CHAPTER OUTLINE

Materials Under Extreme Conditions. http://dx.doi.org/10.1016/B978-0-12-801300-7.00008-5

1. INTRODUCTION

Materials have long been subjected to harsh treatments either under the brunt of Mother Nature or as a part of the technological progress of mankind and made to perform under such conditions. Such extreme conditions include application of high temperature, high pressure, shock wave induced by detonation, corrosive chemical etching, mechanical impact, etc. When two machine parts are subjected to mechanical rubbing during machining, drilling, or otherwise, temperature at the interface shoots up very high, which is why they require frequent cooling during the operation. Alternately, lubricants are employed to reduce chafing between the scrubbing surfaces. At the same time, machine operation induces wear and tear in tools due to constant mechanical action. In this context, hard coatings are provided on tool tips and/or on machine surfaces as protective measure, and at times they serve as solid lubricants too. In this chapter, the mode of deposition and tribological applications of a few hard coatings that have been frequently used for wear resistance are discussed.

1.1 TRIBOLOGY

The term *tribology* was reportedly first coined by Dr. H. Peter Jost of the United Kingdom in 1966, when a committee chaired by him submitted to the UK government a report on the status of industrial lubrication in the country. The term is derived from the Greek words *tribos* meaning "to rub" and *ology* meaning "the study of" [1]. Tribology, therefore, deals with the science and engineering of surfaces undergoing interaction due to the relative motion of one against the other. It is a very important branch of mechanical engineering that involves the principles of friction, lubrication, and wear. In short, tribology involves study of wear and tear arising out of friction between the surfaces and investigation of its remedial measures. Ever since the first mechanical engineer designed his device, issues and challenges pertaining to tribology came into existence. It was there in the earliest of human civilizations in Egypt and Mesopotamia almost 3000 years ago wherein transportation of large stone blocks and carvings were facilitated by use of proper lubrication. They had invented potter's wheel and wheeled carriages that made use of bearings to reduce friction [1]. By the end of the 4th century BC, people in China had been making use of metal ball bearings for war chariots. Evidence of using lathes, levers, gears, and pulleys by the Greeks clearly showed their appreciation for knowledge on tribology. A very significant engineering achievement of the time was the mechanical clock. The concepts of tribology were strongly prevalent in Europe 500 years ago. The brilliant painter and philosopher Leonardo da Vinci is known to have made significant developments in the field of friction, wear, and bearing materials by designing of ingenious rolling element bearings and setting up several innovative testing devices for estimation of friction and other aspects of tribology. This clearly indicates the importance imparted on this branch of mechanical engineering in medieval Europe.

A brief history of the development of the subject is provided elsewhere [1]. It is estimated that in the present world about of one-third of the energy consumed is wasted in the form of friction. However, friction is not always a bane for the engineering industry. It is essential for the operation of devices such as clutches and brakes of automobiles and also traction and friction drives, etc. Today manifestation of tribology exists in every sphere of life be it in (1) industrial manufacturing and processing that involves rolling, turning, grinding, and/polishing activities, (2) mining where heavy drills, pumps, and pulleys are used for excavation and extraction, and (3) natural processes involving erosion/ weathering due to wind and water and tectonic movements. Automobile engines, wristwatch and electric motors used in ceiling fans, grinders, and washing machines are some of the applications of tribology that we come across in our day-to-day lives. Some of the terminologies used in connection with tribology are described in the subsequent sections.

1.2 WEAR

During tribological operation, one of the surfaces undergoes motion relative to the other. As a result of friction between the two surfaces, the softer material tends to get removed by the harder one. Removal of material from either or both the solid surfaces that undergo scratching against each other may be generally termed as wear. Therefore, friction and wear are responses of the same tribological contact process between a pair of given surfaces. Usually, the lower the friction, the lower is the amount of material removed and vice versa [2], but there are contradictory observations as well. There are different underlying mechanisms behind removal of such materials, which include abrasion, adhesion and cohesion, erosion, corrosion, etc. Often more than one mechanism acts at a time for the removal of material [3]. Classification of wear mechanisms has been extensively reviewed in the literature [4,5]. Abrasion is a mechanical process of removing material by way of plastic deformation due to forceful sliding of a considerably harder surface across that of a softer one. The mechanism involves pressing of the hard particles onto the softer surfaces at the contact, thereby forcing it to undergo plastic deformation around the harder particles. The extent of damage, degree of penetration, amount of material removed, and the nature of scratches formed on the surface depend primarily on the geometry of the harder surface and the angle of contact. Application of such kind of abrasive wear is prevalent in mining, metal cutting, and construction. Adhesion and cohesion results from the sticking tendency of different or identical surface materials onto another surface due to different types of intermolecular forces [6,7]. Sometimes during relative motion of surfaces, particles of one of the surfaces get strongly adhered to those of the other. Tangential motion then removes the bulk of the material from the softer material and the process is known as abrasive wear. In this wear mechanism, surface properties of both the materials including impurity phases if any play a very crucial role in removing material. This mechanism is particularly prevalent under vacuum or inert atmosphere or when the flash temperatures of one or both the components are high enough so as to have good adhesion between the surfaces. Erosion occurs as a result of brief interaction between the two colliding surfaces that come in contact with each other for a short time interval. A surface may get eroded when it comes in contact with moving particles or flowing liquid, e.g., wind or river stream. The material is removed in succession by cutting and deformation of the surface. Corrosion is an act of chemical reaction between the corroded surface and the surrounding environment. Chemical reaction when combined with other forms of wear mechanism may produce wear debris. As a result of relative motion between the surfaces, the interfacial temperature at the point of contact rises, which widens the cracks and opens up surfaces for further chemical reaction, which in turn softens the material and makes it vulnerable toward getting removed in the form of debris. A common example of chemical wear is the formation of a native oxide layer on metal surfaces and its removal by abrasive mechanism. Mechanism of chemical wear has been extensively studied by several researchers [8–13] and is nicely illustrated with appropriate diagrams in the book by Holmberg and Matthews [4].

2. GENERAL METHODS OF DEPOSITION

A wide variety of deposition techniques and their recent modifications have made it possible to deposit coatings with amazing tribological properties that were hitherto impossible to achieve. Hard coatings are deposited onto the substrates primarily via gas phase methods or through solution route. Coatings are also produced by molten-state processes and solid-state processes [14]. Usually thicker films are obtained by solid-state route [4]. Two popular gas-phase methods are physical vapor deposition (PVD) and chemical vapor deposition (CVD). Similarly, solution-phase techniques include preparation of coating by spraying slurries, sol–gel method, dip and/or spin coating, and electrochemical processing, etc. A detailed account of deposition techniques is given in the book by Kenneth Holmberg and Allan Matthews and the references therein [4]. In the gas-phase methods, the active species are obtained in the gaseous state before they are made to deposit on the substrates. The active species are activated by employing different methods. The most commonly used techniques are plasma, ion beam, and resistive heating. Plasma-assisted deposition processes, in particular, ensure uniform, clean adherent coatings at a controlled rate over a wide range of substrates and at lower deposition temperatures [4]. In solution-phase methods, deposition takes place primarily in aqueous phase either by virtue of electrochemical or electroless process. Generally, solution-phase deposition techniques are used for thicker films as against the gas-phase methods where the coatings tend to delaminate due to stress generated in them, once thickness reaches a critical value [4].

2.1 PHYSICAL VAPOR DEPOSITION

In physical vapor deposition (PVD) technique, the coating material to be deposited is first made to vaporize from a solid source, and then by using a carrier gas or vacuum system, the vapor is transported to the substrate and made to condense therein in the form of hot atoms that finally nucleate and form a coating of desired thickness. Different modes of activation include resistive heating, induction heating, electron beam, cathodic sputtering, arc, ion beam, plasma, laser beam, etc. In a conventional resistive heating PVD system, the coating material is kept in a container made up of high-melting material such as tungsten, molybdenum, etc. and heated using a resistive heating element. Metals such as gold, silver, aluminum, chromium, copper, lead, etc. are deposited by this technique. Often two or more metals may be deposited sequentially or simultaneously using this technique for obtaining multilayered coating. Induction heating, in contrast to resistive heating, allows direct heating of the coating material as against the holder. Induction heating is particularly useful for deposition of materials that tend to undergo reaction with the holder. Coating materials may be evaporated by producing an electric arc between the two electrodes. Typically it is used to vaporize carbon and other reactive metals. Electron beam is used to deposit high-melting materials. The electron beam ionizes the coating material, which is then deposited on to the substrate placed close to the target. In cathodic sputtering, a target is kept at a negative potential and the argon ions are made to strike the target from which materials are removed and deposited onto the substrate. Therefore, mechanistically it is different from electron beam evaporation where an electron beam is generated from a hot tungsten filament is made to impinge on an anode target kept under high vacuum. Electron beam evaporation is orders of magnitude faster than sputtering. In sputtering, the stoichiometry of the target alloy is maintained in the deposit. In most of the cases laser ablation also involves stoichiometric deposition of the target material at around room temperature. In this method, a pulsed laser beam is made to impinge on the target material (kept under vacuum) from which material is removed under nonequilibrium conditions in the form of a plume which then moves with supersonic speed toward the substrate placed very close to the target and gets deposited there.

2.2 CHEMICAL VAPOR DEPOSITION

In chemical vapor CVD technique, coating is formed by reactive decomposition of the superincumbent atmosphere of the precursor gases and/or volatilized or atomized liquids in the vicinity of the substrate maintained at a suitable temperature. Based on the operational chamber pressure, CVD may be further classified into atmospheric pressure, low pressure, and ultrahigh vacuum (10^{-8} Torr) CVD. Again, depending on the nature of the precursor and their mode of activation, there can be (1) aerosol-assisted CVD in which transportation of the precursor takes place in the form of an aerosol or (2) direct liquid injection CVD (DLICVD) in which a volatile liquid or solid dissolved in an appropriate solvent is first atomized and then the vapors are transported using an inert gas to the deposition chamber. Very high growth rates may be achieved using DLICVD. DC/microwave plasma-assisted CVD (MPCVD) makes use of plasma to deposit coatings at much lower temperature and with increased growth rate. Plasma enhances the reaction kinetics between the precursors. Deposition may be carried out in open atmosphere by flame pyrolysis as well as under high vacuum to form atomic layers. Hot-filament CVD (HFCVD) makes use of a hot filament maintained at a very high temperature to decompose the source gas whereas metal-organic chemical vapor deposition makes use of organic precursors for thin film deposition. It is possible to deposit adherent, uniform and dense coatings by CVD. CVD deposition of coatings has been carried out at 1Pa to 1 atm pressures and from ~30 to 2500°C. The coating texture, composition, and properties may be tailored to a great extent by manipulating different process parameters. For example, grain size of TiN deposited under different conditions may vary from 0.5 to 5 μm [4]. A large number of oxides, oxynitrides, carbo-nitrides, carbides, borides, nitrides may be deposited by this technique.

2.3 ELECTROLESS DEPOSITION

Electroless deposition by solution route proceeds either via direct chemical reaction between the reagents such as metal ion reduction using a suitable reducing agent or upon reaction with the substrate forming its compound or by reactions promoted by catalytic surfaces. Gold, silver or copper layers may be deposited on metal or plastic substrates by in situ reduction of respective metal ions. It is also possible to deposit composite films using this technique [15]. Corrosion-resistant oxide coatings are often obtained on metal surfaces by allowing them to react with phosphoric, chromic, or other mineral acids. The phosphate/chromate layers on top of steel or aluminum surfaces improves lubrication between the moving parts and as a primary coating enhances bonding with the paint given on top of it [4].

2.4 SLURRY METHOD

In the slurry method, a stable dispersion of the material to be coated is prepared using suitable additives and solvent and then brushed and/sprayed onto the substrate, which is then heat-treated at a suitable temperature for drying. Sometimes the slurry is spread in the form of a uniform thin film using a metal vane and the technique is known as doctor's blade technique. The gap between the sweeping vane and the substrate determines the thickness of the film.

2.5 SOL—GEL PROCESS

In sol—gel method, a sol (a dispersion of a solid in liquid) is first prepared by dissolving metal salts or other suitable precursor (viz., alkoxides) of the metal in an appropriate solvent. The sol may be sprayed onto the substrate by means of spin coating or dip coating. In dip coating the sol (coating material in general) is put in a container and the substrate is dipped into it and then pulled out at a very slow speed. Alternately, in spin coating the sol is put dropwise onto the substrate held on a vacuum chuck, which is then rotated at a high speed. Under the influence of the centrifugal force the sol gets spread uniformly over the substrate. In either case, the speed of substrate determines the thickness of the coating. These nanometer-sized particles suspended in liquid and uniformly dispersed on the substrates tend to dry up on the surface, forming a complex network of solvated polymerized macromolecules called gel. The layers are finally cured at different temperature for appropriate applications. Sol—gel method is primarily used for the porous oxide ceramics.

2.6 ELECTROCHEMICAL PROCESSING

In this process, the substrate acts as one of the electrodes inserted in an electrochemical bath and the electrolyte is comprised of dilute acids. Passage of electric current through the electrode brings about the necessary chemical reaction so that the substrate gets coated with the desired material. A large number of metals including iron, cobalt, nickel, copper, zinc, lead, antimony, and arsenic bismuth are deposited by this method. Some metal alloys are also deposited by electrochemical processing. A thick chromium coating deposited electrochemically on other metals is often used as abrasion protective layer. Coatings of nonmetals such as diamond-like carbon have been obtained by decomposition of electrolytes that yielded suitable carbon containing precursors and will be discussed subsequently.

3. CHARACTERIZATION TOOLS

A large number of microscopic and spectroscopic techniques have been extensively used for the characterization of thin films. Whereas microscopy is used for studying grain morphology, orientation and grain boundaries, extent of substrate coverage, film roughness, its thickness, porosity, etc., X-ray diffraction and Raman spectroscopy come handy in analyzing the phase and structure of the coating materials. A brief review of some of these techniques is provided in this section.

3.1 ATOMIC FORCE MICROSCOPY

Atomic force microscopy (AFM) is a widely used nondestructive technique for studying film topography with nanometer resolution [16]. The working principle of AFM is the scanning of the sample surface by a delicate Si/Si_3N_4 microcantilever tip (ca. 100—200 μm in length). A laser light is made to fall on the shining back side of the cantilever, which is then synchronously reflected back to a position-sensitive photodetector that provides feedback signal to piezoelectric scanners that maintain the cantilever tip at constant height or force. Accordingly, information is obtained about the surface roughness or force of interaction between the sample and the tip. Therefore, even a slight movement of the cantilever tip will shift the beam position on the photodetector. AFM measurement may be carried out in three different modes, viz., (1) contact mode, (2) noncontact mode, and (3) tapping mode depending on the type of interaction of the cantilever tip with the surface. A detailed account of the AFM instrumentation is provided in the literature [16,17].

3.2 SCANNING ELECTRON MICROSCOPY

In scanning electron microscopy (SEM) a beam of electrons originating from an electron gun is focused onto the surface of the sample with the help of a set of condenser lens and objective lens. The primary electron beam (~30 kV) is then made to scan the specimen surface, which upon interaction with the material under investigation produces secondary electrons having energy in the range of a few tens of electron volts. The sample chamber is maintained at high vacuum to avoid scattering of electrons. The secondary electrons are deflected using a pair of deflector coils. A scintillation counter detects the escaping secondary electrons. Amplified signal from the counter is fed into a synchronously scanned CRT for display. Diameter of the primary beam ultimately determines the resolution of the image. The extent of magnification is given by the ratio of the CRT raster width to that of the scanning electron probe. The contrast of the image depends on the flux of the secondary electrons emitted and hence depends on the morphology and composition of the samples. Sometimes the samples are coated with a conducting layer to avoid undue charging of the surface.

3.3 X-RAY DIFFRACTION

X-ray diffraction (XRD) is by far the most powerful technique for structural analysis and phase evaluation. When an electromagnetic wave impinges on an obstacle or passes through a slit (which is of the order of its wavelength), it tends to suffer a change in its wave properties, e.g., it tends to undergo a phase shift resulting in constructive or destructive interference. All such phenomena taken together, resulting from the bending of light at the edge of an obstacle may be termed as diffraction. Diffraction of visible light through a grating produces bright and dark interference patterns. Similarly, X-rays may be diffracted using a crystalline lattice of atoms with interplanar spacing similar to that of X-ray wavelength. Classically, the phenomenon of diffraction of X-rays by atomic planes may be described following Bragg's law of diffraction

$$n\lambda = 2d\ \sin\theta \tag{8.1}$$

where λ is the wavelength of incident X-ray, d is the inter planar separation, θ denotes the Bragg angle, and n is the order of diffraction. Fluctuating electric field of X-ray interacts with the electrons of the atoms. The electrons start vibrating and in the process emit radiation in all possible directions. Intensity of the scattered radiation, depends on the number of electrons in the atom and hence on its atomic number (Z). The interplanar lattice spacing may be calculated from X-ray diffraction patterns using Bragg's law wherefrom crystal structure and space group of the materials may be determined precisely. XRD analysis therefore serves as a rapid and efficient fingerprinting characterization technique for crystalline materials. In addition to the routine phase analysis, for nanocrystalline samples, the crystallite size and microstrain in the samples may be estimated from XRD patterns using William Hall plot, wherein full width at half maximum (FWHM) of the XRD peaks (corrected for instrumental broadening) is plotted against Bragg angle according to the equation

$$B\ \cos\theta = (0.9\lambda/D) + 4\varepsilon\ \sin\theta \tag{8.2}$$

where B is the corrected FWHM of the XRD peak, D is the particle size, λ is the wavelength of X-ray, θ denotes Bragg angle, and ε represents the microstrain in the samples [17a].

3.4 RAMAN SPECTROSCOPY

Raman spectroscopy is a nondestructive technique by virtue of which vibrational transitions in a material may be investigated. Different compounds existing in different crystallographic modifications exhibit signature Raman bands. Fingerprinting of the bands forms the basis of material characterization by Raman spectroscopy. Raman spectroscopy not only provides information about the different crystallographic phases but also renders information on local bonding around a central atom. It is particularly useful in studying metal oxide because of the high polarizability of the M—O bonds. It is an excellent tool for studying anionic vacancies and structural defects. Raman spectroscopy studies the inelastic scattering of light by matter. When a laser light is incident on a sample it is not only scattered elastically following classical laws of electromagnetic radiation but a small part is also scattered inelastically, which is then analyzed using a monochromator and finally detected using a photomultiplier or a charge-coupled device. The difference in energy between the incident and the scattered radiation recorded as wave number shift is a measure of the quanta of energy transferred by radiation (for Stokes shift) that induces elementary excitation within the matter in the form of lattice vibrations or phonons (in case of solids) or molecular vibration in case of liquids and gases. Therefore, in Raman spectroscopy, analysis of the wave number change in the scattered radiation yields information about the vibrational modes of the sample.

4. TESTING METHODS

Apart from morphology, structure, and composition, mechanical properties of the coatings are also of supreme importance for tribological applications. The first and foremost criteria of a good coating is that it should be adherent to the substrate. A coating is said to fail if it peels off the substrate. With the advent of modern-day deposition techniques such as physical and chemical vapor deposition, pulsed-laser deposition, etc. more and more adherent coatings are being produced on diverse substrates. Following are a few common methods of adhesion testing of hard coatings. Scotch Tape test is a simple, yet very effective test in which the tape is applied on the coatings and then quickly pulled off. If the coating is removed exposing the substrate, it is said to be nonadherent to the substrate. Therefore, the Scotch Tape test provides a qualitative assessment of adhesion properties of the coatings. Scratch test is an important technique that provides a quantitative measure of adhesion properties of the coatings. In this technique, an indenter is made to scratch across the coating under an increasing normal load [18]. The critical load is adjudged by the damage produced in the coating as viewed under a microscope objective and also from acoustic emission. Historically, the method was used to test adhesion of metal coatings on glass and steel [19—21]. Alternately, a metal cylinder may be fixed with the help of epoxy cement at right angle to the surface of the coating and a gradually increasing load is applied to the bottom of the cylinder parallel to the film surface [22]. The load at which the bonding between the cylinder and the coating breaks and the cylinder starts slipping down is a measure of adhesion of the coating. Fig. 8.1 shows the schematic diagram of the pull-off test for adhesion. Sometimes more than one rod is used to pull against the coatings. The rod may also be fixed at two different positions in the coatings [23]. Loading of the rod introduces either compressive

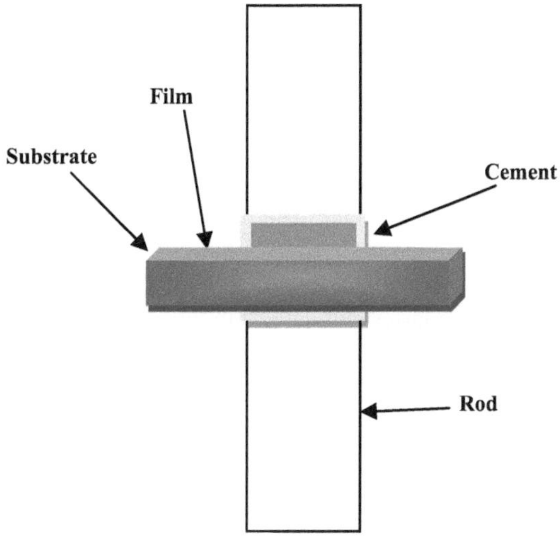

FIGURE 8.1

Schematic representation for the pull-off test for adhesion.

or tensile stress in the coatings forcing it to fail at some point. A detailed review of the adhesion test methods is given in reference [24]. The limiting measurable adhesion by this technique is decided by the bonding strength of the epoxy cement used [4]. Again, a centrifuge rotor may be coated with the material under test and rotated in vacuum at constantly increasing speed [25]. The speed at which the coating fails gives a measure of its adhesion. Similarly, when a coating is subjected to high-frequency ultrasonication, it tends to get delaminated from the substrate and hence ultrasonication forms an important testing method for adhesion [26,27]. Acoustic imaging and laser-based methods are some of the latest developments in the field. Acoustic imaging is a nondestructive method in which a short radio wave generated using a piezo device is focused onto the surface of the coating using a microscope objective. The acoustic wave is then reflected back and eventually converted into an electrical signal, analysis of which provides qualitative information on the adhesion of the coating at that point [28]. By scanning different parts of the coatings, it is possible to get a spatial mapping of the adhesive property of the same. In laser-based technique, a compressive shock wave generated by impinging pulsed laser beam propagates through the back of the substrate right up to the interface between the coating and the substrate and then tends to delaminate the coating if the power of the laser is sufficiently high enough [22]. Therefore, by carefully tuning the laser power, it is possible to get information on the adhesive bonding of the coatings.

Roughness is another important property of coatings. Usually roughness is measured using AFM and/or scanning tunneling microscope (STM) (for relatively smoother films) or by a stylus profilometer for thicker coatings. In AFM or profilometer, either the cantilever or the stylus scans over the surface and maps microundulations on the surface, whereas in STM, the topography is monitored by measuring the tunneling current across the gap between the surface and the STM tip. Here, resolution may be obtained even at the atomic scale depending on the finish of the surface.

Thickness is a quantitative parameter that often decides many important properties of the deposit and is routinely measured for all kinds of deposit be it vapor-deposited or coated by solution route. In PVD, the deposition is usually carried out in presence of a piezoelectric balance, which gives the measure of film thickness. Cross-sectional SEM is another important technique that may be used to measure micron-thick layers. Even AFM and profilometer may be used to measure thickness of coatings, provided a neat step is made at the edge of deposition and measurement is made across the step. Thickness of transparent coatings may be obtained using interference techniques or by using the Beer–Lambert law.

Mechanical evaluation of coatings starts with the measurement of its hardness. An indent is made on the coatings using a hard diamond indenter. The dimension of the indent at a given load gives a measure of the hardness of the coatings. It is a simple, basic, yet very useful technique for measuring hardness of coatings, although measurement may not reflect the intrinsic values and may be seriously limited by the presence of grain boundaries and defects in the film. Based on this simple measurement, it is possible to classify films into ultrasoft, soft, hard, and ultrahard categories, and applications thereof may be envisaged accordingly. Indenters of different shapes are used for hardness measurement, therefore accordingly we have Vickers and Knoop microhardness. The interference due to the substrate may be minimized by carefully choosing the load on the indenter. The Knoop indenter makes a shallower dent on the coating and hence may be used for thin films.

5. HARD COATINGS FOR TRIBOLOGICAL APPLICATIONS

5.1 ALUMINA

Alumina is an amphoteric oxide of the metal aluminum and has the chemical formula Al_2O_3. Alumina occurs in nature primarily as corundum (α-Al_2O_3). Other important naturally occurring sources of alumina include gibbsite ($Al_2O_3 \cdot 3H_2O$), nodstrandite ($Al(OH)_3$), diaspore ($Al_2O_3 \cdot H_2O$), boehmite (γ-AlO(OH)) (obtained as a component of bauxite ore), and bayerite (a polymorph of gibbsite). Bauxite is impure gibbsite. Ruby and sapphire are also naturally occurring forms of alumina. Alumina was first extracted from natural clay in 1754 by Marggraf and later on from alumina-rich earth material named bauxite by Berthier. Manufacturing of alumina started much later in 1860 [29]. Initially, alumina was produced by the reaction of bauxite ore with sodium carbonate and then precipitating out alumina hydrate. Later on, Bayer's method came to be used for extracting alumina from its ores.

5.1.1 Crystal Structure

The thermodynamically stable form of alumina (Al_2O_3) is corundum (α-Al_2O_3). The crystal structure of alumina has been extensively reported in the literature [30–32] and nicely reviewed by Levin and Brandon [33]. The unit cell of corundum exhibits rhombohedral (trigonal) symmetry and belongs to the space group of $R-\overline{3}c$. However, it has been comprehensively described in the literature using a near-closed packed hexagonal sublattice of oxide ion [30] having cell parameters $a = 0.475$ nm and $c = 1.297$ nm and with c/a ratio 2.73. The hexagonal cell may be generated from the rhombohedral cell by a simple transformation with concomitant increase in cell volume. The rhombohedral unit cell of corundum contains 10 atoms, whereas the hexagonal cell contains 30 atoms with six oxide ion layers along its c-axis. Two-thirds of the octahedral interstitial sites are occupied by the aluminum cations. The Wyckoff positions for the oxide anions and the aluminum cations in $R-\overline{3}c$ symmetry are given by 18e and 12c, respectively. Therefore the actual positions of the cations are essentially staggered along the [0001] c-direction. A schematic representation of the crystal structure of α-Al_2O_3 is given in Fig. 8.2. Apart from corundum, there are other metastable polymorphs of alumina with face-

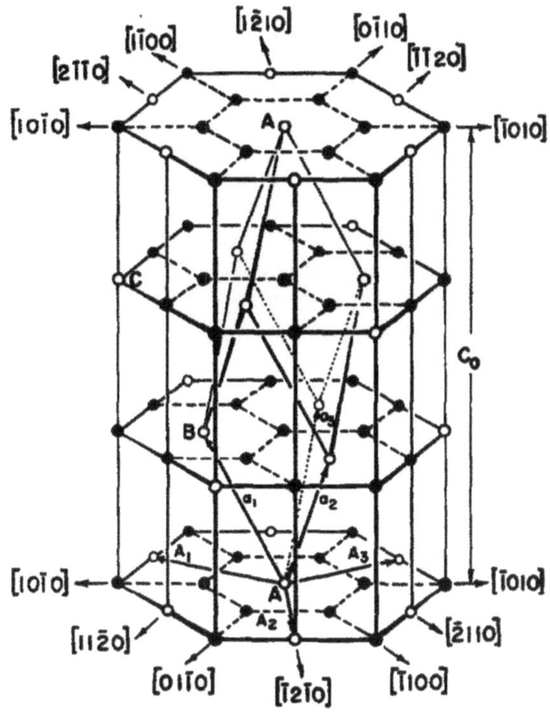

FIGURE 8.2

Crystal structure of α-Al_2O_3.

Adapted from M.L. Kronberg, Plastic deformation of single crystals of sapphire: basal slip and twinning, 5 (1957) 507–524. Reprinted from Acta Metallurgica, Copyright (2016), reproduced with permission from Elsevier.

centered cubic (fcc) or hexagonal close packing (hcp) of oxide ions. The fcc modifications include γ, η (cubic), θ, θ', θ'', λ (monoclinic), and δ (tetrahedral/monoclinic) phases whereas the hcp modifications include α (trigonal), κ (orthorhombic), and χ (hexagonal) phases [33].

5.1.2 Deposition of Alumina

Alumina may be deposited on a large variety of substrates by plasma spraying [35,36], anodizing, physical [37–39] or chemical [40,41] vapor deposition, sol–gel and slurry method. Spray pyrolysis is by far the most popular technique in which plasma of an inert gas is used to heat up the alumina powder, melt it, and spray it to form an adherent coating. Uniform, high-density coatings with particle size ranging from a few nanometers to a few hundred nanometers may be obtained by this process. Coating properties further depend on the processing parameters such as flow rate of plasma gas, nozzle size of the torch, and distance between the torch and substrate. It has been reported that microstructural integrity of the coatings depends significantly on the heat input and size of the particle, which in turn would influence the wear resistance and mechanical performance of the coatings [35]. Further, visible porosity of the coatings decreases and wear resistance and hardness improves with increasing heat input and residence time. Particle size also has a bearing on the formation of different alumina phases since larger particles tend to undergo partial melting at lower heat input resulting in the rapid solidification of the molten mass and hence formation of residual α-Al_2O_3 phases. Tribological properties of alumina coatings may be further improved by addition of secondary ceramic phase such as silicon carbide, zirconia, titania, etc. into the alumina matrix. The major drawback of the conventional plasma spraying is that there is less control over the film thickness and surface finish, which is why post-deposition machining of the surface is often required for sophisticated applications, thereby involving extra processing cost. A variant of the flame spraying technique is the high-velocity oxygen fuel spraying that invokes oxyacetylene flame to melt the alumina powder, which under atomized condition is sprayed onto the substrate in the form of high-velocity jet. Due to higher velocity, denser, smoother, and more adherent coatings are obtained by this technique as compared to the conventional plasma spray. The coatings also exhibit superior wear resistance properties especially at higher temperatures. Similarly, spray coatings may be obtained from liquid feedstock using suspension or precursor solutions; the technique is known as the high-velocity suspension flame spraying (HVSFS). In this process, the nanoparticles in the form of suspension are injected into the gas jet and are almost completely melted, producing very small and flattened lamellae in the range of 100 nm to 1 μm with very few defect inclusions [36]. The coatings thus produced have altogether different morphologies and exhibit better properties. Moreover, a better control over film thickness may be obtained by this approach thereby opening up a wide range of applications such as thermal barrier coatings, etc. Porosity is lower and interlamellar cracks are fewer in number as compared to the conventional spray coatings that improve its wear resistance properties. A recent modification of the technique involves spraying of suspension feedstock with a high-velocity oxygen fuel flame. The coatings are eventually formed with a lesser number of defects. Hardness of the spray coatings decreases with the increase in the indentation load. But the extent of decrease is less for the HVSFS coatings due to a smaller number of interlamellar cracks in them. A similar trend is also exhibited by the elastic modulus of the material [42]. Al_2O_3 coatings are deposited by chemical vapor deposition technique primarily by using a gas mixture of $AlCl_3$, H_2, and CO_2. Other feedstocks have also been tried ([41] and references therein). In CVD, the deposition temperature is usually kept at ~1000°C so that strong metallurgical bonds are formed between the substrate and the coating [43]. However, difference in thermal expansion between the substrate and the coating produces tensile stress that tends to delaminate the coating and form cracks. It has been reported that Al_2O_3 readily nucleates in the gas phase and tends to form upon deposition irregular crystallites, sometimes resulting in poor adherence to the substrate. With the reduction of the feed gas concentration, the rate of deposition decreases significantly (~1 μm/h) resulting in the formation of larger crystallites [41]. Physical vapor deposition may be used as an alternative to CVD for depositing alumina coatings, e.g., electron beam evaporation technique has been employed for depositing alumina coatings on various substrates, viz., SS 304 and 316, Monel 400, molybdenum and copper, etc. [38,44]. Prior to deposition, the substrates were polished, cleaned using an ultrasonic bath, and were vapor degreased. Deposition was typically carried out at 548K and at 10^{-3}–10^{-4} Pa pressure. The rate of deposition was ~0.9 nm/s. An adherent thick coating of alumina for biological applications has been deposited on aluminum alloy using microarc oxidation technique wherein deposition was carried out in a stainless steel bath in the presence of a slightly alkaline electrolyte [45].

5.1.3 Tribological Applications of Alumina

Alumina is an extremely hard, stable, and wear-resistant material and is known for its excellent tribological applications. Hardness and wear resistance arises because of strong ionic bonding between the atoms. Microhardness of an electron beam–deposited alumina coating on Monel 400 has been reported using Jönsson and Hogmark's model as 22.30 GPa [38]. Further, alumina is inexpensive and readily available and therefore promises a wide variety of engineering applications. To date, hard alumina coatings have been used most extensively in industrial applications. The α phase is most desirable for the purpose. The coating minimizes friction and wear properties of the substrate. Hence, it is used as an abrasive as well as on cutting tools. Alumina has been vapor deposited onto carbide tool inserts, thereby largely improving their life and tolerance [4,41]. It is an indispensable component in multilayer coating applied on tool inserts used for machining of ferrous materials [46], carbon, alloys, stainless steel, etc. Alpha alumina is suitable for spacecraft engine and nuclear reactors. A detailed account of the tribological properties and applications of alumina is provided elsewhere [36,47,48]. It has a low coefficient of friction ($\mu \sim 0.2$), comparable to that of graphite ($\mu = 0.1$) and hence may be used as a solid lubricant. Therefore, it finds applications as bearings in high-precision instruments. In one of the studies, it has been reported that the coefficient of friction of alumina coatings against titanium carbide pin is 0.19 and the wear rate is 0.5×10^{-6} mm^3/Nm, respectively. It has been observed that the adhesion of coatings increases with increasing thickness [39]. Thicker films also tend to decrease the wear rate. Scratch-proof alumina coatings are applied on decorative items for their aesthetic looks and also as jewels in wristwatches for their wear-resistant performance. Teflon coatings have low coefficient of friction and are used as nonsticking elements in household utensils. However, these materials wear out easily upon repetitive usage. As an alternative, anodized alumina is provided in household utensils to increase their durability. Moreover, alumina coatings are biocompatible and offer excellent thermal dissipation and protection against corrosion of substrates, which is why they find extensive biomedical applications such as in the field of prosthetics [45]. High-grade γ-alumina is being used in total joint replacement. Alumina is an excellent electrical insulator with very high room temperature resistivity ($\rho_{RT} > 10^{14}$ Ω cm) and possesses high thermal conductivity (c. 30W/m·K). It also possesses a high dielectric constant of 9.5. Alumina is resistant to all strong acids and alkalis even at elevated temperatures and does not get oxidized easily [49]. Hence, it is frequently used as dielectric layers and support for sensors and heating elements [36]. Dielectric alumina coatings are used in aerospace, automotive, and electronic packaging industries. Another important application of alumina is in the form of thermal barrier coating (TBC). Alumina coating in itself may not serve as good thermal barrier coatings primarily used in turbine parts, etc., but when added to yttria-stabilized zirconia (YSZ), it improves upon the hardness of the coating and its resistance toward oxidation without altering significantly the Young's modulus and toughness of the material [50]. Moreover, the coating has a much longer thermal cycle as compared to YSZ alone [51]. The composite coating may be achieved by spray depositing alumina onto YSZ layer. One major disadvantage of using alumina in TBC is that during thermal cycle, alumina undergoes phase change with concomitant change in volume resulting in microcracks in the coating. Such issues may be addressed by doping of alumina with oxides of transition metals. Incorporation of silicon carbide fibers also improves upon mechanical properties of alumina coating. Alumina coatings are also being considered as a potential limiter and liner in Tokamak-type controlled thermonuclear reactors [52,53] that would reduce contamination of plasma and subsequent power loss.

5.2 DIAMOND

Diamond is a crystalline allotropic modification of carbon and primarily consists of cubic arrangement of atoms. It may also exist in hexagonal form. Hexagonal form of diamond is known as "Lonsdaleite." It is rarely found in nature (only in meteorites). Diamond exhibits extreme and unique properties. With intrinsic Vickers hardness ranging between 12,000 and 15,000 kg/mm^2, it is the hardest known material in the bulk and has the highest atom density of $\sim 1.77 \times 10^{23}$/cm^3 and elastic modulus of $\sim 1.2 \times 10^{12}$ N/m^2. Diamond exhibits highest room temperature thermal conductivity of $\sim$20W/cm·K and extremely low coefficient of thermal expansion of $\sim 0.8 \times 10^{-6}$ at 300K [54,55]. Moreover, at room temperature, diamond is inert toward attack of acids and alkalis and is resistant to thermal shock. It is also radiation hard and an excellent electrical insulator ($\rho = \sim 10^{13}$ Ω cm). Topography of diamond surface may be maneuvered from extremely smooth with coefficient of friction as low as 0.1 in air to very rough with protruding edges, depending on the extent of polishing and orientation of the crystallographic planes. Because of these properties, diamond promises tremendous application in the field of tribology especially as a protective coating to cutting/drilling tools and other devices.

5.2.1 Crystal Structure

Space group of cubic diamond lattice is Fd3m. Its primitive (Bravais) cell contains two atoms. Crystal structure of cubic diamond is depicted in Fig. 8.3. It is best described as a face-centered cubic (FCC) lattice caged within another FCC lattice such that they are mutually displaced by (¼, ¼, ¼) part of the lattice constant along the body diagonal. The lattice constant is 3.56 Å. In a diamond lattice, carbon atoms are sp^3 hybridized and connected to the neighboring four carbon atoms arranged tetrahedrally in space through four sigma bonds with bond length of 1.54 Å. The stacking sequence of atoms in a cubic diamond lattice along the (111) direction is given by ABC ABC ABC... pattern as shown in Fig. 8.4. The other isomer is hexagonal diamond or Lonsdaleite. In contrast to the cubic diamond, the stacking

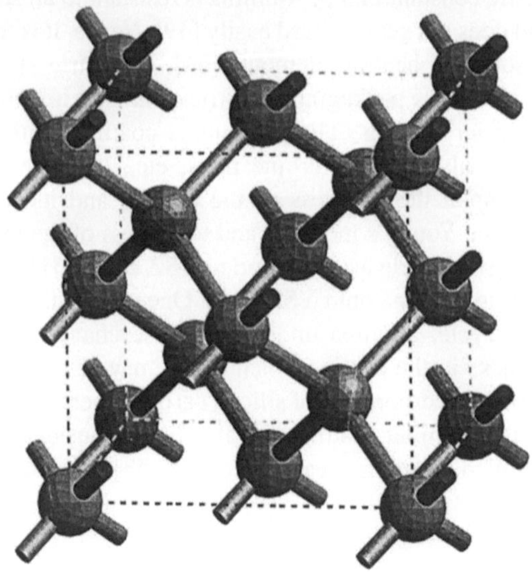

FIGURE 8.3

The crystal structure of cubic diamond.

Adapted from S.-T. Lee, Z. Lin, X. Jiang, CVD diamond films: nucleation and growth, 25 (1999) 123–154. Reprinted from Materials Science and Engineering, Copyright (2016), reproduced with permission from Elsevier.

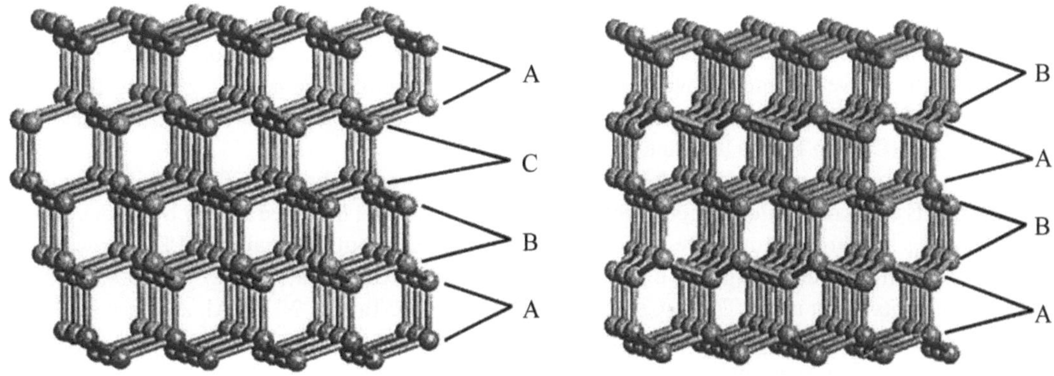

FIGURE 8.4

Crystal structure of cubic and hexagonal diamond with different atomic arrangements.

Adapted from S.-T. Lee, Z. Lin, X. Jiang, CVD diamond films: nucleation and growth, 25 (1999) 123–154. Reprinted from Materials Science and Engineering, Copyright (2016), reproduced with permission from Elsevier.

sequence of atoms in Lonsdaleite follows AB AB AB... pattern (see Fig. 8.4). The lattice constants are given by $a = 2.52$ Å and $c = 4.12$ Å, respectively [55] and bond length of 1.52 Å.

Diamond is metastable at room temperature and under atmospheric pressure and it is evident from the phase diagram (see Fig. 8.5). At ambient pressure and temperature, graphite is thermodynamically

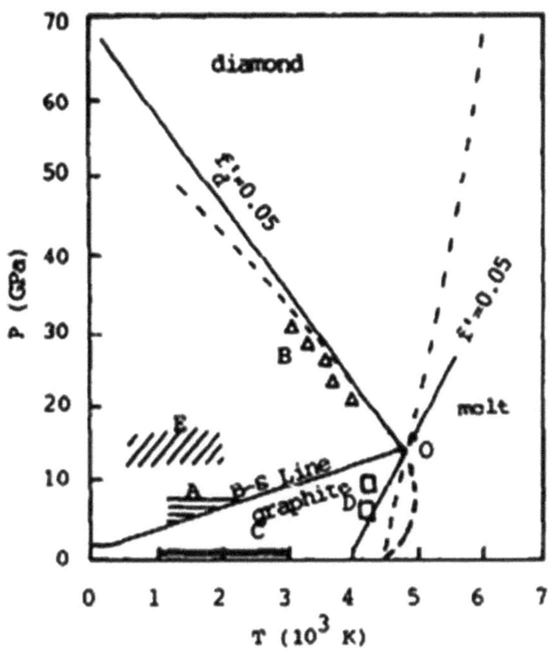

FIGURE 8.5

The phase diagram of carbon.

Adapted from Y. Zhang, F. Zhang, G. Chen, A study of phase transformation between diamond and graphite in P-t diagram of carbon, 32 (1994) 1415–1418. Reprinted from Carbon, Copyright (2016), reproduced with permission from Elsevier.

the most stable form of carbon. Diamond is gradually converted to graphite upon annealing at 1850°C in inert atmosphere and at low pressure (vacuum) [57]. However, in the presence of oxygen, its surface may get oxidized even at much lower temperature with the commencement of graphite formation above 600°C [57]. The rate of graphitization is different for different crystallographic planes. It is slower for <100> than for <110> and <111> planes [58].

Diamond is inert to most of the chemicals at lower temperature (below 600°C). It is resistant to common acids even at elevated temperatures, e.g., graphite may get oxidized slowly by a hot chromic acid cleansing mixture or a mixture of sulfuric and nitric acids but not diamond [57]. However, it reacts weakly with oxidizing agents, e.g., KNO_3, molten hydroxides, salts of oxy-acids and carbide-forming elements (such as Si, Ti, Zr, Cr, Mo, W, Fe, Hf, V, Nb, Ta, Co, Ni, Y, Al, and rare earths) at elevated temperatures. Diamond reacts with water vapor and carbon dioxide above 870K. A beam of atomic oxygen etches away diamond surface [59]. Whereas tantalum, tungsten, titanium, and zirconium form carbides with diamond, it actually gets dissolved in iron, cobalt, nickel, manganese, and chromium, which is why diamond tools may not be used for machining of ferrous metals, including high-speed and hardened steels [55]. As mentioned earlier, an imperfect diamond crystal is one of the hardest materials under ambient conditions. Hardness and wear resistance of diamond increase with increasing impurity content in the material and are believed to be due to hindering of dislocation movement by impurities. Nevertheless, natural diamonds can be cleaved easily along extended planar defects (predominantly along twins) by a simple hammer strike. Cracks also develop along grain boundaries of polycrystalline diamond films. However, it is extremely difficult to polish a piece of diamond. One would take a very long time to polish diamond mechanically using traditional techniques.

But with the help of modern-day laser-based cutting technology, one requires significantly less time. However, due to very high local temperature, the surface of the diamond gets graphitized in the process and needs to be repolished again mechanically. Elastic modulus of diamond is also very high and is highest along <111> direction. Traditionally, natural diamond is used as the abrasive. Nowadays, diamond is being synthesized by a wide range of techniques. Polycrystalline diamond may be deposited onto different substrates by chemical vapor deposition technique, which has opened up altogether new opportunities of using diamond films in the form of wear-resistant coatings. Use of ultrahard coating to protect automobile machinery parts, e.g., gearboxes, engines, etc. might reduce the need of further lubrication without compromising with their lives. Diamond-coated tool bits can work for longer time duration since diamond films are very good thermal conductors as well and would serve as a heat sink, thereby dissipating local heat generated during longer operation of the tools.

5.2.2 Deposition of Diamond Films

The first synthetic diamond was produced by the General Electrical Research Laboratory, New York State, USA [61]. Since then, a large number of techniques have been used to synthesize diamond. Diamond may be synthesized by high-pressure, high-temperature route either using a suitable catalyst (catalyzed synthesis), accompanied by static pressure, or shock produced by explosives (shock synthesis). Diamond films are best deposited by CVD techniques. In the late 1950s, diamond was first deposited under low pressures by CVD method. It was known as cyclic process and was developed by Eversole [62]. Afterward, several researchers [63,64] contributed significantly to the cyclic pyrolysis method, which was primarily restricted to depositing diamond films homoepitaxially onto diamond seeds. The deposition rate was rather slow in this process. It increased by several orders of magnitude after Matsumoto et al. in 1982 [65] started using a hot filament maintained at ~2000°C to activate simultaneously hydrocarbon and hydrogen. By using HFCVD, diamond could be deposited from hydrocarbon precursors onto diamond seed crystals as well as on other suitable nondiamond substrates at subatmospheric pressures where diamond is thermodynamically unstable with respect to graphite. Apart from HFCVD, other means of activation that have been subsequently used for deposition of diamond films include DC, RF, microwave, or electron cyclotron resonance (microwave) plasma, and combustion flame (oxyacetylene or plasma torches). Usually, the substrate is maintained at and above a particular temperature (in the range of 1000–1400K). High substrate temperature is required to ensure sufficient mobility of the hot carbon atoms on the surface of the substrate so that they may get attached to the existing lattice. However, graphitization is also predominant at high temperatures. Atomic hydrogen formed in situ upon activation of the hydrogen gas etches away the graphite so formed thereby ensuring deposition of phase pure diamond films under kinetically controlled conditions. Other etchants for graphite that been used in CVD include OH· radicals, O_2, F_2, O, and F. A few important variants of CVD are discussed in the following sections.

5.2.2.1 Hot Filament-Assisted Chemical Vapor Deposition

Invented and developed by Matsumoto et al. in 1982 [65,66], HFCVD was the earliest technique for the deposition of diamond at sub-atmospheric pressures and continues to remain one of the most popular techniques for deposition of diamond films. The authors demonstrated that using a hot refractory filament maintained at ~2000°C it was possible to dissociate at the same time H_2 gas to atomic hydrogen and also decompose hydrocarbon precursors that served as the source of carbon. Using this approach, it was possible to deposit diamond films onto a nondiamond substrate at an increased rate while substrate temperature could be kept well below 1300°C so as to prevent ubiquitous formation of graphite. Moreover, diamond being very stiff in nature, its sticking coefficient on nondiamond substrates at higher substrate temperatures (TS) is very low that is largely improved upon reducing the substrate temperature. Therefore, different substrates such as tungsten, tantalum, molybdenum, zirconium, silicon, ceramics, etc. could be used as substrates. Tungsten, tantalum, and rhenium, etc. were used as filaments and a large number of carbon sources including methane, propane and ethane, acetone, ethanol, and methanol have been employed [55]. There is usually an increase in the deposition rate at higher pressures (near atmospheric pressures as compared to the sub-atmospheric pressures) but only at the cost of deposition area and input energy. Deposition rate, however, increases with increase in the gas phase temperature [54]. The restriction of small deposition areas may be overcome by using multifilament mesh and convective jet flow of feed gases. A schematic diagram of a HFCVD reactor is shown in Fig. 8.6. It is easy to scale up a HFCVD system for deposition of films simply by increasing the number of filaments in the grid. Power supply and cooling facilities should be adequately modified depending on the load. However, a greater number of filaments invariably increases contamination of diamond films with refractory metal impurities. It is customary to provide the substrates with some pretreatment before they are coated with diamond films. This is done to improve diamond nucleation density on nondiamond substrates. For example, on pristine silicon, nucleation density of diamond is very low (~10^4 particles/cm^2) due to its high surface energy. Such nucleation density may be enhanced by mechanical or ultrasonic abrasion of the surface using diamond seeds, ion implantation, and/or bias application, etc. On applying positive voltage to the substrate, electron bombardment accompanied by desorption of surface hydrogen takes

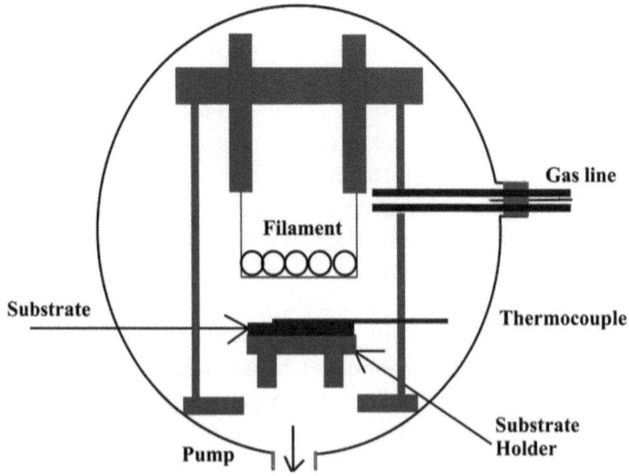

FIGURE 8.6

Schematic diagram of a typical hot filament chemical vapor deposition (HFCVD) reactor.

place, which in turn increases the deposition rate. Upon changing the polarity, bombardment of the substrate with positively charged ions would enhance nucleation density on nondiamond substrates.

Diamond slurry treatment is another convenient technique to improve upon the nucleation density of diamond, and it may be used on all sorts of shapes and complex geometries. A smooth diamond film continuous even at submicron thickness has been deposited on silicon single crystals after mechanical/ultrasonic abrasion of the substrate with diamond and zirconium slurry [68]. Choice of zirconium was based on the fact that it would form cubic zirconium carbide phase with lattice constant close to that of diamond at the interface and hence would reduce the strain in the coating due to lattice mismatch. Moreover, zirconium also has very high heat of hydrogenation and therefore was expected to catalyze diamond nucleation on silicon substrate and would lead to the subimplantation of the diamond seeds for effective nucleation. Very high nucleation density could be achieved for films grown for a short time (~20 min). However, nondiamond carbon content of the films also increased after such slurry pretreatment.

5.2.2.2 DC Plasma-Assisted Chemical Vapor Deposition

Plasma-assisted CVD is another popular technique for deposition of diamond. Here plasma discharge is used to radicalize hydrogen and to decompose the carbon carrier gases (typically methane) for deposition of diamond. In DC plasma-assisted CVD, an electric discharge is produced between the electrodes at low pressures by applying a high-voltage dc supply [69,70]. Fig. 8.7 shows the schematic diagram of a dc plasma-assisted CVD reactor. The substrate is mounted very close to the anode and is kept cool by constant circulation of coolant. Almost 40% of the carbon atoms get converted to diamond during dc plasma-assisted CVD. Hence, the process is much more efficient than hot filament-assisted CVD, which involves only 10% conversion of feed carbon. It is possible to coat large areas by dc plasma-assisted CVD [55]. Moreover, a very high deposition rate may be achieved by this process.

5.2.2.3 Microwave Plasma Chemical Vapor Deposition

MPCVD is another very important method that has been frequently used for diamond deposition. Here, microwave plasma is used to activate the hydrocarbon feed and dissociate molecular hydrogen.

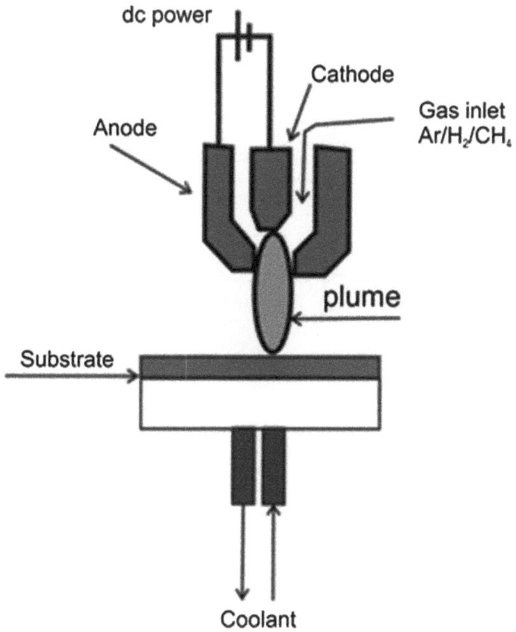

FIGURE 8.7

Schematic diagram of a dc plasma chemical vapour deposition reactor.

Typically, 2.45 GHz is used as excitation source. Microwave plasma oscillates electrons, which in turn produces ions by colliding with gas atoms and molecules. A schematic representation of a microwave plasma reactor manufactured by ASTEX is shown in Fig. 8.8. However, the deposition area is restricted for a microwave CVD reactor. Typical 2–3 cm substrates are used for deposition of diamond in an MPCVD reactor. The size of the plasma ball increases with increase in microwave power. It has been observed that intimate contact of the plasma ball with the substrate is not essential for diamond deposition by MPCVD technique. A substrate of ~10 cm diameter may be coated uniformly with diamond without having the plasma ball in direct contact with the substrate [55].

5.2.2.4 Combustion Flame–Assisted Chemical Vapor Deposition

Diamond may be deposited under ambient conditions using an ordinary welding torch (see Fig. 8.9). This method is known as the combustion flame–assisted CVD and was first proposed by Hirose and coworkers [71]. They used a mixture of roughly 50% acetylene and 50% oxygen to grow diamond. The greatest advantage of this process is its very high deposition rate, simple mode of operation, and low cost of the equipment. Subsequently, oxy-propylene, oxy-butylene, etc. have been used for diamond deposition. Use of propylene in place of acetylene significantly brings down the cost of production. It is possible to deposit diamond on exotic shapes using combustion flame–assisted synthesis. In this method, oxygen acetylene mixture is decomposed at very high temperature (>3000°C) of the flame [71]. Diamond formation takes place at the point of contact of the flame tip and the substrate, which is maintained at ~1000°C. However, because of the fluctuation of the substrate temperature, it may warp

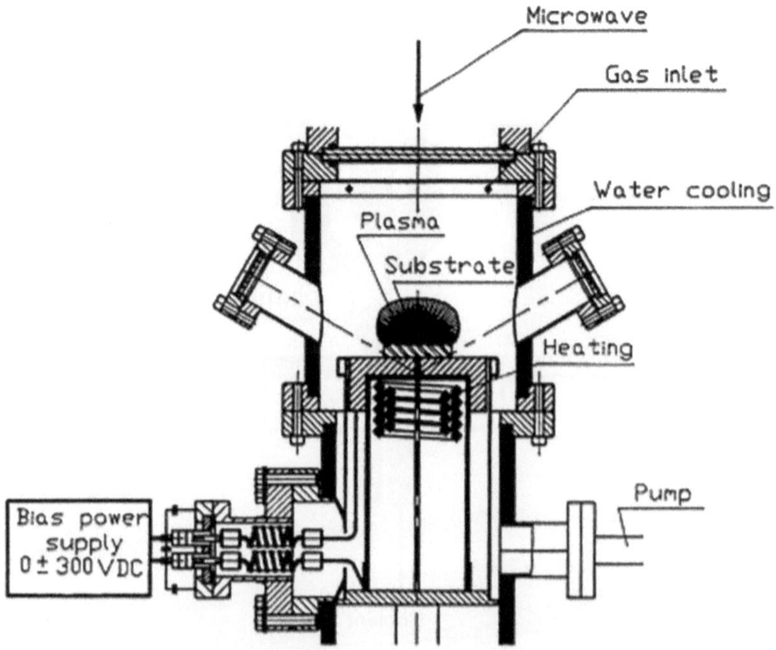

FIGURE 8.8

Schematic representation of a microwave plasma-enhanced chemical vapour deposition reactor manufactured by ASTEX.

Adapted from S.-T. Lee, Z. Lin, X. Jiang, CVD diamond films: nucleation and growth, 25 (1999) 123–154. Reprinted from Materials Science and Engineering, Copyright (2016), reproduced with permission from Elsevier.

or crack thereby delaminating the film. It is also difficult to control the microstructure, composition, and homogeneity of the coating by this method. Further, the deposition area is small and involves huge wastage of feed gases.

5.2.3 Tribological Applications of Diamond Coatings

Diamond is frequently deposited on different substrates as a protective coating. Obviously, being the hardest of materials, it is highly abrasion and wear resistant. However, good adhesion between the substrate and the coating is mandatory for effective tribological applications, which in turn demands an agreeable match between their coefficients of thermal expansion. Diamond coatings have low coefficient of thermal expansion ($\sim 10^{-6} K^{-1}$), which is usually very much lower as compared to that of metals and ceramics. As a result, diamond films on ceramics and metals tend to have residual stress in the films that often leads to spallation of the coating. Tungsten carbide (WC) has coefficient of thermal expansion close to that of diamond, which is why diamond films are deposited on tungsten carbide substrates in a nearly stress-free condition [4,72]. However, there is a serious issue of poor diamond nucleation on the surface of WC tools containing small amount of cobalt due to ready dissolution of carbon into cobalt [73]. On the other hand, formation of WC layer at the substrate-coating interface acts as a barrier against carbon diffusion. Nevertheless, diamond films have been successfully

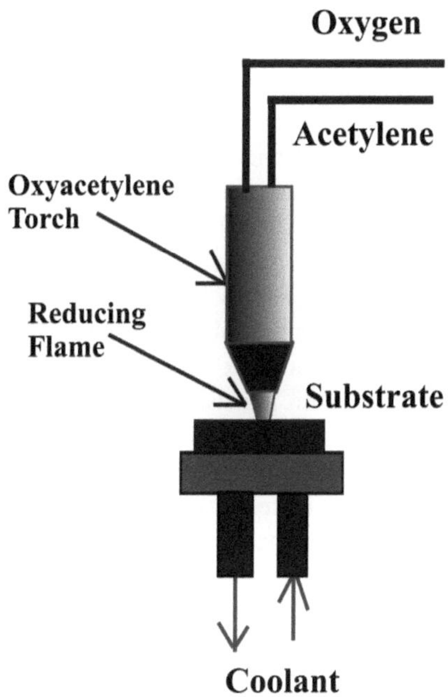

FIGURE 8.9

Schematic diagram of an oxyacetylene flame setup.

deposited on several cobalt-cemented tungsten carbide P-30 tool inserts using both HFCVD and oxyacetylene flame techniques [74]. Diamond coating by oxyacetylene torch took place much faster than that by HFCVD method. Not only were these coated tools better in terms of performance (of cutting and machining of nonferrous metals and alloys) under actual working conditions but also they provided superior surface finish to the products. Therefore, further polishing of the products became minimal implying lowering of processing cost. This would certainly make the use of diamond-coated tool inserts more economically viable and the coating process comparatively environment friendly, since corrosive chemicals used during polishing could be minimized. Diamond has been deposited on a variety of metal and ceramic substrates, such as copper, zirconium, tantalum, niobium, iron, stainless steel, molybdenum, nickel, and alumina, yielding a wide variety of morphology [75]. It has also been deposited on fiber-reinforced polymers and composites [76]. Ferrous materials are difficult to coat with diamond because of its tendency to get solubilized in iron especially at higher temperatures. During machining, temperature at the tool tip increases exorbitantly, initiating dissolution of the diamond coating [77] and eventually wearing out the protective layer. It is also difficult to coat diamond on aluminum metal because of its low melting point. It has been reported that the wear rate of synthetic diamond films is similar to that of natural diamond. Coefficient of friction of diamond sliding on diamond coatings can be as low as 0.03 to 0.5. The lower the surface roughness of the coating and the slider, the lower is the friction [78]. It has been observed that the coefficient of friction and wear rate of diamond coatings is independent of the moisture content of the environment. However, it is very much dependent on the topography and crystallographic orientation of the grains. Plate-like topography is favored for tribological applications because of lower abrasive effects. At higher temperatures, the coefficient of friction of diamond coatings decreases because of partial graphitization of the coating surfaces. Apart from wear resistance and low friction, diamond coatings on tools also provide them with exceptional thermal conductivity. Wear resistance of WC tools is presently improved by depositing an overlayer of titanium carbide and/titanium nitride [73]. It is presumed that with the advent of diamond coating technology, the present-day commercial hard coatings might face tremendous competition in terms of market share in the near future. Tribological application of diamond largely depends on the possibility of its deposition on exotic structures. Phase pure self-standing diamond films have been deposited using a jet flow HFCVD technique [80]. Diamond was coated on all sides of the metal substrates and thereafter the substrates were dissolved using suitable solvent, e.g., H_2O_2 for tungsten substrates. About 200-μm-thick diamond coating has been deposited by the same group on the tapered end of a molybdenum pin using oxyacetylene flame technique [74]. Prior to deposition, the substrate was treated with Murakami's reagent to enhance adhesion between the coating and the substrate. The entire deposition process took nearly 1 h. The diamond-coated molybdenum pin could now be used as a glass cutter with diamond tip. Diamond deposition using jet flow HFCVD has been reported on tantalum substrates. Fig. 8.10 shows the SEM micrograph (adapted from Ref. [80]) of a diamond sieve deposited on tantalum substrate through which holes have been drilled manually [80]. Upon dissolution of the metal and metal carbides formed at the coating interface using a warm mixture of nitric and hydrofluoric acid, self-supporting diamond sieves were formed. Following a similar recipe, polycrystalline diamond was deposited on tungsten wires producing hollow diamond helix and self-supporting diamond microtubes. The tungsten metal wires were dissolved using warm (50°C) H_2O_2 solution.

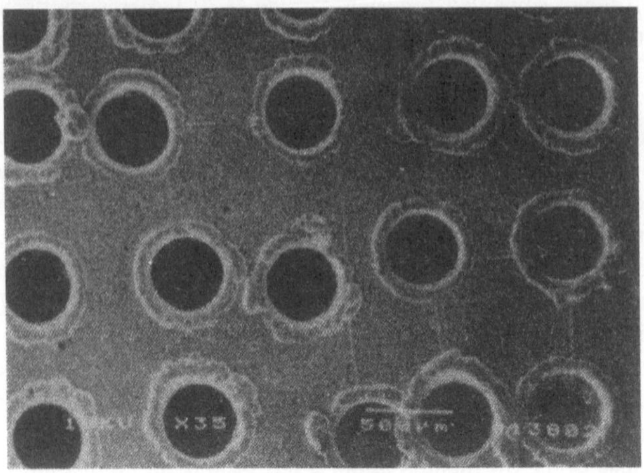

FIGURE 8.10

Scanning electron microscopy pictures of a self-supporting diamond sieve.

Adapted from A.K. Dua, D.D. Pruthi, V.C. George, P. Raj, Formation of self-supporting hollow diamond helix and diamond sieve using "jet-flow" HFCVD, 93 (1995) 759–762. Reprinted from Solid State Communications, Copyright (2016), reproduced with permission from Elsevier.

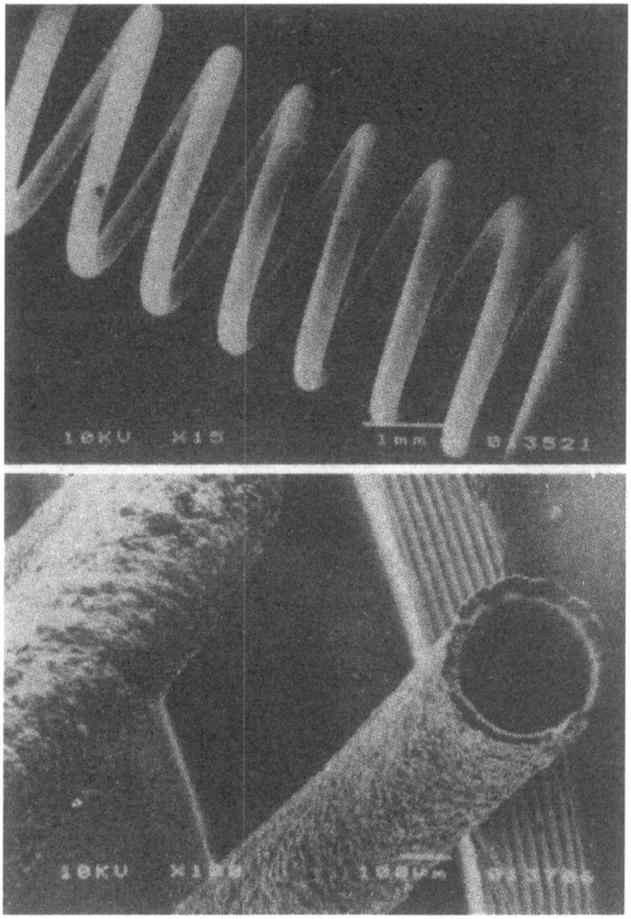

FIGURE 8.11

SEM pictures of a self-supporting diamond helix and tube.

Adapted from A.K. Dua, D.D. Pruthi, V.C. George, P. Raj, Formation of self-supporting hollow diamond helix and diamond sieve using "jet-flow" HFCVD, 93 (1995) 759–762. Reprinted from Solid State Communications, Copyright (2016), reproduced with permission from Elsevier.

Fig. 8.11 shows the SEM micrograph (adapted from Ref. [80]) of a diamond helix and microtube. It is possible to coat practically all kinds of exotic structures using this approach. Diamond-coated metallic wires and/nonmetallic fibers because of their stiffness may find application in aerospace engineering. Free-standing diamond windows have been prepared using CVD technique [73,81]. These windows are transparent almost over the entire spectral range. These low z windows exhibit excellent mechanical properties and hence are likely to substitute conventional beryllium-coated windows in energy dispersive spectrometer [73]. These coatings can withstand hostile environment, which is why they promise tremendous applications in aerospace research as well as masks for X-ray

FIGURE 8.12

Picture of a diamond window assembly.

Adapted from K. Sakamoto, A. Kasugai, M. Tsuneoka, K. Takahashi, T. Imai, T. Kariya, Y. Mitsunaka, High power 170 GHz gyrotron with synthetic diamond window, 70 (1999) 208–212. Reprinted from Review of Scientific Instruments, Copyright (2016), reproduced with permission from AIP Publishing.

lithography. Further, it has been possible to coat large infrared windows with diamond, which not only provides mechanical strength but also protects the windows against aerodynamic erosion by airborne particulates and heating when used in aircrafts. This enables the windows to be used for military purposes [81]. Fig. 8.12 shows the picture (adapted from Ref. [82]) of a diamond window assembly installed on a 170 GHz gyrotron.

5.3 DIAMOND-LIKE CARBON

The term "diamond-like carbon" (DLC) was coined by Aisenberg and Chabot [83] for a class of hard, electrically insulating, and wear-resistant amorphous carbon films deposited from ion beam and having properties close to that of diamond. The term "diamond-like" underlines a set of properties that are similar to that of diamond. In nature, bulk carbon exists in two different crystallographic forms, viz., (1) graphite that consists of a hexagonal lattice of trigonally (π) bonded sp^2 hybridized carbon atoms, and (2) diamond with a cubic lattice of tetragonally bonded sp^3 hybridized carbon atoms. Graphite is the stable allotrope and is soft, opaque, conducts electricity, and is highly reactive, whereas diamond is metastable, hard, electrically insulating, and chemically inert. Moreover, carbon because of its catenation property can undergo different types of hybridization with hydrogen and form polymeric chains of different lengths. Polymers are soft and flexible. DLC encompasses a broad class of amorphous carbon materials that contains all these three forms of carbon but in different proportion as schematically shown in Fig. 8.13. Generally, hydrogen-free amorphous carbon films are termed as a-C/i-C/ta-C and hydrogenated films are called a-C: H. Metal-containing amorphous films are designated as M-C: H and nitrogen-containing DLC is termed as a-C: H: N.

5.3.1 Deposition of Diamond-Like Carbon

Different techniques for deposition of diamond-like carbon have been reviewed in different literature [84,85]. In ion deposition method, diamond-like carbon coatings are produced upon bombardment

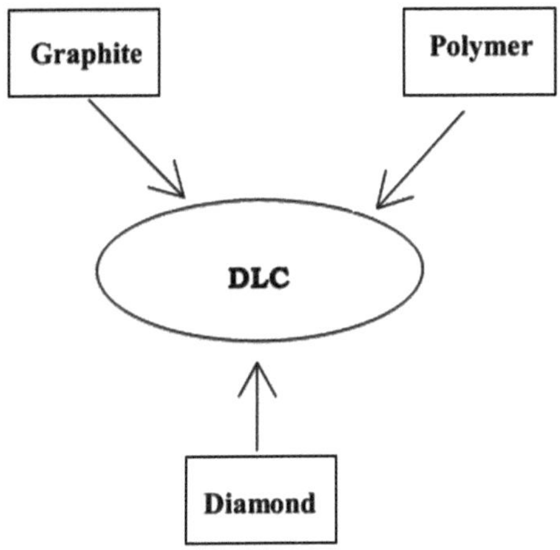

FIGURE 8.13

Schematic composition of diamond-like carbon coatings.

under vacuum of a substrate with highly energetic hydrocarbon ions [83] with energies typically in the range of 100–750 eV and current density of ~2.5 A/cm^2 [86]. The molecule dissociates at the surface depositing a smooth, uniform, and hard (Knoop microhardness ca. 6000 kg/mm^2) coatings of carbon with very low hydrogen content. Typically, the growth rate is ~1 μm/h.

Extremely smooth and practically hydrogen free, superhard coatings of diamond-like carbon may be obtained using a filtered arc deposition method. A cathodic arc is produced by passing current across a graphite cathode and an anode. A radial magnetic field is used to filter out bigger particles, and only filtered carbon ions are used to deposit the film.

The first report of depositing hydrogenated carbon coating from hydrocarbon plasma was from Holland and Ojha [87]. Thereafter, DLC films have been grown on a variety of substrate by a wide range of methods [88–90] including DC or RF PECVD, sputtering, ion beam deposition, pulse laser deposition, etc. Different types of carbon sources have been used. It is observed that properties of DLC largely depend on the technique and nature of feed gas used for deposition. Denser coatings are produced at higher radical energies, whereas amorphous (with carbon source) and polymeric (with hydrocarbon source) carbon deposits are obtained at lower energies. Plasma-enhanced CVD techniques, especially the parallel-plate RF reactors have been extensively used for deposition of DLC coatings. Occasionally, microwave and electron cyclotron resonance plasma reactors are also used. Hydrocarbon sources (be it liquid or gas) having adequate vapor pressure may be used as a feed for PECVD. The list of hydrocarbons that have been used for DLC deposition includes acetylene, ethylene, hexane, benzene, butane, cyclohexane, ethane, isopropane, methane, pentane, propane, and propylene. Sometimes, hydrocarbons are mixed with hydrogen to increase the polymeric content of the film that in turn reduces its coefficient of friction and makes the coating smooth.

Alternately, DLC may be coated on a substrate using a pulsed laser. A high-power laser beam is focused onto a graphite and/polycarbonate target that ablates carbonaceous material from the target and subsequently deposits the same onto the substrate in the form of a thin film [91,92]. Wear rate of such hydrogenated amorphous carbon films is found to be rather low.

Electrodeposition of DLC was first demonstrated by Namba [93]. It is a simple and straightforward technique for deposition of diamond-like films in solution at around room temperature. By this technique, it is possible to coat almost any substrate, viz., aluminum metal which is otherwise difficult to coat using high-temperature processes. Usually a graphite rod acts as an anode and the substrate as cathode. Common organic solvents are used as electrolytes, which under very high voltage (~1000 V) suffer dissociative cleavage, thereby producing precursors suitable for deposition of diamond-like carbon films. A large number of organic solvents including methanol, ethanol, 2-propanol, water-ethylene glycol solution, N-N-dimethyl formamide, acetonitrile, acetone, tetrahydrofuran, acetic acid, formic acid, etc. have been used. It has been observed [94] that the deposition rate and sp^3 carbon content of the films varies with dipole moment and dielectric constant of the electrolyte. Usually films deposited on Si and/SS using alcohols as electrolytes are found to be hydrogen free [94], while those grown on conducting glasses or metal interlayers are found to be hydrogenated in nature. It has been observed that higher voltages favor deposition and a pulsed voltage supply provides better growth rate than continuous dc source [95].

5.3.2 Tribological Applications of Diamond-Like Carbon Film

Hydrogenated amorphous carbon coatings (a-C: H) can be subdivided into two categories based on their tribological properties. They are (1) hard coatings with hardness typically in the range of 10–40 GPa and (2) soft coatings with hardness below 5 GPa. Softer DLC coatings contain higher amount of sp^3 bonding and hydrogen [4,96,97], whereas nonhydrogenated coatings (a-C) contain extremely low hydrogen [4]. Hardness of nonhydrogenated coatings is almost an order of magnitude higher and hence may be aptly termed DLC. Hardness arises from the local tetrahedral bonding between the carbon atoms albeit it does not translate to long-range order [98]. Diamond-like coatings are not only hard but also smooth, and furthermore their tribological properties may be manipulated easily by introducing dopants such as nitrogen, silicon, etc. It has been suggested that while sliding, the outer layers of DLC are converted to graphite that results in the decrease of both friction and wear rates [99]. Coefficient of friction of DLC is ~0.1, which is comparable to that of Teflon. It is even lower for a-C: H and fluorinated DLC [100]. The coefficient of friction is largely affected by the presence of moisture in the atmosphere, since the coatings tend to absorb water vapor, which further reduces the friction due to the coatings. Moreover, the surface of DLC films is generally decorated by a layer of hydrogen. This further reduces the roughness for such films. For a typical film deposited on silicon substrate, the roughness is ~0.05 nm. Thermal conductivity of DLC varies over a wide range from 400 to 1000 $Wm^{-1}K^{-1}$ depending on the deposition process [85]. However, it is difficult to grow thick DLC coatings due to internal stress generated in them as the thickness grows. DLC coatings are known to absorb gases within the voids and microstructures that in turn increase the internal stress in the coating. The coatings tend to delaminate after a critical thickness is reached. Good DLC films with an average thickness of 2μm could be deposited on a variety of substrates [100a]. Extent of adhesion of the coatings depends on the nature of substrates on which DLC has been deposited. Silicon was found to the best substrate whereas it was difficult to deposit adherent coatings of DLC on gold, copper, tungsten, stainless steel, etc. Another great disadvantage of the coatings is that they tend to graphitize above 300°C. Additionally, DLC coatings are biocompatible and hence they find tremendous applications in the fields of biomedical and bioengineering.

Biocompatibility of DLC coatings have been tested successfully on macrophage, fibroblast, human myeloblastic ML-1, and human embryo kidney 293 cells by allowing the cells to grow on DLC coatings and monitoring the rate of their proliferation, differentiation, and other cell responses [101–104]. Moreover, DLC coatings were found to be adherent to the living cells [105]. Viability studies of living cells in contact with DLC have been carried out that show no significant cell damage, indicating that DLC is nontoxic and biocompatible.

DLC forms adherent coatings on metals like titanium, cobalt, plastics, Teflon, and many other materials that are readily used for medical therapeutics. Biomedical applications of DLC have been extensively reviewed elsewhere [84,105–108]. Apart from biomedical applications, DLC finds substantial applications as protective coatings in advanced industries.

5.3.2.1 Orthopedic Applications

DLC coatings are being extensively used in different orthopedic applications. DLC coatings on metals are being preferred over ceramics for total hip and knee replacement. Prosthetic gears made of different metal alloys such Co-Cr, Co-Cr-Mo, alumina, Ti-6Al-4CV, and stainless steel alloy have been coated with DLC, tested under simulated conditions, and found to exhibit lower wear rate and corrosion as compared to the uncoated counterparts [109–113]. It has been observed that DLC-coated metal parts prevented leaching of metallic ions into the body under simulated conditions. A ventricular blood pump device after coating with DLC has been successfully implanted in calves [114]. DLC-coated metals have been used in the fractures of ceramic femoral heads. However, care should be taken to choose the right carbon source for DLC and monitor the roughness of the coatings for optimum biomedical applications.

5.3.2.2 Cardiovascular Applications

Since DLC coatings are hard and nonthrombogenic, DLC-coated NiCr18 stainless steel is being considered as a promising material for artificial heart valve applications [106]. Arterial stents coated with DLC have been found to be better than the uncoated ones in terms of lower platelet activation and initiation of thrombosis. No cytotoxic effects were found in muscle and endothelial cells ([84] and reference therein). Growth of activation-dependent antigens got suppressed and release of toxic metal into blood plasma got reduced upon coating of arterial stent with DLC. Even the blood compatibility of DLC-coated polyurethane was found to be superior to the nonthrombogenic polymer 2-hydroxyethyl methacrylate [115].

5.3.2.3 Other Medical Applications

Catheters are being coated with silver and DLC for better protection against local antibacterial infection. Recently studies showed that many biologically important polymers when coated with DLC exhibit better biocompatibility as compared to the uncoated ones. DLC coatings are also being applied to contact lenses [116] to reduce biofilm formation. Guide wires coated with DLC show more adherence and lower coefficient of friction as compared to those with Teflon coatings [117]. Similar is the case for tools used in advanced surgeries. Performance of needles used in ophthalmic surgery is found to improve upon coating with DLC [106].

5.3.2.4 Industrial and Household Applications

Prospective applications of DLC in today's industries is virtually limitless. Many tribological applications may be addressed with a coating of DLC for better performance. Tools used in everyday life such as knives, scissors, safety razors, etc. are being coated with a protective layer of DLC for retaining their sharpness for an extended period. DLC is coated on industrial tools, drilling bits, gears, weapons, and numerous other items. Fig. 8.14 shows the pictures (adapted from Ref. [118]) of DLC-coated

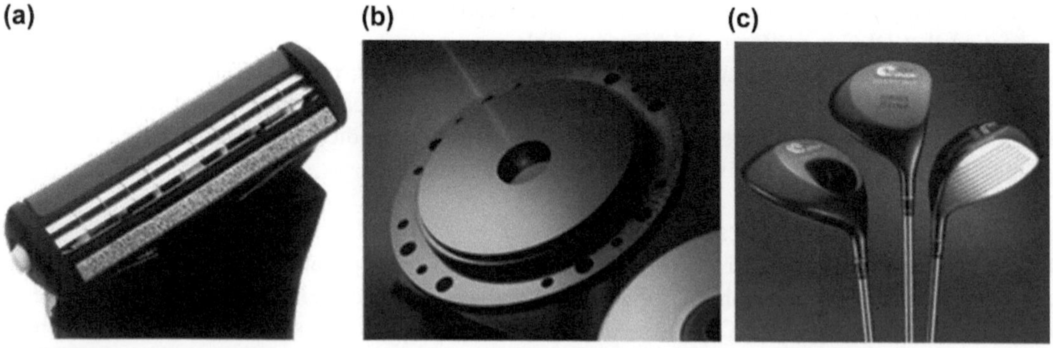

FIGURE 8.14

Diamond-like carbon-coated (a) razor, (b) CD molds, and (c) golf club heads.

Adapted from J.C. Sung, M.-C. Kan, M. Sung, Fluorinated DLC for tribological applications, 27 (2009) 421–426. Reprinted from International Journal of Refractory Metals and Hard Materials, Copyright (2016), reproduced with permission from Elsevier.

razors, CD molds, and golf club heads. Tribological properties of DLC get further enhanced upon fluorination. Not only does it become more inert but also it becomes more lubricating at the same time [100]. Fig. 8.15 shows the pictures (adapted from Ref. [118]) of DLC and fluorinated DLC-coated drills. Memory devices, displays, watches, etc. are also coated with corrosion-resistant and wear-protective coating of DLC. Magnetic discs and heads are coated with DLC to reduce the friction at the point of contact in order to have extended life and reliability of performance. Since DLC may be deposited at very low temperatures, it is possible to coat plastic articles with DLC. Scratchproof polycarbonate sunglasses are usually coated with a protective layer of DLC [119]. Industrial tanks are being coated inside with DLC for preserving corrosive chemicals. Nowadays, drinking water bottles and kitchen containers with an extended shelf life also come with a coating of DLC. Antireflective DLC coatings on

FIGURE 8.15

DLC and fluorinated DLC-coated drills.

Adapted from J.C. Sung, M.-C. Kan, M. Sung, Fluorinated DLC for tribological applications, 27 (2009) 421–426. Reprinted from International Journal of Refractory Metals and Hard Materials, Copyright (2016), reproduced with permission from Elsevier.

IR optics render them scratch and wear resistant without significantly compromising with their transparency. Different motor parts and ancillaries used in the automobile industry such as bearings, pistons, shafts, gears, and pumps are coated with DLC for their better life and improved performance.

6. SUMMARY

It is clear from the discussion that a wide range of hard materials may be deposited in the form of coatings onto different substrates by using a large variety of deposition techniques that tend to protect the underlying substrate against abrasion, wear, and corrosion. Such protective coatings find tremendous tribological application under extreme conditions both in our day-to-day activities and also in industrial production. Newer materials and hence altogether newer applications are being envisaged that promise to change the outlook of modern-day technology.

REFERENCES

[1] J. Williams, Engineering Tribology, Cambridge University Press, 32, Avenue of the Americas, New York, NY 10013-2473, USA: New York, 2008.

[2] N. Saka, Effect of microstructure on friction and wear of metals, in: Fundamentals of Tribology, The MIT Press, London, 1980.

[3] K. Holmberg, Tribological bases for accelerated testing, in: 0 Erational Reliability and Systematic Maintenance, Elsevier London, 1991.

[4] K. Holmberg, A. Matthews, Coating Tribology Properties, Techniques and Applications in Surface Engineering, Elsevier Science B.V., Netherlands, 1998, p. 457.

[5] P.J. Blau, Friction and Wear Transitions of Materials – Break-in, Run-in, Wear-in, Noyes Publications, Park Ridge, New Jirsey, USA, 1989.

[6] K. Kendall, Adhesion: molecules and mechanics, Science 263 (5154) (1994) 1720–1725.

[7] J.N. Israelachvili, Intermolecular and Surface Forces: Revised Third Edition, Academic Press, 2011.

[8] H. Czichos, Tribology – A Systems Approach to the Science and Technology of Friction, Lubrication and Wear, Elsevier, Amsterdam, 1978, p. 400.

[9] D. Buckley, Surface Effects in Adhesion, Friction, Wear, and Lubrication, Elsevier, Amsterdam, 1981, p. 631.

[10] T.F.J. Quinn, Review of oxidational wear. Part I: the origins of oxidational wear, Tribol. Int. 16 (5) (1983) 257–271.

[11] T.F.J. Quinn, W.O. Winer, The thermal aspects of oxidational wear, Wear 102 (1985) 67–80.

[12] J.L. Sullivan, The role of oxides in the protection of tribological surfaces – Part 1, Inst. Mech. Eng. C 143 (87) (1987) 283–291.

[13] J.L. Sullivan, The role of oxides in the protection of tribological surfaces – Part 2, Inst. Mech. Eng. C 144 (87) (1987) 293–301.

[14] D.S. Rickerby, A. Matthews, Advanced Surface Coatings: A Handbook of Surface Engineering, Blackie. Chapman and Hall, 1991.

[15] T.S. Sudarshan, D.G. Bhat, M. Minerals, M. Society, C.A.E.E Microtechnique (Eds.), Surface Modification Technologies III: Proceedings of the Third International Conference Held in Neuchatel, Switzerland, August 28–September 1, 1989, Minerals, Metals & Materials Society, 1990.

[16] P. Russell, D. Batchelor, A.F.M. SEM, Complementary techniques for surface investigations, Microsc. Anal. (2001) 9–12.

[17] E. Meyer, Atomic force microscopy, Prog. Surf. Sci. 41 (1) (1992) 3–49;
[17a] Y.T. Prabhu, K.V. Rao, V.S.S. Kumar, B.S. Kumari, X-Ray analysis by Williamson-Hall and size-strain plot methods of ZnO nanoparticles with fuel variation, World J. Nano Sci. Eng. 4 (2014) 21–28.
[18] T. Sumomogi, K. Kuwahara, H. Fujiyama, Adhesion evaluation of r.f.-sputtered aluminium oxide and titanium carbide thick films from grown carbide tools, Thin Solid Films 79 (1) (1981) 91–100.
[19] O.S. Heavens, L.E. Collins, L'épitaxie dans les lames polycristallines, J. Phys. Radium 13 (12) (1952) 658–660.
[20] P. Benjamin, C. Weaver, The adhesion of evaporated metal films on glass, Proc. R. Soc. London Ser. A Math. Phys. Sci. 261 (1307) (1961) 516–531.
[21] P. Benjamin, C. Weaver, The adhesion of metals to crystal faces, Proc. R. Soc. London Ser. A Math. Phys. Sci. 274 (1357) (1963) 267–273.
[22] K. Mittal, Adhesion measurement: recent progress, unsolved problems, and prospects, in: Adhesion Measurement of Thin Films, Thick Films, and Bulk Coatings, ASTM International, 1978.
[23] D. Butler, A simple film adhesion comparator, J. Phys. E Sci. Instrum. 3 (12) (1970) 979.
[24] J. Valli, A review of adhesion test methods for thin hard coatings, J. Vac. Sci. Technol. A 4 (6) (1986) 3007–3014.
[25] L.I. Maissel, R. Glang, in: L.I. Maissel, R. Glang (Eds.), Handbook of Thin Film Technology, McGraw-Hill, New York, 1970, p. 1.
[26] R. Faure, A. Carlan, J. Crebassa, G. Desrousseaux, B. Robrieux, Modification de la structure des couches minces d'argent soumises a des vibrations mecaniques-mesure de l'adhesion, Thin Solid Films 9 (3) (1972) 329–339.
[27] S. Moses, R.K. Witt, Evaluation of adhesion by ultrasonic vibrations, Ind. Eng. Chem. 41 (10) (1949) 2334–2338.
[28] R.C. Bray, C.F. Quate, J. Calhoun, R. Koch, Film adhesion studies with the acoustic microscope, Thin Solid Films 74 (2) (1980) 295–302.
[29] K. Davis, Material review: alumina (Al_2O_3), Sch. Dr. Stud. Eu. Un. J. 2 (2010).
[30] M.L. Kronberg, Plastic deformation of single crystals of sapphire: basal slip and twinning, Acta Metall. 5 (9) (1957) 507–524.
[31] J.B. Bilde-Sørensen, B.F. Lawlor, T. Geipel, P. Pirouz, A.H. Heuer, K.P.D. Lagerlöf, On basal slip and basal twinning in sapphire (α-Al_2O_3)—I. Basal slip revisited, Acta Mater. 44 (5) (1996) 2145–2152.
[32] W.D. Kaplan, P.R. Kenway, D.G. Brandon, Polymorphic basal twin boundaries and anisotropic growth in α-Al_2O_3, Acta Metall. et Mater. 43 (2) (1995) 835–848.
[33] I. Levin, D. Brandon, Metastable alumina polymorphs: crystal structures and transition sequences, J. Am. Ceram. Soc. 81 (8) (1998) 1995–2012.
[34] M.L. Kronberg, Plastic deformation of single crystals of sapphire: basal slip and twinning, 5 (1957) 507–524. Reprinted from Acta Metallurgica, Copyright (2016), with permission from Elsevier.
[35] L.C. Erickson, T. Troczynski, H.M. Hawthorne, H. Tai, D. Ross, Alumina coatings by plasma spraying of monosize sapphire particles, J. Therm. Spray Technol. 8 (3) (1999) 421–426.
[36] G. Bolelli, J. Rauch, V. Cannillo, A. Killinger, L. Lusvarghi, R. Gadow, Microstructural and tribological investigation of high-velocity suspension flame sprayed (HVSFS) Al_2O_3 coatings, J. Therm. Spray Technol. 18 (1) (2009) 35–49.
[37] H. Fischer, M. Luk, B. Oedekoven, R. Telle, K. Mottaghy, Hemocompatibility of high strength oxide ceramic materials: an in vitro study, J. Biomed. Mater. Res. A 81A (4) (2007) 982–986.
[38] A. Dua, V. George, R. Agarwala, Characterization and microhardness measurement of electron-beam-evaporated alumina coatings, Thin Solid Films 165 (1) (1988) 163–172.
[39] T. Roth, K.H. Kloos, E. Broszeit, Structure, internal stresses, adhesion and wear resistance of sputtered alumina coatings, Thin Solid Films 153 (1–3) (1987) 123–133.

[40] J. Skogsmo, M. Halvarsson, S. Vuorinen, Microstructural study of the κ-Al2O3 → α-Al_2O_3 transformation in CVD κ-Al_2O_3, Surf. Coat. Technol. 54 (1992) 186–192.

[41] B. Lux, C. Colombier, H. Altena, K. Stjernberg, Preparation of alumina coatings by chemical vapour deposition, Thin Solid Films 138 (1) (1986) 49–64.

[42] W.C. Oliver, G.M. Pharr, An improved technique for determining hardness and elastic modulus using load and displacement sensing indentation experiments, J. Mater. Res. 7 (06) (1992) 1564–1583.

[43] A.N. Cloud, A. Aryasomayajula, D.G. Bhat, M.H. Gordon, Determining substrate temperature in an AC inverted cylindrical magnetron sputtering PVD system, Surf. Coat. Technol. 202 (8) (2008) 1564–1567.

[44] A.K. Dua, R. Agarwala, Coatings for reducing plasma contamination and power loss in fusion reactor, Mater. Sci. Forum (1997) 225–238. Trans Tech Publ.

[45] J. Baxi, P. Kar, H. Liang, A. Polat, M. Usta, A.H. Ucisik, Tribological characterization of microarc oxidized alumina coatings for biological applications, Vacuum 83 (1) (2008) 217–222.

[46] A. Aryasomayajula, Low temperature alpha alumina thin film coating for cutting tool application by ac inverted magnetron sputtering technique, ProQuest (2008).

[47] I.-M. Low, Advances in Ceramic Matrix Composites, Woodhead Publishing, 2014.

[48] D. Kopeliovich, Tribological Properties and Applications of Alumina. http://www.substech.com/dokuwiki/doku.php?id=tribological_properties_and_applications_of_alumina&do=recent&DokuWiki=cfdbec5cb9c-7ffb0c3cc71a5a28f10e0. The document: tribological_properties_and_applications_of_alumina.txt. Last modified: 2013/07/16 by dmitri_kopeliovich.

[49] S.F. Hulbert, F.A. Young, R.S. Mathews, J.J. Klawitter, C.D. Talbert, F.H. Stelling, Potential of ceramic materials as permanently implantable skeletal prostheses, J. Biomed. Mater. Res. 4 (3) (1970) 433–456.

[50] X.Q. Cao, R. Vassen, D. Stoever, Ceramic materials for thermal barrier coatings, J. Eur. Ceram. Soc. 24 (1) (2004) 1–10.

[51] H. Xu, H. Guo, F. Liu, S. Gong, Development of gradient thermal barrier coatings and their hot-fatigue behavior, Surf. Coat. Technol. 130 (1) (2000) 133–139.

[52] R. Bardet, M. Bernard, G. Briffod, M. Clement, A. Gauthier, M. Gregoire, P. Grelot, G. Haste, M. Hesse, F. Parlance, D. Pinet, E. Porrot, G. Rey, B. Taquet, J. Weisse, Results with alumina limiter and liner on the tokamak petula, J. Nucl. Mater. 76–77 (1978) 477–480.

[53] R. Bardet, M. Bernard, G. Brifford, M. Clement, R. Frank, A. Gauthier, M. Gregoire, P. Grelot, M. Hesse, S. Gruber, in: The Petula Tokamak Experiment-Effects of Low-Z Materials on Plasma Characteristics, Plasma Physics and Controlled Nuclear Fusion Research 1976, vol. 2, 1977, pp. 259–266.

[54] A. Dua, Trends and future for the emerging diamond deposition technology, Indian J. Eng. Mater. Sci. 7 (5/6) (2000) 237–242.

[55] S.T. Lee, Z. Lin, X. Jiang, CVD diamond films: nucleation and growth, Mater. Sci. Eng. R Rep. 25 (4) (1999) 123–154.

[56] S.-T. Lee, Z. Lin, X. Jiang, CVD diamond films: nucleation and growth, 25 (1999) 123–154. Reprinted from Materials Science and Engineering, Copyright (2016), with permission from Elsevier.

[57] J.E. Field, The Properties of Natural and Synthetic Diamond, Academic Press, 1992.

[58] T. Evans, First International Congress on Diamond in Industry, 1962, p. 275. Paris.

[59] E. Bourdon, A. Raveh, S. Gujrathi, L. Martinu, Etching of a-C: H films by an atomic oxygen beam, J. Vac. Sci. Technol. A 11 (5) (1993) 2530–2535.

[60] Y. Zhang, F. Zhang, G. Chen, A study of phase transformation between diamond and graphite in P-t diagram of carbon, 32 (1994) 1415–1418. Reprinted from Carbon, Copyright (2016), with permission from Elsevier.

[61] H.P. Bovenkerk, F.P. Bundy, H.T. Hall, H.M. Strong, R.H. Wentorf, Preparation of diamond, Nature 184 (4693) (1959) 1094–1098.

[62] W.G. Eversole, Synthesis of Diamond, US Patent 3030188, April 17, 1962.

[63] D.J. Poferl, N.C. Gardner, J.C. Angus, Growth of boron-doped diamond seed crystals by vapor deposition, J. Appl. Phys. 44 (4) (1973) 1428–1434.
[64] B. Spitsyn, L. Bouilov, B. Derjaguin, Vapor growth of diamond on diamond and other surfaces, J. Cryst. Growth 52 (1981) 219–226.
[65] S. Matsumoto, Y. Sato, M. Tsutsumi, N. Setaka, Growth of diamond particles from methane-hydrogen gas, J. Mater. Sci. 17 (11) (1982) 3106–3112.
[66] S. Matsumoto, M. Hino, T. Kobayashi, Synthesis of diamond films in a rf induction thermal plasma, Appl. Phys. Lett. 51 (10) (1987) 737–739.
[67] Deleted in review.
[68] A. Dua, M. Roy, J. Nuwad, V. George, S. Sawant, Enhanced nucleation and post-growth investigations on HFCVD diamond films grown on silicon single crystals pretreated with Zr: diamond mixed slurry, Appl. Surf. Sci. 229 (1) (2004) 254–262.
[69] K. Suzuki, A. Sawabe, H. Yasuda, T. Inuzuka, Growth of diamond thin films by dc plasma chemical vapor deposition, Appl. Phys. Lett. 50 (12) (1987) 728–729.
[70] K. Suzuki, A. Sawabe, T. Inuzuka, Growth of diamond thin films by DC plasma chemical vapor deposition and characteristics of the plasma, Jpn. J. Appl. Phys. 29 (1R) (1990) 153.
[71] Y. Hirose, N. Kondo, 35th Spring Meeting, Jpn. Appl. Phys. Soc., 1988, p. 434.
[72] A. Lindlbauer, R. Haubner, B. Lux, Diamantabscheidung auf keramischen materialien mit der heißdrahtmethode, Wear 159 (1) (1992) 67–77.
[73] K.V. Ravi, Technological applications of CVD diamond, in: K.E. Spear, J.P. Dismukes (Eds.), Synthetic Diamond: Emerging CVD Science and Technology 25, John Wiley & Sons, 1994, pp. 533–579.
[74] A.K. Dua, V.C. George, D.D. Pruthi, P.C.G. Shasikala, P. Raj, Diamond and diamond-like carbon coatings for various applications, in: J. Bhatia, Tribology Society of India (Eds.), Advances in Industrial Tribology, Tata McGraw-Hill Publishing Company, 1998, pp. 75–80.
[75] A. Dua, D. Pruthi, V. George, P. Raj, Preparation of Synthetic Diamond Films at Sub-atmospheric Pressure and Their Characterisation, Bhabha Atomic Research Centre, Bombay, India, 1992.
[76] C. Dold, M. Henerichs, L. Bochmann, K. Wegener, Comparison of ground and laser machined polycrystalline diamond (PCD) tools in cutting carbon fiber reinforced plastics (CFRP) for aircraft structures, Proc. CIRP 1 (2012) 178–183.
[77] Y. Tzeng, M. Yoshikawa, M. Murakawa, A. Feldman, Applications of diamond films and related materials, in: Proceedings of the First International Conference on the Applications of Diamond Films and Related Materials – ADC '91, Auburn, Alabama, USA, August 17–22, 1991, Elsevier Science, 2013.
[78] I. Hayward, I. Singer, L. Seitzman, Effect of roughness on the friction of diamond on CVD diamond coatings, Wear 157 (2) (1992) 215–227.
[79] Deleted in review.
[80] A.K. Dua, D.D. Pruthi, V.C. George, P. Raj, Formation of self-supporting hollow diamond helix and diamond sieve using "jet-flow" HFCVD, 93 (1995) 759–762. Reprinted from Solid State Communications, Copyright (2016), with permission from Elsevier.
[81] A.K. Dua, Recent exciting developments related to diamond coatings, Mater. Sci. Forum (1997) 91–104. Trans Tech Publ.
[82] K. Sakamoto, A. Kasugai, M. Tsuneoka, K. Takahashi, T. Imai, T. Kariya, Y. Mitsunaka, High power 170 GHz gyrotron with synthetic diamond window, 70 (1999) 208–212. Reprinted from Review of Scientific Instruments, Copyright (2016), with the permission of AIP Publishing.
[83] S. Aisenberg, R. Chabot, Ion-beam deposition of thin films of diamondlike carbon, J. Appl. Phys. 42 (7) (1971) 2953–2958.
[84] G. Dearnaley, J.H. Arps, Biomedical applications of diamond-like carbon (DLC) coatings: a review, Surf. Coat. Technol. 200 (7) (2005) 2518–2524.

[85] S.V. Hainsworth, N. Uhure, Diamond like carbon coatings for tribology: production techniques, characterisation methods and applications, Int. Mater. Rev. 52 (3) (2007) 153–174.

[86] Y. Liu, A. Erdemir, E.I. Meletis, A study of the wear mechanism of diamond-like carbon films, Surf. Coat. Technol. 82 (1–2) (1996) 48–56.

[87] L. Holland, S.M. Ojha, Deposition of hard and insulating carbonaceous films on an r.f. target in a butane plasma, Thin Solid Films 38 (2) (1976) L17–L19.

[88] J.C. Angus, P. Koidl, S. Domitz, J. Mort, F. Jansen, Plasma Deposited Thin Films, CRC, Boca Raton, FL, 1986, p. 1564.

[89] C.V. Deshpandey, R.F. Bunshah, Diamond and diamondlike films: deposition processes and properties, J. Vac. Sci. Technol. A 7 (3) (1989) 2294–2302.

[90] Y. Catherine, Preparation techniques for diamond-like carbon, in: R.E. Clausing, L.L. Horton, J.C. Angus, P. Koidl (Eds.), Diamond and Diamond-Like Films and Coatings, Springer, Boston, MA, USA, 1991, pp. 193–227.

[91] A. Voevodin, M. Donley, Preparation of amorphous diamond-like carbon by pulsed laser deposition: a critical review, Surf. Coat. Technol. 82 (3) (1996) 199–213.

[92] A. Voevodin, M. Donley, J. Zabinski, Pulsed laser deposition of diamond-like carbon wear protective coatings: a review, Surf. Coat. Technol. 92 (1) (1997) 42–49.

[93] Y. Namba, Attempt to grow diamond phase carbon films from an organic solution, J. Vac. Sci. Technol. A 10 (5) (1992) 3368–3370.

[94] C. Cao, H. Zhu, H. Wang, Electrodeposition diamond-like carbon films from organic liquids, Thin Solid Films 368 (2) (2000) 203–207.

[95] H. Wang, M.-R. Shen, Z.-Y. Ning, C. Ye, H.-S. Zhu, Pulsed electrodeposition of diamond-like carbon films, J. Mater. Res. 12 (11) (1997) 3102–3105.

[96] M. Ham, K.A. Lou, Diamond-like carbon films grown by a large-scale direct current plasma chemical vapor deposition reactor: system design, film characteristics, and applications, J. Vac. Sci. Technol. A 8 (3) (1990) 2143–2149.

[97] J. Franks, K. Enke, A. Richardt, Diamond-like carbon-properties and applications, Met. Mater. 6 (11) (1990) 695–700.

[98] D.R. McKenzie, Tetrahedral bonding in amorphous carbon, Rep. Prog. Phys. 59 (12) (1996) 1611.

[99] Y. Liu, E.I. Meletis, Evidence of graphitization of diamond-like carbon films during sliding wear, J. Mater. Sci. 32 (13) (1997) 3491–3495.

[100] J.C. Sung, M.-C. Kan, M. Sung, Fluorinated DLC for tribological applications, Int. J. Refract. Met. Hard Mater. 27 (2) (2009) 421–426;
[100a] M. Ham, K.A. Lou, Diamond-like carbon films grown by a large-scale direct current plasma chemical vapor deposition reactor: System design, film characteristics, and applications, J. Vac. Sci. Technol. A 8 (1990) 2143–2149 http://dx.doi.org/10.1116/1.577030.

[101] L.A. Thomson, F.C. Law, N. Rushton, J. Franks, Biocompatibility of diamond-like carbon coating, Biomaterials 12 (1) (1991) 37–40.

[102] M. Allen, F. Law, N. Rushton, The effects of diamond-like carbon coatings on macrophages, fibroblasts and osteoblast-like cells in vitro, Clin. Mater. 17 (1) (1994) 1–10.

[103] L. Lu, M.W. Jones, R.L. Wu, Diamond-like carbon as biological compatible material for cell culture and medical application, Biomed. Mater. Eng. 3 (4) (1993) 223–228.

[104] S. Linder, W. Pinkowski, M. Aepfelbacher, Adhesion, cytoskeletal architecture and activation status of primary human macrophages on a diamond-like carbon coated surface, Biomaterials 23 (3) (2002) 767–773.

[105] F. Cui, D. Li, A review of investigations on biocompatibility of diamond-like carbon and carbon nitride films, Surf. Coat. Technol. 131 (1) (2000) 481–487.

[106] R.S. Butter, A.H. Lettington, Diamond like carbon for biomedical applications. Review, J. Chem. Vap. Deposition (USA) 3 (3) (1995) 182–192.
[107] R. Hauert, A review of modified DLC coatings for biological applications, Diamond Relat. Mater. 12 (3–7) (2003) 583–589.
[108] M. Allen, B. Myer, N. Rushton, In vitro and in vivo investigations into the biocompatibility of diamond-like carbon (DLC) coatings for orthopedic applications, J. Biomed. Mater. Res. 58 (3) (2001) 319–328.
[109] V. Saikko, T. Ahlroos, O. Calonius, J. Keränen, Wear simulation of total hip prostheses with polyethylene against CoCr, alumina and diamond-like carbon, Biomaterials 22 (12) (2001) 1507–1514.
[110] D. Sheeja, B. Tay, S. Lau, L. Nung, Tribological characterisation of diamond-like carbon coatings on Co–Cr–Mo alloy for orthopaedic applications, Surf. Coat. Technol. 146 (2001) 410–416.
[111] D. Sheeja, B. Tay, L. Nung, Tribological characterization of surface modified UHMWPE against DLC-coated Co–Cr–Mo, Surf. Coat. Technol. 190 (2) (2005) 231–237.
[112] T. Xu, L. Pruitt, Diamond-like carbon coatings for orthopaedic applications: an evaluation of tribological performance, J. Mater. Sci. Mater. Med. 10 (2) (1999) 83–90.
[113] J. Onate, M. Comin, I. Braceras, A. Garcia, J. Viviente, M. Brizuela, N. Garagorri, J. Peris, J. Alava, Wear reduction effect on ultra-high-molecular-weight polyethylene by application of hard coatings and ion implantation on cobalt chromium alloy, as measured in a knee wear simulation machine, Surf. Coat. Technol. 142 (2001) 1056–1062.
[114] K. Yamazaki, P. Litwak, O. Tagusari, T. Mori, K. Kono, M. Kameneva, M. Watach, L. Gordon, M. Miyagishima, J. Tomioka, An implantable centrifugal blood pump with a recirculating purge system (Cool-Seal system), Artif. Organs 22 (6) (1998) 466–474.
[115] A. Alanazi, C. Nojiri, T. Noguchi, T. Kido, Y. Komatsu, K. Hirakuri, A. Funakubo, K. Sakai, Y. Fukui, Improved blood compatibility of DLC coated polymeric material, Asaio J. 46 (4) (2000) 440–443.
[116] V. Elinson, V. Sleptsov, A. Laymin, V. Potraysay, L. Kostuychenko, A. Moussina, Barrier properties of carbon films deposited on polymer-based devices in aggressive environments, Diamond Relat. Mater. 8 (12) (1999) 2103–2109.
[117] J. McLaughlin, B. Meenan, P. Maguire, N. Jamieson, Properties of diamond like carbon thin film coatings on stainless steel medical guidewires, Diamond Relat. Mater. 5 (3) (1996) 486–491.
[118] J.C. Sung, M.-C. Kan, M. Sung, Fluorinated DLC for tribological applications, 27 (2009) 421–426. Reprinted from International Journal of Refractory Metals and Hard Materials, Copyright (2016), with permission from Elsevier.
[119] F.M. Kimock, B.J. Knapp, Commercial applications of ion beam deposited diamond-like carbon (DLC) coatings, Surf. Coat. Technol. 56 (3) (1993) 273–279.

CHAPTER

9

INTERMETALLICS AND ALLOYS FOR HIGH TEMPERATURE APPLICATIONS

R. Tewari, N.K. Sarkar, D. Harish, B. Vishwanadh, G.K. Dey, S. Banerjee

Bhabha Atomic Research Centre, Mumbai, India

CHAPTER OUTLINE

Materials Under Extreme Conditions. http://dx.doi.org/10.1016/B978-0-12-801300-7.00009-7

1. INTRODUCTION

With the turn of 20th century, engineers started realizing the shortcomings of steel, the most common structural material. Dramatic effects were noticed during World War II, when many battle machines of Germany failed catastrophically under extreme cold conditions. Similarly, at elevated temperatures sudden softening of steel brought attention of material developers to look out for materials that could perform under these extreme conditions.

In the process of improving the properties of steel under extreme conditions, various alloying additions have been carried out. Out of them, addition of Ni and Cr turn out to be the most rewarding alloying additions. Even though the knowledge of improvement in the corrosion properties of steel by Cr additions has long been known [1,2], detailed mapping of the alloy system was developed after 1920 [3]. Fig. 9.1 provides a section of the ternary phase diagram comprising a Fe-Ni-Cr system. Different phase fields marked in the phase diagram are not in proportion. These are merely used to highlight various commercial alloys developed using this ternary phase diagram as the basic entity. Major commercial alloys developed using these elements as base are superalloys and stainless steels. Nichrome and other nickel chrome alloys were used for various high-temperature applications as components in thermocouples, heating elements, or fixtures in high-temperature furnaces. Development of these alloys pushed the window of operation temperature by a couple of hundred degrees below and few hundred degrees above room temperature. But with the demand of higher operating temperatures in high-tech areas, requirement for newer materials was felt.

During the search of new materials, attractive physical and mechanical properties of ordered intermetallic alloys were recognized. However, initial attempts for developing ordered intermetallic alloys for structural applications could not be successful because of their inadequate ductility or toughness at ambient temperatures and elevated temperature creep strength not matching the expectations. In 1975 an accidental discovery by Aoki and Izumi [4] that small additions of boron dramatically improved ductility in Ni_3Al refocused the attention of materials scientists toward ordered intermetallic phases. The conventional approach of binary combination of metals to obtain an intermetallic phase was replaced by ternary, quaternary, or higher order combination of metals and metalloids (Si, B), which revolutionized the field of intermetallics. New intermetallics were more ductile at ambient temperatures and showed higher creep and oxidation resistance at elevated temperatures [5,6]. One of the reports entitled "Structural Uses for Ductile Ordered Alloys" by the U.S. National

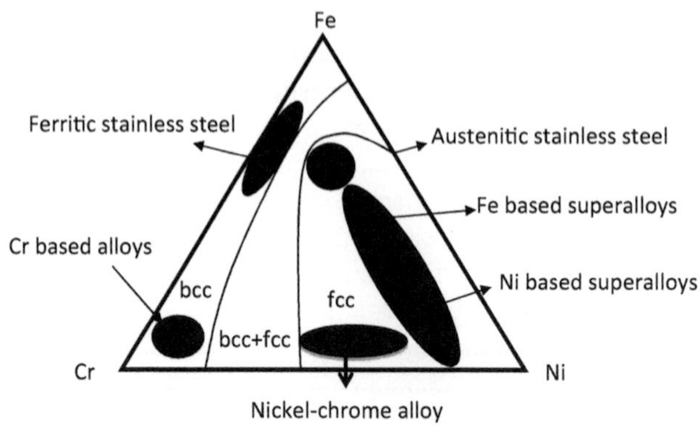

FIGURE 9.1

Schematic representation of the isothermal section of the ternary Fe-Cr-Ni-based system. Phase fields of various commercial alloys are marked in this phase diagram.

Advisory Board identified several applications of intermetallic phases under extreme conditions and proposed a coordinated development program for the new class of intermetallic phases [7]. The Ministry of International Trade and Trade of Japan identified TiAl and Nb_3Al as potential intermetallics for high-temperature applications in a temperature regime, which is beyond application temperature of nickel-based superalloys [8].

Ductilization of Ti_3Al-based intermetallics due to the formation of an orthorhombic phase named as "O" phase by the addition of niobium has opened up yet another possibility [9]. Introduction of a second phase to provide toughness to the base intermetallic phase led to the possibility of ductilization of those intermetallic phases that do not tolerate any stoichiometric variations. Substantial research work at various laboratories and academic institutes led to the development of the γ-TiAl interwoven two-phase microstructure that resulted in many high-temperature attractive properties [10].

As a result of additional activities in the field of intermetallics, other alloy systems also became the active field of research work. One of them was Nb-based alloys. The exploration of Ti-Al-Nb system for high-temperature applications brought the attention of researchers to a Nb-Al-Ti system [11−15]. This alloy system offers many advantages over the Ti-based intermetallic systems. Few of them are in the single-phase field high ductility of niobium and relatively high liquidus point of the alloy systems are some attractive features, while poor high-temperature oxidation properties and low strength remain as major stumbling blocks in the development of these alloys [16]. During the development of the refractory metal alloys another concept of in situ composite was developed. The essential idea behind the development of in situ composite was the distribution of a hard brittle phase in the soft matrix of niobium. Soft matrix would provide necessary ductility at ambient temperatures and hard precipitate phase, generally silicides of niobium, which would provide strength at elevated temperatures ($T > 727$K (1000°C)) [17]. The problem of oxidation resistance was addressed by alloying with elements like Cr, Al, and Ti. In parallel to niobium-based in situ composite development, work on molybdenum-based in situ composite started. In the case of Mo-based alloys, attempts were made to get uniform coating of borosilicate-based glassy phase on the surface, and for this Mo-Si-B-based alloys were developed. In addition to these two refractory metals, exotic alloys like W-Re have also been developed. These alloys have much superior high-temperature properties as compared to any other alloys. However, extremely high density and exorbitant cost do not favor their large-scale application.

The refractory metal beryllides are another attractive class that was examined for possible high-temperature applications because of their high strength-to-weight ratio. $NbBe_{12}$, Nb_2Be_{17}, NbBe, and TiB_{12} are a few intermetallic compounds that were studied extensively. However, absence of ductility even at elevated temperatures is one of the major reasons for reduced active research work in these alloys, which of late surged again once the reason for high brittleness had been identified as the uncontrolled presence of interstitials. Laves phases are another class of alloys that were examined for high-temperature applications. As Cr and refractory metals are the most common elements in the Laves phases, most of the research work was confined to Nb-, Hf-, and Zr-based Laves phases. Though the work later spread to non-Cr Laves phases like HfV_2, successful room temperature rolling of the V-HfV_2 alloy system brought attention back to this field [18].

In spite of the fact that for high-temperature applications Ni-based superalloys still dominate, they have reached nearly the upper limit, as in some applications in aerospace engines the operating temperatures are touching $0.95T_m$, T_m being the melting temperature of the alloy. New materials are, therefore, essential where technology demands still higher operating temperatures. This chapter provides a survey of the other possible candidate materials that have potential to meet this demand.

The first part of the chapter is devoted to various intermetallic phases that are under active consideration, and the second part of the chapter is focused on the development of refractory metal-based alloy and high-entropy alloy systems.

2. INTERMETALLICS

The work on intermetallics was generally focused on two points: one on low-temperature ductility and another on high-temperature strength. Detailed research on various intermetallic systems revealed that brittleness in many of the intermetallic systems was due to environmental effects, which is an extrinsic effect. Therefore ductility can be restored in these intermetallic phases by suitable alloying additions. Examples of such intermetallic phases are Fe_3Al, FeAl, Ni_3Al, and $MoSi_2$ to name a few. In some cases improvement in the properties was achieved by finding out suitable alloying elements [19]. Apart from structural applications many of the intermetallic phases found unique usages due to their unusual physical properties. Superconductivity, hydrogen storage capacity, and hard magnet are a few of the known applications. Table 9.1 summarizes some of the intermetallic phases and their possible applications [20–28]. As exploring the development of intermetallics used for nonstructural applications is beyond the scope of this chapter, work on structural intermetallics will be considered. Most of the intermetallic phases that are useful for structural applications have crystal structures, which are ordered derivatives of the basic metallic crystal structures like fcc, bcc, and hcp. Among this class of intermetallic phases, $L1_2$-based crystal structures are the most studied systems.

2.1 CRYSTAL STRUCTURE OF INTERMETALLIC COMPOUNDS

2.1.1 $L1_2$ Crystal Structure–Based Intermetallics

The $L1_2$ crystal structure is an ordered derivative of the fcc crystal structure. Fig. 9.2a shows the atomic arrangement within the $L1_2$ crystal structure. Stoichiometrically, fully ordered $L1_2$ can be represented by 75:25 atomic distributions. Atomic arrangement in the higher order derivatives DO_{22} and DO_{23} are

Table 9.1 Some Applications of Intermetallics-Based Materials

Intermetallics	Applications/Proposed Applications	Reference(s)
Ni-Al-based intermetallics materials	Tarbocharger rotors, furnace fixtures for heat treatment furnaces, hydroturbines, cutting tools, turbine blades, and valves	[20]
Ti-Al-based intermetallics materials	Tarbocharger rotors, compressor case, piston head for diesel engine	[21]
Fe-Al-based intermetallics materials	Heating elements, hot gas filters, piping, tubing in coal-gasification systems, components in high-temperature sulfidation	[22]
Nb-Al-based intermetallics materials	high critical current density superconductors applications, such as in particle accelerators (LHCs), fusion reactors (ITERs), and nuclear magnetic resonance (NMR)	[23–26]
Laves phases–based intermetallics materials	Hydrogen storage	[27]
Be-based intermetallics materials	Neutron multipliers	[28]

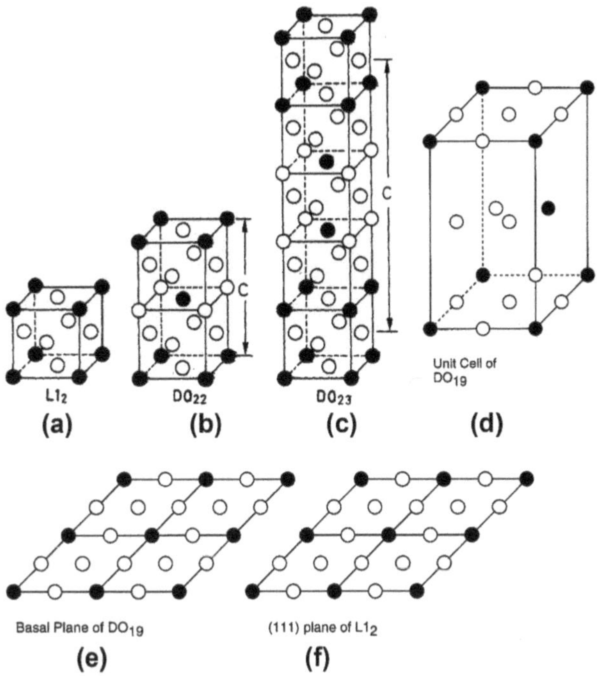

FIGURE 9.2

(a) Unit cells of the $L1_2$ structure (a), its ordered derivatives (b) DO_{22}, (c) DO_{23}, and of the (d) DO_{19} structures are shown. The close-packed planes of the DO_{19} and $L1_2$ are also compared in figure (e) and (f). The isostructural configuration of these close-packed planes can be noticed from the figure. ABAB... stacking sequence of these planes produces the DO_{19} crystal structure whereas ABCABC... stacking sequence produces the $L1_2$ crystal structure.

also shown in Fig. 9.2b and c, respectively. Along with these face-centered-based cubic structures, ordered hexagonal (DO_{19}) is also shown (Fig. 9.2d) as the latter can be created by hexagonal stacking, ABAB... of the basal plane of DO_{19}, which is same as the {111} plane of the $L1_2$ structure (Fig. 9.2e and f). Atomic arrangement of three layers of {111} planes of the $L1_2$ structure is shown in Fig. 9.3.

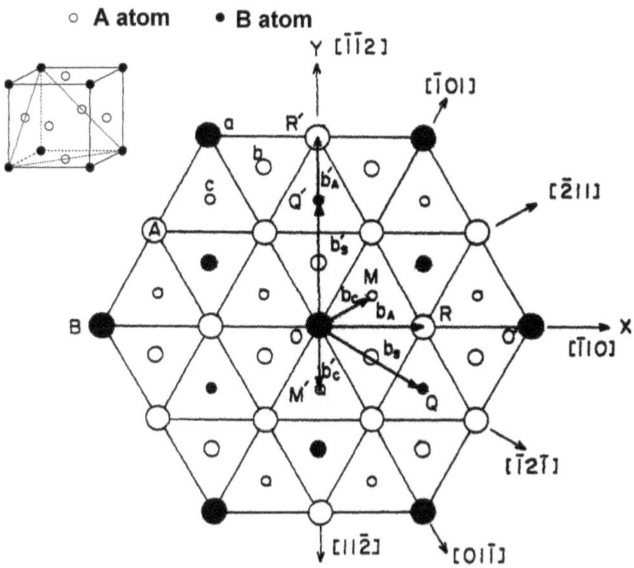

FIGURE 9.3

Atomic arrangement of the {111} plane in the $L1_2$ structure. The a, b, and c layers of the (111) planes are stacked in the sequence abcabc... representing an $L1_2$ lattice shown in inset. The vectors b′A, b′c, and b′s are crystallographic equivalent to bA, bc, and bs, which are the displacement vectors for APB, CSF, and SISF, respectively.

The three layers, marked as a, b, and c, are shown with different dimensions in the figure. On the basis of hard sphere model and geometrical considerations, three types of planar faults on {111} can be considered. These planar faults are the antiphase boundary (APB), superlattice intrinsic stacking fault (SISF), and complex stacking fault (CSF). Whenever atoms on {111} planes are displaced by ½<110> vector, in the A_3B configuration, B atoms are displaced to the antisite, thus generating APB defects. In the case atoms are displaced by 1/3<112> displacement vectors, the B atoms are displaced to the same site but to a position corresponding to a different layer. Therefore, the stacking sequence is altered without the creation of any antisite. This creates intrinsic stacking fault. In the case displacement of 1/6<112> is generated and B atoms move to the antisite and also to the position that corresponds to the different layer. In this process, CSF is created. These displacement vectors are shown as b_A, b_S, and b_C, respectively in Fig. 9.3.

Introduction of parallel APBs with displacement vector ½<110> on every (001) cube plane along a cube axis of an $L1_2$ structure generates long period tetragonal superlattice DO_{22} and the introduction of APB on the alternate (001) planes produces the DO_{23} structure. The DO_{19} structure, shown in Fig. 9.2d, can be obtained by inserting SISFs into the $L1_2$ structure on every alternate {111} planes. Thus, the relative stability of $L1_2$, DO_{22}, DO_{23}, and DO_{19} structures depends on the magnitudes and the temperature dependence of APB and SISF energies. Yamaguchi et al. [29] using six different pairwise interatomic potential sets have shown that SISF is always stable whereas APB-type faults are not always stable. In the case of the Zr_3Al phase, which has $L1_2$ structure, SISF energy is found to be very low: 90 mJ/m^2 at 293K (20°C) [30] and 2 mJ/m^2 at 673K (400°C) [31]. The DO_{19}, therefore, can be produced very easily by splitting of superlattice dislocations into SISF partials 1/3<112> [32]. The estimation of APB energy by interaction parameters has indicated a relatively high value of APB energy in Zr_3Al. Douin [33] has experimentally estimated APB and SISF to be 220 and 15 mJ/m^2, respectively. Thus the splitting of superlattice dislocations into APB partials (½<110>) is energetically not favorable and this fact is supported by the absence of APBs in the microstructure of Zr_3Al-based alloys [31]. Consequently, absence of DO_{22} and DO_{23} phases in Zr_3Al can be attributed to large value of APB energies. These phases could, however, be produced from $L1_2$ by the addition of ternary elements. This is possible if such additions reduce the APB energy relative to the SISF energy.

2.1.2 B2 Crystal Structure—Based Intermetallics

The B2 structure is an ordered bcc structure consisting of two simple cubic interpenetrating sublattices, and stoichiometrically it can be represented by 50:50 atomic distributions. The common ordered derivatives of B2 structures are DO_3 and $L2_1$ structures. These structures along with the B2 structure are shown in Fig. 9.4. In AB compounds with the B2 structure, A atoms occupy α sites and B atoms occupy both β and γ sites. In A_3B compounds with the DO_3 structure, B atoms occupy only β sites and the other two sites are occupied by A atoms. In A_2BC compounds with $L2_1$ structure, A atoms occupy α sites, B atoms β sites, and C atoms occupy γ sites, respectively.

In this class of B2- based intermetallic compounds the most common compounds are NiAl, AgMg, CuZn, NiGa, CoGa, and FeAl. In this group NiAl and FeAl are the most promising materials for structural applications. NiAl has been considered for advanced high-pressure turbine blades. NiAl offers many advantages like low density and high oxidation resistance, but it suffers from poor room temperature ductility, which could not be improved by the addition of boron as was observed in the case of Ni_3Al phase.

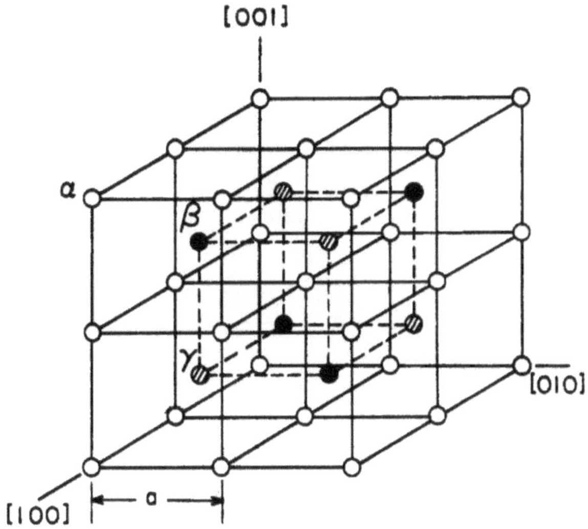

FIGURE 9.4

The atomic arrangement in the bcc-based B2, DO3, and $L2_1$ structures. B2 structure: α site A atom, β and γ sites B atoms; DO_3 structure: α and γ sites A atom, β site B atoms; $L2_1$ structure: α site A atom, β site B atom, and γ site C atom.

Fe_3Al, one of the iron aluminides, has DO_3 structure at low temperature. Upon heating above critical temperature the DO_3 phase of Fe_3Al transforms into the B2 phase. Therefore, above the B2-DO_3 transition temperature antiphase domains of the B2 phase exist, and upon cooling below transition temperature finer domains of the DO_3 appear within a B2 domain. In the case of B2 there is only ½<111> APB displacement vector on {110}, but with DO_3 another fault vector of ½<100> is involved. Therefore, to image both the domains superlattice reflection of (111) is needed whereas to get selective imaging of DO_3 <002> or <222> reflections are used [34,35]. In DO_3 structure, if the same slip system {110}<111> is operative, superlattice dislocations consisting of four partial dislocations are needed to make a perfect lattice translation [36]. Dislocations in DO_3 structure consist of four superpartial ¼<111> dislocations (within the DO_3 structure) connected by two types of APBs. The outer pairs of these four dislocations are connected by ¼<111> APB, which is the same as in B2 APB generated by ½<111> affecting nearest neighbors. The inner pair of dislocations is connected by 1/2<100> APB which is affected by second nearest-neighbor atoms [36]. All the four super partial dislocations are needed to restore the order in the structure. Presence of these four dislocations has also been reported [35]. However, as temperature, composition, or degree of order changes, APBs of the <111> and <100> change, and because of this various other dislocation dissociations are also possible [35].

2.1.3 Intermetallics Having Other Crystal Structures

More complex crystal structures are observed in silicide phases. These are $C11_b$, C40, and C54. Fig. 9.5 shows the crystal structures of these phases. These structures have a similar pattern of stacking as was shown among $L1_2$, DO_{19}, DO_{24}, DO_{22}, and DO_{23} in a previous section. This relationship, known for a long time but with renewed interest in the deformation behavior of the silicide phase, was revisited [37]. There is a difference in the way atomic layers are stacked in silicides and aluminides. In silicides, atoms in the adjacent layer occupy the saddle positions (Fig. 9.6) between two Si atoms, while in aluminides, atoms occupy the pocket formed in the center of three atoms in the neighboring layer as seen in fcc or hcp (Fig. 9.6). This difference in positioning of atoms results in four stacking

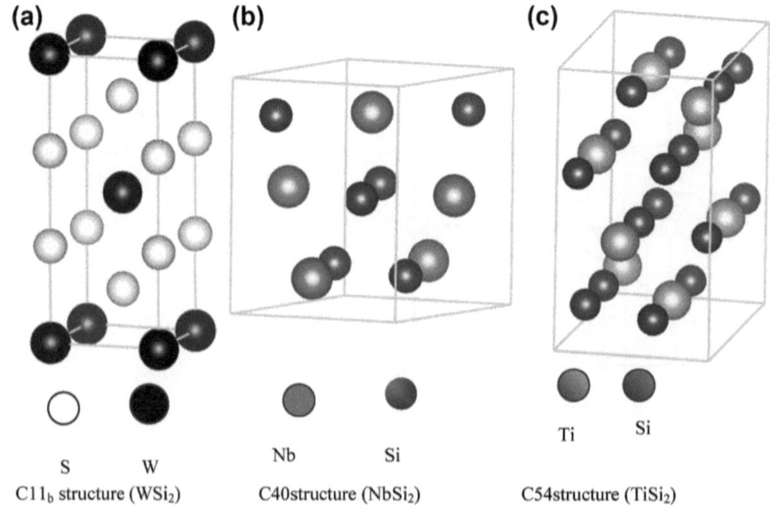

FIGURE 9.5

Showing arrangements of atoms in $C11_b$, C40, and C54 structures. All these structures are observed in silicide phases. These phases bear the same relationship as shown by DO_{22} and DO_{23} with $L1_2$. For details see Fig. 9.6.

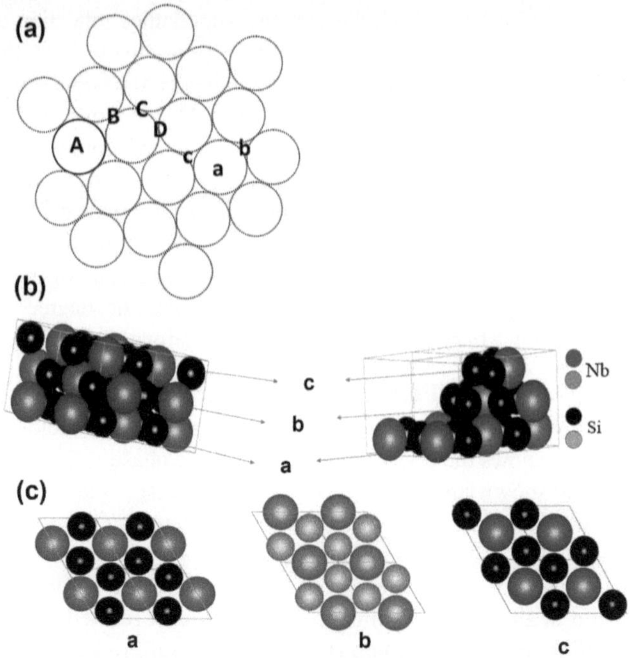

FIGURE 9.6

(a) Shows the densest planes in silicide phases shown in Fig. 9.5. The other layers in silicide phases are marked as A, B, C, and D and for $L1_2$, DO_{19}, DO_{22}, and DO_{23} these atomic plane positions are marked as (a,b,c). (b) Shows atomic arrangements in the four unit cells of C40. These layers are marked as **a**, **b** and **c**. These layers are separately shown in (c).

positions in silicides (marked in the figure as A,B,C,D) but only three in aluminides (a,b,c). With this understanding the $C11_b$ could be reconstructed as AB,AB..., C40 as ABC,ABC..., and C54 as ADBC, ADBC... Among these three structures, in C54 structure the SieSi bonds are interrupted by metallic atoms because of which C54-based silicides show deviation in their behavior as compared to other $C11_b$ or C40 structure-based silicides [38]. Another more subtle relationship exists between C49 ($ZrSi_2$) phase and DO_{22}, where orthorhombic C49 and tetragonal DO_{22} phase are the subgroup of the same point group 4/mmm.

Crystal structures of A15 and $D8_8$ are shown in Fig. 9.7. A15 is the simplest of topologically closed pack structure and therefore, by definition, should not have any octahedral interstitials. $D8_8$ structure, on the other hand, forms connected octahedrons along its *c*-axis, which are generally filled by interstitials. It is a known fact that non-$D8_8$ silicides can be stabilized in $D8_8$ structure by carbon addition.

Laves, which are also topologically closed-packed structures, show three structures: C13, C14, and C36. An excellent review can be found in Ref. [39]. These are shown in Fig. 9.8. Laves phases belong to the class of Frank—Kasper phases showing topologically close-packed structures. They have the general composition AB_2 with the larger A atoms in the center of a 16-atom Frank—Kasper polyhedron and the smaller B atoms in the centers of icosahedra. The coordination number for the A atoms is 16 (4 A and 12 B atoms) and 12 (6 A and 6 B atoms) for the B atoms. The closest packing of hard spheres of types A and B is obtained for the radius ratio $r_A/r_B = \sqrt{(3/2)} \sim 1.225$. The Laves phase can be built up by a particular stacking of atomic layers that are rotated by 180 degrees with axis perpendicular to the layer interface and shifted by $\pm 1/3[-110]$ (Fig. 9.8). This structure could be constructed by stacking four atomic planes parallel to the (001) planes of the hexagonal lattice. The smaller B atoms on these planes form a network commonly known as Kagome net [40]. When this is located at $z = 0$, larger A atoms can be arranged into two planar triangular nets: one below at $z = -3/8$ and another above this B atom plane at $z = 3/8$ positions. The second plane of B atoms located at $z = 1/2$ is the only one where the atom positions change after a rotation by 180 degrees. The thickness of a layer is $(2/3)^{1/2}a_h$ where a_h is the hexagonal unit cell constant in the layer plane. Theoretically, an infinite number of different stacking sequences are possible. However, due to energetic reasons only the shortest possible sequences will be found in real systems. If the starting layer is termed A, the two possible shifted layers are designated as B, in the Ramsdell notation, the given number corresponds to the number of layers of the sequence and the capital letter indicates the crystal system (*H*, hexagonal; *C*, cubic; *R*, trigonal). The letter *h* in the Jagodzinski—Wyckoff notation is assigned to a layer where the directly over- and

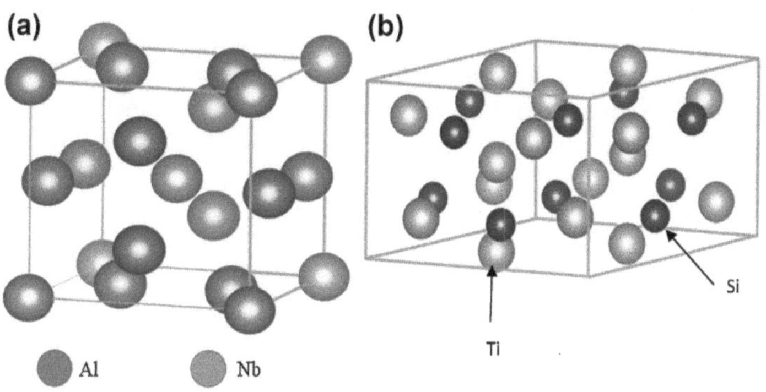

FIGURE 9.7

Atomic arrangements in (a) A15 (Nb_3Al) and (b) $D8_8$ (Ti_5Si_3) unit cells.

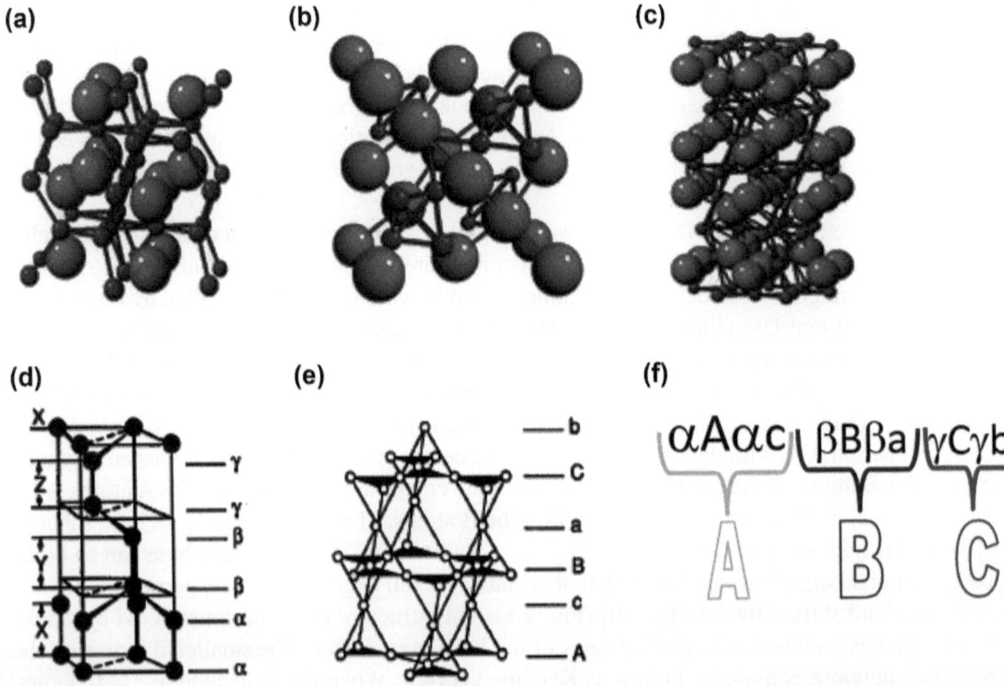

FIGURE 9.8

Atomic arrangements in (a) C14, (b) C15, and (c) C36 structures. As may be noticed from the figure, larger atoms are surrounded by smaller atoms. Upon separating large atoms and assigning the layers name as α, β, γ as shown in (d) and small atoms as A, B, C for base of the triangle and apex of the triangle as a, b, c. The four layers could be bunched as an equivalent to single closest pack planes of fcc and hcp, as shown in (f) the C14 can be constructed as ABAB and C15 as ABCABC... For details, kindly see the text.

underlying layers are displaced in the same direction and the letter C is used if the respective layers are shifted in opposite directions. The hexagonality is defined as the fraction of hexagonal stacking, i.e., as the relative amount of h-type layers in the Jagodzinski–Wyckoff notation. The three generally observed Laves phase types—C14, C15, and C36—correspond to the shortest stacking sequences. However, many of the more complicated stacking sequences have been observed in real systems. For example, Komura et al. [41–43] found long-period poly types with 6-, 8-, 9-, 10-, and even 21-layer structures in Mg-based ternary Laves phase systems, which are ordered along the quasi-binary section according to their hexagonality. A trigonal 12R stacking variant in the Laves phase (Ti0.95 V0.05) Co_2 has also been reported [44].

2.2 NICKEL ALUMINIDES

2.2.1 Ni_3Al

Fig. 9.9 shows the binary phase diagram Ni-Al. As may be noticed from this phase diagram Ni_3Al forms at Al concentrations between 25 and 27 at%. The crystal structure of Ni_3Al is an ordered $L1_2$

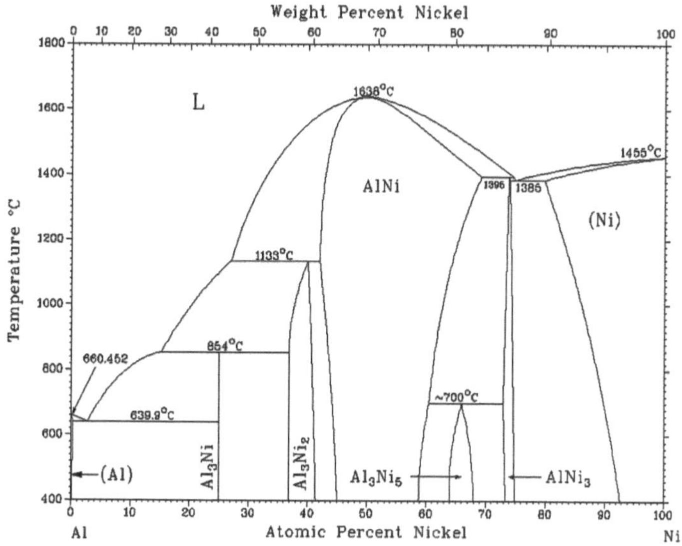

FIGURE 9.9

Binary-phase diagram of Ni-Al system. Phase fields of Ni_3Al and NiAl could be noticed. NiAl has the highest liquidus temperature 1683°C and Ni_3Al has second highest liquidus temperature 1668K (1395°C).

structure. Ni_3Al remains ordered up to very close to its melting point [45]. Cahnet et al. [46] have shown that the ordering temperature of Ni_3Al is 1723K (1450°C). Antiphase domain boundaries very often have been seen in this alloy and are more or less uniform in size except for microstructures generated by rapid solidification. Studies have shown that microsegregation occurs in rapidly solidified alloys. The value of the APB energy evaluated by transmission electron microscopy (TEM) examination [47] is a strong function of solute content and has been found to lie in the range of 85–140 mJ/m^2. Other crystallographic details of the $L1_2$ structures have already been provided in the previous section. As mentioned earlier, after its successful ductilization by microalloying with boron, Ni_3Al has been considered for structural applications also. Additionally, it has often been used as a model material for understanding intermetallic compounds. Commercialization of Ni_3Al for selected applications is underway [20,48,49].

One of the first issues when the properties of the Ni_3Al alloys were investigated at elevated temperatures was the anomalous behavior of the yield strength. This anomalous behavior is identified as increase in yield strength with increasing temperature. Later creep properties also showed inverse creep behavior (also known as sigmodal creep). The inverse creep behavior is identified as increase in the creep rate after primary creep, and the onset of inverse creep usually takes place at very small strains. This strain rate increase may extend over a large strain interval, leading directly to tertiary creep in the absence of a steady-state stage [50–52]. There are many models that could explain this behavior. Prior to delving into these models it is worth noting that in Ni_3Al, similar to disorder fcc, {111} are easy gliding planes but {001} plane could also be thermally activated. This transition of cross-slip of dislocations from octahedral slip to {001} plane could be used to explain these anomalous behaviors.

Experimental evidence has shown that screw-superdislocation cross-slip from the octahedral planes to the cubic planes, where the dislocation core does not completely reside on any one plane leading to the formation of immobilized locks. Based on this observation, macro (or super) kink models have been developed that could reasonably replicate the anomalous yield behavior. In these models mobilization of macrokinks was based on the distribution of heights of kinks. Hemker and his group extended this idea and proposed a model to explain the sigmoidal creep of Ni_3Al. They suggested that octahedral slip during primary creep is exhausted by the formation of these immobilized locks of dislocations due to thermally activated cube cross-slip of the screw segments. Thus, the strain rate is progressively reduced until the cross-slipped segments become thermally activated and are able to bow out and glide on the cube cross-slip plane. Such immobilized locks act as Frank–Read-type dislocation sources for glide on the {001} cube planes. The dislocation generation and subsequent glide on the cube planes leads to an increase in mobile dislocation density and thus to a higher strain rate and a sigmoidal creep process. However, it has also been noticed that octahedral to cubic slip transition was only a necessary but not a sufficient condition as such crossovers were observed even in those temperature ranges where sigmodal creep was not observed. Many such observations are rationalized by invoking a concept of critical length, and if the total length of dislocations cross-slipped on {001} exceeds that length sigmodal creep is observed.

In Ni_3Al alloys, creep attains steady state at a very early stage and extends over a considerable strain range (up to 20%) at high as well as at intermediate temperatures [53,54]. In some strain rate and temperature ranges steady-state creep may be delayed or even absent at intermediate temperatures if sigmoidal creep occurs. The stress exponent, n, in the steady-state condition ranges mostly between 3.2 and 4.4 in both single crystal and polycrystalline alloys. The activation energy for creep ranges between 320 and 380 kJ/mol. This value is higher than the activation energy for diffusion of Ni in Ni_3Al, which varies from 273 to 301 kJ/mol [55]. The diffusivity of Al in NiAl is believed to be higher but it has still not been measured directly. The rate-controlling mechanism of creep of Ni_3Al-based intermetallics is still unclear, and dominance of one single mechanism is doubtful. TEM studies show formation of subgrain boundaries is not very frequent and multiple slip systems remain active during the process. These observations suggested that the interaction between dislocations is an important factor in controlling the creep properties of the alloys. Due to a limited number of studies, the role of grain boundaries is not precisely determined but it has been shown that for small grain size coble creep is the dominating factor. High-temperature performance of off-stoichiometric compounds has shown that creep resistance increases with increasing Ni content. Attempts have also been made to improve the creep strength of Ni_3Al by alloying additions but most of them were not very effective, and many times improvement in creep resistance was at the cost of other properties. Therefore efforts were diverted on the development of multiphase Ni_3Al alloys. Out of them oxide dispersion-strengthened (ODS) Ni_3Al alloys appear to show some promising results. At high temperatures (1273–1523K (1000–1250°C)) high stress exponent, n, 7.2 to 7.8, and high value of activation energy (650–697 kJ/mol) make it an attractive choice. However, there are issues that need to be addressed before giving it active considerations.

One of the possible applications of Ni_3Al is as turbocharger rotors in diesel trucks where it is likely to replace superalloy due to its improved fatigue life and lower cost. Ni_3Al has another application as die material for isothermal forging and mold material due to its excellent strength during deformation at high strain rates. Furnace fixtures, rollers and slabs for heat treatment of auto parts are being replaced by Ni_3Al-based alloys due to its superior strength against carburizing process. Ni_3Al-based alloys have also found potential applications as turbine blades and valves for jet engines [56].

2.2.2 NiAl

NiAl is the ordered intermetallic having simplest order bcc (B2) crystal structure phase with Al atom occupying (x, y, z) position and Ni atom occupying $(x + 1/2, y + 1/2, z + 1/2)$ position (Fig. 9.4). This is a very stable structure that remains ordered until nearly the melting temperature. As illustrated in the phase diagram (Fig. 9.9), NiAl forms at Al concentrations ranging from 40 to about 55 at%. High degree of thermodynamic stability and the existence of wide phase field make its fabrication very easy. Because of this quality, NiAl is developed from very fine powder to single crystal. NiAl alloys are attractive for many applications due to their resistance against oxidation, carburization, and nitriding, as well as their high thermal and electrical conductivity. They are currently used to make metallizations for electronic contacts in advanced semiconductor heterostructures, surface catalysts, and high-current vacuum circuit breakers [57]. Additionally, these alloys are attractive for aerospace structural applications due to their low density (5.98 g/cm^3) and high melting temperature [48]. However, a few major limitations of single-phase NiAl alloys preclude their application as structural materials, namely, poor creep strength at high temperatures, high brittle-ductile transition temperature (~673K (400°C)), and softening at elevated temperatures [58].

Grain boundary brittleness of NiAl appears to be inherent in nature as even in the absence of any harmful impurities any improvement in the fracture strengthening has not been observed [59]. By carrying out detailed atom probe field ion microscopy and high-resolution electron microscopy it was shown that in the vicinity of a random grain boundary, 50% Al site vacancies in NiAl are filled by Ni but Ni site vacancies remained unoccupied, leading to a Ni-rich region [60]. This arrangement may vary with the nature of grain boundary. Comparing the nature of grain boundaries and the propensity for fracture it was shown that random grain boundaries undergo intergranular fracture whereas boundaries like $\Sigma 1$, $\Sigma 3$, or $\Sigma 7$ do not fracture [61], which is commensurate with the findings of others. Some of the simulation studies have shown that boundaries with a surplus of aluminum have appreciably lower cohesive strength than the stoichiometric boundaries or boundaries with a surplus of nickel. The reason for such variation was attributed to the compact structures of boundaries with the stoichiometric configuration or boundaries with a surplus of nickel [62]. Even when the single crystals of NiAl were subjected to deformation, they failed to show very low ductility. The primary reason for the low ductility is that the deformation occurs through {110}<001> slip system, which has only three independent systems and not the five independent systems as needed by von Mises criterion. Brittleness in NiAl has also been traced to dislocation initiation problem as opposed to a dislocation mobility problem [63]. The addition of B in NiAl though changes the intergranular fracture to transgranular fracture, but its ability to provide solid solution strengthening in the alloy more or less offset the beneficial effects of grain boundary strengthening [59,64].

Regarding the high-temperature creep properties are concerned, several strengthening mechanisms have been utilized to improve the creep strength of NiAl alloys. Solid solution of Fe, Nb, Ta, Ti, and Zr produced only limited strengthening [65]. Therefore, solute strengthening is combined with other strengthening mechanisms. One most effective creep resistance was ODS NiAl-Y_2O_3 alloy, which showed high stress exponents ($n = 17$) and activation energies much higher than that of lattice self-diffusion ($Q = 576$ kJ/mol). NiAl matrix reinforcement by larger particles such as TiB_2 and Al_2O_3 or whiskers also shows significant increases in the strength of alloys [66]. Xu and Arsenault [67] have measured stress exponents of NiAl matrix with Al_2O_3 composites ranging from 7.6 to 8.4 and activation energies similar to that of lattice diffusion of Ni in NiAl. They concluded that the rate-controlling deformation mechanism is dislocation climb in NiAl matrix composites during high-temperature deformation.

Although NiAl processing is generally simple and straightforward, reproducibility of mechanical properties of the alloy is an issue. As NiAl is very sensitive to metallurgical variables, most common mechanical properties like yield strength are difficult to reproduce. NiAl alloys have very good environmental durability. Creep strength could be matched with Ni-based superalloys. Inherent ductility and fatigue properties are better than that originally believed. Impact behavior and poor room temperature ductility are a few issues that need attention before realizing the practical application of NiAl alloys [68]. Several industrial applications have already been identified for NiAl-based systems [69].

2.3 TITANIUM ALUMINIDES

2.3.1 Ti_3Al

Most of the alloys developed in this Ti-Al composition are two-phase alloys that contain nearly 23–25 at% Al and 11–18 at% Nb. Addition of Nb appears to distort the ordered hexagonal Ti_3Al (DO_{19} structure) phase and introduces an orthorhombic phase that is known as the "O-phase" [9,70]. This ordered intermetallic phase appears to improve the fracture toughness and creep resistance [71]. Pseudo binary-phase diagram of Ti_3Al-Nb shows that the O-phase has extended phase field with Nb content [72]. It can also dissolve considerable amount of β-stabilizers like V and Mo. The alloys that have been studied generally contain a large amount of Nb and small amount of other β-stabilizers. In this alloy series, alloys like Ti-25Al-10Nb-3V-1Mo and Ti-23Al-17Nb-1Mo have been given active consideration.

Increasing Nb concentration in Ti_3Al upon oil quenching allows retaining high-temperature B2 phase up to room temperature. The retained B2 phase is metastable and contains ω phase, tweed microstructure, and O-phase [73]. Upon aging, ω phase transforms into α_2 phase. In order to have high toughness at room temperature and strength at elevated temperatures, general understanding emerges that [74]:

1. O-phase should be the major constituent.
2. Al content should be greater than 25 at%.
3. 1 at% Mo is necessary for high-temperature properties.
4. Optimized layered structure is needed for high strength, high ductility, and low creep.

Near-stoichiometric orthorhombic alloys based on Ti_2AlNb have exhibited potential for elevated temperature structural applications. These alloys possess good creep resistance as well as reasonable room temperature elongation to failure, which have dependence on the final microstructure [75,76]. Compression creep characteristics of alloys having single O-phase microstructures with compositions as Ti-(24-27)Al-(15-27)Nb at% have been examined and activation energies of creep were estimated between 294 and 376 kJ/mol. This value of activation energy is almost twice that for more conventional Ti_3Al-based alloys such as Ti-24Al-11Nb [77,78]. However, in spite of having active <a> and <c + a> slip [77,79], the O-phase suffers from poor ductility and low-fracture toughness (stress intensity factor (K_{IC}) ~ 6 $MPam^{0.5}$) [80,81]. In this respect the bcc phase plays a critical role in imparting roughness and ductility, as evident by the wavy slip, ductile dimpling, and crack blunting characteristics [77]. In addition, surface slip trace and transmission electron microscopic observations reveal that slip between the O- and bcc phases reduces the tendency for grain boundary cracking, which is prevalent between adjacent O-phase grains [82,83]. Thus, a distribution of bcc phase in the microstructure is essential for practical engineering applications. Fully lath O + bcc microstructures have exhibited good fracture toughness and creep resistance in super bcc-transus-processed and/or heat-treated conditions [71,72]. However, intermediate-to-rapid cooling rates from above the bcc-transus lead to precipitate formation along the prior-bcc grain boundaries, which become preferential sites for cracking [2,7,8]. One possible solution is to introduce thermomechanical processing at subtransus temperatures [84] and tailor the microstructures for obtaining a balance of properties. In particular, an inverse relationship existed between the equiaxed grain size and the secondary creep rates. In addition, the creep resistance of a fully lath O + bcc microstructure was superior to a fully equiaxed microstructure containing an identical bcc-phase volume fraction [85]. Combining these observations, it was felt that fully lath O + bcc microstructures devoid of prior-bcc grain boundaries would provide an attractive combination of tensile and creep behavior. Many studies focused on microstructure—property relationship of O + B2 Ti_2AlNb-based alloy have recently been carried out. In general, these studies can mainly be divided into two categories. In the first category, variations in the mechanical property by adjusting different alloying elements have been investigated [70,72,86,87]. These studies have shown that higher Al/Nb ratio improves oxidation resistance at the expense of the plasticity, and the decrease of Nb concentration reduces the creep resistance of alloys. Since the adjustment of these alloying elements to optimize the mechanical property of the alloy seems to have limitations, addition of other alloy elements (W, Mo, V, Ta) have been attempted. Thus, the other category of the studies is to investigate the effect of the addition of other alloy elements on mechanical properties [86—89]. These studies revealed that addition of Ta to partly substitute Nb can improve the strength and ductility of the orthorhombic alloy [90], whereas Fe and Ni additions strongly degraded creep properties of the alloy [84]. The tensile elongation increases with addition of yttrium as its addition refines the microstructure [91,92]. The addition of boron is beneficial as it reduces prior β-grain size and thereby increases room temperature ductility and creep properties [93]. The addition of W in substitution of Nb was effective in improving the strength, creep resistance, while V was effective in room temperature ductility [94]. In addition, mechanical properties can also be improved via tailoring the microstructure. However, limited work has been done in this direction. Of late it has been realized that the grain size, morphology, and volume fraction are all very important parameters for obtaining the good mechanical properties of the alloy and the thickness of lamellar O plays an important role on these properties [95]. Recently interest in the O-phase has renewed where tensile properties of the alloys have been improved by introducing bimodal distribution of the lamellae of O-phase in the B2 + O phases [96]. This work takes a clue from the fact that coarse lamellar O-phase seems to make a significant contribution for the creep resistance, but for room temperature ductility finer lamellar of O-phase are needed [87]. Thus, by introducing both types of O-phases the superior mechanical properties of the alloy can be induced.

2.3.2 γ-TiAl

Titanium aluminide alloys have been developed with the future possibility of a worthy replacement of heavier materials in high-temperature structural applications such as automotive and aerospace engine components. Titanium aluminide alloys offer many advantages, for example, low density (~4.5 g/cc as compared to 8.3 g/cc of superalloys), high melting temperature, good elevated temperature strength and high modulus, high oxidation resistance, and favorable creep properties [97,98]. In addition, these alloys can also be processed through conventional manufacturing methods such as casting, forging, and machining. In fact, turbocharger turbine wheels made out of γ-TiAl have been used in automobiles [98].

Binary Ti-Al phase diagram (Fig. 9.10) shows presence of ordered intermetallics, like γ-TiAl (ordered face-centered tetragonal, $L1_0$), α_2-Ti_3Al (ordered hexagonal, DO_{19}), along with disordered

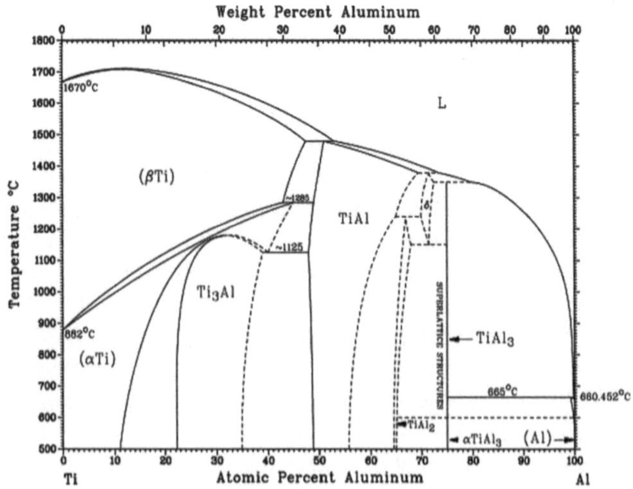

FIGURE 9.10

Binary-phase diagram of Ti-Al.

phases like α-Ti (hcp) and β-Ti (bcc). The three important intermetallic alloy groups that have possible high-temperature applications are γ−TiAl alloys having compositions within 49−66 at% Al, α_2 alloys that contain from 22 at% to approximately 35 at% Al, and two-phase (γ-TiAl + α_2-Ti_3Al) alloys containing Al concentration between 35 and 49 at%. The deformation mechanisms of the two-phase lamellar (γ-TiAl + α_2-Ti_3Al) structure have been studied in detail by Singh et al. [99]. The possibility of transfer of dislocations from one lamella to its neighbor and the dislocation reactions at the interface between them are examined (Fig. 9.11). Overall, two-phase γ-TiAl alloys have more potential for high-temperature applications than α_2 alloys due to their higher oxidation resistance and elastic modulus [100]. However, creep strength of the two-phase γ-TiAl alloys is comparatively lower in comparison to the creep strength of α_2 alloys. Initial work on two-phase γ-TiAl suggested that to get the best creep properties an optimum microstructure consisting of fully lamellar of γ-TiAl + α_2-Ti_3Al is necessary. Modeling of the creep strength in γ-TiAl has become a daunting task as features like grain morphology, lamellar spacing, and precipitate distribution were difficult to incorporate in the conventional creep models and dependence of creep on these parameters was having time and temperature dependence. Most creep studies were performed in the temperature range 949K (676°C)−1150K (877°C) [101] and the stress range 80−500 MPa. These conditions are relevant to the anticipated service conditions [102]. Identifying the rate-controlling creep mechanisms in fully lamellar TiAl alloys is difficult for the following reasons:

1. Unambiguous secondary creep stage was not observed and therefore minimum creep rate could not be obtained.
2. The stress exponent steadily increases with increase in stress [103].
3. Microstructural parameters like lamellar grain size, interface spacing, and irradiated material TEM orientation, precipitate volume fraction, and grain boundary morphology play significant role in creep properties and are difficult to control.

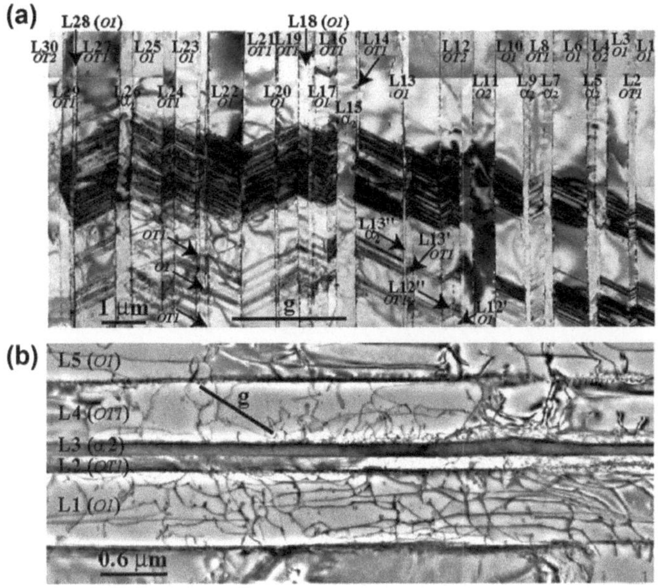

FIGURE 9.11

The deformation microstructure in g-TiAl compressed at room temperature. (a) Image of an area deformed mainly by twinning. Twins have crossed the whole area massively from left to right. This could be inferred with the reducing thickness of twins. a_2 lamellae, marked as L5, L7, L9, L15, and L26 in the figure, were free from dislocations. (b) Shows an area within g-lamellae showing deformation by ordinary dislocations. For further details, see Ref. [99].

Nevertheless, a variety of rate-controlling mechanisms of fully lamellar TiAl alloys have been proposed over the last few years. It has been suggested that the gradual increase in the stress exponent with increasing stress might be due to changes in the creep mechanisms from diffusional creep at low stresses, to dislocation climb as the stress increases, and finally to power-law breakdown at very high stresses [104]. It has been claimed that dislocation climb would most likely be the rate-controlling step during creep at stresses higher than about 200 MPa and temperatures higher than about 973K (700°C). The creep of fully lamellar TiAl alloys appears to be controlled by lattice diffusion of Ti, as estimated activation energy of creep 313 kJ/mol, which is very close to the activation energy for lattice diffusion of Ti in γ-TiAl, which is 291 kJ/mol [105]. Recovery-type dislocation climb mechanism based on microstructural observations of the formation of dislocation arrangements during creep has also been proposed. Several possible explanations have been put forward to reconcile the proposition of dislocation climb and the observation of high stress exponents. For example, it has been suggested [100] that back stresses within lamellar microstructures lead to bowing of dislocations between interfaces, and the estimated shear stress required for bowing a dislocation has been shown as inversely proportional to the lamellar interface spacing. In another proposition, dynamic recrystallization during deformation may contribute to a rise in the strain rate, thus rendering stress exponents with less physical meaning in terms of a single, rate-controlling restoration mechanism [106]. Subgrain size effect where lamellar spacing actually becomes the effective subgrain size has also been proposed. These circumstances are similar to constant structure creep, which is associated with a relatively high stress exponent of 8 or higher.

Based on the dislocation activity and twinning, thinning, and dissolution of α_2-lamellae and coarsening of γ-lamellae in Ti-Al alloys at high stresses, Wang and Nieh [107] proposed a creep model involving movement of ledges at lamellar interfaces to rationalize their observations. This model could be explained by (Fig. 9.12) the growth of the γ-phase at the expense of the α_2-phase occurring by ledge movement. Ledge motion was suggested to involve glide of misfit dislocations and climb of

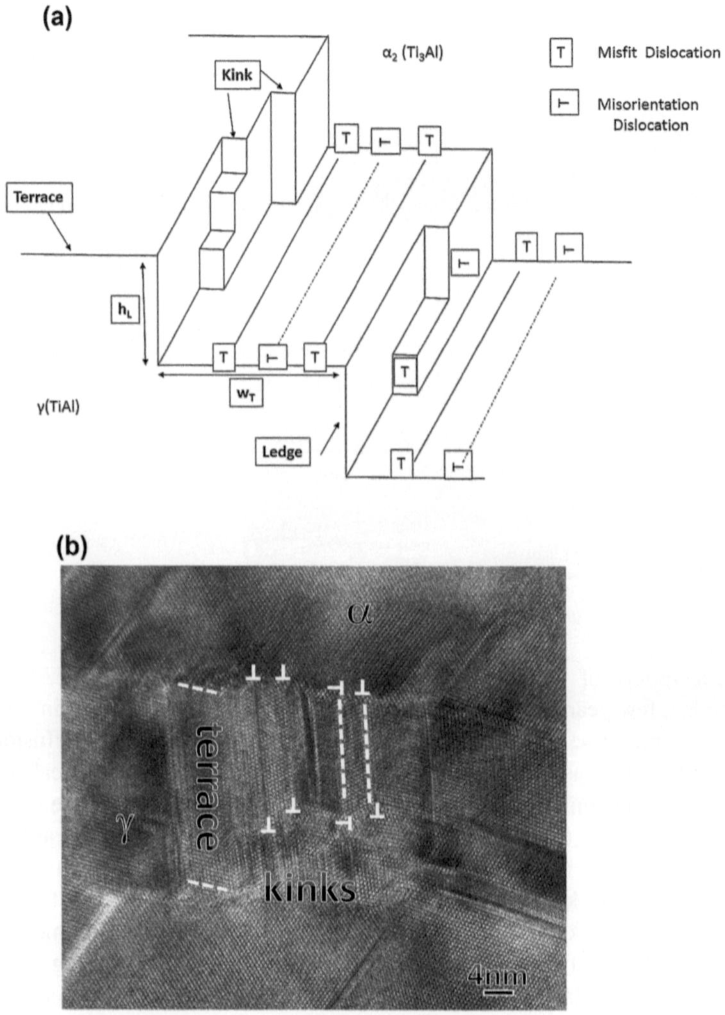

FIGURE 9.12

(a) Schematic presentation of the growth of the g-TiAl in the Ti3Al phase. Diagram is reconstructed based on the mechanism proposed by Wang et al. [107]. (b) High resolution image showing kinks and other defects in γ-TiAl during growth.

misorientation dislocations. Ledge motion leading to the transformation from α_2 to γ may account for a significant amount of the creep deformation, as it involves volume changes during α_2 to γ transformation. At high stresses, multiple ledges are suggested to be able to form and dissolve, and thus deformation may occur. Diffusion of atoms is needed to support the climb of misorientation dislocations to move a ledge. Also, diffusion is needed for the composition change associated with the phase transformation from α_2 to γ. Thus, lattice self-diffusion becomes rate controlling, which is in agreement with previous observations of activation energies.

Alloy additions have been tried to improve creep properties of the alloys. It is a well-known fact that additions of W greatly improve creep resistance [101], which is due to solute hardening within the glide-controlled creep regime and also due to lowering of diffusion rates, thus reducing the dislocation climb rate in the climb-controlled creep regime. Additions of W, Si, C, N, Nb, and Ta improve creep resistance [98,108,109]. Boron surprisingly does not have any influence on minimum creep rates [101].

In summary, the creep behavior of fully lamellar TiAl alloys is influenced by many different microstructural features, and it is difficult to formulate one single model that incorporates all of the relevant variables. An optimum creep behavior may require the following [104]:

1. Smaller lamellar or colony grain size helps to improve creep life by preventing early fracture.
2. A narrow interlamellar spacing reduces the minimum strain rate.
3. Presence of alloying additions such as W, Nb, Mo, V (solution hardening), and C, B, N (precipitation hardening) help in improving creep properties of the alloys.

2.4 IRON ALUMINIDES

Iron aluminides, Fe_3Al and FeAl, have been studied mainly because of their low cost, low density, good wear resistance, ease of fabrication, and resistance to oxidation and corrosion [110,111]. These advantages have led to their potential uses, including heating elements, furnace fixtures, heat-exchanger piping, sintered porous gas-metal filters, automobile and other industrial valve components, catalytic converter substrates, and components for molten salt applications [112]. In addition, Fe_3Al is one of the few structural intermetallics that can be disordered by appropriate treatment. Because of this fact of the disordering temperature being 823K (550°C), it becomes the limiting temperature for structural applications, though it can be raised substantially by suitable alloying [113]. In addition, a few unresolved issues like low ductility and impact resistance at low temperatures and inadequate creep properties need to be addressed before using this alloy for any structural applications.

The excellent corrosion resistance of Fe-Al alloys was first recognized in the 1930s [114], but detailed studies of mechanical behavior commenced with the work of Cahn and his coworkers in the 1960s [115,116]. This work included reports of the creep resistance of Fe_3Al above and below the critical ordering temperature, T_c [115], as well as a description of the effect of long-range order on yielding [117] and mechanical twinning [116]. A discontinuity in the slope of plots of creep rate versus inverse temperature near T_c was attributed to a change in the activation energy for diffusion at this temperature [115]. Later work by Stoloff and Davies dealt with the influence of long-range order on yielding [118] and strain hardening [119] of Fe_3Al. A study of compositional effects on yielding of Fe-Al alloys was reported by Sainfort et al. [120]. Other noteworthy early work by Stoloff and Davies [121] on fracture behavior of Fe_3Al was also done. These studies showed that as aluminum is added to iron, ductility of the alloy drops sharply.

An anomalous variation of the yield stress with temperature has been observed in Fe-Al alloys with an Al concentration range 25–45 at%. A peak in the yield stress appears usually in the intermediate temperature range (673K (400°C)–873K (600°C)) [122–126]. Several mechanisms have been proposed to rationalize the yield-strength peak but the origin of this phenomenon is still not well understood. More comprehensive reviews on this topic as well as critical analyses of the validity of the different strengthening mechanisms could be followed in Refs. [123,127]. Among all, disordering and vacancy-based strengthening appear partially to explain the anomalous yield-strength phenomenon. Loss of order within APBs of mobile <111> superdislocations with increasing temperature leads to structural changes as well as variations in the chemical composition. Thus, the trailing partial of the superdislocation would no longer be able to restore perfect order. A frictional force would therefore be created that will hinder superdislocation movement. With increasing temperature, APB relaxation would be more favored and thus increasing superdislocation pinning would take place, leading to the observed stress peak. Another model that includes roles of constitutional vacancies was proposed [128]. According to this model, with rising temperatures a larger number of vacancies is created. These defects pin superdislocation movement and lead to an increase in yield strength. At temperatures higher than those corresponding to the stress peak, the concentration of thermal vacancies is very large and vacancies are highly mobile. Thus, they may aid dislocation climb processes instead of acting as pinning obstacles for dislocations [129] and softening occurs. This model got some experimental backing [130]. For example, dislocation images show small curvatures, which was taken as evidence for vacancy pinning. Second, it was observed that the stress peak is retained when the samples are quenched from high temperature and tested at room temperature. This is taken as evidence that excess vacancies play a role in the strengthening of the alloy. As strengthening is caused by vacancy concentration, the peak in yield strength becomes independent of strain rates. Experimental results appear to support this argument [131]. However, the vacancy model when extended to Fe_3Al alloys drew some skepticism as vacancy concentration in Fe_3Al is a few orders lower than FeAl. Diffusivity of vacancies is also higher in Fe_3Al, which means they can assist dislocation climb rather than providing strengthening to the alloy. In addition, another shortcoming of the vacancy model was noticed as it could not explain orientation dependence of stress anomaly. Therefore, yield stress anomaly in iron aluminide is still an unresolved issue.

Iron aluminides show superplasticity [132]. Fe-28Al-3Cr alloy showed superplastic behavior in the temperature range of 1123K (850°C)–1273K (1000°C) with a maximum elongation of 210%. The microstructural investigation of deformed samples showed continuous recovery and dynamic recrystallization as the responsible mechanism. Elongations as high as 620% were achieved in a Fe-28 at% Al-2 at% Ti alloy deformed at 1123K (850 C) and the process of continuous recrystallization was attributed to such large deformation. The unusually large starting grain sizes as well as the values of the stress exponents (close to 3) would be consistent with a viscous–drag deformation mechanism. However, significant grain refinement has been observed during deformation [132]. The correlation of grain refinement and high ductility may, in turn, be indicative of the occurrence of grain-boundary sliding.

The low creep strength of iron aluminides was one of the concerns. The decrease in creep strength between 773K (500°C) and 973K (700°C) has been attributed to the rapid increase in diffusivity. Therefore two approaches have been adopted: (1) to alloy with those elements that reduce diffusivity and (2) to alloy with those that retard dislocation motion. Alloying elements such as Zr, Hf, Nb, Ta, and B have been effective in improving creep resistance of FeAl by precipitation hardening. Finally, the most promising strengthening mechanism seems to be the introduction of dispersions of second phases, such as carbides, intermetallic particles, or oxide dispersions [132]. However, it appears that one strengthening mechanism does not provide necessary creep strength, and therefore as suggested by Sundar and Deevi [134], a combination of strengthening mechanisms is perhaps the best way to improve creep resistance of iron aluminides.

2.5 Zr_3Al-BASED ALLOYS

Zr_3Al is, in fact, the first intermetallic that was proposed for any structural application. High strength, low neutron capture cross-section, and good corrosion resistance makes it a suitable candidate for structural application in nuclear reactors [135,136]. Zr_3Al-based materials can be formed into any shapes like strip, rod, and tubing by hot working at temperatures above 1273K (1000°C), which upon subsequent aging at lower temperatures produces a fully ordered $L1_2$ structure exhibiting good combination of strength and ductility [136]. But the Zr_3Al phase, in its binary state, suffers from shortcomings as it shows inadequate control over the distribution of undesirable phases [137], high notch sensitivity [138], and irradiation-induced amorphization [139]. These were the reasons why in spite of having such attractive properties Zr_3Al-based materials were not accepted for structural applications in nuclear reactors [139]. To overcome these shortcomings, addition of niobium in the Zr_3Al phase has been attempted. Addition of Nb allows several advantages over the binary system. As it is a β-stabilizer, it may introduce the β-phase, which could result in a two-phase microstructure allowing greater flexibility for tailoring of microstructure [140]. In case the β-phase is metastable, it would lead to the formation of various other phases generating many alternative routes for the formation of the equilibrium phase. Tailoring the microstructures is, therefore, possible by manipulations of thermomechanical processing for generating a wide range of properties for various applications. Effects of these phases on the mechanical property of the material need to be assessed.

Based on the different phases encountered in the Zr_3Al-Nb alloys, a pseudo phase diagram has been developed (Fig. 9.13). In this phase diagram, size of the $\beta + Zr_2Al$ phase field is of technical importance. This is because, depending on the composition and temperature at which the phase field is obtained, thermomechanical treatments can be designed. As the particle size and the distribution of the Zr_2Al phase control the rate of formation of the equilibrium Zr_3Al phase and relative volume fractions of untransformed phases, final properties and microstructures of the alloys can be controlled by suitable thermomechanical treatments. The morphology of the Zr_2Al particles can be altered from coarse, chunky particles to fine uniformly distributed particles by selecting the appropriate annealing treatments. The optimal temperature region where these alloys could be deformed successfully is between 1050 and 1100K (777–827°C) [141,142]. The material in small quantity has been successfully rolled in this temperature range. The addition of niobium appears to help in increasing the deformability of the Zr_3Al-Nb alloys in two ways: (1) by increasing the stability of the β-phase and (2) by facilitating the dissolution of the intermetallic phases. The microstructural studies of the hot-rolled samples (Fig. 9.14) revealed that the matrix β-phase has undergone dynamic recrystallization and the intermetallic $Zr_2(Al,Nb)$ phase undergoes dissolution and precipitation during rolling (Fig. 9.14).

Variation of the hardness in the Zr_3Al alloys did not show a monotonous fall as observed in many other alloy systems but showed maxima in the intermediate temperature range. A similar phenomenon has been observed in the Zr_3Al-based phases when the yield stress of the material has been plotted as a function of temperature [143].

The microstructural study of the room temperature–deformed sample has shown that the mode of deformation is by glide of dislocation partials with Burgers vectors as a/3 <112> bounding SISFs on {111} planes (Fig. 9.15). The absence of dislocations responsible for the creation of antiphase boundaries clearly indicates that the energy for the formation of APB is much higher than the energy for the formation of SISF. The a/2 <110> type dislocation partials exhibit the cubic slip mode. Between the temperature range of 600–800K (327–527°C) both the modes of deformation occur simultaneously, and above 800K (527°C) dislocations responsible for APB dominate the deformation mechanism. This fact indicates that APB energy is a stronger function of temperature as compared to the SISF energy. Therefore, it is possible that unlike at room temperature, during high-temperature deformation dislocation responsible for SISF further splits into APB dislocation, as has been

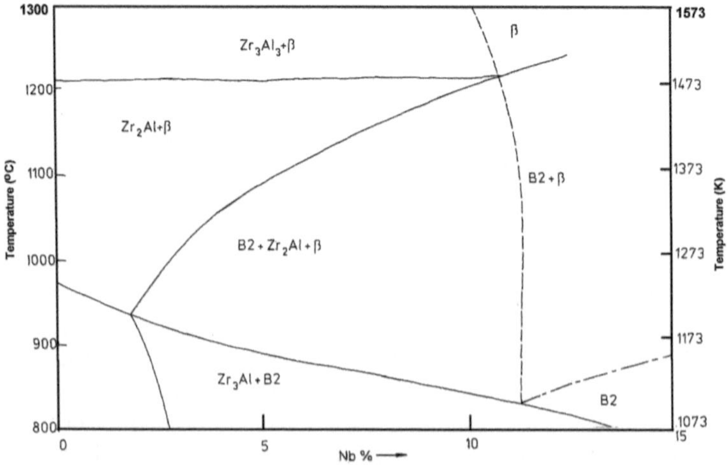

FIGURE 9.13

Schematic pseudo-binary phase diagram of the Zr_3Al-Nb system. Stability of bcc-based structures with the addition of Nb could be noticed.

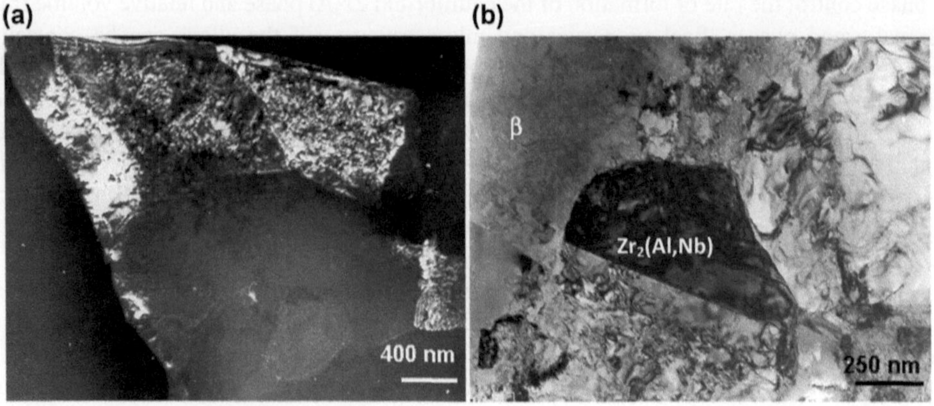

FIGURE 9.14

Showing the deformed microstructures (a) shows dynamic recrystallization of bcc grains and (b) dissolution of intermetallic precipitates during hot working.

observed in the case of other $L1_2$ alloys. This further splitting of dislocation reduces the rate of deformation and can be attributed to the maxima in the observed hardness values, which lies in the intermediate temperature range. This fact needs to be confirmed by further experiments. At temperatures close to melting point, grain sliding dominates other modes of deformation.

For temperatures higher than 800K (527°C), value of stress exponent approaches close to 5, which suggests the onset of the diffusion-assisted phenomena like dislocation climb mechanism. The value of the activation energy of creep in this temperature region falls in the range of 250–350 kJ/mol. On comparison with other Zr-based alloys like zircaloys and the Zr-2.5Nb, the value of the activation

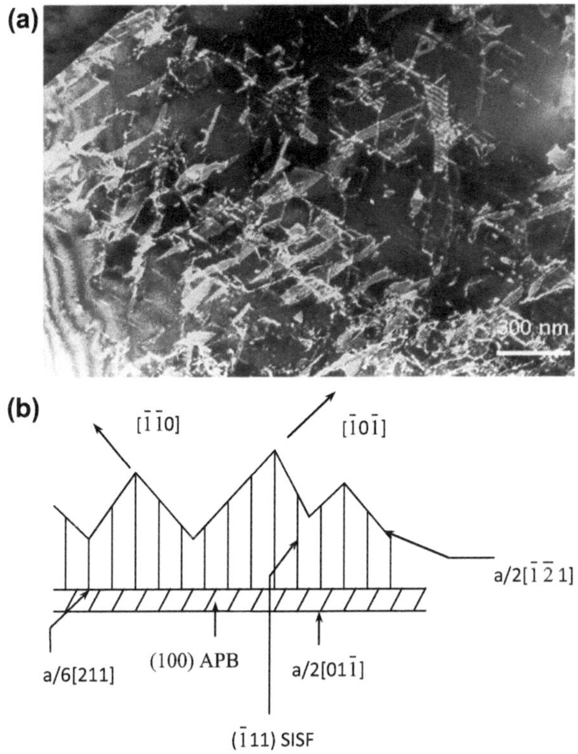

FIGURE 9.15

Showing (a) deformed microstructure of Zr_3Al-Nb alloy. Presence of superlattice intrinsic stacking faults and (b) schematic of SISF showing the partials involved in generating SISF and APB.

energy of creep obtained in the Zr_3Al-based alloys has been found to be higher than that for these alloys. The value of the activation energy of creep in Zr-based alloys is of the order of 230 kJ/mol in the low-temperature range (below the $\beta \rightarrow \alpha$ transition) and 80 kJ/mol in high-temperature ranges. These values correspond to the value of the activation energy required for self-diffusion of zirconium in the α-phase and in the β-phase, respectively. The activation energies of creep in other aluminides such as Ti_3Al, Fe_3Al, etc. also fall in the range 300–400 kJ/mol, which match with the activation energy of self-diffusion of solvent atoms [144–146]. Based on these estimates it was concluded that the activation energy of creep in Zr_3Al-based alloys should also correspond to the self-diffusion of zirconium in the Zr_3Al alloys. The effects of Nb on microstructure Zr_3Al-Nb alloys can be summarized as follows:

1. Addition of Nb to Zr_3Al alloys stabilizes the β-based phases. The cubic phase matrix appears to have a distinct advantage over the hcp alpha-matrix in terms of isotropic deformation behavior and a much higher solubility of hydrogen. On subsequent treatments, the β-based phases undergo transformations where solubility of Nb in these phases is an important factor in the decomposition processes. For example, due to the limited solubility of Nb in the low-temperature phases like α-Zr and $L1_2$ phases, the kinetics of the formation of equilibrium phases is retarded due to the diffusion of Nb away from the advancing front. This results in two-phase microstructure in the Zr_3Al-Nb alloy.
2. Addition of Nb reduces the minimum possible deformation temperature of the alloy, making the alloy more amenable to deformation at low temperatures. Effect of Nb on different parameters

like activation energy of creep, activation area, the liquidus temperature, and Young's modulus has been determined. Considering improvement in these properties, Zr_3Al-Nb-based intermetallics appear more attractive as a structural material compared to presently deployed zirconium alloys.

The evaluation of irradiation-induced degradation of the alloy system is still pending, and only after proper evaluation of the radiation stability of these intermetallics any possible applications in nuclear reactor may be realized.

2.6 TRIALUMINIDES (Al_3X COMPOUNDS)

The quest of developing light weight, oxidation resistance, and good high-temperature properties has led to the development of aluminum-rich intermetallics with the chemistry as Al_3X (where X stands for Group IV and Group V elements), which are named as trialuminides [147]. These phases have DO_{22} or DO_{23} structures. As shown in previous sections, these structures are long periodic derivatives of $L1_2$ structures, which form because of the introduction of periodic antiphase boundaries on the (001) plane with ½<110> displacement vector (Fig. 9.2). The major problem with the most of these compounds is that they are line compounds and do not have any tolerance for chemical variations and because of this it is very difficult to produce them as a single phase. However, single crystals have been successfully grown using the floating zone technique. Another issue with these intermetallic phases is their poor low-temperature ductility. Investigations on Al_3Ti have shown that the major mode of deformation at ambient temperatures is by twinning of the ordered phase resulting from the movement of ½<112] partials on {111} planes. With increasing temperature other modes of deformation, namely, slip along <100] and <110], also start to participate in the deformation process. High anisotropy was also observed in the APB energy [147]. For example, the value of APBs on (001) planes are 25–35 mJ/m^2 as compared to 190–200 mJ/m^2 on {111} planes. In addition, intrinsic stacking fault energy is 100 mJ/m^2. These high values appear to be a reason behind extensive twinning in these materials [147]. To improve low-temperature ductility in these intermetallic compounds it is necessary to enhance the twinning ({111}<112>) and slip activities ((001)<110>)). Therefore, efforts were made to identify those elements that could lower the stacking fault energy and the antiphase boundary energy. This was attempted by making alloys along the tie-line connecting Al_3Ti and Al_3V [148]. However, results were not very encouraging.

By introducing different APBs, DO_{22} structure could be converted into the more ductile $L1_2$ structure. There were efforts to develop this phase by ternary additions and ternary alloy systems like Al-Ti-Ni, Al-Ti-Cr, Al-Ti-Mn, Al-Ti-Ag, and Al-Ti-Fe, which were studied in detail. These compounds ($Al_{66}Ti_{25}Mn_9$ and $Al_{67}Ti_{25}Cr_8$) have shown considerably improved compressive ductility and bend ductility [149,150]. Fracture mode has been found to change from transgranular cleavage at room temperature to intergranular failure at high temperatures. It appears that at high-temperature (~1273K (1000°C)) formation of subgrain boundaries, grain boundary migration and grain boundary sliding help in attaining plastic deformation. Many times serrated grain boundaries have been observed. Formation of the serrated grain boundaries was rationalized in terms of sequence of events involving formation of shear faults at the grain boundaries. Edge dislocations would accumulate and polygonize on these shear faults to form tilt boundaries at which grain boundary would migrate to neutralize interfacial tension and this migration causes grain boundary to assume serrated morphology.

Oxidation resistance of the ternary derivatives of trialuminides was comparable to their binary phases and superior to TiAl phase. Chromium addition appears to promote the formation of Al_2O_3 and suppress the formation of TiO_2, which improves the oxidation properties. In general, Cr-containing trialuminides, which have $L1_2$ structure, show superior resistance as compared to Mn-based trialuminide derivatives [151].

In short, trialuminides offer promising high-temperature properties with low density, better oxidation resistance, and comparable creep properties with respect to γ-TiAl [151]. However, the issue concerning low ductility in the low-temperature region (<873K (600°C)) is one of the major concerns and impeding blocks in realizing full potential of these intermetallic compounds.

2.7 ALUMINIDES WITH A15 STRUCTURES

Several compounds with A15 structure (complex cubic structure; β-W) have been examined for superconducting applications. Of these, Nb_3Al has also been examined for high-temperature applications. Formation of Nb_3Al phase occurs through a peritectic reaction at nearly 2273K (2000°C). Unlike other high-temperature aluminides, Nb_3Al tolerates composition variation. However, detailed microstructural study revealed the presence of other metastable phases [152]. Ternary additions have shown that the phase field could be extended. Mo and Ta turn out to be a most effective alloying addition in this regard. V and Cr appear to give the most tough and useful microstructures. Mechanical properties of two-phase microstructures containing Nb-Nb_3Al show very strong strain rate sensitivity with respect to the volume fraction of the Nb matrix. Increasing Al concentration increases the formation of Nb_2Al, which appears to increase the high-temperature yield strength, whereas addition of Ti reduces strength of the alloy. High-temperature deformation (1473K (1200°C)) shows that the major mode of deformation is <001>{100} slip system but formation of stacking fault has been reported. In steady-state creep region the stress exponent, $n = 2.86$, and activation energy of 367.5 kJ/mol match well with the diffusion model. One of the limiting factors in Nb_3Al is the poor oxidation properties. Experiments have shown that at temperature as low as 1473K (1200°C) Nb_3Al could not survive and crumbled into "yellow oxide flakes" [153].

2.8 LAVES PHASES

Among intermetallics, the Laves phases comprise the single largest structural class. The Laves phases exhibit three crystal structures, namely, the C15($MgCu_2$type) cubic, C14($MgZn_2$-type) hexagonal, and C36($MgNi_2$-type) dihexagonal crystal structures [154]. Unit cells in these complex structures are large resulting in large Burgers vectors, and therefore they show restricted slip activities. This is one possible reason for the excellent high-temperature strength, but also for the absence of ductility at room temperature and the high ductile-to-brittle transition temperatures in Laves phases [155]. In addition, Laves phase in several alloy systems (for example, XCr_2 where X = Ti, Zr and Hf) undergoes polymorphic transformation where high-temperature C14 phase transforms into C15 structure. Many times these transformations result in high density of faults within a system. These transformations are sluggish and involve "synchroshear" partials rather than ordinary Shockley partials. Dislocation glide in these structurally complex systems involves reshuffling of atoms in multiple layers, which is also addressed as "zone glide." Considering inherent difficulty in producing sufficient ductility by slip, it is essential that an alternative deformation mode such as mechanical twinning occurs in order to plastically deform the Laves phases at room temperature [18]. Therefore enhancing the frequency of mechanical twinning is expected to improve room temperature ductility and fracture toughness of Laves phases to make them suitable for high-temperature structural applications. An improvement in cold workability of one C15 cubic Laves compound has been reported, namely the HfV_2 Laves phase with the ternary addition of Nb by Inoue and Tachikawa [156] who suggested that the main deformation mode is {111}<112> twinning. Twinning was later confirmed in the more detailed studies of HfV_2 + Nb ternary alloys having a two-phase C15 + bcc microstructure by Hall and Livingston [157] and Chu and Pope [158,159]. It has been suggested that the positive effect of Nb addition on the HfV_2 alloy is closely related to the substitution behavior of the ternary element on the Hf sublattice. The approach of alloying is to lower the stacking fault energy [160], which may increase twinning density and thereby increase the room temperature ductility in the C15 cubic Laves phases [161]. It was observed in Hf-V-Nb that the density of stacking faults associated with mechanical twinning increases at compositions closer to the C14 phase field in two-phase C15 Laves phase alloys [162]. If the twinning stress in the C15 cubic Laves phase can be reduced below the fracture stress by alloying to reduce stacking fault energy, single-phase Laves compounds may be ductile at room temperature.

2.9 BERYLLIDES

Among all the alloy systems that were explored for high-temperature applications, beryllides are the least-explored systems. Major research work on beryllides started when in 1956 Brush Wellman Inc. under contract with the U.S. Air Force, which was seeking lighter-weight materials with higher-temperature performance than those available, began work on a major R&D effort to investigate a wide range of intermetallic compounds, including beryllides [163]. The main goal of the original work was to identify materials that showed "promise of structural utility in air in the temperature range of 1533K (1260°C)–1923K (1650°C) [163]." Among other materials, beryllides of titanium, tantalum, niobium, and zirconium were also explored. Among these beryllides the other three beryllide families simply outperformed titanium beryllides, and among all beryllides Nb-based beryllides have received the maximum attention. Later on, $TiBe_{12}$ found application as neutron multiplier and renewed interested is gaining momentum with possible application in fusion technology. One of the compounds of Nb-Be, $NbBe_{12}$, with melting point as high as ~3343K (3070°C) and density as low as 2.92 g/cc, was evaluated for it high-temperature capabilities. $NbBe_{12}$ has complex bct structure with Nb occupying corner and body center positions and the remaining 24 atoms of Be distributed in different layers. High-temperature strength and low-temperature brittleness can be expected from the complex crystal structures of the beryllides. In these structures, dislocation motion appears to be limited until temperatures are reached where diffusion processes become rapid. TEM observations of deformed $NbBe_{12}$ have revealed substantial dislocation activity at 1073K (800°C) and above [164]. Most of these dislocations were partials of the type ½<101]{121), ½<101]{101) containing stacking faults within them. Because of the complex bct, crystal structure, faults cannot be created by a simple shift of niobium atoms to beryllium positions and vice versa, since niobium and beryllium atoms do not share equivalent lattice sites. It has been proposed that most of these partials may be related to the $NbBe_{12}$ to Nb_2Be_{17} phase transformation. Therefore, mobility of partials depends upon the stacking fault energy, and elements lowering the stacking fault energy would have the potential to lower the ductile-to-brittle transition temperature.

Two-phase alloys having Be as matrix and titanium or zirconium beryllides as the dispersed phase did not show any improvement in the toughness but suffered from major loss in strength at elevated temperatures [165]. In 2010 attempts have been made to make composites where Nb-Be could be used as the reinforcement material in the NiAl matrix, as the coefficient of thermal expansion of both of them are nearly similar. Beside their brittle nature, toxicity of Be, which demands stringent cautions in processing and characterizing the alloy, made any exploratory work difficult.

3. SILICIDES

One of the first applications of the silicide phase is the use of molybdenum disilicide ($MoSi_2$)-based material as Super Kanthal [166], which has been used along with nickel and chromium as heating elements. Another old application of silicides was as coating materials for protecting refractory metal components [167]. In addition, there is considerable interest in the semiconductor industry to use silicides as diffusion barriers in integrated circuits [168]. Therefore, usage of silicides for high-temperature applications for all practical purposes is a revival where silicides are being developed to exceed the temperature capability of the state-of-the-art single-crystal superalloys. One of the reasons for the current interest is the excellent oxidation resistance of silicide phases, but, at the same time, lack of room temperature ductility and toughness has dampened its progress. To understand the formation of silicide phases, pseudo-quaternary phase diagrams have been developed [37]. Fig. 9.16 shows such phase diagram where Si is sitting at the apex and different ternary elements at the three corners of the base triangle. This diagram does not include all the silicide phases but segregates those that may find potential applications at high temperatures. As summarized in Fig. 9.16, disilicides having C40 and C54b are structurally related. Out of all disilicides, C40 silicides have the widest phase fields and therefore have high possibilities to tailor the microstructures. Among A_5Si_3 (5:3) silicides, phases that have $D8_8$ structures have the highest melting points (liquidus point). One of the $D8_8$ structure base $Zr_3Ti_2Si_3$ compounds has been suggested for ultrahigh-temperature structural materials. High-temperature strength of most of the silicides is within 100–200 MPa at temperatures ranging from 1273K (1000°C)–1473K (1200°C). Ductile-to-brittle transition temperatures of silicides are generally in the range of 1073K (800°C)–1273K (1000°C), which are 673K (400°C)–873K (600°C) above the ductile-to-brittle transition temperatures of NiAl. Currently with no well-defined approach to lower ductile-to-brittle transition temperatures of silicides use of monolithic silicides is not very actively pursued. Fig. 9.17 shows the minimum creep rates of various silicides in comparison to creep rates of single crystals of Ni-based superalloys. It may be noticed that except for Ti_5Si_3, which has comparable creep rates, most of the silicides show inferior creep properties. It has been observed that disilicides generally show high oxidation resistance at elevated temperatures. This high oxidation resistance is due to the formation of adherent, protective silicate scale on the surface. In a few silicides (for example, $MoSi_2$) low-temperature pesting phenomenon is observed due to sluggish kinetics of silicate formation. Most of the silicides show formation of mixed oxide layers that sometimes are not desirable. $MoSi_2$, on the other hand, forms a pure silicate layer and therefore shows good high-temperature oxidation resistance. Fig. 9.18 shows the oxidation of some of the selected alloys, aluminides and silicides. Fig. 9.18 shows significantly improved oxidation resistance of $MoSi_2$ as compared to other aluminides and alloys. Among 5:3 silicides, Ti_5Si_3 shows reasonable oxidation resistance at high temperature but Nb_5Si_3 and Mo_5Si_3 oxidize at a much faster rate. Alloying with Ti or Al shows significant improvement in the oxidation resistance in Nb_5Si_3 whereas alloying with Hf, Zr, or with Mo does not improve oxidation resistance of Nb_5Si_3.

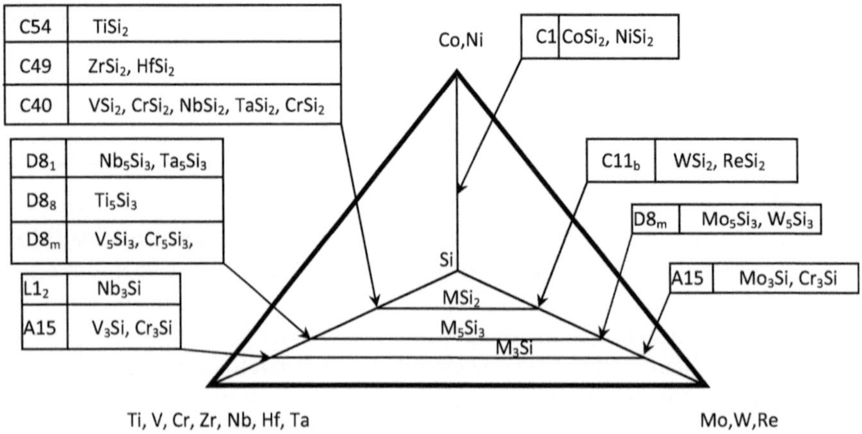

FIGURE 9.16

Projection of a pseudo-quaternary phase diagram summarizing all silicide phases having different crystal structures.

FIGURE 9.17

Minimum creep rates of Ni-based alloys and other silicide phases.

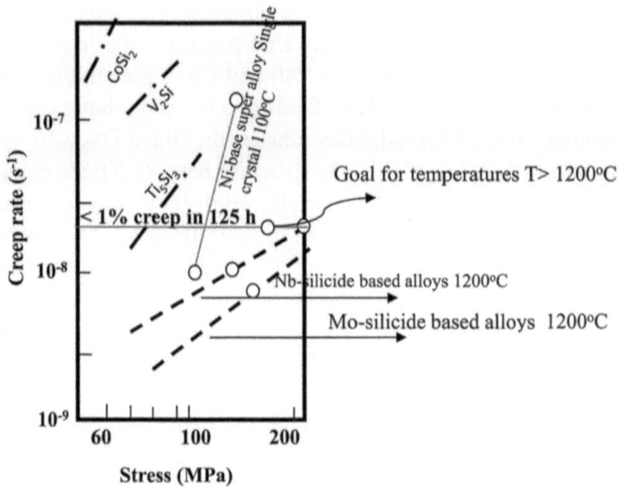

In spite of having very low toughness, silicides were pursued because of their high-temperature oxidation resistance, high thermal conductivity, and ductility at high temperatures. As alloying in silicides could not improve toughness at room temperature as much as it compromised high-temperature properties, use of silicides as monolithic component could not be continued. In the place of a single silicide phase, efforts have been started to develop composites reinforced with fiber reinforcements. These fibers were either a brittle phase (SiC, alumina, etc.) or a ductile phase (Nb, Mo, W, etc.). Work on ductile fiber– reinforced silicides composite is much more exhaustive as compared to the work on brittle-reinforced silicide composites. Phases like SiC and sapphire are chemically compatible with $MoSi_2$. Their thermal expansion coefficients are also similar to the silicide phase.

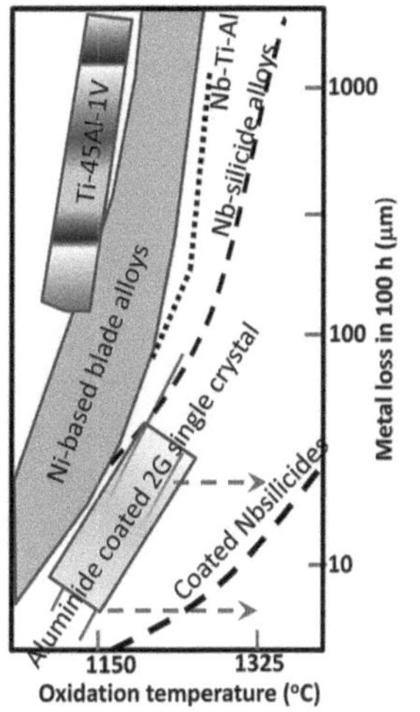

FIGURE 9.18

Oxidation behavior of some high-temperature alloys.

However, improvement in the toughness could not be achieved by silicide composites with brittle reinforcements. Poor resistance to oxidation is one of the major concerns for using ductile reinforcements by refractory metals. In addition, due to interdiffusion between Nb and $MoSi_2$, diffusion barrier coating is needed in Nb-$MoSi_2$ composites. During fabrication of Mo-$MoSi_2$ and W-$MoSi_2$ composite, Mo and W undergo recrystallization and become brittle at room temperatures.

3.1 REFRACTORY METALS-BASED IN SITU COMPOSITES

Refractory metals have been considered periodically for turbine applications operating at temperatures above the present temperature limit [169–171]. Of all the refractory metals, Nb and Mo have been considered as having the greatest potential for turbine applications [172–174]. Both Nb and Mo have varying extents of solubility for other elements, which can improve the desirable properties of the material [172]. Developments based on Nb and Mo alloys have offered significant gains in temperature capability due to their high melting temperatures and relatively low densities compared to other refractory metals. But both Nb and Mo exhibit poor oxidation resistance at elevated temperatures in their monolithic form. Between these two refractory metals, niobium is ductile at room temperature, has wide solubility for a number of strengthening additions, and is relatively low in density (lower than Ni) at 8.56 g/cm^3. Molybdenum, on the other hand, has good solubility for strengtheners, but it has a high density (10.2 g/cm^3) and relatively low ductility at room temperature. It also has substantial limitations with respect to oxidation resistance. The density of Mo is such that unless very thin structures can be produced, the minimum requirements for high-pressure turbine blade wall thickness will yield an increase in the blade weight compared to that of superalloys, and will, therefore, pose a severe penalty on the rotor weight [171]. For this reason, the Nb-based refractory metal-intermetallic composites represent the most promising materials for surpassing superalloys. Extensive investigations have been carried out in the past to find oxidation-resistant alloys, but in Nb most of them were focused on Nb-Al alloys. Of late, alloy developments are focused on the more promising Nb-Si alloys because silica can, in principle, provide a stable surface oxide to higher temperatures than alumina and the lower activation energy for oxygen diffusion in silica may make it more effective than alumina at higher temperatures [175].

3.2 NB-BASED IN SITU COMPOSITE

In this arena, Nb-Si in situ composites that consist of a Nb-based solid solution with Nb_3Si and/or Nb_5Si_3 silicides have shown to have a great potential because of their attractive balance of high- and low-temperature mechanical properties [176–178]. In binary Nb-Si phase diagram, two-phase field comprising $Nb_{ss} + Nb_5Si_3$ phase ranges from 0.6% to 37.5% Si. Therefore it provides the opportunity for manipulating microstructures to a great extent, though for all practical purposes the working range is in between 6% and 18.7%Si [179]. As the Si concentration increases in the binary alloys volume fraction of the Nb_5Si_3 increases, which changes the mode of failure from the ductile to brittle and stress intensity factor drops from 21 $MPa\sqrt{m}$ at Nb-10%Si to 13 $MPa\sqrt{m}$ at Nb-16.5%Si [169]. In this alloy system creep mainly occurs through the matrix Nb phase and follows power law with $n \sim 3$ [171].

Unfortunately, the composites from binary Nb-Si alloys have very poor oxidation resistance because of high reactivity and diffusivity of oxygen in bcc Nb [180]. Therefore, to improve the oxidation resistance, alloy additions are resorted to but most of the alloying elements that improve oxidation resistance bring down the liquidus temperature of the alloy and thereby the maximum operating temperatures [181]. Major alloying additions include elements like Ti, Al, and Hf, all of which form oxides that are thermodynamically more stable than silica and Nb_2O_5. These oxides should have no solubility in the solid state. Under such condition, complex oxide scale formation is expected to occur during oxidation of the multicomponent Nb alloys [169,178,182–189]. Fig. 9.19 shows a tetrahedron showing Nb-Ti-Si-Cr at the apex and the distribution of phases at 1473K (1200°C). As shown the β-phase and Laves remain confined mainly to the Nb-Cr-Ti face of the tetrahedron whereas the silicide phase shows solubility for all the elements. Based on such a diagram relative phase volumes can be roughly estimated. Many these multicomponent alloys showed considerable improvement

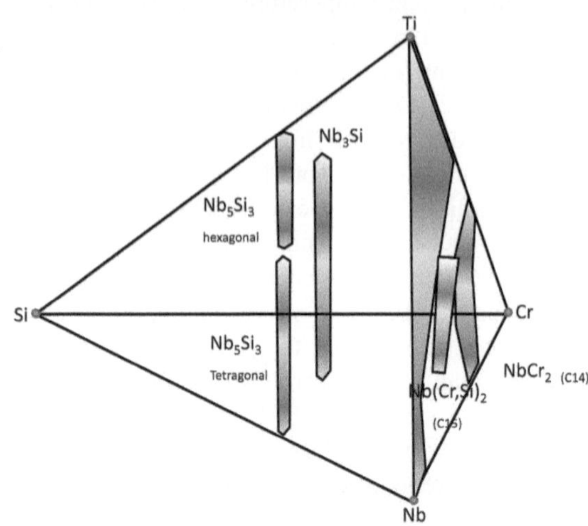

FIGURE 9.19

Schematic representation of various phase fields in quaternary Nb-Cr-Si-Ti-based alloys.

in the high-temperature strength. For example, a directionally solidified Nb-Mo-W-Ti-Si alloy shows compressive strength of 650 MPa at temperature as high as 1771K (1498°C). In addition, it shows a minimum creep of $1.4 \times 10^{-7}\ s^{-1}$ at 1671K (1398°C) [190]. These values of mechanical properties when seen in light of the nominal goal for creep performance in an engine application, which is 1% creep in 125 h at temperatures of 1473K (1200°C) and stresses of 170 MPa, appear to meet the requirement (Fig. 9.17). They have successfully pushed the high-temperature window at least by 473K (200°C). However, oxidation resistance and room temperature ductility are of concern. In order to address high-temperature oxidation resistance these alloys need external protections [169,191]. Refractory metal alloy–based composites are still evolving and probably in the near future may able to replace superalloys in many secondary usages at high temperatures. Regarding replacement of superalloy turbine blade single crystals, the oxidation resistance of coated samples of Nb exhibit properties better than Ni-based super alloys developed for blade material (Fig. 9.18).

3.3 MO-SI-B-BASED ALLOYS

Mo-Si-B alloys, comprising three phases (Mo_5SiB_2, Mo_3Si, and α-Mo) with continuous matrix of α-Mo, have been considered to possess an ideal microstructure for high-temperature applications [192]. In such microstructures, both intermetallic compound phases (Mo_5SiB_2 and Mo_3Si) have superior oxidation resistance due to the formation of SiO_2-B_2O_3 layers, which effectively protect the surface. It is known that the SiO_2-B_2O_3 layer has a much higher fluidity than the SiO_2 layer at high temperatures and can, therefore, quickly cover the virgin surface of the Mo-Si-B alloy during oxidation. While the presence of the α-Mo phase as the continuous matrix imparts low-temperature ductility and fracture toughness, the oxidation resistance arises from a fine and uniform distribution of intermetallic phases such as Mo_5Si_3, T2-phase Mo_5SiB_2, and A15-phase Mo_3Si. Therefore, to simultaneously obtain the combination of good oxidation resistance and mechanical properties, a microstructure consisting of fine intermetallic compound phases uniformly distributed in a continuous matrix of α-Mo phase is desirable [193,194]. Fig. 9.20 shows a schematic representation of the Mo-rich side of the ternary phase diagram. Alloys consisting of Mo_5Si_3, the T2-phase Mo_5SiB_2 and the

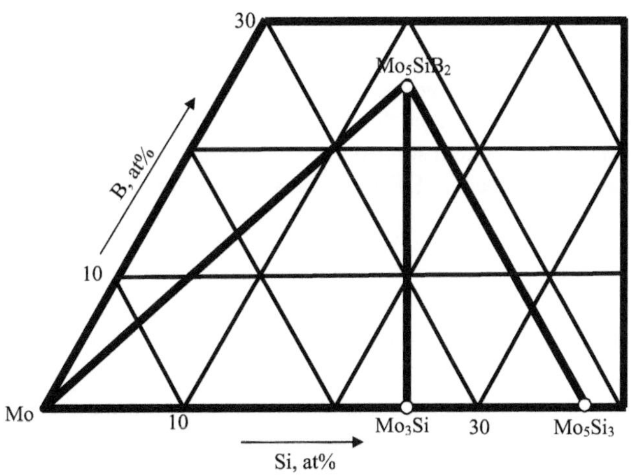

FIGURE 9.20

Schematic presentation of the Mo-rich side of the Mo-Si-B phase diagram.

A15-phase Mo_3Si, exhibit excellent oxidation resistance at elevated temperatures. On the other hand, alloys consisting of α-Mo, Mo_3Si, and T2 are not as oxidation resistant as Mo_5Si_3-T2-Mo_3Si alloys, but alloys show better fracture toughness [194]. Various studies on the control of microstructure have been carried out to date. Currently, techniques used to fabricate Mo-Si-B alloys are mainly divided into the melting-casting and powder metallurgy routes. However, in the former case, the fracture toughness of the fabricated alloy is low as it is difficult to form a microstructure with a continuous matrix phase of α-Mo. Further, there is the possibility of forming a nonequilibrium phase. It is easier to control the microstructure when using the powder metallurgy method. However, the powder metallurgy route has some difficulties in forming a continuous α-Mo phase, which is essential for the desired oxidation resistance. Therefore, there is a necessity to develop a suitable fabrication processing method for production of the desired microstructure. At present, it appears that Nb-based systems can serve up to about 1473K (1200°C) [195], and Mo-based systems can offer capability to above 1573K (1300°C). There have been efforts to develop Nb- and Mo-based alloy systems for industrial applications [195,196]. The refractory metal systems are in the early stages of evolution, but may well enter the forefront of meeting the demands for greater gas turbine efficiency that are driving the need for new high-temperature materials. They may also provide new materials for land-based power-generation turbines and for hypersonic flight for access to space.

4. HIGH-ENTROPY ALLOYS

High-entropy alloys have attracted attention in recent years as new class of high-temperature materials. The basic difference between a high-entropy alloy and a conventional alloy is that conventional alloys have one principal element that forms the base in which a few major or minor alloying elements are added to achieve the desirable combination of mechanical and corrosion properties. In contrast to this, in a high-entropy alloy all constituent elements are nearly in equal atom percentages and therefore the configurational entropy is maximized. The approach of high-entropy alloys is based on the well-known fact that as the number of constituent elements in an alloy system increases, the number of constituent phases does not become numerous. The number of phases formed in such alloys is significantly lower than as estimated by the Gibbs phase rule [197]. This leads to the search for a composition where five or more elements each having an atomic percentage between 5% and 35% in an alloy system form one single phase. Such alloys are commonly addressed as high-entropy alloy (HEA) [198]. Definition of HEA is flexible in a sense that it allows minor elemental additions with atomic percentages even smaller than 5%. However, there is another definition that states that having a configurational entropy larger than 1.5R (where R is the gas constant) in their random solution state is sufficient to incorporate an alloy in the group of HEAs [199]. There are four effects in HEA that make them different from the conventional alloys.

4.1 HIGH-ENTROPY EFFECT

The high-entropy effect means maximization of the configurational entropy in HEAs, which in turn lowers the free energy of solid-solution phases and facilitates their formation, particularly at higher temperatures. Due to this enhanced mutual solubility among constituent elements, the number of phases present in HEAs can be evidently reduced. This means that among the various types of phases in an alloy, those phases with higher entropy have high probability to get stabilized at high

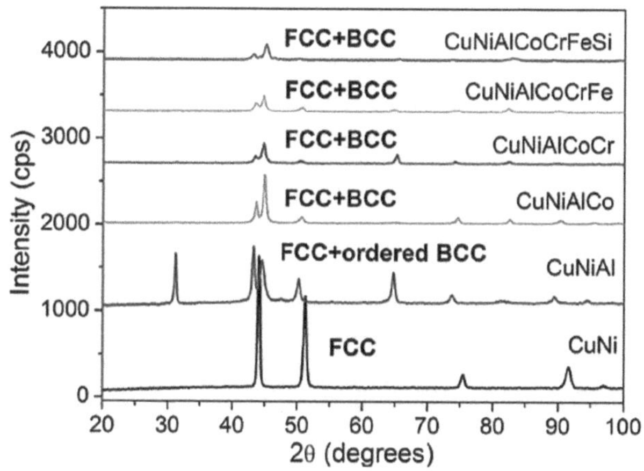

FIGURE 9.21

X-ray diffraction patterns from alloys where number of elements has been systematically increased. Note the reduction in the intensity of the peak with the increase in the number of alloying elements.

temperatures. In general, solid solutions have higher mixing entropy than intermetallic compounds. For HEAs, this difference in entropy between solid solutions and compounds is particularly large owing to the involvement of multiple elements. Thus, it is highly probable that solid-solution phases become the stable phase in HEAs at high temperatures. It is expected that with the increase of temperature, the overall degree of order in HEAs decreases. Thus, even those alloy systems that contain ordered phases in the low-temperature regions can transform to random solid solutions at high temperatures.

There is evidence for the high-entropy effects [200]. Fig. 9.21 shows the X-ray diffraction patterns of a series of binary to septenary alloys. It is noticed from the figure that the structures of phases in quinary, senary, and septenary alloys remain rather simple. Such simple structures contradict conventional expectation—formation of various kinds of binary/ternary compounds. In this regard it should be noted that the formation of a simple structure is not ensured in every case, as the prevailing phase is a consequence of the balance between the enthalpy of formation of ordered phase and the configurational entropy of disordered phases. In cases where the enthalpy of formation of ordered phases is higher, the final microstructure may contain phase mixture of several ordered intermetallics [201].

4.2 SLUGGISH DIFFUSION

Diffusion and phase transformation kinetics in HEAs are slower than those in their conventional counterparts [202]. In HEA the neighboring atoms of each lattice site are often different. The difference in local atomic configuration leads to different bonding and therefore different local energies for each site. When an atom jumps into a low-energy site, it gets "trapped" and the chance to jump out of that site becomes lower. In contrast, if the site is a high-energy site, then the atom has a higher probability to hop back to its original site. Either of these scenarios slows down the diffusion process. Diffusion couple experiments show that the activation energy of diffusion in the Co-Cr-Fe-Mn-Ni alloys is higher than those in other FCC ternary alloys and pure elements [202]. The slow kinetics of atomic movements in HEAs many times leads to formation of nanosized precipitates. Such precipitates contribute to the excellent performance of HEA coatings as diffusion barriers as well as improvement in creep properties.

4.3 LATTICE DISTORTION

A typical lattice in an HEA is composed of several elements (at least five), because of which lattice in HEA is generally distorted. The strain energy associated with lattice distortion raises the overall free energy of the HEA lattice, which affects the properties of HEAs. For example, lattice distortion impedes dislocation movement and improves creep resistance.

4.4 SYNERGISTIC EFFECTS

In HEA, effects of alloying additions are generally in excess to what is predicted from extrapolating effects noticed in their ternary alloys. Thus, the macroscopic properties of HEA not only come from the averaged properties of the component elements but also include the effects from the excess quantities produced by interelemental reactions and lattice distortion. In HEA, beside factors governed by classic Hume−Rothery rules, enthalpy and entropy of mixing become important parameters for HEA phase formation [203]. In order to form simple structures (i.e., fcc, bcc) and their mixtures, including both ordered/disordered cases, the following empirical relationships have been defined: $-22 \leq \Delta H_{mix} \leq 7$ kJ/mol, $\delta \leq 8.5$, and $11 \leq \Delta S_{mix} \leq 19.5$ J(K mol) [203]. Based on the concept of entropy−enthalpy competition, the other criterion to judge whether or not simple phases will form in an HEA is the parameter, $\varepsilon = [T\Delta S_{mix}/\Delta H_{mix}]$ [204]. Thus, larger "ε" suggests a higher possibility of forming simple disordered phases, which agrees well with the analysis [204]. A drawback of the two criteria shown herein is that even if an alloy satisfies both the criteria, it can still have intermetallic phases. Therefore, another criterion that is based on valence electron concentration has been proposed [205]. According to this criterion, if valence electron concentration is above 8, fcc structure is stabilized, and if this value is less than 6.8, bcc structure is stabilized. For in-between values of valence electron concentration, two-phase structure is stabilized.

Refractory metal-based HEA systems have been developed for high-temperature applications [10,206−208]. The major phase in these alloys is typically bcc with high strength ranging from 900 to 1350 MPa [206,209]. Interestingly, such strength is several times higher than that of the weighted average strength of the composing elements. Such high hardness is a result of the solid-solution strengthening effect that originated from the difference in atomic size and modulus among the constituent elements [206]. Room temperature plasticity of refractory HEA is closely related to composition. The plastic strain for fracture under tensile loading is less than 5% for many alloys. However, some alloys can be compressed to 50% without fracture [206,208]. It was suggested that the high plasticity in some alloys is owing to the prevalence of deformation twinning [206−209]. Softening of the refractory metal-based HEA is nearly half of the softening of Inconels [210], which has been attributed to the slower diffusion rate in HEAs. For example, HEA yield strengths of $Nb_{25}Mo_{25}Ta_{25}W_{25}$ and $V_{20}Nb_{20}Mo_{20}Ta_{20}W_{20}$ alloys remain virtually unchanged in the range of 873K (600°C)−1273K (1000°C). Their yield strengths remain above 400 MPa upto 1873K (1600°C). Refractory metal-based HEA shows excellent stable microstructures at elevated temperatures. The reason for such stabilities is low driving force for the elimination of defects. The driving force to eliminate defects originates from the free energy difference between defect-containing and defect-free crystals. However, the defect-free lattice in HEAs is still highly strained owing to the atomic size differences between the elements [210,211]. The activation energy of grain growth for the CoCrFeMnNi alloy (which is nearly a single-phase fcc structure) is 321.7 kJ/mol [212]. This energy is nearly twice the grain growth energy of stainless steel. Another example of phase stability is the alloy NbSiTaTiZr, which resists crystallization up to 800°C [210]. Reasons given for such high stability of amorphous phase in these alloys are lower free energy of highly random amorphous phase and slow kinetics for crystallization [210].

In sum it could be stated that HEA is an emerging field and our understanding of these new materials is still very limited and preliminary. Even though HEA can be produced by conventional casting, wrought working route or powder metallurgy route and even single crystals have also been grown, and exploiting this material for any structural application is a long story. Most of the mechanical properties are assessed from hardness or compression testing. As reasonable ductility is crucial to structural applications, more studies on tensile behaviors are definitely needed. Additionally, except for a few examples [213–215], not much is known about the deformation of HEAs. Detailed studies on dislocation behavior, deformation twinning, deformation microstructures, texture evolution, stacking fault energy, etc. are needed. Therefore in spite of showing early promise, extensive research is needed on HEAs and their real application as high-temperature structural components is a distant dream.

5. SUMMARY

Nickel and, more recently, cobalt-based superalloys dominate the scene of high-temperature structural materials because of some of their basic attributes. Both consist of matrix phases that have close-packed structures (fcc or hcp, respectively), which are associated with inherent ductility at ambient temperatures and a low diffusivity due to their close packing, resulting in good creep resistance. High strength and good creep resistance at elevated temperatures arise due to stronger bonding, which is reflected in high elastic modulus. Superalloys contain one or more dispersed ordered intermetallic phases to varying extents. Generally, the wrought variety has somewhat less volume fraction of intermetallic phases than that of the cast variety. The fundamental point of the design of these alloys is to ensure that the matrix solid solution phase and the dispersed intermetallic phases are coherent with each other and, therefore, resistant to coarsening of the dispersed phase at high operating temperatures. Strengthening is achieved by solid solution and hard particle dispersion and the microstructural stability at elevated operating temperature arises due to the very close matching of the lattices of the matrix and precipitate phases.

As high-tech technologies demand materials for use at still higher temperatures, the first choice falls on intermetallic compounds, which are characterized by stronger bonding and consequent higher melting temperatures. The main limitation of intermetallics is their low ductility at lower temperatures. The choice of intermetallic structures having multiple slip systems and being amenable to plastic accommodation at grain boundaries—essential to impart ductility in polycrystalline materials—has led to the development of only a few structural intermetallics with limited range of applications. Meeting contradictory requirements such as low-temperature ductility, high-temperature strength, creep, and oxidation resistance has remained elusive in single-phase materials. There has been some success in two- or multiphase materials and composites to achieve such combination of properties. The $\gamma + \alpha_2$ structure in Ti-Al system is the best example to illustrate this point.

The use of refractory metals as the base and an appropriate choice of alloy additions giving rise to solid solution strength in the matrix and stable second-phase dispersion for arresting the grain growth yielded some materials for use at temperatures exceeding 1000°C. The high creep rate at elevated temperatures and high ductile-to-brittle transition temperature associated with the bcc structure along with poor oxidation resistance remain as the main obstacles for their extensive applications.

It is natural to select strain-free equilibrium phases for use at elevated temperatures. However, the HEAs with strained lattice and entropy-stabilized phase show considerable promise as high-temperature materials. Because of the cocktail of elements, it may be possible to achieve proper oxidation resistance of HEAs.

REFERENCES

[1] H.M. Cobb, The History of Stainless Steel, ASM International, Materials Park, Ohio, 2010.
[2] Outokumpu Publishers, Handbook of Stainless Steel, Outokumpu Oyj, s.l., 2013.
[3] R. Ruer, Z. Anorg. Chem. (1921) 117.
[4] O. Izumiand, A. Aoki, Phys. Status Solidi (A) 32 (1975) 657–664.
[5] J.H. Fleischer, R.L.Wester, Intermetallic and Compounds-Principal and Practice, John Wiley and Sons, New York, 1994.
[6] D.P. Ezz, S.S. Pope, Int. Met. Rev. 25 (1984) 233.
[7] Structural Uses for Ductile Ordered Alloys: Report, National Research Council, Washington DC, 1984, 0309324521.
[8] A. Sakamoto, M. Tomohiko, Advanced Materials for Extreme Environments, s.l.: 926C3807C Tokyo SENTA ZAIRYO GIJUTSU NO SAISHIN NO SHINPO in Japanese, December 91, 1992.
[9] D. Banerjee, A.K. Gogia, T.K. Nandi, V.A. Joshi, Acta Mater. 36 (1988) 871–882.
[10] Titanium Aluminide Application in High Speed Civil Transport, National Aeronautics and Space Administration, Washington DC, 1999. NASA TM-1999–209071.
[11] F. Ebrahimi, D.T. Hoelzer, J.R. Castillo-Gomez, Mater. Sci. Eng. A 171 (1993) 35–45.
[12] V.K. Sikka, E.A. Loria, Mater. Sci. Eng. A 239–240 (1997) 745–751.
[13] E.S.K. Menon, P.R. Subramanian, D.M. Dimiduk, Metall. Mater. Trans. A 27 (1996) 1647–1659.
[14] F. Ye, C. Mercer, W.O. Soboyejo, Metall. Mater. Trans. A 29 (1998) 2361–2374.
[15] Z. Shi, H. Wei, H. Zhang, D. Wu, T. Jin, X. Sun, Q. Zheng, J. Alloys Compd. 636 (2015) 61–66.
[16] J.R. DiStefano, B.A. Pint, J.H. DeVan, Int. J. Refract. Met. Hard Mater. 18 (2000) 237–243.
[17] K.S. Chan, Mater. Sci. Eng. A 329–331 (2002) 513–522.
[18] W.-Y. Kim, D.E. Luzzi, D.P. Pope, Intermetallics 11 (2003) 257–267.
[19] S. Ochial, Y. Oya, T. Suzuki, Acta Metall. 32 (1984) 289–298.
[20] V.K. Sikka, in: C.T. Liu, D.P. Pope, J.O. Stiegler, S.H. Wang (Eds.), High Tempersture Aluminides and Intermetallics, 1990, pp. 505–520.
[21] Y. Nishiyama, T. Miyashitta, S. Isobe, T. Noda, in: D.P. Pope, J. O. Stiegler, S.H. Wang, C.T. Liu (Eds.), High Temperature Aluminides and Intermetallics, 1990, pp. 557–584.
[22] V.K. Sikka, S. Viswanathan, C.G. McKamey, in: R. Darolia, J.J. Lewandowaski, C.T. Liu, P.L. Martine, D.B. Miracle, M.V. Nathal, Structural Intermetallics, The Materials Society, Warrendale, PA, 1993, pp. 483–491.
[23] T. Takeuchi, Y. Iijima, K. Inoue, H. Wada, B. ten Haken, H.H.J. ten Kate, K. Fukuda, G. Iwaki, S. Sakai, H. Moriai, Strain effects in Nb_3Al multifilamentary conductors prepared by phase transformation from bcc upersaturated-solid solution, Appl. Phys. Lett. 71 (1997) 122–124.
[24] T. Hennmi, N. Koizumi, Y. Nunoya, Y. Okui, K. Matsui, Y. Nabara, T. Isono, Y. Takahashi, K. Okuno, N. Banno, A. Kikuchi, Y. Iijima, T. Takeuchi, Characterization of Nb_3Al strands subjected to an axial-strain for a fusion DEMO reactor, IEEE Trans. Appl. Supercond. 19 (2009) 1540–1543.
[25] K. Tsuchiya, T. Takeuchi, N. Banno, A. Kikuchi, Y. Iijima, S. Nimori, H. Takigawa, A. Terashima, A. Yamamoto, Y. Kuroda, T. Takao, K. Tagawa, K. Tanaka, K. Nakagawa, Study of Nb_3Al wires for high-field accelerator magnet applications, IEEE Trans. Appl. Supercond. 20 (2010) 1416–1419.

[26] T. Takeuchi, N. Banno, K. Tagawa, T. Kobayashi, K. Tsuchiya, N. Tatsumi, T. Fukuzaki, M. Kosuge, H. Wada, Y. Iijima, A. Kikuchi, K. Inoue, Microstructure and n-value of RHQT processed Nb_3Al superconductors for NMR uses, IEEE Trans. Appl. Supercond. 14 (2004) 961−966.
[27] K. Young, J. Nei, D.F. Wong, L. Wang, Structural, hydrogen storage, and electrochemical properties of Laves phase-related body-centered-cubic solid solution metal, Int. J. Hydrogen Energy 39 (2014) 21489−21499.
[28] Y. Mishima, N. Yoshida, H. Takahashi, K. Ishida, H. Kawamura, T. Iwadachi, T. Shibayama, I. Ohnuma, Y. Sato, K. Munakata, H. Iwakiri, M. Uchida, Present status of beryllides for fusion and industrial applications in Japan, Fusion Eng. Des. 1 (2007) 91−97.
[29] M. Yamaguchi, V. Vitek, D.P. Pope, Phil. Mag. A, 43 (1981) 1027.
[30] L.M. Howe, M. Rainville, E.M. Schulson, J. Nucl. Mater. 50 (1974) 139.
[31] P. Staton-Bevan, A.E. Holdway, J. Mater. Sci. (1986) 2843−2849.
[32] P. Mukhopadhyay, V. Raman, S. Banerjee, R. Krishnan, J. Nucl. Mater. 82 (1979) 227.
[33] J. Douin, Philos. Mag. Lett. 63 (1991) 109−116.
[34] W. Liu, H. Rosner, E. Langmaack, A. Gemperle, J. Gemperlova, J. Pesicka, E. Nembach, TEM investigations on the structure of antiphase boundaries in $D0_3$ ordered Fe_3Al, Mater. Sci. Eng. A 258 (1998) 15.
[35] H.Y. Yasuda, K. Nakano, T. Nakajima, M. Ueda, Y. Umakoshi, Effect of ordering process on giant pseudoelasticity in Fe_3Al single crystals, Acta Mater. 51 (2003) 5101−5112.
[36] M.J. Marcinkowski, N. Brown, Theory and direct observation of dislocations in the Fe_3Al superlattices, Acta Metall. 9 (1961) 764−786.
[37] D.M. Shah, D. Berczik, D.L. Anton, R. Hecht, Appraisal of other silicides as structural materials, Mater. Sci. Eng. A 155 (1992) 45−57.
[38] J. Mass, G. Bastin, F. Von Lou, R. Metselaar, Z. Metal. 75 (1984) 140−145.
[39] F. Stein, M. Palm, G. Sauthoff, Structure and stability of Laves phases. Part I. Critical assessment of factors controlling Laves phase stability, Intermetallics 12 (2004) 713−720.
[40] E. Parthe, L. Gelato, B. Chabot, M. Penzo, K. Cenzual, R. Gladyshevskii, Standerized Data and Thecrystal Chemical Characterization of Inorganic Structure Types, in: Gmelin Handbook of Inorganic Chemistry, vol. 1, Springer, s.l., 1993, p. 188.
[41] Y. Komura, Stacking faults and two new modifications of the Laves phase in Mg−Cu−Al system, Acta Crystallogr. 15 (1962) 770−778.
[42] Y. Komura, M. Mitarai, I. Nakatani, H. Iba, T. Shimizu, Structural changes in the alloy systems of Mg-Zn-Cu and Mg-Zn-Ag related to the Friauf-Laves phases, Acta Crystallogr. B26 (1970) 666−668.
[43] Y. Kitano, Y. Komura, Long-period stacking variants and their electron-concentration dependence in the Mg-base Friauf-Laves phases, Acta Crystallogr. B33 (1977) 2496−2501.
[44] Y. Kitano, S. Akitoh, M. Mitarai, K. Ohnishi, K. Noguchi, M. Numata, Microsc. Res. Tech. 40 (1998) 277.
[45] C.T. Stockinger, J.P. Neumann, J. Appl. Crystallogr. 3 (1970) 32.
[46] R.W. Cahn, P.A. Siemers, J.E. Geiger, P. Bardhan, Acta Metall. 35 (1987) 2737−2753.
[47] P. Veyssiere, Philos. Mag. 50A (1984) 189−203.
[48] K.J. Nix, W.D. Hemkerand, Structural Intermetallics, TMS, Warrendale, 1997, p. 21.
[49] D.G. Morris, Intermetallics 6 (1998) 753.
[50] D. Lin, D. Shan, D. Li, Scr. Metall. Mater. 31 (1994) 1455.
[51] S.H. Nieh, T.G. Goods, Scr. Metall. 17 (1983) 23.
[52] D.H. Shah, Scr. Metall. 17 (1983) 997.
[53] D. Lin, D. Li, Y. Liu, Intermetallics 6 (1998) 243.
[54] S. Miura, L.Z. Peng, Y. Mishima, High-Temperature Ordered Intermetallic, 1997.
[55] Y. Zhang, D.L. Lin, High-Temperature Ordered Intermetallic Alloys V, MRS, 1993, p. 611.

[56] V.K. Sikka, J.T. Mavity, K. Anderson, Processing of nickel aluminides and their industrial applications, Mater. Sci. Eng. A 153 (1992) 712–721.

[57] C. Knobloch, V.N. Toloraia, U. Glatzel, Scr. Mater. 37 (1997) 1491.

[58] E.P. George, M. Yamaguchi, K.S. Kumar, C.T. Liu, Ordered intermetallics, Annu. Rev. Mater. Sci. 24 (1994) 409–451.

[59] E.P. George, C.T. Liu, J. Mater. Res. 5 (1990) 754.

[60] P.P. Camus, I. Baker, J.A. Horton, M.K. Miller, J. Phys. Colloques 49 (C6) (November 1988) C6-329–C6-333.

[61] T. Kim, K.T. Hong, K.S. Lee, The relationship between the fracture toughness and grain boundary character distribution in polycrystalline NiAl, Intermetallics 11 (2003) 33–39.

[62] M. Yan, V. Vitek, S.P. Chen, Acta Mater. 44 (1996) 4351–4365.

[63] M.J. Mills, D.B. Miralcle, The structure of a<100> and a<110> dislocation cores in NiAl, Acta Metall. Mater. 41 (1993) 85–95.

[64] E.P. George, C.T. Liu, Alloy Phase Stability and Design, vol. 186, Materials Research Society, s.l., 1991, p. 309.

[65] J. Daniel Whittenberger, L.J. Westfall, M.V. Nathal, Compressive strength of a B2 matrix NiAl-Nb intermetallic at 1200 and 1300 K, Scr. Metall. 23 (1989) 2127–2130.

[66] J. Daniel Whittenberger, S.K. Mannan, K.S. Kumar, 1100 to 1300 K slow plastic compression properties of Ni-38.5Al composites, Scr. Metall. 23 (1989) 2055–2060.

[67] K. Xu, R.J. Arsenault, High temperature deformation of NiAl matrix composites, Acta Mater. 47 (1999) 3023–3040.

[68] D.B. Miracle, Overview No. 104 the physical and mechanical properties of NiAl, Acta Metall. Mater. 41 (1993) 649–684.

[69] R. Darolia, Ductility and fracture toughness issues related to implementation of NiAl for gas turbine applications, Intermetallics 8 (2000) 321–1327.

[70] D. Banerjee, The intermetallic Ti_2AlNb, Prog. Mater. Sci. 42 (1997) 135–158.

[71] C.J. Cowen, C.J. Boehlert, Microstructure, tensile, and creep behavior of Ti-15Al-33Nb (at%) and Ti-21Al-29Nb (at%) Orthorhombic + B, in: Titanium Alloys for High Temperature Applications: A Symposium Dedicated to the Memory of Dr. Martin Blackburn, 2006, p. 93.

[72] D. Banerjee, The evolution of the orthorhombic titanium aluminides, in: Titanium Alloys for High Temperature Applications: A Symposium Dedicated to the Memory of Dr. Martin Blackburn, 2006.

[73] L.A. Bendersky, W.J. Boettinger, F.S. Biancaniello, Intermetallic Ti–Al–Nb alloys based on strengthening of the orthorhombic phase by ω-type phases, in: High Temperature Aluminides and Intermetallics, 1992, pp. 41–47.

[74] R. Boyer, G. Welsch, E.W. Collings, Ti_3Al Alloys. Materials Properties Handbook: Titanium Alloys, 1994, p. 1019.

[75] C.J. Boehlert, B.S. Majumdar, V. Seetharaman, D.B. Miracle, Part I. The microstructural evolution in Ti-Al-Nb O + bcc orthorhombic alloys 30A (1999) 2305–2323.

[76] C.J. Miracle, D.B. Boehlert, Part II. The creep behavior of Ti-Al-Nb O + bcc orthorhombic alloys, Metall. Mater. Trans. A 30A (1999) 2349–2367.

[77] T.K. Nandy, D. Banerjee, Creep of the orthorhombic phase based on the intermetallic Ti_2AlNb, Intermetallics 8 (2000) 915–928.

[78] Creep behaviour of an orthorhombic phase in a Ti-Al-Nb alloy, Scr. Metall. Mater. 28 (1993) 569–574.

[79] T.K. Nandy, D. Banerjee, Deformation mechanisms in the O phase, Intermetallics 8 (2000) 1269–1282.

[80] W. Wei, Z. Weidong, X. Chen, L. Xiaobo, Z. Jianwei, Designed bimodal size lamellar O microstructures in Ti_2AlNb based alloy: microstructural evolution, tensile and creep properties, Mater. Sci. Eng. A 618 (2014) 288–294.

[81] L. Germann, D. Banerjee, J.Y. Guédou, J.-L. Strudel, Effect of composition on the mechanical properties of newly developed Ti_2AlNb-based titanium aluminide, Intermetallics 13 (2005) 920–924.
[82] P.R. Smith, J.A. Graves, C.G. Rhodes, Comparison of orthorhombic and alpha-two titanium aluminides as matrices for continuous SiC reinforced composites, Metall. Mater. Trans. A 25A (1994) 1267–1283.
[83] A. Partridge, E.F.J. Shelton, Processing and mechanical property studies of orthorhombic titanium-aluminide-based alloys, Air Space Eur. 3 (2001) 170–173.
[84] J. Das, A.K. Gogia, D.V.V. Satyanarayana, Effect of iron and nickel impurities on creep and tensile behaviour of Ti–24Al–20Nb–0.5Mo alloy, Mater. Sci. Eng. A 496 (2008) 1–8.
[85] R.G. Rowe, Tri-Titanium Aluminide Alloys Containing At Least Eighteen Atom Percent Niobium, US Patent No. 5032357, July 16, 1991.
[86] A.K. Gogia, T.K. Nandy, D. Banerjee, T. Carisey, J.L. Strudel, J.M. Franchet, Microstructure and mechanical properties of orthorhombic alloys in the Ti-Al-Nb system, Intermetallics 6 (1998) 741–748.
[87] C.J. Cowen, C.J. Boehlert, Microstructure, creep, and tensile behavior of a Ti–21Al–29Nb(at.%) orthorhombic + B2 alloy, Intermetallics 14 (2006) 412–422.
[88] F. Tang, S. Nakazawa, M. Hagiwara, Transient creep of Ti–Al–Nb alloys, Mater. Sci. Eng. A 325 (2002) 194–201.
[89] P.R. Bhowal, W.A. Konkel, H.F. Merrick, Microstructure and properties of fine-grained orthorhombic transformed Ti-25Al-10Nb-3V-1Mo (Super $\alpha 2$) Ti aluminide, Mater. Sci. Eng. A 192–193 (1995) 891–898.
[90] Y. Mao, S. Li, J. Zhang, J. Peng, D. Zou, Z. Zhong, Microstructure and tensile properties of orthorhombic Ti–Al–Nb–Ta alloys, Intermetallics 8 (2000) 659–662.
[91] Y.Y. Chen, Y.F. Si, F.T. Kong, Trans. Nonferrous Met. Soc. China 16 (2006) 316–320.
[92] Y.Y. Chen, Y.F. Si, C.G. Liu, Trans. Nonferrous Met. Soc. China 16 (2006) 849–853.
[93] F. Tang, S. Nakazawa, M. Hagiwara, Mater. Sci. Eng. A 315 (2001) 147–152.
[94] F. Tang, S. Nakazawa, M. Hagiwara, Mater. Sci. Eng. A 329-331 (2002) 492–498.
[95] S. Emura, M. Hagiwara, in: Proceedings of the 9th Conference of Titanium, 7–11, 1999, p. 298.
[96] X. Chen, Z. Weidong, W. Wei, L. Xiaobo, Z. Jianwei, The enhanced tensile property by introducing bimodal size, Mater. Sci. Eng. A 587 (2013) 54–60.
[97] Y.W. Kim, J. Met. 46 (1994) 31.
[98] M. Yamaguchi, H. Inui, K. Ito, High-temperature structural intermetallics, Acta Mater. 48 (2000) 307–322.
[99] J.B. Singh, G. Molenat, M. Sundraraman, S. Banerjee, G. Saada, P. Veyssiere, A. Courete, Philos. Mag. Lett. 86 (2006) 47–60.
[100] Y.W. Kim, J. Met. 41 (1989) 24.
[101] W.J. Zhang, S.C. Deevi, in: D.M. Dimiduk, H. Clemens, R. Darolia, H.J. Inui, M. Larsen, V.K. Sikka, M. Thomas, J.D. Whittenberger, K.J. Hemker (Eds.), Structural Intermetallics, TMS, Warrendale, 2001, p. 699.
[102] L.M. Hsiung, T.G. Nieh, Creep deformation of fully lamellar TiAl controlled by the viscous glide of interfacial dislocations, Intermetallics 7 (1999) 821–827.
[103] J. Beddoes, L. Zhao, P. Au, D. Dudzinsky, J. Triantafillou, in: R. Darolia, C.T. Liu, P.L. Martin, D.B. Miracle, R. Wagner, M. Yamaguchi, M.V. Nathal (Eds.), Structural Intermetallics, TMS, Warrendale, 1997, pp. 109.
[104] J. Beddoes, W. Wallace, L. Zhao, Int. Mater. Rev. 40 (1995) 197.
[105] J.W. Christian, The Theory of Transformations in Metals and Alloys, Pregamon Press, s.l., 1975.
[106] M. Es-souni, A. Bartels, R. Wagner, Creep behaviour of a fully transformed near γ-TiAl alloy Ti-48Al-2Cr 43 (1995) 153–161.
[107] J.N. Wang, T.G. Nieh, The role of ledges in creep of TiAl alloys with fine lamellar structures, Acta Mater. 46 (1998) 1887–1901.

[108] P.-I. Gouma, K. Subramanian, Y.-W. Kim, M.J. Mills, Annealing studies of γ-titanium aluminides alloyed with light elements for creep strengthening, Intermetallics 6 (1998) 689–693.
[109] W.R. Chen, L. Zhao, J. Beddoes, Precipitation hardening of the fully lamellar structure of investment cast Ti-47Al-2Nb-1Mn-0.5Mo-0.5W-0.2Si alloy, Scr. Mater. 41 (1999) 597–603.
[110] C.G. McKamey, J.H. DeVan, P.F. Tortorelli, V.K. Sikka, A review of recent developments in Fe_3Al-based alloys, J. Mater. Res. 6 (1991) 1779–1805.
[111] S.C. Deevi, V.K. Sikka, Nickel and iron aluminides: an overview on properties, processing, and applications, Intermetallics 4 (1996) 357–375.
[112] V.K. Sikka, in: V.K. Sikka, P.J. Maziasz, R.W. Cahn, S.C. Deevi (Eds.), International Symposium on Nickel and Iron Aluminides, ASM International, Materials Park, OH, 1997, p. 361.
[113] N.S. Stoloff, Iron aluminides: present status and future prospects, Mater. Sci. Eng. A 258 (1998) 1–14.
[114] N. Ziegler, Trans. AIME 100 (1932) 267.
[115] A. Lawley, J.A. Coll, R.W. Cahn, Trans. Met. Soc. AIME 218 (1960) 166.
[116] R.W. Cahn, J.A. Coll, Acta Metall. 9 (1961) 138.
[117] A. Lawley, E.A. Vidoz, R.W. Cahn, Acta Metall. 9 (1961) 1287.
[118] N.S. Stoloff, R.G. Davies, Acta Metall. 12 (1964) 473.
[119] R.G. Davies, N.S. Stoloff, Acta Metall. 11 (1963) 1187.
[120] G. Sainfort, P. Mouturat, P. Pein, J. Petit, G. Cabane, M. Salesse, Mem. Sci. Rev. Met. 60 (1963) 125.
[121] N.S. Stoloff, R.G. Davies, Prog. Mater. Sci. 13 (1966) 3.
[122] J.H. Westbrook, R.L. Fleischer (Eds.), Structural Applications of Intermetallic Compounds, Wiley, New York, 2000.
[123] W.R. Chen, L. Zhao, J. Beddoes, Scr. Mater. 41 (1999) 597.
[124] D.Y. Seo, J. Beddoes, L. Zhao, G.A. Botton, Mater. Sci. Eng. A 810 (2002) 329–331.
[125] T.A. Parthasarathy, P.R. Subramanian, M.G. Mendiratta, D.M. Dimiduk, Acta Mater. 48 (2000) 541.
[126] N.S. Davies, R.G. Stoloff, Prog. Mater. Sci. 13 (1966) 1.
[127] W.R. Chen, J. Triantafillou, J. Beddoes, L. Zhao, Intermetallics 7 (1999) 171.
[128] Y. Umakoshi, M. Yamaguchi, Y. Namba, M. Murakami, Acta Metall. 24 (1976) 89.
[129] W. Schroer, C. Hartig, H. Mecking, Z. Metallkd. 84 (1993) 294.
[130] G. Hasemann, J.H. Schneibel, M. Krüger, E.P. George, Vacancy strengthening in Fe_3Al iron aluminides, Intermetallics 54 (2014) 95–103.
[131] J. Beddoes, D.Y. Seo, W.R. Chen, L. Zhao, Intermetallics 9 (2001) 915.
[132] G. Sharma, R. Kishore, M. Sundararaman, R.V. Ramanujan, Superplastic deformation studies in Fe–28 Al–3Cr intermetallic alloy, Mater. Sci. Eng. A 419 (2006) 144–147.
[133] Deleted in review
[134] R.S. Sundar, S.C. Deevi, Mater. Sci. Eng. A 357 (2003) 124.
[135] E.M. Schulson, J. Nucl. Mater. 66 (1977) 322.
[136] E.M. Schulson, Intermetallic Compounds, in: J.H. Westbrook, R.L. Fleischer (Eds.), John Wiley and Sons Ltd., s.l., 1994, p. 133.
[137] E.M. Roy, J.A. Schulson, J. Nucl. Mater. 71 (1977) 124.
[138] H.E. Wilkins, B.J.S. Rosinger, J. Nucl. Mater. 89 (1980) 347.
[139] K. Izumi, O. Aoki, J. Mater. Sci. 14 (1979) 1400.
[140] R. Tewari, G.K. Dey, S. Banerjee, N. Prabhu, Metall. Mater. Trans. 37A (2008) 49–58.
[141] R. Tewari, G.K. Dey, R.K. Fotedar, T.R.G. Kutty, N. Prabhu, Metall. Trans. 35A (2004) 205.
[142] R. Tewari, G.K. Dey, T.R.G. Kutty, A.K. Sengupta, N. Prabhu, S. Banerjee, Metall. Trans. 35A (2004) 189.
[143] R. Tewari, T.R.G. Kutty, G.K. Dey, S. Banerjee, Study on Microstructural Evolution, Phase Transformations and High Temperature Properties on Zr_3Al and Its Ternary Derivatives for Structural Applications in Indian Pressurized Heavy Water Reactors, BARC, s.l., 2009.

[144] A.H. Cottrell, Mater. Sci. Tech. 7 (1991) 981.
[145] C.G. McKamey, P.J. Maziasz, J.W. Jones, J. Mater. Res 7 (1992) 2089.
[146] J. Douinl, F.S. Kumar, P. Veyssiere, Dislocation structure in $L1_2$ Mn-stabilized Al_3Ti deformed between 77 K and 873 K, Mater. Sci. Eng. A 192–193 (1995) 92–96.
[147] R. Morris, D.G. Lerf, Plastic deformation of an iron-modified titanium trialuminide alloy, Acta Metall. Mater. 39 (1991) 2419–2430.
[148] B.C. Muddle, W.-S. Chang, Intermediate phases and phase relations in the composition range Al3(Ti, V) to TiAl in the Al–Ti–V system, Mater. Sci. Eng. A 251 (1998) 232–242.
[149] K.S. Kumar, in: J.J. Lewandowski, C.T. Liu, P.L. Martin, R. Darolia (Eds.), Structural Intermetallics, TMS, Warrandale, 1993, pp. 87–96.
[150] R.A. Varin, M.B. Winnicka, I.S. Virk, in: J.J. Lewandowski, C.T. Liu, P.L. Martin, R. Darolia (Eds.), Structural Intermetallics, TMS, Warendale, 1993, pp. 117–124.
[151] M. Heilmaier, D. Handtrack, R.A. Varin, Creep properties of boron-doped dual-phase Ti-rich $L1_2$-based trialuminide intermetallics, Scr. Mater. 48 (2003) 1409–1414.
[152] K. Yasuda, T. Fujiwada, H. Kodama, M. Suwa, High Temperature Order Intermetallic Alloys IV, MRS, s.l., 1991, pp. 865–870.
[153] D.L. Shah, D.M. Anton, High Temperature Order Intermetallic Alloys IV, vol. 133, MRS, s.l., 1989, pp. 361–371.
[154] P.M. Hazzledine, High temperature ordered intermetallic alloy V, MRS Symposium, 1993, p. 591.
[155] O. Izumi, Mater. Trans. JIM 30 (1989) 639.
[156] K. Inoue, K. Tachikawa, IEEE Trans. Magn. 13 (1977) 840.
[157] J.D. Hall, E.L. Livingston, J. Mater. Res. 5 (1990) 5.
[158] F. Chu, D.P. Pope, High Temperature Ordered Intermetallic VI, MRS Symposium, 1995, p. 1197.
[159] F. Chu, D.P. Pope, Mater. Sci. Eng. A 170 (1993) 39.
[160] A.V. Kazantzis, M. Aindow, I.P. Jones, G.K. Triantafyllidis, J.ThM. De Hosson, Acta Mater. 55 (2007) 1873–1884.
[161] D.E. Luzzi, G. Rao, T.A. Dobbins, D.P. Pope, Acta Mater. 46 (1998) 2913–2927.
[162] C.H. Henager, S.M. Bruemmer, J.P. Hirth, Mater. Sci. Eng. A 170 (1993) 185–197.
[163] C.H. Henager, R.E. Jacobson, S.M. Bruemmer, Mater. Sci. Eng. A (1992) 416–421.
[164] S.M. Bruemmer, L.A. Chariot, J.L. Brimhall, C.H. Henager, J.P. Hirth, Philos. Mag. 65 (1992) 1083–1094.
[165] C.K. Dorn, W.J. Haws, E.E. Vidal, Fusion Eng. Des. 84 (2009) 319–322.
[166] M. Akinc, M.K. Meyer, M.J. Kramer, A.J. Thom, J.J. Huebsch, B. Cook, Boron-doped molybdenum silicides for structural applications, Mater. Sci. Eng. A 261 (1999) 16–23.
[167] K.S. Kumar, Intermetallics: Silicides, in: Encyclopedia of Materials: Science and Technology, second ed., 2001, pp. 4243–4246.
[168] R.W. Mann, L.A. Clevenger, P.D. Agnello, F.R. White, Silicides and local interconnections for high-performance VLSl applications, IBM J. Res. Dev. 39 (July 4, 1995) 403–415.
[169] P.R. Subramanian, M.G. Mendiratta, D.M. Dimiduk, M.A. Stucke, Advanced intermetallic alloys—beyond gamma titanium aluminides, Mater. Sci. Eng. A 239–240 (1997) 1–13.
[170] M.R. Jackson, B.P. Bewlay, R.G. Rowe, D.W. Skelly, H.A. Lipsitt, High-temperature refractory metal-intermetallic composites, J. Met. (January 1996) 39–44.
[171] B.P. Bewlay, M.R. Kackson, M.F.X. Gigliotti, Niobium silicide high temperature in situ composites, in: R.L. Fleischer, J.H. Westbrook (Eds.), Intermetallic Compounds – Principles and Practice: Progress, vol. 3, John Wiley & Sons Ltd., s.l., 2002, pp. 541–560.
[172] J.R. Stephens, D.W. Petrasek, R.H. Titran, Refractory Metal Alloys and Composites for Space Power Systems. Nevada, MRS, NASA, 1988, 20040817 021.
[173] D.M. Berczik Jr., (US Patent 055956616 US), 1997.

[174] J.H. Schneibel, M.J. Kramer, O. Unal, R.N. Wright, Processing and mechanical properties of a molybdenum silicide with the composition Mo−12Si−8.5B (at.%), Intermetallics 9 (2001) 25−31.
[175] R.A. Perkins, K.T. Chiang, G.H. Meier, Effect of Alloying, Rapid Solidification, and Surface Kinetics on the High Temperature Environmental Resistance of Niobium, AFOSR, s.l., 1987. LMSC-F195926(1987).
[176] P.R. Subramanian, M.G. Mendiratta, D.M. Dimiduk, Mater. Res. Soc. Symp. Proc. 322 (1994) 491−502.
[177] D.M. Dimiduk, M.G. Mendiratta, P.R. Subramanian, in: J.J. Lewandowski, C.T. Liu, P.L. Martin, D.B. Miracle, M.V. Nathal, R. Darolia (Eds.), Structural Intermetallics, TMS, Warrendale, PA, 1993, pp. 619−630.
[178] M.G. Dimiduk, D.M. Mendiratta, Strength and toughness of a Nb/Nb_5Si_3 composite, Metall. Mater. Trans. A 24A (1994) 501−504.
[179] B.P. Bewlay, S.D. Sitzman, L.N. Brewer, M.R. Jackson, Analyses of eutectoid phase transformations in Nb−Silicide in situ composites, Microsc. Microanal. 10 (2004) 470−480.
[180] Cyclic-oxidation resistance of niobium-base in situ composites: modeling and experimentation, Oxid. Met. 61 (2004) 165−194.
[181] B.P. Bewlay, M.R. Jackson, H.A. Lipsitt, The balance of mechanical and environmental properties of a multielement niobium-niobium silicide-based in situ composite, Metall. Mater. Trans. A 27A (1996) 3801−3811.
[182] E.S.K. Menon, M.G. Mendiratta, D.M. Dimiduk, in: D.M. Dimiduk, H. Clemens, R. Darolia, H. Inui, J.M. Larsen, V.K. Sikka, M. Thomas, J.D. Whittenberger, J.J. Hemker (Eds.), Structural Intermetallics, TMS, Warrendale, 2001, pp. 591−600.
[183] D.L. Chan, K.S. Davidson, Microstructural and fracture characterization of Nb-Cr-Ti mechanically alloyed materials, Metall. Mater. Trans. A 33A (2002) 401−416.
[184] D.L. Shah, D.M. Anton, in: Materials Research Society Symposia Proceedings, vol. 194, Materials Research Society, s.l., 1990, pp. 45−52.
[185] Z. Li, L.M. Peng, Microstructural and mechanical characterization of Nb-based in situ composites from Nb−Si−Ti ternary system, Acta Mater. 55 (2007) 6573−6585.
[186] D.L. Davidson, K.S. Chan, D.L. Anton, The effects on fracture toughness of ductile-phase composition and morphology in Nb-Cr-Ti and Nb-Si in situ composites, Metall. Mater. Trans. 27A (1996) 3007−3018.
[187] K.S. Chan, D.L. Davidson, D.L. Anton, Fracture toughness and fatigue crack growth in rapidly quenched Nb-Cr-Ti in situ composites, Metall. Mater. Trans. 28A (1997) 1797−1808.
[188] F. Ye, C. Mercer, W.O. Soboyejo, An investigation of the fracture and fatigue crack growth behavior of forged damage-tolerant niobium aluminide intermetallics, Metall. Mater. Trans. 29A (1998) 2361−2374.
[189] R. Tewari, H. Song, G.K. Dey, A. Chatterjee, V.K. Vasudevan, Microstructural evolution in niobium based alloys, Mater. Trans. 39A (2008) 1506.
[190] I. Sha, H. Hirai, T. Tabaru, A. Kitahara, H. Ueno, S. Hanada, Mechanical properties of as-cast and directionally solidified Nb-Mo-W-Ti-Si in-situ composites at high temperatures, Metall. Mater. Trans. A 34 (2003) 85−94.
[191] K. Zelenitsas, P. Tsakiropoulos, Effect of Al, Cr and Ta additions on the oxidation behaviour of Nb−Ti−Si in situ composites at 800°C 416 (2006) 269−280.
[192] J.H. Perepezko, The hotter the engine, the better, Science 326 (2009) 1068−1069.
[193] M.K. Meyer, A.J. Thom, M. Akinc, Oxide scale formation and isothermal oxidation behavior of Mo-Si-B intermetallics at 600−1000°C, Intermetallics 7 (1999) 153−162.
[194] R. Thompson, L.R. Raj, Design of the microstructural scale for optimum toughening in metallic composites, Acta Metall. Mater. 42 (1994) 4135−4142.
[195] B.P. Bewlay, M.R. Jackson, J.-C. Zhao, P.R. Subramanian, M.G. Mendiratta, J.J. Lewandowsk, Ultrahigh-temperature Nb-Silicide-based composites, MRS Bull. (2003) 643−653.

[196] M.R. Jackson, K.D. Jones, in: D.R. Sadoway, R.G. Bautista, K.C. Liddell (Eds.), Mechanical Behavior of Nb − Ti Base Alloys, TMS, Warrendale, 1990, pp. 311−320.
[197] J.W. Yeh, Y.I. Chen, S.J. Lin, S.K. Chen, High-entropy alloys − a new era of exploitation, Mater. Sci. Forum 560 (2007) 1−9.
[198] J.W. Yeh, S.K. Chen, S.J. Lin, J.Y. Gan, T.S. Chin, Nanostructured high entropy alloys with multiple principal elements novel alloys design concepts and outcomes, Adv. Eng. Mater. 6 (2004) 299−303.
[199] Z.P. Lu, H. Wang, M.W. Chen, I. Baker, J.W. Yeh, C.T. Liu, T.G. Nieh, An assessment on the future development of high-entropy alloys:Summary from a recent workshop, Intermetallics 66 (2015) 67−76.
[200] O.N. Senkov, G.B. Wilks, D.B. Miracle, C.P. Chuang, P.K. Laiw, Refractory high entropy alloys, Intermetallics 18 (2010) 1758−1765.
[201] F. Otto, Y. Yang, H. Bei, E.P. George, Relative effects of enthalpy and entropy on the phase stability of equiatomic high entropy alloys, Acta Mater. 61 (2013) 2628−2638.
[202] K.Y. Tsai, M.H. Tsai, J.W. Yeh, Sluggish diffusion in Co−Cr−Fe−Mn−Ni high-entropy alloys, Acta Mater. 61 (2013) 4887−4897.
[203] S. Guo, C.T. Liu, Phase stability in high entropy alloys:formation of solid-solution phase or amorphous phase, Prog. Nat. Sci. Mater. Int. 21 (2011) 433−446.
[204] Y. Zhang, T.T. Zuo, Z. Tang, M.C. Gao, K.A. Dahmen, P.K. Liaw, Z.P. Lu, Microstructures and properties of high-entropy alloys, Prog. Mater. Sci. 61 (2014) 1−93.
[205] X. Yang, Y. Zhang, Prediction of high-entropy stabilized solid-solution in multi-component alloys, Mater. Chem. Phys. 132 (2012) 233−238.
[206] O.N. Senkov, J.M. Scott, S.V. Senkova, D.B. Miracle, C.F. Woodward, Microstructure and room temperature properties of a high-entropy TaNbHfZrTi alloy, J. Alloys Compd. 509 (2011) 6043−6048.
[207] O.N. Senkov, S.V. Senkova, C. Woodward, D.B. Miracle, Low-density, refractory multi-principal element alloys of the Cr−Nb−Ti−V−Zr system: microstructure and phase analysis, Acta Mater. 61 (2013) 1545−1557.
[208] O.N. Senkov, S.V. Senkova, C. Woodward, D.B. Miracle, Mechanical properties of low-density, refractory multiprincipal element alloys of the Cr−Nb−Ti−V−Zr system, Mater. Sci. Eng. A 565 (2013) 51−62.
[209] O.N. Senkov, J.M. Scott, S.V. Senkova, F. Meisenkothen, D.B. Miracle, C. Woodward, Microstructure and elevated temperature properties of a refractory TaNbHfZrTi alloy, J. Mater. Sci. 47 (2012) 4062−4074.
[210] M.-H. Tsai, C.-W. Wang, C.-W. Tsai, W.-J. Shen, J.-W. Yeh, J.-Y. Gan, W.-W. Wu, Thermal stability and performance of NbSiTaTiZr high-entropy alloy barrier for copper metallization, J. Electrochem. Soc. 158 (2011) 1161−1165.
[211] P.K. Huang, J.W. Yeh, Inhibition of grain coarsening up to 1000°C in (AlCrNbSiTiV)N superhard coatings, Scr. Mater. 62 (2010) 105−108.
[212] W.H. Liu, Y. Wu, J.Y. He, T.G. Nieh, Z.P. Lu, Grain growth and the Hall-Petch relationship in a high-entropy FeCrNiCoMn alloy, Scr. Mater. 68 (2013) 526−529.
[213] P.P. Bhattacharjee, G.D. Sathiaraj, M. Zaid, M.J.R. Gatti, C. Lee, C.W. Tsai, J.W. Yeh, Microstructure and texture evolution during annealing of equiatomic CoCrFeMnNi high-entropy alloy, J. Alloys Compd. 587 (2014) 544−552.
[214] F. Otto, A. Dlouhý, C. Somsen, H. Bei, G. Eggeler, E.P. George, The influences of temperature and microstructure on the tensile properties of a CoCrFeMnNi high-entropy alloy, Acta Mater. 61 (2013) 5743−5755.
[215] C. Zhu, Z.P. Lu, T.G. Nieh, Incipient plasticity and dislocation nucleation of FeCoCrNiMn high-entropy alloy, Acta Mater. 61 (2013) 2993−3001.

CHAPTER

SYNTHESIS AND CHARACTERIZATION OF BORIDES, CARBIDES, AND NITRIDES AND THEIR APPLICATIONS

10

R.S. Ningthoujam

Bhabha Atomic Research Centre, Mumbai, India

CHAPTER OUTLINE

Materials Under Extreme Conditions. http://dx.doi.org/10.1016/B978-0-12-801300-7.00010-3

1. INTRODUCTION

Borides, carbides, and nitrides are interesting materials in terms of electronic, optical, thermal, and magnetic properties [1–5]. Some of them show high-superconducting transitions e.g., NbN (T_c ~ 17K), layered $La_3Ni_2B_2N_3$ (12K), Y–Ni–B–C (12–14K), YbPdBC (23K), layered $Li_{0.48}(THF)_yHfNCl$ (26K), MgB_2 (39K) similar to high T_c Cu–O based superconductors. VN is theoretically predicted to show T_c at 30K, but exhibits T_c at 9K (experimentally) due to presence of high electron density states at Fermi level [1,6–14]. Bardeen–Cooper–Schrieffer theory cannot explain many superconducting properties of nitrides, carbides, and nitrides, which is due to unusual chemical bonding and structures. Some of them are metallic/semiconducting/insulating (e.g., Ni_3N (metallic, $\rho_{300K} \sim 152\ \mu\Omega$ cm), TaN (superconducting, $T_c \sim 10.8$K), $TaN_{1.5}$ (insulating, $\rho_{100K} \sim 0.1$ kΩ cm), ScN (semiconducting, $\rho_{300K} \sim 2$ mΩ cm, band gap ~1.3–2.4 eV), BN (insulating, $\rho \sim 10^5\ \Omega$ cm, band gap ~5.9 eV (*h*-phase), 6.1 eV (*c*-phase)), *c*-C_3N_4 (insulating, $\rho \sim 10^{12}\ \Omega$ cm, band gap ~4.75 eV), SiN (insulating, $\rho \sim 10^{14}\ \Omega$ cm, band gap ~4–5 eV) [15–26]. The AlN has band gap of 6.2 eV, which is more than that of any nitrides, carbides, and borides until today [23]. Materials that are electrical insulators but have high thermal conductivity (σ) are Ge_3N_4, Si_3N_4, C_3N_4 (>100 W/m K) and BN (σ = 200–500 W/m K (*h*-phase), 1620 W/m K (nanotubes)), whereas diamond has thermal conductivity ~3300 W/m K [27–30]. Even C_3N_4 is predicted to have very high thermal conductivity of 520 W/m K, which is much higher than that of copper (400 W/m K)

[27]. Among carbon allotropes, the amorphous carbon or diamond-like carbon has lower thermal conductivity than pure diamond. Pure diamond exhibits intrinsic properties such as more reduction in phonon scattering, higher Debye temperature, and purely sp^3 hybridization [29]. Diamond with more amount of ^{12}C isotope has higher thermal conductivity than others [30]. The $c\text{-}C_3N_4$ ($H = 5260$ kgf/mm^2), Si_3N_4 ($H = 43$ GPa), and c-BN ($H = 33-100$ GPa) are the ultra-hard (super-hard) materials comparable with diamond (100 GPa) [22,31–35]. The AlN, GaN, InN, GaAsN, and their mixtures are used as ultraviolet (UV), blue, green, white, and near-infrared (NIR) emitters [36–40]. The diodes operating in the blue and near-UV spectral range are useful in various applications requiring high resolution, such as high-speed laser printers, scanners, and high-density optical data storage. The GaN (blue) and AlN (UV at 210 nm) diodes are some of the examples [37]. Some of them are used as detectors in UV–visible–NIR range [37]. The LaB_6 is used as electron emitter (thermionic materials) in electron microscopy because of lower work function ($\phi = 2.5$ eV) than W (tungsten, $\phi = 4.5$ eV) [41]. In addition, LaB_6 has high conductivity, high melting point, and is chemically stable. Because of these properties, this is used in field ion and field electron emitters [41]. The TiN and ZrN are used as coating substance over the materials, and thus materials are produced as artificial ornaments because the color of TiN or ZrN is similar to gold and they also have high-abrasive resistance [42]. Some of hard materials (WC, BN, TiB_2–TiC, $TiCrB_n$, TiSiBN) are used in cutting and drilling in machines/manufactures [2,3,43–46]. Most of nitrides/carbides/borides (TiN, TaN, WN, SiN, SiC, CN, WC, BN) are refractory materials because of their high melting points and chemical stability [2–4,43–48]. Coating of nitrides and carbides over instruments makes them noncorrosive [2,3] and thus durability increases. Several materials of nitrides, carbides, and borides are used in oil recovery and as catalysts for conversion of alcohol to ketone/aldehydes [2,3,48–50]. Some of them show high magnetization ($Fe_{16}N_{12}$, 260–286 emu/g), which is more than Fe (218 emu/g) [51–55]. The VN is paramagnetic, whereas CrN is antiferromagnetic [56,57]. In FeN_x, as x increases, it becomes ferromagnetic ($x < 0.3$) to antiferromagnetic ($x = 0.5$) and paramagnetic ($x = 1.0$) [58–61]. The BN is used as sample holder in vibrating sample magnetometer for high-temperature measurement because of its diamagnetic nature as well as thermal stability, while SiN_x (diamagnetic) is used as cantilever in atomic force microscopy (AFM) [62]. Most of cutting tools, blades, hypersonic vehicles, missiles, rockets, satellites, and drilling tools are prepared from nitrides/carbides/borides as main components [2–4,63,64]. Sometimes, they are used as insulating layer in electrical purposes [65,66]. Some are highly stable in very high temperature with low thermal expansion [67–70]. Such materials can be used in nuclear reactor [70]. Some of them show negative thermal expansion [70].

The light emitting diodes (LEDs) are the basic component for the electronic lighting technology. LEDs are the semiconducting devices that emit when a sufficient electric current is passed through it. This has p-n junction. The schematic diagram is shown in Fig. 10.1. When a sufficient current is passed through diode, electrons (*es*) in the conduction band of semiconductor flow across the junction from the n-doped side, and holes (*hs*) in the valence band flow from p-side. This results in a band bending. At the p-n junction, *es-hs* recombination takes place and light is emitted with an energy $h\nu \leq E_g$ ($E_g =$ band gap). There are two types of emission/absorption: direct and indirect transitions. In direct transition, energy for transition is almost equal to band gap, whereas in indirect transition, there is photon–phonon interaction and energy of transition is much less than band gap. The transition probability of indirect transition is less than that of direct transition. GaN shows direct band emission [5,71].

High-pressure and temperature treatment can lead to production of new phases of compounds including metastable phase, which cannot be prepared at ambient condition. This will be important in

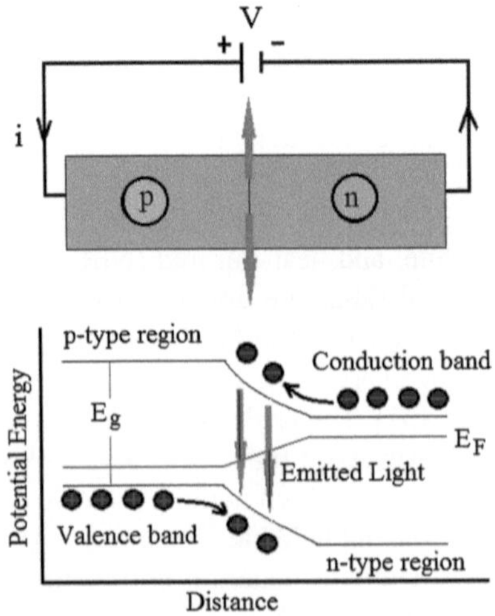

FIGURE 10.1

Schematic diagram of the light emitting diodes (LEDs).

preparation of nitrides, carbides, and borides [72–81]. Even coordination of metal and nonmetal can be increased and, similarly, hybridization (sp^2 to sp^3) can be changed under such conditions. New compounds [IrN_2, OsN_2, PtN_2, ε-Fe_2TMN (TM = Co, Ni)] are coming up.

In the case of binary transition metal nitrides and carbides (i.e., MN, MC), crystal structures are usually simple. The low concentration of N or C atoms can occupy interstitial sites of metal lattice (octahedral voids) [2,3,80,81] and covalent type of chemical bond is formed between *d*- orbital of metal atom M and *p*- orbital of N or C. Volume expansion of metal up to 1–10% occurs with addition of N/C in to M. With further increase on N/C in lattice, crystal structure of metal changes to another depending on synthesis conditions. Determination of crystal structures of binary nitrides of *s*-, *p*-, and *f*-block elements is more complicated as compared to that of *d*-block elements. In case of other ternary/quaternary nitrides, prediction of crystal structure is difficult. On other hand, metal borides (M and B) have many phases (metal rich to boron rich) as compared to nitrides/carbides [4,82,83]. For example, Cr_4B, Cr_2B, Cr_5B_3, CrB, Cr_3B_4, CrB_2, CrB_4, and CrB_6. Volume expansion of M does not usually occur in case of borides after addition of B in M. Instead, it forms another crystal structure contrary to that observed in transition metal nitrides or carbides.

The pseudobinary (substituted ones), ternary, and quaternary borides/carbides/nitrides are also important materials, which have different properties from pure binary ones [1–13]. Some are of layered types in structure. This brings novel properties in terms of electronic, optical, and magnetic properties. They are used as ultrahigh temperature ceramics in extreme conditions (pressure, temperature, radiation, environments, etc.). In generation of advanced nuclear reactors, such materials and their composites with other metals/nonmetals will be useful.

In this chapter, the different techniques of preparation of nitrides existing as of now have been elucidated. Also, a few techniques of preparation of carbides and borides are mentioned. Their chemical bonding and structures are discussed. The electrical resistivity, thermal conductivity, optics, and magnetism of these materials are discussed. Their importance in applications is highlighted.

2. PHENOMENA

2.1 CHEMICAL BONDING

2.1.1 Metal Nitrides and Carbides

Hägg in 1931 suggested empirical rules for prediction of crystal structure of MX, X = N or C or nonmetal atom on basis of atomic radius ratio $r = r_X/r_M$ [2]. When $r < 0.59$, the structure becomes simple such as face-centered cubic (*fcc*), hexagonal closed packed (*hcp*), or simple hexagonal (*sh*), etc. N/C atoms can occupy the interstitial sites of octahedral or cubic prismatic geometry of metal sublattice and cannot occupy the interstitial sites of the tetrahedral geometry since size of tetrahedral site is too small as compared to that of N/C. Usually, binary transition metal nitrides follow Hägg's empirical rules, e.g., stoichiometric VN is *fcc* in which coordinates of V and N are (0, 0, 0), (½, ½, 0) and (½, 0, 0), (½,½,½), respectively [84]. When $r > 0.59$, the complex structure is formed in order to accommodate the larger nonmetal atoms (X) in interstitial sites of metal (M) lattice.

There are many compounds that do not follow Hägg-like empirical rules. In such non-Hägg-like system, C or N atoms do not occupy interstitial sites of metal lattice fully, e.g., ε-TaN [2]. In ternary metal nitrides/carbides ($M_xM_y'X_z$ in which M = Mn, Fe, Co, Ni; M′ = Zn, Cd, Al, Ga; X = N/C and x, y, z are numbers), tetragonal, layered-type and/or antiperovskite structures are found [8,85–87]. Giant negative thermal expansion $\alpha = -25 \times 10^{-6}\ K^{-1}$ in $Mn_{1-x}Ge_xN$ has been reported [86]. The organic/inorganic molecules can occupy the layered/interfacial sites in some cases [12,88].

Metal can form bond with N/C (transition metal nitrides/carbides, e.g., VN, VC [1–3,89], alkali or alkaline earth metal nitrides/carbides, e.g., Li_3N, Mg_3N_2, Li_2C_2, Mg_2C [90–93], *f*-block metal nitrides/carbides, e.g., GdN, GdC [94,95], *p*-block nitrides/carbides, e.g., Si_3N_4, SiC) [25,96]. The M–N or M–C bonds will be of ionic or covalent or mixture type, whereas N–N or C–C bonds show the covalent type. The chemical bonding in MN/MC has been proved by the experimental as well as theoretical results, e.g., ultraviolet photoelectron spectroscopy (UPS) and X-ray photoelectron spectroscopy (XPS) spectrum of VN (stoichiometry) shows a narrow peak closed to E_F (Fermi energy) followed by a broad one at 5 eV binding energy (BE) with two shoulders at 4 and 7 eV [89,97]. The shoulder at 7 eV is attributed to V-3*d*-like states while that at 4 eV is attributed to N 2*p* character. In case of nonstoichiometry ($VN_{0.72}$), an extra peak at 2 eV BE is observed.

2.1.2 Metal Borides

Generally, boron forms covalent bonds with metals in metal borides [98]. The types of bonding are M–B, M–M, and B–B interactions. The structure of metal borides is complex because M can be surrounded by B atoms and B can be surrounded by M atoms. They can form linear, planar, and 3*D* structures. The M/B ratio can change from M_2B, MB, M_3B_4, and MB_2 to MB_{12}. They are either crystalline or amorphous. In planar structure, B contributes sp^2-hybridization, whereas B can have sp^3-hybridization in 3*D*-structure. In some MB_2 systems (e.g., AlB_2), B atoms arrange with graphite-like motif in which metals are electron donor, whereas B atoms are electron acceptor. The borides can be formed with many metals in the Periodic Table [8,99–103].

1. *s*- and *p*-metal borides
   NaB_6, KB_6, Be_5B to BeB_9 (six possible compounds), MgB_2, CaB_6, AlB_2, AlB_{12}, BC, BN.
2. *d*-metal borides
   ScB, TiB, Ti_2B, Ti_3B, Ti_4B, Ti_3B_4, VB, CrB, MoB, FeB, Fe_2B, ZrB_2, TcB_2.
3. *f*-metal borides

ThB_4, ThB_6, UB_2, UB_{12}, PuB, PuB_2, PuB_4, PuB_6

4. Metal-nonmetal-boron ternary system
   a. M-C-B (TiC_5BTi_{12}, $TiC_4B_2Ti_{12}$, LiBC, $Al_8B_4C_7$, UB_2C, $U_5B_2C_7$)
   b. M-N-B (TiB_2−TiN)
   c. M-O-B ($TiB_{0.75}O_{0.25}$, $Li_2B_4O_7$)
   d. M-P-B ($Fe_3B_{0.9}P_{0.1}$)
5. Metal-metal-B ternary system
   $Y_2Fe_{14}B$, $Y_2Co_{14}B$, $LuRh_4B_4$, $Li_3Ni_2B_2$
6. Amorphous metal borides
   Amorphous metal borides are exhibiting as metallic glasses as well as unusual magnetic characteristics.
   Ni_2Fe_2B or $Ni_{40}Fe_{40}B_{20}$
7. Quaternary/higher system of boron
   $YPd_5B_3Co_x$ ($0.3 \leq x \leq 0.4$), $Li_3Ni_2B_2N_3$, $ThPd_3B_2C$, $ThPd_2B_2C$, $ThNi_2B_2C$, $ThPt_2B_2C$

2.2 MICROSTRUCTURES

Density of materials is depending on microstructure evolution under pressure, gas pressure sintering, hot pressing, or hot isostatic pressing. It is also depending on type of additive incorporated during sintering. There are roles of starting materials in getting high density. Generally, the starting materials having spherical particles with uniform size distribution produce high density during sintering. Microstructure can change properties of material in term of mechanical, electrical, optical, and thermal behaviors [104−108]. In this regard, evolution of the different microstructures during sintering of Si_3N_4 has been reported. To get a desired property, it is necessary to have controlled shape, size, and surrounding of particles. Change of microstructure is found when MoO_3 is nitrided in NH_3 to produce Mo_2N at higher temperature. The hardness of WC-10Co composition increases with decrease of grain size of WC.

2.3 CRYSTAL STRUCTURES

Crystal structure of material changes with pressure, temperature, and atmosphere/medium. Hexagonal MoN and cubic Mo_2N can be prepared by pyrolysis of $Mo(NH_2O)_n$ complex at 650 and 800°C, respectively, in NH_3 gas [109]. Many Fe−N phases can be prepared under different conditions (pressure, temperature, quenching) [60,87,110,111]. The IrN_2, OsN_2, and PtN_2 can be prepared under extreme pressures and temperatures [74,76].

The *c*-BN, Si_3N_4, Ge_3N_4, Hf_3N_4, and Fe−Mg−N can be prepared by using different pressure units: simple, 1.2−600 MPa; multianvil apparatus, $P < 25$ GPa; and diamond anvil cell, ~100 GPa [73,112]. Temperature up to 2300K (even more) can be provided using external and internal heating (resistive heater and high-intense laser-based heater). BN can have hexagonal and cubic phases at low and high temperatures/pressures, respectively. Cubic phase is a dense allotrope of boron nitride that has a diamond-like structure. Raman spectroscopy can determine variation in transverse optical phonon of cubic BN under extreme pressure and temperature [113]. Even X-ray diffraction technique can provide the variation of lattice parameters of material with pressure or temperature. Nanocrystalline cubic BN with particle size (10−30 nm) can be prepared from bulk using diamond anvil setup [114]. The phase transition from solid to liquid can be carried out using high pressure/temperature and this transition can also be predicted from the Clausius–Clapeyron equation [115].

3. THEORETICAL CONCEPTS

3.1 ELECTRICAL RESISTIVITY

The electrical resistivity (ρ) of metals comes from scattering of electrons (e) by phonons (ph), other electrons, defects and/or impurities [84,116–118]. The ρ has the relationship with mean free path (l):

$$\rho = \frac{3\pi^2\hbar}{e^2K_F^2 l} \tag{10.1}$$

where K_F is the Fermi wave vector. The mean free path (l) for Cu is large (about 10^2–10^3 Å), which is longer than interatomic distance (d_{hkl}). In temperature range 1–300K, the ρ has two parts: temperature-independent resistivity (like elastic scattering, ρ_e) at very low temperature, and temperature-dependent resistivity (like inelastic scattering, ρ_i) at higher temperature.

$$\rho_T = \rho_e + \rho_i$$

$$\rho_T = \rho_e + \beta T^n \tag{10.2}$$

where β is the temperature coefficient and its value is positive for metallic system. The power "n" varies from 2 to 5 depending on role of *e-e* and *e-ph* scattering in system when $T \ll \theta_D$ (Debye temperature). The "n" becomes fraction (3/2, 5/2, ...) in case of *e-m* (*m*, maganon) scattering. The resistivity increases with temperature in metallic systems. The value of ρ_e is found to increase with increasing impurities/defects present in the system. Beyond critical concentration of disorder/defect/impurity, metallic behavior changes to semiconducting/insulating behavior. In semiconductor, resistivity decreases with rise of temperature. In disordered system, there are two types of electron scattering at low temperature (i.e., weak electron localization and enhanced *e–e* scattering) [119,120]. Weak localization suggests the localized electrons after traveling in both forward and reverse ways. In case of enhanced electron-electron scattering, the electrons undergo intense scattering from the impurities. Their motion becomes diffusive and thus electron mobility reduces as compared to the clean metal. The total resistivity is expressed as:

$$\rho_T = \frac{1}{\sigma_0 + \alpha T^p} + \beta T^n \tag{10.3}$$

In weak electron localization, $p = 1.5$–3, whereas $p = 0.5$ for enhanced *e–e* scattering. This gives rise to minimum in resistivity with temperature at very low temperature.

There is another term called "Kondo effect" observed at very low temperature, which causes resistivity minimum or maximum [121–123]. This occurs when a few ppm (part per million) of transition metals (*d*-block ions) or magnetic impurities are present in noble metals (Ag/Au). Exchange interaction between magnetic spins separated by conduction electrons of noble metals gives rise to antiferromagnetic interactions. The total resistivity is defined by

$$\rho_T = \frac{1}{\sigma_0 + \alpha lnT} + \beta T^n \tag{10.4}$$

Sometimes, resistivity minimum or maximum is found when two ferromagnetic grains are separated by grain boundary or nonmagnetic layer of 1–20 Å [124]. The conduction of electrons occurs through tunneling. By assuming two ferromagnetic grains (i, j) separated by (d_{ij}), degree of spin polarization (P_s) of current carrier is given by

$$P_s = \frac{n\uparrow - n\downarrow}{n\uparrow + n\downarrow} \tag{10.5}$$

where $n\uparrow$ and $n\uparrow$ are the densities of states for spin-up and -down carriers, respectively. There is tunneling resistance (R_{ij}) between grains, which is given by

$$R_{ij} = \frac{r_{ij}}{1 + \varepsilon Cos\theta} \tag{10.6}$$

where $r_{ij} = r_0\exp(2K_Id_{ij})$ and $K_I = \sqrt{(2mU)}/\hbar$ and U is the tunneling barrier height. The ε is the spin-valve coefficient, which is equal to square of degree of spin polarization (i.e., $\varepsilon = P_S^2$). The dip in resistivity minimum decreases with increasing magnetic field.

In semiconductors, conduction mechanism of electrons follows hopping process at Fermi level (E_F) [18,125–127]. The nearest-neighbor hopping process is given by

$$\sigma_{nn} = C_{nn} \exp\left(-\frac{E_{nn}}{k_B T}\right) \tag{10.7}$$

where E_{nn} = activation energy of electrons at the Fermi level. C_{nn} is preexponential function. For the variable range hopping process, conductivity is given by

$$\sigma_{vr} = C_{vr} \exp\left[-\left(\frac{T_0}{T}\right)^{0.25}\right] \tag{10.8}$$

where T_0 is a parameter measured indirectly and is related to the density of centers participating in hopping conduction.

In superconductor, low T_c materials have relationship with Debye temperature (θ_D), Coulombic interaction parameter (μ^*), and electron-phonon coupling constant (λ) [128,129].

$$T_c = \frac{\theta_D}{1.45}\exp\left[-\frac{1.04(1+\lambda)}{\lambda - \mu^*(1+0.62\lambda)}\right] \tag{10.9}$$

The μ^* has relationship with density of states ($N(E_F)$).

$$\mu^* = \frac{0.20N(E_F)}{1 + N(E_F)} \tag{10.10}$$

3.2 THERMAL CONDUCTIVITY

The thermal conductivity (K_{tc}) of a material indicates its ability to conduct heat. In metals, conductivity is primarily due to free electrons. In any material, thermal conductivity is contributed by both electrons and lattice vibrations (phonons). At $T \ll \theta_D$ (Debye temperature), thermal conductivity of nonsuperconducting material follows trend similar to heat capacity (C_p) [2].

$$K_{tc} = \gamma T + eT^n + fT^m (n = 2 \text{ or } 3,\ m = 5-7) \tag{10.11}$$

where γ, e, and f are the temperature coefficients. Overall, total heat/thermal conductivity (K_{tc}) is sum of conductivities from electrons (K_e) and phonons (K_{ph}):

$$K_{tc} = K_e + K_{ph} \tag{10.12}$$

Under steady-state conditions and assuming no radial heat loss or gain, thermal conductivity (K_{tc}) is expressed by one-dimensional Fourier–Biot heat conduction equation [130]:

$$K_{tc} = -Q/A \times (\Delta X/\Delta T) \tag{10.13}$$

Q is the rate of heat flow through specimen having cross-sectional area (A). ΔT is temperature difference for a particular thickness (ΔX) of material.

The materials with good thermal conductivity but poor electrical conductivity are usually used in heat sink applications [131]. Those with good thermal conductivity (K_{tc}) as well as good electrical conductivity are Cu (385 W/m K) [132], silver (427 W/m K) [133], and gold (314 W/m K) [134]. Particularly, Cu (copper) is used as heat sink material because it possesses desirable properties for heat exchangers. In addition, Cu has high resistance to corrosion, biofouling, and microbial activities. Examples of materials having high thermal conductivity (K_{tc}, W/m K) are diamond (2200) [135], graphene (4840) [136], graphite (470) [137], and carbon nanotubes (3180–6600) [138]. Si_3N_4 with $K_{tc} = 200–450$ W/m K was also reported [139].

3.3 OPTICS

Optics is a branch of science that involves interaction of light with matter [140–142]. The semiconducting material absorbs light, which has energy equal to or slightly higher than band gap (ΔE). The distribution of electrons follow Boltzmann's law ($N = N_0\exp(-\Delta E/kT)$). Room temperature thermal energy (0.026 eV) is unable to bridge band gap. In order to bring electrons from the valence band to conduction band of semiconductor, it needs a higher temperature. For example, $\Delta E = 4$ eV of a semiconductor needs 4.6×10^4K, which is extremely difficult to achieve using furnace/source. However, this can be achieved by using photons/lights from suitable source (Xe, W, Hg lamps, laser, or cathode rays). When semiconductor is excited by a suitable light, the electrons (e) leave the holes (h) in the valence band and go to the conduction band. The e–h recombination occurs when excitation source is removed. This gives rise to emission in UV, visible, and NIR regions. Band edge emission is usually sharp, whereas surface emission is broad. Band edge emission of semiconductor occurs at higher energy than that of surface. There is difference in energy between absorption and emission bands due to association of nonradiative process via lattice/defects/impurity. Usually, absorption cross-section of semiconductor is very high due to allowed transition from valence to conduction bands ($\varepsilon_a = 10^{2-5}$ cm^{-1}). However, emission intensity is poor because the excited electrons (energy) undergo relaxation in different paths (nonradiatively). Sources of luminescence quencher are surface of particles, grain boundary, and surrounding environment. In case the surface of particles can be passivated using surfactants, emission intensity can be increased. Enhancement of luminescence with quantum yield up to 90% was observed in quantum dots of CdSe/CdS core–shell [143,144]. As the particle size of semiconductor decreases to nanometers, materials exhibit increase of band gap as compared to bulk. This usually happens only when size of particles is less than the exciton Bohr's radius. The particles whose sizes are less than exciton Bohr's radius are considered as quantum dots (QDs). The exciton Bohr's radius (R_B) is defined by the following relation:

$$R_B = \frac{m_0 \varepsilon_r a_H}{\mu^*}$$

where μ^* is the reduced mass of the effective masses of the electron and hole, respectively (m_e^* and m_h^*). The ε_r and m_0 are the dielectric constant of medium and the mass of free electron, respectively. The a_H is the Bohr's radius of hydrogen. The exciton forms Hydrogen-like atom with diameter significantly larger than the unit cell lengths/parameters (a, b, c). The excitons are the imaginary quasiparticles produced by pairing of electrons (from the conduction band) and holes (from the valence band), due to the Coulombic interaction. QDs show the discrete energy levels due to the quantum confinement. The energy levels can be modeled using a particle in a box. The hole is taken as positive nucleus of hydrogen, whereas electron is taken as negative charge. The band gap (E_g) of QDs is more than the bulk due to the quantum confinement.

$$E_g = E_{Bulk(R \to \infty)} + \frac{h^2}{8R^2}\left(\frac{1}{m_e^*} + \frac{1}{m_h^*}\right)$$

where R is the radius of the particle. QDs confine the electrons and holes in $3D$. When particle changes shape from sphere to rod/sheet and wire (at least one side is in 1–10 nm range), the electrons and holes are confined in $2D$ and $1D$, respectively.

3.4 MAGNETISM

Magnetic moment (μ) of a system arises from spin unbalance between spin-up and -down levels. In general, the magnetism can be divided into five kinds: (1) diamagnetism, (2) paramagnetism, (3) antiferromagnetism, (4) ferromagnetism, and (5) ferrimagnetism. Temperature-dependent susceptibility (χ-T) plot differentiates the five kinds of magnetism [84,123,145]. The χ is defined as ratio of magnetization to the applied magnetic field ($\chi = M/H_a$). The χ for empty space is zero.

The χ for diamagnetic material is negative. Here, induced moment (μ_{ind}) is generated when magnetic field is applied and the direction of μ_{ind} is opposite to applied magnetic field (H_a). Examples are He, NaCl, and superconducting materials below T_c.

The χ for paramagnetic material is positive. H_a does not affect magnetization (remains unchanged). Atomic dipoles do not interact with each other. There are two types of paramagnetism: Langevin (free-spin or Curie-like) paramagnetism in which the χ is related to $1/T$. This is just temperature-dependent susceptibility. Pauli-like paramagnetism is temperature-independent susceptibility. Here, the net effect of spin up and down at the Fermi energy (E_F) does not change with temperature, e.g., VN follows Curie-like behavior at lower temperature and Pauli-like behavior at higher temperature (near room temperature).

The χ for antiferromagnetic material is positive at all temperatures, but χ-T plot shows the maximum. The temperature at which χ shows the maximum is known as Néel temperature (T_N), e.g., MnO has T_N at 122K.

The χ for ferromagnetic material is large and positive. Atomic dipoles have regular pattern in a particular direction (up or down) and have interactions among them to rise to molecular field. Because of molecular fields, these materials have spontaneous magnetization in absence of magnetic field.

Magnetization decreases with temperature and becomes almost zero at a particular temperature called Curie temperature (T_{fp}). Above T_{fp}, the χ follows Curie-like behavior (e.g., χ is related to $1/(T-T_{fp})$). T_{fp} values for Fe, Co, and Ni are 1043, 1400, and 631K, respectively [145].

The χ for ferrimagnetic material is large and positive. The behavior is similar to ferromagnetic material. The only difference is interaction parameter of two planes of spins (e.g., in spinel structure, there are tetrahedral and octahedral planes), which are present in ferrimagnetic structure. The T_{fp} values for Fe_3O_4, $CoFe_2O_4$, and $NiFe_2O_4$ are 858, 793, and 858K, respectively [145]. Mn_4N exhibits ferrimagnetism.

When particle size decreases, ferromagnetic/ferrimagnetic particles become superparamagnetic. The individual particle in superparamagnetic system has magnetic moment ($\sim 10^2-10^4\ \mu_B$). Since particles are so small, spin of particle relaxes in order of 10^{-9} s. Here, particles are assumed to be separated from each other. Overall particles have no magnetic moment in the absence of magnetic field (H_a). There is a blocking temperature (T_B) below which the system behaves as ferromagnetic and above which the system behaves superparamagnetic. Here the magnetic anisotropic energy (KV, K is the anisotropy constant and size of the particles having volume V) is smaller than the thermal energy (kT). T_B is much lower than T_{fp}. For uniaxial particles, T_B is:

$$T_B = KV/25k \tag{10.14}$$

The superparamagnetic relaxation (τ) can be described by the Arrhenius–Néel expression:

$$\tau = \tau_0 \exp(KV/k_BT) \tag{10.15}$$

The value of τ_0 depends on the gyromagnetic precession time and is usually equal to $10^{-10}-10^{-13}$ s.

The crystalline ferromagnetic metals (Fe, Co, Ni) and their alloys, borides, nitrides, and carbides show that T_{fp} decreases with decrease of particle size. An intrinsic property of the material (T_{fp}) depends on the exchange interactions. The particle size dependence (D) of the T_{fp} compared to the bulk value can be explained by using scaling equation as [146]:

$$[T_{fp}(bulk) - (T_{fp}(D)]/T_{fp}(bulk) = \pm(D/D_0)^{-\lambda} \tag{10.16}$$

where $\nu = 1/\lambda =$ correlation length. The D_0 is the order of the characteristic microscopic dimension of the system.

The magnetic susceptibility data can be grouped by the following equations [84]:

$$\chi_g = C/T: \text{(Curie law)}, \tag{10.17}$$

$$\chi_g = \chi_0 + C/T: \text{(Pauli – Curie law)}, \tag{10.18}$$

$$\chi_g = C/(T+\Delta): \text{(Curie – Weiss law)}, \tag{10.19}$$

$$\chi_g = \chi_0 + C/(T+\Delta): \text{(Pauli – Curie – Weiss law)}, \tag{10.20}$$

where C and Δ are the Curie constant and magnetic interaction parameter, respectively; χ_0 is the specific magnetic susceptibility (Pauli-susceptibility) constant, which is independent of temperature. When particle size decreases, change from paramagnetic/diamagnetic to antiferromagnetic/ferromagnetic behavior or from ferromagnetic to antiferromagnetic behavior occurs due to uncompensated spins or/and spin-charge interaction arising in the system. This will give positive or negative exchange parameter in short or long range.

4. SYNTHESIS

4.1 NITRIDES

4.1.1 Binary Nitrides Using Metal and N_2

By direct passing N_2 over metals at high temperature, metal nitrides (M_xN_y) can be produced [2–4].

$$M + N_2 \xrightarrow{\text{heat}} MN$$

for example, VN is prepared at 1200°C from vanadium metal.

4.1.2 Metal Nitrides Using Metal or Metal Precursor and NH_3

In reducing atmosphere (NH_3), metal nitrides are prepared, but formation temperature is lower than that required in N_2 atmosphere [2,3,60,84,147–149]:

1. $M + NH_3 \xrightarrow{\text{heat}} MN$
2. $M_xO_y + NH_3 \xrightarrow{\text{heat}} MN + H_2O$
3. $MCl_x + NH_3 \xrightarrow{\text{heat}} MN + HCl$
4. $MS_x + NH_3 \xrightarrow{\text{heat}} MN + H_2S$

for example, Fe_3N is prepared at 500°C from Fe_2O_3 in NH_3 [60].

4.1.3 Metal Nitrides Using M-N Complex

M–N linkage complex can produce metal nitrides with or without reducing atmosphere [1,109,150–153].

1. $M - N\text{complex} \xrightarrow{\text{heat}} MN + H_2O + CO_2/CO$
2. $M - N\text{complex} + NH_3 \xrightarrow{\text{heat}} MN + H_2O + CO_2/CO$

VN, Ni_3N, and Co-N are prepared from M-N linkage complex in NH_3 atmosphere. The sources for V, Ni, and Co are $[VO(NH_2O)_2Gly].H_2O$, $[Ni(NH_3)_6](NO_3)_2$ and $[Co(NH_3)_6](NO_3)_3$, respectively, in NH_3 atmosphere [1,150,151]. Similarly, MoN and Mo_2N can be prepared from Mo-N complex [109]. Without NH_3/N_2 atmosphere, Ni_3N can be prepared from decomposition of $Ni(NH_2)_2$ at 200°C [152,153].

4.1.4 Single Crystals of Metal Nitrides

Single crystals (0.3–0.7 mm) of GaN can be grown from Ga in sodium melts at 600–800°C and at considerably lower nitrogen pressures of 50–100 bar [154]. Source of nitrogen is from thermal decomposition of sodium azide (Na_3N).

4.1.5 Ternary and Quaternary Nitrides

In general, ternary nitrides are considered as nitrides of two metals. Lithium-based metal nitrides can be prepared from Li_3N and M (M = Al, B, Fe, Si, Ti, Ta) in inert atmosphere and at elevated temperature [154].

$$Li_3N + M = \text{Al, B, Fe, Si, Ti, Ta} \rightarrow Li_xM_yN_z$$

(e.g., $LiSi_2N_3$, Li_2SiN_2, Li_5SiN_3, $Li_{18}Si_3N_{10}$, Li_7TiN_5, Li_7TaN_4)

Ternary nitrides can also be prepared from the respective oxides in NH_3 atmosphere.

$$M^iM^{ii}O_4\ (M^i = Mn, Fe, Co, Ni, M^{ii} = Mo, W) + NH_3 \rightarrow M^iM^{ii}N_2, M^i_3\ M^{ii}{}_3N$$

Also, the halide-based ternary/quaternary nitrides such as lithium nitride halides $Li_{3-2x}N_{1-x}X_x$ (X = Cl, Br), zirconium nitride chlorides A_xZrNCl (A = Li, Na, K), and rare-earth nitride halides (e.g., Ce_2Br_3N) can be prepared.

The barium nitrides-halides such as Ba_2N, Ba_2NF, $Ba_2NCl_{(1-y)}$, Ba_2NBr, and Ba_2NI are prepared at high-temperature reactions under inert atmosphere [155].

4.1.6 Metal Nitrides Using Metathesis Reaction

In metathesis reaction, the cations and anions exchange takes place between two ionic partners in solvent/medium (water). This is simply double-replacement reaction. All ions are surrounded by H_2O molecules in hydrated form.

$$A^+X^- + B^+Y^- \rightarrow BX + AY$$

A^+ ion in water will be as $[A(H_2O)_n]^+$ and X^- as $[X(H_2O)_n]^-$. Similarly, $[B(H_2O)_n]^+$ and $[Y(H_2O)_n]^-$ will be formed. $[B(H_2O)_n]^+$ and $[X(H_2O)_n]^-$ forms BX in water medium, whereas $[A(H_2O)_n]^+$ and $[Y(H_2O)_n]^-$ form AY.

Metal nitrides can be prepared using metathesis route [156]. Here, metal salt such as $MeCl_n$ is reacted with ionic Li_3N, which is source of nitrogen in inert atmosphere and at elevated temperature. The reaction is exothermic and, thus, excess NaCl is added for control. If pressure is provided, the different phases of metal nitrides can be prepared. The final product contains LiCl and NaCl, which can be removed by rinsing with hot water at 70°C.

$$MeCl_n + xLi_3N + yNaCl \rightarrow Me_xN_y + nLiCl + yNaCl$$

For example, ReN_2 with hexagonal structure (lattice parameters $a = 2.806$, $c = 10.747$ Å), which shows the layered type of MoS_2 along c-axis can be prepared by this route. Similarly, other nitrides can be prepared by heating $MeCl_n$ with Ca_3N_2 in inert atmosphere:

$$MeCl_2 + Ca_3N_2 \xrightarrow{heat} MeN + CaCl_2\ + N_2$$

4.1.7 Layered Nitrides

The layered nitrides such as $La_3Ni_2B_2N_3$ and LaNiBN are prepared by arc-melting of precursor in N_2 atmosphere for 3 days above 1000°C [8]. The precursors for $La_3Ni_2B_2N_3$ and LaNiBN are $La_3Ni_2B_2$ and LaNiB, respectively. These precursors are prepared from constituent elements. Both materials crystallize in tetragonal structure. The structure of $La_3Ni_2B_2N_3$ can be considered as a series of three square LaN layers in a rock salt configuration, stacked with single tetrahedral Ni_2B_2 layer, whereas structure of LaNiBN is considered as a series of two LaN layers in a rock salt configuration, stacked with single tetrahedral Ni_2B_2 layer. $La_3Ni_2B_2N_3$ is superconducting material, whereas LaNiBN is nonsuperconducting material. Carbon-doped YNiB having superconducting transition (T_c) above 10K was prepared by Nagarajan and his group for the first time in 1994 [9]. The crystal structure of the material, which is responsible for enhancement of T_c, was shown by Cava and his group for the first time in 1994 [8].

4.1.8 Ceramic Carbides and Nitrides Having High Chemical and Thermal Stability as Well as High Thermal Conductivity

Ceramics are generally used in many applications because they possess one or more of following properties: high hardness, high mechanical strength, high thermal stability, high chemical stability, high thermal conductivity, etc. Nonoxide-based ceramics are carbon nitrides (C_3N_4), boron nitride (BN), silicon carbides (SiC), and nitrides (Si_3N_4) [157]. Bonding in C–N, B–N, Si–C and Si–N is of covalent nature. They possess low diffusion coefficient. For preparation of these materials, high vacuum chamber, high-purity atmosphere (N_2, Ar), other high-purity precursors/materials, and high temperature as well as pressure are needed. Even after preparing samples, it needs sintering at higher temperature and prolongs duration of heat treatment to get high-density materials. This process is not economical. One way to synthesis materials at low temperature is the chemical approach. For example, when metalorganic compound (polysilazane) is heated at 350°C for 100 min in Ar atmosphere, it can produce polymethylsilazane.

$$[-RSiH-NH-]_n \xrightarrow{350^\circ C,Ar} [-RSiH-NH-]_{n-m}[-RSiN-]_m + mH_2$$
$$R{=}CH_3$$

The product polymethylsilazane is of polymeric type. The cylindrical shape of polymeric sample is prepared under pressure with density of 84% and is heated in a quartz tube under Ar or NH_3 at 1000–1400°C. Finally, high-density material (>93%) of Si_3N_4 is obtained.

4.1.9 Stabilization of Different Phases of Metal Nitrides Under High Pressure, Temperature and Quencher/Sudden Cooling Processes

Under high pressure, temperature and sudden cooling/quenching conditions, new phases of nitrides, oxy-carbides/nitrides, carbides, and borides and even metastable compounds can be prepared [73]. The compounds of nitrides (e.g., Zr_3N_4, Si_3N_4, P_3N_5), oxynitrides (e.g., Si_2N_2O, Ge_2N_2O, Ga_3O_3N, $Y_2Si_3O_3N_4$, $Y_5(SiO_4)_3N$, $Si_{3-x}Al_xO_xN_{4-x}$ for $x < 2.1$, Al_3O_3N), nitride-imides (e.g., Si_2CN_4, $PN_x(NH)_y$), and nitride-diazenides (e.g., SrN_2) are prepared under different pressures (even more than 500 GPa) and temperatures (even up to 7000K) in different atmospheres (N_2, H_2, Ar, N_2-O_2).

The surrounding atmosphere of particles can be used as (1) the pressure transmitting medium, (2) protection from other surrounding environment, and (3) source of one of the reactants in formation of compound. N_2/NH_3 gas is used for formation of nitrides, whereas N_2-O_2 mixture is used for the formation of oxynitride. Ar/N_2 gas is used during sintering process. Many techniques are used to synthesize borides, carbides or nitrides under pressure, such as (1) autoclave (1–6000 bar), (2) multianvil device (1–20 GPa), (3) diamond anvil cells (DAC, one few hundred GPa), and (4) shock wave device. There are two ways to provide pressure and heating on a system: static and dynamic. In the static process, both pressure and temperature are slowly applied to the system; and also both cooling and releasing of pressure are operated slowly. The first three techniques can be performed in a static way. For the preparation of large quantity of sample, the first technique is more suitable. Sample up to a few cubic millimeters can be prepared by the second technique, while sample up to a few cubic micrometers can be prepared by the third technique.

In the dynamic process, high-pressure and high-temperature conditions can be achieved by applying projectile impact on the initial sample by explosions or laser shock, or by discharging large quantities of electrical or magnetic energy into the sample chamber. The last/fourth technique can be performed in dynamic way. The fourth technique can prepare sample at very large quantity, but it is very expensive. Since the transition takes place within 10^{-3}–10^{-6} s, it is difficult to understand detailed structural information during compression process.

At high-pressure and high-temperature synthesis, coordination number (CN) of constituent elements of compound can be increased. This can change structure and chemical bonding. Consequently, their electronic, optical, mechanical, and magnetic properties will be different from sample prepared at ambient atmosphere.

In the preparation of CN_x, different precursors are used. The products obtained are usually amorphous. Nitrogen concentration varies depending on precursors, heating conditions (pressure and temperature), and atmosphere. Polycrystalline α-, β-, and c-C_3N_4 phases are prepared. Different precursors can be used for the preparation of amorphous and crystalline CN_x:

$$
\begin{aligned}
&C_3N_3(NH_2)_3 + NH_2NH_2 \xrightarrow{\text{heat}} CN_x(1.14 < x < 1.34)(\text{cryst.})\\
&C_3N_3(NH_2)_3 + C_3N_3Cl_3 \xrightarrow{\text{heat}} CN_x(1.47 < x < 1.50)(\text{partially cryst.})\\
&C_3N_3(NH_2)Cl_2 + NH_2NH_2 \xrightarrow{\text{heat}} CN_x(0.80 < x < 1.46)(\text{cryst.})\\
&C, N, H \xrightarrow{\text{heat}} CN_x(x = 0.25)(\text{cryst.})\\
&C_6Cl_6 + NaN_3 \xrightarrow{\text{heat}} CN_x(0.70 < x < 1.94)(\text{partially cryst.})\\
&C_2N_2 \xrightarrow{\text{heat}} CN_x(0.35 < x < 1.16)(\text{partially cryst.})\\
&C_6N_4 \xrightarrow{\text{heat}} CN_x(0.56 < x < 0.88)(\text{amorphous})\\
&C_6N_4 \xrightarrow{\text{heat}} CN_x(0.56 < x < 0.88)(\text{cryst.})\\
&\text{Graphite} + N_2 \xrightarrow{\text{heat}} CN_x(\text{cryst.})\\
&C_{60} + N_2 \xrightarrow{\text{heat}} CN_x(\text{cryst.})\\
&C_{\text{amorphous}} + N_2 \xrightarrow{\text{heat}} CN_x(\text{cryst.})\\
&C_3N_3Cl_3 \xrightarrow{\text{heat}} CN_x(x = 1.0)(\text{cryst.})\\
&C_3N_4(\text{amorphous}) \xrightarrow{\text{heat}} CN_x(\text{cryst.})\\
&C_3N_4(\text{graphite}) \xrightarrow{\text{heat}} CN_x(x = 1.33)(\text{cryst.})
\end{aligned}
$$

Silicon and germanium prefer four CN (tetrahedral geometry) in compounds prepared at ambient atmosphere. Nitride compounds having spinel structure (cubic, c- or γ-phase) in which one-third of metal (Si, Ge, or Sn) is in tetrahedral coordination (CN) and two-thirds in octahedral CN of metal could be prepared at high temperature and pressure. The spinel c-Si_3N_4, c-Ge_3N_4, and Sn_3N_4 are obtained by means of high-pressure and -temperature syntheses in LH-DAC (laser-heating-based diamond anvil cell), MA (multianvil) and SW (shock wave) experiments.

$$
\begin{aligned}
&Si + N_2 \xrightarrow{\text{heat}} c - Si_3N_4(\text{method} = LH - DAC)\\
&\alpha - Si_3N_4 \xrightarrow{\text{heat}} c - S_3N_4(\text{method} = LH - DAC)\\
&\beta - Si_3N_4 \xrightarrow{\text{heat}} c - S_3N_4(\text{method} = SW)\\
&Si_2N_2NH \xrightarrow{\text{heat}} c - S_3N_4(\text{method} = MA)
\end{aligned}
$$

$$
\begin{aligned}
&Ge + N_2 \xrightarrow{\text{heat}} c - Ge_3N_4(\text{method} = \text{LH} - \text{DAC})\\
&\alpha - Ge_3N_4 + \beta - Ge_3N_4 \xrightarrow{\text{heat}} c - Ge_3N_4(\text{method} = \text{LH} - \text{DAC})\\
&\alpha - Ge_3N_4 + \beta - Ge_3N_4 \xrightarrow{\text{heat}} c - Ge_3N_4(\text{method} = \text{MA})\\
&\beta - Ge_3N_4 \xrightarrow{\text{heat}} c - Ge_3N_4(\text{method} = \text{SW})
\end{aligned}
$$

Depending on the crystal structures, similar compositions show the different properties. Si_3N_4 or Ge_3N_4 or Sn_3N_4 can have three different phases [trigonal (α), hexagonal (β), and cubic (γ, c)]. CN of cation (metal) can be increased from 4 to 6 or 8 with pressures and temperatures during synthesis. Even CNs of anion (N/C) can be obtained as 4 and 6. These can be found in Zr_3N_4 and Hf_3N_4. N can have 5 and 6 CNs in P_3N_5.

Some nitrides in nitrogen/Ar atmosphere are usually formed by heating precursors at high temperature and followed by quenching/sudden cooling to room temperature or liquid nitrogen temperature and then tempering process at low temperature (100–150°C) for long duration up to 10 days in different atmospheres [60,61,111,158,159]. For example, the precursors are α-Fe, γ-Fe, γ-Fe_2O_3, and doped ones. Atmosphere is N_2, Ar, NH_3, or mixture of H_2 and Ar. In phase diagram of Fe-N, the followings phases are reported: (1) α (Fe-bcc, ferromagnetic), (2) γ (Fe-*fcc*, nonmagnetic), (3) ζ (Fe_2N-orthorhombic, antiferromagnetic), (4) ε (Fe_3N-hexagonal, ferromagnetic), (5) γ'(Fe_4N-*fcc*, ferromagnetic), and (6) α''(Fe_8N or $Fe_{16}N_2$-body-centered tetragonal, ferromagnetic (*bct*)). The magnetic moment of α''-$Fe_{16}N_2$ is reported up to 286 emu/g, which is more than that of other nitrides and even α-Fe. Generally, the magnetic moment decreases with increasing nitrogen concentration in nitrides. For the first time, bulk α''-$Fe_{16}N_2$ was prepared by Jack in 1951 with space group I4/mmm (tetragonal, $a = 5.72$ Å and $c = 6.29$ Å). One-sixteenth of the interstitial sites on all the edge centers in the ordered way are occupied by nitrogen atoms in the α-Fe (*bcc*). The unit cell has $2 \times 2 \times 2$ superlattice of α-Fe with expansion of 10% along the c-axis after incorporation of interstitial nitrogen atoms. Similar phase with high magnetic moment was reported in thin films also by Kim and Takahashi in 1972 [158]. However, a pure phase (100%) of α''-$Fe_{16}N_2$ has not been prepared so far. Instead, a mixture of α-Fe and γ'-Fe_4N is associated to α''-$Fe_{16}N_2$.

4.1.10 Reaction Between Lithium Amides and Anhydrous Transition Metal Chlorides

When anhydrous metal chlorides and lithium amide are reacted at 200–400°C, it produces exothermic reaction [160]. The products are lithium chloride and amorphous M-N containing compound. Trituration of the product with tetrahydrofuran removes lithium chloride from the products. Then the remaining product is heated in N_2 gas at 700°C, and producing metal nitride:

$$MCl_n + nLiNH_2 \xrightarrow{200-400^\circ C} M_xN_yClH + nLiCl$$

$$M_xN_yClH + nLiCl \xrightarrow{THF} M_xN_yClH$$

$$M_xN_yClH + N_2 \xrightarrow{700^\circ C} M_xN_y$$

The following compounds TiN, ZrN, HfN, VN, NbN, TaN, Cr_2N, CrN, Mo_2N, W_2N, Mn_3N_2, and Zn_3N_2 are prepared by this method.

4.2 CARBIDES

Metal carbides (MCs) are prepared by direct heating metal (M)/metal hydride (MH)/metal oxide (MO) and carbon in vacuum at higher temperature [2–4].

$$M + C \rightarrow MC$$

$$MH_x + C \rightarrow MC + H_2$$

$$MO + C \rightarrow MC + CO$$

VC is prepared at 1100°C from vanadium metal and carbon in vacuum.

Also, metal carbides are prepared by either metal/metal chloride (MCl_n) with a carburizing gas. When metal carbonyl (M-CO) is reacted with H_2, it forms MC [2,3].

$$M + C_xH_y \rightarrow MC + H_2$$

$$M + CO \rightarrow MC + 1/2\ O_2$$

$$MCl_4 + C_xH_y + H_2 \rightarrow MC + HCl + C_mH_n$$

$$\text{M-Carbonyl} + H_2 \rightarrow MC + CO + CO_2 + H_2 + H_2O$$

Thin films of carbides are prepared by chemical and physical deposition techniques. Single crystals of carbides are prepared by the Verneuil technique, the floating zone technique and other methods of solidification from liquid metals. Ternary chemical bonding among metals and nonmetals (B, C, N, Si) has been explored [161]. Ternary carbides ($Ln_2Mn_{17}C_{3-x}$) are synthesized by reaction of the elemental components in a high-frequency furnace and arc-melting technique. Some of them are exothermic reactions [162]. Synthesis of cemented tungsten carbides and their properties have been explored [163].

4.3 BORIDES

Metal borides are generally prepared by the following procedures [82]:

1. Direct interaction between metal and boron at high temperature and inert atmosphere or vacuum.

$$M + B \rightarrow MB$$

2. Electrolysis of molten boron oxides and salts.
3. Reduction with carbon in a mixture of metal oxide and B_2O_3.
4. Reduction of metal oxides with boron carbide under vacuum.
5. Other synthetic techniques such as chemical and physical vapor depositions.

Different methods of preparation of metal borides are reviewed up to 2013 and their properties and applications are given [164]. Different stoichiometric compositions of transition metal Nb and B are Nb_3B_2, NbB, Nb_5B_6, Nb_3B_4, and NbB_2 [82,83]. Similarly, those of Cr and B are Cr_2B, Cr_5B_3, CrB, Cr_3B_4, CrB_2, CrB_4, and CrB_6. Borides of alkali−earth metal Mg are MgB_2, MgB_4, MgB_6, MgB_7, and MgB_{12}. Borides of alkali metal Li are LiB, Li_5B_4, Li_3B_{14}, $LiB_{12.93}$, $Li_{1.8}B_{14}$, and LiB_2. Borides of *p*-block elements are AlB_2, AlB_{12}, B_4C, SiB_3, SiB_4, SiB_6, and BN. Rare-earth elements form stable hexaborides such as LaB_6, YB_6, EuB_6, and ThB_6. There are the substituted types of compounds (e.g., $Al_{0.5}Mg_{0.5}B_2$). Borides of ternary elements are also formed, e.g., $MgAlB_{14}$, $LiAlB_{14}$, and AlB_3C. Ternary and quaternary borides are prepared, and their thermodynamics and mechanical properties are studied [165].

5. CHARACTERIZATION TECHNIQUES

Characterization of materials (nitrides, carbides, and borides) can be carried out in two ways: physical and chemical approaches. Physical approaches are generally nondestructive. Crystal structures are solved by the X-ray diffraction (XRD) and transmission electron microscopy techniques. The chemical bonding among atoms of systems including electronic states/charge of ions are solved by Fourier transform infrared (FTIR), Raman spectroscopy (RS) and UV/X-ray photoelectron spectroscopy (UPS/XPS), and extended X-ray absorption fine structure (EXAFS) techniques. Since N, C, and B are very light elements, identification of their positions in nitrides, carbides, and borides, respectively, are sometimes difficult by XRD analysis. In this, neutron diffraction (ND) data of these materials or neutron captured by these materials is useful in elucidating their structures. Also, Rutherford backscattering spectroscopy (RBS) and Auger electron spectroscopy (AES) will be useful in identification of such light elements. Microstructure of particles can be investigated by scanning transmission electron microscopy (STEM), transmission electron microscopy (TEM), atomic/magnetic force microscopy (AFM/MFM), scanning electron microscopy (SEM), etc. Chemical approaches are destructive for compositional characterization. Here, a known amount of material is dissolved in acids (HCl and HNO_3, H_2SO_4). Inductively coupled plasma mass spectroscopy (ICP-MS), RBS, energy dispersive analysis of X-rays (EDAX), X-ray fluorescence (XRF), and electron energy loss spectroscopy (EELS) are excellent techniques for study of chemical compositions of nitrides, carbides, and borides. Selected area electron diffraction (SAED) and conversion beam electron diffraction (CBED) studies will help in more depth analysis of phase in nanoscopic scale, but the latter one is better than the former in study of unit cell structure as well as impurities (e.g., carbides in 2-nm size) present in steels. Annular dark field imaging (ADF) and EELS provided in STEM will help in depth analysis of particles (particularly core–shell-type structure). The thermal data (thermogravimetric and differential thermal analyses) will help in understanding thermal stabilities as well as intermediate phase formation. Many metastable phases produced under different conditions can be predicted theoretically.

Generally, nitrides of alkali and alkaline earth metals are very sensitive to environment (ambient atmosphere, H_2, O_2, or any chemical environment). It needs to be protected from such environment during characterization and storing. Such compounds are ionic in nature. e.g., Li_3N, Mg_3N_2. Because of this, they are highly reactive. Some of subnitrides such as Sr_2N, SrN are having ionic, covalent or metallic character.

5.1 CHEMICAL APPROACH

5.1.1 Estimation of Nitrogen in Nitride

Estimation of nitrogen is carried out by Kjeldahl's method [84,166,167]. Initially, Kjeldahl's apparatus has to be tested using standard compound (acetanilide). For analysis, 200 mg of acetanilide or metal nitride sample is taken in a dry Pyrex Kjeldahl flask; 0.5 g of anhydrous copper sulfate and 25 mL of concentrated H_2SO_4 are added to the flask. The purpose of anhydrous copper sulfate is to speed up the digestion process. It is heated in order to get clear solution. The nitrogen present in the compound gets converted to $(NH_4)_2SO_4$ and this can be determined by the addition of aqueous solution (1:1) of caustic alkali (NaOH). The liberated ammonia is distilled off into the standard acid (0.1 N HCl). The amount of acid is determined by acid-base titration. The amount of nitrogen in sample is calculated as: 1 mL of 0.1 N HCl $\equiv$ 1.7034 mg of NH_3.

5.1.2 Estimation of Carbon in Carbide

The carbon content can be determined by heating carbide in pure oxygen atmosphere (1000–2000°C). The CO_2 gas will be evolved during heating. This is passed through known soda lime or soda—lime—asbestos (which can change color after absorbing CO_2 gas). The following equation is useful in determining the amount of carbon present in carbide [166,167].

$$2NaOH + CO_2 \rightarrow NaCO_3 + H_2O$$

5.1.3 Estimation of Boron in Boride

The boron content can be determined by dissolving borides in strong acids. Boron will make complexes with suitable ligands (e.g., salicylic acid). Known amount of salicylic acid can determine amount of boron in solution obtained from borides [166—169].

5.1.4 Estimation of Metal Ions in Nitride, Carbide, or Boride

Metal ions can be determined by analytical methods and ICP-MS, X-ray fluorescence, etc. [166—169]. Also, thermogravimetric analysis can estimate amount of metal(s) present in nitride, carbide, or boride after thermal decomposition.

5.2 PHYSICAL APPROACH

5.2.1 X-Ray Diffraction

The XRD patterns of Mo_2N and MoN are shown in Fig. 10.2a,b [109]. Mo_2N is stable in normal conditions with cubic structure and the lattice parameter is $a = 4.131$ Å. MoN phase prepared in normal condition is unstable as well as disordered due to vacancies in N and Mo sublattices. Under high pressure and high temperature, it becomes stable as well as ordered. MoN can have two phases (cubic as well as hexagonal). The hexagonal phase is written as δ-MoN. The lattice parameters of

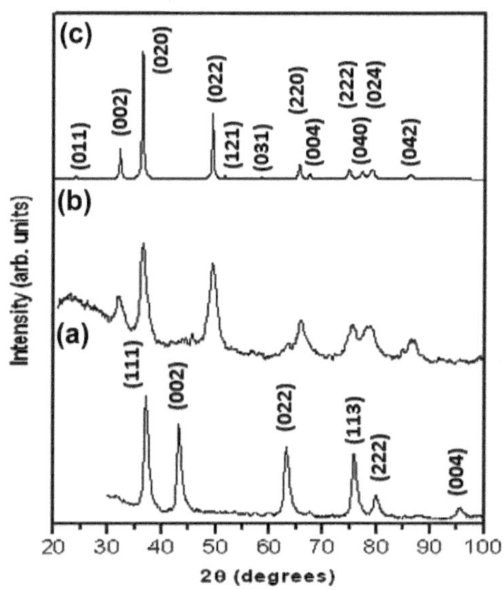

FIGURE 10.2

X-ray diffraction (XRD) patterns of (a) Mo_2N and (b) MoN along with (c) generated XRD pattern of MoN [109].

Reproduced with the permission of John Wiley and Sons Publishers.

δ-MoN are doubled when sufficient pressure and temperature are applied in N_2 atmosphere [170]. δ-MoN is superconducting and its T_c can be increased from 4 to 12K under sufficient pressure and temperature. Among Mo_2N and MoN, XRD pattern of Mo_2N is very sharp, whereas that of MoN is broad. The broadening is due to disorder of N positions in Mo lattice.

It is possible to predict lattice parameters for MoN (cubic and hexagonal phases) [109]. When N atoms occupy the octahedral interstitial sites of Mo lattice, volume expansion occurs. The values of ratio of N to Mo atoms in Mo, Mo_2N, and MoN are 0:1.0, 0.5:1.0, and 1.0:1.0, respectively. The lattice parameters of *bcc* Mo and *fcc* Mo_2N are 3.147 (JCPDF 42-1120) and 4.131 Å, respectively. The unit cell volume of nitride per metal atom is calculated. The straight line ($y = mx + c$) is plotted to data of N/M ratio versus the unit cell volume of nitride per metal atom. There are two points of *bcc* Mo and *fcc* Mo_2N (i.e., $MoN_{0.5}$) in the graph. The values of m and c are found to be 0.243 $(Å^3)^{-1}$ and −3.8, respectively. After substitution of *m* and *c* values, unit cell volume of MoN structure (here one Mo atom and one N atom are considered) per Mo atom is 19.68 $Å^3$. When MoN crystallizes in *fcc* system, there are 4 Mo atoms per unit cell. The unit cell volume of *fcc* MoN will be $V = 4 \times 19.68$ $Å^3$ and thus unit cell parameter will be $a = 4.285$ Å, which is higher than that for *fcc* VN ($a = 4.137$ Å, JCPDF 78-1315). The difference between lattice parameters in VN and MoN ($\Delta a = 0.120$ Å) is due to difference in lattice parameters of metal lattices of *bcc* *V* ($a = 3.027$ Å, JCPDF 22-1058) and *bcc* Mo ($a = 3.147$ Å). This suggests that amount of lattice expansion in metals (having same Bravais lattice in a particular crystal structure, i.e., bcc for Mo and *V*) is the same when the amount of nitrogen is same. This is due to the same number of interstitial sites present in metal lattice. In the case when MoN crystallizes in hexagonal phase, there are 8 Mo atoms and 8 N atoms per unit cell. Since unit cell volume of MoN structure per Mo atom is 19.68 $Å^3$, unit cell volume of hexagonal δ-MoN is $V = 8 \times 19.68$ $Å^3 = 157.44$ $Å^3$. If the ratio of unit cell constants $(c/a) = 0.98$ is taken, then the predicted lattice parameters will be $a = 5.7032$ Å and $c = 5.5891$ Å, which are closed to the other reported values for δ-MoN ($a = 5.725$ and $c = 5.608$ Å; JCPDF 25-1367). Based on space group *P6₃/mmc*, the XRD pattern of hexagonal phase MoN can be generated using lattice parameters $a = 5.703$ Å and $c = 5.589$ Å (Fig. 10.2c).

5.2.2 Neutron Diffraction

XRD cannot find all positions of N atoms in MoN since they are much lighter than Mo atoms. However, neutron diffraction study can determine N positions. Using Rietveld refinement program, observed and calculated profiles for hexagonal MoN are shown in Fig. 10.3 [170]. Here, the space group is $P6_3mc$ (186).

5.2.3 Other Techniques

There are many techniques to characterize nitrides, carbides, and borides such as XPS, Mössbauer spectroscopy, EDAX, XPS, EXAFS, EELS, XFS, FTIR, RBS, AES, AFM/MFM, STEM, etc.

6. PROPERTIES

6.1 ELECTRICAL RESISTIVITY

Fig. 10.4 shows the temperature dependence of the electrical resistivity (ρ) of pure VN, 0.5 at.% B doped VN and 0.5 at.% La doped VN in 4.2−300K, and their superconducting transitions are found to be at 9.2, 8.0, and 7.8K, respectively [1]. The temperature coefficient of resistivity of these samples are positive. In low-temperature range (10−30K), the ρ-T^3 behavior is found. This suggests that the system follows the electron-phonon scattering. Here, Eqs. (10.2) or (10.3) (see Section 3) is used to fit data.

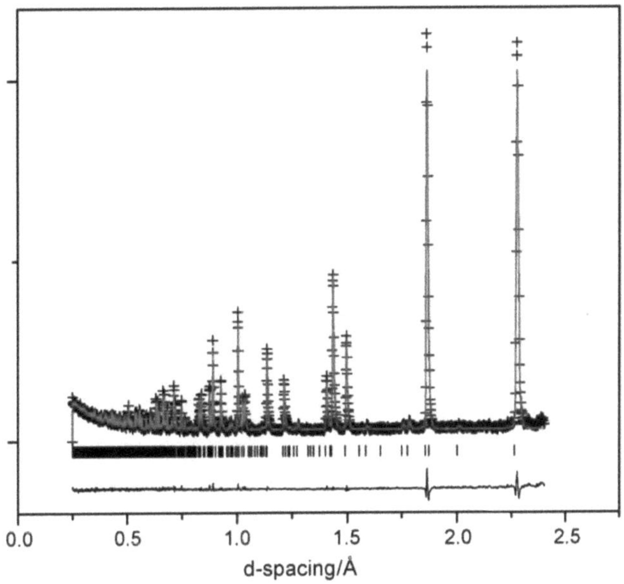

FIGURE 10.3

The neutron observed (+) and calculated (−) profile and difference profiles (bottom trace) for δ-MoN at 290K [170].

Reproduced with the permission of Elsevier Publishers.

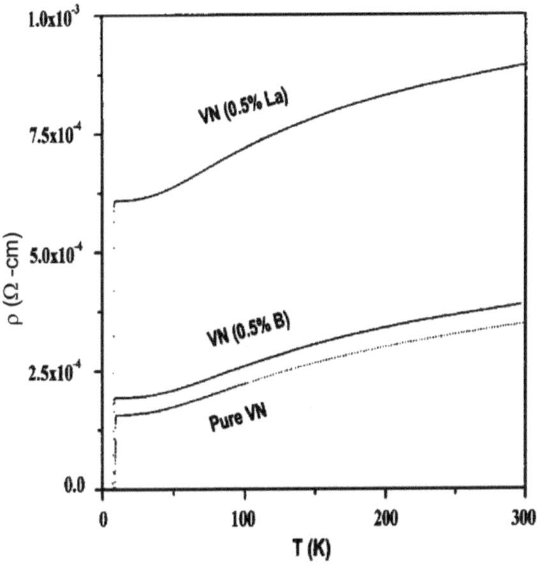

FIGURE 10.4

The variation of resistivity with temperature of pure VN, 0.5 at.% B doped VN and 0.5 at.% La doped VN [1].

Reproduced with the permission of AIP Publishers.

6.2 MAGNETIZATION STUDY

Fig. 10.5 shows the magnetic susceptibility (χ_g) versus temperature for pure VN [56]. Initially, the susceptibility decreases with increase of temperature up to 100K, and then it becomes constant with further increase of temperature. The data are fitted with Eq. (10.20) (see Section 3): The c, θ, and χ_O are found to be 1.0×10^{-5} emu K/g, −2.0K, and 2.21×10^{-6} emu/g, respectively. The density of states ($N(E_F)$) is found to be 1.28 (eV unit cell)$^{-1}$ which is calculated from χ_O. The Coulomb interaction parameter (μ^*) is found to be 0.11. The electron-phonon coupling constant (λ) is calculated using McMillan equation (Eq. (10.9), see Section 3), and it is found to be 0.54. The effective magnetic moment is found to be 0.072 μ_B.

The Fe-N system has the different phases depending on nitrogen concentration in Fe lattice. The phases are α''-$Fe_{16}N_2$ (tetragonal), γ-Fe_4N (fcc cubic), ε-Fe_3N (hexagonal), ζ-Fe_2N (orthorhombic), and FeN (cubic) [51–53,60,61,110,111,158,159,171,172]. α''-$Fe_{16}N_2$ is ferromagnetic and has the highest magnetization among Fe-N system, and its magnetization value is 286 emu/g, which is more than that of *bcc* α-Fe (221 emu/g). γ-Fe_4N has ferromagnetic with magnetization (208 emu/g) [173]. ε-Fe_3N is ferromagnetic. The Fe_3N particles prepared from the acicular-shaped γ-Fe_2O_3 particles in H_2-NH_3 show high magnetization (135 emu/g) and high coercivity (540 Oe), which is shown in Fig. 10.6 [174]. These values are higher than those of γ-Fe_2O_3. The γ-Fe_2O_3 particles have acicular shape along with a few percentages adopting spherical shape. Whereas, ε-Fe_3N particles prepared from α-Fe_2O_3 powder in NH_3 gives smaller spherical particles with 20 nm [60]. These show the superparamagnetic behavior at room temperature. These particles are considered as single domain. Superparamagnetic particles become ferromagnetic below the blocking temperature $T_B = 125$K. The magnetization values measured at 5 and 300K (applied field, $H_a = 4$ T) are found to be 15.2 and 3.9 emu/g, respectively. Fe_2N is antiferromagnetic, whereas FeN is paramagnetic.

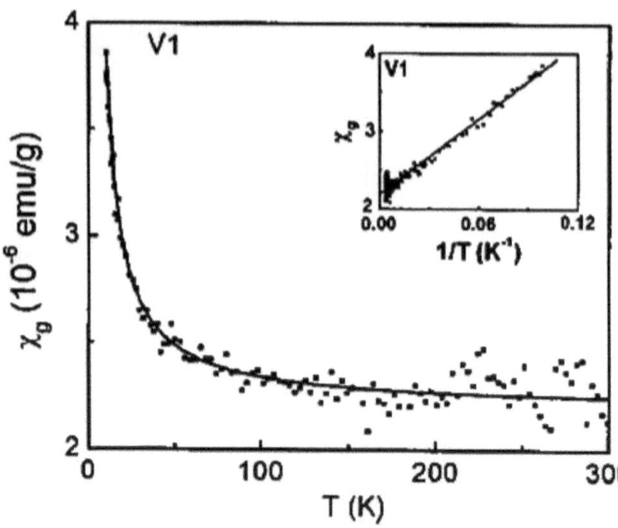

FIGURE 10.5

Magnetic susceptibility (χ_g) versus temperature (K) for pure VN. Data are fitted with Curie-like and Pauli-like behavior. Inset shows its χ_g versus *1/T* below 100K indicating Curie-like behavior [56].

Reproduced with the permission of AIP Publishers.

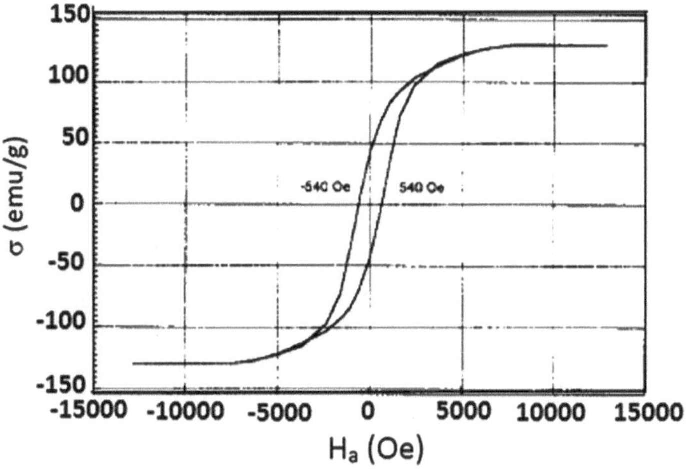

FIGURE 10.6

Magnetization-applied magnetic field (σ-H_a) curve for Fe_3N [174].

Reproduced with the permission of Elsevier Publishers.

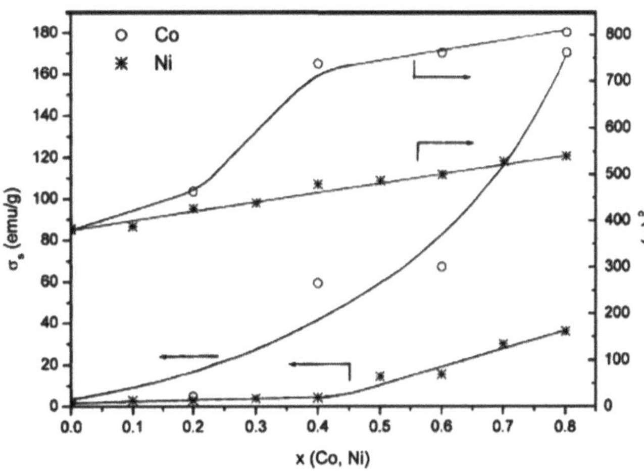

FIGURE 10.7

Variation of saturation magnetization ($M_s = \sigma_s$) and the Curie temperature ($T_c = T_{fp}$) in $Fe_{3-x}M_xN$ (M = Co, Ni) [87].

Reproduced with the permission of Springer Publishers.

When Ni or Co atoms are doped in ε-Fe_3N, both magnetization ($M_s = \sigma_s$) and Curie temperature (T_{fp}, ferromagnetic to paramagnetic transition) increase (Fig. 10.7) [87]. However, rise in M (=σ) and T_{fp} is higher in case of Co doping as compared to Ni doping. Using Mössbauer spectroscopy, magnetism of nitrides, carbides, and borides can be studied extensively.

6.3 HARDNESS

Silicon nitride (Si_3N_4) has two forms (α, β), in which Si is surrounded by 4N (tetrahedron) [175,176]. Each tetrahedron (SiN_4) is linked at each corner to others (like −N−Si−N−Si−N−). These are prepared at normal pressure and high temperature. However, Si_3N_4 prepared at extremely high temperature (2000K) and pressure (15 GPa) can form spinel cubic phase, in which one Si is surrounded by 4N (SiN_4) and another Si is surrounded by 6N (SiN_6). This phase is very hard comparable to diamond and cubic boron nitride.

Nanotwinned cubic boron nitride (*nt-c*-BN) is synthesized from onion-like BN (*o*-BN) precursor under high pressure and temperature [32]. Initially, *o*-BN with grain size (30–150 nm) is prepared by chemical vapor deposition. The *o*-BN particles are put in large volume multianvil system at high pressure, 12–25 GPa, and high temperature, 1600°C. The *nt-c*-BN is formed, and the prepared crystal is 2 mm in diameter and 0.2–0.5 mm in height. Similarly, condition for preparation of single crystal *c*-BN from graphite-like BN precursor needs a very high temperature (2200°C) and high pressure (12–25 GPa).

The *nt-c*-BN sample is optically transparent and its hardness (H_v) is measured with microhardness tester (KB 5 BVZ). The hardness is determined from the asymptotic-hardness region. Fig. 10.8 shows the H_v versus applied load (*F*, expressed in Newton, N). Initially, H_v decreases sharply and becomes constant above 2 N. The asymptotic value is 108 GPa. An optical image of the Vicker indentation with crack produced at a load of 19.6 N is shown as inset of Fig. 10.8 (left). For comparison, H_v-*F* curve of 0.3 mm *c*-BN single crystal sample is shown as inset of Fig. 10.8 (right). It does not show asymptotic up to 4.9 N. The hardness value (108 GPa) of *nt-c*-BN is more than commercial polycrystalline *c*-BN (49 GP with grain size of 1 μm), single crystal *c*-BN (30–43 GPa) and even comparable with synthetic diamond (100 GPa).

The thermal stability in air is performed by thermal techniques. The *nt-c*-BN shows the stability up to 1294°C, which is more than single crystal *c*-BN (1103°C), nanograin *c*-BN (1187°C), and commercially available *c*-BN (1000°C).

Hardness (*H*) of materials is related to its bulk modulus [177]. Bulk modulus (*B*) of ionic compound is given by a general formula: $B \propto (Z_aZ_c/V)$, where Z_a and Z_b are the formal anion and cation charges, respectively; and *V* is the specific volume per ionic pair. For high charges and small volumes,

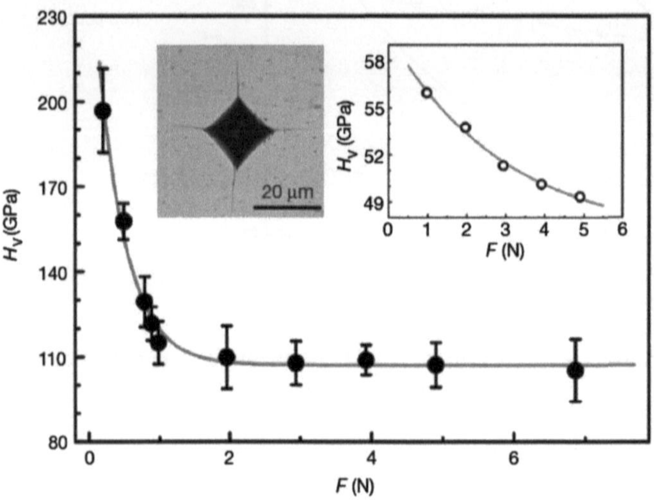

FIGURE 10.8

The H_V of an *nt-c*-BN bulk sample as a function of applied load (*F*). The H_V of the *nt-c*-BN bulk decreases from 196 GPa at 0.2 N to its asymptotic value, 108 GPa, beyond 3 N. Error bars indicate s.d. ($n = 5$). Left inset, an optical micrograph of the Vickers indentation with cracks produced at a load of 19.6 N. Right inset, the H_V–*F* curve of a 0.3-mm *c*-BN single crystal. H_V does not reach its asymptotic value, and the crystal fractures when *F* exceeds 4.9 N [32].

Reproduced with the permission of Nature Publishers.

bulk B will be high. Under pressure, packing efficiency increases at phase transitions at which the CN of cation can increase. In this way, V decreases and thereby B increases. For example, normal glass and quartz contain SiO_2 in which Si^{4+} ion is surrounded by $4O^{2-}$ ions (i.e., CN of Si^{4+} is 4). When quartz is heated at 1000°C under 14 GPa pressure, it retains quartz with $H = 19–33$ GPa. When amorphous silica is heated at 1100°C under 20 GPa, it transforms to polycrystalline stishovite phase in which CN of Si^{4+} is 6 and its average H is 33 GPa with $B = 298$ GPa, which are higher than Al_2O_3 ($H = 21$ GPa, $B = 252$ GPa). The values are comparable with B_4C ($H = 30$ GPa, $B = 200$ GPa), B_6O ($H = 30$ GPa, $B = 200$ GPa), SiC ($H = 29$ GPa, $B = 248$ GPa), c-BN ($H = 32$ GPa, $B = 369$ GPa), but lower than single-crystal diamond ($H = 90$ GPa, $B = 444$ GPa).

6.4 PLASTIC DEFORMATION

Steels are used in manufacture of high-temperature components (such as boiler and turbine) [178]. In this, steel with long plastic deformation/creep is necessary. The plastic deformation is a kind of deformation in materials under low stress. One way is addition of yttrium oxide nanoparticles in steel. However, large-scale production of oxide dispersed steel could not be achieved so far. A second way is addition of carbon nitride (MX) in steel. When 0.002–0.15% C and 0.05% N are present in steel having composition 9%Cr, 0.5% Mo, 1.8% W, 0.2% V, and 0.05% Nb, creep rupture strength can be attained up to 10^5 h at operating temperature (~923K). Here, V and Nb form thermally stable carbonitrides with grain size of 20–30 nm, which act as pinning in grain boundary leading to creep resistant.

6.5 CRYSTAL ELECTROSTATIC POTENTIAL EFFECT

Polarization can be introduced in nitrides when particle shape is changed from spherical to rod or strain is induced from dislocation of atoms or lattice defect or heterointerface between two different nitrides (e.g., GaN and GaN/AlGaN) [179]. Particularly, when crystal structure changes from cubic to hexagonal or tetragonal or orthorhombic, symmetry is reduced. In such case, piezoelectric polarization effect can occur when sample interacts with electromagnetic light or electric field. Crystal electrostatic potential in microscopic region of such material is important in understanding of properties such as luminescence lifetime, *e-h* recombination, etc., and one technique to determine crystal electrostatic potential (V) is the electron hologram.

A coherent electron beam passes through a thin specimen/sample in TEM instrument. Contrast analysis of intensity of electron beam gives the image of particle, but it cannot suggest the V. If the half of electron beam diameter is passed through sample/specimen and another half goes to vacuum, the electrons that travel to sample changes its velocity with respect to that to vacuum. There is gain in phase (i.e., one goes advance). In such situation, constructive interference will not be happen. If one object (i. e., biprism) can bring the constructive interference pattern, the holographic fringes could be seen. A potential (either − or + ve) is applied to biprism to bring constructive interference, and this potential is related to V. The phase difference between sample specimen and vacuum in electron beam ($\Delta\theta$) is given by

$$\Delta\theta = C_E V t \tag{10.21}$$

where C_E is constant and t is the specimen thickness.

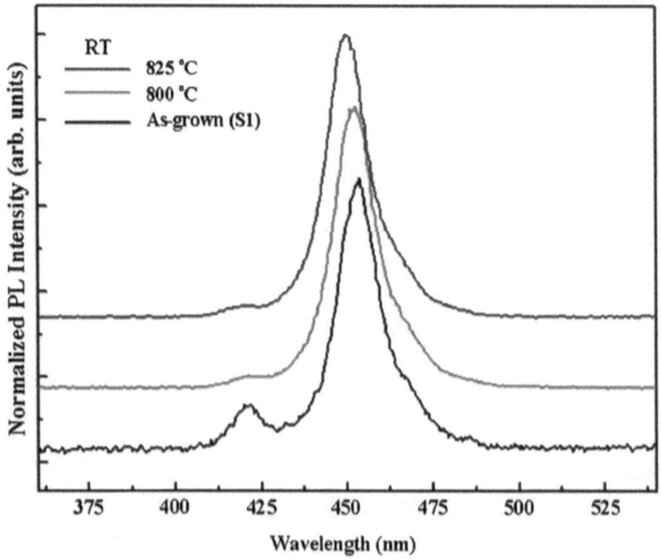

FIGURE 10.9

Photoluminescence of InGaN/GaN QWs on Si at different substrate temperatures [181].

Reproduced with the permission of Elsevier Publishers.

6.6 OPTICS

AlN produces light at 210 nm (UV) under a *dc* bias condition (known as electroluminescence) and excitation (photoluminescence) [38]. The band gaps of GaN and InN are 3.42 and 1.89 eV, respectively [180]. There is a relationship in optical band gap (E_g) when In atoms are doped to GaN (0–20 at.% or $0 < x < 0.20$ in $Ga_{1-x}In_xN$).

$$E_g = 3.42\ eV(1-x) + 1.89\ eV \times x - b \times x(1-x)$$

where $b = 2.6$ eV.

The InGaN/GaN LED samples were grown on the patterned sapphire substrate by using a Thomas Swan Metal–organic chemical-vapor deposition [181]. After the pretreatment, the substrates were set to temperature range between 500 and 600°C at which the deposition of the GaN nucleation layer was introduced. Then, a GaN buffer layer, *n*-type GaN, an $In_yGa_{1-y}N$ underlayer (UD-InGaN), and undoped-GaN layer were deposited. The 6-stacked $In_xGa_{1-x}N$/GaN (InGaN/GaN) quantum wells (QWs) with two different thicknesses (35 and 42 Å) of InGaN were grown on the undoped-GaN layer at the substrate temperature (T_s). Fig. 10.9 shows the photoluminescence spectra of the QWs at different substrate temperatures. Two peaks are observed at 420 and 453 nm. The peak at 420 nm is related to underlayer InGaN and that at 453 nm is related to InGaN/GaN QWs. With increasing substrate temperature, the peak intensity at 420 nm decreases due to improved crystallinity. Slight peak shift for InGaN/GaN QWs is due to interdiffusion of In and Ga atoms in layers. This produces blue light. N- and B-doped SiC stacked with nitrides (AlN–GaN) can provide white light (Fig. 10.10) [182].

7. APPLICATIONS

The borides, carbides, and nitrides are used for many applications due to extraordinary properties as compared to pure metals, alloys, sulfides, and oxides. Nitrides/carbides show higher superconducting

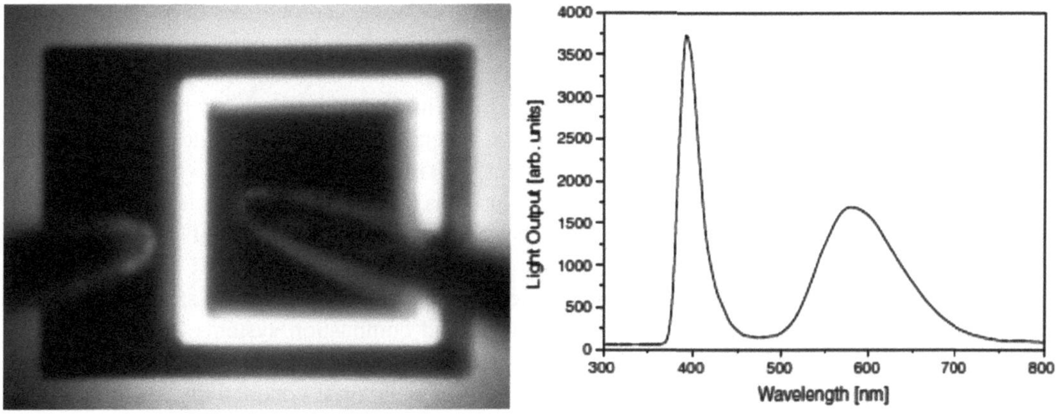

FIGURE 10.10

Photograph of operating warm-white LED (left) and the emission spectrum (right) at a 50 mA input current in N- and B-doped SiC stacked with nitrides (AlN-GaN) [182].

Reproduced with the permission of Elsevier Publishers.

transition (T_c) as compared to pure metals, e.g., V and VN have $T_c = 5.0$ and 9K, respectively [1–3,123]. Ta and TaC have $T_c = 4.5$ and 10.3K, respectively [2,123]. MgB_2 has the highest superconducting transition ($T_c = 39$K) among nitrides, carbides, and borides to date [13]. It is interesting to note that Mg is not a superconductor.

In wireless transmission system, radiofrequency filters are used. These filters have surface acoustic wave (SAW). Nitrides and carbides (e.g., AlN, SiC, AlN/3C-SiC/Si) are used as filters [183]. AlN/3C-SiC/Si multilayer can have phase velocity of 5528 ms^{-1} and an electromechanical coupling coefficient of 0.42%, and this composition is highly stable in terms of thermal and chemical properties. The thermal conductivity of AlN can be increased after covering AlN by Ti, which acts as optical transducer to convert the UV photon flux into IR radiation [184].

The hexagonal wurtzite phase of α-InN, α-GaN, and α-AlN show the direct band gaps of 1.9, 3.4, and 6.2 eV, respectively [36,185,186]. SiC has band gap of 3 eV [187]. AlN, GaN, SiC, InN, InAs, PbSe, GaAs, InAlN, InGaN, and InGaP and their combination can have band gap from very wide to narrow ranges (6–0.3 eV) [36,185–191]. They are used as LEDs, transistors, photovoltaics, solar cells, and detectors (UV-NIR range). The quantum efficiency of luminescence can be increased by fabrication of the structures free of electrostatic fields. The epitaxial growth of GaN/AlGaN quantum wells on tetragonal $LiAlO_2$ is a typical example [39]. Compact disc (CD) technology uses NIR laser diodes to read information because of their long life and reliability. Since size of a focused spot is directly proportional to the square of the wavelength, storage capacity can be enhanced by decreasing wavelength. NIR (780 nm) to red (635) to green (515 nm) lasers were developed to increase storage capacity [5,38,192,193]. GaAs and ZnSe diodes are used for NIR and green emissions. GaN (376 nm) and GaInN (411 nm) quantum wells were developed to get UV and blue lasers.

Semiconducting nitrides (AlN, GaN, InN) are paramagnetic or diamagnetic. However, when they are doped with *d*- and *f*-block elements, the long- or short-range interactions can be introduced. Room temperature ferromagnetic materials can be obtained in some cases. Here, electron-spin interaction occurs within semiconductor. These are used in spintronic applications [190,194]. Spintronic materials can be used in many electronic memory, transistor, and sensor devices. Some carbonitrides of $Sm_2(Fe_{1-x}Co_x)_{17}$ are ferromagnetic materials, which have high Curie temperature (T_{fp}), high coercivity (H_c), and high magnetization (M_s) [3]. Even $Fe_{16}N_2$ has the highest magnetization among nitrides to date [51].

Amongst nitrides, TiN has gold color, high hardness, and high chemical stability [3,195]. These are used as coating materials on windows in warm climates, watches, and jewelry. By addition of carbon in TiN, the color can be varied from red golden to deep purple. This coating helps in corrosion resistant in moisture and air. Some nitrides/carbides/borides are used as barrier between two layers [196–198]. This reduces diffusion among layers and keeps layers stable for long users.Thus, stable microchips and transistors can be fabricated.

High-temperature stable ceramic materials are needed for applications such as structural parts for motor engines, gas turbines, catalytic heat exchangers, and combustion systems [3]. Carbides are usually used as a cutting tools [e.g., WC (Fig. 10.11a), Cr_3C_2 (Fig. 10.11b and c)] since they have great hardness and wear resistance, good thermal shock resistance, thermal conductivity, good oxidation resistance, and their binding properties [2].

Nitrides, carbides, and borides have high surface area. These are used as catalysts. They may take part as both direct contact with reaction medium and support to metal catalyst in many reactions such as hydrogen transfer, isomerization, hydrogenation, hydrosulfurization, oxidation–reduction reactions [3]. A few examples are mentioned here. WN-decorated carbon nanotubes hybrids act as efficient catalyst supports for oxygen reduction reaction [199]. Pt-WN/CNT-M (CNT = carbon nanotubes) is more efficient than Pt/CNT-M catalyst in term of oxygen reduction reaction. This will reduce the price of catalyst cost since a smaller amount of Pt is required. In pharmaceuticals, fine chemicals and fragrance industries, the compounds of α, β-unsaturated aldehydes are used [200]. When the H_2 is passed through the compound (e.g., But-2-enal), many products are obtained. On basis of Gibb's free energy ($\Delta G = -$ve), hydrogenation of C=C is favorable over the C=O (see Eqs. 10.i and 10.ii).

$$CH_3CH{=}CHCHO + H_2 \overset{K_1}{\leftrightarrow} CH_3CH_2CH_2CHO \qquad (10.i)$$

$\Delta G_1 = -138.6$ kJ/mol, $K_1 = 2.0 \times 10^{24}$

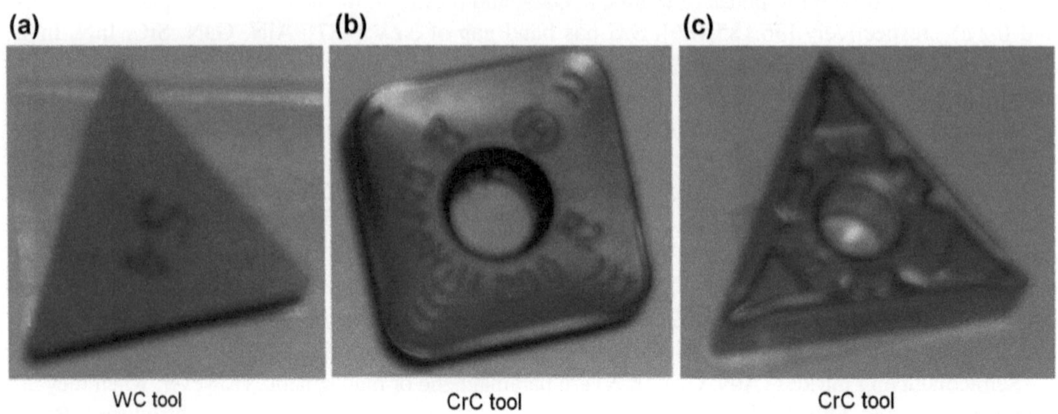

FIGURE 10.11

Cutting tools of (a) WC, (b) CrC, and (c) CrC.

$$CH_3CH{=}CHCHO + H_2 \overset{K_1}{\leftrightarrow} CH_3CH{=}CHCH_2OH \tag{10.ii}$$

$\Delta G_2 = -77.3$ kJ/mol, $K_2 = 3.6 \times 10^{13}$

In order to get site selectivity, it needs the proper catalyst. A good catalyst should have the following characteristics: (1) site selectivity, (2) high yield (conversion), (3) high activity, and (4) economic way (low cost of catalyst, room temperature reactions, etc.). These characteristics can be brought by changes in metal and support sites. In hydrogenation, Pt metal is used and support will be oxides (SiO_2, γ-Al_2O_3, SiO_2-Al_2O_3, and zeolites). Nonoxide-based supports are graphite, carbon nanotubes and graphene, nitrides, carbides, borides and sulfides, etc. Supports possess large surface area, numerous acidic/basic sites, as well as metal-support interaction, which help in changes in catalytic behavior. Fig. 10.12 shows the possible reaction paths of hydrogenation on α, β-unsaturated aldehyde (But-2-enal, a) [200]. Intermediate products are But-2-en-1-ol (b) and Butanal (c). Butanal is more stable than But-2-en-1-ol. On further hydrogenation, both will give Butan-1-ol.

The conversion of But-2-enal and the selectivity of product are calculated using Eqs. (10.22) and (10.23).

$$\text{Conversion}(\%) = [\{C_{\text{But-2-enal}}(\text{Input}) - C_{\text{But-2-enal}}(\text{Output})\} \times 100\%]/[C_{\text{But-2-enal}}(\text{Input})] \tag{10.22}$$

$$\text{Selectivity} = [C_{\text{product}}]/[C_{\text{But-2-enal}}(\text{Input}) - C_{\text{But-2-enal}}(\text{Output})] \tag{10.23}$$

The yield of product (say, But-2-en-1-ol) is calculated by the following:

$$\text{Yield} = \text{Conversion}(\%) \times \text{Selectivity} \tag{10.24}$$

In order to get But-2-en-1-ol (b) at 100°C, $Pt_{1-x}Fe_x/\gamma$-Al_2O_3 is used. However, it gives poor selectivity and low yield, but high activity. On other hand, $Pt_{1-x}Fe_x$/graphite gives good selectivity, high yield, but low activity. $Pt_{1-x}Fe_x$/BN gives good selectivity, high yield, and high activity. Fig. 10.13 shows the effect of Fe loading on the selectivity and yield of products from hydrogenation of But-2-enal at 100°C [200].

Here, H_2/But-2-enal ratio is 59, and this was passed through reactor at rate of 50 mL/min. Upon increasing Fe wt% up to 0.6, selectivity of product (But-2-en-1-ol) increases due to suppression of

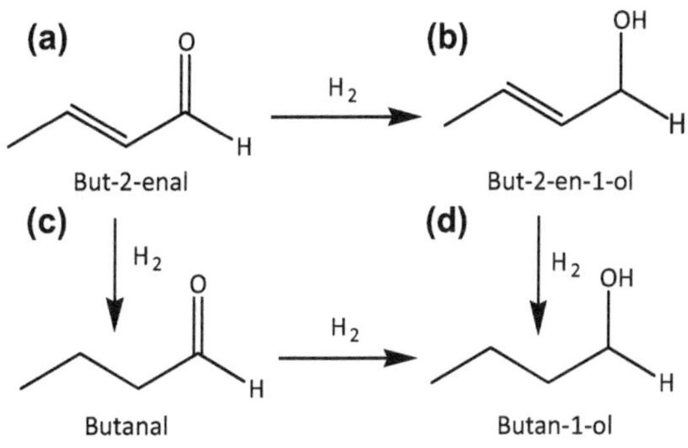

FIGURE 10.12

The possible reaction paths of hydrogenation on α, β-unsaturated aldehyde (But-2-enal, a). Intermediate products are But-2-en-1-ol (b) and Butanal (c), and final product is Butan-1-ol (d).

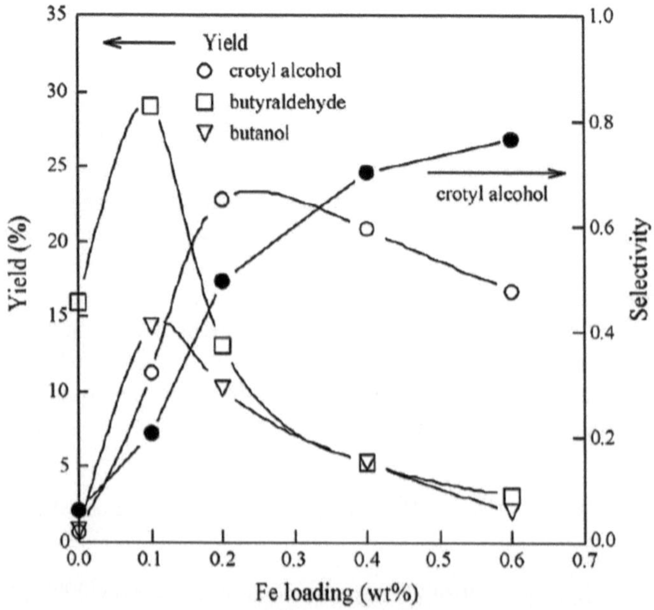

FIGURE 10.13

The effect of Fe loading on the conversion and selectivity on $Pt_{1-x}Fe_x/BN$ catalyst at 100°C. H_2/crotonaldehyde ratio is 59, and this is passed to catalyst (reactor) at the 50 mL/min. Here, crotonaldehyde, crotyl alcohol, butyraldehyde, and butanol are common names, and their names in IUPAC are But-2-enal, Buta-2-en-1-ol, Butanal, and Butan-1-ol, respectively [200].

Reproduced with the permission of Elsevier Publishers.

formation of other products (Butanal and Butan-1-ol). However, the highest yield of product (But-2-en-1-ol) is obtained at 0.2 wt% Fe. Yield of other products (Butanal and Butan-1-ol) is also shown in Fig. 10.13 [200].

The selectivity toward But-2-en-1-ol can be related to the easier formation of PtFe on BN surface than γ-Al_2O_3. The Fe^{2+} on PtFe polarizes C=O bond of But-2-enal and the nearby Pt^o is used to provide the absorbed hydrogen to the C=O bond for hydrogenation, instead of the C=C bond.

Zirconium carbides (ZrC) and silicon carbides (SiC) coating over materials are used as protective layer in nuclear reactor since this is less sensitive to hydrogen (H_2) [201,202]. These are used as cladding materials in nuclear reactors due to high chemical stability, high binding capacity of fission products, as well as low absorption cross-section in thermal neutrons. Boron carbide (B_4C) rods are used in nuclear reactor (light water reactors) as neutron absorber. Now, SiC is used as fuel rod in high-temperature nuclear reactor where oxidation and reduction occur and will be better than ZrC in terms of thermal stability. SiC is used as radiation detector (fast neutrons) for high-temperature applications in harsh radiation environments such as nuclear reactor. A recent review article on the synthesis, characterization, properties of nitrides, and applications will be useful for further understanding of nitrides [203].

8. CONCLUSIONS

Borides, carbides, and nitrides are prepared through chemical routes. Some of them can be prepared only under extreme pressure and temperature. In extreme conditions (temperature, pressure, radiation, environment), some of them show exceptional stability. The nature of chemical bonding is discussed. They can exhibit a range of properties like superconducting, extreme hardness, high chemical stability, high thermal stability, and high magnetic properties. The nitrides, carbides, and borides are used in many applications such as solar cells, coating materials, cutting tools, catalysts, protective layers in nuclear reactors, lasers, and protection from interlayer diffusion due to extraordinary properties as compared to pure metals, alloys, sulfides, and oxides.

REFERENCES

[1] R.S. Ningthoujam, N. Sudhakar, K.P. Rajeev, N.S. Gajbhiye, Electrical resistivity study of La, B doped nanocrystalline superconducting vanadium nitride, J. Appl. Phys. 91 (2002) 6051–6056.

[2] L.E. Toth, Transition Metal Carbides and Nitrides, Academic Press, New York, 1971.

[3] S.T. Oyama, The Chemistry of Transition Metal Carbides and Nitrides, Blackie Academic & Professional, New York, 1996.

[4] L. Xu, S. Li, Y. Zhang, Y. Zhai, Synthesis, properties and applications of nanoscale nitrides, borides and carbides, Nanoscale 4 (2012) 4900–4915.

[5] F.A. Ponce, D.P. Bour, Nitride-based semiconductors for blue and green light-emitting devices, Nature 386 (1997) 351–359.

[6] M. Gurvitch, Universal disordered-induced transition in the resistivity behavior of strongly coupled nitrides, Phys. Rev. Lett. 56 (1986) 647–650.

[7] R.J. Cava, H. Takagi, B. Batlogg, H.W. Zandbergen, J.J. Krajewski, W.F. Peck Jr., R.B. van Dover, J.R. Felder, T. Siegrist, K. Mizuhashi, J.O. Lee, H. Eisaki, S.A. Carter, S. Uchida, Superconductivity at 23K in yttrium palladium boride carbide, Nature 367 (1994) 146–148.

[8] R.J. Cava, H.W. Zandbergen, B. Batlogg, H. Eisaki, H. Takagi, J.J. Krajewski, W.F. Peck Jr., E.M. Gyorgy, S. Uchida, Superconductivity in lanthanum nickel boro-nitride, Nature 372 (1994) 245–247.

[9] R. Nagarajan, C. Mazumdar, Z. Hossain, S.K. Dhar, K.V. Gopalkrishnan, L.C. Gupta, C. Godart, B.D. Padalia, R. Vijayaraghavan, Bulk superconductivity at an elevated temperature ($T_c \sim 12K$) in a nickel containing alloy system Y–Ni–B–C, Phys. Rev. Lett. 72 (1994) 274–277.

[10] H. Eisaki, H. Takagi, R.J. Cava, B. Batlogg, J.J. Krajewski, H.W. Zandbergen, W.F. Peck Jr., K. Mizuhashi, J.O. Lee, S. Uchida, Competition between magnetism and superconductivity in rare-earth nickel boride carbides, Phys. Rev. B 50 (1994) 647–650.

[11] S. Yamanaka, High T_c superconductivity in electron doped layer structured nitrides, Annu. Rev. Mater. Sci. 30 (2000) 53–82.

[12] S. Yamanaka, K. Hotehama, H. Kawaji, Superconductivity at 25.5K in electron-doped layered hafnium nitride, Nature 392 (1998) 580–582.

[13] J. Nagamatsu, N. Nakagawa, T. Muranaka, Y. Zenitani, J. Akimitsu, Superconductivity at 39K in magnesium borides, Nature 410 (2001) 63–64.

[14] H. Rietschel, H. Winter, W. Reichardt, Strong depression of superconductivity in VN by spin fluctuations, Phys. Rev. B 22 (1980) 4284–4292.

[15] R.S. Ningthoujam, N.S. Gajbhiye, Thermal decomposition study of nanocrystalline Ni_3N, Indian J. Phys. 78A (2004) 265–269.

[16] L. Yu, C. Stampfl, D. Marshall, T. Eshrich, V. Narayanan, J.M. Rowell, N. Newman, A.J. Freeman, Mechanism and control of the metal-to-insulator transition in rock salt tantalum nitride, Phys. Rev. B 65 (2002) 245110–245115.

[17] K. Reichelt, W. Nellen, G. Mair, Preparation and compositional analysis of sputtered TaN films, J. Appl. Phys. 49 (1978) 5284–5287.

[18] D. Gall, I. Petrov, J.E. Greene, Epitaxial $Sc_{1-x}Ti_xN$(001): optical and electronic transport properties, J. Appl. Phys. 89 (2001) 401–409.

[19] C. Steinborn, M. Herrmann, U. Keitel, A. Schonecker, J. Rathel, D. Rafaja, Correlation between microstructure and electrical resistivity of hexagonal boron nitride ceramics, J. Eur. Ceram. Soc. 33 (2013) 1225–1235.

[20] Y. Kobayashi, C.-L. Tsai, T. Akasaka, Optical band gap of h-BN epitaxial film grown on *c*-plane sapphire substrate, Phys. Stat. Solidi C 7 (2010) 1906–1908.
[21] X.W. Xhang, Y.J. Zou, B. Wang, X.M. Song, H. Yan, G.H. Chen, S.P. Wang, Optical band gap and refractive index of c-BN thin films synthesized by radio frequency bias sputtering, J. Mater. Sci. 36 (2001) 1957–1961.
[22] X.-M. He, L. Shu, W.-Z. Li, H.-D. Li, Structure and properties of carbon nitride films synthesized by low energy ion bombardment, J. Mater. Res. 12 (1997) 1595–1602.
[23] J. Siwiec, A. Sokolowska, A. Olszyna, R. Dwilinski, M. Kaminska, J. Konwerska-Hrabowska, Photoluminescence properties of nanocrystalline, wide band gap nitrides (C_3N_4, BN, AlN, GaN), Nanostruct. Mater. 10 (1998) 625–634.
[24] D.A. Anderson, W.E. Spear, Electrical and optical properties of amorphous silicon carbide, silicon nitride and germanium carbide prepared by the glow discharge technique, Philos. Mag. 35 (1978) 1–16.
[25] V.Y. Doo, D.R. Nickols, G.A. Silvery, Preparation and properties of pyrolytic silicon nitride, J. Electrochem. Soc. 113 (1966) 1279–1281.
[26] S.V. Deshpande, E. Gulari, S.W. Brown, S.C. Rand, Optical properties of silicon nitride films deposited by hot filament chemical vapor deposition, J. Appl. Phys. 77 (1995) 6534–6541.
[27] D.T. Morelli, J.P. Heremans, Thermal conductivity of germanium, silicon, and carbon nitrides, Appl. Phys. Lett. 81 (2002) 5126–5128.
[28] C.W. Chang, W.-Q. Han, A. Zettle, Thermal conductivity of B–C–N and BN nanotubes, Appl. Phys. Lett. 86 (2005) 173102–173103.
[29] M. Sharmsa, M.L. Liu, A.A. Balandin, C. Casiraghi, W.I. Milne, A.C. Ferrai, Thermal conductivity of diamond-like carbon films, Appl. Phys. Lett. 89 (2006) 161921–161923.
[30] T.R. Anthony, W.F. Banholzer, J.F. Fleischer, L. Wei, P.K. Kuo, R.L. Thomas, R.W. Pryor, Thermal diffusivity of isotopically enriched ^{12}C diamond, Phys. Rev. B 42 (1990) 1104–1111.
[31] I. Tanaka, F. Oba, T. Sakine, E. Ito, A. Kubo, K. Tatsumi, H. Adachi, T. Yamamoto, Hardness of cubic silicon nitride, J. Mater. Res. 17 (2002) 731–733.
[32] Y. Tian, B. Xu, D. Lu, Y. Ma, Y. Wang, Y. Jiang, W. Hu, C. Tang, Y. Gao, K. Luo, Z. Zhao, L.-M. Wang, B. Wen, J. He, Z. Liu, Ultrahard nanotwinned cubic boron nitride, Nature 493 (2013) 385–388.
[33] R.H. Wentorf, R.C. DeVries, R.P. Bundy, Sintered superhard materials, Science 208 (1980) 873–880.
[34] A.R. Krauss, O. Auciello, D.M. Gruen, A. Jayatissa, A. Sumant, J. Tucek, D.C. Mancini, N. Moldovan, A. Erdemir, D. Ersoy, M.N. Gardos, H.G. Busmann, E.M. Meyer, M.Q. Ding, Ultrananocrystalline diamond thin films for MEMS and moving mechanical assembly devices, Diamond Relat. Mater. 10 (2001) 1952–1961.
[35] O.A. Williams, Nanocrystalline diamond, Diamond Relat. Mater. 20 (2011) 621–640.
[36] O. Ambacher, Growth and applications of group iii-nitrides, J. Phys. D Appl. Phys. 31 (1998) 2653–2710.
[37] J.W. Orton, C.T. Foxon, Group III nitride semiconductors for short wavelength light-emitting devices, Rep. Prog. Phys. 61 (1998) 1–75.
[38] Y. Taniyasu, M. Kasu, T. Makimoto, An aluminium nitride light-emitting diode with a wavelength of 210 nanometres, Nature 441 (2006) 325–328.
[39] P. Waltereit, O. Brandt, A. Trampert, H.T. Grahn, J. Menniger, M. Ramsteiner, M. Reiche, K.H. Ploog, Nitride semiconductors free of electrostatic fields for efficient white light-emitting diodes, Nature 406 (2000) 865–868.
[40] V.V. Mamutin, A.Y. Wgorov, N.V. Kryzhanovskaya, Molecular beam epitaxy growth methods of wavelength control for InAs/(In)GaAsN/GaAs heterostructures, Nanotechnology 19 (2008) 445715.
[41] H. Zhang, J. Tang, J. Yuan, J. Ma, N. Shinya, K. Nakajima, H. Murakami, T. Ohkubo, L.-C. Qin, Nanostructured LaB_6 field emitter with lowest apical work function, Nano Lett. 10 (2010) 3539–4544.
[42] S. Niyomsoan, W. Grant, D.L. Olson, B. Mishra, Variation of color in titanium and zirconium nitride decorative thin films, Thin Solids Films 415 (2002) 187–194.
[43] J.A. Arsecularatne, L.C. Zhang, C. Montross, Wear and tool life of tungsten carbide, PCBN and PCD cutting tools, Int. J. Mach. Tools Manuf. 46 (2006) 482–491.

[44] Y.K. Chou, C.J. Eavns, Cubic boron nitride tool wear in interrupted hard cutting, Wear 225–229 (1999) 234–245.

[45] B. Zou, W. Ji, C. Huang, S. Li, J. Wang, Study on microstructure and its formation mechanism, and mechanical properties of TiB_2-TiC laminated $Ti(C_5N_5)$ composite ceramic cutting tool material, Int. J. Refract. Met. Hard Mater. 42 (2014) 169–179.

[46] H. Caliskan, Selection of boron based tribological hard coatings using multi-criteria decision making methods, Mater. Des. 50 (2013) 742–749.

[47] X. Shi, M. Wang, S. Zhang, Q. Zhang, Fabrication and properties of W–20Cu alloy reinforced by titanium nitride coated SiC fibers, Int. J. Refract. Met. Hard Mater. 41 (2013) 60–65.

[48] L. Gladczuk, A. Patel, J.D. Demaree, M. Sosnowski, Sputter deposition of bcc tantalum films with TaN underlayers for protection of steel, Thin Solid Films 476 (2005) 295–302.

[49] B. Ganem, J.O. Osby, Synthetically useful reactions with metal boride and aluminide catalysts, Chem. Rev. 86 (1996) 763–780.

[50] J. Monnier, H. Sulimma, A. Dalai, G. Caravaggio, Hydrodeoxygenation of oleic acid and canola oil over alumina-supported metal nitrides, Appl. Cat. A Gen. 382 (2010) 176–180.

[51] R.M. Metzger, X. Bao, M. Carbucicchio, Magnetism of α''-$Fe_{16}N_2$ (invited), J. Appl. Phys. 76 (1994) 6626–6631.

[52] W.E. Wallace, M.Q. Huang, Enhanced Fe moment in nitrogen martensite and $Fe_{16}N_2$ (invited), J. Appl. Phys. 76 (1994) 6648–6652.

[53] X. Bao, R.M. Metzger, M. Carbucicchio, Synthesis and properties of $Fe_{16}N_2$ in magnetic particles, J. Appl. Phys. 75 (1994) 5870–5872.

[54] S. Kikkawa, A. Yamada, Y. Masubuchi, T. Takeda, Fine $Fe_{16}N_2$ powder prepared by low-temperature nitridation, Mater. Res. Bull. 43 (2008) 3352–3357.

[55] K. Nakajima, S. Okamoto, Large magnetization induced in single crystalline iron films by high-dose nitrogen implantation, Appl. Phys. Lett. 56 (1989) 92–94.

[56] N. Sudhakar, R.S. Ningthoujam, K.P. Rajeev, A.K. Nigam, J. Weissmuller, N.S. Gajbhiye, Effect of La, B doping on the electrical resistivity and magnetic susceptibility of nanocrystalline vanadium nitride, J. Appl. Phys. 96 (2004) 688–695.

[57] L.M. Corliss, N. Elliott, J.M. Hastings, Antiferromagnetic structure of CrN, Phys. Rev. 117 (1960) 929–935.

[58] D.H. Mosca, P.H. Dionisio, W.H. Schreiner, I.J.R. Baurnvol, C. Achete, Compositional and magnetic properties of iron nitride thin films, J. Appl. Phys. 67 (1990) 7514–7519.

[59] T. Himamura, S. Nasu, A study of Fe–N alloy systems, Hyperfine Interact. 111 (1998) 221–226.

[60] R.S. Ningthoujam, N.S. Gajbhiye, Magnetic study of single domain ε-Fe_3N nanoparticles synthesized by precursor techniques, Mater. Res. Bull. 43 (2008) 1079–1085.

[61] N.S. Gajbhiye, R.S. Ningthoujam, J. Weissmular, Mössbauer study of nanocrystalline å-$Fe_{3-x}Co_xN$ system, Hyperfine Interact. 156/157 (2004) 51.

[62] K.B. Gavan, E.M.J.M. van der Drift, W.J. Venstra, M.R. Zuiddam, H.S.J. van der Zant, Effect of undercut on the resonant behaviour of silicon nitride cantilevers, J. Micromech. Microeng. 19 (2009) 035003–035008.

[63] E. Wuchina, E. Opila, M. Opeka, W. Fahrenholtz, I. Talmy, UHTCs: ultra-high temperature ceramic materials for extreme environment applications, Electrochem. Soc. Interface (Winter 2007) 30–36.

[64] J.F. Justin, A. Jakowick, Ultra high temperature ceramics: densification, properties and thermal stability, J. Aerosp. Lab Issue 3 (2011) 11.

[65] K.H. Kim, R.G. Gordon, A. Ritenour, D.A. Antoniadis, Atomic layer deposition of insulating nitride interfacial layers for germanium metal oxide semiconductor field effect transistors with high-κ oxide/tungsten nitride gate stacks, Appl. Phys. Lett. 90 (2007) 212104–212113.

[66] J.C. White, G.L. Harris, D.B. Poker, Semi-insulating layers in 4H and 6H SiC by Si and C ion implantation, Electron. Lett. 38 (2002) 1597–1598.

[67] T.R. Pilladi, G. Panneerselvam, S. Anthonysamy, V. Ganesan, Thermal expansion of nanocrystalline boron carbide, Ceram. Int. 38 (2012) 3723–3728.

[68] T. Yano, Y. Futamura, M. Imai, K. Yoshide, Recovery behavior of neutron-induced damage of AlN irradiated at higher temperatures by thermal annealing, J. Nucl. Mater. 440 (2003) 495–499.

[69] X. Hou, Z. Yu, Y. Li, K.-C. Chou, Preparation and properties of hexagonal boron nitride fibers used as high temperature membrane filter, Mater. Res. Bull. 49 (2014) 39–43.

[70] Z. Chen, R. Huang, X. Chu, X. Wu, Z. Liu, Y. Zhou, L. Li, Negative thermal expansion and nearly zero temperature coefficient of resistivity in anti-perovskite manganese nitride Mn_3CuN co-doped with Ag and Sn, Cryogenics 52 (2012) 629–631.

[71] N. Sarkar, S. Ghosh, Temperature dependent band gap shrinkage in GaN: role of electron-phonon interaction, Solid State Commun. 149 (2009) 1288–1291.

[72] K. Guo, D. Rau, L. Toffoletti, C. Muüller, U. Burkhardt, W. Schnelle, R. Niewa, U. Schwarz, Ternary metastable nitrides ε-Fe_2TMN (TM=Co, Ni): high-pressure, high-temperature synthesis, crystal structure, thermal stability, and magnetic properties, Chem. Mater. 24 (2012) 4600–4606.

[73] E.H. Bordon, R. Riedel, A. Zerr, P.F. McMillan, G. Auffermann, Y. Prots, W. Bronger, R. Kniep, P. Kroll, High-pressure chemistry of nitride-based materials, Chem. Soc. Rev. 35 (2006) 987–1014.

[74] A.F. Young, C. Sanloup, E. Gregoryanz, S. Scandolo, R.J. Hemley, H.-K. Mao, Synthesis of novel transition metal nitrides IrN_2 and OsN_2, Phys. Rev. Lett. 96 (2006) 155501–155504.

[75] E. Gregoryanz, C. Sanloup, M. Somayazulu, J. Badro, G. Fiquet, H.-K. Mao, R.J. Hemley, Synthesis and characterization of a binary noble metal nitride, Nat. Mater. 3 (2004) 294–297.

[76] J.C. Crowhurst, A.F. Goncharov, B. Sadigh, C.L. Evans, P.G. Morrall, J.L. Ferreira, A.J. Nelson, Synthesis and characterization of the nitrides of platinum and iridium, Science 311 (2006) 1275–1278.

[77] V.R. Maslennikova, A.A. Belkina, A.D. Panasyuk, L.I. Struk, V.P. Smirnov, Effect of high-pressure on the structure and properties of materials based on boron carbides, Powder Metall. Met. Ceram. 34 (1995) 491–495.

[78] Y. Kojima, H. Ohfuji, Structure and stability of carbon nitride under high pressure and high temperature up to 125 GPa and 3000K, Diamond Relat. Mater 39 (2013) 1–7.

[79] D. Lahiri, V. Singh, G.R. Rodrigues, T.M.H. Costa, M.R. Gallasc, S.R. Bakshi, S. Seal, A. Agarwal, Ultrahigh-pressure consolidation and deformation of tantalum carbide at ambient and high temperatures, Acta Mater. 61 (2013) 4001–4009.

[80] D.H. Gregory, Nitride chemistry of the s-block elements, Coord. Chem. Rev. 215 (2001) 301–345.

[81] M. Satake, S.A. Iqbal, Chemistry of P-block Elements, Discovery Publishing House, India, New Delhi, 1995.

[82] P. Peshev, L. Leyarovska, G. Bliznakov, On the borothermic preparation of some vanadium, niobium and tantalum borides, J. Less-Common Met. 15 (1968) 259–267.

[83] P. Franke, D. Neuschutz, Binary systems Part 2. Elements and binary systems from B–C to Cr–Zr, Landolt-Bornstein-Group IV Phys. Chem. 19B2 (2004) 1–4.

[84] R.S. Ningthoujam, Synthesis, Solid State Reactivity, Superconducting, Electrical Transport and Magnetic Properties of Nanostructured Metal Nitrides: VN, MoN, Mo_2N, Ni_3N and E-$Fe_{3-x}Co_xN$ ($0.0 \leq X \leq 0.8$) (Ph.D. thesis), IIT Kanpur, India, 2004.

[85] D.S. Bem, C.M. Lampe-Önnerud, H.P. Olsen, H.-C. zur Loye, Synthesis and structure of two new ternary nitrides: $FeWN_2$ and $MnMoN_2$, Inorg. Chem. 35 (1996) 581–585.

[86] K. Takenaka, H. Takagi, Giant negative thermal expansion in Ge-doped anti-perovskite manganese nitrides, Appl. Phys. Lett. 87 (2005) 261902.

[87] N.S. Gajbhiye, R.S. Ningthoujam, S. Bhatacharyya, Magnetic properties of Co and Ni substituted ε-Fe_3N nanoparticles, Hyperfine Interact. 164 (2005) 17–26.

[88] H.W. Zandbergen, J. Jansen, R.J. Cavai, J.J. Krajewskii Jr., W.F. Peck, Structure of the 13-K superconductor $La_3Ni_2B_2N_3$ and the related phase LaNiBN, Nature 372 (1994) 759–761.

[89] P. Marksteiner, P. Weinberger, A. Neckel, R. Zeller, P.H. Dederichs, Electronic structure of substoichiometric carbides and nitrides of titanium and vanadium, Phys. Rev. B 33 (1986) 812–822.
[90] A. Rabenau, H. Schulz, Re-evaluation of the lithium nitride structure, J. Less-Common Met. 50 (1976) 155–159.
[91] D.-W. Kim, T.-H. Kim, H.-W. Park, D.-W. Park, Synthesis of nanocrystalline magnesium nitride (Mg_3N_2) powder using thermal plasma, Appl. Surf. Sci. 257 (2011) 5375–5379.
[92] R. Schmitz, R. Müller, S. Krüger, R.W. Schmitz, S. Nowak, S. Passerini, M. Winter, C. Schreiner, Investigation of lithium carbide contamination in battery grade lithium metal, J. Power Sources 217 (2012) 98–101.
[93] O.O. Kurakevych, T.A. Strobel, D.Y. Kim, G.D. Cody, Synthesis of Mg_2C: a magnesium methanide, Angew. Chem. Int. Ed. 52 (2013) 8930–8933.
[94] J.W. Gerlach, J. Mennig, B. Rauschenbach, Epitaxial gadolinium nitride thin films, Appl. Phys. Lett. 90 (2007) 061919–061923.
[95] P. Jeitschko, A. Simon, R. Ramlau, H. Mattausch, Chemical intergrowth in gadolinium carbide chloride – a HRTEM study, Z. Anorg. Allg. Chem. 623 (1997) 1447–1454.
[96] S. Castelletto, B.C. Johnson, V. Ivády, N. Stavrias, T. Umeda, A. Gali, T. Ohshima, A silicon carbide room-temperature single-photon source, Nat. Mater. 13 (2014) 151–156.
[97] G. Indelkofer, J.M. Mariot, W. Lengauer, E. Beauprez, P. Oelhafen, C.F. Hague, Electronic structure of stoichiometric and substoichiometric vanadium nitride from photoelectron spectroscopy, Solid State Commun. 72 (1989) 419–423.
[98] G.P. Shveikin, A.L. Ivanovskii, The chemical bonding and electronic properties of metal borides, Russ. Chem. Rev. 63 (1994) 711–734.
[99] S. Flandrois, B. Ottaviani, A. Derre, A. Tressaud, Boron-substituted carbons and their intercalation compounds, J. Phys. Chem. Solids 57 (1996) 741–744.
[100] J.L. Sarrao, M.C. de Andrade, J. Herrman, S.H. Han, Z. Fisk, M.B. Maple, R.J. Cava, Superconductivity to 21K in intermetallic thorium-based boride carbides, Phys. C 229 (1994) 65–69.
[101] A.B. Auskern, S. Aronson, Electrical properties of thorium borides, J. Chem. Phys. 49 (1968) 172–176.
[102] P. Rogl, J. Bauer, J. Debuigne, The ternary system uranium-boron-carbon, J. Nucl. Mater 165 (1989) 74–82.
[103] C.H. Liu, Structure and properties of boron carbide with aluminum incorporation, Mater. Sci. Eng. B 72 (2000) 23–26.
[104] C. Schimpf, M. Motylenko, D. Rafaja, Quantitative description of microstructure defects in hexagonal boron nitride using X-ray diffraction analysis, Mater. Charact. 86 (2013) 190–199.
[105] R. Saito, Y. Mosubuchi, T. Motohashi, S. Kikkawa, M. Sasaki, M. Sasaki, Microstructure formation during the annealing of iron nitrides, Mater. Res. Bull. 46 (2011) 547.
[106] V. Tagliazucca, M. Leoni, C. Weidenthaler, Crystal structure and microstructural changes of molybdenum nitrides traced during catalytic reaction by in situ X-ray diffraction studies, Phys. Chem. Chem. Phys. 16 (2014) 6182–6188.
[107] M.J. Hoffmann, G. Petzow, Tailored microstructures of silicon nitride ceramics, Pure Appl. Chem. 66 (1994) 1807–1814.
[108] S.I. Cha, S.H. Hong, G.H. Ha, B.K. Kim, Microstructure and mechanical properties of nanocrystalline WC-10Co cemented carbides, Scr. Mater. 44 (2001) 1535–1539.
[109] N.S. Gajbhiye, R.S. Ningthoujam, Structural, electrical and magnetic studies of nanocrystalline δ-MoN and γ-Mo_2N, Phys. Stat. Sol. C 1 (2004) 3449–3454.
[110] K.H. Jack, The occurrence and the crystal structure of α''-iron nitride: a new type of interstitial alloy formed during the tempering of nitrogen-martensite, Proc. Royal Soc. A 208 (1951) 216–224.
[111] K.H. Jack, The synthesis, structure and characterization of α''-$Fe_{16}N_2$, J. Appl. Phys. 76 (1994) 6620–6625.

[112] G. Serghiou, A.J. McGaff, N. Russell, J.P. Mörniroli, D.J. Frost, N. Odling, R. Boehler, D. Troadec, C. Lathe, Extreme conditions synthesis, processing and characterization of metal nitrides and alloys of mechanical and optoelectronic importance, J. Phys. Conf. Ser. 215 (2010) 012128–012134.
[113] A.F. Goncharov, J.C. Crowhurst, J.K. Dewhurst, S. Sharma, Raman spectroscopy of cubic boron nitride under extreme conditions of high pressure and temperature, Phys. Rev. B 72 (2005) 100104.
[114] V.L. Solozhenko, O.O. Kurakevych, Y.L. Godec, Creation of nanostructures by extreme conditions: high-pressure synthesis of ultrahard nanocrystalline cubic boron nitride, Adv. Mater. 24 (2012) 1540–1544.
[115] A.A. Correa, S.A. Bonev, G. Galli, Carbon under extreme conditions: phase boundaries and electronic properties from first-principles theory, PNAS 103 (2006) 1204–1208.
[116] D. Adler, B.B. Schwartz, M. Silver, Disordered Materials, Plenum Press, New York, 1991.
[117] N.G. Parsonage, L.A.K. Staveley, Disorder in Crystals, Clarendon Press, Oxford, 1978.
[118] O. Gunnarsson, M. Calandra, J.E. Han, Saturation of electrical resistivity, Rev. Mod. Phys. 75 (2003) 1085–1099.
[119] P.A. Lee, T.V. Ramkrishnan, Disordered electronic systems, Rev. Mod. Phys. 57 (1985) 287–337.
[120] B.L. Altshuler, A.G. Aronov, in: A.L. Efros, M. Pollak (Eds.), Electron-electron Interactions in Disordered Systems, North Holland, Amsterdam, 1985.
[121] G.T. Meaden, Electrical Resistance of Metals, Plenum Press, New York, 1965.
[122] D. Kumar, J. Sankar, J. Narayan, R.K. Singh, A.K. Majumdar, Low-temperature resistivity minima in colossal magnetoresistive $La_{0.7}Ca_{0.3}MnO_3$ thin films, Phys. Rev. B 65 (2002) 094407–94416.
[123] C. Kittel, Introduction to Solid State Physics, seventh ed., John Wiley & Sons, Inc., New York, 1995.
[124] E. Rozenberg, M. Auslender, I. Felner, G. Gorodetsky, Low-temperature resistivity minimum in ceramic manganites, J. Appl. Phys. 88 (2000) 2578–2582.
[125] P. Kordos, M. Morvic, J. Betko, J.M.V. Hove, A.M. Wowchak, P.P. Chow, Conductivity and Hall effect characterization of highly resistive molecular-beam epitaxial GaN layers, J. Appl. Phys. 88 (2000) 5821–5826.
[126] N. Savvides, High T_c superconducting B1 phase MoN films prepared by low-energy ion-assisted deposition, J. Appl. Phys. 62 (1987) 600–610.
[127] S. Terada, H. Tanaka, K. Kubota, Heteroepitaxial growth of Cu_3N thin films, J. Cryst. Growth 94 (1989) 567–568.
[128] D.A. Papaconstantopoulos, W.E. Pickett, B.M. Klein, L.L. Boyer, Electronic properties of transition-metal nitrides: the group-V and group-VI nitrides VN, NbN, TaN, CrN, MoN, and WN, Phys. Rev. B 31 (1985) 752–761.
[129] W.L. McMillan, Transition temperature of strong-coupled superconductors, Phys. Rev. 167 (1968) 331–344.
[130] E.M. Sparrow, J.M. Gorman, A. Trawick IV, J.P. Abraham, Novel techniques for measurement of thermal conductivity of both highly and lowly conducting solid media, Int. J. Heat Mass Transf. 55 (2012) 4037–4042.
[131] Y.-J. Park, M.-J. Park, J.-M. Kim, J.-W. Lee, J.-W. Ko, H.-D. Kim, Sintered reaction-bonded silicon nitrides with high thermal conductivity: the effect of the starting Si powder and Si_3N_4 diluents, J. Eur. Ceram. Soc. 34 (2014) 1105–1113.
[132] A.M. Abyzov, S.V. Kidalov, F.M. Shakhov, High thermal conductivity composite of diamond particles with tungsten coating in a copper matrix for heat sink application, Appl. Thermal Eng. 48 (2012) 72–80.
[133] H. Ji, S. Wang, M. Li, J. Kim, Deep crystallization induced high thermal conductivity of low-temperature sintered Ag nanoparticles, Mater. Lett. 116 (2014) 219–222.
[134] B. Feng, Z. Li, X. Zhang, Prediction of size effect on thermal conductivity of nanoscale metallic films, Thin Solid Films 517 (2009) 2803–2807.
[135] D.J. Twitchen, C.S.J. Pickles, S.E. Coe, R.S. Sussmann, C.E. Hall, Thermal conductivity measurements on CVD diamond, Diam. Relat. Mater. 10 (2001) 731–735.

[136] S. Ghosh, I. Calizo, D. Teweldebrhan, E.P. Pokatilov, D.L. Nika, A.A. Balandian, W. Bao, F. Miao, C.N. Lau, Extremely high thermal conductivity of graphene: prospects for thermal management applications in nanoelectronic circuits, Appl. Phys. Lett. 92 (2008) 151911–151913.
[137] B. Wang, X. Wang, W. Lou, J. Hao, Thermal conductivity and rheological properties of graphite/oil nanofluids, Colloids Surf. A Physicochem. Eng. Asp. 414 (2012) 125–131.
[138] J. Hone, Carbon nanotubes: thermal properties, in: Dekker Encyclopedia of Nanoscience & Nanotechnology, 2004, pp. 603–610, http://dx.doi.org/10.1081/E-ENN 120009128.
[139] X.W. Zhou, Y. Sakka, Y. Zhou, K. Hirao, K. Ilatani, A strategy for fabricating textured silicon nitride with enhanced thermal conductivity, J. Eur. Ceram. Soc. 34 (2014) 2585–2589.
[140] R.S. Ningthoujam, Enhancement of luminescence by rare earth ions doping in semiconductor host, in: S.B. Rai, Y. Dwivedi (Eds.), Synthesis, Characterization and Applications of Multifunctional Materials, Nova Science Publishers Inc., Hauppauge, USA, 2012.
[141] D. Curie, Luminescence in Crystals, first ed., John Wiley and Sons Inc., London, 1960.
[142] R.S. Ningthoujam, V. Sudarsan, S.K. Kulshreshtha, SnO_2:Eu nanoparticles dispersed in silica: a low-temperature synthesis and photoluminescence study, J. Lumin. 127 (2007) 747–756.
[143] A.P. Alivistos, Semiconductor clusters, nanocrystals and quantum dots, Science 271 (1996) 933–937.
[144] A.B. Greytak, P.M. Allen, W. Liu, Z. Zhao, E.R. Young, Z. Popovic, B.J. Walker, D.G. Nocera, M.G. Bawendi, Alternative layer addition approach to CdSe/CdS core-shell quantum dots with near-unity quantum yield and high on-time fractions, Chem. Soc. 3 (2012) 2028–2034.
[145] B.D. Cullity, Introduction to Magnetic Materials, Addision-Wesley, Reading, MA, USA, 1972.
[146] J.P. Chen, C.M. Sorensen, K.J. Klabunde, G.C. Hadjipanayis, E. Devlin, A. Kostikas, Size-dependent magnetic properties of $MnFe_2O_4$ fine particles synthesized by coprecipitation, Phys. Rev. B 54 (1996) 9288–9296.
[147] A. Straboni, L. Pichon, T. Girardean, Production of stable and metastable phases of zirconium nitrides by NH_3 plasma nitridation and by double ion beam sputtering zirconium films, Surf. Coat. Technol. 125 (2000) 100–105.
[148] L. Zhang, S.L. Gu, R. Zhang, D.M. Hansen, M.P. Boleslawski, T.F. Kuech, Lateral epitaxial over growth of GaN using diethyl gallium chloride in metal organic vapor phase epitaxy, J. Cryst. Growth 235 (2002) 115–123.
[149] P.S. Herle, M.S. Hegde, N.Y. Vasathacharya, S. Phillip, M.V.R. Rao, T. Sripathi, Synthesis of TiN, VN and CrN from ammonolysis of TiS_2, VS_2 and Cr_2S_3, J. Solid State Chem. 134 (1997) 120–127.
[150] N.S. Gajbhiye, R.S. Ningthoujam, J. Weissmuller, Synthesis and magnetic studies of nanocrystalline nickel nitride material, Phys. Stat. Sol. (a) 189 (2002) 691.
[151] R.S. Ningthoujam, R.N. Panda, N.S. Gajbhiye, Variation of intrinsic magnetic parameters of single domain Co-N interstitial nitrides synthesized via hexa-ammine cobalt nitrate route, Mater. Chem. Phys. 134 (2012) 377–381.
[152] G.W. Watt, D.D. Davis, The catalytic activity of metals produced by the reduction of salts in liquid ammonia I. Nickel, J. Am. Chem. Soc. 70 (1948) 3753–3755.
[153] A. Leineweber, H. Jacobs, S. Hull, Ordering of nitrogen in nickel nitride Ni_3N determined by neutron diffraction, Inorg. Chem. 40 (2001) 5818–5822.
[154] R. Niewa, F.J. DiSalvo, Recent developments in nitride chemistry, Chem. Mater. 10 (1998) 2733–2752.
[155] A.S. Bailey, R.W. Hughes, P. Hubberstey, C. Ritter, R.I. Smith, D.H. Gregory, New ternary and quaternary barium nitride halides: synthesis and crystal chemistry, Inorg. Chem. 50 (2011) 9545–9553.
[156] F. Kawamura, H. Yusa, T. Taniguchi, Synthesis of rhenium nitride crystals with MoS_2 structure, Appl. Phys. Lett. 100 (2012) 251910–251913.
[157] R. Riedel, G. Passing, H. Schonfelder, R.J. Brook, Synthesis of dense silicon-based ceramics at low temperatures, Nature 355 (1992) 714–717.

[158] T.K. Kim, M. Takahashi, New magnetic material having ultrahigh magnetic moment, Appl. Phys. Lett. 20 (1972) 492–494.
[159] K. Yamanaka, Y. Onuma, S. Yamashita, Y. Masubuchi, T. Takeda, S. Kikkawa, Humidity effects in $Fe_{16}N_2$ fine powder preparation by low-temperature nitridation, J. Solid State Chem. 183 (2010) 2236–2241.
[160] I.P. Parkin, A.T. Rowley, Formation of transition-metal nitrides from the reactions of lithium amides and anhydrous transition-metal chlorides, J. Mater. Chem. 5 (1995) 909–912.
[161] B.Y. Kotur, Scandium binary and ternary alloy systems and intermetallic compounds, Croat. Chem. Acta 71 (1998) 635–658.
[162] G. Block, W. Jeitschko, Ternary carbides $Ln_2Mn_{17}C_{3-x}$ (Ln = La, Ce, Pr, Nd, Sm) with filled Th_2Zn_{17} type structure, Inorg. Chem. 25 (1986) 279–282.
[163] G.S. Upadhyaya, Cemented Tungsten Carbides: Production, Properties and Testing, Noyer Publications, USA, 1998.
[164] S. Carenco, D. Portehault, C. Boissiere, N. Mezailles, C. Sanchez, Nanoscaled metal borides and phosphides: recent developments and perspectives, Chem. Rev. 113 (2013) 7981–8065.
[165] L.-M. Pan, Phase Equilibrium and Elastic Moduli of Rapidly Solidified Fe-Cr-Mo-B and Fe-Cr-Ni-B Alloys (Ph.D. thesis), University of Surrey, 1992.
[166] A.I. Vogel, A Text Book of Quantitative Inorganic Analysis, second ed., Longmans, Green and Co Ltd., London, 1960.
[167] H.S. Karp, E.G. Buyok, W.R. Bandi, L.M. Melnick, The determination of metal carbides and metal nitrides in steel by differential thermal analysis-effluent gas analysis, Mater. Res. Bull. 2 (1967) 311–322.
[168] H. Blumenthal, Determination of boron in metal borides, Anal. Chem. 23 (1951) 992–994.
[169] N.N. Greenwood, R.V. Parish, P. Thornton, Metal borides, Q. Rev. Chem. Soc. 20 (1966) 441–464.
[170] C.L. Bull, P.F. McMillan, E. Soignard, K. Leinenweber, Determination of the crystal structure of δ-MoN by neutron diffraction, J. Solid State Chem. 177 (2004) 1488–1492.
[171] K. Suzuki, H. Morita, T. Kaneko, H. Yoshida, H. Fujimori, Crystal structure and magnetic properties of the compound FeN, J. Alloy. Compd. 201 (1993) 11–16.
[172] K.H. Jack, The iron-nitrogen system: the crystal structures of ε-phase iron nitrides, Acta Cryst. 5 (1952) 404–411.
[173] J. Yangy, D. Xuez, F. Li, Local magnetic moments in the $(Fe_{1-x}Ni_x)_4N$ ($0 \leq x \leq 0{:}6$) compounds, J. Phys. Condens. Matter 9 (1997) 2781–2785.
[174] Z.Q. Yu, J.R. Zhang, Y.W. Du, Magnetic properties and preparation of Fe_3N compound, J. Magn. Magn. Mater. 159 (1996) L8–L10.
[175] R.J. Brook, Superhard ceramics, Nature 400 (1999) 312–314.
[176] A. Zerr, G. Miche, G. Serghiou, M. Schwarz, E. Kroke, R. Riedel, H. Fue, P. Kroll, R. Boehler, Synthesis of cubic silicon nitride, Nature 400 (1999) 340–342.
[177] J.M. Leger, J. Hainer, M. Schmidt, J.P. Petitet, A.S. Pereira, J.A.H. da Jornada, Discovery of hardest known oxide, Nature 383 (1996) 401.
[178] M. Taneike, F. Abe, K. Sawada, Creep-strengthening of steel at high temperatures using nano-sized carbonitride dispersions, Nature 424 (2003) 294–296.
[179] F.A. Ponce, Microstructure and polarization fields in nitride semiconductors, J. Phys. Conf. Ser. 326 (2011) 012001–012010.
[180] C. Wetzel, T. Takeuchi, S. Yamaguchi, H. Katoh, H. Amano, I. Akasaki, Optical band gap in $Ga_{1-x}In_xN$ ($0 < x < 0.2$) on GaN by photoreflection spectroscopy, Appl. Phys. Lett. 73 (1998) 1994–1996.
[181] J. Hwang, K. Lee, J.S. Kim, C.-R. Lee, I.-H. Lee, K. Lee, J.H. Lee, J.-Y. Leem, J.S. Kim, J.-H. Ryou, R.D. Dupuis, Manipulation on the optical properties of InGaN/GaN light emitting diodes by adopting InN layer, J. Cryst. Growth 370 (2013) 109–113.
[182] S. Kamiyama, M. Iwaya, T. Takeuchi, I. Akasaki, R. Yakimova, M. Syväjärvi, White light-emitting diode based on fluorescent SiC, Thin Solid Films 522 (2012) 23–25.

[183] C.-M. Lin, Y.-Y. Chen, V.V. Felmetsger, W.-C. Lien, T. Riekkinen, D.G. Senesky, A.P. Pisano, Surface acoustic wave devices on AlN/3C–SiC/Si multilayer structures, J. Micromech. Microeng. 23 (2013) 025019–025028.
[184] K.A. Aissa, N. Semmar, D.D.S. Meneses, L.L. Brizoual, M. Gaillard, A. Petit, P.Y. Jouan, C. Boulmer-Leborgne, M.A. Djouadi, Thermal conductivity measurement of AlN films by fast photothermal method, J. Phys. Conf. Ser. 395 (2012) 012089–012098.
[185] K. Chung, S.I. Park, H. Baek, J.-S. Chung, G.-C. Yi, High-quality GaN films grown on chemical vapor-deposited graphene films, NPG Asia Mater. 4 (2012) 5.
[186] Z. Yan, G. Liu, J.M. Khan, A.A. Balandin, Graphene quilts for thermal management of high-power GaN transistors, Nat. Commun. 3 (May 8, 2012) 827.
[187] Y.-C. Lee, T.-Y. Lu, Y.-H. Lai, H.-L. Chen, D.-L. Ma, C.-C. Lee, S.-C. Cheng, Simulations of light extraction and light propagation properties of light emitting diodes featuring silicon carbide substrates, Opt. Mater. 35 (2013) 1236–1242.
[188] H.J. Haugan, F. Szmulowicz, G.J. Brown, K. Mahalingam, Band gap tuning of InAs/GaSb type-II superlattices, J. Appl. Phys. 96 (2004) 2580–2585.
[189] K.-S. Cho, D.V. Talapin, W. Gaschler, C.B. Murray, Designing PbSe nanowires and nanorings through oriented attachment of nanoparticles, J. Am. Chem. Soc. 127 (2005) 7140–7147.
[190] H. Ohno, A. Shen, F. Matsukura, A. Oiwa, A. Endo, S. Katsumoto, Y. Iye, (Ga,Mn)As: a new diluted magnetic semiconductor based on GaAs, Appl. Phys. Lett. 69 (1996) 363–365.
[191] M.C. Shih, S.C. Wang, C.W. Liang, Fabrication and characterization of InGaAlP/InGaP semiconductor circular ring lasers, J. Vac. Sci. Technol. B 23 (2005) 2180–2183.
[192] T. Someya, K. Tachibana, Y. Arakawa, Blue vertical cavity surface emitting lasers and InGaN quantum dot lasers, in: E.T. Yu, M.O. Manasreh (Eds.), Chapter II, III-V Nitride Semiconductor Applications and Devices, Taylor and Francis Books Inc, New York, 2003.
[193] N. Yamada, M. Otoba, K. Kawahara, N. Miyagawa, H. Ohta, N. Akahira, T. Matsunaga, Phase-change optical disc having a nitride layer, Jpn. J. Appl. Phys. 37 (1998) 2100–2104.
[194] K. O'Donnell, V. Dierolf, Rare-earth Doped III-Nitrides for Optoelectronic and Spintronic Applications, vol. 124, Springer, Netherlands, 2010.
[195] B. Zega, M. Kornmann, J. Amigut, Hard Decorative TiN coating by ion plating, Thin Solid Films 45 (1977) 577–582.
[196] F. Khatkhatay, J. Jian, L. Jiao, Q. Su, J. Gan, J.I. Cole, H. Wang, Diffusion barrier properties of nitride-based coatings on fuel cladding, J. Alloy. Compd. 580 (2013) 442–448.
[197] S.-Y. Chang, M.-K. Chen, High thermal stability of AlCrTaTZr nitride film as diffusion for copper metallization, Thin Solid Films 517 (2009) 4961–4965.
[198] A. Endo, T. Nogushi, M. Kroug, T. Tamuro, H. Inoue, Atomic nitrogen source for the formation of aluminium nitride tunnel barrier SIS junctions, Phys. C 469 (2009) 1589–1592.
[199] S. Jing, L. Luo, S. Yin, F. Huang, Y. Jia, Y. Wei, Z. Sun, Y. Zhao, Tungsten nitride decorated carbon nanotubes hybrid as efficient catalyst supports for oxygen reduction reaction, Appl. Catal. B Environ. 147 (2014) 897–903.
[200] J.C.S. Wu, T.-S. Cheng, C.-L. Lai, Boron nitride supported PtFe catalysts for selective hydrogenation of crotonaldehyde, Appl. Catal. A Gen. 314 (2006) 233–239.
[201] C. Subramanian, A.K. Suri, T.S.R.Ch. Murthy, Development of boron-based materials for nuclear applications, BARC NewsLett. 313 (March–April 2010) 14–22.
[202] E. López-Honorato, J. Tan, P.J. Meadows, G. Marsh, P. Xiao, TRISO coated fuel particles with enhanced SiC properties, J. Nucl. Mater. 392 (2009) 219–224.
[203] R.S. Ningthoujam, N.S. Gajbhiye, Synthesis, electron transport properties of transition metal nitrides and applications, Prog. Mater. Sci. 70 (2015) 50–154.

CHAPTER

HIGH-TEMPERATURE CERAMICS 11

R. Mishra, R.S. Ningthoujam
Bhabha Atomic Research Centre, Mumbai, India

CHAPTER OUTLINE

Materials Under Extreme Conditions. http://dx.doi.org/10.1016/B978-0-12-801300-7.00011-5

1. INTRODUCTION

Refractory ceramics belong to a special class of compounds exhibiting a wide range of properties that are of immense technological importance [1–10]. Large-scale uses of these materials in various branches of science and technology are mainly a result of their excellent physical and chemical properties that meet the criteria needed for industrial applications. The basic functions of refractory materials are to ensure protection of personnel and installations from extreme heat radiating from sources such as molten metals, ingots fed into furnaces, and hot-air flues, and to reduce heat loss. High-temperature ceramics find extensive applications in heavy industries like steel, copper, aluminum, and petrochemical industries and in places such as thermal power plants.

Ceramics refers to material [1] that can sustain extreme environmental conditions of high temperature, pressure, and radiation. According to American Society for Testing and Materials (ASTM-C71) protocol, refractory materials should have thermal stability above 1000°F (538°C). Such materials exhibit high melting points, retain their strength at high temperatures, are resistant to thermal shock, have specific range of thermal conductivity and coefficient of thermal expansion, and most importantly are chemically and physically stable at high temperatures in different operating ambient environments. Several metal oxides meet these requirements. However, there are no ideal oxide materials that satisfy all the above criteria simultaneously.

Many refractory oxides in addition to their mechanical strength and resistance to thermal shock have several other properties useful in various branches of technology. Such oxides can broadly be classified as technical ceramics. Silica is a good refractory and good acid-resistant material (with the exception of hydrogen fluoride). In recent years, silica has been used extensively in optical fiber for transmission of electromagnetic radiation. Many other oxides with wide-ranging properties are used in different branches of technologies. The most interesting properties exhibited by these materials include semiconductivity, superconductivity, electrical insulation, dielectric, piezoelectric and magnetic properties. Many new materials useful for different applications are being invented every year and added to the list of technical ceramics.

High-temperature ceramics have extensive uses in fields such as nuclear industry, solid oxide fuel cells, materials for thermal-to-electric energy conversion, and electronics and communication devices. They are also used as building blocks in smelting furnaces for metallurgical applications, and as oxide ion−conducting electrolytes for power production, gas sensors, and catalysts. High-temperature protective materials are in great demand in the field of power generation and aviation industries. For example, high-temperature oxidation and hot corrosion of combustors, turbine blades, vanes, and outer air seals are minimized by the use of protective oxide-based coating on the surface of the material. Technology related to thermal barrier coatings are extensively used in order to improve the oxidation resistance, particularly at the level of the interphase. Oxidation resistance of super alloys relies on the formation of oxide layer (chromium oxide or aluminum oxide) on the surface to improve durability of components of the structural materials.

Dense ceramic membranes with mixed oxygen ion and electronic conductivity are other applications of high-temperature oxide-based materials receiving considerable attention for their possible uses in oxygen separation, partial oxidation of hydrocarbons, and reduction of organic waste. Perovskite-related oxides are considered to be one of most promising groups of the mixed-conducting membrane materials because of their unique transport properties and stability in reactive atmospheres. Thermoelectric (TE) devices, based on solid-state energy-conversion materials operating at elevated temperatures, have drawn significant interest for their possible application as a new and clean source of energy. The enormous endeavors in the experimental and theoretical works have given a decisive lead for the development of new TE materials of high merit.

In this chapter, recent advances concerning oxide-based high-temperature ceramics are reviewed. The synthesis, nature of bonding, characterization, and high-temperature properties of ceramics will be discussed. Subsequently the trend in the development of materials in high-temperature ceramics and their applications in area of science and technology will be presented.

2. CLASSIFICATION OF CERAMICS

Ceramic materials are generally classified as traditional and advanced ceramics. Traditional ceramics are the silica-based products that are available in nature. Examples of traditional ceramics are clays, bricks and tiles, porcelain, and pottery. These materials usually do not have proper composition and microstructure. Advanced ceramics, on the other hand, are pure and high quality materials with proper microstructures carefully controlled during their processing. Examples of advanced ceramics are Al_2O_3, titania (TiO_2), ZrO_2, CeO_2, MgO, and YPO_4. These materials have high melting points (>1000°C). In recent times, additives in the form of nanoparticles have been added to these materials during processing to improve the quality of materials in terms of their sinterability, heat-resistance properties, melting points, and mechanical strengths.

Ceramic oxides can also be classified based on the nature of their chemical properties. They can be [2] acidic (e.g., silica, aluminosilicates), basic (e.g., magnesite, chrome-magnesite), or neutral (brick, alumina). In addition, they can also be classified according to the nature of their chemical bonding [3–10]. The bonding can be either ionic [MgO, CaO, magnesium aluminate spinel ($MgAl_2O_4$), $Ca_3(PO_4)_2$, YPO_4, ($CaWO_4$)], covalent (SiO_2, GeO_2), or both ionic and covalent (Al_2O_3, $MgSiO_3$). Further, ceramics are classified based on numbers of components present in the ceramics. They can be single component oxides (e.g., BeO, MgO, Al_2O_3, SiO_2, TiO_2, ZrO_2, Nb_2O_3, MnO, MoO_3, SnO_2, HfO_2, Ta_2O_3, CeO_2), two component oxides [e.g., Y_2O_3-doped AO_2 (A = Zr, Hf, Th) and tin (Sn)-doped TiO_2], or multicomponent oxides. On the basis of size of the particles, ceramics can be zero-, one-, two-, or three-dimensional ceramics (0D, 1D, 2D, and 3D). The ceramic semiconductors with particles less than 100 nm in at least one of the dimensions belong to the category of nanomaterials. When the particle size of each material is less than Bohr's exciton radius, it shows the quantum confinement effect [11]. Depending on the size of confinement and shapes of particles, they can be grouped as 0D, 1D, 2D, and 3D ceramics. Examples of 0D, 1D, 2D, and 3D ceramics are nanospheres, nanowires/nanorods, multilayers, and tetrapods/octahedras/snowflakes/triangular shapes, respectively. Particles of different dimensions can be prepared according to a variety of techniques. The particles of same material but with different dimensions can have different electrical, magnetic, optical, and catalytic properties and different chemical and thermal stabilities.

3. CHEMICAL BONDING

Ceramics consist of arrays of interconnected atoms associated with mixed bonding, i.e., combinations of covalent, ionic, and sometimes metallic bonding. Because of strong and long-range coulombic force of attractions, ionic compounds exhibit high melting and boiling temperatures.

In an ionic bond, the ionization energy to form the cation and the electron affinity to form the anion match favorably and the resulting ion pairs have a strong mutual electrostatic attraction, leading to formation of chemical bonding in the molecule. Ionic compounds cannot be considered to be discrete molecules. In ionic crystals there are two forces operating: (1) electrostatic attractive forces acting between the oppositely charged ions, and (2) the interionic repulsive forces (Fig. 11.1). The expression for the attractive energy in ionic bond can be given by the relation shown in Eq. (11.1):

$$E_a = \frac{-|Z_m||Z_x|e^2}{(4\pi\varepsilon_0 r)}, \tag{11.1}$$

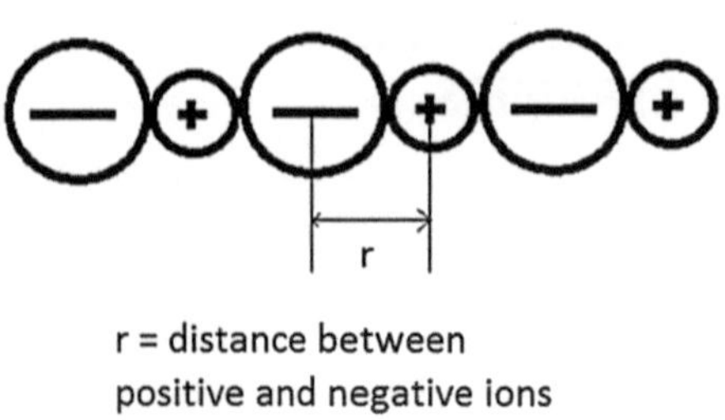

FIGURE 11.1

Linear array of ions with alternate positive and negative charges.

where $Z_m e$ and $Z_x e$ are the charges on the cation (M^{z+}) and anion (X^{z-}), respectively; ε_0 is the dielectric constant of the medium; and r is the distance between the cation and the anion (Fig. 11.2). Under the condition of equilibrium, repulsive forces arising because of strong electron–electron and nucleus–nucleus repulsion energy balances the force of attraction. At equilibrium distance (r_0), the energy becomes minimum, which refers to the lattice energy of the ionic crystal.

The repulsion energy rises rapidly with decreasing distance between the ions. The expression for the repulsive energy can be expressed by Eq. (11.2):

$$E_r = \frac{B}{r^n}, \tag{11.2}$$

where B is a constant and n is the Born exponent. The value B can be derived from compressibility data. The Born exponent depends on the type of ion involved. Larger ions have higher electron densities and hence larger values of n. The total energy of the ion pair can be obtained by adding the attractive and repulsive energies operating on the ion pairs (Eq. 11.3).

$$E = \frac{-|Z_m||Z_x|e^2}{(4\pi\varepsilon_0 r)} + \frac{B}{r^n} \tag{11.3}$$

3.1 MADELUNG CONSTANT

In an ionic crystal lattice, there exists an electrostatic equilibrium between attractive and repulsive forces of ions. The sum of all these interactions is expressed in terms of a constant known as the *Madelung constant* (*A*). 'A' can be defined as the ratio of the coulombic energy of an ion pair in a crystal to the

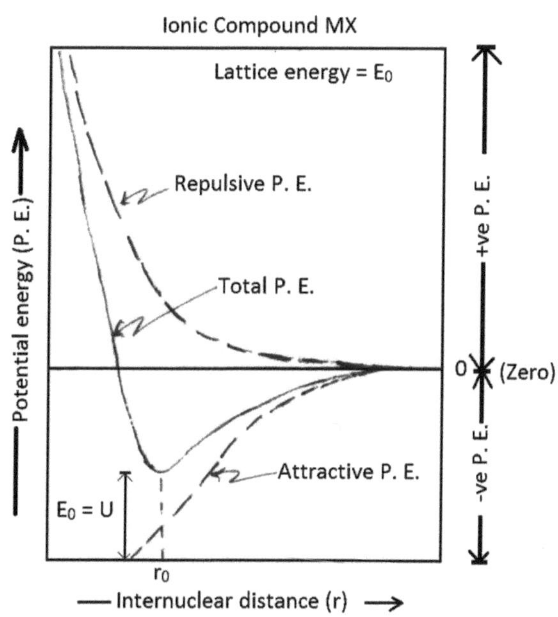

FIGURE 11.2

Potential energy of ionic compound MX versus internuclear distance (*r*) of ions. *PE*, Potential energy; *U*, minimum energy (E_0).

coulombic energy of an isolated ion pair placed at the same distance in vacuum. For a given crystal structure, the value of A is unique. The value of the Madelung constant is determined only by the geometry of the lattice and is independent of the ionic radius and charge. The packing fraction is the most important consideration in determining the structure adopted by the ionic crystals, which in turn influences the value of A.

The energy per ion pair in the crystal involving the Madelung constant is expressed as (Eq. 11.4):

$$E = \frac{-A|Z_m||Z_x|e^2}{(4\pi\varepsilon_0 r)} + \frac{B}{r^n} \quad (11.4)$$

3.2 LATTICE ENERGY

With knowledge of the Madelung constant, the total energy for 1 mol of the crystal lattice containing Avogadro's number (N) of ion pairs is expressed as (Eq. 11.5):

$$E = \frac{-AN|Z_m||Z_x|e^2}{(4\pi\varepsilon_0 r)} + \frac{NB}{r^n} \quad (11.5)$$

The minimum energy, E_0, for this system at distance r_0 can be derived by differentiating Eq. (11.5) with respect to distance r and equating it to zero (Eq. 11.6):

$$\frac{dE}{dr} = 0 = \frac{AN|Z_m||Z_x|e^2}{(4\pi\varepsilon_0 r_0^2)} - \frac{nNB}{r_0^{n+1}} \quad (11.6)$$

The value of B derived from Eq. (11.6) is expressed in Eq. (11.7):

$$B = \frac{A|Z_m||Z_x|e^2 r_0^{n-1}}{(4\pi\varepsilon_0)n} \quad (11.7)$$

The expression for the total binding energy, E_0, can be obtained by putting the value of B in equation Eq. (11.5), as shown in Eq. (11.8):

$$E_0 = \frac{-AN|Z_m||Z_x|e^2}{(4\pi\varepsilon_0 r_0)}\left(1 - \frac{1}{n}\right) \quad (11.8)$$

Eq. (11.8), known as the Born–Lande equation, is quite successful in predicting accurate values of the lattice energy of an ionic compound. Calculation of binding energy of ionic crystals requires knowledge of the crystal structure (to determine A and r_0), which can be readily obtained from X-ray diffraction data. In nature, however, no material exists as a pure ionic compound. Each ionic bonding has some degree of covalent component also.

During melting, the thermal energy of atoms exceeds the binding energy to break the crystal lattice. The higher the bond strength, the greater will be the thermal energy required to break the bonds. In other words strongly bonded materials tend to have higher melting and boiling points. This indicates that ceramics with high melting and boiling points must have high bond energy. The thermal and the electrical conductivities of solids are to a large extent dependent on the presence of free electrons. In ionic compounds, the electron transfer produces a stable inert gas configuration, rendering them with no free electrons. Hence typical ionic materials are good electrical insulators and poor thermal conductors. Similarly the thermal expansion at given temperature tends to be less for strongly bonded material compared with the weakly bonded materials. The mechanical properties of solids are dependent on the strength and directional nature of bonding. Solids with strong and directional bonds tend to be brittle. Ceramics with strong and directional bonding are therefore generally hard and brittle in nature. Typical examples of material with ionic, covalent, and mixed bonding are Li_2O, SiO_2, and Na_2SiO_3, respectively.

4. PHASE TRANSITIONS

The crystal structures of a ceramic material that is based on the effective ionic sizes and the electrostatic forces operating on them are temperature dependent and can change significantly for a given structure to transform to a new structure. The temperature at which both forms are in equilibrium is called a *phase transition temperature.* Phase transitions are often associated with marked changes in properties. During phase transitions the dimensions of crystal lattice change, and this can result in internal stresses, particularly at the crystallite boundaries in a ceramic. This may sometimes lead to internal cracking and a reduction in mechanical strength. The electrical conductivity of a ceramic may also get changed severely by orders of magnitude with change in crystal structure.

Phase transitions like melting, vaporization, and sublimation take place at a given temperature and pressure are associated with changes in volume, enthalpy, and entropy, and can be conveniently described in terms of change in Gibbs function (G) (Eq. 11.9):

$$dG = VdP - SdT \tag{11.9}$$

At constant pressure (Eq. 11.10):

$$\left(\frac{\partial G}{\partial T}\right)_p = -S, \tag{11.10}$$

where V, S, and P are volume, entropy, and pressure, respectively. The change in free energy as a function of temperature, which is continuous, can be discontinuous in the first derivative of $G(T)$ with respect to temperature that represents the change in the entropy at constant pressure. This is a characteristic of a "first-order" transition. In the second-order phase transition, the first derivative of G is continuous but the second derivative is discontinuous. Fig. 11.3 illustrates the nature of free energy temperature dependence of phase transitions. Phase transitions involving ferroelectric and paraelectric states are largely associated with change in crystal symmetry. In the case of magnetic materials, the transition between the spontaneously magnetized and magnetically disordered states that occurs at the Curie or Neel temperature does not involve changes in crystal structure.

5. PHASE DIAGRAM

Phase diagrams are one of the most important sources of information for the development of ceramic materials. A phase diagram can be described as a map of the equilibrium phases associated with various combinations of temperature, pressure, and composition. A phase diagram describes the state of a material system that is in thermodynamic equilibrium under specified conditions of temperature, pressure, and composition. The understanding of phase diagrams and phase equilibrium is crucial for

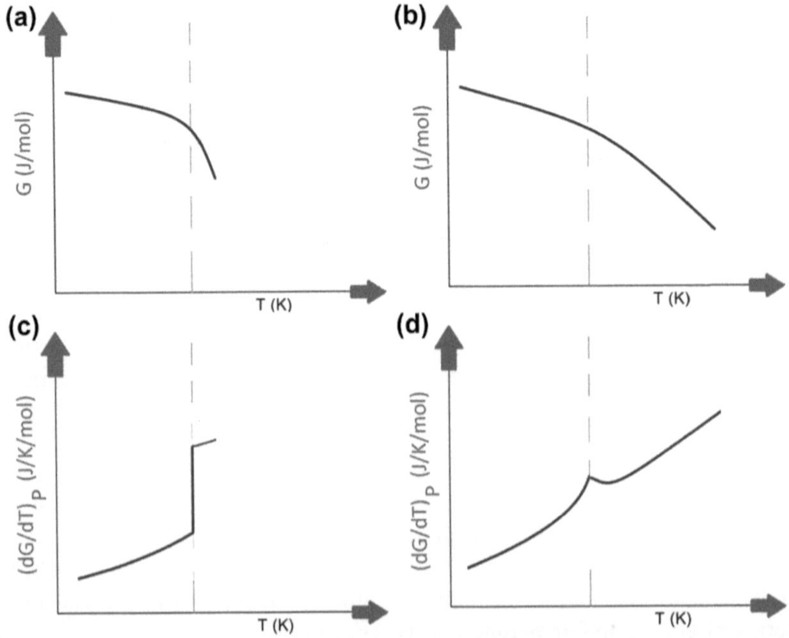

FIGURE 11.3

Nature of free energy temperature dependence of phase transitions: (a) and (c) are Gibbs function (*G*) versus temperature (*T*) curves, and (b) and (d) are derivatives (*dG/dT*) versus *T*.

the insight of processes governing the material's properties and development. Knowledge of phase diagram of given system is quite useful to understand solidification processes, characterization of compounds, and phases, and for understanding the chemical reactions involving multiple components at different temperatures. A phase diagram provides practical insight for the synthesis of compounds in pure form with minimum steps. Further, the physical properties of a material at high temperature governed by its microstructure are in turn influenced by its phase composition, morphology, and synthesis conditions. The study of phase diagrams is therefore very useful to predict the formation of phases under given conditions. Large volumes of literature data are available on principle and applications of phase diagrams [12–19].

Phase diagrams are classified on the basis of the number of the components present in the system. Single-component systems have unary diagrams, two-component systems have binary diagrams, three-component systems give rise to ternary diagrams, and four-component systems have quaternary diagrams. The phase diagram of systems is associated with characteristic properties such as eutectic, peritectic, and monotectic temperatures; solidus–liquidus temperatures; solvus phase boundaries; and phase transition temperatures. Conventional experimental methods such as chemical, microscopic, thermal, and X-ray analyses are generally employed to delineate the phase diagram of a system. Phase diagrams of some of the important high-temperature oxide systems are presented in the following sections.

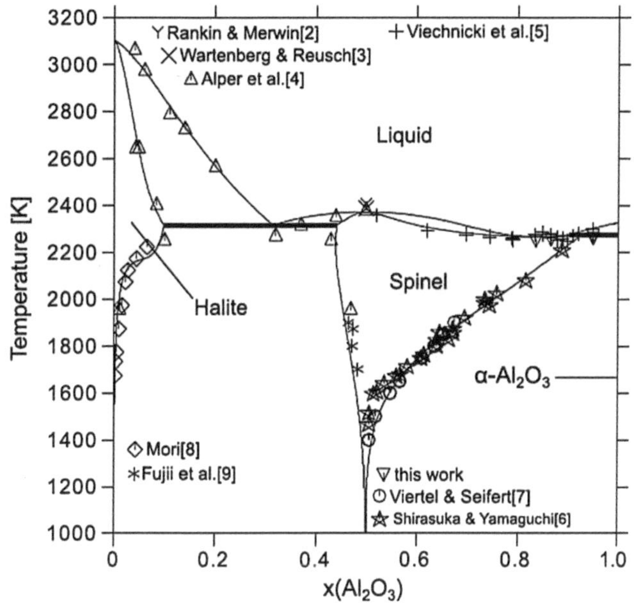

FIGURE 11.4

Phase diagram of the magnesia–alumina *($MgO–Al_2O_3$)* system [20].

Reproduced with the permission of Elsevier publications.

5.1 MAGNESIA–ALUMINA SYSTEM

The phase diagram of the magnesia–alumina ($MgO–Al_2O_3$) system is shown in Fig. 11.4 [20]. It has two eutectic reactions and one intermediate phase: $MgO \cdot Al_2O_3$. There is a significant terminal solubility (~16 mol%) of Al_2O_3 in MgO around 2200°C, whereas the solubility of MgO in Al_2O_3 is negligible. Above 1200°C, the $MgAl_2O_4$ spinel phase exhibits a wide solubility of Al_2O_3. It can be noticed that addition of even a small amount of MgO lowers the melting point of Al_2O_3 appreciably, even though the melting point of MgO is very high. For this reason small quantity of MgO is added to alumina to decrease its melting point and to increase its sintered density. Density and particle size of Al_2O_3 increases with increase in sintering temperature. Scanning electron microscope (SEM) images of Al_2O_3 particles are shown in Fig. 11.5 [21]. The particle sizes are in the range of a few micrometers.

5.2 ALUMINA-SILICA SYSTEM

Mullite with the composition $3Al_2O_3 \cdot SiO_2$ is the only stable crystallized compound in the phase diagram of the alumina–silica ($Al_2O_3–SiO_2$) system (Fig. 11.6) [22]. Mullite crystallizes in the orthorhombic system with space group *Pbam*. The compound incongruently melts around 1828 ± 10°C and exhibits a narrow solubility range below 1753°C. Under metastable conditions, mullite melts congruently at approximately 1880°C. The $Al_2O_3–SiO_2$ system has eutectic composition with melting points 1587 ± 10°C, corresponding to the composition of approximately 6 mol% Al_2O_3.

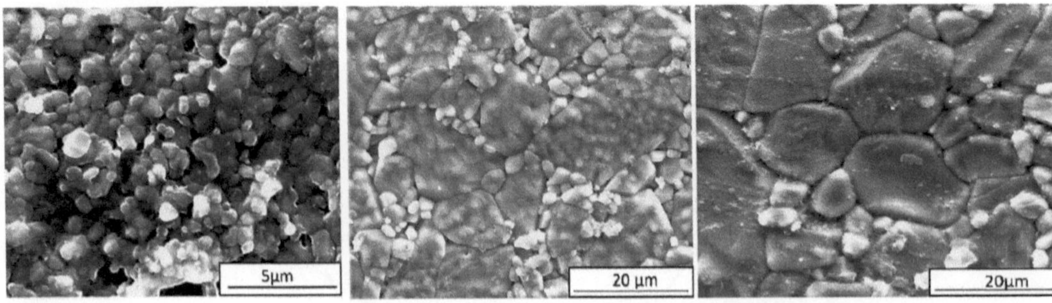

FIGURE 11.5

Micrographs of alumina sintered at 1500, 1600, and 1700°C for 3 h [21].

Reproduced with the permission of Elsevier publications.

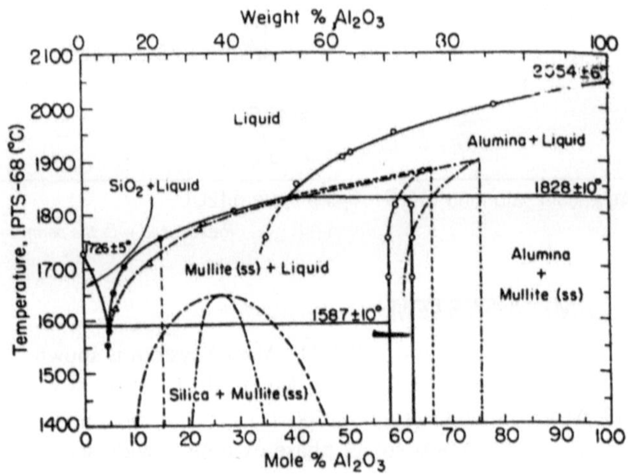

FIGURE 11.6

Phase diagram of the alumina–silica *(SiO_2–Al_2O_3)* system [22]. *IPTS-68*, International Practical Temperature Scale of 1968.

Reproduced with the permission of John Wiley and Sons publishers.

Mullite has a density of 3.2 g/cm^3, hardness of 14 GPa, Young's modulus of 250 GPa, mechanical bending strength of 250 Mpa, and toughness of 2.5 MPa m$^{0.5}$. It has a relatively low thermal expansion coefficient of 6×10^{-6} K^{-1}, which is a very useful property for improving resistance to thermal shocks and mechanical strength [23]. These characteristics make mullite a material of choice for refractory applications, particularly when it is necessary to consider attacks by silica or silicates. The molten silica or silicates react with the alumina component to form refractory mullite. Fig. 11.7 shows pictures of a mullite tube and silica crucible used for high-temperature studies.

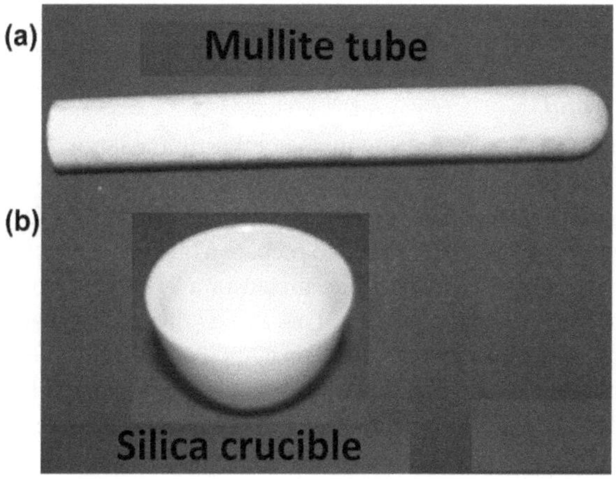

FIGURE 11.7

Mullite tube (a) and silica crucible (b).

6. PREPARATION AND PROCESSING

Refractory ceramic oxides can be synthesized by various physical and chemical methods. Some of the chemical and physical methods of synthesis of ceramic oxides are presented briefly in the following sections.

6.1 CHEMICAL METHODS OF SYNTHESIS

6.1.1 Precipitation Method

Metal oxides can be prepared from metal hydroxides formed by the precipitation of metal ions in the presence of OH^- ions [24–27]. NH_4OH and NaOH are generally used as source of hydroxide ions in this process. The precipitation reaction can be represented by:

$$M^{n+} + nOH^- \rightarrow M(OH)_n \rightarrow M_xO_y \cdot zH_2O \rightarrow M_xO_y + zH_2O$$

$$A^{n+} + B^{m+} + lOH^- \rightarrow A(OH)_n{:}B(OH)_m \rightarrow A_xB_yO_z \cdot pH_2O \rightarrow A_xB_yO_z + qH_2O$$

It is often observed that the prepared metal oxide M_xO_y is associated with adsorbed H_2O on the surface of metal oxide particles. These can be removed by adding acetone or heating the sample up to about 200°C. However, in some cases the coordinated water molecules cannot be removed by heating even up to 600°C. Dehydrated samples in the form of pellets are often sintered at higher temperature (>1000°C) to prepare materials with higher density. For example, sintered SiO_2 and Al_2O_3 can be prepared by heating the precipitates obtained from tetraethyl orthosilicates and aluminum nitrate, respectively. Many times the sintering temperature is lowered by addition of foreign agents (flux) that helps in getting higher crystallization at lower synthesizing temperatures. For example, NaCl is added during coprecipitation to the precursors, which are further heated at 900°C to obtain a high crystalline $(Y_{1-x}Eu_x)_4Al_2O_9$ phase [28]. NaCl is removed by washing powder with hot water.

6.1.2 Combustion Method

In this method the enthalpy change accompanying a combustion reaction is advantageously utilized to prepare a sample at a relatively lower temperatures. For these processes, organic precursors such as citric acid, glucose, polyvinyl alcohol, and glycine are commonly used as fuels [29–34] and nitrate is used as an oxidizing agent. For example, in order to prepare MgO nanoparticles, $Mg(NO_3)_2$ is dissolved in water and glycine is added as a fuel. They are mixed in a 1:1 mol ratio of reductant and oxidant. The mixture is transferred in a 1- or 2-L beaker. This avoids the loss of materials during combustion. The beaker containing materials is heated at 300°C on a hot plate. Within a few seconds, a large volume of white particles is obtained. Instantaneous combustion produces enough heat, which is sufficient to form nanoparticles of MgO powder. However, some organic carbon residue remains after combustion, which can be removed by heating MgO power at a higher temperature such as 600–1000°C.

6.1.3 Sol–Gel Technique

In the sol–gel method metal ions are allowed to react with tertiary butanol or alcohol of higher hydrocarbons [35–39] to form sols of metal alkoxides. These complexes are then heated at 100°C to remove water molecules and form a gel that is a 3D disordered network of M (metal), O, C, OH in a continuous liquid like phase. The gels can be heated at higher temperatures to form uniform particles of desired size. For example, TiO_2 nanoparticles can be prepared by reacting $TiCl_4$ with tertiary butanol through the formation of $Ti(OC(CH_3)_3)_4$ gel. The gel is further heated at 600°C to form TiO_2 nanoparticles.

$$TiCl_4 + 4(CH_3)_3COH \rightarrow Ti(OC(CH_3)_3)_4 + 4HCl$$

$$Ti(OC(CH_3)_3)_4 \rightarrow TiO_2 + CO_2 + CO + H_2O$$

This method can be applied for the synthesis of a large number of ceramic oxides for high-temperature applications.

6.1.4 Solvent-Based Synthesis

In this process, metal precursors (1–2 g) are generally dissolved in HCl or HNO_3. To this metal solution, 10–50 mL of solvents with high boiling points, such as ethylene glycol, polyethylene glycol (PEG), PEG diacid, and polyvinyl alcohol are added. The resulting solution is heated at about 100–200°C for different durations to get the desired size of nanoparticles [40,41]. Urea is generally added to the solution to enhance the formation of hydroxides or oxides. The role of urea is to increases the pH of medium above 7 because of its in situ decomposition to NH_4OH.

6.1.5 Hydrothermal- and Solvothermal-Based Synthesis

In hydrothermal- and solvothermal-based methods, solvents play a big role in the synthesis of materials [42–48]. If the solvent is water, the method is considered a hydrothermal technique, whereas if the solvent is nonaqueous (e.g., alcohol, phenyl ether, benzyl ether, 1-octadecence), the method is considered to be solvothermal. In this method, the precursors are dissolved in suitable solvents and then heated in a closed chamber called an *autoclave* at a high temperature and pressure. The vapor pressure of the solvent is measured by a pressure gauge provided in the system. The method is useful to prepare materials that have different sizes and shapes of particles depending on temperature and the nature of precursors present in the reaction chamber. Some of the metastable phases that could not be prepared by a conventional high-temperature synthesis route can be prepared with this technique. Synthesis of various metals oxides such as ZrO_2, Y_2O_3, CeO_2, SnO_2, TiO_2, SiO_2, ZnO, Fe_2O_3, Fe_3O_4, Al_2O_3, Sb_2O_3, Co_3O_4, HfO_2, MgO, $BaTiO_3$, $LiSiO_3$, Zn_2SnO_4, $MgSiO_3$, $SrTiO_3$, PZT, $PbTiO_3$, $KTaO_3$, (Li, K)NbO_3, ferrites, apatites, tungstates, vanadates, molybdates, zeolites, and even core–shell CdS@ZnO, $CoAl_2O_4$@$ZnAl_2O_4$, and TiO_2@SiO_2 by this method have been reported in the literature.

6.1.6 Microwave Synthesis

In microwave synthesis, heat is produced as a result of absorption of a microwave electromagnetic field by solvents having a polar OH group such as water. Water molecules orient and collide with each other to produce heat to the surrounding area. It takes only a few minutes to reach a temperature as high as 1000°C or higher. This heat is sufficient to produce crystalline products from precursor ions and molecules. Sometimes microwave is provided in an autoclave so that pressure is developed from heat produced by solvents [11,48–52]. This is a widely used technique for the synthesis of many inorganic or organic compounds. A large number of inorganic compounds such as Al_2O_3, $LaMO_3$ ($M^{3+} = Al^{3+}$, Cr^{3+}, Mn^{3+}, Fe^{3+}, Co^{3+}), $RECrO_3$ (where RE stands for *rare earth*), $LaAlO_3$, $LaFeO_3$, $LiNbO_3$, $BaTiO_3$, La_2CuO_4, $CaCu_3Ti_4O_{12}$, $BiFeO_3$, $La_{1-x}Sr_xFeO_{3\pm\delta}$ and $La_{1-x}Sr_xFe_yCo_{1-y}O_{3\pm\delta}$ prepared by either microwave or microwave–hydrothermal techniques have been reported.

6.1.7 Reverse Micelle-Based Synthesis

In the reverse micelle method of synthesis of oxides [53–58], oil (e.g., hexane or isooctane in large quantity), surfactant (e.g., dioctyl sulphosuccinate, cetyl trimethyl ammonium bromide, or oleic acid), and water (a small amount) are taken together in an appropriate ratio to make a reverse micelle. In this method, there is a water-soluble region called a *nanoreactor,* which is surrounded by surfactants. The nanoreactor facilitates the formation of the compound by close confinement of reactants. In this technique, two separately prepared nanoreactors containing two oppositely charged M^{n+} and OH^- ions are brought together to form a fusion nanoreactor in which the metal hydroxides/oxides or mixed hydroxides/oxides are formed. The resulting system is then slightly heated to improve crystallinity of the material. The shape or size of nanoparticles formed by this technique depends on the water-to-surfactant ratio. Some of the examples of the compounds synthesized by this technique are SnO_2, YVO_3@SiO_2 (core–shell), AlOOH, Al_2O_3, and Y_2O_3.

6.1.8 Solid State Synthesis

In solid state synthesis, the precursors in the form of oxides, carbonates, nitrates, oxalates, and acetates are thoroughly mixed in a stoichiometry ratio and heated at the desired temperature under a fixed atmospheric pressure [59–67]. The mixtures are generally reground intermittently for homogenization and pressed to pellets for better contact between the reactants facilitating the diffusion process. Usually, polyvinyl alcohol is added as a binder to the powder for making pellets, which are difficult to compress, and gets removed during sample heating at a high temperature (>500°C). In solid state synthesis route, the size of the product particles formed are generally in the range of a few microns. A large number of oxides including magnesium stannate (Mg_2SnO_4), $BaZrO_3$, $BiTiO_3$, $La_{1.2}Ba_{1.8}Mn_2O_7$, $Y_2Ti_2O_7$, and (Ba/Sr)O–Al_2O_3–$2SiO_2$ have been prepared by this technique.

6.2 OTHER METHODS

6.2.1 Chemical Vapor Deposition Technique

In the chemical vapor deposition (CVD) technique, metal oxides are formed on the heated substrate by the chemical reaction of precursor/s deposited on it via a vapor or gas phase [68,69]. It needs initial activation energy of vaporization of the metal precursor and its subsequent decomposition in the presence of oxygen. The source of heat for the vaporization of the metal precursor can be plasma, laser, or ultraviolet (UV) radiation. In case of chemical vapor condensation, a metalorganic precursor is introduced to the hot zone of the reactor using a mass flow controller followed by condensation of oxide particles. The TiO_2, Al_2O_3, SiO_2, and ZrO_2 nanoparticles are prepared by the CVD method using the following metal precursors: $Ti(oiPr^i)_2(TMHD)_2$ [68], $Al[2\text{-}OC_4H_9]_3$ [69], $Si[OC_2H_5]_4$ [69], and $Zr[3\text{-}OC_4H_9]_4$ [69], respectively.

6.2.2 Physical Vapor Deposition

In the physical vapor deposition (PVD) technique, thin films of materials are deposited on a substrate to functionalize a material or to make it thermally or/and chemically stable. Metal oxides are vaporized using physical processes such as cathodic arc deposition, electron beam PVD, evaporative deposition, pulsed laser deposition, sputter deposition, or sublimation. Semiconducting devices such as thin film solar panels, aluminized film for food packaging, and tools are prepared using the PVD technique. The multilayered zirconia (ZrO_2)-toughened alumina and c-ZrO_2 coating on Ni-based superalloys are carried out by the electron beam PVD method [70].

6.2.3 Ball Milling or Mechanical Attrition

In mechanical attrition, larger particle sizes are put into a ball milling machine in which very hard balls spinning at a high speed collide with the material to produce fine homogeneous powders. In this way, the particle size can be decreased, and even nanomaterials can be prepared [69,71]. The technique is useful to prepare pure phases. For example, $BaZn_2Ti_4O_{11}$ can be prepared by mixing the precursors such as ZnO, $BaCO_3$, and TiO_2 thoroughly with the ball milling process followed by heating at 1250°C [5,71].

6.2.4 Electrolysis

Some of the ceramic oxides can be prepared by using an electrolysis technique [72]. In this technique the metal electrode that is to be oxidized is kept in the electrolyte (H_2O) and maintained at a required pH ($\approx$6–7). An external DC voltage (1–10 V) is applied to the cell for a particular duration. The metal gets oxidized and oxide particles formed on the metal surface are settled in the bottom of the cell. Depending on applied voltage and duration, different shapes of nanoparticles (spherical to wires, tetrapod to cube) can be obtained.

6.3 SINTERING PROCESS

To improve material stability under harsh conditions such as high temperature; variable oxygen partial pressure; moistures; and exposure to radiofrequency, microwave frequency, and radiation, the sample needs to be processed appropriately. Sintering is one of the processes used to increase the density of the material by heating it at high temperatures below its melting point. The sample powders are generally pressed into pellets and then sintered at an appropriate temperature to get high-density material. Samples can also be efficiently sintered by applying both temperature and pressure simultaneously [62–64,73]. This process is also helpful for the synthesis of transparent ceramics [74]. The sintering process increases the grain growth and decreases surface energy and porosity of the material. The density of 3 mol% Y_2O_3 stabilized ZrO_2 can be increased from 70% to 94% of its theoretical density by increasing the sintering temperature from 1473 to 1573K. The hardness and toughness increase with increase in the sintering temperature. $BaTiO_3$ and $BaZrO_3$ with densities of 96–98% can be prepared by sintering the materials above 1500°C.

7. CHARACTERIZATION

A large number of techniques including thermal, diffraction (X-ray or neutron), microscopy, spectroscopy, and chemical analyses are employed for the characterization of ceramic materials.

The phase purity and phase evolution in the ceramic oxides are generally studied with the help of X-ray and neutron diffraction techniques. Complementary techniques to determine the composition of the materials are energy dispersive analysis of X-ray, X-ray fluorescence analysis, inductive coupled plasma–mass spectroscopy, and chemical analysis. High-temperature X-ray diffraction is employed for in situ high-temperature phase analysis. Information on thermal stability, phase transition, glass transition, crystallization, eutectic temperature, peritectic decomposition temperature, melting point, solidus-liquidus temperature, and vaporization behavior of ceramics can be obtained employing thermal techniques like thermogravimetric analysis and differential thermal analysis. For quantitative estimation of energy accompanying the phase transition, heat capacity, melting point, vaporization, or any chemical or physical changes are determined by differential scanning calorimeter and high/low temperature calorimeters. The high-temperature liquidus-solidus phase boundary is determined using the optical spot technique [75]. Thermal expansion behavior, and thermomechanical properties of ceramics can be determined from thermomechanical analysis experiments. The morphology of material down to 10 Å can be studied with the SEM technique and to 1 Å with the transmission electron microscope technique [76].

Information on the nature of the chemical bonding in ceramic materials can be obtained from a variety of experimental techniques including X-ray photoelectron spectroscopy, the extended X-ray absorption fine structure technique, solid state nuclear magnetic resonance, and infrared (IR) and Raman spectroscopy. Knowledge about the physical properties of ceramics such as thermal conductivity, electrical resistivity, dielectric properties, magnetic properties, optical properties, and mechanical properties (hardness, Young's modulus of elasticity, compressibility) is important for the practical deployment of the material at given extreme conditions of temperature, pressure, and chemical environment. The electrical properties of ceramic material such as electronic conductivity, ionic conductivity, dielectric constant, dielectric loss with contribution from grain, and grain boundaries can be conveniently measured by employing an AC frequency analyzer (impedance spectroscopy) measured in the frequency ranging from MHz to mHz at different temperatures. Magnetic properties can be studied with the superconducting quantum interference device and vibrating sample magnetometer techniques.

8. PROPERTIES OF REFRACTORY CERAMICS

Refractory oxides in general are resistant to thermal shock and can have several other properties that are useful in various branches of technology. Such oxides can broadly be classified as *technical ceramics*. Many ceramic oxides have good refractories with high mechanical strength. These are also chemically and physically compatible with a large number of materials at high temperatures. Therefore they are used in the fabrication of structural components for high-temperature applications and liners in high-temperature chemical reactors. Some of the general properties of ceramics are given in the following sections.

8.1 BRITTLENESS

Brittleness is probably the most fundamental property of ceramics, and signifies poor resistance to impact and absence of plasticity. The accidental dropping of a glass beaker or a dinner plate, leading to its fragmentation, is a common experience in everyday life. In ceramics, fracture in the material takes place as soon as the critical threshold stress is applied locally. The fracture takes place without prior plastic deformation as a result of interatomic decoherence. Brittle fracture involves production and propagation of a crack. The fracture occurs by cleavage along defined crystallographic planes. Most ceramics are brittle at room temperature and change into viscous liquid at elevated temperatures. The mixed ionic–covalent nature of bonding in ceramics is the main reason for the brittleness in ceramics.

8.2 TOUGHNESS

The breaking resistance of a structure is referred to as the toughness of the constituent material. This is one of the most important properties of ceramics for many design applications. Ceramics have very low crack-resistance properties, which is a major factor affecting its application in different fields. The low crack resistance is a critical factor for the use of ceramics under extreme mechanical and thermal loads. The predominant method used for the determination of microhardness of ceramic material is the Vickers diamond microhardness indenter technique, which induces radial cracks in the material. The accuracy and reliability of the fracture-toughness data for ceramics have been assessed by international prestandard studies (Round Robin on Fracture Toughness). The toughness of ceramic material lies typically in the range of 0.5–15 MPa $m^{0.5}$, which is about 50–100 times less than that of metals. Sintered alumina with a density of 3.86 g/cm^3 has fracture toughness of 3.5 MPa $m^{0.5}$. Microstructure and addition of a second phase play important roles in the hardness of ceramic materials [77]. As compared with oxide ceramics, carbides and nitrides of tungsten (W), diamond, and CN_x are usually harder. The hardness of oxide ceramic materials can be improved by adding (W, Ti)C as a secondary phase [78]. Hardness values of Al_2O_3, ZrO_2, Al_2O_3-SiO_2, SiC, Si_3N_4, and WC are 2000, 1600, 1450, 2800, 2200, and 1307–2105 kg/mm^2, respectively [79].

8.3 COMPRESSIVE STRENGTH

Compressive strength of a material is its capacity to withstand loads, tending to reduce its size. Ceramics differ significantly in compressive strength compared with metals. Whereas ceramics are stronger in compression, metals exhibit superior tensile and compressive strengths. This difference in properties has a strong effect on load-bearing applications.

8.4 ELECTRICAL CONDUCTIVITY

Ceramic oxides show wide variation in their electrical properties. Oxides like Al_2O_3, SiO_2, MgO, and combinations containing these oxides are the best electrical insulators and are used as structural components in high-voltage transmission lines. Oxides of some elements are also good electronic (semiconductors) and ionic conductors. Rhenium oxide, for example, is more metallic than some metals [80]. The electrical conductivity of some ceramic oxides can be changed either by doping or by applying high pressure. Insulator-to-metal transitions have been observed in several cases employing these methods.

Both intrinsic as well as extrinsic electrical conductions have been observed in metal oxides. It is therefore important to know the amount and type of impurities in these oxides before the measurement of these properties. Impurities can either be electron donors or acceptors. A small impurity of aliovalent ions in an electric insulator can increase its conductivity by several orders of magnitude. The number of charge carriers in an intrinsic semiconductor increases with increasing temperature, whereas the number of charge carriers resulting from the added impurities in an extrinsic semiconductor will be dominant at lower temperatures.

The concentration of the charge carriers is also dependent on the oxygen-to-metal ratio of the oxide under consideration. This is particularly observed in the oxides containing cations with variable oxidation states such as transition metals, rare earth metals, and actinides. Oxides of metals like iron, chromium, manganese, titanium, molybdenum, tungsten, and those of lanthanides and actinides exhibit a wide degree of homogeneity in range of composition. Several properties of such oxides change dramatically with oxygen deficiency or excess and therefore it is extremely important to know the metal-to-oxygen ratio precisely [81].

8.5 VAPORIZATION OF CERAMICS

Many oxides with the metal ion in lower oxidation states oxidize to a higher oxidation state at high temperatures. Such oxides have high vapor pressure in the oxidizing environment. The vaporization behavior of few selected ceramic oxides is discussed in this section. For example, chromium oxides (Cr_2O_3) vaporize in the presence of oxygen at high temperatures to form CrO_3 vapor [82]. The reaction is represented in Eq. (11.11).

$$Cr_2O_3(s) + \frac{3}{2} O_2(g) = 2\ CrO_3(g) \tag{11.11}$$

The depletion of Cr by this reaction in chrome steel subjected to prolonged heating in air at high temperatures could therefore result in a change of its properties. Similarly, oxidation of refractory metals like molybdenum and tungsten on heating in air or oxygen transform to their trioxides, which have high vapor pressures and vaporize congruently to form volatile oxides MoO_3 and WO_3, according to the reaction (Eqs. 11.12 and 11.13) [83].

$$MoO_3(s) = MoO_3(g) \tag{11.12}$$

$$WO_3(s) = WO_3(g) \tag{11.13}$$

Mass spectrometry investigations of the vaporization of these oxides indicated that in addition to these monomers, several polymeric species were also observed in equilibrium vapors [84]. The formation of the trioxide in the vapor phase was also observed in the vaporization of triuranium octoxide (U_3O_8) in air, which is represented in Eq. (11.14):

$$\frac{1}{3} U_3O_8(s) + \frac{1}{6} O_2(g) = UO_3(g) \tag{11.14}$$

The existence of the UO_3 molecule in the vapor phase in equilibrium with the hyper stoichiometric UO_{2+x} was confirmed by mass spectrometry. Several other species like U, UO, and UO_2 were also observed in the vapor phase. The amount of UO_3 in vapor increases with the increasing oxygen content of the condensed UO_{2+x} phase. Such observations are interesting and useful in understanding the fuel redistribution observed in postirradiation studies of nuclear fuel subjected to different burn-up.

Many oxides have been reported to form gaseous hydroxides on interaction with water vapor at high temperatures. The formation of high-temperature gaseous hydroxides is not limited to MO_3-type solids. Oxides of beryllium, aluminum, and boron have been observed to form the gaseous hydroxide species. Thus beryllium oxide, which functions as the best refractory in vacuum, simply evaporates in a moist oxidizing environment [85]. This observation indicates the potential of hydroxide formation in the separation of beryllium from beryl ore.

8.6 THERMODYNAMICS OF NUCLEAR MATERIALS

For long-term use of nuclear fuel inside a reactor, it is necessary to have a thorough analysis of nuclear fuel performance using a reliable database of thermophysical and chemical properties of the fuel in its virgin as well as high burn-up states. Physicochemical databases of fuel and fission products and clad are commonly referenced when planning the fuel cycle program from fuel design and fabrication and reprocessing of irradiated fuel. Knowledge of high-temperature chemistry of the fission products is necessary to predicting the fuel performance. Thermodynamics plays an important role in predicting the behavior of nuclear fuel during irradiation. Formation of a large number of compounds of fission products with the UO_2 fuel matrix, the cladding material, and among each other is extremely important. Many of the compounds formed in this process have structures different from that of the UO_2 fuel matrix and can result in its cracking during power ramping. The cracked pellet facilitates transport of the fission gases, including iodine to the fuel clad gap, which can subsequently result in stress corrosion cracking of the clad and thereby facilitating the ingress of the coolant into the hot irradiated fuel, resulting in transport of radioactivity into the reactor building. Several investigations that enabled prediction of formation of different compounds under changing oxygen potential in the fuel during irradiation were carried out in our laboratory employing vapor pressure measurement and calorimetry techniques. These have been exhaustively reviewed in a recent publication [86] and their role in nuclear technology has been identified.

9. EXAMPLES OF HIGH-TEMPERATURE CERAMICS AND THEIR APPLICATIONS

9.1 MAGNESIA

MgO (face-centered cubic structure, Fig. 11.8) is a refractory ceramic material with a melting point of 2790°C [3,87,88]. The archetype of oxides with ionic bonds, MgO crystallizes in an NaCl-like structure (space group *Fm3m*). Note that all crystal structures shown in this chapter are obtained using diamond software and parameters of crystal structure are obtained from http://www.crystallography.net/cod.

This is a highly stable material that can withstand temperatures higher than 2000°C. The majority of MgO produced is used for refractories in the iron and steel industry as MgO bricks in furnaces. MgO-carbon refractories (about 30% carbon, in various forms) are the reference refractories in steelworks. MgO is also used as a component of Portland cement. Below 350°C, MgO ceramic material has remarkable plasticity, with slips occurring only along (110) <110>, at high temperature (>1700°C). The high reactivity of MgO with water is a disadvantage for its use, but this reactivity decreases if the specific surface area of MgO is decreased. MgO has a band gap of 7.8 eV and

MgO-cubic
Space group: *Fm-33m* (225)
Lattice parameter a = 4.217 Å
Unit cell volume V = 74.99 $Å^3$
Number of Chemical formula per unit cell Z = 4
Atomic positions (x, y, z):
Mg1 (4a) 0, 0, 0
O1 (4b) 0.5, 0.5, 0.5

FIGURE 11.8

Unit cell structure of magnesia.

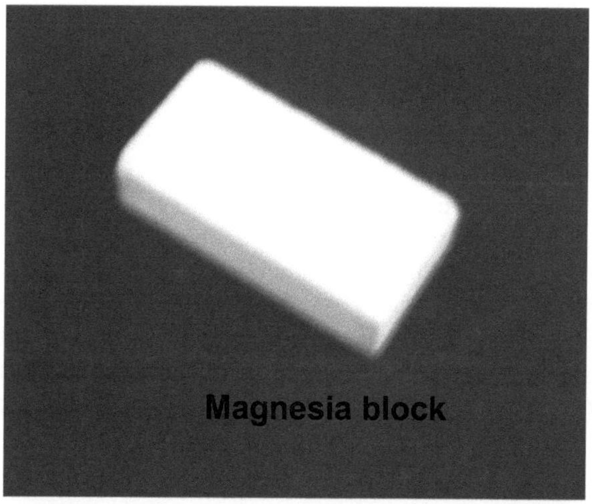

FIGURE 11.9

Magnesia block.

therefore is used as an electrical insulator at high temperatures. Because MgO is transparent to IR radiation, it is used as an optical matrix in many devices. MgO also finds application in wastewater treatment (soil and ground). In medicine, magnesia is given as a dietary supplement. Fig. 11.9 shows the picture of an MgO block, which is used in high-temperature furnaces as a thermal barrier.

9.2 ALUMINA

Al_2O_3 is one of the most commonly used and versatile refractory ceramics with high thermal and chemical (acids and bases) stability [89–91]. Al_2O_3 crystallizes in hexagonal structure (space group *R-3c*). Fig. 11.10 shows the different shapes of Al_2O_3 crucibles used for high-temperature studies.

Al_2O_3 ceramics exhibit high hardness, high moduli of elasticity, satisfactory mechanical strength, wear resistance, and refractory properties. The mechanical properties of Al_2O_3 are sensitive to its

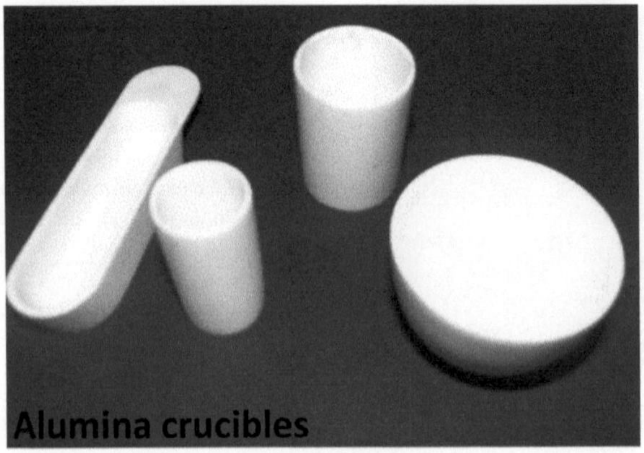

FIGURE 11.10

Different shapes of alumina crucibles.

microstructure and depend highly on the temperature and chemical environment. A dense, fine-grained Al_2O_3 ceramic has Young's modulus of 400 GPa (twice the modulus of steel), Poisson's ratio of 0.25, Vicker's hardness of 20 Gpa, and mechanical bending strength in the range 300–500 GPa. Sintered Al_2O_3 is not prone to corrosion and can be used for high-temperature applications for long duration. Al_2O_3 with toughness 3.5 MPa $m^{0.5}$ at 20°C is certainly not the toughest material, but it is good enough for practical applications. The coefficient of thermal expansion of Al_2O_3 at temperatures of 25–1000°C is $8.5 \times 10^{-6}\, K^{-1}$. Al_2O_3 has very poor thermal shock-absorbing capacity because of unequal thermal expansion coefficients along three crystal axes. Al_2O_3 ceramics are best suited for refractory uses. It is extensively used in many high-temperature furnaces, devices, and instruments. It is also used as a catalyst, a biomaterial for artificial joint replacements and tooth implants, and

FIGURE 11.11

Alumina beads and balls.

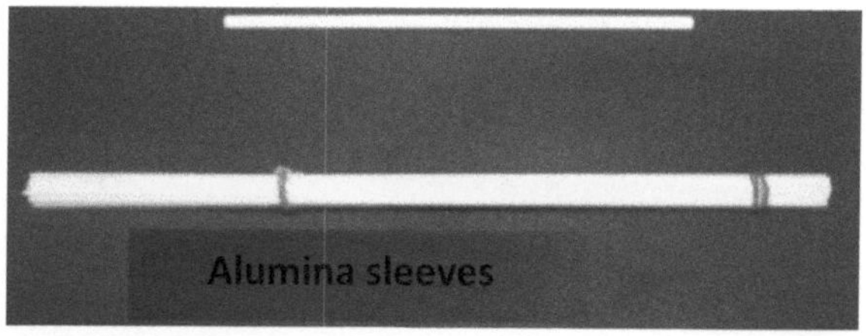

FIGURE 11.12

Alumina sleeves.

polishing materials. Al_2O_3 beads and balls (Fig. 11.11), and sleeves (Fig. 11.12) are used in high-temperature electrical insulation.

9.3 ZIRCONIA

Zirconium dioxide [or zirconia (ZrO_2)] is one of the most important ceramics, with wide-ranging applications in high-temperature thermodynamic measurements [92]. ZrO_2 has a very high melting

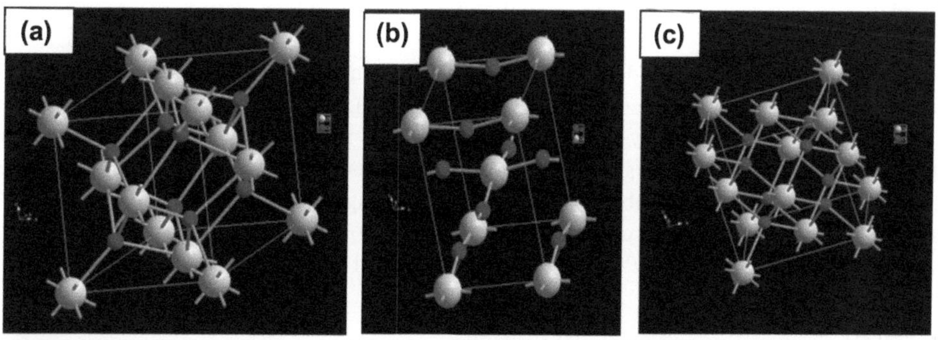

(a) ZrO_2-m,
Space group: *P121/c1* (14)
Lattice parameters a = 5.21, b = 5.26, c = 5.37 Å
V = 145.16 Å^3,
Angle: β = 80.53°,Z = 4
Atomic positions (x, y, z):
Zr1 (2a) 0, 0, 0
Zr2 (2d) 0.5, 0.5, 0
O1 (4e) 0.20, 0.22, 0.21
O2 (4e) 0.288, 0.75, 0.21

(b) ZrO_2-t
Space group:*P 42/n m c* (137)
Lattice parameters a = 3.578, c = 5.162 Å
V = 66.09 Å^3, Z = 2
Atomic positions (x, y, z):
Zr1 (2a) 0,0,0
O1 (4d) 0, 0.5, 0.0484

(c) ZrO_2-c
Space group:*Fm-3m* (225)
Lattice parameters a = 5.09 Å
V = 131.87 Å^3, Z = 4
Atomic positions (x, y, z):
Zr1 (4a) 0, 0, 0
O1 (8c) 0.25, 0.25, 0.25

FIGURE 11.13

Zirconia unit cell structures in (a) monoclinic, (b) tetragonal, and (c) cubic phases.

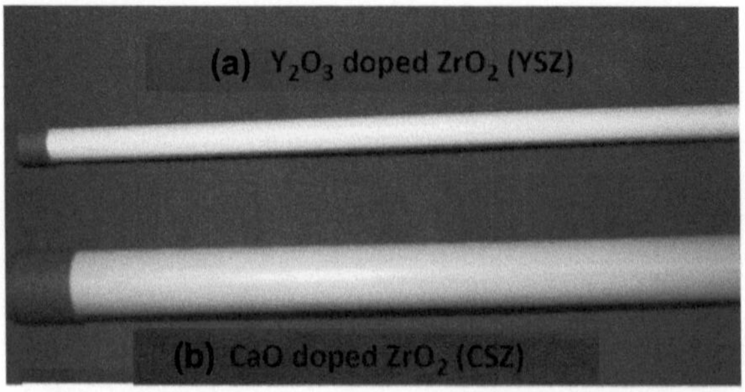

FIGURE 11.14

(a) Yttria-stabilized zirconia *(YSZ;Y_2O-ZrO_2)*. (b) Calcia-stabilized ZrO_2 *(CSZ; CaO-Y_2O)* used as solid oxide electrolytes in galvanic cell measurements.

temperature (2880°C). It undergoes three crystallographic phase transitions: monoclinic (ZrO_2-m, P121/c1) to tetragonal (ZrO_2-t, P42/nmc) at 1170°C and finally to cubic (ZrO_2-c, *Fm-3m*) at 2370°C (Fig. 11.13).

The phase transitions, particularly monoclinic to tetragonal, is accompanied by considerable change in lattice dimensions (increase in volume of ~4%), which causes fragmentation of the material.

The addition of aliovalent oxides stabilizes the high-temperature cubic fluorite phase of ZrO_2. Therefore ZrO_2 is doped with other oxides like calcia (CaO), MgO, or yttria (Y_2O_3) to make CaO-, MgO-, and Y_2O_3-stabilized ZrO_2. Stabilized ZrO_2 is used in the production of monocrystals for jewelry, manufacturing of crucibles and other refractory parts, manufacturing of thermal barriers for the internal protection of the combustion chamber of jet engines, and manufacturing of ionic conductors. Fig. 11.14 show the pictures of Y_2O_3-stabilized ZrO_2 (Fig. 11.14a) and CaO-stabilized ZrO_2 (Fig. 11.14b) used as solid oxide electrolytes in galvanic cell measurements.

9.4 CERIA

Ceria (CeO_2) is one of the most stable rare earth oxides with a +4 oxidation state. Cerium oxide adopts the fluorite structure with space group *Fm-3m* (Fig. 11.15).

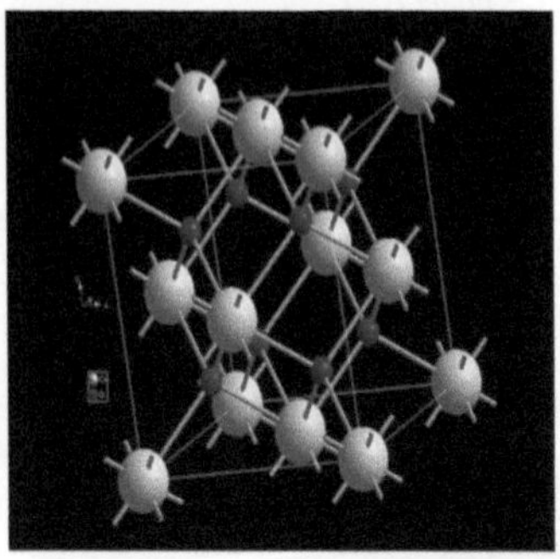

CeO_2-cubic
Space group:*Fm-3m* (225)
Lattice parameter a = 5.409 Å
V = 158.32 Å^3, Z = 4
Atomic positions (x, y, z):
Ce1 (4a) 0, 0, 0
O1(8c) 0.25,0.25,0.25

FIGURE 11.15

Unit cell structure of ceria.

At high temperatures and under reduced oxygen partial pressure, CeO_2 releases oxygen to give a nonstoichiometric anion-deficient compound [93]. The formation of oxygen-deficit species is the basis for use of CeO_2 as a catalyst. The compound is also used as a sensor in catalytic converters in automotive applications, controlling the air-to-exhaust ratio to reduce NO_x and carbon monoxide. CeO_2 is mostly used for polishing purposes, especially in the chemical-mechanical planarization technique. CeO_2 decolorizes glass by converting green-tinted ferrous impurities to nearly colorless ferric oxides. Solid solutions of CeO_2 and ZrO_2 ($Zr_xCe_{1-x}O_2$) have enhanced structural and textural properties, improved thermal stability, and catalytic properties. Addition of SiO_2 as support to $Zr_xCe_{1-x}O_2$ enhances the oxygen storage capacity. CeO_2 may also find application in solid oxide fuel cell (SOFC) technology because of its high oxygen ion conductivity at intermediate temperatures (500–650°C). It has also been proposed to use the cerium (IV) oxide–cerium (III) oxide cycle for a two-step thermochemical water splitting process to produce hydrogen.

9.5 MAGNESIUM ALUMINATE SPINEL

$MgAl_2O_4$ is used as refractory material in glass furnaces, as lining in steel-making furnaces, in burning zones of cement rotatory kilns, and in checker work of glass furnace regenerators [4]. The spinel structure $MgAl_2O_4$ (space group *Fd3m*) is a face-centered cubic stacking of O^{2-} ions containing eight structural units per array. Addition of excess MgO and B_2O_3 during preparation of $MgAl_2O_4$ improves the density of this material to 3.52 g/cm^3, which is about 92% theoretical value.

9.6 MAGNESIUM STANNATE

Mg_2SnO_4 is used in electronic ceramics for manufacturing dielectric plates for phones and high-speed computing devices [65]. This material is also used for capacitors because of its high stability, low permittivity, and low tangent loss.

9.7 ALUMINA-YTTRIA PHASES

$Y_3Al_5O_{12}$ (commonly known as *YAG*), $YAlO_3$, and $Y_4Al_2O_9$ are some of the most important pseudobinary Al_2O_3-Y_2O_3 phases that find wide applications in modern science and technology [28]. They are used as materials for photocatalysts, hosts for phosphors (Nd, Eu), in up-conversions of phosphors and as laser hosts. $(Y_{1-x}Eu_x)_4Al_2O_9$ shows luminescence at the red region. Sintered YAG exhibits 80–90% optical transparency in the 300–1100 nm region [94]. A 1.5-mm thick film of 1 at % Nd-doped YAG can show 83% optical transparency at 1064 nm.

9.8 NiCoCrAlY-$ZrO_2 \cdot Y_2O_3$ BARRIER COATING MATERIALS

NiCoCrAlY-$ZrO_2 \cdot Y_2O_3$ finds application as a thermal barrier coating material in the hot components of gas turbines [95] to increase the durability of Ni-based alloys. In this process, a thin film of NiCoCrAlY is deposited on clean Ni-based alloys (turbine), which is then coated with $ZrO_2 \cdot Y_2O_3$. Bonding between NiCoCrAlY and $ZrO_2 \cdot Y_2O_3$ is an important parameter to get the stable barrier coating. It has been observed that Al_2O_3 and $Y_4Al_2O_9$ phases are formed in the bonding interface. The formation of the $Y_4Al_2O_9$ phase helps preventing high-temperature oxidation of the alloy up to 1100°C.

9.9 METAL TUNGSTATES

Metal tungstates are used in many applications such as laser hosts, phosphor hosts, humidity sensors, light-emitting diodes (LEDs), scintillators, and photocatalytic materials under UV-illumination [52]. $ZnWO_4$ (monoclinic phase, Fig. 11.16), for example, is capable of splitting of water into H_2 and O_2 under UV light and can degrade rhodamine B (RhB) dye up to 98% within 6 hours on exposure to UV irradiation (300 W) (25 mg $ZnWO_4$, RhB 10 mg/L).

9.10 TIN-DOPED INDIUM OXIDE AND $SrCu_2O_2$

Pure indium oxide (In_2O_3) is a transparent n-type semiconductor. The pn-junction involving n-type In_2O_3 and p-type $SrCu_2O_2$ semiconducting materials (tetragonal phase of $SrCu_2O_2$, Fig. 11.17) acts as an excellent luminescent material. Doping of K at the Sr-site of $SrCu_2O_2$ improves optical transparency to 80–90% in the range of 500–3000 nm [44]. The DC electrical conductivity of K-doped $SrCu_2O_2$ at 300K is found to be 4.8×10^{-2} S cm^2 with activation energy of 0.10 eV. At 300K, the hole concentration and mobility are found to be 6.1×10^{17} cm^{-3} and 0.46 cm^2/V s, respectively. In addition, the thin film of In_2O_3 doped with Sn is used in flat panel displays and solar cells because of its controllable conductivity ranging from insulator 10^{-9} to metal 10^4 S cm^2 as well as optical transparency in the visible range.

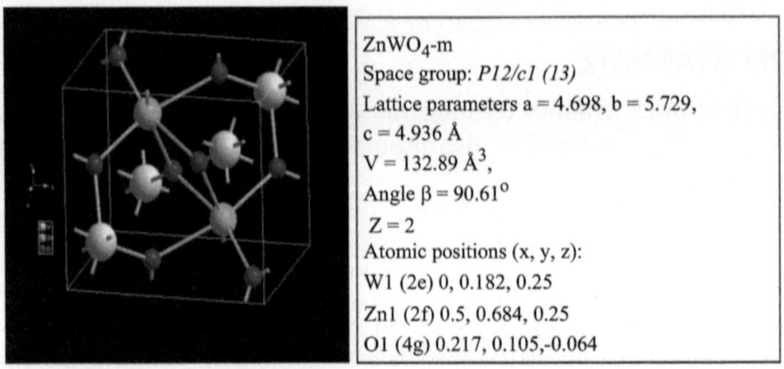

FIGURE 11.16

Unit cell structure of $ZnWO_4$.

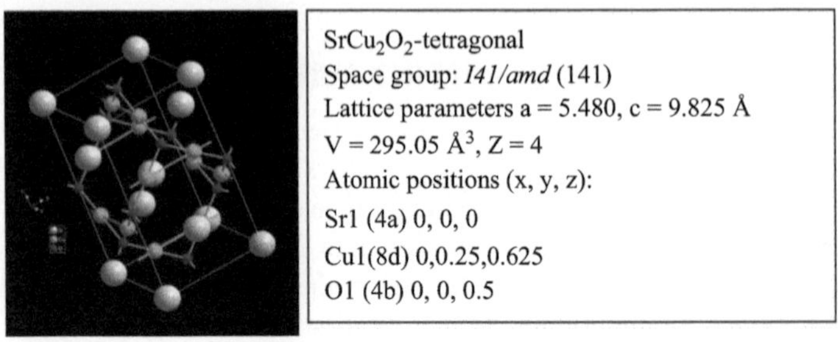

FIGURE 11.17

Unit cell structure of $SrCu_2O_2$.

9.11 $Ba_2Ti_9O_{20}$

Dielectric resonators (DRs) are used in microwave devices to produce microwave radiations of desired frequencies [47,48]. The DRs need to have a dielectric constant ($\varepsilon_r > 20$), a dielectric loss factor ($Q > 2000$), and high microwave stability. $Ba_2Ti_9O_{20}$ and 1 wt% Nb_2O_5-added $Ba_2Ti_9O_{20}$ show excellent DR properties. Whereas $Ba_2Ti_9O_{20}$ has ε_r of 34.6, a temperature coefficient ($\tau = 20.5$ ppm/°C), and $Q = 3820$, 1 wt% Nb_2O_5-added $Ba_2Ti_9O_{20}$ has ε_r of 27.5, a temperature coefficient ($\tau = 10.1$ ppm/°C), and $Q = 2770$. $Co(Ti_{1-x}Zr_x)Nb_2O_8$ is yet another excellent DR material [67].

9.12 AB_2O_4 AND A_2O_3

AB_2O_4-type compounds such as Zn_2SnO_4 show very good photocatalytic activity for the degradation of organic pollutants like phenol [46]. Degradation of 80% of phenol is observed under sunlight within 6 h of exposure. Another AB_2O_4-type compound, $ZnFe_2O_4$, exhibits excellent mechanical strength, high thermal stability, and good catalytic properties. $ZnFe_2O_4$ is used in various applications such as photo-induced electron transfer, photoelectrochemical cells, and photochemical hydrogen production. A_2O_3-type compounds include such materials as sintered aluminum oxynitride (AlON), a transparent material in the visible to near-IR region. The material has excellent mechanical properties, which makes it a useful material for IR windows, visible windows, and missile domes in military aircraft. This material can also be used as a host for LEDs [96]. The highest optical transparency in the material can be achieved by doping AlON with 0.75 wt% Er_2O_3, which increases density of the material. Doped AlON shows up-conversion emission in visible and near-IR regions after excitation at 980 nm. Similarly, Y_2O_3 (Fig. 11.18) is used for IR transmission applications (3–5 μm) at ambient and high-temperatures conditions, luminescent host material, and coating material for hypersonic guided-missile windows [74].

9.13 CERAMICS FOR GAS TURBINES

A gas turbine is a combustion engine at the heart of a power plant that produces electricity. It consists of three components: compressor, combustion chamber, and turbine. An air-fuel mixture produces hot gases that spin a turbine to convert mechanical energy to electricity. In flame combustion, the fuel burns in the presence of air from the compressor, wherein the temperature rises up to 1600–2000°C

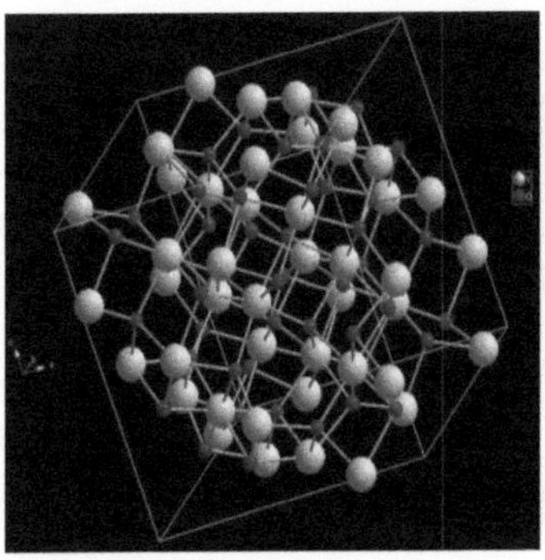

FIGURE 11.18

Unit cell structure of yttria (C-type cubic).

[6]. The turbine, which gets heated by this process, needs to be cooled down using part of the air from compressor. To facilitate the combustion process and to lower the temperature of the system (900−1400°C), catalysts and supported catalysts are used. The support for the catalysts should have the properties like low thermal expansion, high thermal shock resistance, high melting points (>1500°C), and chemical stability in air, NO_x, CO, CO_2, and steam environments. Some of examples of ceramic oxide supports are MgO, Al_2O_3, and ZrO_2. Examples of ceramic catalysts are hexaaluminates, perovskite composites of noble metal/Al_2O_3, spinel, pyrochlore, and zeolite.

9.14 CERAMICS IN SOLID-STATE FUEL CELLS

SOFCs are the electrochemical reactors that convert chemical energy of fuels such as H_2, CH_4, and CH_3OH into electrical energy [32]. The conversion efficiency of such devices is higher than that of the internal combustion engine. The operating temperature of an SOFC is about 600−1000°C. Most high-temperature SOFCs have Y_2O_3-stabilized ZrO_2 (YSZ) as an electrolyte, nickel cermet Ni/YSZ as an anode, and (La, Sr)MnO_3 as a cathode. Interconnections have an important role in the SOFC stacks [29], in which an electrical connection between the anode of one individual cell to the cathode of a neighboring cell is provided. This also acts as mechanical support. Interconnections should possess excellent electrical and thermal conductivity and should be chemically compatible with anode and cathode materials. $LaCrO_3$ and ferritic stainless steels are generally used as interconnecting compounds.

9.15 TITANIA-MODIFIED CEMENTS, TILES, AND BRICKS

The walls of buildings are often fouled or infected by agents like fungus, algae, bacteria, and air pollutants like smoke, NO_x, SO_x, acids, and alkaline ashes [97]. Fungus found in homes, concrete floors, tiles, and lavatories can be avoided using photocatalysts in the presence of sunlight, which reduces cost as compared with other means. One of best choices is the use of the stable semiconducting ceramic materials with a band gap close to the UV-visible range. Some examples of photocatalysts are TiO_2, Sn-doped TiO_2, and In-doped TiO_2. When a small amount of such material is added during preparation of paints, cement, tiles, or bricks as component/binder, it can reduce fouling considerably. Ceramic bricks are extensively used in high-temperature furnaces as a thermal barrier (Fig. 11.19).

9.16 TRANSPARENT MACHINABLE CALCIUM-MICA GLASS-CERAMICS

Machinable ceramics are products that can be cut, drilled, or machined into desired shapes with ordinary tools. They require neither firing nor diamond grinding after machining. The mechanical properties of machinable ceramics tend to be lower than those of common ceramics. Fluormica glass ceramics with general chemical compositions of Ca-Mg-Al-Si-O-F and K-Ca-Mg-Al-Si-O-F are examples of transparent machinable ceramics with high thermal and chemical stability [98,99]. Desired mechanical properties of machinable ceramics can be achieved by sintering them at required temperatures.

9.17 CERAMICS AND PORCELAIN TILES AND BRICKS

Ceramics like porcelain tiles and bricks are generally composed of chemicals such as SiO_2, Al_2O_3, Na_2O, K_2O, MgO, CaO, and Fe_2O_3 [100]. However, their relative concentrations change according to

FIGURE 11.19

Ceramic brick used in high-temperature furnaces as a thermal barrier.

requirements and uses. They are extensively used as construction material for building (floors in rooms such as kitchens and bathrooms, and toilets).

9.18 IMMOBILIZATION OF NUCLEAR WASTES USING CERAMICS

In nuclear power, radioactive waste is produced [78,101]. Immobilization of long-lived high-level radio nuclides with half-lives in the range of 10^3-10^5 years is a challenge. Borosilicate glass matrices have been considered for immobilization of high-level waste. Because of the metastable nature of borosilicate glasses, it is desirable to have alternate matrices for nuclear waste immobilization. Various ceramic-based matrices with mineral structures of monazite, zircon, thorite, britholite, pyrochlore, perovskite, or zirconolite have been proposed for the immobilization of long-lived radioisotopes such as minor actinides [102,103]. The choice of the components is mainly based on their ability to incorporate or immobilize a wider variety of ions simultaneously without further segregation to any other phases, as well as their geothermal and chemical stabilities. Thus the structural and chemical stabilities under diversified external variables, like temperature, pressure, and chemical environment, in particular ground water conditions, are the important information desired for their application. Among these, phosphate-based ceramics like monazites are the most prominent host matrix for immobilization of trivalent and tetravalent actinides. SYNROC is yet another $CaO-ZrO_2-TiO_2$–based low-cost material with chemical stability that is considered for material for disposal of long-lived radioactive waste.

9.19 ACID-RESISTANT CERAMIC MATERIALS

Glasses ($Al_2O_3-CaO-MgO-SiO_2-Sc_2O_3$), Al_2O_3, SnO_2, TiO_2, and CeO_2 are among the high-temperature ceramics stable in acidic medium [104–107]. Al_2O_3 and CeO_2 are among the most stable oxides in acids like HCl and HNO_3, even at higher temperatures. That is why they are considered to be refractory ceramics. However, CeO_2 is dissolved in H_2SO_4 on long heating (1–2 weeks). In ultrasound waves, CeO_2 can be dissolved in HNO_3 as a result of free radicals from cavitation (bubble implosion).

10. CONCLUSIONS

In this chapter, advances in the field of oxide-based high-temperature ceramics are reviewed. General properties of high-temperature ceramic oxides (binary, ternary, and higher members) such as brittleness, toughness, compressive strength, and thermal and electrical insulations are discussed. Classification of ceramics and the nature of their chemical bonding, phase transitions, and phase diagrams are presented. Methods for the synthesis of ceramics using different techniques are described. Applications of some of the important high-temperature ceramics under extreme conditions of temperature, pressure, and chemical environment are presented.

REFERENCES

[1] Moom Township, PA 15108, USA, Harbison-Walker, Handbook of Refractory Practice, 2005.

[2] C.E. Semler, in: J.E. Kogel, N.C. Trivedi, J.M. Barker, S.T. Krukowski (Eds.), Refractories in Industrial Minerals & Rocks, seventh ed., Society for Mining, Metallurgy, and Exploration, Inc., USA, 2006.

[3] K. Byappa, N. Keerthiraj, S.M. Byrappa, Hydrothermal growth of crystals-design and processing (Chapter 14), in: P. Rudolph (Ed.), B Crystal Growth: Basic Technologies and Growth Mechanisms and Dynamics, second ed., Elsevier Pub., USA, 2015.

[4] L.R. Pinga, A.-M. Azadb, T.W. Dung, Magnesium aluminate ($MgAl_2O_4$) spinel produced via self-heat-sustained (SHS) technique, Mater. Res. Bull. 36 (2001) 1417–1430.

[5] W. Wong-Ng, R.S. Roth, T.A. Vanderah, H.F. McMurdie, Phase equilibria and crystallography of ceramic oxides, J. Res. Natl. Inst. Stand. Technol. 106 (2001) 1097–1134.

[6] E.M. Johansson, D. Papadias, P.O. Thevenin, A.G. Ersson, R. Gabrielsson, P.G. Menon, P.H. Bjornbom, S.G. Jaras, Catalysis, vol. 14, The Royal Society of Chemistry, 1999, pp. 183–235.

[7] W.-M. Guo, Z.-L. Zhang, J.-X. Li, Y. You, S.-h. Wu, H.-T. Lin, Improvement of densification and mechanical properties of Al_2O_3–B_4C ceramics, Ceramics Int. 42 (2016) 11486–11489.

[8] J. Guo, Y. Zhang, H. Li, X. Zhang, Effect of the sintering temperature on the microstructure, properties and formation mechanism of ceramic materials obtained from polysiloxane elastomer-based ceramizable composites, J. Alloys Compd. 678 (2016) 499–505.

[9] S. Banijamali, T. Ebadzadeh, Glass-forming ability, sinter-crystallization behavior and microwave dielectric properties of MgO-B_2O_3-(Al_2O_3)-SiO_2 glass-ceramics, J. Non-Cryst. Solids 441 (2016) 34–41.

[10] Y.G. Zhang, H.Y. He, B.C. Pan, Structural features and electronics properties of MgO nanosheets and nanobelts, J. Phys. Chem. 116 (2012) 23130–23135.

[11] M.B. Gawande, S.N. Shelke, R. Zboril, R.S. Varma, Microwave-assisted chemistry: synthetic applications for rapid assembly of nanomaterials and organics, Acc. Chem. Res. 47 (2014) 1338–1348.

[12] B. Predel, Heterogene gleichgewichte: Grundlagen und anwendungen, Steinkopff, Darmstadt, 1982.

[13] B. Predel, M. Hoch, M. Pool, Phase Diagrams and Heterogeneous Equilibria, Springer, Berlin, 2004.

[14] F.N. Rhines, Phase Diagrams in Metallurgy: Their Development and Application, McGraw-Hill, New York, 1956.

[15] P. Gordon, Principles of Phase Diagrams in Materials Systems, McGraw-Hill, New York, 1968.

[16] A. Prince, Alloy Phase Equilibria, Elsevier, Amsterdam, 1966.

[17] J. Zernike, Chemical Phase Theory, N.V. Uitgevers-Maatschappij – AE.E. Kluwer, Deventer, 1955.

[18] T.B. Massalski, et al. (Eds.), Binary Alloy Phase Diagrams, vols. 3, ASM International, Materials Park, OH, 1995.

[19] P. Villars, A. Prince, H. Okamoto (Eds.), Handbook of Ternary Alloy Phase Diagrams, vols. 4, ASM International, Materials Park, OH, 1995.

[20] T. Zienertn, O. Fabrichnaya, Thermodynamic assessment and experiments in the system $MgO-Al_2O_3$, Calphad 40 (2013) 1−9.
[21] K. Maiti, A. Sil, Relationship between fracture toughness characteristics and morphology of sintered Al_2O_3 ceramics, Ceramics Int. 36 (2010) 2337−2347.
[22] A. Aksay, J.A. Pask, Stable and metastable phase equilibria in the system Al_2O_3/SiO_2, J. Am. Ceram. Soc. 58 (1975) 507−512.
[23] P. Boch, T. Chartier, Alumina, mullite and spinel, zirconia (Chapter 6), in: P. Boch, J.-C. Niepce (Eds.), Ceramic Materials: Process, Properties and Application, ISTE Publication, 2007.
[24] S.D. Meetei, S.D. Singh, N.S. Singh, V. Sudarsan, R.S. Ningthoujam, M. Tyagi, S.C. Gadkari, R. Tewari, R.K. Vatsa, Crystal structure and photoluminescence correlations in white emitting nanocrystalline ZrO_2: Eu^{3+} phosphor: effect of doping and annealing, J. Lumin. 132 (2012) 537−544.
[25] A.S. Lanje, S.J. Sharma, R.B. Pode, R.S. Ningthoujam, Synthesis and optical characterization of copper oxide nanoparticles, Adv. Appl. Sci. Res. 1 (2010) 36−40.
[26] A.S. Lanje, R.S. Ningthoujam, S.J. Sharma, R.B. Pode, Luminescence and electrical resistivity properties of cadmium oxide nanoparticles 49 (2011) 234−238.
[27] B.P. Singh, A.K. Parchur, R.S. Ningthoujam, A.A. Ansari, P. Singh, S.B. Rai, Dalton Trans. 43 (2014) 4770−4778.
[28] J. Ma, Z. Yu, D. Zhang, J. Fan, W. Fu, J. Chen, S. Cai, B. Wang, The molten salt synthesis of $(Y_{1-x}Eu_x)_4Al_2O_9$ solid solution powders and their PL properties, J. Am. Ceram. Soc. 98 (2015) 370−373.
[29] X. Liu, J. Wu, C. Johnson, Interconnect materials for SOFC stacks, in: M. Ni, T.S. Zhao (Eds.), RSC Energy and Environment Series No. 7, Solid Oxide Fuel Cells: From Materials to System Modeling, The Royal Society of Chemistry, 2013, pp. 106−134.
[30] S.K. Srivastava, Metal oxide filled micro and nano natural rubber composites, in: S. Thomas, H.J. Maria, J. Joy, C.H. Chan, L.A. Pothen (Eds.), RSC Polymer Chemistry Series No. 8 Natural Rubber Materials, Composites and Nanocomposites, vol. 2, The Royal Society of Chemistry, 2013, pp. 106−134.
[31] P.C.S. Hayfield, Is Ti_4O_7 an optimum ceramic electrode choice?, in: Development of New Materials: Monolithic Ti_4O_7 Ebonex Ceramics, The Royal Society of Chemistry, 2001.
[32] S.P. Jiang, Nano-structured electrodes of solid oxide fuel cells by infiltration, in: M. Ni, T.S. Zhao (Eds.), RSC Energy and Environment Series No. 7, Solid Oxide Fuel Cells: From Materials to System Modeling, The Royal Society of Chemistry, 2013, pp. 106−134.
[33] M. Vallet-Regi, D.A. Navarrete, Synthetic Nanoapatites, RSC Nanoscience & Nanotechnology No. 39, Nanoceramics in Clinical Use: From Materials to Applications, second ed., The Royal Society of Chemistry (2013), 2016, pp. 30−69.
[34] P.C.S. Hayfield, The Properties of Titanium Suboxide & Associated Processes, The Royal Society of Chemistry, 2001, pp. 9−47.
[35] H. Hirashima, E. Onishi, M. Nakagania, Preparation of PZT powders from metal alkoxides, J. Non-Cryst. Solids 121 (1990) 404−406.
[36] A. Suarez-Gomez, R. Sato-Berru, R.A. Toscano, J.M. Saniger-Blesa, F. Calderon-Pinar, On the synthesis and crystallization process of nanocrystalline PZT powders obtained by a hybrid sol−gel alkoxides route, J. Alloys Compd. 450 (2008) 380−386.
[37] M. Hirano, K. Ota, Direct formation and photocatalytic performance of anatase (TiO_2)/silica (SiO_2) composite nanoparticles, J. Am. Ceram. Soc. 87 (2004) 1567−1570.
[38] U.G. Akpan, B.H. Hameed, The advancements in sol−gel method of doped-TiO_2 photocatalysts, Appl. Catal. A 375 (2010) 1−11.
[39] M.A. Willard, L.K. Kurihara, E.E. Carpenter, S. Calvin, V.G. Harris, Chemically prepared magnetic nanoparticles, Int. Mater. Rev. 49 (2004) 125−170.

[40] N.K. Sahu, N.S. Singh, R.S. Ningthoujam, D. Bahadur, Ce^{3+} sensitized $GdPO_4:Tb^{3+}$ nanorods: an investigation on energy transfer, luminescence switching and quantum yield, ACS Photonics 1 (2014) 337–346.

[41] M.N. Luwang, R.S. Ningthoujam, S.K. Srivastava, R.K. Vatsa, Disappearance and recovery of luminescence in Bi^{3+}, Eu^{3+} co-doped YPO_4 nanoparticles due to the presence of water molecules up to 800°C, J. Am. Chem. Soc. 133 (2011) 2998.

[42] T.F.T. Cerqueira, S. Lin, M. Amsler, S. Goedecker, S. Botti, M.A.L. Marques, Identification of novel Cu, Ag, and Au ternary oxides from globalstructural prediction, Chem. Mater. 27 (2015) 4562–4573.

[43] O. Karban, O. Kanunnikova, E. Khazanov, E. Salamatov, O. Khasanov, A. Taranov, Structure and composition of Ba-W-Ti-O ceramics interface regions formed at ultrasonic vibration, ISRN Nanomater. (2012) 1–7.

[44] A. Kudo, H. Yanagi, H. Hosono, H. Kawazoe, $SrCu_2O_2$: a p-type conductive oxide with wide band gap, Appl. Phys. Lett. 73 (1998) 220.

[45] M. Huang, K. Li, H. Li, Q. Fu, Y. Wang, Si-Al-Ir oxidation resistant coating for carbon/carbon composites by slurry dipping, J. Mater. Sci. Technol. 25 (2009) 344–346.

[46] C.G. Anchieta, D. Sallet, E.L. Foletto, S.S. da Silva, O. Chiavone-Filho, C.A.O. do Nascimento, Synthesis of ternary zinc spinel oxides and their application in the photodegradation of organic pollutant, Ceramics Int. 40 (2014) 4173–4178.

[47] A. Rabenau, The role of hydrothermal synthesis in preparative chemistry, Angew. Chem. Int. Ed. Engl. 24 (1985) 11026–11040.

[48] S. Somiya, R. Roy, Hydrothermal synthesis of fine oxide powders, Bull. Mater. Sci. 23 (2000) 453–460.

[49] K.T. Kang, K.S. Park, S.B. Yi, H.-G. Kim, Preparation of ceramic composite membranes by microwave heating, J. Korean Phys. Soc. 45 (2004) 138–140.

[50] T.S.S. Kumar, I. Manjubala, J. Gunasekaran, Synthesis of carbonated calcium phosphate ceramics using microwave irradiation, Biomaterials 21 (2000) 1623–1629.

[51] J. Prado-Gonjal, R. Schmidt, E. Moran, Microwave-assisted synthesis and characterization of perovskite oxides, in: J. Zhang, H. Li (Eds.), Perovskite: Crystallography, Chemistry and Catalytic Performance, Novascience Publishers, Hauppauge, USA, 2013.

[52] J. Yan, Y. Shen, F. Li, T. Li, Synthesis and photocatalytic properties of $ZnWO_4$ nanocrystals via a fast microwave-assisted method, Sci. World J. (2013) 1–8.

[53] M.N. Luwang, R.S. Ningthoujam, N.S. Singh, R. Tewari, S.K. Srivastava, R.K. Vatsa, Surface chemistry of surfactant AOT-stabilized SnO_2 nanoparticles and effect of temperature, J. Colloids Interface Sci. 349 (2010) 27.

[54] M.N. Luwang, R.S. Ningthoujam, S.K. Srivastava, R.K. Vatsa, Preparation of white light emitting YVO_4: Ln^{3+} and silica-coated YVO_4: Ln^{3+} ($Ln^{3+} = Eu^{3+}$, Dy^{3+}, Tm^{3+}) nanoparticles by CTAB/*n*-butanol/hexane/water microemulsion route: energy transfer and site symmetry studies, J. Mater. Chem. 21 (2011) 5326.

[55] M.A. Haider, A.J. Capizzi, M. Murayama, S. McIntosh, Reverse micelle synthesis of perovskite oxide nanoparticles, Solid State Ionics 196 (2011) 65–72.

[56] R. Abazari, S. Sanati, Perovskite $LaFeO_3$ nanoparticles synthesized by the reverse microemulsion nanoreactors in the presence of aerosol-OT: morphology, crystal structure, and their optical properties, Superlattices Microstruct. 64 (2013) 148–157.

[57] O.A. Graeve, H. Fathi, J.P. Kelly, M.S. Saterlie, K. Sinha, G. Rojas-George, R. Kanakala, D.R. Brown, E.A. Lopez, Reverse micelle synthesis of oxide nanopowders: mechanisms of precipitate formation and agglomeration effects, J. Colloid Interface Sci. 407 (2013) 302–309.

[58] S. Vaidya, J. Ahemed, A.K. Ganguli, Controlled synthesis of nanomaterials using reverse micelles, Defence Sci. J. 58 (2008) 531–544.

[59] B.P. Singh, A.K. Parchur, R.S. Ningthoujam, P.V. Ramakrishna, R. Maalej, P. Singh, S.B. Rai, Enhanced upconversion and temperature-sensing behaviour of Er^{3+} and Yb^{3+} co-doped $Y_2Ti_2O_7$ by incorporation of Li^+ ions, Phys. Chem. Chem. Phys. 16 (2014) 22665−22676.

[60] N. Sudhakar, R.S. Ningthoujam, N.S. Gajbhiye, K.P. Rajeev, Structural, magnetic and electron transport studies on nanocrystalline layered manganite $La_{1.2}Ba_{1.8}Mn_2O_7$ system, J. Nanosci. Nanotech. 7 (2007) 965.

[61] Y.-P. Fu, C.-C. Chang, C.-H. Lin, T.-S. Chin, Solid-state synthesis of ceramics in the $BaO-SrO-Al_2O_3-SiO_2$ system, Ceramics Int. 30 (2004) 41−45.

[62] A.-M. Azad, S. Subramanian, Synthesis of $BaZrO_3$ by a solid-state reaction technique using nitrate precursors, Mater. Res. Bull. 37 (2002) 85−97.

[63] M.T. Buscaglia, M. Bassoli, V. Buscaglia, Solid-state synthesis of nanocrystalline $BaTiO_3$: reaction kinetics and powder properties, J. Am. Ceram. Soc. 91 (2008) 2862−2869.

[64] H.L. Chu, C.L. Wang, H.E. Lee, Y.Y. Sie, R.S. Chen, W.S. Hwang, M.C. Wang, Effect of sintering process parameters on the properties of 3YPSZ ceramics, in: IOP Conf. Series: Materials Science and Engineering, vol. 47, 2013, p. 012005.

[65] A.A. Al-Shahrani, Sintering behavior and thermal property of Mg_2SnO_4, J. Mater. Sci. Mater. Electron. 16 (2005) 193−196.

[66] A.-M. Azad, L.J. Min, Mg_2SnO_4 ceramics I. Synthesis processing microstructure correction, Ceramics Int. 27 (2001) 325−334.

[67] Y. Zhang, Y. Zhang, M. Xiang, Crystal structure and microwave dielectric characteristics of Zr-substituted $CoTiNb_2O_8$ ceramics, J. Eur. Ceram. Soc. 36 (2016) 1945−1951.

[68] R. Djenadic, M. Winterer, Chemical vapor synthesis of nano crystalline oxides, in: A. Lorke, M. Winterer, R. Schmechel, C. Schultz (Eds.), Nanoparticles from the Gas Phase, Nano Science and Technology, first ed., Springer-Verlag, Berlin, Heidelberg, 2012, pp. 49−76.

[69] N. Rajput, Methods of preparation of nanoparticles − a review, Int. J. Adv. Eng. Technol. 7 (January 2015) 1806−1811.

[70] E. Roos, S.M. Naga, R.N. Richter, S. Lauf, M. Awaad, A. Lyutovich, M.A. Tameem, Electron beam physical vapour deposition and mechanical properties of c-ZrO_2-ZTA-coatings on alloy 617 substrates, Ceramics Int. 38 (2012) 3317−3326.

[71] N. Obradovic, M.V. Nikolic, N. Nikolic, S. Filipovic, M. Mitric, V. Pavlovic, P.M. Nikolic, A.R. Dordevic, M.M. Ristic, Synthesis of barium-zinc-titanate ceramics, Sci. Sintering 44 (2012) 65−71.

[72] D.P. Singh, O.N. Srivastava, Synthesis and characterization of different metaloxide nanostructures by simple electrolysis based oxidation of metals, J. Nanosci. Nanotechnol. 9 (2009) 5515−5522.

[73] L. Chretien, L. Bonnet, R. Boulesteix, A. Maitre, C. Salle, A. Brenier, Influence of hot isostatic pressing on sintering trajectory and optical properties of transparent Nd:YAG ceramics, J. Eur. Ceram. Soc. 36 (2016) 2035−2042.

[74] X. Li, X. Mao, M. Feng, S. Qi, B. Jiang, L. Zhang, Fabrication of transparent La-doped Y_2O_3 ceramics using different La_2O_3 precursors, J. Eur. Ceram. Soc. 36 (2016) 2549−2553.

[75] R.J. Ackerman, E.G. Rauh, Determination of liquidus curves for the Th-W, Th-Ta, Zr-W, and Hf-W systems: the anomalous behavior of metallic thorium, High Temp. Sci. 4 (1972) 272−282.

[76] T. Nobuo, Present status and future prospects of spherical aberration corrected TEM/STEM for study of nanomaterials, Sci. Technol. Adv. Mater. 9 (2008) 014111.

[77] D. Wang, J. Zhao, C. Xue, Y. Cao, Finite element simulation of Vickers micro-indentation test of micro-nano-composite ceramic tool materials based on microstructure model, Int. J. Refract. Met. Hard Mater. 58 (2016) 34−41.

[78] R.C. Ewing, W.J. Weber, J. Lian, Nuclear waste disposal-pyrochlore ($A_2B_2O_7$): nuclear waste form for the immobilization of plutonium and minor actinides, J. Appl. Phys. 95 (2004) 5949−5971.

[79] Material Properties Charts, Ferro-Ceramic Grinding, Inc., Wakefield, MA. www.ferroceramic.com, (Source).
[80] A. Ferretti, D.B. Rogers, J.B. Goodenough, The relation of the electrical conductivity in single crystals of rhenium trioxide to the conductivities of Sr_2MgReO_6 and $NaxWO_3$, Phys. Chem. Solids 26 (1965) 2007–2011.
[81] C.N.R. Rao, Transition metal oxides, Annu. Rev. Phys. Chem. 40 (1989) 291–326.
[82] R.T. Grimley, R.P. Burns, M.G. Inghram, Thermodynamics of the vaporization of Cr_2O_3: dissociation energies of CrO, CrO_2, and CrO_3, J. Chem. Phys. 34 (1961) 664.
[83] P.E. Blackburn, M. Hoch, H.L. Johnston, The vaporization of molybdenum and tungsten oxides, J. Phys. Chem. 62 (1958) 769–773.
[84] L. Brewer, R.H. Lamoreaux, R. Ferro, R. Marazza, K. Girgis, in: L. Brewer (Ed.), Molybdenum: Physico-Chemical Properties of its Compounds and Alloys, IAEA Review, IAEA, Vienna, 1980. Special Issue No. 7.
[85] N. Jacobsona, D. Myersb, E. Opilaa, E. Coplandc, Interactions of water vapor with oxides at elevated temperatures, J. Phys. Chem. Solids 66 (2005) 471–478.
[86] D. Das, S.R. Bharadwaj, Thoria-based Nuclear Fuels, Green Energy and Technology, Springer-Verlag, London, 2013.
[87] A. Zerr, R. Boehler, Constraints on the melting temperature of the lower mantle from high-pressure experiments on MgO and magnesiowustite, Nature 371 (1994) 506–508.
[88] O.E. Taurian, M. Springborg, N.E. Christensen, Self-consistent electronic structures of MgO and SrO, Solid State Commun. 55 (1985) 351–355.
[89] K. Davis, Material Review: Alumina (Al_2O_3), School of Doctoral Studies (European Union) Journal, 2010, pp. 109–114.
[90] P. Auerkari, Mechanical and Physical Properties of Engineering Alumina Ceramics, Valtion Teknillinen Tutkimuskeskus (VTT), Finland, 1996.
[91] R.F. Geller, P.G. Yavorsky, Melting point of alpha-alumina, J. Res. Natl. Bur. Stand. 34 (1945) 395–401.
[92] K. Kiukkola, C. Wagner, Measurements on galvanic cells involving solid electrolytes, J. Electrochem. Soc. 104 (1957) 379–387.
[93] D. Chen, C. Chen, Z.M. Baiyee, Z. Shao, F. Ciucci, Nonstoichiometric oxides as low-cost and highly-efficient oxygen reduction/evolution catalysts for low-temperature electrochemical devices, Chem. Rev. 115 (2015) 9869–9921.
[94] R. Yin, J. Li, M. Dong, T. Xie, Y. Fu, W. Luo, L. Ge, H. Kou, Y. Pan, J. Guo, Fabrication of Nd:YAG transparent ceramics by non-aqueous gelcasting and vacuum sintering, J. Eur. Ceram. Soc. 36 (2016) 2543–2548.
[95] S. Yang, X. Lan, N. Yuang, Role of $Y_4Al_2O_9$ in high temperature oxidation resistance of NiCoCrAlY-$ZrO_2 \cdot Y_2O_3$ coatings, J. Mater. Sci. Technol. 23 (2007) 568–570.
[96] Y. Wang, X. Xie, J. Qi, S. Wang, N. Wei, Z. Lu, X. Chen, T. Lu, Bifunctional behavior of Er^{3+} ions as the sintering additive and the fluorescent agent in Er^{3+} single doped γ-AlON transparent ceramics, J. Lumin. 175 (2016) 203–206.
[97] V. Augugliaro, V. Loddo, M. Pagliaro, G. Palmisano, L. Palmisano, TiO_2-Modified Cement and Ceramics, Clean by Light Irradiation: Practical Applications of Supported TiO_2, The Royal Society of Chemistry, 2010, pp. 144–167.
[98] M. Goswami, A. Sarkar, P.W. Wagh, K.M. Bandal, V.K. Shrikhande, G.P. Kothiyal, J.V. Yakhmi, V.C. Sahni, Magnesium Aluminum Silicate Machinable Glass Ceramics, BARC, May 2003. News Letter No. 232.
[99] M. Li, N. Zhou, X. Luo, G. Zhang, Z. Xie, L. Xu, P. Liu, Effects of doping $Al_2O_3/2SiO_2$ on the structure and properties of magnesium matrix ceramic, Mater. Chem. Phys. 175 (2016) 6–12.

[100] R. Ji, Z. Zhang, C. Yan, M. Zhu, Z. Li, Preparation of novel ceramic tiles with high Al_2O_3 content derived from coal fly ash, Constr. Build. Mater. 114 (2016) 888−895.

[101] F. Wang, Q. Liao, Y. Dai, H. Zhu, Immobilization of gadolinium in iron borophosphate glasses and iron borophosphate based glass ceramics: implications for the immobilization of plutonium (III), J. Nucl. Mater. 477 (2016) 50−58.

[102] a W. Lutze, R.C. Ewing, Radioactive Waste Forms for the Future, Elsevier Science Publishers B.V., Amsterdam, 1989, pp. 236−335, 495−565;
b Comparison of glass and crystalline nuclear waste forms, in: Material Research Society Symposium Proceedings, vol. 127, 1988, pp. 13−24.

[103] D.G. Brookins, Geochemical Aspects of Radioactive Waste Disposal, Springer-Verlag, New York, 1984, pp. 77−114.

[104] B.L. Krasnyi, V.P. Tarasovskii, E.V. Rakhmanova, V.V. Bondar, Chemical resistance of ceramic materials in acids and alkalis 61 (2004) 337−339.

[105] J.P. Bennett, Corrosion Resistance of Selected Ceramics Materials to Nitric Acid, vol. 8851, United States Department of Interior, Bureau of Mines, 1951, pp. 11−12.

[106] L. Curkovic, S. Kurajica, M.F. Jelaca, M. Marinkovic, Corrosion behaviour of alumina ceramics in aqueous HCl solution, in: Conference on Materials, Processes, Friction and Wear, MATRIB'07, Vela Lika, June 21−23, 2007.

[107] F. Juillet, J.M. Adnet, M. Gasgnier, Ultrasound effects on the dissolution of refractory oxides (CeO_2 and PuO_2) in nitric acid, J. Radioanal. Nucl. Chem. 224 (1997) 137−143.

CHAPTER

COLD PLASMA PROCESSING OF MATERIALS FOR EXTREME CONDITIONS

12

D.S. Patil*, R. Kar, S.S. Chopade

Bhabha Atomic Research Centre, Mumbai, India

CHAPTER OUTLINE

*Department of Metallurgical Engineering and Materials Science, Indian Institute of Technology Bombay, Powai, Mumbai 400076.

Materials Under Extreme Conditions. http://dx.doi.org/10.1016/B978-0-12-801300-7.00012-7

1. INTRODUCTION

Plasmas can be classified according to the method of generation and operating pressure range that is used while generating the plasma. Plasmas generated by high-pressure arc or generated in high-pressure torches have identical temperatures in the different species; these plasmas are classified as thermal plasma. Nonequilibrium plasmas generated at low pressure with constituent species having different *temperatures* can be classified as cold plasmas. Both types can be used for surface modification purposes with their associated advantages and disadvantages. Herein, we are going to focus on the use of cold plasmas in surface modification using a chemical vapor deposition (CVD) process. This process is used for modifying properties of a solid (substrate) by depositing a layer or layers of materials of interest on the substrate through a vapor-phase chemical reaction at elevated temperatures. This process offers many advantages, such as excellent conformal coverage of complex structures, simpler deposition equipment, low-temperature film fabrication, and ease of large-scale deposition. This technique has inherent advantages and it is emerged as a method to prepare thin films for a variety of applications in the electronics and tribology industries.

Three fundamental steps are involved in any CVD process:

1. transportation of the precursor components in vapor phase (with or without carrier gas) to the reaction zone,
2. chemical reaction that takes place in a gaseous medium surrounding the substrate and is generally kept at the high-temperature and
3. removal of the by-products.

The CVD equipment therefore consists of a precursor distribution and mixing system, a reaction chamber (often with a vacuum pumping system if the reactions are to be carried out at low pressure), and a system that will provide energy for the reaction and for heating the substrates, and a neutralization (disposal/scrubber) system for the exhaust gases/products.

The chemical reactions between various precursors occur normally in the vapor phase over the heated substrate, and the film is deposited on the surface. In some cases, however, a film can also be formed from a reaction between the substrate surface and one or more of the constituents of the precursor vapor phase. Therefore the properties of the deposited thin film are dependent on CVD parameters. CVD reactors can be operated at atmospheric pressure and in the low-pressure range. Many different variations of conventional CVD methods have been developed to enhance the efficiency of deposition. This includes metalorganic CVD (MOCVD), laser-induced CVD (LCVD), photo-enhanced CVD (photo CVD), and plasma-enhanced CVD (PECVD). In the MOCVD process, metal organic compounds that are sufficiently volatile and have low decomposition temperatures are used as CVD precursors so that the reactive molecules can be transported in the gas phase in high enough concentrations to allow reasonable film deposition rates and low deposition temperatures.

2. THE ROLE OF PLASMA IN CHEMICAL VAPOR DEPOSITION

In the PECVD mode, the process can be carried out with plasma at atmospheric pressure or at low pressures. Although atmospheric-pressure CVD-based processes have certain advantages like high growth rates, processes using low-pressure plasmas (cold plasmas) are finding a lot of applications in the field of materials processing. This is because of the fact that these plasmas are unique in terms of generation of chemically reactive species at low temperature, and this is because of the nonequilibrium nature of the plasma state. Owing to high internal energy of low-pressure plasma (because of the presence of the charged particles), processes that are thermodynamically allowed but kinetically hindered in a conventional process proceed at a high rate under plasma conditions. This results in a low-deposition temperature and hence technologically a very important process in surface-modification applications, particularly when industrial tool substrates like steel are to be coated. These low pressure plasmas (cold plasmas) are widely used for various industrial processes including thin film deposition by CVD and physical vapor deposition (PVD) such as magnetron sputtering and semiconductor manufacturing.

In the PECVD process, the plasma is generated and controlled by ionizing a gas with the application of a direct current (DC) electric field or radio frequency (rf)/microwave (MW) electromagnetic field of sufficient power. The primary role of the plasma is to promote chemical reactions. Whereas the temperature of the electrons in the bulk gas at room temperature is approximately 0.025 eV, the temperatures of the electrons in the plasma state can be as high as 1–20 eV. This produces an unusual chemically reactive environment at ambient temperatures because of the ability of these highly energetic electrons to ionize and dissociate most types of gas molecules. Although electrons are the ionizing source, collisions involving excited species can also lead to the formation of free radicals and can assist the ionization process. Once ionized, excited gas species react with the substrate surface placed in the discharge plasma, resulting in the modification of the substrate surface. Because of these advantages, the plasma state also enhances the deposition efficiency. Furthermore, high-energy ultraviolet (UV) photons, resulting from the combination and relaxation of dissociated species, may be absorbed by the substrate, creating even more active sites on the substrates. This also assists in reactions when a substance is in the plasma state.

To get a proper quality coating and reproducibility of the deposition process, there are plasma-process parameters that need to be controlled, such as operating pressure and the flow rate of the precursor gases, input power to the plasma discharge, substrate temperature, and the bias on the substrate.

Low-pressure plasma that is used for CVD applications is generally produced in a pressure range of 10^{-5}–10 torr with a continuous gas flow into the reactor. The plasma generated by DC, rf, and MW fields has been successfully used for deposition of various coatings. Each process has its own merits and demerits. Low-pressure plasma sources that are used for CVD application are characterized by a large mean free path of the plasma particles. Therefore these plasmas are suitable for the generation of directed ion currents to the target surface, whereas the formation rate of higher radical densities and reaction products resulting from two-body or even three-body collisions of heavy particles is small.

3. PLASMA SOURCES FOR CHEMICAL VAPOR DEPOSITION APPLICATIONS

Plasma generation results from input of energy into the gas phase that produces a sufficient number of charge carriers (free electrons and positive ions). Plasma sources can be classified according to the nature of energy input to the gas and gas operating pressure. Technological plasmas are mostly generated by electrical discharges in a gas atmosphere. Depending on the temporal behavior of the sustaining electric field, discharges are distinguished as DC and alternating current discharge and work in either continuous or pulsed mode. Most plasma-based CVD systems operate in certain fixed

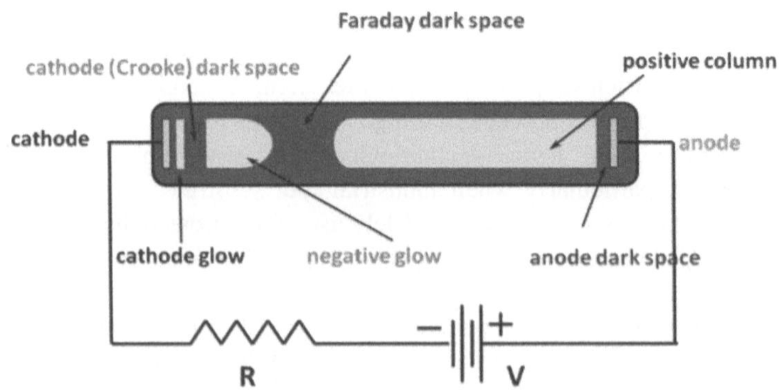

FIGURE 12.1

Schematic of a direct current discharge.

frequencies, including low (50–450 kHZ), rf (13.56 or 27.12 MHZ), and MW (915/2450 MHZ) frequencies. There has been a gradual shift from rf to MW in recent years. MW plasma is widely used because of easy, reliable operation as well as sufficient activity. The details regarding different sources and plasma generation mechanisms are discussed elsewhere [1–4]. However, a brief discussion about these sources is given in the following sections.

3.1 DIRECT CURRENT DISCHARGES

DC plasma is generated in the low-pressure glow discharge ignited in a closed-discharge vessel between inner electrodes (cathode and anode). As shown in Fig. 12.1, the discharge in a long cylindrical vessel shows different characteristic regions: cathode fall, negative glow, Faraday dark space, positive column, and anode fall. The region in the positive column corresponds to plasma. The transition of the glow discharge into an arc is inhibited by putting a current-limiting resistor in the series with the discharge.

3.2 RADIO FREQUENCY DISCHARGES

The rf plasma discharges usually operate in the frequency range of $f = 1-100$ MHz. The corresponding wavelengths ($\lambda = 3-300$ m) are very large in comparison with the reactor dimensions. At lower frequencies, the ions respond to the applied field and collide with the electrode surface and generate secondary electrons. However, at higher frequencies neither the ions nor the electrons can reach the electrode surface during the acceleration phase of the external electric field; therefore charge carriers are generated by electron impact ionization of heavy particles in the plasma volume. In order to have efficient coupling of the applied rf power with the gas/plasma, the rf source is connected to the plasma/gas load through an impedance matching network. The function of this matching network is to match the impedance of the load to that of the generator so that there is maximum power transfer to the load. The rf power to the plasma can be coupled by two different schemes of coupling (inductive and capacitive coupling).

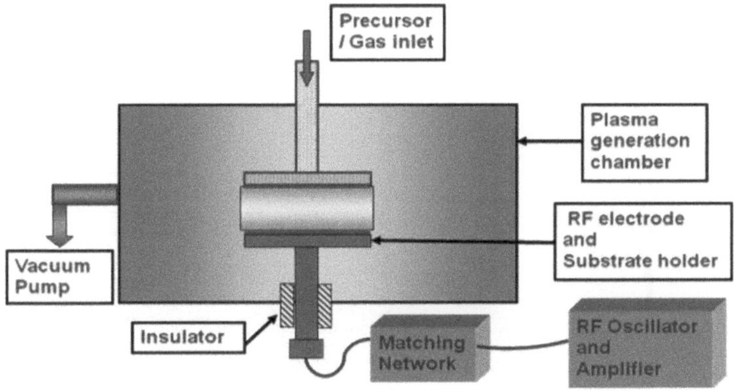

FIGURE 12.2

Capacitively coupled radio frequency discharge.

3.2.1 Capacitively Coupled Radio Frequency Discharges

As shown in Fig. 12.2, the process chamber of a capacitively coupled discharge reactor can have interior circular disc-shaped parallel electrodes, which are separated by a distance of a few centimeters. These electrodes can be in contact with the discharge or they may be insulated from it by a dielectric. Typical gas pressures in which this type of discharge is operating are in the range 10^{-2}–10 torr. The electrodes in the rf discharge are covered by sheath regions; these regions are similar to the cathode dark space in a DC glow discharge. In this configuration, the electrodes form the capacitance and rf power is coupled through the impedance matching network. Using this configuration, plasma electron densities of approximately 0^{9}–10^{10} cm^{-3} and electron temperatures of about 2–5 eV can be obtained. There are also some variations where the power is coupled to the load (plasma) through the dielectric walls. This electrodeless operation is helpful in getting high-purity coatings, because there is no erosion or contamination of the electrode surface.

The configuration of capacitive coupling with inner electrodes also offers a possibility of getting a negative DC self-bias on the electrode when the power electrode is connected through a blocking capacitor. This happens because of the difference in the mobility of the ions and electrons that are reaching the electrode. This self-biasing of the power electrode is particularly very useful in processes where the coating properties are dependent on the energy of the ions that are arriving at the substrate located on the power electrode.

3.2.2 Inductively Coupled Discharges

These discharges are excited by applying an rf electric field of a helical coil surrounding the plasma volume as shown in Fig. 12.3, or of a spiral coil in front of the plasma. Because of the varying magnetic field generated by rf current in the exciting coil, voltage is generated in the plasma volume that is useful in generating the plasma. This coupling between the surrounding coil and the plasma volume can be viewed as a transformer action with several primary windings of the surrounding coil and the plasma as a single winding. This configuration gives higher plasma densities, typically around 10^{11}–10^{12} cm^{-3}, and low electric fields in the plasma (low plasma potential). This low plasma potential capability makes such plasmas suitable for materials processing where low damage to the

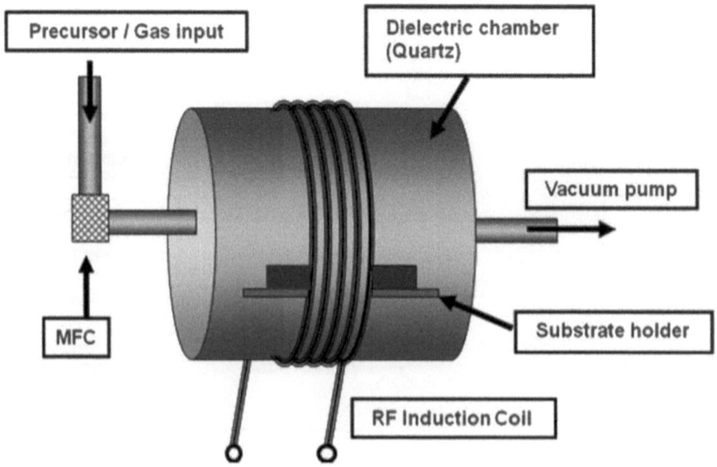

FIGURE 12.3

Inductively coupled radio frequency discharge.

substrates is required. These sources are useful in semiconductor industry and also where heat-sensitive materials are to be treated/coated.

3.3 MICROWAVE DISCHARGES

A typical MW discharge system is shown in Fig. 12.4. It consists of MW components like magnetron that generates MWs, suitable wave guides, and other components like a circulator and water-cooled dummy load that allows forward transmission of the MW power (the reflected power is dumped into a water-cooled dummy load), a cross-coupler that extract signal of the forward and reflected power and is given to the indicating meters after rectification, a stub tuner for impedance matching purpose, a cavity that provides a concentrated MW field for plasma generation, and a sliding short circuit that is used for impedance matching.

Plasma generation using MWs is widely employed in many applications. The characteristic features of MWs are the wavelength that is comparable to the dimensions of the plasma generation system (2.45 GHz: $\lambda = 12.24$ cm), and the short period of the exciting MW field. This results in low amplitude of the oscillations of the electrons in the MW field. For an excitation frequency of $f = 2.45$ GHz and a typical amplitude of $E_0 = 500$ V/cm, it is 3.5×10^{-3} cm. The power absorption in the gas depends on the electron–neutral collision frequency, i.e., on the gas pressure and the gas composition. The absorption efficiency in a 2.45 GHz discharge is high for helium, in the region between 10 and 100 torr, whereas the maximum efficiency for argon is reached around 2 torr. MW discharges can be operated at higher pressures as well (even at atmospheric pressure). The corresponding cut-off density of the electrons at 2.45 GHz is approximately 7×10^{10} cm^{-3}. Waves of this frequency can penetrate into plasmas with higher densities only up to the thickness of the skin sheath, which equals a few centimeters, under these conditions. The MW power absorption inside the skin sheath transfers energy into the plasma via waves with a frequency below the cut-off frequency.

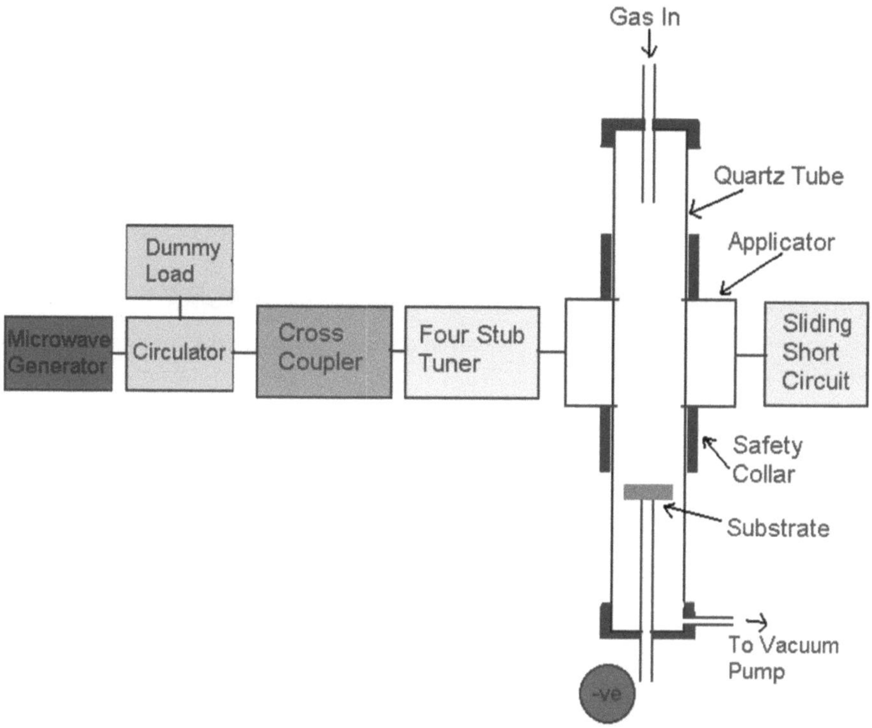

FIGURE 12.4

Schematic diagram of a microwave plasma system.

MW-generated plasmas operated in a low-pressure region have a large number of applications in a variety of fields, including chemical processing and semiconductor manufacturing. The applications are wide ranging and include surface deposition of all types, including diamond/diamondlike carbon (DLC) coatings, etching of semiconductors, promotion of organic reactions, and etching of polymers to improve bonding of the other materials. There is considerable interest in the MW-induced plasma processing of materials because of certain advantages over DC and low-frequency plasmas. Some of the advantages of the MW-excited plasmas are summarized as follows:

1. It produces a much higher degree of ionization and dissociation that commonly gives 10 times higher yield of active species than other types of electrically excited plasma.
2. It is possible to sustain MW plasma over a very wide pressure range, from 10^{-5} torr to several atmospheres. This makes it possible to have more MW plasma applications than with other forms of electrically stimulated plasmas.
3. The electron-to-gas temperature ratio (T_e/T_g) is very high, so the carrier gas and substrate remain moderately cool in the presence of high electron energy. Many reactions can take place in the process with relatively low ion energy. The effects of sputtering or decomposing by the incident ions of the treated surface can be decreased.

4. The absence of internal electrodes removes a source of contamination and makes reaction vessels simpler.
5. The state of the art of MW engineering at high power levels is relatively far advanced.

By using the appropriate MW applicators, MW discharges can be efficiently created and maintained from pressures above several atmospheres to sub-mtorr pressure, resulting in high-pressure discharges with high temperatures (>1000K) for thermal plasma processing applications and low-temperature, low-pressure discharges for applications like thin-film deposition and etching.

3.4 MICROWAVE ELECTRON CYCLOTRON RESONANCE PLASMA

A high-frequency field of amplitude E_0 is present in a volume element that oscillates in x-direction is given in Eq. (12.1).

$$E_x = E_0 e^{i\omega t} \tag{12.1}$$

This field predominantly interacts with light electrons and set the electrons in motion, and is governed by Eq. (12.2).

$$m_e d^2x/dt^2 + \nu\left(m_e dx/dt\right) = -eE_x = -eE_0 e^{i\omega t} \tag{12.2}$$

where ν is the collisional frequency.

The power absorption in the gas is given in Eq. (12.3).

$$P_{\text{abs}} = \left[n_e e^2 E_0^2 /2\, m_e \nu\right]\left[\nu^2/(\nu^2 + \omega^2)\right] \tag{12.3}$$

Eq. (12.4) indicates that the power absorption is reduced in comparison with the DC case by:

$$E_{\text{eff}} = E_0 \left[{}^{1}\!/\left(1 + \left(\omega/\nu\right)^2\right)\right]^{1/2} \tag{12.4}$$

Eq. (12.4) also indicates that there is no power absorption if the electrons experience no collisions and absorption is strongly reduced with the square of frequency if Eq. (12.5):

$$\omega^2 \gg \nu^2 \tag{12.5}$$

However, if a static magnetic field B is applied that is perpendicular to oscillating electric field as shown in Fig. 12.5, then Eq. (12.6):

$$P_{\text{abs}(R+L)} = \left[n_e e^2 E_0^2/4\, m_e \nu\right]\left[\nu^2/\left((\omega - \omega_{\text{ce}})^2 + \nu^2\right) + \nu^2/\left((\omega + \omega_{\text{ce}})^2 + \nu^2\right)\right] \tag{12.6}$$

where $\omega_{\text{ce}} = eB/m_e$, is the electron cyclotron frequency, and R and L are the right and left rotating E-field components.

This equation indicates that the right-hand rotating field will rotate in phase with the gyrating electrons if $\omega = \omega_{\text{ce}}$. This condition is called electron cyclotron resonance (ECR), and in this condition the gyrating electron will experience the DC acceleration from the applied field and continue gaining energy until it collides with another particle. This results in highly energetic electrons in the system.

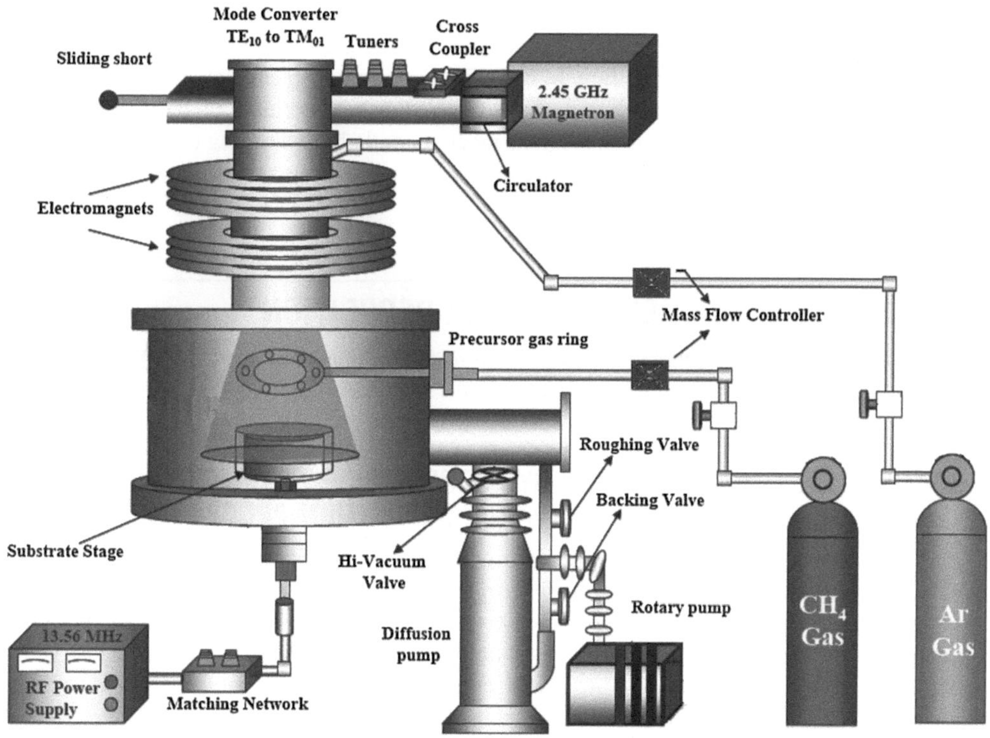

FIGURE 12.5

A typical microwave electron cyclotron resonance discharge system.

In order to have sufficient time for the electrons to gain energy, this phenomenon is effective in the operating pressure range of 10^{-3}–10^{-5} torr. Given a MW frequency of 2.45 GHz, for ECR, the applied magnetic field value is 875 G (Eqs. 12.7 and 12.8):

$$\text{at } \Delta\omega = \omega - \omega_{ce} \rightarrow 0 \tag{12.7}$$

$$E_{\mathrm{eff}(R)} = E_0 \left[1 \Big/ \left(1 + \left(\Delta\omega/\nu \right)^2 \right) \right]^{1/2} \tag{12.8}$$

At resonance (i.e., $\Delta\omega \rightarrow 0$), the frequency term completely drops out, and the absorption corresponds to static case $E_{\mathrm{eff}(R)} = E_0$.

Hence there is a strong improvement of P_{abs} by several orders of magnitude compared with the nonresonant case ($\Delta\omega/\omega = 1$, or no magnetic field) at low collision frequency.

Because of their unique features, MW ECR plasma sources are finding applications in various areas of materials processing. These low-pressure plasmas have a unique place in the development of surface modification technologies and are widely used for various industrial thin film deposition processes, including CVD and PVD. The MW ECR plasma-based CVD process has inherent advantages as compared with other rf and MW sources because of a high degree of ionization of the precursor gases, high density of reactive species, and low pressure operation. This results in high growth rates, efficient gas utilization, and uniformity in deposition. MW ECR sources give typical electron densities of about 10^{12}–10^{13} cm^{-3} and electron temperatures of 4–10 eV. Further, because of the fact that plasma generation is independent of the substrate, it can be independently biased during deposition. Hence it is possible to control the flux and energy of ions impinging on the substrate that is required for deposition of specialized coatings like DLC coatings.

4. PLASMA-ASSISTED CHEMICAL VAPOR DEPOSITION

Although a large variety of coatings can be deposited through the plasma-assisted CVD process, here we will give only few selected examples of depositions.

4.1 DIAMONDLIKE CARBON COATING BY MICROWAVE ELECTRON CYCLOTRON RESONANCE PLASMA CHEMICAL VAPOR DEPOSITION PROCESS

DLC is an amorphous phase consisting of a mixture of sp^3- and sp^2-bonded carbon structures, where sp^2-bonded graphitelike clusters are embedded in an amorphous sp^3-bonded carbon matrix. They can be classified as hydrogenated and hydrogen-free DLCs. In cases where hydrogen atoms are present in the random network as bonded to carbon, they stabilize the sp^3-hybridized carbon bonds, which is responsible for rigidity of the network. They also decrease the connectivity between the sp^3-hybridized carbon–carbon bonds, which influences the mechanical properties of the DLC film. The properties of DLC films depend on the ratio of the carbon bonding types (sp^3/sp^2), bond structure, and hydrogen content. Both hydrogenated and hydrogen-free DLCs can be directly deposited on various substrates such as Ti, W, Si, glass, ceramics, and polymers.

DLC films have attracted a lot of attention because of their unique properties such as high hardness, good wear resistance, excellent chemical inertness, high electrical resistivity, and transparency in the infrared (IR) region. Deposition of these films requires condensation of carbon-containing ions with energy of about 50–200 eV. This can be easily achieved by deposition from any hydrocarbon glow discharges onto the biased substrates. A variety of low-pressure CVD and PVD techniques such as MW and rf-discharge PECVD, pulsed laser deposition (PLD), ion beam-assisted deposition (IBAD), and sputtering have been successfully used for deposition of these coatings [5–7].

It is possible to uniformly deposit these coatings at a low working pressure and low substrate temperature. In comparison with the other processes, an MW ECR plasma-assisted CVD process has a definite advantage because this process offers the possibility of deposition of films on a large area with independent control of the energy of the ions arriving at the substrates. Because of energetic species in the ECR plasma, higher efficiency is achieved by breaking of C–H bonds in the reactant gas mixture. A further advantage of the ECR–CVD process is that it produces ions with a very narrow ion energy distribution (few eV) [8] compared with that of rf-excited plasma, implying that the film properties can be controlled better.

From a large number of literature reports, it is evident that the properties of DLC films are strongly dependent on the deposition process parameters and deposition methods. Substrate bias voltage that controls the energy of the ions impinging on the substrate is known to be an important factor in controlling the properties of the deposited films. The energy of the ions impinging on the substrate can be controlled by biasing the substrate either using a DC voltage or with a DC self-bias voltage (developed at the substrate by application of rf power) and is directly proportional to the bias voltage.

Here we will discuss the deposition and characterization of DLC films on Si (111) substrates that was done in an MW ECR plasma CVD facility set up in our laboratory. These films were deposited under varied influence of rf self-bias (−25 to −200 V) on the substrates during deposition. A schematic of this experimental system is shown in Fig. 12.5. It consists of an MW generator (magnetron) capable of delivering MW power up to 2 kW at 2.45 GHz. The other components in the MW transmission line are a three-port circulator with a water-cooled dummy load (to protect the magnetron from the reflected power), a cross-coupler (that extracts signals for the measurement of forward and reflected power), a three-stub tuner (for impedance matching purposes), a mode converter for TE_{10} to TM_{01}, and a sliding short circuit. The ECR plasma-generation chamber is surrounded by the two water-cooled electromagnets that are powered independently by DC power supplies. The MW hardware is separated from the ECR plasma generation chamber by a quartz window. The current in the coils was set at 130 A (top coil) and 170 A, (bottom coil), respectively. MWs are launched into the plasma-generation chamber axially from the higher magnetic field region, where ω_{ce} (gyration frequency) > ω (MW frequency), before being absorbed at the resonance layer (where $\omega_{ce} \sim \omega$) to get a dense and stable plasma. The divergent magnetic field near the exit of the plasma-generation chamber accelerates the plasma toward the processing chamber where the substrate is located. The plasma processing chamber is evacuated to a base pressure of 2×10^{-6} mbar by a vacuum system consisting of an oil diffusion pump backed by a rotary pump. The substrate is kept on a water-cooled rf electrode that is coupled to the RF generator (13.56 MHz) through an impedance matching unit. The plasma density at the substrate location was approximately $10^{16}/m^3$ and was found to be fairly uniform in the center of the processing chamber, an area of about 15 cm in diameter. Before deposition, the Si substrates were ultrasonically cleaned in isopropyl alcohol. Plasma of argon gas (11.2 SCCM) was generated at an MW power of 360 W and the substrates were sputter cleaned for 10–15 min in this plasma by application of an rf self-bias of about −100 V. This was essential to remove the impurities and oxide from the substrate surfaces. Methane (CH_4, 14.1 SCCM) as a precursor gas was then fed into this chamber through a gas shower ring situated 5 cm above the substrate. The final operating pressure was approximately 4.5×10^{-4} mbar. The DC self-bias to the substrates was varied from −25 to −200 V. Because of low operating pressure the mean free path of ions was much larger compared with the distance between the gas shower ring and the substrate, and therefore the ions reached the substrate without collisions with energy $E_i = e(V_p - V_B)$, which was directly proportional to the substrate bias, where V_p was the plasma potential and V_B was the rf self-bias on the substrates. Depositions were carried out to get DLC films of 0.12–0.69 μm thickness, depending on the substrate bias level during deposition.

The deposited films were characterized by Raman spectroscopy using an excitation wavelength of 532 cm^{-1} and Bomen Fourier transform IR spectroscopy (FTIR) with a resolution of approximately 4 cm^{-1}. Mechanical properties such as hardness and Young's modulus were determined using a nanoindenter (CSM Instruments, Switzerland) with a Berkovich diamond indenter. Scratch resistance of these films was determined using a nanoscratch tester. Complete results of these investigations are reported elsewhere [9]; however, portions of the results are discussed in the following paragraphs.

The DLC films deposited as per the procedure mentioned were smooth, uniform, scratch-free, and adherent to the substrate even at the lowest value of self-bias, i.e., −25 V.

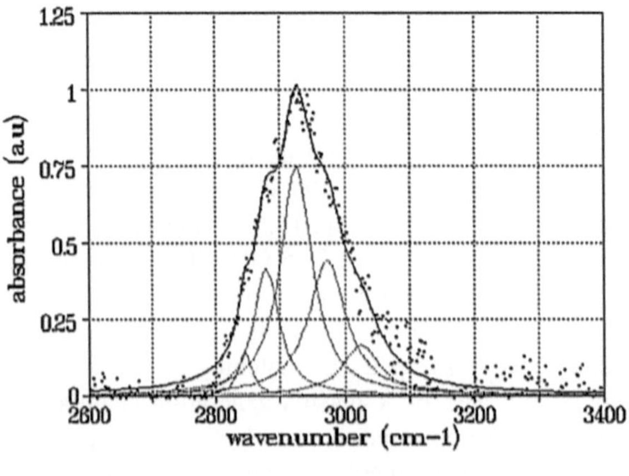

FIGURE 12.6

A typical infrared absorbance spectrum in the range of 2600–3400 cm^{-1} for a diamondlike carbon film deposited at a −150 V bias together with its component peaks. *a.u.*, Arbitrary units.

From S. Singh, M. Pandey, R. Kishore, N. Chand, S. Dash, A. Tyagi, D. Patil, Influence of impinging ion energy on the bonding and mechanical characteristics of diamond like carbon films deposited by microwave ECR plasma CVD, Plasma Process. Polym. 5 (2008) 853–860, reprinted with permission.

IR spectroscopy provided information regarding the bonding characteristics of the DLC films in terms of the different C–H stretching bands. IR absorption spectrums of the coatings recorded in the region of 400–4000 cm^{-1} were used for this purpose. A typical spectra taken on a film deposited at −150 V bias is shown in Fig. 12.6. The details regarding analysis on the FTIR spectra are given elsewhere [9]. From the deconvoluted spectra and the normalized integrated intensities corresponding to various C–H stretching vibrations, it was evident that most of the hydrogen atoms in the DLC films were bonded to the sp^3-hybridized carbon atoms. The average carbon coordination number was determined from these C–H absorption bands by estimating the fraction of sp^3 and sp^2 carbon networks in DLC films deposited at various bias voltages [10]. The carbon coordination number was found to increase with increasing the energy of impinging ions (increasing the substrate bias level) on substrates, reaching a maximum at a −150 V bias before it decreased again.

Raman spectroscopy is widely used to characterize the carbon-based films because of its ability to distinguish different bonding types. Bonding characteristics of the deposited films are investigated using a CCD-based Raman spectrometer (at an excitation wavelength of 532 nm). Fig. 12.7 shows the substrate bias dependence of Raman spectra taken on various DLC films. These spectra are best fitted with two Gaussian functions and a constant background. From the fitting parameters, the peak position, full width at half maximum (FWHM), and the integral intensity ratio (I_D/I_G) of the D and G peaks are obtained. The results are given in Table 12.1.

From this analysis it is seen that in addition to the shifts in the D and G peak positions, at a bias voltage of −150 V, the FWHM of the G peak is highest, and the I_D/I_G is lowest. This decrease in I_D/I_G ratio is related to sp^3 content in the DLC films [11] and the increase in G peak width is related to the hardness and density of the films [12]. A correlation of I_D/I_G ratio and hardness as a function of self-bias (V_{bias}) is shown in Fig. 12.8, which shows that film deposited at a −150 V bias has the highest

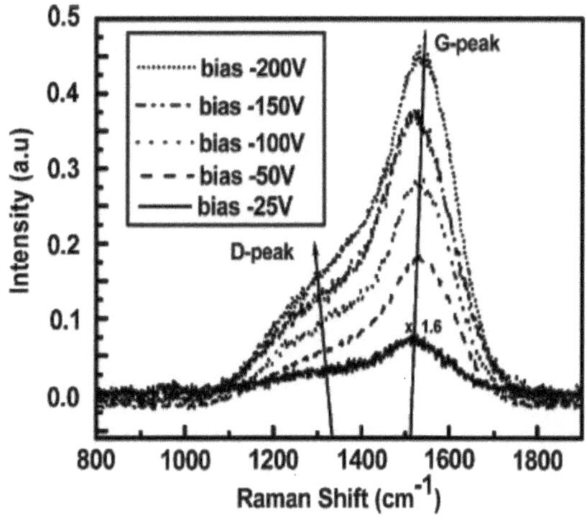

FIGURE 12.7

Raman spectra of diamondlike carbon films deposited at different bias levels. *a.u.*, Arbitrary units.

From S. Singh, M. Pandey, R. Kishore, N. Chand, S. Dash, A. Tyagi, D. Patil, Influence of impinging ion energy on the bonding and mechanical characteristics of diamond like carbon films deposited by microwave ECR plasma CVD, Plasma Process. Polym. 5 (2008) 853–860, reprinted with permission.

Table 12.1 Peak Fitting Details of Raman Spectra

	D Peak		G Peak		
Bias Voltage	Peak Position	FWHM	Peak Position	FWHM	Integrated Area Ratio (I_D/I_G)
V	cm^{-1}	cm^{-1}	cm^{-1}	cm^{-1}	
−200	1358.00	262.78	1546.93	145.77	0.862
−150	1311.73	216.55	1529.00	160.98	0.480
−100	1338.45	230.83	1541.11	146.01	0.657
−50	1344.69	254.00	1549.16	148.54	0.753
−25	1334.65	270.35	1528.93	150.92	0.868

FWHM, *full width at half maximum.*
From S. Singh, M. Pandey, R. Kishore, N. Chand, S. Dash, A. Tyagi, D. Patil, Influence of impinging ion energy on the bonding and mechanical characteristics of diamond like carbon films deposited by microwave ECR plasma CVD, Plasma Process. Polym. 5 (2008) 853–860, reprinted with permission.

hardness and lowest I_D/I_G ratio and therefore should have highest sp^3 content. The load variation of hardness of the films deposited at different bias voltages is shown in Fig. 12.9.

Hardness of the film deposited at −100 V bias is shows a decrease at a higher load (10 mN). This is because the thickness of the film is low for this particular film. Our efforts to deposit films with higher thickness failed; coatings with higher thickness peeled off the substrate, possibly because of higher stress in the film.

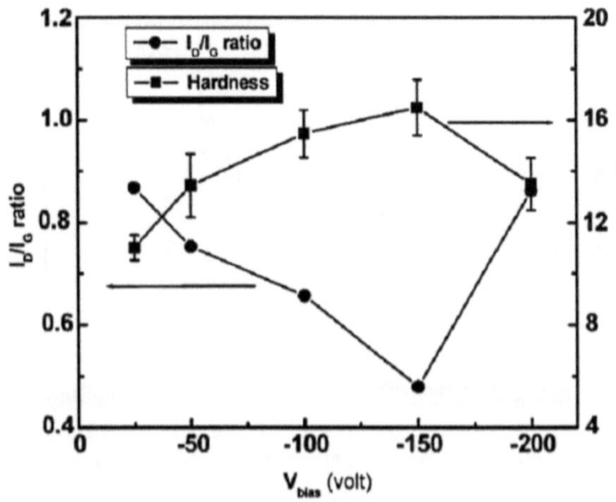

FIGURE 12.8

Variation of integral intensity ratio (I_D/I_G) ratio and hardness of diamondlike carbon films deposited at different bias levels.

From S. Singh, M. Pandey, R. Kishore, N. Chand, S. Dash, A. Tyagi, D. Patil, Influence of impinging ion energy on the bonding and mechanical characteristics of diamond like carbon films deposited by microwave ECR plasma CVD, Plasma Process. Polym. 5 (2008) 853–860, reprinted with permission.

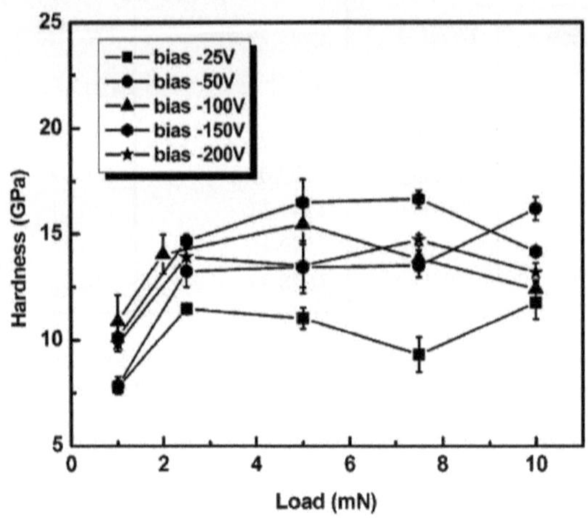

FIGURE 12.9

Variation in the hardness of the deposited coatings with the load.

From S. Singh, M. Pandey, R. Kishore, N. Chand, S. Dash, A. Tyagi, D. Patil, Influence of impinging ion energy on the bonding and mechanical characteristics of diamond like carbon films deposited by microwave ECR plasma CVD, Plasma Process. Polym. 5 (2008) 853–860, reprinted with permission.

From Fig. 12.9, it is very clear that the energy of the ions arriving at the substrate plays a major role with regard to the mechanical properties of these films. The hardness first increases with increasing the bias, reaches maximum, and then decreases. A maximum hardness of approximately16.5 GPa and elastic modulus of about 156 GPa is observed for film deposited at a −150 V bias. This particular film also shows maximum width of the G peak in the Raman spectra, which is in conformity with the observations of Tamor et al. [12]. Mechanical properties of the deposited films are therefore strongly correlated with the observed Raman spectra, indicating its dependence on the bonding characteristics.

Nanoscratch tests on the deposited films were carried out to estimate the critical load (L_c) required for delaminating the deposited DLC films. All nanoscratch tests were performed under progressive loading conditions. A typical nanoscratch curve for the film deposited at a −150 V bias is shown in Fig. 12.10, where the normal force and frictional force are plotted together as a function of position along the scratching line. As the load is increased, the penetration depth increases smoothly until reaching L_C. At this L_C, there is an abrupt increase in the coefficient of friction. This is the point where the film starts to delaminate from the surface. The observed variation of L_c and coefficient of friction with the substrate bias voltage is given in Fig. 12.11. All films show a low coefficient of friction (between ~0.01 and 0.18). It is observed that the load required to delaminate the film deposited at $V_{dc} = -100$ V is lower than that for any other film deposited at other bias voltages. This is possibly because of the higher value of the residual stress in the film. This is also a reason as to why we experienced difficulties in achieving films of higher thickness at this value of bias voltage.

Fig. 12.12 shows the images of the scratched surfaces taken with an optical microscope for films deposited at −25 to −200 V substrate bias. The coating failure that is indicated by initiation of delamination process at the scratched surface for various DLC films is clearly seen from this figure.

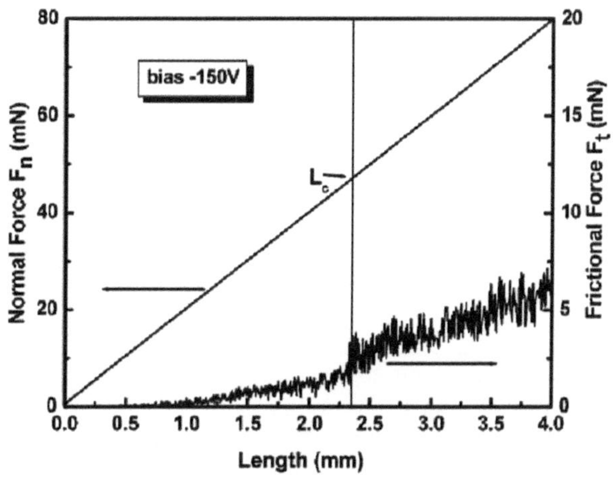

FIGURE 12.10

A typical nanoscratch curve for DLC film deposited at −150 V bias. F_n, normal force; F_t, frictional force; L_c, critical load.

From S. Singh, M. Pandey, R. Kishore, N. Chand, S. Dash, A. Tyagi, D. Patil, Influence of impinging ion energy on the bonding and mechanical characteristics of diamond like carbon films deposited by microwave ECR plasma CVD, Plasma Process. Polym. 5 (2008) 853–860, reprinted with permission.

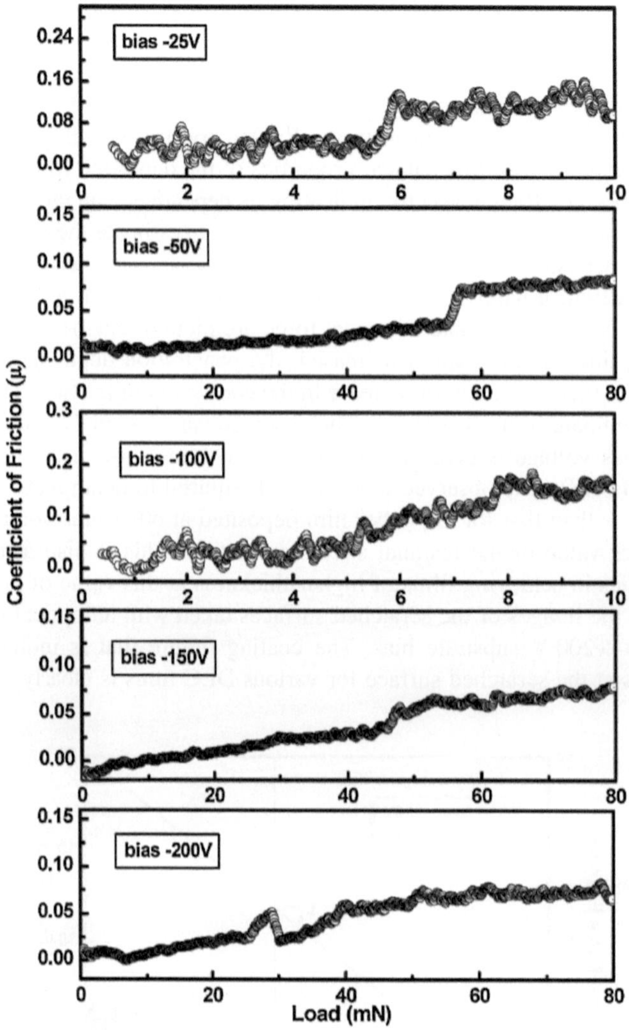

FIGURE 12.11

Variation in the coefficient of friction (μ) versus applied load for diamondlike carbon films deposited at different bias levels between −25 and −200 V.

From S. Singh, M. Pandey, R. Kishore, N. Chand, S. Dash, A. Tyagi, D. Patil, Influence of impinging ion energy on the bonding and mechanical characteristics of diamond like carbon films deposited by microwave ECR plasma CVD, Plasma Process. Polym. 5 (2008) 853–860, reprinted with permission.

Films deposited at low bias, i.e., −25 and −50 V, show a smooth scratch along the length of the surface, whereas film deposited at higher bias voltages shows the spalling of the DLC layers. For films deposited at −25 and −50 V, it is possible that the films have more hydrogen content that lowers the stress in the coating. Compressive stresses are known to make the material harder, but this also results

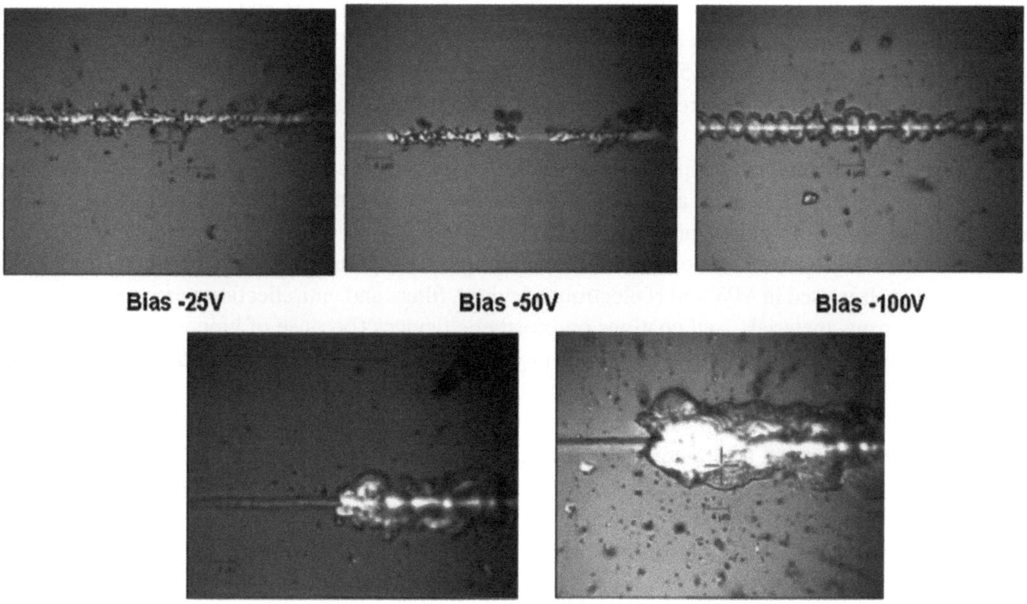

FIGURE 12.12

Optical image of scratched surfaces of diamondlike carbon films deposited at different bias voltages.

From S. Singh, M. Pandey, R. Kishore, N. Chand, S. Dash, A. Tyagi, D. Patil, Influence of impinging ion energy on the bonding and mechanical characteristics of diamond like carbon films deposited by microwave ECR plasma CVD, Plasma Process. Polym. 5 (2008) 853–860, reprinted with permission.

in film spalling and consequently there is associated reduction in the resistance to delamination and increase in the coefficient of friction [13]. From these figures it is clearly evident that L_c is strongly dependent on the energy of the ions that are arriving at the substrates. A more detailed discussion regarding the observed scratch behavior with substrate bias variations are discussed elsewhere [9].

These investigations clearly indicate that the energy of the impinging ions on the substrates play a very crucial role in tuning the mechanical properties and bonding characteristics of DLC films and also indicates that there is an optimum for the energy of impinging ions on the substrate that gives desirable properties.

4.2 MICROWAVE PLASMA-ENHANCED CHEMICAL VAPOR DEPOSITION FOR DEPOSITION OF DIAMOND AND CARBON NANOSTRUCTURES

Carbon is one of the most abundant elements in nature. For ages, people have been using different forms of carbon for various purposes. Diamond, for example, has traditionally been used as a gem, but apart from that, it has many other interesting properties. It is the hardest known naturally occurring material; it is optically transparent from extreme UV light through the visible spectrum to near-IR and for wavelengths larger than 6000 nm (mid-IR and beyond) and it has a refractive index of 2.417. Its thermal conductivity in type-IIa form (20 W/cm K) is the highest of all materials at room temperature, about five times larger than the value for Cu. Electrically, it is an insulator, and the band gap associated with the indirect transition between the valence and conduction bands is as large as 5.47 eV. Its dielectric constant is small ($n = 5.58$). Diamond is chemically inert and resistant to high temperatures, particularly in an oxygenless environment. It is also resistant to X-ray radiation, γ radiation, UV light, and nuclear particles.

Because of these excellent properties it is used as windows in harsh environments such as in free electron lasers, where diamond's ability to dissipate heat rapidly and resist thermal fracture is also important. Free-standing windows are being used in high-power MW devices like gyrotrons. It is also used as heat sinks for high-power electronic devices (a complete integrated circuit can be cooled by a thin diamond layer). It is used in MW and rf electronic devices, filters and antireflection coatings, coatings on tools for nonferrous materials, and coatings on prosthetic devices (because of biocompatibility). It also exhibits excellent corrosion resistance. There is also an interest in developing diamond-based detectors because of their many outstanding properties, such as high energy resolution, high carrier mobility, and high resistivity, that make diamond a very interesting detector for radiation. It is also used as a radiation dosimeter. Indeed, diamond is attractive for the neutron dosimeter because its atomic number ($Z = 6$) is comparable with that of tissues. The low Z-value also ensures low sensitivity to γ rays and this can help in case of mixed neutron–γ radiation fields.

The growth of diamond films from the vapor phase on nondiamond substrates at practical rates is shown by a large number of groups using thermal and plasma-enhanced CVD methods, in which a hydrocarbon gas (usually methane/acetylene) is used as carbon precursor, is mixed with hydrogen, and is energized thermally or in a plasma before contact with a heated substrate. A variety of diamond deposition processes are used by various researchers, including filament-assisted thermal CVD, electron-assisted thermal CVD, laser-assisted thermal CVD, rf plasma CVD, MW plasma CVD, combustion flame–assisted CVD, and DC arc plasma jet CVD.

The various processes appear different in their process details but have a number of common features:

1. A substrate on which diamond can nucleate and grow should be used.
2. A substrate temperature of approximately 700–1000°C is required for growth of diamonds.
3. Atomic hydrogen presence is required at the growth surface.
4. A suitable hydrocarbon species is required at the growth surface.

The deposition technique involves the deposition of carbon atoms that is originated from the dissociation of a carbon-containing gas precursor on a solid substrate. Successful deposition is reported on substrates that are based on pure diamonds as well as nondiamonds. However, deposition of diamond on a nondiamond substrate requires an extra nucleation step in terms of scratching the substrate with a fine-diamond paste and subsequent ultrasonic cleaning.

This scratching provides the nondiamond substrate with the necessary diamond seeds that grow three-dimensionally until the grains coalesce, forming a continuous polycrystalline film with the average grain size increasing with increasing time. The resulting films are nanocrystalline, polycrystalline or ultrananocrystalline, depending on the average grain size, and their macroscopic properties depend on parameters such as grain size as well as the sp^3/sp^2 ratio of carbon bonding in the deposited film.

Although there is as uncertainty regarding the exact mechanisms for the metastable growth of diamond, it is certain that atomic hydrogen plays a crucial role in the process. It has been proposed that atomic hydrogen preferentially attacks graphite if it forms, thermodynamically stabilizes diamond relative to graphite, reacts with other forms of carbon to form diamond nuclei, and plays a vital role in the growth mechanism. It is generally accepted that almost every carbon atom on the growing diamond surface is bonded to a hydrogen atom, yielding a configuration of carbon atoms closely resembling the interior of the crystal lattice.

Thus growth of diamond from carbon-containing molecules diluted in hydrogen involves two processes. The first is carbon deposition, primarily in the form of graphite with a small amount of diamond, and the second is selective etching of graphite by atomic hydrogen. As a result, the various CVD methods for the growth of diamond are optimized to produce atomic hydrogen from molecular hydrogen close to the surface of the growing film. Excellent reviews on diamond thin-film deposition and characterization are available in the literature [14−16].

Apart from diamond there has been a continuous surge in nanotechnology, and it has lead to the popularity of different carbon nanostructures owing to exciting possibilities for their applications. Carbon nanotube (CNT) and graphene nanowall are two such nanostructures. Since the discovery of CNTs in 1991, they are being projected for different applications in diverse fields ranging from electronics to medicine [17−19].

MW PECVD (MPECVD) is one of the most widely used techniques for deposition of carbon-based nanostructures. This is because of the fact that MW plasmas can be generated in a wide pressure range and can be operated for a longer time without many difficulties. Here we are going to discuss the application of MW plasma that is operated at a reasonably high pressure: a few tens of torr. MW plasma operated at this pressure range is really not cold plasma but can be called low-temperature thermal plasma. This plasma can be used for deposition of various carbon nanostructures like CNTs, carbon nanowalls (CNWs), and diamonds (nanocrystalline and bulk). Because of high-pressure operation and efficient power utilization, this plasma gives an abundance of radicals required to grow the carbon nanostructures at a high growth rate. Here we will discuss MPECVD of diamonds and CNWs that are deposited using a MPECVD facility set up in our laboratory. A schematic of the MPECVD system that was used in the growth of CNTs, diamonds, and nanocrystalline diamonds is shown in Fig. 12.13.

The system uses a 2.45-GHz, 2-KW, water-cooled magnetron device for generation of MWs. It also has a circulator with a water-cooled dummy load to take care of the reflected power. MWs generated by the magnetron are launched into a rectangular wave guide (WR340) in TE_{10} mode. In the circuit there are other MW components like three stub tuners that are used for impedance matching purpose. There is also a mode converter in the circuit that converts the MW mode from TE_{10} to TM_{01}. MWs then enter the resonant cavity (an s.s. cylindrical chamber) through a quartz isolation window. The deposition chamber (resonant cavity) has a substrate holder that can be independently heated to 1000°C and also acts as an MW reflector. In order to maintain a vacuum, the chamber is pumped using a rotary vacuum pumping system.

During deposition, maximum power transfer to the plasma load in the process chamber is achieved with the help of three stub tuners and a sliding short circuit. Required gases are fed into the chamber through mass-flow controllers and the deposition chamber is kept at the required pressure by throttling the pumping speed. For the kind of deposition to be discussed here, base vacuum was always kept at close to 0.01 Torr and the operating pressure was chosen at 40−100 Torr, depending on the type of deposition to be done.

Deposition of diamond thin films was carried out on silicon substrates using an MW plasma of gases including hydrogen (H_2, 90 SCCM), argon (Ar, 8 SCCM), and acetylene (C_2H_2, 2 SCCM) at an

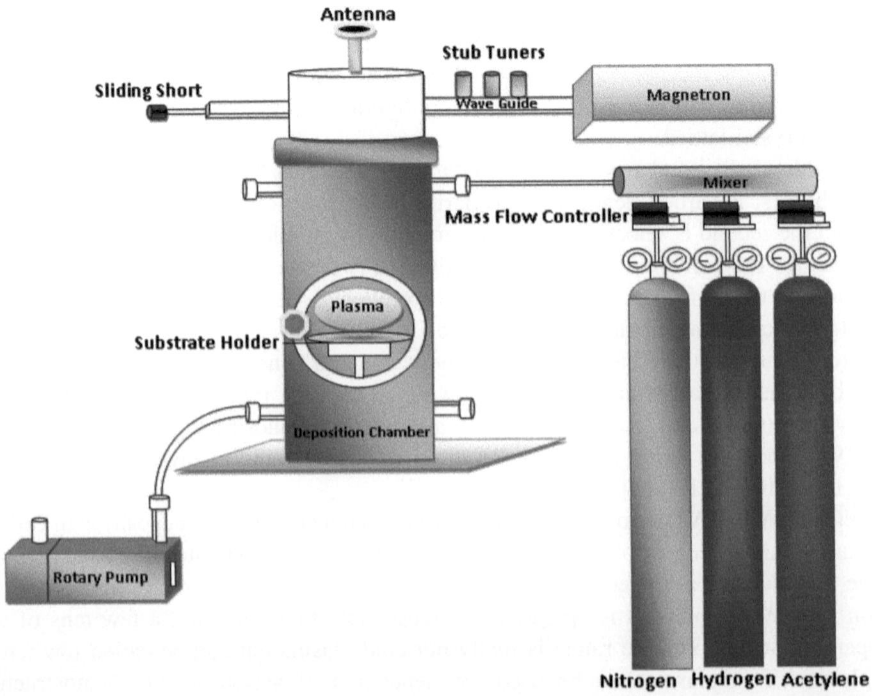

FIGURE 12.13

Schematic diagram of the microwave plasma-enhanced chemical vapor deposition set-up used for the deposition of diamonds and carbon nanotubes.

operating pressure of 73 torr with an incident MW power of 1200 W. The substrate temperature was approximately 950°C and deposition was carried out for 5 h. The Raman spectra taken on a deposited film is shown in Fig. 12.14 and shows deposition of crystalline diamonds on the substrate. The scanning electron microscope (SEM) morphology of the deposited film is shown in Fig. 12.15, and the X-ray diffraction shown in Fig. 12.16 also clearly shows the presence of crystalline diamonds.

We have also grown CNTs directly on Inconel substrates without application of any external catalyst for practical field emission applications like cold cathodes for electron-beam devices. Direct contact of the grown CNTs with the substrate is most important. Direct growth of CNTs on metal surfaces normally involves in situ formation of catalyst nanoparticles on the metallic substrates [20,21], which is normally found to be difficult [22–28]. The experimental details and characterization of the grown CNTs is discussed elsewhere [20].

In this case, for achieving successful and dense growth of CNTs on Inconel substrates, we have attenuated the incident MW power that is on the metal surface. In this scheme, an Si wafer is placed over the Inconel substrate so that MWs are not reflected directly from the metal surface. Reflection of MWs hinders the growth of CNTs on metal surfaces in MPECVD experiments. The advantage of using MW attenuation is discussed elsewhere [29]. It is seen that by introduction of a 0.5 mm thick Si attenuator only about 1.85% of the MW power gets transmitted to the substrate, and it helps stop MWs

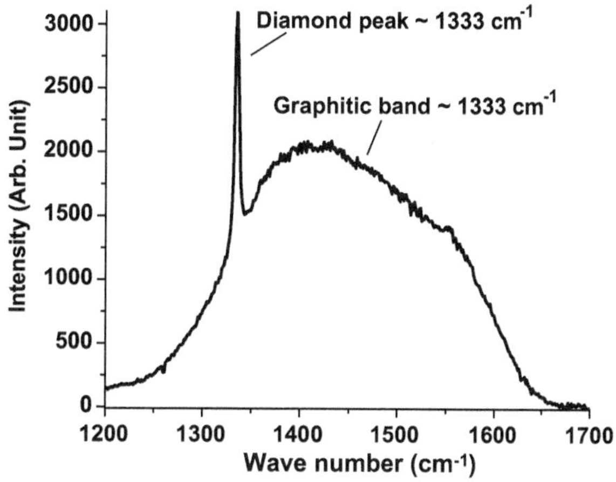

FIGURE 12.14

Raman spectra on diamond thin film. *a.u.*, Arbitrary unit.

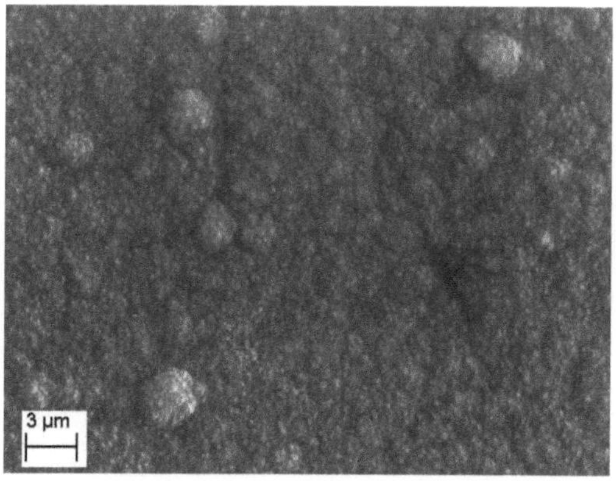

FIGURE 12.15

Scanning electron microscope picture taken on deposited diamond thin film.

from reaching the substrate to the extent that reflection of MWs from the metal surface becomes negligible. Fig. 12.17a shows a photograph of the sample that was loaded in the deposition chamber for an experiment in growing CNTs. Fig. 12.17b shows the actual arrangement inside the PECVD reactor during an experiment. From the photograph it is seen that Si attenuators have become red hot during the experiment, and Inconel substrates are seen beneath them. H_2—C_2H_2 plasma used for generation of CNTs is seen over them.

Before the experiment, Inconel substrates were heated in air to 1100°C. Then they were ultrasonically cleaned with methanol and loaded into the MPECVD chamber. Initially, 100 SCCM H_2 gas was fed into the chamber and plasma was generated for pretreatment of the loaded substrates. The idea behind this was to reduce the oxide on the top surface of the Inconel substrate and to prepare fresh transition metal nanoparticles on the substrate surface itself. After 15 min of this pretreatment

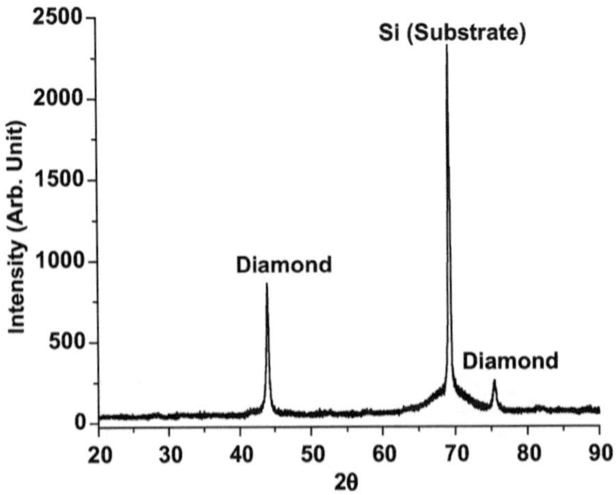

FIGURE 12.16

X-ray diffraction pattern taken on deposited diamond thin film. *a.u.*, Arbitrary unit.

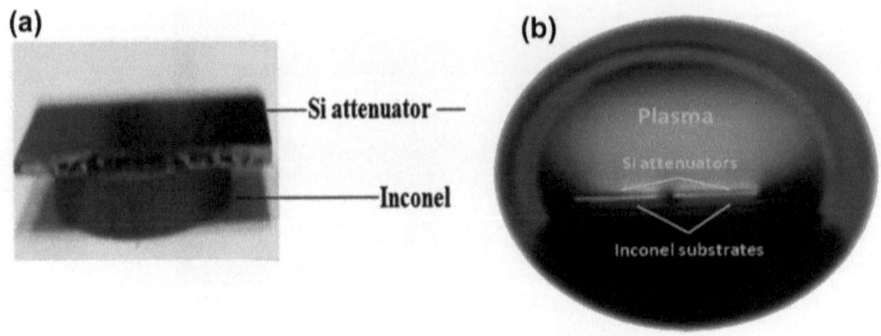

FIGURE 12.17

(a) Photograph of Inconel substrate with Si wafer on top. (b) Photograph showing actual inside of plasma-enhanced chemical vapor deposition chamber during carbon nanotube deposition.

From R. Kar, S.G. Sarkar, C.B. Basak, A. Patsha, S. Dhara, C. Ghosh, D. Ramachandran, N. Chand, S.S. Chopade, D.S. Patil, Effect of substrate heating and microwave attenuation on the catalyst free growth and field emission of carbon nanotubes, Carbon 94 (2015) 256–265, reprinted with permission.

reduction step, 0.5 mm thick Si wafer was kept on the substrate surface. Finally plasma of 100 SCCM hydrogen and 20 SCCM acetylene was used for the growth of CNTs on the substrate. Substrate temperature and deposition time was kept fixed at approximately 900°C and 5 min, respectively.

Fig. 12.18a shows the field emission SEM (FESEM) images of CNTs after deposition. CNTs observed here are long and most of them are well above 10 μm in length. The growth rate in this case is very high, and CNTs grow horizontally without any preferred orientation. The FESEM image also reveals that there is a complete absence of unwanted carbon soot on CNTs. Fig. 12.18b and c show the high-resolution transmission electron microscopy analysis of the sample. It again shows that CNTs

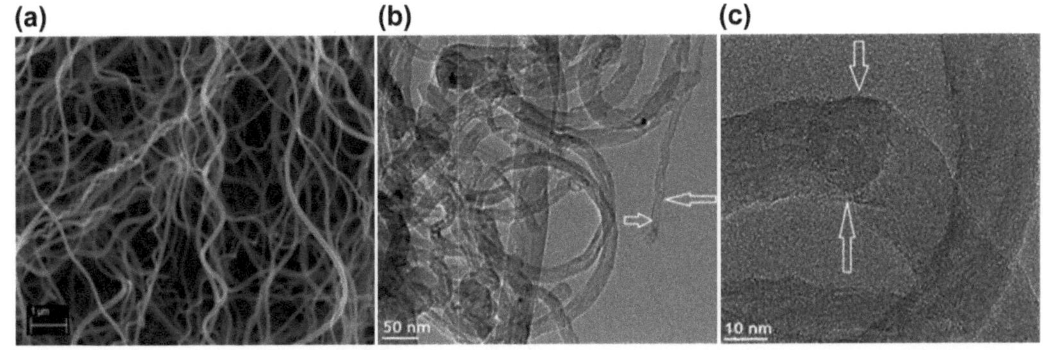

FIGURE 12.18

(a) Field emission scanning electron microscope image of carbon nanotube (CNTs) grown by microwave plasma-enhanced chemical vapor deposition. (b) High-resolution transmission electron microscopy (HRTEM) image showing a large number of CNTs, with bamboolike defects (*arrows*). (c) HRTEM image showing details of bamboolike structure in CNT.

From R. Kar, S.G. Sarkar, C.B. Basak, A. Patsha, S. Dhara, C. Ghosh, D. Ramachandran, N. Chand, S.S. Chopade, D.S. Patil, Effect of substrate heating and microwave attenuation on the catalyst free growth and field emission of carbon nanotubes, Carbon 94 (2015) 256–265, reprinted with permission.

grown here are almost free from unwanted amorphous carbon deposition, and the images also show the presence of favorable bamboolike defects (marked by *arrows*) in the CNTs, which help in the field emission by lowering the turn-on voltage [30–32].

Deposited CNTs were also tested for their field emission characteristics in an indigenously developed set-up shown in Fig. 12.19. It consists of a vacuum chamber, a fixed cathode holder, and a movable anode assembly. The CNT grown substrates were kept on the cathode holder, and the distance between the emitting surface and the anode could be adjusted using a micrometer-controlled translational stage. The measurement chamber was initially evacuated to 1×10^{-6} mbar and all the measurements were done under this condition. A DC voltage power supply was used to bias the cathode and field emission current was measured with microampere accuracy.

Fig. 12.20 shows field emission measurement results on the CNTs deposited samples by MPECVD. It is known that emission follows the Fowler-Nordheim equation as given in Eq. (12.9) [33–35].

$$J = \left(A\gamma^2 E^2/\varphi\right)\exp\left[\left(-B\,\varphi^{3/2}\right)/(\gamma E)\right] \tag{12.9}$$

In Eq. (12.9), J is the emitted current density (A/m^2), φ (eV) is the work function, E is the local electric field, and the constants A and B are 1.541×10^{-6} A eV V^{-2} and 6.831×10^9 eV$^{-3/2}$ Vm^{-1}, respectively. The value of work function, φ, is taken as 4.7 eV [35], γ is called the field enhancement factor (FEF), which decides the quality of the CNTs that are responsible for field emission. From Eq. (12.9) follows Eq. (12.10):

$$\ln J = \ln\left[A\left(\gamma^2 E^2/\varphi\right)\right] - \left(B\,\varphi^{3/2}/\gamma E\right) \tag{12.10}$$

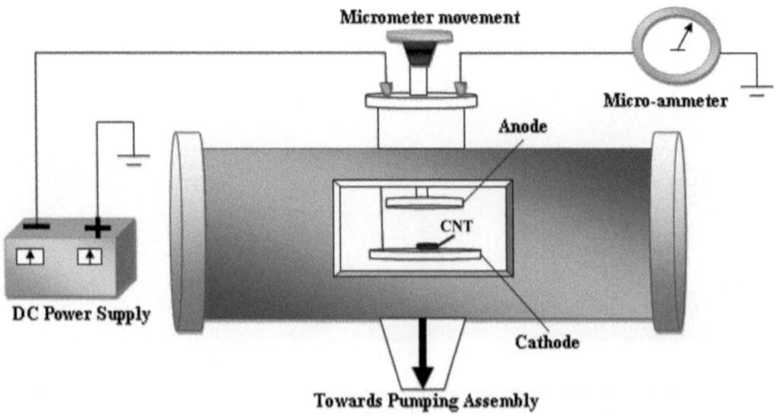

FIGURE 12.19

Schematic diagram of a field emission set-up used in an experiment.

From R. Kar, S.G. Sarkar, C.B. Basak, A. Patsha, S. Dhara, C. Ghosh, D. Ramachandran, N. Chand, S.S. Chopade, D.S. Patil, Effect of substrate heating and microwave attenuation on the catalyst free growth and field emission of carbon nanotubes, Carbon 94 (2015) 256–265, reprinted with permission.

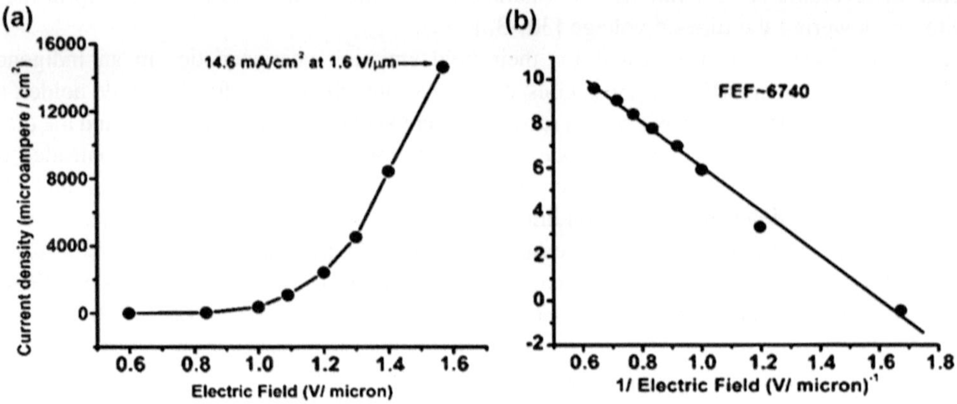

FIGURE 12.20

(a) Current density (J) versus electric field (E) plot for carbon nanotube (CNT) sample deposited by microwave plasma-enhanced chemical vapor deposition (MPECVD), (b) ln J versus 1/E plot CNTs deposited by MPECVD. *FEF*, field enhancement factor.

From R. Kar, S.G. Sarkar, C.B. Basak, A. Patsha, S. Dhara, C. Ghosh, D. Ramachandran, N. Chand, S.S. Chopade, D.S. Patil, Effect of substrate heating and microwave attenuation on the catalyst free growth and field emission of carbon nanotubes, Carbon 94 (2015) 256–265, reprinted with permission.

From Eq. (12.10) it is seen that, a plot of ln J versus $1/E$ should be linear, and FEF can be calculated from the slope of the graph. Fig. 12.20 shows current density versus the electric field plot and it is seen that maximum current density of 14.6 mA/cm^2 is observed at a low electric field of 1.6 V/μm.

In conclusion, it is shown how the MPECVD process can be used for deposition of different carbon nanostructures for use in extreme conditions.

4.3 RADIO FREQUENCY PLASMA-ASSISTED METAL ORGANIC CHEMICAL VAPOR DEPOSITION OF YTTRIUM-OXIDE THIN FILMS

Yttrium oxide (Y_2O_3), commonly known as yttria, is a technologically important refractory ceramic material useful for a large number of applications. It has a high melting point (2410°C), shows chemical inertness, and hence is used as corrosion-resistant coating. Y_2O_3 possesses superior resistance to aggressive chemical attack by molten metals, salts, slag, and glass at high temperatures. It is therefore used for coating crucibles and molds that handle highly reactive molten metals like uranium, titanium, chromium, beryllium, and their alloys. A thin coating of Y_2O_3 is shown to prevent corrosion of the substrate material by molten uranium and other reactive metals. It also has compatibility with graphite up to 1600°C.

In the age of miniaturization, scaling down of microelectronic devices led to a search for materials with a high dielectric constant. Y_2O_3 has been emerging as a suitable alternative to replace SiO_2 as a gate material in microelectronics industry because of its high electrical resistivity ($10^{11}-10^{12}$ Ω cm), high energy band gap (5.5 eV), high dielectric constant (~13), and good lattice constant matching with silicon. It has transparency in the UV and visible range of the electromagnetic spectrum and hence is useful as antireflection and protection layer. Its refractive index is high (~2) and it is well suited for waveguide applications. It also serves as a host matrix for doping of rare-earth ions. The phosphors thus formed find applications in places such as fluorescent lamps, cathode ray tubes, plasma display panels, and field emission displays. It is also used for stabilization of zirconia.

Because of the mentioned technological applications, deposition of Y_2O_3 has been investigated by large number of research groups. A variety of deposition methods have been used to deposit Y_2O_3. Literature indicates that Y_2O_3 has been deposited by spray pyrolysis [36], thermal oxidation [37], thermal CVD in atmospheric pressure [38], low-pressure CVD [39], PECVD [40], MOCVD [41], molecular beam epitaxy [42], atomic layer deposition [43], electron beam evaporation [44], IBAD [45], PLD [46,47], and rf sputtering [48,49].

Here we are going to discuss how cold plasma processing is used for deposition of high-quality coatings of Y_2O_3 on various substrates at low temperatures (350°C). The details regarding the experimental arrangement, deposition process, and characterization are discussed elsewhere [50].

Y_2O_3 coatings using rf plasma-assisted MOCVD is done using metal organic precursors like β-diketonates that are nonhazardous, stable, and safe for handling. These precursors have been used previously for deposition of Y_2O_3 as well as other oxides [51–54]. A significant role is played by rf self-bias on the substrates during deposition of these coatings, because it determines the energy of the ions that are impinging on the substrates. This energy of the ions determines the precursor decomposition characteristics and eventually the properties of the deposited films. In this section, we will discuss the effect of the rf self-bias during deposition on the properties of the deposited films. To investigate the role of rf self-bias on the substrates during deposition, the deposited coatings were characterized using various techniques. The schematic of rf plasma MOCVD deposition system is

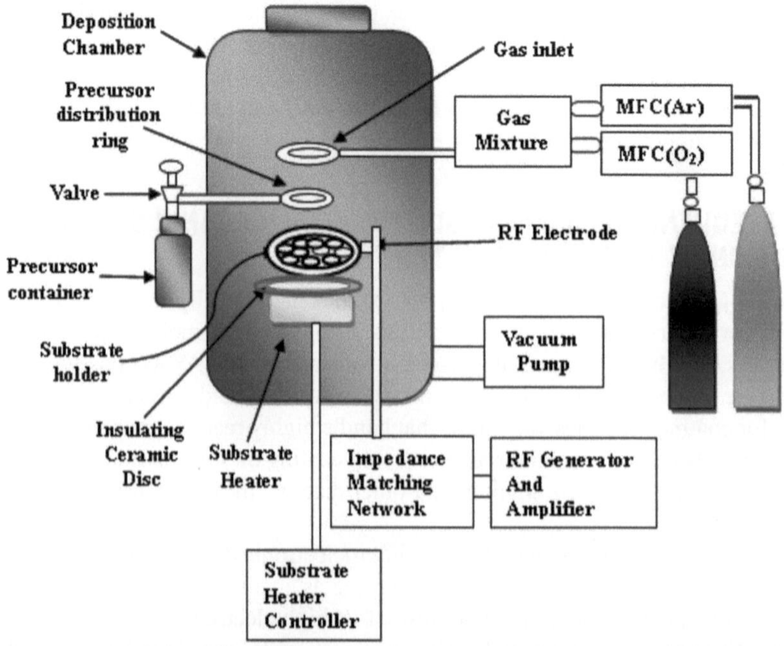

FIGURE 12.21

Schematic of a radio frequency plasma metal organic chemical vapor deposition system used for deposition of yttrium oxide thin films.

From S.S. Chopade, S.A. Barve, K.H.Thulasi Raman, N. Chand, M.N. Deo, A. Biswas, S. Rai, G.S. Lodha, G.M. Rao, D.S. Patil, RF plasma MOCVD of Y_2O_3 thin films: effect of RF self bias on the substrates during deposition, Appl. Surf. Sci. 285P (2013) 524–531, reprinted with permission.

shown in Fig. 12.21. A set of six Y_2O_3 thin films was deposited at different rf self-bias levels on the substrates during deposition, namely, −50 V (film A), −75 V (film B), −100 V (film C), −125 V (film D), −150 V (film E), and −175 V (film F). Plasma of precursor vapors of $Y(thd)_3$ (2,2,6,6–tetramethyl-3,5-heptanedionate) yttrium along with oxygen and argon gases was used for deposition. Silicon (100), stainless steel, quartz, and tantalum substrates were used for deposition. The average roughness values of uncoated substrates were 1.5 nm (silicon), 1.7 nm (quartz), 0.03 μm (stainless steel), and 0.05 μm (tantalum). Experimental details are discussed elsewhere [50].

Thickness measurement results on deposited films by stylus profilometer (model Dektak 150 Veeco) are given in Table 12.2. It clearly shows that film thickness varies with substrate bias during deposition and is maximum for film B. For films deposited for the same duration of 60 min, the thickness of the deposited films was found to vary with the bias voltage. The glancing angle X-ray diffraction (GIXRD) was carried out on the deposited films using a CuK_α radiation source with a Bruker D8 Discover model. The incident (glancing) angle was fixed at 1.8 degrees. The scans were taken in a 2θ range of 10–90 degrees with a scan step of 0.02 degrees per 3 s. These patterns taken on the deposited films are shown in Fig. 12.22. From the X-ray diffraction patterns, it is clearly seen that the films A, B, C, and D show deposition of Y_2O_3 with a body-centered cubic structure. The

Table 12.2 Thickness Measurement Results on Deposited Films at Various Radio Frequency (rf) Self-Bias Voltages

Bias (V)	Thickness (nm)
−50	140
−75	250
−100	65
−125	200
−150	108
−175	75

From S.S. Chopade, S.A. Barve, K.H.Thulasi Raman, N. Chand, M.N. Deo, A. Biswas, S. Rai, G.S. Lodha, G.M. Rao, D.S. Patil, RF plasma MOCVD of Y_2O_3 thin films: effect of RF self bias on the substrates during deposition, Appl. Surf. Sci. 285P (2013) 524–531, reprinted with permission.

diffraction lines corresponding to (222), (400), (440), and (622) reflections, as indicated in JCPDS card No. 79-1716 match well with the observed pattern. Films E and F that show deposits made on the substrate at higher bias levels show no preferred orientation and indicate that they are losing crystalline nature with the increase in the bias level (i.e., with the increase in the energy of the ions that are arriving at the substrates). Film A shows growth in the (111) as well as the (100) directions. Films B and D show textured growth in the (111) direction, whereas film C shows textured growth in the (100) direction.

X-ray photoelectron spectroscopy (XPS) studies were carried out on the deposited films with SPECS spectrometer and a Phoibos 100 MCD energy analyzer. The scans were taken with Al K_α (1486.6 eV energy) with a pass energy of 22 eV. Peaks corresponding to Y 3*d*, O 1*s*, and C 1*s* are scanned in detail with a 0.1 eV scan step. These peaks are deconvoluted by using the XPS peak fitting 4.1 program. The binding energy shifts were corrected using binding energy of physioadsorbed carbon at 284.6 eV, as referenced [55].

Y 3*d* deconvoluted peaks for films deposited at different rf self-bias voltages are shown in Fig. 12.23. The Y 3*d* core level is split into doublet form (Y $3d_{5/2}$ and Y $3d_{3/2}$) as a result of spin orbit coupling. The doublet has an energy separation of 2.1 eV and the Y $3d_{5/2}$-to-Y $3d_{3/2}$ areal intensity ratio is 1.5 [56]. The Y 3*d* peak is deconvoluted with gaussian/lorentzian (G/L) ratio of 5 with FWHM of individual peaks at 1−1.8 eV and a chi squared (X^2) value in the range of 1−2. Three different chemical environments are found for the yttrium atoms in all films. Peak pairs with binding energy of approximately 156.8 and 158.8 eV correspond to stoichiometric Y_2O_3; peak pairs with binding energy of about 157.8 and 160 eV could be either Y−OH or Y−OC bonds. Y−OC bonds will shift the Y 3*d* doublet to a higher binding energy because of the higher electronegativity of carbon [57]. Peak pairs with lower binding energy of about 155.8 and 158 eV indicate the presence of reduced Y_2O_3 [51].

Deconvoluted O 1*s* peaks for the films deposited at different rf self-bias voltages are shown in Fig. 12.24. Deconvolution of an O 1*s* peak is done with a G/L ratio of 15 with FWHM of individual peaks in the range of 1−1.8 eV and a X^2 value in the range of 1−2. The O 1*s* peak is deconvoluted into four subpeaks with binding energies of 529 eV, 530 eV, 531 eV, and 532 eV, corresponding to stoichiometric Y_2O_3, reduced Y_2O_3, the O−C=O bond, and the C=O bond. For film E, the peak

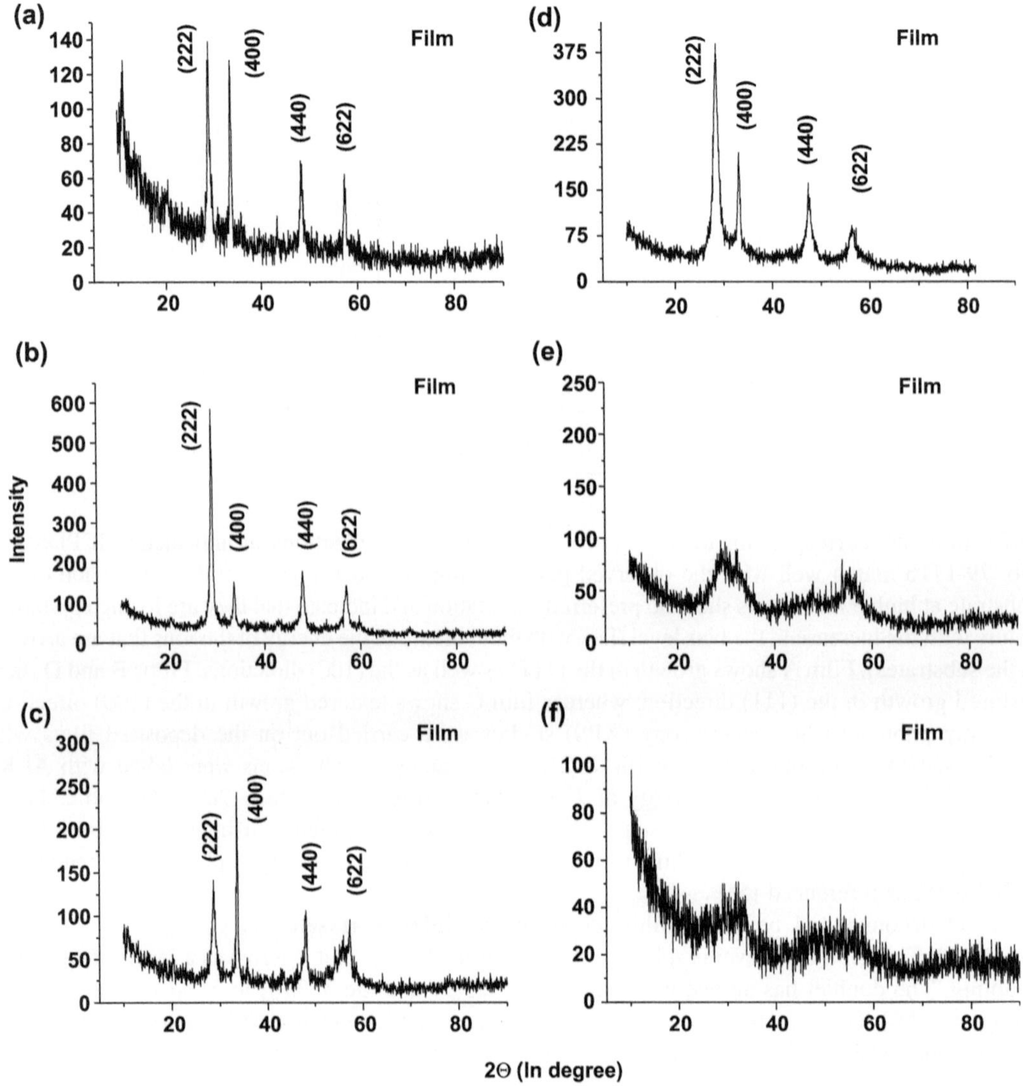

FIGURE 12.22

X-ray diffraction pattern for films deposited at different radio frequency self-bias voltages.

From S.S. Chopade, S.A. Barve, K.H.Thulasi Raman, N. Chand, M.N. Deo, A. Biswas, S. Rai, G.S. Lodha, G.M. Rao, D.S. Patil, RF plasma MOCVD of Y_2O_3 thin films: effect of RF self bias on the substrates during deposition, Appl. Surf. Sci. 285P (2013) 524–531, reprinted with permission.

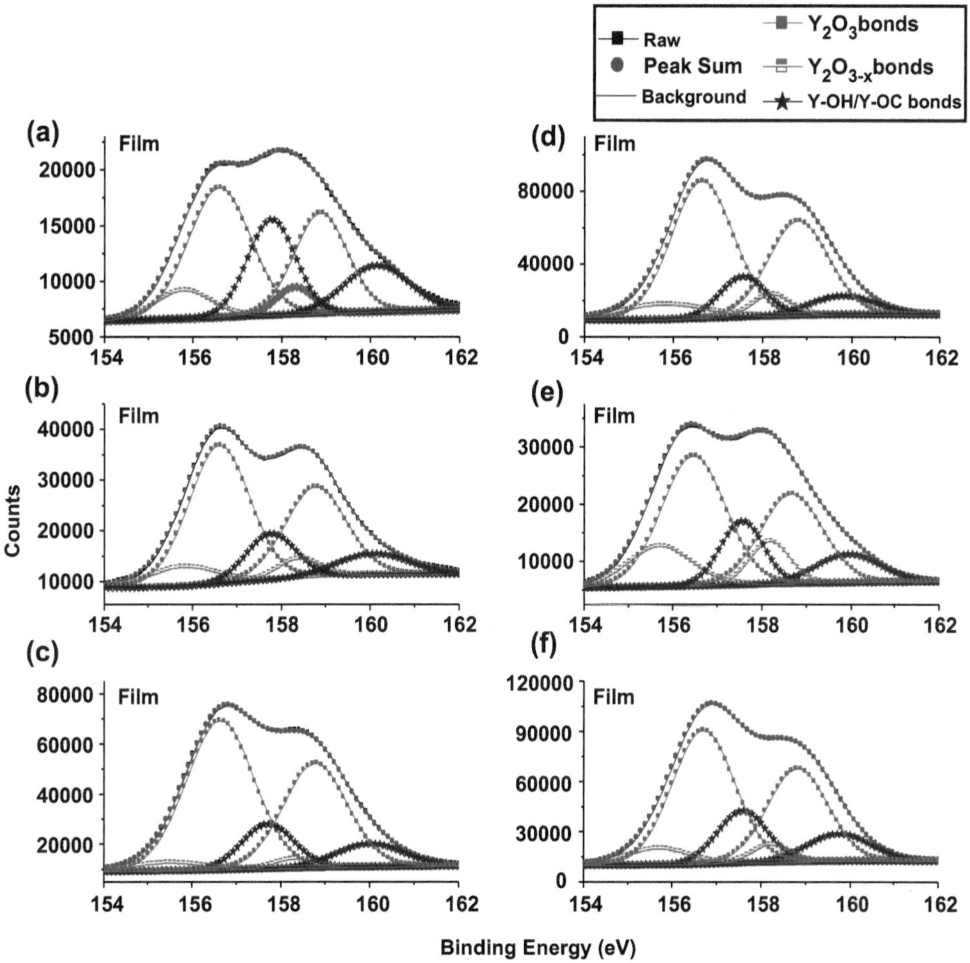

FIGURE 12.23

Y 3*d*-detailed X-ray photoelectron spectroscopy scans for films deposited at different radio frequency self-bias voltages.

From S.S. Chopade, S.A. Barve, K.H.Thulasi Raman, N. Chand, M.N. Deo, A. Biswas, S. Rai, G.S. Lodha, G.M. Rao, D.S. Patil, RF plasma MOCVD of Y_2O_3 thin films: effect of RF self bias on the substrates during deposition, Appl. Surf. Sci. 285P (2013) 524–531, reprinted with permission.

corresponding to binding energy of 528.5 eV corresponds to the Y—OH bond [58]. Because Y_2O_3 is hygroscopic in nature, it absorbs moisture from the atmosphere and reacts to form hydroxide and carbonates. Because in situ sputtering of the sample surfaces before XPS scan is done (to avoid modification of thin film surface), it can be seen that along with stoichiometric Y_2O_3 there exist small amounts of yttrium hydroxide and reduced Y_2O_3 in all films [51]. Because moisture and carbon species

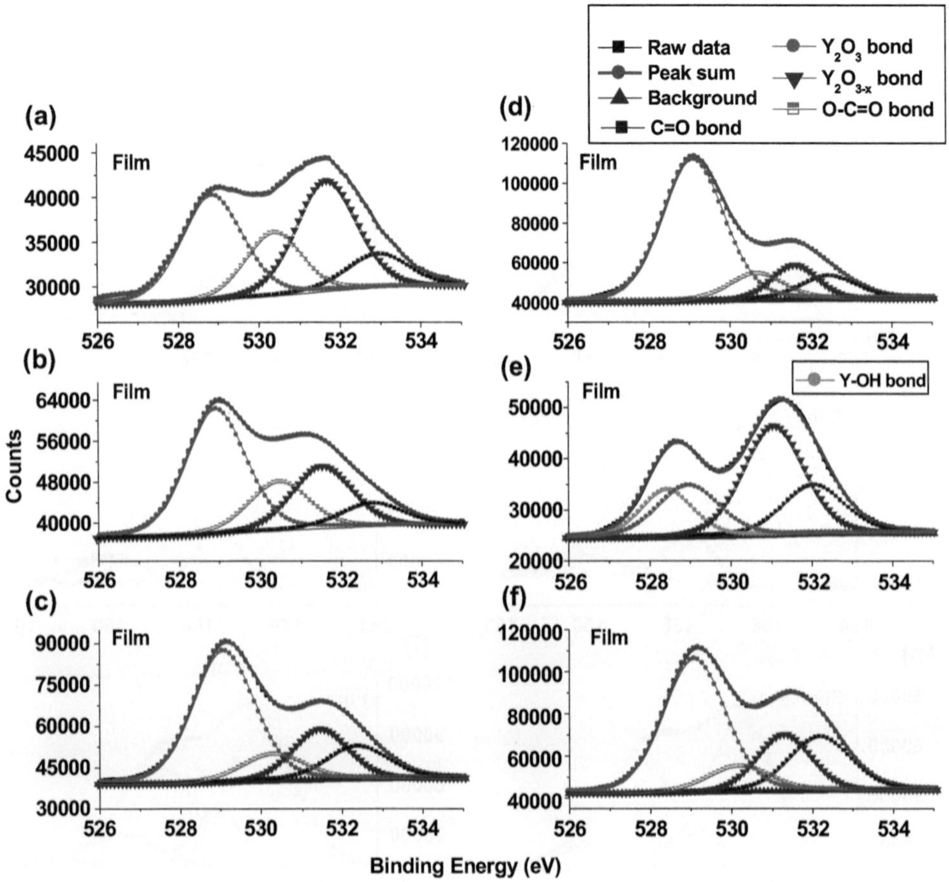

FIGURE 12.24

O 1*s*-detailed X-ray photoelectron spectroscopy scans for films deposited at different radio frequency self-bias voltages.

From S.S. Chopade, S.A. Barve, K.H.Thulasi Raman, N. Chand, M.N. Deo, A. Biswas, S. Rai, G.S. Lodha, G.M. Rao, D.S. Patil, RF plasma MOCVD of Y_2O_3 thin films: effect of RF self bias on the substrates during deposition, Appl. Surf. Sci. 285P (2013) 524–531, reprinted with permission.

are adsorbed on the surface of the film, quantitative analysis based on XPS from O 1*s* and C 1*s* peaks may not represent the actual bulk composition of films. Hence semiquantitative analysis is done only by taking the Y 3*d* peak into consideration [59].

Table 12.3 gives results of a semiquantitative analysis of a Y 3*d* peak and indicates that stoichiometric Y_2O_3 content on the surface increases with bias and the amount of yttrium carbonate decreases. At bias levels above −100 V, there is a decrease in stoichiometric Y_2O_3 content and a slight increase in the amount of yttrium carbonate. However, it is seen that for all films that the amount of reduced Y_2O_3 and yttrium carbonate is very small compared with the quantity of stoichiometric Y_2O_3.

Table 12.3 Semiquantitative Analysis (X-Ray Photoelectron Spectroscopy) of Film Surfaces From a Y 3*d* Peak

Film	% Y_2O_3 Content	% Y_2O_{3-x} Content	%Y–OH/Y–OC Content
A	57.15	10.90	31.80
B	70.29	10.60	19.00
C	77.80	4.10	18.06
D	73.37	10.60	16.03
E	61.02	18.80	20.16
F	69.89	8.33	21.77

From S.S. Chopade, S.A. Barve, K.H.Thulasi Raman, N. Chand, M.N. Deo, A. Biswas, S. Rai, G.S. Lodha, G.M. Rao, D.S. Patil, RF plasma MOCVD of Y_2O_3 thin films: effect of RF self bias on the substrates during deposition, Appl. Surf. Sci. 285P (2013) 524–531, reprinted with permission.

IR spectra were recorded on the deposited films with a Bruker Vertex 80 V FTIR spectrometer configured with a Mylar beam splitter and a deuterated triglycine sulfate (DTGS) detector in the far-IR region at a 4 cm^{-1} spectral resolution. An 11-degree reflection accessory was used for the measurements. Fig. 12.25 gives the far-IR reflectance spectra taken on the films. The triplet consisting of peaks at 305, 330 and 375 cm^{-1} confirm the presence of cubic Y_2O_3 [60]. Additional peaks for cubic Y_2O_3 are seen at approximately 185, 243, 350, 383, 395, 420, 435, 445, and 468 cm^{-1} [61–63]. The triplet is intense for film B and film D compared with the other films. Because IR characterization is a bulk technique, the intensity of IR peaks is directly proportional to the thickness of the film.

The microstructure of the deposited films was studied using the Carl Zeiss FESEM model ULTRA 55. The surface microstructures of the deposited films along with the transverse cross-sections for films A–F as seen by FESEM are given in Fig. 12.26a and b. SEM investigation results indicate that there is a change in the microstructure of the films with applied bias during deposition. It is clearly evident that the deposited films are dense and the grain size changes with bias. For film A, the grain size is approximately 25 nm. For film B, a dense granular structure is visible with a grain size in the range of 40–50 nm. For film D, the grains have grown in elongated shape with grain size of 50–60 nm. Although granular structure is observed, the microstructural features are not clearly seen in cases of films E and F. The cross-sectional images taken on freshly prepared fractured surfaces of all films indicate the deposition of dense films.

For investigating the optical properties of the deposited films, ellipsometric studies were carried out in the range 300–1200 nm with a phase-modulated spectroscopic ellipsometry unit (UVSELTM 460 ISA JOBIN-YVON, SPEX). Variation of the refractive index with wavelength for the films that are deposited at different bias levels is shown in Fig. 12.27a. All films have a fairly constant refractive index in the wavelength region of 600–1200 nm. As wavelength decreases below 600 nm, refractive index increases monotonically as reported in the literature [47]. Film E shows the highest refractive index, whereas other films show a slightly lower refractive index over the range of investigated wavelengths. However, for all the films, the extinction coefficient is zero (Fig. 12.27b) in the wavelength range of 300–1200 nm, indicating that all the films are transparent in the investigated wavelength range.

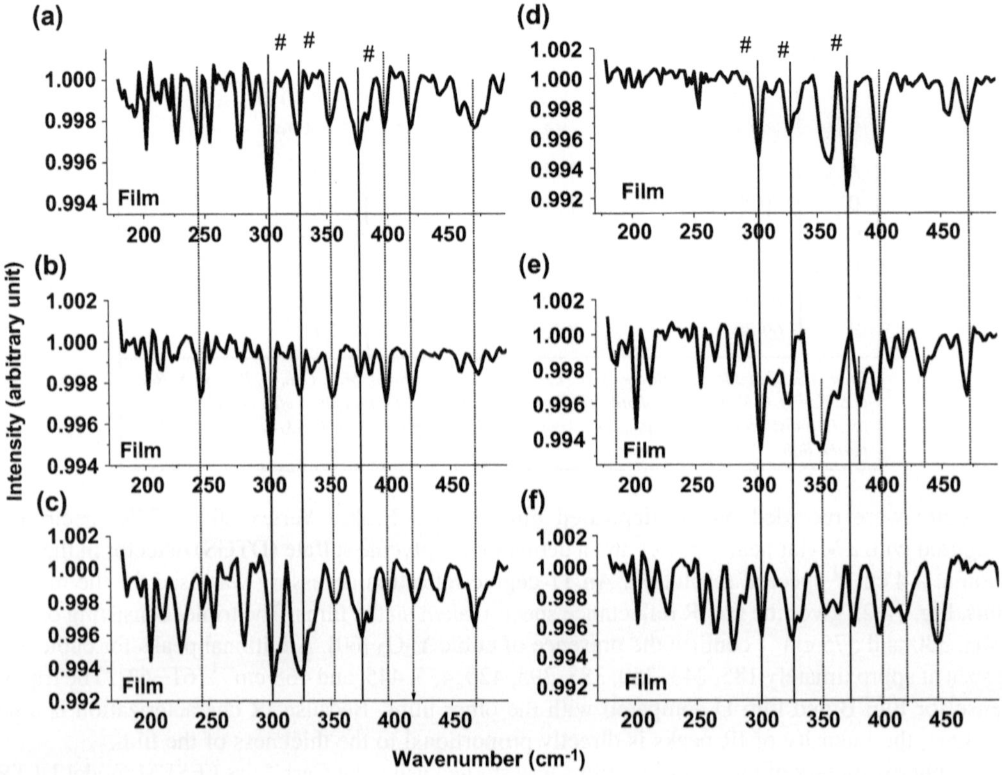

FIGURE 12.25

Far-infrared reflectance spectra of Y_2O_3 thin films deposited at different radio frequency self-bias voltages. *a.u.*, Arbitrary unit.

From S.S. Chopade, S.A. Barve, K.H. Thulasi Raman, N. Chand, M.N. Deo, A. Biswas, S. Rai, G.S. Lodha, G.M. Rao, D.S. Patil, RF plasma MOCVD of Y_2O_3 thin films: effect of RF self bias on the substrates during deposition, Appl. Surf. Sci. 285P (2013) 524–531, reprinted with permission.

The orientation of the films is dependent on the deposition method, the temperature of the substrate, the oxygen partial pressure, and the total gas pressure during the deposition [56]. The interfacial reactions and the thickness of the interface also play a prominent role in deciding the orientation, as reported earlier by Cho and co-workers [64]. During deposition of the investigated films, the substrate temperature, oxygen partial pressure, and total gas pressure are kept same for all films. The initial predeposition treatment conditions are also same. Therefore the change in texture of the deposited films can be clearly attributed to the variation in the rf self-bias level on the substrate during deposition. The net input power to the plasma varies with the bias on the substrates. This power level decides the fragmentation of the precursor and energy of the ions that are impinging on the substrates. This is also evident from the observed deposition rate variation with the bias.

The far-IR spectra show variations in intensity of the peaks with substrate bias. A sharp and intense triplet is obtained for films A, B, C, and D. In the case of films E and F, the triplet at 302, 330, and

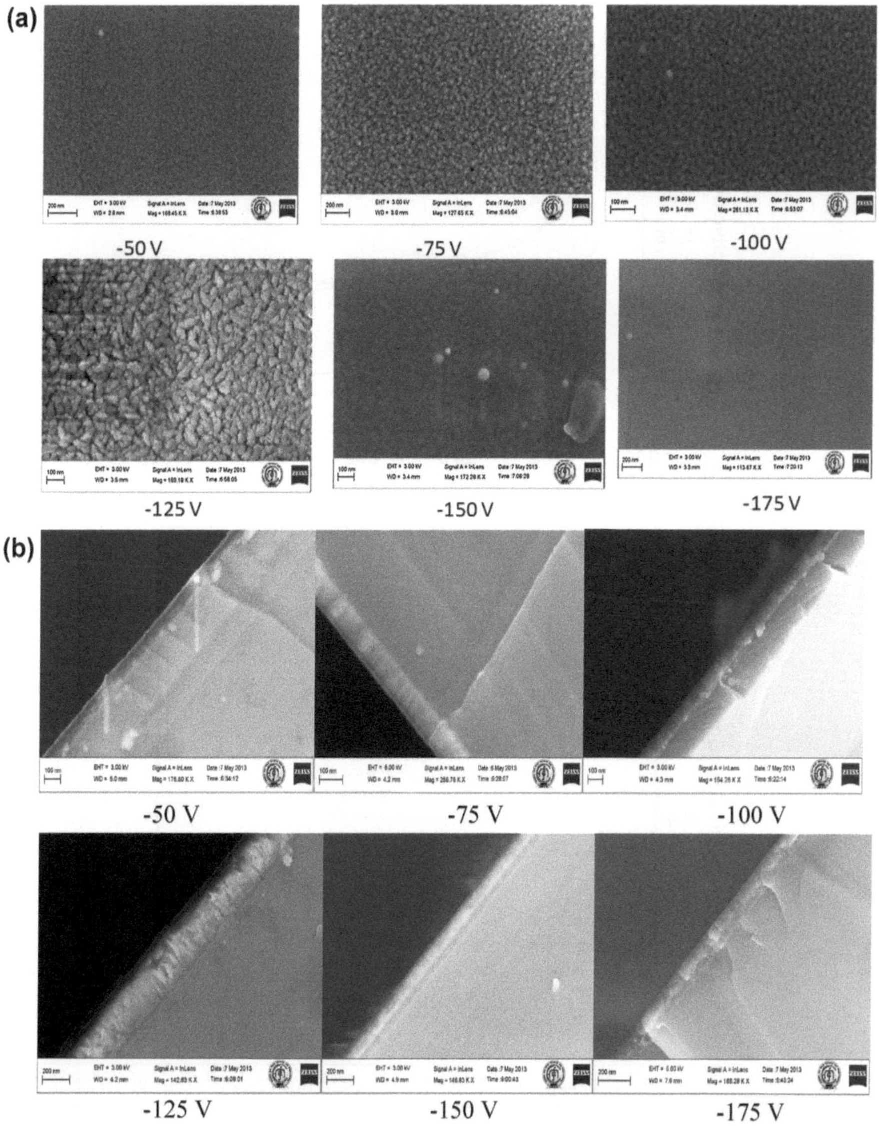

FIGURE 12.26

(a) Scanning electron microscope (SEM) images showing surface microstructures of the deposited films. (b) SEM images showing cross-sectional views of the deposited films.

From S.S. Chopade, S.A. Barve, K.H.Thulasi Raman, N. Chand, M.N. Deo, A. Biswas, S. Rai, G.S. Lodha, G.M. Rao, D.S. Patil, RF plasma MOCVD of Y_2O_3 thin films: effect of RF self bias on the substrates during deposition, Appl. Surf. Sci. 285P (2013) 524–531, reprinted with permission.

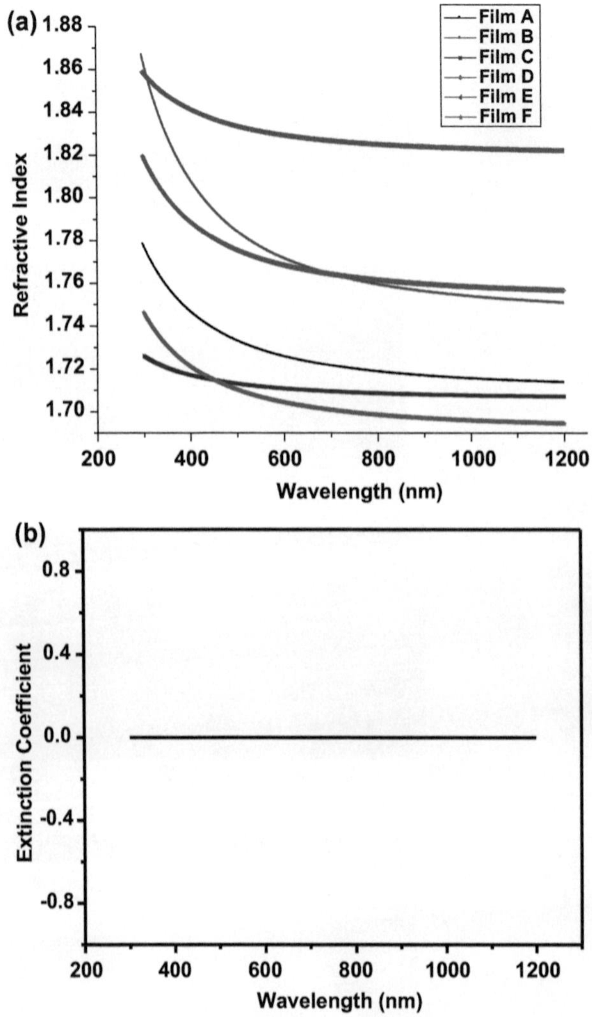

FIGURE 12.27

(a) Variation of refractive index for films deposited at different radio frequency (rf) bias voltages. (b) Variation of the extinction coefficient for films deposited at different rf bias voltages.

From S.S. Chopade, S.A. Barve, K.H.Thulasi Raman, N. Chand, M.N. Deo, A. Biswas, S. Rai, G.S. Lodha, G.M. Rao, D.S. Patil, RF plasma MOCVD of Y_2O_3 thin films: effect of RF self bias on the substrates during deposition, Appl. Surf. Sci. 285P (2013) 524–531, reprinted with permission.

375 cm^{-1} is relatively broader and shifted, indicating degradation in crystalline nature [40]. A peak at 395 cm^{-1} is also present for all films, and a 468 cm^{-1} peak is present in the spectra of films A, B, C, D, and E. A peak at 243 cm^{-1} is observed for films A, B, and C. Peaks at 350 and 418 cm^{-1} are present for films A, B, C, E, and F. Additional peaks at 183, 383, and 435 cm^{-1} are observed for film E. A peak at

445 cm^{-1} is present for film F. Peaks above 300 cm^{-1} are caused by deformation of YO_6 octahedra and the motion of oxygen atoms, and peaks below 300 cm^{-1} are the result of motion of yttrium atoms in their octahedral form [61]. These additional IR absorption peaks (besides the triplets) indicate the presence of a cubic Y_2O_3 phase in the films [61–63]. Thus from IR analysis it is evident that although the GIXRD patterns of films E and F are broad and diffused, these films show distinct peaks for Y–O bonding and hence are not completely amorphous.

For film deposited on the substrate at a −50 V bias (film A), the net power fed in the plasma is relatively less compared with the one at higher bias levels. This affects the degree of ionization in the plasma and also the reactivity of the generated species in the plasma. It is therefore evident that the deposition rate is lower for film A. As the energy of impinging ions is also less, there are voids formed on the surface of film A, resulting in increased moisture absorption and formation of Y–OH and Y–OC bonds [65].

This leads to a decrease in the amount of stoichiometric Y_2O_3, as seen from the XPS analysis (Table 12.3). It is known that Y_2O_3 has a cubic bixbyite structure with ordered oxygen vacancies. The unoccupied oxygen sites form parallel strings along the four (111) directions of the unit cell. These strings help in migration of oxygen through the material as deposition proceeds. Therefore the (111) plane is the lowest surface-energy plane of the bixbyite structure [66]. There is also good lattice constant matching between Si (100) and Y_2O_3 (111), resulting in fewer defects during deposition [47]. For film deposited at −75 V (film B), there is an enhanced degree of ionization of the plasma constituents and the growth is along the (111) direction; hence for these reasons, film B shows a higher deposition rate compared with film A. This trend is supported by XPS analysis, which shows an increase in the amount of stoichiometric Y_2O_3 with an increase in bias up to a bias voltage of −100 V. It is seen that the thickness of film C is reduced drastically. This may be because of the orientation of the film along the (100) direction. Because (111) is the minimum surface energy plane, growth along the (100) direction may produce strain in the film, thereby hampering the deposition rate. However, XPS analyses indicate that film C has the highest amount of stoichiometric Y_2O_3. This suggests that oxygen activity has been enhanced in the case of film C. This high oxygen activity changes the ordering of oxygen vacancies, resulting in the formation of strain. To accommodate strain, growth of film C is along the (100) direction. Film D, which is oriented along (111), also shows an increased deposition rate compared with film C for the reasons mentioned.

During the depositions experiments, argon is used as a plasma generation gas that produces heavy ions in the plasma medium. These ions during collisions can transfer momentum to the other constituents of the plasma, resulting in effective ionization of gases, including a precursor that is used for deposition [67]. However, at higher bias levels as in the case of films E and F, this bombardment by heavy ions results in undesirable effects such as resputtering of the film. This resputtering of the deposited films leads to lower deposition rates, as observed for films E and F. This resputtering also results in an increase in nucleation density. The onset of enhanced nucleation density is evident from elongated grains, as seen in the FESEM image of film D. With further increase in the degree of bombardment, it can be seen that films E and F have transformed to a nanocrystalline phase. Because nanocrystals have increased surface area, there is a tendency to absorb moisture. This is also evident from XPS analysis, which shows an increase in the Y–OH/Y–OC amounts in the films deposited at a bias level above −100 V [68,69].

It is known that optical constants of Y_2O_3 film are highly dependent on its microstructure [70], and even small changes in refractive index values may indicate significant changes in the microstructure of deposited films [71]. The GIXRD pattern for film E shows broader peaks, indicating reduction in grain size from bombardment by energetic ions; also film E exhibits a maximum refractive index. The enhanced refractive index may result from the formation of dense nanocrystals. This dense network of nanocrystals increases the packing density as reported by Ramana and co-workers [70]. Packing density is higher for the (111) plane [58] compared with other planes. The observed high refractive index value for films B and D that are textured along the (111) direction can be explained on this basis. Film A does not show any preferred orientation and hence has an intermediate value of refractive index. The orientation of film C is along the (100) direction, which is not as densely packed as the (111) plane, and hence film C shows a lower refractive index compared with film B. A similar trend was also observed by Kim and co-workers [71]. Film F exhibits the minimum value of refractive index; this may be the result of further degradation in crystallinity of the film from heavy bombardment of highly energetic ions [71].

The extinction coefficient of all the films is zero, suggesting that all films have a good band structure so there is no light absorption [47]. This also indicates that there is sufficient oxygen activity during deposition. The energetic oxygen ions in the plasma can easily react with yttrium to form Y_2O_3 films with maximum transmittance [45]. Although XPS analysis shows that there are indications of formation of reduced Y_2O_3, its quantity is very small compared with the quantity of stoichiometric Y_2O_3.

From the discussion, it is evident that deposition of Y_2O_3 films on silicon substrates is possible at a reasonably low substrate temperature of 350°C using an rf plasma MOCVD process. The films deposited under varying influence of rf self-bias on the substrates show changes in the texture and properties, indicating a strong dependence of the properties on the substrate bias during deposition. Results from different characterization techniques confirm deposition of single-phase cubic Y_2O_3 on the substrates. From the results of these investigations, it is evident that it is important to control the bias level on the substrates during deposition to get the desired properties of the coating.

4.4 RADIO FREQUENCY PLASMA-ASSISTED METAL ORGANIC CHEMICAL VAPOR DEPOSITION OF YTTRIA-STABILIZED ZIRCONIA THIN FILMS

Y_2O_3-stabilized zirconia (YSZ) is one of the most important advanced materials. It has many applications in various fields. It is used as a dielectric gate in the integrated circuits, as a biocompatible and hard material in hip replacement or artificial teeth, as a thermal barrier coating (TBC) and protective coating for its high-temperature superior thermomechanical behavior in turbine blades or other engine components, and as a gas sensor for automotive exhaust applications.

ZrO_2 exists in the monoclinic phase at room temperature, transforms to the tetragonal phase at 1170°C, and to the cubic phase at 2370°C. Change of phase from tetragonal to monoclinic is associated with a 4% volume increase, leading to cracking of the material, which renders pure ZrO_2 useless from an application point of view.

YSZ has an important role in the development of solid oxide fuel cells (SOFCs) as electrolytes. An electrolyte of SOFC must have a high ionic conductivity and a high electronic resistivity. In addition, it must be chemically stable at high temperatures in oxidizing and reducing atmospheres, chemically and mechanically compatible with the other cell components, and gas tight. Whereas the bulk tetragonal zirconia is a poor ionic conductor, the high-temperature cubic phase exhibits a very high ionic conductivity.

Zirconia can be stabilized at room temperature through the doping of oxides such as MgO, CaO, and Y_2O_3. The cubic phase can be stabilized at low temperatures by doping with these oxides, and the obtained solid solution in cubic form induces the creation of O^{2-} vacancies because the substitution of Zr^{4+} for Y^{3+} in the cationic network. As reported in the literature, approximately 3–5 mol% of Y_2O_3 is required for the formation of partially stabilized zirconia (PSZ), whereas 8 mol% Y_2O_3 stabilizes zirconia in the cubic phase (YSZ) [72,73].

For bulk material, the ionic conductivity reaches a maximum for Y_2O_3 content in the ZrO_2:Y_2O_3 solid solution between 8 and 12 mol% (8–12YSZ). PSZ has excellent wear resistance and fracture toughness. YSZ has low thermal conductivity and good erosion resistance and hence is used as a TBC in gas turbines. Because of its high refractive index and low absorption over a wide range of the spectral region, YSZ is also used as an optical coating in high-reflectivity mirrors and optical filters. For many of these applications, it is essential to synthesize high-quality YSZ thin films.

A large variety of methods used for the deposition of YSZ have been reported in the literature, namely DC magnetron sputtering [74], PLD [75], coprecipitation [76], plasma CVD [77], electron beam evaporation [78], and MOCVD [79].

Here, we will confine the discussion to the deposition and characterization of YSZ thin films deposited by rf plasma-enhanced MOCVD using octanedionate (tod) precursors. The details regarding the deposition process and characterization of the deposited coatings are reported elsewhere [80]. In case of rf PECVD, DC self-bias on the substrate acts as an accelerating energy source for the ions and plays an important role in deciding the properties of the deposited films. The process of MOCVD utilizes metal organic precursors that are volatile and for which the flow rate of the precursor can be controlled by adjusting the temperature at which the precursor is heated. The MOCVD technique has various advantages like low deposition temperature, good conformal coverage, uniformity of coating, possibility of large scale production, and ease of compositional control in complex films [81]. In the present study, tod precursors $Y(tod)_3$ and $Zr(tod)_4$ are used for deposition of these coatings. Advantages of tod precursors are enhanced volatility, evaporation without leaving any residue, and reduced carbon contamination in the deposited films.

Three different films with variable content of Y_2O_3 (4, 5, and 9 mol%) have been deposited by varying the temperature of the $Y(tod)_3$ precursor (110, 130, and 160°C) and keeping the temperature of $Zr(tod)_4$ constant at 220°C at a substrate bias level of −100 V. To study the variation in properties of the deposited films with Y_2O_3 content, different characterization techniques are used.

As expected, there is a change of phase with the variation of Y_2O_3 content in the film. Because the properties of the deposited films are dependent on the structure and composition of the films, systematic investigations are carried out on the deposited films and efforts are made to correlate the observed experimental data with the structure and composition of the films.

The schematic of the rf plasma-assisted CVD system is shown in Fig. 12.28. The deposition system consists of a stainless steel chamber with an rf electrode and a heater arrangement. There is an insulating ceramic ring in between the rf electrode and the heater. The rf electrode is connected through an impedance matching network to an rf generator (13.56 MHz). $Zr(tod)_4$ and the $Y(tod)_3$ precursors in pellet form were placed in two different stainless steel containers. The containers are provided with gate valves to regulate the flow of individual precursor. The flow rate of the precursors can be varied by changing the temperature of the precursor container. The precursor vapors were fed into the chamber through a common delivery line (stainless steel tube), which is connected to a perforated circular ring for uniform distribution of precursor vapors (precursor distributor ring).

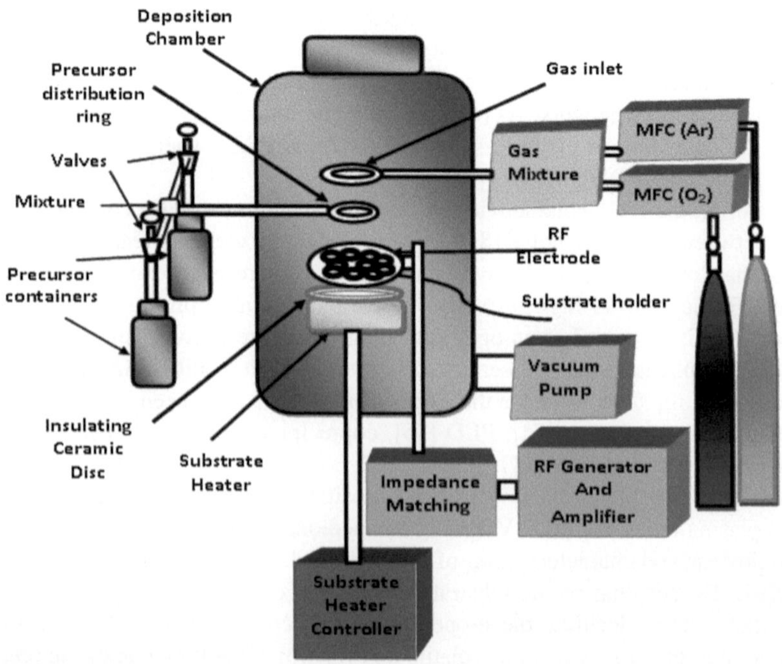

FIGURE 12.28

Schematic of a radio frequency (rf) plasma-assisted metal organic chemical vapor deposition system.

From S.S. Chopade, C. Nayak, D. Bhattacharyya, S.N. Jha, R.B. Tokas, N.K. Sahoo, M.N. Deo, A. Biswas, S. Rai, K.H. Thulasi Raman, G.M. Rao, N.K. Pandey, D.S. Patil, RF plasma enhanced MOCVD of yttria stabillized zirconia thin films using octanedionate precursors and their characterization, Appl. Surf. Sci. 355 (2015) 82–92, reprinted with permission.

The precursor delivery line (190°C) and the precursor distribution ring (200°C) were kept heated during the deposition process. Argon and oxygen gases were fed separately into the chamber through another stainless steel tube. The flow of these gases was controlled through mass flow controllers. Silicon substrates were cleaned by standard RCA procedure.

After placing the cleaned substrates on the electrode, the base vacuum ($\sim 6.9 \times 10^{-2}$ mbar) was achieved with a rotary pump. Substrates were then heated to a temperature of 400°C and subsequently sputtered by argon gas plasma for 20 min. Plasma of argon, oxygen , and vapors of $Zr(tod)_4$ and $Y(tod)_3$ was used for the deposition. The temperature of the $Zr(tod)_4$ container was maintained at 220°C and three different films, namely A, B, and C, were deposited by varying the temperature of the $Y(tod)_3$ container to 110, 130, and 160°C, respectively. For all depositions, the rf self-bias on the substrates was maintained at -100 V and depositions were carried out for 60 min.

Thickness of the films and Y_2O_3 content [estimated from Energy-dispersive X-ray spectroscopy (EDAX) measurement] increase with the increase in the temperature of $Y(tod)_3$ container. EDAX was done using a Bruker Quantax X-ray detector. Earlier, Pasko et al. made a comparative study of properties of monoclinic ZrO_2 films deposited using $Zr(tod)_4$ and $Zr(thd)_4$ precursors and reported that tod precursors are able to produce higher growth rate at lower temperature compared with heptanedionate precursors [52].

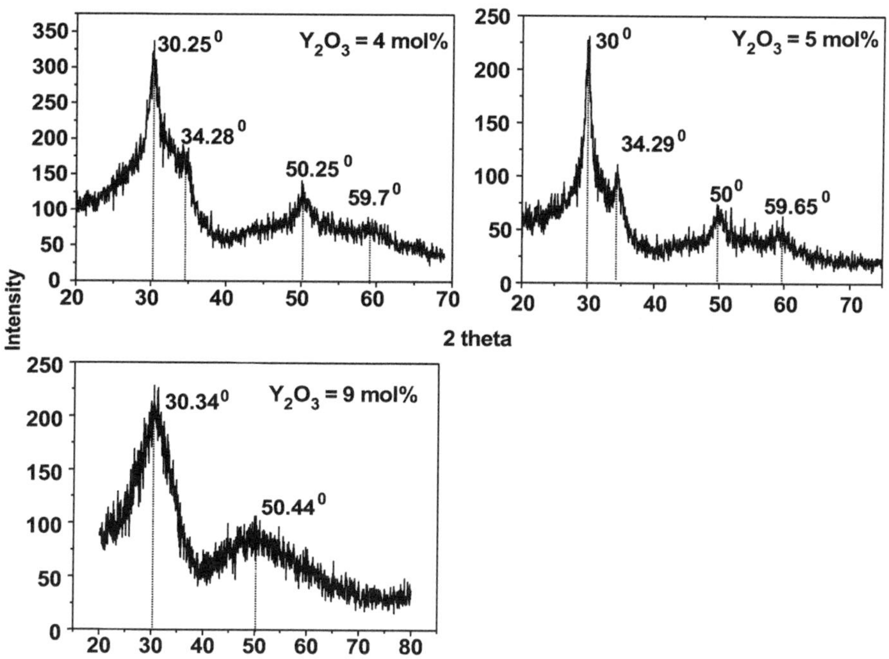

FIGURE 12.29

Glancing angle X-ray diffraction patterns for films deposited with different yttria (Y_2O_3) content.

From S.S. Chopade, C. Nayak, D. Bhattacharyya, S.N. Jha, R.B. Tokas, N.K. Sahoo, M.N. Deo, A. Biswas, S. Rai, K.H. Thulasi Raman, G.M. Rao, N.K. Pandey, D.S. Patil, RF plasma enhanced MOCVD of yttria stabillized zirconia thin films using octanedionate precursors and their characterization, Appl. Surf. Sci. 355 (2015) 82–92, reprinted with permission.

The GIXRD study was carried out on the deposited films with a CuK_α radiation source using the Bruker D8 Discover model. The scans were recorded in 2θ range of 10–70 degrees with scan step of 0.02 degrees per 3 s, keeping the angle of incidence fixed at 1.8 degrees. From the GIXRD pattern shown in Fig. 12.29, it is seen that the diffraction peaks are broad for all three deposited films.

The positions of diffraction lines for cubic and tetragonal phases of YSZ are close and there could be coexistence of other phases, making phase identification difficult. The broadening of diffraction peaks indicates small crystallite size in the deposited films. It is seen that the crystallite size is minimum for film C and maximum for film B. In the case of film C, two broad diffraction peaks without any sharp features indicate the formation of the nanocrystalline phase. The diffraction pattern for film B indicates dominance of the tetragonal phase. The diffraction lines corresponding to reflections (101), (110), (200), and (211) as indicated in JCPDS card 82-1241, are in good agreement with observed diffraction patterns for film B.

The IR spectra were recorded in the far-IR region at 4 cm^{-1} spectral resolution using a Bruker Vertex 80 V FTIR spectrometer configured with Mylar beam splitter and a DTGS detector. An 11-degree reflection accessory was used for the measurements.

FTIR spectra (Fig. 12.30) are useful in identifying the phases present in the deposited films. In the FTIR spectrum for film A, peaks at 228, 334, 380, 419, 504, and 594 cm^{-1} indicate presence of the

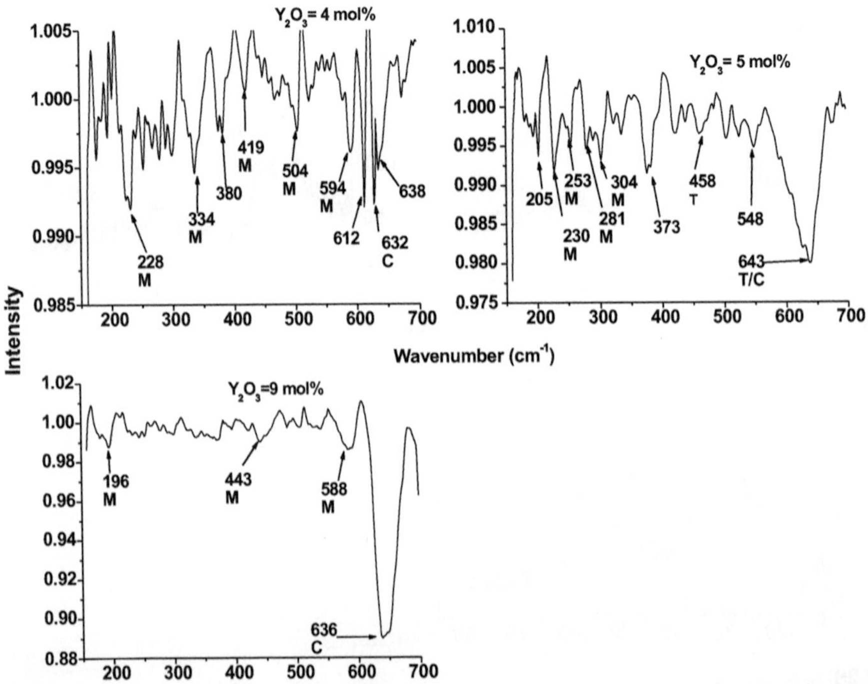

FIGURE 12.30

Far-infrared reflectance spectra for films deposited with different yttria (Y_2O_3) content. *C*, cubic phase; *M*, monoclinic phase; *T*, tetragonal phase.

From S.S. Chopade, C. Nayak, D. Bhattacharyya, S.N. Jha, R.B. Tokas, N.K. Sahoo, M.N. Deo, A. Biswas, S. Rai, K.H. Thulasi Raman, G.M. Rao, N.K. Pandey, D.S. Patil, RF plasma enhanced MOCVD of yttria stabillized zirconia thin films using octanedionate precursors and their characterization, Appl. Surf. Sci. 355 (2015) 82–92, reprinted with permission.

monoclinic phase, whereas the peak at 632 cm^{-1} corresponds to the presence of the tetragonal phase. Thus it is seen that the low Y_2O_3 content in film A shows a mixture of monoclinic and tetragonal phases. In the case of film B, peaks at 230, 253, and 281 cm^{-1} indicate the presence of the monoclinic phase, whereas peaks at 458 and 643 cm^{-1} and the broad band at about 550–650 cm^{-1} indicate the presence of the tetragonal phase, suggesting that as Y_2O_3 content increases in film B, along with monoclinic phase, a dominant tetragonal phase appears, as indicated by a broad band at approximately 550–650 cm^{-1}. With 9 mol% of Y_2O_3 in film C, the prominent peak at close to 636 cm^{-1} indicates the formation of the cubic phase [82–84].

Earlier, Holzschuh and Suhr obtained a fully stabilized cubic structure over a wide range of Y_2O_3 content (from 3.5 to 80 mol%) by using rf PECVD and heptanedionate precursors [85]. However, in the present case with the increase in the Y_2O_3 content, the FTIR spectra show presence of the tetragonal phase and with further increase in Y_2O_3 content, the fully stabilized cubic phase is formed, as observed in case of film C.

Ellipsometric studies were carried out with a phase-modulated spectroscopic unit (UVSELTM 460 ISA JOBIN-YVON, SPEX) in the range of 300–1200 nm. The variation of refractive index with

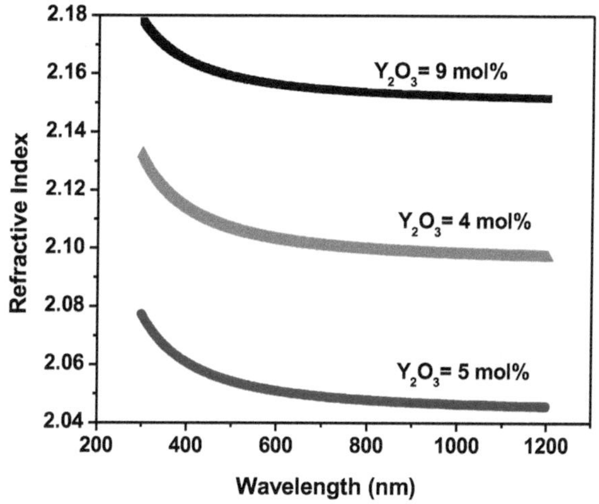

FIGURE 12.31

Variation of the refractive index with wavelength for films deposited with different yttria (Y_2O_3) content.

From S.S. Chopade, C. Nayak, D. Bhattacharyya, S.N. Jha, R.B. Tokas, N.K. Sahoo, M.N. Deo, A. Biswas, S. Rai, K.H. Thulasi Raman, G.M. Rao, N.K. Pandey, D.S. Patil, RF plasma enhanced MOCVD of yttria stabillized zirconia thin films using octanedionate precursors and their characterization, Appl. Surf. Sci. 355 (2015) 82–92, reprinted with permission.

wavelength for the three films is shown in Fig. 12.31. The refractive index remains fairly constant in the wavelength range of 600–1200 nm and increases as wavelength decreases below 600 nm. This is in agreement with the values reported in the literature [86]. The refractive index is highest for film C and lowers for film B. Here we do not observe a particular trend in the variation of refractive index with the Y_2O_3 content. As reported by Heiroth et al., crystallite size has a strong influence on optical properties of zirconia-based oxide thin films [87]. The optical properties degrade in the case of amorphous thin films compared with the crystalline ones. The band gap energy enhances for the nanocrystalline coatings compared with the microcrystalline thin films [87]. Ramana et al. have reported that the nanocrystalline nature increases the density of the film, resulting in an increase in the refractive index value [70]. This is observed for film C, which demonstrates an increase in the value of the refractive index resulting from the smaller crystallite size. Similarly, film A shows a higher value of refractive index because of the smaller crystallite size compared with film B. The extinction the coefficient value for all films is zero within the measurement limit, pointing to the fact that the films are transparent within the investigated spectral range (300–1200 nm) with no interband absorption. Transparency of Y_2O_3-doped films along with good mechanical properties makes them useful for such applications as view ports of scientific instruments and protective optical layers [88]. The transparency of the films can be attributed to the deposition process of rf plasma-enhanced MOCVD and tod precursors. The fragmentation of precursors can be efficiently achieved because of the high internal energy of the plasma medium. There is uniform mixing of precursors, and the reactivity of precursors is enhanced by plasma, resulting in dense and uniform coatings. An advantage of sputtering deposition techniques in enhancing the packing density has been demonstrated earlier [89]. One such technique is rf PECVD, which combines the merits of CVD and the sputtering technique. The rf self-bias on the substrates decides the energy of ions that are bombarding the substrate and therefore plays an important role in determining the composition and structure of the deposited films [50]. The bombarding ions remove the loosely bonded atoms and thus also help in reducing the amount of contamination in the deposited films.

The advantages of tod precursors used in this deposition are that they are volatile as compared with heptanedionate precursors and result in less carbon incorporation in the deposited films, as reported by Pasko et al. [52] and as is evident from the observed transparency of Y_2O_3-doped films in our case. Carbon contamination in films using heptanedionate precursors has been earlier reported. Chevalier et al. have stated that carbon could not be completely removed even at a deposition temperature of 750°C [70], and Varanasi reported blackish-brown coatings caused by carbon incorporation, which became off-white in color after calcination [90]. Carbon codeposition reduces the optical band gap and transmittance of the thin films, as reported by Cho et al. [91]. As in this case, the deposited Y_2O_3-doped films are transparent, and it is evident that carbon contamination in the films is negligible. Transparency of coatings also indicates enhanced density [88].

FESEM (Zeiss Supra) was used to study the morphology of the films. The dense and uniform nature of the coatings is also evident from FESEM images shown in Fig. 12.32.

The chemical environment of the elements in deposited films was studied using XPS. XPS measurement was done with a SPECS spectrometer and a Phoibos 100 MCD energy analyzer. Scans were taken using Al K_α (1486.6 eV energy) with a pass energy of 22 eV. A detailed scan of peaks corresponding to Zr 3*d*, Y 3*d*, and O 1*s* was done with a scan step of 0.1 eV. The XPS peak fitting 4.1 program was used to deconvolute the peaks. The binding energy of physioadsorbed carbon at 284.6 eV was used as a reference for correcting the shifts in binding energy.

Sputtering of the film surfaces before the XPS scan was not done in order to avoid any structural change [92]. Hence carbon and moisture contamination are observed on the surface of the films. The Zr 3*d*-deconvoluted peaks for films deposited with different levels of Y_2O_3 concentration are shown in Fig. 12.33a. The Zr 3*d* core level is split into a doublet (Zr 3*d* $_{5/2}$ and Zr 3*d* $_{3/2}$) because of spin orbit coupling. The energy separation of the doublet is 2.4 eV, and the Zr 3*d* $_{5/2}$-to-Zr 3*d* $_{3/2}$ intensity ratio is 1.5 [92].

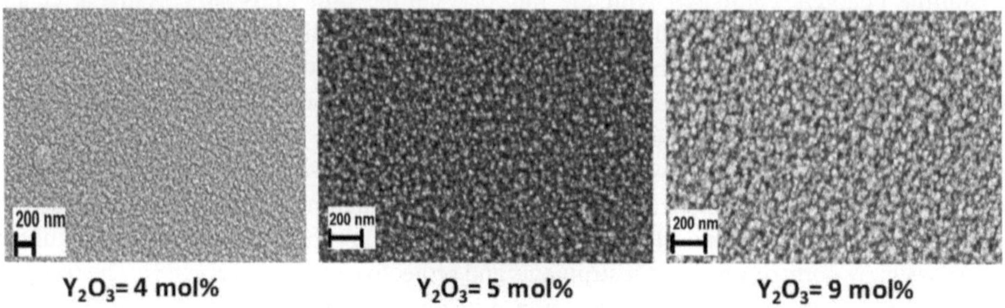

FIGURE 12.32

Field emission scanning electron microscope images for films deposited with different yttria (Y_2O_3) content.

From S.S. Chopade, C. Nayak, D. Bhattacharyya, S.N. Jha, R.B. Tokas, N.K. Sahoo, M.N. Deo, A. Biswas, S. Rai, K.H. Thulasi Raman, G.M. Rao, N.K. Pandey, D.S. Patil, RF plasma enhanced MOCVD of yttria stabillized zirconia thin films using octanedionate precursors and their characterization, Appl. Surf. Sci. 355 (2015) 82–92, reprinted with permission.

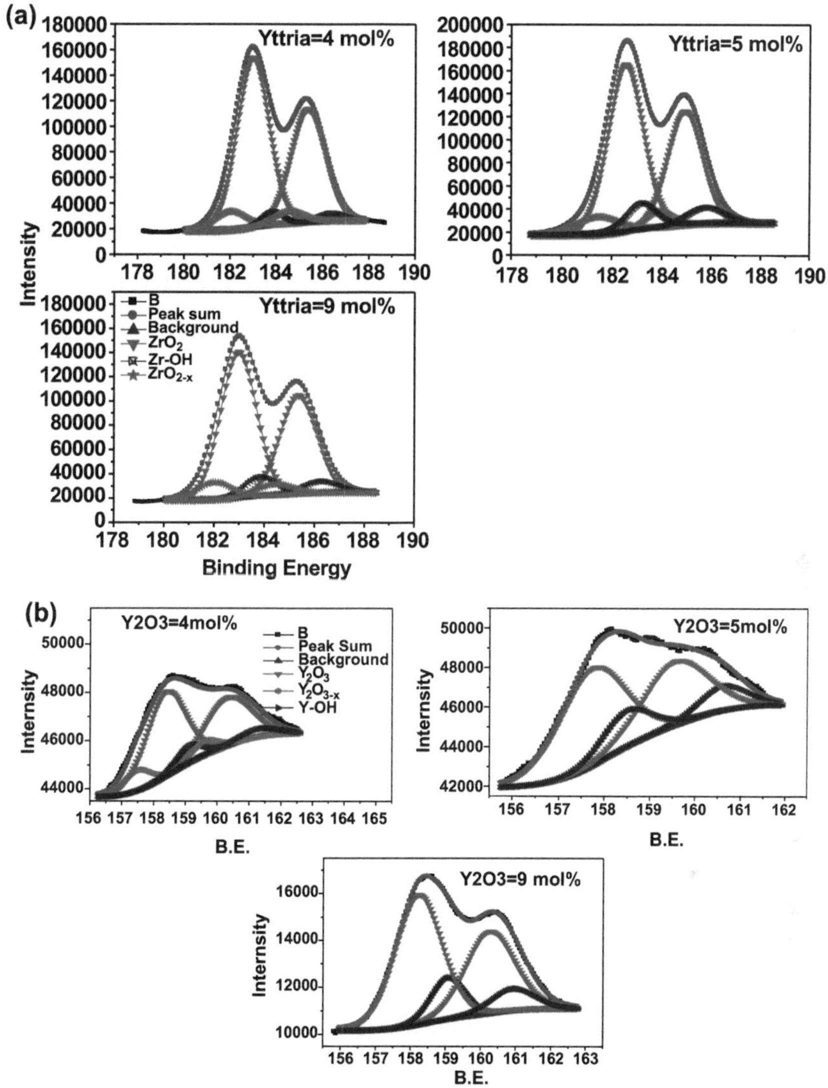

FIGURE 12.33

(a) Zr 3*d*-detailed X-ray photoelectron spectroscopy (XPS) scans for films deposited with different yttria (Y_2O_3) content. (b) Y 3*d*-detailed XPS scans for films deposited with different Y_2O_3 content.

From S.S. Chopade, C. Nayak, D. Bhattacharyya, S.N. Jha, R.B. Tokas, N.K. Sahoo, M.N. Deo, A. Biswas, S. Rai, K.H. Thulasi Raman, G.M. Rao, N.K. Pandey, D.S. Patil, RF plasma enhanced MOCVD of yttria stabillized zirconia thin films using octanedionate precursors and their characterization, Appl. Surf. Sci. 355 (2015) 82–92, reprinted with permission.

Table 12.4 Semiquantitative Analysis (by X-Ray Photoelectron Spectroscopy) of Film Surfaces From the Zr 3*d* Peak

Film	ZrO_2%	ZrO_{2-x}%	Zr–OH%
A	82.2	9	8.9
B	78.9	9.4	11.7
C	86.9	8.9	4.2

From S.S. Chopade, C. Nayak, D. Bhattacharyya, S.N. Jha, R.B. Tokas, N.K. Sahoo, M.N. Deo, A. Biswas, S. Rai, K.H. Thulasi Raman, G.M. Rao, N.K. Pandey, D.S. Patil, RF plasma enhanced MOCVD of yttria stabillized zirconia thin films using octanedionate precursors and their characterization, Appl. Surf. Sci. 355 (2015) 82–92, reprinted with permission.

Three different chemical environments for the zirconium atoms in the deposited films are seen in XPS analyses. The peak pair at binding energies 182.4 and 184.8 eV corresponds to the Zr^{4+} oxidation state. There is some degree of nonstoichiometry on the surface of the films, indicated by the shifting of the peak pair to lower binding energies at 181.5 and 183.9 eV [93]. Zr–OH bonding, on the other hand, shifts binding energies to the higher values of 183.4 and 185.8 eV [94]. Small amounts of Zr–OH may be present on the surface of films because of moisture adsorption, because the films were not sputtered before the XPS scan. Semiquantitative analysis results showing the percentage content of Zr–OH, stoichiometric ZrO_2, and reduced ZrO_{2-x} from the Zr 3*d* peak for the three films are shown in Table 12.4.

The Y 3*d*-deconvoluted peaks for deposited films are shown in Fig. 12.33b. The Y 3*d* core level splits into a doublet (Y $3d_{5/2}$ and Y $3d_{3/2}$) because of spin orbit coupling. The doublet has an energy separation of 2.1 eV and a Y $3d_{5/2}$-to-Y $3d_{3/2}$ intensity ratio of 1.5. The peak corresponding to the Y 3*d* core level is deconvoluted into three peak pairs that correspond to three different chemical environments. Peak pair at binding energies of 156.2 and 158.3 eV correspond to stoichiometric Y_2O_3, whereas the peak pair at 157.4 and 159.4 eV corresponds to Y–OH [95] and the peak pair at 155.2 and 157.2 eV corresponds to reduced Y_2O_3 [50]. For the same reason stated previously, it can be seen from Y 3*d* peak deconvolution that there are peaks corresponding to the Y–OH bond in addition to the peak pair corresponding to stoichiometric Y_2O_3.

From Table 12.4, it is seen that Zr–OH content is higher in film B compared with the other two films.

Tribological studies were done on these coatings. The linear reciprocating mode of a ball on a disk microtribometer (CSM Instruments, Switzerland) was used to carry out friction measurements. In all these measurements, the normal loads, sliding speeds, and sliding distance were kept constant at 1 N, 4 cm/s, and 100 m, respectively. The diameter of the sliding Al_2O_3 ball was 6 mm with a surface roughness of 0.04 μm. Tribology experiments were conducted in ambient atmospheric and non-lubricated conditions at room temperature where the relative humidity level was 75%. FESEM was used to examine the deformation and morphology of the wear tracks. A plot showing variation of the friction coefficient with distance for the three films is shown in Fig. 12.34.

Observed values of the friction coefficient for these films were less than 0.04, which clearly indicates superior tribological properties. In our case, the films were smooth and dense, resulting in enhanced tribological properties. The friction coefficient value for film A was less than 0.02, which is quite low compared with values observed for films B and C. The trend of variation in the friction

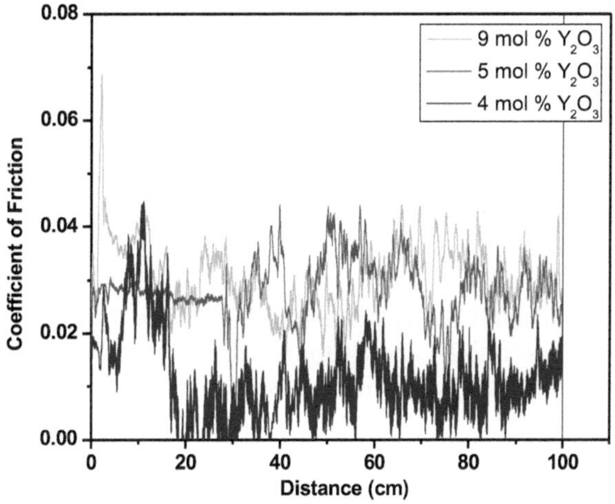

FIGURE 12.34

The coefficient of friction versus distance.

From S.S. Chopade, C. Nayak, D. Bhattacharyya, S.N. Jha, R.B. Tokas, N.K. Sahoo, M.N. Deo, A. Biswas, S. Rai, K.H. Thulasi Raman, G.M. Rao, N.K. Pandey, D.S. Patil, RF plasma enhanced MOCVD of yttria stabillized zirconia thin films using octanedionate precursors and their characterization, Appl. Surf. Sci. 355 (2015) 82–92, reprinted with permission.

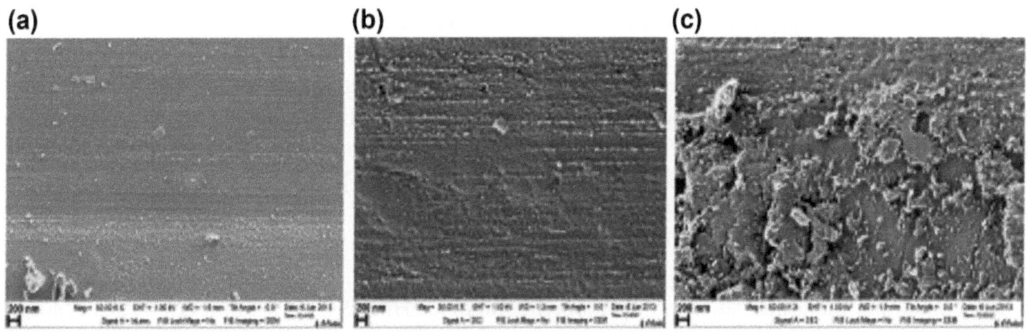

FIGURE 12.35

(a–c) Scanning electron microscope images of the middle portion of the wear track of films A, B, and C.

From S.S. Chopade, C. Nayak, D. Bhattacharyya, S.N. Jha, R.B. Tokas, N.K. Sahoo, M.N. Deo, A. Biswas, S. Rai, K.H. Thulasi Raman, G.M. Rao, N.K. Pandey, D.S. Patil, RF plasma enhanced MOCVD of yttria stabillized zirconia thin films using octanedionate precursors and their characterization, Appl. Surf. Sci. 355 (2015) 82–92, reprinted with permission.

coefficient for films B and C was same. The SEM images showing wear tracks on films A–C are shown in Fig. 12.35.

The enhancement of tribological properties of films A and B in comparison with film C can be attributed to the phenomena of martensitic transformation of the metastable tetragonal phase to the monoclinic phase. There is a 4–5% volume increase associated with this martensitic transformation, leading to compressive stress that arrests crack propagation and imparts fracture toughness to these films [92]. Film C shows the cubic phase and hence is less resistant to fracture compared with the other two films. It has been reported by Sakuma et al. [95] that stability of the tetragonal phase plays an important role in strength and toughness of PSZ. At low Y_2O_3 content, the tetragonal phase is metastable. The metastable tetragonal phase can transform to the monoclinic phase under applied stress. As the Y_2O_3 content increases, the stability of the tetragonal phase increases; the stable tetragonal phase will not undergo stress-induced transformation, leading to a decrease in fracture toughness of the material [95]. This may be the reason for the decrease in wear resistance of film B compared with film A.

In summary, dense Y_2O_3-stabilized zirconia thin films were deposited by rf plasma-enhanced MOCVD using tod precursors $Y(tod)_3$ and $Zr(tod)_4$. Characterization results indicate that growth of these films with improved optical properties is possible through an rf plasma-enhanced MOCVD process using tod precursors. All the films are transparent from the UV to the near-IR range and have nanocrystalline structure. FTIR analysis confirms that the films with 4, 5, and 9 mol% Y_2O_3 are characterized by the presence of predominant monoclinic, tetragonal, and cubic phases, respectively, and mechanical property measurements confirm superior wear resistance of the film with 4 mol % Y_2O_3.

4.5 RADIO FREQUENCY PLASMA-ASSISTED CHEMICAL VAPOR DEPOSITION (FOR SIO_X-BASED CORROSION-RESISTANT COATINGS ON BELL METAL SURFACES

Plasma-based polymerization of organosilicon thin films has received a lot of attention because of various technological applications. Various plasma sources are used for thin film deposition by the plasma-assisted CVD (PACVD) process, and it is evident that rf plasma sources have a special place with regard to the deposition of the dielectric coatings, and in particular, to deposition of polymer-based coatings. This is because PACVD produces highly cross-linked and adherent organosilicon films with a unique combination of properties such as high hardness [96–98], optical transparency [99,100], corrosion resistance [101,102], and low gas permeability [103]. Because of these properties, plasma-polymerized organosilicon films are considered potential candidates for applications in the field of hard and antiscratch coatings [96–98,104], optical filters [105], corrosion resistance layers [97,99,102], gas barriers [106,107], and humidity sensors [108]. For deposition of these films, a silicon-containing monomer such as hexamethyldisiloxane (HMDSO) is commonly used because of its high organic nature and high vapor pressure at room temperature [97]. It is also nontoxic, nonflammable, and cheaper. The quality of the coatings deposited by the rf PACVD process is dependent on the plasma process parameters. To get the desired quality of the coatings, it is essential to optimize the plasma process parameters. The plasma process parameters are rf power, monomer/gas flow rate, working pressure, deposition time, and geometry of the reactor. The rf power, when coupled to the plasma in a capacitively coupled mode, also controls the rf self-bias on the substrates that decides the energy of the ions impinging on the substrates during deposition.

Here we are including an example of the rf PACVD process that is used for deposition of corrosion-resistant and hard plasma-polymerized coatings using an HMDSO precursor. The details regarding this process and characterization of the coatings are given elsewhere [109]. The plasma-polymerized HMDSO (pp-HMDSO) coatings are deposited on bell metal substrates [alloy of copper (78%) and tin (22%) substrates] using a plasma of argon and oxygen gases and vapors of HMDSO. Bell metal is

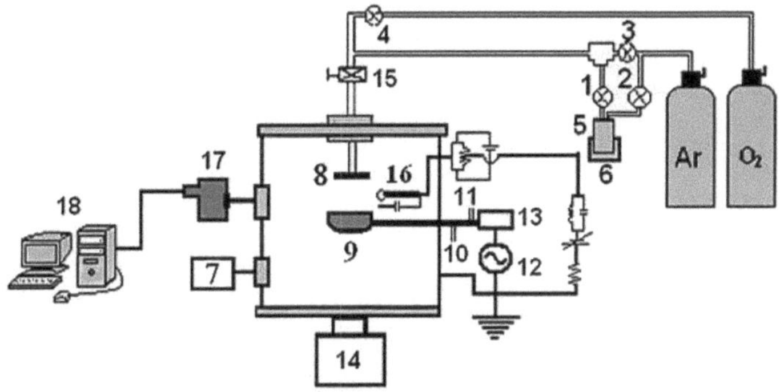

1, 2, 3, 4 = Needle valve	5 = Monomer flask	6 = Constant temperature bath
7 = Vacumn gauge	8 = Gas shower plate	9 = RF electrode
10 = Water in	11 = Water out	12 = RF power source
13 = Matching network	14 = Vacumn system	15 = Stop valve
16 = Self-compensated emissive probe		17 = OES setup
18 = Personal computer		

FIGURE 12.36

Experimental setup of the radio frequency (rf) plasma-polymerization system.

From A.J. Choudhury, J. Chutia, H. Kakati, S.A. Barve, A.R. Pal, N.S. Sarma, D. Chowdhury, D.S. Patil, Studies of radiofrequency plasma deposition of hexamethyldisiloxane films and their thermal stability and corrosion resistance behaviour, Vacuum 84 (2010) 1327–1333, reprinted with permission.

commonly used in the northeastern part of India for making utensils, idols, and various decorative items. Bell metal is susceptible to corrosion on exposure to air and moisture for a long period. The corrosion protection of bell metal components is therefore very important if the items made from this metal are to be used for the said purposes.

The pp-HMDSO films are characterized by step profilometry, FTIR spectroscopy, atomic force microscopy, thermogravimetric analysis (TGA), and the copper acetic acid salt (CASS) test.

A schematic of the capacitively coupled rf plasma system used for deposition of the plasma-polymerized coatings is shown in Fig. 12.36. The experimental chamber used here has a diameter of 400 mm and a height of 450 mm. The chamber is initially evacuated to a base pressure of 6×10^{-6} mbar with a vacuum system consisting of a diffusion pump backed by a rotary pumping system. Argon is used as a carrier gas and bubbled through a container that contains HMDSO (purity >99%, Acros Organics) liquid precursor. The temperature of the container is kept constant at 40°C. The argon gas is then fed through a perforated shower plate so that it enters into the plasma zone that is generated on the substrate holder (10 cm diameter) that is powered by an rf generator (13.56 MHz, 0–500 W) through an impedance matching network. The distance between the rf electrode and shower plate is kept constant at 5 cm. The substrates are kept on this water cooled rf electrode. Because of the capacitively coupled nature of the plasma, an rf self-bias voltage is generated on the rf electrodes and therefore the substrates kept on this electrode are biased to a negative DC self-bias voltage (V_b) that is dependent on the applied rf power. Because the plasma potential (V_p) is positive, the ions are impinging on the substrates with energy that is equal to $q(V_p + V_b)$, where q is the charge of the ion.

Table 12.5 Effect of Radio Frequency Power on the Thickness and Deposition Rate of Plasma-Polymerized Hexamethyldisiloxane Films

Radio Frequency Power (W)	Deposition Time (min)	Film Thickness (nm)	Deposition Rate (nm/min)
20	30	282	9.4
40	30	334	11.1
60	30	551	18.3
80	30	662	22.0
100	30	929	31.0

From A.J. Choudhury, J. Chutia, H. Kakati, S.A. Barve, A.R. Pal, N.S. Sarma, D. Chowdhury, D.S. Patil, Studies of radiofrequency plasma deposition of hexamethyldisiloxane films and their thermal stability and corrosion resistance behaviour, Vacuum 84 (2010) 1327–1333, reprinted with permission.

Because the bias voltage controls the rf self-bias, it also controls the energy of the ions impinging on the substrates. Before the deposition, the substrates are polished and cleaned with acetone, isopropyl alcohol, and distilled water. The substrates are then subjected to an rf argon gas plasma treatment (flow rate: 5 SCCM) at an rf power of 40 W (−55 V bias) for 10 min in order to remove the native oxide layer present on the surfaces. The deposition is then carried out for 30 min at an operating pressure of 1.2×10^{-1} mbar under varying influences of rf power (20–100 W). Both argon and oxygen gas flow rates are kept at 5 SCCM, and the HMDSO monomer flow rate is maintained at 35 SCCM. After the deposition, the rf power is switched off and substrates are cooled in flowing oxygen gas (35 SCCM) for 15 min.

The thickness of the deposited films was measured using a profilometer (Veeco, Dektak-150, USA) with a vertical resolution of 1 A (maximum). The rf bias level on the substrate was in the range of 25–125 V, depending on the power (20–100 W). The effect of rf power on the thickness and deposition rate of pp-HMDSO films is given in Table 12.5. From this data, it is apparent that the deposition rate of the pp-HMDSO film increases with rf power. This is attributed to the increase in electron density with enough available energy, which can be used for plasma polymerization of HMDSO monomer.

FTIR spectra taken on the deposited films to investigate the internal chemical structures of the pp-HMDSO films are shown in Fig. 12.37a. The assignments of the peaks are listed in Table 12.6. As seen from the spectra contained in Fig. 12.37a, the characteristic features of Si–O–Si groups are evident from the presence of the bands at 1053, 800, and 450 cm^{-1}, which signify the asymmetric stretching, bending, and rocking modes, respectively [110–112]. The absorption band at 1260 cm^{-1} relative to the bending mode of the methyl group in $Si(CH_3)_x$ indicates the presence of organic content in the films [111,112]. However, the spectra does not indicate a band corresponding to the $-CH_3$ mode, which is one of the main functional groups of the HMDSO monomer [110,111]. A weak peak at 1620 cm^{-1} signifying the −C=O mode is seen in the spectra above an rf power of 20 W and is indicated in Fig. 12.37a [112,113]. The band corresponding to the broad signal in the region of about 3700–3000 cm^{-1} can be attributed to the −OH stretching mode in H-bonded silanol. For convenience, the FTIR spectra of the pp-HMDSO films in the spectral range (1500–400 cm^{-1}) of interest are shown in Fig. 12.37b.

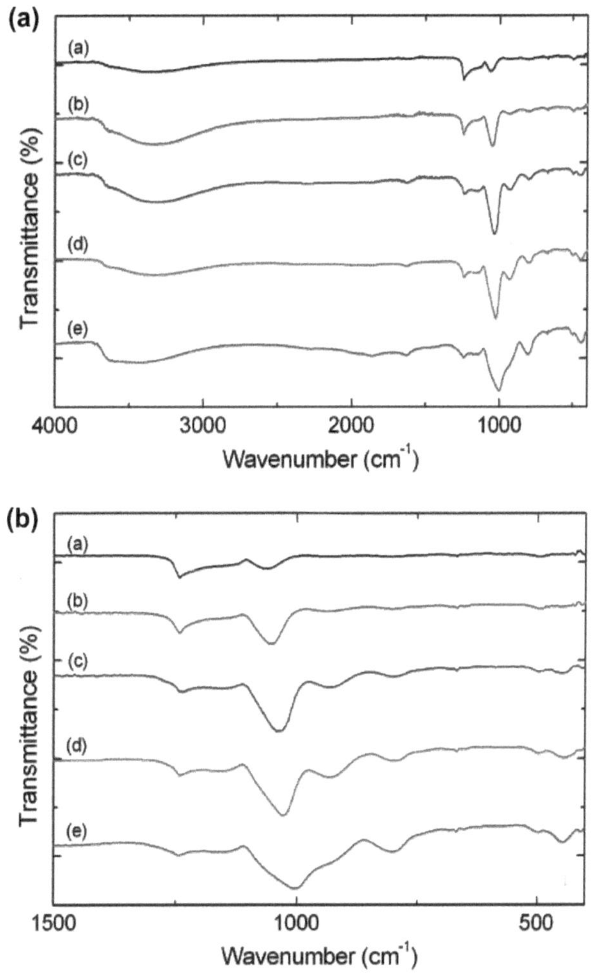

FIGURE 12.37

(a) Fourier transform infrared spectroscopy (FTIR) spectra of plasma-polymerized hexamethyldisiloxane (pp-HMDSO) films prepared at radio frequency (rf) powers of (*a*) 20, (*b*) 40, (*c*) 60, (*d*) 80, and (*e*) 100 W. (b) FTIR spectra of pp-HMDSO films prepared at rf powers of (*a*) 20, (*b*) 40, (*c*) 60, (*d*) 80, and (*e*) 100 W in the spectral range of interest (1500–400 cm^{-1}).

From A.J. Choudhury, J. Chutia, H. Kakati, S.A. Barve, A.R. Pal, N.S. Sarma, D. Chowdhury, D.S. Patil, Studies of radiofrequency plasma deposition of hexamethyldisiloxane films and their thermal stability and corrosion resistance behaviour, Vacuum 84 (2010) 1327–1333, reprinted with permission.

Table 12.6 Assignment of Peaks Detected in Fourier Transform Infrared Spectroscopy Spectra of Plasma-Polymerized Hexamethyldisiloxane Films Prepared in the Radio Frequency Power Range of 20–100 W

Wave Number (cm^{-1})	Assignments
3700–3000	–OH stretching mode
1620	–C=O mode
1253	$Si(CH_3)_x$ symmetrical bending mode
1052	Si–O–Si asymmetric stretching mode
930	Si–OH bending mode
800	Si–O–Si bending mode
450	Si–O–Si rocking mode

From A.J. Choudhury, J. Chutia, H. Kakati, S.A. Barve, A.R. Pal, N.S. Sarma, D. Chowdhury, D.S. Patil, Studies of radiofrequency plasma deposition of hexamethyldisiloxane films and their thermal stability and corrosion resistance behaviour, Vacuum 84 (2010) 1327–1333, reprinted with permission.

Spectra as shown in Fig. 12.37b indicate that:

1. the intensity of bands relative to Si–O–Si (1053 and 800 cm^{-1}) increases;
2. the Si–O–Si stretching peak shifts to a lower wave number region (1053–1040 cm^{-1});
3. the Si–OH (930 cm^{-1}) band intensity increases in the rf power range of 20–80 W, and at 100 W the intensity of the band significantly decreases;
4. the intensity of band resulting from the $Si(CH_3)_x$ mode decreases, and;
5. a new band assigned to the Si–O–Si (450 cm^{-1}) mode appears in the spectra above an rf power of 40 W and subsequently the intensity of the band increases.

The shifting of band for the Si–O–Si mode has previously been reported [107,114,115]. It is now well accepted that shifts in the Si–O–Si stretching wave number toward lower values indicate an increase in the film density [116,117]. The decrease in the intensity of the band relative to the $Si(CH_3)_x$ mode may indicate the enhanced character of the cross-linked structure of the films [110].

From this discussion, it is also evident that pp-HMDSO film chemistry tends to be more inorganic in nature with increasing rf power. From the TGA, it is observed that the deposited films are stable up to 185°C, and about 58% of its total weight loss occurred near 610°C. A typical TGA plot on the film deposited at 60 W rf power is shown in Fig. 12.38.

In order to study the corrosion behavior of the deposited films, a standard CASS test (ASTM-B-368) [118,119] was performed on the deposited pp-HMDSO films. A homogeneous solution prepared from 5% NaCl and 0.026% $CuCl_2$ solutions with pH of the solution maintained at 3.1 by adding the required amount of glacial acetic acid was used for these studies. One-third of each film was dipped inside this solution for a total observation time of 72 h. A virgin bell metal sample (not pp-HMDSO–coated) dipped inside the solution served as a reference. The virgin bell metal sample and the films were observed visually as well as with an optical microscope (Metzer, MBE 9995, resolution × 15) every 12 h. After 24 h of observation, the portion of the virgin bell metal surface

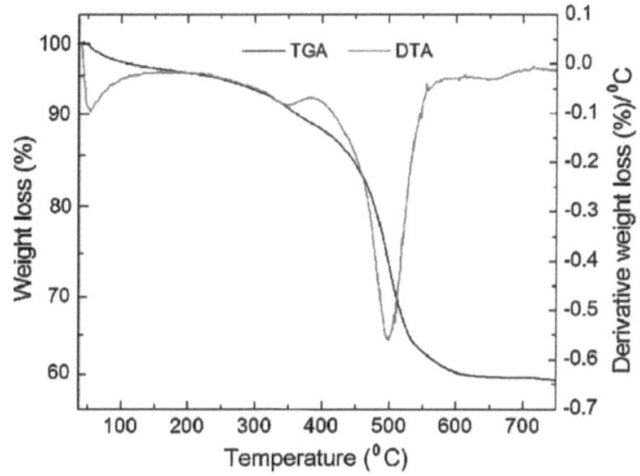

FIGURE 12.38

A typical thermogravimetric analysis (TGA) plot on the film deposited at 60 W radio frequency power. *DTA*, differential thermographic analysis.

From A.J. Choudhury, J. Chutia, H. Kakati, S.A. Barve, A.R. Pal, N.S. Sarma, D. Chowdhury, D.S. Patil, Studies of radiofrequency plasma deposition of hexamethyldisiloxane films and their thermal stability and corrosion resistance behaviour, Vacuum 84 (2010) 1327–1333, reprinted with permission.

dipped inside the solution showed a change in color from bright golden to blackish, thereby showing corrosion of the surface. Fig. 12.39a shows the corrosion of virgin bell metal. The inset of Fig. 12.39a is the microscopic image of the corroded part of the virgin bell metal. During the same observation time (24 h), the pp-HMDSO–coated films showed no sign of cracks or spots on their surfaces. After 48 h, the portions of the films prepared at rf powers of 20 and 40 W, respectively, and dipped inside the solution were peeled off because of permeation of salt solution into the film surfaces. The other films showed no change in their surface morphologies as observed visually, as well as with the optical microscope. Regarding the film prepared at an rf power of 60 W, it was found that at an observation time of 60 h, the portion of its surface dipped inside the solution was peeled off, and minute cracks were observed on the surface of the film prepared at an rf power of 80 W. Fig. 12.39b shows the surface of the film prepared at an rf power of 80 W and the inset shows the microscopic image of the corroded film surface. The film deposited at an rf power of 100 W showed no sign of cracks (Fig. 12.39c), even after the total observation time of 72 h. The inset of Fig. 12.39c shows the microscopic image of the film surface dipped inside the salt solution. The better corrosion resistance behavior of the film deposited at an rf power of 100 W in comparison with the films prepared at lower powers (20–80 W) may be attributed to the relatively high contents of inorganic complexes such as Si–O–Si groups present in the film, as revealed by FTIR analyses.

From the discussion, it is clear that the properties of the deposited pp-HMDSO films are strongly dependent on the plasma process parameters. The impinging ion energies play an important role in determining the properties of the deposited films. It was observed that the pp-HMDSO films tend to be more inorganic in nature with increasing rf power. Film (prepared at 100 W) exhibits better chemical stability in comparison with the films deposited at lower powers (20–80 W). The results also indicate

FIGURE 12.39

Images showing visually observed corrosion tested surfaces of a virgin bell metal sample (a) and plasma-polymerized hexamethyldisiloxane films prepared at radio frequency powers of (b) 80 Wand (c) 100 W, respectively. The insets show the microscopic images of the virgin bell metal and film surfaces kept under the salt solution.

From A.J. Choudhury, J. Chutia, H. Kakati, S.A. Barve, A.R. Pal, N.S. Sarma, D. Chowdhury, D.S. Patil, Studies of radiofrequency plasma deposition of hexamethyldisiloxane films and their thermal stability and corrosion resistance behaviour, Vacuum 84 (2010) 1327–1333, reprinted with permission.

that for corrosion-resistant coating applications, films deposited at 100 W (self bias: −115 V) are the potential candidates to be used on bell metal surfaces against a severe and corrosive environment with good thermal stability.

5. CONCLUSIONS

A detailed discussion regarding cold plasma-based processing of materials that are required for extreme conditions is given here. The chapter has given an in-depth discussion of different types of low-pressure plasma sources that are being used for plasma-assisted CVD of materials for extreme conditions. Importance of plasma in processing the thin films of materials like diamond, DLC, CNTs, Y_2O_3, Y_2O_3-stabilized zirconia, and SiO_x is evident from the detailed discussion regarding the preparation of these materials and their characterization through various tools.

ACKNOWLEDGMENTS

The authors would like to acknowledge the contributions from all coauthors of their various publications from which input was taken in preparation of this manuscript.

REFERENCES

[1] A. Grill, Cold Plasma Materials Fabrication: From Fundamentals to Applications, Wiley-IEEE Press, 1994.
[2] H. Conrads, M. Schmidt, Plasma generation and plasma sources, Plasma Sources Sci. Technol. 9 (2000) 441–454.
[3] A. Ganguli, R.D. Tarey, Understanding plasma sources, Curr. Sci. 83 (3) (2002) 279–290.
[4] D.S. Patil, K. Ramachandran, A.L. Bhide, N. Venkatramani, An Experimental Facility for Microwave Induced Plasma Processing of Materials Report BARC/1997/E/025.
[5] L. Martinu, A. Raveh, A. Domingue, L. Bertrand, J.E. Klemberg-Sapieha, S.C. Gujrathi, M.R. Wertheimer, Hard carbon films deposited under high ion flux, Thin Solid Films 208 (1) (1992) 42–47.
[6] A.A. Voevodin, S.J.P. Laube, S.D. Walck, J.S. Solomon, M.S. Donley, J.S. Zabinski, Pulsed laser deposition of diamond-like amorphous carbon films from graphite and polycarbonate targets, J. Appl. Phys. 78 (1995) 4123–4130.
[7] Z. Has, S. Mitura, Nucleation of allotropic carbon in an external electric field, Thin Solid Films 128 (3–4) (June 28, 1985) 353–360.
[8] P. Reinke, S. Schelz, W. Jacob, W. Moller, Influence of a direct current bias on the energy of ions from an electron cyclotron resonance plasma, J. Vac. Sci. Technol. A 10 (1992) 434–438.
[9] S. Singh, M. Pandey, R. Kishore, N. Chand, S. Dash, A. Tyagi, D. Patil, Influence of impinging ion energy on the bonding and mechanical characteristics of diamond like carbon films deposited by microwave ECR plasma CVD, Plasma Process. Polym. 5 (2008) 853–860.
[10] B. Discchler, A. Bubenzer, P. Koidl, Bonding in hydrogenated hard carbon studied by optical spectroscopy, Solid State Commun. 48 (1983) 105–108.
[11] M. Yoshikawa, G. Katagiri, H. Ishida, A. Ishitani, Raman spectra of diamondlike amorphous carbon films, J. Appl. Phys. 64 (1988) 6464–6468.
[12] M.A. Tamor, W.C. Vassell, Raman "fingerprinting" of amorphous carbon films, J. Appl. Phys. 76 (1994) 3823–3830.

[13] R. Wang, C. Mercer, A.G. Evans, C.V. Cooper, H.K. Yoon, Delamination and spalling of diamond-like-carbon tribological surfaces, Diam. Relat. Mater. 11 (2002) 1797–1803.
[14] J.J. Gracio, Q.H. Fan, J.C. Madaleno, Diamond growth by chemical vapour deposition, J. Phys. D Appl. Phys. 43 (2010) 374017.
[15] E. Karl, Spear diamond -ceramic coating of the future, J. Am. Ceram. Soc. 72 (121) (1989) 171–191.
[16] W. Paul, May Diamond thin films: a 21st-century material, Philos. Trans. R. Soc. Lond. A 358 (2000) 473–495.
[17] S. Iijima, Helical microtubules of graphitic carbon, Nature 354 (1991) 56–58.
[18] P. Jarillo-Herrero, J.A. Van Dam, L.P. Kouwenhoven, Quantum supercurrent transistors in carbon nanotubes, Nature 439 (2006) 953–956.
[19] T.T. Baby, R.B. Rakhi, N. Ravi, S. Ramaprabhu, Cerium oxide dispersed multi-walled carbon nanotubes as cathode material for flexible field emitters, J. Nanosci. Nanotechnol. 12 (2012) 6718–6723.
[20] W. Yi, Q. Yang, CVD growth and field electron emission from of aligned carbon nanotubes on oxidized Inconel plates without addition of catalyst, Diam. Relat. Mater. 19 (2010) 870–874.
[21] W. Yi, Q. Yang, CVD growth of secondary carbon nanotubes and its effect on field electron emission, Diam. Relat. Mater. 19 (2010) 923–926.
[22] W. Yi, Q. Yang, Aligned growth and alignment mechanism of carbon nanotubes by hot filament chemical vapor deposition, Appl. Phys. A 98 (2010) 659–669.
[23] Y.S. Li, A. Hirose, Controlled synthesis of diamond and carbon nanotubes on Ni-base alloy, Appl. Surf. Sci. 255 (2008) 2251–2255.
[24] C. Masarapu, B. Wei, Direct growth of aligned multiwalled carbon nanotubes on treated stainless steel substrates, Langmuir 23 (2007) 9046–9049.
[25] C.E. Baddour, F. Fadlallah, D. Nasuhoglu, R. Mitra, L. Vandsburger, J. Meunier, A simple thermal CVD method for carbon nanotube synthesis on stainless steel 304 without the addition of an external catalyst, Carbon 47 (2008) 313–347.
[26] J. Ting, W. Lin, Unprecedented re-growth of carbon nanotubes on in-situ re-activated catalyst, Nanotechnology 20 (2009) 025608-1–025608-5.
[27] S. Sridhar, L. Ge, C.S. Tiwary, A.C. Hart, S. Ozden, K. Kalaga, S. Lei, S.V. Sridhar, R.K. Sinha, H. Harsh, K. Kordas, P.M. Ajayan, R. Vajtai, Enhanced field emission properties from CNT arrays synthesized on Inconel superalloy, ACS Appl. Mater. Interfaces 6 (2014) 1986–1991.
[28] E. Shin, G. Jeong, Enhancement of carbon nanotube growth yield on Inconel 600 substrates through the surface pretreatments combining thermal annealing and plasma ion bombardment, Curr. Appl. Phys. 14 (2014) 8–12.
[29] R. Kar, S.G. Sarkar, C.B. Basak, A. Patsha, S. Dhara, C. Ghosh, D. Ramachandran, N. Chand, S.S. Chopade, D.S. Patil, Effect of substrate heating and microwave attenuation on the catalyst free growth and field emission of carbon nanotubes, Carbon 94 (2015) 256–265.
[30] V. Kaushik, H. Sharma, A.K. Shukla, V.D. Vankar, Modification in surface morphology and enhanced field emission properties of pristine carbon nanotubes by introducing nitrogen gas, AIP Conf. Proc. 1451 (2012) 148–150.
[31] S.K. Srivastava, V.D. Vankar, D.V. Sridhar Rao, V. Kumar, Enhanced field emission characteristics of nitrogen-doped carbon nanotube films grown by microwave plasma enhanced chemical vapor deposition process, Thin Solid Films 515 (2006) 1851–1856.
[32] B. Padya, D. Kalita, P.K. Jain, G. Padmanabham, M. Ravi, K.S. Bhat, Self-organized growth of bamboo-like carbon nanotube arrays for field emission properties, Appl. Nanosci. 2 (2012) 253–259.
[33] K. Hii, R.R. Vallance, S.B. Chikkamaranahalli, M.P. Mengüç, A.M. Rao, Characterizing field emission from individual carbon nanotubes at small distances, J. Vac. Sci. Technol. B 24 (2006) 1081–1087.

[34] M. Kumari, S. Gautam, P.V. Shah, S. Pal, U.S. Ojha, A. Kumar, A.A. Naik, J.S. Rawat, P.K. Chaudhury, Harsh, R.P. Tandon, Improving the field emission of carbon nanotubes by lanthanum-hexaboride nanoparticles decoration, Appl. Phys. Lett. 101 (2012) 123116-1−123116-5.
[35] R.P. Gao, Z. Pan, Z. Wang, Work function at the tips of multiwalled carbon nanotubes, Appl. Phys. Lett. 78 (2001) 1757-1−1757-3.
[36] S.Y. Wang, Zu H. Lu, Preparation of Y_2O_3 thin films deposited by pulse ultrasonic spray pyrolysis, Mater. Chem. Phys. 78 (2002) 542−545.
[37] M. Gurvitch, L. Manchanda, J.M. Gibson, Study of thermally oxidized yttrium films on silicon, Appl. Phys. Lett. 51 (1987) 919−921.
[38] V.V. Bakovets, T.M. Levashova, V.T. Ratushnyak, L.F. Bakhturova, Chemical vapor deposition of Y_2O_3 films using $Y(dpm)_3$, Inorg. Mater. 38 (4) (2002) 371−373.
[39] R.N. Sharma, A.C. Rastogi, Structure and composition of interfacial silicon oxide layer in chemical vapor deposited Y_2O_3-SiO_2 bilayer dielectrics for metal-insulator-semiconductor devices, J. Appl. Phys. 76 (1994) 4215−4224.
[40] D. Niu, R.W. Ashcraft, Z. Chen, S. Stemmer, G.N. Parsons, Chemical, physical and electrical characterizations of oxygen plasma assisted chemical vapor deposited yttrium oxide on silicon, J. Electrochem. Soc. 150 (5) (2003) F102−F109.
[41] W.J. Varhue, M. Massimo, J.M. Carrulli, V. Baranauskas, E. Adams, Deposition of Y_2O_3 by plasma enhanced organometallic chemical vapor deposition using an electron cyclotron resonance source, J. Vac. Sci. Technol. A 11 (1993) 1870−1874.
[42] G. Apostolopoulous, G. Vellianitis, A. Dimoulas, High epitaxial quality Y_2O_3 high-κ dielectric on vicinal Si (001) surfaces, Appl. Phys. Lett. 81 (19) (2002) 3549−3551.
[43] P. de Rouffignac, J.-S. Park, R.G. Gordon, Atomic layer deposition of Y_2O_3 thin films from yttrium Tris(N,N-diisopropylacetamidinate) and water, Chem. Mater. 17 (2005) 4808−4814.
[44] Y.J. Lee, W.C. Lee, C.W. Nieh, Z.K. Yang, A.R. Kortan, M. Hong, J. Kwo, C.-H. Hsu, High-quality nanothick single-crystal Y_2O_3 films epitaxially grown on Si(111):Growth and structural characteristics, J. Vac. Sci. Technol. B 26 (3) (2008) 1124−1126.
[45] J. Leng, Z. Yu, Y. Li, D. Zhang, X. Liao, W. Xue, Optical and electrical properties of Y_2O_3 thin films prepared by ion beam assisted deposition, Appl. Surf. Sci. 256 (2010) 5832−5836.
[46] R.J. Gaboriaud, F. Pailloux, J. Perriere, Pulsed laser deposition of Y_2O_3 thin films on MgO, Appl. Surf. Sci. 186 (2002) 477−482.
[47] S. Zhang, R. Xiao, Yttrium oxide films prepared by pulsed laser deposition, J. Appl. Phys. 83 (7) (1998) 3842−3848.
[48] J.J. Araiza, M.A. Aguilar-Frutis, C. Falcony, Optical, electrical and structural characteristics of yttrium oxide films deposited on plasma etched silicon substrates, J. Vac. Sci. Technol. B 19 (2001) 2206−2211.
[49] W.M. Cranton, D.M. Spink, R. Stevens, C.B. Thomas, Growth and dielectric characterization of yttrium oxide thin films deposited on Si by r.f.-magnetron sputtering, Thin Solid Films 226 (1993) 156−160.
[50] S.S. Chopade, S.A. Barve, K.H. Thulasi Raman, N. Chand, M.N. Deo, A. Biswas, S. Rai, G.S. Lodha, G.M. Rao, D.S. Patil, RF plasma MOCVD of Y_2O_3 thin films: effect of RF self bias on the substrates during deposition, Appl. Surf. Sci. 285P (2013) 524−531.
[51] S.A. Barve, Jagannath, N. Mithal, M.N. Deo, N. Chand, B.M. Bhanage, L.M. Gantayet, D.S. Patil, Microwave ECR plasma CVD of cubic Y_2O_3 coatings and their characterization, Surf. Coat. Technol. 204 (2010) 3167−3172.
[52] S.V. Pasko, L.G. Hubert-Pfalzgraf, A. Abrutis, P. Richard, A. Burtastyte, V. Kazlauskiene, New sterically hindered Hf, Zr and Y β-diketonates as MOCVD precursors for oxide films, J. Mater. Chem. 14 (2004) 1245−1251.

[53] H. Zama, S. Oda, Properties of metalorganic precursors for chemical vapor deposition of oxide superconductors, Jpn. J. Appl. Phys. 29 (1990) L1072–L1074.

[54] A.C. Jones, P.R. Chalker, Some recent developments in the chemical vapour deposition of electroceramic oxides, J. Phys. D Appl. Phys. 36 (2003) R80–R95.

[55] M.-H. Cho, D.-H. Ko, K. Jeong, S.W. Whangbo, C.N. Whang, S.C. Choi, S.J. Cho, Structural transition of crystalline Y_2O_3 film on Si(111) with substrate temperature, Thin Solid Films 349 (1999) 266–269.

[56] J. Zhu, Y. Zhu, W. Shen, Y. Wang, J. Han, G. Tian, P. Lei, B. Dai, Growth and characterization of yttrium oxide films by reactive magnetron sputtering, Thin Solid Films 519 (2011) 4894–4898.

[57] C. Durand, C. Vallée, C. Dubourdieu, E. Gautier, M. Bonvalot, O. Joubert, Interface formation during the yttrium oxide deposition on Si by pulsed liquid-injection plasma enhanced metal-organic chemical vapor deposition, J. Vac. Sci. Technol. A 22 (6) (2004) 2490–2499.

[58] V. Craciun, E.S. Lambers, N.D. Bassim, R.K. Singh, Characteristics of ultraviolet-assisted pulsed-laser-deposited Y_2O_3 thin films, J. Mater. Res. 15 (2000) 488–494.

[59] S.A. Barve, Jagannath, N. Mithal, M.N. Deo, A. Biswas, R. Mishra, R. Kishore, B.M. Bhanage, L.M. Gantayet, D.S. Patil, Effects of precursor evaporation temperature on the properties of the yttrium oxide thin films deposited by microwave electron cyclotron resonance plasma assisted metal organic chemical vapor deposition, Thin Solid Films 519 (2011) 3011–3020.

[60] T. Gougousi, Z. Chen, Deposition of yttrium oxide thin films in supercritical carbon dioxide, Thin Solid Films 516 (2008) 6197–6204.

[61] Y. Repelin, C. Proust, E. Husson, J.M. Beny, Vibrational spectroscopy of the C-Form of yttrium sesquioxide, J. Solid State Chem. 118 (1995) 163–169.

[62] Y. Nigara, M. Ishigame, T. Sakurai, Infrared properties of yttrium oxide, J. Phys. Soc. Jpn. 30 (1971) 453–458.

[63] N.T. McDevitt, A.D. Davidson, Infrared lattice spectra of cubic rare earth oxides in the region 700 to 50 cm^{-1}, J. Opt. Soc. Am. 56 (1966) 636–638.

[64] M.-H. Cho, D.-H. Ko, K. Jeong, S.W. Whangbo, C.N. Whang, S.C. Choi, S.J. Cho, Growth stage of crystalline Y_2O_3 film on Si(100) grown by ionized cluster beam deposition, J. Appl. Phys. 85 (1999) 2909–2914.

[65] S. Barve, M. Deo, R. Kar, N. Sreenivasan, R. Kishore, A. Biswas, B. Bhanage, M. Rao, L.M. Gantayet, D. Patil, Microwave ECR plasma assisted MOCVD of Y_2O_3 thin films using $Y(tod)_3$ precursor and their characterization, Plasma Process. Polym. 8 (2011) 740–749.

[66] R.J. Gaboriaud, F. Pailloux, F. Paumier, Characterisation of Y_2O_3 thin films deposited by laser ablation on MgO: why a biaxial epitaxy, Appl. Surf. Sci. 188 (2002) 29–35.

[67] S.A. Barve, Jagannath, M.N. Deo, R. Kishore, A. Biswas, L.M. Gantayet, D.S. Patil, Effect of argon ion activity on the properties of Y_2O_3 thin films deposited by low pressure PACVD, Appl. Surf. Sci. 257 (2010) 215–221.

[68] C. Durand, C. Dubourdieu, C. Vallée, V. Loup, M. Bonvalot, O. Joubert, H. Roussel, O. Renault, Microstructure and electrical characterizations of yttrium oxide and yttrium silicate thin films deposited by pulsed liquid injection plasma enhanced metal organic chemical vapor deposition, J. Appl. Phys. 96 (2004) 1719–1729.

[69] Y.H. Lee, A role of energetic ions in RF-biased PECVD of TiO_2, Vacuum 51 (1998) 503–509.

[70] C.V. Ramana, V.H. Mudavakkat, K.K. Bharthi, V.V. Atuchin, L.D. Pokrovsky, V.N. Kruchinin, Enhanced optical constants of nanocrystalline yttrium oxide thin films, Appl. Phys. Lett. 98 (2011) 031905(1)–031905(3).

[71] Y.S. Kim, W.H. Kim, Optical loss mechanism in yttria thin film waveguides, Opt. Mater. 14 (2000) 229–234.

[72] M.S. Miller, W.L. Skeicher, P.B. Narayan, G.T. Swami, J. Crist, R.L. Peterson, R.R. Thomas, H.F. Erskine, The effect of deposition parameters on the wear properties of yttria-stabilized zirconia overcoats, J. Appl. Phys. 67 (9) (1990) 4872–4874.

[73] Q. Xiao, H. He, S. Shao, J. Shao, Z. Fan, Influences of deposition rate and oxygen partial pressure on residual stress and microstructure of YSZ thin films, Thin Solid Films 517 (2009) 4295–4298.

[74] M. Sillassen, P. Eklund, M. Sridharan, N. Pryds, N. Bonanos, J. Bottiger, Ionic conductivity and thermal stability of magnetron-sputtered nanocrystalline yttria-stabilized zirconia, J. Appl. Phys. 105 (2009) 104907(1)–104907(6).

[75] S. Heiroth, R. Frison, J. Rupp, T. Lippert, E. Meier, E. Gubler, M. Dobeli, K. Conder, A. Wokaun, L. Gauckler, Crystallization and grain growth characteristics of yttria-stabilized zirconia thin films grown by pulsed laser deposition, Solid State Ionics 191 (2011) 12–23.

[76] X. Guo, On the degradation of zirconia ceramics during low temperature annealing in water or water vapor, J. Phys. Chem. Solids 60 (1999) 539–546.

[77] C.-B. Cao, J.-T. Wang, W.-J. Yu, D.-K. Peng, G.-Y. Meng, Research on YSZ thin films prepared by plasma-CVD process, Thin Solid Films 249 (1994) 163–167.

[78] M. Hartmanova, I. Thurzo, M. Jergel, J. Bartos, F. Kadlec, V. Zelezny, D. Tunega, F. Kundracik, S. Chromik, M. Brunel, Characterization of yttria-stabilized zirconia thin films deposited by electron beam evaporation on silicon substrates, J. Mater. Sci. 33 (1998) 969–975.

[79] S. Chevalier, M. Kilo, G. Borchardt, J. Larpin, MOCVD deposition of YSZ on stainless steels, Appl. Surf. Sci. 205 (2003) 188–195.

[80] S.S. Chopade, C. Nayak, D. Bhattacharyya, S.N. Jha, R.B. Tokas, N.K. Sahoo, M.N. Deo, A. Biswas, S. Rai, K.H. Thulasi Raman, G.M. Rao, N.K. Pandey, D.S. Patil, RF Plasma enhanced MOCVD of Yttria stabilized Zirconia thin films using octanedionate precursors and their characterization, Appl. Surf. Sci. 355 (2015) 82–92.

[81] F. Maury, Recent trends in the selection of metal-organic precursors for MOCVD process, J. Phys. IV 5 (1995). C5-449–C5-463.

[82] T. Hirata, E. Asari, M. Kitajima, Infrared and Raman spectroscopic studies of ZrO_2 polymorphs doped with Y_2O_3 or $C{-}O_2$, J. Solid State Chem. 110 (1994) 201–207.

[83] D. Liu, C. Perry, R. Ingel, Infrared spectra in non stoichiometric yttria stabilized zirconia mixed crystals at elevated temperatures, J. Appl. Phys. 64 (1988) 1413–1417.

[84] H. Zhang, Y. Liu, K. Zhu, G. Siu, Y. Xiong, C. Xiong, Infrared spectra of nanometer granular zirconia, J. Phys. Condens. Matter 11 (1999) 2035–2042.

[85] H. Holzschuh, H. Suhr, Textured (100) yttria-stabilized zirconia thin films deposited by plasma enhanced chemical vapor deposition, Appl. Phys. Lett. 59 (1991) 470–472.

[86] M. Boulouz, F. Tcheliebou, A. Boyer, Electrical and optical properties of magnetron sputtered Y_2O_3 stabilized ZrO_2 thin films, J. Eur. Ceram. Soc. 17 (1997) 1741–1748.

[87] S. Heiroth, R. Ghisleni, T. Lippert, J. Michler, A. Wokaun, Optical and mechanical properties of amorphous and crystalline yttria-Stabilized zirconia thin films prepared by pulsed laser deposition, Acta Mater. 59 (2011) 2330–2340.

[88] S.R. Casolco, J. Xu, J.E. Garay, Transparent/translucent polycrystalline nanostructured yttria stabilized zirconia with varying colors, Scr. Mater. 58 (2008) 516–519.

[89] M. Boulouz, A. Boulouz, A. Giani, A. Boyer, Influence of substrate temperature and target composition on properties of yttria-stabilized zirconia thin films by r.f. reactive magnetron sputtering, Thin Solid Films 323 (1998) 85–92.

[90] V.G. Varanasi, T.M. Besmann, R.L. Hyde, E.A. Payzant, T.J. Anderson, MOCVD of YSZ coatings using β-diketonate precursors, J. Alloys Compd. 470 (2009) 354–359.

[91] S.-H. Cho, Z.–T. Park, J.–G. Kim, J.–H. Boo, Physical and optical properties of plasma polymerized thin films deposited by PECVD method, Surf. Coat. Technol. 174–175 (2003) 1111–1115.
[92] D. Majumdar, D. Chatterjee, X-ray photoelectron spectroscopic studies on yttria, zirconia and yttria-stabilized zirconia, J. Appl. Phys. 70 (1991) 988–992.
[93] G.M. Ingo, Origin of darkening in 8 wt% yttria-zirconia plasma sprayed thermal barrier coatings, J. Am. Ceram. Soc. 74 (1991) 381–386.
[94] A.M. Korduban, I.A. Yashchishyn, T.E. Konstantinova, I.A. Danilenko, G.K. Volkova, V.A. Glazunova, X-ray photoelectron spectroscopy of nanopowders of ZrO_2-Y_2O_3-Cr_2O_3 compounds, Funct. Mater. 14 (2007) 454–458.
[95] T. Sakuma, Y.-I. Yoshizawa, H. Suto, The microstructure and mechanical properties of yttria-stabilized zirconia prepared by arc-melting, J. Mater. Sci. 20 (1985) 2399–2407.
[96] A.M. Worbel, M.R. Wertheimer, Plasma-Polymerized Organosilicons and Organometallics Plasma Deposition, Treatment, and Etching of Polymers, Academic Press, London, 1990, pp. 163–268.
[97] L. Zajíčková, V. Buršíková, Z. Kučerová, D. Franta, P. Dvořák, R. Šmíd, V. Peřina, A. Macková, Deposition of protective coatings in rf organosilicon discharges, Plasma Sources Sci. Technol. 16 (2007) S123–S132.
[98] L. Zajíková, V. Buríková, V. Peina, A. Macková, J. Jana, Correlation between SiO_x content and properties of DLC: SiO_x films prepared by PECVD, Surf. Coat. Technol. 174 (2003) 281–285.
[99] M. Naddaf, S. Saloum, H. Hamadeh, Photoluminescence from PP-HMDSO thin films deposited using a remote plasma of 13.56 MHz hollow cathode discharge, J. Phys. D Appl. Phys. 40 (2007) 4015–4020.
[100] S. Saloum, M. Naddaf, B. Alkhaled, Properties of thin films deposited from HMDSO/O_2 induced remote plasma: effect of oxygen fraction, Vacuum 82 (2008) 742–747.
[101] E. Vassallo, A. Cremona, L. Laguardia, E. Mesto, Preparation of plasma-polymerized SiO_x-like thin films from a mixture of hexamethyldisiloxane and oxygen to improve the corrosion behaviour, Surf. Coat. Technol. 200 (2006) 3035–3040.
[102] C. Vautrin-Ul, F. Roux, C. Boisse-Laporte, J.L. Pastol, A. Chausse, Hexamethyldisiloxane (HMDSO)-plasma-polymerised coatings as primer for iron corrosion protection: influence of RF bias, J. Mater. Chem. 12 (2002) 2318–2324.
[103] A. Grüniger, A. Bieder, A. Sonnenfeld, P.R. Roh, U. Müller, R. Hauert, Influence of film structure and composition on diffusion barrier performance of SiO_x thin films deposited by PECVD, Surf. Coat. Technol. 200 (2006) 4564–4571.
[104] R. Morent, N.D. Geyter, S.V. Vlierberghe, P. Dubruel, C. Leys, E. Schacht, Organic–inorganic behaviour of HMDSO films plasma-polymerized at atmospheric pressure, Surf. Coat. Technol. 203 (2009) 1366–1372.
[105] S. Gaur, G. Vergason, Plasma polymerization: theory and practice, in: 43rd Annual Technical Conference Proceedings—Denver, April 15–20, 2000, pp. 267–271.
[106] A. Bieder, A. Gruniger, R. von Rohr, Deposition of SiO_x diffusion barriers on flexible packaging materials by PECVD, Surf. Coat. Technol. 200 (2005) 928–931.
[107] M. Creatore, F. Palumbo, R. d'Agostino, P. Fayet, RF plasma deposition of SiO_2-like films: plasma phase diagnostics and gas barrier film properties optimization, Surf. Coat. Technol. 142 (2001) 163–168.
[108] N. Guermat, A. Bellel, S. Sahli, Y. Segui, P. Raynaud, Thin plasma-polymerized layers of hexamethyldisiloxane for humidity sensor development, Thin Solid Films 517 (2009) 4455–4460.
[109] A.J. Choudhury, J. Chutia, H. Kakati, S.A. Barve, A.R. Pal, N.S. Sarma, D. Chowdhury, D.S. Patil, Studies of radiofrequency plasma deposition of hexamethyldisiloxane films and their thermal stability and corrosion resistance behavior, Vacuum 84 (2010) 1327–1333.
[110] A. Pfuch, A. Heft, R. Weidl, K. Lang, Characterization of SiO_2 thin films prepared by plasma-activated chemical vapour deposition, Surf. Coat. Technol. 201 (2006) 189–196.

[111] J. Schwarz, M. Schmidt, A. Ohl, Synthesis of plasma-polymerized hexamethyldisiloxane (HMDSO) films by microwave discharge, Surf. Coat. Technol. 98 (1998) 859–864.
[112] G. Czeremuszkin, M. Latrèche, M. Wertheimer, A. da-Silva Sobrinho, Ultrathin silicon-compound barrier coatings for polymeric packaging materials: an industrial perspective, Plasmas Polym. 6 (2001) 107–120.
[113] J.R. Dyer, Applications of Absorption Spectroscopy of Organic Compounds, Prentice-Hall International, Inc., New Jersey, 1978, pp. 33–47.
[114] N. Benissad, K. Aumaille, A. Granier, A. Goullet, Structure and properties of silicon oxide films deposited in a dual microwave-rf plasma reactor, Thin Solid Films 384 (2001) 230–235.
[115] R.A.B. Devine, Densification-induced infrared and Raman spectra variations of amorphous SiO_2, J. Vac. Sci. Technol. A6 (1988) 3154–3156.
[116] K. Kim, D. Kwon, G.S. Lee, Properties of low dielectric constant fluorinated silicon oxide films prepared by plasma enhanced chemical vapor deposition, Thin Solid Films 332 (1998) 369–374.
[117] P. Raynaud, B. Despax, Y. Segui, H. Caquineau, Ftir plasma phase analysis of hexamethyldisiloxane discharge in microwave multipolar plasma at different electrical powers, Plasma Process. Polym. 2 (2005) 45–52.
[118] H. Kakati, A.R. Pal, H. Bailung, J. Chutia, Deposition of nanostructured crystalline and corrosion resistant alumina film on bell metal at low temperature by rf magnetron sputtering, Appl. Surf. Sci. 255 (2009) 7403–7407.
[119] S.M. Borah, A.R. Pal, H. Bailung, J. Chutia, Optimization of plasma parameters for high rate deposition of titanium nitride films as protective coating on bell-metal by reactive sputtering in cylindrical magnetron device, Appl. Surf. Sci. 254 (2008) 5760–5765.

CHAPTER

TAILORED THERMAL EXPANSION MATERIAL FOR HIGH-TEMPERATURE APPLICATIONS

13

A.K. Tyagi, S.N. Achary

Bhabha Atomic Research Centre, Mumbai, India

CHAPTER OUTLINE

1. INTRODUCTION

Thermal expansion behavior of materials has been of importance since the inception of mankind and evolved chronologically in both understanding fundamentals and using suitable materials for application in various nonambient temperature, and is of paramount importance when a given structural material experiences temperature gradient [1,2]. Thus the thermal behaviors of materials, like metals, ceramics, cooking ware components, and high-end engineering structures such as buildings, bridges,

Materials Under Extreme Conditions. http://dx.doi.org/10.1016/B978-0-12-801300-7.00013-9

air/spacecraft bodies, and furnaces for metal industries need consideration. Some of the conventional metals like iron, copper, and aluminum exhibit a high degree of thermal expansion on heating. Thus alloying them with suitable other elements is one of the earliest methods to tune thermal expansion, as well as to attain net thermal expansion of desired value. The discovery of invariable thermal expansion in Invar (an alloy of 35% Ni and 65% Fe) revolutionized controlled thermal expansion material for practical applications, and the discovery led to the Nobel Prize in 1920. This discovery also paved the way for a range of high-precision mechanical instruments, which further culminated in modern technologies like high accuracy clocks and shadow masks in television screens [3]. Compared with metals and alloys, ceramics in general have low thermal expansion. Fused silica is a typical example of a material with low thermal expansion. It is also possible to tailor thermal expansion behavior in composite materials such as glass-ceramics, graded ceramics that find applications in optical instruments, optical mirror substrates, and glass–metal junctions [4,5]. The knowledge of thermal expansion behavior is required for upcoming technologies like solid oxide fuel cells. Likewise, it is an important property required for design of the nuclear fuel-pin assembly. Thermal expansion behavior of electronic materials can be a limiting factor because it causes their dimensional instability as a result of locally generated temperatures, which may lead to circuit failure. In such aspects, several Invar-type alloys and metal matrix composites have been developed [6]. The fundamental understanding of physical origins and subsequent research and development has led to a large number of crystalline materials with low thermal expansion [7,8]. This chapter will primarily focus on materials with tailored thermal expansion behavior such as negative, low, or zero thermal expansion behavior.

2. THERMAL EXPANSION OF MATERIALS: SOME FUNDAMENTAL ASPECTS

The thermal expansion coefficient (TEC) of any material is defined as a relative change in dimension per unit rise in temperature. There are two types of coefficients of thermal expansions: bulk and lattice TECs. The bulk thermal expansion is measured along the direction of a particular shape or aggregates. Thus it is a cumulative effect of both macroscopic (extrinsic) and microscopic (intrinsic) temperature effects. Typical extrinsic contributions to bulk thermal expansion can be grain boundary, micro/macro cracks, voids, and defects in the samples [9]. The change in dimension, i.e., length, is in general termed a *bulk TEC* (α_l) and it is calculated by using Eq. (13.1).

$$\alpha_l = \frac{l_T - l_o}{l_o(T - T_o)} \tag{13.1}$$

where l_T, length of the materials (at the temperature T), l_o, reference length (at reference temperature T_o), α_l, TEC (°C^{-1} or K^{-1}).

Most commonly, the variation of length with temperature is nonlinear and can be represented by polynomial equation (Eq. 13.2):

$$L_T = L_o + a(T - T_o) + b(T - T_o)^2 + c(T - T_o)^3 + d(T - T_o)^4 + \ldots \tag{13.2}$$

The TEC is calculated with respect to the reference temperature. Thus the coefficient of thermal expansion can be calculated with Eq. (13.3):

$$\alpha_l = \frac{1}{L_o}\left(\frac{dL_T}{dT}\right) \tag{13.3}$$

The basic understanding of thermal expansion of solids has revealed its close correlation with the internal structure and thermal vibrations of atoms or ions [10]. As mentioned, the extrinsic contributions of thermal effects are added to this intrinsic expansion from chemical bonds. Often these contributions contribute negatively and thus mask the actual thermal expansion behavior. Hence the bulk thermal expansion in general has lower values then the actual ones. But the formation of point defects may have a larger value compared with lattice expansions. If the sample is free from the extrinsic contributions, like a defect-free single crystal, a comparative value as the intrinsic unit cell expansion is expected. The intrinsic unit cell expansion, commonly referred to as *lattice thermal expansion*, is measured using X-ray or neutron diffraction data as a function of temperature. The volume TEC (α_V in °C^{-1} or K^{-1}) is defined as (Eq. 13.4):

$$\alpha_V = \frac{1}{V_o}\left(\frac{dV}{dT}\right) \tag{13.4}$$

The average volume of thermal expansion of a material is calculated using Eq. (13.5):

$$\alpha_v = \frac{V_T - V_o}{V_o(T - T_o)} \tag{13.5}$$

where V_T, unit cell volume of the materials at the temperature T, V_o, reference volume; i.e., volume of material at reference temperature T_o.

2.1 ORIGIN OF THERMAL EXPANSION

From the atomistic correlation of thermal expansion, it be easily understood that the thermal expansion of solids is the result of the expansion in bond lengths, and thus its origin remains at the anharmonicity of the interatomic potential energy. Thus the thermal expansion of a solid can be explained easily by the potential-well model [11,12]. The potential energy (U) of two interacting atoms depends on the separation between them and shows an asymmetric well-like feature, as shown in Fig. 13.1. It can be seen from Fig. 13.1 that the width of asymmetry increases on increasing temperature because of longitudinal vibration, which leads to an increase in the equilibrium separation of the bonded pair of atoms by δ_r value. The mean separation (r) and vibration amplitude in turn depend on the force constant of the chemical bond between two atoms. Thus it can be inferred that the thermal expansion of solids is nothing but expansion of chemical bonds. In other words, thermal expansion of solids is related to their chemical bonding strength, which in turn is related to the lattice energy. The crystal structure also has a dominant role in governing thermal expansion behavior, which will be discussed in a subsequent section of this chapter. In summary, the thermal expansion of a solid can be correlated with the nature and strength of the chemical bond, mass of the vibrating atoms, melting point, and crystal structure (e.g., packing fraction). It may be added here that all of these factors that influence thermal expansion behavior are also interrelated.

3. EXPERIMENTAL TECHNIQUES FOR THERMAL EXPANSION MEASUREMENTS

The instrumentation for measurement of thermal expansion of solids has a rich history. In an early stage, marker comparison methods were used for this purpose, which were rather crude and were

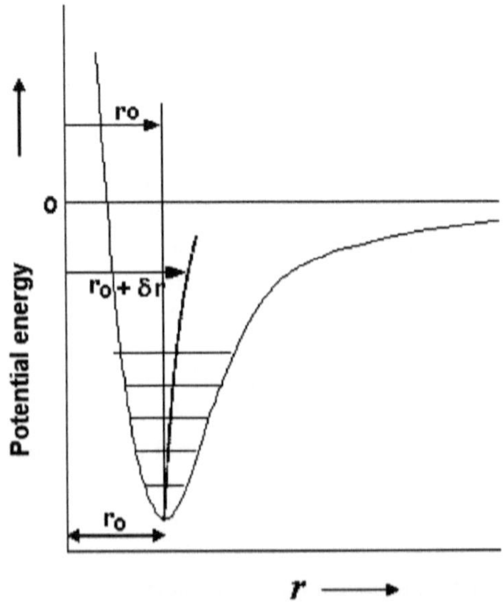

FIGURE 13.1

Interatomic potential energy (*U*) as a function of separation of a pair of atoms (*r*).

applicable for materials exhibiting larger expansion. Presently, sophisticated and automated thermal expansion measurement instruments like the thermodilatometer and interferometers are being employed. The principles and procedures of measurement of thermal expansion of solids have been elaborated extensively [13–17]. As discussed earlier, thermal expansion can be classified as bulk and lattice thermal expansion. The measurement of bulk thermal expansion involves a direct measurement of the change in dimension as a function of temperature, which can be done by a thermodilatometer or interferometer. As far as the lattice thermal expansion is concerned, diffraction methods like variable temperature X-ray or neutron diffraction are used. Usually thermal expansion of a material at a given temperature is measured with respect to a reference temperature, which is often room temperature. In this chapter, mainly dilatometer and diffraction methods for thermal expansion measurement will be discussed.

3.1 DILATOMETER

The most commonly used technique for thermal expansion measurement is push-rod–type dilatometer [2,3,17]. In this technique, the dimensions of the sample in pellet form are measured as per predetermined heating or cooling. The well-sintered pellet is mounted in a furnace and a push-rod is placed in direct contact with it. The expansion or contraction of the pellet moves the push-rod upward or downward. The push-rod is kept in contact with a suitable transducer that quantifies the extent of movement. The quantifiable output of the transducer can be electrical, mechanical, or optical signals. The accuracy of the dilatometer is governed by the sensitivity of the transducer. The linear variable differential transformer (LVDT) is one of the most common transducers used in dilatometers, and measures the expansion or contraction by producing a proportional quantity of electrical signal, i.e. emf in Volts. Also, the dilatometer can have either a vertical or a horizontal configuration. The nature of the push-rod decides the upper temperature limit of the measurement using push-rod dilatometers. Fused silica rods can be used safely below 1000°C because above that temperature, fused silica has tendency toward devitrification. For high-temperature measurements up 1700°C, it is advisable to use high-purity, well-sintered alumina rods, in view of their high melting temperature and superior mechanical strength. Since alumina itself has a high TEC, a blank run of data is always needed to get the exact thermal expansion behavior of a material from the overall expansion (sample + alumina push-rod). Dual push-rod type dilatometers have self-correcting advantages for expansion caused by the push-rod. In case of measurements at still higher temperatures, one can use abrasive metals like Mo and Ta (up to 2000°C) and graphite rods (up to 3000°C, but in a vacuum or inert atmosphere). A typical block diagram of a single push-rod dilatometer is shown in Fig. 13.2. The LVDT attached at the top of the push rod converts the dilation of the material under study to an equivalent electrical (emf) signal and that can be directly read through a digital converter.

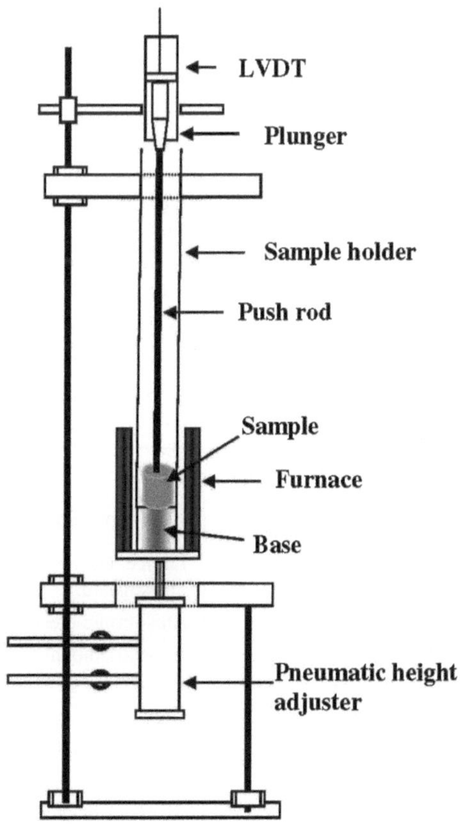

FIGURE 13.2

Schematic diagram of a thermodilatometer.

3.2 DIFFRACTION METHODS FOR THERMAL EXPANSION STUDIES

Diffraction methods, in particular X-ray diffraction (XRD), have been used for accurate measurement of thermal expansion of solids [18,19]. In such measurements, the diffraction data of the sample is recorded as a function of temperature. Most commonly a sample stage-cum-heater is used for this purpose, which often is made of platinum or Pt-Rh strip and can be used at temperatures up to 1600°C. Thermal expansion data at still higher temperatures can be obtained by employing W or Ta heaters. For such transition metals or graphite sample holders or heaters, an inert atmospheres or a vacuum is essential to prevent oxidation. Apart from the temperature requirement, the selection of heaters in diffraction experiments depends on their compatibility with sample. For example, it is not advisable to use platinum-based heaters for Bi- or Pb-containing compounds. On the other hand, platinum-based heaters are highly compatible with fluoride-based samples. These are just a few typical examples to emphasize the criterion of compatibility. It is necessary to evaluate the compatibility on a case-by-case basis. The diffraction data can also be recorded at low temperatures by using a cryostat. There are also cases where an external heater is used and heat is transferred to the sample by radiation or convection. Hot or cold gas guns are also used for heating the sample externally during measurements of diffraction data. The design and heating procedures for diffraction experiments are usually custom made to suit the requirements. Because the measured quantities in diffraction techniques are unit cell parameters, they provide information on lattice expansion.

In summary, thermodilatometer and high-temperature XRD (HT-XRD) provide information on bulk and lattice thermal expansion behaviors, respectively. The diffraction methods have several advantageous over dilatometry. The merits and demerits of diffraction methods and dilatometry are given in Table 13.1. The most important advantage of diffraction methods is their ability to measure anisotropic thermal expansion of materials. It is possible to extract thermal expansion data of a sample of interest even from a multiphasic sample, as well expansion of all phases present in it. Such isolable thermal expansion measurements are not possible by dilatometry. On the other hand, the major disadvantage of diffraction methods is that they cannot be used to study thermal expansion behavior of amorphous materials, which is possible by dilatometry. Thus in order to decipher the intricacies of thermal expansion behavior of a material, often both HT-XRD and dilatometry are used in conjunction.

Table 13.1 Merits and Demerits of High-Temperature X-Ray Diffraction and Dilatometry

Thermodilatometry	High-Temperature X-Ray Diffraction
Only bulk thermal expansion can be investigated.	Lattice and volume thermal expansion can be investigated.
Anisotropic thermal expansion cannot be investigated.	Anisotropic thermal expansion can be investigated.
Usually a large amount of sample is required.	A small amount of powder sample is sufficient.
High-density pellets are preferred.	There is no such requirement.
Any solid material can be investigated.	Only crystalline solids can be investigated.
Sensitivity is good.	Sensitivity is very good.
Sinterability studies can be performed.	Sinterability studies cannot be performed.
Only first-order phase transitions can be investigated.	All types of phase transitions can be investigated.
It does not give any information on crystallographic origin of thermal expansion behavior.	It gives information on crystallographic origin of thermal expansion behavior.

4. CORRELATION OF THE THERMAL EXPANSION COEFFICIENT WITH OTHER PHYSICAL PROPERTIES OF MATERIALS

A number of physical properties have strong influence on the magnitude and behavior of the thermal expansion of materials. These physical properties are melting point, nature and strength of bonds, packing efficiency of the lattice, and defects or impurities. Even phase transition can affect thermal expansion behavior. This section will briefly cover this aspect of thermal expansion behavior. Various factors affecting thermal expansion behavior are summarized in Fig. 13.3 and they are briefly mentioned in this section.

Melting temperature of a solid has a strong and direct bearing on thermal expansion. This can be easily understood because the melting process involves temperature-induced gradual expansion of bonds, which eventually leads to dissociation. The following empirical relation (Eq. 13.6) shows a correlation between the TEC and melting temperature [20,21]:

$$\alpha_V = \frac{1}{16.6T_m} \tag{13.6}$$

Likewise, the bond valence, which is a measure of strength of bond, is also strongly related with to dilation or compression of materials [22–24]. A higher strength of bonds will give rise to a rigid structure and in turn will lead to low expansion. Megaw [22,23] has proposed the following empirical relation (Eq. 13.7) for expansion of the chemical bond and Pauling bond valence (S), which in general holds for most of the ionic solids:

$$\alpha_{\text{bond}} \propto \frac{1}{S^2} \tag{13.7}$$

where S, bond valence ($S = z/n$; z is the charge of the cation, n is the coordination number of the cation).

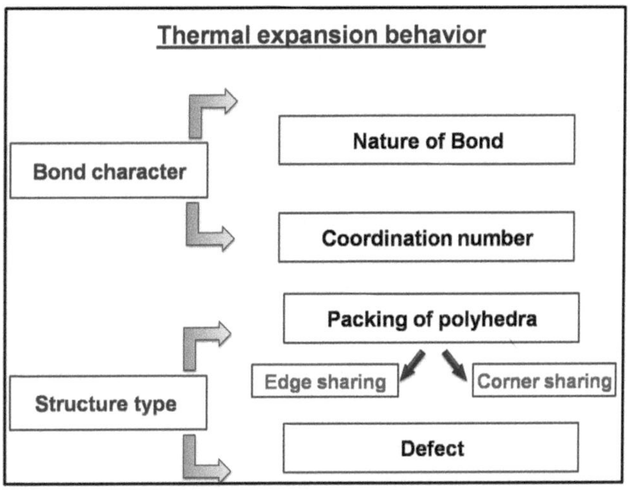

FIGURE 13.3

Scheme depicting various structural factors affecting thermal expansion behavior.

4.1 ROLE OF LATTICE PACKING FRACTION AND COMPRESSIBILITY IN THERMAL EXPANSION

Thermal expansion also bears a close correlation with lattice packing fraction and compressibility of materials. Compressibility in turn has a strong correlation with the nature of chemical bonding and crystallographic structure. The materials that have low packing fractions and high compressibility often show higher thermal expansion behavior, although there could be some deviation from this predicted trend. Typical examples of poor lattice packing and high compressibility are molecular solids [25], which invariably show high thermal expansion. A relationship between compressibility and the TEC can be expressed as $\alpha_l \sim A \times \chi$, where α_l is the TEC, χ = compressibility, and A = a constant [24].

4.2 ROLE OF DEFECTS, IMPURITIES, AND PHASE TRANSITIONS IN THERMAL EXPANSION

The presence of lattice defects like Schottky defects (vacancies) or Frenkel defects (interstitial atoms) has a strong bearing on thermal expansion behavior of a material. Thus the thermal expansion is affected significantly by the nature and kind of defects in the lattice. These defects tend to mask the thermal expansion of a material. In fact, thermal expansion measurements in conjunction with dilatometry and high-temperature XRD are used to study the point defects. Usually the difference between the bulk and lattice TECs gives an estimate of point defects present in the lattice. Likewise, impurities in a material also have an influence on its thermal expansion behavior. Because thermal expansion is an additive property, the presence of any impurity or secondary phase will affect thermal expansion behavior. These have a dominating effect in dilatometric studies. However, they can also influence high-temperature studies when the secondary phase or clusters exert additional strain in the lattice.

There are examples of considerable alteration of electronic structures as a result of solid solution formation, which causes a drastic difference in thermal expansion behavior and cannot be explained just by following the rule of addition. Likewise, phase transitions have a bearing on thermal expansion behavior of a material. In particular, first-order phase transitions are accompanied by an abrupt change in lattice parameters at the transition temperature. The phases above and below phase transition temperatures often exhibit different thermophysical properties owing to the difference in arrangement of atoms and bond distances. Such differences arise because of reconstruction of the structure in transformed phase. HT-XRD is a powerful tool to study the role of phase transition on thermal expansion. Several electroceramics like ferroelectric or relaxor materials are examples of such discontinuity in cell volume as a function of temperature [26–29]. Similarly, materials having electronic phase transition also exhibit different thermal expansion behaviors below and above phase transition temperatures. Typical examples of such materials are Pu, U, Ce, YbGaGe, and $CeAl_3$ [30–33], where anomalous (low or negative) thermal expansion behaviors have been attributed electronic transitions.

5. DIFFERENT CLASSES OF MATERIALS BASED ON THERMAL EXPANSION BEHAVIORS

Materials can be classified into several categories based on their thermal expansion behaviors, including moderate, high, low, or negative thermal expansion material. Materials with moderate thermal expansion have TEC of about 10–30 × 10^{-6}/°C), whereas TEC of about 50–60 × 10^{-6}/°C places materials in the category of high thermal expansion. Typical examples are materials with low melting points such as metals and ionic solids, where the TEC can be as high as about 80–100 × 10^{-6}/°C. On the other hand, materials like silica and diamond, which have strong covalent bonds, are categorized as low TEC materials (TEC about 0.5–5 × 10^{-6}/°C). There are several materials that show negative thermal expansion behavior also. Typical examples of this class of materials are AX_2O_8, AX_2O_7 (A = Zr, Hf; X = W, Mo) [34–38], $M_2(MO_4)_3$ (M = trivalent rare-earth cations like, Sc^{3+}, Y^{3+}, Er^{3+}, and Yb^{3+} or transition metals ions and M = W^{6+} or Mo^{6+}) [39–48], M_2O (M = Cu^+, Ag^+) [49], ABO_2 (A = Cu^+, Ag^+ and B = Al^{3+}, Sc^{3+}, Ln^{3+}) [50]. Likewise, there are materials that exhibit zero thermal expansion behavior.

5.1 TAILORING OF THERMAL EXPANSION BEHAVIOR USING A SOLID SOLUTION FORMATION APPROACH

The discussions covered in the previous sections suggest that the thermal expansion of materials can be tailored in several ways, such as by making appropriate solid solutions, introducing an optimum level of defects, fine tuning crystal structure, tailoring the microstructure, altering the chemical bonding, or designing suitable composites. The subsequent sections of this chapter will discuss strategies to prepare materials with tailored thermal expansion behavior.

Because thermal expansion by and large is an additive property, it follows the rule of addition. Thus it is possible to prepare appropriate solid solutions in such a way so as to obtain a material with desired thermal expansion. Of course, it must be added that there could be other properties also, in addition to thermal expansion, which might govern the composition of solid solutions. The substitutional solid solutions can be of two types: isovalent substituted solid solutions and aliovalent substituted solid solutions. In isovalent solid solutions, the valency of host and guest ions is the same, whereas in case of aliovalent solid solutions, the valency of host and guest ions is different. The aliovalent substitution is always accompanied by cation or anion vacancy, unlike isovalent substitution, which does not introduce any vacancies in cationic or anionic sublattices. It is possible to tailor thermal expansion behavior of materials both by isovalent as well as aliovalent solid solutions. Some typical examples are presented in the following discussions.

Fluorite-type materials with a general composition of MO_2 (M = tetravalent cation) have a range of applications such as nuclear fuels, oxygen storage, and ionic conductivity. Several fluorite-type systems are reported to form isovalent solid solutions either in entire or partial homogeneity, depending upon the processing conditions and relative ionic sizes of host and guest ions. Some of the systems that are well reported in literature are: $Ce_{1-x}Zr_xO_2$ [51,52], $Ce_{1-x}Hf_xO_2$ [53], $Th_{1-x}Ce_xO_2$ [54], $ThO_2-CeO_2-ZrO_2$ [55], CeO_2–YSZ [56], and $Th_{1-x}U_xO_2$ [57,58] systems. Among these systems, ThO_2-CeO_2, ThO_2-UO_2, and ThO_2-PuO_2 form a complete range of solid solution owing to the fact that the end members in these systems are isostructural and the mismatch between host and guest ions is negligible. On the other hand, ZrO_2 or HfO_2 have only a limited solubility (≈5–10 mol %) in CeO_2, ThO_2, or UO_2 because of ionic radii mismatch and also because of differences in their structures.

5.1.1 Isovalent Solid Solution Formation Approach

Thorium has attracted considerable attention because it is expected to play an important role in the third stage of the Indian nuclear power generation program owing to its vast abundance in India.

Thorium is three times more abundant than uranium in earth's crust. ThO_2 is bestowed with excellent properties like low thermal expansion, a high melting point, high thermal conductivity (compared with UO_2), a higher neutron absorption coefficient, and high chemical and radiation stability. These properties make ThO_2 an ideal choice for futuristic nuclear reactors. Advanced heavy water reactors are being contemplated for key role in thorium utilization. Because thorium itself is not a fissile material in the thermal region of neutrons, there is a need to use about 2–6% fissile uranium and plutonium dioxides in the ThO_2 host matrix. In view of this requirement, thermal expansion behavior of a number of thorium-based mixed oxide systems has been investigated [54,59–66]. Because ThO_2 is a high melting solid, its chemical reactivity is poor. In general to prepare thorium-based mixed oxides, a multistep heating protocol is required, which ensures better homogeneity. The lattice thermal expansion behavior of these mixed oxides was investigated by refining the cell parameters at different temperatures using least squares methods.

The thermal expansion behavior of the ThO_2–PuO_2 system is not well reported in literature. The main difficulties in investigating PuO_2-based systems are their high radioactivity and their toxicity. One way to overcome this problem is the use of CeO_2 as a surrogate in place of PuO_2. Both CeO_2 and PuO_2 have quite similar physicochemical properties, such as ionic sizes of Ce^{4+} and Pu^{4+} in octahedral and cubic coordination, melting points, standard enthalpy of formation, and specific heat. Typical physical properties of CeO_2 and ThO_2 are summarized in Table 13.2, which reveals that CeO_2 is an ideal surrogate material to simulate thermophysical properties of PuO_2, and PuO_2-based systems [63]. The gradual decrease in the lattice parameter as a function of Ce^{4+} concentration in a $Th_{1-x}Ce_xO_2$ series can be attributed to the differnces in ionic radii of Th^{4+} and Ce^{4+}, which are 1.05 and 0.90 Å, respectively, in eightfold coordination. It was observed that the substitution of CeO_2 for ThO_2 had a noticeable effect on its thermal expansion behavior. The increase in lattice TEC (α_a) on going from ThO_2 to CeO_2 can be attributed to a relatively higher (α_a)value of CeO_2, which in turn can be correlated to its lower melting point as compared with that of ThO_2. ThO_2 has melting point of 3377°C, whereas CeO_2 melts at 2600°C. The variation of the average lattice TEC as a function of

Table 13.2 Comparison of Properties of Ceria (CeO_2) and Plutonia (PuO_2)

Property	CeO_2	PuO_2
Ionic radius	$Ce^{4+} = 0.87$ Å(CN = 6) $Ce^{4+} = 0.90$ Å(CN = 8)	$Pu^{4+} = 0.86$ Å(CN = 6) $Pu^{4+} = 0.89$ Å(CN = 8)
Structure	Fluorite	Fluorite
Melting point	2873K	2935K
Density	7.30 g/cc	11.44 g/cc
Enthalpy of formation (kJ/mol)	−1090.4	−1055.8
Specific heat (kJ/mol K)	61.63	66.25
Entropy (J/mol K)	62.30	66.13
TEC (α_a) × 10^6 (K^{-1}) (Axial)	11.83	11.61

CN, *coordination number;* TEC, *thermal expansion coefficient.*

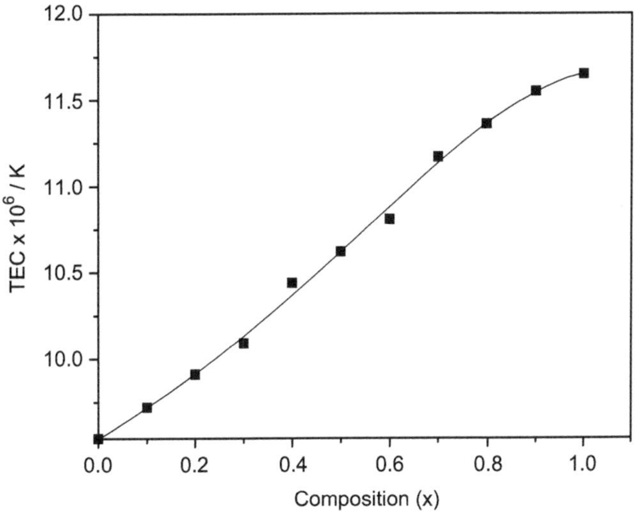

FIGURE 13.4

The variation of the increase in the lattice thermal expansion coefficient (*TEC*) (α_a) as a function of composition (x) in a $Th_{1-x}Ce_xO_2$ series.

composition in a $Th_{1-x}Ce_xO_2$ series is shown in Fig. 13.4 and can also be represented as the following equation (Eq. 13.8).

$$\alpha_a = 9.54 + 1.74x + 0.37x^2 + 1.86x^3 - 1.86x^4 \quad \text{(where } x \text{ is the Ce content)} \tag{13.8}$$

Lattice thermal expansion of several compositions like ThO_2-2, ThO_2-4 and ThO_2-6 wt% UO_2 were studied by powder XRD at higher temperatures [54,61]. At room temperature, the observed cubic unit cell parameters for ThO_2 and ThO_2-2 wt% UO_2, ThO_2-4 wt% UO_2, and ThO_2-6 wt% UO_2 were 5.599, 5.591, 5.589, and 5.585 Å, respectively. The systematic decrease in lattice parameter as a function of U^{4+} content in the solid solutions can be attributed to the smaller ionic size of U^{4+}. This trend in cell parameters indicates the homogeneous incorporation of U^{4+} into the lattice of ThO_2. The coefficients of average lattice thermal expansion (α_a) of ThO_2 and ThO_2-2, ThO_2-4 and ThO_2-6 wt% of UO_2 are 9.67×10^{-6}, 9.82×10^{-6}, 10.09×10^{-6}, and 10.37×10^{-6} K^{-1}, respectively. It can be seen that there is a systematic increase in the average lattice TEC after incorporating UO_2 into ThO_2. This trend can be correlated to the higher melting point of ThO_2 as compared with that of UO_2.

Thus these examples illustrate that how by proper selection of two end members based on their structure and melting points, it is possible to tailor thermal expansion behavior by forming isovalent substitutional solid solutions. In the case of isovalent substation, the average bond lengths and melting points are the guiding factors to effectively tune thermal expansion behavior.

5.1.2 Aliovalent Solid Solution Formation Approach

In addition to the previously mentioned isovalent substituted fluorite-type compounds, the aliovalent substituted products also have importance from the point of view of nuclear reactors because these systems simulate the thermal expansion behavior of nuclear fuels in the presence of fission products.

The major fission products formed in the fission process of U and Pu are alkaline earth elements, rare-earth elements, and platinum group metals. In view of this thermal expansion behavior, a number of aliovalent solid solutions have been investigated [67–74]. In addition, aliovalent solid solutions exhibit enhanced ionic conductivity. The following sections will deal with tailoring thermal expansion behavior by making aliovalent solid solutions. Unlike isovalent solid solutions, the aliovalent solid solutions have point defects and they become additional parameters to influence thermal expansion behavior. The aliovalent substitution is mostly accompanied by anion deficiency or anion excess in the host lattice. Thus the rule of addition to predict thermal expansion behavior based on melting points of two end members or average bond strengths may not holds good for such solid solutions.

5.1.2.1 Thermal Expansion Behavior of $ThO_2-M_2O_3$ Systems ($M = Nd^{3+}$, Eu^{3+}, Gd^{3+}, Dy^{3+})

$ThO_2-M_2O_3$ systems ($M = Nd^{3+}$, Eu^{3+}, Gd^{3+}, Dy^{3+}) have importance for several reasons, including that Nd is used as a burn-up monitor in spent fuel, whereas Eu, Gd, and Dy are commonly used as burnable neutron poisons, which are required for maintaining the reactivity and the controlled fission process in a nuclear reactor. Samples in a $ThO_2-M_2O_3$ system were prepared by a conventional solid state method involving multistep heating, and thermal expansion behaviors have been investigated [68–74]. XRD studies revealed that the cubic fluorite-type solid solutions in this series are retained until the compositions of $Th_{0.50}Nd_{0.50}O_{1.75}$, $Th_{0.50}Eu_{0.50}O_{1.75}$, $Th_{0.60}Gd_{0.40}O_{1.80}$, and $Th_{0.85}Dy_{0.15}O_{1.925}$ in their respective systems. It implies that whereas 50 mol% $NdO_{1.5}$ and $EuO_{1.5}$ can be incorporated into a ThO_2 lattice, only 40 mol% $GdO_{1.5}$ and 15 mol% $DyO_{1.5}$ are soluble in a ThO_2 lattice [69–72]. This trend in solubility behavior of Nd^{3+}, Eu^{3+}, Gd^{3+}, and Dy^{3+} in the lattice of ThO_2 can be explained based on the relative ionic radii of host and guest ions. The ionic radii of Th^{4+}, Nd^{3+}, Eu^{3+}, Gd^{3+}, and Dy^{3+} are 1.05, 1.02, 0.98, 0.97, and 0.94 Å, respectively, in cubic coordination. As the difference between ionic radii of host and guest ions increases, the solubility limit decreases. Therefore the lowest solubility is in the system with the largest difference in ionic radii of host and guest ions, such as with Th^{4+} and Dy^{3+}. The lattice TECs (α_a) of the solid solutions, as evaluated from the HT-XRD data, revealed that the coefficients of average lattice thermal expansion of the solid solutions in the $ThO_2-M_2O_3$ system ($M = Nd^{3+}$, Eu^{3+}, Gd^{3+}, Dy^{3+}) decrease when going from Nd_2O_3 to Dy_2O_3 [69–72]. The observed trend of thermal expansion of these solid solutions appears to be more related to the ionic radii of the cations of the compounds. The thermal expansion characteristics of rare-earth oxides, namely Nd_2O_3, Eu_2O_3, Gd_2O_3, and Dy_2O_3, appear to contribute to the overall thermal expansion behavior of the solid solutions in the $ThO_2-M_2O_3$ system. The lattice TECs of $Th_{1-x}M_xO_{2-x/2}$ ($M = Nd^{3+}$, Eu^{3+}, Gd^{3+} and Dy^{3+}) solid solutions at a given value of x show a decreasing trend when going from Nd^{3+} to Dy^{3+}. The noticeable difference in the thermal expansion behavior of the solid solutions of the systems is observed in thoria–neodymia system. Except for this system, the linear and lattice TECs of the solid solutions of other systems decrease progressively on increasing content of the respective substituent. The decrease in TECs of these solid solutions on increase in the content of the substituent can be explained by the low values of TECs of Eu_2O_3, Gd_2O_3, and Dy_2O_3 compared with that of ThO_2. Another reason could be that as the Eu^{3+}, Gd^{3+}, and Dy^{3+} contents increase, defect concentration (oxygen vacancy) also increases, which can mask thermal expansion to some extent.

5.1.2.2 Thermal Expansion Behavior of Some Ceria-Based Systems

Ceria and substituted ceria ($Ce_{1-x}RE_xO_{2-x/2}$) are important multifunctional materials because they exhibit a wide range of properties such as ionic conductivity, catalysis, and oxygen storage capacitor, just to name a few. They are used as electrolytes in solid oxide fuel cells. In view of these potential applications of ceria and ceria-based systems, thermal expansion behavior needs to be investigated. Ceria, a fluorite-structured material, can be substituted by various aliovalent ions, which alters not only the functional properties but also thermal expansion behavior. This section is devoted to tailor thermal expansion behavior of ceria by an aliovalently substituted solid solution approach. Detailed studies on a $CeO_2-Y_2O_3$ system [71] revealed the formation of a fluorite-type solid solution in CeO_2-rich compositions. On the other hand, C-type (Y_2O_3 group rare-earth oxides) solid solutions were observed in Y_2O_3-rich compositions. As stated earlier, the homogeneity range of F-type and C-type solid solutions depends on ionic radii differences between the rare-earth metal ions and host lattice metal ions (Ce^{4+} in this case). The phase analysis of $Ce_{1-x}Gd_xO_{2-x/2}$ general compositions revealed the existence of F-type solid solutions until the nominal composition $Ce_{0.45}Gd_{0.55}O_{1.45}$, which transformed to C-type solid solutions at and beyond $Ce_{0.40}Gd_{0.60}O_{1.40}$ [69]. The F-type and C-type solid solution phases can have a miscibility gap, depending upon the relative ionic radius of the trivalent rare-earth ion and tetravalent host ion (Ce^{4+}) [51,52,68,70]. Because the aliovalent substitution can lead to either anion-deficiency or anion-excess lattices, its thermal expansion behavior strongly depends on the valence of host and guest ions. Thus in $CeO_2-RE_2O_3$ systems, it is possible to obtain either anion excess C-type or anion-deficient fluorite-type lattice by controlling the stoichiometry of the metal ions. There is another possibility of ordering of these point defects and hence formation of ordered phases. The ordering also has its own effect on thermal expansion behavior. A typical example of tailoring thermal expansion of ceria is taken from a $CeO_2-Ho_2O_3$ system [70]. In this system, the lattice TEC (α_a) increases to $13.0 \times 10^{-6}/°C$ for the composition $Ce_{0.90}Ho_{0.10}O_{1.95}$ compared with $12.6 \times 10^{-6}/°C$ of undoped CeO_2. In addition to the trivalent rare-earth ions, it is also possible to incorporate divalent cations like Sr^{2+} and Ba^{2+} in a ceria lattice. These solid solutions have also been investigated for thermal expansion behavior. It was found that the TEC within the F-type homogeneity range $Ce_{1-x}Sr_xO_{2-x}$ system increases with increasing the Sr^{2+} contents [73,74]. Ceria with about 7.5% Sr^{2+} substitution shows an increase in TEC by about 2.7% in the temperature range of 27–1200°C. This trend was attributed to more ionic nature of Sr–O bonds in the Sr^{2+} substituted ceria.

6. MATERIALS WITH LOW OR NEGATIVE THERMAL EXPANSION BEHAVIOR

This chapter has so far covered examples of materials with positive thermal expansion. However, there are several materials that show contraction on heating. These are called *negative thermal expansion (NTE)* materials [7,8]. Likewise, there are some materials that show low or zero thermal expansion behavior. Of late, low expansion and NTE materials have gained considerable attention in view of their application in high-end technologies as well as from the point of view of fundamental understanding [34,35,75]. Several modifications of silica and compounds with silica-type structure have been reported to possess low thermal expansions [76]. Some of the early examples that showed low thermal expansion are β-eucryptite [77], cordierite ($Mg_2Al_2Si_5O_{18}$) [78], β-spodumene [79] and sodium zirconioum phosphate (NZP) family compounds [80–84]. This behavior has been attributed to highly anisotropic thermal expansion behavior (expansion and contraction along different crystallographic axes), which leads to overall compensation of thermal expansion and in turn low to average TECs are observed. A schematic showing anisotropic behavior of some of these compounds is shown in Fig. 13.5.

In addition, several zeolite-type materials like fauzasite and mesoporous zeolitic materials of Mobil Oil Corporations, named as mobil comopsition material (MCM) are also reported to show unusually

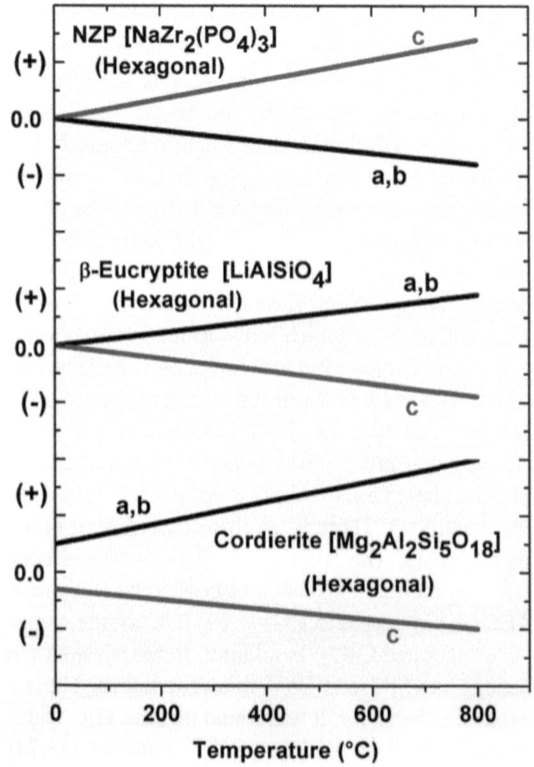

FIGURE 13.5

Some typical examples of anisotropic thermal expansion behavior.

low expansion or NTE behavior [85,86]. There are a few tetrahedrally bonded elements such as Si and Ge that also show NTE at low temperature [87]. Several elements and alloys that exhibit NTE behavior have electronic structure origins. Some typical examples of such electronic effects are f-block elements like Pu and Ce [30,31] and alloys like YbGaGe [32] and $CeAl_3$ [33]. Invar and Kovar alloys were reported to exhibit anomalous expansion behavior [3]. Perovskite-type ferroelectric solid solutions or materials with structural transitions have also been reported to show NTE, albeit over a narrow temperature range [26–29,88,89]. This type of anomalous thermal expansion behavior is intrinsic in nature and often originates as a result of several typical inherent structural features, which will be elaborated in the following sections.

7. TAILORING OF THERMAL EXPANSION OF MATERIALS BASED ON CRYSTALLOGRAPHIC CONCEPTS

So far in this chapter the main emphasis has been on tailoring thermal expansion behavior by rule of addition, by playing with chemical bonding, or by introducing an optimum level of defects. There is another important parameter that influences thermal expansion behavior considerably and that is crystallographic structure. In particular the crystallographic structure plays an important role in designing materials with low, negative, or zero thermal expansion behavior. This section will deal with the rational design of materials with tailored thermal expansion behavior, invoking crystallographic concepts. Several examples will be discussed that will elucidate the vital role of crystal structure in governing thermal expansion behavior of materials. In addition to the nature and type of bonding, the polyhedral building blocks and their packing in the unit cell are also crucial parameters that determine thermal expansion behavior. In addition to the nature of chemical bonds along crystallographic axes, these factors also bestow materials with anisotropic thermal expansion behavior.

The ionic size and charge on central ions govern the coordination number and nature of polyhedral sharing in the unit cell so as to minimize the repulsive forces between the central cations. Thus the polyhedra with the higher charged cations mostly prefer corner sharing, because that ensures that the higher charged cations are separated far apart. The corner sharing of polyhedra leads to appreciably lower packing efficiency of the unit cell and gives rise to the formation of framework solids. These polyhedral units around the central atoms can be tetrahedral, octahedral, or cubic, as governed by the cation-to-anion radius ratio. The corner-shared connectivity of polyhedra in framework solids is shown in Fig. 13.6.

In an ideal and unstrained arrangement of atoms, the farthest separation of the central ions can be achieved if there is a linear linkage between two polyhedra, which leads to an M−X−M bond angle of 180 degrees (M = central ion and X = ligand) and in turn, the separation between two central ions can be represented as 2 × M−X. This arrangement leads to minimum repulsion between two central atoms. It may be noted that corner sharing is most abundant in tetrahedra followed by octahedral polyhedra. However, there are no ionic compounds that have corner-shared cubes. However, in most of the actual situations, owing to considerable strain, most commonly a deviation in the M−X−M angle from 180 degrees is observed in real crystals. This bending is maintained by tilting of polyhedra. For example, silica is a framework solid consisting of corner-sharing SiO_4 tetrahedra, but the Si−O−Si bond angles range from 147 to 152 degrees, which is a considerably lower range than the ideal value of 180 degrees. The tilting of polyhedra leads to lowering of symmetry and also to anisotropic thermal expansion behavior.

Because framework materials have considerable free lattice space and anions in two-fold coordination, the M−X−M bond predominantly shows transverse vibration, which leads to anomalous thermal expansion such as NTE. The increasing transverse vibrations effectively decrease the separation between the next nearest atoms and hence decrease the dimensions of lattice with temperature

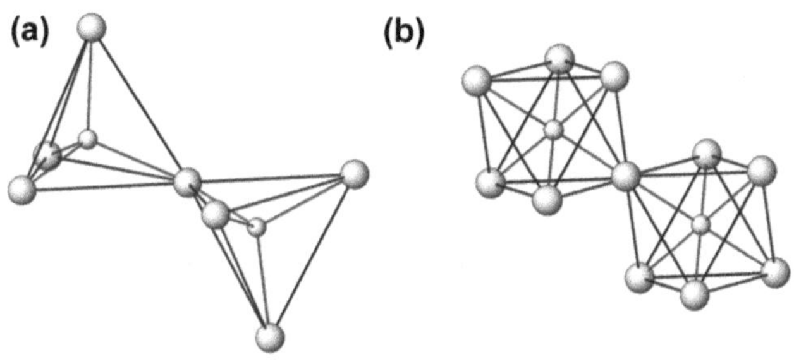

FIGURE 13.6

Typical corner sharing of (a) tetrahedra and (b) octahedra in framework-structured compounds.

[8,10,34,35]. The remarkable observation of NTE behavior exhibited by cubic ZrW_2O_8 and HfW_2O_8 originates from transverse vibrations of M–O–W bonds (M = Zr and Hf), as established by precise temperature-dependent crystal structure analyses [34,35]. In addition to these two tungstates, a wide range of framework materials exhibiting NTE behavior has been reported [7,29,34–41,88–94]. In a later section, some of the framework solids will be discussed more elaborately from the point of view of NTE behavior. There are several recent materials in phosphate, tungstate, and molybdate systems that show unusually low or negative TECs [42–49]. The average low thermal expansion behavior of materials can also be caused by anisotropic unit cell expansion, as depicted in Fig. 13.5. Thermal expansion along one or two axes is compensated by contraction along other axes, which results in net low thermal expansion. A typical example is NZP ($NaZr_2P_3O_{12}$) [54,62–68].

A few cristobalite-type metal phosphates have been reported to show tailored thermal expansion behavior. Relieving the strain of interpolyhedral bond angles has been attributed to lower thermal expansion behavior. Cristobalite is a high-temperature modification of silica. It is possible to isolate it by a prolonged annealing above the cristobalite transition temperature. Several APO_4 (A = trivalent cations) units were prepared in cristobalite phase [95–97]. The structure of these cristobalite-type phosphates consists of AO_4 (A = Al^{3+}, Ga^{3+} and B^{3+}) and PO_4 tetrahedral units that are connected by corner sharing. Thermal expansion behavior as a function of temperature, $AlPO_4$, $GaPO_4$, and a series of solid solutions with general compositions $Al_{1-x}Ga_xPO_4$ have been studied with high-temperature XRD [97]. All the compositions were found to adopt orthorhombic structure with space group $C222_1$, at room temperature. The $Al_{1-x}Ga_xPO_4$ compositions were studied with high-temperature XRD, which revealed a reversible orthorhombic-to-cubic phase transition at higher temperature. The reversible nature of phase transition was also confirmed in differential thermal analysis studies. The phase transition temperature increases as a function of Ga^{3+} content in the $Al_{1-x}Ga_xPO_4$ series. In order to investigate thermal expansion behavior, cell parameters of these compositions in both orthorhombic and cubic modification were determined as a function of temperature. The variation of unit cell parameters for the two end members is shown in Fig. 13.7. The phase transition is accompanied by an abrupt and significant increase in cell volume, which indicates that the packing density of the high-temperature cubic phase in this series of phosphates is low compared with that in the orthorhombic phase. The lattice TECs of both modifications, as determined by HT-XRD are given in Table 13.3. It is interesting to note that the orthorhombic phase shows a considerably higher thermal expansion compared with the cubic phase in all the compositions of the series (Table 13.3). This trend can be correlated to the variation of the M–O–P angle (M = Al^{3+}, Ga^{3+} and ${Al^{3+}}_{1-x}{Ga^{3+}}_x$) as a function of temperature. It was found that a marginal increase or decrease in M–O–P bond angles leads to correspondingly small positive or negative thermal expansion behavior. Thus MPO_4 materials with lower values of bond angles and in turn, higher bond strain, exhibited higher thermal expansion.

Along similar lines, BPO_4 has also been investigated from room temperature up to 900°C [92]. BPO_4 crystallizes in tetragonal lattice with a high cristobalite-type frame. The unit cell parameters were refined at different temperatures, which revealed a strong anisotropic thermal expansion along the *a*-axis compared with the *c*-axis. Thermal expansion behavior of BPO_4 could also be explained from structural parameters such as bond angle and bond length. In this material a gradual increase of B–O–P bond angle as a function of temperature also leads to a corresponding increase in the interpolyhedral separation (B...P nonbonded distance), which causes expansion of the crystallographic axes. In order to arrive at a correlation between interpolyhedral bond angles and TEC, a

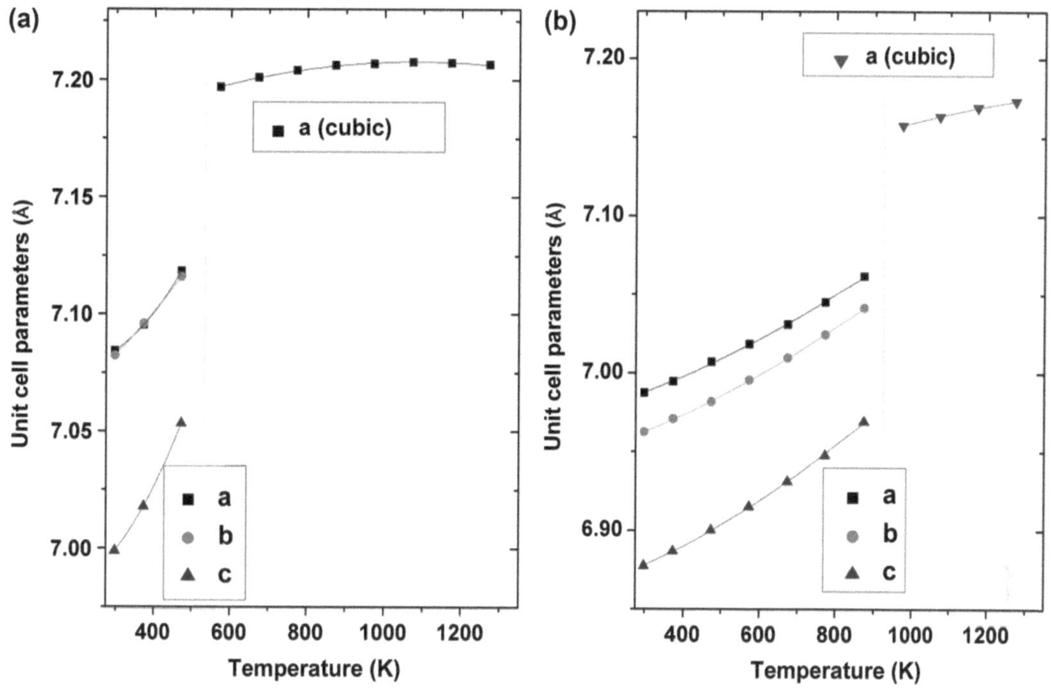

FIGURE 13.7

Variations of unit cell parameters of (a) $AlPO_4$ and (b) $GaPO_4$ with temperature.

Table 13.3 Thermal Expansion Coefficients of Low (*C222₁*) and High (*F-43m*) Cristobalite Forms of $Al_{1-x}Ga_xPO_4$

	Orthorhombic Phase α_a, α_b, α_c, α_V[a]	Phase Transition Temperature (°C)	Cubic Phase α_a, α_v[a]
$AlPO_4$	27.4, 27.2, 44.5, 99.8	202	1.93, 5.75
$Al_{0.8}Ga_{0.2}PO_4$	23.0, 20.3, 35.4, 79.1	273	4.11, 12.3
$Al_{0.5}Ga_{0.5}PO_4$	18.7, 22.4, 30.4, 71.9	340	5.52, 16.6
$Al_{0.2}Ga_{0.8}PO_4$	18.8, 20.3, 25.9, 65.6	506	6.40, 19.3
$GaPO_4$	18.5, 19.8, 23.2, 62.2	605	7.17, 21.4

[a]($\times 10^6$/°C).

comparison of all these high–cristobalite-type MPO_4 ($M = Al^{3+}$, Ga^{3+} and B^{3+}) compounds was made. The coefficient of volume thermal expansion (α_V) and the M–O–P bond angle can be related as: α_V (/°C) $= -191.32 + 4.33 \times [\theta] - 0.02 \times [\theta]^2$, where θ is the M–O–P bond angle in degrees. It is revealed that higher the value of M–O–P bond angle, the lower the TEC. This empirical relation can be used to accurately design cristobalite-type compounds with desired thermal expansion behavior by controlling the M—O—P bond angle. It can be mentioned here that with an appropriately sized M^{3+} ion or a combination of trivalent ions, the M—O—P bond angle can be tuned to the desired value of TECs [37—39,43,44].

7.1 TAILORING OF THERMAL EXPANSION BEHAVIOR IN SOME MOLYBDATES AND TUNGSTATES

A large number of molybdates and tungstates with the general formulas of MX_2O_8 (M = tetravalent ions like Zr^{4+} and Hf^{4+}, and X = W^{6+}, Mo^{6+}) and $M_2(XO_4)_3$ (M = trivalent ions like Y^{3+}, Sc^{3+}, and heavier rare-earth ions Al^{3+}, Cr^{3+}, and Fe^{3+}). These compounds have framework structures consisting of corner-shared octahedral MO_6 or tetrahedral XO_4 units. Several of them exhibit NTE behavior [34—38,98,99]. The size of the counter cation (M^{3+} or M^{4+}) plays a significant role in governing the structure and thermal expansion behavior of these compounds. A few case studies in these systems will be discussed in a later section.

Detailed thermal expansion behavior of $A_2(MoO_4)_3$ compounds (A = heavier rare-earth ions and Cr^{3+}, Fe^{3+}, and Al^{3+}) have been investigated by dilatometry and HT-XRD [43,45—47]. HT-XRD studies revealed that $Al_2(MoO_4)_3$, $Cr_2(MoO_4)_3$, and $Fe_2(MoO_4)_3$ exist in monoclinic ($P2_1/c$) structure at ambient condition and orthorhombic (*Pnca*) structure at higher temperatures. The bulk thermal expansion behavior of these three compounds was studied from 25°C to 850°C. Dilatometric studies revealed a considerable positive expansion in the temperature range of monoclinic phase, whereas there was negative expansion in the temperature ranges of the orthorhombic phase (Fig. 13.8) [45].

These molybdates were also studied with HT-XRD, which revealed a subtle positive thermal expansion in the high-temperature orthorhombic phase of $Cr_2(MoO_4)_3$, unlike NTE as observed with

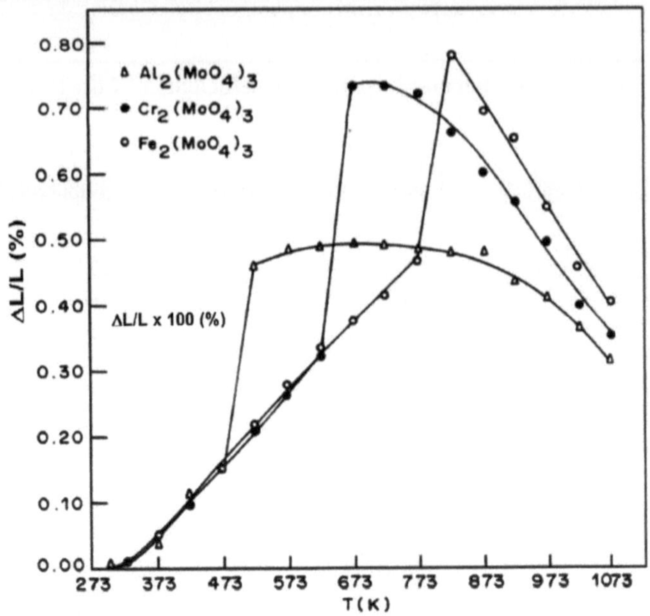

FIGURE 13.8

Bulk thermal expansion (ΔL is change in length with respect to orginal length L) behavior of $A_2(MoO_4)_3$ compounds ($A = Al^{3+}, Cr^{3+}, Fe^{3+}$). *T*, temperature.

Reproduced from A.K. Tyagi, S.N. Achary, M.D. Mathews, J. Alloys Compd. 339 (2002) 207 with permission of Elsevier publishers.

dilatometry. This difference in thermal expansion behavior in dilatometry and HT-XRD indicates that with a complementary use of both techniques, it is possible to obtain intrinsic thermal expansion behavior. Many times, thermal expansion data is underobserved with dilatometry because of low sintered density and microcracks and macrocracks in the pellet. All these factors mask the thermal expansion of the material. However, these factors have no bearing on lattice thermal expansion behavior obtained by HT-XRD. This detailed work on $Al_2(MoO_4)_3$, $Cr_2(MoO_4)_3$, and $Fe_2(MoO_4)_3$ systems opens up another possibility of designing tailor-made thermal expansion materials by preparing the appropriate solid solutions among them. Variable-temperature XRD studies of $Sc_2(MoO_4)_3$ (in the temperature range of 4–300K) have been reported in literature [39], indicating a monoclinic ($\alpha_v = 21.9 \times 10^{-6}$/K) to orthorhombic ($\alpha_v = -6.3 \times 10^{-6}$/K) phase transition at 180K. The monoclinic form shows positive thermal expansion behavior (α_v of 21.9×10^{-6}/K), whereas the orthorhombic form shows NTE behavior ($\alpha_v = -6.3 \times 10^{-6}$/K). $RE_2(MoO_4)_3$ is another class of framework solids that has shown very interesting and tunable thermal expansion behavior. It was revealed through dilatometry and HT-XRD that the ionic radius of RE^{3+} counter cations plays a significant role in thermal expansion behavior of these orthorhombic rare-earth molybdates (Table 13.4). It is evident from this table that the magnitude of negative TEC increases on increasing the ionic radius of trivalent rare-earth counter cations, which facilitates rocking of the MoO_4 tetrahedra, leading to higher NTE behavior. The size of the trivalent counter cation again has a strong bearing on thermal expansion behavior of such molybdates. Thus it is possible to obtain even zero thermal expansion in orthorhombic modification of these molybdates by carefully optimizing the size of trivalent counter cation.

Similar to the $M_2(MoO_4)_3$ molybdates, the corresponding tungstates have been also been investigated for their thermal expansion behavior. This class of materials is also belongs to a framework structure family with corner-sharing MO_6 octahedra and WO_4 tetrahedra. Hence they also exhibit low thermal expansion and/or NTE behavior. Like with molybdates discussed earlier, tungstates with heavier rare-earth or transition metal ions have shown low thermal expansion and/or NTE behavior. A typical example of $Al_2(WO_4)_3$ will be discussed here. Its thermal expansion behavior was investigated with dilatometry and HT-XRD [48,99,100]. The detailed XRD studies revealed $Al_2(WO_4)_3$ to have an orthorhombic structure (space group *Pnca*) comprised of AlO_6 octahedra and WO_4 tetrahedra. The bulk thermal expansion behavior of an $Al_2(WO_4)_3$ pellet of bulk density at about 88% as obtained by dilatometry is shown in Fig. 13.9.

As expected, an NTE behavior was observed for $Al_2(WO_4)_3$ with α_l as -1.5×10^{-6}/°C [100]. As mentioned earlier, it is also important to investigate materials using HT-XRD, which gives intrinsic

Table 13.4 Thermal Expansion Coefficients of a Few Orthorhombic $RE_2(MoO_4)_3$-type Molybdates [75,76]

RE^{3+}	Temperature Range (°C)	$\alpha_a \times 10^6$ (/°C)	$\alpha_b \times 10^6$ (/°C)	$\alpha_c \times 10^6$ (/°C)	$\alpha_l \times 10^6$ (/°C)
Sc^{3+}	−90 to +25	−8.41	10.82	−8.83	−6.30
Y^{3+}	25–800	−11.69	−6.57	−10.04	−9.36
Er^{3+}	25–800	−10.84	−3.34	−8.57	−7.56
Yb^{3+}	25–800	−10.02	−2.99	−5.21	−6.04
Lu^{3+}	25–800	−8.69	−1.64	−7.75	−6.02

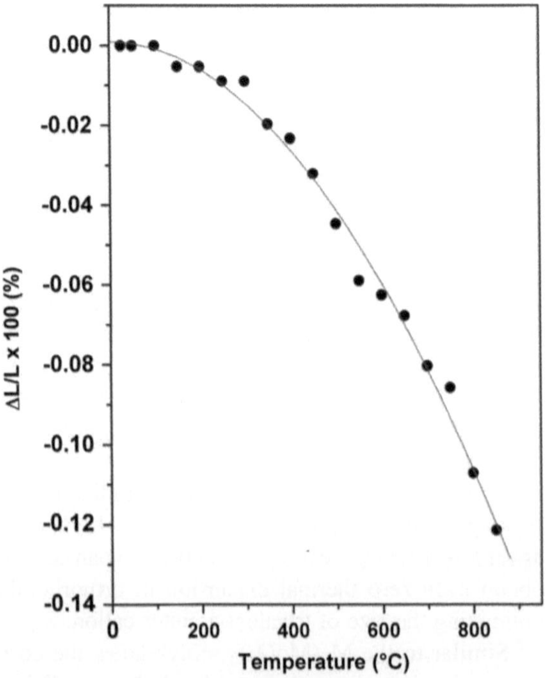

FIGURE 13.9

Variation of percentage of linear thermal expansion (ΔL is change in length with respect to orginal length L) of $Al_2(WO_4)_3$ with temperature.

Reproduced from S.N. Achary, G.D. Mukherjee, A.K. Tyagi, S.N. Vaidya, J. Mater. Sci. 37 (2002) 2501 with permission of Springer.

thermal expansion behavior. This becomes more important in the case of materials for which dilatometry shows feeble NTE behavior, like that shown by $Al_2(WO_4)_3$. The HT-XRD studies of this compound showed an anisotropic thermal expansion with a slight net positive volume expansion, $\alpha_v = +2.2 \times 10^{-6}$/°C [37], $+4.5 \times 10^{-6}$/°C [48], and 3.74×10^{-6}/°C [100]. Thus this example illustrates that how sometimes inferences drawn by dilatometry studies alone can be misleading if the sintered density is poor or the pellet has some microcracks or macrocracks.

As discussed for earlier systems, in the case of tungstates it is also possible to tailor thermal expansion behavior by making solid solutions of tungstates of appropriate rare-earth ions such as monoclinic lighter rare-earth tungstate and orthorhombic heavier rare-earth tungstates. The monoclinic and orthorhombic modifications of rare-earth tungstates show positive thermal expansion and NTE behaviors and in turn, their solid solution can be prepared with desired low or zero thermal expansion behavior, as shown in Fig. 13.10 and Table 13.5.

ZrW_2O_8 and HfW_2O_8, belonging to framework class materials, have corner-sharing AO_6 octahedra and WO_4 tetrahedra, and ZrW_2O_8 has attracted considerable attention [35]. Both of these tungstates adopt cubic lattice ($P2_13$) at ambient temperature, which undergoes an order–disorder transition to *Pa3* lattice upon heating. ZrW_2O_8 has been studied for its thermal expansion behavior extensively, which revealed its negative TECs in two modifications as -9.4×10^{-6}/°C (in −272.7 to 127°C) and -5.5×10^{-6}/°C (in 157–777°C) [34,35]. Other closely related analogues compounds like cubic $ZrMo_2O_8$, $HfMo_2O_8$, ZrP_2O_7, and ZrV_2O_7 have also been explored for thermal expansion behavior. The lattice TECs for cubic $ZrMo_2O_8$ and $HfMo_2O_8$ are -4.9×10^{-6}/°C (in the range 262–300°C) [44] and -4.0×10^{-6}/°C (in the range −196–300°C) [101], respectively.

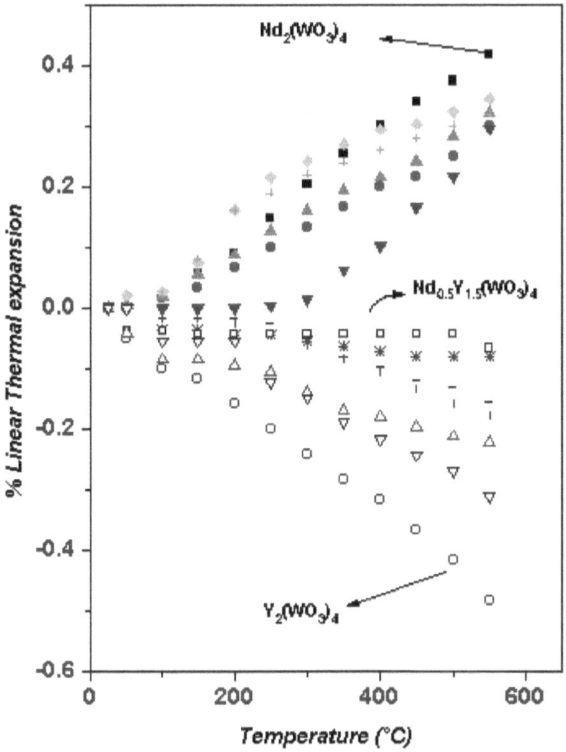

FIGURE 13.10

Materials with tunable thermal expansion behavior in an $Nd_{2-x}Y_x(WO_4)_3$ series.

Table 13.5 Thermal Expansion Coefficients of Materials in an $Nd_{2-x}Y_x(WO_4)_3$ Series

Composition	Linear Thermal Expansion Coefficient ($\alpha_l \times 10^6$ (°C^{-1})
$Nd_2(WO_4)_3$	7.96
$Nd_{1.50}Y_{0.50}(WO_4)_3$	5.63
$Nd_{0.50}Y_{1.50}(WO_4)_3$	−1.26
$Nd_{0.20}Y_{1.50}(WO_4)_3$	−4.26
$Y_2(WO_4)_3$	−5.92

7.2 THERMAL EXPANSION BEHAVIOR OF SCHEELITE- AND ZIRCON-TYPE ABO_4 MATERIALS

ABO_4-type compounds form a very interesting class of wide ranging materials. This class of materials can adopt several different structures including scheelite, monazite, zircon, wolframite, baryte, anhydrite, $MnSO_4$, and $CrPO_4$ structure depending upon ionic radii and charge combinations of the A and B cations [102,103]. Various polymorphs of silica can also lead to ABO_4-type compounds by replacing Si with two different types of cations. However, among all the ABO_4-type compounds, scheelite- and zircon-type lattices are most common. Both of these structures have a tetragonal unit cell consisting of AO_8 polyhedra (bisdisphenoid) and BO_4 (tetrahedra), which are connected differently in these two structures. The scheelite structure is related to the fluorite structure, whereas the zircon structure originates from the rutile structure [104]. The thermal expansion behavior of some scheelite- and zircon-type ABO_4 type compounds will be discussed in this section.

A comparison of thermal expansion behavior of various ABO_4-type compounds reveals that the zircon-type compounds often exhibit low thermal expansion behavior compared with scheelite-type compounds [105−110]. The higher expansion of scheelite type ABO_4 compounds are due to their higher packing efficiency and shorter inetratomic compared to that of zircon type compounds. It has been reported that ABO_4 oxides with zircon-type structure exhibit a range of thermal expansion behavior, which in conjunction depends upon their cation charge and ionic radii of A and B cations [111]. In general silicate based zircon type materials ($B = Si^{4+}$) show much low thermal expansion behavior. A typical example is $ZrSiO_4$ (mineral zircon) which is known to exhibit low thermal expansion behavior [104,110]. Typical variations of unit cell parameters of some zircon and scheelite type compounds are compared in Fig. 13.11. Thermal expansion behavior of several rare-earth phosphates or vanadates, such as $LuVO_4$ and $LuPO_4$, has been investigated with HT-XRD and thermodilatometer, and is well reported in literature [111,112]. It was found in HT-XRD studies that the

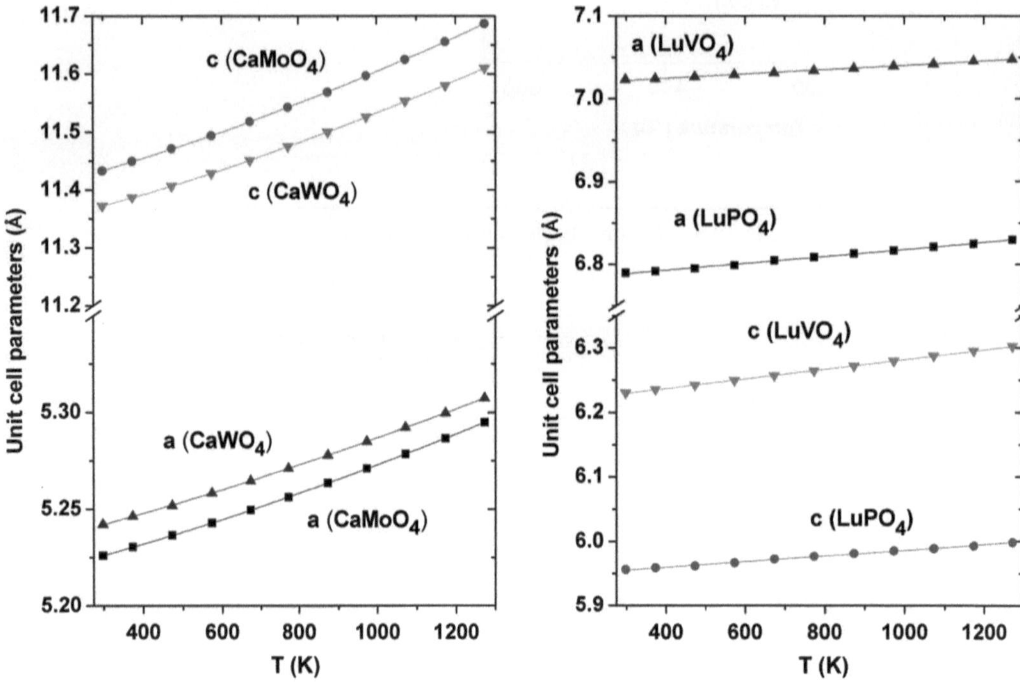

FIGURE 13.11

Variations of unit cell parameters with temperature (*T*) in scheelite-type $CaMoO_4$ and $CaWO_4$ and zircon-type $LuVO_4$, and $LuPO_4$.

coefficients of lattice thermal expansion (25–1000°C) of $LuPO_4$ are $\alpha_a = 6.04 \times 10^{-6}/°C$ and $\alpha_c = 7.25 \times 10^{-6}/°C$, and for $LuVO_4$ they are $\alpha_a = 3.62 \times 10^{-6}/°C$ and $\alpha_c = 11.80 \times 10^{-6}/°C$. Both the compounds show anisotropic lattice thermal expansion behavior, as expected, and the degree of anisotropy is related to the distortion of the polyhedra around Lu^{3+} ions.

Similar HT-XRD studies [110] on scheelite-type $CaMoO_4$ (25–350°C) revealed that the average lattice TECs α_a and α_c are 10.71×10^{-6} and $16.17 \times 10^{-6}/°C$. Likewise, the average lattice TECs for $CaWO_4$ are 13.7×10^{-6} and $21.5 \times 10^{-6}/°C$ (20–1020°C) [110]. The average volume TECs (α_v) for $CaMoO_4$ and $CaWO_4$ are 50.6×10^{-6} and $47.5 \times 10^{-6}/°C$, respectively. The detailed structural analysis revealed that the predominant contribution to thermal expansion behavior of these materials comes from the expansion of Ca–O bonds, given that Mo–O and W–O bonds show only feeble thermal expansion because MoO_4 and WO_4 tetrahedra are relatively rigid.

As discussed earlier, the ABO_4-type scheelite and zircon structures consist of AO_8 polyhedra and BO_4 tetrahedra. Thermal expansion studies on these types of materials revealed that BO_4 polyhedra are thermally rigid because they show negligible or no expansion. The overall positive expansion behavior of ABO_4-type materials primarily originates from thermal expansion of AO_8 polyhedra. Thus the thermal expansion behavior of these materials is primarily dictated by temperature-induced elongation of A–O bonds in AO_8 polyhedra. Among the ABO_4-type scheelite and zircon structures, zircon lattice shows relatively low thermal expansion compared with scheelite lattice. This trend has been attributed to edge-sharing AO_8 polyhedra in addition to the presence of rigid BO_4 tetrahedra.

8. THERMAL EXPANSION BEHAVIOR OF PEROVSKITE-TYPE MATERIALS

Thermal expansion behavior of perovskite (ABO_3)-type materials is influenced by expansion of chemical bonds as well as distortion or rotation of octahedral units. In addition, the temperature-induced phase transitions are controlling factors for their thermal expansion and applications in variable temperatures. In particular, phase transitions showing larger distortion or volume discontinuity are reflected as poor sinterability and growth of poor-quality single crystals as a result of development of microcracks or macrocracks. In absence of any transition, most of the perovskite-type materials show smooth expansion of lattice with increasing temperature. The variation of distortion with temperature or displacement of cations, such as in ferroelectric materials, often shows anomalous or low thermal expansion. Low thermal expansion over a small temperature range has been reported in perovskite-type ferroelectric compounds or solid solutions [29,88,113–115]. Tetragonal $PbTiO_3$ shows negative expansion near ferroelectric transition temperature, which has been attributed to the decreasing tetragonal distortion [114,115]. Similarly, $Pb(Mg_{1/3}Nb_{2/3})O_3$ and $Pb(Zn_{1/3}Nb_{2/3})O_3$ also show low thermal expansion behavior near the phase transition temperature because of distortion [7,113]. Charge ordering as well as temperature-induced valence fluctuation of cations in perovskite-type manganites affect their thermal expansion behavior [116]. Such anomalous thermal expansion behavior accompanied with magnetic or electronic transitions is similar to the magnetostriction or electrostriction effects.

Influence of symmetry and phase transition in some perovskite-type A_2MgWO_6 (A = Ba^{2+}, Sr^{2+}, and Ca^{2+})-type compounds on thermal expansion have been investigated [89,117,118]. The studied compositions, namely, Ba_2MgWO_6, Sr_2MgWO_6, and Ca_2MgWO_6, have cubic (*Fm3m*), tetragonal (*I4/m*), and monoclinic (*P2*$_1$*/n*) lattices, respectively. Temperature-dependent structural studies indicate a structural transition in tetragonal Sr_2MgWO_6 from *I4/m* to cubic *Fm3m* lattice. Anisotropic thermal expansion has been observed in both tetragonal Sr_2MgWO_6 and monoclinic Ca_2MgWO_6 lattices. In fact, tetragonal Sr_2MgWO_6 shows larger expansion along the *a*-axis because of the rotation of octahedral units along the *c*-axis. The feeble negative expansion along the *c*-axis of tetragonal Sr_2MgWO_6 is related to decreasing amplitude of vibration parameter along this axis [89]. It can also be mentioned here that the expansion of the W—O and Mg—O bonds with temperature are much less compared with the A—O bond distances in distorted perovskite systems. Such phenomena are generally reflected as rotational octahedral units and hence anisotropic expansion of Sr_2MgWO_6, as shown in Fig. 13.12. The variation of unit cell parameters of Ca_2MgWO_6 with temperature shows significant differences in TEC along *a*-, *b*-, and *c*- axes [118]. Typical axial TECs of A_2MgWO_6 are given in Table 13.6. It is also observed that the thermal expansion increases with decreasing the ionic radii of the A^{2+} ions. The typical TECs of A_2MgWO_6 can be compared by considering the volume of their corresponding primitive pseudocubic lattices, as shown in Fig. 13.13.

Similar studies on perovskite-type $LaGaO_3$ indicate the ambient-temperature orthorhombic *(Pbnm)* structure transforms to rhombohedral $(R\bar{3}c)$ structure around 150°C, and the changes appear as a discontinuity in unit cell volume [119]. The unit cell parameters of both ambient-temperature orthorhombic and transformed rhombohedral phases can be related to each other by considering the primitive parent cubic perovskite structure. Both of these structures can be derived by tilting of the octahedral units from this parent structure. Typical unit cell relations of pseudocubic (aristotype primitive, *Pm3m*) and orthorhombic or rhombohedral lattices can be given as: $a_o \sim c_o \sim \sqrt{2} \times a_p$; $b_o \sim 2 \times a_p$; $a_h \sim \sqrt{2} \times a_p$ and $c_h \sim 2 \times \sqrt{3} \times a_p$, where a_o, b_o, and c_o are unit cell parameters of the

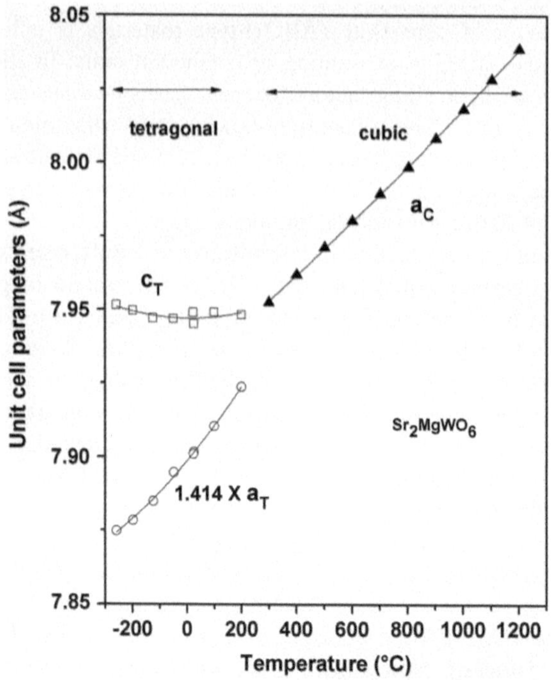

FIGURE 13.12

Variation of unit cell parameters of tetragonal and cubic Sr_2MgWO_6 with temperature (The subscript T and C indicate tetragonal and cubic lattice).

Table 13.6 Unit Cell Parameters and Thermal Expansion Coefficients of A_2MgWO_6

A	Space Group	Z	Unit Cell Parameters (at Room Temperature)	α_a, α_b, α_c, α_V ($\times 10^{-6}$) (K^{-1})	Temperature Range (°C)
Ca	$P2_1/n$	2	$a = 5.4265(2)$ Å $b = 5.5515(1)$ Å $c = 7.7231(2)$ Å $\beta = 89.90(1)$ $V = 232.66(1)$ Å^3	18.9, 4.0, 15.8, 39.1	25–1200
Sr	$Fm3m$	4	$a = 5.5876(1)$ Å $b = 7.9490(3)$ Å $V = 248.18(1)$ Å^3	12.1, 12.1, 12.1, 36.5	300–1200
Ba	$Fm3m$	4	$a = 7.9524(2)$ Å $V = 0.24818(1)$ Å^3	12.0, 12.0, 12.0, 36.4	25–1200

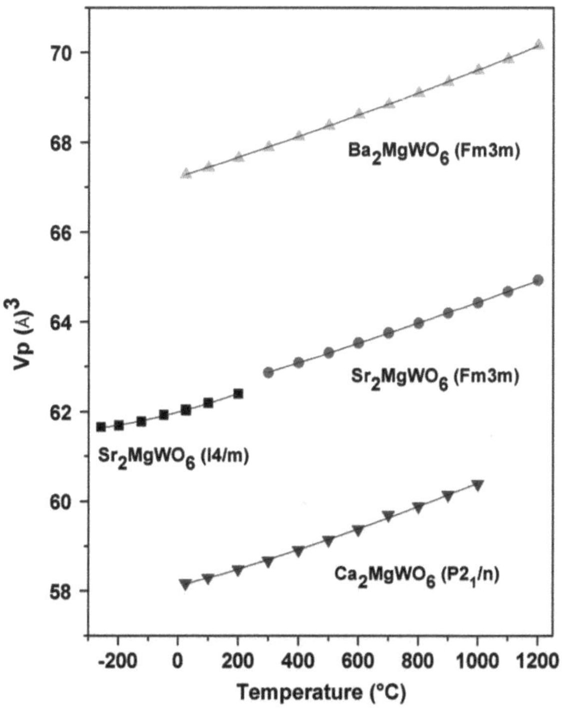

FIGURE 13.13

Variation of molar volume of [volume of primtive psedo cubic lattice (Vp)] A_2MgWO_6 with temperature.

orthorhombic lattice; a_h and c_h are unit cell parameters of the rhombohedral lattice in a hexagonal setting; and a_p is a primitive cubic unit cell parameter. The variation of unit cell parameters with the temperature of $LaGaO_3$ is shown in Fig. 13.14. The transition is observed as a discontinuity in the variation of unit cell volume and is of first-order nature.

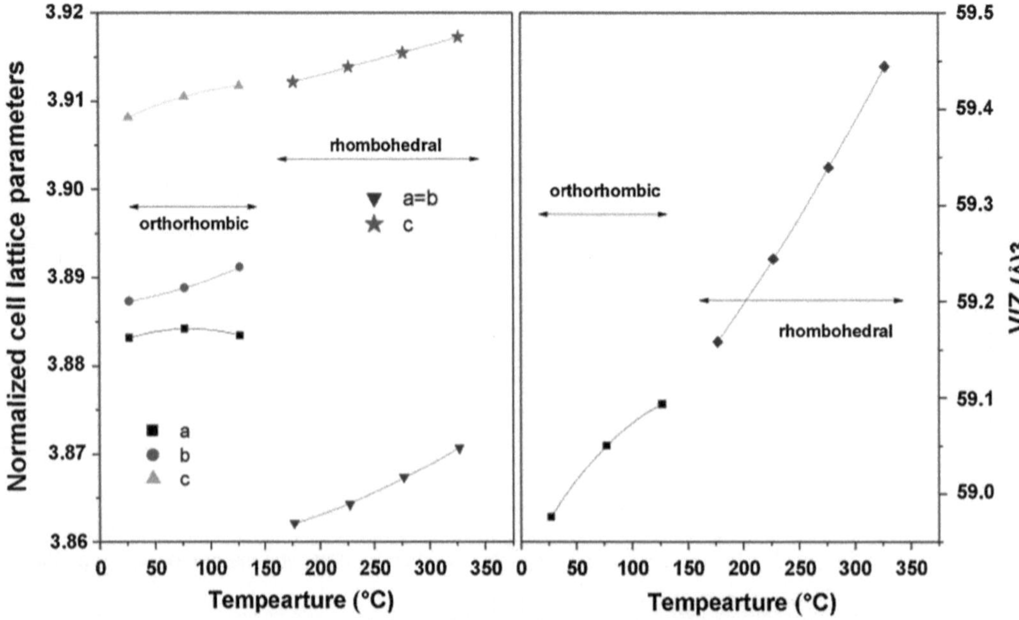

FIGURE 13.14

Variation of normalized unit cell parameters of primitive (P) lattice (left) and molar volume (*V/Z*, right) of orthorhombic and rhombohedral $LaGaO_3$ with temperature.

Because $LaGaO_3$ is a fast oxide ion conductor, its sintering is an important concern owing to the inherent strain and volume anomaly at the transition [120,121]. Literature reveals that the phase transition temperature for orthorhombic $La_{1-x}Ln_xGaO_3$ (Ln = Pr, Nd, or Sm) to rhombohedral *(R-3c)* $La_{1-x}Ln_xGaO_3$ increases on substitution of La^{3+} ions with smaller lanthanide ions [122,123], as well as with a partial substitution of Ga^{3+} ions by trivalent or divalent metal ions [124].

Because the perovskite-type materials exhibit a variety of technologically important properties, the phase transition and anisotropic expansion often hinder the preparation of dense ceramic body, as well as growth of single crystals. Thus tuning of distortion as well as strain at the phase transition is often achieved by suitable substitution. The control of oxygen anion stoichiometry or controlling the tolerance factor by suitable alteration of the A or B site cation are commonly adopted processes to overcome such a drawback.

9. CONCLUSION

In this chapter, attempts were made to explain the origin of thermal expansion behavior of solids. Common techniques of thermodilatometry and HT-XRD were discussed in detail. Various structural parameters that influence thermal expansion behavior of solids were also discussed. In particular, several examples were given, wherein lattice packing efficiency and polyhedral arrangements were shown to considerably affect thermal expansion behavior. The role of chemical bonding in governing thermal expansion behavior was also highlighted. The major emphasis was on materials with low thermal expansion or NTE behavior.

REFERENCES

[1] B. Yates, Thermal Expansion, Plenum Press, New York, 1972.
[2] R.S. Krishnan, R. Srinivasan, S. Devanarayann, Thermal Expansion of Crystal, Pergamon Press, Oxford, 1979.
[3] M. Shiga, Invar alloys, in: R.W. Cahn, P. Haasen, E.J. Kramer (Eds.), Materials Science and Technology, vol. 3B, VHC, Weinheim, 1994, p. 159.
[4] D. Krause, H. Bach, Low Thermal Expansion Glass Ceramics, second ed., Springer, Berllin, 2005.
[5] K.P. Birch, P.T. Wilton, Appl. Opt. 27 (14) (1988) 2813–2815.
[6] H. Holzer, D.C. Dunand, J. Mater. Res. 14 (1999) 780.
[7] R. Roy, D.K. Agrawal, H.A. McKinstry, Annu. Rev. Mater. Sci. 19 (1989) 59.
[8] A.W. Sleight, Annu. Rev. Mater. Sci. 28 (1998) 29.
[9] A. Sher, R. Solomon, K. Lee, M.V. Muller, Phys. Rev. 144 (1966) 593.
[10] S. Stokes, A.J.C. Wilson, Proc. Phys. Soc. 53 (1941) 658.
[11] C. Kittle, Introduction to Solid State Physics, fifth ed., Wiley, New York.
[12] N.W. Ashcroft, N.D. Mermin, Solid State Physics, Holt, Rinehart and Winston, New York, 1976.
[13] G. Rufino, Thermal expansion of solids, in: C.Y. Ho, R.E. Taylor (Eds.), CINDAS Data Series on Material Properties, vol. I-4, ASM International, Materials Park, OH, 1998. Chapter 3.
[14] Y.S. Touloukian, R.K. Kirby, R.E. Taylor, T.Y.R. Lee, Methods for measurement of thermal expansion of solids, in: Thermal Expansion Nonmetallic Solids. Part I, IFL/Plenum Press, New York, 1977.
[15] R.K. Kirby, Methods of measuring thermal expansion, in: D.D. Maglic, A. Cezarirliyan, V.E. Peletsky (Eds.), Compendium of Thermophysical Measurement Methods, vol. II, Plenum Press, New York, 1992.
[16] J.B. Austin, J. Am. Ceram. Soc. 35 (1952) 243.
[17] P.S. Gall, Thermal expansion of solids, in: C.Y. Ho, R.E. Taylor (Eds.), CINDAS Data Series on Material Properties, vol. I-4, ASM International, Materials Park, OH, 1998. Chapter 5.
[18] L.V. Azaroff, M.J. Buerger, The Powder Method in X-ray Crystallography, McGraw-Hill Book Company, NY, 1958.
[19] B.D. Cullity, Elements of X-ray Diffraction, Addison-Wilson Publishing Company, Inc., USA, 1959.
[20] L.G. Van Uitert, H.M. O'Bryon, M.E. Lines, H.G. Guggenheim, G. Zydzik, Mater. Res. Bull. 12 (1977) 261.
[21] S. Yamanaka, T. Maekawa, H. Muta, T. Matsuda, S. Kobayashi, K. Kurosakia, J. Solid State Chem. 177 (2004) 3484.
[22] H.D. Megaw, Z. Kristallogr. 100 (1939) 58.
[23] H.D. Megaw, Mater. Res. Bull. 6 (1971) 1007.
[24] R.M. Hazen, C.T. Prewitt, Am. Mineral. 62 (1977) 309.
[25] A.I. Kitaigorodoski, Molecular Crystals and Molecules, Chapter 5, Academic Press, New York, 1973.
[26] G. Shirane, A. Takeda, J. Phys. Soc. Jpn. 7 (1951) 1.
[27] G. Shirane, S. Hoshina, J. Phys. Soc. Jpn. 6 (1951) 265.
[28] G.A. Smolenskii, J. Phys. Soc. Jpn. 285 (1970) 26.
[29] P.M. Gehring, Z.-G.Y. W Chen, G. Shirane, J. Phys. Condens. Matter 16 (2004) 7113.
[30] A.C. Lawson, J.A. Roberts, B. Martinez, M. Ramos, G. Kotliarz, F.W. Trouw, M.R. Fitzsimmons, M.P. Hehlen, J.C. Lashley, H. Ledbetter, R.J. Mcqueeney, A. Migliori, Philos. Mag. 86 (2006) 2713.
[31] A.C. Lawson, B. Cort, J.A. Roberts, B.I. Bennett, T.O. Brun, R.B. Von Dreele, J.W. Richardson Jr., in: A. Gonis, N. Kioussis, M. Ciftan (Eds.), Lattice Effects in the Light Actinides, Electron Correlation and Materials Properties, Kluwer Academic/Plenum Publishers, New York, 1999, pp. 75–96.

[32] J.R. Salvador, F. Guo, T. Hogan, M.G. Kanatzidis, Nature 425 (2003) 702.
[33] T.K.H. Barron, J. Phys. C: Solid State Phys. 12 (1979) L155.
[34] J.S.O. Evans, T.A. Mary, T. Vogt, M.A. Subramanian, A.W. Sleight, Chem. Mater. 8 (1996) 2809.
[35] T.A. Mary, J.S.O. Evans, A.W. Sleight, T. Vogt, Science 272 (1996) 90.
[36] V. Korthuis, N. Khosrovani, A.W. Sleight, N. Roberts, R. Dupree, W.W. Wareen Jr., Chem. Mater. 7 (1995) 412.
[37] J.S.O. Evans, T.A. Mary, A.W. Sleight, J. Solid State Chem. 133 (1997) 580.
[38] J.Z. Tao, A.W. Sleight, J. Solid State Chem. 173 (2003) 442.
[39] J.S.O. Evans, T.A. Mary, Int. J. Inorg. Mater. 2 (2000) 143.
[40] P.M. Forster, A. Yokochi, A.W. Sleight, J. Solid State Chem. 140 (1998) 157.
[41] T.G. Amos, A.W. Sleight, J. Solid State Chem. 160 (2001) 230.
[42] C. Lind, D.G. Vanderveer, A.P. Wilkinseon, J. Chen, M.T. Vaughan, D.J. Weiner, Chem. Mater. 13 (2001) 487.
[43] J.S.O. Evans, J. Chem. Soc. Dalton Trans. 3317 (1999).
[44] C. Lind, A.P. Wilkinson, Z. Hu, S. Short, J.D. Jorgensen, Chem. Mater. 10 (1998) 2335.
[45] A.K. Tyagi, S.N. Achary, M.D. Mathews, J. Alloys Compd. 339 (2002) 207.
[46] S.C. Abrahams, J.L. Berstein, J. Chem. Phys. 45 (1966) 2475.
[47] S. Sumithra, A.M. Umarji, Solid State Sci. 8 (2006) 1453.
[48] D.A. Woodcock, P. Lightfoot, C. Ritter, J. Solid State Chem. 149 (2000) 92.
[49] W. Tiano, M. Dapiaggi, G. Artioli, J. Appl. Crystallogr. 36 (2003) 1461.
[50] J. Li, A. Yokochi, T.G. Amos, A.W. Sleight, Chem. Mater. 14 (2002) 2602.
[51] V. Grover, A.K. Tyagi, J. Solid State Chem. 177 (2004) 4197.
[52] V. Grover, A.K. Tyagi, J. Nucl. Mater. 305 (2002) 83.
[53] S.V. Chavan, A.K. Tayagi, Mater. Sci. Eng. A 433 (2006) 203.
[54] M.D. Mathews, B.R. Ambekar, A.K. Tyagi, J. Nucl. Mater. 288 (2001) 83.
[55] V. Grover, A.K. Tyagi, Ceram. Int. 3 (2005) 769.
[56] V. Grover, P. Sengupta, A.K. Tyagi, Mater. Sci. Eng. B 138 (2007) 246.
[57] A.K. Tyagi, M.D. Mathews, J. Nucl. Mater. 278 (2000) 123.
[58] A.K. Tyagi, M.D. Mathews, B.R. Ambekar, R. Ramachandran, Thermochim. Acta 421 (2004) 69.
[59] R.D. Purohit, A.K. Tyagi, M.D. Mathews, S. Saha, J. Nucl. Mater. 280 (2000) 51.
[60] M.D. Mathews, B.R. Ambekar, A.K. Tyagi, J. Nucl. Mater. 280 (2000) 246.
[61] R.D. Purohit, S. Saha, A.K. Tyagi, J. Nucl. Mater. 288 (2001) 7.
[62] A.K. Tyagi, M.D. Mathews, R. Ramachandran, J. Nucl. Mater. 294 (2001) 198.
[63] A.K. Tyagi, B.R. Ambekar, M.D. Mathews, J. Alloys Compd. 337 (2002) 277.
[64] R.D. Purohit, S. Saha, A.K. Tyagi, J. Nucl. Mater. 323 (2003) 276.
[65] M.D. Mathews, B.R. Ambekar, A.K. Tyagi, J. Alloys Compd. 386 (2005) 234.
[66] M.D. Mathews, B.R. Ambekar, A.K. Tyagi, J. Nucl. Mater. 341 (2005) 19.
[67] M.D. Mathews, B.R. Ambekar, A.K. Tyagi, Ceram. Int. 32 (2006) 609.
[68] V. Grover, A.K. Tyagi, J. Am. Ceram. Soc. 89 (2006) 2917.
[69] B.P. Mandal, V. Grover, A.K. Tyagi, Mater. Sci. Eng. A 430 (2006) 120.
[70] V. Grover, A.K. tyagi, Mater. Res. Bull. 39 (2004) 859.
[71] S.V. Chavan, M.D. Mathews, A.K. Tyagi, J. Am. Ceram. Soc. 87 (1977) 2004.
[72] S.V. Chavan, A.K. Tyagi, Mater. Sci. Eng. A 404 (2005) 57.
[73] S.V. Chavan, S.J. Patwe, A.K. Tyagi, J. Alloys Compd. 360 (2003) 189.
[74] S.V. Chavan, A.K. Tyagi, Thermochim. Acta 390 (2002) 79.
[75] A.W. Sleight, Endeavour 19 (1995) 64.
[76] D. Taylor, J. Br. Ceram. Trans. 83 (1984) 129.

[77] F.H. Gillery, E.A. Bush, J. Am. Ceram. Soc. 42 (1959) 175.
[78] W. Schreyer, J.F. Schairer, J. Petrol. 2 (1961) 324.
[79] C.-T. Li, D.R. Pecor, Z. Kristallogr. 126 (1968) 46.
[80] J. Alamo, R. Roy, J. Am. Ceram. Soc. 67 (1984) 78.
[81] G. Harshe, D.K. Agrawal, J. Am. Ceram. Soc. 77 (1965) 1994.
[82] S.Y. Limaye, D.K. Agrawal, R. Roy, J. Mater. Sci. 26 (1991) 93.
[83] P.S. Tantri, K. Geetha, A.M. Umarji, S.K. Ramesha, Bull. Mater. Sci. 23 (2000) 49.
[84] D.A. Woodcock, P. Lightfoot, C. Ritter, Chem. Commun. 107 (1998).
[85] P. Lightfoot, D.A. Woodcock, M.J. Maple, L.A. Villaescusa, P.A. Wright, J. Mater. Chem. 11 (2001) 212.
[86] M.P. Attfield, A.W. Sleight, Chem. Commun. 601 (1998).
[87] G.K. White, Contemp. Phys. 34 (1993) 193.
[88] D.K. Agrawal, A. Halliyal, J. Belsick, Mater. Res. Bull. 23 (1988) 159.
[89] S.N. Achary, K.R. Chakraborty, S.J. Patwe, A.B. Shinde, P.S.R. Krishna, A.K. Tyagi, Mater. Res. Bull. 41 (2006) 674.
[90] J. Haines, O. Cambon, G. Fraysse, A. van der Lee, J. Phys. Condens. Matter 17 (2005) 4463.
[91] Y. Zhao, D.J. Weidner, Phys. Chem. Miner. 18 (2004) 294.
[92] S.N. Achary, A.K. Tyagi, J. Solid State Chem. 177 (2004) 3918.
[93] S.N. Achary, O.D. Jayakumar, A.K. Tyagi, S.K. Kulshresththa, J. Solid State Chem. 176 (2003) 37.
[94] K. Robinson, G.V. Gibbs, P.H. Ribe, Science 172 (1971) 567.
[95] F.A. Hummel, J. Am. Ceram. Soc. 32 (1949) 320.
[96] E. Schaffer, R. Roy, J. Am. Ceram. Soc. 39 (1956) 330.
[97] S.N. Achary, R. Mishra, O.D. Jayakumar, S.K. Kulshreshtha, A.K. Tyagi, J. Solid State Chem. 180 (2007) 84.
[98] A.W. Sleight, Inorg. Chem. 37 (1998) 2854.
[99] J.S.O. Evans, T.A. Mary, A.W. Sleight, Physica B 241−243 (1998) 311.
[100] S.N. Achary, G.D. Mukherjee, A.K. Tyagi, S.N. Vaidya, J. Mater. Sci. 37 (2002) 2501.
[101] C.A. Kennedy, M.A. White, A.P. Wilkinson, T. Varga, Appl. Phys. Lett. 90 (2007) 151906.
[102] L.E. Depero, L. Sangalettis, J. Solid State Chem. 129 (1997) 82.
[103] O. Muller, R. Roy, Crystal Chemistry of Non-Metallic Materials, in: The Major Ternary Structural Families, vol. 4, Springer-Verlag, Berlin, 1974.
[104] A. Zalkin, D.H. Templeton, J. Chem. Phys. 40 (1964) 501.
[105] S. Skanthakumar, C.K. Loong, L. Soderholm, J. Nipko, J.W. Richardson Jr., M.M. Abraham, L.A. Boatner, J. Alloys Compd. 225 (1995) 595.
[106] S. Skanthakumar, C.K. Loong, L. Soderholm, J.W. Richardson Jr., M.M. Abraham, L.A. Boatner, Phys. Rev. B 51 (1995) 5644.
[107] R.M. Hazen, L.W. Finger, J.W.E. Mariathasan, J. Phys. Chem. Solids 46 (1985) 253.
[108] E.C. Subbarao, D.K. Agrawal, H.A. McKinstry, C.W. Sallese, R. Roy, J. Am. Ceram. Soc. 73 (1990) 1246.
[109] Z. Mursic, T. Vogt, H. Boysen, F. Frey, J. Appl. Crystallogr. 25 (1992) 519.
[110] S.N. Achary, S.J. Patwe, M.D. Mathews, A.K. Tyagi, J. Phys. Chem. Solids 67 (2006) 774.
[111] H. Li, S. Zhou, S. Zhang, J. Solid State Chem. 80 (2007) 589.
[112] S.J. Patwe, S.N. Achary, A.K. Tyagi, Am. Mineral. 94 (2009) 98.
[113] D.K. Agrawal, R. Roy, H.A. Mckinstry, Mater. Res. Bull. 22 (1987) 83.
[114] J. Chen, X.R. Xing, R.B. Yu, G.R. Liu, Appl. Phys. Lett. 87 (2005) 231915.
[115] A. Chandra, D. Pandey, M.D. Mathews, A.K. Tyagi, J. Mater. Res. 20 (2005) 350.
[116] B.A. Hunter, B.J. Kennedy, T. Vogt, Phys. Rev. B 69 (2004) 020410.
[117] S.J. Patwe, S.N. Achary, M.D. Mathews, A.K. Tyagi, J. Alloys Compd. 390 (2005) 100.
[118] S.J. Patwe, S.N. Achary, M.D. Mathews, A.K. Tyagi, Mater. Chem. Phys. 98 (2006) 486.

[119] P. Dhak, P. Pramanik, S. Bhattacharya, A. Roy, S.N. Achary, A.K. Tyagi, Physica Status Solidi (b) 248 (1884) 2011.
[120] T. Ishihara, H. Matsuda, Y. Takita, J. Am. Chem. Soc. 116 (1994) 3801.
[121] M. Kajitani, M. Matsuda, A. Hoshikawa, S. Harjo, T. Kamiyama, T. Ishigaki, F. Izumi, M. Miyake, Chem. Mater. 17, 4235 (2005).
[122] L. Vasylechko, R. Niewa, H. Borrmann, M. Knapp, D. Savytskii, A. Matkovski, U. Bismayer, M. Berkowski, Solid State Ion. 143 (2001) 219.
[123] M. Berkowski, J. Fink-Finowicki, W. Piekarczyk, L. Perchuc, P. Byszewski, L.O. Vasylechko, D.I. Savytskij, K. Mazur, J. Sass, E. Kowalska, J. Kapusniak, J. Cryst. Growth 209 (2000) 75.
[124] K. Mori, T. Fukunaga, K. Shibata, K. Iwase, S. Harjo, A. Hoshikawa, K. Itoh, T. Kamiyama, T. Ishigaki, Physica B 352 (2004) 147.

CHAPTER

MATERIALS UNDER INTENSE LASER IRRADIATION 14

P.A. Naik, V. Arora, Y.B.S.R. Prasad, J.A. Chakera, P.D. Gupta
Raja Ramanna Centre for Advanced Technology, Indore, India

CHAPTER OUTLINE

Materials Under Extreme Conditions. http://dx.doi.org/10.1016/B978-0-12-801300-7.00014-0

The study of matter at high density and high temperature has been a subject of considerable importance for a variety of research investigations related to the laser-driven inertial confinement fusion (ICF) [1,2], astrophysics [3], and potential application in different branches of applied sciences [4]. Lasers have been used since their invention to create high- temperature, high-density plasmas. The conditions in laser-produced plasmas are extreme for a short time, with density near the solid density, temperature more than one million kelvin, pressure up to several megabars, and magnetic fields larger than several megagauss [5]. The extreme parameters of laser-produced plasmas, similar to those existing in stars, are in fact the main challenge to meet to make ICF using lasers a viable technique for generating controlled fusion energy. In addition, creation of matter at such extreme pressures and temperatures is important for fundamental investigations relevant to laboratory astrophysics, generation of high-energy femtosecond duration X-ray pulses [6], particle acceleration [7], generation of high-energy X-rays [8], energetic ions, protons for cancer therapy [9], laser-produced nuclear reactions [10] like (γ, n) reactions, and photofission of actinides.

In this chapter, the extreme conditions of matter undergoing laser irradiation will be discussed. The chapter is organized as follows: an introduction to high energy density (HED) physics and interaction of intense lasers with matter will be presented in Section 1. Laser-produced plasma is a bright source of X-ray radiation in the subkeV-to-MeV region. At the high densities involved, the opacity of the plasma to generated X-rays becomes an important issue. Section 2 describes X-ray emissions from the laser-produced plasma and opacity of radiation from plasma. Diagnostic techniques used for such plasmas are also described in this section. Ultrashort, ultrahigh-intensity laser—matter interactions have also been exploited for particle generation and acceleration. Section 3 describes electron acceleration based on laser-driven plasma wake fields, ion acceleration, and neutron generation from the ultrahigh-intensity laser—matter interactions. Hot laser-generated plasma ablating rapidly from a solid target surface, as recoil, generates a strong shock wave propagating into the material. This offers the possibility of studying the behavior of matter under extreme high pressures to acquire important experimental data for knowing the "equation of state" (EOS) of the material used as a target. The theoretical concepts of EOS, shock waves, and their diagnostic techniques are presented in Section 4. With the use of ultrashort pulse lasers, extreme material states corresponding to solid densities at several electron volt temperatures, referred to as *warm dense matter (WDM)*, can be created by "isochoric heating" (i.e., heating at a constant volume). The formation of WDM, together with the use of ultrafast X-rays to probe it, is described in Section 5. Radiation and particles emitted from plasmas produced by ultrashort laser pulses are inherently ultrashort in nature (of the order of the laser pulse duration) and are of potential use in time-resolved X-ray diffraction, ion beam cancer therapy, and positron emission tomography (PET). The technological applications of such plasma sources are described in Section 6.

1. HIGH ENERGY DENSITY STATE OF MATTER

An HED state of matter is the state wherein the pressure (i.e., the energy density) inside the matter is greater than 1 Mbar, with an equivalent energy density of 10^{11} J/m^3, and the temperature of the matter is more than 100 million kelvin [11]. Such a condition exists within the interior of stellar objects in the universe and is also produced in nuclear explosions. A similar condition can be recreated in the laboratory using megajoule-class nanosecond duration lasers or petawatt-class ultrashort duration lasers. The various astrophysical phenomena in the HED regime and the parameters of some of the major laser facilities are described in the report of the Committee on High Energy Density Plasma Physics [12].

1.1 HIGH ENERGY DENSITY PHYSICS

The HED state of matter can be produced in a laboratory by focusing a high-energy, high-power laser beam onto a solid target. In this respect, laser technology, from the invention of laser in 1960, has been developed for achieving extremely high fluence (10^8 J/cm^2 at the National Ignition Facility, USA) in a few nanoseconds of pulse duration [13]. This and other high-power laser facilities (like Laser Mega Joule, France) have been designed to compress Deuterium-Tritium (D-T) fuel capsules to high densities and high temperatures, whereby fusion reactions occur, for generating more energy than that put in. The advancements in HED research also owe a lot to technological breakthroughs in the development of ultrashort pulse lasers (few femtoseconds' duration) aimed to achieve higher and higher laser intensity ($\sim 10^{20}$ W/cm^2) in a smaller focal spot. A comparatively recent development in fusion research is the proposed fast-ignition scheme [14], which has relaxed the requirement of driver laser energy and has further rejuvenated research related to laser technology. High-intensity lasers have the capabilities to experimentally investigate a vast field of unexplored phenomena in an HED environment, relevant to both astrophysics and fusion studies.

1.2 INTENSE LASER–MATTER INTERACTION

When the laser light is focused on a target, there is a threshold intensity for the generation of plasma, which roughly depends on the inverse square root of the pulse duration. It is usually 10^{8-9} W-cm^{-2} for few nanoseconds' laser pulses and increases for shorter pulses. The process of plasma production on a solid target by focused laser light (above the threshold intensity) is as follows [5,15,16]: at first, the surface electrons and the electrons produced by multiphoton ionization of target atoms start absorbing laser light. The target material gets ionized and forms plasma. A subsequent part of the laser pulse heats up this plasma and produces more plasma from the target through a cascade process, in which the collisions between free electrons and neighboring atoms facilitate further ionization. This heated plasma, because of its large n_eKT_e pressure, rapidly expands to form an outwardly monotonically decreasing density profile. Rapid expansion is also referred to as *ablation*. In general, the physical structure of the plasma may be described by Fig. 14.1 [17], which shows typical spatial profiles of the electron density (n_e) and the electron temperature (T_e). The laser beam can propagate up to a maximum density called the *critical density* (n_c), given by $n_c = 1.1 \times 10^{21}\ \lambda_L^{-2}$ (μm) cm^{-3}, where λ_L (μm) is the wavelength of the laser in vacuum. This is the density at which the laser frequency equals the local plasma frequency. The laser light cannot propagate in a region of density higher than the critical

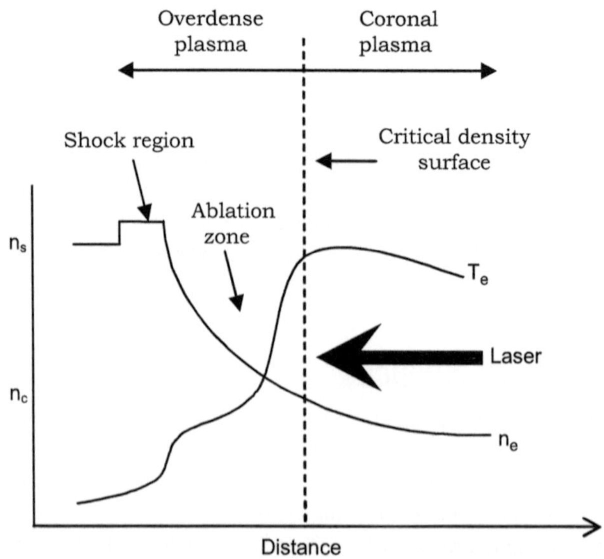

FIGURE 14.1

A schematic diagram of the spatial profiles of electron density and temperature depicting the various zones in laser-produced plasma. n_c, Critical density; n_e, electron density; n_s, solid density; T_e, electron temperature.

density. The region of plasma up to the critical density (underdense region) is thus directly heated by absorption of the laser energy and it is referred to as the *coronal region*. The absorbed laser energy is then transported as heat via thermal conduction and electron transport into the overdense region beyond the critical density (where the laser light cannot penetrate).

Laser energy is absorbed in the plasma by collisional or collisionless mechanisms, depending on the focused intensity, the laser pulse duration, and the laser's wavelength. The two most important absorption mechanisms in the interaction region typical to short and ultrashort laser-produced plasma are: (1) inverse bremsstrahlung (IB), and (2) resonance absorption.

1.2.1 Inverse Bremsstrahlung Absorption

In IB absorption, the electrons oscillating in the electric field of the laser get randomly scattered by collisions with the ions. Thus the laser energy gets converted into thermal energy of the plasma constituents (ions, electrons), resulting in laser light absorption into the plasma. The IB absorption coefficient can be obtained by using the expression for the electron–ion collision time interval (τ_c) [16].

$$\tau_c = \nu_{ei}^{-1} = \frac{3}{4}\frac{(k_B T_e)^{3/2} m_e^{1/2}}{(2\pi)^{1/2} Z_i^2 e^4 n_i \ln \Lambda} = 3.44 \times 10^5 \frac{T_e(eV)^{3/2}}{Z_i^2 n_i \ln \Lambda}[s]; \quad \Lambda \equiv \frac{\lambda_D}{l_{\min}},$$

where n_i is the ion density, Z_i is the average ion charge, T_e is the electron temperature, ln Λ is the Coulomb logarithm, λ_D is the Debye length, and l_{min} is the minimum impact parameter defined by the classical distance of closest approach between the electron and the scattering ion. A significant fraction of absorption takes place in the vicinity of critical density, i.e., when $n_e/n_{cr} \approx 1$.

It should be noted that for an inhomogeneous plasma, the expression for the rate of energy deposition is far more complicated because it depends on the electron density and the plasma temperature, which in turn depend on the time evolution of the expanding plasma. At high laser intensities, the large electric field of the laser can distort the thermal distribution of the electrons, thereby changing the electron–ion collision frequency, ν_{ei}. It follows from the expression for ν_{ei} that the shorter wavelength lasers and high-Z targets result in better absorption of laser light by the process of IB. In fact, IB is the dominant process of laser energy absorption for laser intensity $I_L\lambda^2 \leq 10^{13}$ W-μm^2/cm^2. The collisional absorption process is polarization independent, although the maximum absorption occurs at normal incidence and reduces with increase in the angle of incidence, because the laser light gets reflected from lower and lower density regions. As the collisional absorption increases with increasing density scale length, for the same laser energy the efficiency of collisional absorption decreases with reduction in the laser pulse duration. This is because with the increase in laser intensity, the electron temperature increases and the electron–ion collision rate decreases. Moreover, for sub-nanosecond laser pulses, the density scale length also decreases with pulse duration.

When an ultrashort laser pulse is incident on a solid target, solid density plasma is generated by the initial part of the pulse. This plasma cannot expand significantly during the laser pulse (owing to the very short time involved) and rest of the interaction occurs with the near-solid density plasma, as depicted in Fig. 14.2. The hydrodynamic simulations have shown that the scale length of the plasma is much shorter than the laser wavelength. In such a case, when the laser intensity exceeds 10^{15} W-cm^{-2}, the collisionless processes [18] become the dominant mechanisms of energy absorption. The prominent ones are: (1) resonance absorption, (2) vacuum/Brunel heating, and (3) $j \times B$ heating.

1.2.2 Resonance Absorption

This process occurs when a p-polarized light wave is incident obliquely on plasma with a density gradient. The laser light incident on the plasma at an angle θ is reflected at a lower density given by $n_e = n_c \cos^2(\theta)$. The evanescent wave excited by the reflected wave in the overdense region has longitudinal as well as transverse components, with a frequency equal to the laser light frequency. The longitudinal component of this evanescent wave in the direction of the density gradient resonantly excites the plasma wave at the critical density. The plasma wave can grow resonantly over several laser oscillations and then transfers its energy to the plasma through wave breaking, collisional damping, or

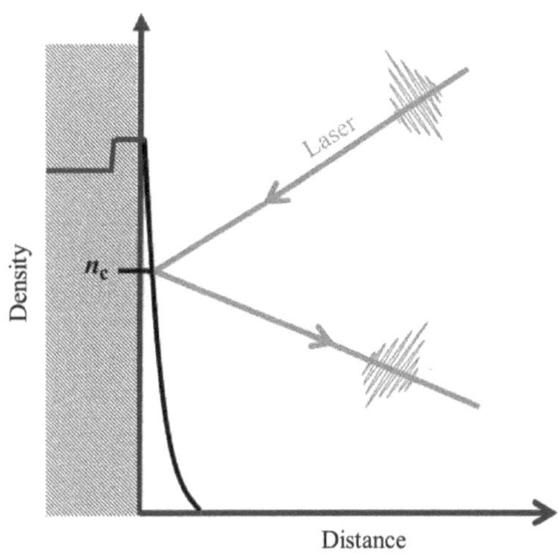

FIGURE 14.2

A schematic representation of the electron density profile in ultrashort laser-produced plasma. n_c, Critical density.

Landau (collisionless) damping. The efficiency of the resonance absorption process depends on the plasma scale length, the angle of incidence, and the polarization of the laser light. It may be noted that this mechanism of absorption leads to absorption of obliquely incident long, as well as short, pulse lasers, although for ultrashort pulses, the mechanisms discussed in the following section become dominant.

1.2.3 Vacuum/Brunel Heating

For scenarios when the density gradient is smaller than λ, which happens in the case of plasma produced by a high-contrast, ultrashort laser pulse, the resonantly driven plasma wave is not able to survive when the amplitude of the electron oscillations exceeds the plasma scale length. In such a case, it was proposed by Brunel [19] that an electron at this sharp plasma–vacuum interface is first accelerated by the laser electric field into the vacuum. When the field of the laser reverses, the electron is pushed back into the plasma. The electron penetrates deeper into the plasma and eventually gets thermalized by collisions. This mechanism is most efficient for ultrashort laser pulses because longer pulses tend to produce plasmas with longer scale-length density profiles. For steep density gradients, the density scale length of the plasma becomes comparable to or even smaller than the skin depth for the laser. Under these conditions, the electrons in the plasma can easily be pulled out in the vacuum by the electric field of the laser and then sent back into the target to undergo a large amount of kinetic energy loss.

1.2.4 j × B Heating

For ultrahigh intensities, the electron motion becomes relativistic and is dominated by the $v \times B$ term of the Lorentz force. The ponderomotive force of the laser [18] can be written as:

$$f_{pond} = -\frac{\partial}{\partial x}\left(\frac{mv_{osc}^2}{2}\frac{4\omega_o^2}{\omega_{pe}^2}\exp^{\frac{2\omega_{pe}}{c}} x\left[\frac{1+\cos 2\omega_o t}{2}\right]\right)$$

Here, the heating results from the component of the $j \times B$ force ($j = n_e\, e\, v$) that causes oscillations of the electrons at the vacuum–plasma interface with a frequency twice the laser frequency. It may be noted that the magnitude of the force is dependent on $\frac{\omega_o^2}{\omega_{pe}^2}$, or equivalently, n_{cr}/n. Thus as the electron density increases, the magnitude of the force goes down, and lesser energy is transferred to the hot electrons. The net result is that the amount of laser light absorbed decreases with increasing density. The second term in the equation leads to the heating of electrons, the first term being the usual ponderomotive force that pushes the electrons out of the high-intensity region. This $j \times B$ term, which is absent only for circular polarization, is most efficient for normal-incidence linearly/elliptically polarized light, and becomes significant at relativistic quiver velocities (when the electron moves in a figure-of-eight trajectory in linearly polarized light). A second interesting effect arises from the fact that the electron exposed to such a high intensity electric field oscillates in the direction of the electric field with very high velocity. The energy associated with this velocity can easily reach megaelectron volt levels, and relativistic effects become important for these interactions.

2. X-RAY EMISSION FROM LASER-PRODUCED PLASMA

Intense laser–matter interaction is a subject of considerable importance for a variety of scientific and technological applications [20], because the plasma is a bright, pulsed source of high-energy X-ray radiation and extremely energetic (MeV) electrons and ions. The duration of the radiation and particles emitted from these plasmas is inherently ultrashort in nature and is of potential use in time-resolved diffraction studies [21]. Proton generation and focusing is one of the new physical phenomena in this field and is of great interest for proton beam cancer therapy [22]. On the other hand, the high-energy electrons can directly create the ignition hot spot required in the fast-ignition concept of ICF [2].

2.1 X-RAY EMISSIONS IN LASER–PLASMA INTERACTIONS

The overall emission spectrum from laser-produced plasma is a superposition of line radiation (bound–bound radiation) on a broad continuum due to free–free and free–bound radiation. The relative yield resulting from these processes depends on the plasma parameters (n_e, T_e) [23]. Excitation and deexcitation of electrons provides the main mechanism responsible for optical and X-ray emissions from a plasma. These take place between bound states, between free and bound states, and between the free states. In the following sections, these processes are referred to as *bound–bound, free–bound*, and *free–free* transitions, respectively. Bound–bound transitions (from a higher to lower energy level) lead to emission of photons with discrete energies determined by the atomic structure of the ion. Free–bound transitions include processes by which a free electron recombines with an ion to one of its bound state. Because the free electron can have any energy, this leads to a continuous emission spectrum. Free–free emission is produced by unbound (free) electrons "colliding" with ions through an interaction between the electron charge and the coulomb field of the ion. Such collisions lead to the continuous acceleration of electrons and the production of bremsstrahlung (free–free) emission radiation. This features a broad range of emitted wavelengths, characteristic of the plasma temperature. The total spectral emission from a plasma has contributions from each of these processes, leading to a mixture of discrete lines riding on continuous emission.

A typical high-resolution X-ray spectrum of magnesium plasma [24], generated using a 2 J, 3 ns laser beam, focused to an intensity of approximately 5×10^{12} W cm^{-2}, is shown in Fig. 14.3. Various prominent lines in the spectral range of 6.9 Å to 9.6 Å are identified as transitions in He-like (Mg^{+10})

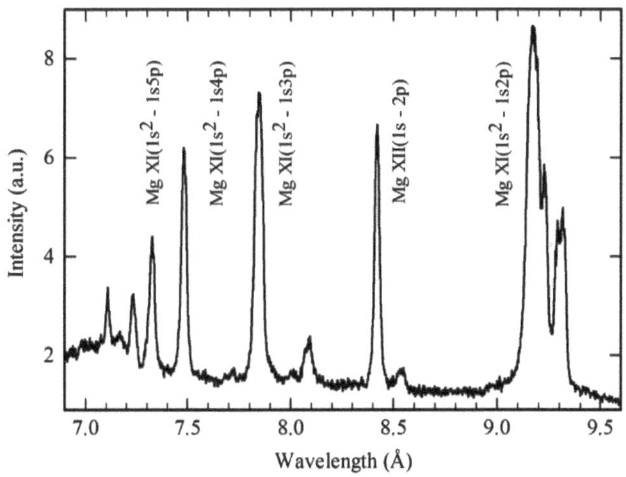

FIGURE 14.3

X-ray spectrum of magnesium plasma produced using nanosecond laser pulses at an intensity of 5×10^{12} W/cm^2. *a.u.*, arbitrary unit.

and H-like (Mg^{+11}) magnesium ions (in increasing order of wavelength): Mg XII 1s − 3p at $\lambda = 7.11$ Å (H-β), Mg XI $1s^2$ − 1s 6p at $\lambda = 7.23$ Å (He-ε), Mg XI $1s^2$ − 1s 5p at $\lambda = 7.31$ Å (He-δ), Mg XI $1s^2$ − 1s 4p at $\lambda = 7.47$ Å (He-γ), Mg XI $1s^2$ − 1s 3p at $\lambda = 7.86$ Å (He-β), Mg XII 1s − 2p at $\lambda = 8.42$ Å (H-α), Mg XI $1s^2$ − 1s 2p (He-α resonance) at $\lambda = 9.17$ Å, and Mg XI $1s^2$ − 1s 2p (He-α intercombination) at $\lambda = 9.23$ Å. The strengths and energies of the bound–bound emission lines are directly related to the electron populations and atomic structure. The plasma temperature and the density influence the rate of collisions, which in turn determine the electron distribution among the bound states. Spectra from plasmas therefore contain information related to both the atomic structure of ions in the plasma and also to the hydrodynamic parameters such as temperature and density. To extract this useful information from a spectrum, theoretical models are required to describe the distribution of bound and free electron populations in the plasma, the interplay between these quantities, and the hydrodynamic properties of the plasma. From these, relationships can be obtained between the atomic populations and emissions that relate directly to the plasma properties.

Highly efficient emissions in the X-ray energy range of 1–2 keV from thermal plasma can be used as back light [25] for high-resolution radiography. However, probing of high–areal density material requires an X-ray source with higher photon energies [26]. The K-α X-ray emission of energy greater than 5 keV can be easily obtained from targets irradiated by an ultrashort pulse laser of moderate focused intensity ($>10^{17}$ Wcm^{-2}). In a simplified description, the K-α line (or L, M shell line) radiation generation is a two-stage process. First, the hot electrons are generated in the plasma, and then they get scattered in the solid target. The K-α yield [27,28] essentially depends on the energy distribution of these electrons and the K-shell ionization cross-section for electron impact. It is important to optimize the energy distribution of the electrons in experimental conditions to achieve high K-α conversion efficiency.

The line radiation is generated by the energetic electrons (called *hot electrons*) produced during the laser irradiation. Under optimized conditions, the X-ray pulse duration is of the order of laser pulse duration [29] because the driving electron pulse is generated only during the ultrashort laser pulse. Further, when the electrons do not penetrate deep inside the bulk solid target or undergo lateral scattering, the source dimension is comparable to the laser focal spot size [30] and therefore it has a high spatial coherence. Such an X-ray source is suitable for time-resolved studies [21], phase-contrast imaging [31], and medical applications [32]. Moreover, the study of this radiation is important because it is related to the presence of hot electrons in ultrashort laser–produced plasma. Fig. 14.4 shows the X-ray spectrum from plasma produced by femtosecond laser pulses. In the spectrum of titanium, both K-α at 4.5 keV and K-β at 4.9 keV can be seen. In the same figure, the copper spectrum shows Cu K-α at 8 keV and the silicon escape peak at ~6.3 keV (because the silicon K-α photon energy is 1.74 keV). The Cu K-β (8.9 keV) is not seen here because of its strong attenuation by a Ni filter used in front of the charged coupled device (CCD) used to record the spectrum. Similarly, the characteristic Zn K-α line at 8.6 keV is seen in the spectrum of zinc.

2.2 OPACITY OF X-RAYS

The hot laser-produced plasma emits radiation covering a wide spectral range from infrared (IR) to X-rays. Therefore it can be considered as a tiny replica of a star/sun in the laboratory. In fact, it has many properties similar to a star, such as very high temperature (much more than that of the sun corona), high density, high radiation opacity, and emission characteristics resembling black body

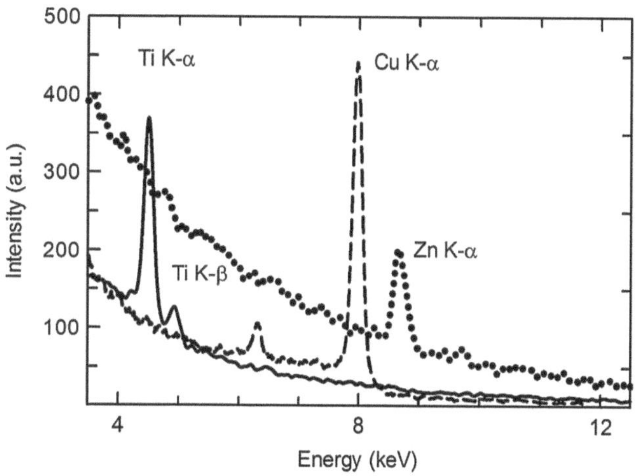

FIGURE 14.4

The K-α X-ray emission spectra of titanium, copper, and zinc from plasma produced by femtosecond laser pulses. *a.u.*, arbitrary unit.

Reproduced with permission from Arora, et al., JINST 8 P01010 (2013). Copyright 2013, IOP Publishing.

emission. The conversion of laser light into soft X-ray radiation [20] during the heating of high-Z material is very important in view of its relevance to the indirect approach to ICF [33], X-ray microscopy [34,35], and X-ray lithography [36]. In such applications, a high-Z material is used as the target because a high-Z plasma is an efficient soft X-ray converter [37,38] and it offers a high opacity to radiation conduction into the cavity wall [39].

A laser-produced plasma may be considered to have two distinct regions [40]: (1) an optically thin region that converts laser light to X-rays (conversion layer), and (2) an optically thick region that absorbs X-rays and remits (reemission zone). The first region is the outer low-density, high-temperature, optically thin plasma in which the laser beam is absorbed up to the critical density and partially converted into X-rays. The X-rays propagating toward the solid target are absorbed in an optically thick, high-density region, which reemits a fraction of the absorbed energy at a lower energy radiation, acting as an optically thick source at a lower temperature. Reemitted radiation intensity from the target is governed by temperature of the optically thick plasma region, which in turn depends on the source flux and the Rosseland mean opacity [39]. The latter comes into play because it determines the ability of the heated material to restrict thermal diffusion loss of radiation into the target, thus making it more like an ideal black body. Rosseland mean opacity (K_R) is defined by the expression [41] $K_R^{-1} = \int_0^\infty K_\nu^{-1}[\delta B\nu/\delta T]d\nu / \int_0^\infty [\delta B\nu/\delta T]d\nu$, where B is the Planck function, T is the plasma temperature, K_ν is the opacity, and $\delta B_\nu/\delta T$ is the weighting function that peaks at roughly 3.8 kT.

Gold is mostly used as an ablator wall material for a hohlraum (hollow cavity) used in the indirect approach to ICF, where the soft X-rays emitted by the cavity walls are used to ablate and compress the *D-T* fuel pellet kept at the center of the hohlraum. Hard X-rays are not desirable here because they penetrate into the *D-T* fuel and preheat it, thereby making it difficult to compress. The gold absorption spectrum at any given temperature, like that of any other element, is not uniform over frequency and has peaks and valleys. The opacity is significantly lower in some frequency regions, as shown in Fig. 14.5 [42]. It has been shown that a mixture of two or more high-Z elements can provide a significantly higher Rosseland mean opacity than pure elements [43–47]. A higher opacity in mix-Z plasma is achieved when the frequency regions of small opacity of one element coincide with the regions of large opacity of another element in the same photon energy band. Because gold has been the

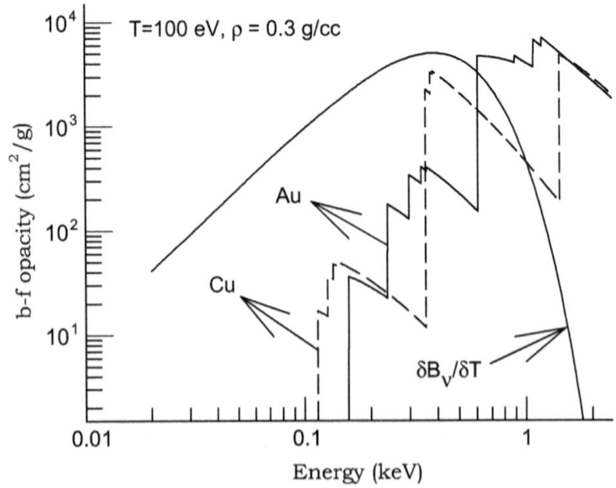

FIGURE 14.5

Bound–free opacity spectrum computed for Cu and Au plasmas.

Reproduced with permission from Chakera, et al., Applied Physics Letters 83 (2003) 27. Copyright 2003, AIP Publishing LLC.

most widely used material for ICF hohlraum fabrication, its mixtures with some other elements, e.g., Au–Gd, Au–Sm [43–45,47], have been theoretically investigated to determine optimum composition to achieve high Rosseland mean opacity. Orzechowski et al. [45] have shown 40% increase in the Rosseland mean opacity at a temperature of 200 eV in Gd mixed with gold compared with pure gold. Colombant et al. [43] have performed computer simulations of Rosseland mean opacity in mixtures of Au with Cu, in view of the latter's advantages of easy availability and alloy formation. It is expected that a wall of optimum atomic composition having higher Rosseland mean opacity will provide a larger conversion of absorbed laser energy into soft X-ray radiation.

Although a large number of theoretical investigations have been carried out on optimization of Rosseland mean opacity using mix-Z targets, experimental studies of X-ray emission from plasma produced from such mix-Z targets are few [48]. Chakera et al. and Naik et al. [42,49] have performed a comparative experimental study of the extreme ultraviolet (XUV)/soft X-ray emission from Au–Cu mix-Z targets of different atomic compositions using a second harmonic of Nd:glass laser at $\lambda_L = 0.53$ μm, focused to a laser intensity of approximately 10^{13} W/cm^2. The radiation intensity in the spectral region of about 15–150 Å and the integrated X-ray yield were found to be maximum for an atomic composition of Au 0.43–Cu 0.57, as shown in Fig. 14.6a [42]. The sub-keV radiation spectrum recorded for this target is shown in Fig. 14.6b [42]. The spectral shape broadly resembles a radiation source of temperature of 90–100 eV under the assumption of an optically thick plasma, i.e., black body (shown as a *dotted line*, Fig. 14.6b). This is in good agreement with a radiation temperature (T_R) of about 100 eV calculated for the laser intensity of approximately 10^{13} W/cm^2 on the target, from the scaling law of T_R (eV) about 130 [I_L (W/cm^2)/10^{14}]$^{0.127}$, as reported by Nishimura et al. [50] for gold plasma heated by 0.53-μm laser pulses. These observations are consistent with the variation of Rosseland mean opacity with different atomic composition calculated using a screened hydrogenic average atom model [44]. Further, keV X-ray yield in the Au–Cu mix-Z plasmas is observed to decrease sharply even for small fraction of the gold in the Au–Cu mix-Z target [51]. This is important for reducing the problem of radiation preheat of the ICF fuel.

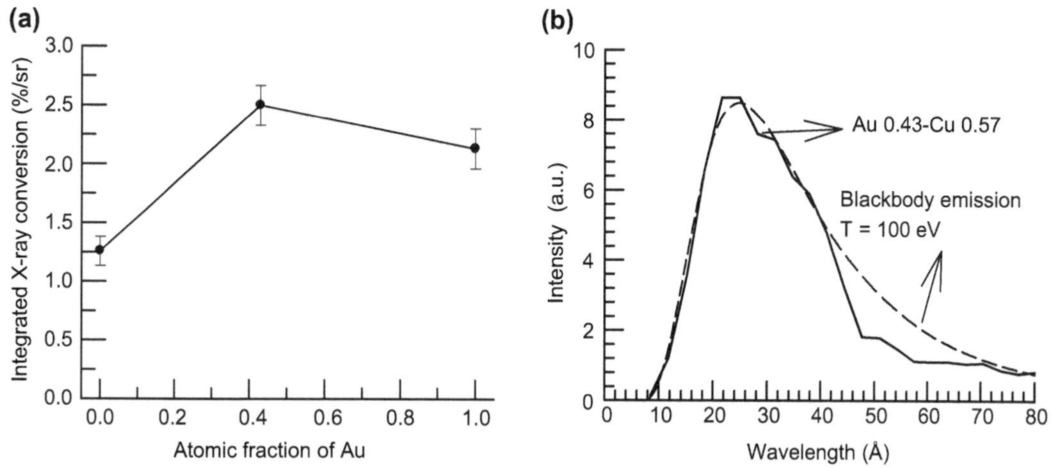

FIGURE 14.6

(a) Integrated X-ray conversion per unit solid angle for Cu, Au, and Au 0.43 –Cu 0.57 mix-Z plasma. (b) Sub-keV emission spectrum for the Au 0.43 –Cu 0.57 mix-Z target. *a.u.*, arbitrary unit.

Reproduced with permission from Chakera, et al., Applied Physics Letter 83, (2003) 27. Copyright 2003, AIP Publishing LLC.

2.3 TECHNIQUES FOR X-RAY DIAGNOSTICS

As mentioned earlier, the radiation emitted from laser-produced plasma covers a wide spectral range from IR, visible, and soft X-rays to very hard X-rays, depending on the laser and target parameters. Of this spectrum, emissions in the X-ray region have been widely used for the investigation of laser–plasma interaction mechanisms [15,20]. The influence of the plasma parameters on the spectral and spatial features of the X-ray emission allows a direct mapping of the laser–plasma interaction processes such as partition of energy between thermal and nonthermal electrons [52] in the plasma or the implosion dynamics of fusion pellets [53]. The time-resolved measurement of X-ray line emissions from highly charged ionic species (H-like, He-like ions) produced by collisional ionization in a high-temperature plasma or by optical field ionization is exploited to derive the temporal evolution of the plasma properties [54]. The diagnostic information on electron temperature and density is obtained by comparing the measured line shapes and the ratio of the intensities with the calculations, assuming an appropriate plasma equilibrium model.

Quantitative interpretation of plasma parameters requires the knowledge of the ionization equilibrium at any time and position inside the plasma. This necessitates the solving of both space- and time-dependent systems of rate equations and equations describing the hydrodynamic motion. Furthermore, the radiative transfer problem, involving the absorption and emission of radiation along a propagation path inside the plasma, must be included because it leads to a significant modification in the intensities and profiles of the lines in the emission spectrum. The spectrum of the X-ray radiation generated by laser-produced plasma contains information about the rapidly evolving high-density high-temperature plasmas. Spectroscopic methods are most widely used for diagnostics in laser plasma on account of the high X-ray flux. These techniques are based on the interpretation of at least one of the following quantities: spectral line width or profiles, relative continuum intensities, and relative line intensities. The spectral lines that are strongly dependent on one of the plasma parameters but weakly dependent on other parameters have to be identified. For example, electron temperature can be determined from the ratio of the intensities of the resonance lines emitted by the plasma.

The basic idea is that the level parameters and the populations in the upper states govern the line intensities, and the relative values of the latter are governed by the plasma temperature. When two transitions arise from the ions of same ionization stage, relative population is simply governed by the Boltzmann distribution. The ratio of line intensities in such a case is given by $I_1/I_2 = (A_1/A_2)(g_1/g_2)(E_1/E_2) \exp[-(E_1-E_2)/kT_e]$, where A_1 and A_2 are the radiative transitions rates of the two upper levels, and E_1 and E_2 are the energy levels with respect to the ground state. It can be seen that the ratio is a strong function of temperature for $|E_1-E_2| > kT_e$. Therefore for sensitive measurements of electron temperature, two lines with their upper level well separated in energy are preferred. From the line intensity ratio, the temperature or density can be calculated using the spectroscopic code RATION [55], which calculates the ratio of various line intensities versus electron density for different electron temperatures or the ratio of intensities versus electron temperature for different values of the electron density.

For low-resolution X-ray spectroscopy, X-ray diodes covered with filters of different materials transmitting X-rays in different spectral region are used. From the output of these diodes, the X-ray spectrum (continuum) is reconstructed and because it has $\exp(-E/kT_e)$ dependence, the temperature can be deduced [56].

For high-resolution spectroscopy, crystals and gratings (transmission or reflective) are used, depending on the spectral region of interest. Crystal spectrographs work on the Bragg diffraction principle and have a very high resolving power. A free-standing transmission-grating spectrograph can be used to cover a large spectral range with moderate resolution. Grazing incidence grating spectrographs are used for a softer X-ray region. Earlier X-ray films or multichannel plates (MCPs) were used as space-resolved X-ray detectors. More recently, X-ray CCD cameras or image plates are used as space-resolved detectors. X-ray diodes can give time-resolved information with low temporal resolution. MCPs can also be used as time-resolved detectors. X-ray streak cameras can give high temporal resolution, along with one-dimensional space resolution. A slit used along with a crystal spectrograph can get energy-resolved images (i.e., two-dimensional space resolution) of the plasma [57]. A pinhole transmission grating can also give the same, albeit with a low radiation throughput [58].

3. PARTICLE ACCELERATION WITH LASERS

There is an ever-increasing demand for using accelerated charged particle energy to explore subatomic particles and to understand the creation of matter. In the conventional particle accelerators based on a radio frequency (rf) accelerating field, the maximum acceleration field is limited to about 50 MV/m, which necessitates building particle accelerators of very large length running in several kilometers to achieve higher and higher particle energy. For example, at CERN in Geneva, the Large Hadron Collider has been set up in a tunnel of about 27 km circumference [59]. Such a huge machine is not only exorbitant in cost but also requires large-scale international collaborative efforts to build and operate such a facility. Ultrashort, ultrahigh-intensity laser—matter interactions have also been used for particle generation and acceleration, and efforts are on to build laser-based TeV electron accelerators.

The basic limit on rf cavity—based accelerating structure comes as a result of the maximum field gradient ($\sim$100 MV/m), which can be sustained by an rf cavity before its breakdown from high voltage. In contrast, laser-based plasma accelerators can sustain huge field gradients ($\sim$100 GV/m; three orders of magnitude higher) and hence there is the prospect of building a high-energy particle accelerator over a small accelerating length of a few centimeters to several meters. As the accelerating field is created in the plasma state, which is already in fully broken down (ionized) state, there is no fear of breakdown of the medium. However, the laser plasma–based accelerators have low average beam current, beam energy stability, and beam pointing stability. With the plasma acceleration technology growing by leaps and bounds, these problems are likely to be sorted out in the near future. The following subsections describe techniques of particle acceleration based on lasers for electrons, protons/ions, and neutron generation.

3.1 PLASMA WAKE-FIELD–BASED ELECTRON ACCELERATION

In 1979, Tajima et al. [60] showed that an ultrashort laser pulse propagating in an underdense plasma can excite an electron plasma wave (also called a *Langmuir wave*), which is an electrostatic longitudinal wave and follows the laser pulse with a phase velocity equal to the group velocity of the laser pulse. Such a longitudinal electrostatic wave generates a very high electric field (called the *wake field* because it is produced in the wake of the laser pulse), analogous to the wake created in water behind a speed boat moving in a river. If electrons are trapped in this field, they can be accelerated to very high energy because of the high electric field gradients of the wake field. The ions, being very heavy, are immobile and form a positive background.

The electron plasma wave is excited in underdense plasma when an ultrashort laser pulse or an electron beam (bunch) propagates through the plasma. The plasma electrons are pushed from the front and as well as from the back side of the laser (or electron beam) by a ponderomotive force, resulting into excitation of the electron plasma wave. The latter can be resonantly excited if the period of plasma wave matches the spatial extent of the driving laser or electron pulse. For an underdense plasma density ($\sim 10^{18}$/cc), the period of plasma wave corresponds to a driver pulse duration of few tens of femtoseconds. Thus a Ti:sapphire laser, providing ultrashort laser pulses in the range of 20 to 50 fs, can be an ideal candidate for electron acceleration based on the described technique.

There are two different types of electron plasma accelerators based on the driver pulse. Electron-beam–driven accelerators are commonly known as *plasma wake-field accelerators* [61], and the laser-beam–driven ones are called *laser wake-field accelerators (LWFAs)* [62]. For a laser pulse duration that is much more than the plasma wave period, the laser pulse undergoes self-modulation in time, resulting in the laser pulse breaking up into multiple pulses, each separated by the plasma wave period, resulting in coherent resonant excitation of the plasma wave to the extent that it can lead to wave-breaking. Such a regime of acceleration is known as a *self-modulated LWFA* [63]. One more type of laser-based acceleration scheme exists, wherein two laser beams of nearly equal frequencies are allowed to beat, resulting into a beat wave. If the plasma density is adjusted so that the beat frequency equals the plasma frequency, the plasma wave can be resonantly excited to accelerate an injected electron bunch. Such an scheme is called a *plasma beat wave accelerator* [64]. The following section describes laser wake-field electron acceleration in more detail.

3.1.1 Principles of Laser Wake-Field Electron Acceleration

When an ultrahigh-intensity, ultra-short pulse laser beam is focused on an He (or any low-Z gas) gas jet (i.e., gas puffed in a vacuum chamber), the leading edge of the laser pulse has sufficient intensity to ionize the gas to form plasma. The ultrahigh-intensity laser pulse, because of the longitudinal

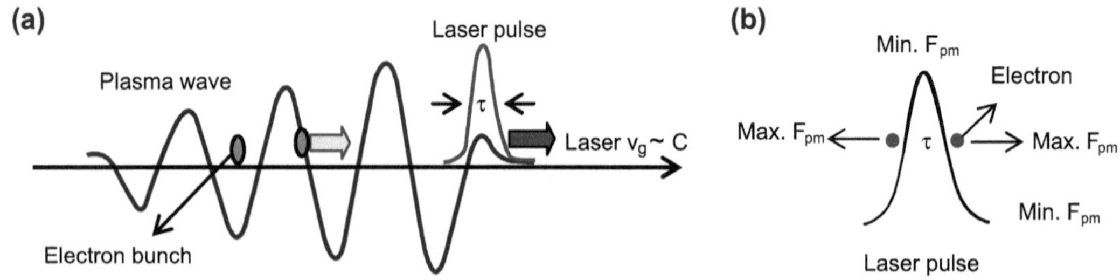

FIGURE 14.7

(a) A schematic representation of laser wake-field acceleration. (b) Ponderomotive force caused by a laser pulse on plasma electrons. τ, Pulse duration ; v_g, group velocity ; F_{pm}, ponderomotive force.

ponderomotive force, pushes the plasma electrons out longitudinally, thereby setting up the plasma oscillations behind the laser pulse propagating through the gas jet. As the plasma wave travels with a phase velocity that is equal to the laser group velocity (close to the velocity of light, c), an electron is trapped in the plasma wake field will be accelerated to the group velocity of the laser pulse, as illustrated in Fig. 14.7a.

The ponderomotive force arises because of a nonuniform space variation of the electric field of the laser beam. It results from the high-order terms of the Lorentz force on the plasma electrons caused by the gradient of the electric field, and is given by $F_p = -\{e^2(1 + \alpha^2)/4m\omega^2\}\ \nabla E^2$, where α is the ellipticity of the laser light ($\alpha = 1$ for circularly polarized light and 0 for linear polarization). Therefore because of the nonuniform intensity profile of a laser pulse, the electrons are pushed away from the region of high intensity to the low-intensity region as shown in Fig. 14.7b. This force displaces the electrons from the uniform plasma (as ions are heavy and immobile), exciting plasma oscillations. The amplitude of plasma wave will grow if it is in resonance with the ponderomotive force on electron. This happens when the half period of plasma oscillations (τ_p) matches with the laser pulse duration (τ_L), i.e., when $\tau_p = 2c\tau_L$.

The longitudinal electric field (E_{max}) created by the plasma Langmuir wave is given by [65] E_{max}(MV/cm) $\cong 2.8 \times 10^{-25}\ \lambda^2(\mu\text{m})\ I\ (\text{W/cm}^2)\ \sqrt{[n_o\ (\text{cm}^{-3})]}$. Here, n_o is the plasma density, and λ is the laser wavelength. For $\lambda = 800$ nm, $n_o = 10^{18}$ cm^{-3}, $I = 10^{18}$ W/cm^2, and $E_{max} = \sim 18$ GV/m, which is orders of magnitude higher than the maximum accelerating field gradient (~ 100 MV/m) achievable in rf-based accelerators.

3.1.2 An Example of Electron Acceleration Using a Laser Wake-Field Accelerator

A typical experimental set-up for a laser wake-field electron accelerator at the Raja Ramanna Centre for Advanced Technology, Indore, is shown in Fig. 14.8. Helium gas jet is used to create plasma by focusing ultrashort, high intensity Ti:sapphire laser pulses. The gas jet is created by puffing the gas through a nozzle by a solenoid valve connected to a high-pressure gas cylinder. The size of the nozzle is typically in the range of a few millimeters. The gas density in the jet is in the range of 10^{17} to 10^{19}/cc. The laser beam is focused just above the nozzle orifice. The laser pulse travels through the gas jet, creating a wake field behind it. The plasma electrons are trapped and accelerated in the wake field behind the laser pulse. The accelerated electron beam comes out in the direction of the laser beam

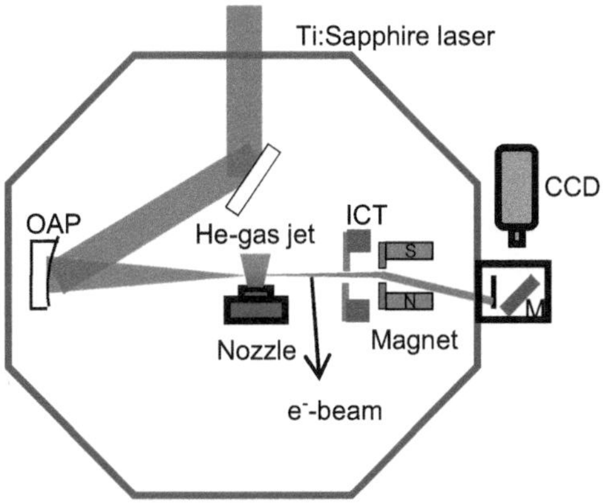

FIGURE 14.8

A typical experimental set-up for electron acceleration using a laser wake-field accelerator scheme. *CCD*, Charge coupled device camera; *ICT*, integrating current transformer; *OAP*, off-axis parabolic mirror.

propagation. A magnetic electron spectrograph placed in the beam path measures the accelerated electron beam energy. The electron beam deflection is recorded on a phosphor screen. The accelerated beam profile and its energy are measured by the deflection on the phosphor screen and its profile in the direction normal to its deflection. The beam current or charge is measured by an integrating current transformer coil kept in the beam bath before the magnetic spectrograph, which gives the total beam current of the accelerated electron beam. Fig. 14.9 shows a typical accelerated electron beam of energy at about 50 MeV recorded on the phosphor screen. For this shot, the He gas density was about 10^{18} cm^{-3}, the laser intensity was approximately 3×10^{18} W/cm^2, and the laser pulse duration was around 45 fs [66].

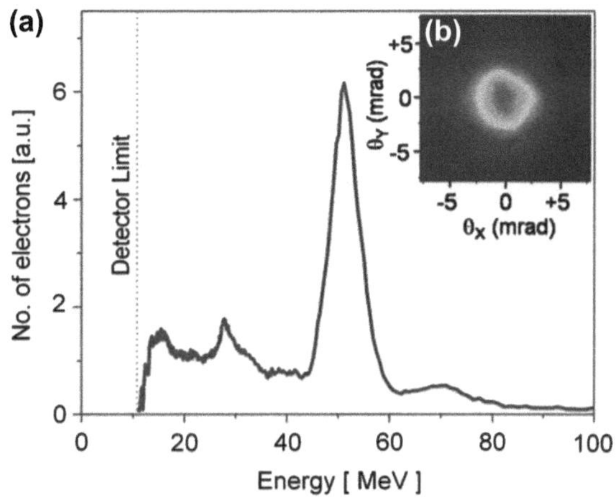

FIGURE 14.9

Accelerated electron beam energy spectrum. The inset shows the electron beam recorded on the phosphor screen. *a.u.*, Arbitrary units.

3.2 PROTON/ION ACCELERATION WITH ULTRAHIGH-INTENSITY LASERS

The interaction of ultrashort, intense laser pulses ($>10^{18}$ W/cm^2) with thin foils offers high brightness, low emittance, and charge-particle sources of high energy on a table top, and thus has the potential to be a viable alternative to the conventional particle accelerators.

3.2.1 The Principle of Proton/Ion Acceleration

The dynamics of ultrashort, high-intensity laser pulse interaction with matter is primarily governed by electrons, which because of their lighter mass, instantly respond to the incident high laser electric field ($>10^9$ V/cm) and gain very high energy (tens of MeV) via various absorption mechanisms. For interaction of high contrast, ultrashort laser pulse with matter, the density gradient of the plasma is very steep. Because the electrons have high excursion lengths at high electric field, during each laser optical cycle they are pulled out of the target into a surrounding vacuum. Also, if the target is very thin (<1 μm), the electrons can emerge out at the rear side of the foil. The energetic electrons, having large penetration depth in the thin foil target, can leave the target, leading to a strong electric field both at the front as well as at the rear of the foil. The high-energy electrons, while emerging out from both sides of the foil, ionize the hydrocarbon layer that always exists as impurity on any foil surface. Therefore protons (H^+) and heavy ions (C^+) are accelerated in the sheath electric field. This process of ion acceleration is known as *target normal sheath acceleration (TNSA)* [7,67] as shown in Fig. 14.10. The sheath field is given by $E_{sh} = KT_h/e\lambda_d$, where T_h is the hot electron temperature, and λ_D is the Debye length (the distance within which a plasma may not be neutral). Thus the higher the hot electron temperature, the stronger is the sheath electric field, and its magnitude is in TV/*m* for a laser intensity close to 10^{18} W/cm^2.

Apart from the TNSA mechanism, there are other proton–ion acceleration mechanisms such as light sail or radiation pressure acceleration (RPA) [68] and relativistic induced-transparency acceleration [69]. These are effective at ultrahigh laser intensities ($>10^{19}$ W/cm^2). In the RPA mechanism, the radiation pressure of the ultrahigh-intensity laser pulse acts as a piston on the thin foil and drives the ionized plasma in the direction of the laser, resulting in compression. The compressed plasma reaches the rear side of the foil and gets expelled out of the foil as accelerated ions. In this case, all the ions have a quasi-monoenergetic energy spectrum because all the ions acquire the same velocity. The monoenergetic ion spectrum is very important for the application of the ion beams in oncology [70]. Protons of energy more than 67 MeV have been demonstrated using a sub–Petawatt (PW)-class laser system [71].

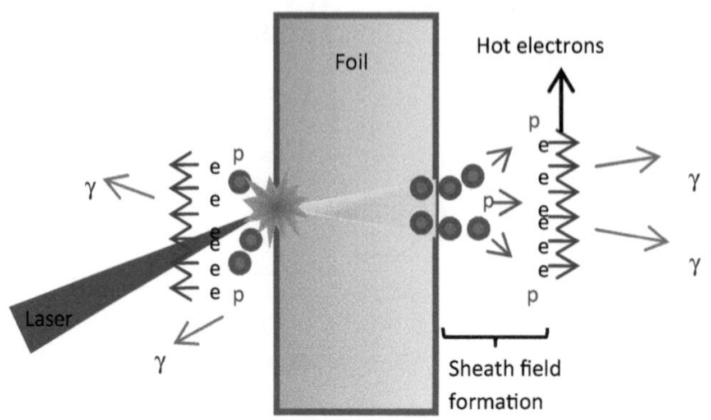

FIGURE 14.10

A schematic diagram illustrating the target normal sheath acceleration mechanism. *e*, Electron; *p*, proton; ν, high energy x-rays.

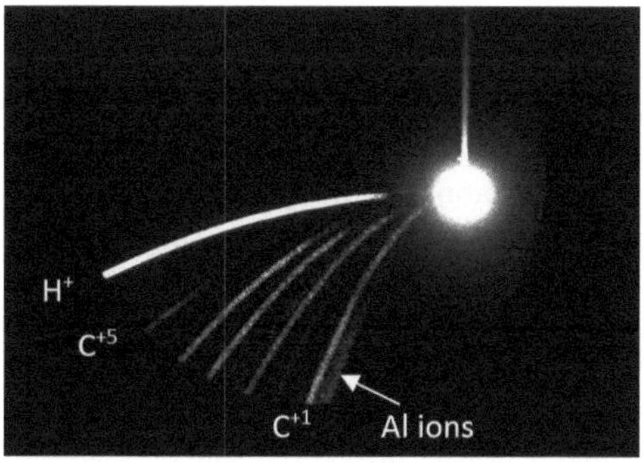

FIGURE 14.11

Typical proton and ion spectra recorded by a Thomson parabola ion spectrograph from an aluminum foil (~12 μm thick).

In TNSA acceleration, the ions have a broad energy spectrum. This is because of nonuniform sheath field formation in space as well as in time, as a result of finite divergence of the hot electrons as they emerge at the rear of the foil. In order to achieve a quasi-monoenergetic proton energy spectrum, mass limited targets or ultrathin foil targets (thickness few tens of nanometers) are being used to achieve uniform sheath field formation at the target rear [72].

3.2.2 An Example of Proton Acceleration

Fig. 14.11 shows the proton and carbon ion spectra recorded in the case of Al foil of 12-μm thickness at the Raja Ramanna Centre for Advanced Technology, Indore. The plasma was formed by irradiating a 45-fs, 10-TW laser beam focused to a spot size of 10 μm on the foil. The ion emission from the freely expanding plasma was monitored by a Thomson parabola ion spectrograph, wherein parallel electric and magnetic fields are applied to disperse the incident multispecies ions according to their charge-to-mass ratio in a parabolic trajectory onto an MCP detector [73]. The maximum proton energy was found to be close to 3 MeV. Spatial profiles of the proton beam recorded on radio chromic films and CR-39 sheet showed that divergence of proton beam >1 MeV was about 10 degrees.

3.3 NEUTRON GENERATION

Plasma created at laser intensities much above 10^{18} W/cm^2 has a dominance of relativistic effect because the plasma electrons have a quiver velocity (oscillatory velocity) nearly approaching the velocity of light. The laser energy is primarily transferred to the plasma electrons and they are accelerated to very high energies [74]. The rapid exit of these hot electrons from the target creates a space—charge field that accelerates protons and ions, as discussed earlier. The loss of electrons makes the remaining plasma highly positively charged, and because of mutual coulomb repulsion between the ions, the plasma explodes. This is called *coulomb explosion* [75]. This results in the production of protons/ions of very high energy.

If deuterated solid target or deuterium gas clusters are irradiated by intense ultrashort laser pulses, it creates deuterium ions of very high energy (>100 keV), and at such high energy that they have sufficiently large cross-section for D—D fusion reaction by the channel $D(d, n)^3$He

$\{{}_1H^2 + {}_1H^2 \rightarrow {}_0n^1 + {}_2He^3\}$, thus resulting in neutron generation of characteristic energy of 2.45 MeV. Further, the propagation of the hot electrons inside the target material produces high-energy bremsstrahlung X-rays that can induce nuclear reactions [76], resulting in generation of neutrons. The technique of neutron generation through a (γ, n) reaction can be used to determine the radiation temperature of the bremsstrahlung photons and the plasma electron temperature [77]. The neutrons, because of their neutral charge and large mean free path, readily carry out information from the plasma without any influence of the electric and/or magnetic fields inside the plasma. The neutrons produced by the $D(d, n)^3He$ reaction in ultra-short laser produced plasma have some unique features like: (1) they are emitted in a very short duration burst (few picoseconds), (2) they have characteristic energy around 2.45 MeV, (3) they have very small source size (of the order of the laser focal spot ~10 μm), and (4) they can be switched on/off at user discretion. The production of a burst of fusion neutrons using a laser could be useful in the study of radiation induced damage in materials [78]. In addition, the short duration inherent in laser driven neutron sources could allow time resolved studies of neutron damage [79]. A laser driven source of neutrons might also serve as a source for fast neutron radiography, where the small source size could lead to high spatial resolution.

An example of neutron generation: An experimental study on the generation of neutrons by irradiating a deuterated polyethylene $(CD_2)_n$ solid target with 10 TW Ti:sapphire laser pulses has also been carried out at the Raja Ramanna Centre for Advanced Technology, Indore [80]. A p-polarized laser beam of 400 mJ energy, 45 fs pulse duration was focused to a focal spot of 10-μm diameter (full width at half maximum, FWHM) with a gold coated *f*/7.5 off-axis parabolic mirror, producing a peak intensity of about 3×10^{18} W/cm^2 on the target. The target was a 1-mm thick sheet of deuterated polyethylene (*D* enrichment over 99%). At an intensity greater than 10^{18} W/cm^2, the electrons were removed from the laser path, leaving behind a positively charged column of deuterium ions that exploded (coulomb explosion) radially, leading to the acceleration of deuterium ions, which in turn, collided with the deuterium ions in the surrounding preplasma, resulting in the release of neutrons by the *D*–*D* fusion reaction. The fusion neutrons were detected using a neutron time of flight (ToF) detector and also confirmed by other detectors like a CR-39 detector, a bubble detector, and a BF_3 counter. The CR-39 sheet showed neutron tracks; by counting these, one could estimate the number of neutrons per shot to be about 10^4. The bubble detectors showed bubble formation. Fig. 14.12 shows the neutron signal recorded with the ToF detector kept at a distance of 1.2 m from the target [80]. The first peak is caused by gamma flash (high-energy bremsstrahlung), and the second peak appears 52 ns after the start of gamma flash and corresponds to the neutron signal. The time of occurrence of the neutron ToF signal closely matches the expected energy of 2.45 MeV.

In the same set-up, changing the target to copper also gave a count of neutrons on the BF_3 detector, although it was much lower. These neutrons were produced by the (γ, n) reactions in the copper target.

4. SHOCK WAVE PROPAGATION

Generating extreme conditions of pressure and temperature in the laboratory and studying the response of different materials under these conditions is of great importance for many scientific applications. One of these reasons is to understand the EOS of the hot, compressed fuel in the *D-T* pellets used in ICF. Issues like reducing the preheat in the fusion fuel and achieving uniform compression requires careful target design with inputs from these studies. In addition, many other areas ranging from

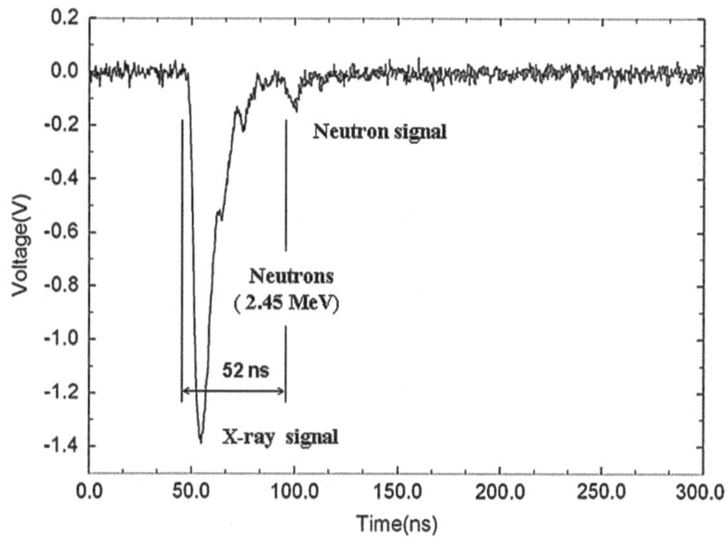

FIGURE 14.12

Neutron signal recorded by a neutron time-of-flight scintillator detector.

astrophysics and geophysics [81,82] to space technology benefit from shock research. These studies also help in designing space crafts that experience impacts at very high velocities (20–40 km/s).

High-pressure research is normally carried out using diamond anvil cells (DACs). However, the pressure generated in a DAC is static in nature. The highest pressure that can be achieved using these cells is limited to a few megabars. With the advent of high-intensity nanosecond-duration laser beams, new techniques have been developed to generate pressures up to several hundreds of megabars. The shock waves generated by the recoil momentum of a rapidly expanding laser-produced plasma coalesce together under optimum conditions to form a single intense shock front propagating at supersonic speed. It is a compression front across which there is a sharp discontinuity of the material properties like density and fluid velocity. The velocity of a shock wave increases with its amplitude. Immediately after the peak of the laser pulse, a rarefaction wave is launched in the compressed medium, which travels faster than the preceding shock wave (in the laboratory frame). Once it reaches the shock front, the amplitude of the shock front rapidly decreases and the shock front decays.

When a shock wave propagates through a material, it can be described by the Rankine–Hugoniot equations which basically express the conservation laws for mass, linear momentum and energy as:

$$\begin{aligned} \rho/\rho_o &= U_s/(U_s - U_p), \\ P - P_o &= \rho_o U_s U_p, \\ E - E_o &= 1/2(P + P_o)(1/\rho - 1/\rho_o), \end{aligned}$$

where ρ_o, ρ, P_o, P, E_o, and E are the initial and final values of mass density, pressure, and internal energy. Here, U_s and U_p are the shock velocity and particle velocity in the medium. Given these three equations and five unknowns U_s, U_p, P, E, and ρ, it follows that experimental determination of any two of them (usually U_s and U_p) are sufficient to define the Hugoniot EOS of the sample if the initial conditions (ρ_o, P_o, and E_o) are known. Among these two (U_s and U_p), mostly U_s is determined experimentally and the value of U_p is used form the standard empirical relation between U_s and U_p, which is valid for most of the elements. However, it is always better to measure experimentally the shock and particle velocities independently, so that other variables can be calculated accurately.

The shock velocities are calculated from the shock luminosity measurements using a fast optical streak camera. The particle velocities are measured by the optical shadowgraphy technique. In order to measure the particle velocity accurately, a velocity interferometer for any reflector (VISAR) [83] is also used. Recent techniques like chirped pulse shadowgraphy [84] are also used to obtain this information. Some details about these techniques are described in the following sections.

4.1 MEASUREMENT OF SHOCK VELOCITY

A schematic diagram of a typical setup for measuring the shock velocity in a material in the form of a thin foil is shown in Fig. 14.13a. An intense laser pulse is focused on the foil. After propagating through the foil, the shock wave reaches the back side of the foil and because of release of shock wave energy, the rear surface of the foil gets heated up and emits visible radiation (referred to as *shock luminosity*). The shock velocity is estimated by measuring the transit time between the arrival of the laser pulse on the front side of the foil and breakout of shock luminosity on the rear side of the foil. Because the material for which the EOS is to be measured is used in the form of a thin foil (thickness ranging from a few hundreds of nanometers to a few microns), and the typical shock speeds are in the range of 10^6 cm/s, the shock transit times in these foils is in tens to hundreds of picoseconds. In order to detect such short durations, optical streak cameras of high temporal accuracy (few picoseconds) are used [85]. Typical streak camera traces of the fiducial (used as zero time marker) and the shock luminosity are shown in Fig. 14.13b.

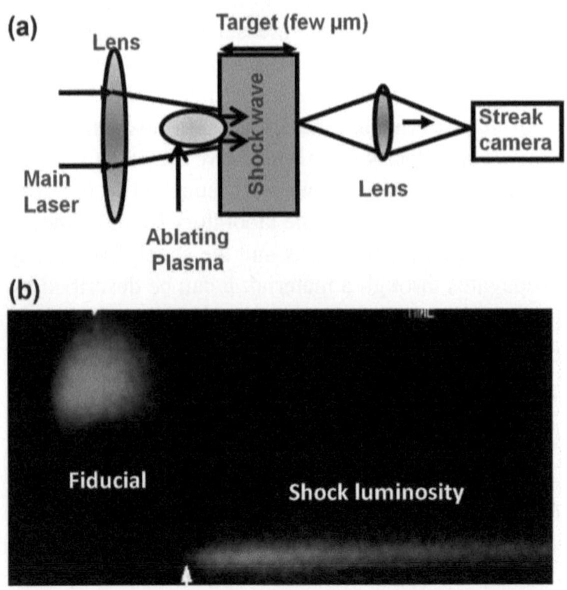

FIGURE 14.13

(a) A schematic set-up for shock velocity measurement. (b) A typical streak image.

4.2 MEASUREMENT OF PARTICLE VELOCITY

As mentioned earlier, the particle velocity can either be estimated from the shock velocity or measured experimentally. The shock velocity and the particle velocity in the media are related as $U_s = a + bU_p$, where the a and b are the coefficients for a given material and can be obtained from the literature [86]. The assumption that the dependence is linear is valid, in general, in a large parameter space. On the other hand, using techniques like optical shadowgraphy or VISAR, one can measure the particle velocity experimentally.

4.2.1 Using Optical Shadowgraphy

In the optical shadowgraphy technique, a short-duration probe laser pulse is passed through the shocked region in a thick transparent material or behind a foil target in the transverse direction as shown in Fig. 14.14a. The shocked region is imaged onto a suitable detector (like a CCD camera) with

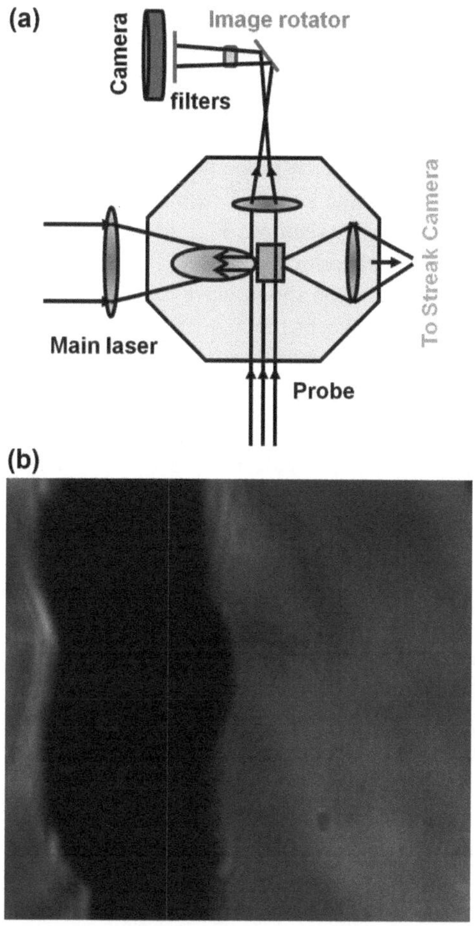

FIGURE 14.14

(a) A schematic set-up for shadowgraphy experiments. (b) A typical shadowgram.

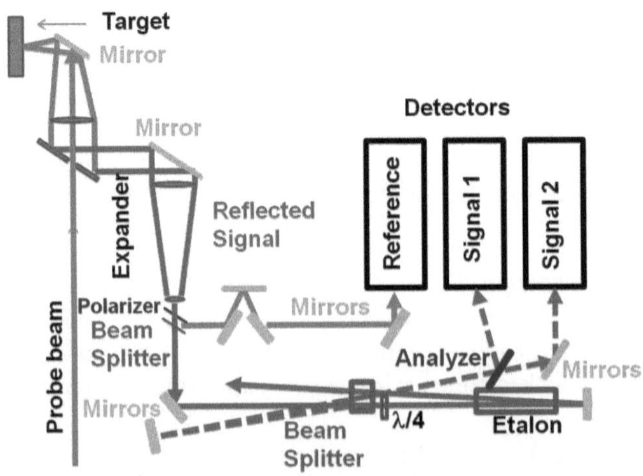

FIGURE 14.15

A schematic set-up for a velocity interferometer for any reflector used by Barker and Hollenbach [83].

appropriate magnification. The duration of the probe beam is also chosen to be as short as possible. In the case of thick transparent material, the shocked region becomes opaque and casts a shadow. By taking shadowgrams (i.e., images of the shadow; Fig. 14.14b) under identical laser conditions for different delays, one can find the shock velocity.

In the case of a thin foil target, once the shock wave reaches the rear side of the foil, the shocked material gets detached from the foil and keeps moving forward (in the direction of the main laser). The expansion velocity of this free surface can be measured by taking shadowgrams for different delays. The particle velocity is twice the free surface expansion velocity.

4.2.2 Using a Velocity Interferometer System for any Reflector

Barker and Hollenbach [83] developed the VISAR technique to measure the particle velocity in a target material in which laser-generated shock waves are propagating. Their earlier technique of velocity interferometer was able to measure the velocity of a moving reflector from the differential Doppler shifts from the expression $U(t) = (\lambda/2\tau)\ F(t)(1 + \Delta\nu/\nu_o)$, where the velocity $U(t)$ at time t was given in terms of fringe count, $F(t)$. However, this technique required the reflector to be a highly polished surface. Practically, as the shock wave reaches the rear side of the foil target, the foil reflectivity reduces and it becomes a diffuse reflector. In order to use the velocity interferometer under these conditions, they introduced an etalon in one of the arms (Fig. 14.15) to obtain the seemingly contradictory requirement of a nonzero time difference with a zero path offset. This relaxes the condition of a spatial coherence requirement so that it could be used for any surface. This VISAR was able to work with any diffused reflecting surface and provided the temporal profile of the particle velocity $U(t-\tau/2)$ in terms of the etalon delay τ and fringe count $F(t)$ as $U(t-\tau/2) = \lambda\ F(t)/[2\tau\ (1 + \Delta\nu/\nu_o)]$.

In order to increase the accuracy of the detecting system, two channels of different polarizations can be used in this system. The VISAR technique has also been successfully used [87] to study the shock wave propagation in transparent materials because the shocked region propagating in time acts like a reflecting surface.

4.2.3 Using Chirped-Pulse Shadowgraphy/Reflectometry

With the advent of short duration (femtosecond) ultraintense lasers, the chirped pulse amplification technique has become standard practice. In order to amplify the femtosecond laser pulses, the pulses are first "chirped" (i.e., stretched in time by dispersing them spectrally through introducing delays in different spectral components). Using these chirped pulses to probe a system, one can get the temporal evolution from the recorded spectrum. By changing the parameters of the pulse stretcher, one can obtain chirped laser pulses of different time durations from a few picoseconds to nanoseconds. This has become a simple probe for laser-produced plasmas as well as laser-generated shocks.

In this technique, a chirped pulse is passed through the shocked region and the spectrum is recorded using a spectrograph. A thin slice of the shocked region is imaged onto the entrance slit of the spectrograph as shown in Fig. 14.16a. Because different wavelengths arrive at different times, the

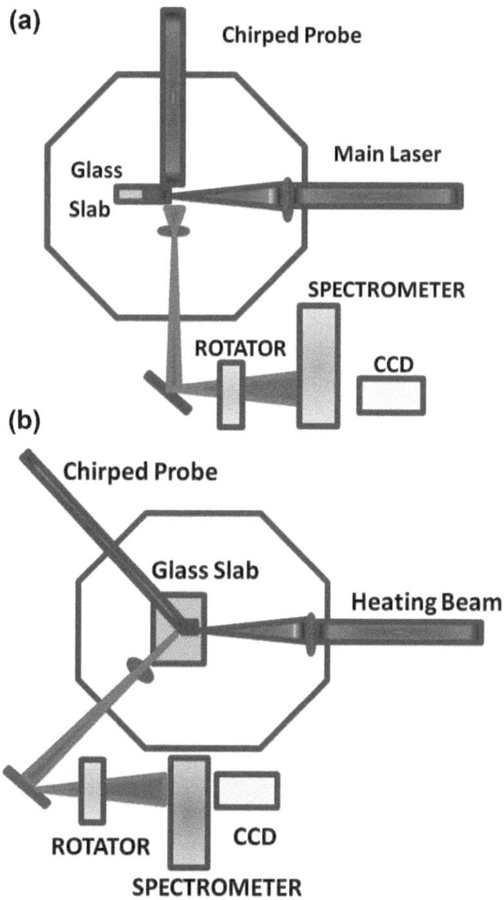

FIGURE 14.16

Schematic set-ups for (a) chirped pulse shadowgraphy, and (b) chirped pulse reflectometry. *CCD*, Charge coupled device.

sequential information is dispersed by the spectrograph and stored in the recorded spectrum. This "chirped pulse shadowgraphy" technique can be used to record in a single shot (unlike multiple shots required in optical shadowgraphy) the movement of the rear side of a foil to provide the particle velocity. It can also be used to get in a single shot the speed of the shadow of the shocked region in a transparent target.

The other configuration in which the probe pulse is obliquely incident on the rear side of the target, shown in Fig. 14.16b, is known as chirped pulse reflectometry. This technique has been used to obtain the exact time of arrival of the shock front at the rear surface of a foil. Benuzzi-Mounaix et al. [88] estimated the shock arrival time by using this technique with a resolution better than 3 ps, which is much better than that obtained with a typical VISAR system. Similarly, they could record the temporal profile of the velocity precisely with approximate picosecond time resolution.

To summarize, the physical parameters of a laser-generated shock wave can be obtained using different optical diagnostics tools like a streak camera, optical shadowgraphy, and VISAR. The recent chirped pulse diagnostics offers a simple alternative to VISAR.

5. WARM DENSE MATTER: AN EXOTIC STATE

WDM is an extreme condition of matter corresponding to near solid densities, temperatures between a few electron volts to tens of electron volts, and pressures of 10–100 Gbars [89]. Creation of materials in these extreme conditions is of fundamental interest because matter in this regime can neither be described as condensed matter nor as being in the plasma state of matter. A WDM region is that area of phase space where the coulomb potential of particles is comparable to the thermal energy, and it is in a sense both coupled and degenerate plasma. It is desirable to obtain the experimental set of data for describing an EOS of such a system [90]. It helps to understand not only the heating processes but also permits the investigation of the subsequent evolution of WDM material for obtaining phase diagrams [91].

The creation and investigation of the WDM state is a challenging task because of high temperature and pressure. Different techniques and methods have been used to produce and measure WDM conditions. The well-developed method of creating WDM is through shock heating. On the other hand, many experiments have been performed using short pulse lasers to isochorically heat solids over small volumes.

5.1 ISOCHORIC HEATING

The well-defined WDM state can be achieved through isochoric heating (i.e., heating without a change in volume) with intense particle or laser beams [92]. It is achieved by delivering the energy (heating) on a time scale faster than the hydrodynamic expansion of the material, so that the heated material does not get time to expand. It implies that $t < d/c_s$, where t is the time over which material is heated, d is the thickness of the material sample, and c_s is the sound speed in the heated material. Taking the thickness of the material sample to be 1 μm and using the sound speed c_s of 10^6 cm/s, the expansion time will be about 1 ps for expanding to a distance equal to 1% of the foil thickness (nearly isochoric). The creation and probing of an extremely fast nonreversible state requires a subpicosecond heating of the sample. The simplest approach is the direct heating of the sample with ultrashort pulsed laser. However, the sample thickness required to maintain the uniform heating should be much smaller than the laser wavelength.

The other alternative for isochoric heating of solids to WDM temperatures is through X-rays or proton beams, both with subpicosecond duration. The ultrashort duration, monochromatic K-α X-ray radiation generated in an ultrashort laser–solid interaction can heat a sample foil to WDM temperatures. The heating is almost uniform when the X-rays have a penetration depth longer than the foil thickness. Moreover, an X-ray heating source of duration shorter than plasma expansion time scale is easy to construct and detect. It also has the advantages over ion heating because it is unaffected by the electromagnetic fields near the target.

More recently, the technique of laser-accelerated proton beams has been used to heat solid density samples up to temperatures of approximately 20 eV before expansion occurs [93]. The proton heating achieved much higher temperatures than X-ray heating because protons deposit most of their energy at the so-called "Bragg peak," whereas the X-ray deposition rate continuously decreases as it propagates into the sample. By focusing the laser-generated proton beams using special target geometries, even higher temperatures may be possible. However, as the proton energies are distributed over a broad energy distribution and vary along their spatial profile, it is difficult to determine the energy deposition with sufficient precision. With this method, different states of matter can be accessed, unlike with the shock-heating technique.

5.2 PROBING OF WARM DENSE MATTER

The creation and subsequent probing of WDM in controlled and well-characterized conditions is a challenging task. The WDM state created with direct heating of thin material foil with an ultrashort optical laser pulse can be probed with a short pulse X-ray source such as an X-ray free-electron laser [94]. The limitation of nonuniform heating of the sample in direct heating with optical laser (owing to small absorption depth) can be overcome by heating with an X-ray laser beam. The subsequent probing of density and temperature by Thomson scattering allows the diagnostics of WDM states. In the shock-generated WDM states, the probing is done by measuring the shock velocity as a function of pressure to obtain the EOS.

Detailed information abut the WDM state can be determined from the characteristic K-α emission from the heated matter. The spectral profile of the emitted K-α X-ray radiation is affected by many processes in the plasma. For instance, the presence of high-energy satellites to the K-α line or the blending of inner-shell radiation from ions is an indicator of the electron temperature of the heated material. Next, the emitted shorter wavelength X-rays with larger escape lengths are able to come out from the deep-solid density plasma and give information about the conditions in the WDM state established in the inner layer of the target. However, detailed information about the WDM state parameters necessitates detailed modeling by considering the several other broadening mechanisms at higher density like self-absorption of radiation, and the strong electric fields in the plasma that may modify the shape of the individual emission lines. This technique has a large advantage for measuring the higher temperatures over time-resolved optical pyrometry. The K-shell emission lines are in the hard X-ray region and a spectrograph with mÅ level resolution is required to resolve the line structure.

6. APPLICATIONS

In this section, we will describe three applications: one for ultrashort X-rays for time-resolved X-ray diffraction studies in materials, a second for application of laser-generated protons to oncology, and third one for use of protons to produce the short-lived isotopes for PET.

6.1 TIME-RESOLVED X-RAY DIFFRACTION

X-rays emitted from ultrashort laser-produced plasmas have been used to study the structural dynamics of a wide range of samples from a single crystal to polycrystalline materials. Measuring ultrafast melting, acoustic phonons, and strain in bulk materials has been an ongoing area of interest [95–98]. Time-resolved studies are performed in a pump–probe scheme where an optical pump (laser) deposits its energy in the sample to modify the lattice. This causes changes in the electron distribution function, leading to changes in the atomic interaction potential so that the matter loses crystalline order and becomes molten on a subpicosecond time scale. The melting can be seen by observing a large decrease of the integrated intensity of the Bragg reflections.

A typical pump–probe set-up [99] with a laser-produced plasma X-ray source, used at the Raja Ramanna Centre for Advanced Technology, is shown in Fig. 14.17a. The pump pulse consists of a 300-

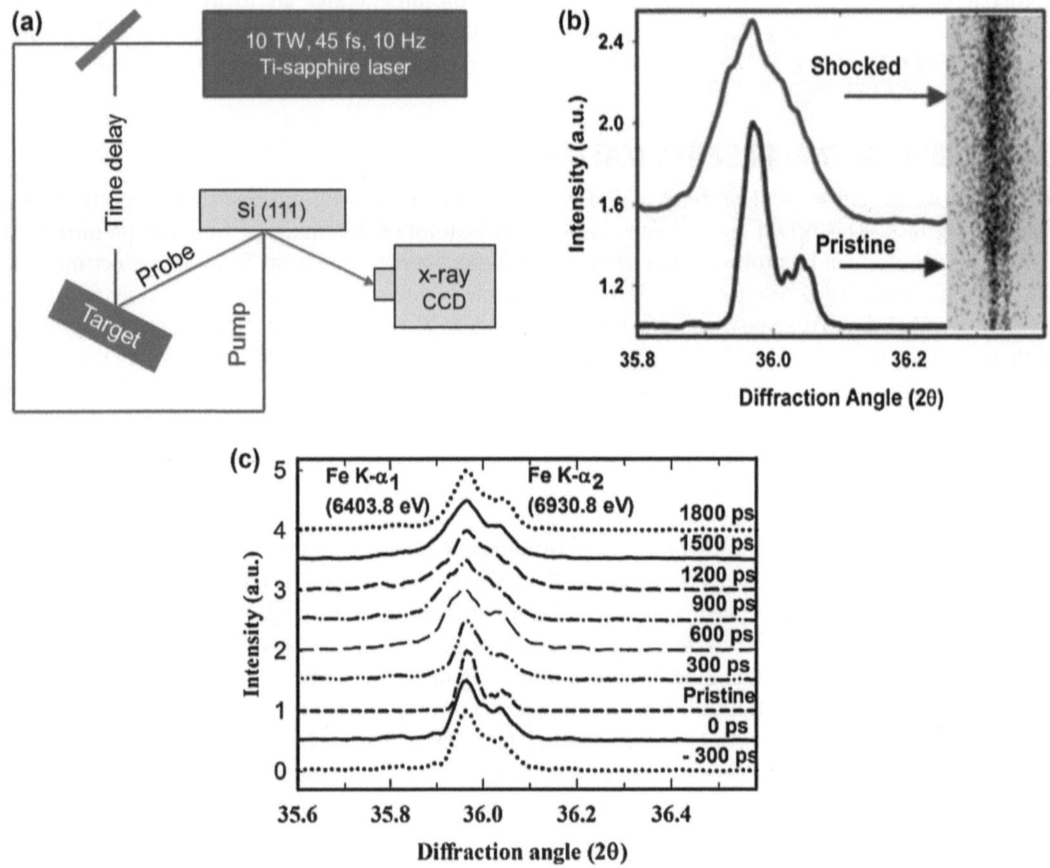

FIGURE 14.17

(a) The experimental set-up, (b) the broadening of the X-ray line, and (C) the observed variation of the full width at half maximum of the Fe K-α_1 line radiation as a function of the delay between the pump and the probe pulse. *a.u.*, arbitrary unit; *CCD*, charge coupled device.

Reproduced with permission from Arora, et al., J. Appl. Phys 114 (2013) 0233302. Copyright 2013, AIP Publishing LLC.

ps laser pulse that is used to modify the lattice structure of the crystalline sample. This pulse is a small fraction of the laser pulse that produces X-rays, which are used as the ultrashort probe. With a variable delay line, the X-ray probe is made to diffract from the perturbed structure at different times after the crystalline lattice is distorted. The X-ray diffraction profile changes as the distance between atomic planes changes, leading to the change in the angle for Bragg reflection. The line shape and position of the reflected line radiation indicate the state of the lattice as transient effects pass through the penetration depth of X-rays. In these experiments, inhomogeneous spacing and lattice expansion are observed. Arora et al. [99] have measured the time-resolved X-ray rocking curves using the setup shown in Fig. 14.17a. Fig. 14.17b shows the image of diffracted X-rays from the pristine and shocked sample. The transient dynamics in the material are resolved by analyzing the shift or broadening of the diffraction signal, as shown in Fig. 14.17c.

Improving the time resolution of X-ray diffraction measurements by moving into the picosecond and subpicosecond time scales is important from the perspective of fundamental physics as well as from the applied technology point of view. The interaction of a material with an ultrashort pulse produces nonequilibrium conditions where various processes may take place, such as carrier excitation, carrier phonon, and carrier–carrier scattering, and carrier recombination. Time-resolved X-ray diffraction can provide information about the evolution of the structural configuration that indeed helps in understanding process development in the transient state. From the applied point of view, interaction of materials with ultrashort laser pulses is the basis for several rapidly developing new technologies such as thin film processing and chemical reaction control. Thus the knowledge of structural dynamics, radiation–matter coupling, and mechanisms of energy transfer will make possible the evaluation of the transient properties of materials and propose the design of new materials suitable for new technological applications.

6.2 USE OF LASER-GENERATED PROTONS IN ONCOLOGY

One of the most promising applications of laser-plasma–accelerated proton/ion beams is in oncology. The main attraction for the use of a proton/light ion beam in tumor therapy comes because of the localized energy deposition of the ion beam to the tumor-affected areas in the human body. This is unlike high-energy X-ray therapy, where the radiation dose deposition is over a large area, leading to unwanted dose delivery to healthy tissues.

Proton and light ion particle beams generated in conventional rf-based accelerators are used to treat cancer patients. Such a facility is very expensive and requires a heavy concrete structure to be built close to the medical facility to provide radiation shielding, because any accelerator facility always has a large radiation dose present around the beam path. In fact, heavy expenses are incurred mainly in the system (gantry) required to deliver the particle beam to the patient. On the other hand, in a laser-plasma–based accelerator, the proton/ion beam can be generated right near the patient, thus avoiding large shielding and heavy gantry issues, because maneuvering a laser beam is much easier and less expensive. Also, a laser-based accelerator is quite cost-effective, even if one uses a petawatt laser facility, compared with the huge installation cost of an rf-based accelerator.

Laser-based proton accelerators are limited to low energy of 30–40 MeV and they have a broad energy spectrum. Therefore these protons cannot be used for treatment of tumors located deep inside the body. With rapid advances in ultrahigh intense laser systems such as PW lasers, it will be possible to achieve proton acceleration up to 60 MeV, which can be used in treatment of eye tumors [100].

Further, with a new scheme of mass-limited target and RPA-based laser accelerators, it will be possible to produce a quasi-monoenergetic proton/ion beam, which is going to be very useful for oncology.

6.3 USE OF LASER-GENERATED PROTONS IN POSITRON EMISSION TOMOGRAPHY

Another important application of the laser-generated protons is to produce short-lived isotopes for PET, which is also being widely used for cancer diagnostics [101]. PET uses very short-lived isotopes that can be injected in the human body for tomography. The unwanted radiation exposure over a long period is avoided because of the short lifetime of the isotopes used. Short-lived isotopes can be produced using conventional accelerators (cyclotron), but such accelerators have to be built close to a hospital because the lifetime of these isotope is only a few minutes. Therefore the other alternative is to produce the very short-lived radio isotopes for use in PET in the hospital itself with a laser-based proton accelerator. Such systems are likely to be a reality in near future.

REFERENCES

[1] S. Atzeni, J. Meyer-Ter-Vehn, The Physics of Inertial Fusion, Oxford University Press, Oxford, 2009.
[2] J.D. Lindl, P. Amendt, R.L. Berger, S.G. Glendinning, S.H. Glenzer, S.W. Haan, et al., The physics basis for ignition using indirect-drive targets on the National Ignition Facility, Phys. Plasmas 11 (2) (2004) 339–491.
[3] L. Stixrude, Structure of iron to 1 Gbar and 40000 K, Phys. Rev. Lett. 108 (5) (2012) 055505, 5.
[4] V. Fortov, Astrophysical Aspects of High Energy Densities in Extreme States of Matter, Springer Berlin Heidelberg, 2011.
[5] S. Eliezer, The Interaction of High-Power Lasers with Plasmas, Institute of Physics Publishing Bristol and Philadelphia, Bristol, 2002.
[6] M.M. Murnane, H.C. Kapteyn, M.D. Rosen, R.W. Falcone, Ultrafast X-ray pulses from laser-produced plasmas, Science 251 (4993) (1991) 531–536.
[7] S.C. Wilks, A.B. Langdon, T.E. Cowan, M. Roth, M. Singh, S. Hatchett, et al., Energetic proton generation in ultra-intense laser–solid interactions, Phys. Plasmas 8 (2) (2001) 542–549.
[8] J.D. Kmetec, C.L. Gordon III, J.J. Macklin, B.E. Lemoff, G.S. Brown, S.E. Harris, MeV X-ray generation with a femtosecond laser, Phys. Rev. Lett. 68 (10) (1992) 1527–1530.
[9] V. Malka, S. Fritzler, E. Lefebvre, E. d'Humières, R. Ferrand, G. Grillon, et al., Practicability of proton therapy using compact laser systems, Med. Phys. 31 (6) (2004) 1587–1592.
[10] F. Ewald, H. Schwoerer, S. Düsterer, R. Sauerbrey, J. Magill, J. Galy, et al., Application of relativistic laser plasmas for the study of nuclear reactions, Plasma Physi. Controlled Fusion 45 (12A) (2003) A83–A91.
[11] R.P. Drake, High-Energy-Density Physics: Fundamentals, Inertial Fusion, and Experimental Astrophysics, Springer, 2006.
[12] R.C. Davidson, D. Arnett, J. Dahlburg, P. Dimotakis, D. Dubin, G. Gabrielse, et al., Frontiers in High Energy Density Physics: The X-games of Contemporary Science, The National Academies, Washington, DC, 2003.
[13] E.I. Moses, R.N. Boyd, B.A. Remington, C.J. Keane, R. Al-Ayat, The national ignition facility: ushering in a new age for high energy density science, Phys. Plasmas 16 (4) (2009) 041006–041013.
[14] M. Tabak, J. Hammer, M.E. Glinsky, W.L. Kruer, S.C. Wilks, J. Woodworth, et al., Ignition and high gain with ultra powerful lasers, Phys. Plasmas 1 (5) (1994) 1626–1634.
[15] D. Giulietti, L.A. Gizzi, X-ray emission from laser-produced plasmas, La Rivista del Nuovo Cimento 21 (10) (1998) 1–93.
[16] W.L. Kruer, The Physics of Laser Plasma Interactions, Addison-Wesley, New York, 1988.

[17] J.A. Chakera, Thermal and Coherent Radiation Emission from Laser Produced Plasmas and Its Applications in X-ray Microscopic Imaging (Ph.D. thesis), Devi Ahilya Vishwa Vidyalaya, Indore, 2005.
[18] P. Gibbon, Short Pulse Laser Interaction with Matter, Imperial College Press, 2005.
[19] F. Brunel, Not-so-resonant, resonant absorption, Phys. Rev. Lett. 59 (1) (1987) 52–55.
[20] I.C.E. Turcu, J.B. Dance, X-Rays from Laser Plasmas: Generation and Applications, Wiley, New York, 1999.
[21] M. Nicoul, U. Shymanovich, A. Tarasevitch, D. von der Linde, K. Sokolowski-Tinten, Picosecond acoustic response of a laser-heated gold-film studied with time-resolved X-ray diffraction, Appl. Phys. Letter 98 (19) (2011) 191902–191903.
[22] A.V. Solovyov, E. Surdutovich, E. Scifoni, I. Mishustin, W. Greiner, Physics of ion beam cancer therapy: a multiscale approach, Phys. Rev. E 79 (1) (2009) 011909.
[23] H.R. Griem, Principles of Plasma Spectroscopy, Cambridge University Press, Cambridge, 1997.
[24] V. Arora, P.A. Naik, B.S. Rao, P.D. Gupta, A comparative study of the ionic keV X-ray line emission from plasma produced by the femtosecond, picosecond and nanosecond duration laser pulses, Pramana 78 (2) (2012) 277–288.
[25] J.A. Cobble, T.E. Tierney, J. Abdallah Jr., Kilovolt radiation from a laser-produced Al plasma, J. Appl. Phys. 102 (2) (2007) 023303–023306.
[26] H.S. Park, et al., High-energy Kα radiography using high-intensity, short-pulse lasers, Phys. Plasmas 13 (5) (2006) 056309–056310.
[27] C. Reich, P. Gibbon, I. Uschmann, E. Förster, Yield optimization and time structure of femtosecond laser plasma K α sources, Phys. Rev. Lett. 84 (21) (2000) 4846–4849.
[28] D. Salzmann, C. Reich, I. Uschmann, E. Förster, P. Gibbon, Theory of K α generation by femtosecond laser-produced hot electrons in thin foils, Phys. Rev. E 65 (3) (2002) 036402–036405.
[29] T. Feurer, A. Morak, I. Uschmann, C. Ziener, H. Schwoerer, E. Förster, et al., An incoherent sub-picosecond X-ray source for time-resolved X-ray-diffraction experiments, Appl. Phys. B 72 (1) (2001) 15–20.
[30] C. Reich, I. Uschmann, F. Ewald, S. Düsterer, A. Lübcke, H. Schwoerer, et al., Spatial characteristics of K α X-ray emission from relativistic femtosecond laser plasmas, Phys. Rev. E 68 (5) (2003) 056408, 08.
[31] R. Toth, S. Fourmaux, T. Ozaki, M. Servol, J.C. Kieffer, R.E. Kincaid Jr., et al., Evaluation of ultrafast laser-based hard X-ray sources for phase-contrast imaging, Phys. Plasmas 14 (5) (2007) 053506–053508.
[32] A. Krol, et al., Laser-based microfocused X-ray source for mammography: feasibility study, Med. Phys. 24 (5) (1997) 725–732.
[33] J. Lindl, Development of the indirect-drive approach to inertial confinement fusion and the target physics basis for ignition and gain, Phys. Plasmas 2 (1995) 3933–4024.
[34] J.A. Chakera, V. Arora, S.R. Kumbhare, P.A. Naik, V. Ganesan, P.D. Gupta, Single shot twin X-ray microscopic imaging using simultaneously produced laser-plasma X-ray sources, Rev. Sci. Instrum. 72 (2) (2001) 1421–1424.
[35] A.D. Stead, R.A. Cotton, J.G. Duckett, J.A. Goode, A.M. Page, T.W. Ford, The use of soft x rays to study the ultrastructure of living biological material, J. X-Ray Sci. Technol. 5 (1) (1995) 52–64.
[36] M. Chaker, et al., Laser plasma sources for proximity printing or projection X-ray lithography, J. Vac. Sci. Technol. B 10 (6) (1992) 3239–3242.
[37] T. Mochizuki, T. Yabe, K. Okada, M. Hamada, N. Ikeda, S. Kiyokawa, et al., Atomic-number dependence of soft-X-ray emission from various targets irradiated by a 0.53-μm-wavelength laser, Phys. Rev. A 33 (1) (1986) 525–539.
[38] R. Popil, P.D. Gupta, R. Fedosejevs, A.A. Offenberger, Measurement of KrF-laser–plasma X-ray radiation from targets with various atomic numbers, Phys. Rev. A 35 (9) (1987) 3874–3882.
[39] G.D. Tsakiris, K. Eidmann, An approximate method for calculating Planck and Rosseland mean opacities in hot, dense plasmas, J. Quant. Spectrosc. Radiat. Transfer 38 (5) (1987) 353–368.

[40] R. Sigel, K. Eidmann, F. Lavarenne, R.F. Schmalz, Conversion of laser light into soft x rays. Part I: dimensional analysis, Phys. Fluids B: Plasma Phys. 2 (1) (1990) 199–207.

[41] I.B. Zel'dovich, Y.B. Zel'dovich, Y.P. Raizer, Physics of Shock Waves and High-Temperature Hydrodynamic Phenomena, Courier Dover Publications, 2002.

[42] J.A. Chakera, et al., Dependence of soft X-ray conversion on atomic composition in laser produced plasma of gold–copper mix-Z targets, Appl. Phys. Lett. 83 (1) (2003) 27–29.

[43] D. Colombant, M. Klapisch, A. Bar-Shalom, Increase in Rosseland mean opacity for inertial fusion hohlraum walls, Phys. Rev. E 57 (3) (1998) 3411–3416.

[44] N.K. Gupta, B.K. Godwal, Effects of various parameters on numerical simulations of inertial confinement fusion hohlraum and radiation hydrodynamics, Laser Part. Beams 19 (02) (2001) 259–265.

[45] T.J. Orzechowski, M.D. Rosen, H.N. Kornblum, J.L. Porter, L.J. Suter, A.R. Thiessen, et al., The Rosseland mean opacity of a mixture of gold and gadolinium at high temperatures, Phys. Rev. Lett. 77 (17) (1996) 3545–3548.

[46] J. Yan, Z.Q. Wu, Theoretical investigation of the increase in the Rosseland mean opacity for hot dense mixtures, Phys. Rev. E 65 (6) (2002) 066401–066406.

[47] P. Wang, J.J. MacFarlane, T.J. Orzechowski, An investigation of the opacity of high-Z mixture and implications for inertial confinement fusion hohlraum design, Rev. Sci. Instrum. 68 (1) (1997) 1107–1110.

[48] H. Nishimura, T. Endo, H. Shiraga, Y. Kato, S. Nakai, X-ray emission from high-Z mixture plasmas generated with intense blue laser light, Appl. Phys. Lett. 62 (12) (1993) 1344–1346.

[49] P.A. Naik, et al., Dependence of soft X-ray conversion on atomic composition in laser produced plasma of copper-gold mix-Z targets, in: ECLIM 2002: 27th European Conference on Laser Interaction with Matter, 2003, pp. 16–20.

[50] H. Nishimura, F. Matsuoka, M. Yagi, K. Yamada, S. Nakai, G.H. McCall, et al., Radiation conversion and related ablation behaviour of a gold-foil target irradiated by 0.35, 0.53, 1.06, and 10.6 μm lasers, Phys. Fluids 26 (6) (1983) 1688–1692.

[51] V. Arora, J.A. Chakera, P.A. Naik, S.R. Kumbhare, P.D. Gupta, N.K. Gupta, Effect of gold on keV X-ray emission yield from laser produced plasma of gold-copper mix-Z targets, J. Appl. Phys. 100 (3) (2006) 033306.

[52] L.L. Ji, A. Pukhov, I.Y. Kostyukov, B.F. Shen, K. Akli, Radiation-reaction trapping of electrons in extreme laser fields, Phys. Rev. Lett. 112 (14) (2014) 145003–145005.

[53] H.S. Park, et al., High-adiabat high-foot inertial confinement fusion implosion experiments on the national ignition facility, Phys. Rev. Lett. 112 (5) (2014) 055001–055005.

[54] K.V. Cone, H.A. Baldis, J. Dunn, M.J. May, M.A. Purvis, M.B. Schneider, et al., Time-resolved soft X-ray spectra from laser-produced Cu plasma, Rev. Sci. Instrum. 83 (10) (2012) 10E318–3.

[55] R.W. Lee, R.E. Strout, R.C. Cauble, L.B. Petway, RATION Spectroscopic Code, Lawrence Livermore National Laboratory, USA, 1990.

[56] E.L. Dewald, Dante soft X-ray power diagnostic for National Ignition Facility, Rev. Sci. Instrum. 75 (10) (2004) 3759–3761.

[57] V. Arora, S.R. Kumbhare, P.A. Naik, P.D. Gupta, A simple high-resolution on-line X-ray imaging crystal spectrograph for laser–plasma interaction studies, Rev. Sci. Instrum. 71 (7) (2000) 2644–2650.

[58] J.F. Pelletier, M. Chaker, J.C. Kieffer, Calibrated time-resolved transmission grating spectrometer for the study of ultrafast X-ray sources, J. X-Ray Sci. Technol. 6 (4) (2012) 359–374.

[59] L. Evans, P. Bryant, LHC machine, J. Instrum. 3 (8) (2008) S08001–S08158.

[60] T. Tajima, J.M. Dawson, Laser electron accelerator, Phys. Rev. Lett. 43 (4) (1979) 267–270.

[61] T. Katsouleas, Physical mechanisms in the plasma wake-field accelerator, Phys. Rev. A 33 (3) (1986) 2056–2064.

[62] J. Faure, et al., A laser–plasma accelerator producing monoenergetic electron beams, Nature 431 (7008) (2004) 541–544.

[63] E. Esarey, C.B. Schroeder, W.P. Leemans, Physics of laser-driven plasma-based electron accelerators, Rev. Mod. Phys. 81 (3) (2009) 1229–1285.

[64] R.J. Noble, Plasma-wave generation in the beat-wave accelerator, Phys. Rev. A 32 (1) (1985) 460–471.

[65] E. Esarey, W.P. Leemans, Scaling laws for laser wakefield accelerators, in: Proceedings of the IEEE Particle Accelerator Conference, vol. 5, 1999, pp. 3699–3701.

[66] B.S. Rao, J.A. Chakera, P.A. Naik, M. Kumar, P.D. Gupta, Laser wake-field acceleration in pre-formed plasma channel created by pre-pulse pedestal of terawatt laser pulse, Phys. Plasmas 18 (9) (2011) 093104–093109.

[67] A. Macchi, M. Borghesi, M. Passoni, Ion acceleration by super intense laser-plasma interaction, Rev. Mod. Phys. 85 (2) (2013) 751.

[68] T. Esirkepov, M. Borghesi, S.V. Bulanov, G. Mourou, T. Tajima, Highly efficient relativistic-ion generation in the laser-piston regime, Phys. Rev. Lett. 92 (17) (2004) 175003–175004.

[69] L. Willingale, et al., Characterization of high-intensity laser propagation in the relativistic transparent regime through measurements of energetic proton beams, Phys. Rev. Lett. 102 (12) (2009) 125002–125004.

[70] S.D. Kraft, et al., Dose-dependent biological damage of tumour cells by laser-accelerated proton beams, New J. Phys. 12 (8) (2010) 085003.

[71] S.A. Gaillard, et al., Increased laser-accelerated proton energies via direct laser-light-pressure acceleration of electrons in microcone targetsa, Phys. Plasmas 18 (5) (2011) 056710–056711.

[72] H. Schwoerer, et al., Laser-plasma acceleration of quasi-monoenergetic protons from microstructured targets, Nature 439 (7075) (2006) 445–448.

[73] M. Tayyab, et al., Role of target material in proton acceleration from thin foils irradiated by ultrashort laser pulses, Phys. Rev. E 90 (2) (2014) 023103.

[74] A. Modena, et al., Electron acceleration from the breaking of relativistic plasma waves, Nature 377 (6550) (1995) 606–608.

[75] T. Ditmire, J. Zweiback, V.P. Yanovsky, T.E. Cowan, G. Hays, K.B. Wharton, Nuclear fusion from explosions of femtosecond laser-heated deuterium clusters, Nature 398 (6727) (1999) 489–492.

[76] T.E. Cowan, et al., High energy electrons, nuclear phenomena and heating in petawatt laser-solid experiments, Laser Part. Beam. 17 (4) (1999) 773–783.

[77] M. Gerbaux, et al., High flux of relativistic electrons produced in femtosecond laser-thin foil target interactions: characterization with nuclear techniques, Rev. Sci. Instrum. 79 (2) (2008) 023504–023508.

[78] R.S. Averback, T. Diaz de La Rubia, Displacement damage in irradiated metals and semiconductors, Solid State Phys. 51 (1997) 281–402.

[79] T. Ditmire, et al., Overview of future directions in high energy-density and high-field science using ultra-intense lasers, Radiat. Phys. Chem. 70 (4) (2004) 535–552.

[80] M. Tayyab, J.A. Chakera, P.A. Naik, M. Kumar, P.D. Gupta, Observation of neutrons in the interaction of high intensity laser pulses with solid targets, in: EPJ Web of Conferences, vol. 59, 2013, p. 17008.

[81] T.J. Ahrens, Shock wave techniques for geophysics and planetary physics (Chapter 6), in: C.G. Sammis, T.L. Henyey (Eds.), Methods of Experimental Physics, vol. 24, Part A, Academic Press, Orlando, 1987, pp. 185–235.

[82] M. Koenig, et al., Recent laboratory astrophysics experiments at LULI, Plasma Fusion Res.: Rev. Articles 4 (044) (2009), 044-1-7.

[83] L.M. Barker, R.E. Hollenbach, Laser interferometer for measuring high velocities of any reflecting surface, J. Appl. Phys. 43 (11) (1970) 4669–4675.

[84] Y.B.S.R. Prasad, S. Barnwal, P.A. Naik, J.A. Chakera, R.A. Khan, P.D. Gupta, Chirped pulse shadowgraphy for single shot time resolved plasma expansion measurements, Appl. Phys. Lett. 96 (22) (2010) 2215031–2215033.
[85] F. Cottet, J.P. Romain, R. Fabbro, B. Faral, Ultra-high-pressure laser-driven shock-wave experiments at 0.26 μm wavelength, Phys. Rev. Lett. 52 (21) (1984) 1884–1886.
[86] S.P. Marsh (Ed.), LASL Shock Hugoniot Data, University of California Press, Berkeley, 1980.
[87] T.R. Boehly, E. Vianello, J.E. Miller, R.S. Craxton, T.J.B. Collins, I. Goncharov, et al., Shock-timing experiments using double-pulse laser irradiation, Phys. Plasmas 13 (5) (2006) 056303–056307.
[88] A. Benuzzi-Mounaix, M. Koenig, J.M. Boudenne, T.A. Hall, D. Batani, F. Scianitti, et al., Chirped pulse reflectivity and frequency domain interferometry in laser driven shock experiments, Phys. Rev. E 60 (3) (1999) R2488–R2491.
[89] R.W. Lee, et al., Plasma-based studies with intense X-ray and particle beam sources, Laser Part. Beams 20 (03) (2002) 527–536.
[90] R. Ernstorfer, M. Harb, C.T. Hebeisen, G. Sciaini, T. Dartigalongue, R.D. Mille, The formation of warm dense matter: experimental evidence for electronic bond hardening in gold, Science 323 (5917) (2009) 1033–1037.
[91] V.E. Fortov, et al., Phase transition in a strongly non-ideal deuterium plasma generated by quasi-isentropical compression at megabar pressures, Phys. Rev. Lett. 99 (18) (2007) 185001–185004.
[92] F. Perez, et al., Enhanced isochoric heating from fast electrons produced by high-contrast, relativistic-intensity laser pulses, Phys. Rev. Lett. 104 (8) (2010) 085001–085004.
[93] G.M. Dyer, et al., Equation-of-state measurement of dense plasmas heated with fast protons, Phys. Rev. Lett. 101 (1) (2008) 015002–015004.
[94] R.R. Fäustlin, et al., Soft X-ray Thomson scattering in warm dense matter at FLASH, in: Ultrafast Phenomena vol. XVI, Springer Berlin Heidelberg, 2009, pp. 241–243.
[95] M. Bargheer, N. Zhavoronkov, Y. Gritsai, J.C. Woo, D.S. Kim, M. Wörner, et al., Coherent atomic motions in a nanostructure studied by femtosecond X-ray diffraction, Science 306 (5702) (2004) 1771–1773.
[96] C.V.K. Schmising, et al., Coupled ultrafast lattice and polarization dynamics in ferroelectric nanolayers, Phys. Rev. Lett. 98 (25) (2007) 257601–257604.
[97] K. Sokolowski-Tinten, C. Blome, J. Blums, A. Cavalleri, C. Dietrich, A. Tarasevitch, et al., Femtosecond X-ray measurement of coherent lattice vibrations near the Lindemann stability limit, Nature 422 (6929) (2003) 287–289.
[98] F. Zamponi, Z. Ansari, M. Wörner, T. Elsässer, Femtosecond powder diffraction with a laser-driven hard X-ray source, Opt. Express 18 (2) (2010) 947–961.
[99] V. Arora, S. Bagchi, M. Gupta, J.A. Chakera, A. Gupta, P.A. Naik, et al., Study of strain propagation in laser irradiated silicon crystal by time-resolved diffraction of K-α X-ray probe of different photon energies, J. Appl. Phys. 114 (2) (2013) 023302–023305.
[100] M.I.K. Santala, et al., Production of radioactive nuclides by energetic protons generated from intense laser-plasma interactions, Appl. Phys. Lett. 78 (1) (2001) 19–21.
[101] R.A. Snavely, et al., Intense high-energy proton beams from petawatt-laser irradiation of solids, Phys. Rev. Lett. 85 (14) (2000) 2945–2948.

CHAPTER

15

LASER-INDUCED VAPORIZATION—MASS SPECTROMETRY STUDIES ON REFRACTORY MATERIALS AT ULTRAHIGH TEMPERATURES

M. Joseph, N. Sivakumar, P. Manoravi

Indira Gandhi Centre for Atomic Research, Kalpakkam, India

CHAPTER OUTLINE

Materials Under Extreme Conditions. http://dx.doi.org/10.1016/B978-0-12-801300-7.00015-2

1. INTRODUCTION

For modern technology, high temperature and high-performance materials are needed in a variety of applications. The term *high temperature* is well defined in the literature [1,2]. From an engineering and technological point of view, high temperatures can be defined as temperatures sufficiently well above normal ambient temperature so as to change significantly the physical and chemical properties of a system. Generally, the vaporization process of these materials becomes highly complex as the temperature is increased, because: (1) the vapor species generated at high temperatures tend to have unusual oxidation numbers and formulae (e.g., barium exhibits +1 and +3 states and aluminum has +1 and +2 states at high temperatures); (2) complex molecular species that are not stable at low temperature have been found under conditions of equilibrium in high-temperature vapors (e.g., molecular species such as $MnFe_2Cl_6$ and Mo_5O_{15} are reported in high temperature vapors); and (3) the relative importance of a species in the vapor phase changes [3,4]. The partial pressures (p) of the various species in the vapor phase is given by Eq. (15.1):

$$\ln p = -\Delta H^\circ/(R * T) + (\Delta S^\circ/R) \tag{15.1}$$

where ΔH° is the enthalpy of vaporization, ΔS° is the entropy of vaporization, and R is the gas constant. Hence a plot of $\ln p$ against $1/T$ will be a straight line with a slope of $\Delta H^\circ/R$. A larger enthalpy of vaporization (ΔH°) means a steeper increase in the pressure with temperature. The species that predominate at low temperatures will have the lowest enthalpy of vaporization, and all other species that are relatively unimportant at low temperatures will have high enthalpy of vaporization. The partial pressure of these minor species will increase more rapidly than the primary species as the temperature is increased. At low temperatures ($T * \Delta S^\circ{}_T$) Eq. (15.1), which gives the difference between $\Delta G^\circ{}_T$ (change in free energy) and $\Delta H^\circ{}_T$, is only a correction. Hence the large $\Delta H^\circ{}_T$ normally determines the sign of $\Delta G^\circ{}_T$, or in other words, the direction of the reaction. But at high temperatures, the entropy term becomes important. This type of behavior is observed in many systems. For example, in the vaporization of SiC below 1800 K, only Si, SiC_2, and Si_2C were observed, but when the temperature was raised to 2200 K, species such as Si_2, Si_3, Si_2C_2, Si_2C_3, and Si_3C were also observed [5]. Similarly, during the vaporization of Si_3N_4 at approximately 5000 K, new species are observed that are not seen below 2500 K, as discussed later in this chapter.

Though uncommon, temperatures above 3000 K are encountered in advanced material processing technologies such as plasma and laser processing, surface modifications, and in rocket plumes [6]. Three specific circumstances where very high temperature (>3000 K) vaporization data required are:

1. In space technology, the aerodynamic frictional heating during reentry of a space vehicle generates excessive temperatures [7]. Usually the heat shields are made out of graphite composites; these are expected to vaporize only minimally under this high heating rate.
2. In nuclear fusion research, the "first wall" material can get exposed to very high temperature [8]. In fusion, one deuteron (D) and one triton (T) fuse to produce a α particle and a neutron. The latter carries 80% of the released energy of 17.6 MeV and can escape unimpeded from the magnetic field containing the D–T plasma because it is electrically neutral. It can penetrate the inner wall of the plasma chamber, the first wall, and be absorbed in the "blanket zone," where its energy is deposited. Generally, the first wall absorbs about 20% of the fusion power from the plasma in the form of radiation or particle fluxes and thereby a very high temperature can be reached in these materials. Some of the candidates for first wall material are SiC, BN, AlN, and Si_3N_4.

3. Safety of nuclear power plants is one of the major concerns in the development of liquid-metal–cooled fast-breeder reactors (LMFBRs). Off-normal events in LMFBRs can result in increase in temperature beyond operating limits. The worst scenario would involve release of a large amount of energy. Such events have the potential of taking the reactor fuel or a portion of it to very high temperatures (up to 5000 K), and can lead to partial or complete disassembly of the core [9]. This type of an event is only postulated and hence called a *hypothetical core disruptive accident (HCDA)*. The analysis of this type of accident requires knowledge of the equation of state of the reactor fuel at high temperatures [10]. The energy release in such accidents depends strongly on how quickly the vapor pressure develops and its magnitude. Hence vapor pressure data on nuclear fuel materials are especially required for such HCDA analysis for LMFBRs. Theoretical approaches to generate vapor pressure of such refractory materials based on the principle of corresponding states (PCS) [11,12], significant structure theory (SST) [13], and molecular dynamics [14] are limited. These models have been applied mostly to organic and simple inorganic systems at temperatures much lower than the region of present interest. They yield parameterized equations with parameters that are obtained by fitting the equations to experimental data. Hence there is the need to generate experimental data on vapor pressure over nuclear fuel materials at very high temperatures.

Conventional experimental techniques that are used for the measurement of thermophysical properties of refractory materials below 2500 K cannot be used at ultrahigh temperatures (>3000 K) because of obvious limitations such as:

1. nonavailability of an inert container material at such elevated temperatures,
2. difficulty in attaining, controlling, and maintaining such high temperatures, and
3. difficulty in making measurement of properties under these conditions. For example, the Knudsen effusion technique is often used to measure vapor pressures of refractory materials at temperatures up to 2500 K. However, this demands that the pressures be in the Knudsen effusion collision-free regime ($<10^{-4}$ atm). Under these conditions it is possible to determine the composition of the vapor phase of the system without much difficulty. But at high temperatures, the pressures will be much higher, of the order of an atmosphere or higher. Apart from the difficulty of sustaining solid–vapor equilibrium under such high-temperature, high-pressure conditions, measurement of equilibrium pressure by sampling the vapor is also difficult.

One way of avoiding these problems is to use a dynamic pulse heating technique, so that there is no need for a container because the sample itself acts as a container and the temperature is generated only for a very short duration (a millisecond or less), thus limiting the energy requirements and the vapor can be sampled easily. The heating is done on a very small sample area and local thermodynamic equilibrium (LTE) is established at that point during the short duration of heating.

1.1 PULSE HEATING METHODS

For the investigation of vaporization behavior, i.e., the determination of vapor pressure of the materials as a function of temperature, methods such as electron pulse heating (EPH), neutron pulse heating (NPH), and laser pulse heating (LPH) [15] are used. The major limitations of EPH are that it can lead to chemical interaction between the sample and the container and also to possible errors in the conversion of the pressure–energy data to pressure–temperature data. The NPH method is limited to materials that can undergo fission and it is not possible to get partial pressure data. Moreover, similar to the EPH method, here too, temperature has to be derived from the energy input.

In the LPH method, the sample is usually heated under vacuum by a pulsed, high-energy laser beam with pulse width ranging from nanoseconds to milliseconds. Using this method, sample properties such as vapor pressure, phase changes, and emissivity can be obtained. The main advantages of this method are: (1) it is applicable to all solid samples and not limited by the property of sample; (2) because it is based on surface heating, the sample itself acts as a container; and (3) it provides better access to the analysis of the gaseous molecules that are formed during laser heating using mass spectrometric and optical spectroscopic techniques. This will lead to a better understanding of the equilibrium conditions.

The main limitations of this method are:

1. Temperature measurement is possible only when the laser pulse is long enough (pulse width of more than a few hundred microseconds).
2. When the laser power density is very high, a variety of physical processes other than simple evaporation, such as particulate ejection and change in isotope distribution in the plume, can occur during the interaction of the laser beam with the solid.

Despite these difficulties, pulsed laser heating has become an important tool in studying the vaporization behavior of materials at temperatures that are unattainable by conventional means, because this method can provide an understanding of the vaporization behavior. Hence of these dynamic pulse heating techniques, the LPH technique was adopted in our study and we have built a laser-induced vaporization–mass spectrometry (LIV-MS) system in our laboratory.

In this chapter, we first discuss the basic processes involved in the laser–solid interaction, with more details on laser-induced vaporization; subsequently, a brief literature survey on laser vaporization methods used to generate vapor pressure data by different groups is given. The relative merits and demerits of these methods are brought out. This is followed by a detailed account on in-house development of the LIV-MS system and the results obtained from a few samples studied, such as graphite, UO_2, UC, ThO_2, and Si_3N_4, will be presented. Further, we will also discuss the possible application of plume characterization using this LIV-MS facility for understanding thin film preparation by the pulsed laser deposition (PLD) method.

2. LASER–SOLID INTERACTION

The first report of laser action in ruby was reported in the year 1960. Soon it was recognized that intense laser beams could be used to heat materials and vaporize them. The relevant property of the laser to be considered for such an application is the low divergence of the laser beams resulting from its high degree of spatial coherence. This property of the laser beam makes it possible to focus it down to very small spot sizes. Effects caused by the absorption of focused laser radiation on the surface of solids are heating, melting, vaporization, and ionization. These effects are used in laser material processing such as surface alloying/cladding, drilling, cutting, micromachining, and welding [16,17]. Further, solid vaporization/ionization by laser are used for sample analysis in methods such as laser-induced breakdown spectroscopy [18] and laser ionization mass spectrometry [19], which are not of interest in the present context. Continuous wave (CW) laser beams are also used as a heat source for heating Knudsen cells to about 3000 K for vapor pressure measurements [20] and toward the determination of melting/freezing temperature of refractory materials such as NpO_2 (3070 ± 62 K) [21] and UO_2 (3130 ± 20) K [22]. However, for vaporization at very high temperatures (>3000 K), a high intensity and short duration (<ms) pulse with a power density between 10^6 and 10^8 W/cm^2 is required to generate a very high temperature, leading to a high rate of material removal, and the heated zone is small. On the other hand, welding or surface modification/annealing processes require the material to be melted but not vaporized. This is best accomplished at lower power densities (between 10^3 and 10^4 W/cm^2) with relatively long pulse durations (~ms) to guarantee sufficient heat transfer into the interior of the solid for melting to occur without the surface temperature becoming too large. The nature of laser–solid interaction depends both on the properties of the materials as well as the parameters of the laser beam. There are many books [23–28] and review articles [29–32] on laser–solid interaction.

2.1 LASER-INDUCED VAPORIZATION

During the interaction of a laser beam with a solid, photons interact with the electrons in the solid and are absorbed. The excited electrons lose their energy through collisions with the lattice. The energy absorbed by the electrons is therefore transferred to the lattice. Thus optical energy is converted into heat instantaneously, in the time of the order of 10^{-13} s in the area where the laser light is absorbed [33]. As the temperature rises, evaporation occurs from the surface. As these processes occur rapidly, one can assume the existence of LTE during LPH [34,35]; i.e., LTE can exist in the phase boundary at the evaporating surface and when LTE prevails, the density distribution in specific quantum states is that of a system in complete thermodynamic equilibrium as defined by the temperature. For the surface with this well-defined temperature, the vaporization rates of all the species are equal to that given by the temperature, which has the same total mass, density, temperature, and chemical composition.

The minimum laser power density (Q_c) required for the evaporation of a solid material can be estimated [30]. As indicated, in vaporization studies the power density required is about 10^6–10^8 W/cm^2 [31]. At these power densities, assuming constant power density (E), a rough estimate of temperature rise can be obtained as (Eq. 15.2) [30,36]:

$$T = T_o + \left[(2E/K)\cdot(\kappa\cdot t/\pi)^{1/2}\right] \tag{15.2}$$

where T is the temperature at the surface of the material; after time t, T_o is the room temperature, K is thermal conductivity, and κ is the thermal diffusivity. Precise calculation of temperature is more complicated [37], owing to temperature-dependent optical and thermal properties of the materials. Data on these properties as a function of temperature are not readily available for most of the materials.

In order to characterize the basic vaporization processes for refractory materials, it is necessary to identify the vapor species. In the vaporization process it is possible to get charged species in addition to neutral species. The production of charged species may be caused by any one or combination of the following processes, namely: thermionic emission, chemical ionization, and photoionization. At the temperature (>3000 K) and pressure of interest (≥1 atm), the laser vaporization processes produce a dynamically stabilized plume. The process of plume formation from an approximate point source (laser focus spot at surface) into a vacuum chamber is similar to that of a free jet expansion through a nozzle [31,38,39]. That is, at the surface, within a few mean free path lengths, vaporization can be modeled as a process with density and composition given by the Langmuir vaporization rate. As the vapor moves away from the surface, further expansion is controlled by gas dynamics. Thus the evaporation process results in the formation of a supersonic gas stream with isentropic flow. In the following sections, we will briefly discuss the nature of Langmuir vaporization, gas dynamic flow, and their implications in laser vaporization studies.

2.2 LANGMUIR VAPORIZATION

The kinetics of vaporization at low surface temperatures has been analyzed by Langmuir [40,41]. According to him, when a condensed phase is in contact with its vapor, a balance is reached between the rate of evaporation and the rate of condensation when equilibrium is attained. In other words, these two processes occur simultaneously at equal rates. At low temperatures, when the vapor pressure of a substance does not exceed 10^{-4} atm, the rate of evaporation of the substance is independent of the presence of vapor around it. That is, the rate of evaporation in a high-vacuum environment is the same as the rate of evaporation in the presence of saturated vapor. Similarly the rate of condensation is determined only by the pressure of the vapor. The rate at which the molecules condense on a surface is given by $\alpha\nu$, where α is the condensation coefficient or sticking coefficient (it represents the ratio between the rate at which the molecules actually condense on the surface and the rate at which they strike the surface) and ν is the number of molecules/second/unit area incident on the surface.

Usually the condensation coefficient is close to unity, unless the vapor species has an appreciably different molecular structure from that of the unit in the condensed phase. For example, both arsenolite and claudite (As_2O_3) consist primarily of As_4O_6 in the gas phase. Arsenolite solid has a molecular lattice of As_4O_6 molecules, whereas claudite does not, and the rate of evaporation of arsenolite is 10^6 times greater than that of claudite [42]. The sluggish rate of atomic rearrangement before evaporation is manifested as a very low evaporation coefficient. The Hertz-Knudsen formula (Eq. 15.3) [41] gives the vaporization flux from the surface as:

$$F = \alpha p(T_s) \Big/ (2\pi MkT_s)^{1/2} \tag{15.3}$$

This is the basic relationship used to measure vapor pressure as a function of temperature by the Langmuir method. In the Eq. (15.3), F is the vaporization flux from the surface; T_s is the surface temperature; M is the mass of evaporating molecule; p is the vapor pressure; and k is the Boltzmann constant. Eq. (15.3) has been modified by a back scattering factor β to account for the retardation caused by collisions in the gas phase as (Eq. 15.4):

$$F = \alpha(1-\beta)p(T_s) \Big/ (2\pi MkT_s)^{1/2} \tag{15.4}$$

If the solid contains more than one element, e.g., A and B, and evaporates as molecules A_iB_j, the flux of a particular species characterized by the stoichiometric numbers i and j is given by Eq. (15.5) [31]:

$$F_{ij} = \alpha_{ij}P_{ij}(T_s, X_s) \Big/ (2\pi M_{ij}kT_s)^{1/2} \tag{15.5}$$

where P_{ij} is the partial pressure of A_iB_j (g) in equilibrium with the condensed phase at the surface temperature T_s and composition X_s. If the composition of the vapor is different from that of the bulk condensed phase, then this results in an incongruent vaporization. Generally pulsed laser evaporation leads to forced congruent vaporization (FCV) on account of its effective surface heating [43]. It is known that the FCV is more favorable at shorter wave lengths and shorter pulse widths [44]. At input power densities of about 10^5–10^6 W/cm^2, only modest surface temperatures are reached. The corresponding vapor pressure of the materials will be low. Because of this, the molecules emitted from the surface do not undergo collisions in the vapor phase (i.e., the free-molecular flow regime). Under this condition, the angular distribution of the vapor species is of a cosine in nature, and the speed distribution corresponds to a Maxwell distribution, characteristic of the surface temperature. The total evaporation rate is given by Hertz-Langmuir formula (Eq. 15.3). At these low-power densities, the interaction has essentially a thermal character; i.e., the material removal rate follows the vapor pressure according to Eq. (15.3) [31].

2.3 GAS DYNAMIC FLOW

If the power density is increased to between 10^6 and 10^8 W/cm^2, collisional effects in the vapor phase become significant and vapor flow changes from free molecular flow to gas dynamic flow. (In molecular flow, the escape probability is unity, i.e., a collision-less flow. The transition from free molecular flow to gas dynamic flow is defined to occur when the escape probability is one-half [31]). In the gas dynamic flow regime, the effect of intermolecular collisions in the vapor will transform random thermal energy of the vapor molecules to translational kinetic energy normal to the surface. This phenomena occurs both in the expansion of the high-pressure gas from an orifice [45,46] or the blow-off from a laser-heated spot on a solid or liquid surface [31,38,39]. However, the analogy between these two situations is not exact, because in the latter a thin transition zone known as the *Knudsen layer* separates the surface and the beginning of the adiabatic expansion. This occurs within a few mean free-path lengths from the surface. For instance, in the case of carbon species, this range is estimated to be of the order of 10 μm at 3500 K and 0.1 μm at 4500 K [47]. In this Knudsen region, vaporization can be modeled as a process that obeys the Langmuir vaporization rate and distribution. The laser-induced vaporization process can be presented as shown in Fig. 15.1.

2.4 KNUDSEN LAYER

The vapor species escaping from a hot surface have a Maxwellian velocity distribution corresponding to the surface temperature, but the velocity vectors all point away from the surface. This anisotropic velocity distribution is transformed into an isotropic one by collisions among the vapor species. This happens within a few mean free paths from the surface, a region known as the *Knudsen layer*, typically on the order of a few micrometers, as indicated previously. In this layer, the translational distribution along the normal direction changes from half Maxwellian at the surface to full Maxwellian on the downstream side of the Knudsen layer [31]. The vapor beyond the Knudsen layer follows an adiabatic gas expansion controlled by gas dynamic parameters. The flow is isentropic, where random thermal energy (characteristic of the surface temperature) is converted to directed translational energy [48]. The adiabatic expansion following the Knudsen layer narrows the velocity distribution. The flow velocity is greater than the most probable velocity of the Maxwellian distribution at the surface temperature by a factor of 1.26–1.64 [49]. The number density of the vapor species decreases as the inverse square of the distance from the target surface. The flow becomes supersonic in the sense that the flow velocity exceeds the local sound velocity, which has decreased because of a drop in temperature. The gas dynamic flow is "cold" because the beam temperature, T_k, is less than surface temperature, T_s, as a result of transformation of thermal motion to directed flow.

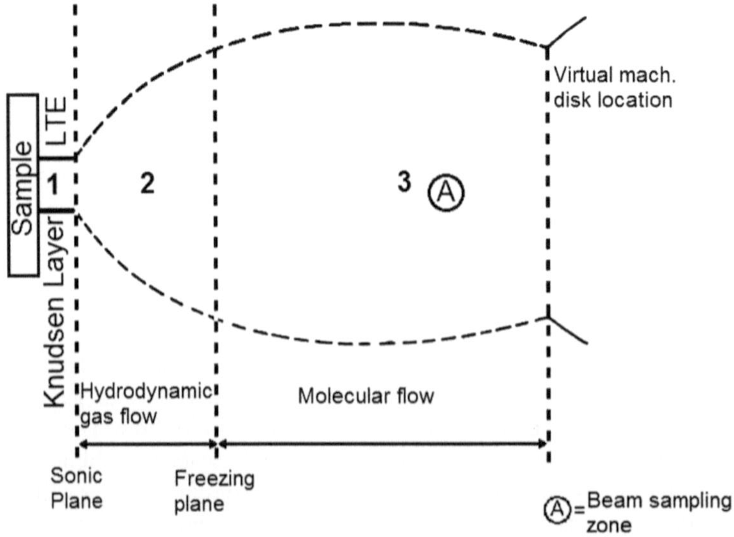

FIGURE 15.1

Schematic of a laser-induced plume expansion showing regions 1, 2, 3. *LTE*, local thermodynamic equilibrium.

The angular distribution of the gas dynamic flow is more sharply peaked about the normal than in the case of molecular flow. The distribution in the case of gas dynamic flow varies as $\cos^n\theta$, where θ is the angle from the surface normal, and $n \geq 2$ [47]. At the freezing plane (as shown in Fig. 15.1), the gas density becomes so low that intermolecular collisions cease. The gas continues to expand in the free-molecular flow regime and its density continues to decrease according to the inverse square of the distance from the target. The time scale for this process of achieving a collision-free (beam) condition is about a microsecond or less at high expansion ratios ($\approx 10^5$) [39]; i.e., the transition from collisional to molecular flow occurs at 1–2 mm from the surface [39]. Hence the chemical composition of the vapor plume is basically the same as the distribution of species at surface temperature. The consequence of the combined effects of the Knudsen layer and gas dynamic expansion is an increase in the axial velocity of the gas stream by 26–64% from its most probable value for $T = T_s$ [49]. In both free-molecular and gas-dynamic expansions, the molecules leave the surface at a rate given by Eq. (15.3), but in the latter case, the following effects in terms of number density seen by the detector quadrupole mass spectrometer (QMS) placed on the downstream of the surface with its axis perpendicular to the molecular axis must be considered [31,39]:

1. Recondensation on the surface that occurs by backscattering of the molecules released from the surface at the outer edge of the Knudsen layer reduces the flux (by 18–25%) [31]. However, the reduction will be very small because the molecule spends very little time in that region.
2. The high flow speed proportionally reduces the residence time in ion source and therefore the sensitivity.

3. Compared with a thermal source, there is an increase in the sensitivity because of the sharp angular distribution of the molecules along the beam axis.

Therefore if a calibration is used at a reference temperature and therefore the corresponding vapor pressure, the difference arising from all these effects is accounted for in the measurement of the pressure from the ion intensity seen by the QMS [50]. Because the density falls rapidly and the vapor species spend only microseconds or so in the collisional region of the beam, the composition of the vapor plume is basically the same as the distribution of species at the surface temperature.

At higher power densities, greater than 10^9 W/cm^2, a variety of other nonequilibrium processes can occur depending on the nature of the material and the angle of the laser beam from the normal to the surface [31]. For example, if the condensed phase does not strongly absorb the incident laser light, in-depth rather than pure surface heating occurs (i.e., the longer the wavelength, the lower the absorption coefficient and the longer the absorption depth) that can be larger than the thermal diffusion length $(\kappa t)^{1/2}$. The energy deposited varies exponentially with depth in the condensed phase (with a characteristic length equal to the reciprocal of the absorption coefficient). The combination of in-depth energy deposition and cooling of the surface owing to the necessity of supplying the heat of vaporization to sustain the initial thermal evaporation results in the temperature maximum being below, rather than at, the surface. The high vapor pressure of the material at the location of the maximum temperature drives nucleation and growth of vapor bubbles, thereby leading to an explosive release of both the gas and entrained liquid. Such droplet emission has been reported on laser irradiation of gold [51]. Emission of chunks of solid can also occur in metals that do not even reach the melting point during the laser pulse. This is because in metals the steep temperature gradient will lead to considerable thermal stress, which exceeds the fracture stress. Such behavior is observed in tungsten [51]. In addition to these effects, it is also observed that at very high power density ($>10^9$ W/cm^2) the plume isotopic composition itself can get altered if the plume contains a large density of ions, thereby causing a plasma centrifuge effect in the expanding plume [52,53].

2.5 LITERATURE SURVEY OF LASER VAPORIZATION STUDIES

Even though a large number of papers have been published on laser–solid interactions, only limited work is reported on vaporization behavior of refractory materials and their vapor pressures at very high temperatures. A summary of the laser-induced vaporization studies concerning measurement of vapor pressure of various materials using different techniques is given in Table 15.1. A brief summary of these is given in the following paragraphs.

Kirillin et al. [54] measured the vapor pressure of graphite using a CW laser beam from a Nd-YAG laser with a power of 500 W as a heat source. The sample was inside a chamber containing inert gas at high pressure. The temperature was measured using a fast micropyrometer to be in the range of 5000 to 7000 K. Ohse et al. started studying the vapor pressures of reactor fuel materials in the mid-1970s, and a summary of all of their studies is presented in the reference list [15]. This group used Nd-YAG lasers of pulse width 1 μs–1 ms (power densities on the order of 10^7–10^8 W/cm^2). Equilibrium vapor pressures were evaluated from the rate of evaporation, obtained by measuring the depth profile of the crater, *d*, formed as a result of vaporization, and this was related to saturated pressure through the relation shown in Eq (15.6) [55]:

$$d = (\alpha \cdot P_s \cdot \tau / \rho) \cdot (M / 2\pi k T)^{1/2}, \tag{15.6}$$

Table 15.1 Summary of Use of Laser Heating Toward Vapor Pressure Measurements

Reference	Material Studied	Laser Pulse Width	Method Adopted	Power Density (W/cm^2)	Temperature Range (K)
Zavitsanos [118]	Graphite (C)	0.5 ms	TOF MS	10^5–10^6	4060–4100 K (cal)
Lincoln et al. [47]	C, Al_2O_3, SiO_2	0.3–0.5 ms	TOF MS	0.5–6×10^4	3200–4200 K (cal)
Asami et al. [65]	UO_2	15 ns	Torsion pendulum	5×10^8–5×10^9	
Kirillin et al. [48]	C	CW	Mass of material evaporated	10^5–10^7	5000–7000 K (meas)
Ohse et al. [15,55,57]	UO_2, UC	1 μs–1 ms	Crater depth, Mach-disk, QMS, Langmuir probe	10^7–10^8	3300–5000 K (meas)
Yano et al. [119]	UO_2	Ruby pulse	Torsion pendulum	–	3460–5880 K (cal)
Bober et al. [59,93]	SS, UO_2	6 ms	Boiling point	$\sim 10^5$	3500–4500 K (meas)
Olander et al. [31,32,62,63]	UO_2, MgO, ZrO_2	0.45–1.1 ms	QMS	10^4–10^6	1900–4200 K (meas)
Joseph et al. (this chapter)	UO_2, UC, ThO_2, C, Si_3N_4	8 ns	QMS	10^6–10^8	4100–5700 K (cal)
Hastie et al. [39,66,68]	C, BN, HfO_2, SiC, ZrO_2, Y_2O_3	7–30 ns	QMS	10^7–10^8	3000–5000 K (cal & meas)
Pflieger et al. [64]	C, UO_2	18–20 m	TOF-MS	$<10^5$	2930–3750 K (meas)

(cal), *calculated;* CW, *continuous wave;* (meas), *measured;* MS, *mass spectrometry;* QMS, *quadrupole mass spectrometer;* ToF, *time of flight.*

where α is the coefficient of evaporation, P_s is the saturated vapor pressure, τ is the laser pulse length, ρ is the density of the condensed phase, M is the mass of the evaporating molecule, and T is the surface temperature. Temperature measurements were made with a specially constructed multiwavelength pyrometer, (allowing measurements at 16 pairs of wavelengths) with a temporal resolution of 10 ns, a spatial resolution of approximately 100 μm, and a temperature resolution of 5 degrees at 5000 K [56]. The main limitation of this method is that Eq. (15.6) assumes that evaporation alone is responsible for the formation of the crater. In addition to the recoil pressure acting on the surface owing to high rates of evaporation, there can be large thermal stress in the planes parallel to the target surface (because of a large temperature gradient) that can affect the morphology of the crater. It is difficult to exactly correct for these factors. In their later studies [57], they used a combination of mass spectrometry, a Langmuir probe, and high-tension diode probes to study the gaseous species (neutrals and ions) that are formed in the vaporization process. Ohse et al. also used the Mach-disk technique to measure the vapor pressure of UO_2 [38]. In this method the analogy of nozzle-type flow sources is applied to that of the pulsed laser-induced vaporization process. In the former (nozzle) type flow, the Mach-disk location, x, is related to the orifice diameter, d; the stagnation pressure in the vessel, p_o; and the external pressure, p_{ext} in Eq. (15.7):

$$x/d = C(\gamma)\cdot(p_o/p_{ext})^{1/2} \tag{15.7}$$

where C is a function of the specific heat ratio, $\gamma = C_p/C_v$, of the expanding gas. Hence a measurement of location of the Mach-disk (using a suitable high-speed camera) allows the determination of the pressure p_o, once d and p_{ext} are known. However, applying this method to the laser-induced vaporization process requires further corrections to the Eq. (15.7), such as expression of p_o in terms of the saturated vapor pressure and suitable correction for d, because the laser pulse gives rise to a nonuniform pressure profile at the orifice. By using this method, the total pressure of UO_2 was reported in the temperature range of 3700–4000 K.

Bober et al. used a recoil momentum measurement technique to evaluate the vapor pressure of UO_2 with a CO_2 laser [58]. The evaporating target is subjected to the recoil force exerted by the evolving vapor. The recoil momentum can be measured by utilizing a target in the form of a small ballistic pendulum. From the measured data on mass of evaporated material, evaporation time, evaporation area, and ballistic and recoil momentum of the sample resulting from the momentum of the vapor released, they determined the vapor flow density, vapor jet velocity, and jet geometry from which the pressure and temperature were derived. The determination of the temperature from the given gas dynamic quantities resulted in considerable uncertainty. Later, this group used a boiling point method [59] that was analogous to the classic boiling point method [60]. In these experiments, the sample is kept in a high-pressure chamber. A definite inert gas pressure of argon or krypton is maintained (between 0.02 and 2 MPa) inside the chamber. A 6-ms wide pulse from a Nd-YAG laser is used to heat the sample at a moderate power density of 10^5 W/cm^2, with a spot diameter of about 1 mm. The increase of central temperature of the irradiation zone is measured with a fast spectral pyrometer. Boiling point is found by correlating the time–temperature curve with the onset of strong evaporation detected by an image converter camera and an ion probe. A heating rate must be chosen so that the surface evaporation will proceed at a condition close to the state of equilibrium. The plateau of the pyrometer signal, which coincides with the rise of the ion probe signal, denotes the onset of boiling. This is a straightforward technique to get vapor pressure data. However, this method requires the emissivity data for the materials in order to measure the temperature when a single wavelength pyrometer is used. (For UO_2 and UC the required emissivities are obtained in separate experiments [61]). Another limitation is that this method gives only the total pressure. Partial pressure information cannot be obtained, which is necessary for a better understanding of the system.

Olander et al. studied the UO_2 system by using a long pulse (450 μs–1.1 ms) from a Nd-glass laser [62]. The peak power density used was about 10^4–10^6 W/cm^2. The vapor species were analyzed by a QMS and the temperature was measured with a fast optical pyrometer focused at the center of the irradiated UO_2 surface. Measurements were reported in the temperature range of 1900 to 4200 K. But they could see the increase in signal only up to 2400 K, and after that there was saturation in the signal. This saturation in the signal was attributed to cluster formation [63]. Here again, because of higher laser pulse width, the evaporated species could have interacted with the laser photons, leading to uncertainties in the chemical composition and in the temperature measurements.

Pflieger et al. [64] also used a millisecond laser (18–20 m) to study the vaporization behavior of graphite and UO_2 at a power density of less than 2×10^5 W/cm^2. They also used a fast pyrometer to measure the surface temperature. Graphite and UO_2 vaporization studies were carried out in the temperature region 3450–3750 K and 2930–3040 K. Asami et al. [65] have studied the laser-induced vaporization of UO_2 by using a ruby laser with pulse width of 15 ns and power densities in the range of 5×10^8 to 5×10^{10} W/cm^2. Pressure was determined by using a torsion pendulum and vapor temperature was determined from the measurement of relative emission intensities of spectral lines of UII. In these measurements, only total pressure was reported. The temperature measured was that of the vapor species and not the surface temperature.

Use of a laser having pulse width of few nanoseconds compared with a relatively long pulse width (microsecond to millisecond) seems to be more advantageous, particularly for heating refractory materials, where the power density required is of the order of 10^6–10^8 W/cm^2 [39,66]. This was demonstrated in the early 1990s, by Hastie et al., from the National Institute of Standards and Technology (NIST, formerly NBS, USA). They have reported vapor pressure measurements on BN [39], graphite [66], SiC, and HfO_2 [67] using a nanosecond pulse from a Nd-YAG laser and they have studied the time-resolved mass spectrum of the species formed by using a QMS. The temperature information was obtained by different indirect methods [39,66]. They also developed the method to measure the surface temperature using spectroscopy [68]. The following part of the chapter presents a summary of data obtained using a LIV-MS system developed in the authors' laboratory.

3. EXPERIMENTAL METHOD

The details of the LIV-MS experimental facility can be found elsewhere [69]. The schematic of the experimental facility is shown in Fig. 15.2a. The photograph of the in-house–developed LIV-MS facility, of which the chambers *A* and *B* form the LIV-MS part, is shown in Fig. 15.2b. The LIV-MS system consists of two ultrahigh-vacuum chambers, differentially pumped. The first chamber is pumped by a 2000-L/s diffusion pump and the second by a 1500-L/s (pumping speed for N_2) cryo pump. The base pressures in the chambers are typically 10^{-6} and 10^{-8} Torr, respectively. The sample is placed on a remotely controlled micropositioner in the first chamber. The QMS is housed in the second chamber in the 90-degrees configuration; i.e., the molecular beam passes through the entrance and exit apertures of the ion source and the ionized molecules are extracted 90 degrees to the molecular beam axis and then mass analyzed. The distance from the sample to the center of the ion source is 56 cm. The neutrals are ionized by electron impact ionization and the energy of the ionizing electrons as well as the emission current can be varied.

A 6-mm aperture is placed (approximately 15 cm from the sample) on the axis of the focal area on the sample to the ion source, for collimation. This also serves as the limiting conductance between the two chambers. This molecular beam axis lies at 30 degrees to the sample surface normal so that the sampling is normally done off-axis to the plume direction, which lies along the surface normal. The second harmonic (532 nm) of a Q-switched Nd:YAG laser is focused on the sample through a quartz viewport by a 25-cm focal length quartz lens. The laser beam axis lies at approximately 37 degrees to the surface, thereby causing an ellipselike focal area. This laser pulse has a duration of approximately 8 ns (full width at half maximum) and a Gaussian spatial profile (99% TEM_{00} mode). The focal areas are measured from magnified optical microscope photographs of the craters. During experiments an He–Ne laser beam is aligned to be along the sampled molecular beam axis through the exit and

(a)

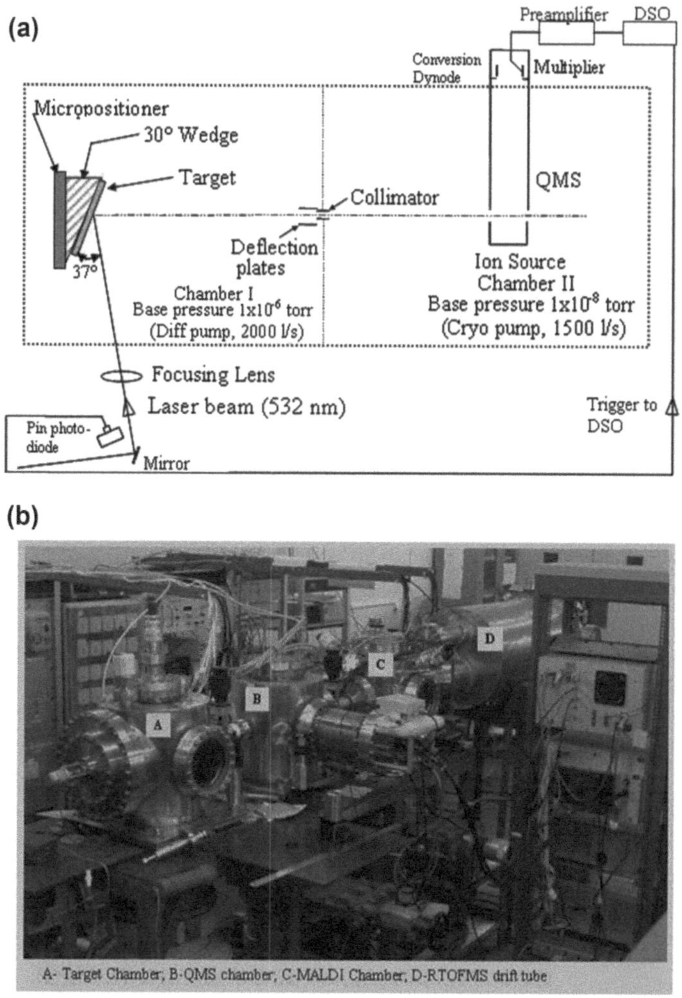

(b)

FIGURE 15.2

(a) Schematic of a laser-induced vaporization–mass spectrometric (LIV-MS) facility [87]. (b) Photograph of an LIV-MS facility. *DSO*, Digital storage oscilloscope; *MALDI*, matrix assisted laser desorption ionization; *QMS*, quadrupole mass spectrometer; *RTOFMS*, reflectron-time of flight mass spectrometer.

(a) Reproduced with permission of Old City Publishing Inc. Publishers.

entrance apertures of the ion source and the collimator. The vaporizing laser beam is focused so that the focal area fully overlaps this He–Ne beam spot on the sample.

The collimator efficiency has been discussed for a similar situation [70], indicating that about two times the dimension of collimating aperture is sampled, although not uniformly over the sampled dimension. A rough estimate for the dimension of the area sampled by the QMS can also be made by considering the geometry. The signal from the electron multiplier of the mass spectrometer (set for the transmission of a particular mass) is preamplified and the waveform is recorded by a digital storage oscilloscope. The oscilloscope is triggered by the pulse from a fast-pin photodiode illuminated by a diffuse reflection of the laser beam, defining the time $t = 0$. The vapor molecules arrive at the ion source after a certain time, depending on their velocity, causing a signal and thus the intensity vs. arrival time is measured, leading to a time of arrival (ToA) profile. Detection of either only the ions in the vapor or the ions together with the neutrals is possible by turning off or on the ion source filament emission.

A pair of deflection plates (5 cm × 5 cm) is placed parallel to the sampled molecular beam axis. These serve to deflect the ions in the vapor away from the sampling axis on application of a suitable voltage. We find that when the density of ions in the vapor is small, it is possible to deflect them off the axis completely. When the densities become large, at least a fraction of the ions still reach the ion source despite the application of a deflection voltage. The largest temperature of measurements is determined by the densities of the ions present for that temperature. That is, on increasing the incident power density and therefore the temperature, we eventually pass to a regime where the ions cannot be deflected completely away from the sampled axis despite the application of the deflecting voltage. In this situation the ions' signal obscures the neutrals' signal and makes measurements impossible.

3.1 EXPERIMENTAL PROCEDURE

The sample is mounted on a kinematic stage and placed in the first chamber. The stage movements are controlled by four direct-current drive motors (two axial translation motion, 25 mm; two tilts, ±7 degrees) through a personal computer. For uniform plume generation, the sample has to be rastered; otherwise a crater forms at the laser focusing point. Typically, the rastering rate is selected such that more than 90% shot-to-shot overlap of the focused spot diameter is achieved for the laser pulse rate of 10 Hz. We find that rastering the sample is necessary for observing a steady signal. The mass spectrometer is tuned to the mass of interest (single ion mode). In order to get a good signal-to-noise ratio, averaging of the signal is done for 100−1000 laser shots. Generally the collection of the signal for about 5 ms from the start of the laser pulse is found to be sufficient to get the ToA profile of each ion mass. By tuning the QMS for different masses of interest, the intensity of each species, which are formed in the laser evaporation, can be obtained. The power density is varied either by changing the laser pulse energy or by changing the area of the laser-focused spot by moving the focusing lens toward and away from the sample surface.

3.2 SAMPLES STUDIED

UO_2 and ThO_2 pellets (15-mm diameter and 3 mm thick) of nuclear grade having 96% of the theoretical density were obtained from the Nuclear Fuel Complex (India). The O/U ratio of UO_2 was about 2.02 and that of ThO_2 was close to stoichiometry. A UC sample in a disc form of 8-mm diameter and 3 mm thick was prepared from UC granules (prepared by carbothermic reduction of the oxide) by the arc-melting process, which gave a guaranteed high and uniform density of a minimum of 95% of the theoretical density. X-ray diffraction examination before and after the experiments confirmed the presence of single-phase UC only, without any UC_2. UC was handled in an inert atmosphere. The Si_3N_4 sample was obtained from M/s Cerac, USA in the form of granules and had a purity of about 98%. This sample was powdered and pellets (12- and 5-mm diameter) were made out of it by cold pressing at 10 MPa. These pellets were sintered at 1500°C in flowing N_2 for about 12 h. These pellets were found to have approximately 70% of the theoretical density. Graphite samples are high-density discs (99.999% purity) of 2.5-cm diameter and 0.5-cm thickness from M/s Leico Industries. These are guaranteed to have uniform and at least 90% of the theoretical density. We find that high- and uniform-density pellets are quite necessary to get steady and reproducible signal as discussed in the following sections. Nonuniform low-density samples can lead to nonuniform temperatures and very irregular morphology with formation of deep craters on the irradiated surface.

4. RESULTS AND DISCUSSION

Graphite is used as a reference material to establish the validity of the LIV-MS facility. This is because it has been relatively well studied compared with any other high-temperature refractory material and also the vapor pressure data reported by the Joint Army Navy Air Force group (JANAF) [71] is expected to be correct, at least up to its proposed melting point (~4000 K).

4.1 VELOCITY DISTRIBUTIONS OF NEUTRALS AND ESTIMATION OF TEMPERATURES

For the determination of partial pressure of a vapor species, we need to define a temperature of vaporization. For this, first we need to show the existence of thermal equilibrium for the vaporization process. If the species formed in the laser vaporization process are in thermal equilibrium with each other, say at temperature *T*, then according to the kinetic theory of gases (Eq. 15.8),

$$1/2\ \mathrm{mv}^2 = 3/2kT \tag{15.8}$$

i.e., $t \propto$ sqrt (m/T) where m is molecular weight of the individual species, k is the Boltzmann constant, and t is the peak position in ToA profile as shown in Fig. 15.3. Hence a plot of ToA as a function of the square root (*sqrt*) of molecular weight should yield a straight line. Fig. 15.4 shows a typical plot obtained for a graphite sample. The straight-line nature of the plot shows that the gas species are in

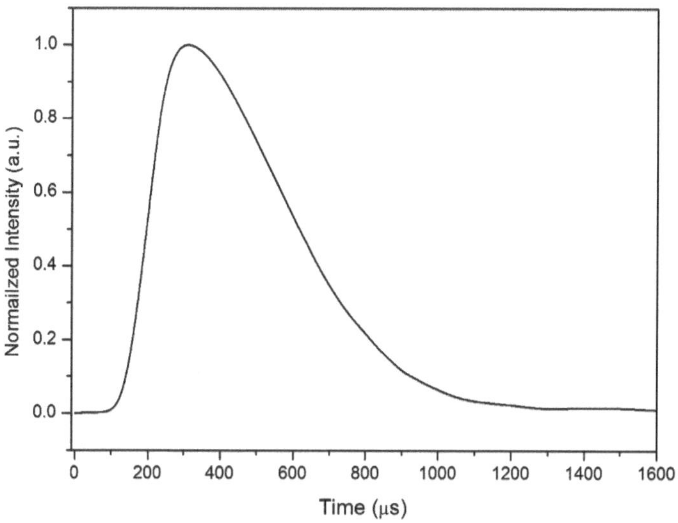

FIGURE 15.3

A typical time of arrival profile of C_3 (each data point is at an interval of 20 ns, containing 50,000 data points). *a.u.*, Arbitrary units.

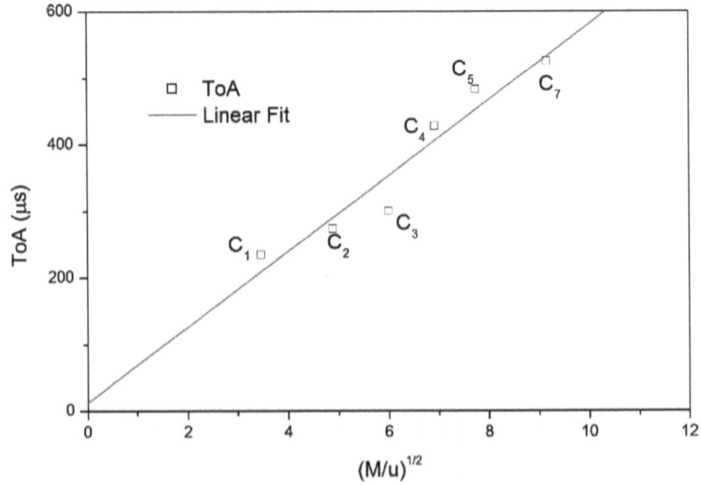

FIGURE 15.4

The plot of time of arrival versus square root of mass corresponding to a laser power density of 8.8×10^7 W/cm^2.

thermal equilibrium with each other, at least with respect to translational energy. Similar behavior was reported [70] for BN, graphite, SiC, and HfO_2. Another study by Sankur et al. [72] on the vaporization of Al_2O_3, SiO_2, and other systems using a 10.6-μm laser pulse of 100 ns pulse width has probed the energy distribution in the vapor by measuring the wavelength-resolved emission intensities. The results of this work also clearly indicated the presence of the local thermal equilibrium (LTE) in nanoseconds of pulsed laser vaporization.

Once the existence of thermal equilibrium is established, it is necessary to measure or estimate the surface temperature. Physical measurement of surface temperature attained during the interaction of laser pulses a few nanoseconds wide is not well established, even though such possibilities are being explored, as can be seen in the literature [68,73]. In our study, an indirect method was used to estimate the surface temperature using the ToA information.

The measured ToAs for C_3 of different power densities are presented in Fig. 15.5a. The same is shown as a bar graph in Fig. 15.5b. It is seen clearly that the maximum of the ToA decreases with increasing power density, and then for the largest power density, it increases again. The decrease in ToA with increasing power density is interpreted as arising from an increase in the surface temperature. The largest power density coincides with that corresponding to large densities of ions in the vapor. The reason for the increase in ToA at this power density corresponding to a large density of ions is not clear. One of the possible reasons is that when the ion and electron densities at the surface are not equal, a space charge field can result close to the surface. Because of this space charge field, a polarized neutral molecule can get aligned in the field and can experience some force directed toward the surface, thereby reducing the velocity.

The nature of the arrival time distributions for laser vaporization has been discussed [74]. The ToA is a convolution of the true speed distribution, creation function, and detector response. The case for vaporization using nanosecond pulse durations has been considered [75]. Because the duration of the temperature transient (i.e., the creation function) is very small compared with the width of the transit time distributions and the time constant of the detector response, the measured profile is simply given by the convolution of the detector response with the speed distribution. The true speed distribution can

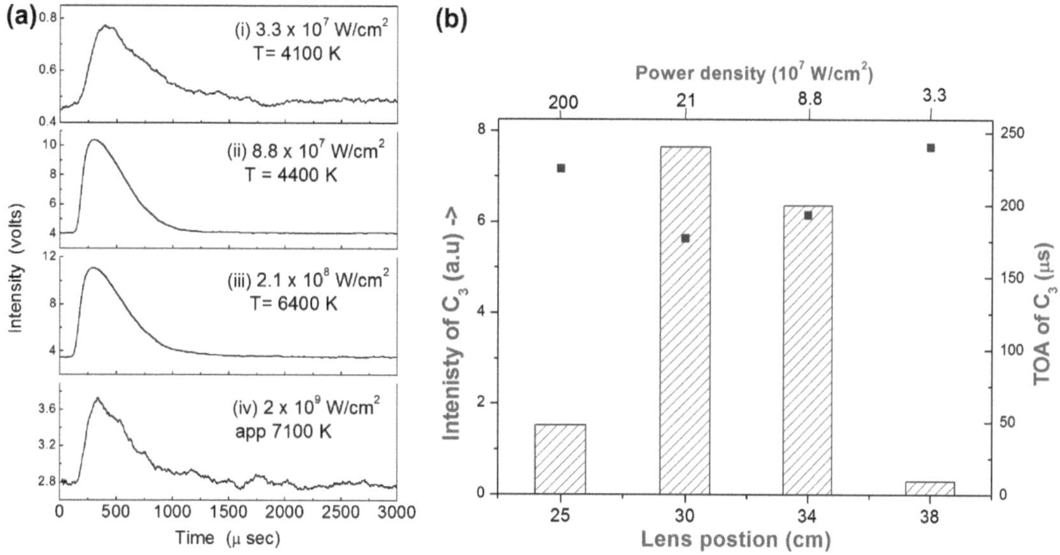

FIGURE 15.5

(a) Time of arrival (ToA) profile of C_3 for incident power density of (i) 3.3×10^7 W/cm^2, (ii) 8.8×10^7 W/cm^2, (iii) 2.1×10^8 W/cm^2, and (iv) 2×10^9 W/cm^2 [86]. (b) Bar graph of ToA and intensity of C_3 for incident power density as shown in (a) [80]. *a.u.*, Arbitrary units.

Reproduced with permission of Elsevier Publishers.

be recovered from the measured profile by deconvoluting the detector response. We do this by using the method suggested in [74,75] and given by Eq. (15.9):

$$J(t)=I(t)+\tau_e\frac{d[I(t)]}{dt} \tag{15.9}$$

where $J(t)$ is the true arrival time profile and $I(t)$ is the measured one. We use a measured time constant τ_e for the preamplifier of 170 μs, which we find by comparison of profiles of Ca^+ ions resulting from vaporization of $CaZrO_3$ using a fast preamplifier and the one used in these measurements. We neglect the transit time through the QMS.

As indicated, for small vaporization rates the gas flow is effusive. On increasing the vaporization rate the gas flow becomes gas dynamic [31]. For this case, the vaporization and the subsequent expansion are understood as arising from: vaporization from the surface with a half-Maxwellian distribution of velocities (only positive components of velocity along the surface normal are present) characterized by a temperature, T_s, followed by a Knudsen layer close to the surface [49,76–78], resulting in a full Maxwellian distribution with a group velocity. Then from the outer boundary of the Knudsen layer: an isentropic expansion of the vapor occurs (similar to supersonic expansions through a nozzle [79]), leading to Mach numbers larger than 1 and cooling. At some distance downstream, the density becomes small enough that collisions cease and a terminal Mach number is reached. Beyond this freezing plane, a molecular flow occurs.

Effectively the observed speed distributions are of the form (for a number density detector) shown in Eq. (15.10)

$$v^2 \exp\left(\frac{-m\left(\frac{L}{t} - v_{gi}\right)^2}{2kT_i}\right) dv \tag{15.10}$$

where v_g is flow velocity and T, the temperature describing the spread in the stream velocity. Conversion of this to an arrival time distribution, using, $v = L/t$ and $dv = -L/_{t^2}\, dt$, gives Eq. (15.11):

$$t^{-4} \exp\left(\frac{-m\left(\frac{L}{t} - v_{gi}\right)^2}{2kT_i}\right) dt \tag{15.11}$$

We find that one distribution is not sufficient to fit the deconvoluted profile, as has been pointed out before [75]. Inclusion of a distribution with a near zero-flow velocity and larger temperature is necessary to get agreement. We fit our deconvoluted profiles to a sum of distributions of this form, as in Eq. (15.12):

$$f(t) = \sum_i A_i t^{-4} exp\left(\frac{-m\left(\frac{L}{t} - v_{gi}\right)^2}{2kT_i}\right) \tag{15.12}$$

comparing the deconvoluted profile by eye to the fitted profile for specified values of the parameters. Table 15.2 gives the parameters of the fits for the carbon molecules at different power densities (the partial pressure reported in this table will be discussed later). Fig. 15.6 shows a representative measured profile, the deconvoluted profile, and the fit for C_3 vapor species.

We have provided a qualitative physical interpretation of these distributions [80] that is in agreement with the earlier observation that this is related to the creation function [75]. The large flow velocity distribution arises from molecules evaporating from the beginning of the temperature transient to on and around peak surface temperature. These molecules expand collisionally and reach large flow velocities. The near-zero flow velocity distribution arises from molecules evaporating at intermediate temperatures on the falling side of the temperature transient. Because the vaporization rates at these times are small and originate at later times, they flow effusively. For nanosecond laser pulses, the peak surface temperatures are reached at the end of the laser pulse. Consideration of the subsequent cooling either because of energy loss from vaporization or conduction (as determined by the thermal diffusion length $(\kappa t)^{1/2}$, where κ is the thermal diffusivity) indicates that the temperature falls rapidly for early times from the peak and the slope becomes progressively smaller at later times. This is in agreement with numerical solutions for the temperature [81]. This means that the sample spends a far larger time at intermediate temperatures on the falling side than it does around the peak surface temperature. This makes this effusive distribution observable despite the fact that the vaporization rate falls exponentially with temperature.

Fig. 15.7 presents the dependence of the flow velocity on the inverse of the square root of mass of the molecule. We include the point corresponding to infinite mass. It is seen that a linear dependence prevails, similar to Fig. 15.4. As discussed, this dependence indicates a common thermal source [75]. This is because collisions in the subsequent expansion can, if anything, only obscure this detail. However, note must be taken that the absorption of the laser beam by the vapor will lead to translational energies larger

Table 15.2 Experimental and Fitted Parameters for Various Carbon Vapor Species [86]

Laser Power Density (W/cm^2)	Estimated Surface Temperature (K)	Species	v_{g1} (cm/s)	T_1 (K)	T_2 (K)	Partial Pressure (atm)
3.3×10^7	4113	C_1	275,000	1500	3500	0.15
		C_2	230,000	500	2500	0.11
		C_3	210,000	1000	2500	0.82[a]
		C_4	145,000	500	2500	0.008
		C_5	130,000	1500	3500	0.03
8.8×10^7	4410	C_1	275,000	1500	3500	2.77
		C_2	250,000	1500	3500	4.32
		C_3	230,000	1500	3500	13.58
		C_4	130,000	1000	2500	0.205
		C_5	130,000	1500	3500	0.59
		C_7	55,000	1500	2500	0.09
9.7×10^7	4979	C_1	290,000	1000	3500	41.29
		C_2	270,000	1000	3500	9.36
		C_3	240,000	1000	3500	49.4
		C_5	120,000	500	3500	0.233
		C_7	–	–	–	0.146
2.1×10^8	6389	C_1	340,000	1250	3500	209.6
		C_2	300,000	1000	3500	31.84
		C_3	275,000	1500	3500	256.5
		C_4	–	–	–	0.152
		C_5	–	–	–	2.03
		C_7	–	–	–	0.92
2.9×10^8	7161	C_1	345,000	1250	3500	296.5
		C_2	300,000	1250	3500	40.66
		C_3	280,000	1250	3500	215
		C_5	90,000	–	–	6.42
		C_7	60,000	–	–	3.712

Ionization cross-section values used (at an electron energy of 28 eV) for C_1 to C_7 (except C_6) are: 1.72, 2.58, 3.15, 3.87, 4.75, and 6.31, and the corresponding ion transmission coefficients are: 0.7, 0.9, 1.0, 1.1, 1.2, and 1.1.
[a]K_{ms} *found using this value of Joint Army Navy Air Force group (JANAF) [71].*

than that given by any reasonable estimate of the surface temperature and can have a similar mass dependence. This we see from the velocities of the ions for the UO_2 system, for example.

In order to estimate the surface temperature (T_s) from the slope of the linear dependence of group velocity on the square root of mass, we need to find the proportionality of the terminal flow velocity to the most probable velocity for that surface temperature, i.e., the value of η in Eq. (15.13).

$$v_g = \eta(2RT_s/m)^{1/2} \tag{15.13}$$

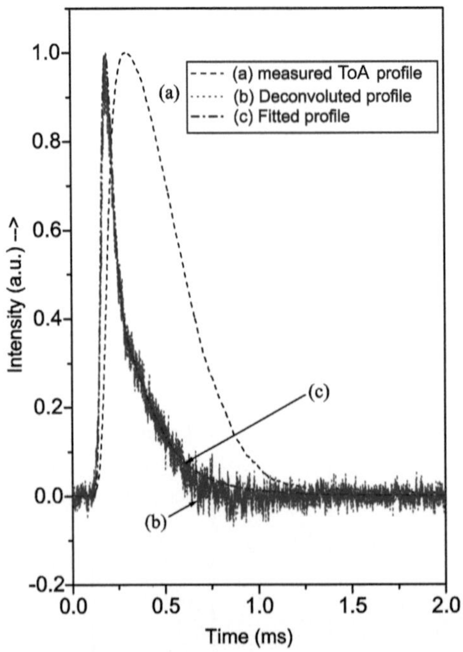

FIGURE 15.6

The fits to the measured arrival time profile of C_3 for an incident power density of 8.8×10^7 W/cm^2: (a) Measured profile, (b) deconvoluted profile, and (c) the fit using the sum of two distributions of $v_{g1} = 230{,}000$ cm/s, $T_1 = 1500$ K, $A_1 = 0.15$ and $v_{g2} = 0$, $T_2 = 3500$ K, $A_2 = 1.0$ [80]. *a.u.*, Arbitrary units; *ToA*, time of arrival.

Reproduced with permission of Elsevier Publishers.

A simple proportionality is indeed expected, as discussed in the literature [82].To find the value of η, we need to consider the effects of the Knudsen layer and the isentropic expansion. Taking the outer boundary of the Knudsen layer as the beginning of the isentropic expansion, one can write Eq. (15.14) [82] for the ratio of the terminal flow velocity (V_{MT}) to the most probable velocity (V_0) as:

$$\frac{V_{MT}}{V_0} = \left[(\gamma/2)\left\{\frac{\left(1+\frac{(\gamma-1)}{2}\right)}{1+\frac{(\gamma-1)M_T^2}{2}}\right\} * T_K\right]^{1/2} * M_T \tag{15.14}$$

where $V_0 =$ the square root of $2RT_s/m$; T_s is the surface temperature; T_K is temperature at the outer boundary of the Knudsen layer, which is given as $T_K = 0.669$ T [76]; M_T is the terminal Mach number. In Eq. (15.14), we have used M_T (instead of Mach number, M) because in our experiments the detector samples the gas stream far after collisions cease (Fig. 15.1), and the value M is the terminal value. The above equation gives the value of η, which depends on the value of M_T used. In the case of laser vaporization of graphite, with a known surface temperature of 4100 K (see the following paragraph) and temperature of the stream velocity distribution (found through fitting) of approximately 1500 K,

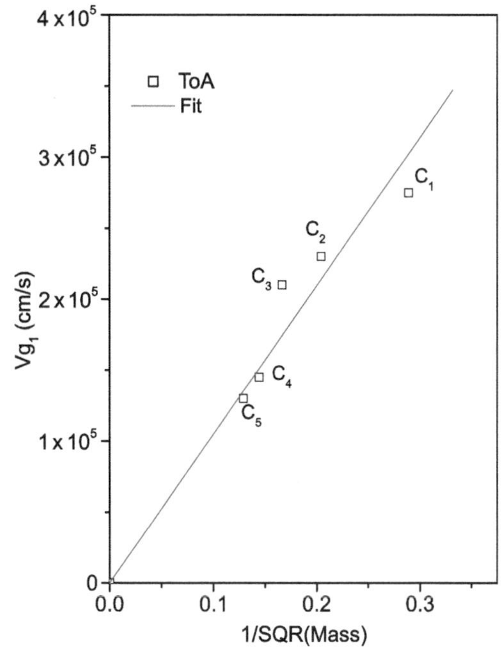

FIGURE 15.7

The plot of group velocity v_g vs. the inverse of the square root of mass for an incident power density of 8.8×10^7 W/cm^2 [86]. *SQR*, square root; *ToA*, time of arrival.

Reproduced with permission of Elsevier Publishers.

a value of 3 for M_T and a value of 1.29 for η are obtained. The value of three for the terminal Mach number, obtained for graphite compares well with those found for the laser vaporization study of lead zirconium titanate [83] and of $YBa_2Cu_3O_7$ [84].

The previous result suggests that the proportionality is rather simple in nature. The actual value obtained for graphite is as follows. In the laser vaporization measurements on graphite, at a power density of 3×10^7 W/cm^2, we have an independent estimate of the surface temperature from the C_5/C_3 ratio that is quite sensitive to temperature below its triple point [66,71]. This leads to a value of 4100 K for the surface temperature at this power density. This is further corroborated by the value of 4100 K from measurements of internal and translational temperatures [85] of the C_2 molecules on vaporization of graphite by a 351-nm laser pulse with the same power density (3×10^7 W/cm^2). Using this value for the surface temperature and the measured v_g values at this power density leads to a value of 1.2 for the proportionality factor, η. We have used this value of η for estimation of surface temperatures for all power densities in experiments on graphite as well as other samples studied, although the value of η may be expected to vary for different source temperatures, source pressures, and chemical compositions. In the limiting case of large values of M_T, the proportionality is indeed the same, but for our studies ($M_T \sim 3$), η can expected to be slightly different. However, in our study the value of 1.2 will be close to the true value because the temperatures describing the stream velocity distributions do not change very rapidly with increasing power densities and with different systems studied. Variation of the velocities from the fitting within reasonable limits leads to a spread of about ±250 K in the surface temperature estimation [86,87].

4.2 DETERMINATION OF PARTIAL PRESSURE

Once the existence of thermal equilibrium is established and the vaporization temperature is defined, then the partial pressure of the vapor species can be determined as given in Eq. (15.15) [88].

Because the mass spectrometer is a number density detector and the residence time of any molecule in the ion source is inversely proportional to its velocity, the observed intensity (I_i^+) will be proportional to the vaporization rate divided by the velocity, leading to $I_i^+ \propto (p_i/T)$. Hence the partial pressure of a particular species (p_i) is related to the intensity by:

$$p_i = K_i I_i^+ T \tag{15.15}$$

where K_i, is a species dependent mass spectrometer constant. A species independent constant is obtained by comparison with the expression in Eq. (15.16):

$$p_i = \left(K_{ms} I_i^+ T\right) / (\sigma_i \tau_i) \tag{15.16}$$

where the subscripts i refer to the ith species: value of intensity (I_i^+), ionization cross section (σ_i), and mass transmission coefficient (τ_i) are used to obtain the partial pressure (p_i) of each species. K_{ms} is the mass spectrometer calibration constant and T is the surface temperature. This will not be true if the expansion modifies the velocity distribution, particularly if the group velocity is not proportional to $(T/M)^{1/2}$. As discussed earlier, in laser evaporation the plume formation can be considered to be similar to the beam formation process of supersonic expansions through a nozzle, with the beam composition after the expansion the same as that of the preexpansion temperature and partial pressure conditions [88]. The high pumping speed of the vacuum system and the pulsed nature of vapor evolution process also ensure the production of high expansion ratios, which is a necessary condition for "freezing in" the species distribution.

We determined the K_{ms} for our experiments, using again, the intensity observed for C_3 at a power density of 3×10^7 W/cm^2, where an independent measure of the surface temperature (4100 K) is available from the C_5/C_3 ratio, as shown in Fig. 15.8. We used the JANAF [71] value for the partial pressure of C_3 of 0.8 atm to determine the constant K_{ms}. (In any set of measurements, we always measure the C_3 intensity at this power density, and in our study of other systems, just before or soon after the measurements). Consequently, any error on this value will be a systematic error in our partial pressures determinations. Therefore the relative values at the different power densities that we have studied will be more meaningful than the absolute values of the partial pressures. Because the measured intensity will be proportional to the total number of molecules evaporated and therefore to the focal area, this is corrected for in the calculation of the partial pressures.

We determined the partial pressures using both the value of the intensity maximum and the integrated area of the ToA profiles. Referring to the discussion in the previous section regarding the presence of two distributions in the deconvoluted profile, it is likely that the pressures derived from the maximum intensity will be more meaningful, because after attaining the maximum temperature (corresponding to the intensity maximum), on the falling side the temperature decreases rather slowly and vaporization will continue to correspond to those lower temperatures [80]. The ionization cross-sections have been taken from Meyer and Lynch [89] using a tabulated value of 1.72 for C_1 for an

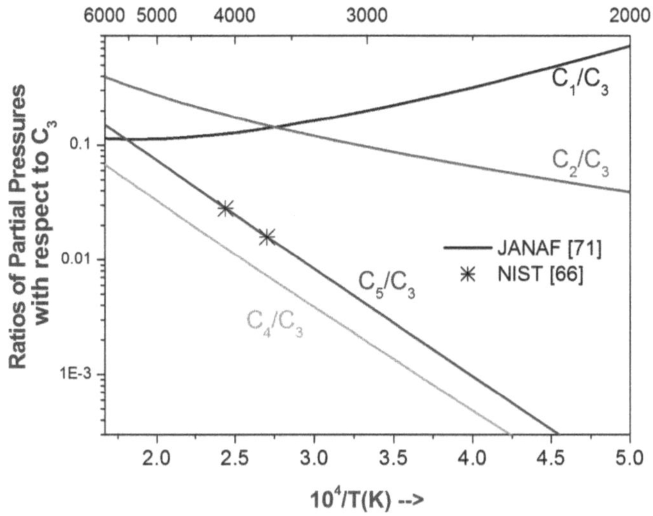

FIGURE 15.8

C_5/C_3 ratio curves are Joint Army Navy Air Force (JANAF)–predicted C_n-species partial pressure ratios with respect to partial pressure of C_3. *, Experimentally observed data for C_5/C_3 ratio by the National Institute of Standards and Technology (*NIST*) (United States) [66].

Reproduced with permission of Elsevier Publishers.

electron energy of 30 eV [90], and the transmission coefficients are those determined [91] for a mass spectrometer of the same model. Because of the use of a conversion dynode in this mass spectrometer, the dependence of the multiplier gain on the mass of the ion has been ignored. In Table 15.2 we give the partial pressures thus obtained using the intensity maximum for the different power densities. The partial pressures and the total pressures obtained are compared with the literature values in Figs. 15.9 and 15.10, respectively.

It is instructive to estimate the maximum pressures that can be measured, having still a defined surface temperature. The rate of transfer of energy from the excited electrons to the lattice is thought to be of the order of 10^{-13} s. As a thermal reservoir is required, we can say that the maximum number of

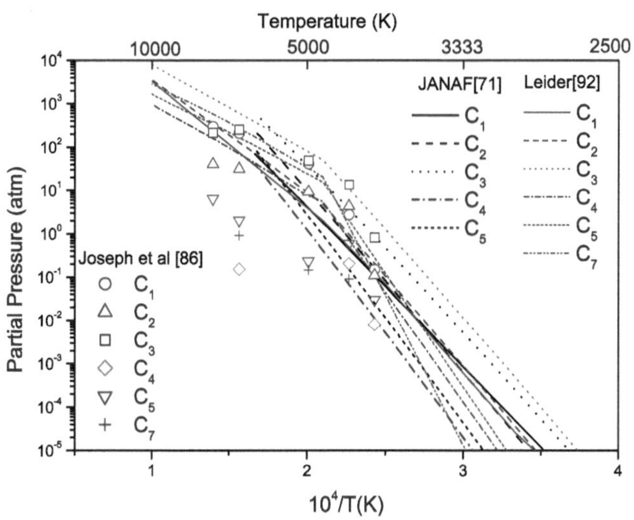

FIGURE 15.9

Partial pressures as reported for various carbon species [86]. Joint Army Navy Air Force (JANAF) [71] and Leider [92].

Reproduced with permission of Elsevier Publishers.

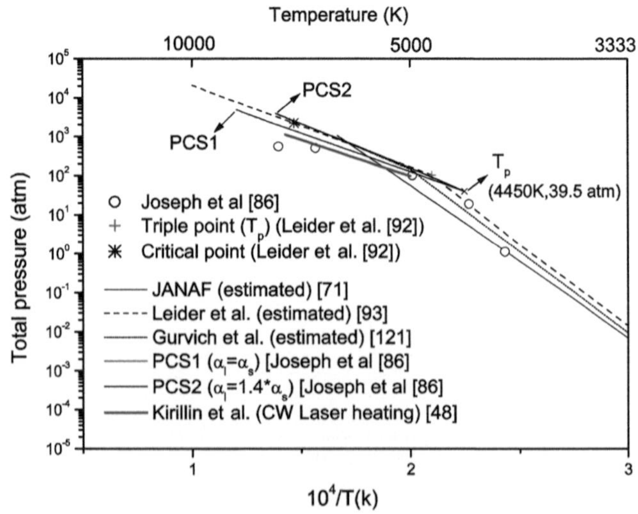

FIGURE 15.10

Total vapor pressure reported for graphite using different methods. *CW*, continuous wave; *JANAF*, Joint Army Navy Air Force; *PCS*, principle of corresponding state. [86,121].

Reproduced with permission of Elsevier Publishers.

molecules (or atoms) vaporized in this time is to be less than (1/100) of the number of atoms present in the volume where the laser beam is absorbed. Assuming a cross-section of 10^{-16} cm^2 and a density of 10^{23}/cm^3, the absorption depth is 10^{-7} cm. This gives the number of atoms to be $A \times 10^{16}$, where A is the focal area. The maximum vaporization rate is then, $A \times 10^{16}$ (1/100)/(10^{-13} A), i.e., 10^{27}/cm^2/s.

4.2.1 Crater Depths

To make an independent assessment of the accuracy of the pressures and temperatures we examined the crater depths for the different cases. The crater profiles are given in Fig. 15.11. It is clear that the craters are not uniform, and in the case of the craters for the two largest power densities used, a ridgelike formation around the crater occurs. This ridge is formed because of the radial movement of the liquid under the pressure exerted by the vaporized molecules leaving the surface.

We estimate the total mass evaporated for one pulse by summing the mass evaporated in discrete time steps. The rising side of the temperature transient is approximated in Eq. (15.17):

$$T = T_S\left(\frac{t}{\tau}\right)^{1/2} \tag{15.17}$$

where T_s is the estimated surface temperature, t is the time (beginning of the laser pulse is time $t = 0$), and τ is the duration of the laser pulse. The falling part (caused by conduction) of the temperature transient after the laser pulse is approximated in Eq. (15.18):

$$T = T_S\left(\frac{\tau}{t}\right)^{1/2} \tag{15.18}$$

From the measured pressures we calculated the vaporization rates (we included only C_1 and C_3 because the partial pressures of the others were far smaller) and therefore the mass vaporized. Then we calculated the crater depth from the total mass evaporated using a value for density of 1.9 g/cm^3. The estimates thus derived are given in Table 15.3. We found indeed that the values were well within a

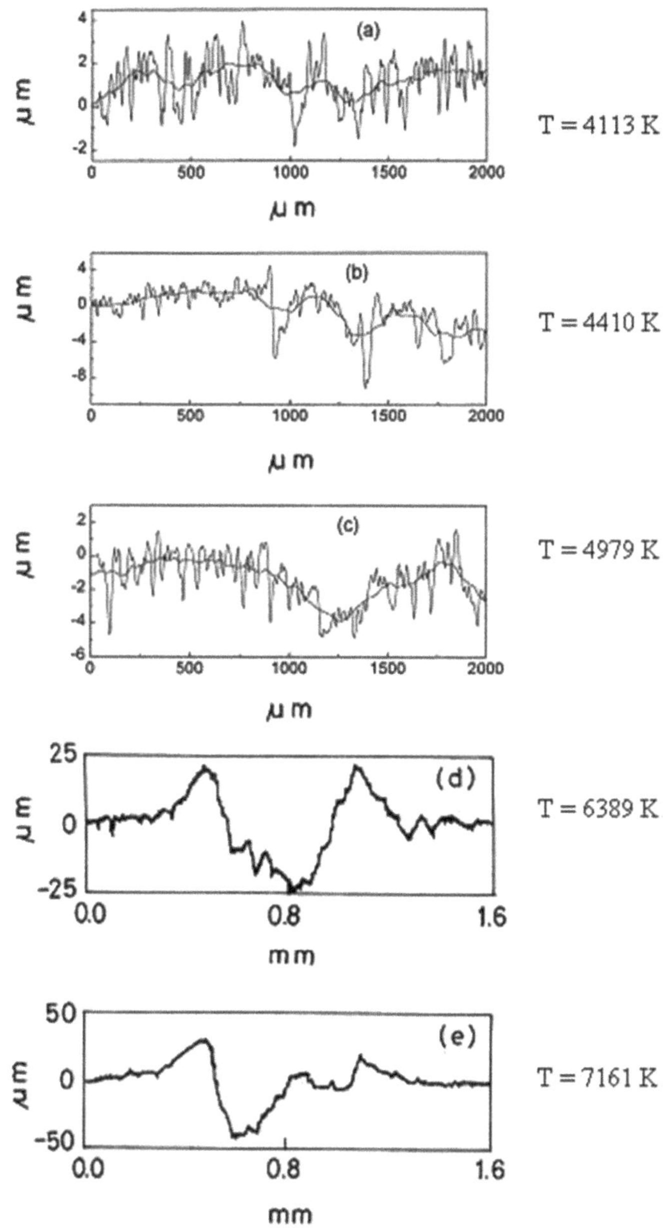

FIGURE 15.11

Crater profiles. (a–c) Measured with profilometer. (d and e) Measured with perthometer. The corresponding peak temperatures are indicated.

factor of three of the measured depths. We neglected the fall in the temperature because of the energy required for vaporization to maintain the corresponding vaporization rates at each temperature. Inclusion of this in the values for temperature vs. time is expected to give better agreement between the calculated and measured values for the crater depths.

Table 15.3 Measured and Calculated Crater Depths on Graphite Target at Different Power Densities

Peak Surface Temperature (K)	Measured Depth for 1800 Pulses (μm)	Calculated Depth for 1800 Pulses (μm)
4113	Within the surface roughness	0.005
4410	1–2	0.5
4979	≈5	6.0
6389	≈30	70
7161	>50	160

4.2.2 UO_2 System

The species observed on $\{UO_2\}$, where { } represents the liquid phase, are U, UO, UO_2, UO_3, and O. At lower laser power densities ($<2.5 \times 10^7/cm^2$), no reproducible signal could be observed. At higher power densities ($>1 \times 10^8/cm^2$), the extent of ion formation was too large, which obscured the detection of neutral species. Hence in our study, the power densities used were $4.8 \times 10^7/cm^2$ and $8.6 \times 10^7/cm^2$. Fig. 15.12 shows a typical ToA profile of UO obtained at a power density of 4.8×10^7 W/cm^2. As can be seen from this Fig. 15.12a, a sharp peak occurs at the shorter time interval (≈300 μs) and is followed by a relatively broader peak at longer times (≈1100 μs). The first sharp peak still appeared even when the ionizer of the QMS was switched off. However, by applying 200 V between the deflection plates, this sharp peak completely disappeared, as shown in Fig. 15.12b. Hence it can be concluded that the sharp peak that occurs at the shorter time interval is the result of UO^+ ions produced from the surface through laser interaction by thermionic and/or photolytic ionization. It is well known that laser vaporization ions have larger velocities compared with neutral species [80]. Once these ions were deflected away from the beam axis through application of suitable electric

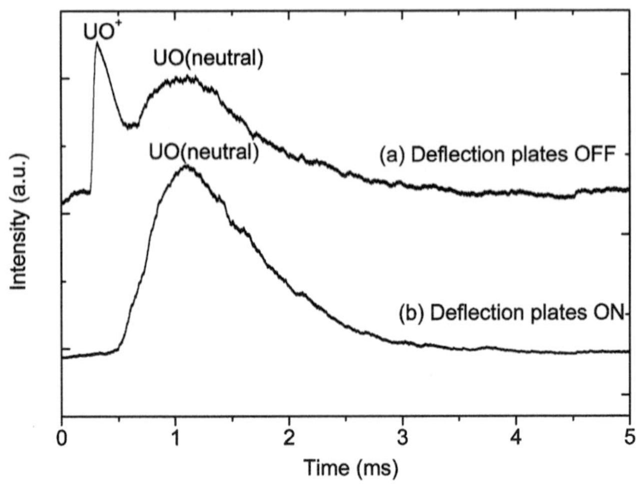

FIGURE 15.12

Typical time of arrival of UO species when the deflection plate is off (a), and on (b) [87].

Reproduced with permission of Old City Publishing Inc. Publishers.

potential to the deflection plates, only the neutral species could reach the ionization source. The broader second peak completely disappeared when the ionizer was switched off. Hence the peak at longer times is a result of the neutral UO species ionized by QMS electron impact ionizer. We have measured similar ToA profiles for all the vapor species.

Compared with the power density of $4.8 \times 10^7/cm^2$, the ToA of all the species decreases when the power density is increased to $8.6 \times 10^7/cm^2$. This decrease indicates the increase in the surface temperature. At these power densities, all the ions could be completely deflected out of the beam by the electric field of 200 V between the deflection plates. Above the power density of $8.6 \times 10^7/cm^2$, even with the application of a higher voltage (500 V) between the deflection plates not all these ions could be deflected completely. This may be either because of the very large number of ions produced or because of their very high velocity, making the electric field insufficient to ensure the total deflection of all the ions. It is to be noted that the ToA of the neutral species increases as the power density incident on the surface increased to 1×10^8 W/cm^2. The neutral beam intensity decreased by a factor of 2, whereas the ion intensity increased by a factor of 6×10^3 compared with that of the value obtained at 8.6×10^7 W/cm^2. Similar behavior was observed in graphite. These results indicate that only a small range of power density can be effectively used to have thermal evaporation as the dominant process; i.e., once the temperature exceeds a certain threshold (corresponding to power density greater than 1×10^8 W/cm^2 in the case of $\{UO_2\}$), a considerable fraction of the vapor is ionized. So once we get to a temperature where a considerable fraction of the vapor is ionized, it is not possible to measure vapor pressures by measuring the intensities of the neutral species alone. It is for this reason that we measured the vapor pressure of UO_2 at only two power densities, namely, 4.8×10^7 and to 8.6×10^7 W/cm^2. Under these conditions, the neutral intensities are large enough to carry out measurements without considerable ionization. Moreover, the partial pressures obtained by measuring the ion intensities of the neutrals represent the true pressure closely.

The intensity of UO_3 was measured at 28 eV, oxygen at 18 eV, and all other species, namely, U, UO, and UO_2 at 12 eV. The measurement of ion intensity at these different electron energies was carried out for the following reasons:

1. The appearance potential of U, UO, UO_2, and UO_3 are 6.1, 5.6, 5.4, and 10.6 eV, respectively [94]. The dissociation energy of U—O bond is about 8 eV [95]. Hence formation of U^+ and UO^+ ions resulting from fragmentation of UO and UO_2, respectively, may not be expected below 13 eV.
2. At 12 eV, the relative ionization efficiency for U, UO, UO_2, and UO_3 are 100:140:35:1.4, respectively [96]. In order to get better sensitivity for UO_3, the ion signal was observed at 28 eV, where it has a maximum ionization probability [94]. Since the appearance potential of oxygen is 13.6 eV [97], the intensity of oxygen was measured at 18 eV. The correction required to account for the contribution of atomic oxygen from the dissociation of UO species is small and can be worked out from the reported fragmentation probabilities of these species [94]. The molecular oxygen could not be detected because of the large background signal present.

The partial pressures obtained at the two power densities (4.8 and 8.6×10^7 W/cm^2) are shown in Table 15.4. The partial pressures obtained are compared in Fig. 15.13 with those of the reported values by Ohse et al. [57] and that of the calculated values by Green et al. [98]. As can be seen form this figure, the partial pressure of UO is higher and those of UO_2 and UO_3 are lower than those of Ohse

Table 15.4 Experimental Results and Parameters Obtained for {UO_2} [87]

Laser Power Density (W/cm^2)	Estimated Surface Temperature T_s, (K)	Vapor Species	v_{g1} (cm/s)	T_1 (K)	T_2 (K)	Intensity (Volts)	Partial Pressure (atm)
4.8×10^7	4715	U	60,000	1500	2500	0.0111	0.47
		UO	47,000	1500	2500	0.0162	0.49
		UO_2	–	–	–	0.0041	0.51
		UO_3	–	–	–	9.54×10^{-4}	1.29
		O	260,000	1500	2500	1.5×10^{-2}	6.95
8.6×10^7	5708	U	70,000	1000	3500	0.00817	0.78
		UO	75,000	1000	3500	0.0137	0.93
		UO_2	40,000	1000	3500	0.0076	2.1
		UO_3	–	–	–	5.05×10^{-3}	15.36
		O	290,000	1500	3500	7.25×10^{-2}	97.87

Ionization cross-section (σ_i) values used for U, UO, UO_2, UO_3, and O are 12.51, 17.71, 4.36, 0.394, and 0.33 [94], and the corresponding ion transmission coefficient (τ_i) values used are 0.3, 0.3, 0.3, 0.3, and 0.8.
U, UO, and UO_2 results correspond to an electron energy of 12 eV.
UO_3 results correspond to electron energy of 28 eV.
O results correspond to electron energy of 18 eV.

et al. and Green et al. The partial pressure of O agrees well with that of Green et al. At 4700 K, UO, UO_2, and UO_3 all have almost equal pressures and the O pressure is comparable to that of U-bearing species. But at 5708 K, O is the dominant vapor species. It is also of interest to note that the total vapor pressure obtained on UO_2 samples ($UO_{2.01}$ and $UO_{2.08}$) using the in-pile measurements did not show any dependence on the stoichiometry [99]. This observation is also in accordance with the calculations, which show that the vapor pressure over $UO_{2\pm x}$ depends only weakly on x above the melting point, and its dependence decreases with increase of temperature [100]. Fig. 15.14 shows the total pressure obtained by the summation of partial pressures in comparison with the literature values. As can be seen, our total pressure is less compared to the literature values (about 20% compared to Ohse et al. [15] and Bober et al. [59], at 4700 K), whereas the total pressure at 5700 K is in reasonable agreement (within 10%). This variation can be attributed to the partial pressure that corresponds to ions of the U-bearing species that will be higher at 5700 K than at 4700 K. However, at 4700 K the contribution of ions (partial pressure caused by ions) is expected to be a considerable fraction compared with the total pressure. In our studies, ions are not accounted for in the pressure measurement. Hence a lower value for the total pressure at 4700 K is expected. At 5700 K, the partial pressure of oxygen is so high that the total pressure is mainly determined by this species. Correspondingly, the contribution from ions of U-bearing species to its total pressure will be less (Table 15.4).

4.2.3 UC System

Data obtained for UC and ThO_2 systems are treated similar to those of UO_2, as discussed previously. In the case of UC, we could observe only U and C species starting from electron energy of 12–28 eV and

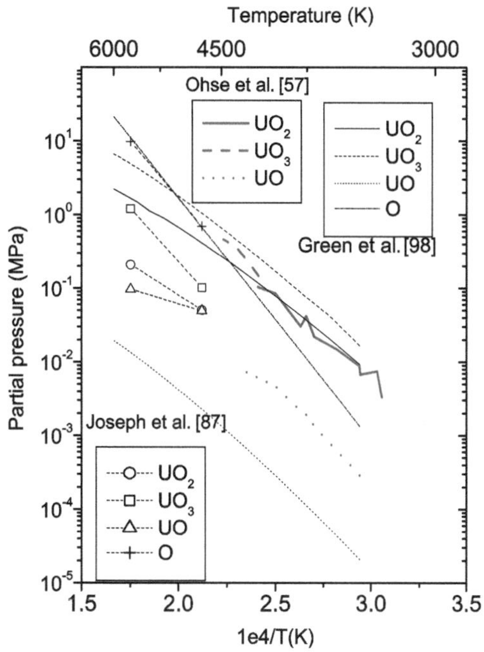

FIGURE 15.13

Partial pressures of various species observed over {UO2} in comparison with the literature values [87].

Reproduced with permission of Old City Publishing Inc. Publishers.

no other molecular species were seen. However, Gupta at al. [101] have identified the molecular species (UC_n, $n = 1-6$) over the U + UC system at a lower temperature (2300–2700 K using Knudsen effusion measurement). It may be of interest to note that similar laser-heating studies by Ohse et al. [55] reported the observation of only U and C species over UC in the temperature range of 6000–7000 K. Because ion intensities are high at 28 eV, the measurements are done at this ionization energy. The surface temperature and partial pressures obtained for the {UC} system are similar to those for {UO_2}. There is no experimental value available in the literature to compare the partial pressures obtained. However, the theoretical values reported by Finn et al. [102] for the partial pressure of U and C are compared with the observed values in Fig. 15.15.

The total pressure obtained by the summation of partial pressures is compared with those of the literature data in Fig. 15.16. As can be seen from this figure, our experimental values are in reasonable agreement with the total pressure reported by Gigli et al. [103] based on SST; less than that of the reported values by Ohse et al. [55], and less than those of PCS [104] by a factor of 2 and 1.4, respectively. However, the present values differ by an order of magnitude compared with values reported by Sheth et al. [105], which were obtained through extrapolating the low-temperature (<melting point) data by correcting for its heat of fusion. Such an extrapolation did not show good agreement in the case of the UO_2 system [106].

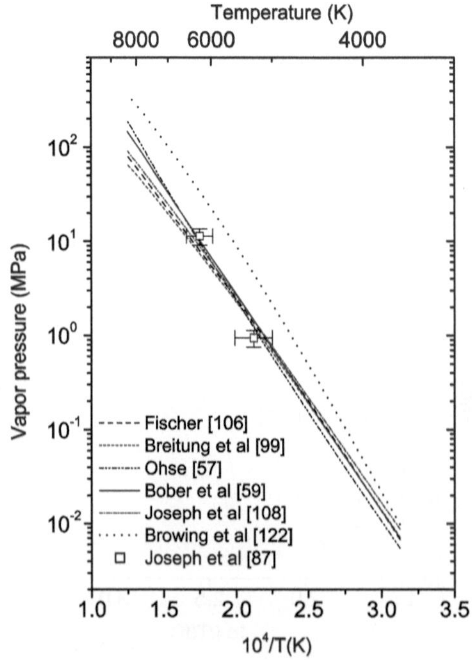

FIGURE 15.14

Total vapor pressure obtained over {UO2} in comparison with the literature values [87].

Reproduced with permission of Old City Publishing Inc. Publishers.

4.2.4 ThO_2 System

In the case of ThO_2, the vapor species observed are Th, ThO, ThO_2, and O. As in the case of UO_2 system, the ion intensities are measured at 12 eV for the Th, ThO, and ThO_2 species because their appearance potentials are 5.9, 6.1, and 8.7 eV, and the dissociation energy of Th—O bond is 9.0 eV [107]. For O, the measurement is done at 18 eV. The partial pressures obtained at two different power densities are shown in Table 15.5. For ThO_2, there is neither experimental nor theoretical data available in the literature for temperature higher than 2800 K. Even basic thermodynamic functions are not available in the high-temperature range in order to make estimation of the pressures, as is done for other systems [67]. The total pressure obtained using the PCS method [87] is compared with that of the experimental values in Fig. 15.17. The total pressure obtained by extrapolating the low temperature data (2400—2800 K) of Ackermann et al. [107] to its melting point (3684 K) is also shown in this figure. As indicated in the introduction, the vapor pressure of different nuclear materials over their liquid region is of interest to reactor safety analysis. The results obtained using LIV-MS are compared with calculated values of vapor pressure in Fig. 15.18 [108]. It may be seen that UC has the lowest vapor pressure, whereas PuN has the highest (for $T > 4000$ K). The mixed nitride, which is a potential fuel, has vapor pressure comparable to that of mixed carbide fuel [109] used in the fast breeder test reactor in India.

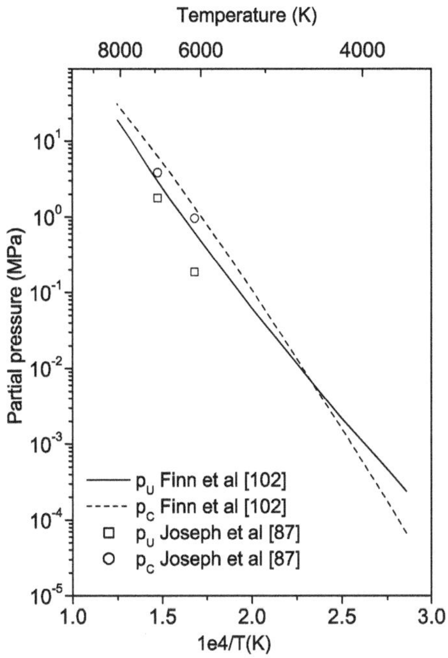

FIGURE 15.15

Partial pressures of U and C species over {UC} [87] compared against literature values [122].

Reproduced with permission of Old City Publishing Inc. Publishers.

4.2.5 Si_3N_4 System

Wang et al. [110] reported laser mass-spectrometric studies on Si_3N_4 and indicated the presence of two high-intensity species at mass 28 u (Si^+) and 70 u (Si_2N^+), and two less intense peaks at a mass to charge ratio (*m*/z) of 42 (SiN^+) and 56 (Si_2^+). Rocabois et al. [111] studied thermal evaporation of Si_3N_4 using a Knudsen cell and observed seven vapor species, namely, N, N_2, Si, SiN, Si_2, Si_2N, and Si_3. The results obtained using our LIV-MS method are presented in detail in another source [112]. Here, we briefly summarize the results. Si_3N_4 is vaporized at two power densities, namely at 3×10^7 and 1×10^8 W/cm^2. At higher power densities ($>1 \times 10^8$ W/cm^2), large amounts of ions are produced that could not be deflected completely by the deflection plates. From the ToA profile, the intensity of each species is obtained. The signal obtained at various m/z values, at power density of 1×10^8 W/cm^2, and ionization energy of 30 eV is shown as a mass spectrum in Fig. 15.19. Except for Si, the most intense isotopic peak of the respective species is shown. For Si, the intensity of isotope 30 is shown. The signals observed at m/z of 14, 30, 42, 56, 70, 84, 98, and 112 were assigned to N, Si, SiN, Si_2, Si_2N, Si_3, Si_3N, and Si_4, on the basis of earlier studies [111,113,114] and from the observed isotopic ratios. The signal at 56 u is assigned to Si_2 on the basis of the isotopic ratio. Even though SiN_2 has been observed by reaction of Si atoms in N_2 matrix [115], no mass spectrometric observation has been reported. The contribution from this species (SiN_2) to the signal at mass 56 may be very small and difficult to confirm from the isotopes ratio. In earlier vaporization (both laser and thermal) studies on

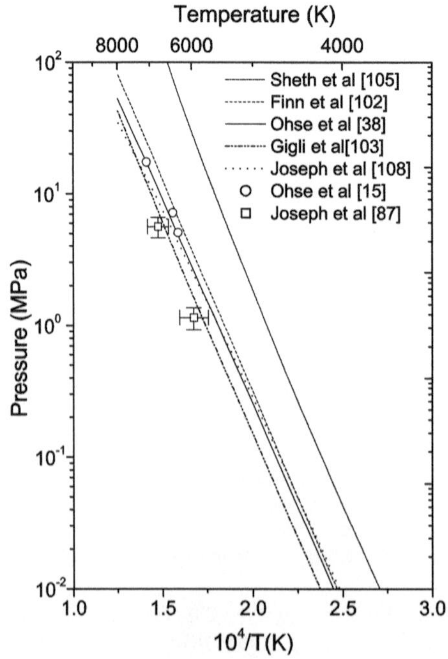

FIGURE 15.16

Total vapor pressure obtained over {UC} in comparison with the literature values [87].

Reproduced with permission of Old City Publishing Inc. Publishers.

Table 15.5 Experimental Results and Parameters Obtained for {ThO_2} [87]

Power Density (W/cm^2)	Estimated Surface Temperature (K)	Vapor Species	v_{g1} (cm/s)	T_1 (K)	T_2 (K)	Intensity (Volts)	Partial Pressure (atm)
4.8×10^7	5146	Th	60,000	1500	3500	7.35×10^{-3}	0.57
		ThO	45,000	1500	3500	2.2×10^{-2}	2.32
		ThO_2	–	–	–	2.17×10^{-3}	1.23
		O	270,000	–	–	3.75×10^{-3}	4.62
1.1×10^8	6233	Th	70,000	1500	3500	2.56×10^{-2}	2.43
		ThO	60,000	1500	3500	4.52×10^{-2}	5.71
		ThO_2	–	–	–	7.93×10^{-3}	5.37
		O	300,000	1500	2500	3.4×10^{-2}	50.72

The ionization cross-section (σ_i) for Th, ThO, ThO_2 are: 13.83, 10.38, and 1.94 [120], and the corresponding transmission coefficients (τ_i) are 0.3, 0 3, 0.3, and 0.8.
Th, ThO, ThO_2 are measured at 12 eV and O at 18 eV.

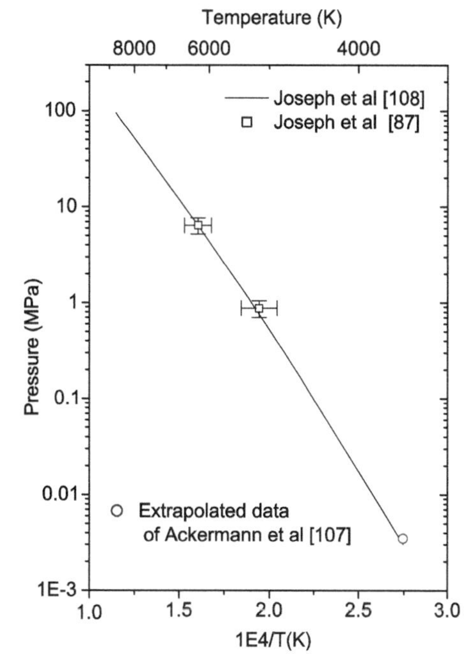

FIGURE 15.17

Total vapor pressure obtained over {ThO_2} in comparison with the literature values [87].

Reproduced with permission of Old City Publishing Inc. Publishers.

Si_3N_4, no species has been observed beyond mass Si_3. But in this study (based on LIV-MS), we observed a clear signal at m/z values of 140, 168, 196, and 280. However, the peak at $m/z = 126$ was seen only at the higher power density (1×10^8 W/cm^2). For each of these high mass peaks, different species can be assigned. For example, for the peak at 168, the possible species that can be assigned are Si_6, Si_4N_4, and Si_3N_6. Similarly, for m/z of 196, the possible species could be Si_7, Si_4N_6, and Si_5N_4. However, the isotopic ratio $[m/(m+1)]/[m/(m+2)]$ expected for each of these possible species will be different. Thus by comparing these expected ratios with those of the observed values, the peaks observed at these three m/z values (140, 168, and 196) can be assigned to Si_3N_4, Si_6, and Si_7, respectively. The peaks obtained at m/z of 126 and 280 are very weak and no definite isotopic ratios could be obtained. However, tentatively the peak at 126 u is assigned to Si_4N. The peak at 280 u can be caused by Si_{10}, the dimer of Si_3N_4, or any other suitable combination of Si and N atoms.

As with the other samples discussed earlier, with Si_3N_4 the main objective of this study was to measure the partial pressures of various species at very high temperatures (3000 K). The surface temperature was determined from the ToA profile as discussed earlier. The linear dependence of the flow velocity versus ($1/\sqrt{\text{mass}}$) indicated that the vaporization is a thermal process and the estimated surface temperature is 4500 ± 300 K for the lower power density (3×10^7 W/cm^2). A similar procedure resulted in a surface temperature of 5000 $\pm$300 K for the higher power density (1×10^8 W/cm^2). As indicated in the experimental section, to perform partial pressure measurements the signal obtained for each pulse should be stable and truly reflect the vaporization rates for that temperature. Unlike the

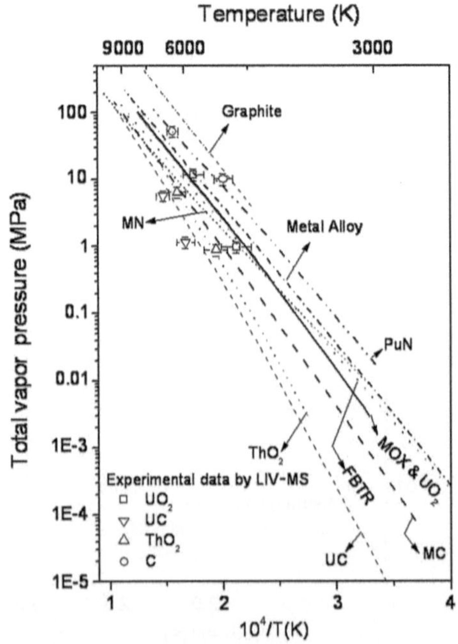

FIGURE 15.18

Vapor pressure for different nuclear materials in their liquid region (line graphs are obtained by the principle of corresponding states-2 method) [108]. *FBTR*, Fast breeder test reactor; *MC*, mixed carbide fuel; *MN*, mixed nitride; *MOX*, mixed oxide; *PuN*, pulutonium nitride.

Reproduced with permission of Elsevier Publishers.

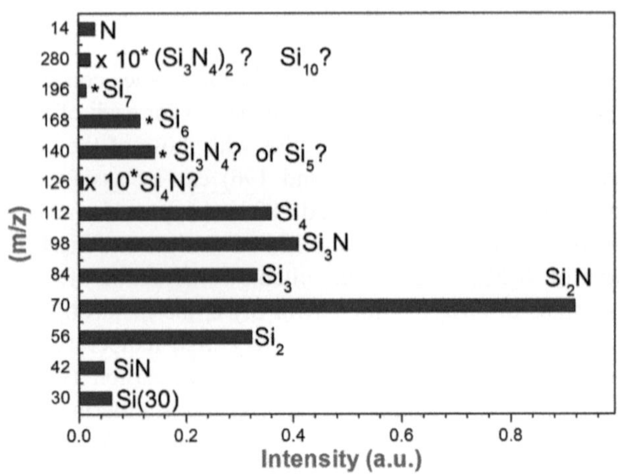

FIGURE 15.19

Mass spectrum of vapor species formed from Si_3N_4 pellets at a power density of 1×10^8 W/cm^2 [112]. *a.u.*, Arbitrary units.

Reproduced with permission of Elsevier Publishers.

other samples studied, even after rastering the sample during the laser irradiation, with 90% shot-to-shot overlap of the irradiated spot, the signal intensity was not steady for Si_3N_4. Typical result is shown in Fig. 15.20 where we compare the UO signal (on vaporization of UO_2) and Si_2N signal (on vaporization of Si_3N_4) at a surface temperature of around 4500 K as a function of number of laser shots that the spot has been exposed to, showing clearly the drop in Si_2N signal. The crater depths for UO_2 and Si_3N_4 in the earlier cases, as a function of laser shots that the spot has been exposed to, are given in Table 15.6. Clearly the reduction in signal correlates with the crater depth. This effect seems to be particularly severe for Si_3N_4, in contrast to experiments on graphite, where we observed a steady signal on rastering. The Si_3N_4 pellets used have only about 70% of the theoretical density, and the density may not be uniform because the sintering is ineffective. We believe that using high and uniform density pellets will be advantageous for such experiments.

4.3 APPLICATION OF LASER-INDUCED VAPORIZATION—MASS SPECTROMETRY IN UNDERSTANDING PULSED LASER DEPOSITION

Preparation of thin films using PLD has become one of the technologically important applications of lasers in materials science. In PLD, the sample is raised to very high temperatures (>3000 K). PLD is surprisingly very simple to do, yet very powerful and flexible [44]. However, the fundamental processes involved in this deposition can be complex and are not well understood. Therefore the

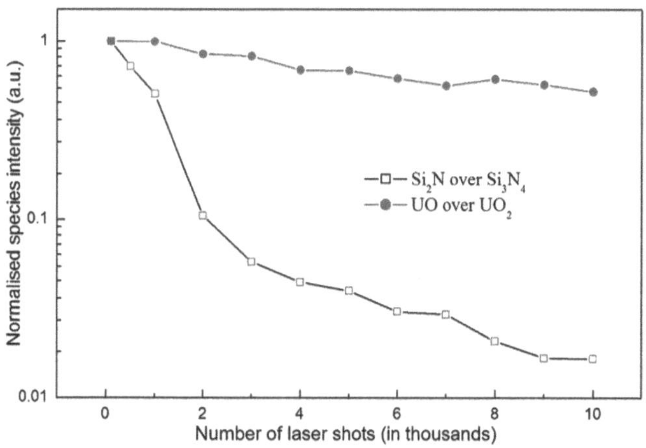

FIGURE 15.20

Change in intensity of Si_2N species over Si_3N_4 and UO species over UO_2 pellet for different numbers of laser shots [112]. *a.u.*, Arbitrary units.

Reproduced with permission of Elsevier Publishers.

Table 15.6 Depth of Crater Measured for Different Number of Laser Shots on Si_3N_4 and UO_2 Pellets for Estimated Surface Temperature of About 4500 K [112]

Pellet	Depth of Crater (μm) for Number of Laser Shots				
Laser shots	100	500	1000	5000	10,000
Si_3N_4	10	17.5	25	30	35
UO_2	<0.1	–	0.2	–	0.5

realization of potential of this technique for the production of optimum quality thin films has been slow.

We have used the LIV-MS facility to characterize the plume formed under PLD conditions to understand the optimal conditions required to get good-quality thin films. We illustrate this aspect in the preparation of two technologically important oxide ion conductors, namely, $CaZrO_3$ [116] and $La_{0.9}Sr_{0.1}Ga_{0.8}Mg_{0.2}O_{2.85}$ (LSGMO) [117]. Both $CaZrO_3$ and LSGMO have potential applications in the form of thin films in the development of oxygen sensors to be used in LMFBRs, i.e., to measure oxygen present as an impurity in liquid sodium, the coolant in LMFBR. In the case of $CaZrO_3$, stoichiometric film was obtained when the deposition was done in an ambient pressure of 150 mTorr of O_2, whereas under vacuum (1×10^{-6} Torr) the film obtained was only calcia-stabilized zirconia (CSZ). On the other hand, the LSGMO, film obtained was stoichiometric only when the deposition was done under vacuum; even at an O_2 background pressure of 5–10 mTorr, only films of mixed oxides resulted. The relationship between film stoichiometry and the mass spectral analysis of the plume obtained in the very high vacuum can explain the observation, as discussed in the following paragraphs (Fig. 15.21).

In the case of $CaZrO_3$, the mass analysis of the plume (Fig. 15.21) reveals clearly a deficiency of CaO^+ ions compared with other ions. The CaO^+ intensity is only about 10^{-3} of the Ca^+ intensity. This suggests that the deficiency of CaO^+ in the plume leads to a film deficient in CaO, namely, CSZ instead of $CaZrO_3$. In the presence of oxygen, the Ca^+ can react to give CaO^+ before deposition, compensating for deficiency of CaO^+ in the plume. Another possibility is that Ca that has already been deposited can react with O_2 to provide the necessary fraction of CaO in the $CaZrO_3$ composition. One would expect Ca^+ in the plume would be more reactive than the Ca^+ or Ca that was already deposited, suggesting that the former process is more likely. It is also of interest to note that in the case of $CaZrO_3$, a large amount of O_2^+ is observed in addition to O and O^+, and the amount of O^+ is greater than that of O. The positive oxygen ionic species may not be available to form required CaO^+ from the available Ca in the plume.

In the case of LSGMO (Fig. 15.21), in oxygen the constituents of the plume react to form oxides, leading to a film of mixed oxides. The reaction can occur before or after deposition, but again one would expect the species in the plume to be more reactive than those deposited already. Furthermore, no O_2 or O_2^+ could be detected and the amount of neutral O is greater than that of O^+. Also, there is no large difference in the elemental ions from oxide-bearing ions, unlike with the $CaZrO_3$. Hence for LSGMO, a high vacuum environment was sufficient to get stoichiometric films because most of the atomic oxygen species formed in the plume would be available to the metal species so as to get them bonded, resulting in stoichiometric films. It is difficult to measure the exact ratio of ions to the total number of neutrals in the plume because the detection of neutrals has to be done by electron impact ionization.

5. CONCLUSION

LIV-MS studies on graphite, UO_2, UC, ThO_2, and Si_3N_4 were carried out at ultrahigh temperatures (>3000 K). The vaporization process at these high temperatures resulted in various vapor species, and the vaporization is regarded as a thermal process. In the case of graphite the various species observed were C_1 to C_7. Similarly, U, UO, UO_2, UO_3, and O over $\{UO_2\}$; U and C over {UC}; and Th, ThO, ThO_2, and O over $\{ThO_2\}$ were observed. In the case of Si_3N_4, we could detect 11 species, namely, Si,

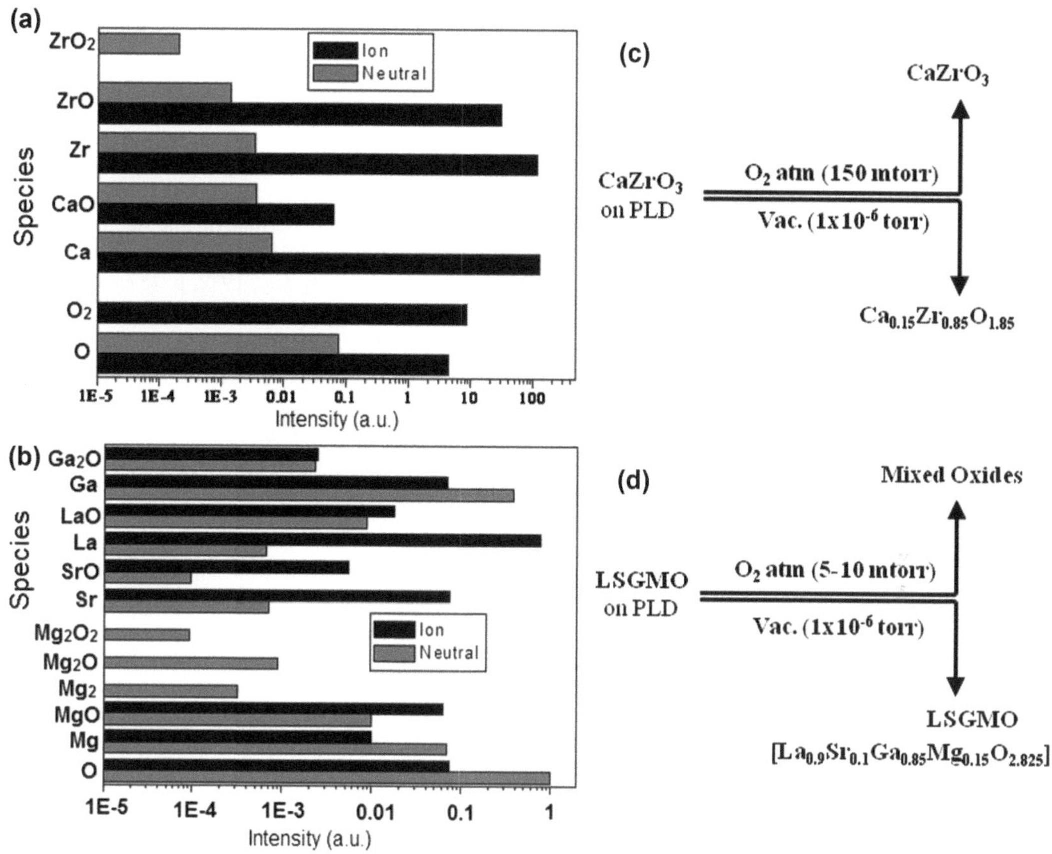

FIGURE 15.21

Typical mass spectrum of species observed on laser vaporization of (a) $CaZrO_3$, and (b) $La_{0.9}Sr_{0.1}Ga_{0.8}Mg_{0.2}O_{2.85}$ (*LSGMO*). (c and d) Stoichiometry of the film obtained for $CaZrO_3$ and LSGMO in an oxygen atmosphere or under vacuum. *a.u.*, Arbitrary units; *PLD*, pulsed laser deposition.

N_2, SiN, Si_2, Si_2N, Si_3, Si_3N, Si_4, Si_3N_4, Si_4N_4, and Si_5N_4, some of which were observed for the first time. Except for Si_3N_4, for all other samples listed, we could measure partial pressure and total vapor pressure. The measured velocity distributions, the analysis, and the reasonable values of temperatures derived there (in comparison with vapor pressures reported) indicate that for the wavelengths and the range of power densities used, the process of vaporization by nanosecond duration pulses is thermal. The present method of determination of surface temperature through ToA information has an uncertainty of about ±250 K. The application of LIV-MS toward the understanding PLD of thin film was also brought out with examples of LSGMO and $CaZrO_3$.

REFERENCES

[1] P. Goldfinger, A definition of high temperature chemistry, in: L. Eyring (Ed.), Advances in High Temperature Chemistry, vol. I, Academic Press, NY, 1967, p. 1.

[2] R.J. Thorn, Some theoretical significance of high temperature chemistry, in: L. Eyring (Ed.), Advances in High Temperature Chemistry vol. 3, Academic Press, NY, 1971, p. 1.

[3] G.D. Maria, Vaporization studies by Knudsen cell-mass spectrometry and thermodynamic stability of high temperature gaseous species, Thermochemistry and its Applications to Chemical and Biochemical Systems, in: M.A.V. Ribeiro da Silva (Ed.), NATO ASI Series, D.Reidel Publishing Co., 1982, p. 157.

[4] L. Brewer, Chemistry and Metallurgy of Miscellaneous Materials: Thermodynamics, 1950.

[5] J. Drowart, G. De Maria, M.G. Inghram, J. Chem. Phys. 29 (1958) 1015.

[6] J.W. Hastie, D.W. Bonnell, in: Proceedings on International Symposium on Thermochemistry and Chemical Processing, 1989, p. 179.

[7] J.I. Slaughter, Performance of re-entry thermal protection materials, in: High Temperature Technology, IUPAC International Symposium on High Temperature Technology, (DC), 1964.

[8] F.W. Clinard, Ceramics Int. 13 (1987) 69.

[9] N.J. Holloway, G.J. Vaughan, M. Mignanelli, J. Havding, P.E. Potter, Science & Technology of Fast Reactor Safety, BNES, London, 1986, pp. 103–109. Paper no. 519.

[10] Summary Report of Specialists Meeting on Equation of State of Materials of Relevance to the Analysis of Hypothetical Fast Reactor Accidents, IAEA-IWGFR/26, IAEA, 1978.

[11] J. Magill, R.W. Ohse, in: R.W. Ohse (Ed.), IUPAC Handbook of Thermodynamic and Transport Properties of Alkali Metals, Blackwell Scientific Publications, Oxford, 1985, p. 86.

[12] R.L. Scott, in: H. Eyring, D. Henderson, W. Jost (Eds.), Physical Chemistry: An Advanced Treatise, vol. 8A, Academic Press, 1972.

[13] E.A. Fischer, Nucl. Sci. 101 (1989) 97.

[14] C. Ronchi, I.L. Iosilevskij, E.S. Yakub, Equation of State of Uranium Dioxide: Data Collection, Verlag Heidelberg, 2004.

[15] R.W. Ohse, J. Chem. Soc. Faraday Trans. 83 (1987) 1235 (and references there in).

[16] D.B. Belforte, M. Levitt (Eds.), The Industrial Laser Annual Handbook, Penn Well Books, Laser Focus, 1986.

[17] M. Bass (Ed.), Laser Material Processing, in: F.Y. Wang (Ed.), Materials Processing Theory and Practice, vol. 3, North Publishing, 1983.

[18] R. Noll, Laser-Induced Breakdown Spectroscopy, Verlag, 2012, http://dx.doi.org/10.1007/978-3-642-20668-9 2.

[19] J.S. Becker, Inorganic Mass Spectrometry Principles and Applications, John Wiley & Sons Ltd., 2007.

[20] J.Y. Colle, F. Capone, Rev. Sci. Inst. 79 (2008) 055105.

[21] R. Bohler, M.J. Welland, F. De Bruycker, K. Boboridis, A. Janssen, R. Eloirdi, R.J.M. Konings, D. Manara, J. Appl. Phys. 111 (2012) 113501.

[22] D. Manara, C. Ronchi, M. Sheindlin, M. Lewis, M. Brykin, J. Nucl. Mater. 342 (148) (2005).

[23] J.F. Ready, Effects of High Power Laser Radiation, Academic Press, 1971.

[24] W.W. Duley, Laser Processing and Analysis of Materials, Plenum Press, 1983.

[25] M. von Allmen, Laser-beam Interactions with Materials Physical Principles and Applications, in: Springer Series in Materials Science, 1987.

[26] L. Moenke-Blankenburg, Laser Microanalysis, Wiley-Interscience Pub., 1989.

[27] J.F. Ready, Industrial Applications of Lasers, Academic Press, 1978.

[28] J.T. Luxson, D.E. Parker, Industrial Lasers and Their Applications, second ed., Prentice-Hall, Inc., 1992.

[29] O.N. Krokhin, Generation of high-temperature vapors and plasmas by laser radiation, in: F.T. Arecchi, E.O. Schulz-Du Bois (Eds.), Laser Handbook, North Holland Publishing Co., 1972, p. p.1371.
[30] J.T. Cheung, H. Sankur, CRC Crit. Rev. Mater. Sci. 15 (1988) 63.
[31] D.R. Olander, High Temp. Sci. 27 (1990) 411.
[32] D.R. Olander, Pure Appl. Chem. 62 (1990) 123.
[33] J.F. Ready, J. Appl. Phys. 36 (1965) 462.
[34] H. Sankur, J.G. Nelson, A.T. Pritt Jr., W.J. Gunning, J. Vac. Sci. Techol. A5 (1987) 15.
[35] H.R. Griem, Plasma Spectroscopy, McGraw-Hill book Co., 1964.
[36] U. Kronert, St Becker, G. Bollen, M.G. Erber, Th Hilbearth, H.-J. Kluge, G. Passler, Nucl. Instrum. Methods Phys. Res. A300 (1991) 522.
[37] R.K. Singh, O.W. Holland, J. Narayan, J. Appl. Phys. 68 (1990) 233.
[38] R.W. Ohse, J.F. Babelot, A. Frezzotti, K.A. Long, J. Magill, High Temp. High Press. 12 (1980) 537.
[39] J.W. Hastie, D.W. Bonnell, P.K. Schenck, High Temp. Sci. 25 (1988) 117.
[40] I. Langmuir, Phys. Rev. 2 (1913) 329.
[41] R.C. Paule, J.L. Margrave, in: J.L. Margrave (Ed.), Free-Evaporation and Effusion Techniques in Characterization of High Temperature Vapors, John Wiley & Sons, Inc., 1967.
[42] L. Brewer, J.S. Kane, J. Phys. Chem. 59 (1955) 105.
[43] W. Breitung, Kernforschungszentrum Karlsruhe, Report no. KfK-2240, 1976.
[44] D.B. Chrisey, G.K. Hubler (Eds.), Pulsed Laser Deposition of Thin Films, Wiley, New York, 1994.
[45] J.B. Anderson (Chapter 1), in: P.P. Wegner (Ed.), Molecular Beams and Low Density Gas Dynamics, Marcel Dekker, 1974.
[46] C. Cerciganani, in: S. Fisher (Ed.), Rarefied Gas Dynamics, AIAA, 1981, p. 305.
[47] M.A. Covington, G.N. Liu, K.A. Lincoln, AIAA J. 15 (1977) 1174.
[48] V.A. Kirillin, V.V. Sychev, A.E. Sheindlin, Engineering Thermodynamics, Mir Publishers, 1976.
[49] R. Kelly, R.W. Dreyfus, Surf. Sci. 198 (1988) 263.
[50] F. Tehranian, D.R. Olander, High Temp. Sci. 25 (1988) 55.
[51] R. Kelly, J.J. Cuomo, P.A. Leary, J.E. Rolhenberg, B.E. Braren, C.F. Aliotta, Nucl. Inst. Method B9 (1985) 329.
[52] M. Joseph, P. Manoravi, Appl. Phys. A 76 (2003) 153.
[53] P.P. Pronko, P.A. VanRompay, Z. Zhang, J.A. Nees, Phys. Rev. Lett. 83 (1999) 2596.
[54] A.V. Krillin, M.D. Kovalenko, M.A. Sheindlin, V.S. Zhivopistsev, Teplofizika Vysokikh Temperatur 23 (1985) 699.
[55] R.W. Ohse, J.F. Babelot, A. Frezzotti, K.A. Long, J. Magill, C. Cercignani, A. Scotti, High Temp. Sci. 13 (1980) 35.
[56] J.P. Hiernaut, R. Beukers, W. Heinz, R. Selfslag, M. Hoch, R.W. Ohse, High Temp. High Press. 18 (1986) 617.
[57] R.W. Ohse, J.F. Babelot, C. Cercignani, J.P. Hiernaut, M. Hoch, G.J. Hyland, J. Magill, J. Nucl. Mater. 130 (1985) 165.
[58] M. Bober, H.V. Karow, K. Schretzmann, in: Thermodynamics of Nuclear Materials, vol. I, IAEA, 1975, p. 295.
[59] M. Bober, J. Singer, M. Trapp, Nucl. Sci. 97 (1987) 344.
[60] P.R. Clopper, R.L. Altman, J.L. Margrave, in: J.L. Margrave (Ed.), Static Techniques, Dew Points, Vapor Densities and Boiling Point Measurements in Characterization of High Temperature Vapors, John Wiley & Sons, Inc., 1967, p. 74.
[61] M. Bober, H.V. Karow, Measurements of spectral emissivity of UO_2 above melting point, in: Proceedings of the 7th Symposium on Thermophysical Properties, ASME, NBS Gaithersburg, MD, USA, 1977, p. 344.
[62] S.K. Yognik, D.R. Olander, Laser Induced Vaporization of Uranium Dioxide, Report, LBL-23391, 1987.
[63] R.J. Tench, M. Balooch, L. Bernardez, M.J. Allen, W.J. Siekhaus, D.R. Olander, W. Wang, J. Vac. Sci. Technol. B9 (2) (1991) 820.

[64] R. Pflieger, M. Sheindlin, J.Y. Colle, Int. J. Thermophys. 26 (2005) 1075–1093.
[65] N. Asami, M. Nishikawa, M. Taguchi, in: Thermodynamics of Nucl. Mater, IAEA, 1975, p. 287.
[66] J.W. Hastie, D.W. Bonnell, P.K. Schenck, High Temp. High Press. 20 (1988) 73.
[67] D.W. Bonnell, P.K. Schenck, J.W. Hastie, M. Joseph, in: W.B. Johnson, R.A. Rapp (Eds.), High Temperature Materials Chemistry – V, The Electrochemical Inc., Pennington, 1990, pp. 156–166.
[68] J.W. Hastie, D.W. Bonnell, P.K. Schenck, J. Nucl. Mater. 294 (2001) 175–178.
[69] M. Joseph, N. Sivakumar, Instrum. Sci. Tech. 26 (1998) 81.
[70] J.W. Hastie, D.W. Bonnell, P.K. Schenck, Pure Appl. Chem. 72 (2000) 2111–2126 (and references there in).
[71] M.W. Chase Jr., J.R. Downey Jr., D.J. Frurip, R.A. McDonald, A.N. Syvernd (Eds.), JANAF Thermochemical Table, third ed., American Institute of Physics, Inc., New York, 1985.
[72] H. Sankur, J.G. Nelson, A.T. Pritt Jr., W.J. Gunning, J. Vac. Sci. Technol. A5 (1987) 15–20.
[73] D. Otte, H. Kleinschmidt, O. Bostanjoglo, Rev. Sci. Instrum. 68 (1997) 1.
[74] W.S. Young, Rev. Sci. Instrum. 44 (1973) 715.
[75] D.W. Bonnell, P.K. Schenck, J.W. Hastie, in: Internatinal Symposium on Laser Process for Microelectronic Applications, 172nd Electrochemical Society Meeting, 1987.
[76] S.I. Anisimov, Soviet Phys. JETP 27 (1968) 182.
[77] C. Cercignani, in: S.S. Fischer (Ed.), Rarefied Gas Dynamics 305, AIAA, NY, 1981.
[78] R. Kelley, R.W. Dreyfus, Nucl. Instrum. Methods Phys. Res. B32 (1988) 341.
[79] J.B. Anderson, J.B. Fenn, Phys. Fluids 8 (1965) 780.
[80] M. Joseph, N. Sivakumar, D. Albert Raj, C.K. Mathews, J. Nucl. Mater. 247 (1997) 21–27.
[81] R.K. Singh, J. Narayan, Mater. Sci. B3 (1989) 217.
[82] H. Pauly, J.P. Toennies, Methods Exp. Phys. 7A (1968) 237.
[83] J.W. Hastie, D.W. Bonnell, A.J. Paul, P.K. Schenck, in: Materials Research Society Symposium Proceedings, vol. 334, 1994, p. 305.
[84] N.H. Cheung, Q.Y. Ying, J.P. Zheng, H.S. Kwok, J. Appl. Phys. 69 (1991) 6349.
[85] R.W. Dreyfus, R. Kelley, R.E. Walkup, Nucl. Instrum. Methods Phys. Res. B23 (1987) 557.
[86] M. Joseph, N. Sivakumar, P. Manoravi, Carbon 40 (2002) 2031–2034.
[87] M. Joseph, N. Sivakumar, P. Manoravi, High Temp. High Press. 34 (2002) 411.
[88] D.W. Bonnell, J.W. Hastie, Characterization of high temperature vapors and gases, in: J.W. Hasite (Ed.), Proceedings of the 10th Materials Research Symposium, NBS SP-561/1,, US Government Printing Office, 1979, p. 537.
[89] R.T. Meyer, A.W. Lynch, High Temp. Sci. 5 (1973) 192.
[90] D.W. Bonnell, Software SIGMA, Version 1.1, 1990, NIST, Gaithersburg, MD, USA.
[91] M. Joseph, (Ph.D thesis), 1995.
[92] H.R. Leider, O.H. Krikorian, D.A. Young, Carbon 11 (1973) 555.
[93] M. Bober, J. Singer, High Temp. Sci. 19 (1985) 329.
[94] P.E. Blackburn, P.M. Danielson, J. Chem. Phys. 54 (1972) 6156–6164.
[95] R.J. Ackermann, E.G. Rauh, M.S. Chandrasekharaiah, Report 'A Thermodynamic Study of the Urania and Uranium System', ANL-7048, 1965.
[96] E.G. Rauh, R.J. Ackermann, J. Chem. Phys. 60 (1974) 1396–1400.
[97] J. Roboz, Introduction to Mass Spectrometry Instrumentation and Techniques, Interscience Publishers, 1968, p. 511.
[98] D.W. Green, L. Leibowitz, Vapor Pressures and Compositions in Equilibrium with Hypostoichiometric Uranium Dioxide at High Temperatures, Report, ANL-CEN-RSD-81–1, 1981.
[99] W. Breitung, K. OReil, Nucl. Sci. Eng. 101 (1989) 26–40.
[100] W. Breitung, Vapor Pressure of Oxide Reactor Fuels above 3000 K: Review and perspective, Report SAND 81-1383, Sandia National Laboratories, Albuquerque, NM, 1981.

[101] S.K. Gupta, K.A. Gingerich, J. Chem. Phys. 71 (1979) 3072–3079.
[102] P.A. Finn, A. Sheth, G. Winslow, L. Leibowitz, in: Proceedings of the Topical Meeting on Advanced LMFBR Fuels, Tucson, USA, 1977 (ANS ERDA, 1977), 1977, p. 189.
[103] C. Gigli, M. Guido, G. De Maria, J. Nucl. Mater. 98 (1981) 35–46.
[104] M. Joseph, N. Sivakumar, C.K. Mathews, Bull. Mater. Sci. 17 (1994) 935–950.
[105] A. Sheth, L. Leibowitz, Equation of State and Transport Properties of Uranium and Plutonium Carbides in the Liquid Region, USAEC report ANL-AFP-11, 1975.
[106] E.A. Fischer, Nucl. Sci. 101 (1989) 97–116.
[107] R.J. Ackermann, E.G. Rauh, High Temp. Sci. 5 (1973) 463–473.
[108] M. Joseph, N. Sivakumar, P. Manoravi, Ann. Nucl. Energy 31 (2004) 1163–1175.
[109] C. Ganguly, C.G. Jain, J.K. Ghosh, P.R. Roy, J. Nucl. Mater. 153 (1988) 178.
[110] S.L. Wang, K.W.D. Ledingham, W.J. Jia, R.P. Singhal, Appl. Surf. Sci. 93 (1996) 205.
[111] P. Rocabois, C. Chatillon, C. Bernard, J. Am, Ceram. Soc. 79 (1996) 1351.
[112] M. Joseph, N. Sivakumar, P. Manoravi, Int. J. Mass Ion Process. 176 (1998) 237–244.
[113] R. Viswanathan, R.W. Schmude Jr., K.A. Gingerich, J. Chem. Thermodyn. 27 (1995) 1303.
[114] K.A. Gingerich, R. Viswanathan, R.W. Schmude Jr., J. Chem. Phys. 106 (1997) 6016.
[115] I.S. Ignatyer, H.F. Schaefer III, J. Phys. Chem. 96 (1992) 7632.
[116] M. Joseph, N. Sivakumar, P. Manoravi, S. Vana Varamban, Ionics 144 (2001) 339–346.
[117] M. Joseph, P. Manoravi, H. Tabata, T. Kawai, J. Appl. Phys. 92 (2002) 997–1001.
[118] P.D. Zavitsanos, G.A. Carlson, J. Chem. Phys. 59 (1973) 2966.
[119] T. Yano, A. Ohtsubo, T. Ishii, J. Nucl. Mater 125 (1984) 71.
[120] D.W. Bonnell, Software 'SIGMA' Version 1.1, NIST, 1990.
[121] L.V. Gurvich, V.S. Iorish, D.V. Chekhovsioi, V.S. Yungman, NIST Special Database 5, "IVTANTHERMO", CRC, Boca Raton, Florida, 1993.
[122] P. Browning, M.J. Gillan, P.E. Potter, The Equation of State of Uranium Dioxide – A Comparison of the Corresponding States and Significant Structure Theories, UKAEA Report AERE-R 8129, 1977.

CHAPTER

16 NANOCLUSTERS UNDER EXTREME IONIZATION CONDITIONS

P. Sharma, R.K. Vatsa
Bhabha Atomic Research Centre, Mumbai, India

CHAPTER OUTLINE

Materials Under Extreme Conditions. http://dx.doi.org/10.1016/B978-0-12-801300-7.00016-4

1. INTRODUCTION TO CLUSTERS

1.1 INTRODUCTION

Advancement of human civilization has gone hand in hand with progress in the field of materials science. The importance of materials in the evolution of human civilization can be realized from the fact that different chronicle landmarks for periodization of human prehistory have been broadly associated with materials, i.e., stone age, bronze age, iron age, etc. [1]. Thus it is certain that advancement in materials science is essential for shaping our society of the future, with development of new and unique materials that have enormous technological potential [2]. As a result, there is always an incessant demand for development of newer advanced materials with novel and exotic properties for diverse applications. In this quest, researchers have focused their attention toward manipulating matter at the nanoscale to craft clusters [3] and nanomaterials/particles [4,5] that exhibit interesting physical and chemical properties altogether different from their bulk counterpart.

1.2 WHAT ARE CLUSTERS AND HOW DO THEY DIFFER FROM NANOPARTICLES?

In general, though clusters and nanoparticles are interrelated, there is a thin line that separates the two. Clusters are defined as aggregate of atoms or molecules that are held together by interactions ranging from weak van der Waals to strong ionic forces [6,7]. In these clusters, unlike liquids or solids, all the constituent atoms/molecules are on or near the surface. Consequently, due to surface effect, properties of a cluster can be significantly varied by loss or addition of a single atom/molecule [8]. This results in size- and shape-dependent alteration in chemical and physical properties [9]. For small clusters, the variation in properties is significant and is not a linear function of size [10,11]. For larger clusters, the dependence on size steadily weakens and the properties of clusters converge toward those of its corresponding bulk phase (Fig. 16.1). These larger clusters, which have dimensions in the nanometer range, are termed as small nanoparticles. In these nanoparticles, the structure of their nucleus is similar to that of their corresponding bulk counterpart, but on the surface they exhibit different structure. Clusters and nanoparticles exhibit distinct properties, which vary spontaneously as a function of their size and shape due to their comparatively higher surface-to-volume ratio. This aspect of engineering materials with desired properties by manipulating the size and shape of matter at nanoscopic level has been the driving force in modern materials science.

1.3 WHY STUDY CLUSTERS UNDER EXTREME LASER IONIZATION CONDITIONS?

The interaction of light with matter is a universal aspect and plays a pivotal role in our day-to-day life. Since the beginning of modern science, light–matter interaction has been a topic of research, and has served as a major scientific tool for systematic exploration of nature. The advent of lasers in 1960 has

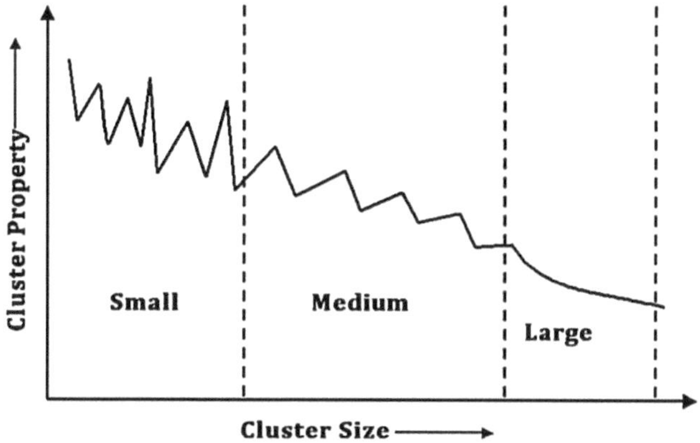

FIGURE 16.1

Representative diagram depicting cluster size–dependent variation in properties of clusters.

From S. Das, P.M. Badani, P. Sharma, R.K. Vatsa. Coulomb explosion phenomenon using gigawatt intensity laser fields: an exotic realm of laser–cluster interaction, Curr. Sci. 100 (2011) 1008–1019. Reproduced with permission of Indian Academy of Sciences.

immensely contributed to fundamental understanding of the underlying basic science of light–matter interaction and has resulted in development of several technological applications [12,13]. The past 50 years have witnessed profound growth in the physics and technology of lasers, leading to development of a wide array of laser systems that operate over a large wavelength range spanning from infrared to X-ray regions, having laser power of petawatt (10^{15} W) order and pulse duration curtailed down to femtosecond (10^{-15} s) timescale. Pulses from such lasers, when focused on a matter, can give rise to intensity exceeding 10^{20} W/cm^2, having electric field of $\sim 10^{11}$ V/cm, magnetic field of $\sim 10^9$ G, temperature of $\sim 10^{10}$K, and pressure of $\sim 10^9$ bar [14]. The enormous pressure generated by such high intense laser pulses has been utilized to compress diamonds up to a pressure of 5×10^{12} Pa, which corresponds to 14 times the pressure experienced at the center of the earth [15]. Materials subjected to such extreme conditions exhibit a wide array of unforeseen and novel effects including material or surface modification [16–18]; generation of multiply charged energetic ions [19], X-rays, and neutrons [20]; and initiation of photonuclear reactions like photofission and photonuclear transmutation, to name a few [21,22].

Understanding such laser–matter interaction at microscopic level is essential for optimizing the desired response of material upon laser irradiation, as well as for unraveling the basic mechanism of laser–matter interaction. In this regard, the matters of interest are atomic and molecular clusters. These clusters are considered as embryos of matter, which act as bridge in understanding the properties of materials as they evolve from atomic/molecular level to bulk (i.e., solid or liquid phase) level [23]. Also, these clusters provide an ideal medium for basic understanding of laser–matter interaction because of their finite size and bulk-like local density [24]. Thus clusters mimic macroscopic phase at microscopic level. In addition, their secluded existence in the gas phase facilitates probing of elementary behavior under well-defined conditions, unperturbed by external factors that reign in the condensed phase [25].

Another aspect that makes laser–cluster interaction fascinating is that clusters are generally metastable and fragile. Here the main attention is devoted to atomic and molecular clusters, which are held together by weak forces (van der Waals and hydrogen bonds) of energy ~ 0.05 eV or even less.

Subjecting such matter to extreme laser ionization conditions would be inquisitive, as for laser intensity $>10^{15}$ W/cm^2, the electric field (E) associated with the laser pulse matches or even exceeds the strength of the electric field experienced by the valence electrons in an atom or molecules [20]. The electric field associated with a laser pulse is given by the expression

$$E(\mathrm{V/cm}) = 27.5\sqrt{I(\mathrm{W/cm^2})} \tag{16.1}$$

Thus a laser pulse of intensity $\sim 10^{15}$ W/cm^2 is associated with an electric field of $\sim 10^9$ V/cm, which is comparable to the Coulombic field within the hydrogen atom. Survival of such facile clusters under such severe laser ionization conditions for adequate laser pulse duration, and subsequently getting transformed into extremely excited state of matter upon efficient energy absorption, is noteworthy.

Since the late 1990s, a large number of studies have been carried out dealing with interaction of clusters with lasers having intensity spanning from $\sim 10^9$ to 10^{20} W/cm^2. In the given intensity region, clusters exhibit several exotic properties arising as a result of efficient laser–cluster interaction resulting in generation of energetic multiply charged atomic ions, higher harmonics, X-rays, and neutrons. This chapter provides the underlying basic principles of cluster generation and characterization, followed by a brief introduction to the fundamentals of laser-cluster interaction. The chapter also highlights a wide array of studies carried out by us, as well as other research groups to probe different aspects of laser-cluster interaction, though the collection of results presented here is by no means complete and we apologize to all those whose work could not be cited. These studies are a stepping-stone for understanding the nitty-gritty of interaction of intense laser with bulk matter having host of potential applications in materials science, accelerator technology, plasma processing, etc., as discussed later.

2. DIFFERENT METHODS FOR PREPARATION OF CLUSTERS

Clusters are finite-size microscopic objects, which can be produced by either aggregating their smaller constituents (atoms and molecules) i.e., top-up approach or by splitting the corresponding larger bulk matter, i.e., bottom-down approach [8]. They can also be synthesized in solution, in the form of colloidal particles [26,27]. However, to carry out studies on pure clusters free from any matrix, it is essential to prepare clusters in gaseous phase. Hence, we will concentrate mainly on the preparation of clusters in gas phase by different methods as outlined below [28–30].

2.1 SUPERSONIC JET SOURCE

The dynamic method, which involves supersonic expansion of gases through a narrow nozzle into vacuum, provides a very suitable means for generating gas phase neutral clusters of atoms and molecules. Here, the basic process comprises adiabatic expansion of a pure gas or a gaseous mixture held at high stagnation pressure (P_0) though a small orifice of diameter d, into an evacuated volume having pressure in the range of 10^{-5}–10^{-6} Torr [31–33]. The two body collisions that transpire during expansion of the gases from higher pressure region to a lower pressure region transform the random motion of the gas into directed flow, thus reducing translation, rotational and vibrational temperature of the gaseous species to a few Kelvins [34–36]. In unison, three body collisions during the course of expansion, coupled with extremely low temperatures and high number density of gas lead to nucleation and condensation of clusters. Three body collisions play a vital role during the initial phase of cluster nucleation and growth, as the process of cluster formation is exoergic in nature and the third body is required for removal of excess energy from the system, in the form of kinetic energy [6]. The process of cluster formation can be schematically represented by the following equations:

$$\mathrm{M + M + X \rightarrow M_2 + X} \quad \text{Cluster nucleation} \tag{16.2}$$

$$\left.\begin{aligned} \mathrm{M_2 + M + X} &\rightarrow \mathrm{M_3 + X} \\ \mathrm{M_{n-1} + M + X} &\rightarrow \mathrm{M_n + X} \end{aligned}\right\} \quad \text{Cluster growth} \tag{16.3}$$

Here M represents the gaseous atom or molecule, while X represents a carrier gas. Generally, less condensable inert gases are used as carrier because they efficiently dispose of the condensation energy from the cluster system via collisions. Subsequently, the dimer acts as site for further growth and condensation of the cluster. Necessity of third-body participation in cluster growth is trivial for the later growth steps (i.e., for larger clusters), as the excess energy can be soaked by the cluster itself, particularly when many intracluster vibrational degrees of freedom are available to act as the sink. Using this method, clusters can be produced as a molecular beam. The degree of cluster formation can be controlled by changing the nozzle diameter (d), stagnation pressure (P_0), and initial gas temperature (T_0) [34]. The rate of three body collisions (Z_3) and in turn the cluster formation probability is related to these parameters by the following equation:

$$Z_3 \propto P_0^2 d / T_0^2 \tag{16.4}$$

Eq. (16.4) implies that cluster formation is facilitated by using large-diameter nozzles and large reservoir pressure held at low temperatures.

2.2 THE PICKUP SOURCE

The pickup technique is a convenient method for generation of embedded heterogeneous clusters, whereby a foreign atom/molecule/cluster is externally attached to the host cluster prepared by supersonic jet [37–39]. Here, the initial clusters generated upon supersonic expansion are crossed perpendicularly by a beam of foreign atom/molecule/cluster, at a certain distance downstream from the nozzle. Collision between the supersonically generated clusters and the externally directed atoms/molecules results in the formation of heterogeneous clusters. This method is widely used for preparation of metal-doped molecular clusters.

2.3 LASER ABLATION AND VAPORIZATION CLUSTER SOURCE

Both the methods mentioned before are well suited for preparation of cluster of volatile and low-melting compounds. However, for preparation of clusters of refractory materials that have low-number density even at higher temperatures, such as metals and ceramics, the direct method of supersonic jet is not suitable. For generating clusters of these refractory materials, the method utilized involves laser vaporization of refractory material coupled with supersonic expansion source [40,41]. Advent of this laser vaporization technique facilitated researchers to produce clusters of virtually any element in the periodic table and in turn ignited extensive interest in the studies of clusters with varying composition,

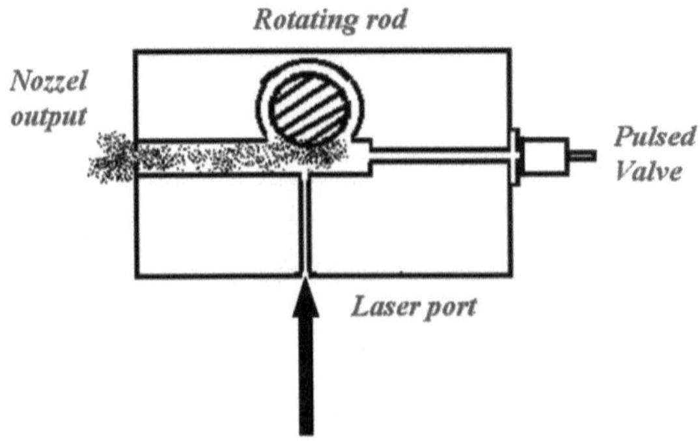

FIGURE 16.2

Schematic diagram of a Smalley-type cluster source.

regardless of their volatility constrain [42–44]. Laser vaporization technique provides high number density of vapors of virtually any material, in a short time interval and in a well-localized volume. Subsequently, by rapid quenching the laser-ablated plasma, using inert carrier gas, clusters can be produced. Hence, laser vaporization method has become one of the most popular techniques for generating cluster beams, especially of refractory materials [45]. In this method a pellet or rod of refractory material whose clusters are to be prepared is ablated with a focused laser pulse. This results in generation of plasma at the target surface having temperature of $\sim 10^4$K [46,47]. The resultant laser-eroded material (atoms/small clusters) is supersonically cooled using monoatomic inert gases, resulting in formation of clusters of refractory materials [48,49].

A similar setup was used by Smalley and coworkers [50] for generation and characterization of carbon cluster, formed by supersonic cooling of plasma produced upon laser ablation of graphite. The resultant carbon clusters produced were detected using time-of-flight mass spectrometry. In their experiment, Smalley et al. observed that under certain expansion conditions the mass peak corresponding to C_{60}^{+} ion was the most dominating peak in the mass spectra, which later led to discovery of Buckminster fullerene having a truncated icosahedral structure made up of 12 pentagonal and 20 hexagonal rings [51]. Hence, such a laser vaporization cluster source is also referred to as Smalley-type source (Fig. 16.2).

3. CHARACTERIZATION OF CLUSTER SOURCE

As discussed before, cluster formation process is statistical in nature. Consequently, in a cluster generation process clusters of a specific size are not produced. Generally, the clusters produced have a wide size distribution, which can be well approximated by a log-normal distribution function (Eq. (16.5)), and the method by which the clusters are produced has little effect on the cluster size distribution [52].

$$g(N) = \frac{1}{N\sqrt{2\pi\sigma^2}} \exp\left(-\frac{\ln^2(N/N_0)}{2\sigma^2} \right) \tag{16.5}$$

In Eq. (16.5), N is the number of atoms in a cluster, σ is the distribution width, and N_0 is the most likely number of atoms. The large width of cluster size distribution in cluster generation process presents a major difficulty when one aims to interpret experimental and theoretical results of a cluster system, as the properties of the cluster depend severely on its size. Hence, a robust theoretical model for predicting the properties of cluster beams (cluster size distribution, structure, etc.), given the source parameters (initial temperature and pressure, nozzle width), is essential. Generally, Raleigh scattering measurements in combination with an empirical scaling law [53] are used to estimate the mean cluster size [54]. Based on the empirical scaling law, the onset of clustering and size of clusters produced can be approximated by an empirical scaling parameter Γ^*, commonly referred to as Hagena parameter [36,55] and is given by:

$$\Gamma^* = k\frac{(d/\tan\alpha)^{0.85}}{T_0^{2.29}}P_0 \tag{16.6}$$

where d is the nozzle diameter (μm),α is the expansion half angle, P_0 is the backing pressure (mbar), T_0 is the preexpansion temperature (Kelvin), and k is a constant related to bond formation. Clustering generally begins for $\Gamma^* > 100-300$, with the number of atoms per cluster (N_c) scaling as $N_c \propto \Gamma^{*2.0-2.5}$ [53,56]. In terms of Hagena parameter, the average size of the cluster, i.e., number of atoms per cluster can be estimated as:

$$N_c \cong 33\left[\frac{\Gamma^*}{1000}\right]^{2.35} \quad (\Gamma^* \le 10^4) \tag{16.7}$$

$$N_c \cong 100\left[\frac{\Gamma^*}{1000}\right]^{1.8} \quad (\Gamma^* > 10^4) \tag{16.8}$$

This empirical procedure provides an estimate of average cluster size and also provides a certain degree of experimental control in regulating the cluster size distribution, by varying the nozzle size and the source parameters, i.e., temperature and pressure.

Electron diffraction is another useful method for characterization of cluster size. Electron diffraction studies also provide significant details regarding the cluster temperature and mean cluster geometry. In a typical electron diffraction experiment, the cluster beam is subjected to well-collimated electron beam with large kinetic energy (25–60 keV). The scattered electrons produce a diffraction pattern similar to that of Debye–Scherrer rings. The shape of the Debye–Scherrer rings provides information regarding the mean cluster size. This method has been utilized for determination of mean size of cluster in supersonic expansion of $(Ar)_n$, $(CO_2)_n$, etc. [57,58].

4. INSTRUMENTATION

Having described different methods for generation of clusters, we would like to mention that production and characterization/analysis of clusters are interrelated and a single cluster setup houses the clusters generation source as well as the analyzer and the detector for characterization of the cluster properties. Fig. 16.3 illustrates a schematic diagram of the cluster setup designed and fabricated

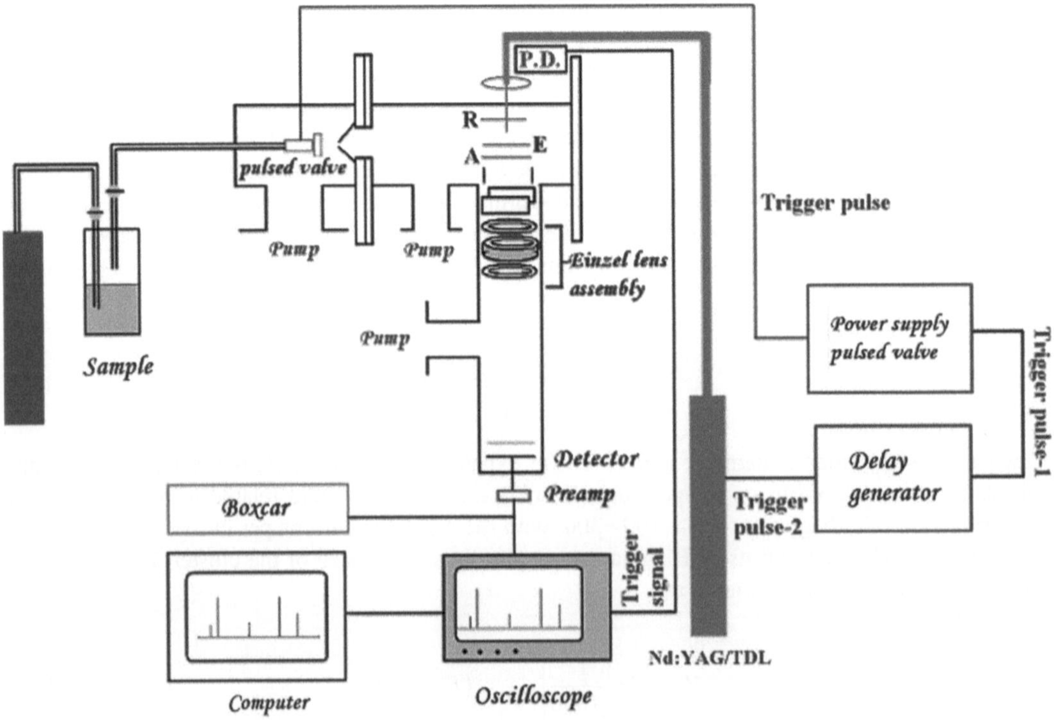

FIGURE 16.3

Schematic layout of cluster setup designed and fabricated at Chemistry Division, Bhabha Atomic Research Center, Mumbai. Here, carrier gas is bubbled through the sample of interest and is supersonically expanded through the pulse valve. The gas pulse is skimmed and at an appropriate delay laser pulse is triggered for ionization of clusters. The ions produced upon laser-cluster interaction are detected using time-of-flight mass spectrometer, and the data is processed using associated electronics.

indigenously at Chemistry Division, Bhabha Atomic Research Center, Mumbai, while Fig. 16.4 gives a pictorial view of the same setup. Here a pulsed valve mounted in the expansion chamber acts as a cluster source for generation of atomic and molecular clusters. In the present setup for formation of clusters, a carrier gas, typically an inert gas (helium or argon) at a stagnation pressure of 1–8 bar is bubbled through the liquid sample. The resultant gas mixture is supersonically expanded through a pulsed nozzle, which results in the formation of clusters having a wide size distribution. The supersonic jet so produced is skimmed with the help of a skimmer (a truncated cone), and a cold beam of clusters is introduced into the analyzer chamber. In the analyzer chamber, these clusters can be investigated by various experimental techniques—as the topic of cluster research is vast and diversified with different scientific and technological significance. Since interest of our group centers around understanding behavior of clusters upon interaction with laser pulses of varying laser intensity and wavelength, we will restrict discussion to this aspect of cluster studies. In our case, we irradiate the clusters with laser photons and analyze the resulting electrons and ions produced. Accordingly in

FIGURE 16.4

Pictorial view of cluster setup designed at Chemistry Division, Bhabha Atomic Research Center, Mumbai.

the present setup, clusters are subjected to focused laser pulses and the resultant cluster photofragments and ions produced upon multiphoton dissociation/ionization of cluster are identified using time-of-flight mass spectrometer [59,60].

In order to appreciate the behavior of clusters upon interaction with lasers, it is pertinent that we first understand how individual atoms and molecules respond when subjected to laser pulses. Upon aggregation, clusters of these atoms and molecules exhibit complex behavior as compared to that exhibited by their individual constituents. In the following section, a brief introduction is provided regarding the basics of interaction of laser with atoms and molecules. This provides us a foundation to understand the mechanism of laser-cluster interaction, as discussed later.

5. CLUSTERS UNDER EXTREME IONIZATION CONDITIONS: INTERACTION WITH LASERS

The process of interaction of intense laser pulses with clusters is a rather complex photophysical/photochemical process. Thus in order to understand how clusters undergo excessive ionization under the influence of laser field (under extreme laser conditions), one should first appreciate the ionization processes that might transpire among individual atoms and molecules, which act as building blocks of cluster.

5.1 BRIEF INTRODUCTION TO INTERACTION OF LASER WITH ATOMS AND MOLECULES

Atoms and molecules, upon interaction with conventional low-intensity ($\sim$mW/cm^2) light sources, may undergo ionization provided the photon energy of the light source exceeds the ionization energy of the

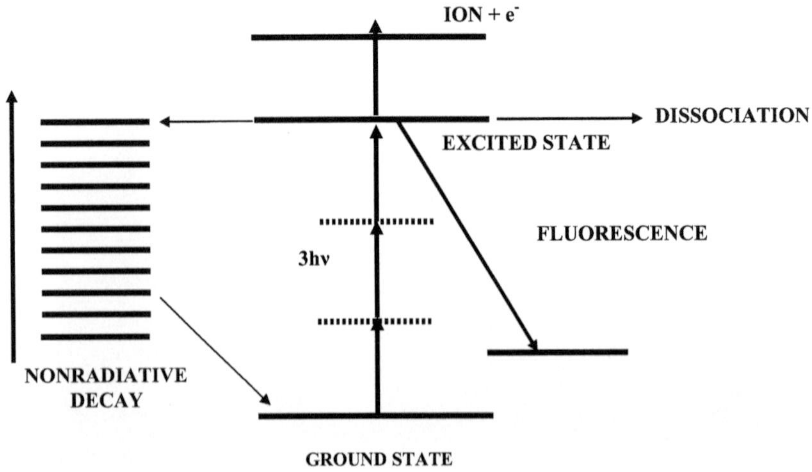

FIGURE 16.5

The different competing channels, i.e., intramolecular relaxation, fluorescence, dissociation, and ionization, which transpire upon laser multiphoton excitation of molecule. Here the molecule depicted exhibits (3 + 1) multiphoton ionization process. Dotted line (......) represents virtual intermediate state.

system under investigation. While in the case of atoms/molecules subjected to intense laser pulses ($\geq 10^7$ W/cm^2), even photons with lower energy may ionize an atom/molecule through a process involving many photons [61–65]. Primarily, interaction of atoms and molecules with laser pulses having photon energy much smaller than the ionization potential leads to their excitation by absorption of one or more photons provided certain optical selection rules are satisfied. These excited atoms or molecules would normally decay back to ground state by emission of radiation or undergo dissociation (in case of molecules). However, if the laser intensity is sufficiently high, then these excited atoms/molecules, before decaying, can absorb additional photons from the laser pulse and reach higher excited states or may even get ionized (Fig. 16.5). This process is called nonlinear ionization, since the ionization of atoms/molecules shows a strong nonlinear dependency on laser intensity. The mode by which these atoms and molecules undergo nonlinear ionization strongly depends on the laser wavelength and intensity. For example, based on the intensity regime, these nonlinear ionization processes have been loosely divided into multiphoton ionization (MPI) and optical field ionization (OFI) regime. The latter is a general term for different types of ionization mechanisms observed at very high laser intensity ($\geq 10^{13}$ W/cm^2), namely, tunneling ionization, above threshold ionization and barrier suppression, which are induced by the electric field of the laser pulse. At low laser intensity, when the electric field associated with the laser pulse is much smaller compared to atomic field strength, atoms and molecules undergo multiphoton ionization via absorption of multiple photons via virtual intermediate excited states. In such cases the ionization rates are significantly enhanced if the multiphoton excitation is facilitated by any intermediate resonant excited state of atom/molecule (Fig. 16.6). Under such laser intensity conditions, the Coulomb field of the system is perturbed negligibly.

While at higher laser intensities and lower frequencies, where the electric field associated with the laser pulse is large enough to distort the Coulombic potential, atoms/molecules undergo optical field

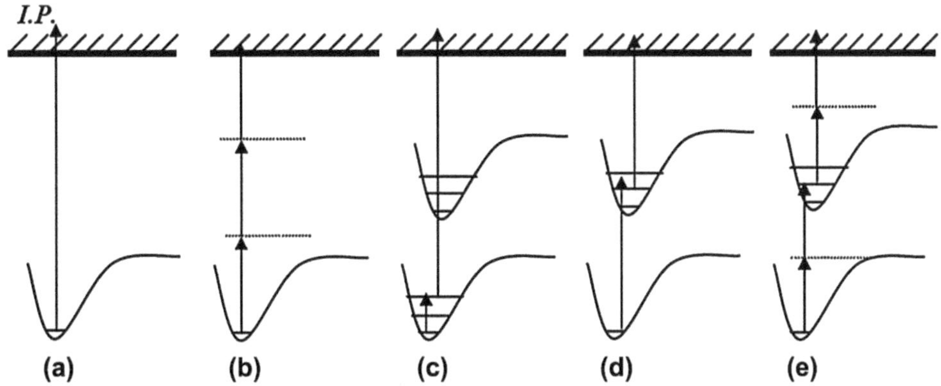

FIGURE 16.6

Different ionization schemes for a molecule. (a) Single photon ionization. (b) Nonresonant multiphoton ionization – three photons are involved. (c) Resonant two-photon ionization via ground state rovibrational level. (d) Resonant two-photon ionization via excited state rovibrational levels. (e) Resonance enhanced multiphoton ionization – (2 + 2) REMPI. Dashed line (......) denotes the virtual intermediate states.

ionization (Fig. 16.7). A question that normally arises is which mechanism is responsible for ionization for a given set of experimental conditions? A useful parameter, γ, the Keldysh adiabaticity parameter, is usually used by experimentalists to separate these nonlinear multiphoton processes into two regimes, MPI and OFI:

$$\gamma = \sqrt{\frac{I.E.}{2U_p}} \tag{16.9}$$

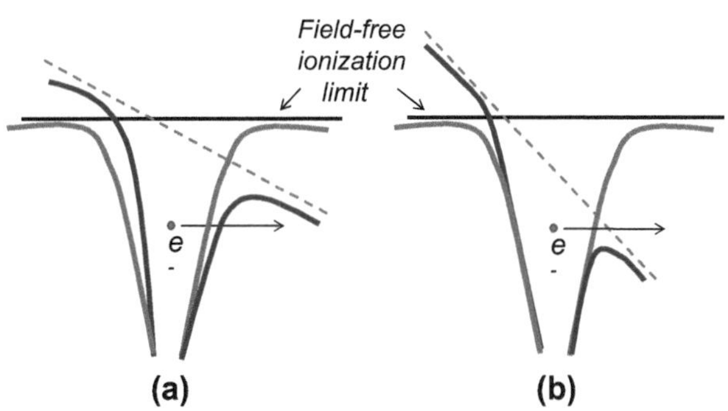

FIGURE 16.7

Illustration depicting field ionization processes, i.e., (a) Tunneling and (b) barrier suppression ionization. Red solid curve (*gray in print version*) represents combined potential of atom and laser field (————). Blue solid line (*light gray in print version*) represents atomic potential under zero field.

where *I.E.* is ionization energy of the species and U_p is the ponderomotive energy, i.e., the cycle-averaged energy of free-electron oscillations. When $\gamma > 1$ the ionization is attributed to MPI processes, while for $\gamma < 1$ ionization is attributed to OFI [66]. However, there is no clear separation between these two regimes as for $\gamma \sim 0.3$ MPI processes have been observed [67] and for $\gamma > 3$ OFI has been reported for benzene molecules [68]. Moreover, for sufficiently higher laser intensity, the maximum of the barrier drops lower than the energy level of the electron. As a result, the bound state for the electron no longer exists and the electron is ionized. This ionization process is called over-the-barrier ionization or barrier-suppression ionization.

5.2 CLUSTERS UNDER INTENSE LASER PULSES

When atoms and molecules aggregate to form clusters, novel properties emerge in relation to their interaction with intense laser pulses. These properties depend not only on the type of atoms and molecules that constitute the cluster but also on the cluster size and laser parameters. Though the concept of interaction of atoms and molecules with laser can be extended to clusters, additional collective behavior of cluster constituents induces complexity in nature of laser-cluster interaction. One such property is the efficient absorption of energy upon intense laser illumination, which ultimately leads to enhanced ionization of the cluster constituents, as compared to their isolated atomic/molecular counterpart and bulk matter. Such effective coupling of laser with the cluster media is facilitated by bulk-like local density of the cluster media coupled with its finite size, which restrains the absorbed laser energy to be dissipated, and is not feasible in case of bulk matter [69]. This absorbed energy leads to several exotic outcomes like generation of multiply charged energetic atomic ions, electrons, and higher-harmonic photons and neutrons. In general, observation of these interesting consequences can be broadly ascribed to efficient coupling of intense laser pulse with the cluster system and the subsequent Coulomb explosion phenomena that transpires [70].

5.2.1 What Is a Coulomb Explosion Phenomenon?

Phenomenon of Coulomb explosion occurs when intense field of laser extricates several electrons from the cluster upon interaction, causing the constituents of the cluster to become extremely ionized [71,72]. The constituents of ionized-charged cluster in turn experience Coulomb repulsion, due to close proximity of ionic charges within the cluster. During the time span of the laser pulse, the ionized cluster endures and further extracts energy from the laser pulse. But once the Coulombic repulsive forces amid the highly charged cluster constituents exceed its total cohesive energy, the cluster explodes (disintegrates), resulting in generation of multiply charged atomic ions associated with large kinetic energy and energetic electrons [19–22].

Coulomb explosion of atomic and molecular clusters has been reported over a broad intensity range varying from 10^9 to 10^{20} W/cm^2 and spanning over wide wavelength range, using laser pulses of varying pulse duration, i.e., nanosecond, picosecond, and femtosecond. However, the potency of Coulomb explosion phenomenon is higher for ultrashort laser pulses of duration $\leq$100 fs, with intensities in the range of 10^{14}–10^{20} W/cm^2. Such behavior of clusters under intense laser conditions, arising due to efficient coupling of laser energy with cluster medium, cannot be explained solely by typical atomic and molecular ionization processes, i.e., multiphoton ionization, optical field ionization, etc., discussed earlier. Besides, further complexity emerges due to cohesive behavior of cluster constituents subjected to laser pulses, having diverse array of wavelength, intensity, and pulse duration [73,74]. As a result, several models have been developed to elucidate behavior of clusters, under the influence of intense laser fields. The next section provides a brief description of the different models proposed to unravel the mechanism of laser-cluster interaction.

5.2.2 Different Models Proposed to Understand Mechanism of Laser-Cluster Interaction

In order to simplify the mechanism of laser-cluster interaction, it is convenient to decouple the process of intense laser-cluster interaction into three stages, i.e., initial inner ionization of the cluster, followed by severe heating of the cluster constituents, and finally explosion of the cluster [75]. In the initial phase, it is considered that the cluster constituents get ionized by either multiphoton or optical field ionization processes at the leading edge of the laser pulse, depending on the laser intensity and wavelength. Consequently, a certain number of ionized electrons produced with sufficient energy leave the cluster. While the ionized electrons that remain confined within the cluster, due to its positive ion background, are called quasi-free electrons as they are free to stray within the cluster. As a result, after first stage, cluster nanoplasma is formed, which constitutes of ions and quasi-free electrons.

In the second stage, the cluster nanoplasma interacts and efficiently absorbs energy from the residual portion of the laser pulse and simultaneously undergoes expansion. Several theoretical models have been proposed to account for efficient absorption of laser energy by the ionized cluster media, at this stage, i.e., ionization ignition model (IIM), coherent electron motion model (CEMM), charge resonance-enhanced ionization (CREI) and the nanoplasma model. Rose-Petruck and coworkers proposed IIM, based on classical trajectory Monte Carlo simulations [76,77]. The model considers that, upon initial ionization of cluster constituents several ionic centers are generated within the cluster, while the ionized electrons are instantly ejected from the cluster by the laser field. As a result, the unshielded ion core in the cluster experiences a very strong and inhomogeneous electric field. This strong electric field coupled with the laser field lowers the ionization barrier of the cluster constituents, facilitating further ionization of cluster constituents. In turn, the effective field increases and further lowers the ionization barrier. Thus the field created by the initial ionization process "ignites" the cluster and facilitates further ionization processes. While CEMM proposes that under the influence of laser field enhanced ionization of cluster occurs through strong electromagnetic coupling between the laser field and clusters, due to coherent motion of the ionized electrons [78–80]. These coherently coupled electrons behave like quasi-particles and get energized by oscillating in the external laser field. These coherently energized electrons initiate subsequent ionization of cluster by stripping electrons, even from the inner shells. On the other hand, CREI model suggests that for sufficiently long laser pulse, upon expansion of cluster, as the internuclear distance between the charged bodies increases, at a critical internuclear separation the frequency of the electron motion in the cluster potential resonates/matches with the frequency of the laser radiation. As a result, the energy of the electron is enhanced through its interaction with the laser light, which subsequently leads to enhanced ionization of the cluster [81–83]. The nanoplasma model considers that solid-like density of cluster facilitates rapid energization of inner-ionized electrons by collisional inverse Bremsstrahlung process, under the influence of laser pulse [84,85], thus causing enhanced ionization of the cluster constituents.

Eventually, in the third and final stage, when the laser pulse has ceased (in case of ultrashort intense laser pulses) or the Coulomb repulsive potential of the ionized cluster overcomes the cohesive energy, the cluster disintegrates producing energetic electrons, multiply charged atomic ions and photons. The recombination of quasi-free electrons with the highly ionized cluster constituents leads to generation

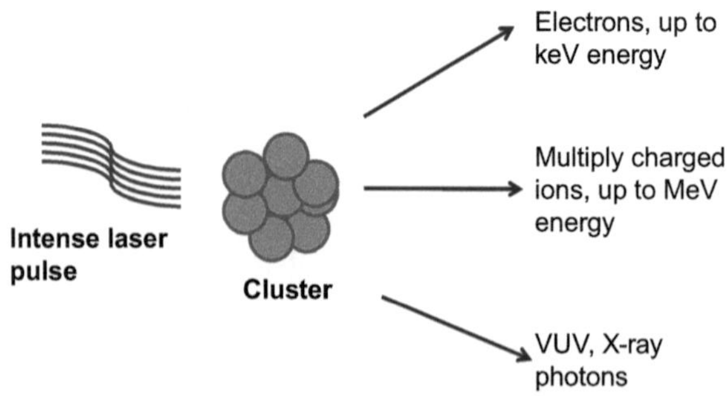

FIGURE 16.8

Illustration depicting different outcomes upon interaction of intense laser pulse with cluster.

of X-rays. In general, all the three stages, i.e., inner ionization, outer ionization, and Coulomb explosion, are interrelated with each other [86] (Fig. 16.8).

Despite the development of several models as discussed earlier, a single model alone could not explain all the aspects of the laser cluster interaction leading to Coulomb explosion and other associated processes. Each of these models explains some part of the overall interaction process, which the cluster undergoes after being irradiated by an intense laser pulse. For example, the IIM explains the generation of multiply charged energetic atomic ions. This model also explains the wavelength dependence of the Coulomb explosion phenomena, as it depends on the ease with which the initial ion core is generated. On the other hand, the coherent electron motion model proposed by Rhodes and coworkers explains the generation of X-rays as a result of interaction of rare gas clusters with intense laser pulses. Generally, it is considered that behavior of small clusters under the influence of strong laser field is governed by a combination of CREI and IIM, while for large cluster systems the nanoplasma model is expected to be a dominant process.

6. DIFFERENT STUDIES RELATED TO INTERACTION OF LASER WITH CLUSTERS

Interaction of laser with clusters depends on several parameters, i.e., cluster size, cluster composition, laser wavelength, laser intensity, laser pulse duration, etc., to name a few. In the following section, we discuss different aspects of laser-cluster interaction categorized as per the interaction of clusters with laser pulses of different pulse regime.

6.1 INTERACTION OF CLUSTERS WITH LASER PULSES OF NANOSECOND PULSE DURATION

Historically, a large number of studies have been carried out wherein interaction of atomic and molecular clusters with nanosecond laser pulses having laser intensity $\sim 10^8 - 10^{11}$ W/cm^2 has been investigated [87–91]. These studies reported regarding generation of different cluster fragment ions following ionization, arising as a result of complex intracluster photochemistry. As discussed earlier, based on Keldysh parameter, the ionization process of clusters under nanosecond laser pulse irradiation falls in the MPI regime ($\gamma \gg 1$). Several studies carried out on different molecular cluster systems have reported intracluster and concerted reactions, which depend on degree of aggregation of the cluster as well as on the mechanism of energy dissipation and transfer. For example, upon multiphoton ionization, ammonia clusters undergo an internal ion–molecule reaction leading to abundance of protonated cluster ions in the mass spectra along with its analogues arising from concomitant evaporative unimolecular dissociation as shown below [92]:

$$\begin{aligned}
(NH_3)_m + nh\nu &\rightarrow NH_3^+(NH_3)_{m-1} + e^- \\
NH_3^+(NH_3)_{m-1} &\rightarrow \left[NH_4^+(NH_3)_{m-2}\right]^* + NH_2 \\
\left[NH_4^+(NH_3)_{m-2}\right]^* &\rightarrow \left[NH_4^+(NH_3)_{m-2-x}\right]^* + xNH_3
\end{aligned} \tag{16.10}$$

Similarly concerted photochemistry has also been reported for molecular cluster systems. For example, in case of methyl iodide clusters, subjected to 266 nm laser pulses, formation of I_2^+ ions has been observed. Generation of I_2^+ ion has been ascribed to concerted reaction within $(CH_3I)_n^+$ cluster [93,94]. Thus clusters exhibit novel reaction channels, which were inaccessible for individual molecules. On the other hand, for studies carried out at longer laser wavelength, methyl iodide clusters exhibited phenomena of Coulomb explosion at laser intensities $\sim 10^9$ W/cm^2, which resulted in generation of energetic multiply charged atomic ions of carbon and iodine up to C^{3+} and I^{3+} (Fig. 16.9) [95,96]. Observation of Coulomb explosion phenomena at such lower laser intensities is rather unusual, as the minimum laser intensity required for the appearance of these multiple charged ions, based on Auguste's formula (Eq. (16.11)), is estimated to be $\sim 10^{14}$–10^{15} W/cm^2, which is at least four orders of magnitude higher than the intensity used in the nanosecond laser experiments.

$$I(\mathrm{W/cm^2}) = 4 \times 10^9 \frac{(I.E.)^4}{Z^2} \tag{16.11}$$

This equation is based on simple Columbic barrier model [98], where *I.E.* is the ionization energy of the multiply charged ion (in eV) and *Z* is charge of the multiply charged ion.

Several other cluster systems were also found to exhibit the phenomenon of Coulomb explosion, as a function of laser wavelength and cluster size at laser intensity of $\sim 10^9$–10^{11} W/cm^2 [19,99–104]. Some of the representative experimental findings are depicted in Figs. 16.10 and 16.11. Based on these results it can be ascertained that the phenomenon of Coulomb explosion is comparatively more pronounced at longer laser wavelengths. As discussed earlier, under laser intensity condition of $\sim 10^9$ W/cm^2 multiphoton ionization process is expected to be the dominant laser absorption process. However, generation of multiply charged atomic ions (Xe^{10+}) requiring *I.E.* as high as 202 eV cannot be explained by multiphoton absorption process alone [105]. As the models discussed in Section 5.2 are mainly applicable for interaction of cluster with ultrashort laser pulses of intensity $\sim 10^{14}$–10^{20} W/cm^2, a variant of nanoplasma model has been proposed to account for generation of multiply charged atomic ions, under nanosecond laser intensity conditions. A brief outline of the proposed model follows.

6.1.1 Three-Stage Cluster Ionization Model

The three-stage cluster ionization model, i.e., "multiphoton ionization ignited-inverse Bremsstrahlung heating–electron impact ionization" has been proposed to explain different aspects of Coulomb

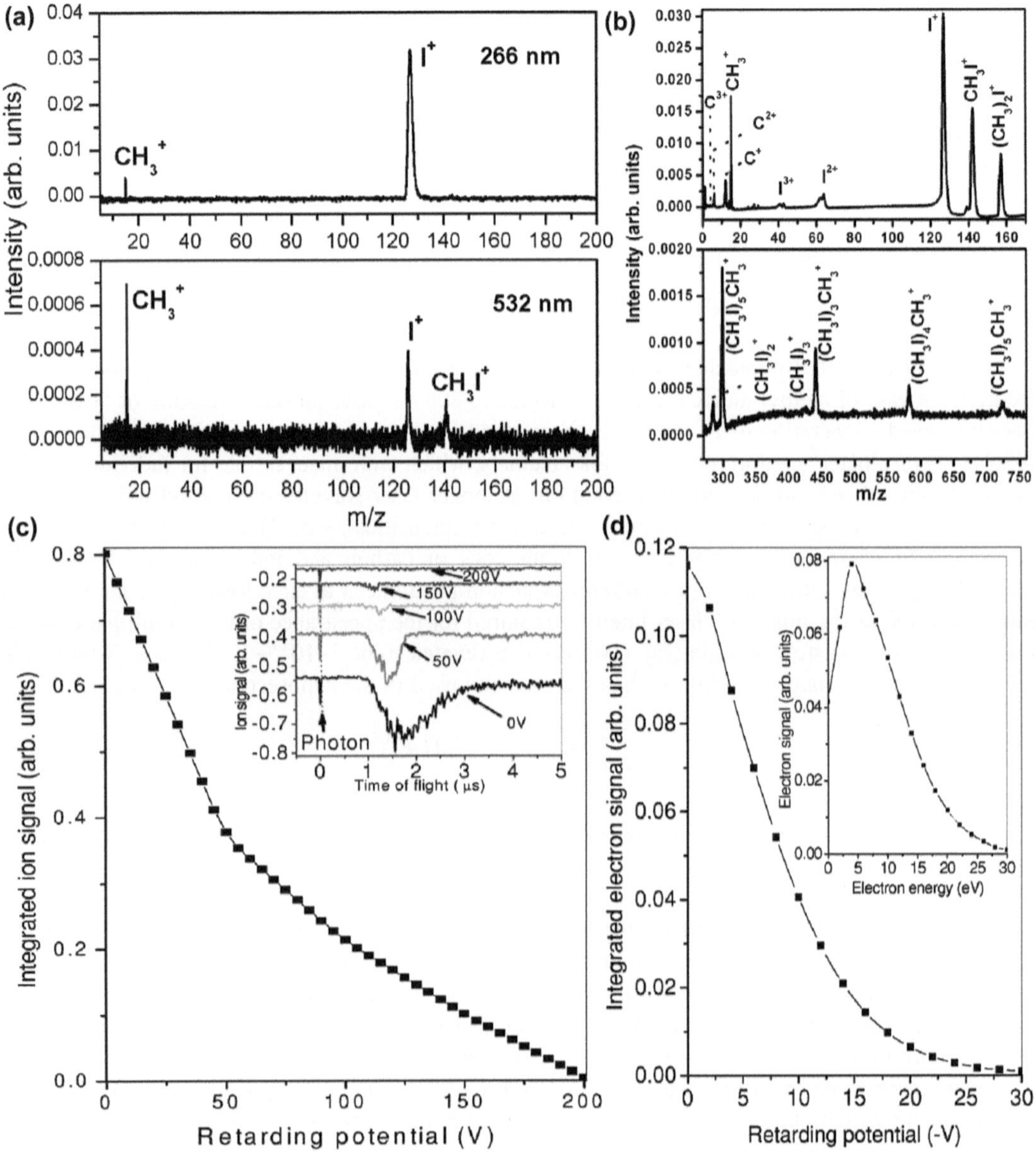

FIGURE 16.9

Upon interaction with laser pulses of nanosecond pulse duration (a) methyl iodide molecules exhibit mutiphoton dissociation/ionization at 266 and 532 nm [97]. While in case of clusters (b) generation of multiply charged atomic ions and clusters fragments has been observed at 532 nm. Additionally, measurement of kinetic energy of (c) ions and (d) electrons carried out using retardation field method reveal regarding generation of energetic ions and electron, upon interaction of $(CH_3I)_n$ clusters with 532 nm laser pulses of intensity $\sim 10^9$ W/cm^2 [96].

(a) Reproduced with permission of John Wiley and Sons. (C) &(d) Reproduced with permission of AIP Publishing LLC.

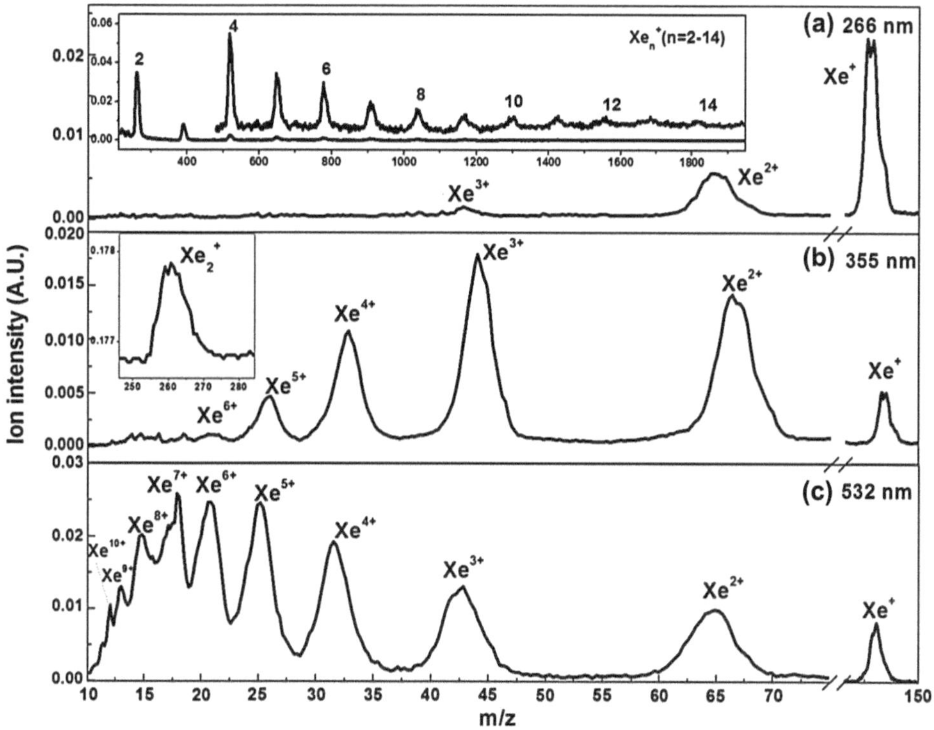

FIGURE 16.10

Time-of-flight mass spectra of xenon clusters subjected to (a) 266 nm, (b) 355 nm, and (c) 532 nm laser pulses of intensity ~5 × 10^9 W/cm^2. Cluster fragments at different wavelengths are shown in the inset [105].

Reproduced with permission of Elsevier.

explosion phenomenon, upon interaction of clusters with nanosecond laser pulses of intensity ~10^9–10^{11} W/cm^2 [106,107]. The model considers that the nanosecond laser-induced Coulomb explosion of cluster is initiated by MPI of the neutral cluster constituents, at the leading edge of the laser pulse. Subsequently, some of the ionized electrons that are caged inside the cluster (quasi-free electrons) extract energy from the laser pulse via inverse bremsstrahlung (IBS) process, i.e., by absorption of energy from the electromagnetic laser field under the influence of neutral and ionic species within the cluster. Once these caged electrons have acquired adequate energy, depending on the energy gained by the electron, they may cause further ionization of the cluster species (i.e., ions and neutrals) via stepwise electron ionization. This leads to efficient coupling of laser energy with the cluster system and causes excessive ionization of the cluster constituents. Eventually, the ionized clusters undergo Coulomb explosion due to strong electrostatic repulsion, resulting in generation of energetic multiply charged atomic ions and electrons. The rate of energy extraction by the confined electrons via IBS process, from the laser pulse, is given by Eq. (16.12) [108,109]:

$$\frac{dE}{dt} = U_p \times \nu \tag{16.12}$$

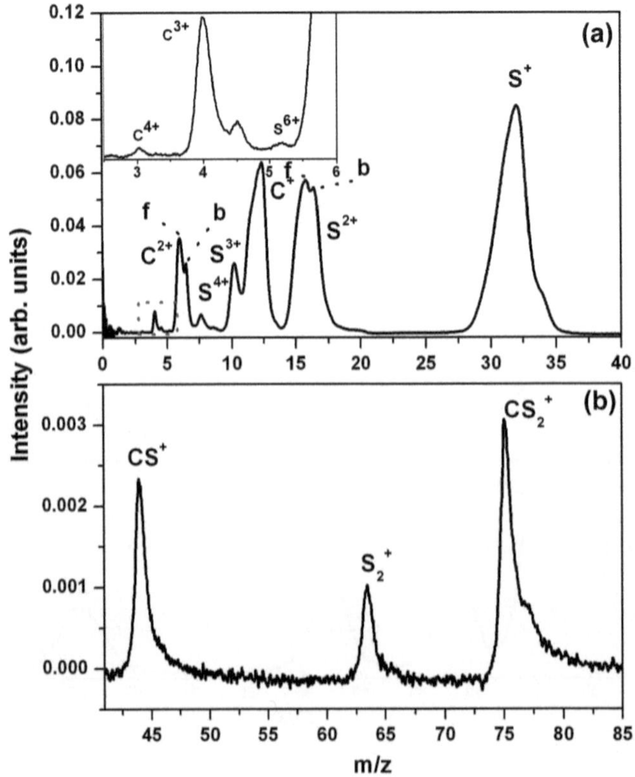

FIGURE 16.11

Time-of-flight mass spectra of CS_2 clusters at 532 nm with laser intensity ~4.6 × 10^9 W/cm^2. The spectra has been divided into two mass regions: (a) m/z 0–40 and (b) m/z 40–85. Inset of (a) shows the enlarged portion in the mass range m/z 2–6. F and b denote the forward and backward component of given mass peak. The separation between the f and b component for a given mass peak can be correlated with the kinetic energy acquired by the ion upon Coulomb explosion [99].

Reproduced with permission of John Wiley and Sons.

Here ν is the collision frequency of the inner-ionized electrons and is of the order of $\sim 10^{14}$–10^{15} Hz (i.e., order of laser frequency) and ponderomotive energy (U_p) is given by Eq. (16.13):

$$U_p = 9.33 \times 10^{-14} I(\mathrm{W/cm^2})\lambda^2(\mu\mathrm{m})^2 \quad (16.13)$$

From Eqs. (16.12) and (16.13), it is obvious that the total energy gained by the inner-ionized electrons, in the total time span starting from initial ionization until the disintegration of cluster, is dictated by the product of ponderomotive energy and the total number of effective electron-ion/neutral collision frequency. In addition, Eq. (16.13) suggests that as the wavelength (λ) increases, ponderomotive energy (U_p) of electrons increases quadratically for a given laser intensity. Hence, higher level of ionization is expected at longer laser wavelengths. Also, it emerges from the model that clusters of sufficiently larger size are essential for confinement of quasi-free electrons, which would facilitate efficient coupling of laser energy with the cluster system. A concise overview of systematic studies carried out by our group that support the proposed three-stage cluster ionization model follows.

6.1.2 Effect of Laser Wavelength on Enhanced Ionization of Cluster Constituents at Laser Intensity of 10^9 W/cm^2

Fig. 16.12 represents mass spectra obtained upon irradiation of $(CH_3I)_n$ cluster with laser pulses of different wavelength, at a laser intensity of $\sim 5 \times 10^9$ W/cm^2. All the studies were carried out under identical expansion conditions, i.e., considering that average cluster size remains the same.

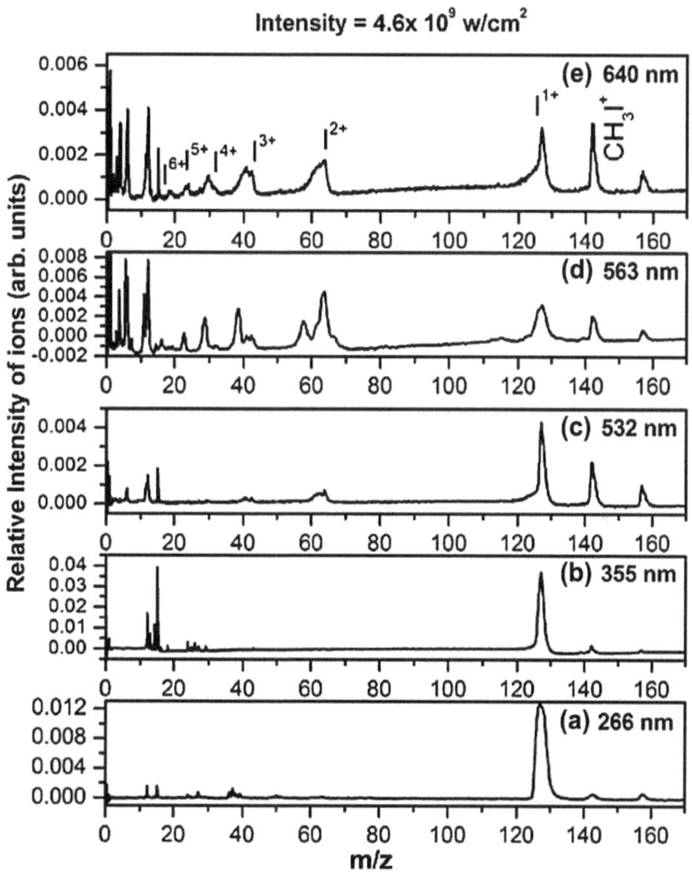

FIGURE 16.12

Time-of-flight mass spectra of methyl iodide clusters exposed to different wavelengths: (a) 266 nm, (b) 355 nm, (c) 532 nm, (d) 563 nm, and (e) 640 nm. The expansion conditions were identical for different sets of experiments [96].

Reproduced with permission of AIP Publishing LLC.

As illustrated in the figure, at 266 and 355 nm $(CH_3I)_n$ clusters undergo usual multiphoton ionization. The mass spectrum exhibits ion signal corresponding to singly charged fragment ions, i.e., CH_3^+, I^+, and CH_3I^+. In addition, cluster fragment ion $CH_3(CH_3I)^+$ is also observed, though no evidence for generation of multiply charged atomic ions of carbon or iodine was observed (Fig. 16.12a and b). At 532 nm, low-intensity peaks corresponding to I^{2+}, I^{3+}, and C^{2+} are observed (Fig. 16.12c), along with other fragment ions. While at 563 nm multiply charged atomic ions up to I^{5+} and C^{4+} are observed Fig. 16.12d. Upon further increasing the laser wavelength to 640 nm, the mass spectrum was similar to that obtained at 563 nm, except that an additional ion signal of I^{6+} (*I.E.*−78.84 eV) also starts appearing (Fig. 16.12e). Thus, these studies demonstrate that Coulomb explosion of $(CH_3I)_n$ clusters under our experimental conditions has an onset laser wavelength and is enhanced at longer wavelengths, in line with the proposed three-stage model (Eq. (16.13)). Similar experimental findings were obtained for studies carried out on xenon clusters that were subjected to different laser wavelengths (Fig. 16.10).

6.1.3 Role of Clusters in the Coulomb Explosion Process

To demonstrate experimentally that clusters of adequate size are essential for observation of multiply charged ions in the time-of-flight mass spectra, due to the ability of large-sized clusters to confine inner-ionized electrons for sufficient duration for extraction of laser energy, studies were carried out on clusters of varying size. As discussed earlier, the average cluster size depends on the diameter of nozzle utilized for supersonic expansion (Eqs. (16.4) and (16.6)). Thus in a set of experiments, studies were carried out on acetone cluster, using nozzles of two different sizes, as larger-sized nozzle would lead to enhancement in the average cluster size of acetone clusters. Figs. 16.13a and b illustrate time-of-flight mass spectrum of acetone clusters subjected to 532-nm laser pulses of intensity $\sim 4.3 \times 10^9$ W/cm^2, using 0.5- and 0.8-mm-diameter nozzle, respectively. As can be inferred from the figure, clusters generated using a 0.5-mm nozzle exhibit usual multiphoton dissociation/ionization behavior resulting in the generation of fragment ions like CH_3CO^+, $(CH_3COCH_3)_n^+$ ($n = 1-4$), $(CH_3COCH_3)_nCH_3CO^+$ ($n = 1-3$), etc. In contrast, acetone clusters generated using a 0.8-mm nozzle exhibited the phenomena of Coulomb explosion resulting in the generation of multiply charged atomic ions of carbon and oxygen, C^{m+}($m = 1-4$) and O^{n+}($n = 1-3$), in addition to other fragment ions [100]. Thus it can be concluded that clusters of larger size are essential for efficient coupling of laser energy with the cluster system, which consequently leads to extreme ionization of cluster constituents.

6.1.4 Role of Inner Ionized Electrons in Enhanced Ionization of Cluster Constituents Under Giga-watt Laser Intensity Pulses

As discussed in above sections, longer laser wavelength and larger cluster size facilitate enhanced ionization of cluster constituents, under gigawatt laser conditions. This qualitatively concurs the three stage cluster ionization model. In order to further substantiate the proposed mechanism, it was pertinent to probe the fate of inner-ionized electrons produced upon subjecting clusters to nanosecond laser pulses. Thus, kinetic energy of electrons ejected upon interaction of methyl iodide clusters with nanosecond laser pulse was measured at different laser wavelengths. As a rule, it is considered that upon simple multiphoton ionization the kinetic energy of electrons (E_e) is estimated by the relation $E_e = nh\nu - I.E.$, where n is the number of photons of laser frequency ν required to cause ionization of the species and *I.E.* is the ionization energy of the species. For methyl iodide clusters ionization energy is expected to be $\leq$9.54 eV (as *I.E.* of methyl iodide clusters is expected to be lower than that of its constituent monomer and the lowering in *I.E.* depends on the degree of clusterization). At 532 nm,

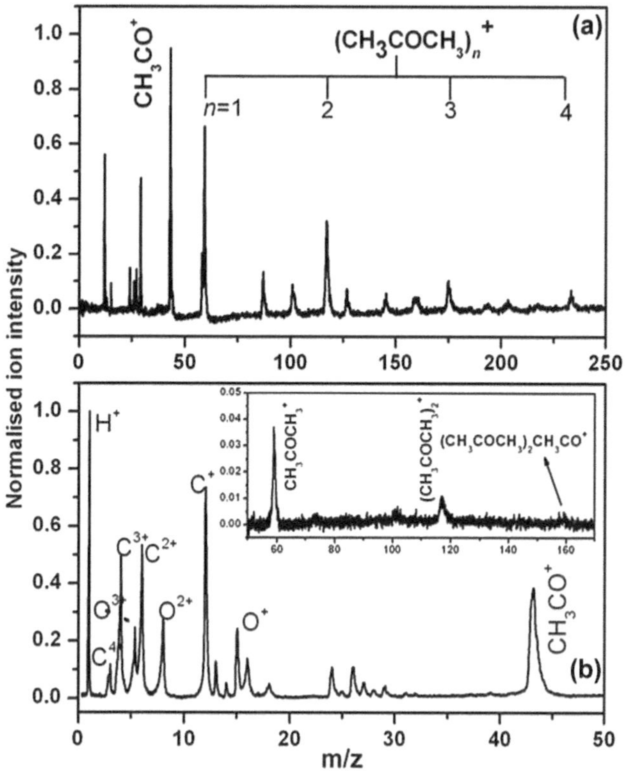

FIGURE 16.13

Time-of-flight mass spectra of acetone clusters subjected to ~4.5 × 10^9 W/cm^2 laser intensity pulses at 532 nm (a) using a 0.5-mm-diameter nozzle and (b) using a 0.8-mm-diameter nozzle at 2-atm helium backup pressure. The inset shows the cluster fragment ions [100].

Reproduced with permission of IOP Publishing.

based on laser power dependency studies the number of photons (n) required to cause multiphoton ionization of methyl iodide clusters has been estimated to be ~5. Thus it was expected that electrons generated upon multiphoton ionization of methyl iodide cluster would acquire kinetic energy ≤2.11 eV. However, kinetic energy distribution measurement studies carried out on methyl iodide clusters using retarding field method reveal that upon interaction with 532-nm laser pulses electrons with kinetic energy up to 30 eV are generated, upon laser-cluster interaction (Fig. 16.9d). Observation of electrons with kinetic energy up to 30 eV provides conclusive proof regarding energization of inner-ionized electrons as proposed in the three-stage cluster ionization model [96]. Further, for studies carried out at different laser wavelengths, researchers have observed enhancement in kinetic energy of electrons at longer laser wavelengths, which further confirms the proposed mechanism [107].

To further cross-check the electron heating mechanism, experiments were also carried out on clusters doped with species having comparatively higher ionization energy. Since ionization by quasi-free electrons is expected to be indiscriminate, it would also facilitate ionization of species having higher ionization energy. Thus studies were carried out on water-doped methyl iodide clusters at 563 nm. Here water was chosen as a dopant because of its higher ionization energy (12.62 eV) as compared to that of methyl iodide (9.54 eV). Also for studies carried out on pure water clusters under our experimental conditions, no ion signal could be observed in the time-of-flight mass spectrum, probably due to lower multiphoton ionization probability of the water. However, when water-doped methyl iodide clusters were subjected to 563-nm laser radiation, in addition to multiply charged atomic ions of carbon and iodine, multiply charged atomic ions of oxygen (up to O^{4+}) were also observed (Fig. 16.14). Thus, observation of multiply charged atomic ions of oxygen clearly demonstrates the significance of inner-ionized electrons in laser-induced ionization of cluster constituents, under nanosecond laser conditions [110]. Thus, the previous set of studies provides concrete evidence regarding role of inner-ionized electrons and their energization under the influence of intense laser pulses.

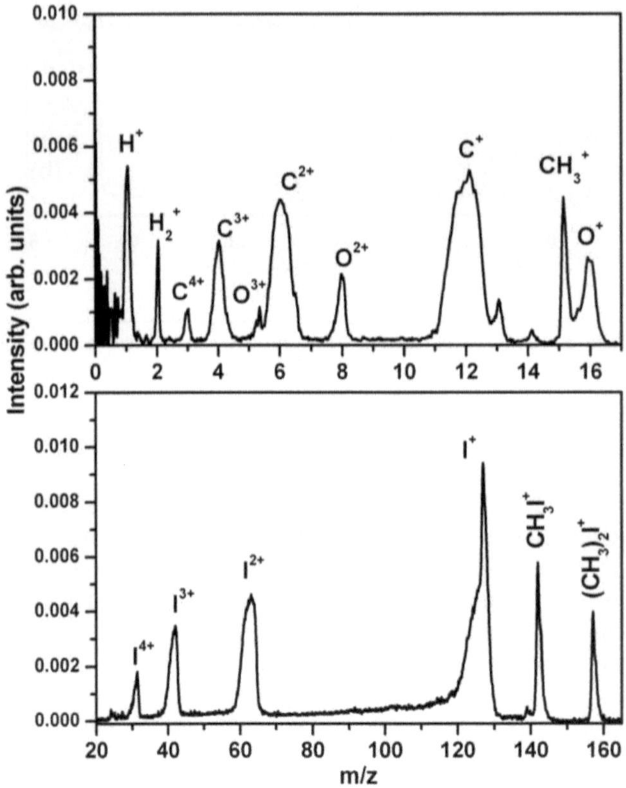

FIGURE 16.14

Time-of-flight mass spectra of $CH_3I–H_2O$ mixed clusters, when subjected to 563-nm laser pulses at an intensity of ~10^9 W/cm^2 [110].

Reproduced with permission of Indian Academy of Sciences.

6.2 INTERACTION WITH PICO- AND FEMTOSECOND LASER PULSES

A large number of studies have been carried out on clusters subjected to picosecond and femtosecond laser pulses. As discusses earlier, clusters exhibit diverse behavior when subjected to laser pulses of varying pulse duration, wavelength, and intensity. In addition, even for fixed laser intensity clusters subjected to laser pulses of varying duration display contrasting behavior. Such distinct behavior upon subjecting clusters to laser pulses of varying duration has been ascribed to temporal intensity profile of a laser pulse, which governs the extent of laser energy deposited within the cluster media prior to its disintegration. For example, it is assumed that upon interaction with femtosecond laser pulses the expansion of the cluster is negligible during the time span of the laser pulse. While the cluster undergoes appreciable expansion and even disintegrates during the time span of picoseconds and nanosecond laser pulse. Consequently, as we will discuss later, the time-dependent evolution of the cluster determines the effectiveness of energy transfer from the laser pulse to the cluster.

Before we discuss response of clusters, it is pertinent to report regarding behavior of atoms and molecules under intense pico- and femtosecond laser pulses, whereby these species have been found to undergo significant deformation and almost instantaneous ionization producing multiply charged atomic ions [111,112]. Historically, C. Manus and coworkers reported generation of multiply charged atomic ions of rare gas atoms up to Ar^{4+} and Xe^{4+} in the laser intensity range of 10^{11}–10^{15} W/cm^2, utilizing 50-ps laser pulses of 1.064 and 0.53 μm, respectively [113,114]. Here generation of these multiply charged atomic ions was ascribed to multiphoton ionization of rare gas atoms. On the contrary, for studies carried out using CO_2 laser, Chin et al. reported tunneling ionization of xenon gas at laser intensity of $\sim 10^{14}$ W/cm^2, producing multiply charged atomic ions up to Xe^{3+} [115], while xenon atoms subjected to X-ray free-electron laser pulse of intensity $\sim 10^{17}$ W/cm^2 (80 fs, 0.83 nm) leads to generation of multiply charged atomic ions up to Xe^{36+} [116]. On the other hand, for simple diatomic and triatomic molecules subjected to intense laser pulses, stepwise ejection of molecular electrons followed by dissociation of transient multicharged molecular ion into multiply charged atomic fragments has been reported. Frasinski et al. observed multiply charged atomic ions up to N^{3+} upon multiple ionization of molecular nitrogen, when irradiated with intense picosecond laser pulses (0.6 ps, 3×10^{15} W/cm^2, 600 nm) [117]. Coulomb explosion of CO_2 molecules into multiply charged atomic fragments, following ejection of up to nine molecular electrons, when subjected to intense femtosecond laser pulses (150 fs, 2×10^{16} W/cm^2, 790 nm) has been reported by Normand and coworkers [118]. Since the pioneering work of Manus and Frasinski, a large number of reports have appeared in literature dealing with extreme ionization of atoms and molecules [119–125]. Having provided glimpses of studies carried out to probe behavior of atoms and molecules under intense laser field, we proceed further to explore the response of cluster, formed upon aggregation of these atoms and molecules, under intense laser field conditions.

Clusters when exposed to intense laser field generated by picoseconds and femtosecond laser pulses undergo complete disintegration into energetic multiply charged atomic ions and electrons. Several experimental observables, like highest charged state of exploding cluster ions [126–130], kinetic energy of ions [131–134] and electrons [135,136], and generation of X-rays [137] and neutrons [138,139] are found in the literature for several atomic and molecular cluster systems. For xenon clusters of $\sim$2500 atoms subjected to 150-fs laser pulse of intensity $\sim 2 \times 10^{16}$ W/cm^2 (800 nm), multiply charged atomic ions of xenon up to Xe^{40+} (*I.E.*–2811 eV) have been observed. In addition, these multiply charged xenon ions produced upon Coulomb explosion of laser-heated xenon clusters are found to be associated with kinetic energy of up to $\sim$1 MeV. These findings suggest that efficient conversion of electromagnetic energy to particle energy can be accomplished under cluster conditions

[140]. Ditmire et al. have measured energy absorption efficiency of high-intensity picosecond laser pulses by low-density gaseous pulse composed of clusters. Their findings suggest that energy absorption efficiency >95% can be realized, which implies deposition of appreciable amount of laser energy per particle within the cluster [141]. For efficient coupling of the laser energy with the clusters media and subsequent ionization, studies have suggested the significant role of laser pulse duration. Various laser pulse–dependency and pump-probe studies [130,142,143] have substantiated the significance of optimal laser pulse duration. For example, clusters of platinum with mean size of Pt_{20} were irradiated using a femtosecond laser of fixed energy (ranging from 1 to 25 mJ), while the laser pulse duration was varied from 140 fs to several ps, providing laser intensity in the ionization region as high as 10^{16} W/cm^2. Under such conditions, the time-of-flight mass spectrum recorded revealed that platinum ions with charge up to Pt^{20+} were generated, in contrast to when single platinum atoms were ionized, charge states of up to Pt^{4+} were observed. An important finding of these studies was that a laser pulse duration of $\tau \sim 600$ fs was found to be optimal for generation of highest charge states, though the laser intensity was comparatively lower than the intensity for the shortest laser pulse by a factor of ~7. Such behavior of platinum clusters was ascribed to resonance absorption, which transpires once the plasmon and laser frequencies coincide during the course of cluster expansion [144]. Similar findings have been reported by various groups, where different experimental observables like absorption of laser energy by the cluster media [145], X-ray emission [146,147], and emission of ions and electrons as a function of laser pulse length [148] have emphasized the role of critical laser pulse duration.

Before concluding this section, it would be desirable to highlight the diverse behavior displayed by clusters subjected to laser pulses of different laser wavelength, intensity, and pulse duration, a summary of studies carried out on two of the most extensively studied cluster systems, i.e., $(CH_3I)_n$ and $(Xe)_n$, is presented. For methyl iodide clusters subjected to nanosecond laser pulses (8–25 ns) of intensity ~10^9–10^{10} W/cm^2, multiply charged atomic ions up to C^{3+} and I^{3+} have been observed at 532 nm, and for studies carried out at 640 nm ions up to C^{4+} and I^{6+} have been reported [95,96]. Meanwhile, Karras and Kosmidis [149] reported multielectron dissociative ionization of methyl iodide clusters, when subjected to 35-ps laser pulses of intensity ~10^{12}–3×10^{13} W/cm^2 at 532 and 1064 nm, producing energetic multiply charged atomic ions up to I^{7+}. For studies carried out using 128-fs laser pulses of intensity ~10^{14}–10^{15} W/cm^2, Ford et al. reported generation of multiply charged atomic ion up to I^{15+} at 795 nm [150]. In contrast, for studies carried out at 397 nm, atomic ions up to I^{7+} were detected, utilizing reflectron time-of-flight mass spectrometer [151]. In addition, the kinetic energy of the multiply charged ions was also found to be influenced by the laser parameters. Table 16.1, summarizes the kinetic energy of multiply charged atomic ions measured for methyl iodide clusters, under different laser excitation conditions.

At laser intensity of ~10^9 W/cm^2, our group has systematically studied size and wavelength response of xenon clusters at 266-, 355-, and 532-nm laser wavelength of 10-ns pulse duration [105]. At 266 nm, xenon clusters were found to exhibit multiphoton ionization behavior, while xenon clusters were found to exhibit the phenomenon of Coulomb explosion, resulting in generation of multiply charged atomic ions up to Xe^{6+} and Xe^{10+} at 355 and 532 nm, respectively (Fig. 16.10). In addition, studies carried out as a function of cluster size suggested that clusters of xenon $\geq Xe_{50}$ in size facilitated enhanced ionization of clusters producing multiply charged atomic ions, while clusters of smaller size undergo multiphoton ionization (Fig. 16.15). While at 10^{11} W/cm^2, generation of ions up to Xe^{11+} has been reported by Luo et al. [152]. For studies carried out using 624-nm laser of pulse duration 350 fs and ~10^{15} W/cm^2 intensity, Snyder et al. reported observation of multiply charged

Table 16.1 Kinetic Energy (eV) of Representative Multiply Charged Carbon and Iodine Ions Measured Under Different Laser Conditions

Wavelength (nm)	Reference	Pulse Width	Intensity (W/cm^2)	I^{1+}	I^{2+}	I^{3+}	C^{1+}	C^{2+}	C^{3+}
532	[123],[a]	35 ps	10^{15}	0.42 ± 0.04	0.42 ± 0.08	0.4 ± 0.1	1.2 ± 0.2	1.7 ± 0.5	
532	[149]	35 ps	3.5×10^{12}		6.1 ± 0.1	48.7 ± 2.1		103 ± 17	
532	[95]	25 ns	10^{11}		9.5	27		55.5	
532	[96]	8 ns	5.1×10^{9}	18 ± 2	114 ± 12	376 ± 25	29 ± 6	115 ± 30	325 ± 26
1064	[149]	35 ps	4×10^{12}		7.5 ± 0.2	82.2 ± 4.3		110 ± 18	
397	[151]	128 fs	10^{15}	12.0 ± 5.4	112 ± 49	334 ± 133	108 ± 38	431 ± 106	714 ± 110
795	[150]	128 fs	10^{14}		24.7 ± 6.6	100 ± 21	39.3 ± 10.2	193 ± 83	310 ± 185

[a]*Ref. [123] is for methyl iodide monomer, while the rest are for studies carried out for methyl iodide clusters.*

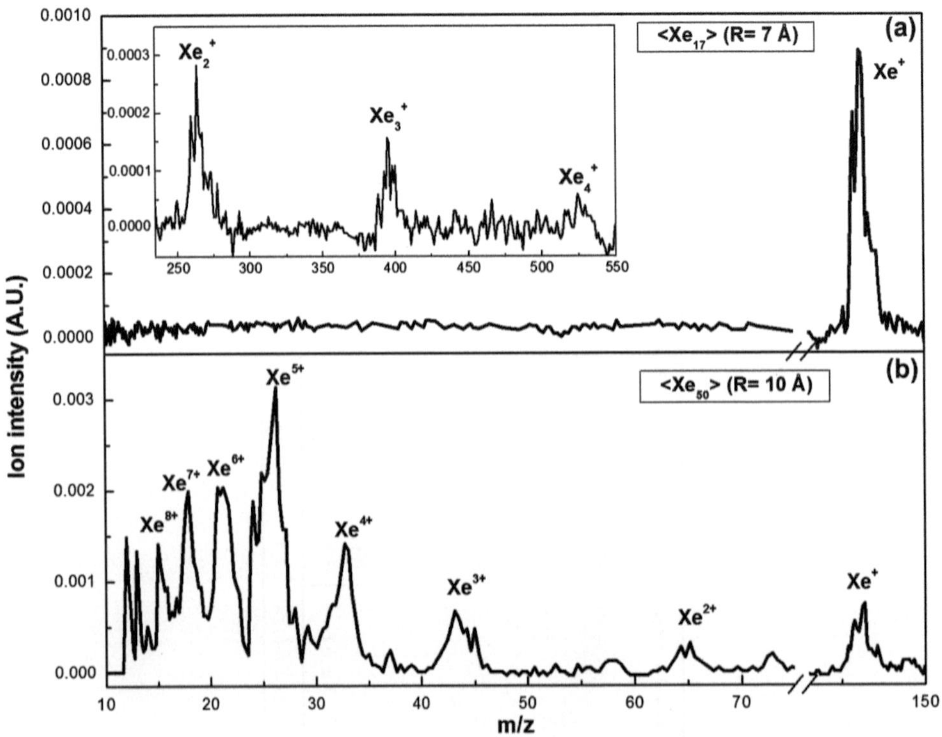

FIGURE 16.15

Xenon clusters of different size subjected to 532-nm laser pulses of intensity ~5 × 10^9 W/cm^2. (a) Cluster of size ~<Xe_{17}> exhibit multiphoton ionization behavior, while (b) clusters of size ~<Xe_{50}> exhibit Coulomb explosion phenomena. Figure depicts that clusters of larger size facilitate efficient coupling of laser energy, producing multiply charged atomic ions [105].

Reproduced with permission of Elsevier.

atomic ions up to Xe^{20+} [128]. Ditmire et al. observed multiply charged atomic ions up to Xe^{40+} associated with kinetic energy up to 1 MeV, upon irradiation of xenon clusters with 150-fs laser pulses of intensity ~2 × 10^{16} W/cm^2, at 780 nm [153]. Striking results have been reported for xenon clusters irradiated with 100 fs intense soft X-ray pulses of 98 nm wavelength from a free-electron laser of up to ~7 × 10^{13} W/cm^2 intensity [154] (Fig. 16.16). Here multiply charged atomic ions of xenon up to 8+ have been observed, arising due to multiple ionization and subsequent Coulomb explosion of xenon clusters. Also it was observed that the yield of different ion states and their kinetic energy depends on the laser intensity. For example, for studies carried out on xenon clusters of size ~<Xe_{1500}> at intensity of 7 × 10^{13} W/cm^2, ions up to Xe^{8+} are observed, while at 2 × 10^{11} W/cm^2 only singly charged atomic ions of xenon are detected.

Thus the above representative experimental findings clearly demonstrate that the extent of extreme ionization of cluster constituents, deduced on the basis of multiply charged ion yield and their kinetic energy, is strongly influenced by cluster size, laser wavelength, laser intensity, and laser-cluster

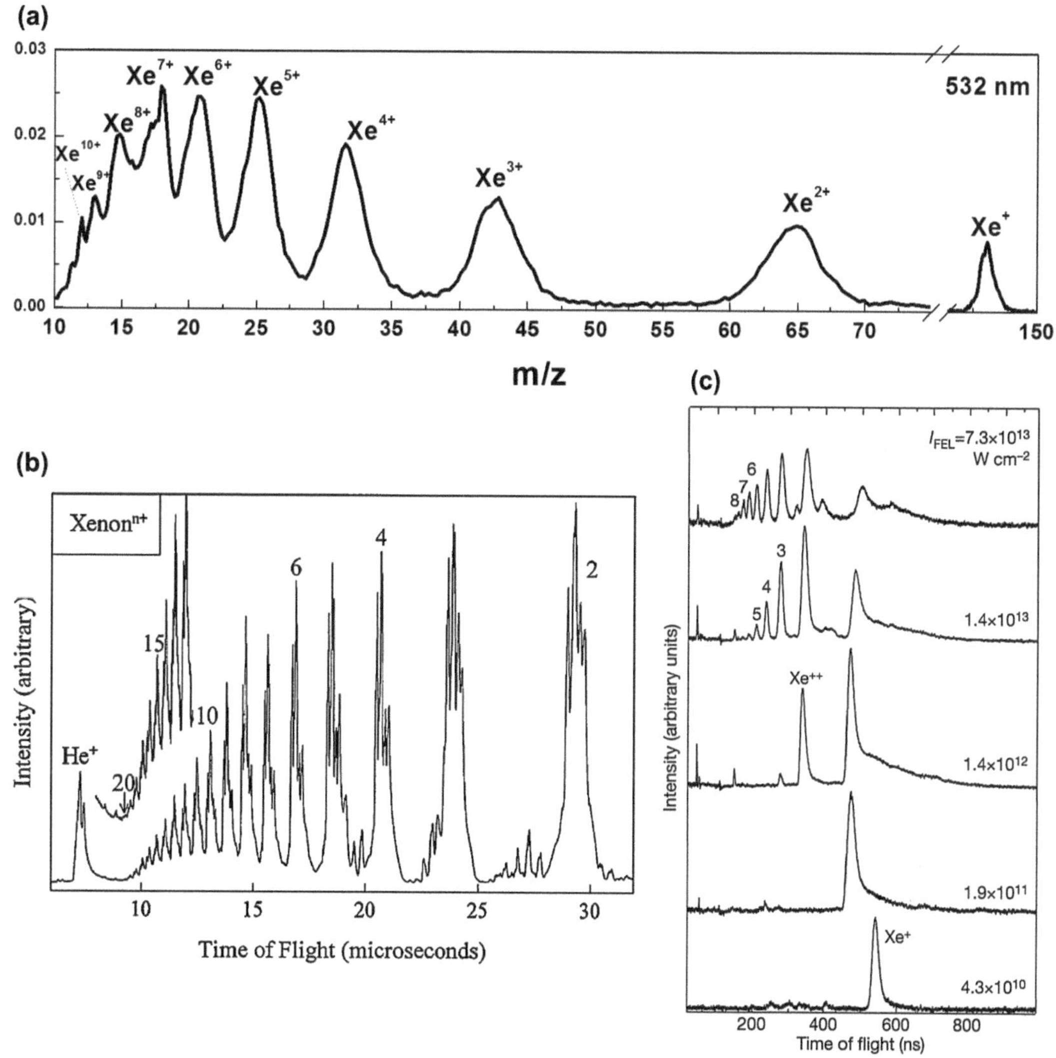

FIGURE 16.16

Xenon clusters subjected to laser pulses of varying laser wavelength and pulse duration. (a) Clusters subjected to 10-ns, 532-nm laser pulses of intensity $\sim 5 \times 10^9$ W/cm^2 [105]. (b) Time-of-flight spectrum obtained upon irradiation with 350-fs, 624-nm laser pulses of intensity $\sim 1 \times 10^{15}$ W/cm^2 [128]. While (c) illustrates spectra obtained upon irradiating xenon cluster of 1500 atoms with ~100-fs-long free-electron laser pulses of 98 nm and intensity $\sim 4.3 \times 10^{10}$ to 7.3×10^{13} W/cm^2 [154].

(a) Reproduced with permission of Elsevier. (b) Reproduced with permission of American Physical Society. (c) Reproduced with permission of Nature Publishing Group.

interaction timescale. To further emphasize the role of laser pulse duration, studies carried out using few-cycle laser pulses are described next.

6.3 INTERACTION WITH FEW-CYCLE LASER PULSES

Laser pulses having pulse duration that is only a small multiple of an optical cycle are described as few-cycle laser pulses. For example, laser pulses of 5 fs generated at 800 nm consist of less than two optical cycles (few cycle) as the laser period of the Ti-Sapphire laser wavelength is ~2.7 fs. These few-cycle pulses are delivered in a diffraction-limited beam and are focused to spot size comparable to the laser wavelength. Thus, due to this extreme temporal and spatial confinement, few-cycle laser pulse of even moderate energy can easily provide peak intensities up to 10^{18} W/cm^2 [155]. Under such few-cycle temporal regime, which is much shorter than the characteristic expansion time of the irradiated cluster, only electron rescattering is considered to be dominant as the ionic degree of freedom is considered to be effectively frozen. Thus, the overall dynamics of clusters subjected to few-cycle laser pulses are expected to be different compared to those subjected to many cycle laser pulses.

Xe_n clusters ($n = 500 - 25{,}000$) have been investigated under few-cycle pulses (12-fs pulse duration, ~4 optical cycles) of 800-nm laser light and peak intensity $\sim 5 \times 10^{14}$ W/cm^2. Under such irradiation conditions, the kinetic energies of ions were measured to be much lower compared to those obtained using longer laser pulses of same intensity. Also, the ion yields were found to be cluster-size independent but were influenced by the polarization of the laser. Fig. 16.17 depicts yield of multiply

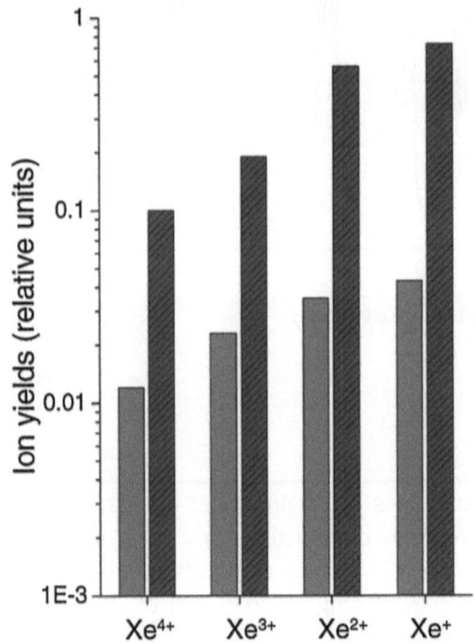

FIGURE 16.17

Yield of Xe ions obtained with 12-fs pulses of peak intensity 5×10^{14} W/cm^2. The hatched bars denote yields obtained when the laser polarization vector was perpendicular to the time-of-flight axis; the unhatched bars are the corresponding yields when the polarization direction was orthogonal.

From D. Mathur, F.A. Rajgara. Communication: ionization and Coulomb explosion of xenon clusters by intense, few-cycle laser pulses, J. Chem. Phys. 133 (2010) 061101 (1–4), reproduced with permission of AIP Publishing LLC.

charged ions as a function of laser polarization [156]. Several studies, experimental and computational, have investigated dynamics of clusters under few-cycle laser conditions [157–160]. These findings further substantiate the decisive role of laser pulse duration for efficient transfer of laser energy to cluster.

Alongside generation of multiply charged atomic ions and energetic electrons upon intense laser-cluster interaction, another important consequence of presence of highly charged ions and energetic electrons in close proximity is recombination in the cluster nanoplasma causing photoemission from bound–bound transition, particularly in the atomic core. For a variety of clusters, characteristic X-ray emission has been observed over wide energy range [161–164]. The X-ray yield shows strong dependency on laser intensity and cluster size [165]. Also for clusters doped with species having lower ionization energies, enhancement in the yield of X-ray generation has been observed. For example, argon clusters doped with water provide enhanced yield of Ar characteristic X-ray compared to pure Ar clusters [166]. Applications and motivations of these studies are discussed in the following section.

7. APPLICATIONS

As discussed in earlier sections, clusters demonstrate an uncanny ability to absorb efficiently the incident laser energy from the intense laser pulse. This absorbed laser energy is subsequently expended into producing energetic multiply charged atomic ions, electrons, high-energy photons, and neutrons. Thus extreme ionization of clusters by intense laser pulses provides a potential scheme for development of tabletop accelerators for transforming laser energy into different particle and optical beams, having the same order of pulse duration as that of the intense laser utilized for cluster irradiation [167]. Though practical realization of a tabletop accelerator based on laser-cluster interaction mechanism is still in its infancy, developments in laser technology and judiciously engineering cluster constituents can significantly enhance the brightness, ion yield, and charge state of the laser-cluster source [166,168]. Currently, availability of lasers having peak intensity of $\sim 10^{22}$ W/cm^2, which corresponds to an electric field of $\sim 10^{12}$ V/cm and effective temperature of $\sim 10^8$K, would further facilitate in the endeavor for development of tabletop accelerators. Some of the applications and motivations that drive and make the field of laser-cluster interaction stimulating are discussed next.

7.1 COULOMB EXPLOSION-DRIVEN NUCLEAR FUSION IN CLUSTERS: GENERATION OF NEUTRONS

Scientists and engineers have striven for several decades to demonstrate tabletop nuclear fusion driven by fusion of deuteron (Eq. (16.14)) to mimic the primary sources of energy from the sun at laboratory scale [169]. In order to overcome the barrier for D–D fusion, it is essential that deuterons acquire kinetic energy of $\sim$MeV range. Such high kinetic energy could be achieved by Coulomb explosion of deuterium clusters exposed to intense laser fields. The first report regarding fusion of $(D_2)_n$ clusters upon their irradiation with 35-fs, 200-mJ laser pulse of intensity $\sim 10^{17}$ W/cm^2 was by Ditmire et al. [138], where deuterons ejected from different clusters collide with adequate kinetic energy to stimulate nuclear fusion.

$$\begin{aligned} D^+ + D^+ &\rightarrow {}^3He^{2+} + n + 3.27\ \text{MeV} \\ D^+ + D^+ &\rightarrow T^+ + H^+ + 4.03\ \text{MeV} \end{aligned} \tag{16.14}$$

Neutron yield (Y) of $\sim 10^3$–10^4 per laser pulse was experimentally reported in these studies [170,171]. In addition, fusion reaction has also been reported for heteronuclear clusters of $(D_2O)_n$

[172], $(CD_4)_n$ [173–175], $(DI)_n$ [176], and $(CD_3I)_n$ [177]. Thus fusion in clusters has facilitated realization of the quest for a tabletop-pulsed source of neutrons having pulse duration <500 ps [170]. Such short-pulsed neutron sources would find potential applications in the field of material characterization, neutron imaging [178], and testing of materials under high neutron flux, subjected to further enhancement in yield of neutrons per laser shot at a higher repetition rate.

7.2 CLUSTER-ASSISTED GENERATION OF ENERGETIC MULTIPLY CHARGED ATOMIC IONS AND ELECTRONS

Interaction of intense laser pulses with cluster has facilitated generation of energetic multiply charged atomic ions and electrons, which are accelerated to MeV energy in few millimeters. Availability of such energetic ultrashort ion and electron pulses has revolutionized the field of particle accelerators. Multiply charged atomic ions hold a large amount of potential energy, which upon collision with solid surface leads to sputtering and surface modification, due to transfer of large amount of energy onto the surface within a short interaction time of few femtoseconds [179]. Unlike kinetic sputtering, where sputtering of the surface is induced by the ions accelerated to very high kinetic energy, here the sputtering is caused by the potential energy contained within the multiply charged ions and is termed as potential sputtering [180]. This soft sputtering method based on multiply charged ions has potential applications in semiconductor industries and nanostructuring of surfaces. In addition, due to electron-deficient nature, multiply charged ions find application in collisional charge transfer [181] and for generation of extreme ultraviolet light [182,183]. Thus, laser-cluster interaction process provides a convenient and compact method for generation of multiply charged atomic ions, which previously could only be generated using electron cyclotron resonance and electron beam ion trap methods, which are bulky and not extensively available [184].

Similarly, energetic electrons liberated upon laser-cluster interactions have been exploited for transmission electron radiography and microscopy [185]. The laser-based electron gun, which involves irradiation of argon clusters with intense femtosecond laser (40 fs, 100 mJ) produces electrons of up to 2 MeV energy, which have been utilized for radiography of organic and inorganic dense objects with a spatial resolution of 60 μm. Similarly, Mangles et al. have reported regarding generation of monoenergetic beams of relativistic electrons of energy ~80 MeV with low divergence [186]. Such monoenergetic electron beams are essential for development of tabletop narrowband ultrashort X-ray sources and free-electron lasers, having widespread applications in materials science and medicine. The ultrashort electron bunches generated upon laser-cluster interaction have been used to carry out time-resolved studies [187] and also to provide an affordable as well as efficient substitute for conventional RF electron guns to carry out electron microscopy and diffraction studies [188,189].

7.3 GENERATION OF INTENSE ULTRASHORT X-RAY PULSES

Generation of X-rays is currently one of the most important outcomes of laser-cluster interaction [137]. Bright betatron X-rays with photon energy >2.4 keV have been produced with an emission yield of 2×10^8 photons/pulse, by irradiating argon clusters with 80-fs laser pulses of intensity $\sim 3 \times 10^{18}$ W/cm^2. Such laser-cluster-driven hard X-ray sources have promising applications in materials and biological sciences, due to their compact size and ultrashort pulse duration. In addition, as the X-rays are perfectly synchronized with the laser pulse, it opens an avenue to perform "single-shot" ultrafast pump probe, X-ray absorption, and diffraction studies without heat-induced blurring [190]. Utilizing gaseous clusters for generation of X-rays also finds potential application in photolithography due to its higher repetition rate and its debris-free nature [165], in contrast to utilization of solid targets for X-ray generation, which has a major drawback of debris generation upon laser ablation, causing damage to expensive X-ray optics [191]. Also, conversion efficiency of up to 10% of the incident laser energy into X-rays has been reported, which is comparable with the X-ray yield obtained from solid targets [109]. Other potential applications include imaging with high spatial resolution utilizing X-ray sources of higher photon flux and higher repetition rate.

ACKNOWLEDGMENTS

The authors acknowledge encouragement and support received from Dr. V.K. Jain, Head, Chemistry Division, and Dr. B.N. Jagatap, Director, Chemistry Group, BARC, Mumbai. The authors also thank Dr. S. Das and Dr. P.M. Badani for their contributions in laser-cluster interaction studies.

REFERENCES

[1] S.L. Sass, The Substance of Civilization: Materials and Human History from the Stone Age to the Age of Silicon, Arcade Publishing, New York, 2011.
[2] B.S. Murty, P. Shankar, B. Raj, B.B. Rath, J. Murday, Nanostructured Materials with High Application Potential, in Textbook of Nanoscience and Nanotechnology, Springer, University Press, 2013.
[3] W. Eberhardt, Clusters as new materials, Surf. Sci. 500 (2002) 242–270.
[4] G. Thornton, Watching nanoparticles grow, Science 300 (2003) 1378–1379.
[5] A.P. Alivisatos, Semiconductor clusters, nanocrystals, and quantum dots, Science 271 (1996) 933–937.
[6] A.W. Castleman Jr., R.G. Keesee, Gas-phase clusters: spanning the states of matter, Science 241 (1988) 36–42.
[7] P. Jena, B.K. Rao, S.N. Khanna (Eds.), Physics and Chemistry of Small Clusters, Plenum, New York, 1987.
[8] P.G. Reinhard, E. Suraud, Introduction to Cluster Dynamics, Wiley-VCH, Weinheim, 2003.
[9] E. Roduner, Size matters:Why nanomaterials are different, Chem. Soc. Rev. 35 (2006) 583–592.
[10] R.L. Johnston, Atomic and Molecular Clusters, Taylor & Francis, 2002.
[11] S. Das, P.M. Badani, P. Sharma, R.K. Vatsa, Coulomb explosion phenomenon using gigawatt intensity laser fields: an exotic realm of laser–cluster interaction, Curr. Sci. 100 (2011) 1008–1019.
[12] T. Maiman, Stimulated optical radiation in ruby, Nature 187 (1960) 493–494.
[13] J. Hecht, D. Teresi, Laser: Light of a Million Uses, Dover Publications, 1998.
[14] C. Yamanaka, S. Sakabe, Prospects of high power laser applications, Rev. Laser Eng. 30 (2002) 185–192.
[15] R.F. Smith, et al., Nature 511 (2014) 330–333.
[16] N. Subramanian, N.V. Chandra Shekar, N.R. Sanjay Kumar, P. Ch Sahu, Curr. Sci. 91 (2006) 175–182.
[17] K. Sugioka, M. Meunier, A. Piqué (Eds.), Laser Precision Microfabrication, Springer, 2010.
[18] J.D. Musgraves, K. Richardson, H. Jain, Laser-induced structural modification, its mechanisms, and applications in glassy optical materials, Opt. Mater. Express 1 (2011) 921–935.
[19] P. Sharma, R.K. Vatsa, Generation of multiply charged atomic ions of halogens using second harmonic of nanosecond Nd:YAG laser, Curr. Appl. Phys. 9 (2009) 140–143.
[20] V.P. Krainov, M.B. Smirnov, Cluster beams in the super-intense femtosecond laser pulse, Phys. Rep. 370 (2002) 237–331.

[21] K.W.D. Ledingham, et al., Laser-driven photo-transmutation of ^{129}I—a long-lived nuclear waste product, J. Phys. D Appl. Phys. 36 (2003) L79–L82.
[22] H. Schwoerer, F. Ewald, R. Sauerbrey, J. Galy, J. Magill, V. Rondinella, et al., Fission of actinides using a tabletop laser, Europhys. Lett. 61 (2003) 47–52.
[23] S. Bj⊘Srnholm, Clusters, condensed matter in embryonic form, Contemp. Phys. 31 (1990) 309–324.
[24] A.W. Castleman Jr., K.H. Bowen Jr., Clusters: Structure, energetics, and dynamics of intermediate states of matter, J. Phys. Chem. 100 (1996) 12911–12944.
[25] Th Fennel, K.-H. Meiwes-Broer, J. Tiggesbaumker, P.-G. Reinhard, P.M. Dinh, E. Suraud, Laser-driven nonlinear cluster dynamics, Rev. Mod. Phys. 82 (2010) 1793–1842.
[26] F. Mafuné, T. Kondow, Formation of small gold clusters in solution by laser excitation of interband transition, Chem. Phys. Lett. 372 (2003) 199–204.
[27] N. Goswami, A. Baksi, A. Giri, P.L. Xavier, G. Basu, T. Pradeep, et al., Luminescent iron clusters in solution, Nanoscale 6 (2014) 1848–1854.
[28] W.A. de Heer, The physics of simple metal clusters: experimental aspects and simple models, Rev. Mod. Phys. 65 (1993) 611–676.
[29] P. Milani, S. Iannotta, Cluster Beam Synthesis of Nanostructured Materials, Springer, Berlin, 1999.
[30] H. Pauly, Atom, Molecule, and Cluster Beams I, Springer, Berlin, 2000.
[31] H. Haberland (Ed.), Clusters of Atoms and Molecules, Springer, Berlin, 1994.
[32] R.A. Larson, S.K. Ncoh, D.R. Herschbach, Seeded supersonic alkali atom beams, Rev. Sci. Instrum. 45 (1984) 1511–1516.
[33] D.H. Levy, Van der Waals molecules, Adv. Chem. Phys. 47 (1981) 323–362.
[34] R. Campargue (Ed.), Atomic and Molecular Beams – the State of the Art 2000, Springer, Berlin, 2001.
[35] G. Scoles (Ed.), Atomic and Molecular Beam Methods, vol. 1, Oxford University Press, New York, 1988.
[36] J. Wörmer, V. Guzielski, J. Stapelfeldt, T. Möller, Fluorescence excitation spectroscopy of xenon clusters in the VUV, Chem. Phys. Lett. 159 (1989) 321–326.
[37] T.E. Gough, M. Mengel, P.A. Rowntree, G. Scoles, Infrared spectroscopy at the surface of clusters: SF_6 on Ar, J. Chem. Phys. 83 (1985) 4958–4961.
[38] D.J. Levandier, S. Goyal, J. McCombie, B. Pate, G. Scoles, Infrared spectroscopy and dimer formation at the surface of medium-large argon clusters, J. Chem. Soc. Faraday Trans. 86 (1990) 2361–2368.
[39] M. Hartmann, R.E. Miller, J.P. Toennies, A. Vilesov, Rotationally resolved spectroscopy of SF_6 in liquid helium clusters: a molecular probe of cluster temperature, Phys. Rev. Lett. 75 (1995) 1566–1569.
[40] E.A. Rohlfing, D.M. Cox, A. Kaldor, Photoionization measurements on isolated iron-atom clusters, Chem. Phys. Lett. 99 (1983) 161–166.
[41] A. Nakajima, K. Hoshino, T. Sugioka, T. Naganuma, T. Taguwa, Y. Yamada, et al., Electronic shell structure of indium-sodium (In_nNa_m) bimetallic clusters examined by their ionization potentials and mass distributions, J. Phys. Chem. 97 (1993) 86–90.
[42] I.G. Dance, K.J. Fisher, G.D. Willett, Molecular manganese sulfide clusters formed by laser ablation, J. Chem. Soc. Dalton Trans. (1997) 2557–2562.
[43] W. Bouwen, P. Thoen, F. Vanhoutte, S. Bouckaert, F. Despa, H. Weidele, et al., Production of bimetallic clusters by a dual-target dual-laser vaporization source, Rev. Sci. Instrum. 71 (2000) 54–58.
[44] T.G. Dietz, M.A. Duncan, D.E. Powers, R.E. Smalley, Laser production of supersonic metal cluster beams, J. Chem. Phys. 74 (1981) 6511–6512.
[45] H. Yu, M.G. Huber, F.W. Froben, Laser ablation of refractory material, cluster formation and deposition, Appl. Surf. Sci. 86 (1995) 74–78.
[46] T. Witke, A. Lenk, B. Schultrich, C. Schultheiss, Investigation of plasma produced by laser and electron pulse ablation, Surf. Coat. Tech. 74–75 (1995) 580–585.
[47] P.R. Willmott, J.R. Huber, Pulsed laser vaporization and deposition, Rev. Mod. Phys. 72 (2000) 315–328.

[48] G. Raina, G.U. Kulkarni, R.T. Yadav, V.S. Ramamurthy, C.N.R. Rao, An indigenous cluster beam apparatus with a reflectron time-of-flight mass spectrometer, Proc. Indian Acad. Sci. (Chem. Sci.) 112 (2000) 83−95.
[49] W.T. Wallace, R.B. Wyrwas, A.J. Leavitt, R.L. Whetten, Adsorption of carbon monoxide on smaller gold-cluster anions in an atmospheric-pressure flow-reactor: temperature and humidity dependence, Phys. Chem. Chem. Phys. 7 (2005) 930−937.
[50] H.W. Kroto, J.R. Heath, S.C. O'Brien, R.F. Curl, R.E. Smalley, C60: Buckminsterfullerene, Nature 318 (1985) 162−163.
[51] Q.L. Zhang, S.C. O'Brien, J.R. Heath, Y. Liu, R.F. Curl, H.W. Kroto, et al., Reactivity of large carbon clusters: spheroidal carbon shells and their possible relevance to the formation and morphology of soot, J. Phys. Chem. 90 (1986) 525−528.
[52] C.G. Granqvist, R.A. Buhrman, Ultrafine metal particles, J. Appl. Phys. 47 (1976) 2200−2219.
[53] O.F. Hagena, W. Obert, Cluster formation in expanding supersonic jets: effect of pressure, temperature, nozzle size, and test gas, J. Chem. Phys. 56 (1972) 1793−1802.
[54] E. Toma, H. Muller, Laser disintegration of Van der Waals clusters of carbon-containing molecules, Phys. Rev. A 66 (2002), 013204 (1−9).
[55] O.F. Hagena, Condensation in free jets: comparison of rare gases and metals, Z. Phys. D 4 (1987) 291−299.
[56] J. Farges, M.F. de Feraudy, B. Raoult, G. Torchet, Noncrystalline structure of argon clusters. II. Multilayer icosahedral structure of Ar_N clusters $50<N<750$, J. Chem. Phys. 84 (1986) 3491−3501.
[57] J. Farges, B. Raoult, G. Torchet, Crystalline and noncrystalline effects in electron diffraction patterns from small clusters in an argon cluster beam, J. Chem. Phys. 59 (1973) 3454−3458.
[58] G. Torchet, M.-F. de Feraudy, A. Boutin, A.H. Fuchs, Structural transformation in $(CO_2)_N$ clusters, N<100, J. Chem. Phys. 105 (1996) 3671−3678.
[59] C. Majumder, O.D. Jayakumar, R.K. Vatsa, S.K. Kulshreshtha, An experimental set-up for generation and characterization of metal clusters, Indian J. Chem. 40 (2001) 577−582.
[60] C. Majumder, O.D. Jayakumar, R.K. Vatsa, S.K. Kulshreshtha, J.P. Mittal, Multiphoton ionisation of acetone at 355 nm: a time-of-flight mass spectrometry study, Chem. Phys. Lett. 304 (1999) 51−59.
[61] S.H. Lin, Y. Fujimura, H.J. Neusser, E.W. Schlag, Multiphoton Spectroscopy of Molecules, Academic Press, Florida, 1984.
[62] S.L. Chin, P. Lambropoulos, Multiphoton Ionization of Atoms, Academic Press, New York, 1984.
[63] D.A. Gobeli, J.J. Yang, M.A. El-Sayed, Laser multiphoton ionization-dissociation mass spectrometry, Chem. Rev. 85 (1985) 529−554.
[64] V.S. Antonov, V.S. Letokhov, Laser multiphoton and multistep photoionization of molecules and mass spectrometry, Appl. Phys. 24 (1981) 89−106.
[65] R.B. Bernstein, Systematics of multiphoton ionization-fragmentation of polyatomic molecules, J. Phys. Chem. 86 (1982) 1178−1184.
[66] L.V. Keldysh, Ionization in the field of a strong electromagnetic wave, Sov. Phys. JETP 20 (1965) 1307−1314.
[67] C. Kosmidis, P. Siozos, S. Kaziannis, L. Robson, K.W.D. Ledingham, P. McKenna, et al., Interaction mechanism of some alkyl iodides with femtosecond laser pulses, J. Phys. Chem. A 109 (2005) 1279−1285.
[68] R.J. Levis, M.J. DeWitt, Photoexcitation, ionization, and dissociation of molecules using intense near-infrared radiation of femtosecond duration, J. Phys. Chem. A 103 (1999) 6493−6507.
[69] Z.B. Walters, R. Santra, C.H. Greene, Interaction of intense VUV radiation with large xenon clusters, Phys. Rev. A 74 (2006), 043204 (1−14).
[70] J. Posthumus (Ed.), Molecules and Clusters in Intense Laser Fields, Cambridge University Press, 2009.
[71] K. Sattler, J. Mühlbach, O. Echt, P. Pfau, E. Recknagel, Evidence for Coulomb explosion of doubly charged microclusters, Phys. Rev. Lett. 47 (1981) 160−163.

[72] W.A. Saunders, N. Dam, Liquid drop model of multiply charged metal cluster fission, Z. Phys. D 20 (1991) 111–113.
[73] T. Desai, High power laser interaction with clusters, Laser Part Beams 19 (2001) 163–168.
[74] U. Saalmann, J.-M. Rost, Ionization of clusters in intense laser pulses through collective electron dynamics, Phys. Rev. Lett. 91 (2003), 223401 (1–4).
[75] U. Saalmann, Ch. Siedschlag, J.M. Rost, Mechanisms of cluster ionization in strong laser pulses, J. Phys. B Mol. Opt. Phys. 39 (2006) R39–R77.
[76] C. Rose-Petruck, K.J. Schafer, C.P.J. Barty, Possible ionization ignition in laser-driven clusters. Application of laser plasma radiation II, SPIE 2523 (1995) 272–282.
[77] C. Rose-Petruck, K.J. Schafer, K.R. Wilson, C.P.J. Barty, Ultrafast electron dynamics and inner-shell ionization in laser driven clusters, Phys. Rev. A 55 (1997) 1182–1190.
[78] A. McPherson, T.S. Luk, B.D. Thompson, K. Boyer, C.K. Rhodes, Multiphoton-induced X-ray emission and amplification from clusters, Appl. Phys. B 57 (1993) 337–347.
[79] A.B. Borisov, J.W. Longworth, A. McPherson, K. Boyer, C.K. Rhodes, Dynamical orbital collapse drives super X-ray emission, J. Phys. B Mol. Opt. Phys. 29 (1996) 247–255.
[80] K. Boyer, B.D. Thompson, A. McPherson, C.K. Rhodes, Evidence for coherent electron motions in multiphoton X-ray production from Kr and Xe clusters, J. Phys. B Mol. Opt. Phys. 27 (1994) 4373–4389.
[81] S. Chelkowski, T. Zuo, O. Atabek, A.D. Bandrauk, Dissociation, ionization, and Coulomb explosion of H_2^+ in an intense laser field by numerical integration of the time-dependent Schrodinger equation, Phys. Rev. A 52 (1995) 2977–2983.
[82] T. Seideman, M.Y. Ivanov, P.B. Corkum, Role of electron localization in intense-field molecular ionization, Phys. Rev. Lett. 75 (1995) 2819–2822.
[83] I. Last, I. Schek, J. Jortner, Energetics and dynamics of Coulomb explosion of highly charged clusters, J. Chem. Phys. 107 (1997) 6685–6692.
[84] T. Ditmire, T. Donnelly, R.W. Falcone, M.D. Perry, Strong X-ray emission from high-temperature plasmas produced by intense irradiation of clusters, Phys. Rev. Lett. 75 (1995) 3122–3125.
[85] P. Hilse, M. Moll, M. Schlanges, Th. Bornath, Laser-cluster-interaction in a nanoplasma-model with inclusion of lowered ionization energies, Laser Phys. 19 (2009) 428–436.
[86] I. Last, J. Jortner, Electron and nuclear dynamics of molecular clusters in ultraintense laser fields. I. Extreme multielectron ionization, J. Chem. Phys. 120 (2004) 1336–1347.
[87] K. Muller-Dethlefs, O. Dopfer, T.G. Wright, Zeke spectroscopy of complexes and clusters, Chem. Rev. 94 (1994) 1845–1871.
[88] Q.Y. Shang, E.R. Bernstein, Energetics, dynamics, and reactions of Rydberg state molecules in van der Waals clusters, Chem. Rev. 94 (1994) 2015–2025.
[89] U. Buck, F. Huisken, Infrared spectroscopy of size-selected water and methanol clusters, Chem. Rev. 100 (2000) 3863–3890.
[90] E.J. Bieske, O. Dopfer, High-resolution spectroscopy of cluster ions, Chem. Rev. 100 (2000) 3963–3998.
[91] C.E.H. Dessent, K. Muller-Dethlefs, Hydrogen-bonding and van der Waals complexes studied by ZEKE and REMPI spectroscopy, Chem. Rev. 100 (2000) 3999–4021.
[92] O. Echt, P.D. Dao, S. Morgan, A.W. Castleman Jr., Multiphoton ionization of ammonia clusters and the dissociation dynamics of protonated cluster ions, J. Chem. Phys. 82 (1985) 4076–4085.
[93] K.V. Vidma, A.V. Baklanov, E.B. Khvorostov, V.N. Ishchenko, S.A. Kochubei, A.T.J.B. Eppink, et al., UV photodissociation of the van der Waals dimer $(CH_3I)_2$ revisited: pathways giving rise to ionic features, J. Chem. Phys. 122 (2005), 204301 (1–11).
[94] Y.-K. Choi, Y.-M. Koo, K.-W. Jung, Multiphoton ionization and fragmentation processes of methyl iodide clusters at 266 and 355 nm, J. Photochem. Photobiol. A Chem. 127 (1999) 1–5.

[95] X. Luo, D. Niu, X. Kong, L. Wen, F. Liang, K. Pei, et al., Cluster-assistant generation of multiply charged atomic ions in nanosecond laser ionization of seeded methyl iodide beam, Chem. Phys. 310 (2005) 17–24.
[96] P. Sharma, R.K. Vatsa, S.K. Kulshreshtha, J. Jha, D. Mathur, M. Krishnamurthy, Energy pooling in multiple ionisation and Coulomb explosion of Clusters by nanosecond long megawatt laser pulses, J. Chem. Phys. 125 (2006), 034304 (1–7).
[97] P. Sharma, R.K. Vatsa, B.N. Rajasekhar, N.C. Das, T.K. Ghanty, S.K. Kulshreshtha, Photoionization of CH_3I mediated by the C state in the visible and ultraviolet regions, Rapid Commun. Mass Spectrom. 19 (2005) 1522–1528.
[98] T. Auguste, P. Monot, G. Mainfray, C. Manus, Multiply charged ions produced in noble gases by a 1 ps laser pulse at lambda = 1053 nm, J. Phys. B Mol. Opt. Phys. 25 (1992) 4181–4194.
[99] P. Sharma, R.K. Vatsa, Wavelength-dependent Coulomb explosion in carbon disulphide (CS_2) clusters: generation of energetic multiply charged carbon and sulphur ions, Rapid Commun. Mass Spectrom. 21 (2007) 2663–2670.
[100] P. Sharma, R.K. Vatsa, Photochemistry of acetone clusters: size dependent observation of Coulomb explosion in the multiphoton ionisation regime, Europhys. Lett. 84 (2008), 43003 (1–6).
[101] P. Sharma, S. Das, R.K. Vatsa, Efficient coupling of nanosecond laser pulses with the cluster medium: generation of hydrogen-like $[C]^{5+}$ atomic ions, Rapid Commun. Mass Spectrom. 29 (2015) 45–53.
[102] X. Kong, X. Luo, D. Niu, H. Li, Cluster assistant generation of C^{2+} and C^{3+} ions in nanosecond laser ionization of seeded benzene beam, Chem. Phys. Lett. 388 (2004) 139–143.
[103] D. Niu, H. Li, W. Wang, X. Xiao, X. Luo, N. Zhang, et al., Cluster-assisted generation of multiply charged ions in nanosecond laser ionization of seeded furan beam at 532 and 1064 nm, Mol. Phys. 106 (2008) 1389–1395.
[104] N. Zhang, W. Wang, W. Zhao, F. Han, H. Li, Multiply ionization of diethyl ether clusters by 532 nm nanosecond laser: the influence of laser intensity and the electron energy distribution, Chem. Phys. 373 (2010) 181–185.
[105] S. Das, P.M. Badani, P. Sharma, R.K. Vatsa, Size and wavelength dependent photoionisation of xenon clusters, Chem. Phys. Lett. 552 (2012) 13–19.
[106] W. Wang, H. Li, D. Niu, L. Wen, N. Zhang, Cluster-assisted multiple-ionization of methyl iodide by a nanosecond laser: wavelength dependence of multiple-charge ions, Chem. Phys. 352 (2008) 111–116.
[107] N. Zhang, W. Wang, H. Cang, H. Wang, H. Li, Multiply ionization of benzene clusters by a nanosecond laser: distributions of the ion charge state and the electron energy, Chem. Phys. Lett. 469 (2009) 14–18.
[108] J.F. Seely, E.G. Harris, Heating of a plasma by multiphoton inverse bremsstrahlung, Phys. Rev. A 7 (1973) 1064–1067.
[109] T. Ditmire, T. Donnelly, A.M. Rubenchik, R.W. Falcone, M.D. Perry, Interaction of intense laser pulses with atomic clusters, Phys. Rev. A 53 (1996) 3379–3402.
[110] S. Das, P. Sharma, R.K. Vatsa, Coulomb explosion of methyl iodide clusters using giga watt laser pulses in the visible region: effect of wavelength, polarisation and doping, J. Chem. Sci. 121 (2009) 965–972.
[111] K. Yamanouchi, The next frontier, Science 295 (2002) 1659–1660.
[112] K. Yamanouchi, Exotic behavior of molecules and clusters in intense laser light fields, in: K. Nakajima, M. Deguchi (Eds.), Science of Superstrong Field Interactions, AIP, 2002.
[113] A. L'Huillier, L.A. Lompre, G. Mainfray, C. Manus, Multiply charged ions formed by multiphoton absorption processes in the continuum, Phys. Rev. Lett. 48 (1982) 1814–1818.
[114] A. L'Huillier, L.A. Lompre, G. Mainfray, C. Manus, Multiply charged ions induced by multiphoton absorption in rare gases at 0.53 μm, Phys. Rev. A 27 (1983) 2503–2512.
[115] S.L. Chin, F. Yergeau, P. Lavigne, Tunnel ionisation of Xe in an ultra-intense CO_2 laser field (10^{14} W cm^{-2}) with multiple charge creation, J. Phys. B 18 (1985) L213–L215.

[116] B. Rudek, S.-K. Son, L.M. Foucar, S.W. Epp, B. Erk, R. Hartmann, et al., Ultra-efficient ionization of heavy atoms by intense X-ray free-electron laser pulses, Nat. Photon. 6 (2012) 858–865.

[117] L.J. Frasinski, K. Codling, P. Hatherly, J. Barr, I.N. Ross, W.T. Toner, Femtosecond dynamics of multielectron dissociative ionization by use of a picosecond laser, Phys. Rev. Lett. 58 (1987) 2424–2427.

[118] C. Cornaggia, M. Schmidt, D. Normand, Coulomb explosion of CO_2 in an intense femtosecond laser field, J. Phys. B Mol. Opt. Phys. 27 (1994) L123–L130.

[119] T.S. Luk, U. Johann, H. Egger, H. Pummer, C.K. Rhodes, Collision-free multiple photon ionization of atoms and molecules at 193 nm, Phys. Rev. A 32 (1985) 214–224.

[120] C. Cornaggia, J. Lavancier, D. Normand, J. Morellec, P. Agostini, J.P. Chambaret, et al., Multielectron dissociative ionization of diatomic molecules in an intense femtosecond laser field, Phys. Rev. A 44 (1991) 4499–4505.

[121] P.A. Hatherly, M. Stankiewicz, K. Codling, L.J. Frasinski, G.M. Cross, The multielectron dissociative ionization of molecular iodine in intense laser fields, J. Phys. B Mol. Opt. Phys. 27 (1994) 2993–3003.

[122] T. Zuo, A.D. Bandrauk, Charge-resonance-enhanced ionization of diatomic molecular ions by intense lasers, Phys. Rev. A 52 (1995) R2511–R2514.

[123] P. Siozos, S. Kaziannis, C. Kosmidis, Multielectron dissociative ionization of CH_3I under strong picosecond laser irradiation, Int. J. Mass Spectrom. 225 (2003) 249–259.

[124] P.B. Corkum, N.H. Burnett, F. Brunel, Multiphoton ionization at large ponderomotive potentials, in: M. Gavrila (Ed.), Atoms in Intense Laser Fields, Academic Press, New York, 1992.

[125] J. Gauer, D. Feldmann, MPI of atoms in intense laser pulses, in: H.G. Muller, M.V. Fedorov (Eds.), Super – Intense Laser – Atom Physics IV, Kluwer Academic Publishers, Netherlands, 1996.

[126] J. Purnell, E.M. Snyder, S. Wei, A.W. Castleman Jr., Ultrafast laser-induced Coulomb explosion of clusters with high charge states, Chem. Phys. Lett. 229 (1994) 333–339.

[127] B.D. Thompson, A. McPherson, K. Boyer, C.K. Rhodes, Multi-electron ejection of inner-shell electrons through multiphoton excitation of clusters, J. Phys. B Mol. Opt. Phys. 27 (1994) 4391–4400.

[128] E.M. Snyder, S.A. Buzza, A.W. Castleman Jr., Intense field-matter interactions: multiple ionization of clusters, Phys. Rev. Lett. 77 (1996) 3347–3350.

[129] M. Lezius, S. Dobosz, D. Normand, M. Schmidt, Explosion dynamics of rare gas clusters in strong laser fields, Phys. Rev. Lett. 80 (1998) 261–264.

[130] S. Zamith, T. Martchenko, Y. Ni, S.A. Aseyev, H.G. Muller, M.J.J. Vrakking, Control of the production of highly charged ions in femtosecond-laser cluster fragmentation, Phys. Rev. A 70 (2004), 011201 (1–4).

[131] E. Springate, N. Hay, J.W.G. Tisch, M.B. Mason, T. Ditmire, J.P. Marangos, et al., Enhanced explosion of atomic clusters irradiated by a sequence of two high-intensity laser pulses, Phys. Rev. A 61 (2000), 044101 (1–4).

[132] V. Kumarappan, M. Krishnamurthy, D. Mathur, Asymmetric high-energy ion emission from argon clusters in intense laser fields, Phys. Rev. Lett. 87 (2001), 085005 (1–4).

[133] Y. Fukuda, K. Yamakawa, Y. Akahane, M. Aoyama, N. Inoue, H. Ueda, et al., Optimized energetic particle emissions from Xe clusters in intense laser fields, Phys. Rev. A 67 (2003), 061201 (1–4).

[134] M. Hirokane, S. Shimizu, M. Hashida, S. Okada, S. Okihara, F. Sato, et al., Energy distributions of ions emitted from argon clusters Coulomb-exploded by intense femtosecond laser pulses, Phys. Rev. A 69 (2004), 063201 (1–4).

[135] Y.L. Shao, T. Ditmire, J.W.G. Tisch, E. Springate, J.P. Marangos, M.H.R. Hutchinson, Multi-keV electron generation in the interaction of intense laser pulses with Xe clusters, Phys. Rev. Lett. 77 (1996) 3343–3346.

[136] L.M. Chen, J.J. Park, K.-H. Hong, J.L. Kim, J. Zhang, C.H. Nam, Emission of a hot electron jet from intense femtosecond-laser–cluster interactions, Phys. Rev. E 66 (2002), 025402 (1–4).

[137] A. McPherson, B.D. Thompson, A.B. Borisov, K. Boyer, C.K. Rhodes, Multiphoton-induced X-ray emission at 4–5 keV from Xe atoms with multiple core vacancies, Nature 370 (1994) 631–634.

[138] T. Ditmire, J. Zweiback, V.P. Yanovsky, T.E. Cowan, G. Hays, K.B. Wharton, Nuclear fusion from explosions of femtosecond laser-heated deuterium clusters, Nature 398 (1999) 489–492.
[139] J. Zweiback, R.A. Smith, T.E. Cowan, G. Hays, K.B. Wharton, V.P. Yanovsky, et al., Nuclear fusion driven by Coulomb explosions of large deuterium clusters, Phys. Rev. Lett. 84 (2000) 2634–2637.
[140] T. Ditmire, J.W.G. Tisch, E. Springate, M.B. Mason, N. Hay, R.A. Smith, et al., High-energy ions produced in explosions of superheated atomic clusters, Nature 386 (1997) 54–56.
[141] T. Ditmire, R.A. Smith, J.W.G. Tisch, M.H.R. Hutchinson, High intensity laser absorption by gases of atomic clusters, Phys. Rev. Lett. 78 (1997) 3121–3124.
[142] T. Döppner, S. Teuber, M. Schumacher, J. Tiggesbäumker, K.H. Meiwes-Broer, Charging dynamics of metal clusters in intense laser fields, Appl. Phys. B 71 (2000) 357–360.
[143] T. Döppner, Th. Fennel, Th. Diederich, J. Tiggesbäumker, K.H. Meiwes-Broer, Controlling the Coulomb explosion of silver clusters by femtosecond dual-pulse laser excitation, Phys. Rev. Lett. 94 (2005), 013401 (1–4).
[144] L. Köller, M. Schumacher, J. Köhn, S. Teuber, J. Tiggesbäumker, K.H. Meiwes-Broer, Plasmon-enhanced multi-ionization of small metal clusters in strong femtosecond laser fields, Phys. Rev. Lett. 82 (1999) 3783–3786.
[145] J. Zweiback, T. Ditmire, M.D. Perry, Femtosecond time-resolved studies of the dynamics of noble-gas cluster explosions, Phys. Rev. A 59 (1999) R3166–R3169.
[146] T. Mocek, C.M. Kim, H.J. Shin, D.G. Lee, Y.H. Cha, K.H. Hong, et al., Soft-X-ray emission from small-sized Ne clusters heated by intense, femtosecond laser pulses, Phys. Rev. E 62 (2000) 4461–4464.
[147] E. Lamour, C. Prigent, J.P. Rozet, D. Vernhet, Physical parameter dependence of the X-ray generation in intense laser–cluster interaction, Nucl. Instrum. Methods Phys. Res. B 235 (2005) 408–413.
[148] V. Kumarappan, M. Krishnamurthy, D. Mathur, Two-dimensional effects in the hydrodynamic expansion of xenon clusters under intense laser irradiation, Phys. Rev. A 66 (2002), 033203 (1–4).
[149] G. Karras, C. Kosmidis, Multielectron dissociative ionization of CH_3I clusters under moderate intensity ps laser irradiation, Int. J. Mass Spectrom. 290 (2010) 133–141.
[150] J.V. Ford, Q. Zhong, L. Poth, A.W. Castleman Jr., Femtosecond laser interactions with methyl iodide clusters. I. Coulomb explosion at 795 nm, J. Chem. Phys. 110 (1999) 6257–6267.
[151] J.V. Ford, L. Poth, Q. Zhong, A.W. Castleman Jr., Femtosecond laser interactions with methyl iodide clusters.: 2. Coulomb explosion at 397 nm, Int. J. Mass Spectrom. 192 (1999) 327–345.
[152] X. Luo, H. Li, D. Niu, L. Wen, F. Liang, B. Wang, et al., Cluster-assisted multiple ionization of xenon and krypton by a nanosecond laser, Phys. Rev. A 72 (2005), 013201 (1–4).
[153] T. Ditmire, J.W.G. Tisch, E. Springate, M.B. Mason, N. Hay, J.P. Marangos, et al., High energy ion explosion of atomic clusters: transition from molecular to plasma behavior, Phys. Rev. Lett. 78 (1997) 2732–2735.
[154] H. Wabnitz, L. Bittner, A.R.B. De Castro, R. Döhrmann, P. Gürtler, T. Laarmann, et al., Multiple ionization of atom clusters by intense soft X-rays from a free-electron laser, Nature 420 (2002) 482–485.
[155] T. Brabec, F. Krausz, Intense few-cycle laser fields: frontiers of nonlinear optics, Rev. Mod. Phys. 72 (2000) 545–591.
[156] D. Mathur, F.A. Rajgara, Communication: ionization and Coulomb explosion of xenon clusters by intense, few-cycle laser pulses, J. Chem. Phys. 133 (2010), 061101 (1–4).
[157] D. Mathur, F.A. Rajgara, Dynamics of atomic clusters in intense optical fields of ultrashort duration, J. Chem. Sci. 124 (2012) 75–81.
[158] F. Greschik, L. Arndt, H.-J. Kull, Absorption of few-cycle laser pulses by outer ionization of atomic clusters, Europhys. Lett. 72 (2005) 376–382.
[159] D. Mathur, F.A. Rajgara, A.R. Holkundkar, N.K. Gupta, Strong-field ionization and Coulomb explosion of argon clusters by few-cycle laser pulses, Phys. Rev. A 82 (2010), 025201 (1–4).

[160] S.R. Krishnan, R. Gopal, R. Rajeev, J. Jha, V. Sharma, M. Mudrich, et al., Photoionization of clusters in intense few-cycle near infrared femtosecond pulses, Phys. Chem. Chem. Phys. 16 (2014) 8721−8730.

[161] V. Kumarappan, M. Krishnamurthy, D. Mathur, L.C. Tribedi, Effect of laser polarization on X-ray emission from Ar_n ($n = 200-10^4$) clusters in intense laser fields, Phys. Rev. A 63 (2001), 023203 (1−6).

[162] C. Deiss, N. Rohringer, J. Burgdörfer, E. Lamour, C. Prigent, J.-P. Rozet, et al., Laser-cluster interaction: X-ray production by short laser pulses, Phys. Rev. Lett. 96 (2006), 013203 (1−4).

[163] J. Jha, M. Krishnamurthy, Collisionless phenomena in heteronuclear clusters, Appl. Phys. Lett. 92 (2008), 191108 (1−3).

[164] H. Singhal, R.A. Ganeev, P.A. Naik, J.A. Chakera, U. Chakravarty, H.S. Vora, et al., High-order harmonic generation in a plasma plume of in situ laser-produced silver nanoparticles, Phys. Rev. A 82 (2010), 043821 (1−4).

[165] T. Ditmire, R.A. Smith, R.S. Marjoribanks, G. Kulcsar, M.H.R. Hutchinson, X-ray yields from Xe clusters heated by short pulse high intensity lasers, Appl. Phys. Lett. 71 (1997) 166−168.

[166] J. Jha, D. Mathur, M. Krishnamurthy, Enhancement of X-ray yields from heteronuclear cluster plasmas irradiated by intense laser light, J. Phys. B Mol. Opt. Phys. 38 (2005) L291−L299.

[167] N. Nakashima, S. Shimizu, T. Yatsuhashi, S. Sakabe, Y. Izawa, Large molecules in high-intensity laser fields, J. Photochem. Photobio. C Photochem. Rev. 1 (2000) 131−143.

[168] J. Jha, D. Mathur, M. Krishnamurthy, Engineering clusters for table-top acceleration of ions, Appl. Phys. Lett. 88 (2006), 041107 (1−3).

[169] J. Jortner, I. Last, Nuclear fusion driven by Coulomb explosion of molecular clusters, ChemPhysChem 3 (2002) 845−848.

[170] J. Zweiback, T.E. Cowan, R.A. Smith, J.H. Hartley, R. Howell, C.A. Steinke, et al., Characterization of fusion burn time in exploding deuterium cluster plasmas, Phys. Rev. Lett. 85 (2000) 3640−3643.

[171] J. Zweiback, T.E. Cowan, J.H. Hartley, R. Howell, K.B. Wharton, J.K. Crane, et al., Detailed study of nuclear fusion from femtosecond laser-driven explosions of deuterium clusters, Phys. Plasmas 9 (2002) 3108−3120.

[172] S. Ter-Avetisyan, M. Schnürer, D. Hilscher, U. Jahnke, S. Busch, P.V. Nickles, et al., Fusion neutron yield from a laser-irradiated heavy-water spray, Phys. Plasmas 12 (2005), 012702 (1−5).

[173] M. Hohenberger, D.R. Symes, K.W. Madison, A. Sumeruk, G. Dyer, A. Edens, et al., Dynamic acceleration effects in explosions of laser-irradiated heteronuclear clusters, Phys. Rev. Lett. 95 (2005), 195003 (1−4).

[174] G. Grillon, Ph. Balcou, J.-P. Chambaret, D. Hulin, J. Martino, S. Moustaizis, et al., Deuterium-deuterium fusion dynamics in low-density molecular-cluster jets irradiated by intense ultrafast laser pulses, Phys. Rev. Lett. 89 (2002), 065005 (1−4).

[175] I. Last, J. Jortner, Nuclear fusion driven by Coulomb explosion of methane clusters, J. Phys. Chem. A 106 (2002) 10877−10885.

[176] I. Last, J. Jortner, Ultrafast high-energy dynamics of thin spherical shells of light ions in the Coulomb explosion of heteroclusters, Phys. Rev. A 71 (2005), 063204 (1−10).

[177] A. Heidenreich, J. Jortner, I. Last, Cluster dynamics transcending chemical dynamics toward nuclear fusion, Proc. Natl. Acad. Sci. USA 103 (2006) 10589−10593.

[178] R. Loveman, J. Bendahan, T. Gozani, J. Stevenson, Time-of-flight fast-neutron radiography, Nucl. Instrum. Methods Phys. Res. B Beam Interact. Mater. Atoms 99 (1995) 765−768.

[179] M. Sporn, G. Libiseller, T. Neidhart, M. Schmid, F. Aumayr, H.P. Winter, et al., Potential sputtering of clean SiO_2 by slow highly charged ions, Phys. Rev. Lett. 79 (1997) 945−948.

[180] T. Neidhart, F. Pichler, F. Aumayr, H.P. Winter, M. Schmid, P. Varga, Potential sputtering of lithium fluoride by slow multicharged ions, Phys. Rev. Lett. 74 (1995) 5280−5283.

[181] N. Nakamura, F.J. Currell, A. Danjo, M. Kimura, A. Matsumoto, S. Ohtani, et al., Target dependence of multi-electron processes in I^{q+} (q = 10, 15) + rare gas (Ne, Ar, Kr and Xe) collisions, J. Phys. B Mol. Opt. Phys. 28 (1995) 2959–2972.

[182] H. Tanuma, H. Ohashi, E. Shibuya, N. Kobayashi, T. Okuno, S. Fujioka, et al., EUV emission spectra from excited multiply charged xenon ions produced in charge-transfer collisions, Nucl. Instrum. Methods Phys. Res. B 235 (2005) 331–336.

[183] F.J. Currell (Ed.), The Physics of Multiply and Highly Charged Ions, vol. 2, Kluwer Academic Publishers Group, 2003.

[184] F.J. Currell (Ed.), The Physics of Multiply and Highly Charged Ions, vol. 1, Kluwer Academic Pub., Dordrecht, 2003.

[185] G.C. Bussolino, A. Faenov, A. Giulietti, D. Giulietti, P. Koester, L. Labate, et al., Electron radiography using a table-top laser-cluster plasma accelerator, J. Phys. D Appl. Phys. 46 (2013), 245501 (1–8).

[186] S.P.D. Mangles, C.D. Murphy, Z. Najmudin, A.G.R. Thomas, J.L. Collier, A.E. Dangor, et al., Monoenergetic beams of relativistic electrons from intense laser–plasma interactions, Nature 431 (2004) 535–538.

[187] W. Schumaker, N. Nakanii, C. McGuffey, C. Zulick, V. Chyvkov, F. Dollar, et al., Ultrafast electron radiography of magnetic fields in high-intensity laser-solid interactions, Phys. Rev. Lett. 110 (2013), 015003 (1–5).

[188] Z.-H. He, A.G.R. Thomas, B. Beaurepaire, J.A. Nees, B. Hou, V. Malka, et al., Electron diffraction using ultrafast electron bunches from a laser-wakefield accelerator at kHz repetition rate, Appl. Phys. Lett. 102 (2013), 064104 (1–4).

[189] J. Faure, C. Rechatin, A. Norlin, A. Lifschitz, Y. Glinec, V. Malka, Controlled injection and acceleration of electrons in plasma wakefields by colliding laser pulses, Nature 444 (2006) 737–739.

[190] L.M. Chen, W.C. Yan, D.Z. Li, Z.D. Hu, L. Zhang, W.M. Wang, et al., Bright betatron X-ray radiation from a laser-driven-clustering gas target, Sci. Rep. 3 (2013), 1912 (1–5).

[191] R. Bobkowski, R. Fedosejevs, Particle emission debris from a KrF laser–plasma X-ray source, J. Vac. Sci. Technol. A 14 (1996) 1973–1980.

CHAPTER

MATERIALS RESPONSE UNDER IRRADIATION

17

M. Vijayalakshmi

Indira Gandhi Centre for Atomic Research, Kalpakkam, India

CHAPTER OUTLINE

1. INTRODUCTION

The principles of alloy design, so far employed for development of aerospace, nuclear, and biomaterials, demonstrated that a given system with desired combination of properties is not the one at its thermodynamic equilibrium. Very often, the required system would belong to one of those lying in suitable local minimum in the free energy landscape. So far, in conventional metallurgy, thermal and thermomechanical routes were adopted to navigate in the free energy landscape to identify the most suitable minimum. Of late, irradiation is being used to energize the system to hop from one free energy minimum to another, until the system reaches the "desirable" local minimum. However, the genesis of "irradiation-engineered" tailoring of properties, an event of serendipity, can be traced [1] back to the discovery of "metamict" 120 years ago—the mineral that was optically isotropic, and the cause was recognized much later as due to radiation-induced amorphization by α-active uranium accidentally present in these minerals.

Materials Under Extreme Conditions. http://dx.doi.org/10.1016/B978-0-12-801300-7.00017-6

The motivation for the vibrant study of materials behavior under irradiation was the necessity to enhance the lifetime of "fuel" in the core of the nuclear reactor. The intense, early efforts of the last century demonstrated [2,3] that degradation of materials under irradiation is the life-limiting factor of the nuclear fuel and not the "fuel" per se. Hence, intense R&D was carried out to understand the behavior of engineering materials under irradiation, leading to the current insight in the field. Recently, the compulsion to meet global energy demand using green technologies has provided strong impetus to revisit nuclear technology. The investigations of the new millennium are directed toward requirements of future Generation IV reactors, life extension of licensed reactors of ongoing Generation III+ systems and fusion technology. Another pertinent area concerning behavior of materials, especially the electronic IC chips, is in aerospace industry [4], due to cosmic ray irradiation, which is not included within the scope of this chapter.

Another technologically important area that motivated study of materials behavior under irradiation, especially ion/electron irradiation, was in the field of semiconductors. Ion and electron irradiation have played [5–20] a pivotal role in engineering the composition and structure of dopants at atomic level to tailor electrical [13–15], optical [6], magnetic [16–18], mechanical [12], and electronic [19,20] properties, in particular, of carbon compounds [7,21]. The "irradiation-mediated engineering" has led [7] to self-assembly or self-organization structures in carbon nanosystems, which have unique ability to reorganize their structure. Many impressive applications include patterning [22] or ordering of magnetic properties of ferromagnetic thin films [23,24], fabrication of nanodots [25], and many more [7]. Morphological instabilities are introduced [26,27] during ion irradiation of thin films or alloys, leading to composition-modulated surface patterns, which open up new possibilities for the growth of three-dimensional, lateral or vertical-ordered nanostructures. Conversion [7] of carbon to diamond under ion irradiation opens up the exciting possibilities of developing diamond-based nanodevices. Electron and ion beams have the potential [7] to be used as "cutting/welding tools," with the atomic precision, laying the foundation for the rapidly emerging "carbon-based electronics."

Materials under irradiation are characterized by [28] significant deviations from its equilibrium structure. The displacement of atoms from their equilibrium positions by the incident beam leads to the formation of point defects, two or three orders of magnitude more than the system under equilibrium at a temperature. The time evolution of these defects leads to two- and three-dimensional defect clusters. The irradiation-induced point defects and their clusters alter the kinetics of precipitation of thermally stable phase, leading to either appearance of phases not expected during thermal exposure or dissolution of thermally stable phases. These events are understood in terms of change in the thermodynamic driving force, by altering the relative stabilities of phases and kinetics, in addition to "ballistic effects" consequent to direct impingement. Hence, it is appropriate to consider the material under irradiation as a "driven" system under extreme conditions.

The irradiation-induced microstructural changes listed previously significantly alter [29–34] the mechanical properties of engineering materials, especially zirconium alloys and steels, which is of concern to the nuclear industry. In steels, irradiation can introduce change in shape and volume by 5–7%, the strength/hardness increase by factors of five or more, ductility is reduced, and there is significant increase in embrittlement. The change in properties of materials under irradiation have been fairly well understood [28–34] based on the irradiation-induced microstructure and microchemistry.

Extensive studies in the last five decades have enabled an in-depth understanding of the irradiation-induced changes in defect dynamics, kinetics, and thermodynamics of phases and their manifestation in properties. Most of these developments are well documented in many books [28,30–34] and reviews [8,31–42]. However, the field continues to remain dynamic, mainly due to the demand from Generation IV reactors [43], fusion technology, life extension of Generation III+ reactors [44], and effort to launch [7] new nanodevice fabrication technologies. Modeling [28,45,46], especially the Monte Carlo [28], molecular dynamics [28,45], and rate theory calculations [28,45], have always been a part of the development of understanding of radiation damage. Despite these efforts, the ability to predict the materials behavior under irradiation for ready use by a nuclear design engineer is still in its infancy.

This chapter attempts to summarize the behavior of materials under extreme conditions of irradiation, which is far different from their behavior under thermal or thermomechanical treatments. The focus of the review is to highlight the "far from equilibrium" behavior of materials under irradiation, rather than assess the field for a specific application. Since there are many new, interesting phenomena exhibited by materials under irradiation, a glimpse of each of these processes, along with basic understanding has been provided, instead of detailed discussion of each process. The review mainly pertains to the progress of the field, to the 21st century. The references are indicative and not exhaustive, in view of the innumerable number of papers, conference proceedings, and books in the field since the 1970s. The review is organized as follows: irradiation as a means of achieving extreme situation, influence of type of irradiation on the damage characteristics, defect dynamics, diffusion, segregation, phase instabilities, and properties. Finally, the future trends are discussed.

2. IRRADIATION – PROCESSES LEADING TO EXTREME SITUATIONS

When a stream of high-energy (keV to MeV) ions or neutrons "bombards" a crystalline lattice, the atoms are "knocked off" from their equilibrium positions, a process that requires ~20–40 eV for most metals. These energetic "primary knock-on" atoms (PKA) generate secondary and tertiary displacements. The process goes on (Table 17.1) until the energy of the knocked atom becomes less than the displacement energy, of ~20–40 eV. Under such circumstances, the small energy that is transferred manifests as increased vibration of the lattice atom about its mean equilibrium position and hence as

Table 17.1 Different Stages of Defect Production (a) Thermal Spike to (d) Annealing Stage

Stage	Time(s)	Event	Result
a	10^{-18}	Energy transfer from incident particles	PKO generated
b	10^{-13}	Displacement of lattice atoms by PKA	Displacement cascade
c	10^{-11}	Energy dissipation, spontaneous recombination, and clustering	Stable Frenkel pairs and defect clusters
d	10^{-8}	Defect reactions by thermal migration	Single interstitial atom and vacancy recombination, clustering, trapping, and defect emission

"heat." At the end of this stage, the lattice ends up with "cascades," wherein vacant volume is generated at the center, with displaced atoms around this volume. This time is not sufficient for the formation of stable lattice defects. The spiked region gets quenched or relaxed, in about 10 ps, leading to formation of stable lattice defects like point defects or the defect clusters. The annealing stage, as the name suggests, is a thermally activated process. Further rearrangement of defects occurs by thermal diffusion of mobile lattice defects, reaching a steady state. The defects produced in the case of ion irradiation have a distinct depth distribution, which does not exist for neutrons.

Unlike interstitials, vacancies are thermodynamically stable defects. Even in a perfect crystal, in the absence of irradiation, presence of vacancies is thermodynamically desirable. It is possible to show that the free energy of a crystal, *F*, is reduced with vacancies.

$$F \sim G = U + pV - TS = H - TS \tag{17.1}$$

where *F* is the free energy of the crystal of volume, *V*, at constant pressure, *p*, at temperature, *T*, *U* the internal energy, *S* the entropy, and *H* the enthalpy of *N* atoms constituting the system.

$$\nabla S_{mix} = k \ln \omega \tag{17.2}$$

where ω is the number of possible configurations.

Consider the introduction of "*n*" defects with *N* available sites in the crystal. The increase in free energy is

$$\nabla G_f = n\nabla H_f - T\nabla S \tag{17.3}$$

∇H_f is the increase in the internal energy due to introduction of defects, and ∇S is the corresponding increase in entropy. It can be shown that

$$w = \{[N(N-1)(N-2)\cdots(N-n+1)]/n!\} \tag{17.4}$$

∇S_{mix} is calculated using the expression for ω in the equation for ∇S_{mix}. The defects introduce vibrational disorder, which can be calculated, enabling estimation of the total ∇S by adding both contributions.

In equilibrium, "*n*" will be such that $dG_f/dn = 0$. This condition can be used in the equation for dG_f to arrive finally at the concentration of "defects" as follows:

$$C = n/N = \exp(-\nabla G_f/kT) \tag{17.5}$$

This general Eq. (17.5) can be now rewritten for the two types of point defects, to get an expression for concentration of the point defects, vacancies, and interstitials.

$$C_v = \exp\left(-E_v^f/kT\right) X \exp(S_v^f/k) \tag{17.6}$$

$$C_i = \exp\left(-E_i^f/kT\right) X \exp(S_i^f/k) \tag{17.7}$$

The concentration of vacancies has two components: thermal and irradiation. At low temperatures, C_vth ~ 0 and the latter is dominant.

The radiation damage until the production of cascade has been shown [28] to be predictable based on computations based on molecular dynamics and kinetic Monte Carlo methods. Routinely used software packages like TRIM/SRIM [47,48], mainly based on Monte Carlo methods, are available for the calculation of defect concentrations, types of defects, and its distribution along the depth for various irradiation species like ions and protons. Rate theory is one of the most powerful techniques [28] for the study of evolution of point defects from the time of birth until steady-state condition is reached.

Conventionally, resistivity is the most commonly used experimental technique for the study of point defects. However, in the later half of last century, many ion beam techniques [42], positron spectroscopy [49], and advanced microscopy methods [50] were routinely used for study of irradiation-induced defects and consequent changes in energetics and mobility. The significant deviations of a material under irradiation from an unirradiated material, in a few typical examples, are illustrated following:

- Normally, the concentration of vacancies produced during irradiation at room temperature is [28] $\sim 10^{-2}$, which is two orders of magnitude higher than thermal equilibrium concentration of vacancies near melting point, of say, copper $\sim 10^{-4}$. The difference becomes significant, especially considering that the temperature dependence of concentration of vacancies is exponential in nature.
- The defect density in any material is increased [28,51] significantly during irradiation. The total defect density in ferritic alloy irradiated using 1 MeV neutrons from BR2/Mol-Belgium to 0.6 dpa at 573K is in the range 10^{19} to 10^{22}, as against 10^6 to 10^7 in annealed ferrite [52].
- Generally, diffusion kinetics is accelerated [53–57], due to the availability of point defects, especially vacancies. Radiation-induced diffusion in copper and its alloys [57] is $D_{rad} \sim 6.5 \times 10^{-21}\ m^2s^{-1}$, at 473K, as against the thermal component of $1.4 \times 10^{-27}\ m^2s^{-1}$. The temperature-dependence studies have revealed the relative contribution of thermal vacancies and irradiation-induced point defects and confirmed that influence of irradiation is dominant at low temperatures.
- Inverse Kirkendall effect, observed [28,58] only under irradiation, and not under thermal exposures, refers to complex interplay between coupling of irradiation-induced point defects with solutes and their migration toward "sinks" in the host lattice. This process alters the "microscopic" chemistry of the host lattice, leading to secondary phases, not predicted by phase diagram, discussed later in detail.
- Normally, the free energy of the lattice with radiation-induced defects, is far higher than the perfect lattice. A typical example is as follows: the free energy of ordered Zr_2Fe, which is -30 kJ/mol, is increased [59] to -20 kJ/mol after irradiation, making it lose its crystallinity.
- Irradiation is found [28] to accelerate formation and dissolution of phases either predicted by phase diagram or entirely new ones, ordering or disordering of the lattice, amorphization and crystallization, depending on the specific interplay of change in energetics, diffusion, and ballistic effects, details of which are explained in Section 7.
- Inverse Ostwald Ripening, characteristic of irradiation, drives [60] the system toward uniform distribution of particles with "single size," due to growth of smaller particles at the expense of larger particles. This phenomenon is exactly opposite of Ostwald ripening, wherein coarse grains/particles grow at the expense of finer grains/particles driven by the difference in solute concentration on the surface of fine/coarse particles. The inverse Ostwald Ripening has been understood in terms of dependence of the kinetics of "recombination" reactions on the size of the second-phase particles, discussed in detail later. These recombination reactions refer to annihilation of irradiation-induced point defects, at the interface between matrix and particles.

The previous typical examples of irradiation-induced processes in materials provide ample scope to support the view that irradiation can drive a system far from equilibrium. Hence, it is appropriate to consider the system under irradiation to be in extreme situation. A system under irradiation is not a "closed" system, since high energy and mass is deposited (ion irradiation) inside a host lattice, making it a "driven" system. The approach of such a system toward equilibrium is driven by the extent of deviation of the system from equilibrium. Under such conditions, very surprising phenomena, unheard of in thermal equilibrium situations, can occur, as discussed next.

3. IRRADIATION USING DIFFERENT INCIDENT BEAMS

The nature and extent of production of defects, discussed previously, depend strongly on the following characteristics of the incident beam: type of incident beam, i.e., neutrons or ions or electrons, incident energy, dose rate (number of incident particles/cm^2/sec.), cumulative dose (total number of displacements per atom, dpa, during the entire exposure of the material to incident beam), and temperature of irradiation. The fundamental difference arises [28] due to the nature of interaction of various incident particles with atoms—hard sphere collisions for neutrons and Coulombic interaction for charged particles. The energy transferred, E_t, by the incident beam to the host lattice atom during every single collision is given following:

$$E_t = \left(4E_0 m_0 M \Big/ (m_0 + M)^2\right) \tag{17.8}$$

depends strongly on the incident energy, E_0, and mass of the incident particle, m_0, and the host lattice atom, M. The dose rate that can be achieved in ion and electron irradiation is orders of magnitude higher (10^{-3} to 10^{-4} dpa/sec), in contrast to extremely low values for neutron, 10^{-6} to 10^{-10} dpa/sec. The damage cross-section is maximum for ions, in the range 10^5 (cm^2) barn, followed by neutrons ~300–800 b, and the least for electrons (~30 b), which follows the same pattern as that of total displacement cross-section.

These fundamental differences in the nature of incident particle and its interaction with the material manifest in differences in the type of radiation damage in the host lattice and depth distribution of the damage. The electron irradiation leads to only formation of Frenkel pairs, and cascades cannot form, while the cascade volume depends on whether the incident beam is ion or neutron or proton, as shown in Fig. 17.1a.

Consequently, the depth of penetration and the depth dependence of total damage caused by the incident beam depend strongly on the nature of incident irradiation beam, which is shown in Fig. 17.1b.

4. DEFECT DYNAMICS IN MATERIALS UNDER IRRADIATION

The energetics and kinetics of evolution of defect clusters in the annealing stage depend [28] on following fundamental parameters: formation and migration energies of vacancies and interstitials, binding energy, bias to various sinks in the host lattice and relaxation volume (i.e., volume of the crystal lattice "distorted" due to formation of a vacancy). It requires much less energy to form vacancies than interstitials ($E_v^f =< 1$ eV and $E_i^f = \sim 4$ eV). The value of C_v is much larger than C_i when the system is at thermal equilibrium. Relative values of both vacancies and interstitials is a matter of concern under irradiation. Vacancies, unlike interstitials, have low formation energy (~1–2 eV), low relaxation volume (0.1–0.5 atomic volume), and high migration energy (>0.5 eV).

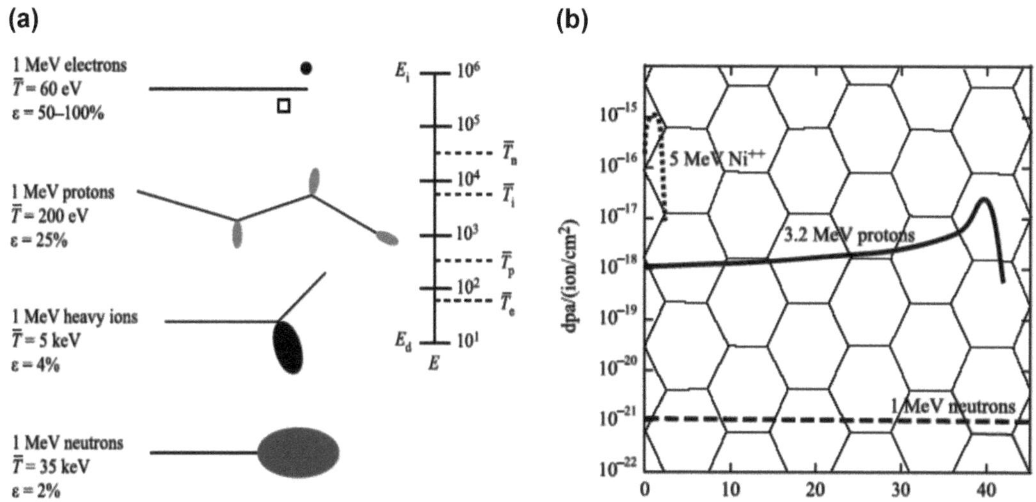

FIGURE 17.1

(a) Difference [28] in the nature of cascade damage produced during irradiation with different incident beams and (b) depth dependence of total damage produced for different incident beams. $\overline{T}$ —average energy transferred, $\overline{T}_{p,i,n,e}$ —energy of incident proton, ion, neutron, and electron, respectively, E_i, E_d—projectile and displacement energy, and ε is the unit charge.

(a) Reproduced with permission from Springer Verlag publisher.

Hence, vacancies are less mobile. On the other hand, interstitials are characterized with a high formation energy (2–4 eV), high relaxation volume (1–2 atomic volume), and low migration energy (0.05–0.1 eV), and, hence, high mobility. Multiple vacancies have less binding energy (0.1 eV) in contrast to high binding energy (1 eV) of multiple interstitials. Hence, vacancy clusters are less stable than clusters of self-interstitial atoms. Vacancies can bind to oversize solute or an oversize impurity atom (B.E = 0.2–1 eV in an fcc lattice), to lower the overall free energy of the system.

The concentration of vacancies and interstitials in a crystal subjected to irradiation is equal in the beginning and dictated by the production rate. However, these values change with time as they evolve, and their evolution is a balance between rate of production and loss of point defects and can be described by "point defect balance" equations. The major defect reactions are recombination reactions and loss to sinks like grain boundaries and defect clusters. The main defect reactions considered are:

$$V + I \rightarrow LS \tag{17.9}$$

where, LS is a lattice site. Point defect reactions with sinks, s, are written as:

$$V + s \rightarrow s \quad \text{or} \quad I + s \rightarrow s \tag{17.10}$$

The chemical rate equations are established, based on the production rate of defects, their recombination rate, and the rate of loss of defects at various types of sinks:

$$dC_v/dt = K_o - K_{iv}C_iC_v - K_{vs}C_vC_s \tag{17.11}$$

$$dC_i/dt = K_o - K_{iv}C_iC_v - K_{is}C_iC_s, \tag{17.12}$$

where C represents the concentration of either vacancy, v, or interstitial, i, or sink, s, and K's represent the rates of either production K_o or K_{iv} or K_{is} the reaction rate between vacancy and interstitial or interstitial/vacancy and sink. Rate theory provides means of calculating the defect density for various combinations of characteristics of incident irradiation and description of the host lattice. The point defects, excess of solubility value, will condense into two- or three-dimensional clusters, constituting the defect structure of materials under irradiation. The defects that are finally introduced in a crystalline lattice during irradiation are as follows: point defects (0D): vacancies and interstitials; line defects – (1D): dislocation lines; planar defects – (2D): dislocation loops; and volume defects (3D): voids and bubbles.

The study of defects under irradiation has been mainly in two classes of materials: neutron irradiation for a number of nuclear structural materials [29–34,38–41], and ion [5–7,37], or electron [7] irradiation in semiconductors [8,35–37], insulators [61,62], and alloys [63,64]. The study of recovery of defects during annealing of irradiated samples has provided [28] rich information regarding production, migration, clustering, and annealing of various defect species, which are well described in excellent reviews and books [5–20,28].

In recent years, considerable interest has been generated in the study of irradiation-induced defect evolution in materials for advanced technologies like fusion [64–73] and accelerator-driven systems [74,75]. In fusion project of ITER, plasma-facing material is exposed to 1500 MW of fusion power, with an average neutron wall loading of ~1 MW/m^2, in addition to thermal and electromagnetic loads, during shutdown/startup of the plasma. The main focus of recent studies has shifted to low-activation, high-temperature materials, like tungsten-based systems [71] and steels [64–70]. Since irradiation facility simulating fusion-like neutron spectrum is not available, as of now, anywhere in the world, most of the studies are confined to irradiation in a fast reactor or with high-energy proton and spallation neutrons [67,68], with helium around 900 appm, at various temperatures. The validity of the extrapolation of inferences from fast reactor irradiation studies to fusion situation has not been confirmed, and the in-service behavior could reveal more surprises.

In austenitic stainless steel, the candidate material for blanket system, the defect structure evolves [76,77] with the temperature. In the temperature range 100–200°C, dominant defect type is the stacking fault tetrahedra, with 20% as interstitial loops. At $T > 250$°C, density of defects decreases steeply, with dominant defect being <111> Frank loops of size 10–20 nm. At higher temperature of 300°C, 2–3 nm sized small cavities (10^{23}/m^3), stabilized by oxygen or helium appear. At still higher temperatures, the number density of cavities reduces due to their conversion to "bias-driven" voids, with reduction in number density of Frank-faulted loops.

The reduced-activation ferritic/martensitic steels (7–10% CrWVTa steels) are attractive [64–73,78–83] for in-vessel components and blanket applications since the martensitic structure with high-defect density-like dislocations, lath boundaries and carbides, provide high "trapping capacity" for the defects and helium ions, offering excellent radiation resistance to ferritic/martensitic steels. This advantage is in addition to the well-established [28] advantages of bcc iron or ferritic steel than fcc iron or austenitic steel: low migration energy (0.5 eV against 1.4 eV in fcc), low relaxation volume, and high binding energy of vacancies to solutes like carbon (0.8 eV against 0.4 eV in fcc). Even at a damage dose level of 200 dpa, the volume increase due to voids in bcc steels is only 1.2% in comparison to ~3–5% at 40–50 dpa for austenitic steels. However, use of ferritic steels is limited to about ~773K, due to deterioration in their high-temperature mechanical properties.

The distinct difference [81] in the defect evolution in fusion steels like F82H F/M steel subjected to a dose of 12–15 dpa, compared to behavior of conventional materials of the last century, has been the copious amounts ($10^{23}/m^3$) of large helium bubbles (1−2 nm) and amorphization of even the most stable carbides like the $M_{23}C_6$. Generally, the defect densities are $\sim 10^{22}/m^3$ of size 3−8 nm.

In order to improve the high-temperature capability of ferritic/martensitic steels, oxide dispersion strengthened (ODS) steel, a ferritic steel with homogeneous distribution of 5 nm yttria (Fig. 17.2), is being developed [29,84−89]. The evolution of defect structure in ODS steel is similar to that of ferritic steels.

The next class of alloys studied more frequently [90−94] is tungsten W and its alloys, for use in first-wall (500−800°C) and divertor (600−1300°C). The high melting point, high thermal conductivity, and good sputtering resistance make W most suitable for plasma-facing components. In addition to displacement damage, (n,γ) reactions promote formation [91,93] of transmutation products like rhenium (Re) and osmium (Os) in W. Hence, irradiation-induced defect structure has been studied in W and its alloys with Re and Os. Voids and dislocation loops are generally the most frequently observed defects. Based on temperature dependence of size and number density of voids, nucleation-dominated region of voids is found to be around 540°C and growth dominated region above 580°C. At high dose values like 0.22 dpa, void lattices have been reported [94], with its axes coincident with the underlying bcc matrix. These studies have enabled the development of "damage structure map" for W and its alloys.

Unlike the evolution of defect structure in bulk materials in nuclear technology, discussed earlier, most of the studies in irradiation-induced defects in nanosystems and thin films [8,35−37], either semiconductors or insulators, are confined to ion/electron irradiation, as stated earlier. The nature of defects is different due to limitation in the size of the system and differences in the nature of interaction: (1) most of the incident beam passes through the nanosystems, with very little deposition of their energy, (2) unlike bulk metals where displacement damage is more dominant with conversion of

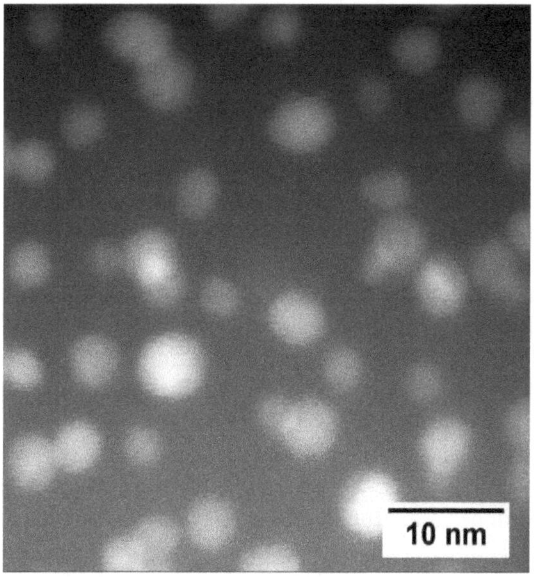

FIGURE 17.2

High-angle annular dark field high-resolution image of 5 nm yttria in ODS-9Cr-1 Mo steel.

Reproduced with permission from Elsevier.

electronic excitement into "heat," insulators experience more damage due to electronic excitement, leading to breaking of bonds, and (3) local temperatures can exceed melting point.

Irradiation-induced defects in semiconductors have opened up exciting possibilities for the development of device technology. High-energy irradiation offers [7] a means for controlled introduction of point defects, essential for improving the yield, performance, and reliability of semiconductor devices. The point defects created by irradiation in semiconductors act as trapping sites for free carriers at deep defect levels introduced into the band gap. Consequently, the resistance of the damaged region, as in zinc oxide [95], a wide band gap semiconductor, increases by orders of magnitude, which is used for isolating adjacent regions in devices. Cavities or open volume defects are very often observed, decorated on its surface with metal impurities, due to solubility differences. Hence, this method is used for removing trace-level, metal impurities in semiconductors.

The most common type of defects in irradiated graphene is shown [7] to be point defect, vacancies and adatoms, in addition to dislocations. However, unlike metals, there is extensive "restructuring" around the vacancy produced by irradiation, leading to entirely different atomic network and bonding. Such an effect in carbon nanotubes results in "healing" effect of "hole" formed during irradiation.

Displacement of several thousands of lattice atoms of silicon and germanium in localized regions around the track of 40 keV heavy ion irradiation has been reported [8,35,37]. The reordering temperature for silicon is 260°C and germanium 180°C for low dose, where there is no overlap of disordered region. At doses higher than 10^{14} ions/cm^2, both show amorphization, suggesting very little dynamic annealing within the cascades. On the contrary, significant dynamic annealing and hence the difficulty in amorphizing gallium nitride under ion irradiation is also reported [96]. Both germanium [97] and silicon [98] recrystallize at much higher temperatures like 510 and 380°C, respectively.

The stability and evolution of defects during annealing in semiconductors are more complex than metals, due to following reasons: (1) different clusters of defects like vacancies, divacancies or trivacancies have different charge states, leading to difference in their migration energy; (2) complex defect interactions due to covalent bonding, which is responsible for the ease of defect trapping at broken bonds and high activation barrier for breakup of defect complex. Most of these aspects of creation, migration, and usefulness of defects are studied using Raman and positron spectroscopy, ion beam analysis, and Hall effect. Detailed reviews are available [6–20] regarding the defect structure in irradiated semiconductors and insulators.

Very recently, "cluster or molecular ion irradiation" has attracted [99–107] a lot of attention, since they could be optimized to obtain "ultrashallow p-n junctions" enabling reduction in size of integrated circuits, a necessity in microelectronics. Under identical conditions of irradiation, cluster ion irradiation produces order of magnitude more defects, in "near-surface" region of the substrate than atomic ion, called the "molecular effect." It is demonstrated that surface charging and beam loss effects in ultralow energy ion implantation are minimized by replacing [108] 500 eV B^+ ion implantation of silicon with 10 keV $B_{18}H_{22}^+$ molecular ion implantation. The "molecular and cluster ion irradiation" has opened up new vistas in the field of ion beam–related phenomena, like interstitial damage, sputtering, secondary electron and ion emission, and ion beam mixing. The sputtering yield during Au cluster irradiation of Au substrates was [103] ~14,000 atoms, the highest ever yield reported in any metal. Broadening due to ion beam mixing during irradiation of tantalum was [109] higher when Al_3 cluster was used instead of aluminum ions, suggesting enhanced energy deposition during cluster ion irradiation. Cluster ion irradiation has provided [110] a new method of synthesis of graphene with reduced defects, by implantation of carbon and its clusters on nickel films.

Atomistic simulations have shown [111–113] that the nature of defects during cluster ion irradiation has been shown to be very different. Unlike point defects and small defect clusters in B1 irradiated silicon, B18 implantation leads [111] to mainly large interstitial clusters, cavities, and amorphous zones. The depth distribution of damage and the type of defect clusters are [107] drastically different during cluster ion irradiation. Molecular or cluster ion irradiation is a [7,114] promising tool for nanoscale surface modifications, such as ultrasmoothing, high-speed etching, and thin film formation required for optoelectronics devices. It is possible to tune surfaces to atomic-level precision, and this is an emerging field of research and technology for optical communication, optical storage, imaging, lighting, sensors, and security.

5. IRRADIATION-ENHANCED DIFFUSION

Diffusion, the random, incessant, movement of atoms in a lattice due to thermal vibrations, is statistical in nature. The vacancies that are in thermal equilibrium with the lattice enable the atoms to diffuse faster. The atoms move in that direction, which eliminates any gradient that may exist in the concentration of defects. Fick's laws, which describe the diffusion kinetics, the microscopic mechanisms of diffusion, the conventional rate laws, self- and interdiffusion coefficients, activation energy analysis, and temperature dependence of diffusion, constitute part of many textbooks [28–30,45].

Diffusion of atoms, A, in a lattice by "vacancy mechanism" is given by

$$D_A^V = f_V \, D_V \, C_V \qquad (17.13)$$

where f is the correlation coefficient, D the diffusion coefficient, and C the concentration. The subscript V refers to vacancies. If other mechanisms operate, the effects are additive in nature. In a pure metal, radiation-induced diffusion is given by

$$D_A^{irr} = D_V C_V + D_i C_i \qquad (17.14)$$

and the process of diffusion is generally accelerated, due to the presence of excess point defects.

Extensive theoretical studies have been developed [28] for predicting many aspects of irradiation-enhanced diffusion: temperature dependence, mechanism, activation energies, and defect dynamics. Useful predictions about migration enthalpy, diffusion coefficients, and activation energies in several systems could be arrived at [28] using several modeling methodologies. Most of the experimental investigations in many iron-, copper-, and silver-based systems use classical methods [28] like resistivity [115], Zener relaxation [116], tracer diffusion [117], Auger spectroscopy [118], and secondary ion mass spectrometry [119]. There are also indirect methods [120,121] of arriving at mechanism of radiation-enhanced diffusion like kinetics of clustering or ordering. Radiation-enhanced diffusion has been studied using electron, ion, and neutron irradiation. Many reviews and books [28,30,72,76,115,121] deal with the subject of irradiation-enhanced diffusion.

In situ transmission electron microscopy studies in electron-irradiated Ag-based alloy (Ag-9%Zn) has shown [122,123] that the activation energy for irradiation-enhanced diffusion is 0.3 and 0.23 eV for low- and high-temperature range, which is half of activation energy for migration of mono-vacancies and interstitials, respectively, for thermal diffusion. It is concluded [122,123] that irradiation-enhanced diffusion at $T < 70$–$80°C$ is vacancy controlled, while at higher temperatures, interstitials dominate the diffusion mechanism, in agreement with theoretical calculations [124].

Ion irradiation and defects recovery studies to evaluate the irradiation-enhanced diffusion have been carried out extensively [125,126]. The influence of irradiation-induced diffusion is more dominant at lower temperatures than higher temperatures. This has been demonstrated in proton-irradiated Ni−Cr alloy [127]. The temperature dependence of radiation-enhanced diffusion is understood in terms of reduction in irradiation-induced vacancies, due to higher recombination reactions at temperatures, where interstitials become more mobile. Positron annihilation spectroscopy has been used [126] for the evaluation of vacancy migration and formation energy in proton-irradiated advanced austenitic stainless steels, D9 with different titanium content. The diffusion coefficient of boron in silicon under irradiation is evaluated [128] to be two orders of magnitude more than thermal diffusion, in the temperature range of 800 to 500°C. Radiation-enhanced diffusion directly is [129] responsible for the modification of creep properties, since the creep behavior depends crucially on diffusion characteristics of the material.

At extremely low doses, ion irradiation has demonstrated [28,130] that "replacement collision sequence" (Fig. 17.3) is one of the major operative diffusion mechanisms and occurs along <100> for most fcc systems. The consequent "dynamic crowdion propagation" destroys order in ordered phases or even dissolves preexisting phases or preexisting defects like twin boundaries.

Another effect often observed due to the above crowdion mechanism is the "ion beam mixing" [131−134] leading to significant differences in the chemistry of implanted region. Many experiments use ion beam mixing of multilayers of semiconductors [135] like AlAs/GaAs, InGaAs/GaAs, and InGaAs/InP to achieve required threshold current density and external quantum efficiency, thus leading to "band gap tuned laser."

An unusual mass transport during ion implantation of isolated nanostructures on suitable substrates has been reported [136]. The mass transport from nanostructures extends beyond the range of the incident ions. This forward mass transport is explained on the basis of thermal effects and creation of rapid diffusion paths at nanoscale regime during the course of ion irradiation. The unusual mass transport is found to be associated with the formation of nanoalloys at subsurfaces. Fig. 17.4 shows a typical example of formation [137] of fcc $Co_xPt_{(1-x)}$ phase, during annealing of silica implanted with Co^+ and Pt^+ ions.

An important consequence of radiation-enhanced diffusion is the segregation of constituent elements in the host lattice, which is discussed next.

6. IRRADIATION-INDUCED SEGREGATION

One of the consequences of radiation-enhanced diffusion is [28] the chemical inhomogeneity introduced in the microscopic regions in the original lattice. The nature and extent of inhomogeneity crucially depends on interaction of point defects with lattice defects and solutes and the mobility of the complexes formed. This phenomenon is referred to as the "radiation-induced segregation," which will be briefly discussed next.

The basic mechanism that gives rise to radiation-induced segregations is the persistent defect fluxes during irradiation and the preferential coupling of solutes to the defect fluxes. Net defect fluxes, which are spatially inhomogeneous, are generated due to following reasons: (1) the defect production during irradiation, especially ion, is spatially nonuniform, and (2) defect annihilation in sinks like grain boundaries introduces flux toward such sinks. The radiation-induced segregation can occur by (1) inverse Kirkendall

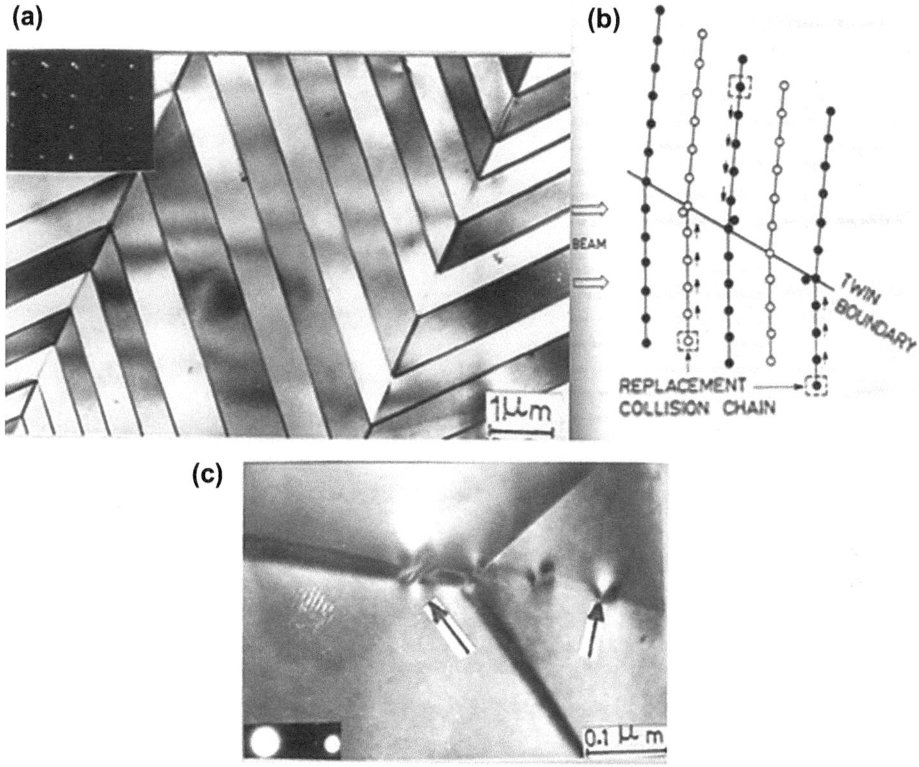

FIGURE 17.3

Replacement collision sequence in ion-irradiated Al_8Mn_5 [130], a rhombohedrally distorted bcc phase in Al-40% Mn alloy, leading to accumulation of defect clusters at the boundary. (a) Transmission electron micrograph of unirradiated Al_8Mn_5, showing crystallographic twins. Inset shows selected area diffraction along <001>. (b) Schematic of mechanism of replacement collision mechanism and (c) formation of defect clusters along the twin interface, after 50 keV helium ion irradiation to a dose of 2.5×10^{15} ions/cm^2. Further increase in dose leads to disappearance of twin interface. Inset shows the two beam condition with [106] spot.

(a) and (c) Reproduced with permission from Elsevier.

effect and (2) defect flux-driven segregation. The final change in microscopic chemistry [138,139] will be the net result of both.

Fig. 17.5 illustrates [140] the segregation of nickel near edge dislocations in EI-847, an austenitic stainless steel irradiated with Cr ions at 625°C. The concentration of nickel in matrix is reduced with increase in dose. Simultaneously, the amount of nickel around loops increases. As a result, there is a loss of beneficial solute nickel in the matrix. Nickel in austenite lattice suppresses void swelling (see Section 8) due to high binding energy with vacancy (0.35 eV). This phenomenon is referred to as "solid solution decay." Consequently, fine voids at smaller dose (Fig. 17.5b) rapidly grow (Fig. 17.5c) at higher dose value.

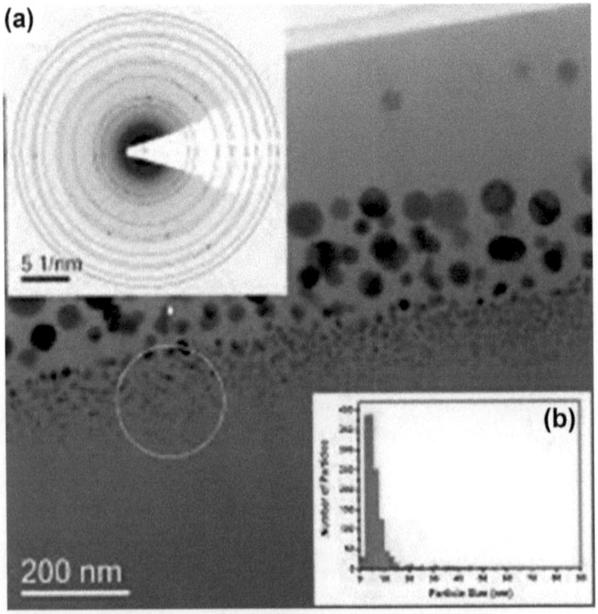

FIGURE 17.4

Cross-section transmission electron micrograph, along with selected area diffraction pattern and histogram of particle size, illustrating [137] the formation of fcc $Co_xPt_{(1-x)}$ phase, after annealing of silica implanted with Co^+ and Pt^+ ions.

Reproduced with permission from Springer-Verlag.

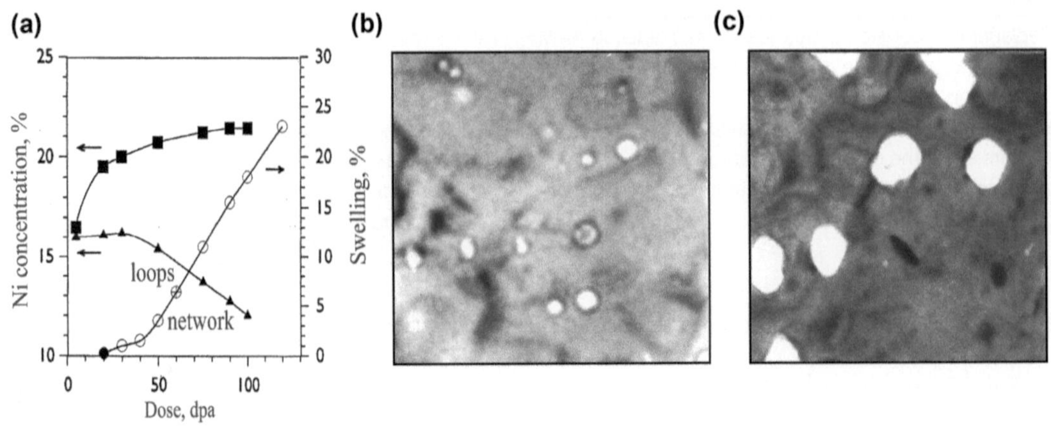

FIGURE 17.5

Radiation-induced segregation of nickel in Cr ion–irradiated ($T = 625°C$) EI-847 austenitic steel. Nickel has segregated toward loops (■) with a concomitant reduction in austenite matrix (▲), leading to solid solution decay and hence more void swelling with increase in dose. (a) Amount of nickel and swelling with dose. (b) Fine voids for 20 dpa and (c) coarse voids after 60 dpa.

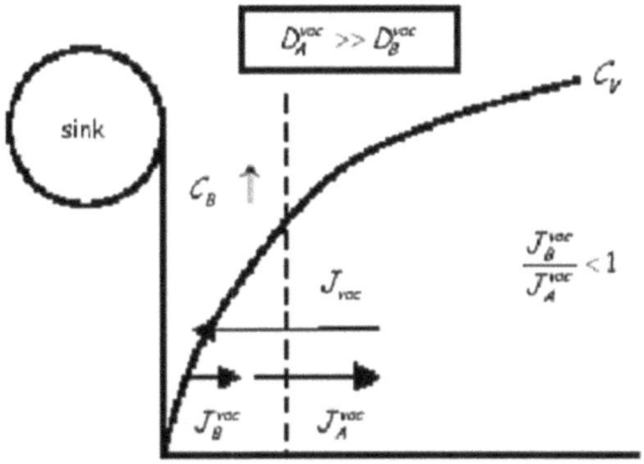

FIGURE 17.6

Radiation-induced segregation altering the original composition of the alloy near strong sinks like grain boundaries.

6.1 INVERSE KIRKENDALL EFFECT

In general, in a binary alloy AB, the vacancy flux J_V toward the sink S gives rise (Fig. 17.6) to atom fluxes J_A and J_B in opposite directions ($J_V = J_A + J_B$). The atomic fluxes, J_A and J_B will be proportional to the concentrations C_A and C_B of the alloy and their diffusion coefficients D_A and D_B. There will be no change in the alloy composition near the sink if $D_A = D_B$. If $D_A \neq D_B$, there is a net flux of faster diffusing element away from regions near the sink, resulting in "depletion" of that component near the sink. The same process will also take place in the case of interstitials, with the only difference that the atom fluxes and the defect fluxes are in the same direction. The two mechanisms, i.e., interstitial flux and vacancy flux, will be operating simultaneously, and the depletion or enrichment observed near the sink will be a net result of the two.

6.2 SOLUTE DEFECT COMPLEXES

In addition to inverse Kirkendall effect, defect flux-driven segregation can also result from the formation of bound defect solute complexes. The defect–solute complexes will be mobile, if

$$E_B^{d-s} + E_m^d < E_m^{d-s} \tag{17.15}$$

where E_B^{d-s} is defect–solute binding energy, E_m^d is migration energy of the defect, and E_m^{d-s} is migration energy of the complex.

The complexes that satisfy the previous relation can be viewed as distinct entities, which can migrate through the crystal lattice. In contrast to inverse Kirkendall effect, both interstitial and vacancy type of complexes will carry the solute toward the sink. Defect–solute interactions become less effective at higher temperatures, and as a result, segregation via defect–solute complexes is expected to dominate at lower temperatures, whereas inverse Kirkendall effect will be more effective at higher temperatures.

For example, the undersized solute elements like Ni, Si, and P, which are present in austenitic stainless steel, move [28,141] toward neutral sinks like grain boundaries and interfaces between precipitates and austenite matrix, while oversized elements like Cr, Mo, and Nb show depletion around these sinks. Nickel and silicon also get enriched near biased sinks like dislocations, while titanium, molybdenum, and chromium move away from such biased sinks. It is also observed that the segregation is less for mobile dislocations and more for conservative Frank loops.

The segregation of chromium and carbon along grain boundaries of heavy ion (Fe)-irradiated ODS reduced activation ferritic martensitic (RAFM) steel has been reported [142] recently, using 3-DAP and STEM.

Nordlund and Averback [58] have shown that ion beam mixing across interfaces of Co/Cu and Ni/Cu favors more copper in cobalt or nickel than the reverse, in addition to more vacancies in copper. This has been understood in terms of inverse Kirkendall effects in agreement with differences in the melting points. Radiation-induced segregation is responsible for the observation [143] of "age hardening" in steels with copper around 0.4%, due to slow nucleation of copper-rich precipitates, with a very large incubation time. This phenomenon assumes relevance since the kinetics is fastest in the range of temperature of operation of pressure vessels in thermal reactors. Another example of industrial relevance of radiation-induced segregation is the faster stress corrosion cracking [144] in austenitic stainless steel, under irradiation, due to segregation of chromium to grain boundaries. Studies to reduce the propensity of radiation-induced segregation have shown [145] that addition of oversized solutes offer strong attractive force to vacancy, increasing recombination rate, thereby reducing radiation-induced segregation.

7. RADIATION-INDUCED/ENHANCED PHASE TRANSFORMATION

Irradiation-induced phase transformations have provided a paradise for physical metallurgists in the last few decades. Inherently, materials under irradiation are in a "far from equilibrium" state. Irradiated systems can shift either far away from thermal equilibrium or approach the equilibrium faster, depending on whether energetics or kinetics is the dominating factor.

The recent surge of activities in nanomaterials has converted [146–153] irradiation-induced phase transformation into a "phase tunable strategy" for tailoring functional materials. Electron/ion beam irradiation is used [6,7] to fine-tune phases of required optical and magnetic properties, for information storage, bar coding, and detection. Complex superlattice of alternate lamellae of oxidic and metallic layers could be fabricated [146,147] using localized phase transformations during low-energy proton irradiation, leading to patterning of devices.

Ion beam mixing [148–153], a phenomenon that has been well studied in recent years, is very often known to lead to solubilities [150,151] higher than thermodynamic predictions. Implantation of substrate of A lattice with B ions of any species like argon or helium leads to mixing of A and B atoms at the interface. It is possible to tune the irradiation parameters to achieve solubilities of B in A lattice higher than thermodynamic solubility or make two immiscible elements mix at the interface region and form a solid solution. In addition to short range of mixing, with displacements <500 Å, involving mixing of 10^3 or more atoms, second-phase formation also is reported [121,148,152,153]. Extensive studies using Rutherford backscattering and glancing angle XRD have reported [154,155] formation of second phase in many systems like formation of Cu_3Si in argon-irradiated Cu/Si bilayer. Many review articles [121,148,152,153] have also compiled all the systems in which such ion beam mixing is reported, along with the evaluation of "mixing efficiency." Ion beam mixing, which is a stochastic process, is dominant in the low-temperature range, where thermal diffusion is low and found to be due to combination of "collision events" and thermal spikes. The operative mechanism also is reported to depend on whether the substrate is refractory or noble. Kim et al. [156], have estimated ion beam mixing parameter experimentally in many refractory and noble metals. The mixing parameter of refractory metals matches well with collisional mixing theory, while in noble metals, it is higher. However, subsequent studies in aluminum alloys support [157] the role of thermal spikes. Theoretical predictions show that thermal spikes are operative mainly in ion beam mixing at low temperatures.

Irradiation-induced phase transformations are caused [158–163] either by introducing (1) relative changes in the stabilities of various phases, or (2) kinetically enhanced by the vacancies generated during irradiation, or (3) "recoil dissolution."

Both allotropic phase transformation in a single element like carbon [7,164,165] to diamond and polymorphic transformation in a compound like zirconium dioxide from most-stable monoclinic to tetragonal are understood [164,165] in terms of destabilization of one phase with respect to the other, due to the irradiation-induced point defects. The crystalline phase transformations in many systems like Y_2O_3 [161] and Al_2O_3 [160], have been attributed to various mechanisms like effect of knock-on collisions, ease of nucleation of extended defects due to presence of structural oxygen vacancies that are sensitive to energetic perturbation, and stress relief effect. The transformation of quasi-crystals to bcc crystals by electron irradiation has also been observed [166], as shown in Fig. 17.7. The order-disorder transformation, ω phase transformation in titanium and zirconium alloys are found to take place in both the directions, depending on the irradiation conditions, which have been discussed in detail elsewhere in this book. The direction of the transformation leading to either ordering or disordering depends on the balance between disorder introduced by irradiation, order reinforced by radiation enhanced diffusion, and ordering energy.

Irradiation-induced amorphization is generally reported in intermetallic compounds [167–170] with either narrow compositional range or line compounds. A typical example is shown (Fig. 17.8) for a line compound of Al_6Mn in an Al–Mn alloy.

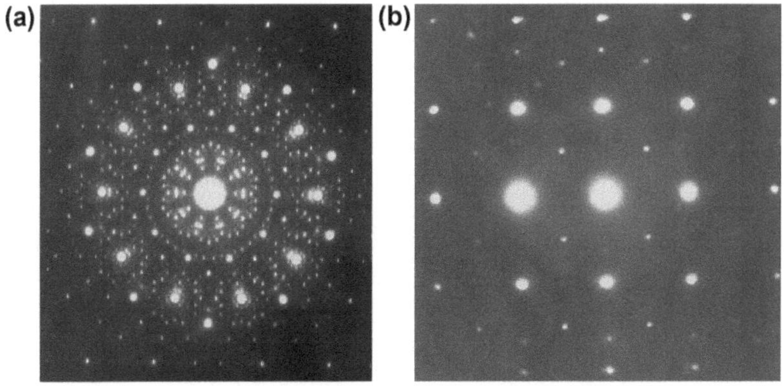

FIGURE 17.7

Irradiation-induced phase transformation in quasi-crystals, studied using in situ transmission electron micrography: (a) Five-fold selected area diffraction pattern of an unirradiated decagonal $Al_{70}Ni_{15}Co_{15}$ system, and (b) B2-type ordered lattice under 400 keV electron beam at 107K up to a dose of 3.6×10^{24} e/m^2 at the dose rate of 4×10^{21} e/m^2/s.

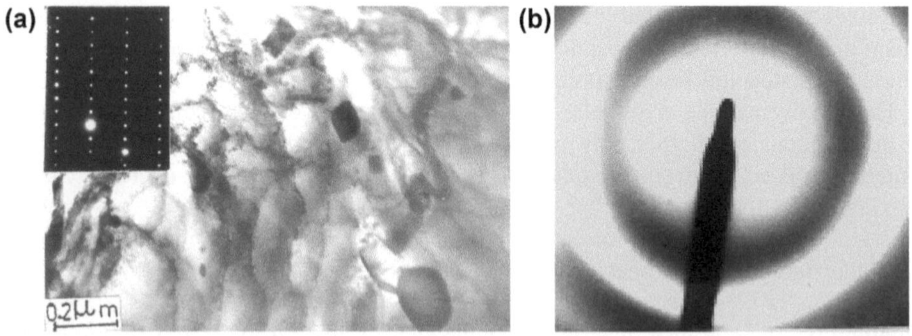

FIGURE 17.8

(a) Transmission electron micrograph of fine Al_6Mn particle in an fcc α matrix of Al-14 a/o Mn alloy. Inset shows the selected area diffraction pattern along <2–20> showing the crystallinity. (b) Diffraction pattern showing evidence for amorphization after irradiation with 30 keV helium ion irradiation to a dose of 4×10^{16} ions/cm^2.

Reproduced with permission from Elsevier.

Similar amorphization has been observed frequently in SiC [171] and other compounds like Fe_3B [172] and B_4C [173]. In complex multicomponent systems, amorphous layer of $Zr(Cr,Fe)_2$ around the core of crystalline $Zr_2(Ni,Fe)$ under neutron irradiation is reported [59], demonstrating the sensitivity of stability on local chemistry. Production of defects in ordered compounds gives rise to local deviations from stoichiometry due to creation of antisite defects. These defects introduce large change in the free energy of the line compounds as compared to those with wide compositional range of existence. Once the increase in the free energy of the crystal with defects exceeds far above the free energy of the amorphous equivalent, amorphization process is initiated. MD simulations have been carried out to predict the dose required for the amorphization of a compound, with reasonably good agreement with experiments in many systems [174].

The critical defect density C_D required for amorphization can be estimated from the difference in free energy, ΔF, between crystalline and amorphous phases,

$$C_D E_F \sim \Delta F \tag{17.16}$$

where E_F is the formation energy of defects. Typical value of ΔF is around 4 kJ/mol and $E_F \sim 4$ eV (interstitials). Therefore, C_D turns out to be 2%, which is far higher than normal steady-state point defect density during irradiation, which is around 10^{-3}. The steady-state values of point defect concentration C_V and C_I can be obtained by solving the coupled differential equations,

$$dC_V/dt = K - \alpha_0(D_I + D_V) - \beta_V D_V C_V \tag{17.17}$$

$$dC_I/dt = K - \alpha_0(D_I + D_V) - \beta_I D_I C_I \tag{17.18}$$

where K is the defect production rate, β the sink density for vacancies and interstitials, α_0 the recombination constant, and D the diffusion constant. High point defect concentrations can be achieved only if their mobility is very low. For typical values of α_0 (5×10^{16}/cm^2) and a dpa rate of 10^{-3}/s to obtain steady-state point defect concentration of 0.02, D_I is 10^{-16} cm^2/s. This value would imply an activation energy for interstitial diffusion around 1 eV—an order of magnitude higher than interstitial migration energies in metals. Though there is not much data on interstitial mobility in intermetallic compounds, activation energies as high as 1−1.3 eV have been reported [175] in systems like FeAl. Such low-defect mobility alone is not enough to accumulate defect densities high enough for amorphization. It is also necessary to prevent their loss by mutual recombination. An additional requisite for amorphization to occur in intermetallic compounds is the low recombination volume. The athermal recombination of point defects due to elastic interaction should be in a volume far below the 100 lattice points, the approximate value for metals.

The second factor, kinetic in nature and favorable for irradiation-induced precipitation, is the increase in mobility due to the point defects introduced in the lattice during irradiation. An example of this mechanism is [176] the recrystallization of amorphous phase like Zr_2Cu, which was a kinetically driven system under irradiation. In ferritic steels like RAFM, irradiation-induced precipitates like α', η or M_6X, χ, σ, M_3P, and MP are reported [177] at neutron doses higher than 30 dpa. Very often, the classical TTT curves of carbides in steels are reported [158,178] to shift toward left, i.e., smaller times, under favorable irradiation conditions. The irradiation-induced point defects react with solute elements in many alloy systems, based on Al, Ni, and Cu resulting in the precipitation of many phases [158] not expected during thermal exposure. The enhanced vacancy concentration many times accelerates the kinetics of precipitation. These observations have been supported [122−128] by rigorous measurement of irradiation-induced diffusion coefficients, which are found to be two orders of magnitude higher than the thermal component. In a pure material, the diffusion coefficient under irradiation is given by

$$D_{irr} = D_v C_v + D_i C_i \tag{17.19}$$

wherein D and C represent the diffusion coefficient and concentration. Since C_v and C_i are two orders of magnitude higher than thermal equilibrium concentration, D_{irr} is far higher than the thermal component.

On the other hand, the technological impact of irradiation-induced phase transformations are too many. Embrittlement of pressure vessel in a reactor is [177,179] due to the irradiation-induced nm-sized Cu precipitates, consequent to radiation-induced segregation stated earlier.

Recoil dissolution refers to the dissolution of thermally stable precipitates, which are destabilized under irradiation, as reported [180] in dissolution of β phase in electron-irradiated Ni-12% Be alloy. Recoil dissolution is generally understood [181] in terms of two possible mechanisms: (1) ejection of atoms by "ballistic" collision, and (2) modification of thermodynamic stability due to increased solubility of solute under enhanced defect concentration. Dissolution of strengthening phases MX, M_2CN in steels, under high-dose irradiation, has resulted [181] in serious impairment of initial properties. Preexisting chromium carbides in RAFM steels, under neutron irradiation, are reported [29] to dissolve and be replaced by newer carbides, with a different crystal structure. In synthetic alloys, the combined effect of recoil dissolution and radiation-enhanced diffusion, has altered [181] the coarsening rate of secondary phases and replaced [180] platelike β particles in Ni-12%Be alloy with superfine Be-rich particles.

The change in energetics of secondary phases or ballistic effects can also lead to dissolution, as in the case of nm-sized precipitates or dispersoids in Zircaloy-4 [158] and ODS ferritic steels [182,183], deteriorating the high-temperature mechanical properties. Extensive research has been recently carried out [86,183,184] in the stability of nm-sized yttria in ODS steels, being developed for high-thermal efficiency. The dispersoids were found to be fairly stable under neutron irradiation up to about 50 dpa, which is understood [182] in terms of loss of point defects at the interface between matrix and the dispersoids. Increasing the temperature to 298 and 723K to a dose of 47 dpa retained the stability of the dispersoids. In contrast, irradiation of yttria dispersed ODS steels with electrons at high temperatures and ions at room temperature, resulting in dissolution of dispersoids [183,184].

Sometimes radiation-induced precipitation can be triggered by inverse Kirkendall effect, explained earlier. The nickel and silicon in stainless steel gets enriched near dislocations and, consequently, a new "deleterious" phase called the *G* phase, nickel silicide (Ni_3Si) [63,185] forms under irradiation, which is not predicted by phase diagrams. Fig. 17.9 shows [186] an example of precipitation of η-(M_6C) carbides, when 20% cold-worked stainless steel was irradiated up to 40 dpa in Fast Breeder Test Reactor, Kalpakkam, India. It is known that M_6C is not expected to form in 316 austenitic stainless steel. Its formation under neutron irradiation in a reactor could only be understood based on irradiation-induced segregation.

As is seen from Fig. 17.9, η-(M_6C) carbides favor the formation of voids at its interface and additionally, it depletes austenite (solid solution decay) of beneficial element like nickel. It is known that nickel with high binding energy with vacancy is efficient in reducing the swelling resistance of the austenite matrix. Hence, the nickel content in next-generation D9 alloys is increased to 15% from 8% of 316 austenitic stainless steel.

Irradiation is reported [28,162,187,188] to enhance spinodal instabilities due to the following reason—recombination of point defects along the interface between coherent particles and matrix is common. In certain situations, flux of self-interstitials to the particles is not balanced by any reverse interstitial flux. As a result, certain component of the system, which occupies interstitial position preferentially, gets pumped into the particle continuously. This leads to spinodal-like compositional instabilities. The most important criterion is the existence of phenomenological diffusion coefficients, relating a flux of one component to concentration gradient in another.

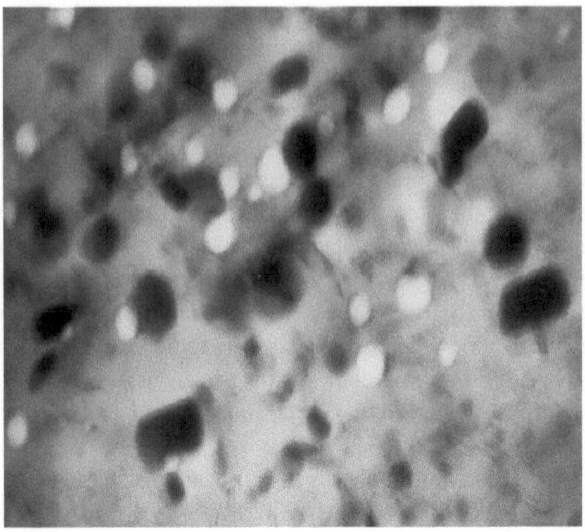

FIGURE 17.9

Precipitation of η-(M_6C) carbides in wrapper, 20% cold-worked austenitic stainless steel, irradiated in Fast Breeder Test Reactor in Kalpakkam, India, to 40 dpa. Association of voids around each carbide confirms the accumulation of vacancies along the interfaces of austenite and carbides.

Reproduced with permission from Universities press.

Another phenomenon, which displays distinct difference under irradiation in contrast to thermal or aging behavior, is the coarsening kinetics. Ostwald ripening [189] is governed by Gibbs–Thomson rule, as stated in any textbook on physical metallurgy. The solubility of solute on the surface of a particle in a matrix is related to the curvature of the particle, leading to higher solubility on the surface of finer particles than coarse particles. The difference in the solubility is known to be the driving force for the Ostwald ripening, leading to growth of coarse particles at the expense of finer particles. Under irradiation, larger particles are found to dissolve [28,58,63], transferring mass from larger to smaller particles, finally leading to "monosized" distribution of fine particles. This phenomenon is termed as "inverse Ostwald ripening," and has been studied using kinetic Monte Carlo methods and experimental methods mainly of Au ion irradiation in SiO_2 [190]. This phenomenon is understood in terms of unusually strong binding of solute B atoms on the surface of particles of A atoms, sometimes leading to negative surface energy, increasing the total surface-to-volume ratio. The total free energy of such system is reduced by transfer of mass from larger particles to smaller ones, exactly opposite to conventional Ostwald ripening.

8. INFLUENCE OF RADIATION-INDUCED MICROSTRUCTURE ON MECHANICAL PROPERTIES

The degradation of materials like steels, exposed to severe irradiation as in a fast reactor [28,39,191], manifest in the following ways: void swelling, irradiation creep, irradiation hardening, irradiation embrittlement, and irradiation growth, which will be briefly described next.

Void swelling stems from the "preferential bias" (2%) of matrix dislocations toward interstitials than vacancies, leaving behind an excess concentration of vacancies. The excess concentration of vacancies condense into "voids," when its amount exceeds the equilibrium concentration of vacancies, at the temperature of irradiation. Void swelling measured in terms of normalized increase in volume ($\nabla V/V$) varies with dose (Fig. 17.10). Region 1 (transient regime) corresponds to the nucleation of stable voids and region 2 (linear regime) to steady-state growth of voids. The threshold dose ϕ_c needs to be maximized to increase the burn up of the fuel in a nuclear reactor.

Void swelling is generally dominant in a specific temperature range for each material, since swelling is less at higher temperatures due to emission of vacancies from voids, while at lower

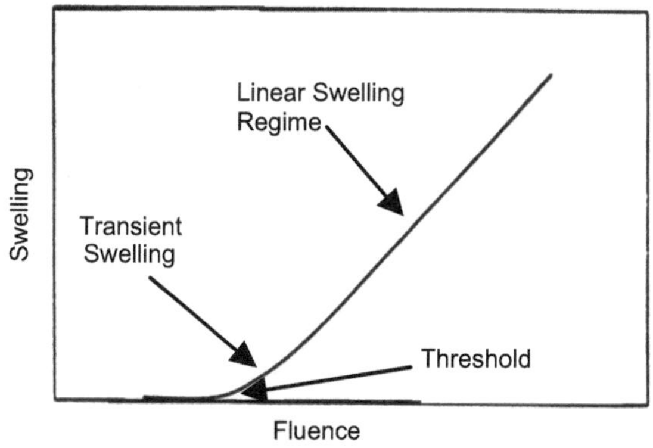

FIGURE 17.10

Schematic of variation of void swelling with dose.

temperatures mobility of vacancies is not sufficient enough to condense into voids. Rate theory models have been developed to predict swelling rates. Void swelling suppression mechanisms mainly focus on [39,190,191] promoting recombination reactions between vacancies and interstitials annihilating each other and some of them are as follows: reduce origin of helium (B.E. with vacancy ~2 eV), which stabilizes voids, or add solute with high binding energy with vacancies or introduce optimum density of dislocations or distribute fine coherent second-phase particles in the matrix, like TiC or Ni_3Si in austenite, since the interface promotes recombination reactions, or choose the system with high binding energy with vacancy, high formation energy for vacancies, and low mobility for interstitials like ferritic steels. Most of these methods are known [39,190,191] to reduce the severity of other degradation mechanisms during irradiation, stated earlier, since all the degradation mechanisms are caused by the irradiation-induced point defects and their interaction with lattice inhomogeneities. Current efforts are [192–194] focused on development of multiscale modeling methods to predict swelling rate for multicomponent, multiphase systems. The main challenge appears to be the development of reliable interatomic potentials for multicomponent systems.

The current interest in RAFM steels for fusion applications has motivated [68,69,195] a large number of studies on various aspects of neutron irradiation in RAFM steel, with respect to defect production, evolution of point defects into clusters, and irradiation-induced precipitation and their effect on void swelling and hardening. Akihiko Kimura [69] has reviewed all these studies and summarized the various events in a comprehensive manner as shown in Fig. 17.11.

"Irradiation enhanced/induced creep" refers to either "augmentation" of thermal creep or introducing creep, even in the absence of thermal creep due to irradiation. It has been demonstrated [196,197] that the irradiation creep damage increases with hoop stress in 20% cold-worked austenitic

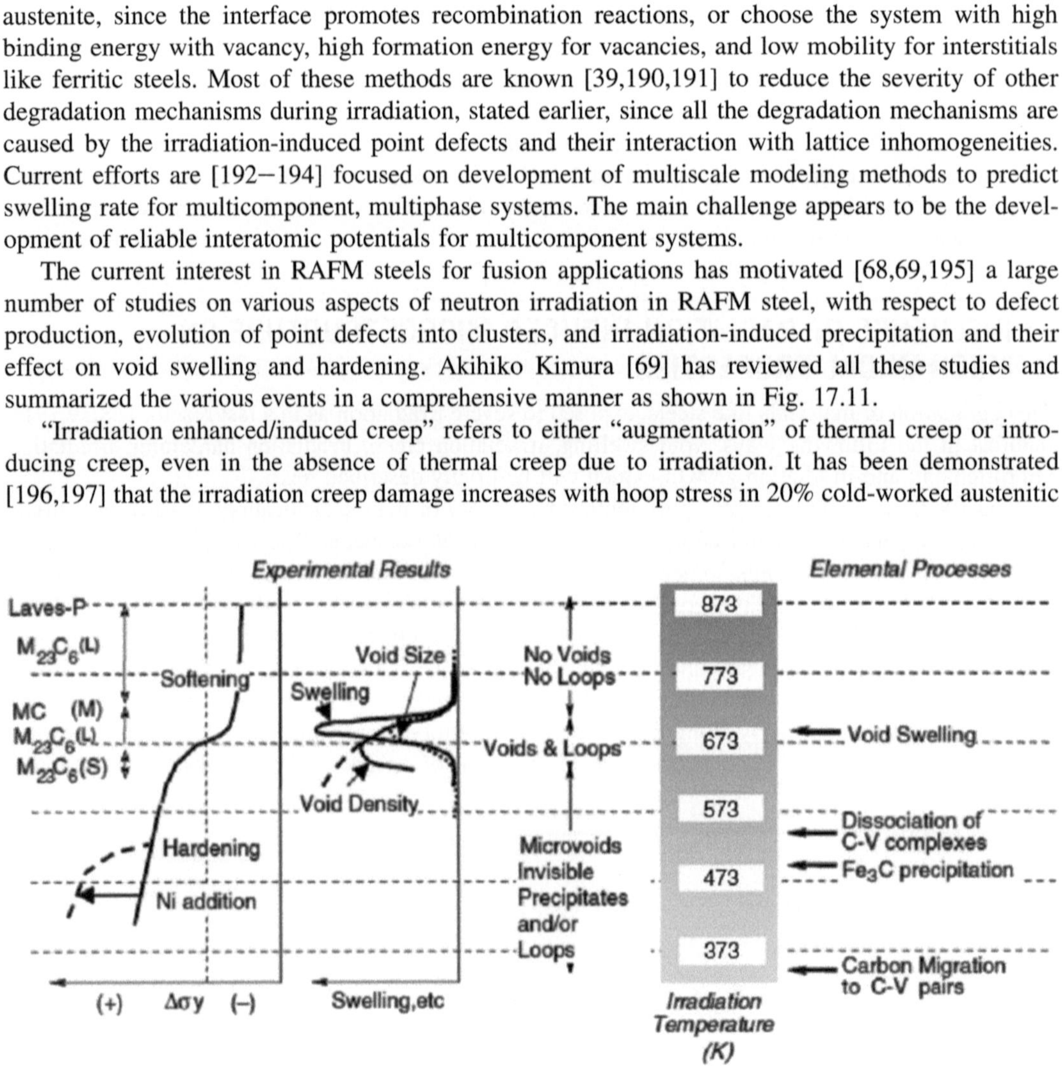

FIGURE 17.11

Comprehensive summary of various events during neutron irradiation at various temperatures in RAFM steels.

Reproduced with permission from Elsevier.

stainless steel exposed in EBR-II reactor. This observation highlights the role of engineering stress on each of the core components of a nuclear reactor.

At $T < 0.3\ T_m$, reduction in irradiation creep rate is due to the difficulty of dislocations to glide in a matrix hardened by the irradiation-induced defect clusters. At $T > 0.3\ T_m$, irradiation increases creep rate, due to enhanced climb-controlled glide by irradiation-induced vacancies. At $T > 0.6\ T_m$, thermal creep becomes dominant. Most often, swelling influences irradiation creep. At very small dose levels, swelling enhances creep rates by pinning effect of fine voids. At higher dose levels, creep reduces due to unfaulting of dislocation loops to increase dislocation density and reduced availability of interstitials to dislocations. At still higher dose, creep disappears, while swelling continues.

Numerous mechanisms have been identified [39,190] to understand the irradiation creep: "stress-induced preferential absorption" (SIPA), stress-induced preferential nucleation (SIPN), and the point defect migration along the grain boundaries. SIPA refers to creation of more new planes of atoms nearly perpendicular to the stress axis due to excess flow of interstitials to dislocations whose extra half planes are perpendicular to the stress axis. SIPN refers to creation of extra planes selectively perpendicular to the tensile axis due to preferential nucleation of dislocation loops in these planes, at high-stress levels. The same occurs for vacancy loops at low values of external stress. At very high temperatures, the grain boundary—aided thermal creep is augmented by irradiation-induced point defects.

Irradiation growth refers to the "dimensional distortion" of a metal in an irradiation field even if stress is absent, retaining the volume constant. The irradiation growth is anisotropic in nature and most frequently observed [198] in hcp systems like zirconium-based alloys, leading to gradual increase in "*a*" and reduction in "*c*" of the unit cell parameters. The distortion is due to the condensation of interstitials preferentially as dislocation loops lying on prism planes of type (11..0) of hcp. Vacancy loops form in the depleted zones on the basal planes (0001). This process is equivalent to transfer of atoms from the basal planes to prism planes, via irradiation-induced point defects resulting dilation of "*a*" and shrinkage of "c."

Irradiation hardening refers to classical understanding of enhanced yield strength, a smaller increase in ultimate tensile strength, a reduction in the work hardening rate, and a reduction in the uniform and total elongation. This is caused by the pinning of mobile dislocations by the large density of dislocation loops. Source hardening is caused by the additional stress required to unpin the dislocations, and friction hardening is due to the defect density, i.e., the nanodispersoids, acting as obstacle to mobile dislocations, increasing the flow stress. However, a more dominating mechanism in ODS steels by the dispersoids is found to be the recombination reactions. The counteracting mechanism is offered by the recombination reactions by trapping of defects at the particle/matrix interface, as shown in Fig. 17.12. Fig. 17.12 demonstrates [199] the irradiation hardening in ODS steels as a function of sink strength, i.e., the density of nanodispersoids.

Irradiation embrittlement is characterized by nearly nil amount of plastic or creep deformation that occurs prior to fracture. Irradiation reduces [200] the ductile—brittle transformation temperature (DBTT) and lowers the upper shelf energy, as shown in Fig. 17.13.

Irradiation induced segregation of embrittling elements like phosphorous, antimony, and tin, called "metalloids," reduces [41,201] fracture stress, causing additional shift in DBTT. Irradiation embrittlement is more significant in ferritic steels than austenitic due to temper embrittlement phenomenon.

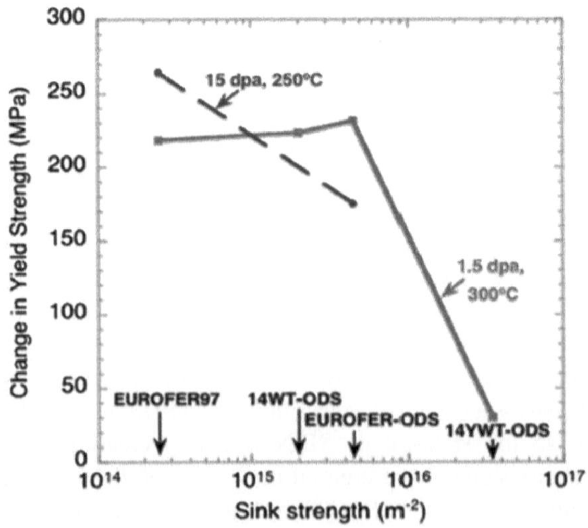

FIGURE 17.12

Irradiation strengthening as a function of sink strength, i.e., density of oxide particles [199] in ODS steel.

Reproduced with permission from IOP Science.

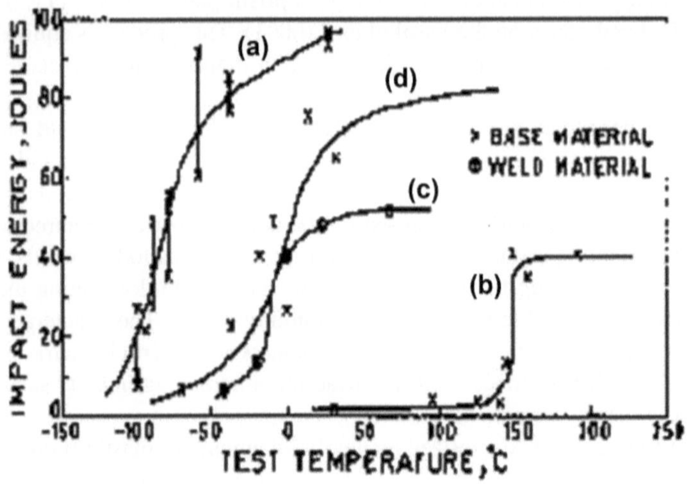

FIGURE 17.13

Variation of impact energy versus test temperature for an ASTM A 203 steel for different dose values and recovery of ductility after annealing of defects: (a) unirradiated; (b) irradiated in RAPS, India, to a fluence of 3.5×10^{19} n/cm^2; (c) irradiated to a fluence of 5×10^{18} n/cm^2; (d) annealed at 300°C for 15 days after irradiation to a fluence of 3.5×10^{19} n/cm^2. Please note the increase in ductile–brittle transformation temperature with dose and reduction in the upper shelf energy, with recovery of damage after annealing.

This is a major concern, especially in view of superior radiation resistance of ferritic steels. Monte Carlo methods, percolation model, and experimental methods have shown that finer grain size could reduce [202] DBTT by 50°C. The high-temperature capability of ferritic steels is improved by dispersing 5 nm yttria particles in ferrite, whose stability under irradiation and manufacturing technology have been established [203]. Currently, long-term in-reactor performance is being evaluated.

In recent years, concerted effort has been directed [68,69] toward the development of reduced activation ferritic steels for fusion applications. In addition to the optimization [80] of composition using unirradiated samples, in-reactor performance has been evaluated [68,69,204,205] for a number of steels like Eurofer [78], F82H [70], and OPTIFER [78]. Shift in DBTT increases by about 200°C for a dose of around 7 dpa at the irradiation temperature of around 250–300°C, wherein the severity is maximum. However, the DBTT values per se are below −30°C and hence are selected for the fusion applications. Helium embrittlement of RAFM steels for fusion applications have also been investigated [206] in recent years by irradiating helium preimplanted steels in high-flux reactor. For helium amounts around 80 wppm, for irradiations at temperature of around 200°C, an additional embrittlement due to hardening of the matrix by helium bubbles was observed. While, for higher amounts of helium, "helium embrittlement" was dominant. The effects reduced at temperatures higher than 450°C. It is believed that the high density of defects in martensitic structure of RAFM promotes [69,207] trapping, absorption and annihilation of point defects, suppressing the formation of helium bubbles, and embrittlement.

Irradiation-induced embrittlement has been studied [29,206,208] extensively in ODS steels with different chromium content and V–Ti–Cr [207] alloys, the earliest choice for fusion applications. Surprisingly, ODS steels were found to be resistant to neutron irradiation embrittlement and hardening without loss of ductility, the main cause for the observed embrittlement. 14 Cr ODS steels were not found to be susceptible to hydrogen and helium embrittlement [69,209]. Vanadium-based alloys were evaluated [210] in the 1980s for fusion applications due to their low activation, excellent compatibility with Pb–Li, low thermal stress, and high heat flux capability. But radiation resistance of the alloys needs further development. Currently, combination of RAFM and ODS steels is the most preferred option for fusion structural materials.

In recent years, a large number of micromechanical testing procedures have been developed [211–214] for the evaluation of degradation of materials under irradiation. These techniques can be comfortably used for studying the effect of ion irradiation or for surveillance of costly neutron-irradiated nuclear materials. The results of such studies can be used for benchmarking the multi-scale models to predict the degradation of materials under irradiation. One of the small specimen testing procedures is the shear punch test, which yields results comparable to standard tensile test. In this, a flat-tipped cylinder shears through center of a disk-shaped specimen and stress–strain plots are plotted. The toughness and DBTT measurements could be demonstrated to be reliable if done with subsize Charpy specimens. Extensive evaluations have been carried out [213,214] to standardize the testing and analysis procedures. Some of these tests are ASTM approved, due to acceptable correlation between standard tests and the small specimen testing methods. These developments would play a crucial role in the development of fusion and ceramic materials, especially if the appropriate irradiation simulation environment close to fusion reactor operating conditions also becomes available, for which intense international efforts are required.

9. MULTISCALE MODELING TO PREDICT IRRADIATION BEHAVIOR OF MATERIALS

The basic understanding of the behavior of materials under irradiation was arrived at by the end of the 1900s. However, it was required to consolidate all the capabilities of different investigative tools, spanning from quantum mechanics to continuum mechanics, to finally predict the behavior of materials under service conditions. Of late, multiscale modeling is being increasingly applied [192–195,215] to examine the predictive capabilities for newly emerging advanced materials like ODS and RAFM steels, for the GEN-IV fission and fusion reactors. The basic concept of multiscale modeling is to combine various levels of well-demonstrated predictive models, finally to predict the behavior of large components. Each level addresses a specific phenomenon, in a specific range of length and time, followed by suitable integration schemes. For instance, the molecular dynamics (MD) includes [28,216,217] role of electrons in bonding between atoms to provide atomic information in primary damage. The kinetic Monte Carlo (KMC) deals with accumulation of defects over different length scales. The rate theory, a stochastic method, addresses [28] the time evolution of irradiation-induced defects, using defect balance equations, to arrive at the best description for the lattice with different defect clusters. The dislocation dynamics (DD) solves [218,219] equations of motion for dislocation lines and simulates collective motion of dislocation interaction with defects. Finally, a suitable model for the appropriate property is chosen, to be analyzed using finite element methods [220,221] based on continuum mechanics, to predict the performance of components. The major challenge arises from the availability of empirical potentials for multicomponent systems in models related to atomic level, appropriate linkages between available powerful models, and ability to handle multicomponent commercial materials.

Embrittlement in reactor pressure vessel (RPV) steels is one of the problems often studied using multiscale modeling. Many experimental and modeling investigations, in a wide range of testing conditions, have been combined to develop a multiscale model [194] for prediction of embrittlement properties in RPV steel. Kwon et al. has [222] integrated MD, KMC, and DD to calculate the critical resolved shear stress (CRSS) for pure iron, which agreed well with experimental estimations. They could also arrive at the general scheme for predicting radiation hardening in RPV steels.

The efforts of multiscale modeling in RAFM steels and other fusion materials are directed [223] toward modeling evolution of helium bubbles in continuous and pulsed mode of irradiation. This stage employs MD, KMC, and rate theory calculations based on nucleation and growth of helium bubbles. The second step involves the calculation of hardening due to the formation of helium bubbles, using dispersion barrier hardening model.

The ODS steels, which are generally fabricated [84,224] using mechanical alloying, have been studied widely, with respect to the stability and defects, during fabrication and irradiation. The multiscale modeling with respect to stability and structure of yttria nanoclusters in the presence of excess vacancies generated during mechanical alloying and titanium has revealed [225] that size of nanodispersoids increased with higher "excess" oxygen. Extensive calculations of binding energy for different cluster configurations have confirmed [183,226] that addition of zirconium and titanium lead to efficient refining of cluster size. Ab initio total energy calculations of defects, in oxygen-enriched nanoclusters in iron, have shown [183] that increase in concentration of vacancies reduces the solubility of oxygen in iron. MD simulation has been carried out [183] for the evolution of cascade during

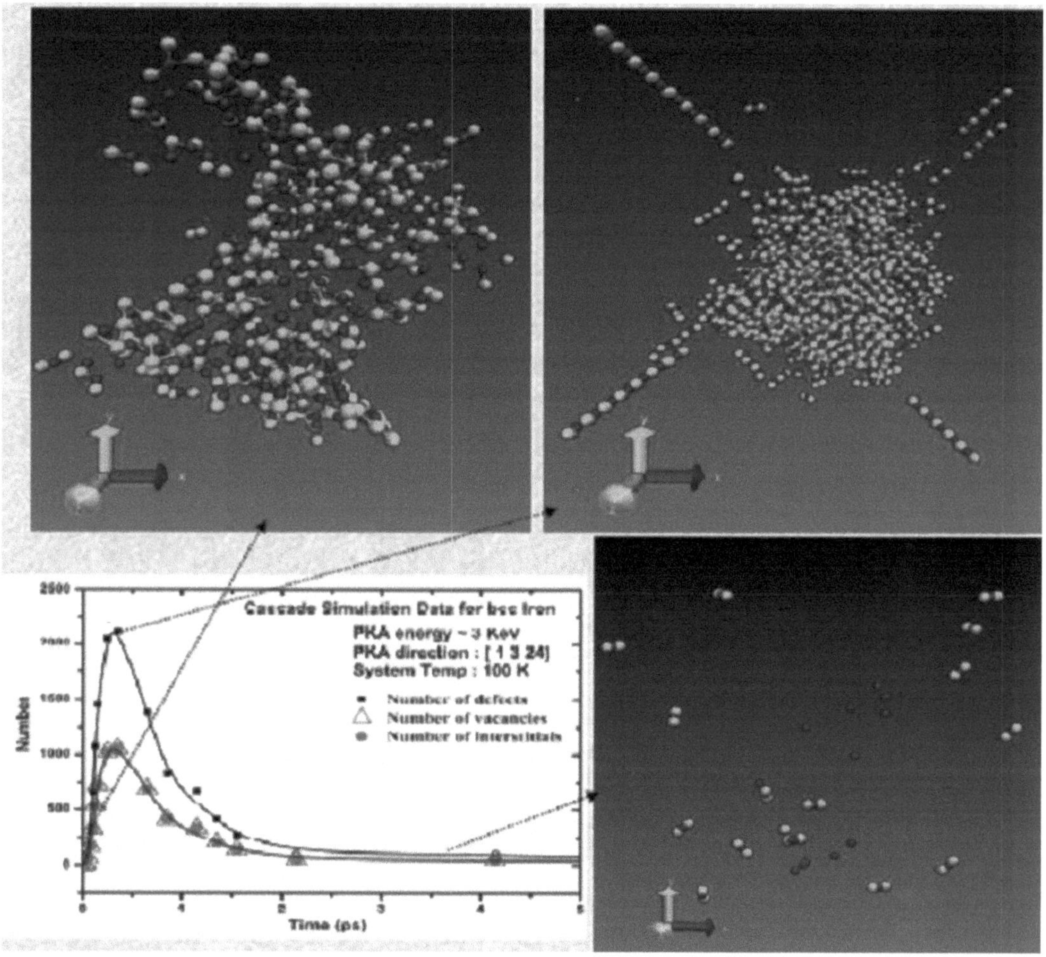

FIGURE 17.14

Molecular dynamics simulation [184] of radiation damage cascade in a nanocluster during initiation, after thermal spike and after annealing stage. Green (*gray in print version*) refers to interstitials and red (*light gray in print version*) to vacancies. The dumbbells with two interstitials single vacancy are the most stable defect cluster. The graph shows the time evolution of vacancies and interstitials.

irradiation of a nanocluster, during initiation, thermal spike, and annealing stage, in addition to quantitative evaluation of point defect concentrations. A typical simulation result of the radiation damage of nanocluster is shown in Fig. 17.14.

There has been an intense effort to develop modeling methodologies for the response of ODS steels for external stress and derive the mechanical properties, the final concern for design of nuclear materials. In ODS steels, multiscale modeling efforts [227–229] could simulate interaction of dislocations with nm-sized yttria particles, using DD and generalized Peierls–Nabaro model, yielding CRSS for the system. Density functional theory has been used to study the evolution of defect clusters in ODS steels. It has been shown that the interface between iron and yttria acts as strong traps for both vacancies and interstitials, through the long-range attraction. This increases the recombination probability, explaining the improved radiation creep and void swelling in ODS steels. Model ODS alloys have also attracted a lot of attention [230] in recent years with respect to stability of yttria dispersoids under ion irradiation and defect cluster evolution. However, a large number of commercial ODS steels have shown [231] the stability of yttria under neutron irradiation and acceptable toughness, encouraging the reactor designers to reach higher temperatures and burnup using ODS steels.

Currently, most of the modeling for fusion materials is being carried out in ferritic iron or iron-based alloys. The calibration of the models and transfer of data set from one length and time scale to the next is being carried out using simulation results employing high-energy protons and spallation neutrons in the Swiss Spallation Neutron Source (SINQ) in Switzerland and the Joint Accelerator for Nanoscience and Nuclear Simulation (JANNUS) in France.

10. FUTURE TRENDS

The extensive R&D carried out to understand the physics and metallurgy of materials under irradiation, in the last century, laid the foundation for the in-depth understanding of the subject. The current effort is threefold: (1) development of materials including alloys and composites, for GEN IV reactor concepts and fusion reactors; (2) consolidation of the existing powerful predictive models to yield a single multiscale model, with appropriate linkages; and (3) calibration of predictive models based on results from simulation irradiation facilities, for which the reliability of small-specimen testing methods needs to reach the level of international acceptance.

The irradiation behavior of semiconductors has progressed rapidly along the directions of development of optoelectronic materials for microelectronics, carbon-based semiconductor industry, and use of irradiation as precise cutting or welding tool to achieve atomic-level precision. These ambitious concepts need to be demonstrated satisfactorily in laboratories and developed to the level of industrial practice.

11. SUMMARY

The review has explained the reasons for considering "irradiation" to be a parameter that can generate extreme environment in materials exposed to energetic neutrons or ions or electrons, based on thermodynamic reasons and defect density. These possibilities have opened up newer opportunities for navigating the free energy landscape in domains that have not so far been explored. Recent ion irradiation studies on graphene, semiconductors, and insulators have led to "irradiation engineered technologies" yielding powerful self-organization structures and cutting/welding tools with atomic precision, in addition to laying foundation for the newly emerging carbon-based electronics. In engineering materials of relevance to nuclear technology, materials degradation under irradiation and mitigation methods have been studied for nearly more than four decades. Hence, the review is focused on recent studies on advanced materials for future fission reactors and fusion reactors. The various phenomena like irradiation-induced defect generation, their evolution, their impact on diffusion, segregation, phase stability, and mechanical properties are briefly presented. The materials development programs for fusion technology with an environment of 14 MeV neutrons with copious amount of helium needs to await the availability of such environments. The extrapolation of materials behavior in fast neutron spectrum (~1 MeV) to fusion environment, at best, is only an "available" nearest best solution. However, the in-service behavior of fusion materials could produce many interesting surprises.

REFERENCES

[1] R.C. Ewing, Nucl. Instrum. Methods Phys. Res. B 91 (1994) 22.
[2] P. Rudling, R. Adamson, B. Cox, F. Garzarolli, A. Strasser, Nucl. Eng. Tech. 40-1 (2008) 1.
[3] T. Allen, J. Busby, M. Meyer, D. Patti, Mater. Today 13 (12) (2010) 14.
[4] R. Gaillard, Single event effects: mechanisms and classification, in soft errors in modern electronic system, in: M. Nicolaidis (Ed.), Frontiers in Electronic Testing, vol. 41, Springer Science + Business Media, LLC, 2011, p. 27, http://dx.doi.org/10.1007/978-1-4419-6993-4_2 (Chapter 2).
[5] B.I. Kharisov, O.V. Kharissova, U.O. Mendez, Radiation Synthesis of Materials and Compounds, Science, CRC Press, 2013, ISBN 9781466505223, p. 586. CAT#K14554.
[6] E. Rimini, Ion Implantation: Basics to Device Fabrication, Kluwer Academic Publishers, Springer Science–Business Media, LLC, 1995, ISBN 978-0-7923-9520-1, http://dx.doi.org/10.0007/978-1-4615-2259-1.
[7] A.V. Krasheninnikov, F. Banhart, Nat. Mater. 6 (2007) 723, http://dx.doi.org/10.1038/nmat1996.
[8] N.T. Fourches, J. Appl. Phys. 77 (1995) 3684, http://dx.doi.org/10.1063/1.358607.
[9] M. Miglierini, M. Hasiak, J. Nanomater. 2015 (2015), 175407, http://dx.doi.org/10.1155/2015/175407.
[10] S. Kuriakose, D.K. Avasthi, S. Mohapatra, Beilstein, J. Nanotechnol. 6 (2015) 928.
[11] K.E. Sickafus, E.A. Kotomin, B.P. Uberuaga, Radiation Effects in Solids, Science, Springer Science & Business Media, 2007, p. 592.
[12] A. Kis, G.C. Sanyi, J.P. Salvetat, T.N. Lee, E. Couteau, A.J. Kulik, et al., Nat. Mater. 3 (2004) 153.
[13] C. Gómez-Navarro, P.J. De Pablo, J. Gómez-Herrero, B. Biel, F.J. Garcia-Vidal, A. Rubio, et al., Nat. Mater. 4 (2005) 534.
[14] H. Stahl, J. Appenzeller, R. Martel, P. Avouris, B. Lengeler, Phys. Rev. Lett. 85 (2000) 5186.
[15] B.Q. Wei, J. D'Arcy-Gall, P. Ajayan, G. Ramanath, Appl. Phys. Lett. 83 (2003) 3581.
[16] S. Talapatra, P.G. Ganesan, T. Kim, R. Vajtai, M. Huang, M. Shima, et al., Phys. Rev. Lett. 95 (2005) 097201.
[17] P. Esquinazi, D. Spearmann, R. Höhne, A. Setzer, T. Butz, Phys. Rev. Lett. 91 (2003) 227201.
[18] D. McGrouther, J.N. Chapman, F.W.M. Vanhelmont, J. Appl. Phys. 95 (12) (2004) 7772.
[19] J.R. Heath, H.W. Kroto, S.C. O'Brien, R.F. Curl, R.E. Smalley, Nature 318 (1985) 162.
[20] S. Iijima, Nature 354 (1991) 56.
[21] F. Giannazzo, S. Sonde, V. Raineri, Carbon 47 (2009) 3201.
[22] O. Bobes, K. Zhang, H. Hofsass, Phys. Rev. B 86 (2012) 235414.
[23] C. Chappert, V. Kottler, T. Devolder, V. Mathet, H. Bernas, Science 280 (1998) 1919.
[24] H. Bernas, J.-P. Attane, K.-H. Heinig, D. Halley, D. Ravelosona, A. Marty, et al., Phys. Rev. Lett. 91 (7) (2003) 077203.
[25] E. Akcöltekin, T. Peters, R. Meyer, A. Duvenbeck, M. Klusmann, I. Monnet, et al., Nat. Nanotechnol. 2 (2007) 290.
[26] P. SanthanaRaman, K.G.M. Nair, M. Kamruddin, A.K. Tyagi, A. Rath, P.V. Satyam, et al., Appl. Surf. Sci. 258 (2012) 4156.
[27] F. Watt, A.A. Bettiol, J.A. Vankan, E.J. Teo, M.B.H. Breese, Int. J. Nanosci. 4 (2005) 269.
[28] G.S. Was, Fundamentals of Radiation Materials Science, Metals and Alloys, second ed., Springer, 2015, p. 433 (Chapter 9).

[29] G.R. Odette, M.J. Alinger, B.D. Wirth, Annu. Rev. Mater. Res. 38 (2008) 471, http://dx.doi.org/10.1146/annurev.matsci.38.060407.130315.
[30] G.S. Was, Radiation effects in solids, Science, in: E. Kurt, E. Sickafus, A. Kotomin, P.U. Blas (Eds.), Springer Science & Business Media, 2007, p. 421.
[31] Kathy Lu John, Materials in Energy Conversion, Harvesting, Storage Wiley & Sons, 2014 (Chapter 3 Technology & Engineering p. 25).
[32] R.J.M. Koning, T.R. Allen, R. Stoller, S. Yamanaks, Comprehensive Nuclear Materials, Published by Elsevier Ltd., 2012, ISBN 978-0-08-056033-5, http://dx.doi.org/10.1016/B978-0-08-056033-5.09002-9.
[33] W.J.S. Yang, Radiation induced changes in microstructure, in: F.A. Garner, N.H. Packan (Eds.), in: A.S. Kumar (Ed.), ASTM International STP, vol. 955, 1987, p. 628.
[34] V.S. Neustroev, V.G. Dvoretzky, Z.E. Ostrovsky, V.K. Shamardin, G.A. Simansky, ASTM STP 1447, in: M.L. Grossbeck, T.R. Allen, R.G. Lott, A.S. Kumar (Eds.), Effects of Radiation on Materials, ASTM Publications, 2004, p. 32.
[35] C. Claeys, E. Simoen, in: C. Claeys, E. Simoen (Eds.), Radiation Effects in Advanced Semi Conductor Materials and Devices, Springer Series in Materials Science, Springer-Verlag Berlin, Heidelberg, 2002, p. 9.
[36] V.S. Vavilov, Radiation Effects in Semiconductors and Semiconductor Devices, Springer Science & Business Media, 2012.
[37] J.S. Williams, S.O. Kucheyev, H.H. Tan, J. Wong-Leung, C. Jaadish, Philos. Mag. 85 (2005) 617.
[38] B. Raj, U. Kamachi Mudali, M. Vijayalakshmi, M.D. Mathew, A.K. Bhaduri, P. Chellapandi, et al., Advanced Materials Research, vol. 794, Trans Tech Publications, Switzerland, 2013, p. 3. http://dx.doi.org/10.4028/www.scientific.net/AMR.794.3.
[39] B. Raj, M. Vijayalakshmi, P.R. Vasudeva Rao, K.B.S. Rao, Challenges in materials research for sustainable nuclear energy, MRS Bull. Spec. Issue Harnessing Mater. Energy 33 (4) (2008) 327.
[40] B. Raj, M. Vijayalakshmi, Nuclear energy encyclopedia: science, technology, applications, in: S.B. Krivit, J.H. Lehr, T.B. Kingery (Eds.), Wiley Series on Energy, fifth ed., John Wiley & Sons, Inc., 2011.
[41] B. Raj, M. Vijayalakshmi, JOM 62 (9) (2010) 75. www.tms.org/jom.html.
[42] S. Picraux, Applications of Ion Beams to Metals, Springer Science & Business Media, 2012 (Technology & Engineering).
[43] www.world-nuclear.org/info/nuclear-fuel-cycle/power-reactors/generation-iv-nuclear-reactors/.
[44] www.world-nuclear.org/info/Nuclear-Fuel-Cycle/Power-Reactors/Advanced-Nuclear-Power-Reactors/.
[45] T. Diaz de la Rubia, Annu. Rev. Mater. Sci. 26 (1996) 513.
[46] K.E. Sickafus, E.A. Kotomin, B.P. Uberuaga, Radiation effects in solids, in: Proc. of NATO Advanced Study Institute on Radiation Effects in Solids, Held in Sicily, Italy, July, 2004, Springer Science & Business Media, Netherlands, 2007.
[47] J.P. Biersack, L.G. Haggmark, Nucl. Instrum. Methods 174 (1980) 257. http://www.srim.org.
[48] R.E. Stoller, M. Toloczko, G.S. Was, F.A. Garner, Nucl. Instrum. Methods Phys. B 310 (2013) 80. http://dx.doi.org/10.106/j.nimb.2013.05.008.montecarlosrim.
[49] A.P. Druzhkove, V.L. Arbuzov, D.A. Perminov, J. Nucl. Mater. 341 (2005) 153, http://dx.doi.org/10.1016/j.jnucmat.2005.01.021.
[50] E. Meslin, M. Lambrecht, M. Hernández-Mayoral, F. Bergner, L. Malerba, P. Pareige, et al., J. Nucl. Mater. 406 (2010) 73.
[51] M. Matijasevic, E. Lucon, A. Almazouzi, J. Nucl. Mater. 377 (2008) 101.
[52] M. Matijasevic, A. Almazouzi, Association of metallurgical engineers of Serbia scientific paper AMES UDC:669.15'24.085.5-198=20, 246, MJoM Metall. − J. Metall.
[53] W. Schule, Radiation Enhanced Diffusion in FCC Alloys, European Commission, L2920, Luxemborg, 1994. Report EUR 16109 EN.
[54] B. Tsuchiya, T. Shikama, S. Nagata, K. Toh, M. Narui, M. Yamazaki, J. Nucl. Mater. 367−370 (2007) 1073.

[55] A.K. Tyagi, M.P. Macht, V. Naundorf, J. Nucl. Mater. 179 (1991) 1026.
[56] G. Arnold, G. Battaglin, G. Della, G. De Marchi, P. Mazzoldi, A. Miotello, Nucl. Instrum. Methods Phys. Res. B Beam Interact. Mater. Atoms 32 (1988) 315.
[57] M.S. Wechsler, Radiation effects on metals and neutron dosimetry, in: Proc. of Symp. on Fourth Pacific Area National Meeting, Los Angeles, ASTM STP, vol. 341(1963), 1962, p. 86.
[58] K. Nordlund, R.S. Averback, Phys. Rev. B 59 (1) (1999) 20.
[59] A.T. Motta, J. Nucl. Mater. 244 (1997) 227.
[60] G.S. Was, T. Allen, Mater. Charact. 32 (1994) 235.
[61] G. Szenes, Phys. Rev. B 51 (1995) 8026.
[62] L. Thome, S. Moll, A. Debelle, F. Garrido, G. Sattonnay, J. Jagielski, Radiation effects in nuclear ceramics, Adv. Mater. Sci. Eng. (2012), 905474, http://dx.doi.org/10.1155/2012/905474.
[63] S.I. Porollo, S.V. Shulepin, Y.U.V. Konobeev, F.A. Garner, J. Nucl. Mater. 378 (2008) 17.
[64] S.J. Zinkle, Phys. Plasmas 12 (2005) 058101.
[65] S.J. Zinkle, L.L. Snead, Annu .Rev. Mater. Res. 44 (2014) 241, http://dx.doi.org/10.1146/annurev-matsci-070813-113627.
[66] S.L. Dudarev, J.L. Boutard, R. Lasser, M.J. Caturla, P.M. Derlet, M. Fivel, et al., J. Nucl. Mater. 386-388 (2009) 1.
[67] L.K. Mansur, A.F. Rowcliffe, R.K. Nanstad, S.J. Zinkle, W.R. Corwin, R.E. Stoller, J. Nucl. Mater. 329 (2004) 166.
[68] E. Gaganidze, H.C. Schneider, B. Dafferner, J. Aktaa, J. Nucl. Mater. 355 (1−3) (2006) 83.
[69] A. Kimura, Mater. Trans. 46 (2005) 394.
[70] A.-A.F. Tavassoli, J. Nucl. Mater. 302 (2−3) (2002) 73.
[71] A. Hasegawa, M. Fukuda, K. Yabuuchi, S. Nogami, J. Nucl. Mater. 471 (2016) 175, http://dx.doi.org/10.1016/j.jnucmat.2015.10.047.
[72] A. Hasegawa, K. Abe, T. Muroga, N. Yamamoto, In effects of radiation on materials, Materials, in: M.L. Grossbeck, T.R. Allen, R.G. Lott, A.S. Kumar (Eds.), ASTM STP, 1447ASTM International, 2004, p. 693.
[73] S.J. Zinkle, N.M. Ghoniem, Fusion Eng. Des. 51−52 (2000) 55.
[74] M.K. Hossain, Md A. Taher, M.K. Das, World J. Nucl. Sci. Technol. 5 (5) (2015) 287, http://dx.doi.org/10.4236/wjnst.2015.54028. Published in Sci Res, http://www.scirp.org/journal/wjnst.
[75] www.neutron.kth.se/publications/reviews/ADS_Status.../STATUS.PDF.
[76] C.W. Barnes, M.A. Bourke, S.A. Maloy, F.G. Mariam, F.E. Merrill, M. Nastasi et al, Radiation Damage from Atomic to Meso-scales in Extreme Environments, FTP, p. 1. https://www.iaea.org/About/Policy/GC/GC51/.../gc51inf-3-att7_en.pdf.
[77] H. Ullmaier, F. Carsughi, Nucl. Instrum. Methods Phys. Res. B 101 (1995) 406.
[78] R.L. Klueh, D.S. Gelles, S. Jitsukawa, A. Kimura, G.R. Odette, B. van der Schaaf, et al., J. Nucl. Mat. 307−311 (2002) 458.
[79] Y. Chen, Nucl. Eng. Technol. 45 (2013) 311.
[80] T. Jayakumar, M.D. Mathew, K. Laha, S.K. Albert, S. Saroja, E. Rajendra Kumar, et al., Fusion Sci. Tech. 65 (2) (2014) 171.
[81] X. Jia, Y. Dai, M. Victoria, J. Nucl. Mater. ISSN: 0022-3115 305 (2002) 1.
[82] K. Laha, S. Saroja, A. Moitra, R. Sandhya, M.D. Mathew, T. Jayakumar, et al., J. Nucl. Mater. 439 (2013) 41.
[83] R. Mythili, J. Ravikirana, J. Vanaja, K. Laha, S. Saroja, T. Jayakumar, et al., Proced. Eng. 55 (2013) 295.
[84] S. Saroja, A. Dasgupta, R. Divakar, S. Raju, E. Mohandas, M. Vijayalakshmi, et al., Nucl. Mater. 409 (2) (2011) 131.

[85] N. Akasaka, S. Yamashita, T. Yoshitake, S. Ukai, A. Kimura, J. Nucl. Mater. 1053–1056 (2004) 329, http://dx.doi.org/10.1016/j.jnucmat.2004.04.133.
[86] S. Ukai, M. Fujiwara, J. Nucl. Mater. 307–311 (2002) 749.
[87] M.S. El-Genk, J.M. Tournier, J. Nucl. Mater. 340 (2005) 93.
[88] G. Sundararajan, R. Vijay, A.V. Reddy, Curr. Sci. 105 (2013) 1100.
[89] A. Dasgupta, R. Divakar, P.K. Parida, C. Ghosh, E. Mohandas, M. Vijayalakshmi, et al., in: T. Bollinghaus, J. Lexow, T. Kishi, M. Kitagawa (Eds.), Materials Challenges and Testing for Supply of Energy and Resources, Springer-Verlag Berlin, Heidelberg, 2012, p. 117, http://dx.doi.org/10.1007/978-3-642-23348-7.
[90] W.L. El-Guebaly, R. Kurtz, M. Rieth, H. Kurishita, A. Robinson, Fusion Sci. Tech. 60 (2011) 185.
[91] A. Hasegawa, M. Fukuda, T. Tanno, S. Nogami, Mater. Trans. 54 (2013) 466.
[92] W.A. Hasegawa, M. Fukuda, S. Nogami, K. Yabuuchi, Nucl. Eng. Des. Fusion 89 (7) (2014), http://dx.doi.org/10.1016/j.fusengdes2014.04.035.
[93] T. Tanno, M. Fukuda, S. Nogami, A. Hasegawa, Mater. Trans. 52 (2011) 1447.
[94] V.K. Sikka, J. Moteff, J. Appl. Phys. 43 (1972) 4942, http://dx.doi.org/10.1063/1.1661050.
[95] S.O. Kucheyev, C. Jagdish, in: J. Stephen (Ed.), Zinc Oxide Bulk, Thin Films and Nanostructures: Processing, Properties, and Applications Chennupati Jagadish, Pearton Elsevier, Oxford, UK, 2011, p. 285 (Chapter 8 Technology & Engineering).
[96] S. Dhara, A. Datta, C.T. Wu, Z.H. Lan, K.H. Chen, Y.L. Wang, et al., App. Phys.Lett. 82 (2003) 3.
[97] T. Som, J. Ghatak, O.P. Sinha, R. Sivakumar, D. Kanjilal, J. App. Phys. 103 (2008) 123532.
[98] L. Pelaz, L.A. Marqúes, J. Barbolla. www.ele.uva.es/~simulacion/update2005/Pelaz_JAP04_m.pdf.
[99] W. Brandt, A. Ratkowski, R.H. Ritchie, Phys. Rev. Lett. 33 (1974) 1325.
[100] S. Della-Negra, J. Depauw, D. Jacquet, Y. Le Beyec, M. Pautrat, K. Baudin, et al., Phys. Rev. A 63 (2001) 022902.
[101] D. Shen, X. Lu, Z. Xia, X. Wang, Q. Zhao, Nucl. Instrum. Methods Phys. Res. B 129 (1995) 392.
[102] H. Rothard, K. Kroneberger, E. Veje, A. Clouvas, J. Kemmler, P. Koschar, et al., Phys. Rev. B 41 (1990) 3959.
[103] S. Bouneau, A. Brunelle, S. Della-Negra, J. Depauw, D. Jacquet, Y. Le Beyec, et al., Phys. Rev. B 65 (2002) 144106.
[104] A.I. Titov, S.O. Kucheyev, V.S. Belyakov, A.Y. Azarov, J. Appl. Phys. 90 (2001) 3867.
[105] H. Yamamoto, T. Saito, H. Asaoka, Appl. Surf. Sci. 178 (2001) 127.
[106] A.Y. Azarov, A. Titov, Semiconductors 41 (2001) 5.
[107] P. Bhattacharya, R. Fornari, H. Kamimura, in: H. Kamimura (Ed.), Comprehensive Semiconductor Science and Technology, Physics and Fundamental Theory, Elsevier Publications, 2011, p. 1.
[108] L.A. Larson, J.M. Williams, M.I. Current, Rev. Accl. Sci. Tech. 4 (2011) 11.
[109] W.H. He, H.S. Kang, D.Y. Shen, R. Nie, Y.X. Shen, F.R. Ding, Nucl. Instrum. Methods B 201 (2003) 449.
[110] R. Zhang, Z. Zhang, Z. Wang, S. Wang, W. Wang, D. Fu, et al., Appl. Phys. Lett. 101 (2012) 011905.
[111] L.A. Marques, L. Pelaz, I. Santos, Phys. Rev. B 74 (2006) 201201R.
[112] M.W. Ullah, A. Kuronen, K. Nordlund, F. Djurabekova, P.A. Karaseov, A.I. Titov, J. Appl. Phys. 112 (2012) 043517.
[113] T. Aoki, T. Seki, J. Matsuo, Z. Insepov, I. Yamada, Nucl. Instrum. Methods Phys. Res. B 153 (1995) 264.
[114] I. Yamada, J. Matsuo, Z. Insepov, T. Aoki, T. Seki, N. Toyoda, Nucl. Instrum. Methods Phys. Res. B 164–165 (2000) 944.
[115] G.J. Dienes, A.C. Damask, J. Appl. Phys. 29 (1958) 1713, http://dx.doi.org/10.1063/1.1723032.
[116] M. Halbwachs, H. Gonzalez, J. Hillairet, Scr. Metall. 9 (1975) 575.
[117] R.S. Averback, H. Hahn, Phys. Rev. B 37 (1988) 6533.
[118] S. Hofmann, Auger- and X-Ray Photoelectron Spectroscopy in Materials Science, Science, Springer Science & Business Media, 2012, p. 327.

[119] V.C. Venezia, L. Pelaz, H.J. Ludwig, T.E. Haynes, Phys. Rev. B 69 (2004) 125215, http://dx.doi.org/10.1103/PhysRevB.69.125215.
[120] R. Sizmann, J. Nucl. Mater. 69 (1978) 386.
[121] S. Takeda, Jpn. J. Appl. Phys. 30 (1991) L639.
[122] M. Halbwachs, N. Yoshida, Phil. Mag. 43 (1981) 1125.
[123] W. Schule, J. Phys. F Metal Phys. 10 (1980) 2345.
[124] Yu.M. Platov, S.V. Simakov, F.K. Obrab, Materials 6 (1988) 5.
[125] J. Arunkumar, S. Abhaya, R. Rajaraman, G. Amarendra, K.G.M. Nair, C.S. Sundar, et al., Nucl. Mater. 384 (2009) 245.
[126] J. Arunkumar, Studies on Defect Recovery and Solute Diffusion in Ti Modified Austenitic Stainless Steel (Ph.D. thesis), University of Madras, 2011.
[127] M. Gieb, J. Heieck, W. Schule, Radiation Enhanced Diffusion in Nickel 10.6% Chromium Alloys, European Commission Eur 15793 EN, 1994.
[128] D.G. Nelson, J.F. Gibbons, W.S. Johnson, Appl. Phys. Lett. 15 (1969) 246, http://dx.doi.org/10.1063/1.1652986.
[129] W.A. Coghlan, Review of recent irradiation creep results, Int. Metals Rev. 31 (1986), http://dx.doi.org/10.1179/imtr.1986.31.1.245.
[130] K.G.M. Nair, K. Krishan, M. Vijayalakshmi, Scr. Metall. Mater. 25 (5) (1991) 1217.
[131] B.M. Paine, R.S. Averback, Nucl. Instrum. Methods Phys. Res. B 7/8 (1985) 666.
[132] E. Ruedl, P. Delavignette, S. Amelincks, J. Nucl. Mater. 6 (1962) 46.
[133] L.E. Rehn, P.R. Okamoto, Nucl. Instrum. Methods Phys. Res. B 39 (1989) 104.
[134] K. Nordlund, M. Ghaly, R.S. Averback, J. Appl. Phys. 83 (1998) 1238.
[135] B.N. Dev, Physics at Surfaces and Interfaces, World Scientific, Singapore, UK and USA, 2003, ISBN 981-238-575-4, p. 161.
[136] J. Ghatak, M.U. Bhatta, B. Sundaravel, K.G.M. Nair, Sz-C. Liou, C.-H. Chen, et al., Nanotechnology 19 (2008) 325602.
[137] S. Balaji, B.K. Panigrahi, S. Amirthapandian, S. Kalavathi, A. Gupta, K.G.M. Nair, Appl. Phys. A 116 (4) (2014) 1595.
[138] R.S. Averback, L.E. Rehn, W. Wagner, H. Wiedersich, P.R. Okamoto, Phys. Rev. B. 28 (1983) 3100.
[139] L.E. Rehn, W. Wagner, H. Wiedersich, Scr. Metall. 15 (1981) 683.
[140] V.N. Voyevodin, V.V. Bryk, A.S. Kalchenko, I.M. Nklyudov, G.D. Tolstolutskaya, UDC 620.187:621.039:531 ISSN 1562-6016 BAHT, No. 4 92, 2014, p. 3.
[141] P.J. Maziasz, C.J. McHargue, Int. Mater. Rev. 32 (1987) 190, http://dx.doi.org/10.1179/095066087790150331.
[142] S. Pasebani, I. Charit, J. Burns, S. Alsagabi, D.P. Butt, I.J. Cole, et al., Nucl. Mater. 462 (2015) 191.
[143] A. Hempel, M. Saneyasu, Z. Tang, M. Hasegawa, G. Brauer, F. Plazaola, et al., in: M. L.Hamilton, A.S. Kumar, S.T. Rosinski, M.L. Grossbeck (Eds.), Effects of Radiation on Materials: 19th International Symposium ASTM, STP 1366, ASTM, West Conshchocken, PA, 2000, p. 560.
[144] G.S. Was, P.L. Andresen, JOM 44 (1992) 8.
[145] M.J. Hackett, J.T. Busby, G.S. Was, Metall. Mater. Trans. 39A (2008) 218.
[146] J. Fassbender, J. McCord, J. Magn. Magn. Mater. 320 (2008) 579.
[147] S. Kim, S. Lee, J. Ko, J. Son, M. Kim, S. Kang, et al., Nat. Nanotechnol. 7 (2012) 567, http://dx.doi.org/10.1038/nnano.2012.125.
[148] M. Nastasi, J.W. Mayer, Mater. Sci. Eng. R Rep. 12 (1994) 1.
[149] S. Amirthapandian, B.K. Panigrahi, A.K. Srivastava, S. Dhara, A. Gupta, V.S. Sastry, et al., J. Appl. Phys. 95 (2004) 5295, http://dx.doi.org/10.1063/1.1687039.
[150] G. Martin, A. Barbu, Nucl. Instrum. Methods Phys. Res. 209 (1983) 203.

[151] J. Schroers, W. Jhonson, Phys. Rev. Lett. 93 (2004) 20.
[152] S. Klaumunzer, G. Schumacher, Phys. Rev. Lett. 51 (1983) 1987.
[153] S. Mayr, Phys. Rev. B 71 (2005) 1.
[154] J. Yang, H.B. Zhang, K. Tao, Y. Fan, Appl. Phys. Lett. 64 (1994) 1800, http://dx.doi.org/10.1063/1.111812.
[155] A. Markwitz, W. Matz, M. Waldschmidt, G. Demortier, Surf. Interface Anal. 26 (1998) 160–174.
[156] S.J. Kim, M.A. Nicolet, R.S. Averback, D. Peak, Phys. Rev. B 37 (1988) 38.
[157] P. Borgesen, D.A. Lilienfield, H.H. Johnson, J. Appl. Phys. 66 (1989) 513, http://dx.doi.org/10.1063/1.343567.
[158] K.C. Russel, Prog. Mater. Sci. 28 (1984) 229.
[159] G.S. Was, Science, in: E. Kurt, E. Sickafus, A. Kotomin, P.U. Blas (Eds.), Radiation Effects in Solids, Springer Science & Business Media, 2007, p. 545 (Chapter11).
[160] C.L. Chen, K. Arakawa, J.G. Lee, H. Mori, Scr. Mater. 63 (2010) 1013.
[161] R.J. Gaboriaud, F. Paumier, M. Jublot, B. Lacroix, Nucl. Instrum. Methods Phys. Res. B Beam Interact. Mater. Atoms 311 (2013) 86.
[162] ASTM STP 955, ASTM, PA, USAF.A. Garner, N.H. Packan, A.S. Kumar (Eds.), Radiation-induced Changes in Microstructure, 1987.
[163] K.G.M Nair, M. Vijayalakshmi, K. Krishan, Scripta Metallurgica et Materialia 25 (5) (1991) 1217.
[164] A. Benyagoub, F. Couvreur, S. Bouffard, F. Levesque, C. Dufour, E. Paumier, Nucl. Instrum. Methods Phys. Res. B Beam Interact. Mater. Atoms 175–177 (2001) 417.
[165] D. Simeone, J.L. Bechader, D. Gosset, A. Chevarier, P. Daniel, H. Pilliaire, et al., J. Nucl. Mater. 281 (2000) 171.
[166] M. Vijayalakshmi, M. Feurbacher, K. Urban, (unpublished results).
[167] K.G.M. Nair, M. Vijayalakshmi, K. Krishan, Scr. Mater. 29 (1993) 913.
[168] R.C. Birtcher, L.M. Wang, Nucl. Instrum. Methods B 59/60 (1991) 966.
[169] L.M. Howe, D.P. McCooeye, M.H. Rainville, J.D. Bonnett, D. Phillips, Nucl. Instrum. Methods B 59/60 (1991) 884.
[170] J.L. Brimhall, H.E. Kissinger, L.A. Charlot, Rad. Eff. 77 (1983) 273.
[171] L.L. Snead, S.J. Zinkle, J.C. Hay, M.C. Oshorne, Nucl. Instrum. Methods Phys. Res. B 141 (1998) 123.
[172] S.M. Cheremisin, A. Yu Dudkin, I.V. Metveau, Nucl. Instrum. Methods Phys. Res. B 84 (1994) 102.
[173] D. Gosset, S. Miro, S. Doriot, G. Victor, V. Motte, Nucl. Instrum. Methods Phys. Res. B 365 (2015) 300.
[174] K. Trachenko, M.T. Dove, E. Artacho, I.T. Todorov, W. Smith, Phys. Rev. B 73 (2006) 74207. http://dx.doi.org/10.1103.
[175] J.P. Riviere, Rad. Eff. 33 (1977) 21.
[176] T. Nagase, Y. Umakoshi, Mater. Trans. 46 (3) (2005) 616.
[177] M.L. Grossbeck, T.R. Allen, R.G. Lott, A.S. Kumar (Eds.), Effects of Radiation on Materials, ASTM STP 1447, ASTM International, 2004, p. 344.
[178] M. Vijayalakshmi, P. Rodriguez, in: P.K. Rao (Ed.), Proc. of Int. Conf. on Stainless Steel, Held in Bombay, India, 1989 (1992) p. 227.
[179] G.R. Odette, G.E. Lucas, Rad. Eff. Defects Solids 144 (1–4) (1998) 189.
[180] Z.G. Liu, T. Al-Kassab, Phys. Stat. Solidi 131 (2) (1992) 429.
[181] R.A. Johnson, A.N. Orlov, Physics of Radiation Effects in Crystals, Science, Elsevier, 2012.
[182] S. Yamashita, N. Akasaka, S. Ukai, S. Ohnuki, J. Nucl. Mater. 367-370 (2007) 202.
[183] C. Robertson, B.K. Panigrahi, S. Balaji, S. Kataria, Y. Serruys, M.H. Mathon, et al., J. Nucl. Mater. 426 (2012) 240.
[184] B.K. Panigrahi, Presented at IAEA-Technical Meeting (622-13-TM-47051) Priorities in Modeling and Simulation for FastNeutron Systems, April 14–16, 2015, Vienna.
[185] P.J. Maziasz, J. Nucl. Mater. 205 (1993) 118.

[186] C.N. Venkiteswaran, P. Parameswaran, V. Anandaraj, V. Karthik, N.G. Muralidharan, V. Venugopal, et al., in: B. Raj, T. Jayakumar, K.B.S. Rao, S. Saroja, P.V. Sivaprasad, P. Shankar (Eds.), Advances in Stainless Steels, MS Universities Press, 2009, pp. 59–66.
[187] J.J. Hoyt, M. Haataja, Phys. Rev. B 83 (2011) 174106.
[188] K. Nakai, C. Kinoshita, J. Nucl. Mater. 169 (1989) 116.
[189] P.W. Voorhees, J. Stat. Phys. 38 (½) (1985) 231.
[190] G. Rizza, M. Strobel, K.H. Heinig, H. Bernas, Nucl. Instrum. Methods Phys. Res. B 178 (2001) 78.
[191] B. Raj, R. Divakar, M. Vijayalakshmi, Trans. Indian Instrum. Metals 62 (2) (2009) 89.
[192] M. Samaras, M. Victoria, Mater. Today 11 (12) (2008) 54.
[193] S.L. Dudarev, Annu. Rev. Mater. Res. 43 (2013) 35.
[194] G.R. Odette, B.D. Wirth, D.J. Bacon, N.M. Ghoniem, MRS Bull. (2001) 176.
[195] S.J. Zinkle, N.M. Ghoniem, J. Nucl. Mater. 417 (2011) 2.
[196] J.L. Seren, V. Levy, P. Dubuisson, D. Gilban, A. Mailard, A. Fissolo, et al., in: R.E. Stoller, A.S. Kumar, D.S. Gilles (Eds.), ASTM STP 1125, ASTM, PA, 1992, p. 1209.
[197] M.B. Tolozko, F.A. Garner, J. ASTM Int. 1 (2004) 4.
[198] A. Rogerson, J. Nucl. Mater. 159 (1988) 43.
[199] S.J. Zinkle, A. Moslang, T. Muroga, H. Tanigawa, Nucl. Fusion 53 (2013) 104024–104037.
[200] S.K. Ray, P.R. Sreenivasan, K.G. Samuel, P. Rodriguez, Int. J. Press. Vessel Pip. 54 (3) (1993) 481.
[201] B. Raj, S. Saroja, K. Laha, T. Karthikeyan, M. Vijayalakshmi, K.B.S. Rao, J. Mater. Sci. 44 (9) (2009) 2239.
[202] T. Karthikeyan, V. Thomas Paul, S. Saroja, A. Moitra, G. Sasikala, M. Vijayalakshmi, J. Nucl. Mater. 419 (1) (2011) 256.
[203] H. Zhu, T. Wei, R. Harrison, L. Edwards, K. Maruyama, in: J. Mathew, L. Ma, A. Tan, M. Weijnen, J. Lee (Eds.), Engineering Asset Management and Infrastructure Sustainability, Springer Verlag London Ltd., 2011, p. 1147.
[204] M.R. Gilbert, S. Zhang, R. Kemp, L.W. Packer, S.L. Dudarev, J. Ch, Sublet Fusion Sci. Tech. 66 (2014) 9.
[205] K. Ehrlich, Fusion Eng. Des. 56–57 (2001) 71.
[206] E. Gaganidze, J. Aktaa, Fusion Eng. Des. 83 (10) (2008) 1498.
[207] D.S. Gelles, in: R.L. Klueh, D.S. Gelles, M. Okada, N.H. Packan (Eds.), Reduced Activation Materials for Fusiona Reactors, 1990, p. 113. ASTM STP 1047, ASTM, PA.
[208] E.A. Marquis, S. Lozano Perez, V. de Castro, J. Nucl. Mater. 417 (1) (2011) 257.
[209] K. Lee, Materials in Energy Conversion, Harvesting and Storage Pub, John Wiley & Sons, 2014 (Chapter 3).
[210] J. Bethin, A. Tobin, J. Nucl. Mater. 122–123 (1984) 864.
[211] G.E. Lucas, Metall. Trans. 21A (1990) 1105.
[212] X. Mao, H. Takahashi, J. Nucl. Mater. 150 (1) (1987) 42.
[213] V. Karthik, K. Laha, P. Parameswaran, K.S. Chandravathi, K.V. Kasiviswanathan, T. Jayakumar, et al., Int. J. Press. Vessel. Pip. 88 (2011) 375.
[214] K.V. Kasiviswanathan, V. Karthik, C.N. Venkiteswaran, T. Johny, N.G. Muralidharan, J. Joseph, Energy Proc. 7 (2011) 129.
[215] E. Lucon, P. Benoit, P. Jacquet, E. Diegele, R. Lasser, A. Alamo, et al., Fusion Eng. Des. 81 (2006) 917.
[216] http://www.tgm.mawi.tu-darmstadt.de/Materials_Modelling_part003.pdf.
[217] J.T. Buchan, M. Robinson, H.J. Christie, D.L. Roach, D.K. Ross, N.A. Marks, J. Appl. Phys. 117 (2015) 245901.
[218] T. Diaz de la Rubia, H.M. Zbib, T.A. Khraishi, B.D. Wirth, M. Victoria, M.J. Caturle, Nature 406 (2000) 871, http://dx.doi.org/10.1038/35022544.
[219] F. Ferroni, E. Tarleton, S. Fitzerald, Mater. Sci. Eng. 22 (2014) 045009, http://dx.doi.org/10.1088/0965-0393/22/4/045009.
[220] M.S. Starvin, S. Babu, S. Aithal, K. Manisekar, P. Chellapandi, Sci. Res. Essays 8 (3) (2013) 266.

[221] D.K.L. Tsang, B.J. Marsden, J. Nucl. Mater. 350 (2006) 208.
[222] J. Kwon, G.G. Lee, C. Shin, Nucl. Eng. Tech. 41 (1) (2009) 11.
[223] C. Dethloff, Modeling of Helium Bubble Nucleation and Growth in Neutron Irradiated RAFM Steels. ISSN: 2192-9963, KIT Scientific Publishing, 2012, ISBN 978-3-86644-901-5.
[224] P. Dubuisson, Y. de Charlan, V. Garat, M. Blat, J. Nucl. Mater. 428 (1–3) (2012) 6.
[225] H. Hu, Z. Zhou, L. Liao, M. Wang, S. Li, J. Phys. Conf. Ser. 419 (2013) 012029, http://dx.doi.org/10.1088/1742-6596/419/012029.
[226] N. Baluc, R. Schaublin, P. Spatig, N. Ilchuk, L. Veleva, Z. Oksiuta, et al., Nucl. Fusion 51 (2011) 113006.
[227] A. Takahashi, Z. Chen, N. Kioussis, N.M. Ghoniem, in: ICFRM-14, 2009.
[228] B.D. Wirth, G.R. Odette, J. Marian, L. Ventelon, A. Young-Vandersall, L.A. Zepeda-Ruiz, Multiscale Modeling of Radiation Damage in Fe-based Alloys in the Fusion Environment, 2009.
[229] N.M. Ghoniem, N. Kioussis, Multiscale Modeling of the Deformation of Advanced Ferritic Steels for Generation IV Nuclear Energy, 2009. www.osti.gov/scitech/servlets/purl/953345.
[230] E.M. Grieveson, S.G. Roberts, Fusion Energy (2013) 1, http://dx.doi.org/10.1109/SOFE, 6635334.
[231] R. Lindau, A. Möslang, M. Rieth, M. Klimiankou, E. Materna-Morris, A. Alamo, et al., Fusion Eng. Des. 75–79 (2005) 989.

CHAPTER

RADIATION—MATTER INTERACTION AND RADIATION-TOLERANT OXIDES

18

V. Grover, M.K. Patel, A.K. Tyagi, S. Banerjee
Bhabha Atomic Research Centre, Mumbai, India

CHAPTER OUTLINE

Materials Under Extreme Conditions. http://dx.doi.org/10.1016/B978-0-12-801300-7.00018-8

1. INTRODUCTION

Research on materials under extreme environments has received significant attention in the context of next-generation energy, defense as well as transportation technologies. These kinds of applications demand materials to perform under extreme conditions of stress, temperature, pressure, radiation dose, corrosive environment, etc. Generation of energy by employing nuclear fission reactors is the most promising source of sustainable energy. Materials used in fission reactors and for storage and retention of nuclear waste encounter a constant influx of radiation that may modify materials properties. The futuristic nuclear applications require the structural materials capable of withstanding elevated temperatures and radiation fluxes for long periods of time.

Thus, new concepts of designing such radiation-resistant materials are at the forefront of research on materials science and engineering [1]. Many studies have established that the structures that lie at the borderline of two stable polymorphs tend to show enhanced radiation tolerance. This has helped in identifying the ceramics that would serve as better host lattices for futuristic requirements. Some very recent studies have thrown light on an interesting observation that nanocrystalline materials show enhanced radiation stability. This has been attributed to a large number of grain boundaries in nanomaterials and their ability to act as sink for irradiation-induced defects. The knowledge that surfaces, grain boundaries, and interphase boundaries act as sinks for radiation-induced point defects and traps is now being utilized to develop nanocomposites with enhanced radiation performance and their performance in bulk. Materials in a reactor can be divided into two categories, structural and fuel materials. There is another class of materials employed for storing spent nuclear fuel (SNF), which are called nuclear waste forms. For the limited scope of this chapter, we will present examples of ceramics, and hence our focus will be on nuclear fuel and waste-form materials.

Various types of radiation include energetic α, β, and γ particles, neutrons, heavy atom recoils generated by neutrons, and the fission fragments (FFs). These energetic radiations deposit energy in the materials creating nonequilibrium conditions resulting in displacement of atoms from their specific lattice sites. Since the structure or the crystallography of materials governs various physicochemical properties, such perturbation may result in drastic change in properties of the material. These modifications may occur on timescales ranging from a few days in fuels to millions of years in waste forms. Thus, it is important to understand and predict the long-term behavior of materials exposed to radiation.

Damage simulation by accelerated ion beams is an efficient and a controlled way to understand the structural modifications due to the energy deposited by the radiations of concern.

This chapter will briefly introduce the sources of neutrons, α, β, γ, and FF (fission fragments) and the generic nature of damages caused by them. It further describes the problems associated with nuclear waste and possible solutions. A description of radiation–matter interaction and its consequence on materials is presented to give insight into the change in material behavior. This is followed by a brief review of ion beam–induced modifications in typical nuclear materials with emphasis on effect of structure and powder properties on radiation tolerance of ceramics.

2. NUCLEAR FISSION

In fission, a relatively heavy fissile atomic nucleus (^{235}U, ^{233}U, ^{239}Pu, ^{241}Pu, etc.) absorbs a neutron and splits into two or more lighter nuclei, releasing kinetic energy, and free neutrons. In a nuclear reactor, a sustained chain reaction is generated when a part of these neutrons is absorbed and sustains the fission process. Such a chain reaction occurring in ^{235}U (Eq. 18.1), results in fission products (FP) X and Y (which are not determined uniquely) and have a typical mass distribution as shown in Fig. 18.1 [2,3].

$$^{235}_{92}\mathrm{U} + \mathrm{n} \rightarrow \mathrm{X} + \mathrm{Y} + k\mathrm{n} + \sim 200\ \mathrm{MeV} \tag{18.1}$$

($k = 2.43 \pm 0.07$ average number of fission neutrons.)

3. TYPICAL NUCLEAR REACTORS

The nuclear reactor works on controlled fission reaction. The reactors may broadly be classified as thermal or fast (with or without breeding). Natural U-based fuel is used in pressurized heavy water reactors (PHWRs) to generate electricity and produce ^{239}Pu. The ^{239}Pu can be used as a fuel material

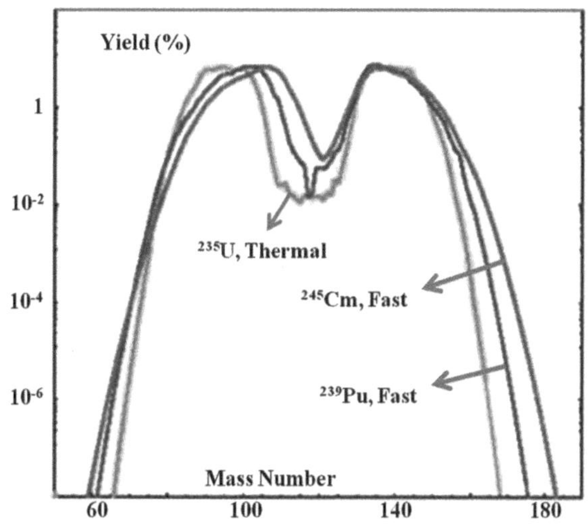

FIGURE 18.1

The mass number distribution of fission products for the thermal fission of ^{235}U, fast fission of ^{239}Pu, and fast fission of ^{245}Cm.

for fast breeder reactor [4]. In the concept of breeder reactors, mixed oxide fuel in conjunction with a blanket material like thoria is used (Eq. 18.2).

$$^{232}_{90}\mathrm{Th} + \mathrm{n} \rightarrow {}^{233}_{90}\mathrm{Th} \xrightarrow{22\ \mathrm{min}} {}^{233}_{91}\mathrm{Pa} + \beta^- + \vartheta \xrightarrow{27\ \mathrm{days}} {}^{233}_{92}\mathrm{U} + \beta^- + \vartheta \quad (18.2)$$

Advanced heavy water reactor (AHWR) is another class of reactors based on thorium-based technology envisaged in Indian nuclear power programme [5]. The futuristic concepts like accelerator-driven systems (ADS) are also being envisaged for generating copious neutrons per single spallation event. These neutrons can augment fission neutrons in sustaining the fission chain reaction in a subcritical reactor, converting fertile to fissile materials and transmuting long-lived radionuclides into short-lived radioisotopes.

4. REACTORS AND RADIOACTIVITY

All the components within the containment structure are susceptible to undergo degradation due to radiation effects.

$$^{238}_{92}\mathrm{U} + \mathrm{n} \rightarrow {}^{239}_{92}\mathrm{U} \xrightarrow{23\ \mathrm{min}} {}^{239}_{93}\mathrm{Np} + \beta^- + \vartheta \xrightarrow{2.4\ \mathrm{days}} {}^{239}_{94}\mathrm{Pu}(2.4 \times 10^4 \mathrm{years}) + \beta^- + \vartheta \quad (18.3)$$

Eq. (18.3) depicts the production of ^{239}Pu from ^{238}U by absorption of neutrons. Similarly, transuranic actinides like americium (Am) and curium (Cm) are also created by absorption of neutrons. Table 18.1 shows half-lives of some of the long-lived nuclides [3,6]. The spent nuclear fuel (SNF) contains minor

Table 18.1 Typical Components of High-Level Nuclear Waste [3]

Parent Isotope	Half-Life (years)	Decay and Energy (MeV)	Daughter Products	Problem
^{137}Cs	30.7	β (1.176)	^{137}Ba	Fission product producing heat by β-decay and displacement damage to a limited scale.
^{134}Cs	2.062	β (2.059)	^{134}Ba	
^{90}Sr	28.5	β (0.54)	^{90}Y	
^{85}Kr	10.72	β (0.687)	^{85}Rb	
^{106}Ru	1.020	β (3.54)	^{106}Ru	
^{144}Ce	1.3	β (0.319)	^{144}Pr	
^{99}Tc	2.13×10^5	β (0.294)	^{99}Ru	
^{129}I	1.57×10^7	β (0.194)	^{129}Xe	
^{243}Cm	28.5	α (6.169)	^{239}Pu	Minor actinides, produce α, causing extensive displacement damage
^{244}Cm	18.1	α (5.902)	^{240}Pu	
^{245}Cm	8500	α (5.623)	^{241}Pu	
^{241}Am	432.7	α (5.638)	^{237}Np	
^{242}Am	141	α (5.637)	^{238}Np	
^{243}Am	7380	α (5.438)	^{239}Np	
^{238}Pu	87.74	α (5.5)	^{234}U	
^{239}Pu	2.411×10^4	α (5.245)	^{235}U	
^{240}Pu	6563	β (0.005)	^{236}U	
^{237}Np	2.140×10^6	α (4.959)	^{233}Pa	

actinides (MAs), unused fuel, and FPs. Materials that have levels of radiation greater than that approved by regulatory authorities require immobilization and shielding to prevent radioactivity from spreading in the biosphere. A typical nuclear power plant generating 1 GW power produces annually ≈30 tons of SNF [3]. Nuclear waste is generated at every step of the nuclear fuel cycle. These nuclides are relatively long lived, emitting α, β, and γ radiations, which need to be managed. Thus materials in a reactor are exposed to energetic neutrons, FF, α-particles and their recoils, and β and γ radiation. These radiations deposit energy in the materials that deteriorates the material properties by causing unwanted structural and chemical changes like swelling, fission gas release, amorphization, segregation of radionuclides, precipitation, cracking, and many other such changes, henceforth referred to as damage.

5. METHODS TO DISPOSE NUCLEAR WASTE AND MATERIAL REQUIREMENTS

5.1 DIRECT DISPOSAL

In the case of direct disposal of nuclear waste [3,6], the strategy adopted all over the world is to convert the liquid waste into some kind of suitable solid form (immobile). The material has to be highly radiation resistant and chemically, mechanically, and thermally stable beyond the lifetime of major long-lived isotopes (Table 18.1). Solidification provides improved safety during handling and shipping. It also circumvents the release of radionuclides by groundwater leaching, a major route by which radioactivity can enter the biosphere. Commonly used waste-form materials are borosilicate glasses and futuristic ceramics (pyrochlores, zirconolites, hollandites, and synthetic rocks) and various ceramic composites.

5.2 PARTITIONING AND TRANSMUTATION

Partitioning and transmutation (P&T) is a concept that will involve separation of long-lived radionuclides from the waste usually by chemical means (partitioning) and then their transmutation in fast reactors or futuristic ADS [3,7]. This requires that the waste be diluted in a material that is inert against formation of new actinides by absorption of neutrons in a reactor or ADS (Eq. 18.3). Such materials are called inert matrix fuels [7]. Transmutation reduces all the long-term risks of storing and securing the nuclear waste. Some transmutation reactions are given as following for I and Tc.

$$^{129}_{53}\mathrm{I} + \mathrm{n} \rightarrow {}^{130}_{53}\mathrm{I} \xrightarrow{12.4\ \mathrm{h}} {}^{130}_{54}\mathrm{Xe(stable)} + \beta^- \tag{18.4}$$

$$^{99}_{43}\mathrm{Tc} + \mathrm{n} \rightarrow {}^{100}_{43}\mathrm{Tc} \xrightarrow{15.8\ \mathrm{s}} {}^{100}_{44}\mathrm{Ru(stable)} + \beta^- \tag{18.5}$$

Two obvious requirements for materials employed for the above-mentioned applications are:

1. They should be able to accommodate various chemical forms of radioactive nuclides (Table 18.1), which is required for their safe retention and dilution.
2. They should maintain their structural integrity on atomic and macroscopic scale under the previously mentioned radiation environment, since energetic radiation degrades material properties by depositing a large amount of energy, which creates highly nonequilibrium thermodynamic conditions.

The next section discusses the damage caused by the various types of radiation.

6. DAMAGE CAUSED BY RADIATION

The types of radiation and the specific damage caused by them in materials are described following.

6.1 NEUTRONS

Energetic neutrons ($E > 1$ MeV) displace atoms from their lattice sites in the target material as their momentum and energy are transferred to the target atoms. These recoil atoms, typically called primary knock-on atoms (PKA), have $\sim$100 keV of energy and hence produce displacement of many more atoms by direct head-on collisions in a cascade of displacements. These PKAs produce point defects and cascades, the latter consisting of vacancy-rich cores with interstitial-rich peripheries. The net result is production of point defects much in excess of their equilibrium concentrations. As operating temperatures of nuclear reactors are high enough for a significant mobility of point defects, various diffusional processes occur. These are annihilation of defects by mutual recombination of vacancies and interstitials, clustering of the same type of defects forming vacancy and interstitial clusters, cascade collapse forming dislocation loops, coupling of fluxes of vacancies and impurity atoms resulting in impurity segregation, and phase transitions including amorphization. At higher doses, they cause localized defect clusters and cascades, which can result in large swelling, amorphization, and even phase transformations [3,8,9].

6.2 β-, γ-DECAYS

Electrons can produce isolated displaced atoms (Frenkel pairs) if their energy is high, typically >1 MeV. Thus, β-decay causes very few isolated point defects. However, ionization processes may produce radiolysis. This may lead to radiation-induced segregation and enhanced diffusion of constituent elements and radionuclides [3,8,9].

6.3 α-DECAY

The α-particles produced during decay, for example, from ^{239}Pu (Fig. 18.2) cause damage mainly by electronic stopping (See Section 11.2 for electronic and nuclear stopping regime), producing about 200 displacements, most of them as largely isolated defects at the end of its range of some 10–20 μm.

Here, the heavy recoil atom ^{235}U receives a recoil energy typically 100 keV and causes damage predominantly by nuclear stopping. This produces a dense collision cascade in a few picoseconds with typically ≈1500 displacements in a short distance of ≈20 nm [3,8,9].

FIGURE 18.2

Schematic of α-decay process.

Adopted from IAEA-TECDOC-1516, Viability of Inert Matrix Fuels in Reducing Plutonium Amounts in Reactors, IAEA, Vienna, 2006.

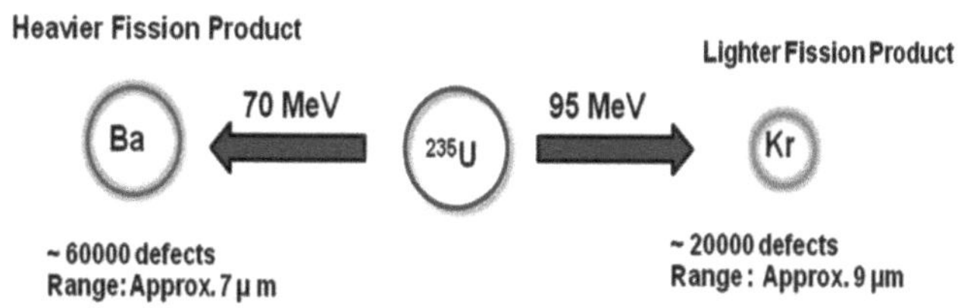

FIGURE 18.3

Schematic of fission-fragmentation process.

Adopted from IAEA-TECDOC-1516, Viability of Inert Matrix Fuels in Reducing Plutonium Amounts in Reactors, IAEA, Vienna, 2006.

6.4 FISSION FRAGMENTS

Fission produces two heavy fragments (plus two to three neutrons) with mass distribution as shown in Fig. 18.1. FFs are elements between Ga and Dy with energies between 70 and 120 MeV, and the process is as shown in Fig. 18.3.

These ions have a range of about ≈10 μm depending on the density of the material. Most of the energy deposited is by electronic energy loss, typically 20–30 keV/nm. This energy causes modifications in small discontinuous linear regions termed as "tracks" of size about 10 nm diameters. In conventional fuels like UO_2 they cause stresses, gas release, phase changes, etc.

7. THE RATE/DURATION OF RADIATION DAMAGE

The rate and duration of radiation damage varies over a large range. For MAs that have a large half-life and chain of unstable daughter products (Table 18.1), the duration is longer. The damage accumulated in the case of fuel stored in a repository is about 100 dpa in 10^5 years [3,6,8]. One dpa is defined as the dose at which, on an average, each atom has been displaced from its lattice position once. Flux of neutrons in the present PHWRs is about 10^{13} n/cm^2/s, which results in an accumulation of 100 displacements per atoms (dpa) in its lifetime. However, the flux is of the order of 10^{16} n/cm^2/s for fast reactors and even more for ADS and fusion reactors. Thus, in these, 100 dpa are expected to be accumulated only within 1–2 years. FFs may cause other changes like high-temperature spikes and shock pressures leading to high density of dislocations. Thus, there is a need to assess materials for their radiation stability and structural modifications within a shorter timescale and in a nonradioactive environment to facilitate detailed studies by employing several powerful analytical techniques. In the following section, the methods employed to test the radiation stability of materials to be used for the herein-mentioned applications are described.

8. METHODS TO EVALUATE DAMAGE

In the following sections, the four major methods used for testing the effects of radiation on materials are briefly discussed.

8.1 IN-PILE IRRADIATION IN REACTOR

In-pile neutron irradiation in a reactor is one of the oldest methods to assess radiation damage in materials to be employed for nuclear waste management [10–16]. These studies give direct assessment of radiation damage and provide the most reliable data. However, the cost of material testing under neutron irradiation is large, and currently the duration of necessary examinations of an improved material for industrial employment in existing reactors takes more than 20 years. Hence it is difficult to conduct direct reactor experiments for development of new materials.

8.2 ANALYSIS OF NATURAL ANALOGUES AND ACTINIDE DOPING

Natural minerals containing about 30 wt% of U and Th can also serve as natural analogues for studying self-irradiation effects [17—19]. They are useful in studying radiation effects in waste forms. However, in mineral samples the dose rates are extremely low, and the thermal and environmental histories of minerals are not known clearly [20]. In order to increase dose rates, the effects of alpha-decay events have been studied by incorporating short-lived actinides in lab, such as ^{238}Pu ($t_{1/2} = 87.7$ years) and ^{244}Cm ($t_{1/2} = 18.1$ years) [21,22]. While studies involving short-lived actinides provide realistic irradiation conditions and bulk damage, they are time-consuming, involving one or more years for completion, and generally provide only limited data under just a few sets of experimental conditions. Furthermore, detailed characterization and study of such highly radioactive samples is nontrivial.

8.3 COMPUTER SIMULATIONS

Computer simulations can be employed to study material properties by empirical potential-based molecular dynamics [23—29], by first principles [30], Monte Carlo simulations, or rate theory calculations [31]. However, these calculations involve many approximations or are limited by availability of correct potentials and many times by the speed and cost of computation. They can simulate only certain type of the damage.

8.4 ION BEAMS FROM ACCELERATORS

Suitably chosen ion beams can be efficiently used to perform accelerated tests of radiation damage caused by all the herein-mentioned sources [8]. As shown in Fig. 18.4, low-energy heavy ions ($\approx$100 keV—1 MeV) can be used to simulate damage by α-recoils and neutrons. High-energy, 70—120 MeV, ions (Ga to Dy) can simulate damage by FFs.

5 MeV helium ions can cause damage similar to α-particles, and 2—3 MeV electrons can efficiently simulate β-particles coming from FPs. Ion beams can cause the same amount of damage in a much shorter time. Thus, they allow assessing the extent of radiation damage in materials in a reasonable duration of time and outside a radioactive environment. Many studies have shown experimentally that the damage by ions is similar to that observed in naturally found minerals containing MAs [32]. The advantages are several, including the following:

1. The damage accumulation takes place within few hours; high-dose damage can be easily achieved.
2. There is control over dose, energy, and ion species.
3. Allows study of kinetics of damage at various temperatures.
4. Several in situ characterization tools can be used to probe the structural evolution.
5. Several ex-situ structural and chemical analyses can also be performed.

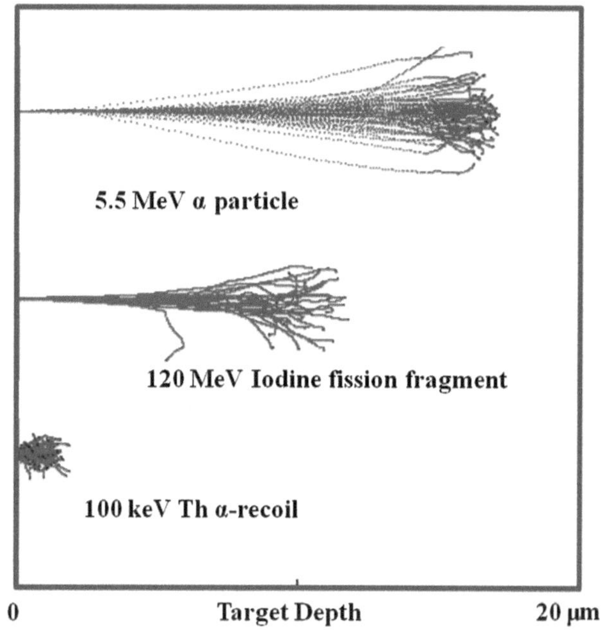

FIGURE 18.4

An illustration how appropriate ions of suitable energy can simulate radiation damage.

6. Most importantly, all this can be done in a laboratory environment unhindered by high radioactivity, and the damage due to individual types of radiation can be isolated by choosing appropriate ions of required energy.

One of the shortcomings of using accelerated ions is that the rate at which they cause damage is very large. The rate effect needs to be studied carefully, and presently there is an intensive research effort on the part of the International Atomic Energy Agency and many countries to demonstrate how experimental accelerator simulation methods and theoretical modeling tools could be applied for investigation of irradiation phenomena.

The following section describes the physics of ion beam interaction.

9. ION–SOLID INTERACTION

Ion–solid interactions have been a subject of study for a century. A detailed description of the stopping theory can be found in several standard books [33–35]; here only important concepts related to ion–solid interactions are described.

Depending on the characteristics of the ion beam, ion–material interactions result in sputtering, X-ray and electron emission, ion scattering and channeling, nuclear reactions, luminescence, formation of defects, and phase transformations/separation.

9.1 ION BEAM CHARACTERISTICS AND ENERGY TRANSFER

Important characteristics of ion beams are ionic mass, energy (MeV/amu), ionic state, and beam current. When an energetic particle penetrates a medium, it loses energy along its path depending on the ion characteristics and material properties. The spatial rate of energy loss is given by [34,35],

$$\frac{dE}{dx} = N\,S(E) \tag{18.6}$$

where $S(E)$ is the total stopping cross-section and N is the atomic density of the target. dE/dx, inside the material, has no spatial or temporal coherence but can be measured and calculated. It is an important parameter on which the effects of irradiation depend and is defined for a given target-ion combination. Large rate of energy deposition dE/dx (20–30 keV/nm) occurs near characteristic energy region where ions have velocity close to Bohr velocity (V_B, velocity of electron in Bohr orbit of the target). The Bohr velocity can be derived from the uncertainty principle as, $V_B \approx \left(\frac{1}{4\pi\varepsilon_0}\right)\left(\frac{e^2}{\hbar}\right) \approx 2.2 \times 10^6\ \text{m/s}$, where $\left(\frac{1}{4\pi\varepsilon_0}\right) = 9.0 \times 10^9\ \text{Nm}^2/\text{C}^2$, ε_0 is the permittivity of constant, e is a unit charge, and $\hbar$ is Planck's constant divided by 2π.

9.2 RANGE

The mean range is another important characteristic of an ion beam. The ions undergo several collisions inside the material and move stochastically until they finally come to rest.

The penetration depth or ion range is defined as the total distance travelled by the ion inside the material on the average and is given by:

$$R = \int_{E_0}^{0} \frac{1}{\left(\frac{dE}{dx}\right)_{total}}\, dE \tag{18.7}$$

where E_0 is the energy of the incident ion and $\left(\frac{dE}{dx}\right)_{total}$ represents the total energy loss in a material that can be partitioned into electronic and nuclear components. Fig. 18.5 shows schematics of the ion range in a solid. The projection of the ion trajectory on the axis of incidence is called the projected range R_p. Since the number of collisions per unit distance and the energy lost per collision are random variables, there is a spatial distribution of implanted ions with the same initial energy inside a material. These statistical fluctuations in the R_p are called the projected straggling σ_p, and similar fluctuations also exist along an axis perpendicular to the axis of incidence, called lateral straggling $\sigma_\perp$. In general, straggling is the spread in R_p.

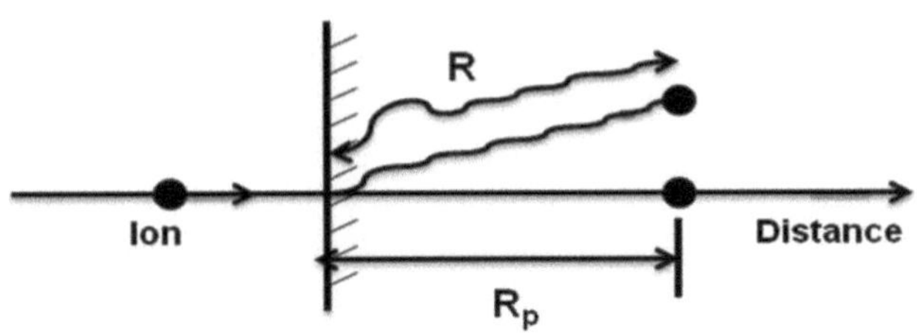

FIGURE 18.5

Total path length R and projected range R_p of the incident ion.

Adopted from J.W. Mayer, L. Eriksson, J.A. Davies, Ion Implantation Is Semiconductors, Academic Press, New York, 1970.

9.3 ENERGY-LOSS MECHANISMS

As the ion penetrates a solid it loses energy first by interacting with target electrons and ionizing them. As it slows down it interacts with screened target nuclei causing recoils or direct displacements.

The latter process is called nuclear energy loss and is governed by nuclear stopping cross-section (S_n), and the former is known as electronic energy loss and is governed by electronic stopping cross-section (S_e). When expressed in units of energy transferred per path length in the target, S_e and S_n are denoted as the electronic and nuclear energy loss or stopping power. Fig. 18.6 shows the variation of S_n of and S_e for the Xe ion in ThO_2. Both S_n and S_e depend upon the kinetic energy of the ion because the velocity determines the interaction time, distance of approach, the type of interaction potential, and charge state of the ion. When the initial energy of ion is in the range <1 keV/amu, S_n is dominant, and when it is in the range >0.1 MeV/amu S_e is dominant (Fig. 18.6). Hence, for a high-energy ion, S_e is the major term at its entry in the solid and S_n is dominant only at the end of the ions' range. An example is shown in Fig. 18.7a where a 100 MeV I (note that iodine (I) serves as a typical FF) penetrates through ThO_2, for up to 6 μm, S_e is the dominant component of energy loss, and as the ion penetrates further, the contribution from S_n increases and finally overtakes S_e.

For a low-energy ion, however, S_n remains the major term. This is shown in Fig. 18.7b for a 100 keV iodine ion penetrating through ThO_2, where from the surface itself the major mode of energy loss is by S_n. Here the low-energy iodine has the same behavior as α-recoil or PKAs generated by neutrons.

9.4 COMPUTATION OF NUCLEAR ENERGY LOSS

S_n arises from the collisions between the energetic particles and the target nuclei, which cause atomic displacements and excitation of phonons. Nuclear stopping becomes important when an ion slows down well below the V_B (Bohr velocity). Displacement occurs when the colliding particle imparts an energy greater than the displacement threshold energy (E_d) to the target atom so that it leaves its lattice position. If the energy is not greater than E_d, then the same is dissipated in causing atomic vibrations or

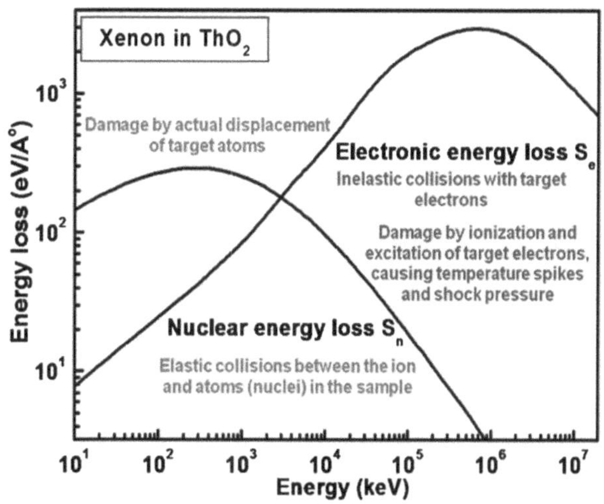

FIGURE 18.6

Variation of electronic and nuclear energy loss as function of energy for Xe ions in ThO_2 [35].

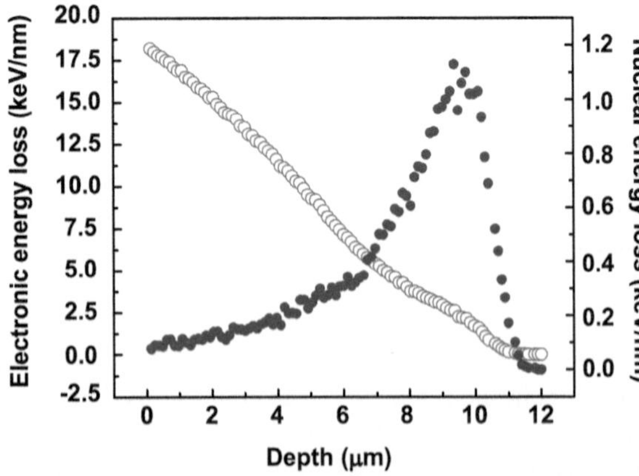

FIGURE 18.7A

Variation of electronic and nuclear energy loss as function of depth for 100 MeV I ions in ThO_2 [35]. [*Red color curve* (gray in print versions) denotes electronic energy loss and *green color curve* (dark gray in print versions) denotes nuclear energy loss.)

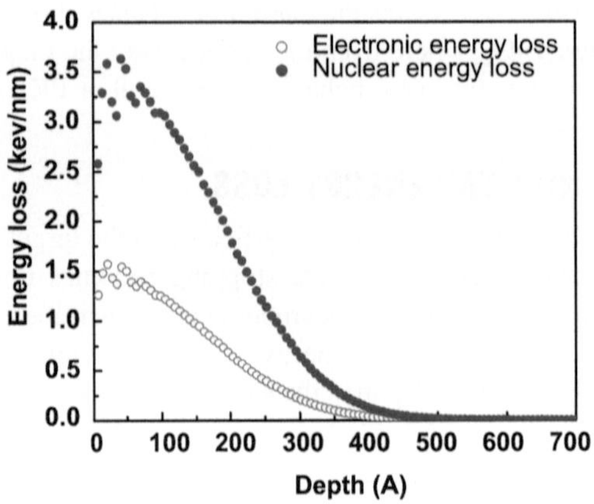

FIGURE 18.7B

Variation of electronic and nuclear energy loss as function of depth for 100 keV U ions in ThO_2 [35].

phonons. E_d depends on target Z, crystallographic location, and type of bonding in a material and will be defined later.

The nuclear stopping cross-section is given by,

$$S_{\text{nuclear}} = \int_0^{\infty} T(E_0, p) 2\pi p dp \tag{18.8}$$

where E_0 is the initial kinetic energy of the projectile, "p" is the impact parameter, which depends on the particle velocity and the interatomic potential between the two colliding particles, and $T = \frac{4M_0M_1}{(M_0M_1)^2} E \sin^2(\frac{\theta}{2})$ is the transferred energy, E is the kinetic energy of the incident atom, θ is the scattering angle in the center of mass system, and M_0 and M_1 are the mass numbers of the projectile and target atoms, respectively. θ and r are two statistical parameters that are directly correlated for a given collision event. The calculation of differential cross-section for energy transfer requires the knowledge of interatomic potential $V(r)$ between ion and the target atom, "r" being the distance between the nuclei and ion. In principle it is a repulsive Coulomb potential due to the positive charges of the cores. This Coulomb potential is screened by the surrounding electrons. The effect of the electrons can be described by assuming a dimensionless screening function $\Phi(r)$ to the Coulomb potential.

$$V(r) = \frac{Z_0 Z_1 e^2}{r} \Phi(r) \tag{18.9}$$

where Z_0 and Z_1 are the atomic masses of the projectile and the target atoms and "e" is the electronic charge. A large number of different repulsive potentials and screening functions have been proposed over the years, some determined semiempirically, others from theoretical calculations. The Ziegler, Biersack, and Littmark (ZBL) repulsive potential is mostly used. It has been constructed by fitting a universal screening function to theoretically obtained potentials calculated for a large variety of atom pairs. The ZBL screening function has the form

$$\Phi(r) = 0.1818e^{-3.2x} + 0.5099e^{-0.9423x} + 0.2802e^{-0.4029x} + 0.02817e^{-0.2016x} \tag{18.10}$$

where $x = r/a_U$ is the reduced distance, a_U the screening distance given by,

$$a_U = \frac{0.8854a_0}{Z_0^{0.23} + Z_1^{0.23}} \tag{18.11}$$

For practical calculations, ZBL nuclear stopping power for an ion with energy E_0 and the reduced nuclear cross-section ($S_n(\varepsilon)$) is given by the following equation,

$$S_n(E_0) = \frac{8.462 \times 10^{-15} Z_0 Z_1 S_n(\varepsilon)}{(M_0 + M_1)\left(Z_0^{0.23} + Z_1^{0.23}\right)} \frac{\text{eV} - \text{cm}^2}{\text{atom}} \tag{18.12}$$

with the reduced energy ε given as,

$$\varepsilon = \frac{32.53 M_1 E_0}{Z_0 Z_1 (M_0 + M_1)\left(Z_0^{0.23} + Z_1^{0.23}\right)} \tag{18.13}$$

and the reduced nuclear stopping is calculated as:

$$\text{For } \varepsilon \leq 30: \quad S_n(\varepsilon) = \frac{\ln(1 + 1.1383\varepsilon)}{2\left[\varepsilon + 0.01321\varepsilon^{0.21226} + 0.19593\varepsilon^{0.5}\right]} \tag{18.14}$$

$$\text{For } \varepsilon > 30: \quad S_n(\varepsilon) = \frac{\ln(\varepsilon)}{2\varepsilon} \tag{18.15}$$

9.5 COMPUTATION OF ELECTRONIC ENERGY LOSS

S_e arises from electromagnetic interaction between the positively charged ion and the target electrons. In this case, only a small momentum is transferred to the target atoms; most of the energy of the ions is lost by inelastic scattering of the ions by target electrons resulting in electronic excitation and/or ionization of the target atoms. In both the cases excited electrons (plasmons) eventually transfer energy to lattice and they thermalize. Theoretical and experimental evidence suggests that approximately one-half of the electronic energy loss is due to electronic excitations and the other half due to ionizations. This is often referred to as the equipartition principle [37,38]. Electronic stopping is determined mainly by the charge state of the ion and its velocity. It is derived for elemental targets and then applied to target consisting of different atoms by employing Bragg formula/approximation (Net S_e is equated to weighted sum of energy loss due to each target element).

9.5.1 Analytical Expression for Electronic Energy Loss

When an ion passes through a medium, its orbital electrons are stripped off to various extents depending upon the ion velocity (V_{ion}). The effective charge on a positive ion is given, in terms of V_{ion} and V_B, by Northcliffe [39] as,

$$Z^*_{eff} = Z_0\left[1 - a\exp\left(-b\frac{V_{ion}}{V_B}Z_0^{-2/3}\right)\right] \tag{18.16}$$

where Z_0 is the atomic number and a and b are fitting constants.

A mathematical expression for the electronic stopping cross-section is given by,

$$S_{electronic} = \int I(V_{ion}, \rho)\left(Z_1^*(V_{ion})\right)^2 \rho dV \tag{18.17}$$

This stopping power is calculated for a particle in the plasma of each volume element's density, and the final expression is computed by averaging over these values weighted by their distribution in the solid. Here, $I(V_{ion}, \rho)$ is the stopping interaction function and it determines the interaction of the particle with velocity v in a plasma of density ρ. Z_1^* is the charge of the proton, the asterisk indicates that this value may be different from the atomic number because the ion may not be fully stripped; dv is a volume element. Integrating over ρdv weighs each density interaction by the probability of that density occurring in the solid. $I(V_{ion}, \rho)$ varies with the particle velocity and plasma oscillation frequencies of the electrons in the stopping medium. Fast particles ($V_{ion} \gg V_B$) have little interaction time and so electronic stopping decreases with increasing velocity beyond this velocity. Slow particles ($V_{ion} < V_B$) pick up electrons, so have a lower effective charge. Therefore, there is a broad maximum in the electronic stopping and it occurs near the Bohr velocity.

At ion velocities where $V_{ion} < V_B Z^{2/3}$, the widely used form of S_e is given by Lindhard and Scharff [40]. In this region, the S_e is given by:

$$S_e(E) = 3.83\frac{Z_0^{7/6}Z_1}{\left(Z_0^{2/3} + Z_1^{2/3}\right)^{3/2}}\left(\frac{E}{M_0}\right)^{1/2} = K_L E^{1/2} \tag{18.18}$$

where $K_L = 3.83\frac{Z_0^{7/6}Z_1}{M_0^{1/2}\left(Z_0^{2/3}+Z_1^{2/3}\right)^{3/2}}$ and $S_e(E)$ is in units of 10^{-15} eV/cm^2/atom for ion energies E given in keV and M_0 in atomic mass units.

For higher ion velocities $V_{ion} \gg V_B Z^{2/3}$, Bethe and Bloch provide a good estimate assuming that the interaction is distant enough so that the dipole approximation applies [41–43].

$$S_e = \frac{4\pi e_0^2 e^2}{mV_{ion}^2}Z_2 \ln\left(\frac{2mV_{ion}^2}{I}\right) \tag{18.19}$$

For $M_0 \gg M_1$ the mean logarithmic ionization energy I is defined by,

$$\ln I = \sum_j f_{j0} \ln(\varepsilon_j - \varepsilon_0) \tag{18.20}$$

where $e_0 = Z_0 e$ is the projectile charge and e is elementary charge.

9.5.2 Relativistic Form

The relativistic form of the S_e proposed by Bohr, Bethe, and Fano [44,45] is

$$S_e = \frac{4\pi e_0^2 e^2}{m V_{ion}^2}\left[\ln\left(\frac{C m V_{ion}^3}{e_1 e \omega_0}\right) - \ln\left(1 - \frac{V_{ion}^2}{c^2}\right) - \frac{V_{ion}^2}{2c^2}\right] \tag{18.21}$$

This expression was proposed by Bohr and the term $-\frac{V_{ion}^2}{2c^2}$ is for distant interactions. The relativistic Bethe formula for a heavy projectile is,

$$-\frac{dE}{dx} = \frac{4\pi e_0^2 e^2 Z_1 N}{m V_{ion}^2} \sum_j f_{j0}\left[\ln\left(\frac{2 m V_{ion}^2}{\hbar \omega_{j0}}\right) - \ln\left(1 - \frac{V_{ion}^2}{c^2}\right) - \frac{V_{ion}^2}{c^2}\right] \tag{18.22}$$

This equation differs from the one proposed by Bohr by the last term $-\frac{V_{ion}^2}{c^2}$, which arises from close collisions. Here, C is coefficient in Bohr's stopping formula $2\exp - \gamma = 1.1229$, $\gamma = 0.5572$ is Euler's constant ω_0 and $\omega_{j0} = (\varepsilon_j - \varepsilon_0)$ is the resonance frequency, c is the speed of light, N is the number of atoms per unit volume, Z_1 is the atomic number of the target nucleus, and f_{j0} is dipole oscillator strength.

The effects of ion irradiation depend primarily on S_n and S_e and may be estimated for a target-ion beam pair by a Monte Carlo program which is called Stopping and Range of Ions in Matter (SRIM) [35] that employ expressions discussed before. The effect may be defect formation by displacement of atoms by momentum transfer, when S_e is low. When S_e exceeds a material-dependent critical value of electronic energy loss (S_{eth}), it results in the formation of a spike in excitations in both electronic and atomic systems within a cylindrical volume surrounding the ion path. Values of this track radius can be calculated from the nature of transformation that occurs within the track as described in later sections.

10. DAMAGE IN NUCLEAR-STOPPING REGIME

Generally, damage by S_n is assessed in terms of dpa. One dpa is the dose at which, on an average, each atom has been displaced from its lattice position once. The expression of displacement dose in dpa as per NRT theory [46] is given by

$$\text{dpa} = \frac{0.8}{2E_d(\text{eV})} S_n(\text{eV/cm}) \times \frac{\text{ion fluence}(\text{ions/cm}^2)}{\text{target atomic density}(\text{atoms/cm}^3)} \tag{18.23}$$

where E_d is determined experimentally or calculated and S_n is calculated from SRIM. The term *dose* is expressed often in terms of the energy deposited per unit mass of material, generally in units of eV/kg or Gy (=100 Rad). The term *fluence* indicates the number of ions injected per unit area of target material (ions/cm^2). In the conversion from fluence to dose, the SI unit of Gray (Gy) is obtained by multiplying the fluence in (ions/m^2) by the energy loss (eV/nm) and dividing by the density of the target material (kg/m^3) and employing the conversion factor of 1 Gy = 6.24×10^{18} eV/kg [47].

11. DAMAGE IN ELECTRONIC-STOPPING REGIME: TRACK FORMATION

Ions that deposit energy in the electronic-stopping regime are generally termed as swift heavy ions (SHI), since the ions are with high velocity and high Z. When an SHI passes through matter, it strips off the target electrons in a cylindrical region thus ionizing the atoms. The mutual repulsion of positively charged target ions results in atomic motion. All these phenomena take place within a timescale of the order of 10^{-17} to 10^{-13} s. At higher timescales, the electronic excitations lead to the formation of hot plasma via electron-phonon coupling, which leads to spike in the temperatures of the order of several thousands of K followed by fast quenching rates [48]. In nuclear materials, FFs create such spikes and cause structural modifications in materials like amorphization or other phase transformations, polygonization, etc. All these various phenomena are reviewed in the next sections.

11.1 IDENTIFICATION OF TRACKS

Tracks were first observed [49] in an electron microscope image in 1959. Chemical etching [50,51] and optical microscopy can also be used to detect tracks produced by FF (Fig. 18.8).

Currently morphology of the track is studied by various methods, and this reveals spherical discontinuous and cylindrical continuous regions of structural disturbances. Naturally occurring tracks are found in minerals containing uranium. Here, the α-decay of ^{238}U leads to a recoiling daughter nucleus ^{234}Th capable of creating nuclear track [52,53]. Recoil tracks have important implications in the field of nuclear power generation. A detailed review of nuclear tracks, their use, and production, is available elsewhere in literature [54].

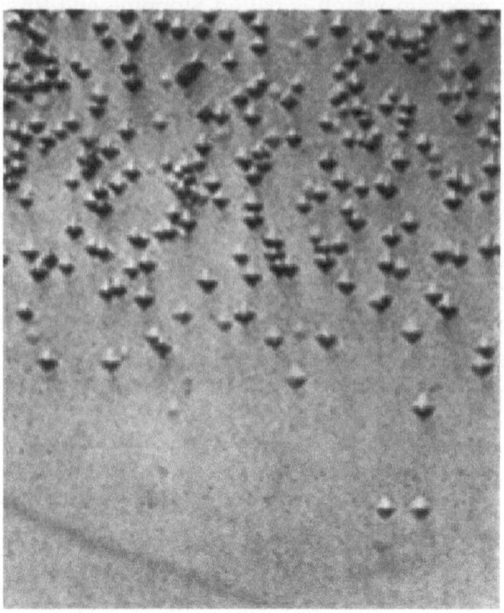

FIGURE 18.8

Etched fission tracks in LiF [50].

Reproduced with permission from Nature Publishing Group.

11.2 HIGH PRESSURE–TEMPERATURE AND SWIFT HEAVY IONS

The injection of energy by SHI is faster than any of the relaxation mechanisms in the materials and results in the transformations that are similar to those occurring under the influence of shock wave. However, simulations indicate that the temperature may attain much higher values than those obtained along the Hugoniot [55]. Nevertheless, very often material inside the tracks will have close similarity/relation with the high-pressure phases of the material or may become amorphous because of the very fast quenching.

12. MODELS OF TRACK FORMATION

The effects of SHI interaction with a material are normally described by the Coulomb explosion and the thermal spike models, and these are described in the following sections.

12.1 COULOMB EXPLOSION

The coulomb explosion model is the first to have been proposed to explain the phase changes, and it was initially assumed that the model is applicable to insulators [56,57]. According to this model, a positively charged region of a few nanometers is created by ionization of target atoms, after the passage of SHI (Fig. 18.9). The coulomb repulsion between these ions in the small volume converts the potential energy of target ions into the kinetic energy of atoms and causes very fast motion of atoms. This leads to radiolysis, atomic disordering, and effects similar to those caused by shock waves. This model has been applied to explain track formation in polymers, mica, cellulose nitrate, zircon, quartz, and glasses. The tracks are formed only after a critical S_e (S_{eth}) is reached. This is consistent with the fact that after SHI irradiation, materials undergo structural modifications similar to those observed after shock compression and a threshold pressure is needed for introducing such transitions. The phase transformation of hcp (α) to hexagonal (ω) transition in titanium irradiated with 2.2 GeV U, Fig. 18.10, is an example of such structural transformation [58]. The model is applicable to metallic systems as well. However, there is not much quantitative development in this model.

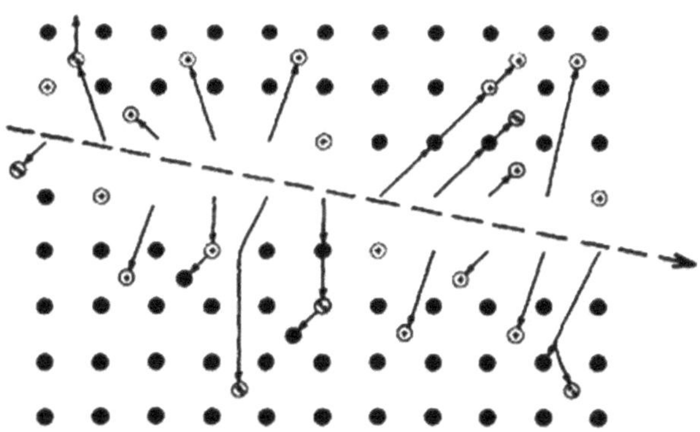

FIGURE 18.9

Coulomb explosion in the wake of the SHI [56,57].

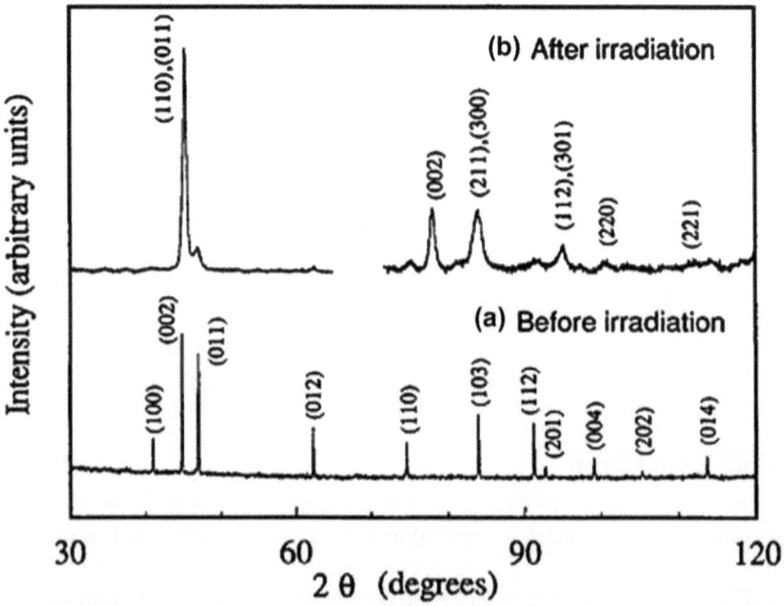

FIGURE 18.10

Phase transformation of hcp-α to hexagonal-ω in titanium irradiated with 2.2 GeV U ions [58]. (a) Before irradiation. (b) After irradiation.

Reproduced with permission from Taylor & Francis.

12.1.1 Thermal Spike Model

F. Desauer first pointed out that the electronic excitation in the wake of the charged particle could result in the transfer of energy of the ion from the electronic to the vibrational system of the lattice through electron-phonon coupling [59]. The first explanation of phase transformation in materials was then formulated by Seitz and Koehler based on the herein-mentioned hypothesis [60]. Extending this proposal, the energy transferred by SHI to the lattice results in the rise in temperatures of the order of thousands of K in a few picoseconds followed by fast quenching. This model is commonly known as "thermal spike." Detailed calculations based on the thermal spike model have been developed by Toulemonde et al. [61–63] and Szenes [64].

The following sections discuss some of the effects observed in the post-irradiated materials.

13. PHASE TRANSFORMATIONS BY SHI IRRADIATION

The SHI causes structural phase transformations in materials, which can be related to creation of a thermal spike; for example, the monoclinic to tetragonal transformation observed in ZrO_2 doped with fissionable impurities upon neutron irradiation [65]. Later, the kinetics of the same were studied in detail by Benyagoub in HfO_2 and ZrO_2 irradiated with SHI [66]. Similarly, a cubic to high pressure monoclinic phase is observed in the case of Y_2O_3 [67,68].

14. ION HAMMERING AND TEXTURE CHANGES

In addition to what was just discussed, amorphous alloys like some metallic glasses show dimensional changes due to mass transport when irradiated with SHI, known as "ion hammering." Ion hammering denotes the phenomenon that amorphous (metallic glass) or amorphizing solids exhibit irreversible deformation when subject to bombardment with track-generating ions [69,70]. In a thin sample of metallic glass, the dimensions perpendicular to the ion beam increase linearly with fluence while the mass density remains virtually constant, implying that the dimension parallel to the ion beam shrinks [71–73]. Recently, SHI-induced texture changes in nanocrystalline titanium layers have been observed. During off-normal irradiation, the nanocrystals undergo rotation up to 90 degrees, and a reversal of the ion incidence angle leads to a back rotation of the grains. At the highest ion fluence, the whole layer exhibits shear flow [74].

15. IRRADIATION EFFECTS IN SOME REPRESENTATIVE NUCLEAR MATERIALS

Ceramics with proposed applications in hostile environments such as front end and back end of the nuclear cycle (nuclear reactors and waste immobilization) must be radiation resistant. They should be resistant to radiation-induced amorphization or swelling. The radiation performance of the material depends, in a large part, on its ability to accommodate lattice defects and in turn would be governed by its structure. Over the last few years, this has been a fertile area for materials scientists and engineers and a lot of work has been devoted to identifying radiation-tolerant materials and understanding their structural rationale so that this knowledge can be utilized to design tailor-made materials with improved radiation resistance. Herein, the trends in radiation damage behavior in some classes of materials have been elucidated based on their crystal structure. The radiation-induced damage may manifest in many undesirable effects in lattices such as amorphization, swelling induced by clustering of vacancies, interstitial point-defect clustering, and defect-induced precipitation of new crystalline phases. The most important structural class of compounds pertaining to nuclear materials are fluorites and fluorite-related compounds such as pyrochlores ($A_2B_2O_7$), $A_3B_4O_{12}$, etc. The ideal fluorite structure is adopted by materials of nuclear significance such as UO_2, PuO_2, ThO_2, CeO_2, HfO_2, and higher actinides [75].

Binary metal oxides possessing fluorite structure are known to exhibit a very high radiation stability [28,75]. This property was well recognized many years ago and was one of the major factors leading to early use of uranium dioxide UO_2 (urania) as nuclear fuel instead of metallic uranium [76]. There have been many studies wherein the radiation stability of fluorite-type materials has been investigated under both low- and high-energy irradiation and over a range of fluences. Low-energy irradiation mimics the damage caused by alpha recoils or self-irradiation in the oxides employed at both front end or as a waste immobilization form. Similarly, structural modifications caused by SHI (swift heavy ions) simulate the effect of fission fragments. Such studies have been extensively carried out in UO_2 (as fuel materials) and CeO_2, yttria-stabilized zirconia (YSZ), pyrochlores such as $Gd_2Zr_2O_7$ (as proposed matrices for "inert matrix fuel" concept). In a detailed study [77], UO_2 was irradiated with various SHI at a wide range of energies corresponding to typical FFs up to a maximum fluence of about 10^{17} ions/cm^2. S_{eth} was found to be between 22 and 29 keV/nm [77,78]. SHI caused lattice parameter to increase, fission gas bubble formation, resolution of fission gas from bubbles, and fission-enhanced diffusion. Ion irradiation experiments performed using ^{85}Kr as tracer and 72 MeV iodine ions as FF showed an athermal release of Kr between ambient temperature and 500°C. No amorphization could be observed, however, with increase in fluence an increase in the loop density was observed. The rim effect in UO_2 has also been studied extensively [79]. It is an interesting physical phenomenon that has been observed in UO_2 at extended lifetimes (burnups). The cold outer part of UO_2 pellets used as fuel experiences more radiation damage and contains more fission damage as compared to bulk of pellet. This happens due to neutron resonance capture by U-238 in this rim region to form fissile Pu-239. The consequence is the polygonization or grain subdivision of sintered UO_2 into smaller grains, which has implications on properties like thermal conductivity, mechanical properties, fission gas release, etc. [79]. Irradiation studies on YSZ were performed with 940 MeV Pb ions at various fluences. The macroscopic stresses developed were measured using X-ray diffraction (XRD) employing the $\sin^2\psi$ method [80]. The state of stress in the irradiated layer can be described by a combination of hydrostatic stress caused by the formation of damaged tracks leading to swelling and biaxial stress imposed by the bulk undamaged material, which controls the lateral expansion of the surface-damaged layer [80]. This stress created by SHI resulted in an increase in the hardness and fracture toughness of YSZ. The improvement of the mechanical properties of ceramics by ion irradiation could be due to the fact that the residual compressive stresses are introduced into the surface layer of samples [81]. Various techniques have been employed to characterize the damage in the materials. Rutherford backscattering with ion channeling analysis was employed to study the damage depth distributions in SHI-irradiated YSZ [82,83]. Steady values of the defect concentrations in both the cationic and anionic sublattices are obtained at depths higher than 500 nm, with depletion in the surface region. The S_{eth} was found to be 20–25 keV/nm, above which the defect concentration was found to increase strongly with a more pronounced effect in the oxygen sublattice as compared to the Zr. No amorphization, however, could be observed, even at S_e of 50 keV/nm [72,83].

Ceria (CeO_2), is also an important nuclear material. It is used as a surrogate for UO_2, PuO_2, etc. for studying the material properties of the fuel oxides since ceria possesses identical structure and has similar thermophysical properties. Further, zirconia and ceria are the possible components of inert matrix fuels such as rock-like oxides [84] or CERMET fuels, etc. Considering this, quite a few studies have been performed on the radiation stability of ceria under both low-energy ions as well as swift heavy-ion irradiation. In a study performed by Sonoda et al. [85], irradiation of CeO_2 with 210 MeV Xe ions led to the formation of ion tracks wherein the crystalline lattice structure inside the ion track was maintained. Ishikawa et al. [86] showed detailed structural investigations using XRD thin films irradiated with 150 and 200 MeV ^{197}Au ions and found that these ion tracks possess modified crystalline lattice with larger lattice parameters.

It must be noted that other than structural factors, the powder properties of the materials such as their particle size, densities, microstructures, etc. also have been recently shown to exert noticeable influence on the radiation stability. The following examples will show the dependence of radiation stability of materials on structure as well as microstructure of complex oxides of nuclear significance.

15.1 RADIATION STABILITY AND CRYSTAL STRUCTURE

Pyrochlores ($A_2B_2O_7$) hold a special importance in terms of their proposed applications as one of the candidates for host lattices for MAs transmutation as well as waste immobilization because of their structural compatibility with the larger radionuclide species [28]. In view of this, understanding their radiation stability is of paramount importance. A lot of research has thus been dedicated to explore their behavior under both low- and high-energy irradiation. Interestingly, they are also excellent candidates for studying the effect of structure on irradiation behavior. Among all the pyrochlores, $Gd_2Zr_2O_7$ has been considered to be the most radiation-stable material because of its unique crystal structure.

The pyrochlore structure $A_2B_2O_7$ is a superstructure of fluorite-type structure AO_2 (Fm-3m). In general, the introduction of trivalent B_2O_3 in fluorite-type AO_2 will lead to formation of a defect fluorite, wherein an oxygen vacancy accompanies every two incumbent B^{3+} cations for the charge compensation. In a defect fluorite structure, A^{4+} and B^{3+} cations are randomized over the 4a site (cationic site in fluorite structure) while the oxide ions and vacancies are randomized over the 8c sites (occupied by oxygen anions in ideal fluorite structure). However, in certain cases, depending on relative cationic radii of A^{4+} and B^{3+}, there occurs an ordering of A^{4+} and B^{3+} and vacancy occupies a distinct position (8a) [28] rather than getting randomized with oxide ion position, and this results in pyrochlore structure. The pyrochlore structure is observed when r_A/r_B lies between 1.46 and 1.75, known as the radius ratio rule for pyrochlores. For the radius ratio lower than 1.46, defect fluorite is observed while for ratios higher than 1.75 a monoclinic structure is obtained. It could also simply be understood as follows—the closer are the ionic radii of A and B cations, the higher would be their tendency to randomize and they would crystallize as defect fluorite. The more the dissimilarity (up to a certain extent, in accordance with radius ratio rule), the higher is their inherent tendency to order up and yield pyrochlore structure. The schematic diagrams of pyrochlore and fluorite are depicted in Fig. 18.11.

The radiation, in general, causes the formation of Frenkel defects wherein each Frenkel pair consists of atomic interstitial and a vacancy. The irradiation thus leads to the formation of excess Frenkel pairs in the lattice. In pyrochlore lattice, the formation of Frenkel defects occurs on both cationic and anionic sublattices. The Frenkel defects on cationic sublattice lead to antisite defects, which means that A- and B-cations exchange lattice sites. Sickafus et al. [75] described the radiation damage effects in an ordered complex oxide as follows:

$$A_A + B_B \rightleftharpoons B_A + A_B \tag{18.24}$$

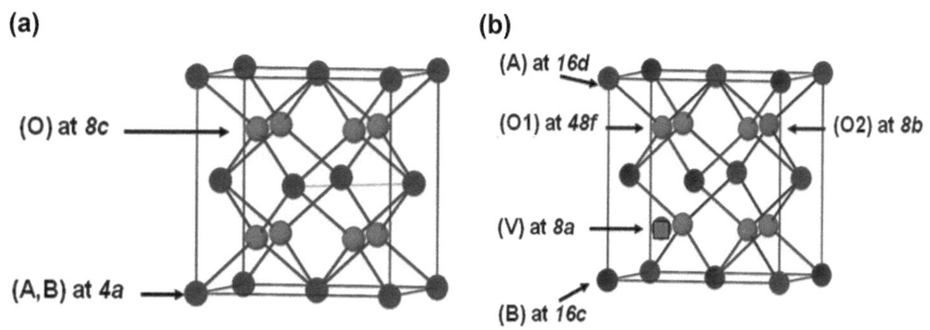

FIGURE 18.11

Schematic diagram of fluorite (a) and pyrochlore (b) structures.

$$O_O \rightleftharpoons V_O + O_i \quad (18.25)$$

Eq. (18.24) represents cation antisite defect formation and Eq. (18.25) represents oxygen anion sublattice interstitial—vacancy pair. Interestingly, these are the same defects that occur in an order-disorder phase transformation accompanying pyrochlore to defect-fluorite phase transitions. Thus, the irradiation of pyrochlore lattice should also lead to defect fluorite phase. However, it was explained in the preceding discussion that not all the $A_2B_2O_7$ compounds form pyrochlore structure. It has been lucidly explained in literature that the damage evolution or radiation tolerance of the complex oxides would depend on two factors: (1) the extent of lattice stability upon accumulation of defects; and (2) the fate of accumulated interstitials and vacancies. Therefore, in light of this, it may be safely predicted that the $A_2B_2O_7$ pyrochlore composition, which can transform to the defect fluorite structure subsequent to formation of defects (cation antisites and oxygen-vacancy interstitials), would show enhanced stability under irradiation. Otherwise, there would be cluster formation of interstitials and vacancies leading to formation of dislocation loops, void swelling, and even amorphization. Thus, from a structural point of view, the closer the r_A/r_B ratio, toward the lower end of the stability regime of pyrochlores, the better would be the chances of annihilation of defects leading to formation of defect fluorites and hence the higher would be the radiation tolerance.

From structural considerations, this implies that the $A_2B_2O_7$ compositions closer to borderline radius ratio of defect fluorite-pyrochlore would be better candidates for radiation-stable materials. On the other hand, in case of highly ordered pyrochlores, there is no way of annihilating the Frenkel defects (since the vast difference in sizes of A^{3+} and B^{4+} would not favor cation antisites) and hence they amorphize readily. An extensive study to establish the radiation tolerance of the pyrochlores with r_A/r_B ration was performed by Patel et al. wherein they chose three pyrochlores, viz., $Gd_2Ti_2O_7$, $Nd_2Zr_2O_7$, and $Gd_2Zr_2O_7$ with radius ratios as 1.46, 1.54, and 1.75, respectively [87]. The irradiation was performed with 120 meV Au, 90 MeV I, and 70 MeV Ni ions in order to mimic typical fission fragment energies. Under any particular energy it was observed that the radiation stability indeed varied with the radius ratios, and whereas $Gd_2Ti_2O_7$ amorphized completely, $Gd_2Zr_2O_7$ did not amorphize and rather showed an order-disorder (O-D) transformation. Fig. 18.12 shows the XRD patterns of the three pyrochlores under investigation under the irradiation of 90 MeV I and at the fluence of 1×10^{13} ions/cm^2. This was well corroborated by the antisite formation energies calculated by Sickafus et al. [28], which showed highest value for $Gd_2Ti_2O_7$ and lowest for $Gd_2Zr_2O_7$ among these three. They have, in fact, concluded that in $A_2B_2O_7$ compounds, isolated cation antisite defect energy is the function of A and B cation radii. Typically, a low energy for fluorite disorder defect pair implies that material can accommodate disorder more easily. Patel et al. [87] have also attributed this O-D transformation in $Gd_2Zr_2O_7$ to thermal spike phenomenon wherein the SHI passing through the material cause a large temperature rise in localized regions due to electron-phonon coupling followed by fast quenching. The comparison of Raman spectra of these disordered fluorites match well with those obtained by melt quenching of $Gd_2Zr_2O_7$, thus supporting this observation. Sattonay et al. [88] have studied the irradiation behavior of $Gd_2(Ti_{2-x}Zr_x)O_7$ system under both low and high energy ranging from few MeV to few GeV to compare the effects of nuclear collisions and electronic excitations. They have also observed decrease in radiation stability with increase in Ti content (or decrease in r_A/r_B), which is in agreement with several other researchers [89—91]. In addition, they observed that at higher energies (1.5 GeV Xe or 2.6 GeV U), similar structural changes (as under low-energy irradiation) were observed at much lower fluences. An interesting observation put forth by Devanathan et al. on the basis of molecular dynamic simulations of radiation damage events in these

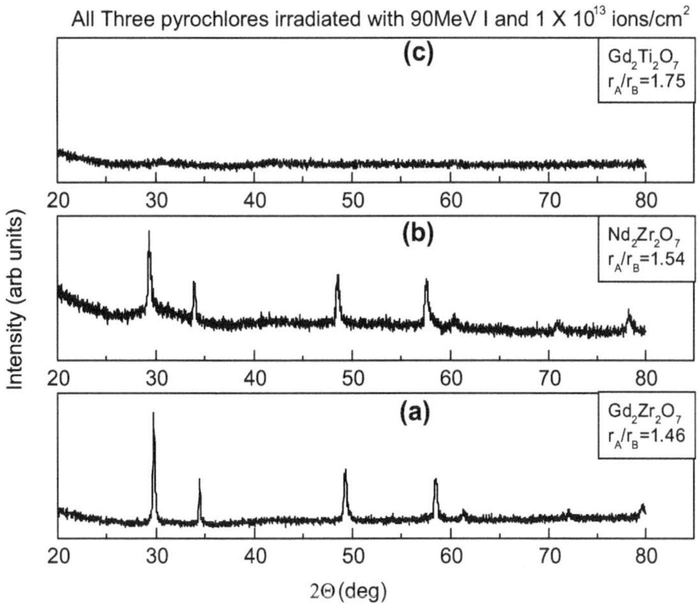

FIGURE 18.12

X-Ray Diffraction (XRD) patterns of (a) $Gd_2Zr_2O_7$, (b) $Nd_2Zr_2O_7$, and (c) $Gd_2Ti_2O_7$ under 90 MeV I irradiation and at the fluence of 1×10^{13} ions/cm^2.

pyrochlores suggested that the enhanced point defect clustering within the cascades in $Gd_2Ti_2O_7$ as compared to $Gd_2Zr_2O_7$ may facilitate amorphization in the former under irradiation. Lang et al. [92] observed that structure of the ion tracks created by irradiation (with 1.43 GeV Xe ions) depended on the composition of the pyrochlore, and the damage cross-section increased with increase in Ti content in the $Gd_2(Ti_{2-x}Zr_x)O_7$ $(0 \leq x \leq 1)$.

The radiation stability of these materials as a function of temperature has been studied by Kulriya et al. by comparing the effect of SHI (120 MeV Au ions) on $Gd_2Ti_2O_7$ at both room temperature and 1000K [93]. An in situ high-temperature (1000K) setup has been designed and installed in the materials science beamline of superconducting linear accelerator at the Inter-University Accelerator Centre (IUAC) for temperature dependent ion irradiation studies on the materials exposed with SHI irradiation, details of which can be found elsewhere [93]. The schematic diagram of the high-temperature target ladder developed at IUAC, Delhi, India, is shown in Fig. 18.13. The temperature-dependent ion irradiation studies showed that the ion fluence required to cause amorphization at 1000K irradiation is significantly higher than that required at room temperature, which can be attributed to a higher defect annihilation probability with increase in temperature.

Thus, as explained by Sickafus et al., the formation of O-D defect reaction pair (antisite defect in cation position and vacancy at oxygen position, with oxide interstitial) more completely defines the disordering process. The compounds possessing lower O-D defect pair energies will also exhibit lower propensity for cluster or cavity formation, in addition to showing tolerance to radiation-induced amorphization. If the energy requirement is not much for the formation of oxygen Frenkel pair, the driving force for the oxygen interstitials to join a cluster or the vacancy to adsorb on the growing voids will be less. Similarly, a lower energy for the formation of cation antisite would be efficient in

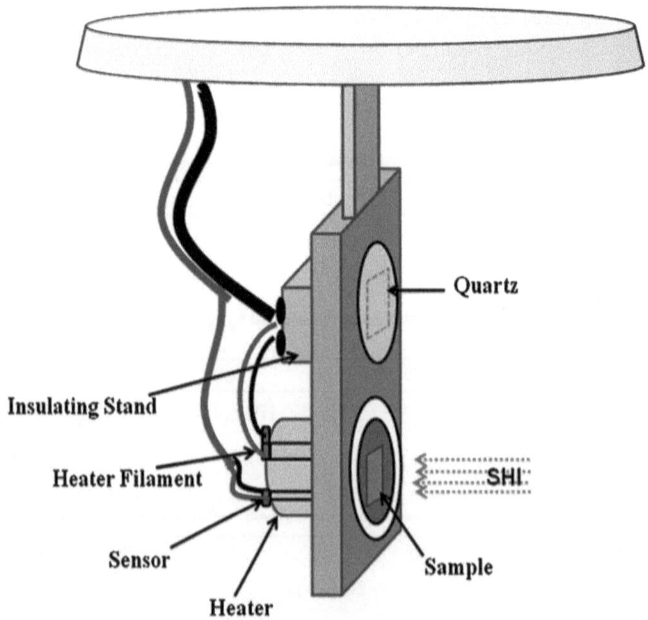

FIGURE 18.13

Schematic diagram of high-temperature target ladder designed at IUAC, New Delhi, India.

Reproduced with permission from Elsevier.

annihilating freely migrating cations and prevent their contribution to the growth of undesirable defect structures such as clusters and voids. It must, however, be noted that there are certain other factors as well that contribute to the overall radiation tolerance of complex oxides. Simple parameters like bond ionicities, cationic size, electronic configuration, and oxygen positional parameters have been explained as a few such factors that can be used to attribute radiation response of a wide variety of materials [89].

15.2 EFFECT OF POWDER PROPERTIES

It has been mentioned in the literature that the ability of the material to resist radiation damage also depends largely on the microstructure, i.e., how well it can accommodate and dissipate the vacancies and interstitials in equal numbers [94]. In this regard, nanomaterials have been shown to possess excellent radiation resistance in certain cases. They exhibit better retention of properties such as ductility and exhibit lower swelling. Bai et al. [95] have done extensive simulation studies on Cu that suggest that nanomaterials in general might possess exceptional radiation resistance due to annihilation of defects at the grain boundaries (GBs). According to a mechanism proposed by Bai et al., the interstitial defects created during damage events are absorbed or loaded into the grain boundaries and are reemitted into the bulk material to annihilate vacancies leading to the healing effect, and this promotes radiation tolerance. Substantially high interfacial areas occurring in nanocrystalline samples results in shorter diffusion distances for the defect migration. This promotes the healing process and consequently the enhanced radiation tolerance is enhanced. The mechanism is pictorially depicted in Fig. 18.14. Nanocrystalline partially inverse spinel $MgGa_2O_4$ was observed to tolerate doses up to 96

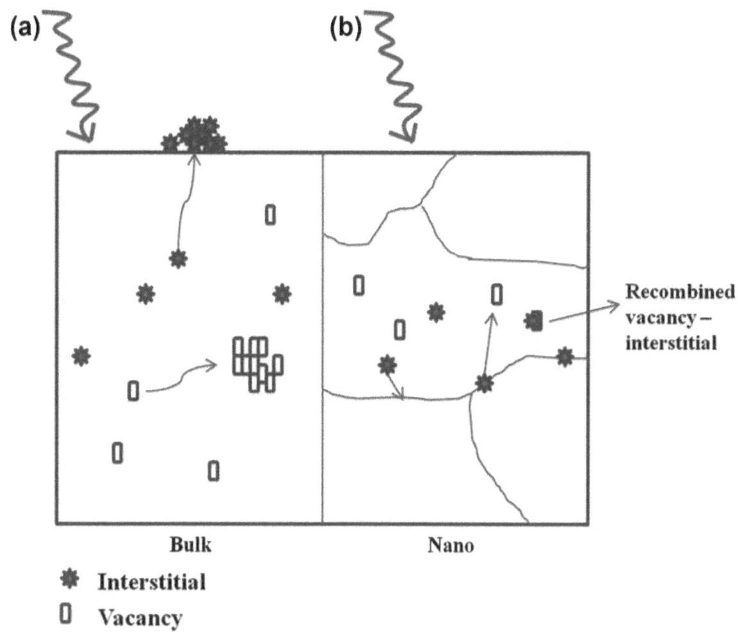

FIGURE 18.14

Schematics of annihilation in bulk and nanomaterials. (a) In bulk materials interstitials migrate to surface causing swelling while vacancies aggregate to form voids. (b) In nanostructured materials, interstitials are absorbed by grain boundaries and then reemitted back, and they recombine with vacancies leaving the healed region.

displacements per atom (dpa) as compared to its bulk counterpart, which was amorphized by a dose of just 12 dpa [96].

Swift heavy ions have been shown to cause annealing of defects in the carbon nanostructure like carbon nanotubes, fullerenes, and graphene [97–100]. Improved radiation resistance has been reported recently in several nanocrystalline materials containing a large fraction of GBs like nickel [101], copper [101], gold [102], palladium [103], ZrO_2 [103], and $MgGa_2O_4$ [96] as compared to their polycrystalline counterparts. There are contrary views as well on the same, e.g., it can be argued that the scattering of the electron produced by SHIs from the grain boundaries result in confinement of the energy within the grain boundaries, thereby increasing the transient lattice temperature, and this transient temperature rise will be reasonably higher in samples with smaller grain size [104–106]. This can cause higher radiation damage. Further, since grain boundaries and interfaces have been known to act as sinks for defects, this has been considered detrimental to materials performance, due to grain boundary embrittlement by elements such as helium, sulfur, etc. [94]. An additional difficulty that is encountered with nanomaterials is their tendency to recrystallize (grain growth) under high localized temperatures generated under irradiation, and during this phenomenon there would obviously be the loss of properties associated with nanoscale materials. However, the proposed mechanism is general and also applies to interface between two phases in the composite. A study carried out by Misra et al. [107] on Nb-Cu composites showed enhanced tolerance to radiation damage, and, further because these elements are immiscible, the nanostructure is largely retained. Similarly, ferritic steel dispersed with nanoparticles of Y_2O_3 was shown to display enhanced radiation tolerance [108].

The formation of continuous ion tracks along ion paths is a major characteristic of the electronic excitation effects under SHI irradiation [62]. It is reasonable to assume that the formation of these ion tracks would be critically dependent on the density and packing of the lattice or, in other words, the microstructure and powder properties of the material should have tremendous bearing on the radiation damage brought about by high-energy irradiation. Considering the nuclear importance of ceria-based material in conjunction with curiosity of exploring the dependence of radiation stability on microstructure, in particular with the grain size, detailed investigations were done by Grover et al. [109]. They have attempted to relate the irradiation behavior of various sintered ceria samples to their microstructure. The ceria nanopowders were synthesized and annealed in pellet form under various conditions. Three different particle sizes 25 nm, 50 nm, and 4–5 μm with three different densities and morphologies were obtained. These samples were then irradiated with 100-MeV silver ions (SHI) at different fluences and were studied by XRD, Raman spectroscopy, scanning electron microscopy and X-ray photoelectron spectroscopy. The XRD patterns of samples with smallest and largest particle sizes (in this study) as a function of fluence are depicted in Fig. 18.15 (a and b). Inset shows their corresponding scanning electron microscope (SEM) of the unirradiated samples. None of the samples were found to amorphize up to the highest fluence of 1×10^{14} ions/cm^2 employed. However, the sample with lowest grain size suffered maximum damage. The sample with largest grain size was most stable and showed little change in crystallinity. Defect characterization by Raman spectroscopy indicated the enhanced formation of Ce^{3+} and related defects upon irradiation in sample with larger grain size. The most intriguing result was the absence of Ce^{3+}-related defects in sample with lowest

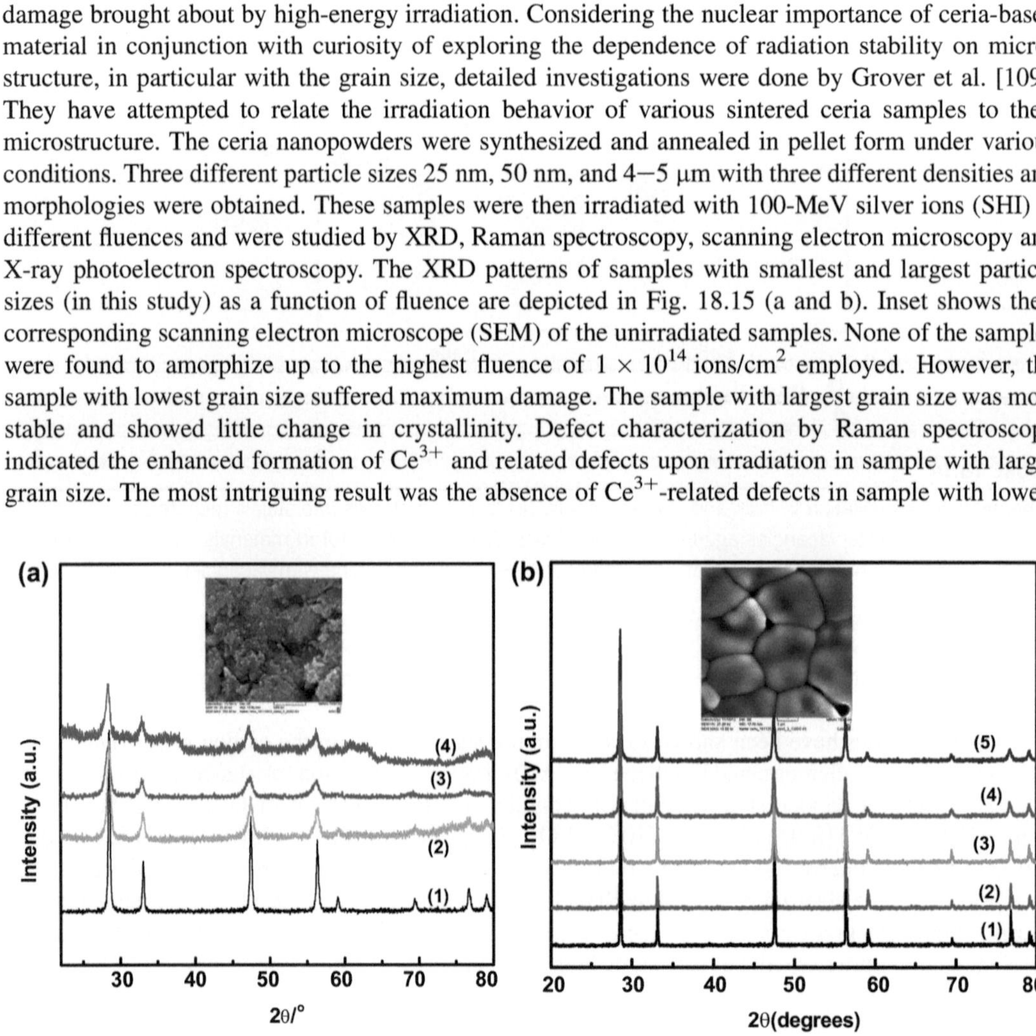

FIGURE 18.15

(a) The in situ XRD patterns of sample (particle size ~25 nm) at various fluences: (1) pristine, (2) 3×10^{12} ions/cm^2, (3) 3×10^{13} ions/cm^2, (4) 1×10^{14} ions/cm^2; (b) The in situ XRD patterns of sample (particle size ~ 4–5 μm) at various fluences: (1) pristine, (2) 1×10^{12} ions/cm^2, (3) 1×10^{13} ions/cm^2, (4) 5×10^{13} ions/cm^2, (5) 1×10^{14} ions/cm^2; Insets show the SEM of the corresponding unirradiated samples.

grain size, which actually showed maximum damage upon irradiation. The previous two observations have been explained as follows. As mentioned before, the grain boundaries act as a sink for defects and interstitial being more mobile move to grain boundaries faster. In the smallest grain-sized samples, faster removal of interstitials leaves the interior of the grains highly rich in vacancies. In case of larger-sized samples, the diffusion of defects is much slower and hence the vacancies and interstitials have a higher probability of recombination and hence less damage. The higher propensity of Ce^{3+}-related defects in larger grain-sized samples was proved by XPS studies. Similarly, a recent study carried out by Dey et al. [110] on irradiation of nanocrystalline YSZ by 400 keV Kr ions revealed the effect of grain boundaries on damage evolution. None of the samples under study showed amorphization. However, the highest concentration of defects was observed in large-grained sample (220 nm) relative to nanograined ones. Extensive scanning transmission electron microscope images showed that in large-grained samples, both interstitial clusters and vacancies were observed whereas nanograined samples depicted the propensity of only vacancy clusters, thus indicating that interstitial can migrate to grain boundaries in the nanosized samples. Grain growth was observed in nanosized samples, which could be attributed to the coherent migration of radiation-induced defects to the boundaries causing an increase in grain boundary mobility. Absence of grain growth in larger-grained samples obviously indicated that the diffusion length of defects is much smaller compared to grain size. In another work by Edmondson et al. [111] thin films of nanocrystalline ceria grown on silicon substrates were irradiated with 3 MeV Au^{+} ions at two different temperatures of 300 and 400K. The materials were found to be phase stable with lattice contraction at low doses of irradiation, which was attributed to stress relief. The grains are observed to undergo a temperature-dependent, defect-stimulated growth.

In the field of radiation-stable nanomaterials, a lot is desired to be done but the mechanism proposed behind the additional radiation tolerance seems to be highly promising. However, much still needs to be done in order to identify the best structure with superior morphology for applications in nuclear reactors.

16. CONCLUSIONS

Understanding of radiation-induced modifications is crucial for the efficient use of nuclear reactors and management of nuclear waste. Thus, in order to give an introduction to a beginner in the field, we have summarized the various types of radiation existing in a nuclear reactor and waste environment and how these radiations interact with the materials to modify their physical and chemical properties of materials. Insight has been provided into the effect of radiation on materials on nuclear importance. Specific examples have been cited to show how radiation tolerance depends on basic properties like structure, microstructure, and powder properties. The radiation response of a material depends on a wide variety of factors like bond ionicities, cationic size, electronic configuration, and oxygen positional parameters (in case of oxides), etc. It has also been discussed in detail that the compounds that possess borderline structures and/or wide-ranging nonstoichiometry are likely to exhibit enhanced radiation tolerance. Despite a lot of work directed toward the search for the best candidate materials with optimized structure and superior microstructure from the point of view of radiation response, the quest is far from over. These insights may serve as vital guidelines to design and envisage novel radiation-tolerant materials.

REFERENCES

[1] A. Misra, L. Thilly, MRS Bull. 35 (December 12, 2010) 965.
[2] B.R. Martin, Nuclear and Particle Physics, John Wiley & Sons, Ltd., 2006.
[3] M.I. Ojovan, W.E. Lee, An Introduction to Nuclear Waste Immobilization, Elsevier, 2005.
[4] S.C. Chetal, V. Balasubramaniyan, P. Chellapandi, P. Mohanakrishnan, P. Puthiyavinayagam, C.P. Pillai, S. Raghupathy, T.K. Shanmugham, C. Sivathanu Pillai, Nucl. Eng. Des. 236 (2006) 852.
[5] R.K. Sinha, A. Kakodkar, Nucl. Eng. Des. 236 (2006) 683.
[6] R.C. Ewing, W.J. Weber, F.W. Clinard Jr., Prog. Nucl. Energy 29 (1995) 63.
[7] IAEA-TECDOC-1516, Viability of Inert Matrix Fuels in Reducing Plutonium Amounts in Reactors, IAEA, Vienna, 2006.
[8] K.E. Sickafus, E.A. Kotomin, B.P. Uberuaga, Radiation effects in solids, in: Proceedings of the NATO Advanced Study Institute, 2004.
[9] G.S. Was, Fundamentals of Radiation Materials Science – Metals and Alloys, Springer, 2007.
[10] J.-F. Babelot, R. Conrad, R.J.M. Konings, G. Mühling, M. Salvatores, G. Vambenepe, J. Alloys Compd. 271–273 (1998) 606.
[11] T. Yamashita, K. Kuramoto, N. Nitani, Y. Nakano, H. Akie, H. Nagashima, Y. Kimura, T. Ohmichi, J. Nucl. Mater. 320 (2003) 126.
[12] E.A.C. Neeft, K. Bakker, R.P.C. Schram, R. Conard, R.J.M. Konings, J. Nucl. Mater. 320 (2003) 106.
[13] T. Wiss, R.J.M. Konings, C.T. Walker, H. Thiele, J. Nucl. Mater. 320 (2003) 85.
[14] T. Yamashita, N. Nitani, H. Kanazawa, M. Magara, T. Ohmichi, H. Takano, T. Muromura, J. Nucl. Mater. 274 (1999) 98.
[15] R.J.M. Konings, R. Conard, J. Nucl. Mater. 274 (1999) 336.
[16] E.A.C. Neeft, R.J.M. Konings, K. Bakker, J.G. Boshoven, H. Hein, R.P.C. Schram, A. van Veen, R. Conard, J. Nucl. Mater. 274 (1999) 78.
[17] G.R. Lumpkin, J. Nucl. Mater. 289 (2001) 136.
[18] G.R. Lumpkin, R.C. Ewing, Phys. Chem. Miner. 16 (1988) 2.
[19] R.C. Ewing, B.C. Chakoumakos, G.R. Lumpkin, T. Murakami, R.B. Greegor, F.W. Lytle, Nucl. Instrum. Methods Phys. Res. B 32 (1988) 487.
[20] G.R. Lumpkin, R.C. Ewing, B.C. Chakoumakos, R.B. Greegor, F.W. Lytle, E.M. Foltyn, F.W. Clinard Jr., L.A. Boatner, M.M. Abraham, J. Mater. Res. 1 (1986) 564.
[21] W.J. Weber, J.W. Wald, H. Matzke, Mater. Lett. 3 (1985) 173.
[22] W.J. Weber, J.W. Wald, H. Matzke, J. Nucl. Mater. 138 (1986) 196.
[23] A.R. Cleave, Atomistic Scale Simulations for Waste Form Applications, Imperial College of Science, 2006 (thesis).
[24] K.J.W. Atkinson, Atomic Scale Simulations of Defects in Bulk Materials and Monolayer Surfaces, Imperial College of Science, 2002 (thesis).
[25] J.A. Ball, Computer Simulations of Disorder in Ceramic Materials, Imperial College of Science, 2006 (thesis).
[26] J.A. Purton, N.L. Allan, J. Mater. Sci. 12 (2002) 2923.
[27] A. Chartier, C. Meis, J.P. Crocombette, W.J. Weber, L.R. Corrales, Phys. Rev. Lett. 94 (2005) 0225505.
[28] K.E. Sickafus, L. Minervini, R.W. Grimes, J.A. Valdez, M. Ishimaru, F. Li, K.J. McClellan, T. Hartman, Science 289 (2000) 748.
[29] I.T. Todorov, N.L. Allan, J.A. Purton, M.T. Dove, W. Smith, J. Mater. Sci. 42 (2007) 1920.
[30] P. Nerikar, T. Watanabe, J.S. Tulenko, S.R. Phillpot, S.B. Sinnott, J. Nucl. Mater. 384 (2009) 61.
[31] M.J. Caturla, N. Soneda, T. Diaz de la Rubia, M. Fluss, J. Nucl. Mater. 351 (2009) 78.

[32] W.J. Weber, R.C. Ewing, L.M. Wang, J. Mater. Res. 9 (1994) 688.

[33] P. Sigmund, Particle Penetration and Radiation Effects General Aspects and Stopping of Swift Point Charges, Springer-Verlag, Berlin, Heidelberg, 2006.

[34] M. Nastasi, J.W. Mayer, J.K. Hirvonen, Ion-Solid Interactions: Fundamentals and Applications, Cambridge University Press, 1996.

[35] J.F. Ziegler, J.P. Biersack, U. Littmark, The Stopping and Range of Ions in Solids, vol. 1, Pergamon Press, 1985. The program is downloadable from: www.srim.org.

[36] J.W. Mayer, L. Eriksson, J.A. Davies, Ion Implantation Is Semiconductors, Academic Press, New York, 1970.

[37] A. Chatterjee, H.D. Maccabee, C.A. Tobias, Radiat. Res. 54 (1973) 479.

[38] L.C. Feldman, J.W. Mayer, S.T. Picraux, Materials Analysis by Ion Channeling (Chapter 2), Academic Press, New York, 1982.

[39] L.C. Northcliffe, Phys. Rev. 120 (1960) 1744.

[40] J. Lindhard, M. Scharff, Mat. Fys. Medd. Dan. Vid. Selsk. 27 (1953) 1–31.

[41] R.M. Barnett, et al., Phys. Rev. D 54 (1996) 1.

[42] H. Bethe, Zur Theorie des Durchgangs schneller Korpuskularstrahlendurch Materie, Ann. Phys. 5 (1930) 324.

[43] F. Bloch, Bremsvermögen von Atomen mit mehreren Elektronen, Z. Phys. 81 (1933) 363.

[44] N. Bohr, Philos. Mag. 30 (1915) 581.

[45] [a] H. Bethe, Bremsformel für Elektronen relativistischer Geschwindigkeit, Z. Phys. 76 (1932) 293; [b] U. Fano, Ann. Rev. Nucl. Sci. 13 (1963) 1–66.

[46] M.J. Norgett, M.T. Robinson, I.M. Torrens, Nucl. Eng. Des. 33 (1975) 50.

[47] E.H. Lee, Nucl. Instrum. Methods Phys. Res. B 151 (1999) 29.

[48] G. Schiwietz, K. Czerski, M. Roth, F. Staufenbiel, P.L. Grande, Nucl. Instrum. Methods Phys. Res. B 255 (2004) 4.

[49] E.C.M. Silk, R.S. Barnes, Philos. Mag. 7 (1959) 970.

[50] D.A. Young, Nature 182 (1958) 375.

[51] P.B. Price, R.M. Walker, Phys. Rev. Lett. 8 (1962) 271.

[52] R.L. Fleischer, Science 207 (1980) 979.

[53] W.H. Hung, R.M. Walker, Science 155 (1967) 1103.

[54] B.E. Fischer, R. Spohr, Rev. Mod. Phys. 55 (1983) 907.

[55] V.E. Fortov, D.H. Hoffmann, B.Y. Sharkov, Phys. Usp. 51 (2008) 109.

[56] R.L. Fleischer, P.B. Price, R.M. Walker, J. Appl. Phys. 36 (1965) 645.

[57] R.L. Fleischer, P.B. Price, R.M. Walker, E.L. Hubbard, Phys. Rev. 156 (1967) 353.

[58] H. Dhammak, A. Barbu, A. Dunlop, D. Lesueur, N. Lorenzelli, Philos. Mag. Lett. 67 (1993) 253.

[59] F. Desaures, Z. Phys. 38 (1923) 12.

[60] F. Seitz, J.S. Koehler, Solid State Phys. 2 (1956) 305.

[61] A. Meftah, F. Brisard, J.M. Constantini, E. Dooryhee, M. Hage-Ali, M. Hervieu, J.P. Stoquert, F. Studer, M. Toulemonde, Phys. Rev. B 49 (1994) 12457.

[62] M. Toulemonde, W. Assmann, C. Dufour, A. Meftah, F. Studer, C. Trautmann, Mat. fys. Medd. 52 (2006) 263.

[63] Z.G. Wang, C. Dufour, E. Paumier, M. Toulemonde, J. Phys. Condens. Matter 6 (1994) 6733.

[64] G. Szenes, Phys. Rev. B 51 (1995) 8026.

[65] M.C. Wittels, F.A. Sherrill, Phys. Rev. Lett. 3 (1959) 176.

[66] A. Benyagoub, Phys. Rev. B 72 (2005) 094114.

[67] S. Hémon, V. Chailley, E. Dooryhée, C. Dufour, F. Gourbilleau, F. Levesque, E. Paumier, Nucl. Instrum. Methods Phys. Res. B 122 (1997) 563–565.

[68] S. Hémon, C. Dufour, F. Gourbilleau, E. Paumier, E. Dooryhée, S. Bégin-Colin, Nucl. Instrum. Methods Phys. Res. B 146 (1998) 443.
[69] S. Klaumünzer, C.L. Li, S. Löffler, M. Rammensee, G. Schumacher, H.C. Neitzert, Radiat. Eff. Defects Solids 108 (1989) 131.
[70] M.D. Hou, S. Klaumüunzer, G. Schumacher, Phys. Rev. B 41 (1990) 1144.
[71] A.I. Ryazanov, A.E. Volkov, S. Klaumünzer, Phys. Rev. B 51 (1995) 12107.
[72] S. Klaumünzer, M.-D. Hou, G. Schumacher, Phys. Rev. Lett. 57 (1990) 850.
[73] C. Trautmann, S. Klaumünzer, H. Trinkaus, Phys. Rev. Lett. 85 (2000) 3648.
[74] I. Zizak, N. Darowski, S. Klaumünzer, G. Schumacher, J.W. Gerlach, W. Assmann, Phys. Rev. Lett. 101 (2008) 65503.
[75] K.E. Sickafus, R.W. Grimes, J.A. Valdez, A. Cleave, M. Tang, M. Ishimaru, S.M. Corish, C.R. Stanek, B.P. Uberuaga, Nat. Mater. 6 (2007) 217.
[76] L.R. Morss, N.M. Edelstein, J. Fuger, J.J. Katz (Eds.), The Chemistry of the Actinide and Transactinide Elements, Springer, Dordrecht, 2006.
[77] H. Matzke, P.G. Lucuta, T. Wiss, Nucl. Instrum. Methods Phys. Res. 166–167 (2000) 920.
[78] T. Wiss, H. Matzke, C. Trautmann, M. Toulemonde, S. Klaumünzer, Nucl. Instrum. Methods Phys. Res. B 122 (1997) 583.
[79] H. Matzke, Nucl. Instrum. Methods Phys. Res. B 116 (1996) 121.
[80] G. Sattonnay, M. Lahrichi, M. Herbst-Ghysel, F. Garrido, L. Thomé, J. Appl. Phys. 101 (2007) 103516.
[81] V. Menvie Bekale, G. Sattonnay, C. Legros, A.M. Huntz, S. Poissonnet, L. Thomé, J. Nucl. Mater. 384 (2009) 70.
[82] A. Gentils, L. Thomé, J. Jagielski, L. Nowicki, S. Klaumünzer, F. Garrido, M. Beauvy, Nucl. Instrum. Methods Phys. Res. B 218 (2004) 457.
[83] J. Jagielski, A. Gentils, L. Thomé, L. Nowicki, F. Garrido, S. Klaumünzer, Nucl. Instrum. Methods Phys. Res. B 219–220 (2004) 626.
[84] N. Nitani, K. Kuramoto, T. Yamashita, Y. Nihei, Y. Kimura, J. Nucl. Mater. 319 (2003) 102.
[85] T. Sonoda, M. Kinoshita, Y. Chimi, N. Ishikawa, M. Sataka, A. Iwase, Nucl. Instrum. Methods Phys. Res. B 250 (2006) 254.
[86] N. Ishikawa, Y. Chimi, O. Michikami, Y. Ohta, K. Ohhara, M. Lang, R. Neumann, Nucl. Instrum. Methods Phys. Res. B 266 (2008) 3033.
[87] M.K. Patel, V. Vijaykumar, D.K. Avasthi, S. Kailas, J.C. Pivin, V. Grover, B.P. Mandal, A.K. Tyagi, Nucl. Instrum. Methods Phys. Res. B 266 (2008) 2898.
[88] G. Sattonnay, S. Moll, L. Thome, C. Legros, M. Herbst-Ghysel, F. Garrido, J.M. Costantini, C. Trautmann, Nucl. Instrum. Methods Phys. Res. B 266 (2008) 3043.
[89] R.C. Ewing, W.J. Weber, J. Lian, J. Appl. Phys. 95 (2004) 5949.
[90] S.X. Wang, B.D. Begg, L.M. Wang, R.C. Ewing, W.J. Weber, K.V.G. Kutty, J. Mater. Res. 14 (1999) 4470.
[91] B.D. Begg, N.J. Hess, D.E. McCready, S. Thevuthasam, W.J. Weber, J. Nucl. Mater. 289 (2001) 188.
[92] M. Lang, J. Lian, J. Zhang, F. Zhang, W.J. Weber, C. Trautmann, R.C. Ewing, Phys. Rev. B 79 (2009) 224105.
[93] P.K. Kulriya, R. Kumari, R. Kumar, V. Grover, R. Shukla, A.K. Tyagi, D.K. Avasthi, Nucl. Instrum. Methods Phys. Res. B 342 (2015) 98.
[94] G. Ackland, Science 327 (2010) 1587.
[95] X.M. Bai, A.F. Voter, R.G. Hoagland, M. Nastasi, B.P. Uberuaga, Science 327 (2010) 1631.
[96] T.D. Shen, S. Feng, M. Tang, J.A. Valdez, Y. Wang, K.E. Sickafus, Appl. Phys. Lett. 90 (2007) 263115.
[97] N. Bajwa, A. Ingale, D.K. Avasthi, R. Kumar, A. Tripathi, J. Appl. Phys. 104 (2008) 054306.
[98] K. Jeet, V.K. Jindal, L.M. Bharadwaj, D.K. Avasthi, K. Dharamvir, J. Appl. Phys. 108 (2010) 034302.
[99] A. Kumar, D.K. Avasthi, J.C. Pivin, P.M. Koinkar, Appl. Phys. Lett. 92 (2008) 221904.

[100] S. Kumar, A. Tripathi, F. Singh, S.A. Khan, V. Baranwal, D.K. Avasthi, Nanoscale Res. Lett. 9 (2014) 126.
[101] N. Nita, R. Schaeublin, M. Victoria, J. Nucl. Mater. 329–333 (2004) 953.
[102] Y. Chimi, A. Iwase, N. Ishikawa, M. Kobiyama, T. Inami, S. Okuda, J. Nucl. Mater. 297 (2001) 355–357.
[103] M. Rose, A.G. Balogh, H. Hahn, Nucl. Instrum. Methods Phys. Res. B 127–128 (1997) 119–122.
[104] A. Gupta, D.K. Avasthi, Phys. Rev. B 64 (2001) 155407.
[105] D.K. Avasthi, S. Ghosh, S.K. Srivastava, W. Assmann, Nucl. Instrum. Methods Phys. Res. B 219 (2004) 206.
[106] M. Kumar, S.A. Khan, P. Rajput, F. Singh, A. Tripathi, D.K. Avasthi, A.C. Pandey, J. Appl. Phys. 102 (2007) 083510.
[107] A. Misra, M.J. Demkowicz, X. Zhang, R.G. Hoagland, J. Mater. 59 (2007) 62.
[108] S. Ukai, M. Fujiwara, J. Nucl. Mater. 307–311 (2002) 749.
[109] V. Grover, R. Shukla, R. Kumari, B.P. Mandal, P.K. Kulriya, S.K. Srivastava, S. Ghosh, A.K. Tyagi, D.K. Avasthi, Phys. Chem. Chem. Phys. 16 (2014) 27065.
[110] S. Dey, J.W. Drazin, Y. Wang, J.A. Valdez, T.G. Holesinger, B.P. Uberuaga, R.H.R. Castro, Sci. Rep. 5 (2015) 7746.
[111] P.D. Edmondson, Y. Zhang, S. Moll, F. Namavar, W.J. Weber, Acta Mater. 60 (2012) 5408.

CHAPTER

DIAMOND-BASED RADIATION DETECTORS FOR APPLICATIONS IN HIGHLY CORROSIVE SOLUTIONS AND HIGH-RADIATION FIELDS

19

D. Jain, J. Nuwad, N. Manoj, V. Sudarsan
Bhabha Atomic Research Centre, Mumbai, India

CHAPTER OUTLINE

Materials Under Extreme Conditions. http://dx.doi.org/10.1016/B978-0-12-801300-7.00019-X

1. INTRODUCTION

The ever-increasing curiosity of understanding Mother Nature has greeted humankind with continuing discoveries. However, more important has been the innovative approaches added to these discoveries that have led to new inventions and versatile applications. Such rapid evolution has only been possible by understanding the fundamental behavior of materials existing around us. It is not an exaggeration to say that development of materials and their unique applications is happening at an unprecedented pace at present. In particular, the present era of technologically important materials includes an important share coming from carbon-based materials. Here we restrict ourselves to materials almost exclusively composed of carbon, otherwise we enter ourselves into a gigantic world of carbon with its presence in almost every nook and corner of the existence of life. Research in the past few decades has discovered many novel forms of carbon other than conventionally known diamond and graphite. These include fullerenes (C_{60}, C_{70}, C_{80}, etc.), carbon nanotubes (CNTs), carbon fibers, amorphous carbon, diamond-like carbon (DLC), and most recently graphene. These materials, with their unique structure–property correlation, are highly promising for several applications and in fact have already found their place in various products/processes. Among all these materials, diamond is one form of carbon, perhaps the oldest known to mankind and still continuing to strengthen its unique place. Discovered first in nature and generally found deep beneath the Earth's surface or in seabeds (to a lesser extent), diamond has been long cherished by the rich and wealthier classes as a precious gemstone or the most confined way to store wealth. Earliest synthetic efforts to make it in the lab were also guided by the objective to conquer its rapidly increasing prices. However, with time, as it was examined more from the point of view of a fundamental scientific understanding, its unique properties were unraveled and interest in diamond kept growing for several technological/industrial applications other than as a gemstone. This was followed by efforts to synthesize diamond in lab, and few methods were successfully invented, the most exciting being the chemical vapor deposition (CVD) route. Like any material, applications of diamond also originate from its physicochemical properties. One of the important applications is radiation detection by diamond under off-normal or extreme conditions, as is the theme of this book. This chapter is aimed at reviewing the subject of diamond-based particle radiation detectors for use under extreme conditions such as corrosive chemical environments or intense radiation fields. Some of our results on development of CVD-grown diamond film-based α-particle detectors for estimation of actinides in acidic solutions will also be presented.

2. STRUCTURE AND PROPERTIES OF DIAMOND

Compositionally, diamond is the pure elemental form of carbon, which is thermodynamically stable in high-temperature, high-pressure (HTHP) region of carbon's equilibrium phase diagram [1]. Local structure of ideal diamond crystal can be viewed as a gigantic molecule of carbon atoms composed of covalently bonded sp^3-hybridized atoms (C–C distance ~ 1.54 Å) and oriented in a regular tetrahedral geometry. This structure is extended in three dimensions (due to catenation property of carbon) to give a network of strongly bonded carbon atoms. The crystal lattice of diamond is face-centered cubic (FCC, Space group: $Fd3m$ (No.: 227), lattice parameter $a = 3.54$ Å) [2] and structure can be visualized in terms of carbon atoms occupying all lattice sites of FCC lattice along with occupation of half of the tetrahedral void positions by carbon in alternate order such that the closest distance between occupied tetrahedral voids remains half the length of face diagonal <110> in its unit cell. Another polymorph of diamond, which is usually detected as inclusions in meteorite samples recovered from Earth [3] or is sometimes partially formed during laboratory synthesis of diamond [4], is lonsdaleite or hexagonal diamond (Space group: P6₃mc (No.: 186), lattice parameters: $a = 2.49$ Å and $c = 4.14$ Å) [5]. Both cubic and hexagonal polymorphs consist of sp^3-bonded network of carbon atoms but differ in terms of their local microstructure. The cubic form can be visualized as planes of six-member nonplanar (puckered) carbon rings stacked together in ABC ABC ... pattern (with all rings in chair conformation) when viewed along the direction of body diagonal <111>. On the other hand, the lonsdaleite polymorph is formed by stacking of similar six-member carbon rings in AB AB ... pattern (in-plane chair conformation and out-of-plane boat conformation) along the same viewing direction [6]. Both polymorphs have nearly the same theoretical density (~3.5 g/cc at ambient conditions) with very low packing fraction (~34%) indicating that a large space in diamond structure is void. This has important implications toward transport properties (heat transport and charge carrier movement) in diamond crystal.

Properties of bulk diamond (natural/synthetic) are well known and qualitatively mentioned here. All properties stem from its unique electronic structure and nature of bonding. Although a metastable form of carbon under ambient conditions of temperature and pressure, its conversion rate to thermodynamically stable graphite is negligibly small due to high activation barrier required for the transition. It therefore remains stable with all properties intact under ambient conditions for practically unlimited duration. Bulk diamond has the highest hardness among all known materials with very high wear resistance and least friction coefficient, making it useful for tribological applications as well as life-enhancing coating for machining tools. Electrically, pure bulk diamond is insulating in nature with large band gap (~5.5 eV). However, its band structure can be modified to behave as semiconductor (both p- and n-types) and even semimetal by appropriate doping. Low-temperature superconductivity is also known in heavily boron-doped p-type diamond synthesized by HTHP [7] as well as CVD approach [8], and transition temperatures up to 11 K have been reported [9]. Tunability of electrical properties becomes useful while employing diamond-based materials for radiation detection and electrochemical applications. Room temperature thermal conductivity of diamond is highest among all known bulk materials and, in combination with high-temperature chemical as well as thermo-mechanical stability, it provides avenues for its use as a heat sink medium. Diamond is highly transparent to a variety of electromagnetic radiations and therefore is very useful for optical windows, if used in the form of thin films/plates. Diamond has excellent chemical stability in a wide range of chemical environments (solution/vapors) including mineral acids, alkalis, fluorinated acids, organic solvents, ionic liquids, etc. However, chemical stability of diamond in high-temperature molten salts is yet to be fully established. A very important attribute of diamond is its low atomic number ($z = 6$) and chemical nontoxicity. Both make it a fully biocompatible material (effective atomic number of biological tissues: $Z_{\text{effective}} \sim 7.4$), which can be safely placed inside the human body (for medical dosimetry applications). Another important property of diamond is tunability of its surface chemical structure. Surface properties can be modified by terminating it with oxygen, carbon, hydrogen, fluorine, and even functional organic molecules. This opens up an arena of several applications based on surface interactions such as gas sensing, biosensors, etc. The previous discussion on properties of diamond clearly indicates its unique place among all carbon-based materials and its potential for numerous applications. The following sections will now cover issues specific to radiation detection by diamond under extreme operational conditions.

3. DIAMOND AS A HOST FOR RADIATION SENSING

For a material to be used as host for radiation sensing application, one must look for radiation interaction with that material and the resulting response generated, which can be detected by suitable electronics. The simplest and widely occurring effect is creation of free charge carriers in a material as a result of radiation interaction, which are then collected in external circuit by applying a suitable electric field. A material's electronic properties therefore play an important role in assessing its potential as radiation detector. To understand diamond's potential as a host matrix for radiation detection, one must look into its electronic properties. Further, being a solid-state candidate, its properties must be compared with those of the best-known alternate that is in use on an industrially acceptable and economically viable scale. Silicon is the only choice as an alternate material with which we shall compare diamond for the following reasons: (1) silicon is highly abundant and technology to extract silicon from the Earth's crust and convert it to high-purity and large-size single crystals is well established, and (2) silicon-based solid-state radiation detectors are being successfully used for many applications and vast experience with them already exists. With this background, we examine the electronic properties of diamond and silicon as shown in Table 19.1, to assess if diamond holds some superlative attributes.

Table 19.1 Properties of Diamond and Silicon at 300K (in Reference to Radiation Detection Application)

Parameter	Diamond	Silicon
Lattice parameter (Å)	3.54	5.43
Density (g/cc)	3.50	2.33
Band gap at (eV)	5.5	1.1
Coefficient of thermal expansion (10^6 K^{-1})	0.9–1.5	3.6
Maximum operating temperature (K)	1400	600
Melting point (K)	Phase changes to graphite	1690
Physical stability	Very good	Good
Chemical stability	Excellent	Poor in acids/alkalis
Radiation hardness	Excellent	Moderate
Electron mobility (cm^2/Vs)	~2200	1400
Hole mobility (cm^2/Vs)	~1600–1800	500–600
Breakdown voltage (10^6 V/cm)	10	0.3
Electrical resistivity (Ω cm)	$>10^{12}$	$\sim 2 \times 10^5$
Thermal conductivity (W/cm K)	20	1.5
Saturation electron drift velocity (10^7 cm/s)	2–3	1
Static dielectric constant	5.5–5.8	11.8–11.9
Energy to form electron–hole pair (eV)	~13	~3.6

From Table 19.1 it is clear that diamond is superior to silicon in basic properties required for a radiation detector except the limitation of lower sensitivity of diamond (three to four times less sensitive than Si) due to its large band gap. Since intrinsic diamond has a band gap of 5.5 eV, any electromagnetic radiation having energy higher than that [ultraviolet (UV) rays, X-rays, γ-rays] can be detected in principle. In addition, particle radiation (α-particles, β-particles, protons, neutrons, metal ion, fission fragments, pions, etc.) can also be detected. With large band gap and high resistivity ($\sim 10^{13}-10^{14}$ Ω-cm at room temperature), diamond radiation detectors would show very low background noise (leakage currents). As a result, device construction becomes quite simple with diamond separated by two metallic electrodes in ohmic contact configuration. Contrary to that, silicon-based detectors (low band gap and lower resistivity $\sim 10^5$ Ω-cm at room temperature) have appreciable background noise under ambient conditions. This makes it compulsory to configure the electrodes in reversed bias p-n junction or ac-coupling oxide layer (at amplifier input configuration) to obtain noise-free operation. Another advantage of diamond over silicon is the possibility to operate diamond detectors in both positive and negative bias with stable performance. Also, due to high breakdown voltage, diamond detectors can be operated at higher bias voltages (up to 10,000 V cm). Lower dielectric constant of diamond imparts very low capacitance to diamond-based detectors as compared to silicon, and a simple metal-insulator-metal (MIM) structure is sufficient for radiation detection purpose. Diamond has high-saturated carrier velocity (for both electrons and holes), which indicates its potential use for radiation detection in very high event rate conditions as well as fast (low time constant) counting device.

Principle of radiation detection by diamond usually remains the same irrespective of the nature of radiation or manner of detector operation, i.e., generation of free charge carriers and their subsequent collection in external circuit, as mentioned before. Detectors can be operated either in (1) pulse counting mode or (2) current mode. A simplified explanation of basic principle of radiation detection in diamond is given by Kania et al. [10], and a schematic representation is shown in Fig. 19.1.

In spite of inherent advantages of diamond, the dominance of silicon in the area of solid-state radiation detectors is not surprising for the simple reason of limited availability of electronic

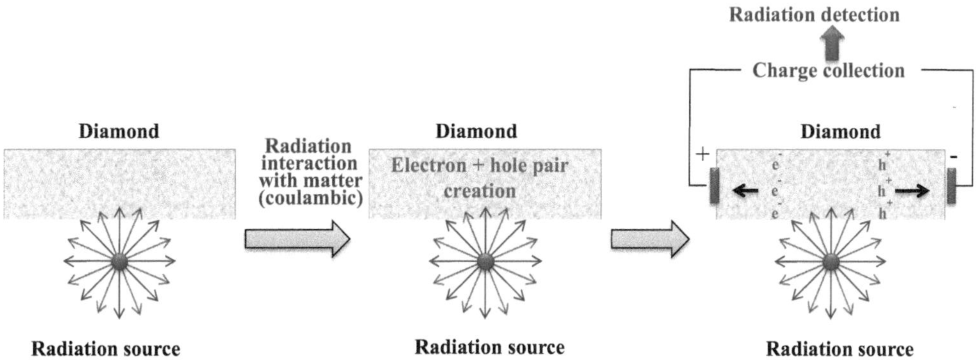

FIGURE 19.1

Schematic representation of radiation detection mechanism in diamond.

quality pure diamond crystals and large-area films/wafers. In fact, availability of diamond films/wafers was out of the question for a long period until the invention of CVD route of diamond synthesis using simple gaseous precursors [6,11,12]. From the previous discussion, it is seen that diamond is an excellent host for radiation detection applications. In the next section, we outline the requirements of radiation detection under extreme operating conditions followed by presenting the case of diamond as a unique choice for such applications.

3.1 DIAMOND FOR RADIATION DETECTION UNDER EXTREME CONDITIONS

In simple definition, the term "extreme condition" can be referred to higher deviations from routine ambient conditions. They are also referred to as conditions far away from thermodynamic equilibrium. The higher the extent of deviation, the more we approach toward extremities. For example, thermal extremities correspond to both very high and very low temperatures. Similarly, pressure extremities can be defined as high/ultrahigh pressures (shock waves, compressed fluids, deep geological conditions, etc.) on one side while low pressures (vacuum/ultrahigh vacuum) on other side. Mechanical extremities can also be referred to as a state of material under intense compressive/tensile load. Concentration extremities can be referred to as saturated/super saturated solutions on one end and infinitely dilute solutions on the other. For radiation fields, extremities can be referred to in terms of energy of radiation with highly ionizing radiations (gamma rays, cosmic rays) on one end and weakly ionizing/nonionizing radiations (radio waves) on the other hand. Radiation extremities can also be defined in terms of net radiation intensities quantifiable by radiation flux/fluence. Chemical extremities can be referred to as highly reactive materials at one end while feebly reactive/inert materials on other end for a given chemical reaction (such as redox reaction, combustion, dissolution, etc.). Under such extremities, materials behave differently or deviate from what is observed under ambient conditions. Therefore, it can be said that understanding a material's behavior or a given phenomenon under extreme conditions would require knowledge of a system's physicochemical behavior under that extremity. Here 'system' is referred to as a combination of material and the given phenomenon.

With this background of material behavior under extreme conditions, we now take up the case of diamond as radiation detector under extreme conditions. It is clear that diamond has all the essential attributes required for detection of a wide variety of radiations except a lower sensitivity than silicon. A further look at its properties (Table 19.1) explains why it is unique and an excellent host under extreme conditions. With a large band gap and near complete transmission of visible light, diamond-based devices can be used as solar-blind detectors. This is essential for on-field, windowless radiation detection applications (radiation surveillance, material movement tracking, etc.) where a detector needs to be operated in open condition under visible light and the active surface also needs to be exposed to a radiation source. It is known that exposure of diamond detectors to strong visible light is even beneficial as it minimizes the polarization-induced deterioration in detector performance. For such applications, Si-based detectors cannot be used for their small band gap causes higher background noise; making it necessary to use these detectors in dark or cooled conditions with suitable windows. Another attribute of large band gap and low-leakage currents is that detectors can be operated at higher temperatures without any appreciable deterioration in signal-to-noise ratio (S/N). Such applications are required in space technology where radiation exposure to the outer surface of spacecrafts/satellites, which suffer steep temperature gradients, is to be measured. Research carried out in this area has revealed that natural diamonds neither show any appreciable increase in leakage background current nor cause excess polarization up to 500°C operational temperature [13].

Diamond is highly stable under hostile chemical environments (concentrated acids/alkalies/fumes/organic reagents/fluoroacids) whereas silicon is moderately/severely damaged depending on chemical aggressiveness of the medium. Therefore diamond is a safe and reliable option for radiation detection under hostile chemical environments. Such applications arise in nuclear industry, particularly in spent fuel reprocessing and waste management cycles. During these stages, large volumes of acidic/aqueous/organic liquids containing alpha-emitting actinides (Th, U, Pu, Am, etc.) and other radionuclides are processed at industrial scale. Efficient operation and control of these process plants depends upon fast, reliable, and real-time assay of actinides in various process streams. Presently, it is done by conventional alpha spectrometry using solid-state detectors such as silicon detectors, ZnS(Ag) detectors, etc. The first step involves sampling of solutions from different process streams followed by appropriate separation/partitioning of actinides in desired form/solvent. Solutions are diluted as required to minimize radiation exposure to analysts. A known small volume of solution is thereafter placed on a heated metallic disk for solvent evaporation followed by high-temperature firing inside Bunsen flame for fixing the activity. The entire operation is carried out in fume hoods or glove boxes. The solidified activity is thereafter transferred to a radiometric counting lab and measured by conventional alpha spectrometry in air/vacuum to estimate the actinide concentration in sampled solution. The entire process is time-consuming, involves errors at individual steps (sampling, separation, dilution, solvent evaporation, firing, counting), as well as exposes analysts to radiation. Apart from these factors, information obtained is not in real time and therefore is less useful for online process control. Such processes ideally require detectors that can be placed in contact with process solutions and estimate actinide concentration directly. Silicon-based detectors are not successful due to severe chemical degradation in corrosive chemicals. Attempts have been made in the past to use conventional solid-state alpha detectors (Si, ZnS) with thin protective coatings (PTFE, Mylar, or silicon oxide) [14–16]. However, none of them resulted in reliable and reproducible measurements over continuous long operation, especially in corrosive solutions (nitrates, fluorides, etc.). Detector sensitivity is affected as few low-energy α-particles entering into the protective layer get absorbed within the layer and are not registered as an event. Further, since thin protective coatings (submicron thickness) are difficult to make, the detector's active surface degrades over a period of time influencing its performance and finally damaging it permanently. Most attempts with Si-based detectors were faced with degradation of host matrix or poor detector performance. In contrast to silicon, chemical inertness and high-temperature stability of diamond, coupled with its excellent detector properties, makes it an ideal choice for alpha detection in chemically aggressive solutions. Such detectors can be kept in direct contact with solutions for practically an unlimited time, as far as diamond's stability is concerned. Development of such detectors can be traced back to the early 1970s when natural diamond crystals were used as host matrix [17]. Significant research has been carried out since then using both natural as well as synthetic diamond, and successful results have been reported with such detectors employed at a French nuclear fuel reprocessing center for performance evaluation [10,18–20]. Research activities are presently underway in our group to synthesize high-quality CVD-grown polycrystalline diamond films and develop α-detectors based on these films for use in corrosive acidic solutions [21]. The objective is to aid the nuclear fuel reprocessing industry; a few results will be presented in later sections.

Other important areas for utilization of diamond-based radiation detectors under extreme conditions are (1) high-energy physics research and (2) fusion energy research. While the former is directed toward fundamental understanding of matter interactions at extremely high energies in order to improve our understanding on the formation of universe, the latter is directed toward finding a nearly inexhaustible and sustainable energy source for future generations. Several megascale research programs are presently underway/planned in these areas including the Large Hadron Collider (LHC and SuperLHC) at CERN, Geneva; Facility for Antiproton and Ion Research (FAIR) at Darmstadt; International Linear Collider (ILC) and International Thermonuclear Experimental Reactor (ITER). In case of high-energy accelerators such as LHC, ions beams (light to heavier ions) accelerated to extremely high kinetic energies ($\sim$1AMeV to 1ATeV) are made to collide and resulting reaction products (numbering from two to even thousands) are analyzed. The area around the collision zone (known as quark-gluon plasma) is full of high-energy ionizing radiation with unprecedented intensities ($>10^{16}$ particles/cm^2) and detection of such radiation requires extremely radiation hard, fast detectors capable to continuously operate at higher temperatures for long durations (from several months to even years). With its superior radiation hardness, high-temperature stability and fast response, diamond-based detectors are the most preferred choice for these applications. On the other hand, in fusion research experiments, such as ITER, fusion plasma is a complex high-temperature mix of neutrons, protons, hydrogen isotopes and other radiation. Continuous monitoring of plasma is essential and diamond-based particle detectors are promising candidates for this purpose. Apart from particle detectors, diamond-based X-ray and vacuum ultraviolet (VUV) detectors are also useful in nuclear fusion experiments. A substantial amount of work is presently being pursued on development of diamond-based beam monitoring arrays and neutron monitors. Results have shown promising performance and have further excited the research community in this direction [22–26].

Yet another case of not so extreme but crucially sensitive operational conditions is the requirement of radiation detection for biological systems and medical applications (particularly radiation dosimetry). Equivalence of atomic number of diamond ($Z = 6$) to that of human body tissues ($Z_{\text{effective}} = 7.4$) and nontoxic nature coupled with high S/N and radiation hardness makes it ideal to use diamond-based radiation dosimeters for in vivo as well as in vitro applications. Recent progress in use of diamond detectors for clinical dosimetry applications such as high-energy photon beam therapy, electron beam radiotherapy [27] and proton beam radiotherapy [28] have shown encouraging results with high-response function and spatial resolution, especially when single-crystal CVD diamonds were used.

From the previous discussion, it is clear that diamond's potential as a radiation detector under extreme conditions is remarkable and indispensable. For a long time, due to limited availability of electronic quality diamonds (natural/synthetic), the scientific and industrial communities were of the view that diamond-based detectors would be confined to special applications where Si-based detectors remained inferior for technical reasons. The envisaged uses were high-temperature operation, intense radiation fields, fast radiation events, and hostile chemical environments. Silicon seemed dominating in the field of solid-state semiconducting radiation detectors until the following advancements occurred: (1) remarkable development in CVD route of diamond synthesis making it possible to produce high-quality diamond in desired form (single crystals/polycrystals), size/shape (thin films, plates, etc.) and purity (doped/undoped) in an economically acceptable manner; and (2) expansion in application areas for diamond-based radiation detectors including UV/X-ray detection, radiation dosimetry, thermoluminescence (TL) detectors, neutron beam monitoring, fission fragment detection, etc. Except for synthesizing large single-crystal films (>10 mm $\times$ 10 mm area), CVD methods have reasonably matured for routine synthesis of high-quality polycrystalline diamond films (up to 8 inch diameter), small single crystals (<10 mm × 10 mm), gem-quality single crystals, doped diamonds and composites of diamond with other carbon allotropes. This is evidenced by the presence of several industrial players commercially competing in the field of synthetic diamonds (CVD/HTHP) for industrial/technological applications [29–31]. As CVD methods are continuously evolving and quality of CVD diamonds is moving from better to the best, diamond-based radiation detection devices are likely to be a major player in the future.

3.2 NATURAL DIAMONDS FOR RADIATION DETECTION

One of the oldest known materials, almost exclusively treasured as precious gemstones, natural diamonds are formed very slowly deep down in the Earth's crust where temperature and pressure conditions mimic its thermodynamic stability regime. These are usually brought to shallower regions during volcanic eruptions and subsequently mined after prolonged cooling. These diamond mines (mainly found in South Africa, Russia, and Latin America) contain diamonds with several other constituents, referred to as geological 'plum pudding'. Natural diamonds differ widely in terms of size, shape, color (colorless, blue, yellow, pink, brown, etc.), crystal quality and inclusions present in them. Overall quality of these stones differs distinctly from one region to other depending upon prevailing geothermal chemistry. Diamonds mined from the same region, which appear similar in terms of size, shape and color, may differ distinctly in microscopic properties (crystal quality, impurities, defects and defect distribution). Further, the occurrence of large diamonds (>1 cm^3 volume) is rare, making them highly expensive. One can therefore understand that obtaining high-purity crystals is difficult. In contrast, strict control and detailed knowledge of microscopic properties of diamond is required for radiation detection purposes. It is also impractical and uneconomical to characterize a large number of natural crystals in order to obtain a handful of specimens that fit the quality index for radiation detection. From the previous discussion, it turns out that while pioneering work on radiation detection indeed involved natural diamonds and promising potential was identified [32,33], the research suffered from highly nonreproducible and nonquantitative results obtained over a large number of samples. This hampered further research in this area until the CVD route for laboratory synthesis of diamond was invented [11,12]. Before discussing CVD diamond-based radiation detectors, it is therefore pertinent to give a brief account of laboratory synthesis of diamond.

4. LABORATORY SYNTHESIS OF DIAMOND

For earliest attempts of diamond synthesis the guiding factor was equilibrium phase diagram of carbon [1], which shows diamond in metastable form under ambient conditions and graphite as a thermodynamically stable phase by a small enthalpy margin (~2.9 kJ/mol). Higher stability and negligible diamond-to-graphite conversion rate indicated a high-activation barrier for diamond-to-graphite conversion under ambient conditions. Further, while diamond can easily be converted to graphite at higher temperatures (>1000°C) at ambient pressure, the reverse transformation has been a technological challenge. Therefore laboratory synthesis of diamond either mimics the temperature-pressure conditions of its thermodynamic stability region (known as equilibrium synthesis methods) or preferentially crystallizes diamond in metastable state by kinetic control of reaction (known as nonequilibrium synthesis methods). In equilibrium methods, diamond is crystallized from (1) carbon present in molten metals/alloys on a small diamond seed crystal under HTHP method [34,35], (2) detonation of explosive mixtures (e.g., TNT + RDX, etc.) in controlled manner (shock wave method) [36] or (3) electron beam irradiation of fullerene-based structures where nanodiamond crystals nucleate due to pressure exerted at the central region of fullerenes [37].

In case of nonequilibrium methods, diamond is kinetically stabilized under low-pressure and high-temperature conditions using simple gaseous precursors such as hydrogen and hydrocarbons [11,12]. The present discussion is only focused on CVD methods as the advent of this method has brought a revolution in development of diamond-based radiation detectors. Since its invention, CVD methods have advanced significantly [6,38] and led to both homo- and heteroepitaxial growth of diamond (pure/doped) as thin films, plates, coatings, crystals, nanocrystalline diamonds, oriented/random films, etc. CVD methods can further be classified in terms of manner of activation of gaseous precursor to dissociate them into smaller fragments, which deposit over the substrate to form diamond. CVD synthesis is usually carried out under highly reducing conditions (i.e., H_2-rich atmosphere) for preferred crystallization of diamond over graphite. Different CVD methods are (1) hot filament CVD (HFCVD), (2) plasma-assisted CVD [microwave plasma-CVD (MPCVD) and radio frequency plasma CVD (RFCVD)] and (3) DC-arc jet CVD and their several variants based on conditions introduced in these CVD reactors. Examples of such variants include electron-assisted CVD, bias-enhanced CVD, CVD with separate substrate heating stage, magneto-MPCVD, etc. Detailed discussion of these methods is beyond the scope of this chapter.

Heteroepitaxial growth is carried out on a nondiamond substrate (Si, Mo, SiC, etc.) and results into polycrystalline films. Single crystals are not formed mainly due to large mismatch in the lattice parameter of diamond and substrate material. The main advantage of heteroepitaxial CVD is that large-area uniform films can be synthesized in view of easy availability of high-purity large-area single crystal silicon substrates. Another advantage is synthesis of thick films (up to several hundred microns) of desired quality by optimizing the growth parameters (gas composition, pressure, temperature, etc.) and increasing the deposition rate as well as deposition time. Finally, the substrate can be leached out using mineral acids (HF solution or fluorinated aqua-regia) and free-standing diamond films/plates can be obtained. These films, when polished from both flat surfaces, give highly transparent diamond films and find application in several high-end technologies such as optical windows for high-power applications. A major disadvantage of heteroepitaxial growth is that polycrystalline films have highly defective diamond–substrate interface and large fraction of grain boundaries, both of which are detrimental to film's electronic properties and thereby influence radiation detection performance.

Homoepitaxial CVD growth of diamond is carried out on single-crystal diamond substrates to grow single-crystal diamonds. As-grown crystals can be separated from seed crystal by "lift-off" technique. The advantage of this method is that single crystals with minimum defects and nearly no grain boundaries are produced, which are ideal for diamond-based electronic devices. Future improvement may be accomplished by combining the advantages of homo- and heteroepitaxial methods. In this approach, optically smooth polycrystalline diamond plates processed from heteroepitaxially grown films can be used as seed for further homoepitaxial growth. Films resulting from such an approach are likely to be improved with larger grain sizes and would be better for electronic applications compared to heteroepitaxially grown films.

CVD methods have now reasonably matured and, with control of growth conditions (deposition parameters, feed gas purity, and substrate characteristics), high-quality diamonds can be routinely synthesized. Ample literature is available on recipes for synthesizing diamond with desired purity (undoped/doped), crystallinity and crystallographic orientation, morphology, uniformity, film thickness, surface termination, nanostructured diamonds, etc. Among CVD methods, MPCVD route has been the most successful to synthesize high-quality electronic-grade films. Other routes, namely, HFCVD, RFCVD, and their hybrid variants, have their own importance for given product quality. We have used both HFCVD and MPCVD routes to prepare high-quality polycrystalline diamond films on single-crystal silicon substrates. Deposition parameters were systematically optimized to ensure minimization of nitrogen and nondiamond impurities and grow films with good electronic properties for detector applications. Details of our results on CVD synthesis of diamond films are reported elsewhere [39]. Improvements in CVD route have evolved the field from small-size HTHP-grown crystals (<few mm^2) to large-area films and coatings on complex three-dimensional surfaces (machine tools). The challenge for the scientific community now is to achieve the synthesis of large-area (>1 inch diameter) single-crystal diamond by homoepitaxial growth, which at present is limited for the want of large single-crystal diamond seeds (substrate). Another area where progress is required is synthesis of diamonds with tunable intrinsic defect density.

In summary, development of diamond synthesis methods has expanded the applications of diamond out from gemstones and hard abrasives to a large number of technological/industrial applications (hard coatings, heat sinks, optical devices, radiation detectors, tribological applications, biosensors, chemical sensors, etc.). These developments have transformed the diamond industry. At present, synthetic diamonds are a major share in the diamond economy with ~600 metric tons annual production for industry, jewelry and other technological applications in comparison to ~25 metric tons of natural diamonds mined annually, which are exclusively used for gemstone purposes [40].

4.1 CHARACTERIZATION OF CVD DIAMONDS FOR RADIATION DETECTION

To evaluate the quality of a CVD diamond for radiation detection applications, its detailed characterization is essential. Several techniques are used to evaluate the bulk and microscopic properties of CVD diamonds. X-ray diffraction (XRD) and vibration spectroscopy (IR and Raman) are preliminary tools to identify the bulk structure. Raman spectroscopy, in particular, is very useful for diamond as well as other carbon-based materials. It gives detailed information about phase identification, structural stress, defects at grain level, etc. Microscopic techniques (optical microscopy/electron microscopy/atomic force microscopy) are used to characterize surface structure, topography, and morphology of CVD diamond films. Luminescence measurements such as photoluminescence (PL), cathodoluminescence (CL), and thermoluminescence (TL) are useful in identifying the nature of defect centers in diamond and therefore the type of diamond. Low-temperature physical property measurements (thermal conductivity and electrical resistivity) are major tools to characterize the purity and electronic quality. Current-voltage (I-V) characterization is carried out to estimate the average resistivity of films to qualify it for detector application. Secondary ionization mass spectroscopy (SIMS) is useful in finding the nitrogen impurity and nitrogen-vacancy defect centers. While other techniques can also be used, those just mentioned are adequate for characterizing CVD diamond for radiation detection purpose.

4.2 POSTGROWTH TREATMENT OF CVD DIAMOND FOR DETECTOR APPLICATION

For radiation detection applications, extreme care must be taken while depositing diamond films. One should ensure leak tightness of CVD reactor, high purity of feed gases, desired gas mixing ratios, substrate pretreatment, growth rate, etc., so that high quality films are produced. In spite of such care, as-grown films still contain certain impurities and defects, which act as trap centers for charge carriers during radiation detection. To minimize these impurities/defects, pretreatment of CVD-grown diamonds is required before using them for detector fabrication. Carbon-rich nondiamond species are usually deposited while terminating CVD growth thereby covering the film surface. Since graphitic carbon is preferentially reduced over diamond at high temperatures in highly reducing atmosphere, nondiamond carbon impurities from the growth surface can be minimized by continuing the CVD process in pure hydrogen flow for some time at the end of film growth. Up to 99.9% surface graphitic phase can be removed by heating the films in hydrogen ~1000°C with negligible loss (~0.2%) of diamond phase [11]. Such films have hydrogen-terminated surface and uniform polycrystalline morphology with average grain size approximately ~10% of the film thickness [19]. Such films are then subjected to a sequence of physicochemical treatment, which includes (1) vacuum annealing of as-grown films at 800–1000°C to relax internal strain and minimize structural defects associated with thermally induced atomic movements, (2) chemical treatment in strong oxidizing reagents ($H_2O_2 + H_2SO_4$ mixture/freshly prepared chromic acid) to further minimize graphitic carbon impurities and (3) chemical treatment in hydrofluoric acid solution to leach out silicon substrate, in case a detector is to be made from free-standing film. It is essential to control temperature and oxidizing conditions during chemical treatment while using chromic acid so that microprecipitation of chromium oxide and related impurities does not occur [39]. Finally, films are washed several times in warm deionized water and dried in vacuum before deposition of electrodes and mounting in detector.

5. HISTORY OF DIAMOND-BASED RADIATION DETECTORS

A review of literature on interaction of radiation with diamond takes us to the beginning of the 20th century when fluorescence or coloration of transparent natural diamonds was reported upon exposure to alpha-emitting radioactive salts/vapors [32,41,42]. For quite a long time, these effects were debated in terms of the dominating effect of alpha, beta or gamma radiation. However, nothing could be definitively concluded due to very little characterization details of studied crystals. Since it is known that, apart from chemical doping (in synthetic diamonds) or ion implantation (in natural diamonds), electron/gamma irradiation is an industrially practiced method for diamond coloration by creating color centers in diamond lattice, the same might have been the reason for those early observations. The potential of diamond to be used as a solid-state counter for alpha radiation was observed in the late 1940s when conductivity pulses were recorded in natural diamond crystals exposed to radium and subjected to electrical potential across the opposite crystal surface [33,43]. These results were significant in a way that basic requirements and key issues in utilizing diamond as a detector host for charge particle radiations were brought out. This indicated that diamond-based radiation counters may have advantageous features like small size, room temperature operation, fast response and higher material density compared to large-size gas-filled ionization chambers. Requirement of minimum carrier trap centers in crystal was also brought out in these studies. The fact that only a few crystals gave conductivity pulse among several visibly similar crystals indicated wide variation in electronic properties of natural diamonds. Saturation in photoconductivity was also observed with increasing bias potential. Most importantly, a very important issue of diamond polarization due to space-charge buildup was highlighted as a major challenge in diamond-based radiation detection technology [44].

In summary, while successful radiation detection in diamond kept the scientific community interested in this research, the overall efforts were hampered due to extremely poor consistency of results with one or two positive results among hundreds of apparently colorless and transparent diamonds. It was even believed that diamonds capable of radiation counting are 'rare'.

Research efforts in this area remained sluggish until notable progress occurred in CVD methods of diamond synthesis. Even the HTHP method, which was invented earlier than the CVD route, did not bring much progress in this area as very small crystals could be grown and were usually not of highest electronic quality. Based on the classification of natural diamonds [45], it was proposed that only type-II diamonds (transparent to UV light down to 225 nm) exclusively show radiation detection properties. Later in the 1970s, notable work mainly from Russian scientists (then the USSR) demonstrated the application of carefully selected natural diamond crystals for alpha-particle detection in air/vacuum and corrosive liquid solutions [17,18,46,47]. With the present understanding of physics of radiation detection in diamond, one can say that inconsistencies in early results were due to (1) vast variation in quality of natural/HTHP diamonds and (2) lack of characterization of studied specimens. Research on CVD diamond–based radiation detectors picked up in the mid-1990s [10,48] and since then has continued to rapidly advance.

5.1 DESIGN OF DIAMOND-BASED ALPHA-RADIATION DETECTORS

Similar to conventional solid-state devices, diamond-based radiation detectors can be operated in charge-pulse mode or current mode, depending upon the manner in which free carriers (electrons and holes) are collected by external electronics. While several designs are reported in literature, all can be grouped into two types: (1) in-plane electrode configuration (Fig. 19.2), in which electrodes are present on the same surface of diamond crystal/film/plate separated by a finite distance (*L*); and (2) out-of-plane electrode configuration (sandwich configuration) shown in Fig. 19.3, where electrodes are made across the diamond crystal/film separated by thickness (*d*) of the crystal. In-plane design may include simple spot pattern (Fig. 19.2) or interdigitated electrodes deposited on diamond (Fig. 19.4). In principle, either design can be adopted for radiation detection. The selection depends upon nature and energy of radiation, mode of interaction, average range of radiation in diamond, crystal quality, detection environment, etc.

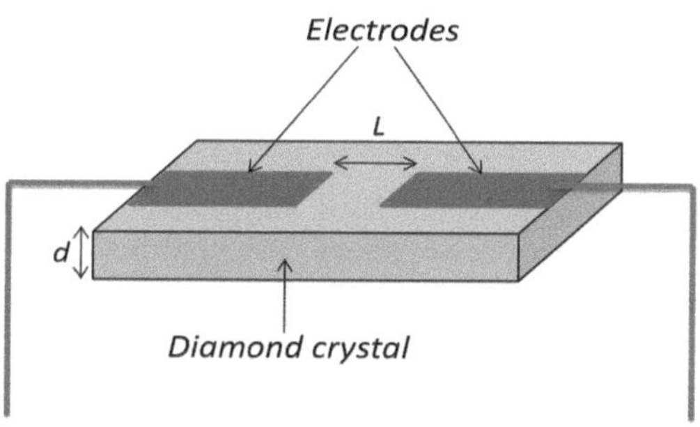

FIGURE 19.2

In-plane design of diamond detector.

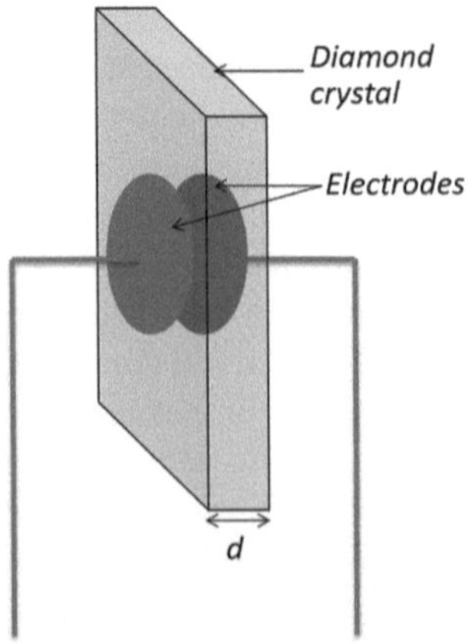

FIGURE 19.3

Out-of-plane (sandwich) design of diamond detector.

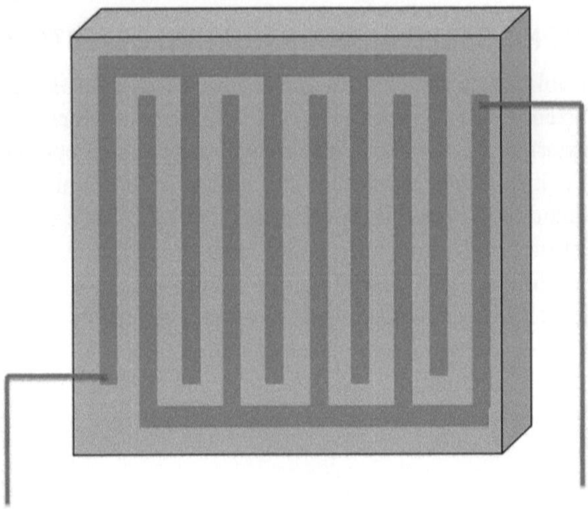

FIGURE 19.4

In-plane design with interdigitated electrodes.

For penetrating radiations/particles, which have longer range (UV rays, X-rays, gamma-rays, neutrons, high-energy electrons, pions, and sometimes even protons), thick diamond films (up to few hundreds of microns) are required and detectors are designed in typical out-of-plate MIM heterostructure. These detectors can be operated in positive as well as negative bias conditions. On the other hand, for charge particle detection (α-particles, metal ions, fission fragments, low-energy protons, etc.), which have higher specific ionization and limited range, detector design requires thin CVD films (few microns to few tens of microns) deposited on a suitable substrate (Si, Mo, SiC) and can be configured as (1) in-plane MIM heterostructure or (2) out-of-plane metal-insulator-semiconductor

(MIS) or metal-insulator-semiconductor-metal (MISM) heterostructure. Proper adherence of diamond film over the substrate (chemically bonded interface) is required for stable and reproducible performance of detector in MIS/MISM configuration. In-plane MIM configuration can also be used with thick diamond films (>100 μm) for charge particle radiation detectors such as ~5 MeV α-particles, generally emitted from actinides with limited range (~15 μm) in diamond. It is essential to deposit electrodes on growth surface in in-plane MIM configuration as crystalline quality of films is best on the growth surface with large crystallites (up to several tens of microns) resulting in the best charge collection signal. The substrate–diamond interface, in contrast, is highly impure, near amorphous, and possesses high defect density and therefore is not usually recommended for electrode deposition with in-plane MIM configuration.

One of the main parameters that governs the performance of diamond α-detectors is speed and efficiency of charge collection, once produced by radiation–matter interaction in the detector. Providing electrode contacts therefore is crucial, and the basic requirement is formation of good ohmic contacts. Electrode formation on diamond-based detectors is well established and can be carried out in multiple ways such as (1) physical vapor deposition, (2) sputtering, (3) injecting electrodes by thermal diffusion/ion implantation and (4) doping of diamond surface itself. Example of doping diamond surface for electrode formation is growing a thin layer of boron-doped diamond (p-type) on the growth surface by introducing boron precursor in CVD reactor after required thickness of pure diamond has already grown. Ion beam injection of elements like boron on diamond surface to create heavily B-doped semiconducting electrode (known as three-dimensional electrodes) is another method. Further, in case of metallic electrodes, use of pure metals coating (Au, Ag, Pt, etc.) or layers of more than one metal (Au-Cr, Au-Pt, Au-Mo, Ti-Cr-Au, etc.) are commonly found in literature. High-temperature vacuum annealing is sometimes followed for ensuring ohmic nature of electrodes. In our lab, we use PVD for Au evaporation on CVD diamond films followed by high-temperature vacuum annealing for this purpose.

5.2 DESIGN OF DIAMOND α-DETECTORS FOR USE IN CORROSIVE LIQUID MEDIA

Use of diamond for detection of alpha-emitting radio nuclides in corrosive liquids has been a long-sought application, particularly for nuclear technology. Therefore, design of such detectors requires special consideration. Two types of detectors can be developed: (1) for off-line measurements where a small volume of liquid is brought in contact with detector's active area, and (2) for online measurements where a detector's active volume is to be immersed into radioactive liquid. Both these detectors have unique application. The former is used as an off-line analytical tool to estimate actinide concentration in radioactive solutions. Such detectors are useful for direct analysis of liquid samples and to minimize the analysis time, radiation exposure to analyst and errors present in conventional radiometric method as discussed previously. The latter design of immersible detector can be used for online real-time determination of alpha activity (actinide concentration) in a process stream and thereby control the process on real-time scale. One example of such application is various process streams involved in nuclear fuel reprocessing by PUREX process, where different solutions (aqueous/organic) are encountered during separation and partitioning of actinides from spent nuclear fuel. Our group is engaged in developing both types of detectors for specific use in acidic solutions. Figs. 19.5 and 19.6 represent schematic designs of two types of detectors [49] while Figs. 19.7 and 19.8 show as-developed detectors for use in air and off-line liquid solution operation [21], respectively. Importantly,

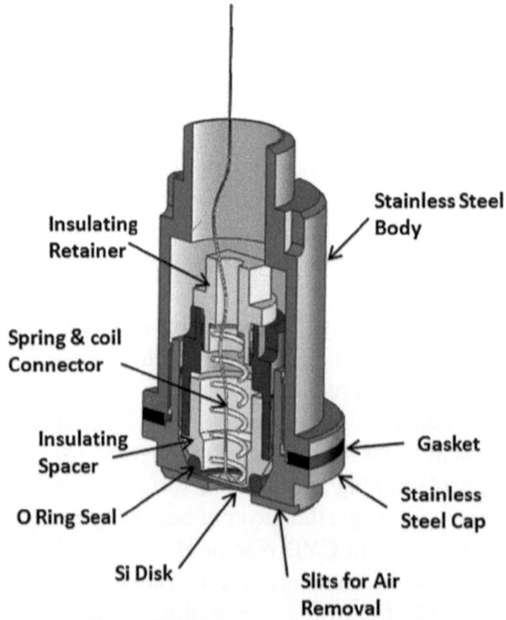

FIGURE 19.5

Schematic of diamond α-detector for online liquid source monitoring [49].

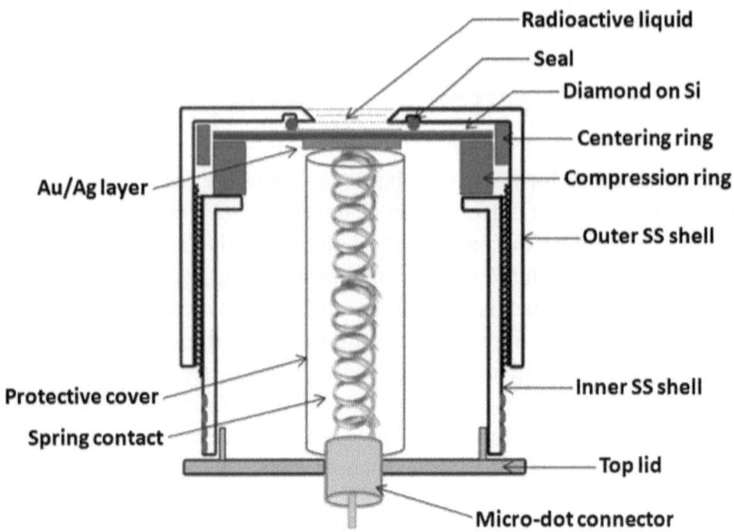

FIGURE 19.6

Schematic of diamond α-detector for off-line liquid source monitoring [49].

FIGURE 19.7

As-developed diamond α-detectors for air/vacuum operation [50].

FIGURE 19.8

As-developed diamond α-detector for off-line liquid source monitoring [21].

in case of liquid source monitoring, the solution in contact with the diamond surface acts as a highly conducting medium. Therefore, the detector shell itself acts as a perfectly grounded electrical contact for completion of circuit and no separate electrode from the front side of diamond surface (exposed to liquid) is needed.

While chemical inertness of diamond enables direct contact of diamond surface to corrosive solutions, detector failure can occur if electrode contacts or detector shell gets damaged. If chemically stable detector shell is used and electrical contacts are secured, such detectors can be used continuously for a fairly long period of time without significant deterioration in performance. This requirement can be met with availability of (1) acid-resistant stainless steels for fabrication of detector shell and (2) acid-resistant polymers such as polytetrafluoroethylene (PTFE), resins and sealants for securing electrode contacts, silicon substrate and lead wires from acidic solutions/fumes. We have used stainless steel (SS-304) for fabrication of detector shell and PTFE and acid-resistant sealants for isolating the active diamond surface from the rest of detector. Care is taken to ensure firm electrode contact at rear surface by using enameled copper wire coil soldered to gold-plated copper disk, which remains glued on the rear surface of diamond film (or Si substrate) using electronic-grade silver paste [49]. The rear electrode contact is soldered to a microdot connector and further secured by coating with insulating varnish followed by covering with PTFE tape. Detectors developed for off-line operation have ~6 mm diameter of active film area with ~50 μL cavity volume. Such detectors could be installed in a plutonium-active fume hood and continuously operated for off-line liquid source monitoring for several months without significant deterioration in performance. Detectors made from free-standing diamond films were delicate in handling due to the fragile nature of film. While their performance in air was satisfactory, they were not tested for liquid source monitoring.

5.3 ELECTRONICS AND DETECTION GEOMETRY FOR CVD DIAMOND α-DETECTORS

In order to obtain good performance of diamond-based radiation detectors, development of fast, dedicated and noise-free electronics is essential apart from fabrication of a good-quality detector. Capacitance of as-fabricated detector becomes crucial in deciding the nature of electronics. Ideally, it should be kept as low as possible for measurements involving weak signals (~2 fC/particle). Diamond detectors are advantageous being inherently low-capacitance detectors due to low dielectric constant of diamond. Depending on the nature of signal generated inside the detector and desired accuracy of measurements, low-to-moderate noise circuits can be used. For charge particle detection such as α-particles, a charge-sensitive preamplifier coupled with an amplifier, shaper and multichannel analyzer/counter is employed to record the pulse height spectra. Oscilloscopes are used to observe the pulse profile, pulse heating and other detection parameters (gain, bias voltage, shaping time, etc.) while optimizing the operating conditions for a given detector.

Electronics associated with diamond α-detectors developed in our lab consist of detector head with spring-loaded back electrode connected through microdot miniature connector. The detector is connected to a charge-sensitive preamplifier (Model: PA4850, ECIL, India) through a minimum-length shielded cable. Smaller length of this cable is important as it minimizes the signal loss coming from detector to preamplifier. Preamplifier output is fed to a spectroscopy amplifier (Model: PA4903N, M/s. ECIL, India) having a gain range adjustable between 2.5 and 1500 (<5.0 μV RMS noise with 3 μs shaping time). The amplifier output is given to a compact 8K multichannel pulse height analyzer and associated window-based software (Model: USB-MCA, M/s. ECIL India) with 0–10 V signal output range. Amplifier output is also fed to a counter (Model: EC 5104N, M/s. ECIL India) for operating the detector in gross counting mode as well as to an oscilloscope. Bias is applied at the detector back contact using a high-voltage bias supply unit (Model: HV4828N, M/s. ECIL India). A schematic diagram of an alpha spectrometer assembly is shown in Fig. 19.9.

Each detector needs to be tested in air before liquid source monitoring. For testing in air, detector was kept with exposed diamond film facing downward. Electro-deposited actinide source (^{239}Pu, ^{241}Am) with activity ranging over ~60 pCi to 90 nCi were kept at ~8 mm distance from detector shell. Bias voltage was varied between 25 and 40 V (electric field strength: ~8.3 to ~13 kV/cm) for detectors made with thin polycrystalline (~20–25 μm average thickness) films. Initially, detection parameters were optimized for better S/N and counting was continuously monitored for several hours to check detector stability and count rate reproducibility. For liquid source monitoring, detector shell and preamplifier assembly are covered by multiple layers of parafilm and PTFE tape. Cables required for connecting bias supply and amplifier are enclosed in polythene tubes taken inside plutonium-active fume hood such that detector civility is kept upward where finite volume of analyte solution (up to 50 μL) can be poured using a micropipette. The rest of the electronics are kept in an inactive area outside the fume hood. Counting is carried out for up to 1 h for a given solution to check count rate

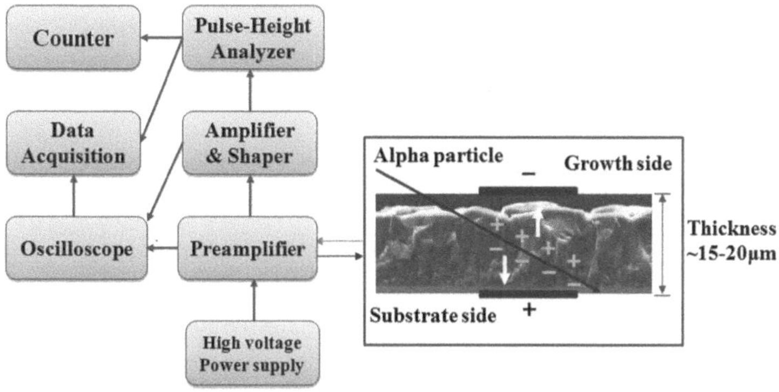

FIGURE 19.9

Schematic diagram of alpha spectrometer setup for diamond α-detector.

reproducibility. After each sample, detector civility is rinsed with actinide-free solvent (nitric acid, oxalic acid, hydrochloric acid, EDTA solution) to bring the baseline count rate nearly equal to virgin detector [50].

6. PERFORMANCE OF CVD DIAMOND α-DETECTORS

A review of the literature reveals that research in CVD diamond-based α-detectors has resulted in development of stable, reproducible, and quantitative detectors for (1) gross α-counting using polycrystalline films [10,20,51] and (2) high-resolution α-spectroscopy applications using single-crystal films/plates [52,53] in air/vacuum operation. Further, it has been demonstrated that these detectors are ideal for determination of actinide concentration in corrosive liquid solutions [19,54]. Even newer methods based on coating of boron-doped nanocrystalline diamond on conventional silicon detectors have been proposed for trace detection of actinides in dilute aqueous solutions [55]. A few results obtained on α-detectors developed in our lab using MPCVD-grown polycrystalline diamond thin films are presented here as an example. Unlike Si-based detectors, diamond α-detectors can be operated in both positive- and negative-bias conditions with respect to diamond surface exposed to radiation in air/liquid medium. Results obtained in our lab also confirmed the same. Before employing the detectors for liquid source monitoring, these detectors were thoroughly tested in air using electrodeposited plutonium sources. Fig. 19.10 shows as-measured pulse height spectra for different plutonium sources (bias = 24 V, amplified gain ~1k, shaping time ~3 μs) in air.

Fig. 19.11 shows the calibration plot for these measurements in comparison to calibration plot obtained by using a commercial ZnS(Ag) detector in air under standard operating conditions. It can be seen that while sensitivity of diamond detectors is significantly lower than ZnS(Ag) detector, the performance is reasonably linear [50]. Linearity of these detectors is also shown in Fig. 19.12 where peak counts are plotted as a function of time for a representative detector.

After thorough testing in air, these detectors were tested in liquid solutions for estimation of (1) Pu (1−10,000 μg/L) in 3 M HNO_3, (2) Pu (1−200 μg/L) in (0.1 M oxalic acid + 2 M HNO_3), and (3) ^{233}U (40−1500 μg/L) in 3 M HNO_3. Figs. 19.13 and 19.14 show as-measured pulse height spectra and

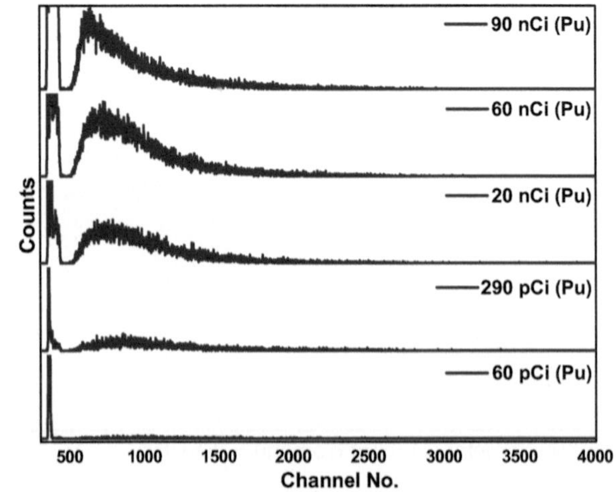

FIGURE 19.10

As-measured pulse height spectra on electrodeposited Pu samples (60 pCi–90 nCi). Vertical axis in each curve has same magnitude [50].

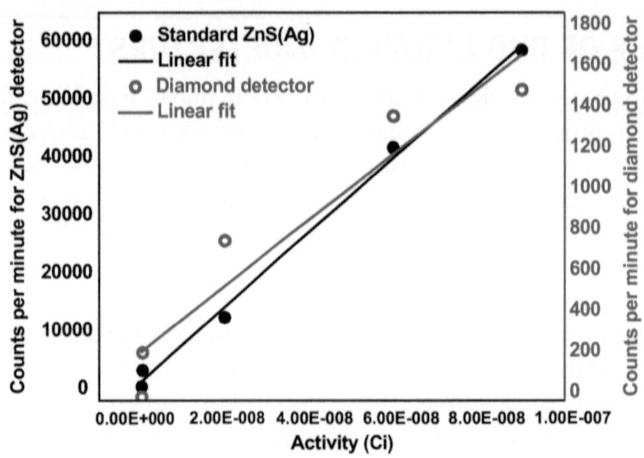

FIGURE 19.11

Calibration plot for diamond detector in air. Calibration plot for ZnS(Ag) detector is also shown for the purpose of comparison [50].

calibration plot for estimation of plutonium in 3 M HNO_3 solution, respectively. Our results confirmed that these detectors are stable and perform linearly as well as reproducibly over concentration range from 1 to 1000 μg/L. Similar results were obtained for estimation of Pu in (0.1 M oxalic acid + 2 M HNO_3) and ^{233}U in 3 M HNO_3 solution [21,50].

Our results confirmed the suitability of polycrystalline CVD diamond film-based α-detectors for off-line gross estimation of actinides directly in a wide variety of corrosive chemical environments.

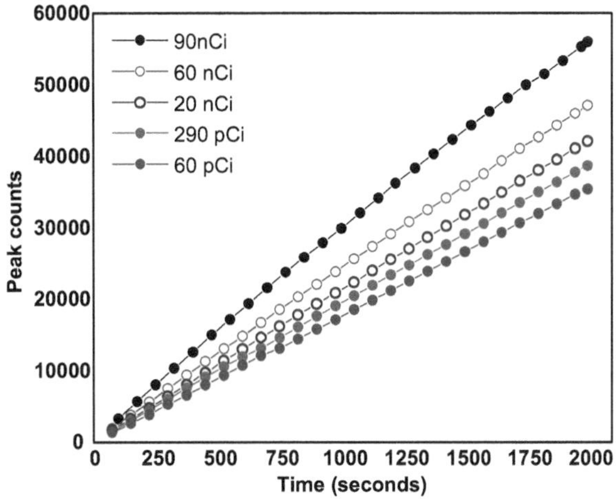

FIGURE 19.12

Peak counts as a function of time, indicating linear behavior of detector [50].

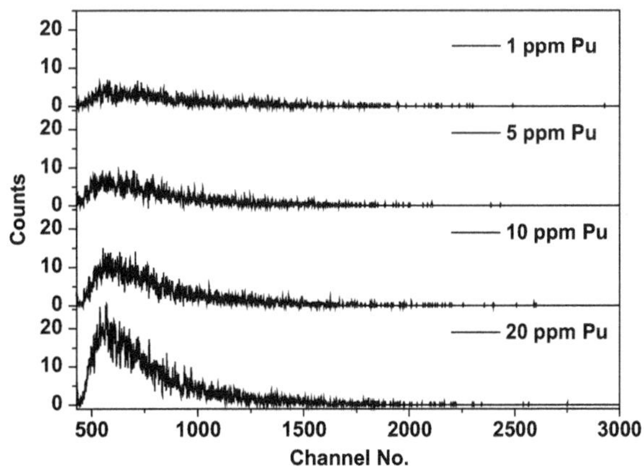

FIGURE 19.13

As-measured pulse height spectra recorded with Pu in 3 M HNO_3 [21].

Present efforts are directed toward (1) developing such detectors for organic liquid source monitoring, (2) detectors made of free-standing, optically polished, polycrystalline thick CVD films (50–100 μm), and (3) online detectors for real-time measurements in process streams.

An important feature of CVD-grown polycrystalline diamond film-based α-detectors is their broad spectral resolution. As observed in Figs. 19.10 and 19.13, the pulse height spectra obtained from actinide source in air or liquid solutions are quite broad. The broadening increases in liquid medium due to relatively higher self-absorption of alpha energies within the liquid. The broadening comes

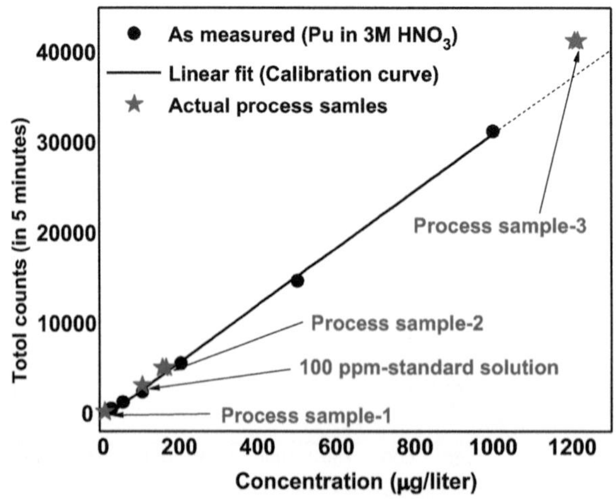

FIGURE 19.14

Calibration plot for diamond detector for liquid source monitoring (Pu in 3 M HNO_3). Performance of standard solution as well as actual reprocessing plant solutions is shown by stars [50].

because polycrystalline films have large fraction of grain boundaries along with other crystal defects/impurities that act as trap centers for charge carriers. A notable contribution to spectral broadening also comes from polarization phenomenon in diamond due to space charge buildup. The polarization problem in diamond-based particle detectors is discussed in detail in a later section of this chapter. As a result, energy resolution remains poor for polycrystalline diamond film-based α-detectors. These detectors can therefore be used only in counting mode for measurement of gross alpha activity and thereby estimate the total actinide concentration. However, given the advantages of (1) direct analysis of liquid solutions, (2) applicability in corrosive media, (3) fast counting, (4) no tedious sample preparation, and (5) minimization of radiation exposure to the analyst, these detectors hold excellent potential especially in the nuclear industry toward the back-end of the fuel cycle. At the maximum, sample dilution may be required, in case actinide concentration is beyond the linear working range of such detectors. It would be interesting to develop α-detectors using highly polished polycrystalline films (~50–100 μm) and see if the spectral broadening is reduced. This is envisaged for modified microstructure of diamond, involving larger grains exposed to alpha particles. So far, we have not been able to locate convincing literature on this aspect. Ultimately, best performance is expected from α-detectors developed using homoepitaxially grown CVD single-crystal diamond films/plates. This would be similar to natural diamond crystals that were used initially for α-detection in liquid solutions [17,18]. However, unlike natural diamonds, CVD-grown single crystals have the added advantage of controllable and reproducible crystal properties. Technology for single-crystal diamond-based α-detectors with spectroscopic resolution is already showing promise for air/vacuum operation [52,56] and such detectors are likely to work excellently with improved energy resolution for actinide estimation in corrosive liquids.

An important feature of these detectors is their electrochemical inertness during measurements in corrosive liquid solutions. This comes from high electrical resistivity of CVD-grown diamond films

($\sim 10^{13}-10^{14}$ Ω cm at room temperature), which allows only a negligibly small current to flow between the two electrodes (thereby through the liquid solution) to cause any significant electrochemical reaction to take place in the solution.

7. FACTORS GOVERNING THE PERFORMANCE OF CVD DIAMOND α-DETECTORS

The nature of radiation interaction with diamond is the basis to assess the performance of a diamond radiation detector. Therefore, one can look into this issue in two parts: (1) related to the quality of the CVD diamond and (2) related to the nature of radiation and medium in which the radiation is present. Regarding the quality of CVD diamond, it is mainly governed by the nature of trap centers (impurities, defects, vacancies, grain boundaries, etc.) and their distribution in it. Therefore, detailed understanding of them is essential to predict the performance of CVD diamond α-detectors. Detailed information about trap centers would help in modeling the band structure of given CVD diamond and thereby predict the resulting detector's behavior. Fortunately, several experimental tools are available to gain insight into trap center behavior. Some of them are PL, CL, photoabsorption, electron paramagnetic spectroscopy (EPR), electron field emission, carrier lifetime measurements, carrier mobility measurements, etc. The second factor related to the nature of radiation and medium in which it is present is also essential, as this would influence the extent of interaction with diamond. Once these aspects are understood, other parameters such as detection efficiency, stability, counting/spectrometry application, high-temperature operation, etc., can be predicted/estimated. Some of the important parameters on which performance of CVD diamond α-detectors depends are (1) carrier mobility in diamond crystal and factors influencing it, (2) mean free path, (3) charge collection efficiency (CCE), (4) charge collection distance (CCD) and (5) polarization phenomenon in diamond.

7.1 EFFECT OF BIAS POTENTIAL

Let us consider the ideal case of a defect-free, pure diamond crystal that is configured in a basic MIM sandwich-type detector structure. Once charged particle radiation interacts in the active volume of the detector, charge carriers (electrons and holes) are produced. Average energy required for liberation of one e-h pair in diamond lattice is taken as 13 eV, however, different values are reported in the literature ranging over 8.6–17 eV per electron–hole pair [57]. If no bias potential is applied across the electrodes, these charge carriers tend to recombine to undergo radiative or thermal de-excitation within their lifetime in diamond lattice. If a nonzero bias is applied, the charge carriers migrate toward oppositely charged electrodes and their migration competes with the recombination process. In this case, some charge pulse (conductivity pulse) is recorded in the circuit indicating the qualitative detection of radiation. If bias is slowly increased, tendency/efficiency of charge carrier migration toward the electrode surface increases and charge pulse with higher amplitude is recorded. In technical terms, it is said that with increasing bias the charge collection efficiency increases. With continuous increase in bias, there comes a limiting situation where charge pulse amplitude saturates. This is identified as state of maximum charge collection efficiency for a given detector. Fig. 19.15 shows a schematic plot of variation of measured pulse amplitude (a measure of charge collection efficiency) with applied bias voltage.

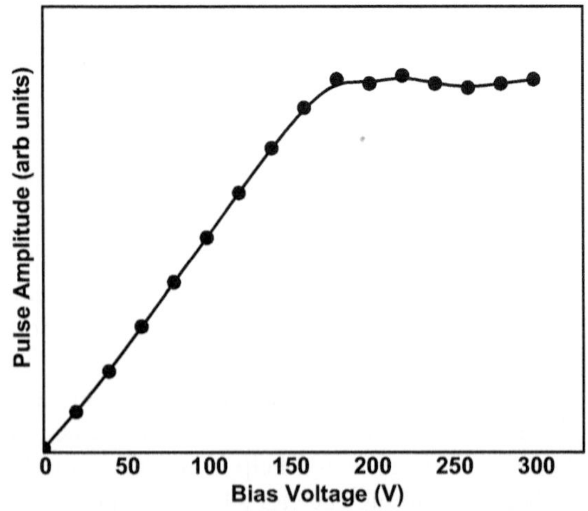

FIGURE 19.15

Schematic representation of saturation of charge collection as a function of applied bias voltage for an ideal diamond crystal.

Previous discussion was based on an ideal defect-free, isotopically pure diamond crystal where no trap centers (impurities/defects/grain boundaries) were present. In such case, application of saturation bias potential leads to near complete charge collection (charge collection efficiency; CCE ~100%). In actual crystals, however, such purity or defect-free quality is not possible (even in completely pure and defect-free diamond, there is ~1% ^{13}C atoms, which act as isotopic impurity centers). In such cases, another phenomenon of charge carrier entrapment by trap centers becomes operative as soon as charge carriers are produced. As a result, net CCE decreases (<100%), however, the nature and shape of curve for CCE versus applied bias remains similar to that shown in Fig. 19.15. The extent of decrease in CCE mainly depends upon the quality of diamond and experimental conditions of measurements. For optimum performance, effect of bias needs to be studied during the calibration of detector so that it is operated with the highest possible CCE.

7.2 CHARGE COLLECTION EFFICIENCY

CCE is one of the most important parameters governing the performance of CVD diamond-based radiation detectors. It is defined as:

$$\text{CCE} = \frac{Q_m}{Q_g} \tag{19.1}$$

Here, 'Q_m' and 'Q_g' are quantity of charge measured by counting circuit and charge generated due to energy deposition by ionizing radiation in the bulk of diamond lattice, respectively. CCE and background current (noise) are important parameters to be evaluated for any detector before it can be used for quantitative measurements. Background current (noise) is defined as signal measured in electronic circuit when detector is biased but no radiation enters into the active volume at a given temperature. Q_m can be evaluated by taking the integral of signal amplitude (pulse height spectrum) as a function of time and corrected with (multiplied by) detector calibration factor to convert count rate (cps) to disintegration rate (dps). In this manner, total counts can be converted into number of events (charge pulses generated).

Q_g can be obtained by performing a standard measurement using ^{90}Sr source (pure beta emitter) and measuring the energy deposited into the detector by an electron (considering ~13 eV be the average energy required to create an electron–hole pair in diamond lattice). There are, however, reports that use range of average work function (~13–18 eV) for detector calibration and estimation of Q_g.

In most cases, CCE initially increases with bias voltage and gets saturated. Onset of saturation voltage can be taken as operating voltage for a given detector. Further increase in bias voltage does not benefit in terms of counting efficiency but may contribute to enhancement of background noise. CCE can easily reach up to 90–95% for type IIa natural diamond crystals. For homoepitaxially grown single-crystal CVD diamonds, near-complete CCE (~100%) is also reported [52]. However, it is quite lower in polycrystalline CVD diamond films (~20–40%, depending upon film quality) due to presence of grain boundaries as inherent charge trap centers. Improvement in CCE of polycrystalline CVD diamond films is an area where significant research has been carried out and there is further scope for more work to be done.

Factors responsible for incomplete charge collection in CVD-grown diamond mainly originate from (1) film quality, (2) dimension of device, (3) nature of radiation interaction, (4) radiation type, and (5) charge collection circuit.

A simplified expression for calculation of CCE, which can be applied for diamond-based α-detectors, is given by Hecht formula [58]. As per this expression, CCE for a sandwich-detector configuration is given as:

$$\frac{Q_m(t)}{Q_g} = \frac{\mu\tau E}{L}\left(1 - e^{-\frac{L}{\mu\tau E}}\right) \tag{19.2}$$

Here, '$Q_m(t)$' and 'Q_g' are as defined previously, 'μ' is carrier mobility, 'τ' is carrier lifetime, 'E' is average applied electric field across the electrodes, and 'L' is the thickness of the detector. One can see from the expression that by limiting the detector thickness, higher values of CCE can be attained. However, the minimum thickness 'L_{min}' required for a given detector system is limited by average range of radiation in diamond. Thin crystals add further advantage in terms of through passage of low-ionizing radiations (beta, gamma, and X-rays) so that selective and exclusive detection of heavy charged particles (α-particles, heavy ions, fission fragments, protons, etc.) can be ensured. Usual thickness employed for α-detectors based on silicon-supported polycrystalline CVD diamond films ranges between 20 and 30 μm. Here it is to be noted that for detection of different types of radiation (UV, X-rays, gamma rays, neutrons, etc.) by diamond, crystal thickness is selected based on the range of these radiations in diamond and the nature of radiation interaction, which gives detectable response signal.

7.3 CHARGE COLLECTION DISTANCE

CCD is another essential parameter that defines electrical quality of a diamond sample. When charge carriers drift inside diamond crystal under the influence of external bias, they may get trapped at defect centers/impurity centers and thus not reach the electrode at all. CCD is defined as the average distance by which electrons and holes are separated before they get trapped. Average CCD is given by the following expression:

$$\mathrm{CCD} = \mu.\tau.E = \frac{Q_m}{Q_g/L}, \quad \text{for CCD} < L \tag{19.3}$$

$$Q_g = \frac{e.W}{\varepsilon} \tag{19.4}$$

Also, average CCD is given as

$$\text{CCD}_{\text{avg}} = \mu.\tau.E = (\mu_e.\tau_e. + \mu_h.\tau_h).E \tag{19.5}$$

Here 'μ' is average carrier mobility (cm^2/V s), 'τ' is the average carrier lifetime(s), 'E' is the applied electric field (V/cm), 'L' is the detector thickness (cm), 'Q_m' is the measured charge in the circuit (Coulombs), and 'Q_g' is the total charge generated in the detector (Coulombs) due to radiation interaction with diamond, 'e' is the elementary charge on electron (1.6×10^{-19} C), 'W' is energy deposited by incident radiation and 'ε' is work function or average energy required for creation of an electron–hole pair in diamond lattice (~13 eV), 'μ_e' and 'μ_h' are mobilities, while 'τ_e' and 'τ_h' are average lifetime of electrons and holes, respectively. Depending upon the bias condition [i.e., which of the carriers (*e* or *h*) is closer to one of the electrodes], mobility and lifetime of dominating carrier is used for evaluation of CCD. It is important to note that CCD of natural diamonds varies over several orders of magnitude from few millimeters (type IIa crystals) to few microns (crystals with higher impurities/defect densities) and is not controllable from sample to sample. On the other hand, consistent values of CCD can be reproducibly obtained in carefully synthesized CVD diamonds with maximum value around ~250 μm. However, large inhomogeneity is found in terms of distribution, due to presence of grain boundaries [59].

7.4 POLARIZATION IN DIAMOND-BASED DETECTORS

Usually observed in solid-state semiconductor detectors that have high intrinsic resistivity (CdTe, HgI_2, etc.), bulk polarization is also a challenge for diamond-based devices (natural/synthetic/CVD). In brief, this phenomenon can be explained as follows. During radiation interaction (charged particles/photons), free charge carriers are generated. As these carriers drift toward oppositely charged electrodes under applied bias condition, they may get trapped by inherent trap centers (defects, vacancies, impurities, grain boundaries, etc.) and do not reach to the electrode surface. Such trapped charges create their own local electric field, also known as space charge buildup effect, which is usually in the opposite direction to the applied bias and therefore lowers the effective bias field. As a result, charge collection efficiency reduces with time and is reflected in terms of lowering of observed pulse height amplitude as well as pulse counts in the detection circuit with time. This leads to unstable detector performance over a long period of operation and is therefore undesirable for several applications. Typical indications of polarization effect are (1) gradual shift of pulse height spectrum toward lower channel numbers (i.e., toward low energy) and/or (2) gradual increase in electronic noise counts. This is shown in Fig. 19.16 for a typical detector tested in our lab for direct estimation of plutonium in 3 M HNO_3 solution for spectra collected after each half hour [50].

As seen in Fig. 19.16, polarization effect was observed for a few detectors developed in our group. A successive decrease in pulse amplitude was observed during oscilloscope monitoring. Another observation was a sudden burst of counts at random periods during counting experiments. This could be due to the saturation of trap centers by charge carriers. Once trap centers are saturated, there is no further trapping possible. At that instant, space charge buildup is disturbed and all the charge come as a sudden burst of pulses, which can be seen with the oscilloscope. These pulses, however, do not significantly influence the counting characteristics as the burst of pulses usually contribute to electronic noise peak. Another observation is negative pulses seen in the oscilloscope once bias is made

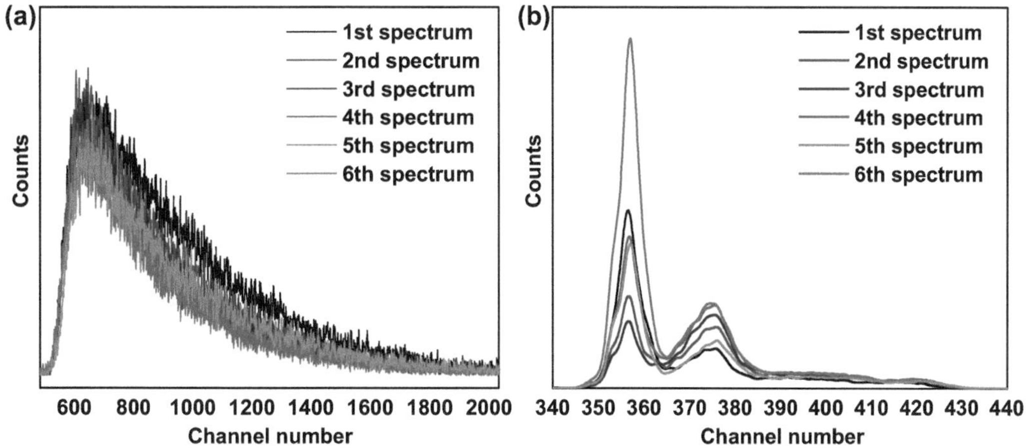

FIGURE 19.16

Pulse height spectra successively recorded on a polycrystalline CVD diamond α-detector using plutonium in 3 M HNO_3 solution every spectrum is recorded for 30 min. (a) Peak response due to plutonium, (b) electronic noise peak [50].

zero. The longer the period of detector operation, the more persistent can be such negative pulses. This could be explained as follows. As space charge buildup usually acts opposite to the applied bias direction, the opposite field slowly contributes to pulse collection on the opposite electrodes than expected and gives rise to negative pulses. These pulses also do not affect the nature of alpha spectrum during the counting time (under biased conditions). However, they get subtracted from the expected counts and thereby influence the detector's sensitivity and efficiency.

Knowing the origin of polarization effect, it is possible to say that its minimization requires good control over trap centers (impurities and defects) present in diamond. This can be accomplished almost exclusively (except inherent point defects, dislocations and isotopic defects) by homoepitaxial growth of CVD single-crystal diamond films/plates grown under highly controlled atmosphere. However, at present it is not easy to grow large-area ($>$10 mm $\times$ 10 mm) single-crystal diamond plates due to limited availability of large-area high-quality single-crystal diamond seeds. Polycrystalline films therefore still remain the material of choice for several radiation detection applications. These films always have grain boundaries and a highly defected substrate–diamond interface. The top surface (growth surface) remains highly crystalline, pure but quite rough with distribution of diamond crystallites. While one cannot get away with grain boundaries in polycrystalline films, a few strategies can be practiced to minimize defect densities (and therefore polarization) in these films. These are (1) accurate control of growth chemistry to ensure minimum ingress of nitrogen and other impurities during film synthesis, (2) minimization of surface roughness (growth surface) by lapping/polishing to optical level smoothness, (3) removal of substrate–diamond interface by leaching out the substrate and thereafter removing defect-prone region of diamond films by polishing, (4) chemical processing of as-grown films to get rid of surface impurities and nondiamond contaminants (graphite and other carbon structures), (5) annealing of films ($\sim$1400°C in vacuum or 700–900°C in air) to minimize residual structural stress, (6) priming of as-grown films by X-ray or electron irradiation and (7) applying an oscillating electrical bias potential to minimize the effect of polarization on detector performance, etc. While few strategies can be applied to all types of films (growth chemistry, thermal annealing, priming, etc.), some are not practically suitable for thin films (lapping, polishing, leaching, etc.). These thin films are required for specific applications such as detection of fission fragments, heavy charge particles, etc. In such cases, one has to minimize polarization by other possible methods or employ single-crystal diamonds, wherever possible. Importantly, control over polarization effect is essential if one needs to use the detector for quantitative measurements. If polarization persists, count rate would decrease with time and detector calibration would not be possible.

Polarization effects are more pronounced with intense radiation sources. This puts a limiting condition to linear operational range of diamond α-detector for estimation of concentration. Beyond the limiting upper concentration, the event rate in the bulk matrix exceeds the detector's dead time and therefore the detector is not able to register a separate count for an individual event. These may therefore get registered as combined counts (shifting the pulse height spectra toward higher channel numbers) and detector behavior becomes nonlinear for higher concentrations. Similar effects are seen in our detectors, where good linearity was observed over the concentration range from 1 to 1000 μg/L for plutonium or ^{233}U in 3 M HNO_3 solution. Beyond this, detector performance was nonlinear as well as nonreproducible. Note that linear range of a CVD diamond α-detector would vary with film quality and likely to improve with single-crystal CVD films.

7.5 POLARIZATION QUENCHING

Minimization of polarization effect in diamond α-detectors is referred to as polarization quenching. Several aspects have been mentioned in the previous section to control/minimize polarization. Early work on addressing this issue reported use of injecting electrode contacts in natural diamonds [60]. In many reports, it is demonstrated that bulk polarization can be effectively stopped (or minimized) in a reasonable period of time if the detector is illuminated by strong visible light source (such as incandescent lamp) continuously during the operation. Thereafter, the detector can be calibrated and used for quantification of unknown actinide sources in both air as well as liquid solutions [61]. Similar to visible light illumination, exposure to X-rays, UV rays and soft gamma rays has also been reported to minimize polarization-induced deterioration of charge collection properties in diamond detectors. Another approach useful for polarization quenching is to heat the detector during measurements. This is only possible with diamond as large band gap, high-temperature stability and excellent thermal conductivity allow detector operation at higher temperature without significant increase in background noise. Laser light illumination is another possible method for polarization quenching. For intense source (high specific activity), one needs to ensure that rate of polarization quenching by thermal or optical excitation is higher than the rate of trapping so that polarization quenching can be made independent of source strength. Selection of specific quenching methodology depends upon this aspect.

8. NEUTRON DETECTION IN DIAMOND

Only a brief introduction is presented here for neutron detection by CVD diamonds. Neutrons being neutral particles do not interact the way charged particles (alpha, protons, fission fragments) interact with diamond. They may (1) pass through the diamond crystal without losing much of energy or (2) get scattered resulting into displacing an atom from its original site, or (3) get absorbed by carbon atoms through a nuclear reaction (activation). The probability of any interaction is essentially governed by incident neutron energy and interaction cross-section (expressed in barns; $1\ b = 10^{-24}\ cm^2$). For example, the capture cross-section for low-energy neutrons ($E < 5$ MeV) is only 0.6 b while collision cross-section for 1 MeV neutrons is ~2.6 b. The cross-section increases with energy and shows resonance maxima at energies >8 MeV. Only if neutrons cause atomic displacement or get absorbed, the displaced atom or ejectile emitted from the activation product can cause further ionization in diamond matrix resulting into formation of charge carriers. These charge carriers are then collected by suitable electrodes in the same manner as found in diamond-based charged particle detectors. Hence, neutrons are always detected indirectly in diamond detectors. Neutron detection by CVD diamond detectors has very important use in fusion research, beam monitoring in spallation sources and radiation surveillance.

9. CHALLENGES AND FUTURE PROSPECTIVE IN DIAMOND RADIATION DETECTORS

Given the fact that applications of CVD-grown diamond radiation detectors are increasing and feedback obtained from performance of these detectors is guiding future improvements, sustainable growth in this area is inevitable. This will be further supported by increasing interest and applications in areas other than radiation detection, which include optical, thermal, biological, medical and mechanical applications. All these applications require further advancements in CVD methods for synthesizing high-quality diamonds in desired shape and form. For example, a near-term realizable goal is to increase the area and thickness of heteroepitaxially grown electronic-grade polycrystalline diamond films/plates. This can be achieved by suitable modification in CVD reactor design and routine availability of large-area (>8 inch diameter) single-crystal silicon wafers. Another advancement is awaited in terms of method development for synthesis of large-area homoepitaxial single-crystal diamond films/plates. Experimental efforts need to be directed toward growth of CVD diamond on optically polished large-area polycrystalline diamond plates. This may innovate a novel method to prepare large-area single-crystal films using such quasi-homoepitaxial growth methodology. Given the success of the first run of LHC experiments and experiments planned in future such as super LHC, high-energy accelerator experiments, fusion plasma experiments, etc., diamond-based radiation detectors would play a very important role. In developing countries like India, where large-scale expansion of nuclear technology with a closed fuel cycle is planned, use of diamond detectors for online real-time control of chemical processes involving actinides is likely to take a giant leap. Another possible application of CVD diamonds in nuclear technology could be deposition of diamond-based coatings on large three-dimensional surfaces to improve the corrosion resistance of metallic containers used for immobilization of high-level radioactive waste in glass form while imparting excellent heat dissipation path. Continued innovations and sustained efforts should remain the prime objectives in the pursuit of future development of diamond radiation detection devices.

10. CONCLUSIONS

From its physicochemical attributes, diamond is undoubtedly an excellent host for radiation detection applications, particularly under extreme operative conditions such as corrosive chemical environments, high temperatures, intense radiation fields, etc. Its unique attributes offer simple design of devices with robust performance under such conditions. Remarkable progress in CVD methods of diamond synthesis has made it possible to prepare diamonds suitable for radiation detection with tailored properties on routine basis. Continuing improvements in CVD methods offer exciting future advancements such as heteroepitaxially grown polycrystalline films deposited on complex three-dimensional surfaces and large-area homoepitaxially grown single-crystal diamond films/plates. Growing applications and promising results shown by diamond-based radiation detection devices have accelerated R&D efforts in several areas including nuclear technology, fusion energy science, high-energy physics research, medical dosimetry, etc. This brief review has highlighted the subject in reference to principles of radiation detection in diamond, its suitability under extreme conditions, historical evolution of such devices, design and performance of diamond-based α-detectors for use in corrosive liquid solutions, and factors governing their performance. Continued innovation and sustained efforts would guide further growth and future developments in this area.

ACKNOWLEDGMENT

The authors thank their former colleagues Dr. A. K. Dua and Dr. C. G. S. Pillai for initiating research on CVD methods of diamond synthesis and their applications for radiation detection, respectively. Sincere thanks are due to Dr. S. Panja and Dr. P. S. Dhami, Nuclear Recycle Group, BARC, for carrying out experiments in corrosive radioactive solutions. The authors also thank Dr. V. K. Jain, Head, Chemistry Division, BARC; Dr. D. Das, former Head, Chemistry Division, BARC; Dr. B. N. Jagatap, Director, Chemistry Group, BARC; Shri P. K. Wattal, former Director, Nuclear Recycle Group, BARC; and Dr. T. Mukherjee, former Director, Chemistry group, BARC, for their support and encouragement.

REFERENCES

[1] A. Savvatimskiy, Book chapter entitled: evolution of experimental carbon phase diagram, in: Carbon at High Temperatures, vol. 134 (8), Springer Series in Materials Science, 2015, pp. 183–211, http://dx.doi.org/10.1007/978-3-319-21350-7.

[2] JCPDS Card No.: 75-0410 PCPDF Version 2.0, ICDD Database 1998.

[3] C. Frondel, U.B. Marvin, Lonsdaleite, a hexagonal polymorph of diamond, Nature 214 (1967) 587–589.

[4] F.P. Bundy, J.S. Kasper, Hexagonal diamond – a new form of carbon, J. Chem. Phys. 46 (9) (1967) 3437–3446.

[5] JCPDS Card No.: 80-0004 PCPDF Version 2.0, ICDD Database 1998.

[6] J.C. Angus, C.C. Hayman, Low-pressure metastable growth of diamond and "diamondlike" phases, Science 241 (1988) 913–921.

[7] E.A. Ekimov, V.A. Sidorov, E.D. Bauer, N.N. Mel'nik, N.J. Curro, J.D. Thompson, S.M. Stishov, Superconductivity in diamond, Nature 428 (2004) 542–545.

[8] Y. Takano, M. Nagao, I. Sakaguchi, M. Tachiki, T. Hatano, K. Kobayashi, H. Umezawa, H. Kawarada, Superconductivity in diamond thin films well above liquid helium temperature, Appl. Phys. Lett. 85 (2004) 2851–2853.

[9] Y. Takano, T. Takenouchi, S. Ishii, S. Ueda, T. Okutsu, I. Sakaguchi, H. Umezawa, H. Kawarada, M. Tachiki, Superconducting properties of homoepitaxial CVD diamond, Diam. Relat. Mater. 16 (4–7) (2007) 911–914.

[10] D.R. Kania, M.I. Landstrass, M.A. Plano, L.S. Pan, S. Han, Diamond radiation detectors, Diam. Relat. Mater. 2 (1993) 1012–1019.
[11] J.C. Angus, H.A. Will, W.S. Stanko, Growth of diamond seed crystals by vapor deposition, J. Appl. Phys. 39 (6) (1968) 2915–2921.
[12] W.G. Eversole, N.Y. Kenmore, Synthesis of Diamond, US Patent No.: 3030187 & 3030188, 1962.
[13] F. Nava, C. Canali, M. Artuso, E. Gatti, P.F. Manfredi, S.F. Kozlov, Transport properties of natural diamond as nuclear particle detector for a wide temperature range, IEEE Trans. Nucl. Sci. NS-26 (1) (1979) 308–315.
[14] J.T. Byrne, G.A. Rost, Alpha counter for the direct determination of plutonium in solution, Anal. Chem. 33 (6) (1961) 758–761.
[15] G. Amsel, Oxide stabilized surface barrier detectors, Rev. Sci. Instrum. 32 (1961) 1253.
[16] L. Cathey, W.J. Jenkins, The operation of solid-state alpha counters in chemical process streams, IRE Trans. Nucl. Sci. 9 (3) (1962) 193–199.
[17] S.F. Kozlov, E.A. Konorova, M.I. Krapivin, V.A. Nadein, V.G. Yudina, Using diamond detectors as immersed α-counters, Soviat At. Energy 40 (6) (1976) 574–575.
[18] S.F. Kozlov, Preparation and characteristics of natural diamond nuclear radiation detectors, IEEE Trans. Nucl. Sci. 22 (1975) 160–170.
[19] P. Bergonzo, F. Foulon, A. Brambilla, D. Tromson, C. Jany, S. Haan, Corrosion hard CVD diamond alpha particle detectors for nuclear liquid source monitoring, Diam. Relat. Mater. 9 (2000) 1003–1007.
[20] P. Bergonzo, A. Brambilla, D. Tromson, C. Mer, B. Guizard, R.D. Marshall, F. Foulon, CVD diamond for nuclear detection applications, Nucl. Instrum. Methods Phys. Res. A 476 (2002) 694–700.
[21] J. Nuwad, D. Jain, N. Manoj, S. Panja, P.S. Dhami, V. Sudarsan, Development of diamond thin film-based alpha particle detectors for online assay of plutonium content in corrosive liquid medium, in: Proceedings of the DAE-BRNS Fifth Interdisciplinary Symposium on Materials Chemistry ISMC-2014, 2014.
[22] F. Bachmair on behalf of the RD42-Collaboration, Diamond sensors for future high energy experiments, Nucl. Instrum. Methods Phys. Res. A (2016). http://dx.doi.org/10.1016/j.nima.2016.03.039i.
[23] X. Xie, X. Yuan, X. Zhang, Z. Chen, X. Peng, T. Du, T. Li, Z. Hu, Z. Cui, J. Chen, X. Li, G. Zhang, T. Fan, G. Yuan, J. Yang, Q. Yang, Application of a single crystal chemical vapor deposition diamond detector for deuteron plasma neutron measurement, Nucl. Instrum. Methods Phys. Res. A 761 (2014) 28–33.
[24] E. Berdermann, M. Pomorski, W. de Boer, M. Ciobanu, S. Dunst, C. Grah, M. Kiš, W. Koenig, W. Lange, W. Lohmann, R. Lovrinčić, P. Moritz, J. Morse, S. Mueller, A. Pucci, M. Schreck, S. Rahman, M. Trager, Diamond detectors for hadron physics research, Diam. Relat. Mater. 19 (2010) 358–367.
[25] M. Artuso, et al., A 3D diamond detector for particle tracking, Nucl. Instrum. Methods Phys. Res. A 824 (2016) 402–405.
[26] A. Balducci, M. Marinelli, E. Milani, M.E. Morgada, A. Tucciarone, G. Verona-Rinati, M. Angelone, M. Pillon, Extreme ultraviolet single-crystal diamond detectors by chemical vapor deposition, Appl. Phys. Lett. 86 (2005) 193509-1–193509-3.
[27] N. Ade, T.L. Nam, T.E. Derry, S.H. Mhlang, The dose rate dependence of synthetic diamond detectors in the relative dosimetry of high-energy electron therapy beams, Radiat. Phys. Chem. 98 (2014) 155–162.
[28] M. Marinelli, F. Pompili, G. Prestopino, C. Verona, G. Verona-Rinati, G.A.P. Cirrone, G. Cuttone, R.M. LaRosa, L. Raffaele, F. Romano, C. Tuvè, Dosimetric characterization of a synthetic single crystal diamond detector in a clinical 62 MeV ocular therapy proton beam, Nucl. Instrum. Methods Phys. Res. A 767 (2014) 310–317.
[29] http://2atechnologies.com.
[30] http://www.e6.com.
[31] http://www.diamond-materials.com.
[32] R.C.W., The action of α-radiation on diamond, Nature 74 (1906) 271.

[33] D.E. Wooldridge, A.J. Ahearn, J.A. Burton, Conductivity pulses induced in diamond by alpha-particles, Phys. Rev. 71 (12) (1947) 913.
[34] J.B. Hannay, Artificial diamonds, Nature 22 (1880) 255–257.
[35] F.P. Bundy, H.T. Hall, H.M. Strong, R.H. Wentorf, Man-made diamonds, Nature 176 (1955) 51–55.
[36] V.Y. Dolmatov, Detonation synthesis ultradispersed nanodiamonds: properties and applications, Rus. Chem. Rev. 70 (7) (2001) 607.
[37] F. Banhart, P.M. Ajayan, Carbon onions as nanoscopic pressure cells for diamond formation, Nature 382 (1996) 433–435.
[38] W.A. Yarbrough, R. Messier, Current issues and problems in the chemical vapor deposition of diamond, Science 247 (1990) 688–692.
[39] J. Nuwad, D. Jain, C.G.S. Pillai, V. Sudarsan, Chemical vapor deposited diamond thin films: a brief overview on synthesis and characterization, SMC Bull. (Soc. Mater. Chem.) 5 (1) (2014) 15–25.
[40] http://www.manmadediamondinfo.com.
[41] W. Crookes, On the action of radium emanations on diamond, Proc. R. Soc. Lond. 74 (1904) 47–49.
[42] S.C. Lind, D.C. Bardwell, The coloring of the diamond by radium radiation, J. Frankl. Inst. 196 (4) (1923) 521–528.
[43] A.J. Ahearn, Conductivity induced in diamond by alpha-particle bombardment and its variation among specimens, Phys. Rev. 73 (1948) 1113.
[44] F.C. Champion, Electrical counting properties of diamonds, Proc. Phys. Soc. Sect. B 65 (7) (1952) 465–472.
[45] R. Robertson, T.J. Fox, A.E. Martin, Two types of diamonds, Philos. Trans. R. Soc. Lond. A 453 232 (1934) 463–535.
[46] S.F. Kozlov, E. Belcarz, M. Hage-Ali, R. Stuck, P. Siffert, Diamond nuclear radiation detectors, Nucl. Instrum. Methods 117 (1974) 277–283.
[47] C. Canali, E. Gatti, S.F. Kozlov, P.F. Manfredi, C. Manfredotti, F. Nana, A. Quirini, Historical properties and performance of natural diamond nuclear radiation detectors, Nucl. Instrum. Methods 160 (1979) 73–77.
[48] P. Bergonzo, F. Foulon, R.D. Marshall, C. Jany, A. Brambilla, R.D. McKeag, R.B. Jackman, High Collection efficiency CVD diamond alpha detectors, IEEE Trans. (1998) 30–33.
[49] J. Nuwad, D. Jain, N. Manoj, V. Sudarsan, Unpublished results.
[50] J. Nuwad, D. Jain, N. Manoj, S. Panja, P.S. Dhami, V. Sudarsan, Unpublished results.
[51] I. Nasiekaa, V. Strelchuka, M. Boykoa, V. Voevodinb, A. Vierovkinb, A. Rybkab, V. Kutniyb, S. Dudnikb, V. Gritsinab, O. Opalevb, V. Strel'nitskij, Raman and photoluminescence characterization of diamond films for radiation detectors, Sens. Actuators A 223 (2015) 18–23.
[52] Y. Sato, H. Murakami, T. Shimaoka, M. Tsubota, J.H. Kaneko, Charge-collection efficiency and long-term stability of single-crystal CVD diamond detector under different carrier-drift conditions, Jpn. J. Appl. Phys. 55 (2016), 046401.1–046104.5.
[53] M. Rebai, A. Fazzi, C. Cazzaniga, G. Croci, M. Tardocchi, E. Perelli Cippo, C.D. Frost, D. Zaccagnino, V. Varoli, G. Gorini, Time-stability of a single-crystal diamond detector for fast neutron beam diagnostic under alpha and neutron irradiation, Diam. Relat. Mater. 61 (2016) 1–6.
[54] R.P. Taleyarkhan, J. Lapinskas, B. Archambault, J.A. Webster, T.F. Grimes, A. Hagen, K. Fisher, S. McDeavitt, W. Charlton, Real-time monitoring of actinides in chemical nuclear fuel reprocessing plants, Chem. Eng. Res. Des. 91 (2013) 688–702.
[55] M. Pomorski, C. Mer, J. de Sanoit, T. Quang Tran, P. Bergonzo, Diamond for actinide traces detection and spectrometry in liquids, IEEE Trans. (2013) 1–6, http://dx.doi.org/10.1109/ANIMMA.2013.6727986.
[56] Y. Sato, H. Murakami, T. Shimaoka, M. Tsubota, J.H. Kaneko, Single-crystal CVD diamond detector for high-resolution particle spectrometry, EPL 108 (2014) 42001, http://dx.doi.org/10.1209/0295-5075/108/42001.

[57] G.P. Freeman, H.A. Van Der Velden, Some aspects of the counting properties of diamond, Physica 17 (6) (1951) 565–572.
[58] K. Hecht, On the Mechanism of the photovoltaic primary stream in insulating crystals, Z. Phys. 77 (1932) 208–216 (Original article in German language).
[59] A. Mainwood, Recent developments of diamond detectors for particle and UV radiation, Semicond. Sci. Technol. 15 (2000) R55–R63 (and references therein).
[60] E.A. Konorova, S.F. Kozlov, Soc. Phys.Semi. Cond. 4 (1971) 1600.
[61] E.-K. Souw, R.J. Meilunas, Response of CVD diamond detectors to alpha radiations, Nucl. Instrum. Methods Phys. Res. Sect. A: Accel. Spectrom. Detect. Assoc. Equip. 400 (1) (1997) 69–86.

CHAPTER

SEVERE PLASTIC DEFORMATION OF MATERIALS 20

R. Kapoor
Bhabha Atomic Research Centre, Mumbai, India

CHAPTER OUTLINE

1. INTRODUCTION

Application of sufficient stress on metallic materials inevitably results in plastic deformation. As the capacity for plastic deformation of most materials is limited, attempts to severely deform materials usually lead to their failure. However, if during plastic deformation restoration mechanisms prevail (mechanisms where crystalline defect density rearranges or annihilates) the ability of a material to

Materials Under Extreme Conditions. http://dx.doi.org/10.1016/B978-0-12-801300-7.00020-6

plastically deform can be extended. The restoration mechanisms in metals are either dynamic recovery or dynamic recrystallization. These help to modify the microstructure by arranging line defects (dislocations) into surface defects (boundaries) thus ineffect reducing their spacing, and refining the microstructure. The ability of a single-phase material to undergo large plastic deformation is dependent on its crystal structure. Crystal structures that allow the activation of a large number of slip systems have the ability to undergo large plastic deformation, whereas those with limited slip systems have difficulty in deforming. In general, metallic materials with fcc and bcc crystal structures have sufficient slip systems and can be deformed to a large extent at ambient temperatures, whereas those with hcp structures have limited slip systems and require to be deformed at higher temperatures at which additional slip systems become active.

The refinement of grains through severe plastic deformation (SPD) is thought to take place by a variant of a recovery process referred to as continuous dynamic recrystallization (cDRX) or through a process of geometric dynamic recrystallization (gDRX). In cDRX the grains divide into cells and subgrains, which eventually develop into high-angle grain boundaries. In gDRX the large deformation strains produce serrated grain boundaries that meet and pinch off, resulting in finer grains. The descriptions of different dynamic recrystallization processes are given in Ref. [1,2]. As the microstructure of a material changes (defect density, grain size, etc.) so do its mechanical properties. It is well established through the Hall–Petch relation [3,4] that the strength varies with $d^{-0.5}$ where d is the grain size. Thus an important technique of achieving high strength for a given alloy system is to refine its grain size. One such technique of refining grain size is to subject the material to a thermomechanical treatment where the process of dynamic recrystallization would result in fine grains [5]. However, grain sizes obtained by this process are usually in the 1–10 μm range and are still not sufficient to increase the strength substantially. A more recent technique of refining the grains to submicron levels is by subjecting the material to severe plastic deformation (when effective plastic strains exceed about 2). This is a top down approach to producing ultrafine or nanograined materials where the initial coarse-grained (>100 μm) microstructure is refined through a series of plastic deformation steps to ultrafine-grained (UFG) microstructure, where the average grain size is less than 1 μm.

Severe plastic deformation is achieved when a material can be repeatedly plastically deformed without a net change of shape, i.e., the overall shape at the beginning of the deformation is the same as that at the end. The process is repeated and the plastic strain in effect gets added with each cycle resulting in a large achieved strain. Materials that have undergone severe plastic deformation exhibit interesting and unique properties not usually observed in conventional coarse-grained (CG) and mildly deformed materials as will be seen in this chapter. There are several techniques of achieving SPD, the most popular being that of equal channel angular pressing (ECAP) pioneered by Segal and his group [6–9] and further developed by Valiev [10–12]. A general review of severe plastic deformation is given in Ref. [13] and a review more specifically on equal channel angular extrusion in Ref. [14]. Section 2 of this chapter discusses the theory and phenomena of severe plastic deformation, where the ability of a material to undergo SPD is discussed and the mechanism of microstructural refinement is explained. Section 3 discusses the various methods by which severe plastic deformation can be imparted to materials. Section 4 discusses the characterization of the microstructure after SPD and its subsequent mechanical properties. Lastly, Section 5 discusses current and future applications of materials processed by severe plastic deformation.

2. THEORY AND PHENOMENA OF SEVERE PLASTIC DEFORMATION

2.1 QUANTIFYING PLASTIC DEFORMATION

Plastic deformation in metallic materials takes place through the movement of defects, either point defects or line defects depending on the temperature, stress, and rate of deformation. At low to warm temperatures (less than half its melting point on absolute scale)—the temperature range where SPD is carried out—it is primarily the movement of dislocations on specific slip planes that result in shear strains. The movement of dislocations in the reverse direction adds up in terms of effective plastic strain, ε_{eff}, determined as [15],

$$\varepsilon_{eff} = \left\{\frac{2}{9}\left[(\varepsilon_{11} - \varepsilon_{22})^2 + (\varepsilon_{22} - \varepsilon_{33})^2 + (\varepsilon_{33} - \varepsilon_{11})^2\right] + \frac{1}{3}\left(\gamma_{12}^2 + \gamma_{23}^2 + \gamma_{31}^2\right)\right\}^{0.5} \tag{20.1}$$

Where ε_{ij} and γ_{ij} are the normal and shear components of the strain tensor. Using Eq. (20.1) the effective strain for any plastic deformation process can be determined. Table 20.1 gives a few examples of the effective strains achieved during uniaxial compression and plane strain compression. True strain values are considered while calculating the effective strain. It is seen that for a 90% reduction in height during rolling the effective strain is 2.66. In terms of effective strain achieved, for the same height reduction plane strain compression is more effective as compared to uniaxial compression. However it is seen that even a 90% reduction during rolling leads to ε_{eff} of 2.66 and thus it would be difficult to achieve higher plastic strains using this process. As will be discussed later, during SPD the material commonly experiences effective strains of about 8–10, and in some cases up to 20, and most importantly these are achieved without a net change of shape.

2.2 MECHANISM OF MICROSTRUCTURAL REFINEMENT DURING SPD

At low homologous temperatures metallic material undergoes plastic deformation by the movement and interaction of dislocations that with increasing strain form a substructure of configuration of dislocations arranged in cells. The walls of the cells contain dislocations whereas the interior of the cell

Table 20.1 Comparison of Effective Strain for Different Percentages of Uniaxial and Plane Strain Compression

Process	Extent of Compression	Components of Strain Tensor	ε_{eff}
Uniaxial compression	50% reduction in height	$\varepsilon_{11} = \ln(0.5)$, $\varepsilon_{22} = \varepsilon_{33} = -0.5\varepsilon_{11}$, $\gamma_{ij} = 0$	0.69
	90% reduction in height	$\varepsilon_{11} = \ln(0.1)$, $\varepsilon_{22} = \varepsilon_{33} = -0.5\varepsilon_{11}$, $\gamma_{ij} = 0$	2.30
Plane strain compression (rolling)	50% reduction in height	$\varepsilon_{11} = \ln(0.5)$, $\varepsilon_{22} = 0$, $\varepsilon_{33} = -\varepsilon_{11}$, $\gamma_{ij} = 0$	0.80
	90% reduction in height	$\varepsilon_{11} = \ln(0.1)$, $\varepsilon_{22} = 0$, $\varepsilon_{33} = -\varepsilon_{11}$, $\gamma_{ij} = 0$	2.66

has low dislocation density. This mechanism of cell formation at intermediate strains is a well-known phenomenon [16,17], the classical understanding being that with increasing strain the cells transform to subgrains and eventually to high-angle boundaries. The evolution of microstructure with increasing effective strain has been seen by many researchers [18–26]. One of the earlier descriptions of this process suggested that severe plastic deformation (carried out in Mg) results in a fragmented microstructure with highly nonequilibrium grain boundaries [10,27]. Upon subsequent heating to warm temperatures the dislocations in the boundary recover and convert the nonequilibrium boundaries to equilibrium high-angle grain boundaries. Iwahashi et al. [18] explained the formation of ultrafine grains in pure Al due to intersecting deformation bands during ECAP. The first grains that pass through ECAP die (shown in Section 3.1), which resulted in elongated grains with a high dislocation density within those grains. The second pass created deformation bands, which intersect the earlier elongated grains and fragment them into smaller subgrains. With each subsequent pass, the grains further fragment and reach an equilibrium size based on the force response of the material.

One recent study by Kawasaki et al. showed the evolution of microstructure with increasing severity of deformation through an ECAP die [26]. Here annealed pure Al with large grains of about 200 μm (Fig. 20.1a) was subjected to SPD through ECAP. After the first pass (Fig. 20.1b) elongated and fragmented grains were observed with a high fraction of low-angle boundaries (boundaries with misorientation angle $\varphi < 15$ degrees). After two passes (Fig. 20.1c), subgrains were distinctly visible, however, they were grouped together in similar orientations (represented by colors according to stereographic projection) suggesting that they are separated by low-angle boundaries. With increasing number of passes (8 passes Fig. 20.1c) the grains become well defined and equiaxed and share boundaries with grains of different orientations (represented here as colors as depicted in stereographic triangle) suggesting an increased fraction of high-angle boundaries. In most cases it is observed that the maximum refinement is done in the first few deformation steps after which the size of grains (or subgrains) tends to saturate as seen in Fig. 20.2 for Al and its alloys. By an effective strain of four, the final grain size has been nearly achieved.

The evolution of the microstructure during SPD can be traced back to the inverse relation between the steady-state cell or subgrain size d achieved and the square root of total dislocation density ρ of the material.

$$d = \frac{K}{\sqrt{\rho}} \tag{20.2}$$

Further, using the relation between stress σ and the dislocation density ρ: $\sigma = M\alpha\mu b\sqrt{\rho}$, where M is the Taylor factor, α is a constant of value around 0.5, and μ is the shear modulus, Eq. (20.2) can be written as,

$$d = \frac{KM\alpha\mu b}{\sigma} \tag{20.3}$$

Eq. (20.3) states that higher the stress response of the material, the smaller will be the evolved grain size. Thus using Eq. (20.3) an estimate of the grain size can be made. For $K \approx 10$, $\alpha = 0.5$, and $M = 3$, and taking the maximum available stress as 10% of shear modulus ($\sigma/\mu \approx 0.1$), the cell size obtained $d = 150b \approx 500$ nm. This is about the typical value of grain size achieved during severe plastic deformation. During deformation, which is a thermally activated process, a higher strain rate and lower

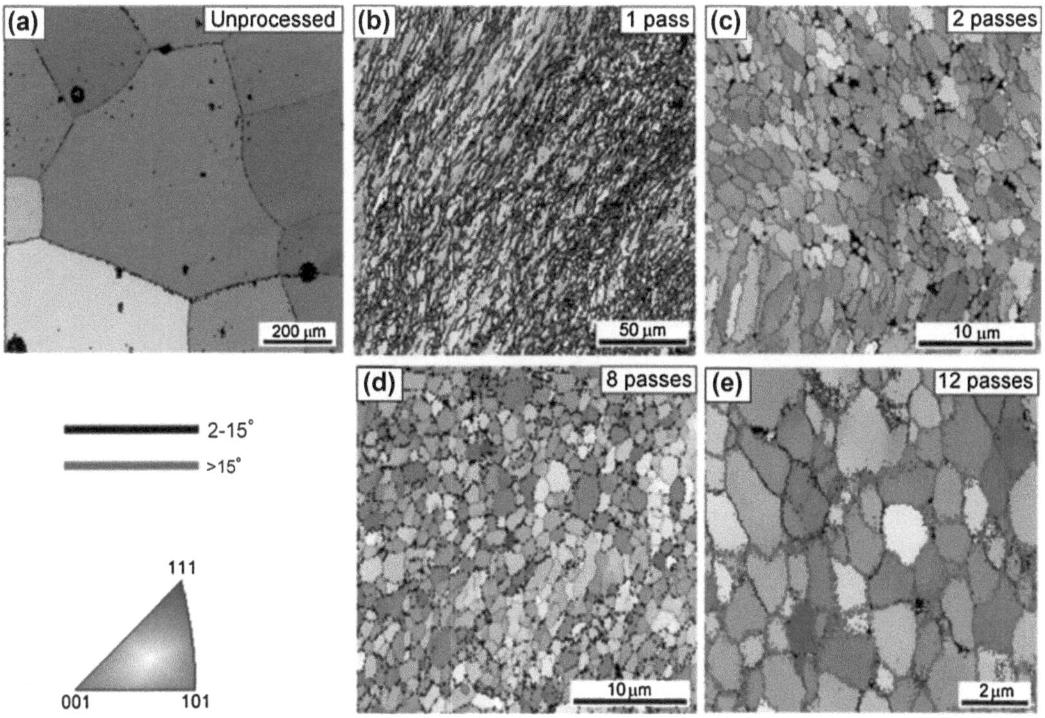

FIGURE 20.1

Electron backscatter diffraction (EBSD) map for high-purity aluminum in (a) the initial unpressed condition and after equal channel angular pressing (ECAP) at room temperature through (b) 1, (c) 2, (d) 8, (e) 12 passes using route B_C. *Black lines* represent boundaries with 2 degrees $< \varphi <$ 15 degrees and *red lines* (gray in print version) as $\varphi >$ 15 degrees. The grain colors correspond to the orientations in the unit triangle.

Adapted from M. Kawasaki, Z. Horita, T.G. Langdon, Mater. Sci. Eng. A 524 (2009) 143–150 with permission from publisher.

temperature will increase the stress and thus result in a smaller grain size. This "trick" of deforming materials dynamically and in the cryotemperature range has been used by researchers [28–34] to refine the grain size below that obtainable by deformation at room temperature and nominal strain rates.

Another proposed mechanism of the breakup of large grains to smaller grains during SPD is through the formation of microshear bands across the sample. This was shown in a multiaxially pressed ferritic steel [35] and AA7075 [36]. Here initial pressings created microshear bands within grains. Repeating the deformation in another direction produced intersecting microshear bands. New grains with high-angle boundaries were seen to form near these intersections.

In many studies it is not usually specified whether the term "grain size" implies only high-angle boundaries or all boundaries. High-angle boundaries are defined as those boundaries whose misorientation angle (φ) across the boundary are greater than or equal to 15 degrees, i.e., $\varphi \geq$ 15 degrees. During SPD, the grain size considering all boundaries (i.e., $\varphi >$ 2 degrees) saturates at low values of effective strain, whereas the fraction of high-angle boundaries ($\varphi >$ 15 degrees) increases with increasing effective strain. This has been brought out well in Fig. 20.3. Different studies have shown

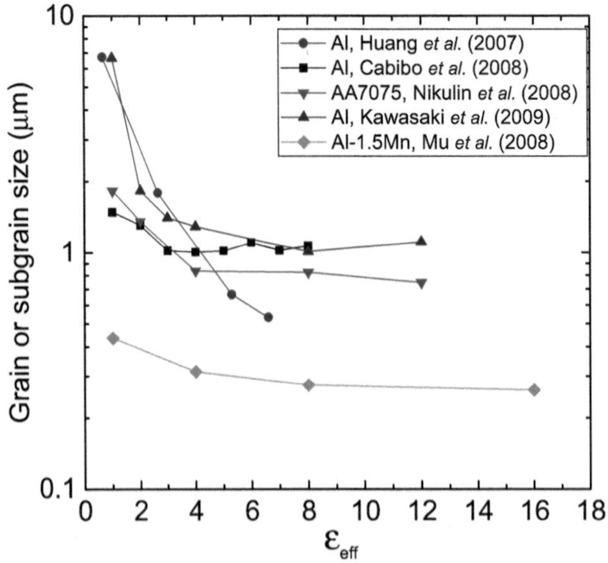

FIGURE 20.2

Grain (or subgrain) size as function of the effective plastic strain for Al alloys.

Data replotted from Y. Huang, P.B. Prangnell, Scr. Mater. 56 (2007) 333–336; M. Cabibbo, W. Blum, E. Evangelista, M.E. Kassner, M.A. Meyers, Metall. Mater. Trans. A 39 (2008) 181–189; I. Nikulin, Y. Motohashi, R. Kaibyshev, Mater. Sci. Forum 584–586 (2008) 691–696; M. Kawasaki, Z. Horita, T.G. Langdon, Mater. Sci. Eng. A 524 (2009) 143–150; S. Mu, W. Hua, G. Gottstein, Mater. Sci. Forum 584–586 (2008) 697–702.

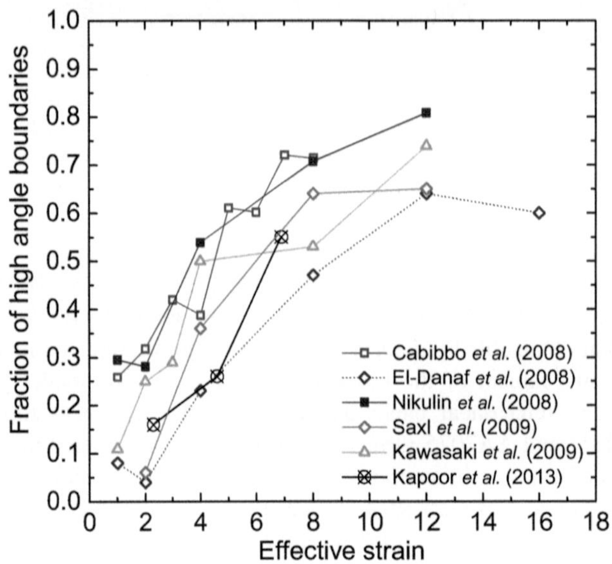

FIGURE 20.3

Plot of fraction of high-angle grain boundaries versus effective strain for different studies of severe plastic deformation. The fraction of high-angle boundaries appears to saturate at about 70–80%.

Data replotted from M. Cabibbo, W. Blum, E. Evangelista, M.E. Kassner, M.A. Meyers, Metall. Mater. Trans. A 39 (2008) 181–189; E.A. El-Danaf, Mater. Sci. Eng. A 487 (2008) 189–200; I. Nikulin, Y. Motohashi, R. Kaibyshev, Mater. Sci. Forum 584–586 (2008) 691–696; I. Saxl, A. Kalousová, L. Ilucová, V. Sklenicka, Mater. Charact. 60 (2009) 1163–1167; M. Kawasaki, Z. Horita, T.G. Langdon, Mater. Sci. Eng. A 524 (2009) 143–150; R. Kapoor, A. Sarkar, R. Yogi, S.K. Shekhawat, I. Samajdar, J.K. Chakravartty, Mater. Sci. Eng. A 560 (2013) 404–412.

that the fraction of high-angle boundaries f_{hab} increases with effective strain. Most studies show that after an effective strain of about 8–10 the fraction of high-angle boundaries tends to saturate between 70% and 80%. Thus after severe plastic deformation there still exists about 20–30% of low-angle boundaries.

3. METHODS TO IMPART SEVERE PLASTIC DEFORMATION

It is a challenge to deform metallic materials to very large plastic strains without introducing volumetric defects such as cracks. Normal working processes such as forging, rolling, extrusion, etc. are usually carried out at warm to hot temperatures if large strains are to be imparted. Imparting large strains uniaxially at ambient temperatures will certainly introduce some defects. Thus severe plastic deformation is imparted in manner such that compressive hydrostatic stresses are also present along with the shear stresses required for plastic deformation. The hydrostatic stresses prevent the sample from cracking, while the shear stresses impart the required plastic strains. In addition, the process should be such that at the end of any "pass" (or the beginning of the next pass) the overall shape remains the same. This ensures that the process can be repeated again, thereby accumulating further plastic strain. Some of the common and popular methods to impart severe plastic deformation are described next.

3.1 EQUAL CHANNEL ANGULAR PRESSING

ECAP is the most popular and widely studied SPD technique. Fig. 20.4 schematically describes the ECAP process. As the name suggests, a sample with square (or round) cross-section is pressed into a

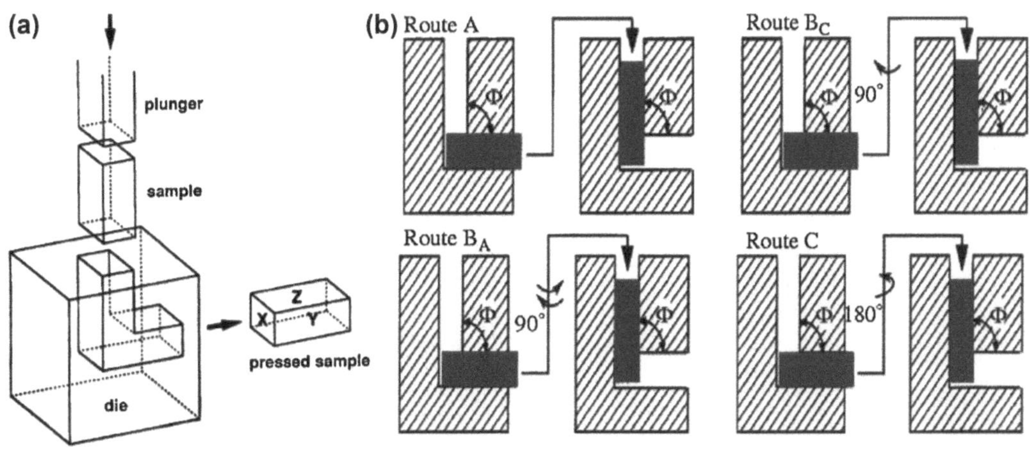

FIGURE 20.4

(a) Schematic of the equal channel angular pressing (ECAP) process showing the sample on top before insertion into die and the pressed sample horizontally exiting from the die [37]. (b) Different techniques in which the sample after exit can be reentered into the die. Route A entry without any change in orientation, route B rotation by 90 degrees, and route C rotation by 180 degrees [38].

(a) Reproduced with permission from publisher.

die with two intersecting channels of equal cross-section area. As the sample is pressed using a plunger, it shears at the intersection and exits through the second channel. All of the material passes through the intersecting section and experiences shear strain. As the cross-section is symmetric (i.e., either square or circular) the sample can be rotated before reentry into the die. If the sample is reintroduced without rotation the route is called route-A, if it is rotated by 180 degrees it is called route-C, if it is rotated by 90 degrees in the same direction every pass it is called route B_C, and if it is rotated by 90 degrees but alternatively in different sense it is referred to as route B_A (Fig. 20.4b). Passing sample through an ECAP die produces a simple shear deformation as shown schematically in Fig. 20.5. Here an element *abcd* gets transformed to *a'b'c'd'*. From geometry, Iwahashi et al. [39] calculated the shear strain experienced by such an element as,

$$\varepsilon_{\text{eff}} = \frac{1}{\sqrt{3}}\left[2\cot\left(\frac{\Phi+\Psi}{2}\right) + \Psi\operatorname{cosec}\left(\frac{\Phi+\Psi}{2}\right)\right] \tag{20.4}$$

where Φ and Ψ are described in Fig. 20.5. It is seen that for $\Phi = 90$ degrees and $\Psi = 0$ degree $\varepsilon_{\text{eff}} = 2/\sqrt{3} = 1.15$. For more common angle of $\Phi = 90$ degrees and $\Psi = 20$ degrees $\varepsilon_{\text{eff}} \approx 1$. Thus each pass through a 90 degrees ECAP die will result in an effective strain of 1 and *N* such passes will result in an effective strain of *N*. In the design of the die it should be kept in mind that the channels should be "very smooth" so as to reduce friction between die and sample. It is also essential that the exit channel should be very slightly smaller (within tolerances) than the entrance channel for easy reentry for the next pass.

There is a large amount of literature on use of ECAP to impart severe plastic deformation and it is not possible to exhaustively list out all references here, however, there are review articles that contain exhaustive lists, including for example Refs. [13,14,40,41].

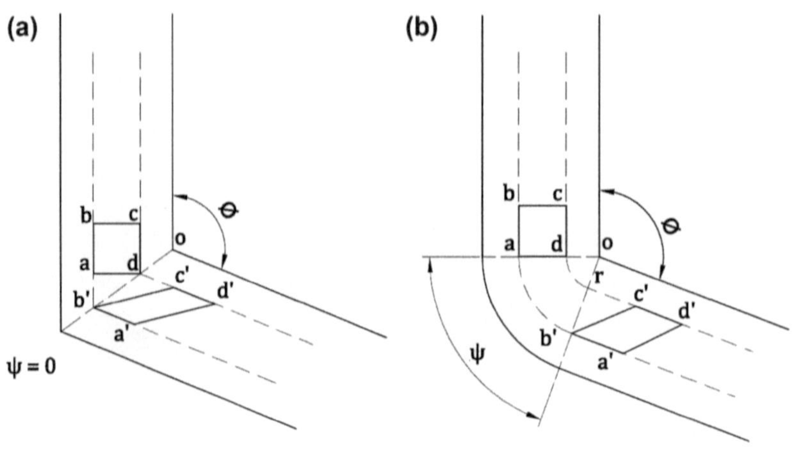

FIGURE 20.5

Schematic of the shear strain experienced by a material element as it passes through the channels of an ECAP die. Φ is the angle the channels meet at, Ψ is the angle of curvature at the outer point of contact of the channels. (a) For $\Psi = 0$, (b) for $\Psi > 0$ [39].

Reproduced with permission from publisher.

3.2 HIGH PRESSURE TORSION

High pressure torsion (HPT) is a technique where thin disks of a material are subjected to simultaneous torsion and high pressure as shown in Fig. 20.6 [42]. This is a variant of the technique first used by Bridgman [43]. The dimension of the sample is usually about 20 mm in diameter and about 0.2–0.5 mm in thickness. A pressure in the GPa range (around 5 GPa) is applied vertically to press down the sample on the support. This setup constrains the sample on all sides. The plunger is then rotated, thereby applying a torsional strain on the sample. The frictional force between the plunger and the sample, and the sample and the anvil, helps shear the material by the rotating plunger. For N rotations the shear strain γ achieved is,

$$\gamma = \frac{2\pi r N}{l}, \tag{20.5}$$

where r is the radius and l is the thickness of the disk. The effective strain using Eq. (20.1) is $\varepsilon_{\text{eff}} = \gamma/\sqrt{3}$. From Eq. (20.5) it is seen that the shear strain increases with increasing radius, thus making the deformation inhomogeneous. However, it has been observed that with increasing number of rotations the microstructure does tend to homogenize, as inferred from microhardness measurements across the disk and its correlation with microstructure. Ref. [44] is a comprehensive review of the HPT process. It is routine for HPT to result in grains in the sub-100-nm range and with a significant fraction of high-angle boundaries.

3.3 MULTIAXIAL FORGING

Multiaxial forging (MAF) is a technique in which the sample is compressed sequentially along three orthogonal axes, by 90 degrees rotation after each pressing (or forging). This technique was developed

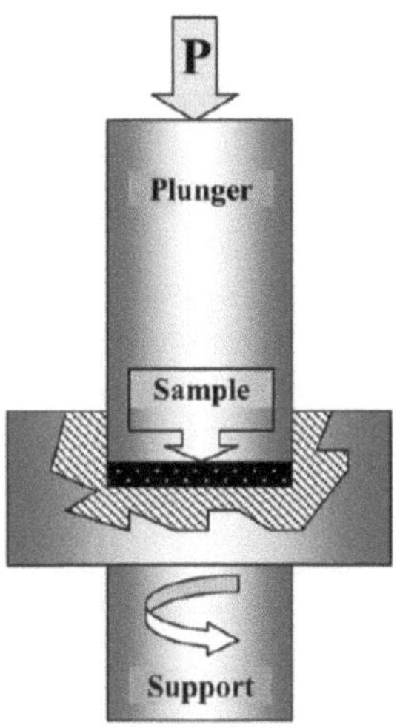

FIGURE 20.6

Schematic diagram of the high-pressure torsion setup [42].

Reproduced with permission from publisher.

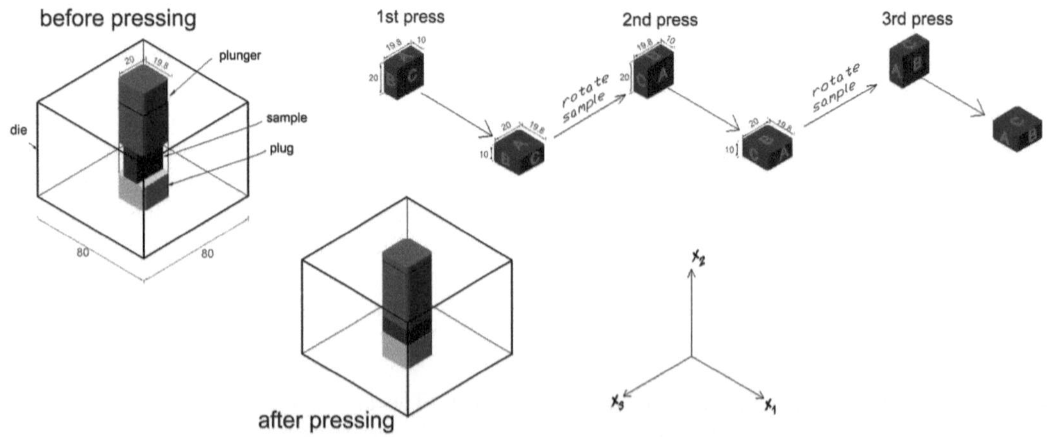

FIGURE 20.7

Schematic of a confined channel die showing the pressing of sample along three orthogonal axes. This is an example where the sample is reduced to 50% of its height in each pass ($\varepsilon_{eff} = 0.78$) [47].

Reproduced with permission from publisher.

and patented in the late 1980s [45] and has been further developed and used [46–51]. A simple variation of this technique is to uniaxially compress the sample repeatedly on three orthogonal axes [35,52–54]. However, this repeated uniaxial compression would require the sample to be ground after each pass to produce flat faces in order to compress it again. This process of grinding can be avoided if a channel die is used, either as open die [49] or confined as seen in Fig. 20.7 [47]. In the example shown here the die is so designed that if the sample is compressed by 50% it can be rotated and reinserted into the die and pressed again, repeating the process for all three orthogonal directions. The channel die provides a plane strain compression to the sample, which for the same reduction in height produces a higher effective strain compared to uniaxial compression (from Eq. 20.1). Fig. 20.7 shows the sequence of pressing. For 50% reduction in height $\varepsilon_{eff} = 0.78$. This process is repeated in all three orthogonal directions. The open channel die has a disadvantage that one pair of faces bulges at the end of a cycle leading to nonuniform deformation and early cracking. The confined channel die ensures that after each press the sample takes the shape of the die, and the face that would have bulged now becomes flat, ensuring more uniform deformation.

3.4 ACCUMULATIVE ROLL BONDING

Here the sample is rolled to half the initial thickness, which makes its length double. The sample is then cut in half, the surfaces cleaned, and the halves stacked together and rolled again to the same reduction (Fig. 20.8). This process was pioneered by Saito et al. [55] and further developed and used primarily for Al alloys. The rolling is usually carried out at warm conditions to ensure bonding the cut surfaces. Before stacking the sheets together, the surfaces are degreased and cleaned with wire brush to remove the oxide layer to facilitate the joining of the sheets during subsequent rolling.

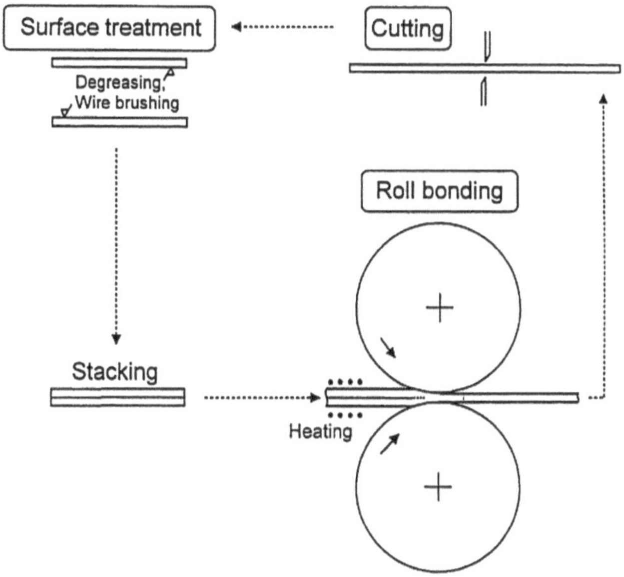

FIGURE 20.8

Schematic of the accumulative roll bonding process [55].

Reproduced with permission from publisher.

3.5 OTHER SPD PROCESSES

Other less common severe plastic deformation processing techniques exist. Some of these are twist extrusion, cyclic extrusion compression, repetitive corrugation and straightening, parallel channel ECAP, rotary die ECAP, and integrated extrusion and ECAP (Ref. [40] gives brief summaries of these). These severe plastic deformation techniques are all batch processes like the earlier processes of ECAP, HPT, MAF, and ARB discussed in previous sections. Some continuous SPD processes are discussed in the next section.

3.6 CONTINUOUS SPD PROCESSES

All of the SPD processes discussed before are discontinuous processes where specific short-dimension specimens are used. If the SPD technique is to be made attractive to industries, it should be a continuous rather than a batch process. Table 20.2 lists out and briefly describes SPD techniques in which sheets and rods can be continuously passed through a die.

4. CHARACTERIZATION AND PROPERTIES

4.1 MICROSTRUCTURE

It was seen in an earlier section that severe plastic deformation produces an ultrafine-grained microstructure. The primary mechanism of this refinement in grain size is thought to be through the process of continuous dynamic recrystallization, which is in essence a recovery-based process [53]. In this process with increasing strain the grain boundaries change from being low angle to high angle. Almost all studies on SPD characterize the subsequent microstructure. This list is exhaustive and is not

Table 20.2 Continuous SPD Techniques

Process	Schematic Diagram
ECAP conform [56] The workpiece is fed into a groove on a rotating shaft. Frictional forces on the contact surface of the groove drive the workpiece forward. A stationary constraint die forces the workpiece to bend 90 degrees with a shearing action like an ECAP die.	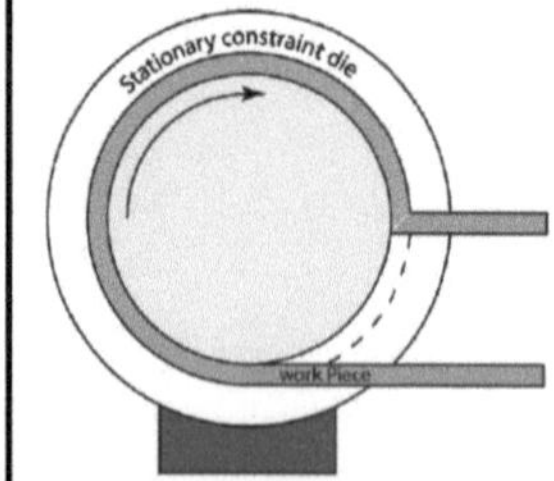
Continuous frictional angular extrusion [20] Sheets pass through the frictional roles and enter an ECAP-like die to exit at 90 degrees. (1) Driving roll; (2) Sheetwork piece; (3) Support block; (4) Die assembly; (5) Entrance channel; (6) Exit channel.	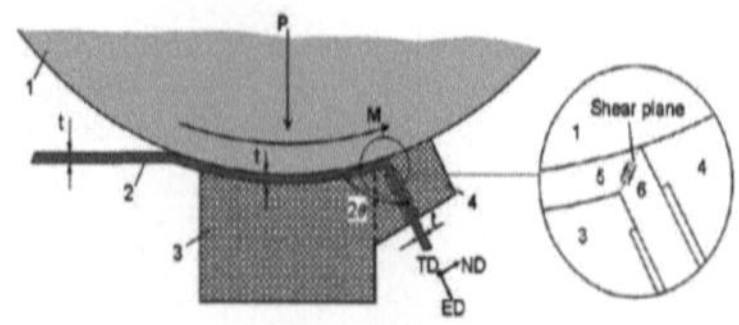
Conshearing[57] The material is passed over a large central roll guided by smaller satellite rolls. The strip is extruded through an ECAP-like die abutting the large roll.	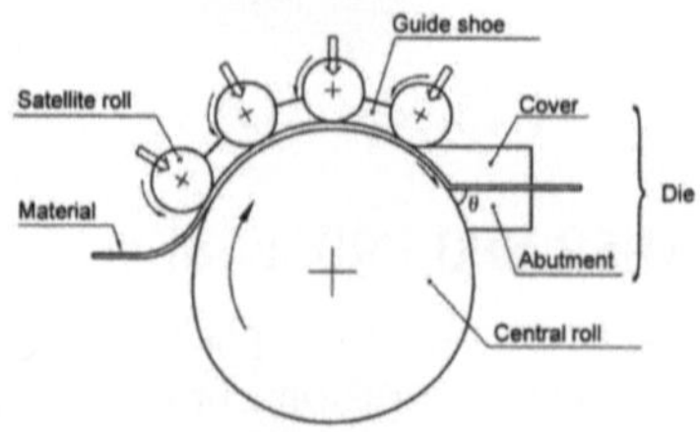
Constrained strip shearing [58] Twin rolls with nulls feed the workpiece into an ECAP-like die with channel intersection angle of 120 degrees.	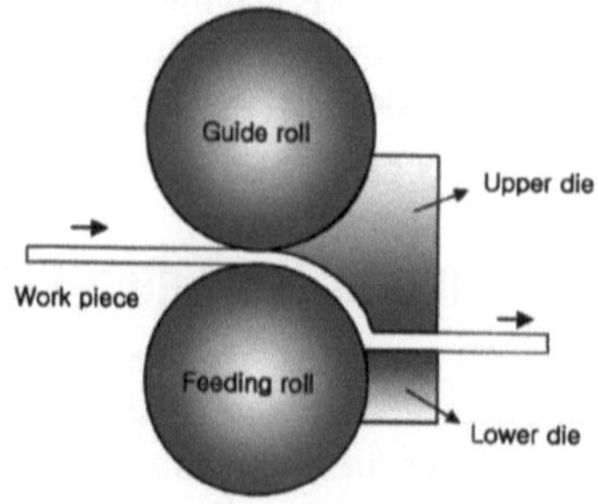

All figures are reproduced with permission from corresponding publisher.

provided here, however, a few salient features of the microstructure after SPD are brought out. The first is that after most SPD processes (ECAP and its variants, multiaxial forging, accumulative roll bonding, etc.) the length scale of microstructure is in the submicron level typically ranging from 0.2 to 1 μm. In one severe plastic deformation method, the high-pressure torsion technique, the grains can be refined to below 100 nm, making the microstructure truly nanosized [11]. There are numerous studies showing the microstructure of SPD materials, and shown here are few examples of the microstructure of fcc, bcc, and hcp materials after SPD. Fig. 20.9 shows the microstructure of pure Al after pressing through

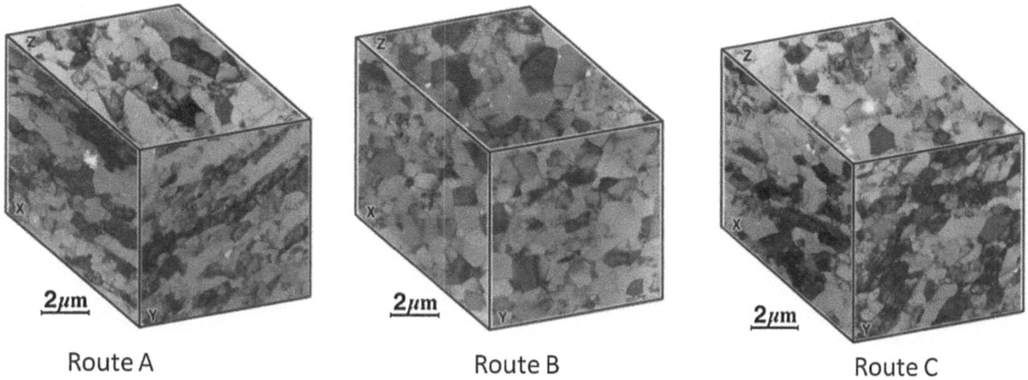

FIGURE 20.9

Microstructure of pure Al after 4 passes through an ECAP die for different routes [59]. The faces *x*, *y*, and *z* are according to that shown in Fig.20.4.

Reproduced with permission from publisher.

different routes in an ECAP die [59]. The resulting grain size varied from about 0.5 to 1.5 μm depending on the route and the face observed. Route B showed a microstructure with equiaxed grains on all three faces, unlike route A where significant elongated grains were observed. From these and other earlier works it was established that route B (in particular B_C) was the most desirable in producing equiaxed grains; this is followed by route C, which shows elongated grains on one face and finally route A, which shows elongated grains on two faces [59,60].

Fig. 20.10 shows some examples of transmission electron microscope (TEM) images and electron backscatter diffraction (EBSD) maps of Al and Al-Mg alloy [47,61,62]. ECAP of Al-1.5Mg after effective strain of 12 showed an ultrafine-grained microstructure (Fig. 20.10a). In Fig. 20.10b and c the colors on the EBSD plot represent the orientation of the normal plane with respect to stereographic triangle shown in the figure. In comparison to TEM images, EBSD maps can better distinguish between subgrains and grains and provide better statistics in terms of grain-size estimation. Fig. 20.10d–f are the corresponding distribution of boundary spacings of all boundaries with misorientation greater than 2 degrees. It is seen that the boundary spacings range from 0.2 to 10 μm and with alloying the spacings get shifted to smaller values also seen in Refs. [18,63]. There are many examples of other fcc metals in which the microstructure has been refined to ultrafine subgrains/grains after severe plastic deformation, e.g., Cu [21,29,64] and Ni [42], some of which are shown in Fig. 20.11. HPT is known to produce the finest microstructure as can be seen in Ni after undergoing HPT. It shows grains as low as 100 nm and averaging about 200 nm. Cu severely deformed under cryoconditions not only produces UFG microstructure but also results in twins [29,65]. Fig. 20.11c shows twins of nanowidth in Cu dynamically deformed at cryotemperatures. The widths of the twins were not observed to change with increasing strain. In addition, this cryodeformed Cu showed high-dislocation activity near the boundaries of the deformation twins.

Severe plastic deformation of bcc materials also results in ultrafine grains; one example is shown in Fig. 20.12 of ECAP of Nb [66,67]. The TEM image shows subgrains/grains of about 200 nm size with the grain size distribution as computed from EBSD scans showing most of the distribution falling

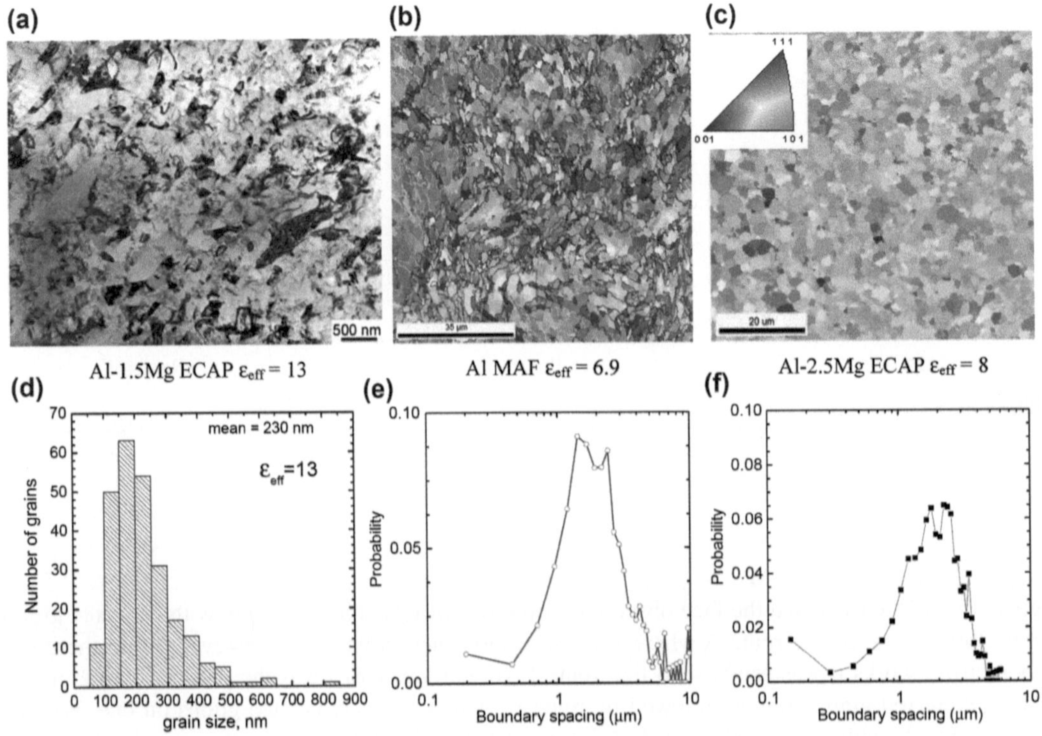

FIGURE 20.10

(a) Transmission electron microscope (TEM) image of Al-1.5Mg after ECAP at room temperature to $\varepsilon_{\mathrm{eff}} = 12$ [61], EBSD inverse pole figure maps of (b) Al after multiaxial forging to $\varepsilon_{\mathrm{eff}} = 6.9$ [47], (c) AA5052 after ECAP to $\varepsilon_{\mathrm{eff}} = 8$ [62], and (d)–(f) the corresponding distribution of boundary spacings of (a)–(c).

(a)–(c) reproduced with permission from publisher.

FIGURE 20.11

TEM images of (a) Cu after ECAP of $\varepsilon_{\mathrm{eff}} = 8$ [21], (b) Ni after HPT at 9 GPa and 5 turns [42], (c) Cu dynamically deformed at cryotemperatures revealing twins of nanothickness [29].

Reproduced with permission from publishers.

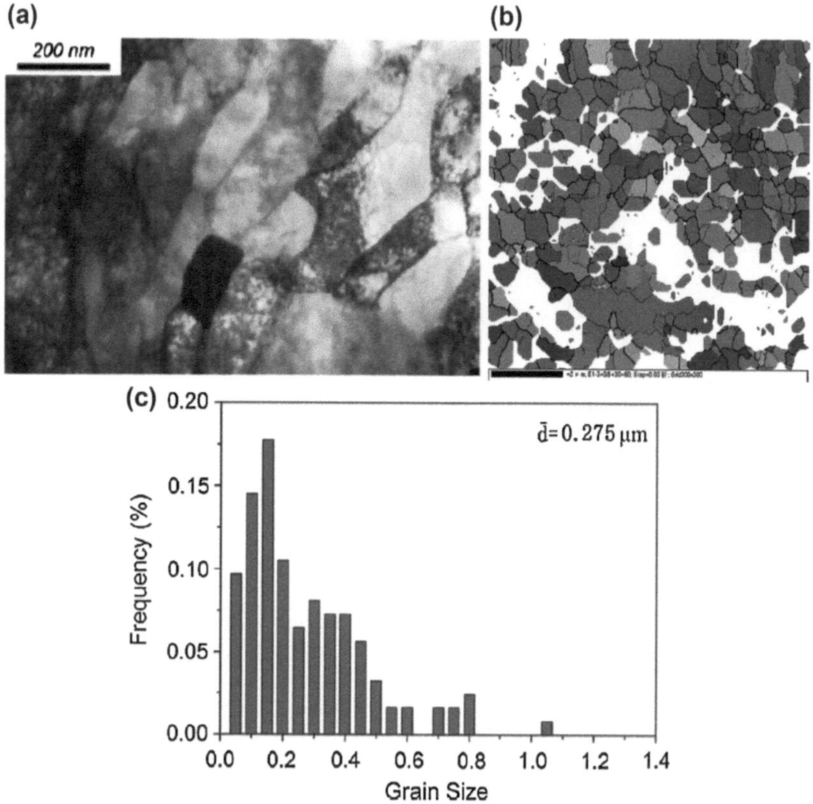

FIGURE 20.12

(a) TEM image of Nb after ECAP to $\varepsilon_{eff} = 18$ showing subgrains with high dislocation density within [66], (b) EBSD map of Nb after ECAP to $\varepsilon_{eff} = 26$ [67], and (c) the grain size distribution [67].

Reproduced with permission from publishers.

below 0.8 μm. HCP materials too can be subjected to severe plastic deformation, however, their deformation at cold temperatures is difficult and they need to be deformed at warm temperatures to sustain severe deformation, e.g., SPD in Mg is carried out at around 200−300°C, Ti at around 400−450 °C [68−70]. The grain sizes of hcp materials after SPD are also slightly larger having a mean of about 1−3 μm. Two examples of the microstructure of Mg [69] and Zr [70] after severe plastic deformation are shown in Fig. 20.13.

As the grain boundaries produced during SPD are through the process of continuous DRX (evolution of cell boundaries to subgrain boundaries to grain boundaries) it is expected that the interior of subgrains and grains would have a high density of dislocations. This can be seen from one such example of Al-1.5Mg after ECAP at room temperature as shown in Fig. 20.14a. Further, in this process the boundaries themselves retain dislocations making the boundary "nonequilibrium" as seen in Fig. 20.14b [71]. High-resolution microscopy of severely plastically deformed materials shows the existence of dislocations in the close vicinity to boundaries [72,73]. These dislocations would interact

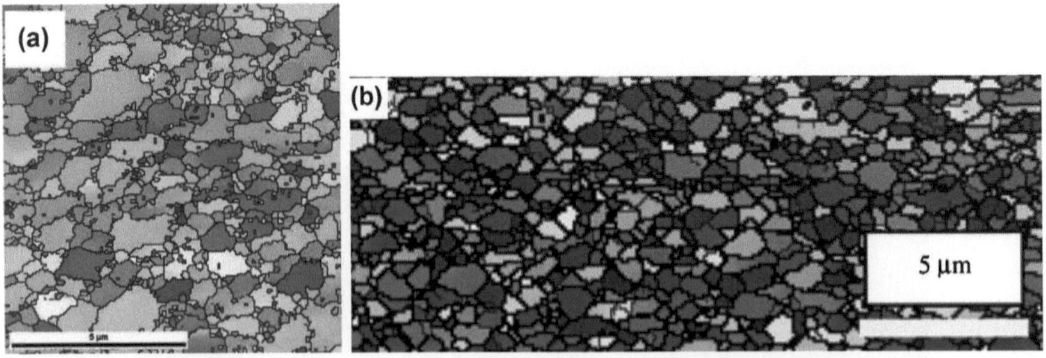

FIGURE 20.13

(a) EBSD IPF map of normal direction of Mg after ECAP to $\varepsilon_{eff} = 8$ [69]. (b) EBSD map of pure Zr after ARB to $\varepsilon_{eff} = 5.25$ [70].

Reproduced with permission from publishers.

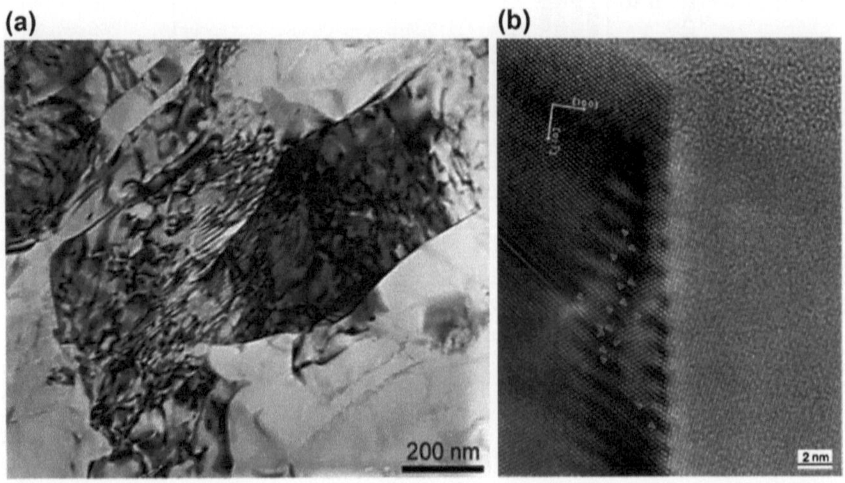

FIGURE 20.14

TEM images of (a) Al-1.5Mg after ECAP to $\varepsilon_{eff} = 12$ showing high-dislocation density within submicron grains (unpublished work of author), (b) high-resolution TEM of Al after HPT [71].

(b) Reproduced with permission from publisher.

with boundaries making the boundaries highly nonequilibrium. As these nonequilibrium boundaries have higher energy and long-range stress fields, these are unstable. Thermal energy would result in the migration of these boundaries leading to grain growth.

The average equilibrium grain size as a function of annealing temperature after severe plastic deformation for Al and some of its alloys [74–80] is shown in Fig. 20.15. It is seen that the average

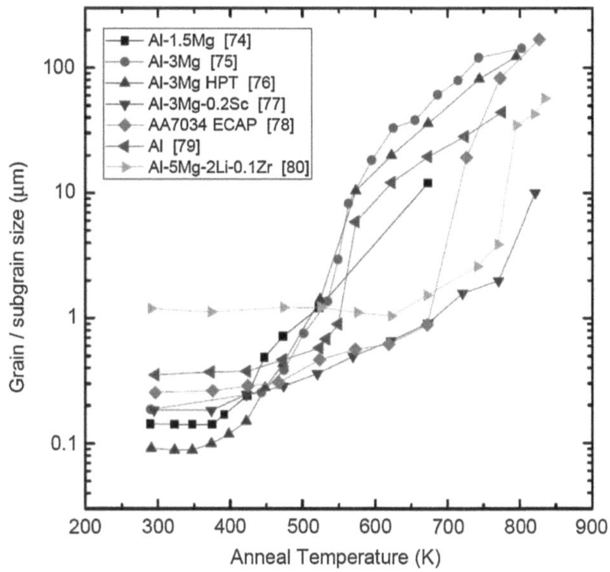

FIGURE 20.15

Average grain size as a function of annealing temperature after severe plastic deformation for Al-1.5Mg [74], Al-3Mg [75], Al-3Mg HPT [76], Al-3Mg-0.2Sc ECAP [77], AA7034 ECAP [78], Al [79], and Al-5Mg-2Li-0.1Zr [80].

grain size for Al and Al-Mg alloys increases with increasing temperature. However, the extent of increase is different for single-phase Al alloys and those with second-phase precipitates. For single-phase Al alloys the grains remain submicron only up to 500K, whereas for those alloys that have second-phase precipitates such as Al-Mg-Sc alloy, Al-Mg-Li-Zr alloy, and AA7034 the grains remain submicron up to a higher temperature of 650–700K. Second-phase intermetallics help in preventing grain growth and are hence desirable for warm temperature use of UFG Al alloys processed by SPD. The desirable alloy would be one where the grains remain submicron up to higher temperatures, i.e., the graph shifts to lower grain sizes and the transition temperature shifts to higher temperatures. There have been attempts to stabilize the ultrafine grains by heating to temperatures where dislocation density within grains is reduced but temperature is not high enough for grain growth. The attempt is to recover the excess dislocation density and transform the nonequilibrium grain boundaries to equilibrium high-angle grain boundaries, thus in effect reducing the driving force for grain growth. These would result in the microstructure remaining UFG with higher thermal stability. However, such attempts have had limited success, e.g., in Al-Mg system [81,82].

4.2 MECHANICAL BEHAVIOR

4.2.1 Stress–Strain Behavior

As discussed earlier one of the primary benefits of subjecting a material to severe plastic deformation is the refinement of its microstructure, which in turn enhances strength—a fundamental principle in metallic materials. Severe plastic deformation of a material can increase its yield strength (YS) anywhere from three to five times as compared to an annealed one. As SPD reduces the work-hardening ability, the increase in the ultimate tensile strength (UTS) is usually not that high. Fig. 20.16 shows some examples of the tensile stress–strain curves for Al [83], Cu [84,85], austenitic and ferritic steel [86,87], Ti [88], and a β Ti alloy [89] tested at ambient temperature and quasistatic

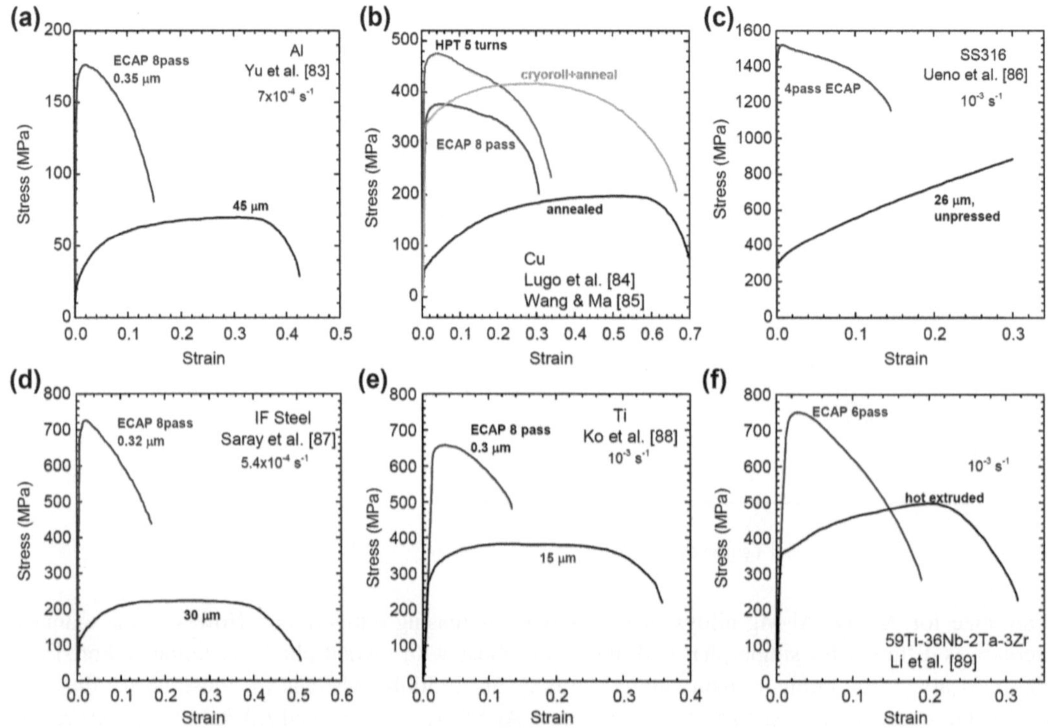

FIGURE 20.16

Tensile engineering stress versus strain curves at ambient temperature and strain rates as mentioned in plots for (a) Al [83], (b) Cu [84,85], (c) austenitic stainless steel SS316 [86], (d) ferritic interstitial free steel [87], (e) Ti [88], and (f) Ti alloy [89]. The strain rates for each of the tests are mentioned in plot.

strain rates before and after SPD. In all cases the YS and UTS increase and the total elongation to failure (ductility) decreases. Of the total elongation to failure, the uniform elongation is smaller than the nonuniform elongation. In one study by Wang et al. [90], Cu cryorolled to produce a UFG microstructure was given an annealing treatment that resulted in a bimodal grain size distribution. This Cu with bimodal distribution showed a significant uniform elongation and a total elongation matching that of annealed Cu (shown in Fig. 20.16). It was inferred that the small grains impart the high strength and the inhomogeneous grain size distribution enhances the work-hardening rate, which increase the uniform elongation and thus the overall ductility.

Metallic materials show an inverse relation between the yield stress and ductility (represented here as the % elongation for tensile tests) shown as a solid line in Fig. 20.17a representing the upper bound of the data of pure metals (data extracted from Valiev et al. [91]). Superimposed on the graph are some tensile data for Al [92,93], Cu [84,94–102], Fe [103], Ti [104], and V [105] before and after SPD. It is seen that SPD increased the yield stress and lowered the ductility. The large variation in the data of Cu is because of the different SPD processing techniques used. For Cu deformed at cryotemperatures or using dynamic deformation, the ductility appears to be higher than that after SPD carried out at

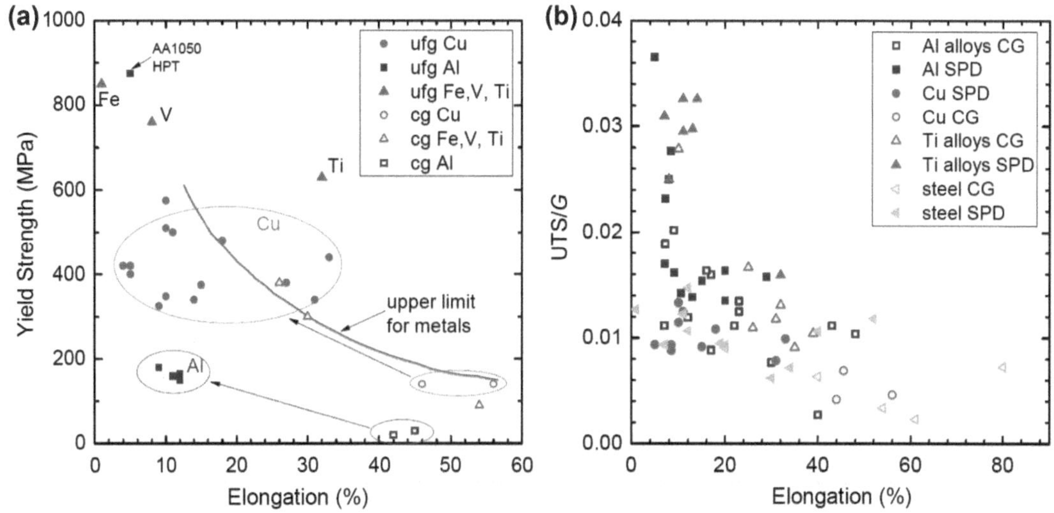

FIGURE 20.17

(a) Yield strength versus elongation for pure metals before and after undergoing severe plastic deformation. The *gray line* is from Ref. [91], which shows the boundary of the data of pure metals ranging from Al to W. All data points lie below that curve. Data after SPD are shown as solid symbols for Al [92,93], Cu [84,94–102], Fe [103], Ti [104], and V [105]. The *hollow symbols* represent the corresponding undeformed metals with *arrows* indicating the shift in position before and after SPD. (b) UTS normalized with shear modulus (*G*) versus elongation for different alloys of Al [40,81,106–113], Cu [84,94–102], steel [114–116], and Ti alloys [88,89,104,117–121]. The *open and filled symbols* represent non-SPD and SPD materials, respectively.

ambient temperatures. SPD of Cu at cryotemperatures results in twin boundaries that appear to have a beneficial effect on ductility. In a related study where nanocrystalline Cu with nanosized twins was produced by electropulsed-deposition technique [122,123] the ductility showed improvement over nanosized Cu without twins. Nanosized twins have partial dislocations on their boundaries, which interact with dislocations and take up strain, thereby increasing ductility. Further, of the various processing techniques HPT too results in a higher strength and some ductility. In one study it was shown that an Al alloy subjected to HPT increased its yield strength to near 850 MPa retaining about 3% elongation [123], in comparison to a yield strength of about 200–300 MPa obtained after other SPD techniques. In order to compare alloys of different metals, the UTS was normalized with the shear modulus and plotted with % elongation (Fig. 20.17b) for alloys and pure metals before and after SPD. It is seen that (1) in general with increasing strength the ductility falls, and (2) that the SPD materials are shifted to the upper-left direction in the graph (higher strength and lower ductility). The condition for stability during a tensile test depends on both strain hardening and strain rate hardening according to Hart's criteria [124].

$$\gamma = 1 - m, \qquad (20.6\text{a}-\text{c})$$

where $\gamma = \frac{1}{\sigma}\frac{\partial\sigma}{\partial\varepsilon}$ and $m = \frac{\partial \ln\sigma}{\partial \ln\dot{\varepsilon}}$.

Here γ is the normalized work-hardening rate and m is the strain rate sensitivity for true stress σ, true strain ε, and true strain rate $\dot{\varepsilon}$. The high dislocation density and ultrafine-grained microstructure after SPD leave the material with a very limited ability to further increase the dislocation density, hence to strain harden. Materials after SPD have a high strain hardening rate at low strains, but with increasing strain this rapidly decreases to very low values [125]. This limited strain-hardening capacity results in necking at early strains (from Eq. 20.6). The subsequent nonuniform deformation also depends on γ and m according to Refs. [126,127] and recently shown for UFG Al alloys [128]. The total elongation to failure E_{tot} is,

$$E_{tot} = \eta^{m/(1-\gamma)} - 1 \tag{20.7}$$

where $\eta = (mY'/Y_o)$ and Y_o is the initial value of the normalized area gradient $Y = (1/A)(\delta A/\delta x)_{time}$ and $Y' = \partial Y/\partial\varepsilon$ takes on very large values, meaning that the gradient $\delta \ln A/\delta\, x$ takes on very large values with strain. Eq. (20.7) says that the ductility would increase with increase in both normalized strain-hardening rate γ and strain-rate sensitivity m. A plot showing the increase in ductility with increasing m according to Eq. (20.7) is shown in Refs. [127,128].

4.2.2 Stress—Grain Size Relation

The strength of metallic materials increases with a reduction in grain size as established by the empirical Hall—Petch relation for grain sizes in the micrometer to millimeter range [3,4],

$$\sigma_y = \sigma_o + k_{HP} d^{-1/2}, \tag{20.8}$$

where d is grain size, σ_y is yield strength, and k_{HP} is a material-dependent Hall—Petch constant. This equation is theoretically explained based on dislocations piling up on grain boundaries, which act as barriers to dislocation motion. The pileups cause stress concentration on the boundary. The smaller the grain size the less the pileup and the higher the stress required for the dislocation to penetrate the boundary. With the grain size reducing to the submicron and nanometer scale, dislocation pileups may no longer be able to form, and this physical model of pileup at grain boundaries may break down. Fig. 20.18 shows the yield strength versus grain size data for Al, Al-Mg alloys, and Cu. Fig. 20.18a shows Al data [47,113,129—133] and data for Al-Mg alloys [134,135]. The conventional H-P equation for Al is shown by solid line taken from Ref. [140]. At values less than 10 μm, the YS data start appearing higher than the H-P trend for coarse-grain Al. This has been explained on the basis that the strength depends not only on the grain size but also on the dislocation density within grains by [129,141],

$$\sigma_y = \sigma_o + k_{HP} d^{-1/2} + \alpha M G b \sqrt{\rho}, \tag{20.9}$$

where $\alpha \approx 0.25$, M is the Taylor factor, G is the shear modulus, b is the magnitude of the Burgers vector, and ρ is the total dislocation density. Thus, SPD materials would be expected to be stronger than the expected Hall—Petch hardening alone due to the extra dislocations within grains as shown for Al in Fig. 20.18a. Infact it has been shown that in severely deformed metals the dislocation hardening term (second term) has a higher contribution to strength than the grain boundary strengthening term [47,129]. Thus, SPD materials that have higher dislocation density result in a positive deviation from the Hall—Petch relation (larger than H-P prediction). Most other data too appear above the line. For the Al-7.5Mg alloy, as the grain size reduced below 0.5 μm a reduced H-P slope k_{HP} [134] was seen as compared to the coarse-grained data of Ref. [140]. This reduction in H-P slope is also seen in Al

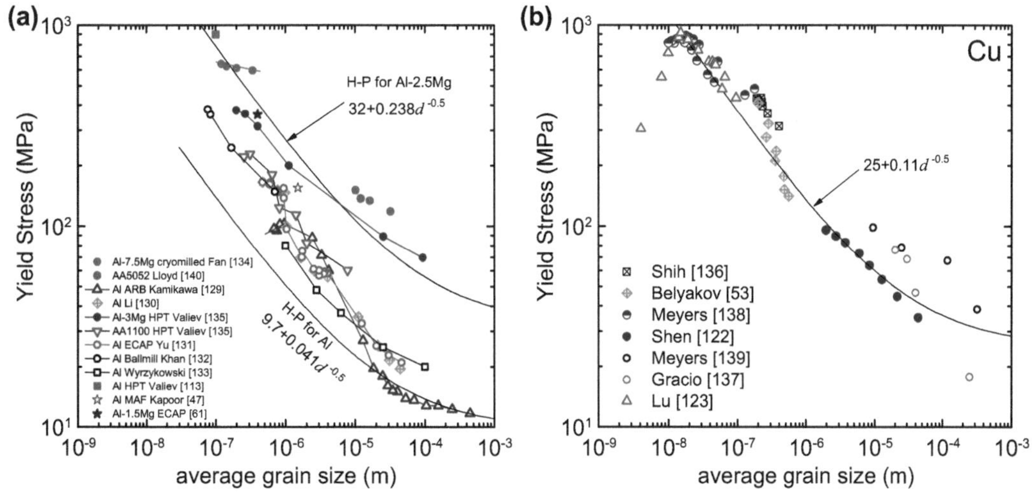

FIGURE 20.18

Yield stress vs average grain size on a log–log plot for (a) Al [47,113,129–133] and Al-Mg alloys [134,135] and (b) Cu [53,122,136–139]. Shown in each plot are lines for the Hall–Petch relations as determined from coarse-grained data. Note that this is a log–log plot and the H-P line will be curved due to the additive term (Eq. 20.8).

[129,131,135]. A similar plot is shown for Cu in Fig. 20.18b for data from literature [53,122,136–139]. It is seen that UFG Cu does appear to follow the Hall–Petch slope as extrapolated from the micrometer regime. Here too the strength through ECAP [136] and MAF [53] appeared higher than the H-P relation. The data by Meyers et al. [138] does follow the H-P relation, but although not mentioned in their review paper it does not appear to be SPD data and appears like data for pulsed-electrodeposited material. In comparison to SPD data, that of nanotwinned Cu with twin boundary spacing lower than ~20 nm produced through electropulsed deposition [123] showed that the strength decreased with further grain size decrease—a negative H-P relation. It should be mentioned that such fine microstructures with boundary spacing in the 10s of nm range are not produced by SPD.

In summary it can be said that in general the Hall–Petch relation established for micrometer grain sizes does not extrapolate to finer-grain sizes—with decreasing grain size through SPD, there is a hardening above the H-P relation due to the excessive dislocation density, and with further decrease in grain sizes to the nanometer range a decrease in the H-P slope occurs. At such reduced grain sizes the usual dislocation–dislocation interaction and pileup of dislocations present in micrometer-sized grains most likely do not exist. It is likely that grain boundary–mediated activities start, however, there is still a lack of clarity in the understanding of the strengthening mechanisms at these length scales. The strain rate sensitivities are still not high enough to establish a grain boundary–mediated deformation process. The discussion on strain rate sensitivity follows.

4.2.3 Strain Rate Sensitivity and Activation Volume

As discussed earlier the ability of a material to sustain large uniform elongation depends on a macroscopic quantity, the strain rate sensitivity m defined in Eq. (20.6). Strain rate sensitivity is usually determined by performing strain rate change tests during either uniaxial tension or compression tests. Severe plastic deformation is known to increase the strain rate sensitivity of fcc materials, an example of which is shown in Fig. 20.19 for Al-1.5Mg before and after room temperature ECAP [82]. Coarse-grained Al-1.5Mg (annealed) showed serrations in the stress–strain curve at 100°C (inset) representative of dynamic strain aging. For this condition, the strain rate sensitivity was negative (as seen in the inset), where a reduction in strain rate resulted in an increase in stress (inset showing serrations). After 12 passes in a 120 degrees ECAP die ($\varepsilon_{eff} = 8$) strain rate change tests showed a positive strain rate sensitivity and an absence of serrations. This is a clear demonstration of the increase in m after SPD. As the process of SPD also results in a UFG microstructure, it is expected that m shows an increase with decreasing grain size. Such a plot for fcc and bcc metals is shown in Fig. 20.20. It is seen that Al and Cu (representative of fcc materials) show an increase in m with decreasing grain size. In case of bcc metals, the opposite trend is seen where a decrease in grain size results in a decrease in m (Fig. 20.20). At very low values the strain rate sensitivity of bcc metals appears to show an increase, although there is not sufficient data to support this trend. This difference in behavior of m with grain size for fcc and bcc materials can be understood using the concept of activation volume.

Strain rate sensitivity is a macroscopic engineering parameter but it is related to a more fundamental thermodynamic parameter, the activation volume V. Physically V is the product of the activation area (area swept out by the dislocation during its thermally activated motion) and the magnitude of the Burgers vector b, i.e., $V = Ab$. It is also thermodynamically expressed as [151,152],

$$V = -\frac{d(\Delta G)}{d\tau^*} = kT\left(\frac{\partial \ln \dot{\gamma}_o}{\partial \tau^*} + \frac{\partial \ln \dot{\gamma}}{\partial \tau^*}\right), \tag{20.10}$$

where τ^* is the thermal component of shear stress, $\dot{\gamma}$ is the shear strain rate, k is Boltzmann constant, T is temperature, and ΔG is the free energy of thermal activation during deformation and is related to the shear strain rate as $\dot{\gamma} = \dot{\gamma}_o \exp(-\Delta G/kT)$ where $\dot{\gamma}_o$ is the preexponential factor, a product of the dislocation

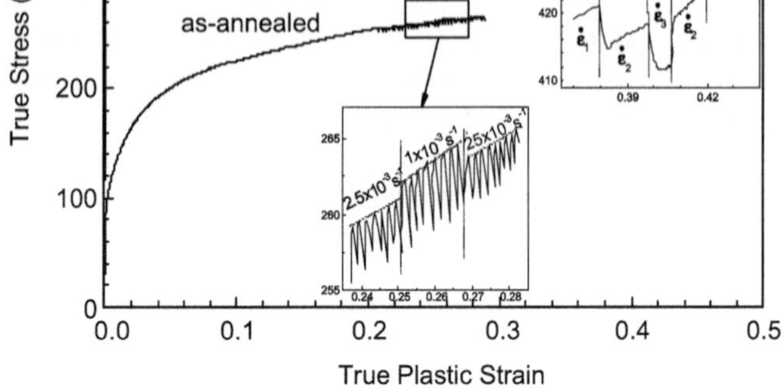

FIGURE 20.19

Compression true stress versus true strain plots of Al-1.5Mg before and after ECAP. ECAP was done at room temperature and compression testing at 100°C, showing strain rate changes from strain rates as indicated [82].

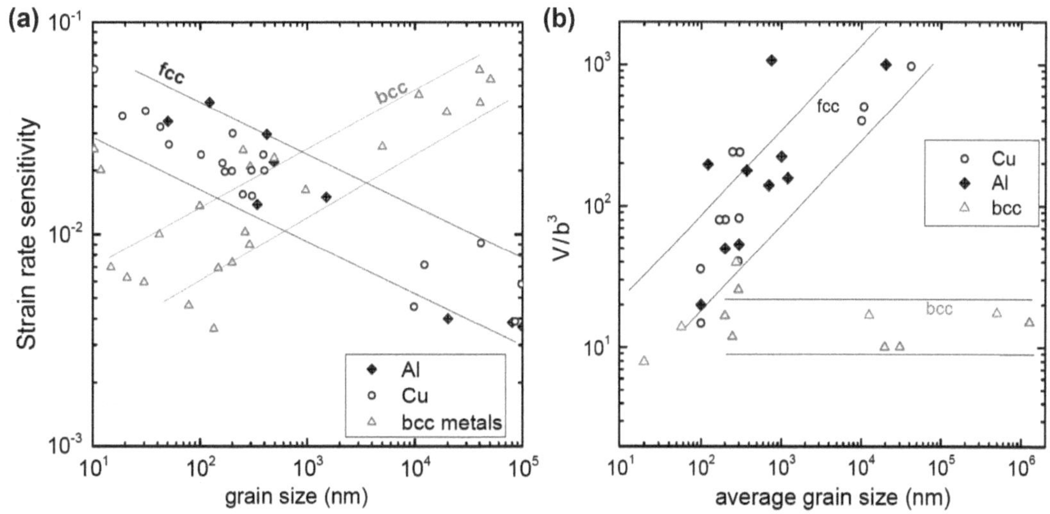

FIGURE 20.20

(a) Strain rate sensitivity plotted versus grain size for fcc metals Cu and Al [47,49,142] and for different bcc metals Fe, W, Ta, and Cr using data extracted from Ref. [142]. (b) Apparent activation volume versus grain size for fcc Cu [49,85,143,144], Al [47,51,135,145–149], and bcc metals [143] Nb [67] and Cr [150].

density, the thermal vibration frequency, the Burgers vector, and activation distance. The activation volume is determined by performing sudden changes of either strain rate or stress, in which case the microstructure is assumed to be constant. The athermal part does not change, thus $\partial\tau^* = \partial\tau$ and Eq. (20.10) can be written accordingly. Further, for a polycrystalline material the shear stress is converted to uniaxial stress through the Taylor factor M, $\sigma = M\tau$. When changes in $\dot{\gamma}_0$ with τ are not considered an apparent activation volume V_a is obtained. Using this, Eqs. (20.10) and (20.6c), and the expression for m is,

$$m = \frac{MkT}{\sigma V_a} \tag{20.11}$$

Eq. (20.11) says that at a given temperature the strain rate sensitivity is inversely related to the stress and the apparent activation volume. The variation of apparent activation volume is inherently different for fcc and bcc materials. In bcc materials the short-range barrier is the Peierls stress, which does not change with strain and hence the thermal component of stress is independent of strain. The activation volume can be expressed as $Ab = \xi lb$ where ξ is the distance of activation, l is the dislocation length that is activated. For bcc materials, l and ξ are determined by the short-range barrier and do not change with strain, and hence V_a does not change with strain. It also does not change with a reduction in grain size as seen from Fig. 20.20b. However, with increasing severity of deformation the grain size decreases and the stress increases, which according to Eq. (20.11) decreases m. Thus for bcc materials the strain rate sensitivity decreases with decreasing grain size. In an exception, the refinement of Ta to nanosized grains via HPT showed a higher m than the general trend for bcc metals [153].

For fcc materials the short-range barrier is the dislocation–dislocation intersection, and thus the spacing of the barriers changes with strain (increase in dislocation density). During thermal activation the dislocation length that is activated and the distance between obstacles are proportional to $1/\sqrt{\rho}$, thus activation volume would be $\xi lb = ab/\rho$ for some proportional constant a. Using this and Eq. (20.9), Eq. (20.11) can be written as,

$$m = \frac{MkT}{ab} \frac{\rho}{\left(k_{\mathrm{HP}} d^{-1/2} + \alpha MGb\sqrt{\rho}\right)}. \tag{20.12}$$

The numerator has a dependence of ρ whereas the denominator has dependence of $\sqrt{\rho}$, thus with increasing dislocation density m would increase. For larger grain sizes the grain size term would have weak influence. As the grain size decreases in the 100 nm range, it is likely that both l and ξ are controlled by and proportional to the grain size d. The strain rate sensitivity can be expressed as [154],

$$m = \frac{MkT}{ab\xi} \frac{1}{\left(k_{\mathrm{HP}}\sqrt{d} + \alpha MGb\sqrt{\rho}\, d\right)} \tag{20.13}$$

This equation states that in the nanoregime the strain rate sensitivity should increase with decreasing d; $\sqrt{\rho}$ has a weak dependence. The previous expressions for m are not rigorous and are only a semiquantitative attempt to explain the increase of m with severe plastic deformation.

4.2.4 Fatigue

Cyclic deformation studies are generally separated into two regimes, the low-cycle fatigue (LCF) and high-cycle fatigue (HCF) regime based on the number of cycles to failure: LCF for less than 10^4 cycles and HCF for higher than 10^4 cycles to failure. In terms of strain amplitudes, LCF occurs if the macroscopic plastic strain amplitudes (usually $>10^{-3}$) are greater than the elastic strain amplitudes. During HCF the macroplastic strain amplitudes are very small (usually in the 10^{-4} to 10^{-5} range) and smaller than the elastic strain amplitudes. However, in either case when looked at from the microscopic level the damage evolution mechanisms are similar and the final failure is due to repeated micro-yielding at local regions. For review on fatigue behavior of materials, see Ref. [155]. HCF testing is carried out in stress-control mode whereas LCF is usually carried out in strain-control mode with a measureable plastic strain amplitude. During HCF testing the stress amplitude at $R = -1$ is sufficiently below the yield stress, and thus the majority of life is governed by the nucleation of a crack—a sort of probe for the resistance to crack initiation. In LCF tests as there is a measureable plastic strain amplitude the nucleation of the crack is rapid and failure is controlled by its propagation. From HCF tests a stress amplitude versus number of cycles to failure curve is established (the "S–N" curve) from which the fatigue limit (stress amplitude to 10^7 cycles or more) can be determined. The fatigue limit is known to be proportional to the UTS, i.e., the stronger a material the higher is its fatigue limit. Materials after SPD also follow this trend as seen in Fig. 20.21. The data of materials after SPD and data of materials not having undergone SPD have the same trend, the slope of the best-fit line being about 0.54. As materials after SPD have higher UTS, they would also have a higher fatigue limit (Fig. 20.21).

Thus it is expected that as the UTS increases after SPD, so will the overall HCF life. Examples of this are seen in Fig. 20.22 for different metals and alloys. Fig. 20.22a shows that with increasing severity of deformation the fatigue life increases for an Al-0.5Mg alloy [156], being highest after an effective strain of 12. Fig. 20.22b shows the S–N curves before and after ECAP for Ti [157] and Ti-6Al-4V [117]. In both cases the fatigue strength increases after severe plastic deformation. Ti-6Al-4V is stronger than Ti and hence has higher fatigue limit. Similar improvement of fatigue life after SPD is seen in Mg alloys too, e.g., Ref. [160]. Fig. 20.22c shows the S–N curve for Cu from different sources

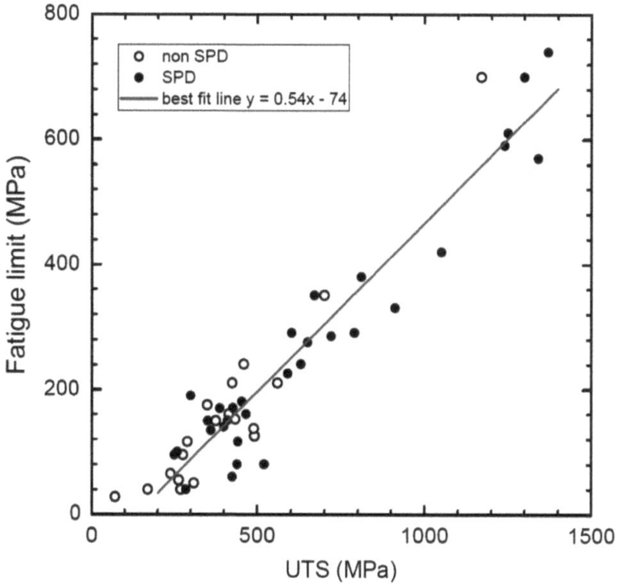

FIGURE 20.21

Fatigue limit versus ultimate tensile strength (UTS) for Mg, Al, Cu, Ti alloys, and steel.

Data taken from Y. Estrin, A. Vinogradov, Acta Mater. 61 (2013) 782–817.

before and after SPD [156–159]. The plot shows data up to the ultrahigh cycle fatigue regime (10^{10} cycles). In the lower cycle range, CG Cu has a lower fatigue life than UFG Cu produced by SPD, but at 10^7 cycles the fatigue lives get similar as seen in recent work by Goto et al. [159] for high-purity Cu. A lower-purity Cu (99.9%) showed a significantly higher fatigue life and was higher than the CG Cu up to the ultrahigh cycle fatigue regime.

Strain-controlled LCF tests show a reduction in fatigue life after grain refinement through SPD as shown in Fig. 20.22d for AA5056 and Cu [95]. The UFG curves are shifted to the left (lower cycles to failure). Giving a 170°C treatment for 120 min to ECAP, Cu increased its fatigue life to better than that of CG Cu. A similar treatment of Al does increase the fatigue life to better than as ECAP but does not increase it better than CG Al. The annealing treatment to ECAP Cu produces a bimodal microstructure, which increases ductility and hence enhances the LCF life. However, similar types of heat treatments to Al and its alloys have failed to produce a bimodal microstructure and hence failed to improve the LCF life beyond the CG condition. The ability to produce a bimodal microstructure depends on the ability of a material to readily recrystallize, which Cu does but Al does not.

The increase in HCF life and decrease in LCF life after SPD can be explained on the basis of the combined Coffin–Manson and Basquin equations [95]. The total strain amplitude $\Delta\varepsilon_{tot}$ is composed of the elastic strain amplitude $\Delta\varepsilon_{el}$ and the plastic strain amplitude $\Delta\varepsilon_{pl}$.

$$\frac{\Delta\varepsilon_{tot}}{2} = \frac{\Delta\varepsilon_{el}}{2} + \frac{\Delta\varepsilon_{pl}}{2} \tag{20.14}$$

The expression for the elastic part is the Basquin equation and for the plastic part is the Coffin–Manson equation. Eq. (20.14) can be written in terms of the total life N_f (total reversals is $2N_f$) as,

$$\frac{\Delta\varepsilon_{tot}}{2} = \frac{\sigma'_f}{E}(2N_f)^b + \frac{\varepsilon'_f}{2}(2N_f)^c. \tag{20.15}$$

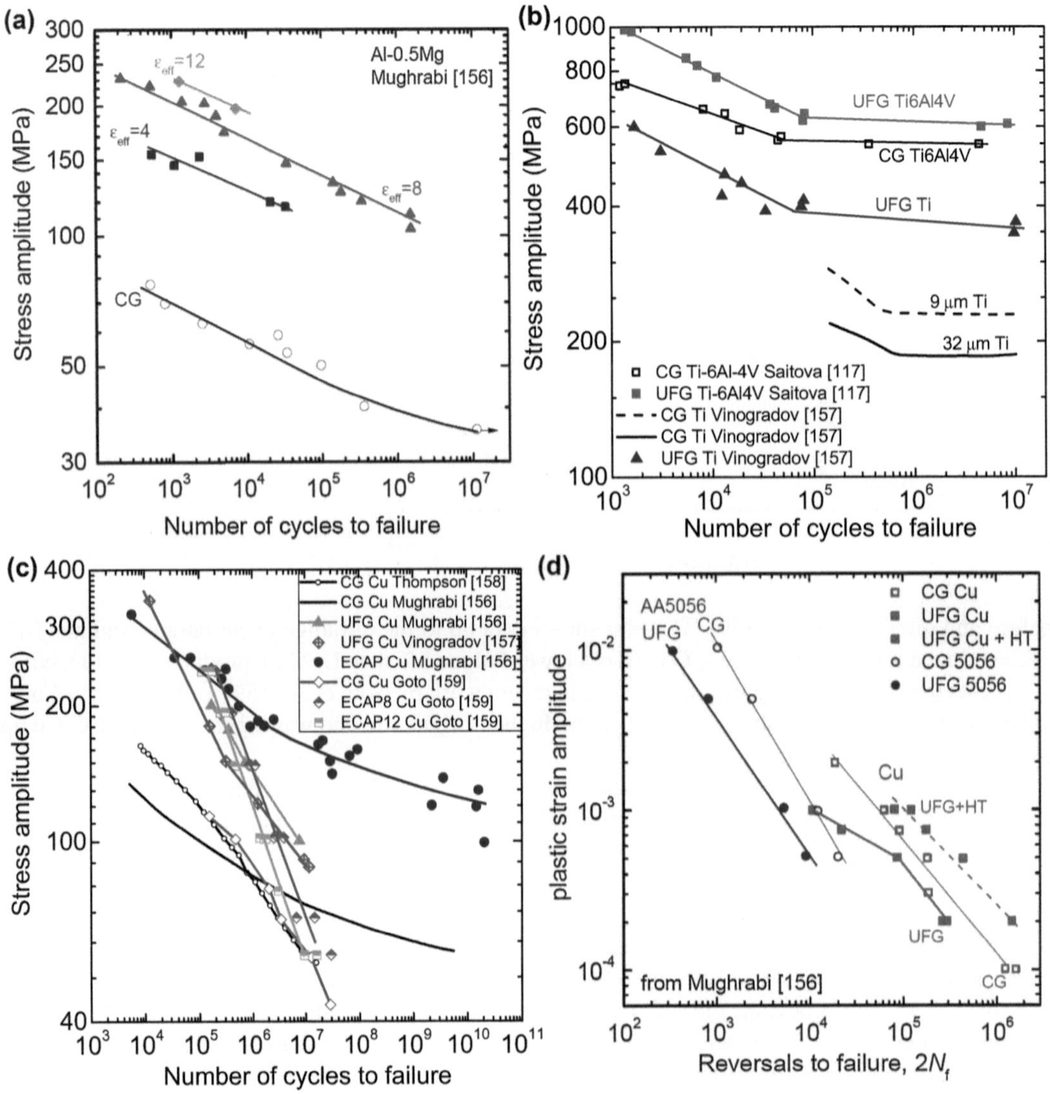

FIGURE 20.22

Stress amplitude versus number of cycles to failure for (a) Al-0.5Mg [156], (b) Ti [157], and Ti-6Al-4V [117] where 9 and 32 μm represent the average grain sizes and (c) Cu [156–159]. (d) Plastic strain amplitude versus cycles to failure in plastic strain-controlled tests for AA5056 and Cu [95]. *CG*, coarse grained; *ECAP8 and 12*, represent 8 and 12 ECAP passes; *HT*, heat treated 170°C for 20 min; *UFG*, ultrafine grained.

where σ'_f and ε'_f are the fatigue strength and ductility coefficient. respectively, b and c are the corresponding exponents, and E is the Young's modulus. A value of $N_f = ½$ implies a monotonic test and hence σ'_f and ε'_f should represent (or scale with) the fracture stress and strain. A schematic plot of Eq. (20.15) is shown in Fig. 20.23 showing the shift in the curves from CG to UFG microstructure. Monotonic tests of materials after SPD are known to show an increased fracture stress and a decreased ductility, which implies an increase in σ'_f and a reduction in ε'_f. In the LCF regime, where the plastic

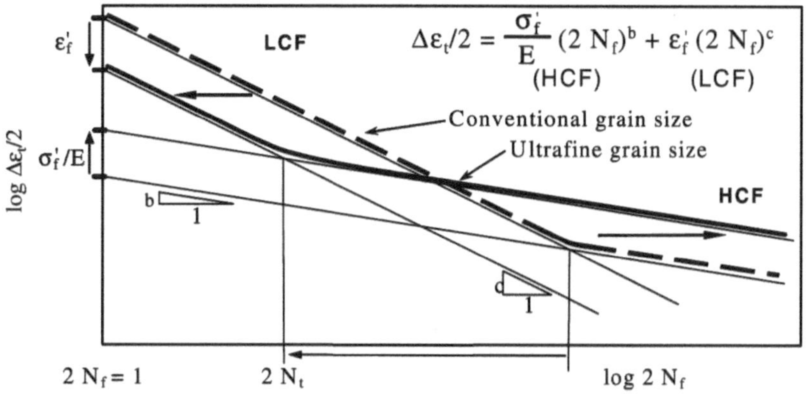

FIGURE 20.23

Schematic of the total strain fatigue life diagram for coarse-grain conventional materials and after severe plastic deformation resulting in ultrafine grains [95].

Reproduced with permission from publisher.

strain dominates (Coffin–Manson equation), ε'_f is reduced after SPD and hence the curve shifts downward. In the HCF regime, where the elastic part dominates (Basquin equation), σ'_f is increased after SPD and hence the curve shifts upward. This explains the increase in HCF life and reduction in LCF life of UFG alloys produced by SPD. Due to the high plastic strain per cycle in LCF regime, the number of cycles to initiate a crack is low and is usually an insignificant part of the total life, thus it is crack propagation that dominates. Conversely, in the HCF regime the number of cycles for crack nucleation is substantial and hence life is governed by crack initiation rather than propagation.

Fatigue crack growth behavior of materials after severe plastic deformation has the same regimes as those observed in coarse-grained materials—the threshold of crack propagation, the Paris regime where da/dN is power law of ΔK, and the failure regime. However, the fatigue crack growth rate is higher for UFG materials produced by severe plastic deformation as compared to that of the undeformed condition (CG microstructure). Examples of this are seen in Fig. 20.24 for Cu [40], Ti [161], and an Al-Mg alloy AA5056 [162]. In Fig. 20.24a, with increasing severity of deformation of Cu (from 1 pass to 8 pass of ECAP) the fatigue crack growth rate da/dN increases and the threshold for fatigue crack growth decreases. Similar increase in fatigue crack growth rate and decrease in threshold are seen for Ti (Fig. 20.24b) and AA5056 (Fig. 20.24c). It has been suggested that the higher crack growth rate in UFG microstructure could be due to less out-of-plane movement of the crack as compared to a coarser microstructure where deviations of crack path could be substantial. Thus although UFG materials produced by SPD have a higher resistance to crack initiation (better HCF life), they have a poorer resistance to crack propagation.

4.2.5 Superplasticity

The ability of a material to exhibit very large elongations to failure (usually taken as being greater than around 300%) is referred to as superplasticity. Such large elongations take place when tested at elevated temperatures where diffusion and/or nonconservative dislocation motion takes place and

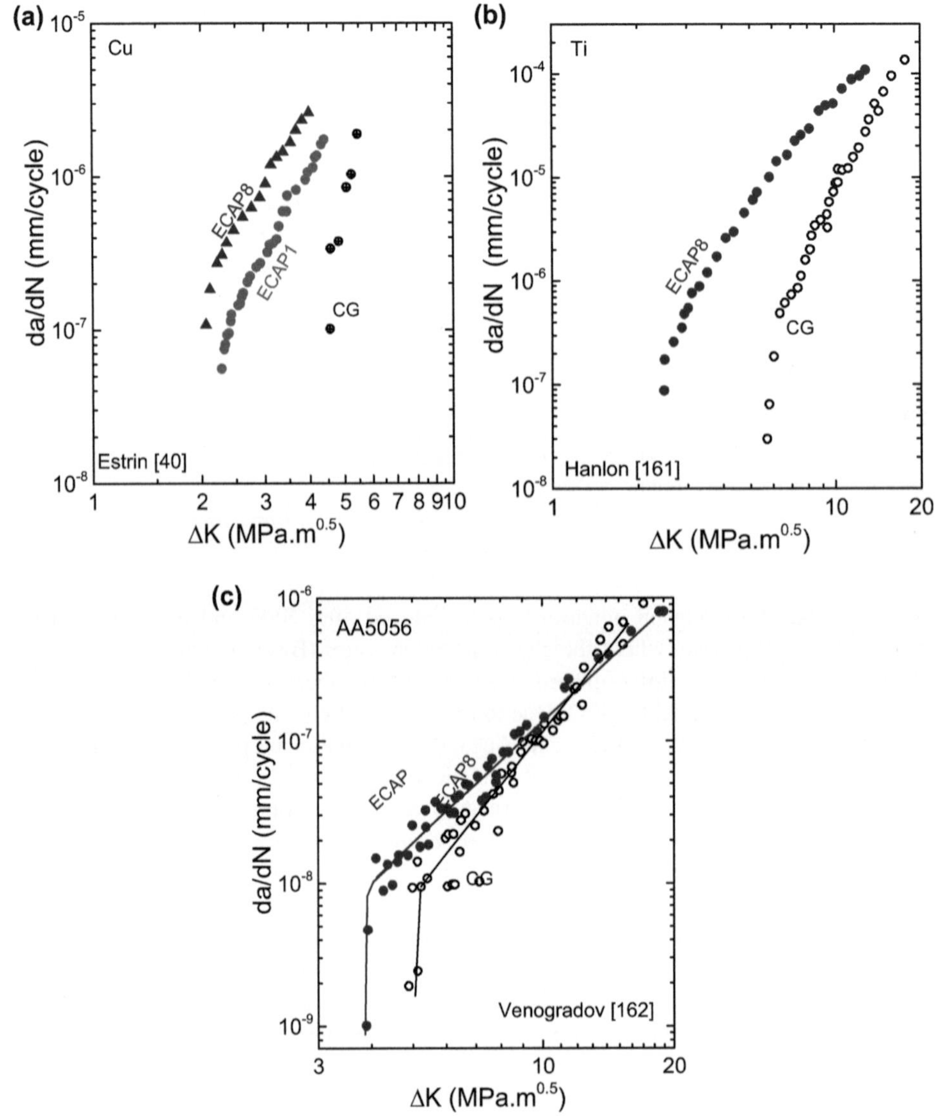

FIGURE 20.24

Log–log plots of the crack growth rate *da/dN* versus the applied stress intensity factor range ΔK for (a) Cu [40], (b) Ti [161], and (c) AA5056[162].

where grain boundaries play an active role in deformation [163]. It is imperative that if grain boundaries are the primary participants in high-temperature deformation, superplasticity would be aided by smaller grain sizes. The grain size dependence on the strain rate of deformation at higher temperatures is [164],

$$\dot{\varepsilon} = A\left(\frac{DGb}{kT}\right)\left(\frac{b}{d}\right)^p\left(\frac{\sigma}{G}\right)^n$$
$$= A\left(\frac{D_0Gb}{kT}\right)\left(\frac{b}{d}\right)^p\left(\frac{\sigma}{G}\right)^n \exp\left(\frac{-Q}{RT}\right). \quad (20.16)$$

Here n is the stress exponent, p is the grain size exponent, b is the magnitude of the Burgers vector, G is the shear modulus, D is the diffusion coefficient $= D_0\exp(-Q/RT)$, D_0 the frequency parameter, Q the activation energy of an appropriate diffusional process, A is a dimensionless parameter, and k is the Boltzmann constant. For a grain boundary deformation process (grain boundary sliding) $p = 2$ and $n = 2$. Eq. (20.16) states that at a given temperature and stress, as the grain size decreases the strain rate increases, i.e., the material gets softer. Note the difference in the role grain boundaries play at low and high temperatures. At low temperatures where dislocation glide is the major carrier of deformation and grain boundaries act as obstacles, the Hall–Petch equation is valid. At high temperatures the grain boundaries deform and material gets softer with decreasing grain size (Eq. 20.16). In the former, as grain size decreases stress increases, thereby making the material stronger; however, at higher temperatures when grain boundaries start deforming, as the grain size decreases, the material gets softer. The value of $n = 2$ corresponds to a strain rate sensitivity of $m = 0.5$, which leads to superplastic deformation. Smaller grain-sized materials would shift the superplasticity domain to lower temperatures and higher strain rates as seen from Eq. (20.16).

Materials subjected to SPD techniques result in a UFG microstructure and hence have the potential for superplasticity. However, as discussed in Section 4.1, single-phase UFG materials have a tendency for grain growth at higher temperatures, thereby negating the advantage of a UFG microstructure. Thus it is unlikely that single-phase UFG metals would result in superplasticity. The key to achieving superplasticity would be to prevent grain growth of UFG metals at high temperatures. This is achieved by producing UFG materials with second-phase precipitates that pin grain boundaries and prevent grain growth. Superplasticity has been observed in severely plastically deformed Al alloys, Mg alloys, and Zn alloys; for a more exhaustive list, refer to Refs. [165,166]. Fig. 20.25 shows some examples of superplasticity for two-phase alloys of Al-5.5%Mg-2.2%Li-0.12%Zr [167], Mg-5.5%Zn-0.5%Zr alloys [168], and Zn-22%Al [169]. The advantage of a stable UFG microstructure produced through SPD is that the grains remain fine enough to shift the superplastic domain to higher strain rates (typically $>10^{-2}\ s^{-1}$), thus making it attractive for industrial processing. For example, such increases in strain rates corresponding to superplastic deformation were observed in many Al alloys as listed in Ref. [165].

5. APPLICATIONS

It has been discussed in this review that severe plastic deformation carried out by repetitive plastic deformation steps results in an ultrafine-grained microstructure with increased strength and high cycle fatigue limit. The next logical step would be to use it for engineering and technological applications, which would require industrial production of UFG materials through the use of severe plastic deformation. The most common and well-studied SPD techniques are batch processes, be it ECAP, HPT, ARB, or multiaxial forging as discussed earlier. There have been few innovative techniques where SPD of long shapes (rods, bars, sheets, etc.) were made possible, such as ECAP conform, conshearing, etc. as listed in Table 20.2. However, the adoption of any of the SPD processes by the

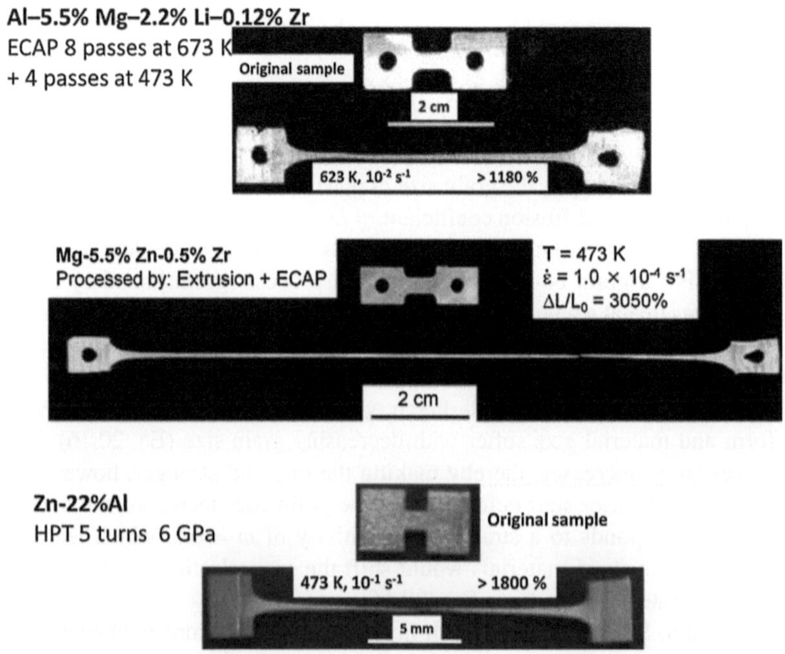

FIGURE 20.25

Examples of superplasticity in ultrafine-grained alloys processed using severe plastic deformation as seen in Al-5.5%Mg-2.2%Li-0.12%Zr [167], Mg-5.5%Zn-0.5%Zr [168], and Zn-22%Al[169].

Reproduced with permission from publishers.

industry has been slow, primarily due to cost factor, the lack of proper market for such materials, and finally lack of awareness and an understanding of the process despite the large amounts of lab-scale production and testing. In spite of this drawback its use in making small components for specific applications is still viable. The engineering aspects of the use of severe plastic deformation techniques have been considered in Refs. [170,171].

As possibly the first commercial industrial application, Honeywell Electronics Materials developed sputtering targets of UFG Al and Cu (Fig. 20.26 top left) using ECAP on a 4000-ton-capacity press [172]. These targets were of monolithic Al-0.5Cu alloy with average grain size of 0.5 μm. The traditional sputtering targets are diffusion-bonded pure metal with a stronger alloy as backing. As UFG materials have higher strength, the targets can be made of monolithic UFG metals thus avoiding the diffusion-bonding process and hence increasing the life of targets. In addition to this, Honeywell Electronics also developed a turbocharger component from ECAP AA2618 Al alloy for aerospace and transportation industry applications, and landing gear components from AA7xxx Al alloy for use in airplanes.

Apart from the herein-mentioned industrial products through SPD, there are studies that attempt to demonstrate the feasibility of manufacturing engineering components with UFG microstructure using the SPD techniques [170,171,173]. The SPD processes have been used to make small components for structural engineering applications, e.g., microbolts from UFG carbon steel processed by ECAP [174],

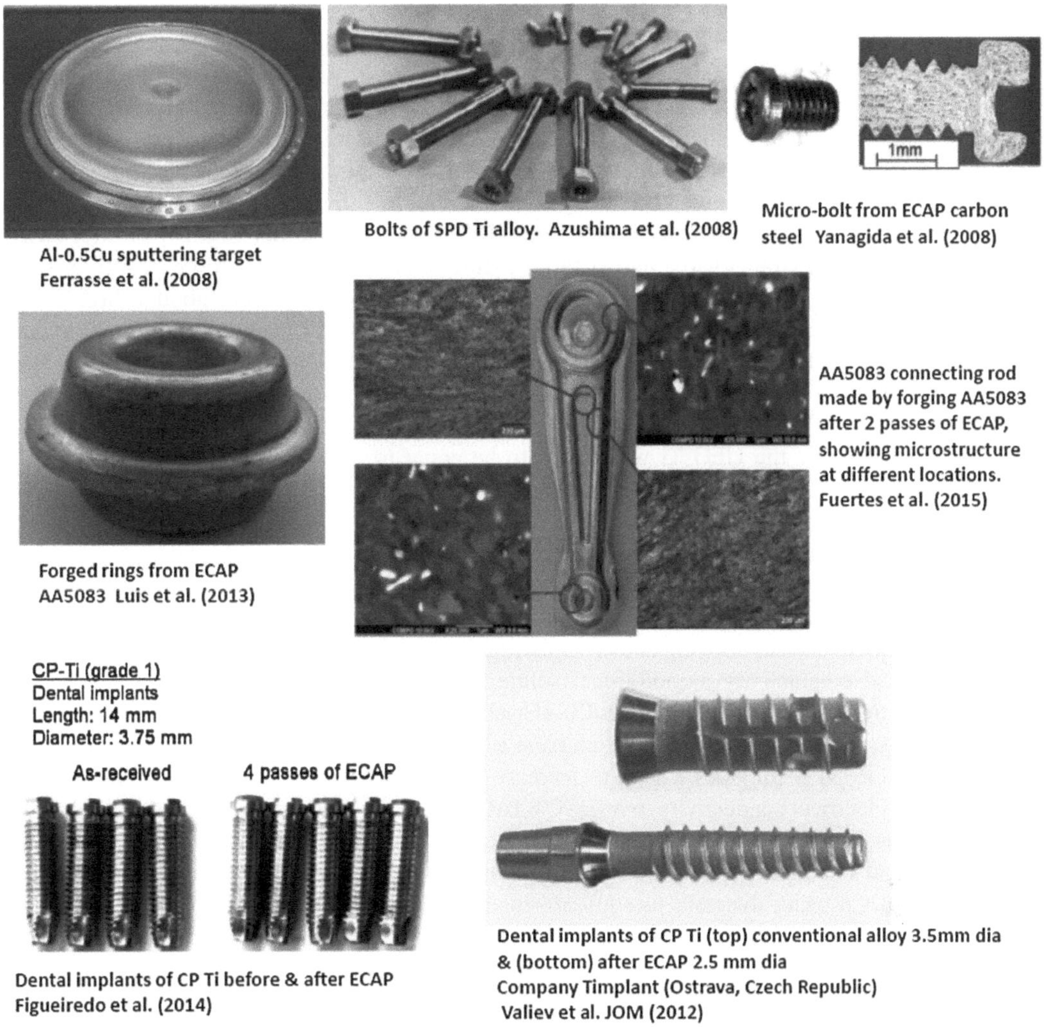

FIGURE 20.26

Example of products made using severe plastic deformation taken from Ferrasse et al. [172], Azushima et al. [173], Yanagida et al. [174], Luis et al. [175], Fuertes et al. [176], Figueiredo et al. [177], and Valiev et al. [178].

Reproduced with permission from publishers.

connecting rod of AA1050 and AA5083 after ECAP and forging [176], small turbine blades of AA6063 by ECAP and hot forging [179], forged rings from ECAP AA5083 [175], high-strength bolts of AA6061 from hybrid process of drawing-ECAP-rolling [180], high-strength bolts of AA6061 from ECAP [181], and superplastic forming and diffusion bonding of UFG Ti-6Al-4V [182]. Some pictures of these applications are shown in Fig. 20.26. The use of the batch process of SPD would also be useful in making miniature components for small devices such as microgears [183] and components for microelectromechanical systems (MEMS) [184]. The advantage of making very small components with UFG materials is that the component would still contain significant number of grains within, thus ensuring relative microstructural uniformity and hence more reliability of the component.

Another important application is that for bioimplants. Of the various metals, pure Ti has very good biocompatibility and hence is widely used as bioimplants. Since coarse-grained pure Ti is not strong enough, it needs to be alloyed for strength. However, the alloying elements make the product less biocompatible. In a study by Medvedev et al. [185] it was shown that Grade 4 Ti after ECAP and appropriate heat treatment had a higher fatigue limit than that of CG Ti6Al4V. A similar study on Grade 1 Ti also showed increased fatigue limit after ECAP, however, not as good as high grades of Ti [177]. Thus a stronger UFG Ti of higher grade having a higher fatigue limit would make a better bioimplant material than conventional grained Ti6Al4V. Timplant company has been manufacturing dental implants with conventional Ti alloys and has also manufactured implants with UFG Ti from SPD under the name Nanoimplant [178] as shown on bottom right of Fig. 20.26. The osseointegration of the UFG Ti was found to be equal to or slightly better than that of CG commercially pure Ti grade 2 [186]. Thus the use of UFG pure Ti for bioimplants is a very promising application.

Another application is for electrical-conducting materials such as Cu and Al. For achieving high electrical conductivity it is desired to keep purity levels high, however, this reduces the strength of materials and makes it difficult to use. Usually this increase in strength is achieved by conventional hardening techniques such as alloying and precipitation hardening. The drawback of such hardening is reduced electrical conductivity as additional solute, precipitates, and defects cause scattering of conduction electrons. One such example is Cu-1Cr alloys used in electrical conductor parts such as cable connectors, circuit breakers, etc., which have a 60–70% International Annealed Copper Standard (IACS) value. SPD provides a way to increase strength of pure Cu to levels of alloys while maintaining the electrical conductivity to near 90% IACS. Grain boundaries also degrade the electrical conductivity, however, special boundaries like coincident site lattice boundaries have lower energy and degrade electrical conductivity only marginally. Zhang et al. [187] introduced nanoscale twins in ultrafine-grained Cu using dynamic plastic deformation at cryogenic temperatures and showed an increased strength with 95% IACS conductivity. Al is also extensively used in overhead electrical transmission line cables. This use requires both high conductivity and reasonable strength. CG Al has an electrical conductivity of about 60% IACS, and alloying to increase strength reduces the conductivity further. The technique to improve conductivity while maintaining strength in Al alloys is to subject them to SPD and then age them so as to precipitate out the solutes from the matrix [188,189]. Precipitation reduces the negative effect of solute atoms on conductivity, and overaging would enhance the conductivity further with little loss in strength as that comes from the reduced microstructural scales due to SPD.

6. SUMMARY

The technique of severe plastic deformation to refine the microstructural lengthscales to submicron levels (ultrafine grained) in materials is now well established. All processes of SPD require that in each "pass" the overall shape of the sample does not change. This ensures that repetitive plastic deformation can be performed leading to the accumulation of large plastic strain. Severe plastic deformation refines the microstructure by the process of continuous dynamic recrystallization, where dislocations first arrange in cells, which then form subgrains, and eventually with further strain transform to high-angle boundaries. The typical grain size after SPD ranges from 200 nm to 1 μm with around 70% fraction of high-angle boundaries. This high-angle boundary fraction can be further increased by appropriate annealing treatments that would transform the low-angle boundaries to high-angle ones. High-pressure torsion is one SPD process where the grain size can be reduced to below 100 nm. However, for all SPD processes the microstructure produced is in a nonequilibrium state, making the thermal stability of the grains a concern. This is more so in single-phase materials where grain growth is unhindered. Thus for any high-temperature application single-phase materials should be avoided. In contrast, two-phase materials with precipitates help pin boundaries and prevent grain growth and are thermally stable up to higher temperatures. These are thus desirable for high-temperature applications. As materials with UFG microstructure have significantly large grain boundary area, grain boundary deformation takes place in high-temperature regimes, thereby enhancing the deformation rates. This brings out an opportunity for achieving high–strain rate superplasticity in UFG materials. Two-phase alloys with thermally stable UFG microstructures are potential candidates for use in superplastic forming. The UFG microstructure allows the superplastic domain to shift to lower temperatures and higher strain rates, thus enabling feasible superplastic forming.

The first and most obvious advantage of severe plastic deformation is its increased room temperature strength due to the reduced grain size (Hall–Petch effect) and increased dislocation density. However, in most cases the increased strength corresponds to a reduced ductility. In some cases, such as Cu, the ductility can be increased by introducing twin boundaries in the microstructure. A bimodal microstructure (UFG and coarse grains) in Cu was also seen to increase the ductility. Another important engineering parameter is the strain rate sensitivity, which influences the uniform elongation and ductility and general formability. Severe plastic deformation increases the strain rate sensitivity in fcc materials but decreases it in bcc materials. This is related to the activation volumes that decrease with decreasing grain size in case of fcc materials, and remain unchanged with grain size for bcc materials. Another important mechanical property of a structural material is its fatigue behavior. Severe plastic deformation enhances the high-cycle fatigue life but reduces the low-cycle fatigue life, these being related to the increase in strength and reduction in ductility, respectively.

The last two decades have seen considerable development of the field of severe plastic deformation in terms of developing processes, understanding microstructure and its evolution, and establishing mechanical properties. However, the use of severe plastic deformation by industry has been slow. Despite this the feasibility of manufacture of small components made through SPD has been demonstrated, such as sputtering targets, bolts, small turbine blades, bioimplants etc. The fabrication of small components for MEMS is one promising area where the batch-processing limitation of SPD technique is not of concern. The use of UFG Ti for bioimplants has been demonstrated and shows promise for wider use. The use of grain boundary engineering along with severe plastic deformation is a new and challenging area of work. The introduction of nanotwins in a UFG microstructure increases strength without any significant loss in electrical conductivity making it a potential replacement to copper alloys used for electrical applications that require strength. Severe plastic deformation can be considered as a tool for microstructurally engineering materials for required properties. Although a fair amount of understanding has been developed in the microstructure–mechanical property correlation of severely deformed materials, other multifunctional properties such as corrosion, diffusion, electrical and magnetic properties, radiation resistance, etc. are yet to be understood. As this understanding builds up, the use of severe plastic deformation for applications will follow.

REFERENCES

[1] H.J. McQueen, Mater. Sci. Eng. A 387–389 (2004) 203–208.
[2] M.E. Kassner, S.R. Barrabes, Mater. Sci. Eng. A 410–411 (2005) 152–155.
[3] E.O. Hall, Proc. Phys. Soc. B64 (1951) 747.
[4] N.J. Petch, J. Iron Steel Inst. Lond. 173 (1953) 25.
[5] T. Sakai, J.J. Jonas, Acta Metall. 32 (1984) 189–209.
[6] V.M. Segal, USSR patent number 575892. (1977).
[7] V.M. Segal, V.I. Reznikov, A.E. Drobyshevskiy, V.I. Kopylov, Russ. Metall. 1 (1981) 115.
[8] V.M. Segal, Mater. Sci. Eng. A 197 (1995) 157–164.
[9] V.M. Segal, Mater. Sci. Eng. A 271 (1999) 322–333.
[10] R.Z. Valiev, N.A. Krasilnikov, N.K. Tsenev, Mater. Sci. Eng. A 137 (1991) 35–40.
[11] R.Z. Valiev, A.V. Korznikov, R.R. Mulyukov, Mater. Sci. Eng. A 168 (1993) 141–148.
[12] R.Z. Valiev, Mater. Sci. Eng. A 234–236 (1997) 59–66.
[13] R.Z. Valiev, R.K. Islamgaliev, I.V. Alexandrov, Prog. Mater. Sci. 45 (2000) 103–189.
[14] R.Z. Valiev, T.G. Langdon, Prog. Mater. Sci. 51 (2006) 881–981.
[15] C.N. Reid, Deformation Geometry for Materials Scientist, Pergamon Press, Oxford, 1973.
[16] R.D. Doherty, D.A. Hughes, F.J. Humphreys, J.J. Jonas, D.J. Jensen, M.E. Kassner, W.E. King, T.R. McNelley, H.J. McQueen, A.D. Rollett, Mater. Sci. Eng. A 238 (1997) 219–274.
[17] B. Verlinden, J. Driver, I. Samajdar, R.D. Doherty, Thermo-Mechanical Processing of Metallic Materials, Pergamon Press, Oxford, 2007.
[18] Y. Iwahashi, Z. Horito, M. Nemoto, T.G. Langdon, Metall. Mater. Trans. A 29 (1998) 2503–2510.
[19] W.Q. Cao, A. Godfrey, Q. Liu, Mater. Sci. Eng. A 361 (2003) 9–14.
[20] Y. Huang, P.B. Prangnell, Scr. Mater. 56 (2007) 333–336.
[21] A. Mishra, B.K. Kad, F. Gregori, M.A. Meyers, Acta Mater. 55 (2007) 13–28.
[22] M. Cabibbo, W. Blum, E. Evangelista, M.E. Kassner, M.A. Meyers, Metall. Mater. Trans. A 39 (2008) 181–189.
[23] E.A. El-Danaf, Mater. Sci. Eng. A 487 (2008) 189–200.
[24] I. Nikulin, Y. Motohashi, R. Kaibyshev, Mater. Sci. Forum 584–586 (2008) 691–696.
[25] I. Saxl, A. Kalousová, L. Ilucová, V. Sklenicka, Mater. Charact. 60 (2009) 1163–1167.
[26] M. Kawasaki, Z. Horita, T.G. Langdon, Mater. Sci. Eng. A 524 (2009) 143–150.
[27] R.Z. Abdulov, R.Z. Valiev, N.A. Krasilnikov, J. Mater. Sci. Lett. 9 (1990) 1445–1447.
[28] B.K. Kad, J.M. Gebert, M.T. Perez-Prado, M.E. Kassner, M.A. Meyers, Acta Mater. 54 (2006) 4111–4127.
[29] Y.S. Li, N.R. Tao, K. Lu, Acta Mater. 56 (2008) 230–241.
[30] F. Tang, J.M. Schoenung, Mater. Sci. Eng. A 493 (2008) 101–103.
[31] L. Tang, Z. Chen, C. Zhan, X. Yang, C. Liu, Metall. Mater. Trans. A 44 (2013) 793–805.
[32] V.V. Popov, E.N. Popova, D.D. Kuznetsov, A.V. Stolbovsky, E.V. Shorohov, P.A. Nasonov, K.A. Gaan, G. Reglitz, S.V. Divinski, G. Wilde, Mater. Sci. Eng. A 585 (2013) 281–291.
[33] A. Chatterjee, G. Sharma, A. Sarkar, J.B. Singh, J.K. Chakravartty, Mater. Sci. Eng. A 556 (2012) 653–657.
[34] J.B. Singh, R. Kapoor, A. Durga Prasad, J.K. Chakravartty, Mater. Sci. Eng. A 581 (2014) 26–30.
[35] T. Sakai, A. Belyakov, H. Miura, Metall. Mater. Trans. A 39 (2008) 2206–2214.
[36] O. Sitdikov, T. Sakai, A. Goloborodko, H. Miura, R. Kaibyshev, Philosophical Magazine 85 (2005) 1159–1175.

[37] P.B. Berbon, M. Furukawa, Z. Horita, M. Nemoto, T.G. Langdon, Metall. Mater. Trans. A 30 (1999) 1989–1997.
[38] V. Sklenicka, J. Dvorak, M. Svoboda, P. Kral, M. Kvapilova, Equal-channel angular pressing and creep in ultrafine-grained aluminium and its alloys, in: Z. Ahmad (Ed.), Aluminium Alloys – New Trends in Fabrication and Applications, InTech, 2013, http://dx.doi.org/10.5772/51242 (Open Access).
[39] Y. Iwahashi, J. Wang, Z. Horita, M. Nemoto, T.G. Langdon, Scr. Mater. 35 (1996) 143–146.
[40] Y. Estrin, A. Vinogradov, Acta Mater. 61 (2013) 782–817.
[41] I. Sabirov, M.Y. Murashkin, R.Z. Valiev, Mater. Sci. Eng. A 560 (2013) 1–24.
[42] A.P. Zhilyaev, G.V. Nurislamova, B.K. Kim, M.D. Baró, J.A. Szpunar, T.G. Langdon, Acta Mater. 51 (2003) 753–765.
[43] P.W. Bridgman, J. Appl. Phys. (1937) 328–336.
[44] A.P. Zhilyaev, T.G. Langdon, Prog. Mater. Sci. 53 (2008) 893–979.
[45] A.K. Ghosh, US Patent, Method for Producing a Fine Grain Aluminium Alloy Using Three Axis Deformation, Patent number 4721,537, (1988).
[46] S. Mu, W. Hua, G. Gottstein, Mater. Sci. Forum 584–586 (2008) 697–702.
[47] R. Kapoor, A. Sarkar, R. Yogi, S.K. Shekhawat, I. Samajdar, J.K. Chakravartty, Mater. Sci. Eng. A 560 (2013) 404–412.
[48] A. Vorhauer, R. Pippen, in: R.Z.V.M. Zehetbauer (Ed.), Nanomaterials by Severe Plastic Deformation – NANOSPD2, Wiley-VCH, Vienna, Austria, 2002, p. 691.
[49] A. Kundu, R. Kapoor, R. Tewari, J.K. Chakravartty, Scr. Mater. 58 (2008) 235–238.
[50] T. Sakai, H. Miura, X. Yang, Mater. Sci. Eng. A 499 (2009) 2–6.
[51] S. Zhang, W. Hu, R. Berghammer, G. Gottstein, Acta Mater. 58 (2010) 6695–6705.
[52] A. Belyakov, W. Gao, H. Miura, T. Sakai, Metall. Mater. Trans. A 29 (1998) 2957–2965.
[53] A. Belyakov, T. Sakai, H. Miura, K. Tsuzaki, Philos. Mag. A 81 (2001) 2629–2643.
[54] Y.J. Li, W. Blum, Phys. Stat. Sol. 202 (2005) R119.
[55] Y. Saito, H. Utsunomiya, N. Tsuji, T. Sakai, Acta Mater. 47 (1999) 579–583.
[56] G.J. Raab, R.Z. Valiev, T.C. Lowe, Y.T. Zhu, Mater. Sci. Eng. A 382 (2004) 30–34.
[57] Y. Saito, H. Utsunomiya, H. Suzuki, T. Sakai, Scr. Mater. 42 (2000) 1139–1144.
[58] J.-C. Lee, H.-K. Seok, J.-H. Han, Y.-H. Chung, Mater. Res. Bull. 36 (2001) 997–1004.
[59] Y. Iwahashi, Z. Horita, M. Nemoto, T.G. Langdon, Acta Mater. 46 (1998) 3317–3331.
[60] Y. Iwahashi, Z. Horito, M. Nemoto, T.G. Langdon, Acta Mater. 45 (1997) 4733–4741.
[61] R. Kapoor, J.K. Chakravartty, Acta Mater. 55 (2007) 5408–5418.
[62] R. Kapoor, S.K. Shekhawat, I. Samajdar, Mater. Sci. Eng. A 611 (2014) 114–122.
[63] H. Hasegawa, S. Komura, A. Utsunomiya, Z. Horita, M. Furukawa, M. Nemoto, T.G. Langdon, Mater. Sci. Eng. A 265 (1999) 188–196.
[64] J.T. Wang, Y. Zhang, J.Q. Liu, Mater. Sci. Forum 584–586 (2008) 929–937.
[65] F. Dalla Torre, A.A. Gazder, G.F. Gu, C.H.J. Davies, E.V. Pereloma, Metall. Mater. Trans. A 38A (2007) 1080–1095.
[66] V.V. Popov, E.N. Popova, A.V. Stolbovskiy, Mater. Sci. Eng. A 539 (2012) 22–29.
[67] Z. Pan, F. Xu, S.N. Mathaudhu, L.J. Kecskes, W.H. Yin, X.Y. Zhang, K.T. Hartwig, Q. Wei, Acta Mater. 60 (2012) 2310–2323.
[68] S. Suwas, B. Beausir, L.S. Tóth, J.J. Fundenberger, G. Gottstein, Acta Mater. 59 (2011) 1121–1133.
[69] S. Biswas, S. Singh Dhinwal, S. Suwas, Acta Mater. 58 (2010) 3247–3261.
[70] L. Jiang, M.T. Pérez-Prado, P.A. Gruber, E. Arzt, O.A. Ruano, M.E. Kassner, Acta Mater. 56 (2008) 1228–1242.
[71] Z. Horita, D.J. Smith, M. Furukawa, M. Nemoto, R.Z. Valiev, T.G. Langdon, J. Mater. Res. 11 (1996) 1880–1890.

[72] K. Ikeda, K. Yamada, N. Takata, F. Yoshida, H. Nakashima, N. Tsuji, Mater. Trans. 49 (2008) 24–30.
[73] S. Ii, M. Hishida, N. Takata, K. Ikeda, H. Nakashima, N. Tsuji, Mater. Sci. Forum 584–586 (2008) 716–721.
[74] R.Z. Valiev, F. Chmelik, F. Bordeaux, G. Kapelski, B. Baudelet, Scr. Metall. Mater. 27 (1992) 855–860.
[75] J. Wang, Y. Iwahashi, Z. Horita, M. Furukawa, M. Nemoto, R.Z. Valiev, T.G. Langdon, Acta Mater. 44 (1996) 2973–2982.
[76] M. Furukawa, Z. Horita, M. Nemoto, R.Z. Valiev, T.G. Langdon, Acta Mater. 44 (1996) 4619–4629.
[77] P.B. Berbon, S. Komura, A. Utsunomiya, Z. Horita, M. Furukawa, M. Nemoto, T.G. Langdon, Mater. Trans. JIM 40 (1999) 772–778.
[78] C. Xu, M. Furukawa, Z. Horita, T.G. Langdon, Acta Mater. 51 (2003) 6139–6149.
[79] C.Y. Yu, P.L. Sun, P.W. Kao, C.P. Chang, Mater. Sci. Eng. A 366 (2004) 310–317.
[80] M. Furukawa, Y. Iwahashi, Z. Horita, M. Nemoto, N.K. Tsenev, R.Z. Valiev, T.G. Langdon, Acta Mater. 45 (1997) 4751–4757.
[81] H.W. Höppel, R.Z. Valiev, Z. Metallakunde 93 (2002) 641–648.
[82] R. Kapoor, C. Gupta, G. Sharma, J.K. Chakravartty, Scr. Mater. 53 (2005) 1389–1393.
[83] C.Y. Yu, P.L. Sun, P.W. Kao, C.P. Chang, Scr. Mater. 52 (2005) 359–363.
[84] N. Lugo, N. Llorca, J.M. Cabrera, Z. Horita, Mater. Sci. Eng. A 477 (2008) 366–371.
[85] Y.M. Wang, E. Ma, Appl. Phys. Lett. 83 (2003) 3165–3167.
[86] H. Ueno, K. Kakihata, Y. Kaneko, S. Hashimoto, A. Vinogradov, Acta Mater. 59 (2011) 7060–7069.
[87] O. Saray, G. Purcek, I. Karaman, H. Maier, Metall. Mater. Trans. A 43 (2012) 4320–4330.
[88] Y.G. Ko, D.H. Shin, K.-T. Park, C.S. Lee, Scr. Mater. 54 (2006) 1785–1789.
[89] Z. Li, B. Zheng, Y. Wang, T. Topping, Y. Zhou, R. Valiev, A. Shan, E. Lavernia, J. Mater. Sci. 49 (2014) 6656–6666.
[90] Y. Wang, M. Chen, F. Zhou, E. Ma, Nature 419 (2002) 912–915.
[91] R.Z. Valiev, I.V. Alexandrov, Y.T. Zhu, T.C. Lowe, J. Mater. Res. 17 (2002) 5–8.
[92] G. Horii, D. Terada, N. Tsuji, J. Phys. 240 (2010) 012114.
[93] P.L. Sun, E.K. Cerreta, J.F. Bingert, G.T. Gray Iii, M.F. Hundley, Mater. Sci. Eng. A 464 (2007) 343–350.
[94] R. Pippan, S. Scheriau, A. Taylor, M. Hafok, A. Hohenwarter, A. Bachmaier, Annu. Rev. Mater. Res. 40 (2010) 319–343.
[95] H. Mughrabi, H.W. Höppel, M. Kautz, Scr. Mater. 51 (2004) 807–812.
[96] I. Semenova, G. Salimgareeva, G. Da Costa, W. Lefebvre, R. Valiev, Adv. Eng. Mater. 12 (2010) 803–807.
[97] F. Dalla Torre, R. Lapovok, J. Sandlin, P.F. Thomson, C.H.J. Davies, E.V. Pereloma, Acta Mater. 52 (2004) 4819–4832.
[98] W.S. Zhao, N.R. Tao, J.Y. Guo, Q.H. Lu, K. Lu, Scr. Mater. 53 (2005) 745–749.
[99] Y.M. Wang, E. Ma, M.W. Chen, Appl. Phys. Lett. 80 (2002) 2395–2397.
[100] Y.H. Zhao, J.F. Bingert, X.Z. Liao, B.Z. Cui, K. Han, A.V. Sergueeva, A.K. Mukherjee, R.Z. Valiev, T.G. Langdon, Y.T. Zhu, Adv. Mater. 18 (2006) 2949–2953.
[101] Y.H. Zhao, J.F. Bingert, Y.T. Zhu, X.Z. Liao, R.Z. Valiev, Z. Horita, T.G. Langdon, Y.Z. Zhou, E.J. Lavernia, Appl. Phys. Lett. 92 (2008) 081903.
[102] A. Vinogradov, T. Suzuki, S. Hashimoto, K. Kitagawa, A. Kuznetsov, S. Dobatkin, Structure and mechanical properties of submicrocrystalline copper produced by ECAP to very high strains, Mater. Sci. Forum 503–504 (2006) 971–976.
[103] V.M. Segal, V.I. Reznikov, V.I. Kopylov, D.A. Pavlik, V.F. Malyshev, Processy Plasticheskogo Structyroob Razovania Metallov, Sci Eng, Minsk, 1994.
[104] W.J. Kim, C.Y. Hyun, H.K. Kim, Scr. Mater. 54 (2006) 1745–1750.
[105] Y.B. Chun, S.H. Ahn, D.H. Shin, S.K. Hwang, Mater. Sci. Eng. A 508 (2009) 253–258.
[106] C.S. Chung, J.K. Kim, H.K. Kim, W.J. Kim, Mater. Sci. Eng. A 337 (2002) 39–44.
[107] R. Lapovok, C. Loader, F.H.D. Torre, S.L. Semiatin, Mater. Sci. Eng. A 425 (2006) 36–46.

[108] J. May, M. Dinkel, D. Amberger, H.W. Höppel, M. Göken, Metall. Mater. Trans. A 38 (2007) 1941–1945.
[109] T.L. Tsai, P.L. Sun, P.W. Kao, C.P. Chang, Mater. Sci. Eng. A 342 (2003) 144–151.
[110] Y.H. Zhao, X.Z. Liao, Z. Jin, R.Z. Valiev, Y.T. Zhu, Acta Mater. 52 (2004) 4589–4599.
[111] P. Cavaliere, M. Cabibbo, Mater. Charact. 59 (2008) 197–203.
[112] N. Tsuji, Y. Ito, Y. Saito, Y. Minamino, Scr. Mater. 47 (2002) 893–899.
[113] R. Valiev, M. Murashkin, I. Semenova, Metall. Mater. Trans. A 41 (2010) 816–822.
[114] Y. Fukuda, K. Oh-Ishi, Z. Horita, T.G. Langdon, Acta Mater. 50 (2002) 1359–1368.
[115] K.-T. Park, D. Shin, Metall. Mater. Trans. A 33 (2002) 705–707.
[116] S. Qu, C.X. Huang, Y.L. Gao, G. Yang, S.D. Wu, Q.S. Zang, Z.F. Zhang, Mater. Sci. Eng. A 475 (2008) 207–216.
[117] L.R. Saitova, H.W. Höppel, M. Göken, I.P. Semenova, R.Z. Valiev, Int. J. Fatigue 31 (2009) 322–331.
[118] I.P. Semenova, R.Z. Valiev, E.B. Yakushina, G.H. Salimgareeva, T.C. Lowe, J. Mater. Sci. 43 (2008) 7354–7359.
[119] R.Z. Valiev, I.V. Alexandrov, N.A. Enikeev, M.Y. Murashkin, I.P. Semenova, Rev. Adv. Mater. Sci. 25 (2010) 1–10.
[120] G.G. Yapici, I. Karaman, H.J. Maier, Mater. Sci. Eng. A 434 (2006) 294–302.
[121] S. Zherebtsov, G. Salishchev, R. Galeyev, K. Maekawa, Mater. Trans. 46 (2005) 2020–2025.
[122] Y.F. Shen, L. Lu, Q.H. Lu, Z.H. Jin, K. Lu, Scr. Mater. 52 (2005) 989–994.
[123] L. Lu, X. Chen, X. Huang, K. Lu, Science 323 (2009) 607–610.
[124] E.W. Hart, Acta Metall. 15 (1967) 351.
[125] R. Kapoor, N. Kumar, R.S. Mishra, C.S. Huskamp, K.K. Sankaran, Mater. Sci. Eng. A 527 (2010) 5246–5254.
[126] U.F. Kocks, J.J. Jonas, H. Mecking, Acta Metall. 27 (1979) 419–432.
[127] I.H. Lin, J.P. Hirth, E.W. Hart, Acta Metall. 29 (1981) 819–827.
[128] R. Kapoor, P.S. De, R.S. Mishra, An analysis of strength and ductility of ultrafine grained Al alloys, Mater. Sci. Forum 633–634 (2010) 165–177.
[129] N. Kamikawa, X. Huang, N. Tsuji, N. Hansen, Acta Mater. 57 (2009) 4198–4208.
[130] S.X. Li, G.R. Cui, J. Appl. Phys. 101 (2007) 083525–083526.
[131] C.Y. Yu, P.W. Kao, C.P. Chang, Acta Mater. 53 (2005) 4019–4028.
[132] A.S. Khan, B. Farrokh, L. Takacs, Mater. Sci. Eng. A 489 (2008) 77–84.
[133] J.W. Wyrzykowski, M.W. Grabski, Philos. Mag. A 53 (1986) 505–520.
[134] G.J. Fan, H. Choo, P.K. Liaw, E.J. Lavernia, Acta Mater. 54 (2006) 1759–1766.
[135] R.Z. Valiev, N.A. Enikeev, M.Y. Murashkin, V.U. Kazykhanov, X. Sauvage, Scr. Mater. 63 (2010) 949–952.
[136] M.H. Shih, C.Y. Yu, P.W. Kao, C.P. Chang, Scr. Mater. 45 (2001) 793–799.
[137] J.J. Gracio, J.V. Fernandes, J.H. Schmitt, Mater. Sci. Eng. A 118 (1989) 97–105.
[138] M.A. Meyers, A. Mishra, D.J. Benson, Prog. Mater. Sci. 51 (2006) 427–556.
[139] M.A. Meyers, U. Andrade, A. Chokshi, Metall. Mater. Trans. A 26 (1995) 2881–2893.
[140] D.J. LLoyd, S.A. Court, Mater. Sci. Technol. 19 (2003) 1349–1354.
[141] N. Hansen, Scr. Mater. 51 (2004) 801–806.
[142] Q. Wei, J. Mater. Sci. 42 (2007) 1709–1727.
[143] Y. Wang, Y. Liu, J.T. Wang, Mater. Sci. Eng. A 635 (2015) 86–93.
[144] F.H. Dalla Torre, E.V. Pereloma, C.H.J. Davies, Acta Mater. 54 (2006) 1135–1146.
[145] N.Q. Chinh, P. Szommer, T. Csanádi, T.G. Langdon, Mater. Sci. Eng. A 434 (2006) 326–334.
[146] R.W. Hayes, D. Witkin, F. Zhou, E.J. Lavernia, Acta Mater. 52 (2004) 4259–4271.
[147] M. Hokka, J. Kokkonen, J. Seidt, T. Matrka, A. Gilat, V.T. Kuokkala, Exp. Mech. 52 (2012) 195–203.
[148] H. Miyamoto, K. Ota, T. Mimaki, Scr. Mater. 54 (2006) 1721–1725.

[149] I. Sabirov, M.R. Barnett, Y. Estrin, P.D. Hodgson, Scr. Mater. 61 (2009) 181–184.
[150] D. Wu, X.L. Wang, T.G. Nieh, J. Phys. D 47 (2014) 175303.
[151] H. Conrad, J. Met. (July 1964) 582–588.
[152] U.F. Kocks, A.S. Argon, M.F. Ashby, Prog. Mater. Sci. 19 (1975) 1.
[153] Q. Wei, Z.L. Pan, X.L. Wu, B.E. Schuster, L.J. Kecskes, R.Z. Valiev, Acta Mater. 59 (2011) 2423–2436.
[154] Q. Wei, S. Cheng, K.T. Ramesh, E. Ma, Mater. Sci. Eng. A 381 (2004) 71–79.
[155] S. Suresh, Fatigue of Materials, Cambridge University Press, Cambridge, 1998.
[156] H. Mughrabi, H.W. Höppel, Int. J. Fatigue 32 (2010) 1413–1427.
[157] A. Vinogradov, S. Hashimoto, Mater. Trans. 42 (2001) 74–84.
[158] A.W. Thompson, W.A. Backofen, Acta Metall. 19 (1971) 597–606.
[159] M. Goto, S.Z. Han, J. Kitamura, T. Yakushiji, J.H. Ahn, S.S. Kim, M. Baba, T. Yamamoto, J. Lee, Int. J. Fatigue 73 (2015) 98–109.
[160] O. Kulyasova, R. Islamgaliev, B. Mingler, M. Zehetbauer, Mater. Sci. Eng. A 503 (2009) 176–180.
[161] T. Hanlon, E.D. Tabachnikova, S. Suresh, Int. J. Fatigue 27 (2005) 1147–1158.
[162] A. Vinogradov, S. Nagasaki, V. Patlan, K. Kitagawa, M. Kawazoe, Nanostruct. Mater. 11 (1999) 925–934.
[163] T.G. Langdon, J. Mater. Sci. 44 (2009) 5998–6010.
[164] A.K. Mukherjee, J.E. Bird, J.E. Dorn, Trans. ASM 62 (1969) 155.
[165] M. Kawasaki, T.G. Langdon, J. Mater. Sci. 42 (2007) 1782–1796.
[166] T.G. Langdon, Mech. Mater. 67 (2013) 2–8.
[167] R.Z. Valiev, D.A. Salimonenko, N.K. Tsenev, P.B. Berbon, T.G. Langdon, Scr. Mater. 37 (1997) 1945–1950.
[168] R.B. Figueiredo, T.G. Langdon, Adv. Eng. Mater. 10 (2008) 37–40.
[169] M. Kawasaki, T.G. Langdon, Mater. Sci. Eng. A 528 (2011) 6140–6145.
[170] V.M. Segal, Mater. Sci. Eng. A 386 (2004) 269–276.
[171] R. Georgy, IOP Conf. Ser. 63 (2014) 012009.
[172] S. Ferrasse, V.M. Segal, F. Alford, J. Kardokus, S. Strothers, Mater. Sci. Eng. A 493 (2008) 130–140.
[173] A. Azushima, R. Kopp, A. Korhonen, D.Y. Yang, F. Micari, G.D. Lahoti, P. Groche, J. Yanagimoto, N. Tsuji, A. Rosochowski, A. Yanagida, CIRP Ann. Manuf. Technol. 57 (2008) 716–735.
[174] A. Yanagida, K. Joko, A. Azushima, J. Mater. Process. Technol. 201 (2008) 390–394.
[175] C.J. Luis, D. Salcedo, J. Leon, I. Puertas, J.P. Fuertes, R. Luri, J. Nanomater. 2013 (2013) 14.
[176] J.P. Fuertes, J. Leon, C.J. Luis, D. Salcedo, I. Puertas, R. Luri, J. Nanomater. 2015 (2015).
[177] R.B. Figueiredo, E.R. de, C. Barbosa, X. Zhao, X. Yang, X. Liu, P.R. Cetlin, T.G. Langdon, Mater. Sci. Eng. A 619 (2014) 312–318.
[178] R.Z. Valiev, I. Sabirov, A.P. Zhilyaev, T.G. Langdon, JOM 64 (2012) 1134–1142.
[179] D. Salcedo, C.J. Luis, J. León, I. Puertas, J.P. Fuertes, R. Luri, J. Manuf. Sci. Eng. 136 (2013) 011009.
[180] J.H. Kim, S.K. Hwang, Y.-T. Im, I.-H. Son, C.M. Bae, Mater. Sci. Eng. A 552 (2012) 316–322.
[181] Y.G. Jin, H.M. Baek, S.K. Hwang, Y.-T. Im, B.C. Jeon, J. Mater. Process. Technol. 212 (2012) 848–855.
[182] O.A. Kaibyshev, R.V. Safiullin, R.Y. Lutfullin, O.R. Valiakhmetov, R.M. Galeyev, A. Dutta, T. Raghu, G.G. Saha, Mater. Sci. Technol. 22 (2006) 343–348.
[183] W.J. Kim, Y.K. Sa, Scr. Mater. 54 (2006) 1391–1395.
[184] Y. Estrin, M. Janecek, G.I. Raab, R.Z. Valiev, A. Zi, Metall. Mater. Trans. A 38 (2007) 1906–1909.
[185] A. Medvedev, H.P. Ng, R. Lapovok, Y. Estrin, T.C. Lowe, V.N. Anumalasetty, Mater. Lett. 145 (2015) 308–311.
[186] C.N. Elias, M.A. Meyers, R.Z. Valiev, S.N. Monteiro, J. Mater. Res. Technol. 2 (2013) 340–350.
[187] Y. Zhang, Y.S. Li, N.R. Tao, K. Lu, Appl. Phys. Lett. 91 (2007) 211901.
[188] E.V. Bobruk, M.Y. Murashkin, V.U. Kazykhanov, R.Z. Valiev, Rev. Adv. Mater. Sci. 31 (2012) 109–115.
[189] M.Y. Murashkin, I. Sabirov, V.U. Kazykhanov, E.V. Bobruk, A.A. Dubravina, R.Z. Valiev, J. Mater. Sci. 48 (2013) 4501–4509.

CHAPTER

MATERIALS IN A HIGH MAGNETIC FIELD

21

S.B. Roy

Raja Ramanna Centre for Advanced Technology, Indore, India

CHAPTER OUTLINE

Materials Under Extreme Conditions. http://dx.doi.org/10.1016/B978-0-12-801300-7.00021-8

1. INTRODUCTION

1.1 TRADITIONAL AND NEWER DEFINITIONS OF HIGH MAGNETIC FIELD

Physical and chemical properties of a material are mainly determined by the dynamics of electrons, which in a material are arranged in the atomic orbitals. In the case of metals, the electrons in the outermost orbitals are collectivized and shared between all the atoms. An externally applied magnetic field interacts with the electrons in a material through the spin- and orbital-magnetic moments or via the Lorentz force, causing cyclotron motion. Hence understandably an applied magnetic field can be an important probe for investigating the microscopic material properties. However, in comparison with energy density of an applied magnetic field, the interaction energy (Zeeman energy) between a material and the magnetic field can be quite small. For example, a 10 T field corresponds to 4×10^7 Jm^{-3}, whereas the interaction energy at 10 T is only about 2×10^{-22} J [1]; so the effect of magnetic field on materials is relatively small in comparison with the effects caused by other external parameters like temperature and pressure.

In 1980s a 10 T magnetic field was considered to be a very strong field. Nowadays magnetic fields up to 15 T are quite common in the laboratories worldwide and can be easily obtained by conventional superconducting magnets. In specialized high magnetic-field laboratories worldwide like the National High Magnetic Field Laboratory (NHMFL) at Tallahassee, Florida, United States; the Nijmegen Magnet Laboratory, Holland; the Grenoble High Magnetic Field Laboratory, France; and the National Institute of Materials Science, Tsukuba, Japan, water-cooled magnet in the range of 30 T and hybrid magnets (combination of superconducting and resistive magnets) in the range of 40 T are now regularly operational. A magnetic field is considered steady when the timescale of the change in magnetic field is in tens of minutes. On the other hand, a magnetic field is called a *pulsed field* when the changes take place in a very short time, like milliseconds. A nondestructive pulsed magnetic field up to 70 T can be generated with coils, which can work for up to hundreds of pulses. However, such nondestructive pulsed magnetic fields cannot be generated in the megagauss limit, because the coils get destroyed as a result of the combined action of Joule heating and Maxwell stress. A "single-turn coil method" can provide a magnetic field up to 300 T, but the coil gets destroyed after the peak magnetic field. To achieve magnetic fields higher than 300 T, an extreme method of magnetic flux compression is used.

The quest to achieve higher and higher magnetic fields has been mainly attributed to the traditional belief that a very high magnetic field is required in order to get more detailed information on the microscopic properties of materials via Zeeman energy. The interactions increase strongly with increasing magnetic field, and a magnetic field stronger than the interaction energy of a material is likely to reveal its magnetic and electronic states more accurately and with a higher degree of resolution. As a result, the term *high magnetic field* was mainly associated with the field regime above 10 T, which was generally suitable for a meaningful materials characterization and where magnetic field–induced transitions were observed in many materials. However, many new magnetic materials have been discovered, in which a magnetic field of relatively low strength ($<$10 T) induces a first-order magnetostructural phase transition that is often accompanied by the interesting functional properties of a large magnetocaloric effect (MCE), magnetoresistance, and magnetostriction. The magnetic field in these materials facilitates a coupling between the spin and lattice degree of freedom, which is a key to the understanding and subsequent tuning of the observed multifunctional properties. On the other hand, in a type-II superconductor the magnetic field penetrates in the form of quantized magnetic flux lines or vortices, which arrange themselves in some regular fashion similar to the structural lattice of a real material, popularly known as *vortex matter.* The density of the vortex matter changes with an applied magnetic field of relatively low strength, and the vortex matter can undergo various kinds of magnetic field–induced phase transitions, including a vortex solid–vortex liquid transition. Understanding the magnetic field response of vortex matter is absolutely essential for the tuning the dissipationless current-carrying properties of a superconductor. The lower limit of a "high" magnetic field in the case of biological materials is usually taken to be lower by one or two orders of magnitude than that for nonbiological materials, 0.1–1 T. Hence the required strength of the magnetic field to investigate the properties of materials does actually depend on the type of material and/or material's properties under investigation, and often newer and interesting physical phenomena are observed in even in magnetic fields less than 10 T.

1.2 WHY THE STUDY OF MATERIALS IN A MAGNETIC FIELD IS IMPORTANT

To start with, high magnetic fields can be used for detailed characterization of materials in a broad sense. For example, quantum electronic structures of materials can be investigated through the quantum Hall effect, cyclotron resonance, and the de Haas–van Alphen (dHvA) effect. On the other hand, measurements of the critical magnetic field and critical current density of superconductors are very important for basic characterization of superconductors as well as for the assessment of their suitability for technological applications. Furthermore, a variety of phase transitions can be induced using an applied magnetic field, and there is always an expectation that new phenomena may also be waiting to be discovered in the high magnetic-field regime; the discovery of the quantum fluid in a high magnetic field is an example in this direction. If a strong magnetic field is applied during the solidification process of a crystal, it can strongly influence the morphology of the crystal through the orientation effect. This effect leads to new methods of magnetic processing of high superconducting transition temperature (T_C) superconductors, permanent magnets, and biological materials. In the low field (<10 T) regime, the interest is mainly focused on magnetic field–induced magnetostructural phase transition in newer classes of materials, leading to interesting sensor and actuator applications. One classic example is the giant magnetoresistance (GMR)–based sensors/read heads, which along with the newer magnetic materials for data storage has increased hard disk capacity in computers dramatically since the mid-1990s. Another area of interest is, where the influence of a relatively low magnetic field correlates with the biological functions associated with a material.

2. GENERATION OF HIGH MAGNETIC FIELDS

If the timescale of the increasing (or decreasing) magnetic field during an experiment is a few minutes or a few tens of minutes, the field is called a *steady field.* On the other hand, if the magnetic field changes in a very short time, like milliseconds, it is called a *pulsed field* [1]. There is, however, no strict distinction between a steady field and a pulsed field regarding the timescale. It depends on the field sweep time, which is faster or slower than the thermal relaxation time of the coil producing the magnetic field. Pulsed fields with a duration of several hundred milliseconds up to several seconds are sometimes called *long pulse fields* [1].

2.1 STEADY FIELDS

Superconducting magnets are usually used in laboratories worldwide to produce fields up to 20 T. A magnetic field of 7 T using an Nb_3Sn-based superconducting magnet was first generated in 1961 by Kunzler et al. [2], and the superconducting magnets that produced over 10 T became commercially available in the 1970s [1]. However, quenching of the superconducting magnets remained a problem in the field region greater than 10 T, and it was only in the late 1980s that a 20 T superconducting magnet was finally built successfully.

Special techniques are needed for generating a magnetic field above 20 T; the maximum field limit is around 23 T, using a high T_C superconductor [3]. There is an ongoing project at NHMFL at Florida State University (Tallahassee, Florida, United States) for construction of a 32 T superconducting magnet consisting of a state-of-the-art 15 T low-temperature superconductor outer magnet and a transformational 17 T high-temperature superconductor inner coil set using a high-T_C yttrium barium copper oxide (YBCO)−coated conductor. Superconducting magnets consume lots of liquid helium, and hence they are not quite user-friendly. Cryogen-free superconducting magnets have been developed, but the available field for real experiments is still limited to around 15 T [1].

High-power water-cooled magnets called *Bitter magnets* or *hybrid magnets*, which are composed of a Bitter magnet and a large-bore superconducting magnet, need to be used to generate steady fields above 20 T. The important points in the design of Bitter magnets (named after Francis Bitter who designed this type of magnet at the Massachusetts Institute of Technology, United States) are "how to remove the huge amount of heat" and "how to support the strong magnetic forces in the radial and axial directions" [1]. In order to solve these two problems simultaneously, Bitter made an ingenious design with a stack of plates, which could support both axial and radial strong forces and remove heat by water flowing along the axial direction in the holes [1]. Bitter magnets are still used to produce fields of up to 33 T at NHMFL at Florida State University, Tallahassee, Florida, United States.

A hybrid magnet, which is a combination of a superconducting magnet and a Bitter magnet, is used to obtain a higher steady field than that obtained by a Bitter magnet. Such a 45 T hybrid magnet is regularly used for experiments at NHMFL. This magnet combines a superconducting magnet of 11.5 T with a resistive Bitter magnet of 33.5 T. Such magnets consume electricity at high power and also require a huge amount of liquid helium for the large-bore superconducting magnet installed outside the Bitter magnet.

2.2 PULSED FIELDS

In a pulsed field magnet, the power consumption is much smaller than that of a steady field magnet; hence the technical problems in the generation of static and pulsed magnetic fields are of a different nature [1]. Kapitza at the Cavendish Laboratory was the first to build large pulsed magnets of field strength 35−50 T [4,5], and used these magnets for many pioneering experimental studies on magnetoresistance, magnetization, magnetostriction, the Zeeman effect, and the study of electron orbits. The pulsed field is quite economical as compared with steady fields, and it is relatively easy to generate a pulsed field on the order 50 T for about 10 ms in a 20-mm bore [1]. The highest nondestructive pulsed field obtained so far is 100.75 T and the highest field for a long-pulsed magnet is 60 T (http://www.magnet.fsu.edu/mediacenter/news/pressreleases/2012/100tshot.html., 2012).

In the methods to obtain magnetic fields higher than 100 T, destruction of the magnet is unavoidable. A metal will explode violently when it is exposed to such a high magnetic field pulse because of the combined effect of Joule heating and Maxwell stress [6]. This explosion starts as a result of the skin effect at the surface of the inner wall of the magnet coil, and is accompanied by shock waves compressing the magnetic coil material by pushing the surface away from the magnetic field. In the "single turn coil method," a large current is supplied to a small coil for a few microseconds and the destruction of the coil occurs after the current reaches the peak field, as a result of the inertia of the coil material. However, because the coil actually expands during the experiment, the sample kept inside the coil survives. A magnetic field up to 300 T can be obtained using this method [1]. An extreme method of magnetic flux compression is employed to generate fields higher than 300 T. In this method, a metallic cylinder called a *liner* with magnetic flux inside is imploded uniformly from outside while cylindrical symmetry is maintained. As a result, the magnetic flux gets compressed according to the ratio of the initial area and the final area of the cross-section inside the cylinder. Using the method of electromagnetic flux, compression a field up to 600 T can be generated. Even higher fields up to 1000 T can be achieved by compressing the magnetic flux with a high explosive [1]. Such experiments involve currents of the order of several megaamperes and the pulse duration is about 10–50 μs. However, both the magnet coil and the sample are completely destroyed during these experiments.

3. RESPONSE OF MATERIALS IN A MAGNETIC FIELD

3.1 METALS AND SEMICONDUCTORS

The subject matter and the discussion in this section will be closely in line with the textbooks by Ziman [7], Ashcroft and Mermin [8], and Mizutani [9].

3.1.1 Orbiting Electrons and Quantum Oscillations

The concept of the Fermi surface is very useful in condensed matter physics for predicting the thermal, electrical, magnetic, and optical properties of metals, semimetals, and doped semiconductors. The Fermi surface is an abstract boundary in a reciprocal space, and its shape is derived from the periodicity and symmetry of the crystalline lattice and from the occupation of electronic energy bands. In materials with complex crystal structures, the Fermi surface can be quite intricate.

The conduction electrons moving with the velocity v in the presence of electric field E and magnetic field $B = \mu_0 H$ experience the Lorentz force given in Eq. (21.1):

$$F = -e(E + v \times B) \tag{21.1}$$

where $-e$ is the electronic charge. Considering the effect of the magnetic field alone on the motion of the conduction electron, Eq. (21.1) indicates that the electron will rotate in a closed orbit if the velocity v has no component along the magnetic field B. A nonzero component of the velocity in the direction of B will cause the electron to move in a helical path. The magnetic field changes the direction of velocity but not its magnitude; hence the energy of the electron remains unchanged. For free electrons, the frequency of this periodic motion is $w_C = eB/m$.

It is natural to expect that such periodic motion of the electron could be detected by resonance with an electromagnetic field of suitable frequency. The condition of such resonance is that the electron should complete at least one circuit of the orbit before being scattered [7]. A cyclotron resonance experiment can directly determine the effective mass of the charge career in semiconductors and metals. Observation of the surface resistance, reflecting power, absorption, and other properties of the materials will show a cyclotron resonance line of width $1/\tau$ at the frequency $w = w_C$. A high value of $w_C\,\tau$ is required for the electrons to undergo periodic motion; the resonance condition $w = w_C$ is satisfied at microwave frequencies. In a semiconductor, the electron density is relatively small and a microwave field can penetrate the semiconductors reasonably well. The carriers in a semiconductor, whether electrons or holes, occupy little pockets of the energy $E(k)$ at minima or maxima of the band. The energy surfaces are usually ellipsoid in these neighborhoods [7]. Cyclotron resonance frequency w_C varies with orientation and thus provides information about the axes of the ellipsoid. In metals the electrons can absorb energy only when they are within a skin depth of the surface. At microwave frequencies where the skin depth is quite small compared with the electronic mean free path, one is in the extreme anomalous regime.

With metals, one needs to go above the plasma frequency to see the same sort of resonance line as may be readily detected in a semiconductor [7]. Also in order to resolve the resonance line, one needs to work at high frequencies with $w\tau >> 1$. In this condition in a metal, the skin depth is very small; the radius of the electron orbit in the magnetic field in real space will be considerably larger than the penetration distance of the electric field. In such situations, Azbel and Kaner suggested placing the magnetic field parallel to the surface of the metal [8]. An electron can resonantly absorb energy from an electric field if it experiences the electric field of the same phase each time it enters the skin depth. This will only be possible if the applied field completes an integral number of periods, T_E, each time the electron returns to the surface, i.e., $T = nT_E$, where T is the period of cyclotron motion and n is an integer. This leads to the relation $w = nw_C$. With frequency kept fixed, if the absorption is plotted as a function of $(1/B)$, the resonant peaks because a given cyclotron period will be uniformly spaced. The quantity actually measured experimentally is the derivative of the real part of the surface impedance dR/dH. Such an absorption derivative curve for elemental copper [10] is shown in Fig. 21.1.

In a solid, the electrons are represented by the Bloch wave function, and the force F exerted on the Bloch electron of wave vector k is generally expressed as $F = (h/2\pi)dk/dt$, where h is the Planck constant. Eq. (21.1) now takes the form of Eq. (21.2):

$$\left(\frac{h}{2\pi}\right)\frac{dk}{dt} = -e(v \times B) \tag{21.2}$$

Hence in the reciprocal space the motion of an electron of wave vector k can be described as rotating in a closed orbit on a constant energy surface normal to the magnetic field B. Integration of Eq. (21.2) yields Eq. (21.3).

$$(k - k_0) = -\left(\frac{2\pi e}{h}\right)(r \times B) \tag{21.3}$$

where r is the position vector of the electron and k_0 is a constant. The trajectory of the electron in reciprocal space has the same shape as that in real space but is rotated by 90 degrees and scaled by the factor $(-2\pi e/h)B$.

In the presence of the magnetic field, the closed orbit of the Bloch electron is quantized in the same manner as the electron orbit of a free atom. Following the semiclassical approach of Onsager and Lifshitz, the Bohr–Sommerfeld quantization rule for the motion of the Bloch electron in the magnetic field is written as Eq. (21.4) [9]:

$$\int\left(\left(\frac{h}{2\pi}\right)k - eA\right)ds = \left(n + \frac{1}{2}\right)\left(\frac{h}{2\pi}\right) \tag{21.4}$$

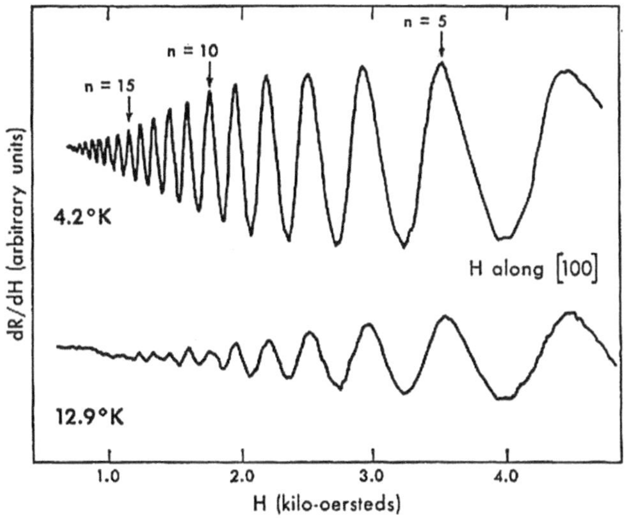

FIGURE 21.1

Absorption derivative curves for copper belly orbits at 4.2 and 12.9 K. *a.u.*, Arbitrary unit; *dR/dH*, surface impedance, *H*, magnetic field.

P. Haussler, S.J.Welles, 1962. Phys. Rev. 152 675, http://www.magnet.fsu.edu/mediacenter/news/pressreleases/2012/100tshot.html 2012. Copyright 1962, Reproduced with the permission of American Physical Society.

where n is a positive integer including zero and A is the vector potential defined as $B = \nabla \times A$, and the integral is carried out over the closed orbit of the electron in real space.

Using Eqs. (21.3) and (21.4), Eq. (21.5) can be derived [9]:

$$eBS = \left(n + \frac{1}{2}\right)\left(\frac{h}{2\pi}\right) \tag{21.5}$$

where S is the area of the closed orbit in real space. The area $A(E)$ of the close orbit in reciprocal space is related to the area S of the closed orbit in real space through the relation in Eq. (21.6):

$$A(\varepsilon) = \left(\frac{2\pi eB}{h}\right)^2 S \tag{21.6}$$

Eq. (21.6) can be rewritten with the help of Eq. (21.5) as Eq. (21.7):

$$A_n(\varepsilon) = \left(\frac{4\pi^2 eB}{h}\right)\left(n + \frac{1}{2}\right) \tag{21.7}$$

Eq. (21.7) indicates that the area enclosed by the electron orbit is quantized and takes only discrete sets of values, which correspond to the principal quantum number n. The enclosed area differs by $(4\pi^2 eB/h)$, when the electron jumps from a given orbit to another. When a magnetic field is applied along the z-axis, the Bloch electron circulates in the k_xk_y plane normal to the magnetic field in reciprocal space, and only a single parameter n is needed to describe its stationary circular motion. Three independent quantum numbers k_x, k_y, and k_z, of the wave vector k are then reduced to two quantum numbers, n and k_z, in the presence of a magnetic field along the z-axis. A set of concentric cylinders with their axes along the z-direction replace the spherical Fermi surface of the free electron

(Fig. 21.2a). A projection of the Fermi surface of the concentric cylinders onto the $k_x k_y$ plane obtains the concentric circles specified by the principal quantum number $n = 0, 1, 2, 3, \ldots$ (Fig. 21.2b). The energy eigenvalue of the free electron in the magnetic field is given by Eq. (21.8) [9]:

$$\varepsilon = \left(n + \frac{1}{2}\right)\left(\frac{h\omega C}{2\pi}\right) + \left(\frac{h^2 k_z^2}{8m\pi^2}\right); \quad (n = 0, 1, 2) \tag{21.8}$$

where n is the principal quantum number and $w_C = eB/m$.

The corresponding ε–k_z relations consist of a discrete set of parabolas with different integers, which are called the *Landau levels* (Fig. 21.3). Here, each electronic state is specified by a set of two quantum numbers n and k_z. The separation of the adjacent Landau levels is widened with increasing magnetic field because w_C in Eq. (21.8) is proportional to the magnetic field. The wider the energy separation, the more electrons should be accommodated in a given level. The internal energy U of the conduction electrons depends critically on the closeness of the highest occupied Landau level to the Fermi level E_F. The internal energy is minimized at the magnetic field, where the highest occupied Landau level has just passed E_F and is emptied. These electrons must be absorbed in the next highest

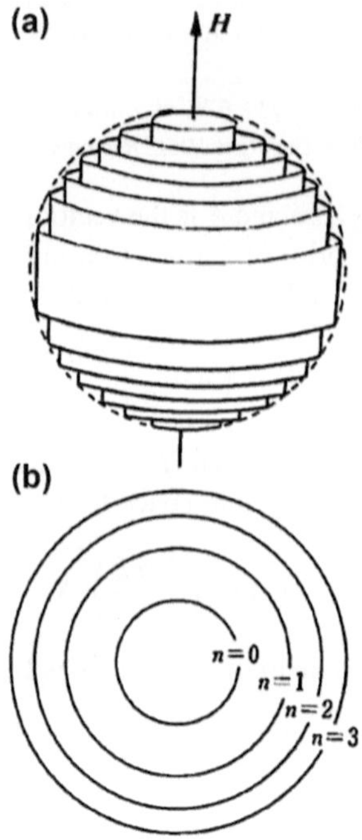

FIGURE 21.2

(a) The free-electron Fermi surface in the presence of a magnetic field along the z-axis. (b) Its projection onto the $k_x k_y$ plane. H, Magnetic field.

U. Mizutani, Introduction to the Electron Theory of Metals, Cambridge University Press, 2001. Copyright 2001, Reproduced with the permission of Cambridge University Press.

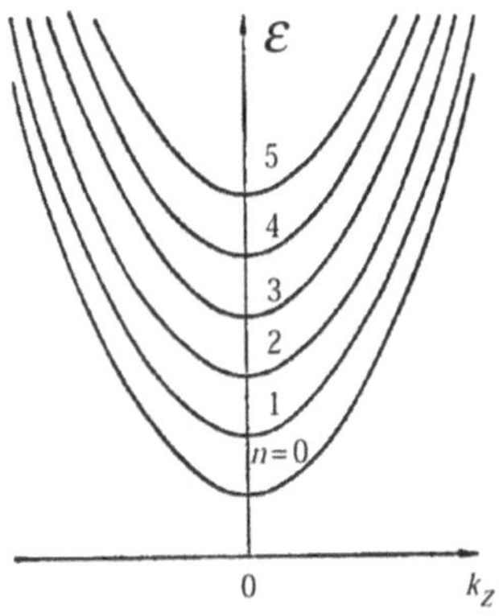

FIGURE 21.3

ε–k_z Relation given by Eq. (21.8). The Landau level refers to each parabola specified by *n*.

U. Mizutani, Introduction to the Electron Theory of Metals, Cambridge University Press, 2001. Copyright 2001, Reproduced with the permission of Cambridge University Press.

Landau level below E_F. This process is repeated every time the Landau level passes the Fermi level, resulting in oscillations of the internal energy of the conduction electrons as a function of the magnetic field.

3.1.2 Experimental Studies of the Fermi Surface

The effect of oscillations in the internal energy of the conduction electron system is reflected in various physical properties of metals and semimetals. The differentiation of the internal energy with respect to the magnetic field leads to the experimentally measureable quantity of magnetic susceptibility, which oscillates as a function of the magnetic field. This is called the dHvA *effect.* Denoting magnetic fields as B_1 and B_2, where the two successive Landau levels n and $n-1$ pass the Fermi level, from Eq. (21.7) one can derive Eq. (21.9):

$$\Delta\left(\frac{1}{B}\right) = \left(\frac{1}{B_1} - \frac{1}{B_2}\right) = \frac{4\pi^2 e}{hA(\varepsilon)} \tag{21.9}$$

Plotting the dHvA signal as a function of an inverse of the applied magnetic field, from Eq. (21.9), an external cross-section of the Fermi surface $A(E)$ normal to the applied magnetic field can be deduced from the period of oscillations. Fig. 21.4 shows the dHvA oscillations of pure Zn measured by the torque method [11]. In a metal or semimetal, the Fermi surface topology can be determined by measuring the quantum oscillations in magnetic fields applied along different directions relative to the principal crystal axes of a single crystal. An accurate measurement of the period of the oscillations can be made, even if the signal itself is rather weak. The Fermi radius, for example, in pure Cu, can be determined with an accuracy of 0.001% [9].

The specific heat of a material is derived from the differentiation of the internal energy with respect to temperature. Hence the specific heat also oscillates with increasing magnetic field. This is known as *magnetothermal oscillation.* On the other hand, the de Haas–Shubnikov effect refers to the oscillations of the electrical resistivity as a function of a magnetic field. Similarly, an oscillatory behavior can be observed in the Hall coefficient and the thermoelectric power [9].

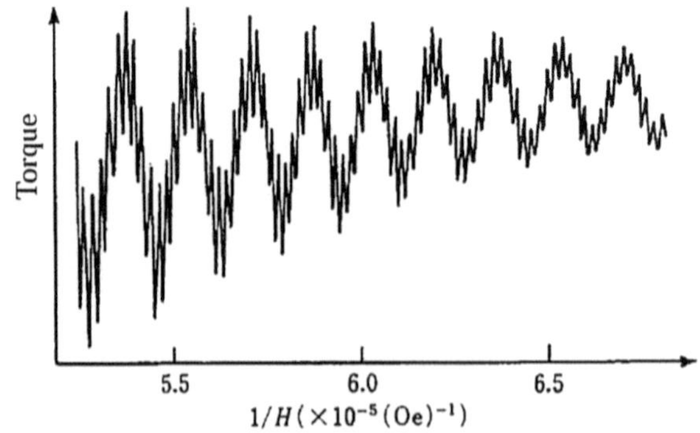

FIGURE 21.4

dHvA oscillations of pure Zn measured by the torque method.

U. Mizutani, Introduction to the Electron Theory of Metals, Cambridge University Press, 2001. Copyright 2001, Reproduced with the permission of Cambridge University Press.

Useful information about the Fermi surface geometry can be extracted by measuring attenuation of sound waves in a metal as they propagate perpendicular to a uniform magnetic field [8]. This is particularly so if the wave is carried by displacement of ions that are both perpendicular to its direction of propagation and to the magnetic field. Such a sound wave is accompanied by an electric field of the same frequency and wave vector because the ions are electrically charged. The electrons in a metal interact with the sound wave through this electric field, and support or hinder the sound wave propagation. If the condition $w_C \tau >> 1$ for permitting the electrons in the magnetic to complete many orbits between the collisions is satisfied, then the sound attenuation can depend on wavelength in a manner reflecting the Fermi surface geometry. An electron will be weakly coupled to the wave when its orbit diameter is an integer number of wavelengths. On the other hand, it can be strongly coupled when the orbit diameter differs from an integer number of wavelength by half a wavelength. For a more detailed discussion, readers are referred to the book by Ashcroft and Mermin [8]. The sound attenuation displays a periodic variation with inverse wavelength, in which the period is equal to the inverse of the extremal diameter of the Fermi surface along the direction of sound propagation. The shape of the Fermi surface can be deduced from the observed structure in the experimentally determined sound attenuation by varying the direction of sound propagation and the direction of the applied magnetic field. For a more detailed discussion on the experimental studies of Fermi surface using high magnetic field, readers are referred to the books by Mizutani [9], Ziman [7], and Ashcroft and Mermin [8].

In all these cases discussed in the previous paragraph, the magnetic field effect is related to the cyclotron frequency (or Landau energy) defined as $w_C = eB/m$. If this energy is higher than the Fermi energy, the condition is called *quantum limit*, and the corresponding field may be called a *high magnetic field* [1]. This magnetic field depends on the carrier density of the material and can be extremely high for metals, whereas it may be quite moderate for semiconductors with a low carrier density.

In real materials, the formation of the Landau level, which arises because of the circular motion of the conduction electron caused by the Lorentz force in the magnetic field, is greatly affected by the scattering of the electrons by impurities and phonons. The quantum oscillations will disappear if the electron is scattered before the completion of a revolution [9]. This necessitates that the electron mean free path in the materials is long enough for completion of a revolving motion in the magnetic field. This in turn demands that the measurement be made on high-quality single crystals and at very low temperatures under high magnetic fields. The requirement of a high-purity single crystal certainly limits the materials that can be studied, and the dHvA effect has been measured mostly on elemental metals and intermetallic compounds with a stoichiometric composition and with long electron mean free path at low temperatures. Between 1960 and 1970, the Fermi surface of pure elements was well investigated through extensive studies of the quantum oscillation. In comparison, the Fermi surface studies of alloys and intermetallic compounds are relatively more difficult. However, the study of quantum oscillations has revealed many important features of three classes of interesting materials, namely, organic conductors and superconductors, heavy fermion systems, and high T_C superconductors. A detailed discussion on the high magnetic field effects in these interesting materials is beyond the scope of this chapter. However, we provide a brief discussion in the following paragraphs.

Before going into the discussions on these rather exotic materials, we would like to point out interest in bismuth (Bi), which was the first element to have its Fermi surface experimentally identified [12]. In fact, experimental studies of Bi first led to the discovery of quantum oscillations in both magnetization [13] and resistivity [14]. This semimetal Bi is distinguished from other metals by an extremely small Fermi surface occupying 10^{-5} of the Brillouin zone [15] and a very long mean free path exceeding 2 mm at room temperature [16]. The quantum limit in Bi can be reached by the application of a magnetic field of modest strength (9 T, along the trigonal axis) because of the low carrier density. In this limit, electrons are all pushed to the lowest Landau level and the radius of the lowest-energy quantized isolated electron orbit in a magnetic field becomes shorter than the Fermi wavelength [1,17]. Behnia et al. studied transport coefficients of a single crystal of Bi up to 33 T, which is deep inside the ultraquantum limit [17]. They observed three unexpected maxima in the Nernst coefficient that are related to the quasiplateaus in the Hall coefficient. Their results suggest that bulk Bi may host an exotic quantum fluid similar to the one associated with the fractional quantum Hall effect (FQHE) (to be discussed in detail later in this chapter).

The organic conductors have generally complicated crystalline structures with low symmetry and large unit cells involving hundreds of atoms. The band structure calculations, however, predict a multiband quasi two-dimensional electronic structure, giving rise to a relatively simple Fermi surface in most of these organic conductors. Experimental studies on numerous organic conductors (e.g., θ-$(ET)_4CoBr_4(C_6H_4Cl_2)$, β^t-$(ET)_4NH_4[Fe(C_2O_4)_3(C_3H_7NO)$, $(ET)_8[Hg_4Cl_{12}(C_6H_5Br)_2$, $(TMTSF)_2NO_3$, and $(BEDO)_4Ni(CN)_44CH_3CN)$ revealed Fermi surfaces that can be regarded as model systems for the study of quantum oscillations in networks of orbits coupled by magnetic breakdown [18]. Two types of networks have been distinguished, namely, the linear chain of coupled orbits and the networks of compensated orbits. In addition there are additional anomalous behaviors in the observed quantum oscillations in various organic conductors, which of course need further attention and experiments in order to get a better insight into these interesting materials.

Heavy-fermion systems are intermetallic compounds based on the rare-earth elements Ce and Yb or actinide elements U, Np, and Pu. In these intermetallic compounds 4f- or 5f- electron orbitals are extended in nature and strongly hybridized with the conduction electron bands. The resultant strong electronic correlations in turn give rise to the formation of heavy quasiparticles with a strongly enhanced effective mass m_{eff}, which is 100–1000 times the value of the free-electron mass m_0, hence the name *heavy fermion systems*. The f-electron magnetic moments in heavy fermion systems tend to get screened by the conduction electrons in heavy fermion systems via the Kondo interaction. On the other hand, the conduction electrons can also mediate Ruderman–Kittel–Kasuya–Yosida (RRKY) indirect exchange between the f-electron magnetic moments, which promotes long-range magnetic ordering. A subtle competition between Kondo and RKKY interactions govern the interesting properties of heavy fermion systems. The characteristic energy scale of Kondo and RKKY interaction is represented by $T_K \propto (-1/Jn\ (E_F))$ and $T_{RKKY} \propto J^2n\ (E_F)$, respectively, where J is the exchange interaction between the f-electron and the conduction electron and $n(E_F)$ is the density of state at the Fermi surface [19]. The parameter $Jn\ (E_F)$ can be influenced by applying either hydrostatic or chemical pressure and in many heavy fermion systems, it is possible to tune the magnetic ordering temperature to zero by pressure or chemical substitutions, giving rise to a quantum phase transition between antiferromagnetic (AFM) and nonmagnetic ground states. The amplitude of quantum magnetic fluctuations grows rapidly in the vicinity of the quantum phase transition, and this can lead to two distinct phenomena: non-Fermi liquid behavior and unconventional superconductivity. It is conjectured that in contrast to phonons in conventional superconductors, electron pairing in heavy fermion superconductors is probably the result of quantum critical magnetic fluctuations where the intensity is maximum at the quantum phase transition. The exact nature of the quantum critical fluctuations, however, still remains a matter of debate. A proposal based on the critical fluctuations of the magnetic order parameter has failed to describe the non-Fermi liquid behavior observed in many heavy fermion systems. In this regard, an unconventional framework based on wave vector q-independent local critical magnetic fluctuations is doing its round [19]. The observation of the enhancement of AFM fluctuations (but not of the local magnetic fluctuations) at the quantum phase transition in the heavy-fermion system $Ce_{1-x}La_xRu_2Si_2$, however, supports the scenario based on quantum critical fluctuations of the order parameter. Another interesting question is whether the f-electrons are itinerant or localized on both sides of the quantum phase transition, and this can be answered by comparing the results of the theoretical band structure calculations with the experimental results of dHvA effect. Such dHvA experiments can also probe the effective masses for each band, and in many heavy-fermion compounds dHvA measurements have identified quasiparticles with masses of up to 100 times the bare electron mass [19,20], thus providing a direct evidence of a large quasiparticle density of states at the Fermi level. A high magnetic field is expected to quench the Kondo effect in heavy fermion systems, leading to a localization of the f-electrons. It is an interesting question as to whether the magnetic field–induced localization of f-electrons occurs suddenly at a field-induced transition or progressively develops in a magnetic field. This has also triggered many high magnetic field experimental studies of heavy-fermion systems [19].

The hole-doped high-T_C copper-oxide superconductors show various interesting properties as a function of hole-doping. The normal state properties of these copper oxide superconductors on the overdoped side can be understood within the framework of a generalized Fermi-liquid theory. On the other hand, several types of experiments, including nuclear magnetic resonance (NMR), angle-resolved photoemission spectroscopy, polarized [21] elastic neutron diffraction, the polar Kerr effect, scanning tunneling spectroscopy, resonance ultrasound spectroscopy, and the Nernst effect revealed the presence of a mysterious pseudogap in the underdoped side of the phase diagram. The exact nature of this pseudogap, whether this a precursor effect to superconductivity or a distinct phase competing with superconductivity, is still not quite clear. The quantum oscillations have been observed in a high magnetic field in underdoped $YBa_2Cu_3O_y$ and overdoped $Tl_2Ba_2CuO_{6+\delta}$ (Fig. 21.5) which

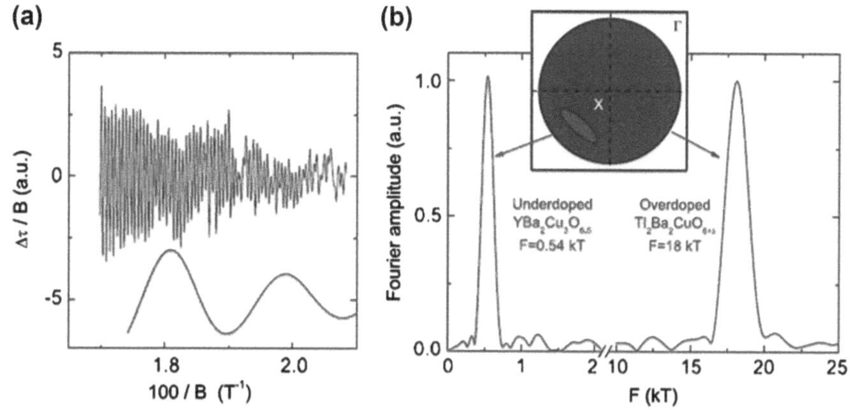

FIGURE 21.5

(a) Oscillatory part of the magnetization (raw data) versus $1/B$ of $Tl_2Ba_2CuO_{6+\delta}$ (*magenta*, dark gray in print versions) and $YBa_2Cu_3O_{6.5}$ (*red*, light gray in print versions). (b) Fourier analysis of the raw data shown in (a). The inset is a sketch of the size of the Fermi pocket in the first Brillouin zone deduced from the frequency of the quantum oscillations for $Tl_2Ba_2CuO_{6+\delta}$ (*magenta*, dark gray in print versions) and $YBa_2Cu_3O_{6.5}$ (*red*, light gray in print versions). *a.u.*, Arbitrary units.

B. Vignolle, D. Vignolles, M. Julien, C. Proust, C. R. Phys. 14 (2013) 39. Copyright 2013, Reproduced with the permission of Elsevier.

proved the existence of a Fermi surface in the high T_C copper oxide compounds both in the underdoped and overdoped regions.

The Fermi surface of the overdoped $Tl_2Ba_2CuO_{6+\delta}$ was described by a large quasi–two-dimensional Fermi cylinder covering 65% of the first Brillouin zone. The observation of quantum oscillations implies that in spite of strong electron–electron interactions in the overdoped $Tl_2Ba_2CuO_{6+\delta}$, quasiparticles exist at all the points of the Fermi surface. Measurements of the dHvA effect in the underdoped $YBa_2Cu_3O_{6.5}$ suggested the existence of a closed and coherent Fermi surface. However, the quantum oscillation frequency was found to be significantly smaller than that observed in overdoped compounds, which pointed toward a relatively small external area representing only about 2% of the first Brillouin zone. All these experimental studies suggest that below a critical hole-doping level, copper-oxide compounds undergo a Fermi surface reconstruction, where the large holelike Fermi surface gets reconstructed into small pockets. Whereas the results of band structure calculations agree well with the experimental results of overdoped $Tl_2Ba_2CuO_{6+\delta}$, the Fermi surface deduced from band structure calculations in $YBa_2Cu_3O_{6.5}$ suggest two large holelike tubular CuO_2 sheets, plus three quasi–one-dimensional sheets. The observation of negative Hall and Seebeck coefficients in the underdoped copper-oxide compounds [21], along with the results of dHvA effect studies, however, suggest that the large Fermi surface suggested by the band structure calculation undergoes a reconstruction in the high field and low-temperature limits. This in turn suggests the possibility of a translational symmetry breaking as a result of some form of spin, charge, or orbital order. NMR studies have indeed revealed a magnetic field–induced charge order in the copper-oxide planes of the underdoped copper-oxide compounds as soon as superconductivity is sufficiently weakened by the magnetic field [21].

3.2 CLASSICAL AND QUANTUM HALL EFFECT

We shall now consider the effect of an electric field E_x applied along the x direction, which is in addition to the perpendicular magnetic field (Fig. 21.6). In this discussion, we shall closely follow the treatments presented in the book by Kaxiras [22]. The applied electric field will induce a current $j_x = (-e)nv_x$ in the same direction, where v_x is the velocity and n is the density of electrons in the plane. The moving electrons will experience a force in the y direction because of the presence of the magnetic field, which is given in Eq. (21.10):

$$F_y = (-e)\left(\frac{v_x}{c}\right)H \tag{21.10}$$

Using the expression for the current, and the fact that the motion in the y direction can be associated with an effective electric field E_y, we obtain Eq. (21.11):

$$E_y = \frac{F_y}{-e} = \left(\frac{j_x}{(-e)nc}\right)H \tag{21.11}$$

or Eq. (21.12):

$$\frac{J_x}{E_y} = \frac{(-e)nc}{H} \tag{21.12}$$

Here, n is the density of the electrons. Eq. (21.12) is defined as the Hall conductivity, σ_H. It is the ratio of the current in the x direction to the effective electric field E_y in the perpendicular direction. More generally, this ratio can be viewed as one of the off-diagonal components of the conductivity tensor σ_{ij} $(i, j = x, y)$, which relates the current to the electric field (Eqs. 12.13 and 21.14):

$$j_x = \sigma_{xx}E_x + \sigma_{xy}E_y \tag{21.13}$$

$$j_y = \sigma_{yx}E_x + \sigma_{yy}E_y \tag{21.14}$$

The electric field can be expressed in terms of the current using the resistivity tensor (Eqs. 21.15–21.17):

$$\rho_{ij}(i,j = x, y) \tag{21.15}$$

$$E_x = \rho_{xx}j_x + \rho_{xy}j_y \tag{21.16}$$

$$E_y = \rho_{yx}j_x + \rho_{yy}j_y \tag{21.17}$$

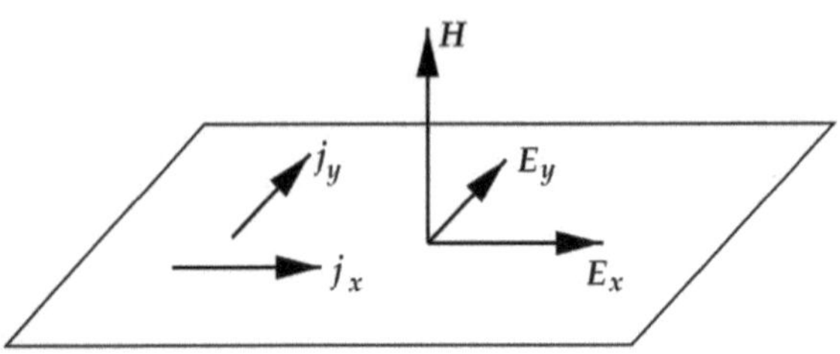

FIGURE 21.6

Geometry of the two-dimensional electrons in the Hall effect, with the external magnetic field H, the x, y components of the electric field E_x, E_y, and the current j_x, j_y.

E. Kaxiras, Atomic and Electronic Structure of Solids, Cambridge University Press, 2003. Copyright 2003, Reproduced with the permission of Cambridge University Press.

which is the inverse of the conductivity tensor. Assuming the case of an isotropic solid, i.e., $\rho_{yx} = -\rho_{xy}$ and $\rho_{xx} = \rho_{yy}$, the relationship between the two tensors can be found through Eqs. (21.18) and (21.19):

$$\sigma_{xx} = \frac{\rho_{xx}}{\rho_{xy}^2 + \rho_x^2} = \sigma_{yy} \tag{21.18}$$

$$\sigma_{xy} = \frac{-\rho_{xy}}{\rho_{xy}^2 + \rho_x^2} = -\sigma_{yx} \tag{21.19}$$

It is assumed that the conditions in the system of confined electrons are such that $\sigma_{xx} = 0 \Rightarrow \rho_{xx} = 0$.

We then find for the Hall conductivity $\sigma H = \sigma_{xy} = -1/\rho_{xy} = j_x/E_y$. Now, if the field H fulfills the condition shown in Eq. (21.20),

$$H = \frac{nch}{e(l+1)} \tag{21.20}$$

exactly $l+1$ Landau levels will be filled with index numbers 0, 1,...,l. If the electrons cannot be accommodated exactly by an integer number of Landau levels, then levels up to index l will be completely filled and the level with index $l+1$ will be partially filled.

With $j_x/E_y = (-enc)/H$ one can now write (Eq. 21.21):

$$\sigma_H = -\frac{enc}{H} = -\frac{e^2}{h}(l+1) \tag{21.21}$$

where $l+1$ is the number of filled Landau levels (up to index l). Equivalently, we can obtain the Hall resistivity (the off-diagonal component of the resistivity tensor) as Eq. (21.22):

$$\rho_{xy} = -\frac{1}{\sigma_{xy}} = \frac{h}{e^2}\left(\frac{1}{l+1}\right) \tag{21.22}$$

Eqs. (21.21) and (21.22) suggest that the Hall conductivity and resistivity are quantized. The important assumption here is that the value of the magnetic field and the density of electrons in the plane are such that exactly $l+1$ Landau levels are filled. In 1980, experiments by von Klitzing and co-workers [23] indeed showed that for very high magnetic fields and low temperatures, the Hall resistivity ρ_{xy} as a function of the field H has wide plateaus that correspond to integer values of $l+1$ in Eq. (21.22). The diagonal resistivity ρ_{xx} vanishes when ρ_{xy} has the quantized values (as suggested by Eq. 21.22). This behavior is shown schematically in Fig. 21.7. This striking observation is now called the integer quantum Hall effect. The existence of plateaus in the value of ρ_{xy} is because of the fact that not all states in the system are extended Bloch states. The Landau levels are not δ-functions in energy in a real system, but are broadened because of the presence of impurities. The states with energy very close to the ideal Landau levels only are extended and can carry current, whereas the states lying between ideal Landau levels are localized and cannot carry current. As the magnetic field is varied and localized states with energy away from an ideal Landau level are being filled, there can be no change in the conductivity of the system, which remains stuck to its quantized value corresponding to the ideal Landau level. This produces the plateaus in the Hall resistivity ρ_{xy}. This striking phenomenon demonstrates the importance of defects in the crystal, which can dominate the electronic behavior. It should also be noted that the plateaus in Hall resistivity can be measured with unprecedented precision, reaching one part in 10^7. Because the ratio h/e^2 appears in several

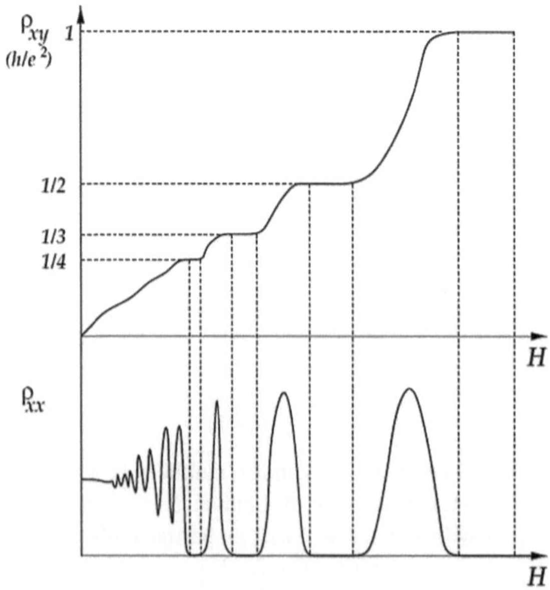

FIGURE 21.7

Schematic representation of the integral quantum Hall effect. The plateaus in the Hall resistivity ρ_{xy} as a function of the magnetic field H, where its values are given by $(h/e^2)\ (1/(l+1))$ with $l+1$ an integer, are identified; the diagonal resistivity ρ_{xx} vanishes for the values of H corresponding to the plateaus.

E. Kaxiras, Atomic and Electronic Structure of Solids, Cambridge University Press, 2003. Copyright 2003, Reproduced with the permission of Cambridge University Press.

fundamental units, this has great significance in metrology. A couple of years after the experiments of von Klitzing and co-workers, experiments by Tsui, Stormer, and Gossard [24,25] on samples of very high purity at high magnetic fields and low temperatures revealed that the plateaus in the Hall resistivity can also occur for values larger than h/e^2. In particular, they occur at values 3, 3/2, 5, 5/2, 5/3,…, in units of h/e^2. This is known as the *FQHE*, and Laughlin, Tsui, and Stormer were awarded the 1998 Nobel Prize for Physics for their work on the FQHE.

3.3 DIAMAGNETIC, PARAMAGNETIC, AND FERROMAGNETIC MATERIALS

In this section we shall closely follow the book by Blundell [26], and readers are referred to this book for further details.

In a diamagnetic material, a magnetic field induces a magnetic moment that opposes the very magnetic field that actually causes it. The diamagnetism is a quantum mechanical phenomenon, and a classical explanation of it in terms of Lenz's law is not quite correct. For a material of volume V and N ions (each with Z number of electrons of mass m_e and ionic radius r) with all shells filled, the diamagnetic susceptibility is expressed as Eq. (21.23) [26]:

$$\chi = \frac{Ne^2\mu_0}{V6m_e}\sum_{i=1}^{Z}\langle r_i^2\rangle \tag{21.23}$$

Diamagnetic susceptibility is largely temperature independent. It is present in all materials and is associated with the closed atomic shells of the materials, but is usually much weaker than the other magnetic effects. Materials with delocalized π electrons such as naphthalene and graphite show large diamagnetic susceptibilities. The effective ring diameter in molecular naphthalene and the loosely bound sheets of hexagonal layers in graphite is several times larger than an atomic diameter, thus producing a large diamagnetic susceptibility in such materials.

In paramagnetic materials, an applied magnetic field induces magnetization, which tends to align with the magnetic field. In contrast with the diamagnetic materials with closed atomic shells and hence no magnetic moment, the atoms in paramagnetic materials possess a nonzero moment arising out of unpaired electrons. This nonzero magnetic moment of an atom is associated with its total angular momentum, which is a sum of the orbital angular momentum L and the spin angular momentum S. These magnetic moments associated with the atoms, however, point at random directions because of very weak interactions among themselves, and for any practical purpose they are considered to be independent. The magnetic moments in the paramagnetic materials tend to get aligned with the application of an external magnetic field, and the degree of alignment depends on the strength of the magnetic field. On the other hand, an increase of temperature in the materials will randomize the moments; hence the magnetization of a paramagnetic materials will depend on the ratio of the external magnetic field B and temperature T, i.e., B/T.

In the general case of a paramagnetic material, where total angular momentum J can take any integer or half-integer value, the induced magnetization in an applied magnetic field can be expressed as Eq. (21.24) [26]:

$$M = M_S B_J(y) \tag{21.24}$$

where Eq. (21.25):

$$y = \frac{g_J \mu_B J B}{k_B T} \tag{21.25}$$

and Eq. (21.26):

$$M_S = n g_J \mu_B J \tag{21.26}$$

and $B_J(y)$ is the Brillouin function given by Eq. (21.27):

$$B_J(y) = \frac{(2J+1)}{2J}\coth\left(\frac{2J+1}{2J}y\right) - \frac{1}{2J}\coth\left(\frac{y}{2J}\right) \tag{21.27}$$

Except at very low temperature and/or in very high magnetic fields, the experimental situation corresponds to $y << 1$, which leads to Eq. (21.28):

$$B_J(y) = \frac{(J+1)y}{3J} + O(y^3) \tag{21.28}$$

Thus for low magnetic fields the susceptibility in a paramagnetic materials is given by Eq. (21.29):

$$\xi \approx \frac{\mu_0 M}{B} = \frac{\eta \mu_0 \mu_{eff}^2}{3 k_B T} \tag{21.29}$$

which is the Curie–Weiss law. A measurement of susceptibility in paramagnetic materials therefore allows one to deduce the value of effective magnetic moment μ_{eff} (Eq. 21.30):

$$\mu_{eff} = g_J \mu_B \sqrt{J(J+1)} \tag{21.30}$$

where Eq. (21.31):

$$g_J = \frac{3}{2} + \frac{S(S+1) - L(L+1)}{2J(J+1)} \tag{21.31}$$

and g_J is known as the *Lande-g-factor.*

In ferromagnetic (FM) materials a spontaneous magnetization exists even in the absence of an applied magnetic field. This effect arises because of the exchange interaction (which is quantum mechanical in origin and much stronger than the dipolar interaction between magnetic moments) between the moments at the atomic sites. One of the earliest models to explain ferromagnetism is the Weiss model, where the net effect of the exchange interaction is represented by an effective molecular field, as shown in Eq. (21.32):

$$B_{mf} = -\frac{2}{g\mu_B} \sum_j J_{ij} S_j \tag{21.32}$$

Here it is assumed that one is dealing with a system where zero angular momentum, i.e., $L = 0$ and $J = S$, and all the magnetic atoms experience the same molecular field. This molecular field B_{mf} is a measure of the effect of ordering of the system, hence one can write Eq. (21.33):

$$B_{mf} = \lambda M \tag{21.33}$$

One can now proceed to investigate the response of a FM material in the same way as in the case of a paramagnetic material, which is placed in a magnetic field $B + B_{mf}$, where B is the externally applied magnetic field. At low enough temperatures, the internal molecular field B_{mf} can align the magnetic

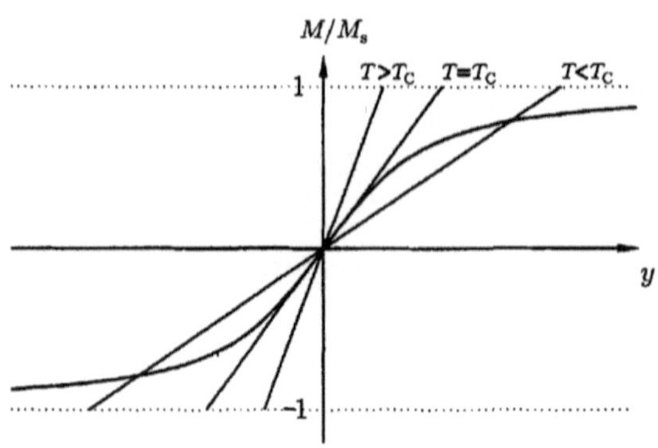

FIGURE 21.8

The graphical solutions of Eqs. (21.34) and (21.35).

S.J. Blundell, Magnetism in Condensed Matter, Cambridge University Press, 2001. Copyright 2001, Reproduced with the permission of Oxford University Press.

moments even in the absence of an applied magnetic field. To find a solution to the Weiss model the following equations are solved graphically [26]:

$$\frac{M}{M_S} = B_J(y) \tag{21.34}$$

and

$$y = \frac{g_J \mu_B J(B + \lambda M)}{k_B T} \tag{21.35}$$

This treatment in the absence of the term λM would be identical to the treatment of a paramagnetic material (described earlier in this section). Restricting to the case of $B = 0$, hence $M = k_B\, Ty/g_J\mu_B J\lambda$, the straight line produced by plotting M against y has a gradient that is proportional to temperature T. There is no simultaneous solution of the previous equations except at the origin, where $y = 0$ and $M_S = 0$ (Fig. 21.8). The situation, however, changes below a critical temperature T_C, when the gradient of the M versus y line is less than the Brillouin function $B_J\,(y)$ at the origin. At temperatures $T < T_C$, there are three solutions: 1 at $M_S = 0$ and another two for M_S at $\pm$ some nonzero value (Fig. 21.8). It is found that the nonzero solutions are stable, whereas the zero solution is unstable. Thus nonzero spontaneous magnetization appears in FM materials below a critical temperature T_C, even in the absence of an external magnetic field, and this spontaneous magnetization grows further with the decrease in temperature. Above this critical temperature T_C, the FM materials behave like a paramagnet.

This critical temperature in the FM materials is known as the *Curie temperature*, and can be estimated by finding when the gradients of the line $M = k_B Ty/g_J\mu_B J\lambda M_S$ and the curve $M = M_S B_J(y)$ are equal at the origin [26]. For small values of y, $B_J(y) = (J + 1)y/3J + O(y^3)$ and the Curie temperature T_C is then defined as Eq. (21.36):

$$T_C = \frac{g_J \mu_B (J + 1) M_S}{3k_B} = \frac{n\lambda \mu_{eff}^2}{3k_B} \tag{21.36}$$

The molecular field is given by Eq. (21.37):

$$B_{mf} = \lambda M_S = \frac{3k_B T_C}{g_J \mu_B (J + 1)} \tag{21.37}$$

For a ferromagnet with $J = 1/2$ and $T_C \approx$ 1000 K, B_{mf} is estimated to be about 1500 T [26]. This effective field is an enormous magnetic field, which reflects the strength of the exchange interaction.

The effect of an applied magnetic field in FM materials is to shift the straight line in the graphical solution of the equations to the right, which results in a nonzero solution for M for all temperatures. From energetic point of view there is always an advantage in FM materials in an applied magnetic field to have a nonzero magnetization with moments lining up along the magnetic field.

3.4 MULTIFERROIC MATERIALS

Multiferroic materials display magnetic and electric or elastic order simultaneously [27]. They exhibit an interesting magnetoelectric effect, where magnetization can be induced by the electric field and electric polarization by the magnetic field [28]. The surge of interest in this field arose mainly because of two developments that gave rise to the possibility of various interesting technological applications involving such materials [29]. First, the observation of significant enhancement of multiferroic properties in the thin film form of $BiFeO_3$ [30], and second, the discovery of a novel class of multiferroics, namely, $TbMnO_3$ [31] and $TbMn_2O_5$ [32], in which magnetism and ferroelectricity do not just coexist, but in which magnetism causes ferroelectricity. The multiferroics can be classified into two groups: type I and type II multiferroics. The type I multiferroics include those materials in which magnetism and ferroelectricity originate from different sources but with some coupling between them. In these materials, ferroelectricity with large spontaneous polarization usually appears at higher temperatures than magnetism. There are several subclasses of type I multiferroics, depending on the mechanism of ferroelectricity in them. $BiFeO_3$ ($T_{FE} = 1100$ K, $T_N = 643$ K) and $YMnO_3$ ($T_{FE} = 914$ K, $T_N = 76$ K) are typical examples of type I multiferroics.

The type II multiferroics show a strong coupling between ferroelectricity and magnetism and the ferroelectricity exists only in a magnetically ordered state. A typical example of type II multiferroics is $TbMnO_3$, where magnetic ordering appears at $T_{N1} = 41$ K. The magnetic structure changes at a lower temperature, $T_{N2} = 28$ K, and a nonzero electric polarization appears in this low-temperature magnetic state. This electric polarization can be influenced strongly by the application of an external magnetic field, and when a critical magnetic field is applied along a certain direction the polarization is rotated by 90 degrees [31]. A stronger magnetic field–induced effect on electrical polarization has been observed in $TbMn_2O_5$. In this compound, the electrical polarization changes sign with field, and oscillations in the polarization have been observed, with a magnetic field alternating between $+1.5$ and -1.5 T.

3.5 BIOLOGICAL MATERIALS

A biological material differs from a nonbiological material because its molecular description includes the notion of biological function. According to Hagen [33], a definition of biological function is the rate of a chemical reaction and its underlying molecular mechanism in the presence of a biocatalyst, usually an enzyme. A biological system is not different from a nonbiological system (e.g., condensed matter and complex fluids) if the influence of the magnetic field does not correlate to biological function. The lower limit of a "high" magnetic field in the case of biological materials is usually taken to be lower by one or two orders of magnitude than that for nonbiological materials, and lies in the range of 0.1 to 1 T. In rest of this section we shall closely follow the rather comprehensive review of Hagen [33] on this subject, and the readers are referred to this article for further details. Hagen [33] introduced a subdivision of classes of increasing complexity for the biological systems: (1) a "component," which is the smallest unit consisting of a single biomacromolecule or a set of chemically identical biomacromolecules; (2) a "single activity," which can be a reaction catalyzed by a single enzyme, or an interaction event between two molecules such as the binding of a substrate to an enzyme, the binding of a hormone to a receptor, or the interaction between two different biomacromolecules; a process is an extension of a single activity; and (3) a "cell," which is the minimal unit of life capable of proliferation. A complex system is an extension of the cell.

Complete orientation of noninteracting molecules in aqueous solution cannot be obtained in the present-day laboratory magnetic fields. Interacting biomacromolecules can, however, form larger structures like liquid crystalline phases, membranes, and crystals, and any object of sufficient size and with anisotropic diamagnetism can be partially or completely oriented in a homogeneous magnetic field. A small degree of orientation of biomacromolecules may influence the averaging required for high-resolution liquid-state NMR studies. It also appears that such weak ordering may even be exploited to study protein structure and dynamics in a hybrid solution and solid-state NMR. The fact that biomolecular structures and protein crystals can be partially oriented in magnetic fields suggests that the process of biomacromolecular crystallization may be influenced directly by the magnetic field [33]. It has indeed been observed that in a homogenous magnetic field of 10 T, a supersaturated solution of the protein lysosome forms a smaller number of crystals, but the crystals are of larger macrodimensions than those formed in the absence of the magnetic field [34,35].

Ataka and Wakayama [36] have suggested a possible explanation of the effect of a magnetic field on protein crystallization. A common growth mechanism for biomacromolecules is three-dimensional nucleation, where microcrystals tend to attach to a larger crystal surface and eventually join to form one crystal. A preorientation of the microcrystals in a magnetic field apparently creates a favorable condition. The apparent viscosity of the supersaturated protein solutions may also increase in the presence of a magnetic field, thus resulting in reduced convection [37,38]. The majority of the studies on biomolecular crystallization in magnetic fields have been pursued in hen egg white lysosome, although other proteins occasionally studied typically include those that are monomeric, globular, well soluble, and of small to intermediate size. Other systems like multimeric proteins, complex enzymes, oxygen-sensitive proteins, membrane proteins, paramagnetic proteins, and nucleic acids are yet to be studied [33].

The interaction between individual molecules and magnetic fields typically produced in the laboratory are much smaller than thermal energy at ambient temperatures. As a result, a magnetic field can affect the course of a biochemical reaction only indirectly via the perturbation of an intermediate quantum mechanical process. A possible mechanism is the modulation produced by an external magnetic field in the rate of intersystem crossing between singlet and triplet states of two interacting radicals created at the same time [33]. An increased activity of the enzyme catalase in fields up to 6 T has been reported in the literature [39]. However, similar investigations of magnetic field–induced change in activity have turned out to be negative in a few other systems. Intersystem crossing is an isoenergetic radiationless transition between electronic states with different multiplicities. The triplet and singlet germinate pair normally will not be isoenergetic, but they may become so by mixing of states through a variety of perturbations, namely, spin–orbit coupling, spin–spin interaction, hyperfine interaction, spin–lattice relaxation, and exchange interaction [33]. An external magnetic field of increasing strength will then change the relative weight of these interactions by increasing the Zeeman interaction and interfere with the intersystem crossing. According to Hagen [33], when combining these considerations, one would tend to conclude that the exploration of enzyme activities in a high magnetic field has been far from exhaustive on two counts: (1) the number of systems studied so far is limited, and (2) the explored magnetic fields are of relatively modest strength, typically 0–0.5 T. Many enzymes are metalloproteins with the pair made up of an organic radical and a transition metal ion radical, where the influence of the latter through spin–orbit coupling may demand much higher magnetic fields to modulate intersystem crossing rates.

The effects of magnetic fields (ranging from very weak to very strong) on complex subcellular, cellular, and supercellular processes are yet to be investigated in a systematic manner. There are many open questions that need more consistent experimental results along with convincing theories/models to have proper answers.

4. SUPERCONDUCTORS IN MAGNETIC FIELD

4.1 FLUX-LINE LATTICE OR VORTEX MATTER IN TYPE II SUPERCONDUCTORS

A superconductor is characterized by its perfectly diamagnetic property known as the *Meissner effect*, apart from, of course, the zero direct current resistance. However, it is also well known that in certain superconductors the magnetic field can partly enter the superconductors above a critical field H_{C1}, below which the material remains in the Meissner state. Above H_{C1}, however, the material continues to be in the superconducting state until an upper critical field H_{C2}, where it turns into a normal conducting state. Abrikosv in 1957 provided an elegant explanation of such partial magnetic flux penetration in terms of a periodic solution of the Ginzburg and Landau phenomenological theory of superconductivity [40]. Abrikosov's solution suggested that in the magnetic fields above H_{C1}, the magnetic field penetration in the superconductors gives rise to a regular triangular (hexagonal) lattice of parallel flux lines, also variously known as *flux tubes*, *fluxons*, or *Abrikosov vortex lines*. Each of these flux lines carries a quantum of magnetic flux $\Phi_0 = h/2e = 2.07 \times 10^{-15}$ Tm^2. The superconducting order parameter $\Psi(r)$ is suppressed to zero at the center of a flux line from its value $\Psi = 1$ that it attains in the Meissner state. The line $\Psi = 0$ is surrounded by a tube of radius ξ, the vortex core, which in turn is surrounded by a circulating supercurrent $J(r)$, which generates the magnetic field $B(r)$ of the flux line. In bulk superconductors, the vortex current and field are confined to a tubular region of radius λ; at large distances $r >> \lambda$, current and field of an isolated vortex decay as $\exp(-r/\lambda)$. The ratio $\kappa = \lambda/\xi$ is called the *Ginzburg–Landau parameter* of a superconductor. The flux-line lattice, or Abrikosov lattice, exists only in superconductors with $\kappa > 1/\sqrt{2}$, which are known as type II superconductors. In contrast, the type I superconductors have $\kappa < 1/\sqrt{2}$, and they exist only in complete diamagnetic or Meissner state.

4.2 VORTEX MATTER AND ITS PHASE TRANSITIONS

Thermal fluctuations play a very important role in determining the properties of flux-line lattice. The importance of thermal effects for a particular superconductor can be gauged through the estimation of the Ginzburg number, and the higher the value of the Ginzburg number, the more important thermal fluctuations become for the physical properties of that superconductor [41,42].

The Ginzburg number is quite small (of the order of 10^{-8} for the conventional metallic superconductors with low superconducting transition temperatures). The effect of thermal fluctuations is considered to be small in these superconductors and the flux-line lattice can be considered as a system of rigid rods arranged periodically in a triangular (hexagonal) lattice. On the other hand, in the layered copper-oxide superconductors, conjunction of extreme physical properties such as high transition temperatures, short coherence length, and large anisotropy together enhance the effect of thermal fluctuations. In such materials, flux-line lattice is considered to be a system of flexible interacting line objects. The flux-line lattice in the presence of pinning and thermal fluctuations is now commonly referred to as *vortex matter* [41]. In rest of this section, we shall closely adhere to the rather comprehensive reviews of Blatter on this subject, and readers are referred to these articles for further details [41,42].

In vortex matter, two structural elements play very important roles. The circular current flow in the scale λ produces a line elasticity $\varepsilon_1 \propto \varepsilon_0$, where $\varepsilon_0 = (\Phi_0/4\pi\lambda^2)$ is the basic energy scale of the problem. In an isotropic superconductor, the line elasticity is presented through the increase in length of the vortex, whereas in anisotropic high-T_C superconductors, the anisotropic constant strongly reduces the line elasticity. The vortex core, on the other hand, plays an important role in the dynamic behavior of the individual flux line, and also adds a correction in the line energy and a term in the line elasticity. The normal electrons (or quasiparticles) trapped in the vortex core region can give rise to a dissipative response, which limits the technological applicability of the type II superconductors. This actually motivates the quest for the so-called "hard" type II superconductors, where the vortices are pinned by material defects.

There is an important difference between ordinary and vortex matter. The interaction potential between atoms/ions in ordinary matter has an attractive component in $V(r)$, whereas in vortex matter, the vortex–vortex interaction is purely repulsive. As a result, the density of the vortex matter is determined by the external magnetic field and there is no generic density of the vortex matter. The interaction between flux lines wins over the entropy at low temperatures and, in general, a regular triangular (hexagonal) lattice is formed. The thermal fluctuations in vortex matter increase with the increase in temperature, and this ultimately leads to a first-order melting transition from a vortex–solid phase to a vortex–liquid phase. In conventional low-temperature superconductors, this vortex–solid melting transition takes place very near to the upper critical field $H_{C2}(T)$ line, and it is quite difficult to detect with standard experimental methods. However, in high-T_C copper-oxide superconductors, the enhanced thermal fluctuations bring the melting line well below the H_{C2}. An interesting feature of this melting line H_M (T) is that it has a negative slope, i.e., $dH_M\,(T)/dT < 1$. This is similar to what is observed during the melting of ice, and implies that the density of vortex–liquid is higher than that of vortex–solid. The first order nature of the vortex–solid melting transition in high-T_C copper-oxide superconductors has been demonstrated through various experimental techniques. A schematic vortex matter H–T phase diagram is shown in Fig. 21.9. In the Bi-based copper-oxide compound BSSCO, the vortex–solid melts at very low field values $B < \Phi_0/\lambda^2$. At such low field values, the shear modulus in the vortex–solid decreases exponentially as the circular current flow of the individual vortex tends to vanish, and the melting line shows a reentrant behavior resulting in a dilute vortex–liquid phase (see Fig. 21.9).

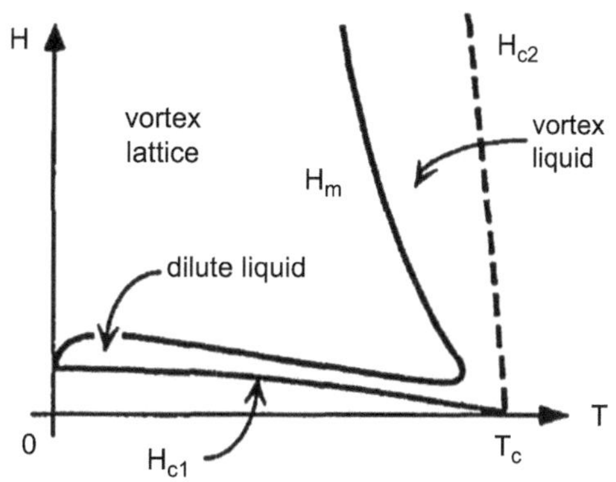

FIGURE 21.9

Phase diagram of an anisotropic high-T_C superconductor with a large Ginzburg number. H, Magnetic field; H_C, lower critical field; H_m, field for vortex solid–vortex liquid melting; T, temperature; T_C, superconducting transition temperature.

G. Blatter, Phys. C 282–287 (1997) 19. Copyright 1997, Reproduced with the permission of Elsevier.

4.3 QUENCHED DISORDER IN VORTEX MATTER AND CRITICAL CURRENT-CARRYING CAPACITY OF TYPE II SUPERCONDUCTORS

The physics of the vortex matter in the presence of quenched disorder or material defects is of considerable interest, especially because of the technological importance of the critical current (J_C) carrying-capacity of the concerned superconductor. Such material defects prevent the flux lines from moving under the action of the Lorentz force when an external current is applied to the superconductor and hence enable the superconductor to carry a true dissipationless supercurrent. The quenched disorder can be: (1) uncorrelated, e.g., point defects; and (2) correlated, which can be provided by twin boundaries in high-T_C copper-oxide compounds like YBCO or artificially generated by heavy ion irradiation, producing columnar defects. The random disorder potential caused by point defects acting on the flux lines can destroy the long-range translational order of the flux-line lattice and transforms the system to a quasi—long-range order glasslike phase, termed *Bragg glass*. This Bragg-glass phase can be transformed into a more strongly disordered vortex—glass phase with the increasing amount of quenched disorder. The first-order nature of the melting transition still exists for a weak enough disorder, but the experimental results suggest that at high field/low temperature, the first-order line terminates in a critical point with the emergence of an additional second-order transition (Fig. 21.10). The strongly correlated disorder-like columnar defects can completely destroy the original lattice order, and the new emerging phase is termed *Bose glass*. In contrast with the first-order nature of the melting transition in the defect-free or weakly disordered superconductor, the transition from Bose glass to a disordered state is a second-order transition. This Bose-glass transition is also shifted to higher temperatures and magnetic fields, and thus enhances the region of the H—T phase space where the superconductor can be used for technological applications. The Bragg, vortex, and Bose glass features also have significant influence in the reduction of flux creep and flux-flow behavior observed in the superconductors with weak disorder.

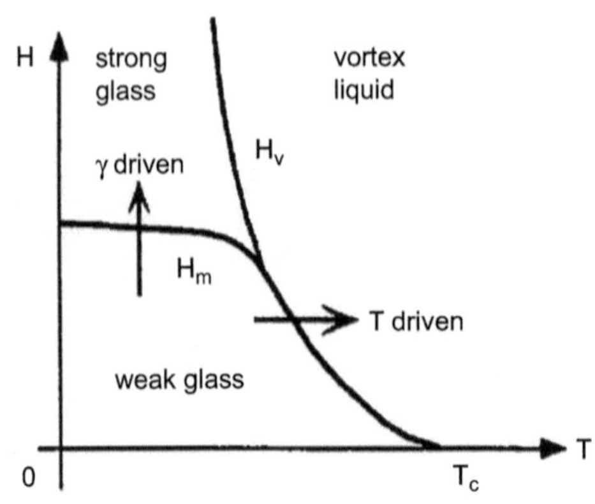

FIGURE 21.10

Possible phase diagram indicating the positions of the thermal- and disorder-driven melting transitions in the vortex lattice. H, Magnetic field; H_m, field for vortex solid—vortex liquid transition; H_v, field for vortex glass—vortex liquid transition; T, temperature ; T_C, superconducting transition temperature.

G. Blatter, Phys. C 282–287 (1997) 19. Copyright 1997, Reproduced with the permission of Elsevier.

5. MAGNETOELASTIC MATERIALS WITH CURRENT TECHNOLOGICAL INTEREST

The existence of a magnetic field–induced first-order phase transition (FOPT) involving a change in the lattice and spin degrees of freedom is fairly well known in various classes of solids [43]. It has become apparent that such FOPTs can actually give rise to interesting functional properties of technological importance. Three classes of magnetic materials have emerged with much promise for technological applications. These are: (1) magnetic shape memory alloys (SMAs), (2) magnetocaloric materials, and (3) GMR and colossal magnetoresistance (CMR) materials.

5.1 MAGNETOSTRICTIVE MATERIALS AND MAGNETIC SHAPE MEMORY ALLOYS

A magnetostrictive material, when magnetized by the application of an external magnetic field, exhibits a change in length per unit length. This phenomenon was first observed by James P. Joule in the 1840s in iron samples. This effect is known as *Joule magnetostriction*, and it is the most common magnetostrictive mechanism. The change in length produced by the Joule effect is accompanied by a transverse change in dimensions. Magnetostriction is an intrinsic effect of FM materials, and nearly all FM materials exhibit a measureable change in shape resulting from the application of a magnetic field and the subsequent change in magnetization. Magnetostrictive materials thus can convert magnetic energy to mechanical energy and vice versa. However, in most common FM materials like nickel, iron, and cobalt, the change in length is of the order of a few parts per million, whereas the change in volume is relatively small. Thus for any practical usage of these materials, one needs a high magnetic and cryogenic temperature environment, which limits their usage in engineering applications. In the 1940s during the Second World War, sonar devices were contemplated for use with the magnetostrictive effect, but were mostly left unexplored mainly because of the advances in piezoelectric materials.

The interest in magnetostrictive materials was revived in 1960s owing to the discovery of relatively large magnetostriction (1000s of ppm) in rare-earth alloys by Clark [44]. This renewed research activity in magnetostrictive materials led to the development of the engineering alloy Terfenol-D ($Tb_{0.27}Dy_{0.73}Fe_2$), a specially formulated alloy of terbium, dysprosium, and iron, which exhibited a large magnetostriction at room temperature and in relatively small applied magnetic fields. Initial usages of Terfenol-D were mainly confined to sonar devices for military applications. However, many commercial applications involving Terfenol-D, including magnetostrictive linear motors and flat-board loudspeakers have been realized.

SMAs have the ability to return back to a predetermined shape after being bent, stretched, or otherwise mechanically deformed [45,46]. A "thermoelastic martensitic transition," a reversible transition between two different crystal microstructures in the concerned material's system, is at the root of the shape memory effect. Discovery of the shape memory effect in equiatomic nickel–titanium alloys in the early 1960s gave a big impetus to the research in SMAs [47]. The conventional SMAs are rather slow to respond because they rely on variations in temperature and the flow of heat. However, a new class of materials was discovered in the late 1990s, which showed a magnetic field–induced large reversible deformations. These materials are now known as *magnetic SMAs* (*MSMAs*) [48]. The magnetic control offers a faster response in MSMAs in comparison with ordinary SMAs.

The first observation of the magnetic field–induced shape memory effect was reported in the stoichiometric Heusler alloy Ni_2MnGa by Ullakko et al. [49] in the mid-1990s. The authors observed a reversible deformation of 0.2% with a moderate field of about 1 T. This large magnetostriction in Ni_2MnGa originates from a magnetic field–induced twin-boundary motion leading to a twin-related variant reorientation. This mechanism produces considerably larger deformations than the usual magnetostriction mechanism arising from the magnetic field–induced change of interatomic distances. The magnetic field–induced deformation in Ni–Mn–Ga alloys is about two orders of magnitude higher than that observed in the commercial magnetostrictive material Terfenol-D. Large deformations can also arise from a magnetic field–induced martensite-to-austenite phase transition. If this magnetic field–induced transition is reversible in nature and involves a large change in magnetization, then it may give rise to magnetic superelasticity. In contrast, the systems displaying a field-induced variant reorientation are characterized by a high magnetocrystalline anisotropy in the martensite phase. In addition to large deformations, the superelastic deformation mechanism also provides large work output. In Ni–Mn–Ga alloys, a large magnetic field over 10 T is required to induce a martensite–austenite phase transition. However, a large magnetic field–induced strain associated with martensite–austenite phase transition has been reported in $Ni_{45}Co_5Mn_{36.7}In_{13.3}$ [50], $Ni_{50}Mn_{34}In_{16}$ [51], and NiCuMnIn alloys [52]. A similar large magnetostrictive effect associated with a magnetic field–induced first-order magnetostructural transition has also been observed in various other classes of magnetic materials, including the large magnetocaloric materials like Gd_5Ge_4 [53] and FeRh [54].

5.2 MAGNETOCALORIC AND MAGNETORESISTIVE MATERIALS

Cooling or heating of magnetic materials with the variation of an externally applied magnetic field is generally defined as the MCE. The disorder of the spin system of the concerned materials will get reduced with the isothermal application of an external magnetic field, which in turn will lower the magnetic part of the total entropy by the amount ΔS_{Mag}. An isothermal withdrawal of the external magnetic field will restore the zero-field magnetic entropy of the system, changing it by ΔS_{Mag}, subject to the condition that there are no hysteresis losses during the process of demagnetization. The total entropy of the system will, however, remain constant if the external field is reduced to zero under adiabatic conditions. The lattice contribution (and electronic contributions in case of metals) to entropy of the material in this situation needs to change by an amount opposite of ΔS_{Mag}. As a result of this adiabatic demagnetization (or magnetization) of a magnetic material, the temperature of the system will then decrease (increase) by an amount $\Delta T_{Adiabatic}$. These experimentally measurable quantities ΔS_{Mag} and $\Delta T_{Adiabatic}$ are used to express MCE in a material [55].

In 1881, Warburg [56] first discovered the changes of temperature in an iron sample by a few millikelvin when it was moved in and out of a magnetic field. The first explanation of MCE was provided by Weiss and Piccard [57]. Thereafter Debye [58] and Giauque [59] independently put forward an explanation of MCE based on thermodynamics, which in turn suggested the possibility of cooling by adiabatic demagnetization. Giauque and MacDougall [60] indeed constructed an adiabatic demagnetization refrigerator, and using a rare-earth–based paramagnetic salt $Gd_2(SO_4)_{3.8}H_2O$ and an external magnetic field of 0.8 T, they could reach 0.53, 0.34, and 0.25 K starting at 3.4, 2.0, and 1.5 K, respectively.

The change in electrical resistance in a material with an applied magnetic field is commonly termed *magnetoresistance*. This phenomenon finds applications as sensors in several areas of technology. For quite some time, personal computers have been using GMR sensors as a read head of the memory storage hard disc. The GMR device is basically a three-layer sandwich of a magnetic metal such as cobalt or iron with a nonmagnetic metal filling such as silver or platinum [61]. The interesting GMR properties in a layered magnetic structure were first reported in the mid-1980s. This discovery naturally spurred more research activities, and in early 1990s a class of rare-earth manganese-oxide materials (commonly termed *manganites*) were identified with CMR properties [62]. Manganites show exotic physical properties in the form of a metal–insulator transition and a variety of magnetic, charge, and orbital ordering dictated by strong electron–electron interactions and electron–lattice interactions [62].

A large MCE is expected in materials in the temperature region where the derivative of magnetization with respect to temperature is large. This usually happens when magnetic ordering takes place in a paramagnetic solid as a function of temperature and the bulk magnetization undergoes a large change in a relatively narrow temperature region around the transition temperature. In this direction, various rare-earth and transition metals and alloys have been investigated in some detail from 1950 to the early 1990s, with preference being given to rare-earth–based systems because of their relatively large (theoretically available) magnetic entropy. For a detailed review of these works, readers are referred to review articles [63] by Gschneidner and Pecharsky [63], Pecharsky and Gschneidner [64], and Tishin [65]. The MCE in elemental gadolinium (Gd) is often considered to be a benchmark for near room-temperature magnetic refrigeration. However, the maximum MCE of Gd ($\approx$16 K) even in relatively strong magnetic fields of 7 T remains considerably smaller than the temperature span of 40–50 K, which is usually required for many practical applications. Also, the MCE in Gd reduces rapidly as the operating temperature deviates from the temperature regime of paramagnetic to FM transition. Hence in the path of realization of magnetic refrigeration into a competing and commercially viable technology, one of the core issues remained in finding materials with tunable magnetocaloric properties better than Gd in a wide temperature span. The discovery of $Gd_5(Si_2Ge_2)$ compounds with giant MCEs [66] was a breakthrough in this direction. The MCE of this material is about 50% larger than elemental Gd, albeit with a lower magnetic ordering temperature of 270 K. Substitution of a small amount of Ga in place of Si and Ge could increase the magnetic ordering temperature of $Gd_5(Si_2Ge_2)$ to 285 K without much loss of the MCE [67]. The magnetic ordering temperature of the $Gd_5(Si,Ge)_2$ alloys could be varied from 40 to 270 K by changing the Si/Ge ratio but keeping the MCE same or even increasing it [68].

It was generally accepted until late the 1990s that the magnetic entropy change ΔS_{Mag} was maximum in the region of second-order phase transition from the paramagnetic to the FM state. Elemental Gd, for example, shows maximum $\Delta S_{Mag} \approx 4.5$ J/kg K in an applied magnetic field of 20 kOe at the paramagnetic to FM transition temperature of 294 K. As a result, the search for new MCE materials was mainly directed toward identifying magnetic materials with magnetic ordering transitions in the temperature region of interest [65,69]. It was realized soon after the discovery of the giant MCE in $Gd_5(Si,Ge)_2$ that this giant MCE was associated with a first-order magnetostructural phase transition [70]. Subsequently, it was proposed that the magnetic field–induced total entropy change ΔS_M consisted of the conventional magnetic entropy change ΔS_{Mag} and entropy differences ΔS_{St} of the two crystallographic polymorphs of $Gd_5(Si,Ge)_2$ [66]. In the low applied magnetic fields (<20 kOe), the structural part of the entropy ΔS_{St} accounted for more than 50% of the total entropy [66]. All these findings, apart from providing a new direction to the search of MCE materials, also generated much interest in the condensed matter physics community toward understanding the underlying physics of magnetostructural phase transition in the concerned solid, which is a result of the interaction between spin and lattice degrees of freedom.

Since early 2000, new classes of materials have been discovered showing large MCEs associated with a first-order magnetostructural phase transition. The same first-order magnetostructural phase transition also gives rise to large magnetostriction and large magnetoresistance in many of these materials. Often all three functional properties are observed in the same material. A framework of disorder-influenced FOPT provides the key to understanding and tuning the observed functional properties [43]. We shall highlight this generality with the case study of an NiMnIn-based SMA, which shows a large MCE, large magnetoresistance, and large magnetostriction driven by a temperature- and magnetic field induced—first order magnetostructural phase transition. A brief comparison will be made with other classes of magnetic materials including giant magnetocaloric materials and CMR manganites.

The off-stoichiometric Ni_2MnIn alloy with the composition $Ni_{50}Mn_{34}In_{16}$ belongs to the same family of Heusler alloys as that of the well-known magnetic SMA Ni_2MnGa, which undergoes a martensitic transition (an FOPT) from high-temperature austenite phase to low-temperature martensite phase. These Heusler alloys have an $L2_1$ cubic crystal structure in the austenite state, with different types of structure possible in the martensite state, the most prevalent one being the tetragonal $L1_0$ structure. Fig. 21.11 presents a magnetization versus temperature plot for the $Ni_{50}Mn_{34}In_{16}$ alloy. A large change in magnetization is observed across the austenite-to-martensite phase transition taking place around 240 K [71]. A distinct thermal hysteresis is observed in the temperature region of phase transition, which highlights the first-order nature of this transition [71]. In many off-stoichiometric compositions of Ni_2MnIn, Heusler alloys (including the present $Ni_{50}Mn_{34}In_{16}$ alloy), a magnetic field—induced transition from a low-magnetization martensite state to a higher-magnetization austenite state is observed in the low-temperature region. Fig. 21.12 presents the typical isothermal magnetization versus magnetic field curves showing magnetic field—induced martensite-to-austenite

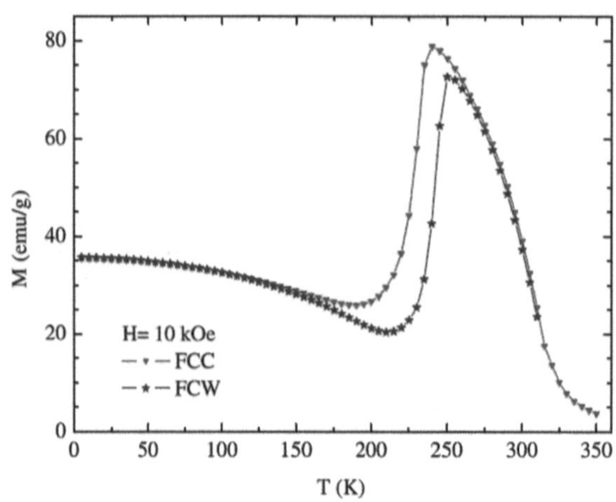

FIGURE 21.11

Magnetization (*M*) versus temperature (*T*) plot for $Ni_{50}Mn_{34}In_{16}$ alloy obtained while cooling in a magnetic field of 10 kOe [(field cooled cooling) *FCC* mode] and warming in the same field [(field cooled warming) *FCW* mode].

V.K. Sharma, M.K. Chattopadhyay, S.B. Roy, J. Phys. Condens. Matter 20 (2008) 425210. Copyright 2008, Reproduced with the permission of IOP Publishing.

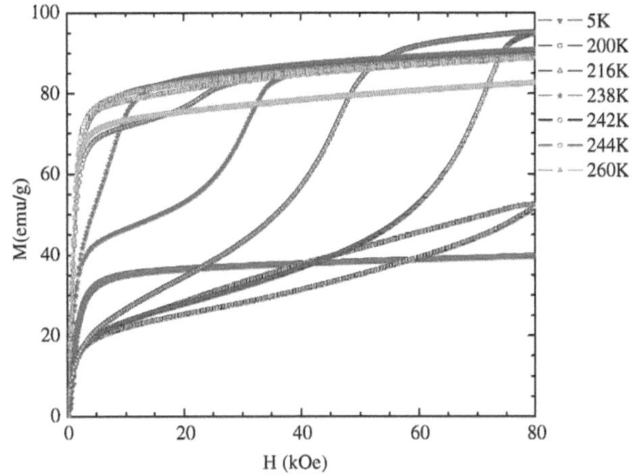

FIGURE 21.12

Isothermal magnetization (M) versus field (H) curves of $Ni_{50}Mn_{34}In_{16}$ alloy at representative temperatures. The M–H curves at $T = 200$, 216, 238, 242, and 244 K show the signature of a field-induced transition from martensite to austenite phase.

V.K. Sharma, M.K. Chattopadhyay, S.B. Roy, J. Phys. Condens. Matter 20 (2008) 425210. Copyright 2008, Reproduced with the permission of IOP Publishing.

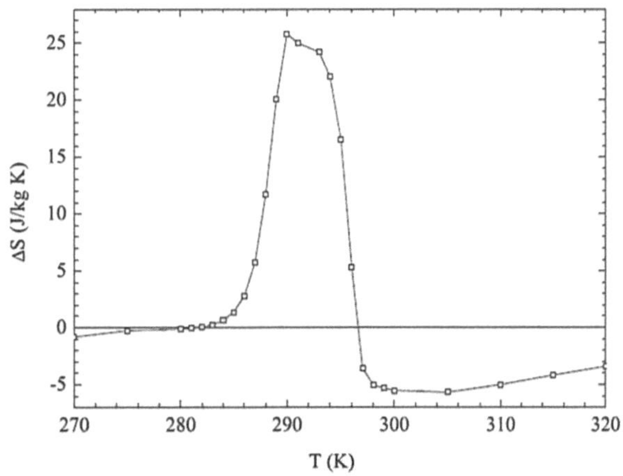

FIGURE 21.13

MCE (represented by isothermal entropy change, ΔS) versus temperature plot in a 1% Cu-doped $Ni_{50}Mn_{34}In_{16}$ alloy for a magnetic field variation of 45 kOe.

V.K. Sharma, M.K. Chattopadhyay, A. Khandelwal, S.B. Roy, Phys. Rev. B 82 (2010) 172411. Copyright 2010, Reproduced with the permission of American Physical Society.

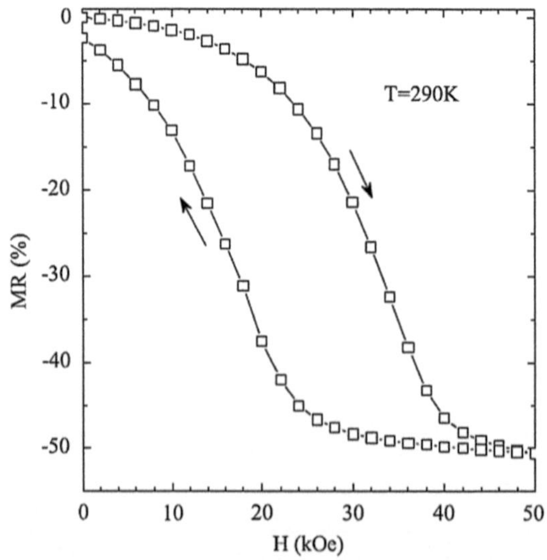

FIGURE 21.14

Isothermal magnetoresistance (*MR*) as a function of a magnetic field (*H*) in 1% Cu-doped $Ni_{50}Mn_{34}In_{16}$ alloy at a temperature (*T*) of 290 K.

V.K. Sharma, M.K. Chattopadhyay, A. Khandelwal, S.B. Roy, Phys. Rev. B 82 (2010) 172411. Copyright 2010, Reproduced with the permission of American Physical Society.

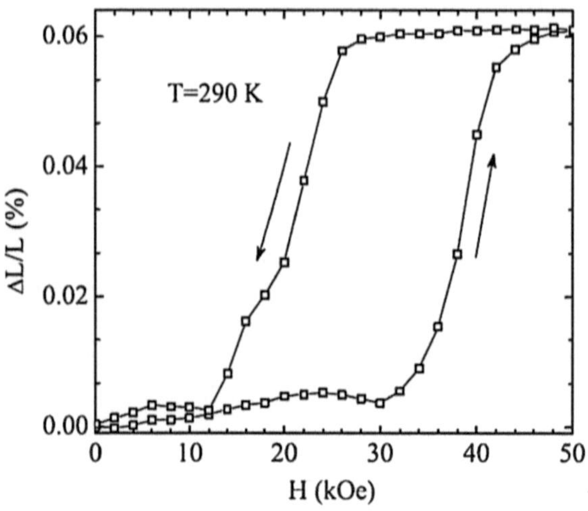

FIGURE 21.15

Isothermal strain as a function of magnetic field (*H*) in 1% Cu-doped $Ni_{50}Mn_{34}In_{16}$ alloy at a temperature (*T*) of 290 K. $\Delta L/L$, change in length/length.

V.K. Sharma, M.K. Chattopadhyay, A. Khandelwal, S.B. Roy, Phys. Rev. B 82 (2010) 172411. Copyright 2010, Reproduced with the permission of American Physical Society.

phase transition in the $Ni_{50}Mn_{34}In_{16}$ alloy at various temperatures. Various characteristic features of such *M-H* curves can be explained within the framework of a disorder-broadened FOPT [71].

Scanning Hall-probe imaging experiments performed across the magnetic field–induced martensite-to-austenite phase transition in the $Ni_{50}Mn_{34}In_{16}$ alloy provide snapshots of the coexistence of the martensite phase (with a low value of magnetization) and the austenite phase (with a high value of magnetization) on a length scale of tens of micrometers, extended over a wide magnetic field regime [72]. The characteristics exhibited by the local Hall voltage loops also indicate the presence of a landscape of critical magnetic fields for the martensitic phase transition across the sample, which in turn confirms that this magnetic field–induced FOPT in the $Ni_{50}Mn_{34}In_{16}$ alloy is indeed influenced by disorder. This kind of disorder-induced landscape of transition temperatures and magnetic fields along with phase-coexistence and metastability had previously been observed in doped-$CeFe_2$ [73] alloys [which belong to the same family as that of the well-known commercial magnetostrictive material Terfenol-(Tb,Dy)Fe_2] and FeRh alloys [74] with multifunctional properties [43].

The observed large drop in magnetization across the FOPT in the $Ni_{50}Mn_{34}In_{16}$ alloy (Fig. 21.11), along with the magnetic field dependence of the transition temperature lead to a large MCE of about 18 J/kg-K for a magnetic field variation of 50 kOe [75]. Similar large MCEs associated with the FOPT have been observed in other compositions of Ni–Mn–In Huesler alloys [76]. The lattice entropy change across the FOPT accounts for about one-half of the total entropy change in these Ni–Mn–In alloys, which points to the important role of magnetostructural coupling in these materials [77]. The same temperature and magnetic field–induced FOPT also gave rise to GMR [78] and superelasticity [51] in the $Ni_{50}Mn_{34}In_{16}$ alloy. The martensitic transition temperature of $Ni_{50}Mn_{24}In_{16}$ and the associated functional properties (e.g., MCE, GMR and magnetostriction) can be shifted toward room temperature with Cr-doping at Mn sites and Cu-doping at the Ni sites [52]. Figs. 21.13–21.15 present examples of multifunctional properties in a 1% Cu-doped $Ni_{50}Mn_{34}In_{16}$ near room temperature.

The thermomagnetic hysteresis, phase coexistence, and metastability discussed for $Ni_{50}Mn_{34}In_{16}$ are the characteristic features of a disorder-influenced FOPT, and the same characteristic features have been observed in various other classes of functional magnetic materials [43]. We present here a brief discussion on the giant MCE material Gd_5Ge_4, CMR manganites, and equiatomic FeRh alloys to highlight the role of disorder-influenced FOPTs in the multifunctional properties of these materials. The generality of this observation can be extended to many other materials, including giant magnetocaloric materials $LaFe_{13-x}Si_x$ and $MnCoGeB_x$; for a detailed discussion, readers are referred to some recent review articles [55,79].

The giant MCE material Gd_5Ge_4 undergoes a low-temperature first-order AFM-to-FM transition with the application of a magnetic field. Thermomagnetic hysteresis is one of the numerous evidences of the first order nature of this AFM–FM phase transition in Gd_5Ge_4 [80]. The AFM–FM transition in Gd_5Ge_4 has been imaged with scanning micro-Hall probe technique, which clearly revealed the coexistence of FM and AFM phase in micrometer scale in a finite magnetic field–temperature regime around the FOPT line [81]. These results, in combination with dimensional analysis and numerical simulation, further pointed out how this AFM–FM phase coexistence and the associated dynamics (nucleation and growth) could provide useful information for tuning the functional properties of Gd_5Ge_4. The unique layered structure of Gd_5Ge_4 is at the root of the observed temperature- and magnetic field–induced first order magnetostructural transition from a monoclinic–paramagnetic to an orthorhombic–FM structure, which in turn gives rise to the giant MCE [66]. The same first-order AFM–FM transition in Gd_5Ge_4 is also accompanied by large magnetostriction [53] and magnetoresistance [82].

In one of the earliest works on CMR manganites, an FOPT from a metallic FM state to an insulating charge-ordered AFM state in $La_{0.5}Ca_{0.5}MnO_3$ marked by sharp drop in magnetization (M) and change in the resistivity (ρ) by several orders of magnitude was highlighted [83]. In the phase transition region a distinct thermal hysteresis was observed both in $M(T)$ and $\rho(T)$, which was taken as a signature of the FOPT [84]. A drastic change of lattice constants was also observed in the vicinity of the AFM−FM transition in $La_{0.5}Ca_{0.5}MnO_3$ [84]. A very similar first-order FM−AFM transition as a function of temperature was reported in $Nd_{0.5}Sr_{0.5}MnO_3$ [85]. An isothermal AFM−FM transformation could be induced inside the AFM state with the application of an external magnetic field, and that was clearly marked with sharp changes in physical properties and the accompanying thermomagnetic hysteresis [85]. The CMR observed in manganites has been explained in terms of a percolation path involving the metallic FM phase in the phase coexistence regime across the first order AFM−FM transition, which could be manipulated by an applied magnetic field [62]. The existence of such a percolative conducting path in the phase transition region along with phase coexistence on a micrometer scale was confirmed experimentally [86,87]. A substantial MCE associated with such first-order AFM−FM transitions has been reported in several CMR manganites [88,89]. The magnetic field−induced variation in the configurational entropy associated with the structural transition and the magnetic inhomogeneity arising from the AFM−FM phase coexistence across the disorder-influenced FOPT are responsible for the large observed MCE.

The equiatomic FeRh alloy, which forms in the cubic CsCl-type structure, undergoes an FOPT from an FM to an AFM state around 330 K. This FOPT in the FeRh alloy gives rise to GMR, a giant MCE, a giant elastocaloric effect, and giant magnetostriction close to room temperature. The FOPT in the Fe−Rh alloy is again accompanied by the characteristic thermomagnetic history effects, phase coexistence, and the landscape of transition temperatures, which have been studied in detail through magnetization, magnetic force microscopy, and linear strain measurements [54,74,90,91]. An MCE of about 12 J/kg-K reproducible under multiple field cycles could be achieved in FeRh-based alloys by properly choosing a combination of isothermal and adiabatic field variation cycles in the magnetic field−temperature phase space [92]. The commonality in the various features associated with the magnetic field− and temperature-induced first-order AFM-FM transition in CMR manganite systems, giant magnetocaloric materials including $Gd_5(Ge, Si)_4$, and doped-$CeFe_2$ alloys has been highlighted [80].

6. FUTURE OUTLOOK

Summarizing the discussions presented in the earlier sections, we say that in a material the applied magnetic field couples most strongly to the electrons in two possible ways: (1) by acting on electrical currents generated from quantum mechanical motion of the electrons, or (2) by coupling to the magnetic moments arising from the intrinsic spin of the electrons. Thus magnetic field has been an important tool in the investigation on the electronic structure of materials. In many cases the applied magnetic field has a relatively small effect on the electronic structure; nevertheless such effect leads to the development of experimental techniques revealing properties related to the electronic structure of the materials, which otherwise would have been inaccessible. In some other cases, the applied magnetic field can be strong enough to change the magnetostructural properties, revealing new and novel magnetic field−induced phenomena. The discovery of such new phenomena subsequently leads to the development of new devices and products of significant technological and societal importance. Most of the discussions were mostly confined to bulk materials and two-dimensional thin solid film structures. However, in recent times more novel structures and materials like surfaces of solids and interfaces between solids, small particles of a solid, wires with nanometer scale diameters, and two-dimensional materials such as grapheme (which is just a single layer of carbon atoms) are under scrutiny, with significant promises for new physics and device applications.

We have also not discussed the relatively weaker coupling of magnetic fields to the magnetic moments of nuclear spins, which forms the physical basis for NMR techniques. This is a quite mature field with widespread applications in the areas starting from basic science research to technology and health care. For a recent study of NMR in materials at very high fields, the readers are referred to the review article by Moulton and Reys [93].

A relatively new direction of high magnetic field research is in the area of biological materials with mostly diamagnetic response, with a promise of newer applications. In gestalt, the field of materials research in a high magnetic field is far from being saturated; rather it is opening up. With a steady increase in the high magnetic field facilities all over the world there is a considerable anticipation of the discovery of newer and exotic magnetic field–induced phenomena and the associated technological and societal applications.

REFERENCES

[1] M. Motokawa, Rep. Prog. Phys. 67 (2004) 1995.
[2] J.E. Kunzler, E. Buehler, F.S. Hsu, J.H. Wernick, Phys. Rev. Lett. 6 (1961) 89.
[3] T.E.A. Kiyoshi, IEEE Trans. Appl. Supercond. 10 (2000) 472.
[4] P.L. Kapitza, Proc. R. Soc. A 105 (1924) 691.
[5] P.L. Kapitza, Proc. R. Soc. A 115 (1927) 658.
[6] N. Miura, F. Herlach, in: F. Herlach, N. Miura (Eds.), High Magnetic Fields: Science and Technology, vol. 1, World Scientific, 2003, p. 1.
[7] J.M. Ziman, Principles of the Theory of Solids, Cambridge University Press, 2011.
[8] N.W. Ashcroft, N.D. Mermin, Solid State Physics, Brooks/Cole: Cengage Learning, 2009.
[9] U. Mizutani, Introduction to the Electron Theory of Metals, Cambridge University Press, 2001.
[10] P. Haussler, S.J. Welles, Phys. Rev. 152 (1962) 675.
[11] A.S. Joseph, W.L. Gordon, Phys. Rev. 126 (1962) 489.
[12] D. Shoenberg, Proc. Roy. Soc. A170 (1939) 341.
[13] W.Y. de Haas, P.M. van Alphen, Proc. Acad. Sci. Amsterdam 33 (1930) 1106.
[14] L. Shubnikov, W.Y. de Haas, Comm. Phys. Lab. Univ. Leiden 207a (1930).
[15] V.S. Edelman, Adv. Phys. 25 (1976) 555.
[16] A.B. Pippard, R.G. Chambers, Proc. Phys. Soc. A65 (1952) 955.
[17] K. Behnia, L. Balicas, Y. Kopelevich, Science 317 (2007) 1729.
[18] A. Audouard, J.Y. Fortin, C. R. Phys. 14 (2013) 15.
[19] D. Aoki, W. Knafo, I. Sheikin, C. R. Phys. 14 (2013) 53.
[20] M. Springford, Phys. B 171 (1991) 151.
[21] B. Vignolle, D. Vignolles, M. Julien, C. Proust, C. R. Phys. 14 (2013) 39.
[22] E. Kaxiras, Atomic and Electronic Structure of Solids, Cambridge University Press, 2003.
[23] K. von Klitzing, G. Dorda, M. Pepper, Phys. Rev. Lett. 45 (1980) 494.
[24] D.C. Tsui, H.L. Stormer, A.C. Gossard, Phys. Rev. Lett. 48 (1982) 1559.
[25] D.C. Tsui, H.L. Stormer, A.C. Gossard, Phys. Rev. B 25 (1982) 1405.
[26] S.J. Blundell, Magnetism in Condensed Matter, Cambridge University Press, 2001.

[27] H. Schmid, Ferroelectrics 162 (1994) 317.
[28] M. Fiebig, T. Lottermoser, D. Frohlich, A.V. Goltsev, R.V. Pisarev, Nature 419 (2002) 818.
[29] D.I. Khomskii, Physics 2 (2009) 20.
[30] J.E.A. Wang, Science 299 (2003) 1719.
[31] T.E.A. Kimura, Nature 426 (2003) 55.
[32] N. Hur, et al., Nature 429 (2004) 392.
[33] W.R. Hagen, in: F. Herlach, N. Miura (Eds.), High Magnetic Fields: Science and Technology, vol. 1, World Scientific, 2006, p. 207.
[34] G. Sazaki, J. Cryst. Growth 173 (1997) 231.
[35] G. Sazaki, A. Moreno, K. Nakjima, J. Cryst. Growth 262 (2004) 499.
[36] M. Ataka, N.I. Wakayama, Acta Cryst. D58 (2002) 1708.
[37] N. Ramachandran, F.W. Leslie, J. Cryst. Growth 274 (2005) 297.
[38] L. Wang, C. Zhong, N.I. Wakayama, J. Cryst. Growth 237 (2002) 312.
[39] W. Haberdutzl, Nature 213 (1967) 72.
[40] J.F. Annett, Superconductivity, Superfluids and Condensates, Oxford University Press, 2008.
[41] G. Blatter, Phys. C 282–287 (1997) 19.
[42] G. Blatter, M. Feigel'man, V. Geshkenbein, A. Larkin, V. Vinokur, Rev. Mod. Phys. 66 (1994) 1125.
[43] S.B. Roy, J. Phys. Condens. Matter 25 (2013) 183201.
[44] A.E. Clark, in: E.P. Wohlfarth (Ed.), Handbook of Magnetic Materials, vol. 1, 1980, p. 531 (North-Holland, Amsterdam).
[45] K. Otsuka, T. Kakeshita, MRS Bull. (February 2002) 91.
[46] K. Otsuka, C.M. Wayman, Shape Memory Materials, Cambridge University Press, 1998.
[47] W.J. Buehler, J.W. Gilfrich, R.C. Wiley, J. Appl. Phys. 34 (1963) 1475.
[48] M. Acet, L. Manosa, A. Planes, in: K.H.J. Buschow (Ed.), Handbook of Magnetic Materials, vol. 19, Elsevier, 2011.
[49] K. Ullakko, J.K. Huang, V.V. Kokorin, R.C. O'Handley, Scr. Mater. 36 (1997) 1133.
[50] R. Kainuma, et al., Nature 439 (2006) 957.
[51] T. Krenke, E. Duman, M. Acet, E.F. Wassermann, X. Moya, L. Manosa, A. Planes, E. Suard, B. Ouladdiaf, Phys. Rev. B 75 (2007) 104414.
[52] V.K. Sharma, M.K. Chattopadhyay, A. Khandelwal, S.B. Roy, Phys. Rev. B 82 (2010) 172411.
[53] C. Magen, L. Morellon, P.A. Algarabel, C. Marquina, M.R. Ibarra, J. Phys. Condens. Matter 15 (2003) 2389.
[54] M. Manekar, V.K. Sharma, S.B. Roy, J. Phys. Condens. Matter 24 (2012) 216004.
[55] S.B. Roy, in: K.H.J. Buschow (Ed.), Handbook of Magnetic Materials, vol. 23, Elsevier, 2013.
[56] E.G. Warburg, Ann. Phys. 13 (1881) 141.
[57] P. Weiss, A. Piccard, C. R. Acad. Sci. 166 (1918) 352.
[58] P. Debye, Ann. Phys. 81 (1926) 1154.
[59] W.F. Giauque, J. Am. Chem. Soc. 49 (1927) 1864.
[60] W.F. Giauque, D.P. MacDougall, Phys. Rev. 43 (1933) 76.
[61] G. Prinz, Science 282 (1998) 1660.
[62] E. Dagotto, Nanoscale Phase Separation and Colossal Magneto- Resistance, Springer, 2002.
[63] K. Gschneidner Jr., V.K. Pecharsky, Annu. Rev. Matr. Sci. 30 (2000) 387.
[64] V.K. Pecharsky, K. Gschneidner Jr., J. Magn. Magn. Mater. 200 (1999) 44.
[65] A.M. Tishin, in: Handbook of Magnetic Materials, vol. 12, Elsevier, 1999, p. 395 (Chapter 4).
[66] V.K. Pecharsky, K. Gschneidner Jr., Phys. Rev. Lett. 78 (1997) 4494.
[67] V.K. Pecharsky, K. Gschneidner Jr., J. Magn. Magn. Mater. 167 (1997). L–179.
[68] V.K. Pecharsky, K. Gschneidner Jr., Appl. Phys. Lett. 70 (1997) 3299.
[69] J.A. Barclay, J. Alloys. Comp. 207–208 (1994) 355.

[70] W. Choe, V.K. Pecharsky, A.O. Pecharsky, K.A. Gschneidner Jr., V.G. Young Jr., G.J. Miller, Phys. Rev. Lett. 84 (2000) 4617.
[71] V.K. Sharma, M.K. Chattopadhyay, S.B. Roy, J. Phys. Condens. Matter 20 (2008) 425210.
[72] V.K. Sharma, J.D. Moore, M.K. Chattopadhyay, K. Morrison, L.F. Cohen, S.B. Roy, J. Phys. Condens. Matter 22 (2010) 016008.
[73] S.B. Roy, G.K. Perkins, M.K. Chattopadhyay, K.J.S. Sokhey, A.K. Nigam, P. Chaddah, A.D. Caplin, L.F. Cohen, Phys. Rev. Lett. 92 (2004) 147203.
[74] M.A. Manekar, C. Mukherjee, S.B. Roy, Euro. Phys. Lett. 80 (2007) 17004.
[75] V.K. Sharma, M.K. Chattopadhyay, S.B. Roy, J. Phys. D Appl. Phys. 40 (2007) 1869.
[76] A. Planes, J. Phys. Condens. Matter 21 (2009) 233201.
[77] B. Li, W.J. Ren, Q. Zhang, X.V. Lv, X.G. Liu, H. Meng, J. Li, D. Li, Z.D. Zhang, Appl. Phys. Lett. 95 (2009) 172506.
[78] V.K. Sharma, M.K. Chattopadhyay, K.H.B. Shaeb, A. Chouhan, S.B. Roy, Appl. Phys. Lett. 89 (2006) 222509.
[79] K.A.J. Gschneidner, Y. Mudryk, V.K. Pecharsky, Scr. Mater. 67 (2012) 572.
[80] S.B. Roy, P. Chaddah, V.K. Pecharsky, K.A. Gschneidner Jr., Acta Mater. 56 (2008) 5895.
[81] G.K. Perkins, J.D. Moore, M.K. Chattopadhyay, S.B. Roy, P. Chaddah, V.K. Pecharsky, K.A. Gschneidner Jr., L.F. Cohen, J. Phys. Condens. Matter 19 (2007) 176213.
[82] E.M. Levin, V.K. Pecharsky, K.A. Gschneidner Jr., G.J. Miller, Phys. Rev. B 64 (2001) 235108.
[83] P. Schiffer, A. Ramirez, O. Bao, S.W. Cheong, Phys. Rev. Lett. 75 (1995) 3336.
[84] P.G. Radaeli, D.E. Cox, M. Marezio, S.W. Cheong, P. Schiffer, A.P. Ramirez, Phys. Rev. Lett. 75 (1995) 4488.
[85] H. Kuwahara, Y. Tomioka, A. Asamitsu, Y. Moritomo, Y. Tokura, Science 270 (1995) 961.
[86] Y. Murakami, J.H. Yoo, D. Shindo, T. Atou, M. Kikuchi, Nature 423 (2003) 965.
[87] L. Zhang, C. Israel, A. Biswas, R. Greene, A. Lozanne, Science 298 (2002) 805.
[88] K.A. Gschneidner Jr., V.K. Pecharsky, A.O. Tsokol, Rep. Prog. Phys. 68 (2005) 1479.
[89] M.H. Phan, S.C. Yu, J. Magn. Magn. Mater. 308 (2007) 325.
[90] M.A. Manekar, S.B. Roy, Euro. Phys. J. B 64 (2008) 19.
[91] M.A. Manekar, S.B. Roy, J. Phys. Condens. Matter 20 (2008) 325208.
[92] M.A. Manekar, S.B. Roy, J. Phys. D Appl. Phys. 44 (2011) 242001.
[93] W.G. Moulton, A. Reys, in: F. Herlach, N. Miura (Eds.), High Magnetic Fields: Science and Technology, vol. 3, World Scientific, 2006, p. 185.

CHAPTER

PROPERTIES OF MATERIALS UNDER HIGH ELECTRIC FIELD

22

S.N. Achary, A.K. Tyagi

Bhabha Atomic Research Centre, Mumbai, India

CHAPTER OUTLINE

1. INTRODUCTION

The behavior and stability of materials in high electric and electromagnetic field govern the final reliability and quality of electrical appliances and their dimensions. In general, the behavior of materials depends on their atomic constituents and electronic structures [1,2]. Based on the electronic structure, electrical materials can be grouped as conductors and insulators (wide, narrow, or medium-band gap). The properties of these materials, in particular insulating or conducting properties, also depend on atomic and nanoscale defects as well as structures and microstructures. Wider tunability and control on these properties have been demonstrated by controlled chemistry, structure, and microstructure [3−10]. Such multifaceted approach from different disciplines of science and engineering led to dramatically enhanced capacity of known materials and discovery of newer materials. For example, the current carrying capacity of a conductor is the backbone for civilization in terms of producing current inside a power generator and transmitting power from generator to the end use, as well as governing the speed and size of motors, etc. In addition to these normal electrical applications, the size and performance of electronic components in integrated circuits are also based on the materials' properties.

Materials Under Extreme Conditions. http://dx.doi.org/10.1016/B978-0-12-801300-7.00022-X

The electrical properties of materials are characterized in a number of ways, and most commonly explained terminology is based on resistance (R), which means the extent of opposition a material exhibits to the flow of electrical current, i.e., the flow of charge carrier, which may be electron (e), hole (h), or ions under the influence of electrical field. The fundamental quantity of material, defined as a term resistivity, $\rho = R \times (l/A)$, where R, l, and A are resistance, length, and area of cross-section of the conductor, is in general used to explain its electrical property. Broadly, materials exhibiting conductivity (defined as $\sigma = 1/\rho$), 10^4–10^6 S/cm ($\rho = 1$–10^2 μΩ·cm) are classed as *conductor*, while those having conductivity ($<10^{-15}$ S/cm) are classed as insulators. The term *superconductors* is used for materials having zero resistivity. Materials that are intermediate-level conductors are generally called *semiconductors*. The conductivity of materials depends on number and charge of the carrier of current as well as their mobility carrier ($\sigma = n \cdot e \cdot \mu$, where $n =$ number of carriers, i.e., carrier density, $e =$ charge of the carrier, and μ is their mobility). Thus, all sorts of materials can be classed either as an insulator, semiconductor, or conductor. A more appropriate mode of classification of the electrical properties is based on the temperature dependence of electrical resistance/resistivity or their reciprocals, conductance/conductivity. It has been realized that different types of materials have some relevant applications and they either may be as current conductor, or current blocker (insulator), or as controlled conduction as in electronic circuits [3–9].

The conducing, insulating, or semiconducting properties of materials under the influence of electrical field are the consequence of their electronic band structure, and thus more specifically the electrical properties are explained by electronic band model. According to the Pauli exclusion principle, no two electrons are occupied in the same energy level. In other words, electrons have quantized momentum states in an atom. Two electrons having opposite spins are paired and they are occupied in one electron orbital. Thus at maximum two electrons are occupied in an orbital of an atom. If more than one atom comes to the proximity, as in a bonded system or periodically arranged atoms as in crystal, states of neighboring atoms overlap and thus the energy levels split to bonding and antibonding orbital with a discrete energy separation. In a periodic system, the interactions of bonding or antibonding orbital result in the formation of delocalized orbitals called bands. A schematic for the overlapping of the orbitals to form bands is shown in Fig. 22.1. The uppermost energy level formed by antibonding orbitals is termed a *conduction band* while those formed by the bonding orbitals are termed as *valence band*. The *electronic band* represents a path where the electrons can move freely. However, the change of state from one to another band needs to have a change in energy. Typical valence and conduction band of different types of materials are depicted in Fig. 22.2. In case of

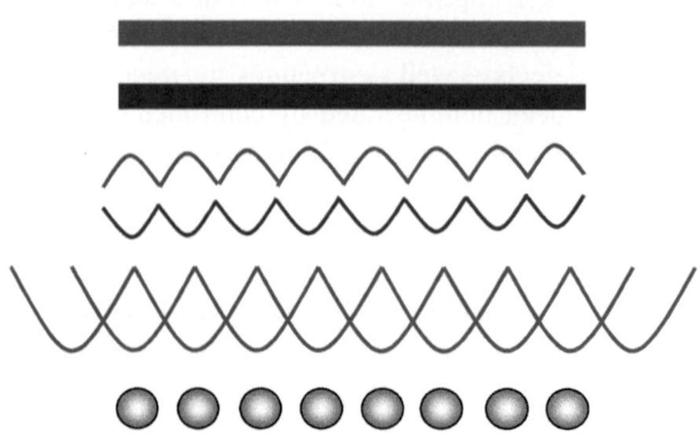

FIGURE 22.1

Electronic bands in periodic atomic systems.

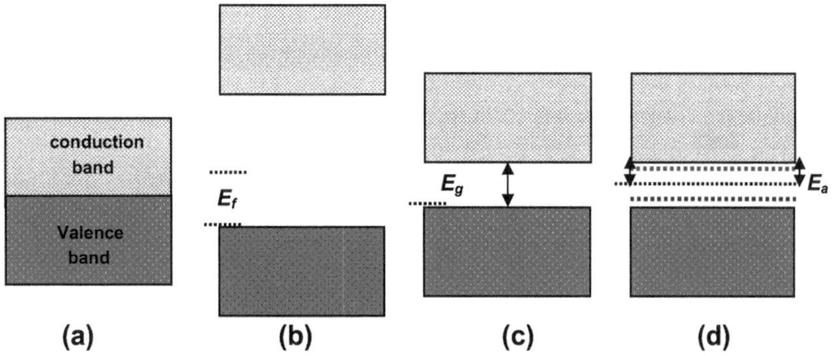

FIGURE 22.2

Electronic band structure in metal (a), insulator (b), and semiconductor (c). The donor and acceptor defect levels in extrinsic semiconductor are shown (d).

conductors, as in case of metals, the valence band and conduction band overlap and thus the electron is mobile. In case of insulators, the valence band is completely filled and is well separated from the conduction band. A partial overlap and lower gap between the valence and conduction band are characteristics of semiconducting materials.

2. INTERACTION OF MATTER WITH ELECTRIC AND ELECTROMAGNETIC FIELD

As mentioned earlier, the responses of insulators, semiconductors, or conductors toward the electric field are different and they are a consequence of their physical states as well as internal atomic and electronic structures. Depending on their response, they can be conductor or insulator in a broader sense. In case of conductor transportation the charge depends on the direction and nature of applied field, while in the insulator a feeble displacement of charges or rearrangements of charges occur without any long range transportation of charges. The change bears some fundamental characteristics, like charges are additive and quantized as well as follow the law of conservation. The rearrangements of charges are reflected in several microscopic or macroscopic levels depending on the nature of effects. In this section, the effects of electric field in conductors and insulators are briefly explained. More detailed explanations are found in fundamental textbooks of solid state physics [1,2,10,11].

It is well known that a charge in a conductor remains uniformly distributed due to the higher and faster mode of equilibration. Also, a charge experience is a force from the field and direction of force depends on the nature and magnitude of the charge as well as the direction of field. It may be noted that charge distributes over the surface and creates a field around it but not inside it. The electric field lines remain perpendicular to the surface of the conductors. The strength of electric field (E) at the surface of the conductors is related to surface charge density and the permittivity of the medium in which the conductor is placed. The electric field due to a point charge (q) at a distance r is given as

$$E(r) = \frac{1}{4\pi\varepsilon_0} \frac{q}{r^2} \tag{22.1}$$

In case of vacuum the permittivity is defined as permittivity of free space "ε_0" and the magnitude is calculated as $\varepsilon_0 = 1/c^2\mu_0^2$, where c = velocity of light and μ_0 is permeability of free space. The value of ε_0 for free space is $8{:}85 \times 10^{-12}$ F/m. For any other medium, the permittivity is given as relative to ε_0 and called relative permittivity (ε_r), where $\varepsilon = \varepsilon_r \cdot \varepsilon_0 = (1 + \chi)\, \varepsilon_0$,and χ is the electric susceptibility of the material. The electric susceptibility of a material is obtained from the electric field and electric polarization ($E = \varepsilon_0 \cdot \chi \cdot P$). The electric field at a point due to several charges is given by the addition of fields of the individual charges at the measured point. The force exerted by the charge q on another charge (q') placed at a distance r is defined as the product of electric field at r ($E(r)$) and q' and thus can be given as:

$$F(r) = \frac{1}{4\pi\varepsilon_0}\frac{q}{r^2}\cdot q' = E(r)\cdot q' \tag{22.2}$$

In a conductor, the electric field causes displacement of carrier, and that in turn results as a flow of current and a voltage drop across the conductor. The relation between voltage (electric potential) drop and net current is related by Ohm's law ($V = IR$, where I is current though the conductors and R is resistance of the conductor; $V = E \times L$, where E = electric field strength, L = length of the conductor). Alternatively, the electric potential is defined as net work done in bringing a unit test charge from infinite distance to a point with field $E(r)$ and thus the potential at point due to charge q is given as, $V(r) = \frac{1}{4\pi\varepsilon_0}\frac{q}{r}$. In case of alternating current, the term *resistance* is replaced by *impedance*, which represents the cumulative effect of resistance, inductance, and capacitance. The mobility of carrier is defined as the drift velocity per unit field strength.

2.1 CONDUCTORS

In any conductors, there are a number of electrons that are free to move. In the case of metal, the valence electrons form a sea of electrons where the positively charged kernels, i.e., atomic part without the valence electrons, are distributed to maintain charge neutrality. Under the influence of electric field all the electrons accelerate in a direction opposite to that of the field, by virtue of their negative charge. In principle these electrons do not have any interaction with the atoms in the lattice, and thus they do not have any hindrance from the atoms in the lattice sites. The acceleration of the electrons depends on the electric field. If V is potential across the conductor of length l, the electric field, $E = V/l$, and thus the acceleration of the electron is given as

$$a = \frac{e}{m}E \tag{22.3}$$

where e and m are charge and mass of electron.

Though the electrons do not have any retarding interaction from the positively charged kernels at the lattice site of solid conductors, they experience frictional forces from the imperfection due to impurity atoms, vacancies, interstitial atoms, dislocations, as well as thermal vibrations of electrons and atoms. Thus the electrons lose their kinetic energy and change the direction of acceleration and thus feel a net frictional force in their motion. This oppositional force is the earlier mentioned term resistance (R). If τ is the average time between collisions, also known as relaxation time, the average velocity of electron in the direction of field (v_d) is given as

$$v_d = a\tau = \frac{e}{m}\tau E \tag{22.4}$$

where e and m are charge and mass of electron, $E =$ strength of electric field.

Assuming all the electrons in the solid have the same drift velocity, the current density (J), i.e., the net charge transported across unit area per second in the direction of field direction is given as

$$J = Nev_d = Ne\left(\frac{e}{m}\tau\right)E = Ne(\mu_e)E = \sigma E \tag{22.5}$$

where N is the number of free electrons per unit volume and $\sigma =$ conductivity. The term μ_e is electron mobility (defined by $e\tau/m$). This direct relation is the general expression for Ohm's law. This expression indicates that the net current depends on the number of free electrons and drift velocity, which in turn depends on the impurity and thermal effect on the atoms and electrons. As the metals have a large number of free electron density ($\sim 10^{22}/\text{cm}^3$), they exhibit large electrical conductivity ($\sim 10^7\ \Omega^{-1}\text{m}^{-1}$). The Ohm's law further implies that the conductivity (σ) is a constant, and hence the relaxation time is independent of electric field. However, it is reasonable to assume that mean free path is independent of electric field than the relaxation time. Since the drift velocity depends on the field, τ needs to vary with field. Thus the validation of Ohm's law strictly remains only in metals where thermal velocity of electron $\left(\frac{1}{2}mv_{thermal} = \frac{3}{2}kT\right)$ is much higher than the drift velocity. Also it is important to note that the resistivity $\rho(=1/\sigma)$ is dependent on the intrinsic thermal scattering centers as well as the defects or imperfections like impurity or dislocations, etc. The total resistivity is thus a sum of these contributions, which is known as Matthiessen's additive rule ($\rho_{total} = \rho_{temp.} + \rho_{defects\ or\ impurities}$). Also, as the temperature increases, the vibrations of atoms or electrons also increase and they noticeably affect the mobility of electron. Thus the temperature-dependent resistivity is explained as $\rho_{temp.} = \rho_0 + \alpha T$, where ρ_0 and α are characteristics of materials. At low temperature, the electron–electron scattering becomes the dominating effect, and thus they follow a larger temperature dependency and they can be explained with higher power of temperature $\left(\rho_{temp.} = \rho_0 + \alpha' T^5\right)$.

As in the case of constant electric field, the time-dependent electric field has closely similar effect on the motion of charge carrier in the materials. In such case the current density becomes time dependent. Replacing the average electron velocity with the time-dependent term, the current density can be expressed as

$$J = N\ e\frac{p(t)}{m} \tag{22.6}$$

where the $p(t)$ is momentum of electron at time t and m is mass of electron.

The electron will acquire additional momentum due to the force on it by the electric and magnetic fields around it. If the time-dependent electric field is explained by sinusoidal relations with ω as the frequency of oscillation, the current density can be written as

$$J(\omega) = Ne\frac{p(\omega)}{m} = \frac{Ne\left(\frac{e}{m}\right)}{\frac{1}{\tau} - i\omega}E(\omega) = \sigma(\omega)E(\omega) \tag{22.7}$$

where $\sigma(\omega) = \frac{N\ e^2\ \tau}{m(1-i\omega\tau)}$ is called as *ac* or *frequency dependent* conductivity. Comparing this relation with that of time-independent field relation, it can be written as: $\sigma(\omega) = \frac{\sigma_0}{(1-i\omega\tau)}$, where $\sigma_0 = \frac{Ne^2\tau}{m}$ is the conductivity at zero frequency.

As the time-dependent current creates a magnetic field, the motion of electron or carrier is influenced by this magnetic field and also it further affects the motion of other carriers within the conductor as well as those of nearby conductors [12]. Thus, within a conductor the distribution of current is not evenly distributed. Hence, any conductor can be considered as one or more concentric circular elements, and the carriers at the center portion of the conductor will be covered by larger magnetic flux than those on the outside. Thus the induced back-*emf* will be more toward the center of the conductor, which will result in lower current density at the center than at the surface. Such variation of current density is called *skin effect* and becomes dominating in conductors carrying high alternating current and larger diameter of the conductors. Similarly, the high alternating current conductors in closer proximity affect the current density of each other. The magnetic flux created by one deflects the carriers either closer to surface or farther from the surface depending on the direction of current. Thus the carriers feel extra resistance in their flow, and this effect is known as proximity effect [12–15]. The distribution of carrier or current density due to skin and proximity effects can be visualized from Fig. 22.3. Due to skin effect the current density at the center of the conductor of circular cross-section is almost zero (as in Fig. 22.3a), while due to proximity effect the carriers are pushed to or away from the surface depending on the direction of current of adjacent conductor (Fig. 22.3b).

The additional resistances observed in the conductor carrying alternating current due to skin and proximity effects are generally given by empirical relations [13–15]. For the copper conductors, the resistances due to skin and proximity effects are given as following. The numerals of such relations depend on the characteristics of materials and shape of the conductor.

$$R_s = \frac{\left(\frac{8\pi f 10^{-7} k_s}{R_{dc}}\right)^2}{192 + \left(\frac{8\pi f 10^{-7} k_s}{R_{dc}}\right)^2} \tag{22.8}$$

$$R_p = \frac{\left(\frac{8\pi f 10^{-7} k_p}{R_{dc}}\right)^2}{\left(192 + 0.8\left(\frac{8\pi f 10^{-7} k_p}{R_{dc}}\right)^2\right) \times \left(\frac{d}{S}\right) \times 2.9} \tag{22.9}$$

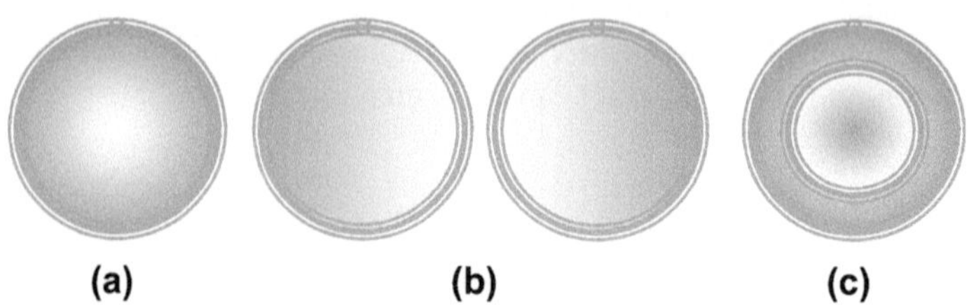

FIGURE 22.3

Schematics of current density across a circular conductor(s) due to skin (a), proximity effect (b), and both (c). The increasing color intensity indicates increasing current density across the cross-section.

where f = frequency in Hz; k_s and k_p are shape factor for the conductor, usually 1 for circular wires; d = diameters of the conductor, in mm; S = separation between conductors; R_{dc} = resistance of the conductors for direct current.

Thus the total resistances of the conductor are given as

$$R_{ac} = R_{dc} + R_s + R_p \tag{22.10}$$

The cables to carry high current-carrying conductors are thus designed suitably to minimize these resistances and often used as concentric or stranded to optimize the current density. The current density due to skin effect is essentially lower compared to that at the surface of the conductor, and the current density at a depth d from the surface is given as: $j = j_s e^{-d/\delta}$, where δ is skin depth, defined as the depth from the surface of conductors where the current decreases to $1/e$ times of that at surface, and d is distance from the surface. The skin depth (δ), depends on the conductivity, magnetic permeability, and frequency of alternating current. The variation of current density due to skin effect for cylindrical Cu and Fe rods are shown in Fig. 22.4. With increasing frequency the skin depth decreases very fast. However, if an additional conductor exists nearby, due to proximity effect the carriers are pushed to the center as shown in Fig. 22.3c. Thus the diameter of strand and material for high-tension cables are governed by these effects.

In the general power electric and electronic industry, the transportation of high current without loss and interference from the neighboring conductors not only focuses on the nature of conducting material but also on the insulating materials separating them. In simple power transmission line, the large loss is also minimized by selection of appropriate insulator and conductor carrying the current. Common metals show low resistivity, which is in the range of 1.5−150 μΩ·cm. The elemental metallic conductors like Cu, Ag, Au, Al, and Pt show low resistivity and higher melting points, and they are the prime choice for transmission of electricity [16]. Typical properties of some common elemental conductors are given in Table 22.1. The selection of material depends on the end use as well as cost-effectiveness and stability in harsh environment. Though the noble metals like Ag, Au, and Pt

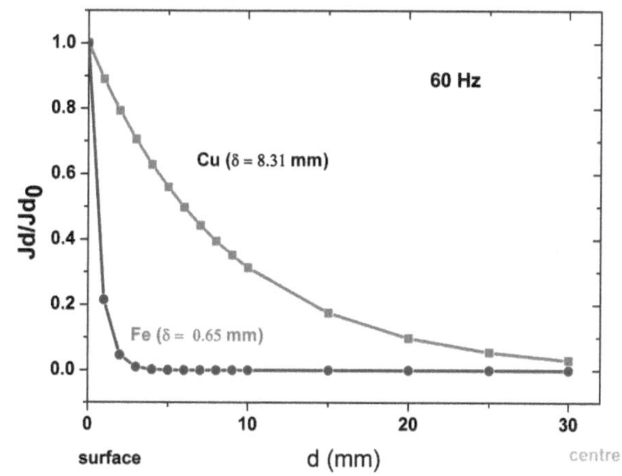

FIGURE 22.4

Caparison of current density across the depth of cylindrical wires of Cu and Fe.

Parameters were taken from Z. Popovic, B.D. Popovic, Introductory Electromagnetic, Prentice Hall, Inc. Ver-12, 2012.

Table 22.1 Physical Properties of Some Practical Transmission Cable Materials [16,19]

Element	Melting Point (°C)	ρ (μΩ·cm)	α ppm (Ω/Ω × °C)
Cu	1085	1.68	4300
Al	660	2.65	4500
Ag	962	1.59	4100
Au	1064	2.21	3500
W	3422	5.3	4500
Mo	2623	5.7	5300
Rh	1964	4.51	4500
Ir	2446	5.3	3930
Pt	1768	10.59	3927

show corrosion-resistive properties under harsh environment, their large-scale utilization is not cost-effective. Thus the transmission cable relies on Fe, Cu, and Al, even though they show corrosion in air or water.

Due to nonavailability of appropriate materials, the loss of power in the electric transmission is tolerated to about 20–30% against conceptually achievable loss of about 2–6% [18]. In particular, extremely high voltage transmission line suffers serious losses, due to the energy dissipation in the conductor and joule heating (long line resistance and so high I^2R losses) of the conductors. The discharge of dielectric gas molecules around the conductor as well as leakage currents though the insulations are also causes of transmission losses [18–20]. The large-scale integrated circuits suffer the dimensional limitations owing to the same reason. Thus appropriate insulating materials are essential to confine the electricity in high-current wires or nanodimensional semiconductor in electronic devices. Over the years the ceramic insulators, like porcelain, Bakelite, etc., have been used for separating high-tension cables in power transmission, while the electronic industries rely on the SiO_2 layer over silicon. Also numbers of polymers and rubbers have been evolved for creating an insulating barrier for high-current and high-voltage cables. But still the losses and limitations of operating conditions are yet to be overcome. Thus continuous research and development is being pursued for suitable dielectric material. The main considered property of chosen material for such high–electric field applications is conductivity if the material is supposed to carry current and breakdown voltage if the material is supposed to behave as insulator.

2.2 SEMICONDUCTORS

Besides conductors, the materials exhibiting intermediate range of resistivities but appreciable conduction compared to the insulators belong to the category of semiconductor [8–11]. The classification of materials as semiconductor was initially based on the temperature effect of resistivity, wherein the resistivity decreases with the increasing temperature. In the past few decades these materials have dominated the electronic industry in the form of integrated circuits for regulated flow of charge. Further, the high-speed electronic circuits in the electronic industry rely on the semiconducting materials as well as on the passive separating insulating materials, discussed later in this chapter. From the band structure of semiconductors, it is known that the valence and conduction bands are separated by a small gap. As the materials may contain some defects, an appreciable amount of carriers are available for charge transport and they can be easily increased by increasing temperature. Thus under normal condition, semiconductors are poor conductors, but the conductivity can be enhanced by suitable doping. An unbalanced carrier introduced by such process makes them conducting, and the conduction is either through one type of carrier, i.e., electron or hole, or both as in ambipolar semiconductor. Typical variation of conductivity of a semiconductor with temperature or defect is shown in Fig. 22.5. Early applications for regulated power transmission as well as rectification have been explored using appropriate carrier-conducting semiconductors, like p–n junction. Semiconducting junctions for current regulation, control as well as switch and light emission are the examples known in modern electronic industry. The voltage and current follow the linear relation in one direction while in opposite direction they do not, and thus the reverse biased junctions are used as voltage regulator. The variation of current with voltage in a p–n junction shown in Fig. 22.6 indicates their usability as a constant current or constant voltage source.

Transport properties of semiconductors in high field are important as they are the guiding factor for miniaturization of electronic devices. The linear relation of V and I or mobility fails at higher field [21–23]. In the extremely high electric field the semiconducting materials exhibit high carrier mobility, but in addition several other effects, like impact ionization, generation of hot carriers, breakdown, etc., are also observed [24–27]. The transport properties of semiconducting materials in extremely high electric field are crucial for several electronic applications, like field-effect transistors [27], photodiodes [28,29], and fast switches [30]. The linear relation of current and voltage breaks in semiconductors at high applied field. At higher potential the mobility as well as energy of the carriers increase, but beyond certain limit they may overcome the optical phonon energy and thus they emit an optical phonon. It may also happen that the kinetic energy of carrier, electron may exceed the binding energy, thus leading to an emission of electron, as in field emission. In case of activation of optical phonon, the carrier velocity may saturate with further increasing electric field. In common

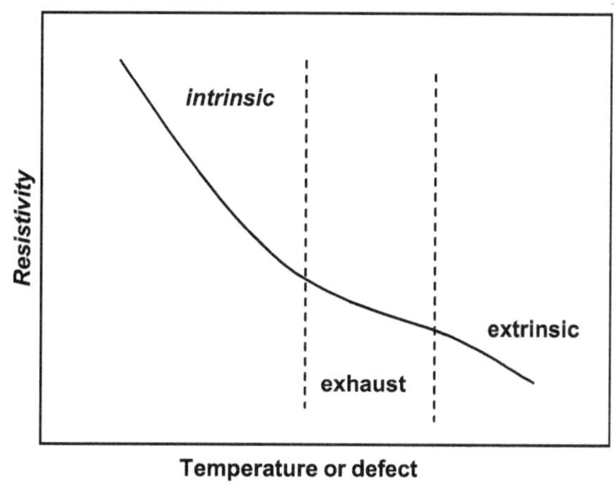

FIGURE 22.5

Variation of resistivity of semiconductor with defect concentration or temperature.

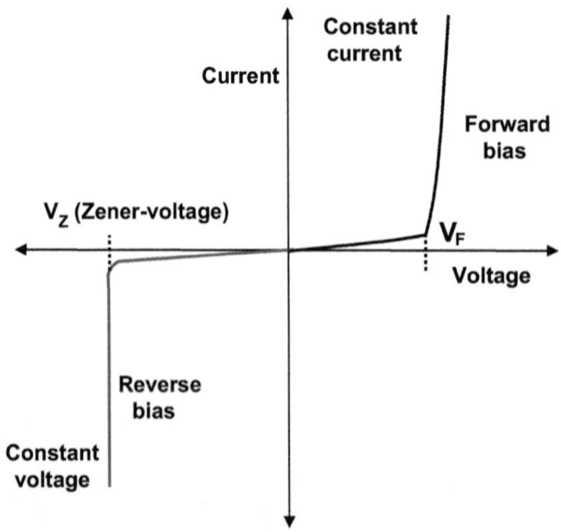

FIGURE 22.6

Voltage and current regulation in a p–n junction.

semiconductors, like Si or Ge, such saturation levels are not attained and show monotonous increase in current or carrier mobility with voltage due to inaccessible alternate modes. However, a drastically decreasing trend in dc electrical conductivity of silicon at higher field has been reported [21]. This trend has been attributed to the shortening of collision time and mobility of electron due to quantum emission of additional electrons. However, in case of GaAs-like semiconductors, a drop in current at higher field is also noticed [31]. The microwave oscillation of current in n-type semiconductor like GaAs and InP in high electric field was initially observed by Gunn, and the phenomena is known as Gunn effect [31]. The devices based on this effect are used for high-frequency oscillator for microwave generation, where an n-type semiconductor with two heavily doped terminals compared to the middle one is subject to dc electric field and the current voltage character shows a sudden drop in current followed by an oscillating pattern. The rendered negative differential resistivity is due to the accumulation of charge carrier at the central region of the doped semiconductor. Such devices are practically used to generate frequency from 1 to 100 GHz [31,32].

The effects arising from thermal and mechanical stress induced by high electric field are also important contributors for the failure of semiconductor devices. Electric overload, either as voltage or current, in a semiconductor may also lead to an increase in temperature due to I^2R effect at the junction. The conductivity of high-voltage semiconductor devices, like bipolar transistors and gated circuits, is affected by overvoltage. The overvoltage-induced current abnormality in oxide semiconductor gates may also be due to the structural failure as well as field-induced ion transportation. The avalanche of hot (high-energy) carriers in semiconductors at high electric field is also an important contribution for the failure of the device [33]. Besides, the additional surrounding components are also affected by such high-field semiconductors, which may be due to leakage current or induced electric or magnetic field or static charges. Attempts to minimize these effects are carried out by the selection of appropriate insulating or dielectric materials, and they are discussed subsequently in this chapter.

Additionally the multilayers of differently conducting materials are affected in high electric field. Fundamental aspects of such semiconductor multilayers have been explored by various researchers. It has been observed that the carrier dynamics are influenced due to the electron–electron scattering as well as electron-phonon scattering and also accumulation of carriers at the interface of such multilayer. The charge separation introduced at the interface alters the initial band alignment and hence alters the transport behavior [34]. Due to such charge accumulation process at the interface, such materials become a subject of study for second-harmonic generation and to study the carrier dynamics at semiconductor interfaces [35,36]. Glinka et al. have investigated the carrier dynamics in the GaAs/GaSb and GaAs/GaSb/InAs heterostructure through the second-harmonic signal generated by electric field–induced charge redistribution at the interface [37].

Semiconducting materials used for stimulated optical emission are also influenced by high electric field due to the alternation of optical absorption by impurity atoms in the semiconductors [38]. Such phenomena are initially observed by the influence of strong electromagnetic field of laser on the hydrogenous impurities materials [39,40]. Probability of ionization of atoms and solid bodies in the field of a strong electromagnetic wave whose frequency is lower than the ionization potential has been explained by Keldysh [41]. This ionization process is explained by the tunneling autoionization process in low-frequency region or multiphoton absorption in optical frequency region of electromagnetic field.

2.3 INSULATORS

Insulators are the class of materials where no or almost negligible conduction is observed. According to the electronic structure, they have completely filled valence band and are separated from the conduction band with a large gap. Insulators are practically used to prevent or block the flow of charge from the source to surroundings. In principle no material is a perfect insulator as in some condition, either by substitution of impurities or by the effect of external parameters like temperature, pressure, or humidity, nonnegligible conductions are observed in most of the insulators. The practical requirements made the insulators to evolve from jute, paper to glass, ceramics and cross-linked polymeric materials or ceramic, polymers, and ceramic-polymer composites. Widely used insulating materials for high-voltage transmission are glass or composite ceramics like porcelain, a composite of clay and quartz or alumina or feldspar, etc. The water-retarding properties of such ceramic porcelain are generally achieved by smoothening the surface of the materials. In practical materials the dielectric strength, mechanical strength, and temperature of operation are the important parameters that need to be considered. The voltage difference across the insulators may lead to catastrophic breakdown, like discharge, electrolysis, or other chemical reactions. Thus the class of insulators for low or high or ultrahigh voltage are different. The electric field effect on insulating materials is thus prior hand investigation for their utilization. Under the high-tension influence the electric field is distributed inside the insulator by either a slight reorientation of the molecules or ions in the material [42–44].

Ideally, the insulator should not show any flow of charge regardless of electric field. But if the field is having sufficiently higher strength, the flow of charge occurs in them and the process is known as breakdown. The charging and discharging of such insulators may also result in electron pinning and lead to the deterioration in the chemical or microstructure integrity of the insulators or secondary-field emission electron and/or due to surface charging [45–48]. Often they are inferred by studying the electron impact effect on insulators [49,50]. More details of the breakdown of such insulating materials are given in the section for dielectrics and dielectric materials.

In the opposite sense of superconductors, the terminology *superinsulator* is also used for some materials, like titanium nitride, which under certain conditions at low temperature exhibit infinite resistance [51]. Such materials are of significant interest to fundamental physics and the phenomenon is accepted as an electronic state and is sensitive to the temperature, electric and magnetic field. It is proposed that the electron pair, analogous to Cooper pair, repel each other, which does not allow them to flow under the field influence at least up to certain strength. But they all lose their insulating property beyond a certain electrical field.

2.4 DIELECTRIC

Practically dielectric materials are insulators, which do not show any conductivity but exhibit a property of charge storage due to internal rearrangement of charges. For several applications like energy storage and information processing, dielectric materials have evolved rapidly in the last few decades. Most important applications of dielectric materials are the information storage and insulation of electronic devices. The degree of miniaturization in memory devices based on capacitive components, such a static and dynamic random access memories, depends on the dielectric constant of the material. As the dielectric materials are poor conductors of electricity, they isolate two different electric potentials and store the electrical energy imposed on them. Thus they are also useful as capacitors for high-voltage sources. In principle the storage property of the dielectric materials is characterized in the terms (a) *capacitance*, which in principle governs the efficiency of the materials to keep away the conducting plates from coming in contact and (b) *breakdown voltage*, which governs the maximum voltage that can be applied on it without sparking. The capacitance of a dielectric material is defined as

$$C = \varepsilon_0 \varepsilon_r \frac{A}{d} \tag{22.11}$$

where ε_0 is the dielectric constant of the free space (8.854×10^{-12} F/m), d is distance between the plate, A is area of the dielectric materials, and ε_r is the dielectric constant of the dielectric layer.

The net charge stored in a dielectric material is given as

$$Q = CV \tag{22.12}$$

where C is capacitance and V is potential across the capacitor.

The energy stored in the capacitor may be considered to be the same as the work done in moving a charge from low to high potential, and it can be written as

$$U = \frac{Q^2}{2C} = \frac{C^2V^2}{2C} = \frac{1}{2}CV^2 \tag{22.13}$$

where Q is charge on the metallic foils and $V =$ potential difference between the plates.

The phenomenon by which the charge is stored in a dielectric material of a capacitor is explained by different types of polarization that occur in it. In general, on application of electric field at the two opposite ends of dielectric material, the opposite charges are displaced, thus creating dipoles inside the bulk matrix of it. This is measured as the extent of polarization, which is defined as net dipole moments per unit volume ($P = \sum \mu_i / V$).

The polarization in a dielectric material can be of different origins, like electronic, ionic, dipolar, and interfacial polarization (Fig. 22.7), which are activated at different frequencies of alternating electric or electromagnetic field. Electronic polarization occurs due to displacement of the center of negatively charged electron cloud relative to the positive nucleus of atom by the electric field. Ionic

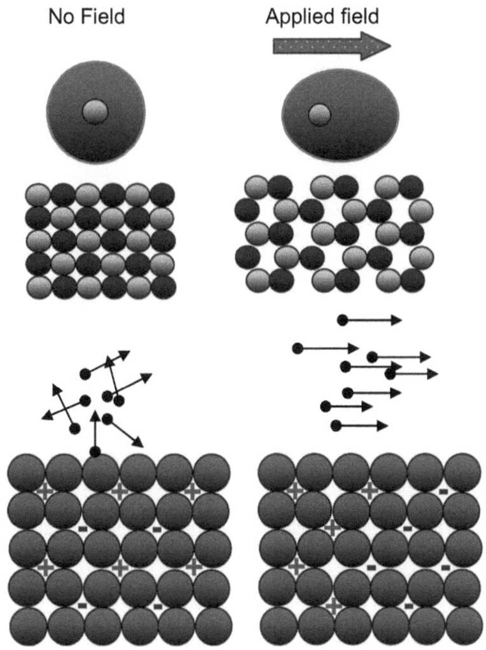

FIGURE 22.7

Polarization mechanism of dielectric materials under applied electric field. (a) Electronic polarization, (b) ionic polarization, (c) orientation polarization, and (d) space charge polarization (from top to bottom).

polarization occurs in ionic materials. In an ionic lattice, the positive ions are displaced in the direction of applied electric field, whereas negative ions get displaced in the opposite direction to the applied electric field, giving rise to net dipole moment. If the molecules possess permanent dipole moments, the rotation of these dipoles under an electric field will produce the dipolar polarization. Therefore, the dipolar polarization is strongly dependent on the frequency of the applied electric field and on the temperature. By the application of low-frequency field some of these bound charge carriers migrate over a distance. The translating charge carriers are accelerated by an applied field until they are opposed by or trapped at the physical barriers in these heterogeneous systems. This buildup of charges causes the polarization of the material. Space charge or interfacial polarization occurs in heterogeneous systems such as multicomponent materials or incompatible chemical substance-containing materials. Space charge polarization is the governing factor of dielectric properties at lower frequency. Electronic, ionic, and orientation polarization arises when charges are locally bound in atoms, molecules, or structures of solids or liquids. The total contribution of polarization to the dielectric constant is therefore a summation of the just-discussed individual polarizations.

As mentioned, the polarizations' effects originating from the different types of dipoles respond to alternating field at different frequencies (Fig. 22.8) due to differences in their order of time reversal. When the frequency of the applied field is sufficiently low, all types of polarization can be observed. Electronic displacement responds rapidly to the field reversals in the frequency range of 10^{17} Hz and thus are only controlled in the optical region of electromagnetic wave. The typical frequencies for the

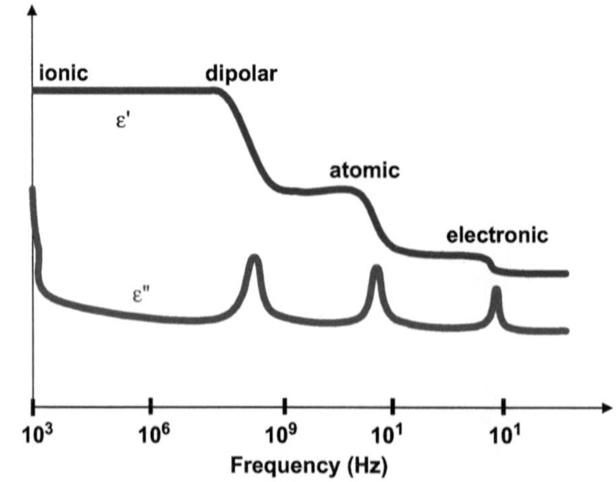

FIGURE 22.8

Response of real (ε') and imaginary (ε'') part of relative permittivity due to different polarization effects to frequency.

ionic displacement, oriental motion of permanent dipoles are in the frequency range of GHz to MHz region of electromagnetic spectra. At low frequency the ionic movements, in terms of ionic conductivity, appeared, and at sufficiently low frequency they appeared as high capacitance due to blocking electrode effect known as space charge polarization. Each dielectric mechanism has a characteristic "cutoff frequency." As frequency increases, the slow mechanisms drop out in turn, leaving the faster one contribute to ε_r. As expected, capacitance value, i.e., dielectric constant, always decreases with increasing frequency.

Though the dielectric materials retain the charge, in real situation the phase lag between the input and output field resulted as an energy loss and the loss is represented as: $\tan\delta = \varepsilon''/\varepsilon'$, where ε' and ε'' are real (relative permittivity or dielectric constant) and imaginary part (loss part) of relative permittivity of the dielectric medium. For a good dielectric material, $\tan\delta$ is significantly low and insensitive to the frequency of the applied field. The dielectric constant is one of the basic electrical properties of solids, which are studied with frequency and temperature to understand polarization mechanism [52–56], the process of electrical conduction, defect concentration and nature of bonding, etc. [57–59]. Depending on the response of the dielectric materials to the electric field or stress or temperature and then nature of dielectric constants, the dielectric materials can be grouped into different classes, namely ferroelectric/antiferroelectric, piezoelectric or pyroelectric, and their corresponding analogs nonpolar. The details of the materials are not covered in this chapter and can be found in other literature [60].

3. DIELECTRIC MATERIALS FOR HIGH-SPEED ELECTRONIC CIRCUITS

The dielectric materials are often used as substrate and separator for electronic circuits. The thermal conductivity and strength of the dielectric materials are the important parameters considered for such applications. In high-speed electronic processing units, in particular with operating frequency over 1 GHz range, the aforementioned skin and proximity effects as well as surface roughness and other electromechanical properties are of concern. The dissipation of signal to the dielectric elements is also dependent on the permittivity and loss factor of the dielectric medium. The signal loss from the element to these passive elements limits the operation of circuit elements and leads to propagation delay, which is expressed as, $t_{pd} = 1.017 \cdot \varepsilon_{eff}$, where t_{pd} = propagation delay in nsec/foot and ε_{eff} = effective relative dielectric constant of material. The signal loss due to the loss tangent is given as: $\alpha = 2.3\, f \tan\delta \sqrt{e_{eff}}$, where α = attenuation in dB/in, f = frequency component in GHz, $\tan\delta$ = material dissipation factor, and ε_{eff} = effective relative dielectric constant of the material [61,62]. Thus, the materials with relatively lower dielectric constant are preferred for such application [63]. Several magnetic and nonmagnetic oxides, like SiO_2, ZnO, Bi-doped rare-earth iron garnets, and tantalates, are considered for such applications [63]. The point defects present in such materials often play a crucial role for their usage in the high-frequency region.

4. DIELECTRICS UNDER EXTREME POTENTIAL (DIELECTRIC BREAKDOWN)

Insulators and dielectric materials at certain electric field become conducting instantaneously, which is called *breakdown phenomena*. This limits the material's practical use for either insulation or energy storage. The breakdown phenomena in solid, liquid, as well as gaseous dielectric materials has been known for a long time and is an important property considered for their safe operation limit. The mechanisms of initiation of breakdown phenomena are different in different materials, but they eventually create conducting channels between two plates of electrode of capacitor or source and ground (surrounding) of insulators.

In general, the electric breakdown may occur due to a number of processes due to degradation of material by evaporation, melting, or chemical or electrochemical changes or destruction of mechanical integrity by change of microstructure in itself. More importantly, the self-heating in the materials due to the action leakage current rapidly increases the temperature, which changes the structure, microstructure, or carrier concentration. These changes are giving way to the flow of current in the material. The chemical and electrochemical processes are slowly developed in materials over time and thus over a period of time the resistive properties of dielectric materials decrease considerably. The field strength that a material can withstand without undergoing breakdown is known as *dielectric strength*. In semiconducting materials, the breakdown fields are in the range of 10^5–10^6 V/cm, while in insulating materials they are in a higher range ($\sim 10^7$ V/cm). In soft polymeric dielectric materials this can be at around 10^6 V/cm [64]. The field-induced electrochemical reactions are infrequent but can occur due to many factors, like cyclic charging-discharging, trapped impurities, or contaminations. In order to understand the failure mechanism of the dielectric or semiconducting materials, a large number of studies were performed many years ago, and they are reviewed in several reports [64–71]. Next, we briefly explain the typical breakdown phenomena in various potential dielectrics for high-voltage applications.

The mechanisms of breakdown in semiconductors are mainly by impact ionization, tunneling, thermal instability, etc. [64–69]. The breakdown of interconnects in high-voltage terminal is a limitation in the integrated circuits, due to large accumulation of charges at the semiconductor surface

[72]. This redistributes the field in the semiconductor itself or creates hot carriers and reduces the breakdown voltage of the barrier dielectric [72–74]. Alternate approaches for reducing the charge accumulation are considered by the usage of suitable alloys of Si–Ge or Si–C [75,76], which reduces the external resistance. A number of metal-rich silicides of Ni or Co as well as other metals like Yb, Er, and Pt have also been investigated for such problems [77–82]. The use of high-dielectric materials in the place of currently used SiO_2 are also considered as solutions for breakdown problems. The high electric field–related Gunn effects and unregulated current flow in semiconductors were mentioned earlier. The breakdown of phenomena in Zener diode, a reverse bias diode, is utilized to regulate power supply. Impact ionization, field emission, double injection, light activation, and space charge–limited currents at the trap are also the possible processes for breakdown of semiconducting materials.

The electric breakdown of dielectric gases in general occurs due to the partial ionization of gas molecules under the influences of electric field. The dielectric strength of air is about 3×10^6 V/cm and beyond it is partially ionized and forms an electric arc between the gap of electrodes or a spark by recombination of charges [83]. The breakdown voltages of gas dielectric materials depend on the nature of gas as well as the electrode configurations and gas pressure. Paschen's law empirically explains the breakdown voltage for various gases under influence of only electric field and is given as

$$V_{bd} = \frac{a \times p \times d}{\ln(p \times d) + b} \tag{22.14}$$

where V_{bd} is breakdown voltage, a and b are characteristic constants, p is gas pressure, and d is distance between two electrodes. The breakdown of gas dielectric occurs through a nonlinear current voltage relation (Fig. 22.9), viz., initial acceleration of ions by voltage followed by constant current.

At higher field, the energetic ions or electrons collide and cause further ionization very rapidly, often called avalanche ionization or Townsend discharge where ionization increases exponentially. In addition, the cathode materials and external ionizing radiations, like electron or photon, cosmic or earth radiation also affect the breakdown phenomena of gas, due to their contribution to initialize the

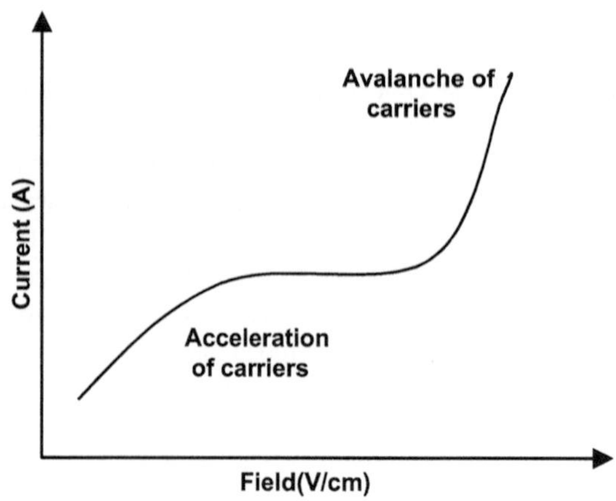

FIGURE 22.9

The voltage current relation of gaseous dielectrics. In between the acceleration and avalanche, a constant current region is observed due to free and fast transportation of carrier.

ionization process. The field emission, photoelectronic emission, and thermoemission of electrons are also initiation of dielectric breakdown of any media.

An enormous amount of research efforts on insulating gas medium have been carried out to find better insulating materials with higher breakdown field. The gaseous insulators with the most potential for high-voltage application are fluorocarbon or fluorosulfur. The fast recombination of the decomposed products makes them sufficiently stable, but most of them have large ozone depletion and global warming potentials. Practically, SF_6 is the most widely used gaseous insulator in high-voltage systems, like accelerator and high-voltage electron microscopes, despite its ozone and global warming potential, which possesses about three more dielectric strength compared to air and is an excellent quencher for discharge due to recombination of radicals or ions produced by disintegration [84–88]. In order to avoid SF_6, research attention is being focused to find newer dielectric gas or modification of common gases like N_2 for practical applications. Literature shows that the dielectric strength of N_2 is enhanced significantly by mixing it with small amounts ($\sim$5%) of SF_6 [89,90]. Alternate molecular gas like NF_3, which has dielectric strength a little lower than SF_6, but appreciably higher strength compared to air, is also proposed to replace SF_6 [91,92] but could not gain technical success due to chemical reactivity at relatively lower temperature.

The breakdown in liquid dielectrics used for voltage insulation as well as cooling in power systems deteriorates their performance. But the complex structure and better self-healing process as well as better thermal conductivity of liquid dielectric compared to solid dielectrics make them better and cheaper materials for the just-discussed applications. Water, however, has high dielectric constant ($\sim$78) and high cooling efficiency is not often suitable for insulation purposes due to high conductivity due to trapped ions. Thus, the normal mineral oils or saturated hydrocarbons like hexane or heptanes are used even though their dielectric constants are low ($\sim$1) and highly flammable. Silicone oils find wider choice due to their higher dielectric constant and strength, but they are expensive and hence preferred for specific applications. Often the mineral oils are used as liquid dielectric used in transformer due to cost considerations. The failure of such dielectrics can be catastrophic from several aspects, like instrumental and environmental. The liquid dielectrics are also used in electrorheological separation and machining applications, which are not covered in this chapter. The electric field effect on the dielectric properties of liquids and dielectric properties are also concerned in synthetic chemistry either as a solvent or reaction medium [93,94]. In the liquid dielectrics, the classical Clausius–Mossotti relation equating the dielectric constant to electrical polarizabilities and permanent dipoles is not satisfactory. The following empirical relation proposed by van Arkel and Snoek [95] can satisfactorily explain the dielectric constant and electric field dependency of pure polar dielectric liquids:

$$\frac{\varepsilon - 1}{\varepsilon + 2} = \frac{4\pi N}{3}\left(\alpha + \frac{\mu^2}{3kT + CN\mu^2}\right) \tag{22.15}$$

where $\frac{\varepsilon + 2}{3}E$ is defined as Clausius–Mossotti electrical field. N = No of molecules per cm^3; α = polarizability of the molecule; μ = permanent dipole moment of the molecule.

C is an empirical constant proposed by van Arkel and Snoek for polar molecules and is slightly temperature dependent. Later, Symth [96] and Muller [97] replaced the values of C by $3C/4\pi$. More details of electric properties of polar organic molecules can be referred to in the literature [93,98–102]. The liquid dielectric power system not only acts as insulator but also acts as a coolant and arc arrestor by rapidly recombining the ions produced by discharge [102]. The basic property of such liquids should be high dielectric strength and thermal conductivity as well as chemically inert to other electrical and container material. However, the chemical degradation due to contamination, absorption, or trapping of gas molecules or contamination with decomposed products by electric discharge reduces their dielectric strength.

Due to large practical requirements, the effects of electric field on dielectric liquids have been studied since the early days of electrical engineering. However, the characterizations of these were done by simple analytical or mechanical methods and thus not much information on mechanism of electrical effects was known. The mechanisms of breakdown of liquid dielectric are by the ionization of the liquid or by ionization of gas bubble of the liquid. In the latter case, the low densities of gas in the bubble ionize faster by field ionization and propagate further into the liquid. These two mechanisms are explained by avalanche–streamer and bubble theory [103–105]. In dielectric liquids, the coexistences of both phenomena are also supported in these references. Similar to gas dielectrics, the microstructure and morphology of electrode surfaces are also controlling factors for the dielectric breakdown strength of liquids. In nonscratched smooth electrode the breakdown strength increases [106,107].

The solid dielectrics under the influence of higher electric field also undergo breakdown, and the breakdown is permanent and not retrieved by itself. The breakdown of solid dielectric is a challenge for the electronic circuit and is a serious limitation in scaling the integrated circuits. The electronic industries as well as energy storage applications rely not only on high dielectric constant materials but also materials with low loss. As mentioned earlier, the polarization process in the materials is frequency dependent and thus the complex permittivity depends on frequency and that represents the characteristics of the dielectric loss in ac field. The loss tangent defined as, $\tan\delta = \varepsilon'/\varepsilon''$, where the ε'' and ε' are imaginary and real part of the complex permittivity, governs the dielectric strength of the materials.

In order to develop low-loss materials, continuous research efforts have been made by selecting a material or altering the size, porosity, morphology, and thickness, adding selected additives, etc. The loss tangents for practically used dielectrics are in the range of 0.0001–0.001. The dielectric strengths of the dielectric films used in conventional ICs are in between 10^6 and 10^7 Vcm^{-1}. The power loss in dielectric capacitors depends on the loss tangent and also electric field.

The mechanism of breakdown in solid has been reported in a large number of research articles [64–69]. Breakdown in solid dielectrics can be intrinsic or extrinsic. The intrinsic breakdown of a dielectric is related to the free carriers and defects. Under high electric field, emission of electron from atom or by breaking a bond is also encountered. Also, under the influence of field, the electrons can accelerate and collide with other atoms to ionize them. The processes of primary and secondary ionization lead to an avalanche of carriers, which create an easy current conduction path in insulator. This process destroys the insulating character of the insulator. Thus defect-free materials with high-ionic characters show higher breakdown strength. The trapped gas molecules, in particular in porous ceramics, show lower dielectric field strength due to faster ionization of the gas molecules in the trap.

In addition, alternate breakdown mechanisms like thermal and electrochemical process are common in solid dielectric. The solid dielectric shows feeble but nonnegligible current due to leakage and thus causes heating by joule process. Unless the heat is removed, they increase conductivity of the dielectric leading to failure. In addition, the thermal effect produces local heating zones, known as hot spots, where the temperature can increase until melting or even to evaporation and induce erosion or reaction with additional components. This is also reflected in the change in microstructure-related degradation. Besides, electrochemical breakdown also occurs in the dielectric due to the field-induced movement of ions to the respective electrodes. Electrofracture and deformation also occur in solid dielectrics due to the electrostatic field and ion movements. The mechanical failure due to electrostriction in response to alternate field also occurs in dielectric and more commonly in piezoelectric materials [108–114] used as actuators. In general, dielectric strength decreases with increasing sample thickness and temperature due to defect contributions.

Dielectric breakdowns in solids relating to the microstructure in bulk and at the interface have been investigated in a number of studies [115–132]. Since the breakdowns are a consequence of trapping of charges and scattering of carriers, they are greatly influenced by the microstructures [121–123]. Studies on the relation between the microstructure and breakdown voltages for alumina indicate that the diffusion of injected charge by suitable additive or sintering aid increases the breakdown voltage. It is also observed that the dissolution of Si^{4+} ion in Al_2O_3 increases the cation vacancy with negative charge and they act as positron trap site, which hinders the breakdown [120,124–126]. Studies on breakdown strength of Y_2O_3-stabilized ZrO_2 indicate an increasing trend of breakdown strength with the size while there is a decreasing trend with the size of pores. The breakdown strength increases to maximum value (of about 2 MV/cm) at 10% Y_2O_3 [127].

5. CASE STUDIES ON HIGH AND LOW DIELECTRIC CONSTANT MATERIALS

The dielectric materials of both high and low dielectric constant have applications in electrical and electronic engineering. Microelectronics use both of these materials, as gate or interlayer insulating materials. As SiO_2 ($\varepsilon' = 3.9$) forms an adherent film on the Si surface, it has been successfully used as gate dielectric in integrated circuit. The breakdown of SiO_2 layer based on tunneling current is about 1.2 nm. The success of very large-scale integrated circuit has been attained near to this limit [128,129]. However, suitable dielectric material with high dielectric constant and high dielectric strength is a problem for the futuristic electronics. Though several high dielectric materials have been developed, the low breakdown potential kept them aside. For example, for $SrTiO_3$ the dielectric constant is 200, but low breakdown potential (3.3 MV/cm) makes it unsuitable for application as a gate dielectric [130–133]. The prospects and suitability of the dielectric materials for application in conjunction with Si in integrated circuits have been explained by Hall et al. [134]. The difficulty to replace SiO_2 (often as SiO_x) with other high dielectric materials arises from the stringent requirements, like thermodynamic stability of itself as well as in contact with Si, low cost, high dielectric, high band gap, etc. [134–136]. The potential high dielectric constant materials, like SiO_2, Al_2O_3, $ZrSiO_4$, ZrO_2, and HfO_2 are considered to be promising due to their wide band gap ($\geq$5 eV). The materials possessing higher dielectric constants, like Ta_2O_5, are more suitable for capacitor-based energy storage or memory applications but not for gates for integrated circuits due to their reactivity with Si [137,138]. TiO_2 also has limitations for such application due to higher leakage currents arising from the smaller band gap, formation of multiple oxidation states during the preparation and operation conditions [139–142]. Research also indicates the increasing stacking of SiO_x layers increases dielectric constant as well-reduced leakage current [143], but this is not so in most other cases.

In recent years, HfO_2 has been considered as a more promising gate dielectric, and its chemistry and electronics properties have been explored extensively in literature. Though ZrO_2 and HfO_2 have almost similar physical and electronic properties, lower chemical reactivity of HfO_2 with Si compared to ZrO_2 favors the former for gate dielectric. Chio et al. [138] have analyzed a large number of preparation methods for HfO_2 thin film as well as its growth on a number of semiconducting materials' surfaces and revealed high dielectric constant of about 15–25. But diversified breakdown voltages depending on the methods of preparation are shown in this study. Additionally, HfO_2 derivatives, like composites of HfO_2–SiO_2 with suitable combination of merits of both oxides, are also considered for such applications [136,137,144–149].

Rutile (TiO_2) and compositions with related rutile structure form the most widely studied group of materials for the dielectric properties. Rutile is well known for its incipient ferroelectric property arising from the distortion in TiO_6 octahedra at low temperature as well as for its large frequency-dependent static dielectric permittivity [150]. A large number of solid solutions with rutile structure or ordered rutile-related structures are formed in composition $M_{1-x}M'_xO_2$, where M and M′ are the cations suitable for the formation of octahedral coordination and charge neutrality. Often the substituted cations are disordered for a wide range of compositions and retain the parent tetragonal lattice and symmetry. Several metal oxides with composition MO_2, $MM'O_4$, and MM'_2O_6 stoichiometry adopt either rutile-type tetragonal or related structure types. Recent studies on such materials indicate large permittivity and relaxor-type ferroelectric properties. Subsequently, a number of studies have been available in literature indicating their dielectric properties. Mani et al. have systematically investigated a number of rutile-related compositions in $MM'O_4$ and $MM'M''O_6$ -type compositions and revealed rutile-type tetragonal structure for most of them and large dielectric permittivity in $FeTiTaO_6$ [151,152].

The structural studies of Mani et al. on $MM'M''O_6$, (M = Fe, Cr, Al, Ga, M′ = Ti and M″ = Nb, Ta, Sb)-type materials indicated abnormality in residuals of refinements, due to some deviation of intensity as well as unindexed peaks [151]. The phase pure samples of such compositions, viz., $FeTiTaO_6$, $CrTiTaO_6$, and $AlTiTaO_6$, prepared by soft chemical methods also reproduce the high dielectric constants [153,154]. The dielectric properties investigated by both the groups indicate large permittivity for $FeTiTaO_6$ compositions only. Among these substituted rutile-type materials, only $FeTiTaO_6$ has larger frequency dispersion and large permittivity but also shows larger loss. Typical permittivities of $FeTiTaO_6$ and $FeTiNbO_6$ at different temperatures are shown in Fig. 22.10. Also, it has been concluded

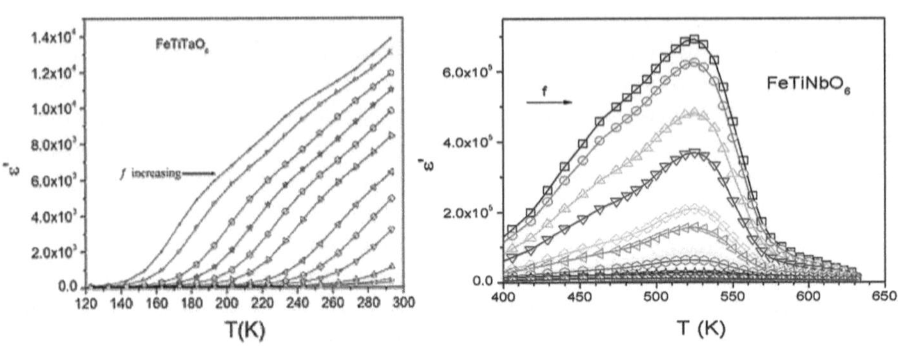

FIGURE 22.10

Temperature dependences of the dielectric constant of $FeTiTaO_6$ and $FeTiNbO_6$ at different frequencies.

that the compositions containing Fe as the trivalent ion show larger permittivity compared to others [151,152]. From low-temperature and high-temperature dielectric studies the dielectric loss has been attributed to the polaronic conduction and point defects due to variable oxidation state of Fe as in them [151,152,155]. It has been reported that the polaronic conduction is significant in the case of Fe-containing compositions, which can be explained by the feeble deviation of oxidation state of Fe^{3+} in them. A decrease in loss but still appreciably higher values ($\tan\delta \sim 50-250$) is observed even in the denser ($\sim$97%) sintered sample [156]. Later studies on $FeTiTaO_6$ and its W^{6+} substituted compositions reproduce the high dielectric constant of such materials [157,158]. But the authors have attributed the high dielectric constant to the polar nanodomains in heterogeneously distributed cations and possible oxygen vacancies as well as electrode effects [157,158].

Since rutile is an incipient ferroelectric material, its dielectric properties are sensitive to the substitution. The permittivity of TiO_2 is enhanced significantly by substitution of Ti^{4+} with other cations. Colossal permittivity of the order of 10^4-10^5 is observed by cosubstituting Co and Nb in TiO_2 [159]. There have been studies on (Nb + In) co-doped TiO_2 exhibiting temperature- and frequency-independent colossal permittivity ($>10^4$), and low loss (<0.05) has been attributed to defect dipoles formed by alliovalent cations in the lattice [160]. The metal-doped enhancement of the dielectric constant in perovskite-type $BaTiO_3$ has been practically used for several applications. Larger dielectric permittivities are also reported in Pt metal particle dispersed $SrTiO_3$ [161,162]. Within the percolative threshold limit, the metal particles inside the dielectric make them multilayer capacitors or assembly of microcapacitors and thus enhance the net capacitance [159,160].

Among lead-free ferroelectric ceramics, several studies on perovskite and double perovskites are available in literature. Discovery of ferroelectricity and subsequent explanation of origin ferroelectricity of $BaTiO_3$, materials with perovskites have been explored extensively for high dielectric constant materials. The perovskite-type titanates are formed with divalent cations, like alkaline-earth, Pb^{2+}, Cd^{2+}, etc. Only ideal $SrTiO_3$ has an ideal primitive cubic structure at ambient temperature while distorted structures are formed with cations larger or smaller than the Sr^{2+} ions. At ambient temperature $BaTiO_3$ has a tetragonal structure and with increasing temperature the dielectric constant increases and reaches to a maximum value of about 10^3 at the transition temperature [163,164]. Despite the ferroelectric nature, the dielectric constant of $BaTiO_3$ is only 20–30 at ambient condition. Thus to increase the dielectric constant, doping in structure or making composite with other materials, in particular conducting metallics like Cu, Ni, Ag, etc., have been explored in literature [165–168]. The dielectric constants of about 10^4 have been reported in platinum, tin, carbon, etc., containing composites [163]. Similarly, the dielectric properties of $SrTiO_3$ have also been investigated, in particular the low-loss dielectric constant of $SrTiO_3$ and incipient ferroelectric behavior at lower temperature [168–172]. Though the $SrTiO_3$ lattice is expected to show ferroelectric distortion, the ferroelectric transition has never been observed in it. At lower temperature, $SrTiO_3$ shows cubic to tetragonal structural transition, which is reflected as an anomaly in the temperature-dependent dielectric constant. Quantum ferroelectric behavior in $SrTiO_3$ at low temperature is reported at lower temperature ($\sim$30K) and that gets diffused on substitution by other cations both at Sr^{2+} or Ti^{4+} sites. This has been attributed to the formation of local polar clusters and internal lattice strain, but the dielectric constant of about 1000–2000 has been reported for ion-doped $SrTiO_3$ [173–175].

$PbTiO_3$ and its complex solid solution compositions are widely used electroceramics for piezoelectric and actuator application due to large electromechanical coupling coefficients [176,177]. The structural variation in the solid solutions has been a subject of study itself due to their complex phases diagram and coexistence of two or more closely related phases in a range of compositions. The morphotropic phase boundary compositions of such solid solutions often show relaxor-like dielectric properties with large dielectric constant over a wider frequency and temperature range instead of narrow temperature of ferroelectric materials [176–179]. Thus relaxor ferroelectrics with morphotropic phase boundary compositions have been explored in recent years. The polarization and stabilities of such phases are strongly influenced by temperature, electric field as well as mechanical stress. Low-loss high dielectric constant relaxor like ferroelectric properties has been reported on single crystal sample with composition $[Pb(Sc_{1/2}Nb_{1/2})O_3]_{0.58}$–$[PbTiO_3]_{0.42}$ [180,181] The temperature-dependent dielectric constant and loss tangents of this sample measured at some frequencies are shown in Fig. 22.11. The ferroelectric to paraelectric transition is observed as a single peak in temperature peak around 234°C while the loss peak indicates multiple peaks, characteristics of multidomain relaxor ferroelectric phase [181]. On increasing temperature, a sequential phase transition from ferroelectric and paraelectric phase has been observed by polarizing microscope where the electric field of light interacts differently with different polar domains. Typical domains observed by polarized optical microscope at various temperatures are shown in Fig. 22.12.

A large number of studies on electrical properties have also been performed on doped $PbTiO_3$, in particular the transition metal ion-doped $PbTiO_3$ in the aim to control magnetic properties by electric field. Such materials are considered as multiferroic and magnetoelectric or magnetodielectric materials and is a subject itself [182]. The electric field often enhances the leakage current, in particular, the transition metal substituted ferroelectric materials. Several alternate approaches, like suitably modifying the composition or controlling the defects by the control of anion stoichiometry, are being explored. Chemical substitution by appropriate elements for charge balance is one of the common approaches for this. The enhanced ferroelectric character as well as a considerable decrease in leakage current in Gd and Fe cosubstituted $PbTiO_3$ was achieved by such charge balance concept [183].

Electric field effect on $BiFeO_3$ has also been investigated in several reports for controlling the electric and magnetic polarizations and understanding the mechanism. Evolution of large spontaneous polarization in $BiFeO_3$ under the influence of cyclic electric field, a large electric polarization (about 100 $\mu C/cm^2$) in $BiFeO_3$, has been reported by Lebeugle et al. [184]. Such high polarization has been

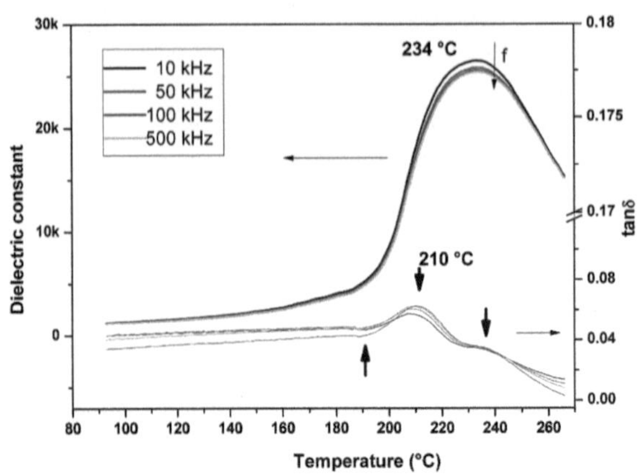

FIGURE 22.11

Temperature-dependent dielectric constant and loss factor of $[Pb(Sc_{1/2}Nb_{1/2})O_3]_{0.58}$–$[PbTiO_3]_{0.42}$ single crystal measure across the 001 planes.

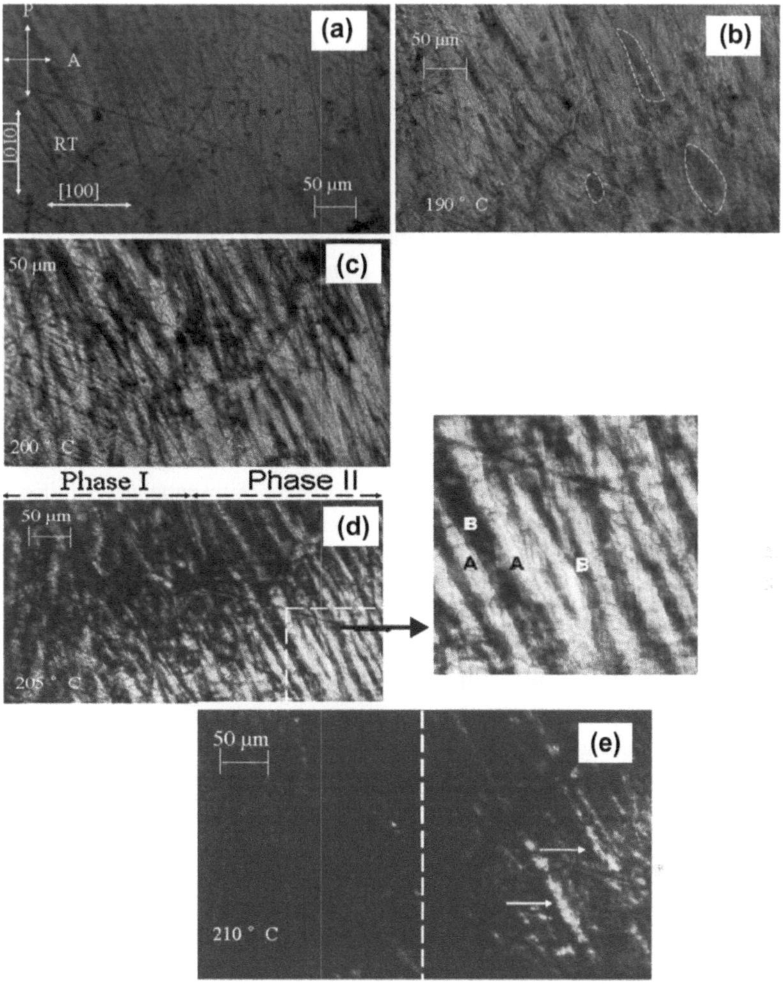

FIGURE 22.12

Domain patterns of (001) $[Pb(Sc_{1/2}Nb_{1/2})O_3]_{0.58}$–$[PbTiO_3]_{0.42}$ single crystal at various temperatures while heating under crossed polarizer at $\delta = 90$ degrees. The changes in orientation of polar domains interact differently with the electric field of light and they are reflected as contrast variation.

Reproduced from S.V. Rajasekaran, S.N. Achary, S.J. Patwe, R. Jayavel, G. Mangamma, A.K. Tyagi, J. Mater. Res. 29 (2014) 1054 with the permission of Cambridge University Press.

predicted only from structural parameters and theoretical calculation and never observed experimentally on bulk sample [185,186]. In strained thin films, such high polarization under electric fields has been reported by Wang et al. [187].

The mixed-cation perovskite-type solid solutions like $AAl_{0.5}Nb_{0.5}O_3$–$ATiO_3$, A = Ca, Sr and Ba [188] systems show larger variation of dielectric constants with compositions. The d^0 cations, like

Ti^{4+}, Nb^{5+}, Ta^{5+}, W^{6+}, Mo^{6+}, etc., often show second-order Jahn–Teller distortion, which is reflected as ferroelectric or relaxor-like dielectric properties [189–191]. Recently, tunable dielectric properties that vary from incipient ferroelectric to ferroelectric to relaxor-like properties have been observed in $SrTiO_3$–$SrFe_{0.5}Ta_{0.5}O_3$ system [192]. With increasing the Fe^{3+} and Ta^{5+} contents at the Ti^{4+} site of $SrTi_{1-2x}Fe_xTa_xO_3$ composition, the structure transforms from cubic $SrTiO_3$-type to orthorhombic $SrFe_{0.5}Ta_{0.5}O_3$-type structures through a biphasic region. The dielectric constants vary appreciably (Fig. 22.13) with composition of this system. In the cubic solid solution compositions, maximum dielectric constants (~8000) are observed at higher temperature while lower values are observed in biphasic compositions. All the compositions show significant loss while the losses are less in the biphasic compositions.

Similarly, this system shows a systematic variation in electric field–induced polarization (Fig. 22.14) with composition. It may be noted here that a feeble substitution at the Ti^{4+} site of $SrTiO_3$ introduces ferroelectric-like behavior in it.

Large dielectric permittivity ($\varepsilon_0 \sim 10^5$) has been reported for a cation-ordered perovskite-type compound, namely $CaCu_3Ti_4O_{12}$ [193]. The colossal permittivity is reported to be quite stable in a wider range of temperature as well as frequency up to MHz. Subsequently, similar orders of permittivities have been reported in a number of perovskite-type compounds with or without cation ordering [168,169,194,195]. However, studies also indicate that the colossal dielectric properties of $CaCu_3Ti_4O_{12}$ are of extrinsic microstructure origin and mainly due to interfacial polarization at grain boundary [196,197]. Recently a large colossal breakdown strength (~21 kV/cm) was observed in Al^{3+}-doped $CaCu_3Ti_4O_{12}$ sample [198] prepared by liquid-phase doping of Al_2O_3 in $CaCu_3Ti_4O_{12}$. The authors emphasized that there was a drastic increase in resistance (from 0.47 to 13.55 MΩ) at the grain boundary. This suggests that appropriate modification of the microstructure or composition is promising for practical dielectric of storage as well as gate applications.

The monoclinic perovskite-type titanates, like $Ln_2Ti_2O_7$, Ln = rare-earth ions, have also been considered as promising electronics materials due to their high Curie temperature ferroelectric and piezoelectric properties, high dielectric constants, and nonlinear optical properties [199–203]. The ferroelectric to paraelectric transition temperatures of $La_2Ti_2O_7$ and $Nd_2Ti_2O_7$-like perovskites are above 1073K [199]. Studies on optical and dielectric properties of $Ln_2Ti_2O_7$ (Ln = La, Nd, and Pr) indicate larger nonlinear optical activity and dielectric constant of about 10–40 [204,205]. The typical dielectric constant of $Pr_2Ti_2O_7$ at different temperatures and frequencies are shown in Fig. 22.15.

Literature also shows the ferroelectric transition of $Pr_2Ti_2O_7$ depends on the nature of sample preparation, viz., ferroelectric transition around 740°C has been reported in the spark plasma-sintered sample of $Pr_2Ti_2O_7$ [202] while no transition is reported up to 1200°C in other studies [199,205]. The spontaneous polarizations as calculated from the shift of the ions are in the range of 7.4–8.3 $\mu C/cm^2$ in such layered perovskites [205–207]. However, the practically observed polarizations are higher in $La_2Ti_2O_7$ and $Nd_2Ti_2O_7$ compared to $Pr_2Ti_2O_7$ due to conduction contributions from defects and oxidation states of praseodymium [205,208–210]. The experimentally measured polarization versus voltage properties of bulk polycrystalline $Pr_2Ti_2O_7$ are shown in Fig. 22.16.

In view of lead-free materials for high Curie temperature ferroelectric and piezoelectric applications, compounds with Aurivillius-type structures have also been considered [211,212]. Bismuth titanate ($Bi_4Ti_3O_{12}$) of this class has drawn considerable interest for dielectric and ferroelectric application due to low leakage and fatigue free and high-remanent polarization as well as high-ferroelectric to paraelectric-phase transition temperature ($T_c \sim 675°C$) [212,213]. A number of

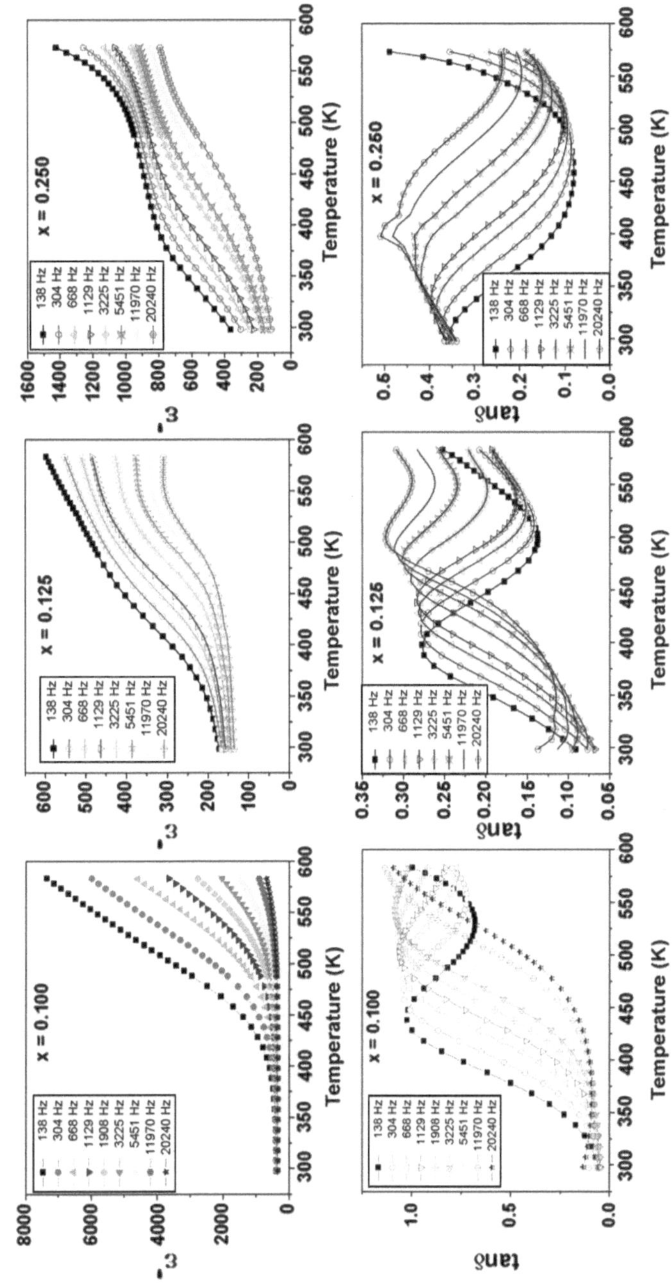

FIGURE 22.13

Dielectric constant and loss factor of compositions in $SrTi_{1-2x}Fe_xTa_xO_3$ [$x = 0.100$ (cubic); $x = 0.125$ (cubic + orthorhombic); $x = 0.250$ (orthorhombic)].

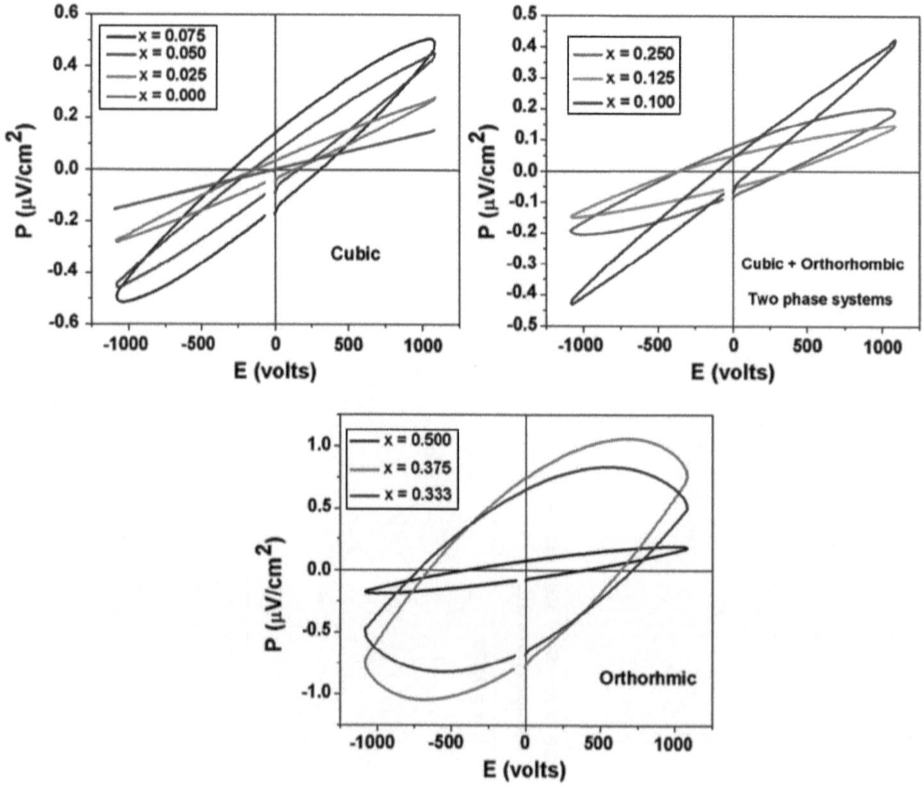

FIGURE 22.14

P versus E loops of $(0.00 \leq x \leq 0.50)$ compositions in $SrTi_{1-2x}Fe_xTa_xO_3$ compositions.

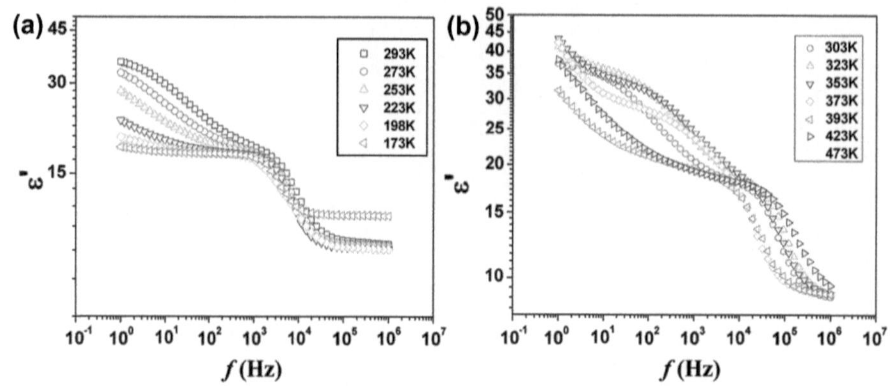

FIGURE 22.15

The temperature and frequency-dependent dielectric constants of $Pr_2Ti_2O_7$.

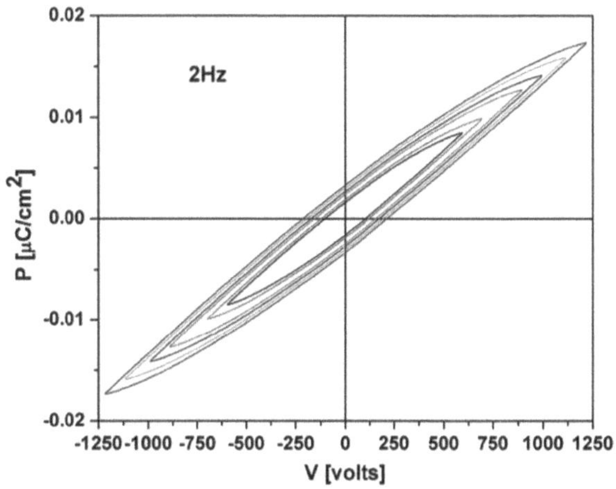

FIGURE 22.16

The variation of polarization versus voltage of pellets of bulk polycrystalline $Pr_2Ti_2O_7$ [205].

studies in bulk and thin films of $Bi_4Ti_3O_{12}$ prepared by various chemical and physical methods indicate large dielectric constant (~30–60) for it [214]. However, due to its lower chemical stabilities and compatibilities with other components, a number of research efforts have been paid to analogous phases, like $SrBi_2Ta_2O_7$, $SrBi_4Ti_4O_{15}$, $SrBi_2Nb_2O_9$, etc. [213,215–219]. Research on structural and ferroelectric properties of the under electric field revealed the tantalates and niobates are fatigue free ($>10^{12}$ cycles) and are more promising for practical applications [213,218]. Thin films of $SrBi_2Ta_2O_9$ find practical applications as nonvolatile computer memory chips [220]. Further research on these materials to prepare thin film and composition modifications to increase the cyclic operation is underway.

Besides the herein-mentioned examples, a large number of ceramics as well as organic and polymeric materials have been investigated for understanding their electrical properties. The ferroelectrics and relaxors have emerged extensively as materials for high dielectric property applications. However, in this chapter we have not included them to focus on the high-field effect on the materials. Also, we have not included other applications, like electrorheology, electrosynthesis, processing, and analytical as well as separation techniques as they themselves are complete research areas. Broadly it can be suggested the potential materials to circumvent the breakdown problem of gate dielectric is a change for future research. Similarly, the development of materials for high-speed electronics as well as high-voltage application with environmentally friendly nonlead materials are current research areas of electrical materials.

6. SUMMARY AND CONCLUSIONS

In summary, this chapter presents a brief introduction to electrical properties of materials, electric field, and behavior of materials under electric field. The effect of electric field on conductors, semiconductors, and insulators and their consequences are explained. The failure of the electrical and electronic materials under high electric field are discussed. Also, the effects of high electric field in controlling the properties are also touched upon. A more detailed description of dielectric breakdown and current problems is presented. The technological challenges and development of some dielectric materials are presented at the end. It is concluded that the development of dielectric materials for replacement of SiO_2 and high dielectric constant in free materials needs more research attention for future technologies.

REFERENCES

[1] N.W. Ashcroft, N.D. Mermin, Solid State Physics, Holt, Rinehart and Winston, New York, 1976.
[2] L. Solymar, D. Walsh, Electrical Properties of Materials, seventh ed., Oxford University Press, 2004.
[3] L.C. Sengupta, S. Sengupta, Mater. Res. Innov. 2 (1999) 278.
[4] L.B. Kong, S. Li, T.S. Zhang, J.W. Zhai, F.Y.C. Boey, J. Ma, Prog. Mater. Sci. 55 (2010) 840.
[5] B.D. Hatton, K. Landskron, W.J. Hunks, M.R. Bennett, D. Shukaris, D.D. Perovic, et al., Mater. Today 9 (2006) 22.
[6] P. Mavinakuli, S. Wei, Q. Wang, A.B. Karki, S. Dhage, Z. Wang, et al., J. Phys. Chem. C 114 (2010) 3874.
[7] L.J. Gibson, M.F. Ashby, Cellular Solids, Cambridge University Press, Cambridge, UK, 1997.
[8] L. Łukasiak, A. Jakubowski, J. Telecommun. Inf. Technol. 1 (2010) 3.
[9] G. Busch, Eur. J. Phys. 10 (1989) 254.
[10] D.A. Neamen, Semiconductor Physics and Devices, Basic Principles, third ed., The McGraw-Hill Companies, Inc., New York, 2003.
[11] R.E. Hummel, Electronic Properties of Materials, third ed., Springer—Verlag, New York, 2001. Inc.
[12] Z. Popovic, B.D. Popovic, Introductory Electromagnetic, Prentice Hall, Inc. Ver-12, 2012.
[13] F.E. Terman, Radio Engineers' Handbook, McGraw-Hill, 1943.
[14] D.G. Fink, H.W. Beatty, Standard Handbook for Electrical Engineers, eleventh ed., Mc-Graw Hill, 1978.
[15] A. Ducluzaux, ECT 83: Extra Losses Caused in High Current Conductors by Skin and Proximity Effects, Cahier Technique Schneider Electric No. 83, 1983, p. 3.
[16] G. Matijasevic, L. Brandt, Conductive material, wires and cables, in: Electrical Engineering, Encyclopedia of life support systems (EOLSS), vol. II, 1995.
[17] CRC Handbook of Chemistry and Physics, CRC Press, 2005.
[18] Total Losses in Power Distribution and Transmission Lines (1). http://electrical-engineering-portal.com/total-losses-in-power-distribution-and- transmission-lines-1.
[19] J.P. Holtzhausen, W. L Vosloo, High Voltage Engineering: Practice and Theory, 2014.
[20] L. Paris, G.Z. Valtorta, G. Manzoni, A. Invernizzi, N. de Franco, A. Vian, et al., Present limits of very long distance transmission systems, article 37-12, in: Int. Conf. on Large High Voltage Electric Systems, 1984. Paris.
[21] T. Getinet, Indian J. Pure Appl. Phys. 48 (2010) 192.
[22] S.M. Komirenko, K.W. Kim, V.A. Kochelap, M.A. Stroscio, Int. J. High Speed Electron. Sys. 12 (2002) 1057.
[23] K.D. Yong, K.C. Sub, B.C. Eu, J. Korean Phys. Soc. 38 (2001) 712.
[24] S.G. Liu, Appl. Phys. Lett. 9 (1966) 79.
[25] B. Gelmont, M. Shur, J. Phys. D 6 (1973) 842.
[26] P.N. Butcher, W. Fawcett, C. Hilsum, Br. J. Appl. Phys. 17 (1966) 841.
[27] M.S. Shur, T.A. Fjeldly, Compound semiconductor field effect transistors, in: S.M. Sze (Ed.), Modern Semiconductor Device Physics, Wiley-Interscience Publication, New York, 1998.
[28] D.S. Ong, G.J. Rees, J.P.R. David, J. Appl. Phys. 93 (2003) 4232.
[29] D.S. Ong, K.F. Li, S.A. Plimmer, G.J. Rees, J.P.R. David, P.N. Robson, J. Appl. Phys. 87 (2000) 7885.
[30] M. Levinshtein, J. Kostamovaara, S. Vainshtein, Breakdown Phenomena in Semiconductors and Semiconductor Devices, World Scientific, Singapore, 2005.

[31] J.B. Gunn, Solid State Commun. 1 (1963) 88.
[32] S. Vainshtein, V. Yuferev, V. Palankovski, D.S. Ong, J. Kostamovaara, Appl. Phys. Lett. 92 (2008) 062114.
[33] Technical Note, Panasonic T04007BE-3, 4 2009 3−2.
[34] S. Koshihara, A. Oiwa, M. Hirasawa, S. Katsumoto, Y. Iye, C. Urano, et al., Phys. Rev. Lett. 78 (1997) 4617.
[35] G. Lupke, Surf. Sci. Rep. 35 (1999) 75.
[36] Y.D. Glinka, W. Wang, S.K. Singh, Z. Marka, S.N. Rashkeev, Y. Rogachyova, et al., Phys. Rev. B 65 (2002) 193103.
[37] Y.D. Glinka, T.V. Shahbazyan, I.E. Perakis, N.H. Tolk, X. Liu, Y. Sasaki, et al., Appl. Phys. Lett. 81 (2002) 3717.
[38] C. Nunes, High electric field effects on the impurity optical absorption coefficient in semiconductors, in: D.S. Critin (Ed.), Radiative Processes and Dephasing in Semiconductors 18, Optical Society of America, TOPS 1, 1998.
[39] Q. Fanyao, A.L.A. Fonseca, O.A.C. Nunes, Phys. Rev. B 54 (1996) 16405.
[40] Q. Fanyao, A.L.A. Fonseca, O.A.C. Nunes, J. Appl. Phys. 82 (1997) 1236.
[41] L.V. Keldysh, J. Exp. Theor. Phys. 47 (1964) 1945.
[42] K. Eleperuma, T.P. Saha, T. Gillespie, Aust. J. Electr. Electron. Eng. 4 (2008) 239.
[43] J.S. Forrest, J. Inst. Electr. Eng. − Part II 89 (1942) 60.
[44] R. Krishnan, S. Samanta, R. Sudha, K. Govardhan, Int. J. Comput. Sci. Inf. 1 (2012) 2231.
[45] A. Neuber, M. Butcher, L.L. Hatfield, H. Krompholz, J. Appl. Phys. 85 (1999) 3084.
[46] Y.S. Choi, H.J. Kim, B.J. Shin, IEEE Trans. Electron Devices 51 (2004) 1241.
[47] K.G. Balmain, W. Hirt, IEEE Trans. Electr. Insul. 18 (1983) 498.
[48] C. Le Gressus, G. Blaise, IEEE Trans. Electr. Insul. 27 (1992) 472.
[49] Y. Yamano, A. Ohashi, K. Kato, H. Okubo, Y. Hakamata, IEEE Trans. Dielectr. Electr. Insul. 6 (1999) 464.
[50] Y.S. Liu, G.J. Zhang, W.B. Zhao, Z. Yan, Appl. Surf. Sci. 230 (2004) 12.
[51] V.M. Vinokur, T.I. Baturina, M.V. Fistul, A.Y. Mironov, M.R. Baklanov, C. Strunk, Nature 452 (2008) 613.
[52] G.A. Samara, Phys. Rev. 13 (1976) 4529.
[53] W. Kleemann, J. Dec, P. Lehnen, R. Blinc, B. Zalar, R. Pankrath, Eur. Phys. Lett. 57 (2002) 14.
[54] M. Bouziane, A. Benabad, J.C. Niepce, B. Elouadi, Phys. Chem. News 44 (2008) 133.
[55] W.H. Huang, D. Viehland, R.R. Neurgaonkar, J. Appl. Phys. 76 (1994) 490.
[56] S.P. Gaikwad, V. Samuel, R. Pasricha, V. Ravi, Mater. Lett. 58 (2004) 3700.
[57] C. Ramasastry, Y.V.G.S. Murthi, J. Phys. Chem. Solids 24 (1963) 384.
[58] G.A. Samara, J. Phys, Chem. Solids 40 (1979) 509.
[59] L.W. Barr, A.B. Lidiard, Defects in ionic crystals, in: vol. 10 of Physical Chemistry an Advanced Treatise, Academic Press, 1971.
[60] L.E. Cross, Ferroic materials and composites: past, present and future, Ch. 5, in: S. Somiya (Ed.), Advanced Ceramic III, Elsevier Applied Science, New York, 1998.
[61] R. Hartley, Base Materials for High Speed, High Frequency PC Boards, PCB&A, March 2002.
[62] E. Bogatin, Technical Note Materials Requirements for High Performance Digital Systems, Gigatest Labs, 2003.
[63] T. Fujii, A. Ando, Y. Sakabe, J. Eur. Ceram. Soc. 26 (2006) 1857.
[64] N. Klein, Thin Solid Films 50 (1978) 223.
[65] J.F. Verweij, J.H. Klootwijk, Microelectron. J. 27 (1996) 611.
[66] J.H. Klootwijk, J.F. Verweij, J.B. Rem, S. Bijlsma, Microelectron. J. 27 (1996) 623.
[67] K.G. McKay, K.B. McAfee, Phys. Rev. 91 (1953) 1079.
[68] A.G. Chynoweth, W.L. Feldmann, C.A. Lee, R.A. Logan, G.L. Pearson, P. Aigrain, Phys. Rev. 118 (1960) 425.
[69] S. Knight, Proc. IEEE 54 (1966) 1004.
[70] A.M. Barnett, A.G. Milnes, J. Appl. Phys. 37 (1966) 4215.

[71] N. Klein, Adv. Electron. Electron Phys. 26 (1969) 309.
[72] M. Qiao, X. Zhang, S. Wen, B. Zhang, Z. Li, Microelectron. Reliab. 54 (2014) 2704.
[73] Y. Sugawara, T. Kamei, IEEE Trans. Electr. Devices 34 (1987) 1816.
[74] E. Falck, W. Gerlach, J. Korec, IEEE Trans. Electron. Devices 40 (1993) 439.
[75] M. Ieong, V. Narayanan, D. Singh, A. Topol, V. Chan, Z. Ren, Mater. Today 9 (2006) 26.
[76] M.C. Ozturk, J. Liu, H. Mo, IEDM Technical Digest (IEEE) 03(2003), 2003, p. 20.5.1.
[77] J. Kedzierski, D. Boyd, P. Ronsheim, S. Zafar, J. Newbury, J. Ott, et al., IEDM Tech. Dig. IEEE (2003) 13.3.1.
[78] K. Takahashi, K. Manabe, T. Ikarashi, N. Ikarashi, T. Hase, T. Yoshihara, et al., IEDM Tech. Dig. IEEE (2004) 91.
[79] Y.H. Kim, C. Cabral, E.P. Gusev, IEDM Tech. Dig. IEEE (2005) 657.
[80] S. Zhu, J. Chen, M.-F. Li, S.J. Lee, J. Singh, C.X. Zhu, et al., IEEE Electron Device Lett. 25 (2004) 565.
[81] S. Zhu, H.Y. Yu, S.J. Whang, J.H. Chen, C. Shen, C. Zhu, et al., IEEE Electron Device Lett. 25 (2004) 268.
[82] M. Jang, J. Oh, S. Maeng, W. Cho, S. Lee, K. Kang, et al., Appl. Phys. Lett. 83 (2003) 2611.
[83] A. Hong, Dielectric Strength of Air, The Physics Factbook, 2000.
[84] F. Jakob, N. Perjanik, Sulfur Hexafluoride, a Unique Dielectric, Analytical Chem. Tech International, Inc. 2011.
[85] Thornson-Houston Co., Ltd., (Brit. Patent 532670) January 29, 1941.
[86] G. Gamilli, General Electric Review, vol. 59, 1956, p. 41.
[87] F.S. Cooper, General Electric Co. (US. Patent 2221671) November 12, 1940.
[88] J.F. Gall, Ind. Eng. Chem. 49 (1957) 1331.
[89] A. Pedersen, W. McAllister, G.C. Crichton, S. Vibholm, Formulation of the streamer breakdown criterion and its application to strongly electronegative gases and gas mixtures, Arch. Elektrotechn. 67 (1984). Springer−Verlag.
[90] B. Novak, Comparison of heat transfer in air and SF_6 in high voltage GIS, in: Electrical Insulation Conference (EIC), 2013, p. 85.
[91] R.L. Jaw, M.C. Miller, J. Phys. Chem. 60 (1956) 1412.
[92] H.H. Rogers, J. Chem. Eng. Data 2 (1961) 250.
[93] J.N. Wilson, Chem. Rev. 25 (1939) 377.
[94] Y. Yamini, S. Seidi, M. Rezazadeh, Anal. Chim. Acta 814 (2014) 1.
[95] A.E. Vanarkel, J. L Snoek, Z. Phys. 33 (1932) 662, 36 (1934) 187.
[96] P. Smyth, J. Phys. Chem. 43 (1939) 131.
[97] F.H. Muller, Z. Phys. 38 (1937) 498.
[98] R. Stanway, Mater. Sci. Technol. 20 (2004) 931.
[99] A.A. Maryott, E.R. Smith, Table of Dielectric Constants of Pure Liquids, National Bureau of Standards Circular 514, August 10, 1951.
[100] R.C. Chu, Trans. ASME 126 (2004) 491.
[101] A.E. Bergles, in: ASME HTD 57, 1986, p. 1.
[102] S. Naidu, V. Kamaraju, High Voltage Engineering, Tata McGraw Hill Education Private Limited, 2009.
[103] H. Yamashita, H. Amano, T. Mori, J. Phys. D 10 (1977) 1753.
[104] E.O. Forster, Research in the dynamics of electrical breakdown in liquid dielectrics, in: US Japan Seminar on Electrical Conduction and Breakdown in Dielectrics Gainsville, Florida, October 30, 1979.
[105] P.K. Watson, A.H. Sharbaugh, J. Electrochem. Soc. 107 (1960) 516.
[106] Y. Inuishi, J. Electrostat. 7 (1979) 1.
[107] A.H. Sharbaugh, J.C. Devins, S.J. Rzad, Trans. Electr. Insul. 13 (1978) 249.
[108] H. Kahn, A.H. Heuer, Mater. Today 2 (1999) 3.
[109] P. E van Kessel, L.J. Hornbeck, L.E. Meier, M.I. L Douglass, Proc. IEEE 186 (1998) 1687.

[110] I.L. Ballarini, R. L Mullen, H. Kahn, A.H. Heuer, MRS Symp. Proc. 518 (1998) 137.
[111] S.B. Brown, W. van Arsddl, C. L Muhlstein, in: Proceedings of the IEEE International Conference on Solid-State Sensors and Actuators, Chicago, Illinois, vol. 1, 1997, pp. 591–593.
[112] J.A. Connally, S.B. Brown, Science 256 (1992) 1537.
[113] G. Cornelia, R.P. Vinci, R.S. Iyer, R.H. Dauskardt, J.C. Bravman, MRS Symp. Proc. 518 (1999) 81.
[114] C.-H. Ho, S.Y. Kim, K. Roy, Microelectron. Reliab. 55 (2015) 308.
[115] R. Gerson, T. Marshall, J. Appl. Phys. 30 (1959) 1650.
[116] T. Tunkasiri, G. Rujijanagul, J. Mater. Sci. Lett. 15 (1996) 1767.
[117] T. Nagaya, Y. Isibashi, Jpn. J. Appl. Phys. Part I 36 (1997) 6136.
[118] J.S. Lee, U. Anselmi-Tamburini, A.A. Munir, S. Kim, Electrochem. Solid-State Lett. 9 (2006) J34.
[119] J. Liébault, J. Vallayer, D. Goeuriot, D. Tréheux, F. Thévenot, J. Eur. Ceram. Soc. 21 (2001) 389.
[120] A. Si Ahmed, J. Kansky, K. Zarbout, G. Moya, J. Liébault, D. Goeuriot, J. Eur. Ceram. Soc. 25 (2005) 2813.
[121] G. Blaise, C. Le Gressus, C.R. Acad. Sci. Paris 314 (1992) 1017.
[122] F.G. Will, H.G. Delorenzi, K.H. Janora, J. Am. Ceram. Soc. 75 (1992) 295.
[123] J. Kolk, E.L. Heasell, Solid-State Electron. 23 (1980) 101.
[124] M. Touzin, D. Goeuriot, H.-J. Fitting, C. Guerret-Piécourt, D. Juve, D. Treheux, J. Eur. Ceram. Soc. 27 (2007) 1193.
[125] M. Touzin, D. Goeuriot, C. Guerret-Piécourt, D. Juvé, H.-J. Fitting, J. Eur. Ceram. Soc. 30 (2010) 805.
[126] M. Fusi, L. Lamagna, S. Spiga, M. Fanciulli, G. Brammertz, C. Merckling, et al., Microelectron. Eng. 88 (2010) 435.
[127] O. Jongprateep, V. Petrovsky, F. Dogan, J. Met. Mater. Miner. 18 (2008) 9.
[128] G.D. Wilk, R.M. Wallace, J.M. Anthony, J. Appl. Phys. 89 (2001) 5243.
[129] S.O. Kasap, S.O. Kasap, Principles of Electronic Materials and Devices, McGraw-Hill, Boston, 2002.
[130] A. Kawamoto, K.J. Cho, R. Dutton, J. Comput. Aided Mater. Des. 8 (2001) 39.
[131] E. Gerritsen, N. Emonet, C. Caillat, N. Jourdan, M. Piazza, D. Fraboulet, et al., Solid State Electron 49 (2005) 1767.
[132] C. Kwon, Q.X. Jia, Y. Fan, M.F. Hundley, D.W. Reagor, J. Appl. Phys. 83 (1998) 7052.
[133] F. Amy, A.S. Wan, A. Kahn, F.J. Walker, R.A. McKee, J. Appl. Phys. 96 (2004) 1635.
[134] S. Hall, O. Buiu, I.Z. Mitrovic, Y. Lu, W.M. Davey, J. Telecommun. Inf. Technol. 2 (2007) 33.
[135] J. Robertson, Solid-State Electron. 49 (2005) 283.
[136] M. Houssa, L. Pantisano, L.-Å. Ragnarsson, R. Degraeve, T. Schram, G. Pourtois, et al., Mater. Sci. Eng. R 51 (2006) 37.
[137] H.T. Johnson-Steigelman, A.V. Brinck, J.P. Chang, P.F. Lyman, J. Vac. Sci. Technol. A 24 (2006) 1218.
[138] J.H. Choi, Y. Mao, J.P. Chang, Mater. Sci. Eng. R 72 (2011) 97.
[139] S.A. Campbell, D.C. Gilmer, X.C. Wang, M.T. Hsieh, H.S. Kim, W.L. Gladfelter, et al., IEEE Trans. Electron Devices 44 (1997) 104.
[140] R.C. Smith, T.Z. Ma, N. Hoilien, L.Y. Tsung, M.J. Bevan, L. Colombo, et al., Adv. Mater. Opt. Electron. 10 (2000) 105.
[141] M.K. Bera, C. Mahata, C.X. Maiti, Thin Solid Films 517 (2008) 27.
[142] K. Yokota, Y. Yano, K. Nakamura, M. Ohnishi, F. Miyashita, Nucl. Instrum. Methods B 242 (2006) 393.
[143] J.-P. Locquet, C. Marchiori, M. Sousa, J. Fompeyrine, J.W. Seo, J. Appl. Phys. 100 (2006) 051610.
[144] Y.Y. Lebedinskii, A. Zenkevich, E.P. Gusev, M. Gribelyuk, Appl. Phys. Lett. 86 (2005) 191904.
[145] M.R. Visokay, J.J. Chambers, A.L.P. Rotondaro, A. Shanware, L. Colombo, Appl. Phys. Lett. 80 (2002) 3183.
[146] J. Morais, L. Miotti, G.V. Soares, S.R. Teixeira, R. Pezzi, K.P. Bastos, et al., Appl. Phys. Lett. 81 (2002) 2995.

[147] M. Koyama, Y. Kamimuta, T. Ino, A. Kaneko, S. Inumiya, K. Eguchi, et al., IEDM Tech. Dig. IEEE (2004) 499.
[148] T. Watanabe, M. Takayanagi, K. Kojima, K. Sekine, H. Yamasaki, K. Eguchi, et al., IEDM Tech. Dig. IEEE (2004) 507.
[149] M.A. Quevedo-Lopez, J.J. Chambers, M.R. Visokay, A. Shanware, L. Colombo, Appl. Phys. Lett. 87 (2005) 012902.
[150] C. Lee, P. Ghosez, X. Gonze, Phys. Rev. B 50 (1994) 13379.
[151] R. Mani, S.N. Achary, K.R. Chakraborty, S.K. Deshpande, J.E. Joy, A. Nag, et al., Adv. Mater. 20 (2008) 1348.
[152] R. Mani, S.N. Achary, K.R. Chakraborty, S.K. Deshpande, J.E. Joy, A. Nag, et al., J. Solid State Chem. 183 (2010) 1380.
[153] Y.-D. Hou, Y. Shi, H.-Y. Ge, M.-K. Zhu, H. Yan, Mater. Res. Bull. 47 (2012) 184.
[154] S. Wang, Y.-D. Hou, M.-P. Zheng, H.-Y. Ge, M.-K. Zhu, H. Yan, Mater. Res. Bull. 48 (2013) 3098.
[155] S.K. Deshpande, S.N. Achary, R. Mani, J. Gopalakrishnan, Tyagi, Phys. Rev. B 84 (2011) 064301.
[156] Y. Shi, Y.-D. Hou, C. Wang, H.-Y. Ge, M.-K. Zhu, J. Am. Ceram. Soc. 93 (2010) 2491.
[157] K.T. Fehr, R. Hochleitner, E. Schmidbauer, J. Electroceram. 29 (2012) 240.
[158] K.T. Fehr, R. Hochleitner, E. Schmidbauer, J. Electroceram. 30 (2013) 191.
[159] H. Du, X. Lin, H. Zheng, B. Qu, Y. Huang, D. Chu, J. Alloy. Comp. 663 (2016) 848.
[160] W. Hu, Y. Liu, R.L. Withers, T.J. Frankcombe, L. Norén, A. Snashall, M. Kitchin, et al., Nat. Mater. 12 (2013) 821.
[161] W.W. Cho, I. Kagomiya, K.-I. Kakimoto, H. Ohsato, Appl. Phys. Lett. 89 (2006) 152905.
[162] W.W. Cho, I. Kagomiya, K.-I. Kakimoto, O. Hitoshi, J. Eur. Ceram. Soc. 27 (2007) 2907.
[163] M. Kahn, D.P. Burks, I. Burn, W.A. Schulze, in: L.M. Levinson (Ed.), Electronic Ceramics: Properties, Devices and Applications, Marcel Dekker, New York, 1988, p. 202.
[164] G. Goodman, in: R.C. Buchanan (Ed.), Ceramic Materials for Electronic: Processing, Properties and Applications, Marcel Dekker, New York, 1986, p. 131.
[165] X. Ning, P. Yongping, W. Bo, W. Haidong, C. Kai, Ceram. Int. 38 (2012) 141.
[166] W.-H. Tuan, Y.-C. Huang, Mater. Chem. Phys. 118 (2009) 187.
[167] M. Saleem, J.S. Song, I. Kim, M.-S. Kim, S.M. Jang, U. Farooq, et al., J. Nanoelectron. Optoelectron. 10 (2015) 386.
[168] T.A. Vanderah, Science 298 (2002) 1182.
[169] L.A. Benderskya, J.J. Krajewski, R.J. Cava, J. Eur. Ceram. Soc. 21 (2001) 2653.
[170] W. Wersing, Mater. Process. Wirel. Commun. 1 (1996) 715.
[171] R.A. Cowley, Phys. Rev. Lett. 9 (1962) 159.
[172] D.A. Bruce, W.G. Stirling, J. Phys. C 16 (1983) 841.
[173] C. Hubert, J. Levy, A.C. Carter, W. Chang, S.W. Kiechoefer, J.S. Horwitz, et al., Appl. Phys. Lett. 71 (1997) 3353.
[174] C. Ang, Z. Yu, P.M. Vilarinho, J.L. Baptista, Phys. Rev. B 57 (1998) 4703.
[175] J.H. Haeni, P. Irvin, W. Chang, R. Uecker, P. Reiche, Y.L. Li, et al., Nature 430 (2004) 758.
[176] E. Park, W. Hackenberger, Curr. Opin. Solid State Mater. Sci. 6 (2002) 11.
[177] A.A. Bokov, X. Long, Z.-G. Ye, Phys. Rev. B 81 (2010) 172103.
[178] M. Davis, D. Damjanvoic, N. Setter, Phys. Rev. B 73 (2006) 014115.
[179] S.-E. Park, T.R. Shrout, J. Appl. Phys. 82 (1997) 1804.
[180] R. Haumont, A. Al-Barakaty, B. Dkhil, J.M. Kiat, L. Bellaiche, Phys. Rev. B 71 (2005) 104106.
[181] S.V. Rajasekaran, S.N. Achary, S.J. Patwe, R. Jayavel, G. Mangamma, A.K. Tyagi, J. Mater. Res. 29 (2014) 1054.
[182] S.N. Achary, O.D. Jayakumar, A.K. Tyag, Multiferroic materials, in: S. Bannerjee, A.K. Tyagi (Eds.), Functional Material, Elsevier Inc., 2012.

[183] D.P. Dutta, B.P. Mandal, E. Abdelhamid, R. Naik, A.K. Tyagi, Dalton Trans. 44 (2015) 11388.
[184] D. Lebeugle, D. Colson, A. Forget, M. Viret, Arxiv: 0706, 1681.
[185] O.D. Jayakumar, S.N. Achary, K.G. Girija, A.K. Tyagi, C. Sudakar, G. Lawes, et al., Appl. Phys. Lett. 96 (2010) 32903.
[186] C. Tabares-Munoz, J.-P. Rivera, A. Bezinges, A. Monnier, H. Schmid, Jpn. J. Appl. Phys. 24 (Suppl. 2) (1985) 1051.
[187] J. Wang, J.B. Neaton, H. Zheng, V. Nagarajan, S.B. Ogale, B. Liu, et al., Science 299 (2003) 1719.
[188] I. Levin, J.Y. Chan, J.E. Maslar, T.A. Vanderah, J. Appl. Phys. 90 (2001) 904.
[189] R.E. Cohen, Nature 358 (1992) 136.
[190] P.S. Halasyamani, K.R. Poeppelmeier, Chem. Mater. 10 (1998) 2753.
[191] N.S.P. Bhuvanesh, J. Gopalakrishnan, J. Mater. Chem. 7 (1997) 2297.
[192] R. shukla, S.J. Patwe, S.K. Deshpande, S.N. Achary, P.S.R. Krishna, A.B. Shinde, et al., Nat. Sci. Rep. 6 (2016) 23400.
[193] M.A. Subramanian, D. Li, N. Duan, B.A. Reisner, A.W. Sleight, J. Solid State Chem. 151 (2000) 32.
[194] C.C. Homes, T. Vogt, S.M. Shapiro, S. Wakimoto, A.P. Ramirez, Science 293 (2001) 673.
[195] L. Singh, U.S. Rai, K.D. Mandal, N.B. Singh, Prog. Cryst. Growth Charact. Mater. 60 (2014) 15.
[196] T.-T. Fang, C.P. Liu, Chem. Mater. 17 (2005) 5167.
[197] P. Lunkenheimer, R. Fichtl, S.G. Ebbinghaus, A. Loidl, Phys. Rev. B 70 (2004) 172102.
[198] R. Jia, X. Zhao, J. Li, X. Tang, Mater. Sci. Eng. B 185 (2014) 79.
[199] H. Yan, H. Ning, Y. Kan, P. Wang, M. Reece, J. Am. Ceram. Soc. 92 (2009) 2270.
[200] T.-M. Pan, M.-D. Huang, C.-W. Lin, M.-H. Wu, Sens. Actuator B 144 (2010) 139.
[201] S. Nanamatsu, M. Kimura, K. Doi, S. Matsushita, N. Yamada, Ferroelectrics 8 (1974) 511.
[202] Z.P. Gao, H.X. Yan, H.P. Ning, M.J. Reece, Adv. Appl. Ceram. 112 (2013) 69.
[203] M. Kimura, S. Nakamatsu, T. Kawamura, S. Matsushita, Jpn. J. Appl. Phys. 13 (1974) 1473.
[204] C. Wenger, G. Lupina, M. Lukosius, O. Seifarth, H.-J. Mussig, S. Pasko, et al., J. Appl. Phys. 103 (2008) 104103.
[205] S.J. Patwe, V. Katari, N.P. Salke, S.K. Deshpande, R. Rao, M.K. Gupta, et al., J. Mater. Chem. C 3 (2015) 4570.
[206] E. Bruyer, A. Sayede, J. Appl. Phys. 108 (2010) 053705.
[207] J. Lopez-Perez, J. Iniguez, Phys. Rev. B 84 (2001) 075121.
[208] W.S. Kim, S.-M. Ha, S. Yun, H.-H. Park, Thin Solid Films 420–421 (2002) 575.
[209] A. Bayart, S. Saitzek, A. Ferri, R. Pouhet, M.-H. Chambrier, P. Roussel, et al., Thin Solid Films 553 (2014) 71.
[210] L. Sun, L. Ju, H. Qin, M. Zhao, W. Su, J. Hu, Physica B 431 (2010) 49.
[211] R.E. Newnham, R.W. Wolfe, F. Dorrian, Mater. Res. Bull. 6 (1971) 1029.
[212] J.F. Dorian, R.E. Newnham, D.K. Smith, Ferroelectrics 3 (1971) 17.
[213] B.H. Park, B.S. Kang, S.D. Bu, T.W. Noh, J. Lee, W. Jo, Nature 401 (1999) 682.
[214] T. Bertaud, C. Bermond, S. Blonkowski, C. Valle, T. Lacrevaz, A. Farcy, et al., IEEE Trans. Compon. Packag. Manuf. Technol. 2 (2012) 502.
[215] V.V. Sidsky, A.V. Semchenko, A.G. Rybakov, V.V. Kolos, A.S. Turtsevich, A.N. Asadchyi, et al., J. Rare Earths 32 (2014) 277.
[216] Y. Shimakawa, Y. Kubo, Y. Nakagawa, T. Kamiyama, H. Asano, F. Izumi, Appl. Phys. Lett. 74 (1999) 1904.
[217] Y. Noguchi, M. Miyayama, T. Kudo, Phys. Rev. B 63 (2001) 214102.
[218] Y. Noguchi, M. Miyayama, T. Kudo, Appl. Phys. Lett. 77 (2000) 3639.
[219] Y.-H. Wu, C.-C. Lin, L.-L. Chen, Y.-C. Ju, J.-R. Wu, Appl. Phys. Lett. 98 (2011) 013506.
[220] C.A. Araujo, J.D. Cuchlaro, L.D. McMillan, M.C. Scott, J.F. Scott, Nature 374 (1995) 627.

Index

'*Note*: Page numbers followed by "f" indicate figures and "t" indicate tables.'

E

O

P

Q

R

Kolibri 360 GmbH
Ferdinand-Jühlke-Straße 7
99095 Erfurt, Deutschland
produktsicherheit@kolibri360.de